WELDING HANDBOOK

Eighth Edition

Volume 2

WELDING PROCESSES

The Three Volumes of the Welding Handbook, Eighth Edition

1) WELDING TECHNOLOGY

2) WELDING PROCESSES

3) MATERIALS AND APPLICATIONS

WELDING HANDBOOK

Eighth Edition

Volume 2

WELDING PROCESSES

R. L. O'Brien
Editor

AMERICAN WELDING SOCIETY
550 N.W. LEJEUNE ROAD
P.O. BOX 351040
MIAMI, FL 33135

Library of Congress Number: 90-085465
International Standard Book Number: 0-87171-354-3

American Welding Society, 550 N.W. LeJeune Rd., P.O. Box 351040,
Miami, FL 33135

THE WELDING HANDBOOK is a collective effort of many volunteer technical specialists to provide information to assist with the design and application of welding and allied processes.

Reasonable care is taken in the compilation and publication of the Welding Handbook to insure authenticity of the contents. No representation or warranty is made as to the accuracy or reliability of this information.

The information contained in the Welding Handbook shall not be construed as a grant of any right of manufacture, sale, use, or reproduction in connection with any method, process, apparatus, product, composition, or system, which is covered by patent, copyright, or trademark. Also, it shall not be construed as a defense against any liability for such infringement. No effort has been made to determine whether any information in the Handbook is covered by any patent, copyright, or trademark, and no research has been conducted to determine whether an infringement would occur.

Printed in the United States of America

CONTENTS

CONTENTS

CHAPTER 26, DIFFUSION WELDING AND BRAZING 813

CHAPTER 27, ADHESIVE BONDING OF METALS 839

CHAPTER 28, THERMAL SPRAYING 863

WELDING HANDBOOK COMMITTEE

May 31, 1990

M. J. Tomsic, Chairman	Plastronic, Incorporated
C. W. Case, 1st Vice Chairman	Inco Alloys International
D. R. Amos, 2nd Vice Chairman	Westinghouse Turbine Plant
R. L. O'Brien, Secretary	American Welding Society
J. R. Condra	E. I. duPont de Nemours and Company
G. N. Fischer	Fischer Engineering Company
J. R. Hannahs	Midmark Corporation
A. F. Manz	A. F. Manz Associates
J. C. Papritan	Ohio State University
L. J. Privoznik	Westinghouse Electric Corporation
E. G. Shifrin	Detroit Edison (Retired)
B. R. Somers	Consultant
P. I. Temple	Detroit Edison
R. M. Walkosak	Westinghouse Electric Corporation

WELDING HANDBOOK COMMITTEE

June 1, 1990

C. W. Case, Chairman	Inco Alloys International
D. R. Amos, 1st Vice Chairman	Westinghouse Turbine Plant
B. R. Somers, 2nd Vice Chairman	Consultant
R. L. O'Brien, Secretary	American Welding Society
J. R. Condra	E. I. duPont de Nemours and Company
J. Feldstein	Teledyne McKay
J. R. Hannahs	PMI Food Equipment Group
L. Heckendorn	Toledo Scale Company
A. F. Manz	A. F. Manz Associates
J. C. Papritan	Ohio State University
P. I. Temple	Detroit Edison
M. J. Tomsic	Plastronic, Incorporated
R. M. Walkosak	Westinghouse Electric Corporation

FOREWORD

This is the 53rd anniversary of the first AWS *Welding Handbook*. Beginning with that start in 1938, the Welding Handbook Committee has been a dedicated body of Volunteers willing to donate personal time and effort to the production of the Handbook.

Some of the processes described in the First Edition have changed little in the ensuing years. Other processes in this edition, such as plasma arc, laser beam, and electron beam, are beyond the imagination of that first Handbook Committee. We pause to consider what processes our successors will be describing fifty years hence, and the mode of presentation.

Considerable time goes into selecting experts in various fields to represent equipment manufacturers, users, and general interest groups to balance Handbook Chapter Committees. This procedure ensures that the Handbook Chapters contain the latest data and present the material without bias.

R. L. O'Brien, Editor
Welding Handbook

PREFACE

This is the second volume of the Eighth Edition of the *Welding Handbook*. This volume, Welding Processes, covers the material previously presented in Volumes Two and Three of the Seventh Edition.

Volume Three will address materials, their weldability, and applications. It will update the content of Volume Four of the Seventh Edition and Volume Five of the Sixth Edition.

The authors of this volume have upgraded the material from the Seventh Edition to reflect the latest technology. They have also increased the number of applications to make the process descriptions more meaningful and have made more extensive use of color illustrations. Several sections on Safety have been added.

The Welding Handbook Committee and the Chapter Committee Members have put in thousands of hours of personal time to produce this volume. We have recognized their contributions by listing them on the title pages of the respective chapters. We wish to thank them for their generous contributions of time and talent, and we extend our appreciation to their employers for supporting this work.

The Welding Handbook Committee expresses appreciation to Alexander Lesnewich, Hallock C. Campbell, and Leonard P. Connor for their editorial overview, to Deborah Givens for editorial assistance, and to Linda Williams for word processing assistance.

Your comments on the Handbook are welcome. Please address them to the Editor, Welding Handbook, American Welding Society, P.O. Box 351040, Miami, FL 33135.

M. J. Tomsic, Chairman
Welding Handbook Committee
1987-1990

R. L. O'Brien, Editor
Welding Handbook

ARC WELDING POWER SOURCES

PREPARED BY A COMMITTEE CONSISTING OF:

M. J. Tomsic, Chairman
Plastronic, Inc.

N. Crump
Hobart Brothers Co.

J. F. Grist
Miller Electric Mfg. Co.

W. T. Rankin
Pow Con Incorporated

J. M. Thommes
Pow Con Incorporated

J. L. Winn
L-Tec Welding and Cutting Systems

WELDING HANDBOOK COMMITTEE MEMBER:
M. J. Tomsic
Plastronic, Inc.

ARC WELDING POWER SOURCES

INTRODUCTION

MANY TYPES OF power sources are necessary to meet the unique electrical requirements of the various arc welding processes. The arc welding power sources described in this chapter include those for shielded metal arc (SMAW), gas metal arc (GMAW), flux cored arc (FCAW), gas tungsten arc (GTAW), submerged arc (SAW), electroslag (ESW), electrogas (EGW), plasma arc (PAW), and arc stud welding (ASW). These power sources include both pulsed and nonpulsed configurations that are either manually or automatically controlled.

This chapter is intended to be a guide to understanding and selecting the proper power source. The applications outlined are typical and serve only to illustrate and explain the relationship of power source to process.

Selecting the correct power source, of course, depends upon process requirements. The first step is to determine the electrical requirements of the welding process with which it will be used. Other factors to consider include such things as future requirements, maintenance, economic considerations, portability, environment, available skills, safety, manufacturer's support, code compliance, and standardization. This chapter, however, will concentrate only on the technical aspects of power.

GENERAL

THE VOLTAGE SUPPLIED by power companies for industrial purposes is too high to use directly in arc welding. Therefore, the first function of arc welding power sources is to reduce the high input or line voltage to a suitable output voltage range [usually 20 to 80 volts (V)]. Either a transformer, solid-state inverter, or a motor-generator can be used to reduce the 120, 240, or 480 V utility power to the rated terminal or open circuit voltage appropriate for arc welding. Alternatively, a power source for arc welding may derive its power from a prime mover such as an internal combustion engine. Sources deriving power from internal combustion engines must use rotating generators or alternators as the electrical source.

The same device (transformer or motor generator) also provides a high welding current, generally ranging from 30 to 1500 amperes (A). The typical output of a power source may be alternating current (ac), direct current (dc), or both. It may be either constant-current, constant-voltage, or both. It may also provide a pulsing output mode.

Some power source configurations deliver only certain types of current. For example, transformer type power sources will deliver only alternating current. Transformer-rectifier power sources may deliver either ac or dc. Electric motor-generator power sources usually deliver direct current output. A motor-alternator will deliver ac or, when equipped with rectifiers, dc.

Power sources can also be classified by subcategories. For example, a gas tungsten arc welding power source might be identified as transformer-rectifier, constant current, ac/dc. A more complete description should include welding current rating, duty cycle rating, service classification, and input power requirements. Special features can also be included, such as remote control, high frequency stabilization, current pulsing capability, starting and finishing current versus time programming, wave balancing capabilities, and line voltage compensation. Current or voltage control might also be included. Typical conventional controls are moveable shunts, saturable reactors, magnetic

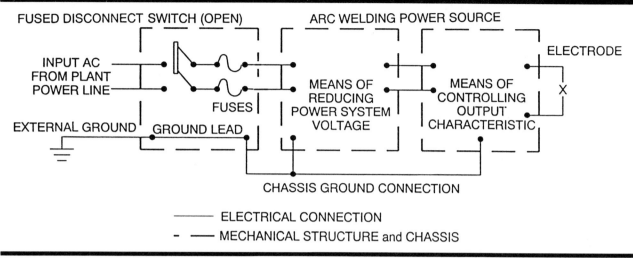

Figure 1.1–Basic Elements of an Arc Welding Power Source

amplifiers, series impedance, or tapped windings. Solid-state electronic controls may be phase control or inverter control semiconductors; these elements may be controlled by a microprocessor.

Figure 1.1 shows the basic elements of a welding power source supplied from utility lines. The arc welding power source itself doesn't usually include the fused disconnect switch; however, this is a necessary protective element. An engine-driven power supply would require elements different from those shown in Figure 1.1. It would require an engine, an engine speed regulator, an alternator with or without a rectifier, or a generator, and an output control.

Until the advent of welding processes that use pulsed current, welding power sources were commonly classified as constant current or constant voltage. Such classifications are based on the static volt-ampere characteristics of the power supply, not the dynamic characteristics. Generally, the word *constant* is true only to a degree. *Constant-voltage* power sources are usually much closer to constant-voltage output than *constant-current* sources are to constant-current output. In either case, specialized power sources are available that will hold output voltage or current truly constant. Constant-current power sources are also referred to as *variable-voltage* power sources, and constant-voltage power sources are often called *constant-potential* power sources. The fast response solid-state sources made available in recent years can provide power in pulses ranging over a broad range of frequencies.

CONSTANT-CURRENT MACHINES

THE NATIONAL ELECTRICAL Manufacturers Association (NEMA) Publication EW-1, *Electric Arc Welding Power Sources*, defines a constant-current arc welding machine as one " . . . which has means for adjusting the load current and which has a static volt-ampere curve that tends to pro-

duce a relatively constant load current. The load voltage, at a given load current, is responsive to the rate at which a consumable electrode is fed into the arc, except that, when a nonconsumable electrode is used, the load voltage is responsive to the electrode-to-work distance."

These characteristics are such that if the arc length varies because of external influences which result in slight changes in arc voltage, the welding current remains substantially constant. Each current setting yields a separate volt-ampere curve when tested under steady conditions as with a resistive load. In the vicinity of the operating point, the percent change in current is less than the percent change in voltage.

The no-load or open circuit voltage of constant-current arc welding power sources is considerably higher than the arc voltage.

These power sources are generally used for manual welding with a covered electrode or a tungsten electrode, where variations in arc length are unavoidable because of the human element.

When used in a semiautomated or automated application where constant arc length is required, external control devices are necessary. For example, an arc voltage sensing wirefeeder can be used to maintain constant arc length for gas metal arc welding (GMAW) or flux cored arc welding (FCAW). With gas tungsten arc welding (GTAW), the arc voltage is monitored and, via a closed loop feedback, used to regulate a motorized slide which positions the torch to maintain a constant arc length (voltage).

CONSTANT VOLTAGE MACHINES

THE NEMA STANDARD defines a constant voltage source as follows: "A constant-voltage arc welding power source is a power source which has means for adjusting the load voltage and which has a static volt-ampere curve that tends to produce a relatively constant load voltage. The load current, at a given load voltage, is responsive to the rate at which a consumable electrode is fed into the arc." A constant-voltage arc welding machine is usually used with welding processes that employ a continuously fed consumable electrode, generally in the form of wire.

A welding arc powered by a constant-voltage source, using a consumable electrode and a constant-speed wire feed, is essentially a self-regulating system. It tends to stabilize the arc length despite momentary changes in the torch position. The arc current will be approximately proportional to wire feed for all wire sizes.

CONSTANT-CURRENT/ CONSTANT-VOLTAGE MACHINES

A POWER SOURCE that provides both constant current and constant voltage is defined by NEMA as: "A constant-current/constant-voltage arc welding machine power source is a power source which has the selectable characteristics of a constant-current arc welding power source and a constant-voltage arc welding power source."

Additionally, some designs feature an automatic change from constant current to constant voltage (arc force control for SMAW) or constant voltage to constant current (current limit control for constant-voltage power source).

PRINCIPLES OF OPERATION

ARC WELDING INVOLVES low-voltage, high-current arcs between an electrode and the work piece. The means of reducing power system voltage in Figure 1.1 may be a transformer or an electric generator or alternator driven by an electric motor.

Electric generators built for arc welding usually are designed for dc welding only. In this case, the electromagnetic means of controlling the volt-ampere characteristic of the arc welding power source is usually an integral part of the generator and not a separate element, as shown in Figure 1.1. Unlike generators, alternators provide ac output which must be rectified to provide a dc output. Various configurations are employed in the construction of dc generators. They may use a separate exciter and either differential or cumulative compounding for controlling and selecting volt-ampere output characteristics.

WELDING TRANSFORMER

FIGURE 1.2 SHOWS the basic elements of a welding transformer and associated components. For a transformer, the significant relationships between winding turns and input and output voltages and currents are as follows:

$$\frac{N_1}{N_2} = \frac{E_1}{E_2} = \frac{I_2}{I_1} \qquad (1.1)$$

where

N_1 = the number of turns on the primary winding of the transformer
N_2 = the number of turns on the secondary winding
E_1 = the input voltage
E_2 = the output voltage
I_1 = the input current
I_2 = the output (load) current.

Taps in the transformer secondary winding may be used to change the number of turns in the secondary, as shown in Figure 1.3, to vary the open circuit (no-load) output voltage. In this case, the tapped transformer permits the selection of the number of turns, N_2, in the secondary winding of the transformer. When the number of turns is decreased on the secondary, output voltage is lowered because a smaller proportion of the transformer secondary winding is in use. The tap selection, therefore, controls open circuit voltage. As shown by the equation, the primary-secondary current ratio is inversely proportional to the primary-secondary voltage ratio. Thus, large secondary (welding) currents can be obtained from relatively low line currents.

A transformer may be designed so that the tap selection will directly adjust the output volt-ampere slope characteristics for a proper welding condition. More often, however, an impedance source is inserted in series with the transformer secondary windings to provide this characteristic, as shown in Figure 1.4. Some types of power sources use a combination of these arrangements, with the taps adjusting the open circuit (or no-load) voltage of the welding machine and the impedance providing the desired volt-ampere slope characteristics.

In constant-current power supplies, the voltage drop, E_x, across the impedance shown in Figure 1.4 increases greatly as the load current is increased. The increase in voltage drop, E_x, causes a large reduction in the arc voltage, E_A. Adjustment of the value of the series impedance controls its voltage drop and the relation of load current to load voltage. This is called *current control* or, in some cases, *slope control*. Voltage E_O essentially equals the no-load (open circuit) voltage of the power supply.

In constant-voltage power sources, the output voltage is very close to that required by the arc. The voltage drop E_x, across the impedance (reactor) increases only slightly as the load current increases. The reduction in load voltage is small. Adjustment in the value of reactance gives slight control of the relation of load current to load voltage.

This method of slope control with simple reactors also serves as a method to control voltage with saturable reactors or magnetic amplifiers. Figure 1.5 shows an ideal vector relationship of the alternating voltages for the circuit of Fig-

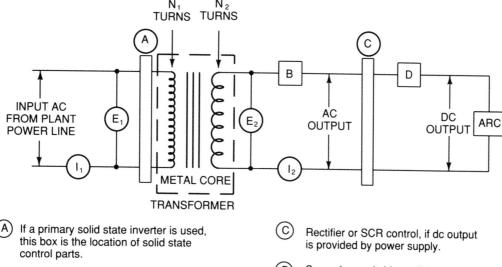

N_1 TURNS N_2 TURNS

INPUT AC FROM PLANT POWER LINE E_1 I_1 METAL CORE   TRANSFORMER E_2 I_2 AC OUTPUT   DC OUTPUT   ARC

(A) If a primary solid state inverter is used, this box is the location of solid state control parts.

(B) Location of series control components if used.

(C) Rectifier or SCR control, if dc output is provided by power supply.

(D) Secondary switching solid state device for chopper type control if used. Also location of slope resistor if used, or inductor for ac circuit.

Figure 1.2–Principal Electrical Elements of a Transformer Power Supply

ure 1.4 when a reactor is used as an impedance device. The voltage drop across the impedance plus the load voltage equals the no-load voltage only when added vectorially. In the example pictured, the open circuit voltage of the transformer is 80 V; the voltage drop across the reactor is about 69 V when the load (equivalent to a resistor) voltage is 40 V.

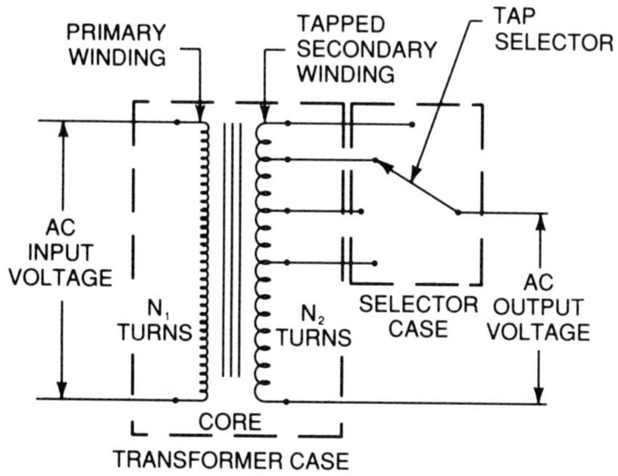

The vectorial addition is necessary because the alternating load and impedance voltages are not in time phase.

The voltage drop across a series impedance in an ac circuit is added vectorially to the load voltage to equal the transformer secondary voltage. By varying the voltage drop across the impedance, the load voltage may be changed. This peculiar characteristic (vectorial addition) of impedance voltage in ac circuits is related directly to the fact that both reactors and resistances may be used to produce a drooping voltage characteristic. An advantage of a reactor is that it consumes little or no power, even though a current flows through it and a voltage can be measured across it.

When resistors are used, power is lost and temperature rises. Theoretically, in a purely resistive circuit (no reactance), the voltage drop across the resistor could be added arithmetically to the load voltage to equal the output voltage of the transformer. For example, a welding machine with an approximately constant-current characteristic, an 80 V open circuit, and powering a 25 V, 200 A arc would need to dissipate 55 V x 200 A or 11 000 watts (W) in the resistor to supply 5000 W to the arc. The reason for this is that in the resistive circuit, the voltage and current are in phase. In the reactor circuit, phase shift accounts for the greatly reduced power loss. In the reactor circuit, there are only the iron and copper losses, which are very small by comparison.

Variable inductive reactance or variable mutual inductance may be used to control the volt-ampere characteristics in typical transformer or transformer-rectifier arc welding power sources. The equivalent impedance of a

PRIMARY WINDING   TAPPED SECONDARY WINDING   TAP SELECTOR

AC INPUT VOLTAGE N_1 TURNS N_2 TURNS   SELECTOR CASE   AC OUTPUT VOLTAGE

CORE   TRANSFORMER CASE

Figure 1.3–Welding Transformer With Tapped Secondary Winding

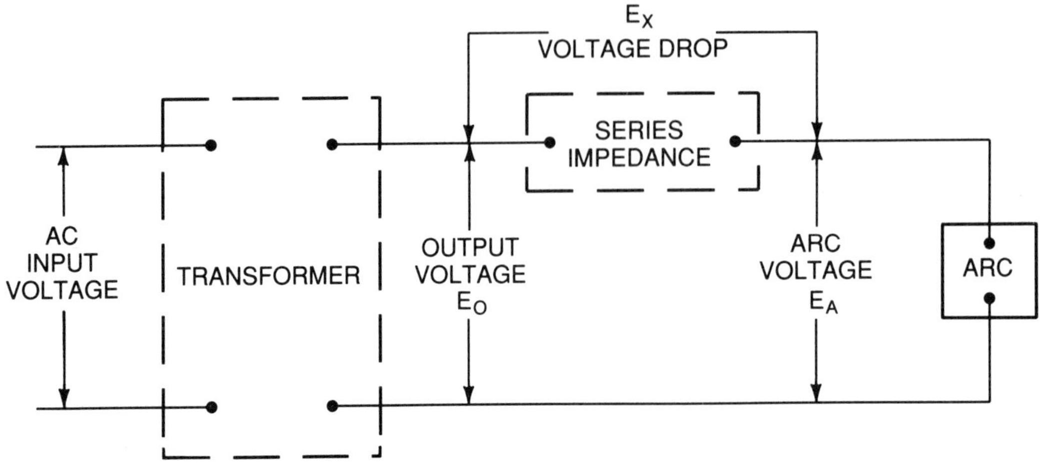

Figure 1.4—Typical Series Impedance Control of Output Current

variable inductive reactance or mutual inductance is located in the ac electrical circuit of the power source in series with the secondary circuit of the transformer, as shown in Figure 1.4. Another major advantage of inductive reactance is that the phase shift produced in the alternating current by the reactor improves ac arc stability for a given open circuit voltage. This is an advantage with gas tungsten arc and shielded metal arc welding processes.

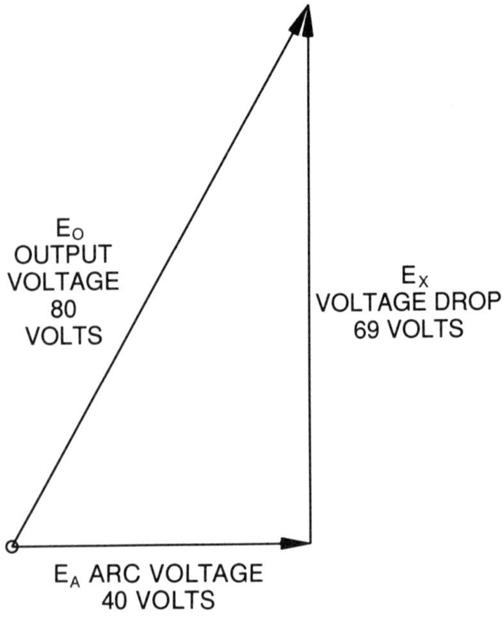

Figure 1.5—Ideal Vector Relationship of the Alternating Voltage Output Using Reactor Control

The reactance of a reactor can be varied by several means. One way is by changing taps on a coil or by other electrical/mechanical schemes discussed later. Varying the reactance alters the voltage drop across the reactor. Thus, for any given value of inductive reactance, a specific volt-ampere curve can be plotted. This creates the dominant control feature of such power supplies.

In addition to adjusting reactance, mutual inductance between the primary and secondary coils can also be adjusted. This can be done by moving the coils relative to each other or by using a movable shunt that can be inserted or withdrawn from the transformer. These methods change the magnetic coupling of the coils to produce adjustable mutual inductance.

In ac-dc welding power sources incorporating a transformer and a rectifier, the rectifier is located between the adjustable impedance or transformer taps and the output terminal. In addition, the transformer-rectifier type of arc welding power supply usually incorporates a stabilizing inductance or choke, located in the dc welding circuit, to improve arc stability.

GENERATOR AND ALTERNATOR

ROTATING MACHINERY IS also used as a source of power for arc welding. These machines are divided into two types; generators which produce direct current and alternators which produce alternating current.

The no-load output voltage of a dc generator may be controlled with a relatively small variable current in the main or shunt field winding. This current controls the output of the dc generator series or bucking field winding that supplies the welding current. Polarity can be reversed by changing the interconnection between the exciter and the main field. An inductor or filter reactor is not usually needed to improve arc stability with this type of welding equipment. Instead, the several turns of series winding on the field poles of the rotating generator provide more than enough inductance to ensure satisfactory arc stability. Such units are described in greater detail later in this chapter.

An alternator power source (a rotating-type power source in which ac is produced and either used directly or rectified into dc) can use a combination of the means of adjustment mentioned above. A tapped reactor can be employed for gross adjustment of the welding output, and the field strength can be controlled for fine adjustment.

SOLID-STATE DIODES

Solid-state derives its name from solid state physics: the science of the crystalline solid. Methods have been developed for treating certain materials in order to modify their electrical properties. The most important of these materials is silicon.

Transformer-rectifier or alternator-rectifier power sources rely on rectifiers to convert ac to dc. Early welding machines used selenium rectifiers. Today, most rectifiers are made of silicon for reasons of economy, current-carrying capacity, reliability, and efficiency.

A single rectifying element is called a *diode*, which is a one-way electrical valve. When placed in an electric circuit, a diode allows current to flow in one direction only: when the anode of the diode is positive with respect to the cathode. Using a proper arrangement of diodes, it is possible to convert ac to dc.

The resistance to current flow through a diode results in a voltage drop across the component and generates heat within the diode. Unless this heat is dissipated, the diode temperature can increase enough to cause failure. Therefore, diodes are normally mounted on heat sinks (aluminum plates) to remove that heat.

Diodes have limits as to the amount of voltage they can block in the reverse direction (anode negative and cathode positive). This is expressed as the voltage rating of the device. Welding power source diodes are usually selected with a blocking rating at least twice the open circuit voltage in order to provide a safe operating margin.

A diode can accommodate current peaks well beyond its normal steady state rating, but a high, reverse-voltage transient could damage it. Most rectifier power sources have a resistor, capacitor, or other electronic device to suppress voltage transients that could damage the rectifiers.

SILICON CONTROLLED RECTIFIER (SCR)— THYRISTOR

SOLID-STATE DEVICES with special characteristics also can be used to directly control welding power by altering the welding current or voltage wave form. Such solid-state devices have now replaced saturable reactors, moving shunts, moving coils, etc., formerly used to control the output of

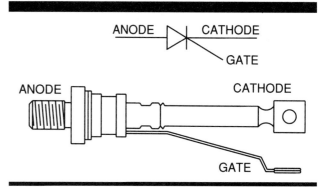

Figure 1.6—Silicon Controlled Rectifier

welding transformers. One of the most important of these devices is the silicon controlled rectifier (SCR), sometimes called a *thyristor*.

The SCR is a diode variation with a trigger called a *gate*, as shown in Figure 1.6. An SCR is nonconducting until a positive electrical signal is applied to the gate. When this happens, the device becomes a diode, and will conduct current as long as the anode is positive with respect to the cathode. However, once it conducts, the current cannot be turned off by a signal to the gate. Conduction will stop only if the voltage applied to the anode becomes negative with respect to the cathode. Conduction will not take place again until a positive voltage is applied to the anode and another gate signal is received.

There are two main uses for SCRs: in inverter configurations, and in phase control mode with transformers. Using the action of a gate signal to selectively turn on the SCR, the output of a welding power source can be controlled. A typical phase control SCR circuit is shown in Figure 1.7.

In Figure 1.7, during the time that point B is positive with resjpective to point E, no current will flow until both

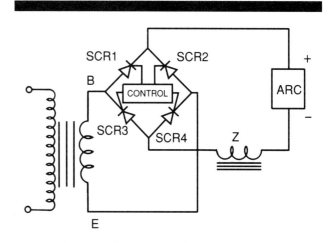

Figure 1.7—Single-Phase DC Power Source Using an SCR Bridge for Control

SCR 1 and SCR 4 receive a gate signal to turn on. At that instant, current will flow through the load. At the end of that half-cycle, when the polarity of B and E reverses, a negative voltage will be impressed across SCR 1 and SCR 4, and they will turn off. With point E positive with respect to point B, a gate signal applied to SCR 2 and SCR 3 by the control will cause these two to conduct, again applying power to the load circuit. To adjust power in the load, it is necessary to precisely time where, in any given half-cycle, conduction is to initiate.

When high power is required, conduction is started early in the half-cycle. Should low power be required, conduction is delayed until later in a half-cycle. This is known as *phase control*. The result is shown in Figure 1.8. The resulting power is supplied in pulses to the load, and is proportional to the shaded area under the wave form envelope. Figure 1.8 shows that significant intervals may exist when no power is supplied to the load. This can cause arc outages, especially at low-power levels. Therefore, wave filtering is required.

Figure 1.7 shows a large inductance, Z, in the load circuit. For a single-phase circuit to operate over a significant range of control, Z must be very large to smooth out the pulses enough to increase the conductance times. If, however, SCRs are used in a three-phase circuit, the nonconducting intervals would be reduced significantly. The inductance (Z) would be sized accordingly. For this reason three-phase SCR systems are more practical for welding power sources, unless the output is ac/dc and wave shaping is employed.

Timing of the gate signals must be precisely controlled. This is another function of the control shown in Figure 1.7. To adapt the system satisfactorily for welding service, another feature, feedback, is necessary. The nature of the feedback depends on the parameter to be controlled and the degree of control required. To provide constant-voltage characteristics, the feedback must consist of some signal that is proportional to arc voltage. That signal controls the precise arc voltage at any instant so that the control can properly time and sequence the initiation of the SCR to hold the preset voltage. The same effect is achieved with constant current by using a current reference.

Most commercial SCR phase controlled welding power supplies are three-phase machines; either the constant-current or constant-voltage type. Such power supplies have distinct features because the output characteristics are controlled electronically. For example, automatic line-voltage compensation is very easily accomplished, allowing welding power to be held precisely as set, even if the input line voltage varies. Also, volt-ampere curves can be shaped and tailored for a particular welding process or its application. Such machines can adapt their static characteristic to any welding process from one approaching a truly constant voltage to one having a virtually constant current. Other capabilities are pulsing, controlled current with respect to arc voltage, controlled arc voltage with respect to current, and a high initial current or voltage pulse at the start of the weld.

An SCR can also serve as a secondary contactor, allowing welding current to flow only when the control allows the SCRs to conduct. This is a useful feature in rapid cycling operations, such as spot and tack welding. However, an SCR contactor does not provide the electrical isolation that a mechanical contactor or switch provides. Therefore, a primary circuit breaker, or some other device, is required to provide isolation for electrical safety.

Several SCR configurations can be used for arc welding. Figure 1.9 shows a three-phase bridge with six SCR de-

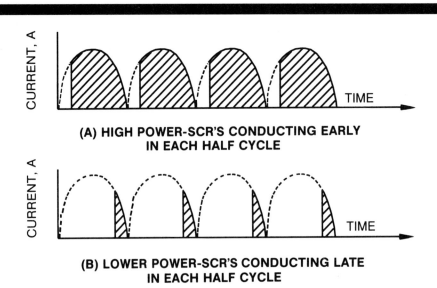

(A) HIGH POWER-SCR'S CONDUCTING EARLY
IN EACH HALF CYCLE

(B) LOWER POWER-SCR'S CONDUCTING LATE
IN EACH HALF CYCLE

Figure 1.8—Phase Control Using an SCR Bridge

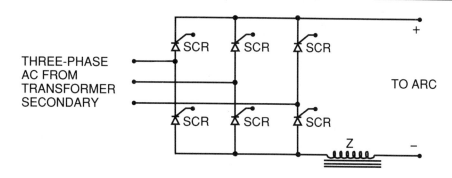

Figure 1.9–Three-Phase Bridge Using Six SCRs (Full Wave Control)

vices. With a 60 Hz main frequency this arrangement produces a 360 Hz ripple frequency under load. It also provides precise control and quick response; in fact, each half-cycle of the three-phase output is controlled separately. Dynamic response is enhanced because it reduces the size of the inductor needed to smooth out the welding current.

Figure 1.10 diagrams a three-phase bridge rectifier with three diodes and three SCRs. Because of greater current ripple this configuration requires a larger inductor than the six SCR unit. For that reason it has a slower dynamic response. A fourth diode, called a *freewheeling diode*, can be added to recirculate the inductive currents from the inductor so that the SCRs will turn off, i.e., commutate. This offers greater economy over the six SCR unit because it uses fewer SCRs and a lower cost control unit.

TRANSISTORS

THE TRANSISTOR IS another solid-state device which is used in welding power supplies. Because of their cost, the use of transistors is limited to power supplies requiring precise control of a number of variables. The transistor differs from the SCR in several ways. One is that conduction through the device is proportional to the control signal applied. With no signal, there is no conduction. When a small signal is applied, there is a corresponding small conduction; with a large signal, there is a correspondingly large conduction. Unlike the SCR, the control can turn off the device without waiting for polarity reversal or an "off" time. Since transistors do not have the current-carrying capacity of SCRs, several may be required to yield the output of one SCR.

Several methods can be used to take advantage of transistors in welding power supplies, such as frequency modulation and pulse width modulation. With frequency modulation, the welding current is controlled by varying the frequency supplied to the main transformer. Since the frequency is changing, the response time varies also. The size of the transformer and inductor must be optimized for the lowest operating frequency.

With pulse width modulation, welding output is controlled by varying the conduction time of the switching device. Since the frequency is constant, the response time is constant and the magnetics can be optimized for one operating frequency.

SOLID-STATE INVERTER

THE PRIMARY CONTRIBUTORS to weight/mass in any power source are the magnetics (main transformer and fil-

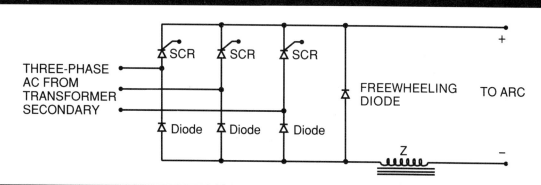

Figure 1.10–Three-Phase Hybrid Bridge Using Three SCRs and Four Diodes (Half-Wave Control)

ter inductor). Various attempts have been made to reduce their weight and size, for example, the substitution of aluminum windings for copper. The use of an inverter circuit can produce significant reductions in size and weight of these components as well as decrease their electrical losses. An inverter-based power source is smaller, more compact, requires less electricity than conventional welding power sources, and offers a faster response time.

An inverter is a circuit which uses solid-state devices (SCR's or transistors) to convert DC into high-frequency AC, usually in the range of 1 kHz to 50 kHz. Conventional welding power sources use transformers operating from a line frequency of 50 or 60 Hz. Since transformer size is inversely proportional to line or applied frequency, reductions of up to 75 percent in power source size and weight are possible using inverter circuits.

Inverter circuits control the output power using the principle of time ratio control (TRC). The solid-state devices (semiconductors) in an inverter act as switches; they are either switched "on" and conducting, or they are switched "off" and blocking. This operation of switching "on" and "off" is sometimes referred to as *switch mode operation*. TRC is the regulation of the "on" and "off" times of the switches to control the output. Figure 1.11 illustrates a simple TRC circuit which controls the output to a load such as a welding arc.

When the switch is on, the voltage out (V_{out}) equals voltage in (V_{in}); when the switch is off, $V_{out} = 0$. The average value of V_{out} is as follows:

$$V_{out} = \frac{t_{on} \cdot V_{in} + 0 \cdot t_{off}}{t_{on} + t_{off}} \text{ or } \frac{V_{in} \cdot t_{on}}{t_{on} + t_{off}} \qquad (1.2)$$

Thus: $V_{out} = V_{in} \cdot \dfrac{t_{on}}{t_p}$

where
$$t_{on} = \text{on time (conducting)}$$
$$t_{off} = \text{off time (blocking)}$$
$$t_p = t_{on} + t_{off} \text{ or time of 1 cycle}$$

V_{out} is controlled by regulating the time ratio t_{on} / t_p.

Since the on/off cycle is repeated for every t_p interval, the frequency (f) of the on/off cycles is defined as:

$$f = \frac{1}{t_p} \qquad (1.3)$$

Thus, the TRC formula can now be written as:

$$V_{out} = V_{in} \cdot t_{on} \cdot f$$

The TRC formula written in this manner points to two methods of controlling an inverter welding power source. By varying t_{on} the inverter uses pulse-width modulated TRC. Another method of inverter control called *frequency modulation TRC* varies f. Both frequency modulation and pulse width modulation have been used in commercially available welding inverters.

Figure 1.12 is a block diagram of an inverter used for dc welding. Incoming three-phase or single-phase 50/60 HZ power is converted to dc by a full wave rectifier. This dc is applied to the inverter which, using semiconductor switches, inverts it into high-frequency square wave ac. In another variation used for welding, the inverter produces sine waves in a resonant technology, with frequency modulation control. The switching of the semiconductors takes place between 1 kHz and 50 kHz depending on the component used and method of control.

This high-frequency voltage allows the use of a smaller step down transformer. After being transformed, the ac is

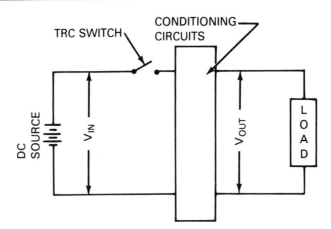

Figure 1.11–Simplified Diagram of an Inverter Circuit Used to Demonstrate the Principle of Time Ratio Control.(Note that conditioning circuits include components such as transformer, rectifier and inductor shown in Figure 1.7.)

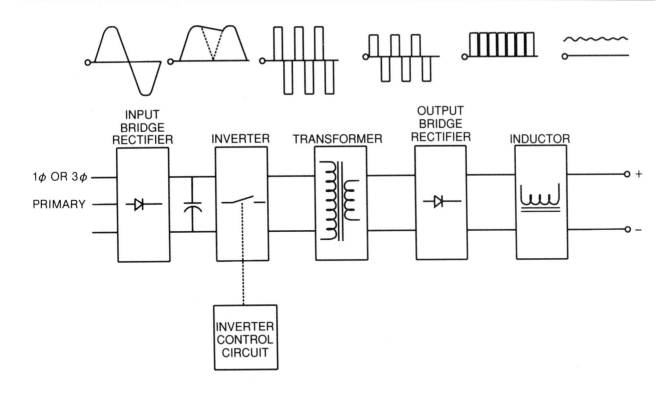

Figure 1.12–Inverter Diagram Showing Power Supply Sections and Voltage Wave Forms

rectified to dc for welding. Solid-state controls enable the operator to select either constant-current or constant-voltage output and, with appropriate options, these sources can provide pulsed outputs.

The capabilities of the semiconductors and particular circuit topology determine response time and switching frequency. Faster response times are generally associated with the higher switching and control frequencies, resulting in more stable arcs and superior arc performance. However, other variables, such as length of weld cables, must be considered since they may affect the power supply performance. Table 1.1 compares inverter switching devices and the frequency applied to the transformer.

Inverter technology can also be used to produce enhanced performance ac welding power sources. Another application is dc constant-current power sources used for plasma cutting.

Table 1.1
Types of Interver Switching Devices and the Frequency Applied to the Transformer

Switching Device	Frequency Range
SCR-type Devices	1 KHz to 10 KHz
Transistor-type Devices	10 KHz to 100 KHz

VOLT-AMPERE CHARACTERISTICS

THE EFFECTIVENESS OF all welding power sources is determined by two kinds of operating characteristics, each of which affects welding performance differently. They are defined as the static and dynamic characteristics. Both affect arc stability, but in different ways depending on the welding process.

Static output characteristics are readily measured under steady-state conditions by conventional test procedures using resistive loads. A set of output-voltage versus output-current characteristic curves (volt-ampere curves) are usually used to describe the static characteristics.

The dynamic characteristic of an arc welding power source is determined by measuring the transient variations in output current and voltage that appear in the arc. Dynamic characteristics describe instantaneous variations or those that occur during very short intervals, such as 0.001 second.

Most welding arcs operate in continually changing conditions. In particular, transients occur (1) during the striking of the arc, (2) during rapid changes in arc length, (3) during the transfer of metal across the arc, and (4) in the case of ac welding, during arc extinction and reignition at each half-cycle.

These arc transients can occur in 0.001 second, the time interval during which a significant change in ionization of the arc column occurs. The power source must respond rapidly to these demands. For this reason, it is important to control the dynamic characteristics of an arc welding power source. The steady-state or static volt-ampere characteristics have little significance in determining dynamic characteristics of an arc welding system.

Among the arc welding power supply design features that do have an effect on dynamic characteristics are those that provide:

(1) Local transient energy storage, such as parallel capacitance circuits or dc series inductance
(2) Feedback controls in automatically regulated systems
(3) Modifications of waveform or circuit-operating frequencies

An improvement in arc stability is typically the goal for modifying or controlling these characteristics. Beneficial results include:

(1) Improvement in the uniformity of metal transfer
(2) Reduction in metal spatter
(3) Reduced weld pool turbulence

Static volt-ampere characteristics are generally published by the power supply manufacturer. There is no universally recognized method by which dynamic characteristics are specified. The user should obtain assurance from the manufacturer that both the static and dynamic characteristics of the power supply are acceptable for the intended application.

CONSTANT CURRENT

TYPICAL VOLT-AMPERE (V-A) output curves for a conventional constant-current power source are shown in Figure 1.13. It is sometimes called a *drooper* because of the substantial downward (negative) slope of the curves. The power source might have open circuit voltage adjustment in addition to output current control. A change in either control will change the slope of the volt-ampere curve.

The effect of the slope of the V-A curve on power output is shown in Figure 1.13. With curve A, which has an 80 V open circuit, a steady increase in arc voltage from 20 to 25 V (25 percent) would result in a decrease in current from 123 to 115 A (6.5 percent). The change in current is relatively small. Therefore, with a consumable electrode welding process, electrode melting rate would remain fairly constant with a slight change in arc length.

Setting the power source for 50 V open circuit and more shallow slope intercepting the same 20 V, 123 A position will give volt-ampere curve B. In this case, the same increase in arc voltage from 20 to 25 V would decrease the current from 123 to 100 A (19 percent), a significantly greater change. In manual welding, the flatter V-A curve would give a skilled welder the opportunity to substantially vary the current by changing the arc length. This could be useful for out-of-position welding because a welder could control the electrode melting rate and molten pool size. Generally, however, less skilled welders would prefer the current to stay constant should the arc length change.

Current control is used to provide lower output. It would result in volt-ampere curves with greater slope, as illustrated by curves C and D. They offer the advantage of more nearly constant-current output, allowing greater changes in voltage with minor changes in current.

CONSTANT VOLTAGE

A TYPICAL VOLT-AMPERE curve for a conventional constant-voltage power source is shown in Figure 1.14. This power source does not have true constant-voltage output. It has a slightly downward (negative) slope because internal electrical impedance in the welding circuit causes a minor voltage droop in the output. Changing that impedance will alter the slope of the volt-ampere curve.

Starting at point B in Figure 1.14, the diagram shows that an increase or decrease in voltage to A or C (5 V or 25 percent) produces a large change in amperage (100 A or 50 percent). This V-A characteristic is suitable for constant feed electrode processes, such as gas metal arc, submerged arc, and flux cored arc welding, in order to maintain a constant arc length. A slight change in arc length (voltage) will cause a fairly large change in welding current. This will automatically increase or decrease the electrode melting rate to regain the desired arc length (voltage). This effect has been called *self regulation*. Adjustments are sometimes provided with constant-voltage power sources to change or modify the slope or shape of the V-A curve. If done with inductive devices, the dynamic characteristics will also change.

The curve shown in Figure 1.14 can also be used to explain the difference between static and dynamic power supply characteristics. For example, during GMAW short circuiting transfer, the welding electrode tip touches the weld pool, causing a short circuit. At this point, the arc voltage approaches zero and only the circuit resistance or inductance limits the rapid increase of current. If the power supply responded instantly, very high current would immediately flow through the welding circuit,

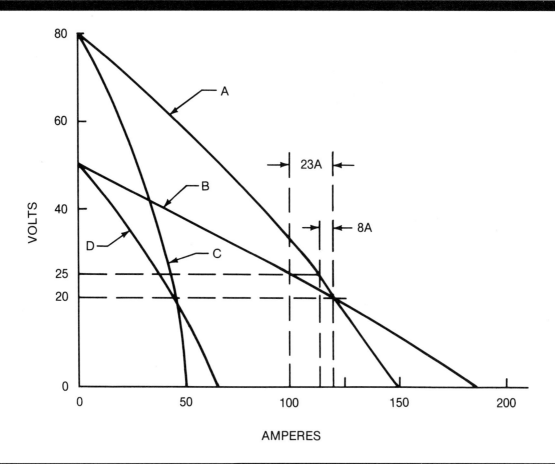

Figure 1.13–Typical Volt-Ampere characteristics of a "Drooping" Power Source With Adjustable Open Circuit Voltage

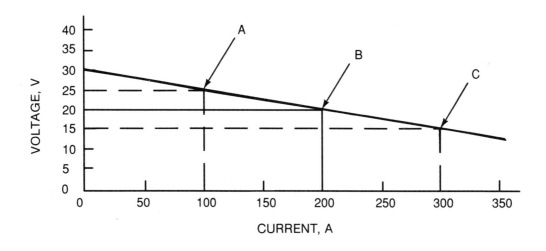

Figure 1.14–Volt-Ampere Output Relationship for a Constant Voltage Power Source

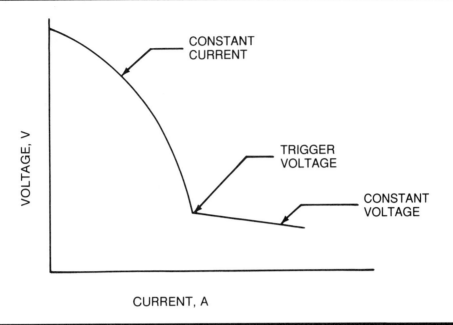

Figure 1.15–Combination Volt-Ampere Curve

quickly melting the short-circuited electrode and freeing it with an explosive force, dispelling the weld metal as spatter. Dynamic characteristics designed into this power source compensate for this action by limiting the rate of current change — thereby decreasing the explosive force.

COMBINED CONSTANT-CURRENT AND CONSTANT-VOLTAGE CHARACTERISTICS

ELECTRONIC CONTROLS CAN be designed to provide either a constant-voltage or a constant-current output from a single power source. Such a power supply can be used for a variety of welding processes.

Electronically controlled outputs can also provide output curves that are a combination of constant current and constant voltage as shown in Figure 1.15. The top part of the curve is essentially constant current; below a certain trigger voltage, however, the curve switches to constant voltage. This type of curve is beneficial for SMAW to assist starting and to avoid electrode stubbing (sticking in the puddle) should a welder use too short an arc length.

DUTY CYCLE

INTERNAL COMPONENTS OF a welding power supply tend to heat up as welding current flows through the unit. The amount of heat tolerated is determined by (1) the breakdown temperature of the electrical components, and (2) the media used to insulate the transformer windings and other components. These maximum temperatures are specified by component manufacturers and organizations involved with standards in the field of electrical insulation.

Fundamentally, the *duty cycle* is a ratio of the load-on time allowed to a specified test interval time. Observing this ratio is important in order that the internal windings and components and their electrical insulation system will not be heated above their rated temperature. These maxi-

mum temperature criteria do not change with the duty cycle or current rating of the power supply.

Duty cycle expresses, as a percentage, the maximum time that the power supply can deliver its rated output during each of a number of successive test intervals without its temperature exceeding a predetermined limit. In the United States, NEMA duty cycles are based on a test interval of 10 minutes. Some agencies and manufacturers in other countries use shorter test intervals such as 5 minutes. Thus, a 60 percent NEMA duty cycle (a standard industrial rating) means that the power supply can deliver its rated output for 6 out of every 10 minutes without overheating. (Note that uninterrupted operation at "rated load" for 36 minutes out of one hour does not constitute a 60 percent duty cycle but 100 percent). A 100 percent duty-cycle power supply is designed to produce its rated output continuously without exceeding the prescribed temperature limits of its components.

In the past, for very high-current power supplies (750 A and higher), a one-hour duty-cycle rating was also used. Some manufacturers still provide power supplies with a one-hour duty rating. To determine the rated output of these machines, they are loaded for one hour at rated output. Then the output is reduced immediately to 75 percent of the rated current value, and sustained for an additional three hours. Component temperatures are measured at the end of the first one-hour period and at the conclusion of the test. These temperatures have to be within established allowable limits.

Duty cycle is a major factor in determining the type of service for which a power supply is designed. Industrial units designed for manual welding normally are rated at 60 percent duty cycle. For automatic and semiautomatic processes, the rating is usually 100 percent duty cycle. Light-duty power supplies are usually 20 percent duty cycle. Ratings at other duty-cycle values are available from the manufacturers.

An important point is that the duty cycle of a power supply is based on the output current and not on a kilovolt-ampere or kilowatt rating. Manufacturers perform duty-cycle tests under what NEMA defines as usual service conditions. Caution should be observed in basing operation on other than usual service conditions. Factors which contribute to lower than tested or calculated performance include high ambient temperatures, insufficient cooling air quantity, and low line voltage.

The following formulas are given for estimating the duty cycle at other than rated output (1.4), and for estimating

other than rated output current at a specified duty cycle (1.5):

$$T_a = \left(\frac{I}{I_a}\right)^2 \times T \tag{1.4}$$

$$I_a = I \times \left(\frac{T}{T_a}\right)^{1/2} \tag{1.5}$$

where

T = rated duty cycle in percent
T_a = the required duty cycle in percent
I = rated current at rated duty cycle
I_a = the maximum current at required duty cycle.

The power supply should never be operated above its rated current or duty cycle unless approved by the manufacturer.

Example: At what duty cycle can a 200 A power supply rated at 60 percent duty cycle be operated at 250 A output? Using equation (1.4):

$$T_a = \left(\frac{200}{250}\right)^2 \times 60\% = (.8)^2 \times .6 = 38\% \tag{1.6}$$

Therefore, this unit must not be operated more than 3.8 minutes out of each 10 minute period at 250 A. If used in this way, welding at 250 A should not exceed the current rating of any power supply component.

Example: The aforesaid power supply is to be operated continuously (100 percent duty cycle). What output current must not be exceeded? Using equation (1.5):

$$I_a = 200 \times \left(\frac{60}{100}\right)^{1/2} = 200 \times .775 = 155 \text{ amps} \tag{1.7}$$

If operated continuously, the current should be limited to 155 A output.

OPEN CIRCUIT VOLTAGE

OPEN CIRCUIT VOLTAGE is the voltage at the output terminals of a welding power source when it is energized, but current is not being drawn. Open circuit voltage is one of the design factors influencing the performance of all welding power sources. In a transformer, open circuit voltage is a function of the primary input voltage and the ratio of primary-to-secondary coils. Although a high open circuit voltage may be desirable from the standpoint of arc initiation and stability, the electrical hazard factors associated with high voltages preclude values higher than those which are used. Cost is another but secondary factor.

The open circuit voltage of generators or alternators is related to design features such as the strength of the magnetic field, the speed of rotation, the number of turns in the load coils, etc. These power sources generally have controls with which the open circuit voltage can be varied.

NEMA Standard EW-1 contains specific requirements for maximum open circuit voltage. When the rated line voltage is applied to the primary winding of a transformer, or when a generator-type arc welding machine is operating at maximum-rated no-load speed, the open circuit voltages are limited to the levels shown in Table 1.2.

NEMA Class I and Class II power sources normally have open circuit voltage at, or close to, the maximum specified. Class III arc power sources frequently provide two or more open circuit voltages. One arrangement is to have a high and low range of amperage output from the power source. The low range normally has approximately 80 V open circuit, with the high range somewhat lower. Another arrangement is the tapped secondary coil method, described earlier, in which, at each current setting, the open circuit voltage changes about 2 to 4 volts.

Table 1.2
Maximum Open Circuit Voltages for Various Types of Arc Welding Power Sources

For Manual and Semiautomatic Applications	
Alternating current	80 V rms
Direct current—over 10% ripple voltage*	80 V rms
Direct current—10% or less ripple voltage*	100 V avg.
For Automatic Applications	
Alternating current	100 V rms
Direct current—over 10% ripple voltage*	100 V rms
Direct current—10% or less ripple voltage	100 V avg

$$* \text{ Ripple voltage, } \% = \frac{\text{ripple voltage, rms}}{\text{avg. total voltage}}$$

In this country, both transformer and alternator ac power sources produce reverses in the direction of current flow each 1/120 second (60 Hz). Typical sine wave forms of a dual-range machine with open circuit voltages of 80 and 55 V rms are diagrammed in Figure 1.16.

Since the current must change direction after each half-cycle, it is apparent that, for an instant, at the point at which the current wave form crosses the zero line, the current flow in the arc ceases. An instant later, the current must reverse its direction of flow. However, during the period in which current decreases and reaches zero, the arc plasma cools, reducing ionization of the arc stream.

Welding current cannot be re-established in the opposite direction unless ionization within the arc gap is either maintained or quickly reinitiated. With conventional power supplies, ionization may not be sustained depending on the welding process and electrode being used. Its reinitiation is improved by providing an appropriately high voltage across the arc gap: the recovery voltage. The greater this recovery voltage, the shorter the period during which the arc is extinguished. If insufficient, the arc cannot be re-established without shorting the electrode.

Figure 1.16 shows the phase relations between open circuit voltage and equal currents for two different open circuit voltages, assuming the same arc voltage (not shown) in each case. As can be seen in this figure, the available peak voltage of 113 V is greater with 80 V (rms) open circuit. Peak voltage of 78 V available with 55 V (rms) open circuit may not be sufficient to sustain a stable arc. The greater phase shift shown for the low-range condition causes a current reversal at a higher recovery voltage because it is near the peak of the open circuit voltage wave form, the best condition for reignition. Resistance is not used to regulate alternating welding current because the power source voltage and current would be in phase. Since the recovery voltage would be zero during current reversal, it would be impossible to maintain a stable arc.

For SMAW with low open circuit voltage machines, it is necessary to incorporate ingredients in electrode coverings which help maintain ionization and provide favorable

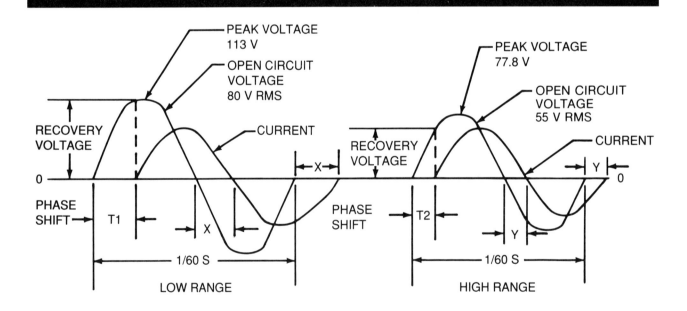

NOTE: T1 IS GREATER THAN T2

Figure 1.16–Typical Voltage and Current Waves of a Dual Range AC Power Source

metal transfer characteristics to prevent sudden, gross increases in the arc length.

In a dc system, once the arc is established, the welding current does not pass through zero. Thus, rapid voltage increase is not critical; resistors are suitable current controls for dc machines. However, with some processes, direct current power sources must function in much the same way with respect to the need to provide open circuit voltage when the arc length changes abruptly. Often reactance or inductance is built into these power sources to enhance this effect.

NEMA POWER SOURCE REQUIREMENTS

THE NATIONAL ELECTRICAL Manufacturers Association (NEMA) publication EW-1 (latest revision) covers the requirements for electric arc welding apparatus including power sources.

NEMA CLASSIFICATIONS

NEMA CLASSIFIES ARC welding power sources primarily on the basis of duty cycle. There are three classes:

(**1**) "A NEMA Class I arc welding machine is characterized by its ability to deliver rated output at duty cycles of 60, 80, or 100 percent. If a power source is manufactured in accordance with the applicable standards for Class I machines, it shall be marked 'NEMA Class I (60)', 'NEMA Class I (80)', or 'NEMA Class I (100)'."

(**2**) "A NEMA Class II arc welding machine is characterized by its ability to deliver rated output at duty cycles of 30, 40, or 50 percent. If a machine is manufactured in accordance with the applicable standards for Class II machines, it shall be marked 'NEMA Class II (30)', 'NEMA Class II (40)', or 'NEMA Class II (50)'."

(**3**) "A NEMA Class III arc welding machine is characterized by its ability to deliver rated output at a duty cycle of 20 percent. If a machine is manufactured in accordance with the applicable standards for Class III machines, it shall be marked 'NEMA Class III (20)'."

NEMA Class I and II power supplies are further defined as completely assembled arc welding power supplies which are comprised of the characteristics of the following:

(**1**) A constant-current, a constant-voltage, or a constant-current/constant-voltage machine
(**2**) A single-operator machine
(**3**) One of the following:
(**a**) DC generator arc welding power source
(**b**) AC generator arc welding power source
(**c**) DC generator-rectifier arc welding power source
(**d**) AC/DC generator-rectifier arc welding power source
(**e**) AC transformer arc welding power source
(**f**) DC transformer-rectifier arc welding power source

(**g**) AC/DC transformer-rectifier arc welding power source

OUTPUT AND INPUT REQUIREMENTS

IN ADDITION TO duty cycle, NEMA specifies the output ratings and performance capabilities of power supplies of each class. Table 1.3 gives the output current ratings (size) for Class I, II, and III arc welding machines. The rated load volts (E) for Class I and II machines under 500 A can be calculated using $E = 20 + 0.04I$, where I is the rated load current. For sizes of 600 A and higher ratings, the rated load voltage is 44. The output ratings in amperes and load volts and also the minimum and maximum output currents and load volts for power supplies are given in NEMA publication EW-1 (latest edition).

The electrical input requirements of NEMA Class I and II transformer arc welding machines for 50 and 60 Hz are the following:

60 Hz: 200, 230, 460, and 575 V
50 Hz: 220, 380, and 440 V

For NEMA Class III transformer-type arc welding machines, the electrical input requirement is 60 Hz, 230 V. The transformer primary windings are usually tapped to permit selection of two or three alternate voltage supplies, such as 200, 230, and 460 V.

Table 1.3
NEMA-Rated Output Current for Arc Welding Machines

Rated Output current, A		
Class I	Class II	Class III
200	150	180-230
250	175	235-295
300	200	
400	225	
500	250	
600	300	
800	350	
1000		
1200		
1500		

Table 1.4
Typical Nameplate Specifications for an AC–DC Arc Welding Power Source

Model	Welding Current, A				Open Circuit Voltage AC & DC	Amperes Input at Rated Load Output—60 Hz Single Phase			
	AC		DC			Amperes			
	Gas Tungsten Arc	Shielded Metal Arc	Gas Tungsten Arc	Shielded Metal Arc		230 V	460 V	kVA	kW
300 Ampere	5-48 20-230 190-435	5-48 20-245 200-460	5-60 20-250 200-460	5-45 16-200 150-350	80	104	52	23.9	21.8

Table 1.5
Typical Primary Conductor and Fuse Size Recommendations

Model	Input Wire Size, AWG (a)				Fuse Size in Amperes			
	200 V	230 V	460 V	575 V	200 V	230 V	460 V	575 V
300 A	No. 2 (No. 6)(b)	No. 2 (No. 6)(b)	No. 8 (No. 8)(b)	No. 8 (No. 8)(b)	200	175	90	70

a. American wire gage

b. Indicates ground conductor size

The voltage and frequency standards for welding generator drive motors are the same as for NEMA Class I and II transformer primaries.

NAMEPLATE DATA

THE MINIMUM NAMEPLATE data for an arc welding power supply specified in NEMA publication EW-1 is the following:

(1) Manufacturer's type designation or identification number, or both
(2) NEMA class designation
(3) Maximum open circuit voltage (OCV)
(4) Rated load volts
(5) Rated load amperes
(6) Duty cycle at rated load
(7) Maximum speed in rpm at no-load (generator or alternator)
(8) Frequency of power source
(9) Number of phases of power supply
(10) Input voltage(s) of power supply
(11) Amperes input at rated load output

The instruction book supplied with each power source is the prime source of data concerning electrical input requirements. General data is also given on the power source nameplate, usually in tabular form along with other pertinent data that might apply to the particular unit. Table 1.4 shows typical information for a NEMA-rated 300 A GTAW power supply. The welding current ranges are given with respect to welding process. The power source may use one of two input voltages with the corresponding input current listed for each voltage when the machine is producing its rated load. The kilovolt-ampere (kVA) and kilowatt (kW) input data are also listed. The power factor, pf, can be calculated as follows:

$$pf = \frac{kW}{kVA} \qquad (1.8)$$

The manufacturer will also provide other useful data concerning input requirements, such as primary conductor size and recommended fuse size. Power sources cannot be protected with fuses of equal value to their primary current demand. If this is done, nuisance blowing of the fuses or tripping of circuit breakers will result. Table 1.5 shows typical fuse and input wire sizes for the 300 A power source of Table 1.4. All pertinent codes should be consulted in addition to these recommendations.

ALTERNATING-CURRENT POWER SOURCES

TRANSFORMER POWER SOURCES

ALTERNATING-CURRENT POWER sources normally are single-phase transformers that connect to ac building power lines and transform the input voltage and amperage to levels suitable for arc welding. The transformers also serve to isolate the welding circuits from the plant power lines. Because various welding applications have different welding power requirements, means for control of welding current or arc voltage, or both, must be incorporated within the welding transformer power source. The methods commonly used to control the welding circuit output are described in the following sections.

Movable-Coil Control

A MOVABLE-COIL transformer consists essentially of an elongated core on which are located primary and secondary coils. Either the primary coil or the secondary coil may be movable, while the other one is fixed in position. Most ac transformers of this design have a fixed-position secondary coil. The primary coil is normally attached to a lead screw and, as the screw is turned, the coil moves closer to or farther from the secondary coil.

The varying distance between the two coils regulates the inductive coupling of the magnetic lines of force between them. The farther the two coils are apart, the more vertical is the volt-ampere output curve and the lower the maximum short-circuit current value. Conversely, when the two coils are closer together, the maximum short-circuit current is higher and the slope of the volt-ampere output curve is less steep.

Figure 1.17A shows one form of a movable-coil transformer with the coils far apart for minimum output and a steep slope of the volt-ampere curve. Figure 1.17B shows the coils as close together as possible. The volt-ampere curve is indicated at maximum output with less slope than the curve of Figure 1.17A.

Another form of moveable coil employs a pivot motion. When the two coils are at a right angle to each other, output is at a minimum. When the coils are aligned with one coil nested inside the other, output is at maximum.

Movable-Shunt Control

IN THIS DESIGN, the primary coils and the secondary coils are fixed in position. Control is obtained with a laminated iron core shunt that is moved between the primary and secondary coils. It is made of the same material as that used for the transformer core.

As the shunt is moved into position between the primary and secondary coils, as shown in Figure 1.18A, some magnetic lines of force are diverted through the iron shunt rather than to the secondary coils. With the iron shunt between the primary and secondary coils, the slope of the volt-ampere curve increases and the available welding current is decreased. Minimum current output is obtained when the shunt is fully in place.

As illustrated in Figure 1.18B, the arrangement of the magnetic lines of force, or magnetic flux, is unobstructed when the iron shunt is separated from the primary and secondary coils. Here the output current is at its maximum.

Tapped Secondary Coil Control

A TAPPED SECONDARY coil may be used for control of the volt-ampere output of a transformer as shown in Figure 1.3. This method of adjustment is often used with NEMA Class III power sources, commonly called *farm welders*. They are the least expensive and most universally used of all welding power supplies. Basic construction is somewhat similar to the movable-shunt type, except that the shunt is permanently located inside the main core and the secondary coils are tapped to permit adjustment of the number of turns. Decreasing secondary turns reduces open circuit voltage and, also, the inductance of the transformer, causing welding current to increase.

Movable-Core Reactor

THE MOVABLE-CORE reactor type of ac welding machine consists of a constant-voltage transformer and a reactor in series. The inductance of the reactor is varied by mechanically moving a section of its iron core. The machine is diagramed in Figure 1.19. When the movable section of the core is in a withdrawn position, the permeability of the magnetic path is very low due to the air gap. The result is a low inductive reactance that permits a large welding current to flow. When the movable-core section is advanced into the stationary core, as shown by broken lines in Figure 1.19, the increase in permeability causes an increase in inductive reactance and, thus, welding current is reduced.

Saturable Reactor Control

A SATURABLE REACTOR control is an electrical control which uses a low voltage, low amperage dc circuit to change the effective magnetic characteristics of reactor cores. A self-saturating saturable reactor is referred to as a *magnetic amplifier* because a relatively small control power change will produce a sizeable output power change. This type of control circuit makes remote control of output from the power source relatively easy, and it normally requires less maintenance than the mechanical controls.

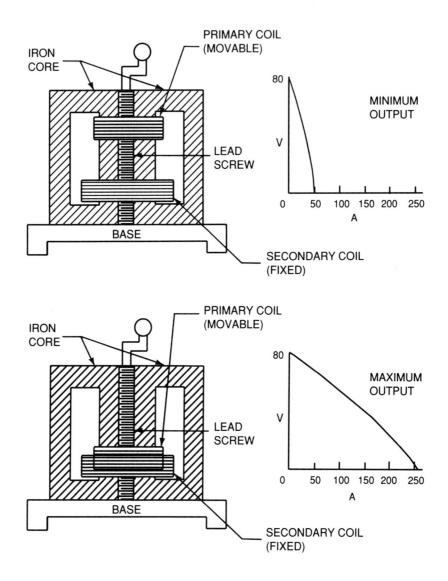

Figure 1.17—Movable-Coil AC Power Source

With this construction, the main transformer has no moving parts. The volt-ampere characteristics are determined by the transformer and the saturable-reactor configurations. The dc control circuit to the reactor system allows adjusting the output volt-ampere curve from minimum to maximum.

A simple, saturable-reactor power source is diagramed in Figure 1.20. The reactor coils are connected in opposition to the dc control coils. If this were not done, transformer action would cause high circulating currents to be present in the control circuit. With the opposing connection, the instantaneous voltages and currents tend to can-

cel out. Saturable reactors tend to cause severe distortion of the sine wave supplied by the transformer. This is not desireable for gas tungsten arc welding because the wave form for that process is important. One method of reducing this distortion is by introducing an air gap in the reactor core. Another is to insert a large choke in the dc control circuit. Either method, or a combination of both, will produce desirable results.

The amount of current adjustment in a saturable reactor is based on the ampere-turns of the various coils. The term *ampere-turns* is defined as the number of turns in the coil multiplied by the current in amperes flowing through the

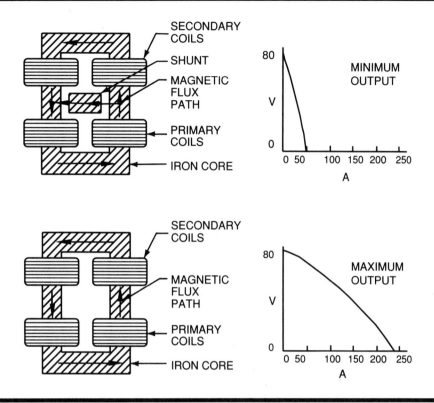

Figure 1.18–Movable-Shunt AC Power Source

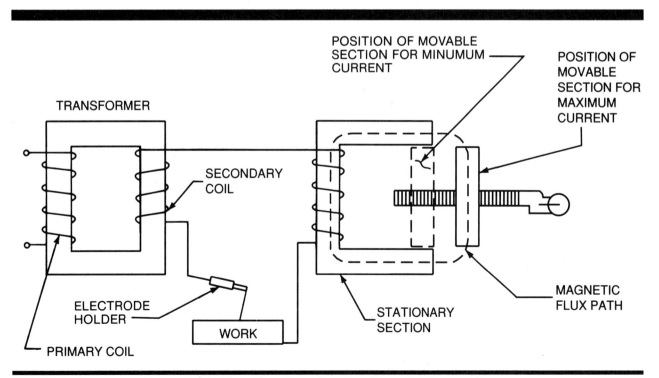

Figure 1.19–Movable-Core Reactor Type AC Power Source

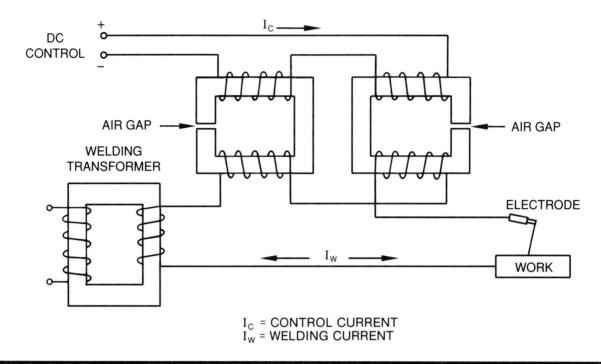

I_C = CONTROL CURRENT
I_W = WELDING CURRENT

Figure 1.20–Saturable Reactor Type AC Welding Power Source

coil. In the basic saturable reactor, the law of equal ampere-turns applies. To increase output in the welding circuit, a current must be made to flow in the control circuit. The amount of change can be approximated with the following equation:

$$I_w = \frac{I_c N_c}{N_w} \tag{1.9}$$

where

I_w = change in welding current, A
I_c = change in current, A, in the control circuit
N_c = number of turns in the control circuit
N_w = number of turns in the welding current circuit

The minimum current of the power source is established by the number of turns in the welding current reactor coils and the amount of iron in the reactor core. For a low minimum current, either a large amount of iron or a relatively large number of turns, or both, are required. If a large number of turns are used, then either a large number of control turns or a high control current, or both, are necessary. To reduce the requirement for large control coils, large amounts of iron, or high control currents, the saturable reactors often employ taps on the welding current coils, creating multi-range machines. The higher ranges would have fewer turns in these windings and, thus, correspondingly higher minimum currents.

Magnetic Amplifier Control

TECHNICALLY, THE MAGNETIC amplifier is a self-saturating saturable reactor. It is called a *magnetic amplifier* because it uses the output current of the power supply to provide additional magnetization of the reactors. In this way, the control currents can be reduced and control coils can be smaller. While magnetic amplifier machines often are multi-range, the ranges of control can be much broader than those possible with an ordinary saturable reactor control.

Referring to Figure 1.21, it can be seen that by using a different connection for the welding current coils and rectifying diodes in series with the coils, the load ampere-turns are used to assist the control ampere-turns in magnetizing the cores. A smaller amount of control ampere-turns will cause a correspondingly larger welding current to flow because the welding current will essentially "turn itself on." The control windings are polarity sensitive.

Power Factor

CONSTANT-CURRENT AC power sources are characterized by low power factors due to their large inductive reactance. This is often objectionable because the line currents are high with heavy loads and rate penalties are charged to most industrial users for low power factor. Power factor

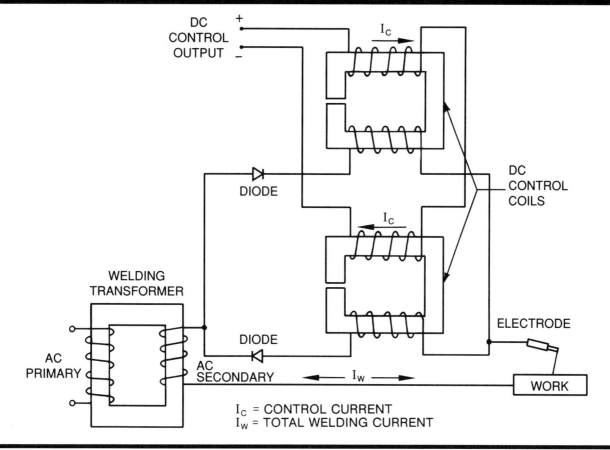

DC CONTROL OUTPUT +

I_C

DC CONTROL COILS

DIODE

I_C

WELDING TRANSFORMER

ELECTRODE

AC PRIMARY

DIODE

AC SECONDARY

I_W

WORK

I_C = CONTROL CURRENT
I_W = TOTAL WELDING CURRENT

Figure 1.21–Magnetic Amplifier Welding Current Control

may be improved by adding capacitors to the primary circuit of inductive loads such as welding transformers. This reduces the primary current from power lines while welding is being performed. Unfortunately, the current draw under light or no-load conditions will increase.

Large alternating-current-transformer power sources may be equipped with capacitors for power-factor correction to approximately 75 percent at rated load. At lower-than-rated load-current settings, the power factor may have a leading characteristic. When the transformer is operating at no-load or very light loads, the capacitors are drawing their full corrective kVA, thus contributing power-factor correction to the remainder of the load on the total electrical system.

When several transformer type welding sources are operating at light loads, care should be taken that the combined power-factor correction capacitance will not upset the voltage stability of the line. If three-phase primary power is used, the load on each phase of the primary system should be balanced for best performance. Power-factor correction, under normal conditions, has no bearing on welding performance.

Auxiliary Features

CONSTANT-CURRENT AC power sources are available in many configurations with respect to their auxiliary features. Generally, these features are incorporated to better adapt the unit to a specific process or application, or to make it more convenient to operate. The manufacturer should be consulted for available features when considering such power sources.

Primary contactors or manually operated power switches are usually included in ac power sources to turn the unit on and off. Most NEMA Class I and Class II units are furnished with a terminal board or other means for connection of various rated primary-line voltages. Input supply cords are not normally supplied with NEMA Class I and Class II welding power sources. The smaller NEMA Class III power sources are generally equipped with a manually operated primary switch and an input supply cord.

Some ac power sources incorporate a system for supplying a higher-than-normal current to the arc for a fraction of a second at the start of a weld. This "hot start" feature provides starting surge characteristics, similar to those of motor-generator sets, to assist in initiating the arc, particularly at current levels under 100 A. Other power sources may be equipped with a start control to provide an adjustable "soft" start or reduced current start, to minimize transfer of tungsten from the electrode with the GTAW process.

Equipment designed for the gas tungsten arc welding process usually incorporates electrically operated valves and timers as well to control the flow of shielding gas and coolant to the electrode holder. High-frequency units may be added to assist in starting and stabilizing the ac arc.

NEMA Class I and Class II power sources may be provided with means for remote adjustment of output power. This may consist of a motor-driven device for use with crank-adjusted units or a hand control at the work station when an electrically adjusted power supply is being used. When a weldment requires frequent changes of amperage or when welding must be performed in an inconvenient location, remote control adjustments can be very helpful. Foot-operated remote controls free the operator's hands and permit gradual increase and reduction of welding current. This is of great assistance in crater filling for gas tungsten arc welding.

Safety voltage controls are available to reduce the open circuit voltage of ac arc welding power sources. They reduce the open circuit voltage at the electrode holder to about 30 V. Voltage reducers may consist of relays and contactors that either reconnect the secondary winding of the main transformer for a lower voltage, or switch the welding load from the main transformer to an auxiliary transformer with a lower no-load voltage.

ALTERNATING-CURRENT POWER SOURCES—ALTERNATORS

ANOTHER SOURCE OF ac welding power is an alternator (often called an *ac generator*), which converts mechanical energy into electrical power suitable for arc welding. The mechanical power may be obtained from various sources, such as an internal combustion engine or an electric motor. They differ from standard dc generators in that the alternator rotor assembly contains the magnetic field coils (see Figure 1.22) instead of the stator coils as in other generators (see Figure 1.23). Slip rings are used to conduct low dc power into the rotating member to produce a rotating magnetic field. This configuration precludes the necessity of the commutator and the brushes used with dc output generators. The stator (stationary portion) has the welding current coils wound in slots in the iron core. The rotation of the field generates ac welding power in these coils.

The frequency of the output welding current is controlled by the speed of rotation of the rotor assembly and by the number of poles in the alternator design. A two-pole alternator must operate at 3600 rpm to produce 60 Hz current, whereas a four-pole alternator design must operate at 1800 rpm to produce 60 Hz current.

Saturable reactors and moving-core reactors may be used for output control of these units. However, the normal method is to provide a tapped reactor for broad control of current ranges, in combination with control of the alternator magnetic field to produce fine control within these ranges. These controls are shown in Figure 1.24.

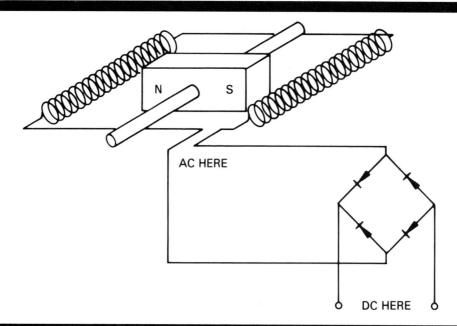

Figure 1.22–Schematic View of an Alternator Showing the Magnetic Field Contained in the Rotor Assembly

SQUARE-WAVE ALTERNATING-CURRENT POWER SOURCES

AC WELDING MACHINES for the SMAW, GTAW, PAW and SAW process traditionally have been based on methods for regulating their fields: (1) moving coils, (2) moving shunt, and (3) saturable reactors. The need for wider current ranges and remote current control led to the development of magnetic amplifiers with silicon diodes. While this technology has served the industry well, the industry needed machines with which higher weld quality and improved reliability could be obtained. With the growing acceptance and reliability of power semiconductors, a new generation of welding machines has been developed.

With 60 Hz alternating current, the welding current is reversed 120 times per second. With magnetic-type supplies, this reversal occurs slowly, hampering reignition of the next half-wave. Even though auxiliary means can be used to provide a high ionizing voltage, such as superimposed high-frequency energy for GTAW and PAW, often the instantaneously available current is too low to assure reliable arc ignition.

This problem can be avoided by using a current with a square-wave form as diagramed in Figure 1.25. With its rapid zero-crossing, deionization may not occur or, at the very least, arc reignition is enhanced to the extent that high-frequency (hf) reignition systems are unnecessary.

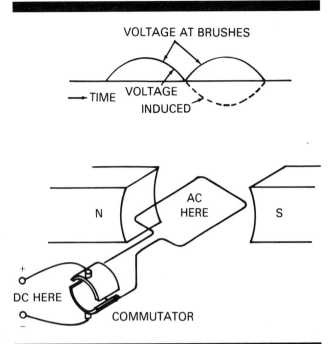

Figure 1.23–Schematic View of a Generator Showing the Magnetic Field Contained in the Stator Assembly

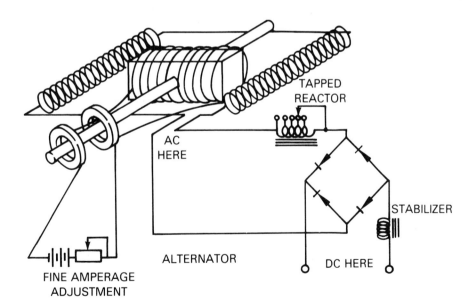

Figure 1.24–Schematic View of an Alternator Type Power Supply Showing a Tapped Reactor for Coarse Current Control and Adjustable Magnetic Field Amperage for Fine Output Current Control

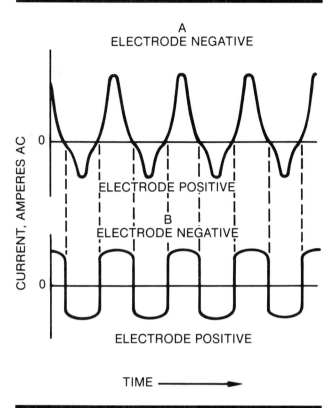

Figure 1.25–Comparison of Arc Current Waveforms of (A) Magnetic Amplifier and (B) Square Wave Power Source at Same Average Current Level

The square shape of the wave form trailing edge keeps the cover gas ionized and the electrode tip hot longer in preparation for reignition at the opposite polarity. These features are important in those installations where it becomes desirable to eliminate hf because (1) hf may cause radio or television interference, (2) etching of the work in the immediate vicinity of the weld may be cosmetically undesirable, (3) hf leakage may bother the operator, and (4) peripheral equipment may be damaged by hf.

Various design approaches have been used to produce square ac wave forms. Some machines use single-phase and some use three-phase input. Two common approaches are the use of a memory core and inverter circuits.

Memory Core

A MEMORY CORE is a magnetic device that keeps the current flowing at a constant value (a kind of electric fly-wheel). In conjunction with a set of four power SCRs, it can be used to develop a square-wave ac current. Current forcing is employed, wherein the memory core stores energy in proportion to the previous half-cycle current, then pumps that same amount of current to the arc at the beginning of each new opposite polarity half-cycle. Only one value of the current can be assumed once the gas has been reionized. The value is the "remembered" multicycle average-current value maintained by the memory core device. The transition time from one polarity to the other is very short, in the range of 80 microseconds.

A sensor placed in the memory core current path produces a voltage signal which is proportional to the ac current output. That current signal is compared with the desired current reference signal at a regulator amplifier. The resulting actuating error signal is processed to phase fire four SCRs in the proper sequence to bring the output to the proper level. Consequently, the weld current is held within ±1 percent for line voltage variations of ±10 percent. Response time is fast, thus lending itself to pulsed ac GTAW operations.

Another feature designed into this type of power source is a variable asymmetric wave shape. This enables the operator to obtain balanced current or various degrees of controlled imbalance of DCEN (direct current electrode negative or straight polarity) versus DCEP (direct current electrode positive or reverse polarity). Such capability provides a powerful tool for arc control. The main reason for using ac with the GTAW process is that it provides a cleaning action. This is especially important when welding aluminum. During DCEP cycles, the oxides on the surface of the workpiece are thrown off by emissive electron forces, exposing clean metal to be welded. Tests with various asymmetrical power sources established that only a small amount of DCEP current is required. Amounts as low as 10 percent would be adequate with the exception of cases in which hydrocarbons may be introduced by the filler wire.

Balance is set with a single knob, from balanced current to 70-80 percent DCEN, 20-30 percent DCEP at one extreme, and 35-45 percent DCEN, 55-65 percent DCEP at the other (see Figure 1.26). The balance control in effect adjusts the width of each polarity, without changing current amplitude or frequency. The regulating system holds the selected balance ratio constant as other amperage values are selected.

The usefulness of such a balance control becomes obvious. With a reasonably clean workpiece, the operator can dial for a low percentage of cleaning action (DCEP). With the resulting high percent of DCEN wave form, the heat balance approaches that of DCEN, providing more heat into the work, less arc wander, and a narrower bead width. Considering that GTAW is often selected because of its concentrated arc, this allows greatest utilization of its best characteristic.

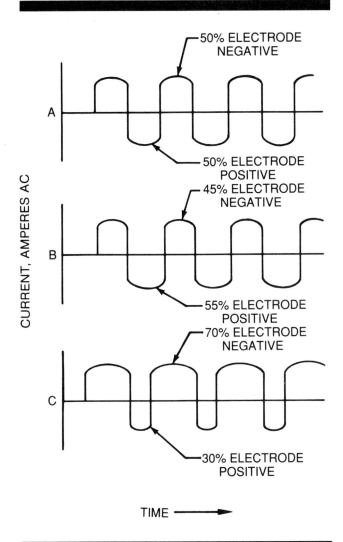

Figure 1.26–Typical Waveforms Produced by Square Wave Power Balance Control

The asymmetrical wave, with less DCEP time, allows the operator to use smaller diameter electrodes without the danger of high temperatures eroding the tip. In effect, it allows a higher current density. This results in a smaller diameter arc cone and better heat concentration. With a smaller gas cup, it often allows the operator to get into tighter joint configurations.

Inverter With AC Output

ANOTHER APPROACH TO achieve a square-wave ac output is to use inverter circuits. Several systems are used with the inverter approach to achieve a square-wave ac output with

rapid zero crossover. They are (1) dual sources with inverter switching, (2) single source with inverter switching, and (3) synchronous rectifier inverter.

The dual source with inverter switching approach makes use of solid-state SCR technology. It combines two three-phase adjustable-current dc power supplies. One that provides the main weld current is SCR controlled and is rated typically for 300 A, 50 V dc output. It supplies current during both DCEN (straight polarity) and DCEP (reverse polarity) phases of operation. The other power supply is a conventional reactor-controlled power supply typically rated at 5 to 100 A, 50 V dc output. Its function is to provide higher current during the DCEP (reverse polarity) phases of operation so that the cleaning is improved. Tests have shown that the most effective etching is obtained when DCEP current is higher, but applied for much shorter time than the DCEN current. Both supplies must provide 50 V output to ensure good current regulation when welding since the actual arc voltage during the DCEP phase may approach this voltage.

The switching and combining of the current from the two power supplies is controlled by five SCRs. See Figure 1.27. Four of these SCRs are part of an inverter circuit which switches the polarity of the current supplied to the arc. These four SCRs are arranged to operate in pairs.

One pair (SCR1 and SCR4) is switched on to provide current from the main supply during the DCEN portions of the square wave. The other pair (SCR2 and SCR3) is switched on to provide current from both power supplies during DCEP portions of the square wave. A shorting SCR (SCR5) is used with a blocking diode to bypass current from the second power supply around the inverter circuit during the DCEN portion of the cycle, thereby preventing its addition to the welding current.

The SCRs are turned on by a gating circuit, which includes timing provisions for adjustment of the DCEN and DCEP portions of the square-wave output. The DCEN time can typically be adjusted from 5 to 100 milliseconds (msec), and the DCEP time from 1 to 100 milliseconds (msec). A typical time setting for welding thick aluminum might be 19 msec DCEN and 3 msec DCEP time. The SCRs are turned off by individual commutation circuits. The current waveform is shown in Figure 1.28.

The single source with inverter switching is a much simpler and less bulky approach than the dual-source system. With the single-source approach, a single dc constant-current power source is used. Figure 1.29 shows a single-source ac square-wave inverter using transistors instead of SCRs. The operation of this source is very similar to that of the dual source. The four transistors are arranged to operate in pairs. Since there is not an additional reverse-current source, a fifth transistor and blocking diode are not necessary. AC balance can be controlled like the memory-core source and the dual-source inverter. However, the reverse-polarity current must be the same in amplitude as the straight-polarity current and cannot be increased as with

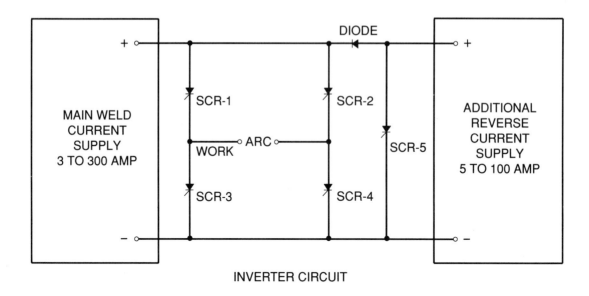

INVERTER CIRCUIT

Figure 1.27–Inverter Circuit Used With Dual DC Power Supplies for Controlling Heat Balance in Gas Tungsten Arc Welding

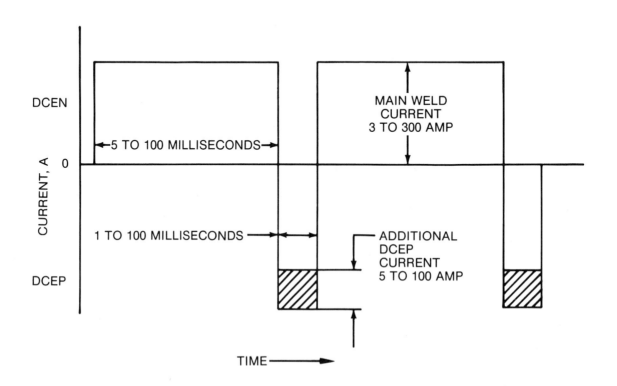

Figure 1.28–Typical Arc Current Waveform for Dual Adjustable Balance Inverter Power Supply

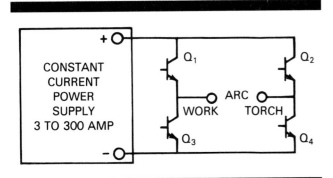

Figure 1.29–AC Inverter Circuit (Single Source)

the dual-source inverter. Both the single-source and dual-source inverters can vary the frequency of the ac square-wave output whereas the memory-core source must operate at line frequency (50 Hz or 60 Hz).

A third approach utilizes a device called a *synchronous rectifier*. Here, one starts with a power source with an inverter in the primary, producing a high-frequency ac output. This high-frequency ac is applied to the synchronous rectifier circuit which, upon command, rectifies the high-frequency ac into either DCEN or DCEP output. By switching the synchronous rectifier alternately between DCEN and DCEP, a synthesized lower frequency ac output can be created.

DIRECT CURRENT POWER SUPPLIES

CONSTANT VOLTAGE

CONSTANT-VOLTAGE (POTENTIAL) power sources are commonly used for GMAW, FCAW, and SAW. They are either rotating, transformer-rectifier, or inverter-type machines. Generators that can supply constant-voltage welding power are normally the separately excited, modified compound-wound type. The compounding of constant-voltage units differs from that of constant-current units to produce flat volt-ampere output characteristics. These machines may have solid-state devices in the excitation circuit to optimize performance and to provide remote-control capability. Various types of electronic circuits are used for this purpose, such as phase-angle controlled SCRs and inverter circuits.

Transformer-rectifier and inverter-type constant-voltage power sources are normally three-phase units. However, small single-phase units, usually rated 200 A or below, are marketed for light-duty applications.

Electrical Characteristics

CONSTANT-POTENTIAL POWER sources are characterized by their typically flat volt-ampere curves. A negative slope of 1 or 2 V per 100 A is not uncommon. This means that the maximum short circuiting current is usually very high, sometimes in the range of thousands of amperes. Machines with volt-ampere curves having slopes of up to 8 V per 100 A are still referred to as constant-voltage power supplies.

There are many varieties and combinations of constant-voltage power sources. A fixed slope may be built into the power supply, or the unit may have an adjustment to tailor the slope of the volt-ampere curve to the welding process.

The dynamic characteristics of these power supplies are of prime importance. Should inductance be used to adjust the slope, it will change not only the static but also the dynamic characteristics of the machine. In some cases, adjustable inductors are used in the dc portion of the circuit to obtain separate control of the static and dynamic features. The dc inductor will not alter the static characteristics but will affect the dynamic characteristics, which are very important for short circuiting transfer in gas metal arc welding.

General Design

MANY DESIGNS OF constant-voltage machines are available. The advantage of any particular type is related to the application and to the expectations of the user.

Open Circuit Voltage. The open circuit voltage of some transformer-rectifier machines is adjusted by changing taps on the transformer. Another type of machine controls the open circuit voltage with secondary coils wound in such fashion that carbon brushes, driven by a lead screw, slide along the secondary coil conductors. A second control is often provided with them to adjust the volt-ampere characteristics to provide the requirements of the welding process. Because of the additional effect on the volt-ampere output curve, this is called *slope control*.

Constant-voltage power sources have a wide range of open circuit voltages. Electrically controlled machines may have as high as 75 V open circuit. Tapped or adjustable-transformer types have open circuit voltage which may be varied from 30 to 50 V maximum to 10 V minimum.

Slope. Slope control is generally obtained by changing taps on reactors in series with the ac portion of the circuit. Slope control may be provided by carbon brushes, attached to a lead screw, contacting the reactor turns. This variable reactor provides continuous adjustment of slope. Another method of control uses magnetic amplifiers or solid-state devices to electrically regulate output voltage. These machines may have either voltage taps or slope taps in addition to electrical control.

Some advantages of electrical controls are the ease of adjustment, the ability to use remote control, and the absence of moving parts. Also, some electrically controlled machines permit adjustment of output during welding. This is helpful when "crater filling" or changing welding conditions. The combination of taps with electrical control to give fine output adjustment between taps is a suitable arrangement in a service application where the machine requires little attention during welding. Fully electrically controlled machines are easier to set up and readjust when welding requirements change rapidly. Slope can also be controlled electronically by circuitry in most phase-angle controlled SCR and inverter power supplies.

Electrically controlled machines often do not have separate slope controls. A fixed, all-purpose slope is designed into them.

On constant-voltage generators, slope control is usually provided by a tapped resistor in the welding circuit. This is desirable because of the inherent slow dynamic response of the generator to changing arc conditions. Resistance-type slope controls limit maximum short circuit current. Reactor slope control will also limit maximum short circuiting current. However, it will slow the rate of response of the power source to changing arc conditions more than resistive type slope control.

While there is no fixed rule for volt-ampere slope in the welding range, most machines have slopes of from 1 to 3 V per 100 A.

Inductance. Gas metal arc machines designed for short circuiting transfer generally incorporate additional dc inductance to improve performance by providing the dynamic characteristics required. Such inductance can be variable or fixed.

Ripple. Single-phase power sources generally require some type of ripple filter arrangement in the welding circuit. Usually this filter is a bank of electrolytic capacitors across the rectifier output. The purpose is to provide a smooth dc output, capable of clearing a short circuit. An inductor is used to control the output of the capacitors. Without some inductance, the discharge of the capacitors through a short circuit would be much too violent for good welding operation.

Control Devices

CONSTANT-VOLTAGE POWER sources are usually equipped with primary contactors. Electrically controlled models usually have remote voltage-control capabilities. Other features available on certain machines are line-voltage compensation and accessories to interface with wire feed equipment to change both feed rate and welding current.

Electrical Rating

PRIMARY RATINGS ARE similar to those discussed earlier. Constant-voltage machines usually have a more favorable power factor than constant-current machines and do not require power factor correction. Open circuit voltage, while subject to NEMA specification, is usually well below the established maximum. Current ratings of NEMA Class I machines range from 200 to 1500 A.

Constant-voltage power sources are normally classified as NEMA Class I or Class II. It is the usual practice to rate them at 100 percent duty cycle, except for some of the light-duty units of 200 A and under, which may be rated as low as 20 percent duty cycle.

CONSTANT CURRENT

WELDING POWER SOURCES that are called *constant-current machines* are typically used for SMAW, GTAW, PAC, PAW, and SAW. These machines can be inverters, transformer-rectifiers, or generators. The transformer-rectifier and inverter machines are static, transforming AC to DC power. Generators convert mechanical energy of rotation to electrical power.

The open circuit voltages of constant current, rectifier-type power sources vary depending on the intended welding application. They range from 50 to 100 V. Most NEMA Class I and Class II machines are usually fixed in the 70 and 80 V range.

Electrical Characteristics

AN IMPORTANT ELECTRICAL characteristic is the relation of the output current to the output voltage. Both a static (steady-state) relationship and a dynamic (transient) relationship are of special interest. The static relationship is usually shown by volt-ampere curves, such as those in Figures 1.13 and 1.14. The curves usually represent the maximum and minimum for each current range setting. As discussed in a previous section, the dynamic relationship is difficult to define and measure for all load conditions. The dynamic characteristics determine the stability of the arc under actual welding conditions. They are influenced by circuit design and control.

General Design

THE USUAL VOLTAGES of the ac supply mains in the United States are nominally 208, 240, 480, and 600 with a frequency of 60 Hz. Transformers are designed to work on these voltages. This is done by arranging the primary coils in sections with taps. In that way, the leads from each section can be connected in series or parallel with other sections to suitably match the incoming line voltage. On three-phase machines, the primary can be connected in delta or wye configuration. The secondary is frequently connected in delta; a delta connection is preferred for low voltage and high current.

The current is usually controlled in the section of the machine between the transformer and the rectifiers. Current control employs the principle of variable inductance or impedance. The following are methods for varying the impedance for current control:

(1) Moving coil
(2) Moving shunt
(3) Saturable reactor or magnetic amplifiers
(4) Tapped reactor
(5) Moving reactor core
(6) Solid-state

In addition to these six control systems, there is a type that employs resistors in series with the dc portion of the welding circuit. Methods (1), (2), and (5) are classed as mechanical controls; methods (3) and (6), as electric control; method (4) and the external resistor type, as tap controls. These same methods are also used for controlling constant-current transformer sources.

An inductance is usually used in the dc welding circuit to control excessive surges in load current. These current surges may occur due to dynamic changes in arc load. It is also used to reduce the inherent ripple found after rectifying the alternating current. A three-phase rectifier produces relatively little ripple; therefore, the size of its inductor is determined primarily by the need to control arc load surges. A high ripple is associated with single-phase rectification. The size of inductors for single-phase machines is determined by the need to reduce ripple. Therefore, they are larger than those of three-phase machines of the same rating. Power sources of this type usually have a switch on the dc output so that the polarity of the voltage at the machine terminals can be reversed without reversing the welding cables.

Auxiliary Features

THE AUXILIARY FEATURES are similar to those available for constant-current ac power sources, although all features are not available on all power sources. The manufacturer can supply complete information.

In addition to previously listed features, current-pulsing capabilities are available with many dc power sources as standard or optional equipment. Pulse power sources are capable of alternately switching from high to low welding current repetitively. Normally, high- and low-current values, pulse duration, and pulse repetition rate are independently adjustable. This feature is useful for out-of-position welding and critical gas tungsten arc welding applications.

INVERTER POWER SUPPLY

AN INVERTER MACHINE is different from a transformer-rectifier type in that the inverter will rectify 60 Hz ac line current, use a chopper circuit to produce a high-frequency ac, reduce that voltage with an ac transformer, and rectify that to obtain the required dc current output. Changing the ac frequency to a much higher frequency allows a greatly reduced size of transformer and reduced transformer losses as well.

GENERATORS AND ALTERNATORS— MOTOR AND ENGINE DRIVEN

GENERATOR-TYPE POWER sources convert mechanical energy into electrical power suitable for arc welding. The mechanical power can be obtained from an internal combustion engine, an electric motor, or from a power take-off from other equipment. For welding, two basic types of rotating power sources are used; the generator and the alternator. Both have a rotating member, called a *rotor* or an *armature*, and a stationary member, called a *stator*. A system of excitation is needed for both types.

The principle of any rotating power source is that current is produced in electrical conductors when they are moved through a magnetic field. Physically, it makes no difference whether the magnetic field moves or the conductor moves, just so the coil experiences a changing magnetic intensity. In actual practice, a generator has a stationary field and moving conductors, and an alternator has a moving field and stationary conductors.

The dc generator has a commutator-brush arrangement for changing ac to dc welding power. Normally, the dc generator is a three-phase electrical device. Three-phase systems provide the smoothest welding power of any of the electromechanical welding power sources.

A dc generator consists of a rotor and stator. The rotor assembly is comprised of (1) a through shaft, (2) two end bearings to support the rotor and shaft load, (3) an armature which includes the laminated armature iron core and the current-carrying armature coils, and (4) a commutator. It is in the armature coils that welding power is generated.

The stator is the stationary portion of the generator within which the rotor assembly turns. It holds the magnetic field coils of the generator. The magnetic field coils conduct a small amount of dc to maintain the necessary continuous magnetic field required for power generation. The dc amperage is normally no more than 10 to 15 A and very often is less.

In electric power generation, there must be relative motion between a magnetic field and a current-carrying conductor. In the dc generator, it is the armature that is the current-carrying conductor. The magnetic field coils are located in the stator. The armature turns within the stator and its magnetic field system, and welding current is generated.

The armature conductors of a welding generator are relatively heavy because they carry the welding current. The commutator is located at one end of the armature. It is a group of conducting bars arranged parallel to the rotating shaft to make switching contact with a set of stationary carbon brushes. These bars are connected to the armature conductors. The whole arrangement is constructed in proper synchronization with the magnetic field so that, as the armature rotates, the commutator performs the function of mechanical rectification.

An alternator power source is very similar, except that generally the magnetic field coils are wound on the rotor, and the heavy welding current winding is wound into the stator. These machines are also called *revolving* or *rotating field machines*.

The ac voltage produced by the armature coils moving through the magnetic field of the stator is carried to copper commutator bars through electrical conductors from the armature coils. The conductors are soft-soldered to the individual commutator bars. The latter may be considered as terminals, or "collector bars", for the alternating current generated from the armature.

The commutator is a system of copper bars mounted on the rotor shaft. Each copper bar has a machined and polished top surface. Contact brushes ride on that top surface to pick up each half-cycle of the generated alternating current. The purpose of the commutator is to carry both half-cycles of the generated ac sine wave, but on separate copper commutator bars. Each of the copper commutator bars is insulated from all the other copper bars.

The carbon contact brushes pick up each half-cycle of generated alternating current and direct it into a conductor as direct current. It may be said that the brush-commutator arrangement is a type of mechanical rectifier since it does change the generated alternating current (ac) to direct current (dc). Most of the brushes used are an alloy of carbon, graphite, and small copper flakes.

Placing the heavy conductors in the stator eliminates the need for carbon brushes and a commutator to carry high current. The output, however, is ac, which requires external rectification for dc application. Rectification is usually done with a bridge, using silicon diodes. An alternator usually has brushes and slip rings to provide the low dc power to the field coils. It is not usual practice in alternators to feed back a portion of the welding current to the field circuit. Both single- and three-phase alternators are available to supply ac to the necessary rectifier system. The dc characteristics are similar to those of single- and three-phase transformer-rectifier units.

An alternator or generator may be either self-excited or separately-excited, depending on the source of the field power. Either may use a small auxiliary alternator or generator, with the rotor on the same shaft as the main rotor, to provide exciting power. On many engine-driven units, a portion of exciter field power is available to operate tools or lights necessary to the welding operation. In the case of a generator, this auxiliary power is usually 115 V dc. With an alternator-type power source, 120 or 120/240 V ac is usually available. Voltage frequency depends on the engine speed.

Output Characteristics

BOTH GENERATOR- AND alternator-type power sources generally provide welding current adjustment in broad steps called *ranges*. A rheostat or other control is usually placed in the field circuit to adjust the internal magnetic field strength for fine adjustment of power output. The fine adjustment, because it regulates the strength of the magnetic field, will also change the open circuit voltage. When adjusted near the bottom of the range, the open circuit voltage will normally be substantially lower than at the high end of the range.

Figure 1.30 shows a family of volt-ampere curve characteristics for either an alternator- or generator-type power supply.

With many alternator power supplies, broad ranges are obtained from taps on a reactor in the ac portion of the circuit. As such, the basic machine does not often have the dynamic response required for shielded metal arc welding. Thus, a suitable inductor is generally inserted in series connection in one leg of the dc output from the rectifier. Welding generators do not normally require an inductor.

There is a limited range of overlap normally associated with rotating equipment where the desired welding current can be obtained over a range of open circuit voltages. If welding is done in this area, welders have the opportunity to better tailor the power supply to the job. With lower open circuit voltage, the slope of the curve is less. This allows the welder to regulate the welding current to some degree by varying the arc length. This can assist in weld-pool control, particularly for out-of-position work.

Some welding generators carry this feature beyond the limited steps described above. Generators that are compound wound with separate and continuous current and voltage controls can provide the operator with a selection of volt-ampere curves at nearly any amperage capability within the total range of machine. Thus, the welder can set the desired arc voltage with one control and the arc current with another. This adjusts the generator power source to provide a static volt-ampere characteristic that can be "tailored" to the job throughout most of its range. The volt-ampere curves that result when each control is changed independently are shown in Figures 1.31 and 1.32.

Welding power sources are available that produce both constant current and constant voltage. These units are used for field applications where both are needed at the job site and utility power is not available. Also, many new designs use electronic solid-state circuitry to obtain a variety of volt-ampere characteristics.

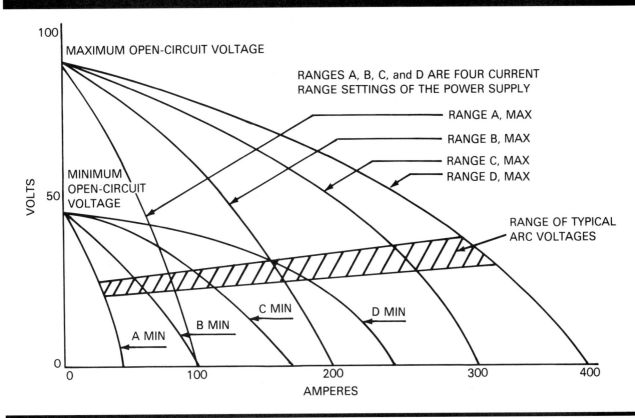

Figure 1.30–Volt-Ampere Relationship for a Typical Constant-Current Rotating Type Power Source

Sources of Mechanical Power

GENERATORS ARE AVAILABLE with ac drive motors of several voltage and frequency ratings and also with dc motors. Welding generators are usually single units with the drive motor and generator assembled on the same shaft.

Induction motor-driven welding generators are normally available for 200, 240, 480, and 600 V three-phase, 60 Hz input. Other standard input requirements are 220, 380, and 440 V 50 Hz. Few are made with single-phase motors, since transformer type welding power supplies usually fill the need for single-phase operation. The most commonly used driving motor is the 230/460 V, three-phase, 60 Hz induction motor.

Figure 1.33 summarizes some of the electrical characteristics of a typical 230/460 V, three-phase, 60 Hz induction motor-generator set: overall efficiency, power factor, and current input. The motors of dc welding generators usually have a good power factor (80 to 90 percent) when under load, and from 30 to 40 percent lagging power factor at no-load. No-load power input can range from 2 to 5 kW, depending upon the rating of the motor-generator set. The power factor of induction motor-driven welding generators may be improved by the use of static capacitors similar to those used on welding transformers. Welding generators have been built with synchronous motor drives in order to correct the low power factor.

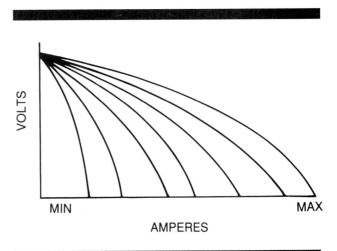

Figure 1.31–Effect of Current Control Variations on Generator Output

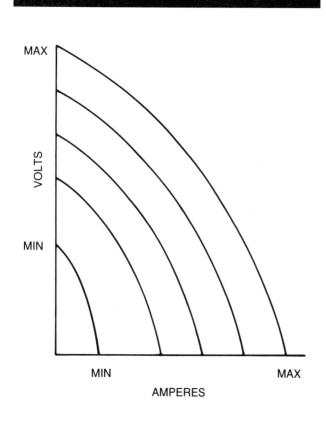

Figure 1.32—Effect of Voltage Control Variations on Generator Output

Rotating-type power supplies are used for field erection work when no electric power is available. For this use, a wide variety of internal combustion engines is available. Both liquid-cooled and air-cooled engines are used, depending on specific power source applications.

In the United States, gasoline is the most popular fuel because of price and availability. Diesel fuel is popular because of its high flashpoint. Also, some Federal laws will permit only diesel fuel for engines used in specific applications. A good example is the use of diesel engines for welding power sources on off-shore drilling rigs. Propane is used in some applications because it is cleaner burning than gasoline. However, it does require a special carburation system.

Parallel Operation

INCREASED CURRENT OUTPUT can be obtained by connecting welding generators in parallel. However, parallel connection is *not* advised unless the manufacturer's specific instructions are followed. Such caution is necessary because successful paralleling depends upon matching the output voltage, output setting, and polarity of each ma-

chine. In the case of self-excited generators, the problem is further complicated by the necessity to equalize the excitation between the generators.

The blocking nature of the rectifiers makes dc alternator units easy to operate in parallel. Care should be taken to ensure that connections are the same polarity. All units paralleled must be set to deliver equal outputs.

Auxiliary Features

ROTATING TYPE POWER sources are available with many auxiliary features. Units may be equipped with a remote control attachment. It may be either a hand or foot control that the operator takes to the work station to make power source adjustments while welding.

Gas engines are often equipped with idling devices to save fuel. These devices are automatic in that the engine will run at some idle speed until the electrode is touched to the work. Under idle, the open circuit voltage of the generator is low. Touching the electrode to the work energizes a sensing circuit that automatically accelerates the engine to operating speed. When the arc is broken, the engine will return to idle after a set time.

Engine-driven generators are often equipped with a provision for auxiliary electric power. This power is available at all times on some units and only during idle time on other units. Other auxiliary features that can often be obtained on engine-driven welding machines are polarity switches (to easily change from DCEN to DCEP), running-hour meters, fuel gauges, battery charger, high-frequency arc starters, tachometers, and output meters. Some larger units are equipped with air compressors.

PULSED AND SYNERGIC PULSED POWER SUPPLIES

PULSED CURRENT POWER supplies have been used for SMAW, GTAW, GMAW, FCAW, and SAW. These power supplies first appeared on the market in the 1960's. The most common are pulsed power supplies for GMAW and GTAW.

Pulsed Gas Metal Arc Welding Power Supplies

PULSED POWER SUPPLIES are used with the GMAW process to reduce the arc power and wire deposition rate while retaining the desirable spray transfer. The concept is based on the fact that metal transfers from the wire electrode at two different rates depending on the welding current; one approximating a few hundred drops per second (the spray arc mode) when the current exceeds a critical level, and one that is less than 10 drops per second (the globular mode) when the welding current is below the critical level. The critical current is called the *transition current*. By pulsing the current between these two regions, it is possible to

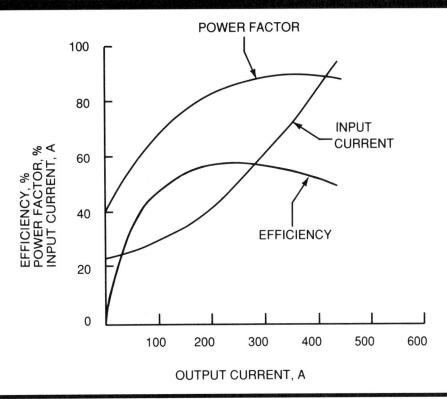

Figure 1.33–Typical Characteristic Curves of a 300 Ampere DC Motor-Generator Power Supply

obtain the desirable qualities of spray transfer while reducing the average current and deposition rates significantly. This allows the GMAW process to be used in all positions and for welding sheet metal.

In practice, the current level during the globular interval is kept sufficiently low to prevent any metal from being transferred but high enough to sustain ionization in the arc regions. For that reason, it has been called the "keep alive" current, but more commonly the *background current*. During the spray interval, the current is raised above the transition current for a sufficient time to allow a drop or two to transfer; this is the pulse current. Power supplies have been designed with the necessary controls to produce the controlled output for pulsed GMAW.

The first pulsed GMAW power sources were fixed frequency machines. Still being used, they consist of a constant-voltage power supply for background current and a half-wave rectified power supply for pulses of current. See Figure 1.34.

The next improved version provided pulses at either 60 or 120 Hz; it used a constant-current power supply as background current and a phase-angle controlled SCR power source for adjusting the peak current. See Figure 1.35.

With the appearance of transistorized power sources in the late 1960's and early 1970's, research was conducted with pulsed GMAW using variable-frequency power supplies. With solid-state power sources such as inverters, it became possible to make independent settings of all the pulse variables: peak current, background current, peak current time (pulse width), and background current time. See Figure 1.36. By controlling these variables singly or by causing them to interact, it is possible to gain almost total control of metal transfer, allowing only single drops to transfer per pulse, while retaining control of the average current.

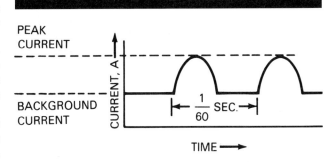

Figure 1.34–Current Output From Pulsed GMAW Power Supply. Constant Voltage Power Supply for Background Current and Half-Wave Rectified Power Source for Pulses of Current

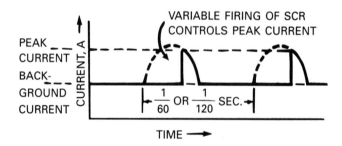

Figure 1.35–Current Output From Pulsed GMAW Power Supply. Constant Current Power Supply for Background Current and Phase Angle Controlled SCR Power Supply for Variable Peak Current

When all of the pulsed conditions are set, it may be difficult to reset some of the more sophisticated power supplies when, for example, the wire feed speed must be changed, since a number of variables then need to be reset. Such pulsed GMAW power supplies can be made simpler to operate by using electronic and microprocessor controls to set the optimum pulsed conditions, based on the wire feed speed setting. Typical circuit elements for this type power supply are diagrammed in Figure 1.37.

The variables used and the controls exercised depend on the objectives of their designers. For example, to change deposition rate it is possible to make proportional changes in the pulse frequency while fixing the other variables. Instead of changing frequency, the pulse width could be varied in proportion to the demands of the arc. Some power supplies are designed to change both the pulsing frequency and the background current, depending on the wire feed speed.

The word *synergic* (several things acting as one) is used to describe pulsed GMAW machines that set the pulsed GMAW variables based on the wire feed speed. Through the use of electronic controls, it is possible to choose a variety of synergic curves to satisfy particular applications.

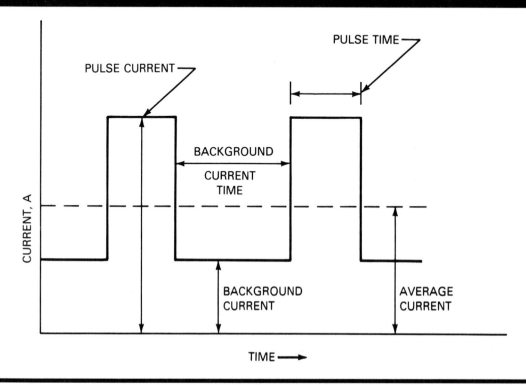

Figure 1.36–Pulsed Gas Metal Arc Welding Variables

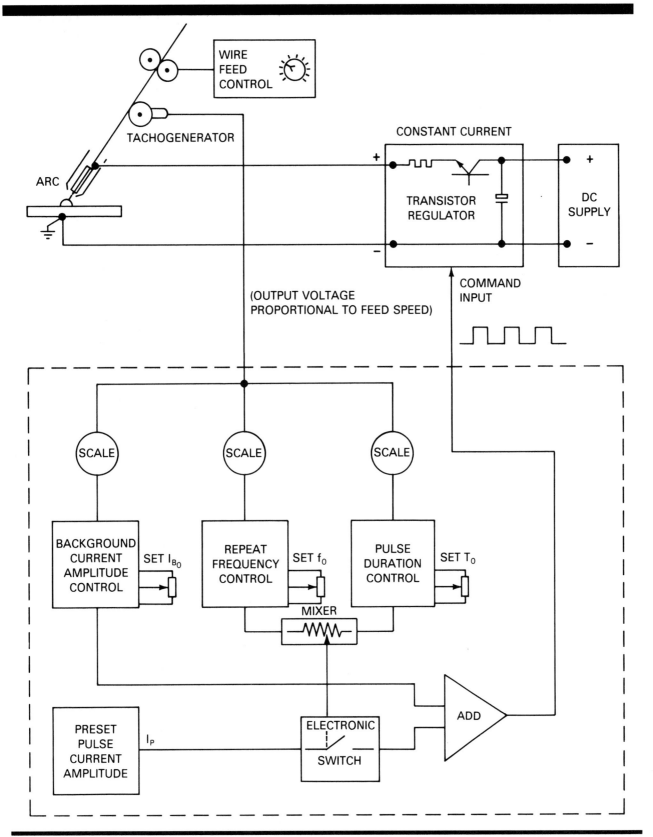

Figure 1.37–Basic Circuit for Synergic Pulse Operation

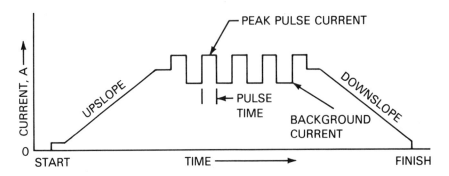

Figure 1.38—Typical Pulsed GTAW Program Showing Upslope and Downslope

In the 1960's, pure transistorized power supplies were fairly expensive. With the advent of high power and fast response semiconductors, the price of variable frequency pulsed power supplies has been greatly reduced. Most inverter welding power sources have the capability to be used for pulsed GMAW. Response times, frequency, pulse width, peak or background levels are all adjustable with solid-state inverters. Some synergic power supplies are designed to assure the transfer of only single drops with each pulse by making instantaneous adjustments of the pulse frequency and width depending on the voltage sensed across the arc. Some machines vary either pulse frequency, peak current time (pulse width), or wire feed speed as stickout changes. Pulsed GMAW power sources typically range up to 500 A peak current, and frequency varies from 60 to 200 Hz.

Pulsed Gas Tungsten Arc Power Supplies

THE CONCEPT OF pulsing current also has been used with considerable effectiveness with the GTAW process. The frequencies differ from those found using the GMAW process, generally ranging from 2 seconds per pulse to 10 pulses per second. The lower frequencies are more commonly used. Because of the electrical sophistication required for pulsing, it is simple to provide other desireable features such as starting circuits and controlled current slopes. A diagram showing the changes of current possible in programming a weld with such machines is illustrated in Figure 1.38.

Pulsed GTAW is characterized by a repetitive variation in arc current from a background (low) value to a peak (high) value. Both peak and background current levels are adjustable over a wide range. Peak current time plus background current time constitute the duration of one pulse cycle. The peak current duration and background current duration also are independently adjustable.

The purpose of pulsing is to alternately heat and cool the molten weld metal. The heating cycle (peak current) is based on achieving a suitable molten weld pool size during the peak pulse without excessive side wall fusion or melt-through, depending on the joint being welded. Background current and duration are determined by the desired cooling of the weld pool. The purpose of the cooling (background current) portion of the cycle is to speed up the solidification rate and reduce the size of the molten pool without interrupting the arc. Thus, pulsing allows alternately increasing and decreasing the size of the molten pool.

The fluctuation in molten pool size and penetration is related to the pulsing variables, travel speed, the type, thickness, and mass of the base metal, filler metal size, and position of welding. Because the size of the molten pool is partially controlled by the current pulsing action, the need for arc manipulation to control the molten pool is reduced or eliminated. Thus, pulsed current is a useful tool for manual out-of-position GTAW, such as in-place pipe joints, and for automatic butt welding of thin wall tubing.

SPECIAL POWER SOURCES

MULTIPLE-OPERATOR WELDING

MULTIPLE-OPERATOR WELDING equipment is economical for shops having a number of welding stations in a small area. Most such equipment has been used to advantage in shipbuilding and construction shops that rely on the shielded metal arc welding process.

A multiple-operator (MO) installation typically consists of one large constant-potential power supply with an open circuit voltage of 70 to 80 V feeding eight or 10 welding stations.

Commercially available units vary from 500 A to 2500 A for motor-generator units, from 500 A to 1500 A for transformer-rectifier type installations, and from 400 A to 2000 A for transformers. Overload devices and circuit breakers protect the equipment from damage. The electrode holder at each welding station is connected to the power supply through a variable resistor or grid. This provides a drooping, essentially constant-current characteristic at the welding arc. A common work lead is used for all welding stations.

Individual Modules

ONE TYPE OF multiple-operator power source consists of a bank of individual power modules that, housed in a common cabinet, provide remotely controlled dc welding power from the main unit to individual stations at distances up to 200 ft (60 m). Where the output of an individual module is not sufficient for a particular welding job, two or more modules may be paralleled. A common work lead connection for all modules is provided. Each welder can select either polarity; each module is isolated, and each can be individually controlled.

Design Concept

MULTIPLE-OPERATOR EQUIPMENT is effective when operating duty cycles of each manual welding station is quite low (20 to 25 percent). Such conditions prevail where welders must change electrodes frequently, check fit-up, change position, and chip slag. Thus, their actual arc time will be relatively short in relation to their total work period. Multiple-operator equipment takes advantage of this low duty cycle to reduce welding equipment cost and increase flexibility and portability.

For example, if the average arc current required is 160 A and the average 10 minute duty cycle of a welding machine is 25 percent, the average current per welding machine is 160 x 0.25 or 40 A. Thus, in any 10 minute period an 800 A power source could supply an average 40 A to about 20 welding machines on the job site.

Features

IN LARGE INSTALLATIONS, the use of multiple-operator power supplies usually reduces the cost of capital equipment and its installation. Maintenance costs are also reduced, since only one power supply must be maintained instead of many. The resistor banks or reactor panels can usually be located close to each individual welder to enable current adjustments to be made conveniently.

SUBMERGED ARC WELDING

THE POWER SOURCE used for submerged arc welding (SAW) may be ac or dc constant current, or dc constant voltage. Direct current power supplies may be either motor-generators or transformer-rectifiers. Constant-current power sources are often used in combination with arc voltage controls for the electrode feed rate. Constant-voltage power sources are set to provide the required arc length, and the electrode feed speed is used to regulate the welding current. Submerged arc welding generally is done at high currents (350 to 1200 A) so the power source must have a high current rating at high duty cycle.

A standard NEMA-rated transformer-rectifier type or motor-generator type dc power source can be used for submerged arc welding if rated adequately for this application. Power supplies may be paralleled according to the manufacturer's instructions to obtain the necessary welding current capacity. Duplex units are available, consisting of two single-operator units assembled and connected for single or parallel operation. The use of duplex machines or single units with high-current rating is preferred to standard power sources connected in parallel.

Constant-voltage dc power sources used for submerged arc welding should have an open circuit voltage in the 50 V range and an adequate current rating for the application at hand. The welding current is automatically controlled by the feed rate of the electrode wire. One of the advantages of this method is that a simple control system provides a uniformly stable arc voltage, which is of particular advantage for high-speed, light-gage welding and also more consistent arc starting because of the high initial-current surge.

With motor-generator power sources, the high-power requirements of some submerged arc applications may cause excessive loading of the drive motor. At a given current, the input to the generator is roughly proportional to its load voltage. Therefore, care should be taken to select units having adequate motor-power rating. This precaution also applies to some transformer-rectifier type machines when the actual arc voltage exceeds the rated output voltage.

The flow of welding current from a generator may be started and stopped by a magnetic contactor in the welding circuit, or by a relay in the generator field circuit, depending upon its design and characteristics. Transformer-type power sources control the flow of current by means of a contactor in the primary line of the machine.

Magnetic deflection of the arc (arc blow), a characteristic of direct current welding, usually limits the magnitude of dc which can be used in submerged arc welding. Alternating current minimizes arc blow.

Transformers with ratings up to and including 2000 A are available with special features for adapting them to submerged arc welding application. Open circuit voltage should be at least 80 V, but preferably 85 to 100.

Welding current for multiple arc welding systems may be supplied in a number of different ways. For parallel arc welding, a power supply may be connected in the conventional single electrode manner. Then two or more welding electrodes may be fed by a single-drive head through a common contact nozzle or jaw.

In a series arc system, a single transformer or dc power source may be used for supplying two independent electrodes feeding into the same weld pool. The output terminals of the power supply are connected separately to one of the two welding heads, not to the work, so the workpiece is not part of the electrical circuit. This is called a *series arc system*. It requires a power supply with a high open circuit voltage.

High-speed tandem welding generally uses two independent welding heads. They are supplied by multiple transformer units connected to a three-phase line using either a closed delta connection or a Scott connection. Since heavy currents are frequently used in tandem welding, these systems distribute the power load on the three phases.

The closed delta system requires the use of three transformers with separate current control reactors. The transformer secondaries are connected in closed delta ahead of the reactors, as shown in Figure 1.39. This system provides for the adjustment of welding currents in the two arcs, the ground current, and the phase-angle displacement between the three currents. Adjustment of these conditions is important to obtain desired arc deflection (magnetically), weld penetration, and weld contour. Arc currents cannot be independently adjusted. A change in one will change the other two because of the phase angles.

The Scott-connected system uses two transformers. At least one of them should be specifically designed for Scott connection, with the primaries and secondaries connected as shown in Figure 1.40. The units should have 85 to 100 V open circuit. This system overcomes the interrelated current adjustment inherent in the closed delta system and provides independent control of arc currents.

STUD ARC WELDING

STUD ARC WELDING (SW) must be done with a dc power source. The process requires higher capacity, better consistency of operation, and better dynamic current control than is normally available with conventional power sources. The general characteristics desired in a stud arc welding power source are the following:

(1) High open circuit voltage, in the range of 70 to 100 V
(2) A drooping volt-ampere characteristic such that 25 to 35 V appears across the arc at maximum load
(3) A rapid rate of current rise
(4) High-current output for a relatively low duty cycle

Each of the various types of special power sources available has its own characteristics. Consequently, it is difficult to compare these types with other sources of power. One method of comparison would be to evaluate each

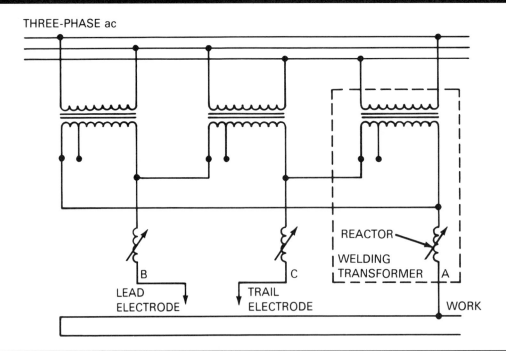

Figure 1.39—Arc Welding Transformer Connections for Three-Phase Delta System. This Method Equalizes Line Loadings.

power source in terms of current output and stud base diameter. The power source should be capable of delivering the required welding current for the size of stud to be welded.

ELECTROSLAG AND ELECTROGAS WELDING

THE EQUIPMENT USED for electroslag and electrogas welding is very similar to that required for submerged arc or flux cored arc welding. The same power sources can be used for either process, with one exception: ac power supplies are not used with the electrogas process. Both ac and dc power supplies are used with the electroslag process. Standard power sources used for either process should have an open circuit voltage up to 80 V and be capable of delivering 600 A continuously (100 percent duty cycle). The power supplies should be equipped with remote controls. The number of power supplies required depends on the number of welding electrodes being used to fill the joint. One power supply is required for each welding electrode.

Special constant-voltage dc power supplies designed for electroslag and electrogas welding are available. Typical power supplies are transformer-rectifiers having 74 V open circuit and rated at 750 A at 50 V output, 100 percent duty cycle. The primary input is 60 Hz, three-phase, 230/460 V.

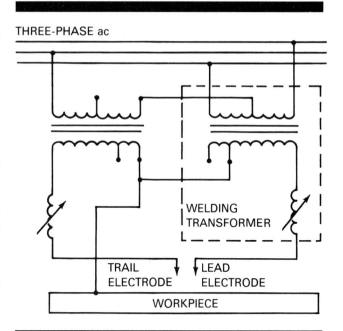

Figure 1.40—Arc Welding Transformer Connections for Scott-Connected System

SUPPLEMENTARY READING LIST

Amin, M. and Watkins, P.V.C. "Synergic pulse MIG welding." Document 46. The Welding Institute, August 1977.

Amin, M. "Microcomputer control of synergic pulsed MIG welding." Document 166/1981. The Welding Institute, December 1981.

Bailey, K. and Richardson, R. "A microprocessor-based SCR type arc welding power supply." Technical Report 529501-83-14. Columbus, OH: Center for Welding Research.

Brosilow, R. "The new GMAW power supplies." *Welding Design and Fabrication.* 22-28, June 1987.

Correy, T. B., Atteridge, D. G., Page, R. E., and Wismer, M. C. "Arc starting in gas tungsten arc welding, radio frequency free." *Welding Journal* 65(2): 33-41; February 1986.

Cullison, A., and Newton Montiel, B. "Changes are coming for welding power sources." *Welding Journal* 69(5): 37; May 1990.

Frederick, J. E., et al. "Solid state remote controlled welding." *Welding Journal* 57(8): 32-39; August 1978.

Gasawara, T. O., Maruyama, T, Saito, T., Sato, M., and Hida, Y. "A power source for gas shielded arc welding with new current waveforms." *Welding Journal* 66(3): 57-63; March 1987.

Grist, F. J. "Improved, lower cost aluminum welding with solid state power source." *Welding Journal* 54(5): 348-357; May 1975.

Grist, F. J., and Armstrong, F. W. "A new AC constant potential power source for heavy plate, deep groove welding." *Welding Journal* 59(6): 35-35; June 1980.

Kashima, T. and Yamanaha. "Development of the inverter controlled DC TIG arc welding power source." Document IIW-XII-878-85. England: International Institute of Welding, June 1985.

Kyselica, S. "High-frequency reversing arc switch for plasma arc welding of aluminum." *Welding Journal* 66 (1): 31-35; January 1987.

Lesnewich, A. "MIG welding with pulsed power." Bulletin 170. New York: Welding Research Council, Febrary 1972.

Lucas, W. "A review of recent advancements in arc welding power sources and welding processes in Japan." Document 199/1982. Abington, U.K.: Welding Institute, November 1982.

Malinowski-Brodnicka, M. et al. "Effect of electromagnetic stirring on GTA welds in austenitic stainless steel." *Welding Journal* 69(2): 525; February 1990.

Manz, A. F. *Welding power handbook*. New York: Union Carbide Corporation, 1973.

Needham, J. C. "Review of new designs of power sources for arc welding processes." Document XII-F-217-80. England: International Institute of Welding.

National Electrical Manufacturers Association. *Electric arc welding power sources*, EW1-1988. National Electrical Manufacturers Association: Washington, DC, 1988.

Pierre, E. R. *Welding processes and power sources*, Third edition, 1985.

Rankin T. "New power source design breaks with tradition." *Welding Journal* 69(5): 30, May 1990.

Schiedermayer, M. "The inverter power source." *Welding Design and Fabrication*, 30-33, June 1987.

Shira, C. "Converter power supplies - more options for arc welding." *Welding Design and Fabrication*, June 1985.

Spicer, R. A. "Elemental effects on GTA spot weld penetration in cast alloy 718." *Welding Journal* 69(8): 285s-288s; August 1990.

Tomsic, M. J., Barhorst, S. E., and Cary, H. B. "Welding of aluminum with variable polarity power." Document No. XII 83984. England: International Institute of Welding.

Villafuerte, J. C. and Kerr, H. W. "Electromagnetic stirring and grain refinement in stainless steel GTA welds." *Welding Journal* 69(1): 1s; January 1990.

Xiao, Y. H., and Van Ouden G. "A study of GTA weld pool oscillation." *Welding Journal* 69(8): 289s-293s; August 1990.

SHIELDED METAL ARC WELDING

PREPARED BY A COMMITTEE CONSISTING OF:

D. R. Amos, Chairman
Westinghouse Electric Corp.

D. A. Fink
Lincoln Electric Co.

J. R. Hannahs
Midmark Corporation

R. W. Heid
Newport News Shipbuilding

A. R. Hollins
Duke Power Co.

J. E. Mathers
Welding Consultants, Inc.

L. C. Northard*
Tennessee Valley Authority

M. Parekh
Hobart Brothers Co.

A. Pollack
Consultant

M. S. Sierdzinski
Alloy Rods Co.

M. J. Tomsic
Plastronic Inc.

WELDING HANDBOOK COMMITTEE MEMBER:

D. R. Amos
Westinghouse Electric Corp.

* *Deceased*

CHAPTER **2**

SHIELDED METAL ARC WELDING

FUNDAMENTALS OF THE PROCESS

DEFINITION AND GENERAL DESCRIPTION

SHIELDED METAL ARC welding (SMAW) is an arc welding process in which coalescence of metals is produced by heat from an electric arc that is maintained between the tip of a covered electrode and the surface of the base metal in the joint being welded.

The core of the covered electrode consists of either a solid metal rod of drawn or cast material or one fabricated by encasing metal powders in a metallic sheath. The core rod conducts the electric current to the arc and provides filler metal for the joint. The primary functions of the electrode covering are to provide arc stability and to shield the molten metal from the atmosphere with gases created as the coating decomposes from the heat of the arc.

The shielding employed, along with other ingredients in the covering and the core wire, largely controls the mechanical properties, chemical composition, and metallurgical structure of the weld metal, as well as the arc characteristics of the electrode. The composition of the electrode covering varies according to the type of electrode.

PRINCIPLES OF OPERATION

SHIELDED METAL ARC welding is by far the most widely used of the various arc welding processes. It employs the heat of the arc to melt the base metal and the tip of a consumable covered electrode. The electrode and the work are part of an electric circuit illustrated in Figure 2.1. This circuit begins with the electric power source and includes the welding cables, an electrode holder, a workpiece connection, the workpiece (weldment), and an arc welding electrode. One of the two cables from the power source is attached to the work. The other is attached to the electrode holder.

Welding commences when an electric arc is struck between the tip of the electrode and the work. The intense heat of the arc melts the tip of the electrode and the surface of the the work close to the arc. Tiny globules of molten metal rapidly form on the tip of the electrode, then transfer through the arc stream into the molten weld pool.[1] In this manner, filler metal is deposited as the electrode is progressively consumed. The arc is moved over the work at an appropriate arc length and travel speed, melting and fusing a portion of the base metal and continuously adding filler metal. Since the arc is one of the hottest of the commercial sources of heat [temperatures above 9000°F (5000°C) have been measured at its center], melting of the base metal takes place almost instantaneously upon arc initiation. If welds are made in either the flat or the horizontal position, metal transfer is induced by the force of gravity, gas expansion, electric and electromagnetic forces, and surface tension. For welds in other positions, gravity works against the other forces.

The process requires sufficient electric current to melt both the electrode and a proper amount of base metal. It also requires an appropriate gap between the tip of the electrode and the base metal or the molten weld pool. These requirements are necessary to set the stage for coalescence. The sizes and types of electrodes for shielded metal arc welding define the arc voltage requirements (within the overall range of 16 to 40 V) and the amperage requirements (within the overall range of 20 to 550 A). The current may be either alternating or direct, depending upon the electrode being used, but the power source must be able to control the level of current within a reasonable range in order to respond to the complex variables of the welding process itself.

---

1. Metal transfer across the welding arc is described in Chapter 2, "Physics of Welding," *Welding Handbook*, Vol. 1, 8th ed., pp. 50–54.

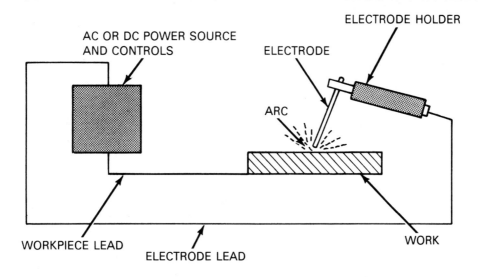

Figure 2.1—Elements of a Typical Welding Circuit for Shielded Metal Arc Welding

Covered Electrodes

IN ADDITION TO establishing the arc and supplying filler metal for the weld deposit, the electrode introduces other materials into or around the arc, or both. Depending upon the type of electrode being used, the covering performs one or more of the following functions:

(1) Provides a gas to shield the arc and prevent excessive atmospheric contamination of the molten filler metal.

(2) Provides scavengers, deoxidizers, and fluxing agents to cleanse the weld and prevent excessive grain growth in the weld metal.

(3) Establishes the electrical characteristics of the electrode.

(4) Provides a slag blanket to protect the hot weld metal from the air and enhance the mechanical properties, bead shape, and surface cleanliness of the weld metal.

(5) Provides a means of adding alloying elements to change the mechanical properties of the weld metal.

Functions 1 and 4 prevent the pickup of oxygen and nitrogen from the air by the molten filler metal in the arc stream and by the weld metal as it solidifies and cools.

The covering on shielded metal arc electrodes is applied by either the extrusion or the dipping process. Extrusion is much more widely used. The dipping process is used primarily for cast and some fabricated core rods. In either case, the covering contains most of the shielding, scavenging, and deoxidizing materials. Most SMAW electrodes have a solid metal core. Some are made with a fabricated or composite core consisting of metal powders encased in a metallic sheath. In this latter case, the purpose of some or even all of the metal powders is to produce an alloy weld deposit.

In addition to improving the mechanical properties of the weld metal, electrode coverings can be designed for welding with alternating current (ac). With ac, the welding arc goes out and is reestablished each time the current reverses its direction. For good arc stability, it is necessary to have a gas in the arc stream that will remain ionized during each reversal of the current. This ionized gas makes possible the reignition of the arc. Gases that readily ionize are available from a variety of compounds, including those that contain potassium. It is the incorporation of these compounds in the electrode covering that enables the electrode to operate on ac.

To increase the deposition rate, the coverings of some carbon and low alloy steel electrodes contain iron powder. The iron powder is another source of metal available for deposition, in addition to that obtained from the core of the electrode. The presence of iron powder in the covering also makes more efficient use of the arc energy. Metal powders other than iron are frequently used to alter the mechanical properties of the weld metal.

The thick coverings on electrodes with relatively large amounts of iron powder increase the depth of the crucible at the tip of the electrode. This deep crucible helps to contain the heat of the arc and permits the use of the drag technique (described in the next paragraph) to maintain a constant arc length. When iron or other metal powders are added in relatively large amounts, the deposition rate and welding speed usually increase.

Iron powder electrodes with thick coverings reduce the level of skill needed to weld. The tip of the electrode can be dragged along the surface of the work while maintaining a welding arc. For this reason, heavy iron powder electrodes frequently are called *drag electrodes*. Deposition rates are high, but, because slag solidification is slow, these electrodes are not suitable for out-of-position use.

Arc Shielding

THE ARC SHIELDING action, illustrated in Figure 2.2, is essentially the same for all electrodes, but the specific method of shielding and the volume of slag produced vary from type to type. The bulk of the covering materials on some electrodes is converted to gas by the heat of the arc, and only a small amount of slag is produced. This type of electrode depends largely upon a gaseous shield to prevent atmospheric contamination. Weld metal from such electrodes can be identified by the incomplete or light layer of slag which covers the bead.

For electrodes at the other extreme, the bulk of the covering is converted to slag by the heat of the arc, and only a small volume of shielding gas is produced. The tiny globules of metal being transferred across the arc are entirely coated with a thin film of molten slag. This molten slag floats to the surface of the weld puddle because it is lighter than the metal. The slag solidifies after the weld metal has solidified. Welds made with these electrodes are identified by the heavy slag deposits that completely cover the weld beads. Between these extremes is a wide variety of electrode types, each with a different combination of gas and slag shielding.

Variations in the amount of slag and gas shielding also influence the welding characteristics of covered electrodes. Electrodes which produce a heavy slag can carry high amperage and provide high deposition rates, making them ideal for heavy weldments in the flat position. Electrodes which produce a light slag layer are used with lower amperage and provide lower deposition rates. These electrodes produce a smaller weld pool and are suitable for making welds in all positions. Because of the differences in their welding characteristics, one type of covered electrode usually will be best suited for a given application.

PROCESS CAPABILITIES AND LIMITATIONS

SHIELDED METAL ARC welding is one of the most widely used processes, particularly for short welds in production, maintenance and repair work, and for field construction. The following are advantages of this process:

(1) The equipment is relatively simple, inexpensive, and portable.

(2) The filler metal, and the means of protecting it and the weld metal from harmful oxidation during welding, are provided by the covered electrode.

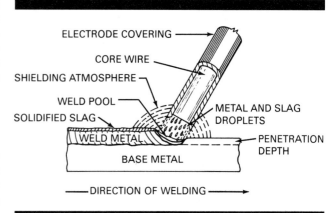

Figure 2.2–Shielded Metal Arc Welding

(3) Auxiliary gas shielding or granular flux is not required.

(4) The process is less sensitive to wind and draft than gas shielded arc welding processes.

(5) It can be used in areas of limited access.

(6) The process is suitable for most of the commonly used metals and alloys.

SMAW electrodes are available to weld carbon and low alloy steels, stainless steels, cast irons, copper, and nickel and their alloys, and for some aluminum applications. Low melting metals, such as lead, tin, and zinc, and their alloys, are not welded with SMAW because the intense heat of the arc is too high for them. SMAW is not suitable for reactive metals such as titanium, zirconium, tantalum, and columbium because the shielding provided is inadequate to prevent oxygen contamination of the weld.

Covered electrodes are produced in lengths of 9 to 18 in. (230 to 460 mm). As the arc is first struck, the current flows the entire length of the electrode. The amount of current that can be used, therefore, is limited by the electrical resistance of the core wire. Excessive amperage overheats the electrode and breaks down the covering. This, in turn, changes the arc characteristics and the shielding that is obtained. Because of this limitation, deposition rates are generally lower than for a welding process such as GMAW.

Operator duty cycle and overall deposition rates for covered electrodes are usually less than provided with a continuous electrode process such as FCAW. This is because electrodes can be consumed only to some certain minimum length. When that length has been reached, the welder must discard the unconsumed electrode stub and insert a new electrode into the holder. In addition, slag usually must be removed at starts and stops and before depositing a weld bead next to or onto a previously deposited bead.

EQUIPMENT

POWER SOURCES

Type of Output Current

EITHER ALTERNATING CURRENT (ac) or direct current (dc) may be employed for shielded metal arc welding, depending upon the current supplied by the power source and the electrode selected. The specific type of current employed influences the performance of the electrode. Each current type has its advantages and limitations, and these must be considered when selecting the type of current for a specific application. Factors which need to be considered are as follows:

Voltage Drop. Voltage drop in the welding cables is lower with ac. This makes ac more suitable if the welding is to be done at long distances from the power supply. However, long cables which carry ac should not be coiled because the inductive losses encountered in such cases can be substantial.

Low Current. With small diameter electrodes and low welding currents, dc provides better operating characteristics and a more stable arc.

Arc Starting. Striking the arc is generally easier with dc, particularly if small diameter electrodes are used. With ac, the welding current passes through zero each half cycle, and this presents problems for arc starting and arc stability.

Arc Length. Welding with a short arc length (low arc voltage) is easier with dc than with ac. This is an important consideration, except for the heavy iron powder electrodes. With those electrodes, the deep crucible formed by the heavy covering automatically maintains the proper arc length when the electrode tip is dragged on the surface of the joint.

Arc Blow. Alternating current rarely presents a problem with arc blow because the magnetic field is constantly reversing (120 times per second). Arc blow can be a significant problem with dc welding of ferritic steel because of unbalanced magnetic fields around the arc.[2]

Welding Position. Direct current is somewhat better than ac for vertical and overhead welds because lower amperage can be used. With suitable electrodes, however, satisfactory welds can be made in all positions with ac.

---

2. The influence of magnetic fields on arcs and arc blow is discussed in Chapter 2, "Physics of Welding," *Welding Handbook*, Vol. 1, 8th ed., pp. 47–49.

Metal Thickness. Both sheet metal and heavy sections can be welded using dc. The welding of sheet metal with ac is less desirable than with dc. Arc conditions at low current levels required for thin materials are less stable on ac power than on dc power.

Review of a welding application will generally indicate whether alternating or direct current is most suitable. Power sources are available as dc, ac, or combination ac/dc units. The power source for the SMAW process must be a constant-current type rather than a constant-voltage type, because it is difficult for a welder to hold the constant arc length required with constant-voltage power sources.

Significance of the Volt-Ampere Curve

FIGURE 2.3 SHOWS typical volt-ampere output characteristics for both ac and dc power sources. Constant-voltage power sources are not suitable for SMAW because with their flat volt-ampere curve, even a small change in arc length (voltage) produces a relatively large change in amperage. A constant-current power source is preferred for manual welding, because the steeper the slope of the volt-ampere curve (within the welding range), the smaller the change in current for a given change in arc voltage (arc length).

For applications that involve large diameter electrodes and high welding currents, a steep volt-ampere curve is desirable.

Where more precise control of the size of the molten pool is required (out-of-position welds and root passes of joints with varying fit-up, for example), a flatter volt-ampere curve is desirable. This enables the welder to change the welding current within a specific range simply by changing arc length. In this manner, the welder has some control over the amount of filler metal that is being deposited. Figure 2.4 portrays these different volt-ampere curves for a typical welding power source. Even though the difference in the slope of the various curves is substantial, the power source is still considered a constant-current power source. The changes shown in the volt-ampere curve are accomplished by adjusting both the open circuit voltage (OCV) and the current settings on the power source.

Open Circuit Voltage

OPEN CIRCUIT VOLTAGE, WHICH is the voltage set on the power source, does not refer to arc voltage. Arc voltage is the voltage between the electrode and the work during welding and is determined by arc length for any given electrode. Open circuit voltage, on the other hand, is the voltage generated by the welding machine when no welding is being done. Open circuit voltages generally run between

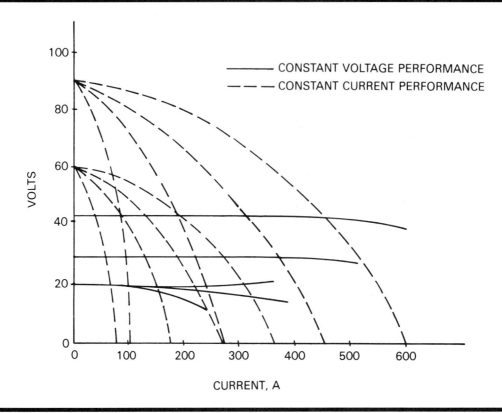

Figure 2.3–Typical Volt-Ampere Curves for Constant Current and Constant Voltage Power Sources

50 and 100 V, whereas arc voltages are between 17 and 40 V. The open circuit voltage drops to the arc voltage when the arc is struck and the welding load comes on the machine. The arc length and the type of electrode being used determine just what this arc voltage will be. If the arc is lengthened, the arc voltage will increase and the welding current will decrease. The change in amperage which a change in arc length produces is determined by the slope of the volt-ampere curve within the welding range.

Some power sources do not provide for control of the open circuit voltage because this control is not needed for all welding processes. It is a useful feature for SMAW, yet it is not necessary for all applications of the process.

Power Source Selection

SEVERAL FACTORS NEED to be considered when a power source for SMAW is selected:

(1) The type of welding current required
(2) The amperage range required
(3) The positions in which welding will be done
(4) The primary power available at the work station

Selection of the type of current, ac, dc, or both, will be based largely on the types of electrodes to be used and the kind of welds to be made. For ac, a transformer or an alternator type of power source may be used. For dc, transformer-rectifier or motor-generator power sources are available. When both ac and dc will be needed, a single-phase transformer-rectifier or an alternator-rectifier power source may be used. Otherwise, two welding machines will be required, one for ac and one for dc.

The amperage requirements will be determined by the sizes and types of electrodes to be used. When a variety will be encountered, the power supply must be capable of providing the amperage range needed. The duty cycle must be adequate.[3]

The positions in which welding will be done should also be considered. If vertical and overhead welding are planned, adjustment of the slope of the V-A curve probably will be desirable (see Figure 2.4). If so, the power supply must provide this feature. This usually requires controls for both the output voltage and the current.

---

3. See Chapter 1, pp. 14–15, for an explanation of duty cycle.

A supply of primary power is needed. If line power is available, it should be determined whether the power is single-phase or three-phase. The welding power source must be designed for either single- or three-phase power, and it must be used with the one it was designed for. If line power is not available, an engine-driven generator or alternator must be used.

ACCESSORY EQUIPMENT

Electrode Holder

AN ELECTRODE HOLDER is a clamping device which allows the welder to hold and control the electrode, as shown in Figure 2.5. It also serves as a device for conducting the

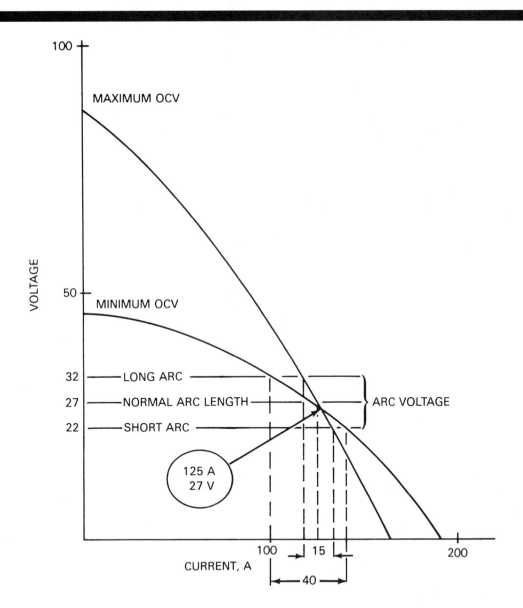

NOTE: LOWER SLOPE GIVES A GREATER CHANGE IN WELDING CURRENT FOR A GIVEN CHANGE IN ARC VOLTAGE.

Figure 2.4–The Effect of Volt-Ampere Curve Slope on Current Output With a Change in Arc Voltage

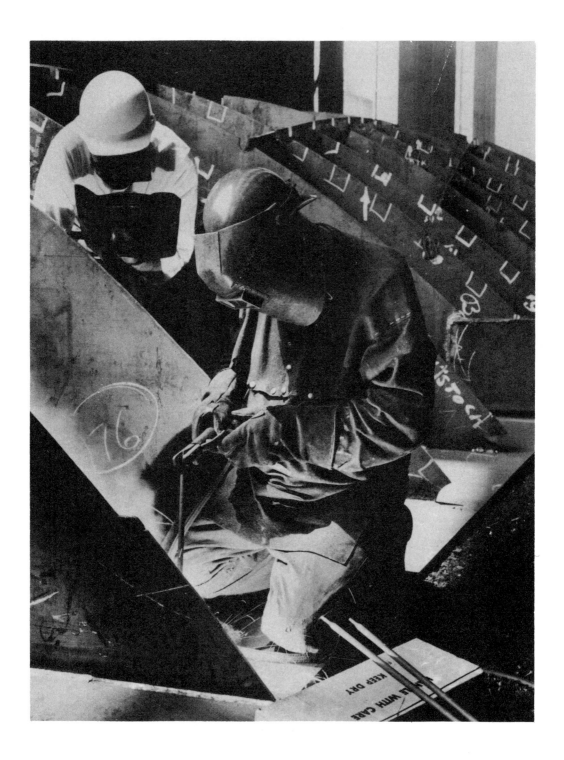

Figure 2.5–Welding a Structure With the Shielded Metal Arc Welding Process

welding current from the welding cable to the electrode. An insulated handle on the holder separates the welder's hand from the welding circuit. The current is transferred to the electrode through the jaws of the holder. To assure minimum contact resistance and to avoid overheating of the holder, the jaws must be kept in good condition. Overheating of the holder not only makes it uncomfortable for the welder, but also it can cause excessive voltage drop in the welding circuit. Both can impair the welder's performance and reduce the quality of the weld.

The holder must grip the electrode securely and hold it in position with good electrical contact. Installation of the electrode must be quick and easy. The holder needs to be light in weight and easy to handle, yet it must be sturdy enough to withstand rough use. Most holders have insulating material around the jaws to prevent grounding of the jaws to the work.

Electrode holders are produced in sizes to accommodate a range of standard electrode diameters. Each size of holder is designed to carry the current required for the largest diameter electrode that it will hold. The smallest size holder that can be used without overheating is the best one for the job. It will be the lightest, and it will provide the best operator comfort.

Workpiece Connection

A WORKPIECE CONNECTION is a device for connecting the workpiece lead to the workpiece. It should produce a strong connection, yet be able to be attached quickly and easily to the work. For light duty, a spring-loaded clamp may be suitable. For high currents, however, a screw clamp may be needed to provide a good connection without overheating the clamp.

Welding Cables

WELDING CABLES ARE used to connect the electrode holder and the ground clamp to the power source. They are part of the welding circuit (see Figure 2.1). The cable is constructed for maximum flexibility to permit easy manipulation, particularly of the electrode holder. It also must be wear and abrasion resistant.

Welding cable consists of many fine copper or aluminum wires stranded together and enclosed in a flexible, insulating jacket. The jacket is made of synthetic rubber or of a plastic that has good toughness, high electrical resistance, and good heat resistance. A protective wrapping is placed between the stranded conductor wires and the insulating jacket to permit some movement between them and provide maximum flexibility.

Welding cable is produced in a range of sizes (from about AWG 6 to 4/0[4]). The size of the cable required for a particular application depends on the maximum amperage to be used for welding, the length of the welding circuit (welding and work cables combined), and the duty cycle of the welding machine. Table 2.1 shows the recommended size of copper welding cable for various power sources and circuit lengths. When aluminum cable is used, it should be two AWG sizes larger than copper cable for the application. Cable sizes are increased as the length of the welding circuit increases to keep the voltage drop and the attendant power loss in the cable at acceptable levels.

If long cables are necessary, short sections can be joined by suitable cable connectors. The connectors must provide good electrical contact with low resistance, and their insulation must be equivalent to that of the cable. Lugs, at the end of each cable, are used to connect the cables to the power source. The connection between the cable and a connector or lug must be strong with low electrical resistance. Soldered joints and mechanical connections are

---

4. American wire gage sizes.

Table 2.1
Recommended Copper Welding Cable Sizes

Power Source		Awg Cable Size for Combined Length of Electrode and Ground Cables				
Size in Amperes	Duty Cycle, %	0 to 50 ft (0 to 15 m)	50 to 100 ft (15 to 30 m)	100 to 150 ft (30 to 46 m)	150 to 200 ft (46 to 61 m)	200 to 250 ft (61 to 76 m)
100	20	6	4	3	2	1
180	20-30	4	4	3	2	1
200	60	2	2	2	1	1/0
200	50	3	3	2	1	1/0
250	30	3	3	2	1	1/0
300	60	1/0	1/0	1/0	2/0	3/0
400	60	2/0	2/0	2/0	3/0	4/0
500	60	2/0	2/0	3/0	3/0	4/0
600	60	2/0	2/0	3/0	4/0	*

* Use two 3/0 cables in parallel.

used. Aluminum cable requires a good mechanical connection to avoid overheating. Oxidation of the aluminum significantly increases the electrical resistance of the connection. This of course, can lead to overheating, excessive power loss, and cable failure.

Care must be taken to avoid damage to the jacket of the cable, particularly for the electrode cable. Contact with hot metal or sharp edges may penetrate the jacket and ground the cable.

Helmet

THE PURPOSE OF the helmet is to protect the welder's eyes, face, forehead, neck, and ears from the direct rays of the arc and from flying sparks and spatter. Some helmets have an optional "flip lid" which permits the dark filter plate over the opening in the shield to be flipped up so the welder can see while the slag is being chipped from the weld. This protects the welder's face and eyes from flying slag. Slag can cause serious injury if it strikes a person, particularly while it is hot. It can be harmful to the eyes whether it is hot or cold.

Helmets are generally constructed of pressed fiber or fiberglass insulating material. A helmet should be light in weight and should be designed to give the welder the greatest possible comfort. The welder in Figure 2.5 has a helmet on. The observer is using a hand shield.

Miscellaneous Equipment

CLEANLINESS IS IMPORTANT in welding. The surfaces of the workpieces and the previously deposited weld metal must be cleaned of dirt, slag, and any other foreign matter that would interfere with welding. To accomplish this, the welder should have a steel wire brush, a hammer, a chisel, and a chipping hammer. These tools are used to remove dirt and rust from the base metal, cut tack welds, and chip slag from the weld bead.

The joint to be welded may require backing to support the molten weld pool during deposition of the first layer of weld metal. Backing strips or nonmetallic backing materials are sometimes used, particularly for joints which are accessible from only one side.

MATERIALS

BASE METALS

THE SMAW PROCESS is used in joining and surfacing applications on a variety of base metals. The suitability of the process for any specific base metal depends on the availability of a covered electrode whose weld metal has the required composition and properties. Electrodes are available for the following base metals:

(1) Carbon steels
(2) Low alloy steels
(3) Corrosion resisting steels
(4) Cast irons (ductile and gray)
(5) Aluminum and aluminum alloys
(6) Copper and copper alloys
(7) Nickel and nickel alloys

Electrodes are available for application of wear, impact, or corrosion resistant surfaces to these same base metals.

COVERED ELECTRODES

COVERED ELECTRODES ARE classified according to the requirements of specifications issued by the American Welding Society. Certain agencies of the Department of Defense also issue specifications for covered electrodes. The AWS specification numbers and their electrode classifications

are given in Table 2.2. The electrodes are classified on the basis of the chemical composition or mechanical properties, or both, of their undiluted weld metal. Carbon steel, low alloy steel, and stainless steel electrodes are also classified according to the type of welding current they are suited for and sometimes according to the positions of welding that they can be used in.

Carbon Steel Electrodes

IN ANSI/AWS A5.1, *Specification for Covered Carbon Steel Arc Welding Electrodes,* a simple numbering system is used for electrode classification. In E6010, for example,

Table 2.2
AWS Specifications for Covered Electrodes

Type of Electrode	AWS Specification
Carbon steel	A5.1
Low alloy steel	A5.5
Corrosion resistant steel	A5.4
Cast iron	A5.15
Aluminum and aluminum alloys	A5.3
Copper and copper alloys	A5.6
Nickel and nickel alloys	A5.11
Surfacing	A5.13 and A5.21

the E designates an electrode. The first two digits (60) signify the minimum tensile strength of the undiluted weld metal in ksi, in the as-welded condition. The third digit represents the welding position (1, in this case, refers to all positions). The last digit refers to the covering type and type of current with which the electrode can be used.

There are two strength levels of carbon steel electrodes: the 60 series and the 70 series. The minimum allowable tensile strength of the weld metal for the 60 series is 62 ksi (427 MPa), although additional elongation may allow some of these to go as low as 60 ksi (414 MPa). For the 70 series, it is 72 ksi (496 MPa) and, again, some of these may go as low as 70 ksi (483 MPa), with additional elongation. Maximum chemical composition limits for significant elements are provided within the applicable AWS specifications for most electrode classifications. Charpy V-notch impact requirements are given for some electrodes in both series.

Certain of the carbon steel electrodes are designed to operate only on dc. Others are for either ac or dc. Polarity on dc usually is reverse (electrode positive), although a few of the electrodes are intended for straight polarity. Some of these may be used with either polarity.

Most electrodes are designed for welding in all positions. However, those which contain large amounts of iron powder or iron oxide in the coatings are generally restricted to groove welds in the flat position and horizontal fillet welds. The coverings on these electrodes are very heavy, which precludes their operation in the vertical and overhead positions.

Several electrodes of the 70 series are low hydrogen type. Their coatings are formulated with ingredients that are low in moisture and cellulose and, hence, in hydrogen content. Hydrogen is responsible for low ductility and for underbead cracking sometimes encountered in highly restrained welds. For this reason, low hydrogen electrodes are used to weld hardenable steels. They are also used for high sulfur steels and to provide weld metal having good low temperature notch toughness.

The specification does not set a limit on the moisture content of these electrodes, but less than 0.6 percent is recommended. To control moisture, proper storage and handling are required. Typical storage and baking conditions are given in ANSI/AWS A5.1.

Low Alloy Steel Electrodes

ANSI/AWS A5.5, *Specification for Low Alloy Steel Covered Arc Welding Electrodes*, classifies low alloy steel covered electrodes according to a numbering system which is similar to that just described for carbon steel electrodes. It uses, in addition, a suffix such as A1 to designate the chemical composition (alloy system) of the weld metal. Thus, a complete electrode classification is E7010-A1. Another is E8016-C2. Alloy systems into which the electrodes fall are carbon-molybdenum steel, chromium-molybdenum steel, nickel steel, and manganese-molybdenum steel. Weld

metal strength levels range from 70 to 120 ksi (480 to 830 MPa) minimum tensile strength, in 10 ksi (70 MPa) increments. In this specification, weld metals that are commonly used in the as-welded condition are classified on the basis of their properties in that condition. Similarly, those that are commonly used in the stress-relieved condition are classified on the basis of their properties after a stress relief heat treatment.

In this connection, it should be noted that the stress relief called for in ANSI/AWS A5.5 consists of holding the test assembly at temperature for one hour. Fabricators using holding times that are significantly different from one hour at temperature may have to be more selective in the electrodes they use, and may be required to run tests to demonstrate that the selected weld metal mechanical properties will be adequate after a specific heat treatment temperature and holding time. Radiographic quality standards for deposited weld metal and notch toughness requirements are included for many SMAW electrode classifications.

The military specifications for low alloy steel electrodes sometimes use designations that are similar to those in the AWS specification. Also, some electrodes are produced that are not classified in AWS specifications, but which are designed for specific materials or which broadly match standard AISI low alloy steel base metal compositions, such as 4130. The A5.5 specification sets limits on the moisture content of low hydrogen electrodes packaged in hermetically sealed containers. These limits range from 0.2 percent to 0.6 percent by weight, depending on the classification of the electrode. The higher the strength level, the lower the limit on the moisture content. The reason for this is that moisture is a primary source of hydrogen, and hydrogen can produce cracking in most low alloy steels, unless high preheats and long, slow cooling cycles are employed. The higher the strength of the weld and the base metal, the greater the need for low moisture levels to avoid cracking. Exposure to high humidity (in the range of 70 percent relative humidity or higher) may increase the moisture content of the electrode in only a few hours.

Corrosion Resisting Steel Electrodes

COVERED ELECTRODES FOR welding corrosion resisting steels are classified in ANSI/AWS A5.4, *Specification for Covered Corrosion-Resisting Chromium and Chromium-Nickel Steel Welding Electrodes*. Classification in this specification is based on the chemical composition of the undiluted weld metal, the positions of welding, and the type of welding current for which the electrodes are suitable. The classification system is similar to the one used for carbon and low alloy steel electrodes. Taking E310-15 and E310-16 as examples, the prefix E indicates an electrode. The first three digits refer to the alloy type (with respect to chemical composition). They may be followed by a letter or letters to indicate modification, such as E310Mo-15. The last two digits refer to the position of welding and the type of current for which the electrodes are suitable. The

-1 indicates that the electrodes are usable in all positions through 5/32 in. (4 mm) diameter. The number 5 indicates that the electrodes are suitable for use with dcrp (electrode positive). The number 6 means that electrodes are suitable for either ac or dcrp (electrode positive). Electrodes over 5/32 in. (4 mm) diameter are for use in the flat and horizontal positions.

The specification does not describe the covering ingredients, but -15 coverings usually contain a large proportion of limestone (calcium carbonate). This ingredient provides the CO and CO_2 that are used to shield the arc. The binder which holds the ingredients together in this case is sodium silicate. The -16 covering also contains limestone for arc shielding. In addition, it usually contains considerable titania (titanium dioxide) for arc stability. The binder in this case is likely to be potassium silicate.

Differences in the proportion of these ingredients result in differences in arc characteristics. The -15 electrodes (lime type coverings) tend to provide a more penetrating arc and to produce a more convex and coarsely rippled bead. The slag solidifies relatively rapidly so that these electrodes often are preferred for out of position work, such as pipe welding. On the other hand, the -16 coverings (titania type) produce a smoother arc, less spatter, and a more uniform, finely rippled bead. The slag, however, is more fluid, and the electrode usually is difficult to handle in out-of-position work.

Stainless steels can be separated into three basic types: austenitic, martensitic, and ferritic. The austenitic group (2XX and 3XX) is, by far, the largest one. Normally, the composition of the weld metal from a stainless steel electrode is similar to the base metal composition that the electrode is designed to weld.

For austenitic stainless steels, the composition of weld metal differs from the base metal compositions slightly in order to produce a weld deposit which contains ferrite (i.e., which is not fully austenitic) to prevent fissuring or hot cracking of the weld metal. The amount of ferrite common to the various welding electrodes is discussed in ANSI/AWS A5.4 in some detail. In general, a minimum ferrite content in the range of 3 to 5 ferrite number is sufficient to prevent cracking. Ferrite content as high as 20 FN may be acceptable for some welds when no postweld heat treatment is employed. The Schaeffler diagram, or the DeLong modification of a portion of that diagram, can be used to predict the ferrite content of stainless steel weld metals. Magnetic instruments are available to make direct measurements of the ferrite content of deposited weld metal. (See ANSI/AWS A4.2, *Standard Procedures for Calibrating Magnetic Instruments to Measure Delta Ferrite Content of Austenitic Stainless Steel Weld Metal.*)

Certain austenitic stainless steel weld metals (Types 310, 320, and 330, for example) do not form ferrite because their nickel content is too high. For these materials, the phosphorus, sulfur, and silicon content of the weld metal is limited or the carbon content increased as a means of minimizing fissuring and cracking.

Appropriate welding procedures can also be used to reduce fissuring and cracking. Low amperage, for example, is beneficial. Some small amount of weaving, as a means of promoting cellular grain growth, may be helpful. Proper procedures in terminating the arc should be used to avoid crater cracks.

ANSI/AWS A5.4 contains two covered electrode classifications for the straight chromium stainless steels (4XX series). One contains 11 to 13.5 percent chromium, the other, 15 to 18 percent. The carbon content of both is 0.1 percent maximum. Both weld metals are air hardening and their weldments require preheat and postheat treatment to provide the ductility which is necessary in most engineering applications.

The specification also contains three electrode classifications that are used to weld the four to ten percent chromium-molybdenum steels. These materials, too, are air hardening, and preheat and postheat treatment are required for sound, serviceable joints.

Nickel and Nickel Alloy Electrodes

COVERED ELECTRODES FOR shielded metal arc welding of nickel and its alloys have compositions which are generally similar to those of the base metals they are used to join, and some have additions of elements such as titanium, manganese, and columbium to deoxidize the weld metal and thereby prevent cracking.

ANSI/AWS A5.11, *Specification for Nickel and Nickel Alloy Covered Welding Electrodes*, classifies the electrodes in groups according to their principal alloying elements. The letter "E" at the beginning indicates an electrode, and the chemical symbol "Ni" identifies the weld metals as nickel base alloys. Other chemical symbols are added to show the principal alloying elements. Then, successive numbers are added to identify each classification within its group. ENiCrFe-1, for example, contains significant chromium and iron in addition to nickel.

Most of the electrodes are intended for use with dcrp (electrode positive). Some are also capable of operating with ac, to overcome problems that may be encountered with arc blow (when nine percent nickel steel is welded, for example). Most electrodes can be used in all positions, but best results for out-of-position welding are achieved using electrodes of 1/8 in. (3.2 mm) diameter and smaller.

The electrical resistivity of the core wire in these electrodes is exceptionally high. For this reason, excessive amperage will overheat the electrode and damage the covering, causing arc instability and unacceptable amounts of spatter. Each classification and size of electrode has an optimum amperage range.

Aluminum and Aluminum Alloy Electrodes

ANSI/AWS A5.3, *Specification for Aluminum and Aluminum Alloy Electrodes for Shielded Metal Arc Welding*, contains two classifications of covered electrodes for the welding of aluminum base metals. These classifications are based on the mechanical properties of the weld metal in the as-welded condition and the chemical composition of the core wire. One core wire is commercially pure aluminum (1100), and the other is an aluminum-five percent silicon alloy (4043). Both electrodes are used with dcrp (electrode positive).

The covering on these electrodes has three functions. It provides a gas to shield the arc, a flux to dissolve the aluminum oxide, and a protective slag to cover the weld bead. Because the slag can be very corrosive to aluminum, it is important that all of it be removed upon completion of the weld.

The presence of moisture in the covering of these electrodes is a major source of porosity in the weld metal. To avoid this porosity, the electrodes should be stored in a heated cabinet until they are to be used. Those electrodes that have been exposed to moisture should be reconditioned (baked) before use, or discarded.

One difficulty which may occur in welding is the fusing of slag over the end of the electrode if the arc is broken. In order to restrike the arc, this fused slag must be removed.

Covered aluminum electrodes are used primarily for noncritical welding and repair applications. They should be used only on base metals for which either the 1100 or 4043 filler metals are recommended. These weld metals do not respond to precipitation hardening heat treatments. If they are used for such material, each application should be carefully evaluated.

Copper and Copper Alloy Electrodes

ANSI/AWS A5.6, *Specification for Covered Copper and Copper Alloy Arc Welding Electrodes*, classifies copper and copper alloy electrodes on the basis of their all-weld-metal properties and the chemical composition of their undiluted weld metal. The designation system is similar to that used for nickel alloy electrodes. The major difference is that each individual classification within a group is identified by a letter. This letter is sometimes followed by a number, as in ECuAl-A2, for example. The groups are: CuSi for silicon bronze, CuSn for phosphor bronze, CuNi for copper-nickel, and CuAl for aluminum bronze. These electrodes, generally, are used with dcrp (electrode positive).

Copper electrodes are used to weld unalloyed copper and to repair copper cladding on steel or cast iron. Silicon bronze electrodes are used to weld copper-zinc alloys, copper, and some iron base materials. They are also used for surfacing to provide corrosion resistance.

Phosphor bronze and brass base metals are welded with phosphor bronze electrodes. These electrodes are also used to braze weld copper alloys to steel and cast iron. The phosphor bronzes are rather viscous when molten, but their fluidity is improved by preheating to about 400°F (200°C). The electrodes and the work must be dry.

Copper-nickel electrodes are used to weld a wide range of copper-nickel alloys and also copper-nickel

cladding on steel. In general, no preheat is necessary for these materials.

Aluminum bronze electrodes have broad use for welding copper base alloys and some dissimilar metal combinations. They are used to braze weld many ferrous metals and to apply wear and corrosion resistant bearing surfaces. Welding is usually done in the flat position with some preheat.

Electrodes for Cast Iron

ANSI/AWS A5.15, *Specification for Welding Electrodes and Rods for Cast Iron*, classifies covered electrodes for welding cast iron. The electrodes classified in A5.15 are nickel, nickel-iron, nickel-copper alloys, and one steel alloy. Preheat is recommended when welding iron castings, particularly if the steel electrode is used. The specific temperature depends on the size and complexity of the casting and the machinability requirements. Small pits and cracks can be welded without preheat, but the weld will not be machinable. Welding is done with low amperage dcrp (electrode positive) to minimize dilution with the base metal. Preheating is not used in this case, except to minimize the residual stresses in other parts of the casting.

Proprietary nickel and nickel alloy electrodes may also be used to repair castings and to join the various types of cast iron to themselves and to other metals. The hardness of the weld metal depends on the amount of base metal dilution.

Phosphor bronze and aluminum bronze electrodes are used to braze weld cast iron. The melting temperature of their weld metals is below that of cast iron. The casting should be preheated to about 400°F (200°C), and welding should be done with dcrp (electrode positive), using the lowest amperage that will produce good bonding between the weld metal and groove faces. The cast iron surfaces should not be melted.

Surfacing Electrodes

MOST HARD SURFACING electrodes are designed to meet ANSI/AWS A5.13, *Specification for Solid Surfacing Welding Rods and Electrodes*, or ANSI/AWS A5.21, *Specification for Composit Surfacing Welding Rods and Electrodes*. A wide range of SMAW electrodes is available (under these and other AWS filler metal specifications) to provide wear, impact, heat, or corrosion resistant layers on a variety of base metals. All of the covered electrodes specified in A5.13 have a solid core wire. Those specified in A5.21 have a composite core. The electrode designation system in both specifications is similar to that used for copper alloy electrodes with the exception of tungsten carbide electrodes. The E in the designation for these electrodes is followed by WC. The mesh size limits for the tungsten carbide granules in the core follow these to complete the designation. The core, in this case, consists of a steel tube filled with the tungsten carbide granules.

Surfacing with covered electrodes is used for cladding, buttering, buildup, and hard surfacing. The weld deposit in these applications is intended to provide one or more of the following for the surfaces to which they are applied:

(**1**) Corrosion resistance
(**2**) Metallurgical control
(**3**) Dimensional control
(**4**) Wear resistance
(**5**) Impact resistance

Covered electrodes for a particular surfacing application should be selected after a careful review of the required properties of the weld metal when it is applied to a specific base metal.

Electrode Conditioning

SMAW ELECTRODE COVERINGS are hygroscopic (they readily absorb and retain moisture). Some coverings are more hygroscopic than others. The moisture they pick up on exposure to a humid atmosphere dissociates to form hydrogen and oxygen during welding. The atoms of hydrogen dissolve in the weld and the heat-affected zone and may cause cold cracking. This type of crack is more prevalent in hardenable steel base metals and high strength steel weld metals. Excessive moisture in electrode coverings can cause porosity in the deposited weld metal.

To minimize moisture problems, particularly for low hydrogen electrodes, they must be properly packaged, stored, and handled. Such control is critical for electrodes which are to be used to weld hardenable base metals. Control of moisture becomes increasingly important as the strength of the weld metal or the base metal increases. Holding ovens are used for low hydrogen electrodes once those electrodes have been removed from their sealed container and have not been used within a certain period of time. This period varies from as little as half an hour to as much as eight hours depending on the strength of the electrode, the humidity during exposure, and even the specific covering on the electrode. The time which an electrode can be kept out of an oven or rod warmer is reduced as the humidity increases.

The temperature of the holding oven should be within the range of 150 to 300°F (65 to 150°C). Electrodes that have been exposed too long require baking at a substantially higher temperature to drive off the absorbed moisture. The specific recommendations of the manufacturer of the electrode need to be followed because the time and temperature limitations can vary from manufacturer to manufacturer, even for electrodes within a given classification. Excessive heating can damage the covering on an electrode.

APPLICATIONS

MATERIALS

THE SMAW PROCESS can be used to join most of the common metals and alloys. The list includes the carbon steels, the low alloy steels, the stainless steels, and cast iron, as well as copper, nickel, and aluminum and their alloys. Shielded metal arc welding is also used to join a wide range of chemically dissimilar materials.

The process is not used for materials for which shielding of the arc by the gaseous products of an electrode covering is unsatisfactory. The reactive (Ti, Zr) and refractory (Cb, Ta, Mo) metals fall into this group.

THICKNESSES

THE SHIELDED METAL arc process is adaptable to any material thickness within certain practical and economic limitations. For material thicknesses less than about 1/16 in. (1.6 mm), the base metal will melt through and the molten metal will fall away before a common weld pool can be established, unless special fixturing and welding procedures are employed. There is no upper limit on thickness, but other processes such as SAW or FCAW are capable of providing higher deposition rates and economies for most applications involving thicknesses exceeding 1-1/2 in. (38 mm). Most of the SMAW applications are on thicknesses between 1/8 and 1-1/2 in (3 and 38 mm), except where irregular configurations are encountered. Such configurations put an automated welding process at an economic disadvantage. In such instances, the shielded metal arc process is commonly used to weld materials as thick as 10 in. (250 mm).

POSITION OF WELDING

ONE OF THE major advantages of SMAW is that welding can be done in any position on most of the materials for which the process is suitable. This makes the process useful on joints that cannot be placed in the flat position. Despite this advantage, welding should be done in the flat position whenever practical because less skill is required, and larger electrodes with correspondingly higher deposition rates can be used. Welds in the vertical and overhead positions require more skill on the welder's part and are performed using smaller diameter electrodes. Joint designs for vertical and overhead welding may be different from those suitable for flat position welding.

LOCATION OF WELDING

THE SIMPLICITY OF the equipment makes SMAW an extremely versatile process with respect to the location and environment of the operation. Welding can be done indoors or outdoors, on a production line, a ship, a bridge, a building framework, an oil refinery, a cross-country pipeline, or any such types of work. No gas or water hoses are needed and the welding cables can be extended quite some distance from the power source. In remote areas, gasoline or diesel powered units can be used. Despite this versatility, the process should always be used in an environment which shelters it from the wind, rain, and snow.

JOINT DESIGN AND PREPARATION

TYPES OF WELDS

WELDED JOINTS ARE designed primarily on the basis of the strength and safety required of the weldment under the service conditions imposed on it. The manner in which the service stresses will be applied and the temperature of the weldment in service must always be considered. A joint required for dynamic loading may be quite different from one permitted in static loading.

Dynamic loading requires consideration of fatigue strength and resistance to brittle fracture. These, among other things, require that the joints be designed to reduce or eliminate points of stress concentration. The design should also balance the residual stresses and obtain as low a residual stress level as possible. The weld must produce adequate joint strength.

In addition to service requirements, weld joints need to be designed to provide economy and accessibility for the welder during fabrication. Joint accessibility can improve the ability of the welder to meet good workmanship and quality requirements, and can assist in control of distortion and reduction of welding costs. The effect of joint design on some of these considerations is discussed below.

Groove Welds

GROOVE WELD JOINT designs of different types are used. Selection of the most appropriate design for a specific application is influenced by the following:

(1) Suitability for the structure under consideration
(2) Accessibility to the joint for welding
(3) Cost of welding
(4) Position in which welding is to be done

A square groove is the most economical to prepare. It only requires squaring-off of the edge of each member. This type of joint is limited to those thicknesses with which satisfactory strength and soundness can be obtained. For SMAW, that thickness is usually not greater than about 1/4 in. (6 mm) and then only when the joint is to be welded in the flat position from both sides. The type of material to be welded is also a consideration.

When thicker members are to be welded, the edge of each member must be prepared to a contour that will permit the arc to be directed to the point where the weld metal must be deposited. This is necessary to provide fusion to whatever depth is required.

For economy as well as to reduce distortion and residual stresses, the joint design should have a root opening and a groove angle that will provide adequate strength and soundness with the deposition of the least amount of filler metal. The key to soundness is accessibility to the root and sidewalls of the joint. J-groove and U-groove joints are desirable for thick sections. In very thick sections, the savings in filler metal and welding time alone are sufficient to more than offset the added cost of this joint preparation. The angle of the sidewalls must be large enough to prevent slag entrapment.

Fillet Welds

WHERE THE SERVICE requirements of the weldment permit, fillet welds frequently are used in preference to groove welds. Fillet welds require little or no joint preparation, although groove welds sometimes require less welding. Intermittent fillet welding may be used when a continuous weld would provide more strength than is required to carry the load.

A fillet weld is often combined with a groove weld to provide the required strength and reduce the stress concentration at the joint. Minimum stress concentration at the toes of the weld is obtained with concave fillets.

WELD BACKING

WHEN FULL PENETRATION welds are required and welding is done from one side of the joint, weld backing may be required. Its purpose is to provide something on which to deposit the first layer of metal and thereby prevent the molten metal in that layer from escaping through the root of the joint.

Four types of backing are commonly used:

(1) Backing strip
(2) Backing weld
(3) Copper backing bar
(4) Nonmetallic backing

Backing Strip

A BACKING STRIP is a strip of metal placed on the back of the joint, as shown in Figure 2.6(A). The first weld pass ties both members of the joint together and to the backing strip. The strip may be left in place if it will not interfere with the serviceability of the joint. Otherwise, it should be removed, in which case the back side of the joint must be accessible. If the back side is not accessible, some other means of obtaining a proper root pass must be used.

The backing strip must always be made of a material that is metallurgically compatible with the base metal and the welding electrode to be used. Where design permits, another member of the structure may serve as backing for the weld. Figure 2.6(B) provides an example of this. In all cases, it is important that the backing strip as well as the surfaces of the joint be clean to avoid porosity and slag inclusions in the weld. It is also important that the backing strip fit properly. Otherwise, the molten weld metal can run out through any gap between the strip and the base metal at the root of the joint.

Copper Backing Bar

A COPPER BAR is sometimes used as a means of supporting the molten weld pool at the root of the joint. Copper is used because of its high thermal conductivity. This high conductivity helps prevent the weld metal from fusing to the backing bar. Despite this, the copper bar must have sufficient mass to avoid melting during deposition of the first weld pass. In high production use, water can be passed through holes in the bar to remove the heat that accumulates during continuous welding. Regardless of the method of cooling, the arc should not be allowed to impinge on the copper bar, for if any copper melts, the weld metal can become contaminated with copper. The copper bar may be grooved to provide the desired root surface contour and reinforcement.

Nonmetallic Backing

NONMETALLIC BACKING OF either granular flux or refractory material is also a method that is used to produce a sound first pass. The flux is used primarily to support the weld metal and to shape the root surface. A granular flux layer is supported against the back side of the weld by some method such as a pressurized fire hose. A system of this type is generally used for production line work, although it is not widely used for SMAW.

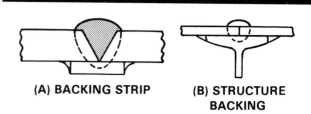

(A) BACKING STRIP   **(B) STRUCTURE BACKING**

Figure 2.6–Fusible Metal Backing for a Weld

Refractory type backing consists of a flexible, shaped form that is held on the back side of the joint by clamps or by pressure sensitive tape. This type of backing is sometimes used with the SMAW process, although special welding techniques are required to consistently produce good results. The recommendations of the manufacturer of the backing should be followed.

Backing Weld

A BACKING WELD is one or more backing passes in a single groove weld joint. This weld is deposited on the back side of the joint before the first pass is deposited on the face side. The concept is illustrated in Figure 2.7. After the backing weld, all subsequent passes are made in the groove from the face side. The root of the joint may be ground or gouged after the backing weld is made to produce sound, clean metal on which to deposit the first pass on the face side of the joint.

The backing weld can be made with the same process or with a different process from that to be used for welding the groove. If the same process is used, the electrodes should be of the same classification as those to be used for welding the groove. If a different process such as gas tungsten arc welding is used, the welding rods should deposit weld metal having composition and properties similar to

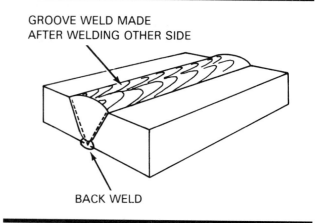

GROOVE WELD MADE
AFTER WELDING OTHER SIDE

BACK WELD

Figure 2.7–A Typical Backing Weld

those of the SMAW weld metal. The backing weld must be large enough to support any load that is placed on it. This is especially important when the weldment must be repositioned after the backing weld has been deposited and before the groove weld is made.

FIT-UP

JOINT FIT-UP INVOLVES the positioning of the members of the joint to provide the specified groove dimensions and alignment. The points of concern are the root opening and the alignment of the members along the root of the weld. Both of these have an important influence on the quality of the weld and the economics of the process. After the joint has been properly aligned throughout its length, the position of the members should be maintained by clamps or tack welds. Finger bars or U-shaped bridges can be placed across the joint and tack welded to each of the members.

If the root opening is not uniform, the amount of weld metal will vary from location to location along the joint. As a result, shrinkage, and, hence, distortion, will not be uniform. This can cause problems when the finished dimensions have been predicated on the basis of uniform, controlled shrinkage.

Misalignment along the root of the weld may cause lack of penetration in some areas or poor root surface contour, or both. Inadequate root opening can cause lack of complete joint penetration. Too wide a root opening makes welding difficult and requires more weld metal to fill the joint. This, of course, also involves additional cost. In thin members, an excessive root opening may cause excessive melt-through on the back side. It may even cause the edge of one or both members to melt away.

TYPICAL JOINT GEOMETRIES

THE WELD GROOVES shown in Figure 2.8 illustrate typical designs and dimensions of joints for shielded metal arc welding of steel. These joints are generally suitable for economically achieving sound welds. Other joint designs or changes in suggested dimensions may be required for special applications.

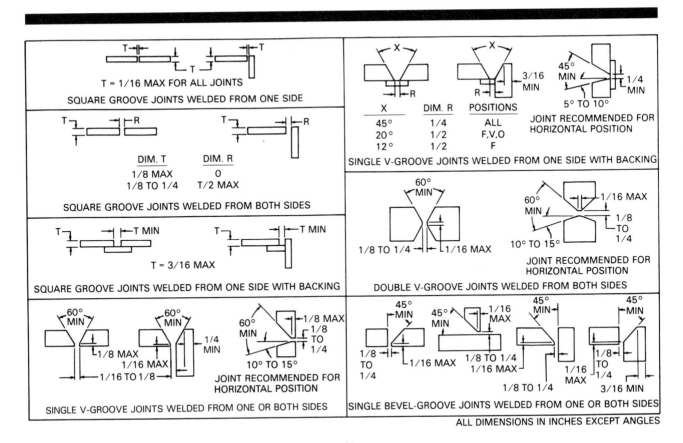

(A)

Figure 2.8–Typical Joint Geometries for Shielded Metal Arc Welding of Steel

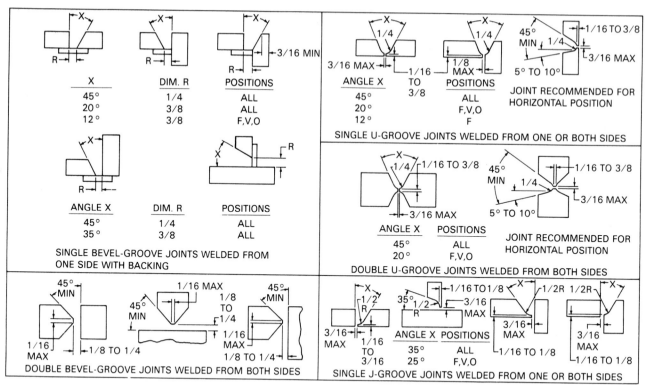

(B)

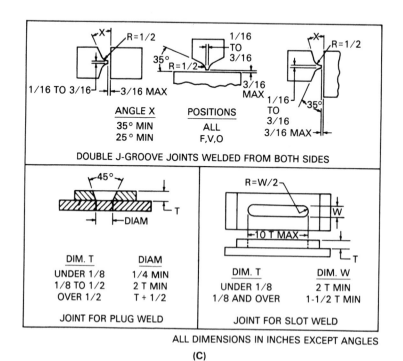

ALL DIMENSIONS IN INCHES EXCEPT ANGLES

(C)

Figure 2.8–(Continued)–Typical Joint Geometries for Shielded Metal Arc Welding of Steel

RUNOFF TABS

IN SOME APPLICATIONS, it is necessary to completely fill out the groove right to the very ends of the joint. In such cases, runoff tabs are used. They, in effect, extend the groove beyond the ends of the members to be welded. The weld is carried over into the tabs. This assures that the entire length of the joint is filled to the necessary depth with sound weld metal. A typical runoff tab is shown in Figure 2.9. Runoff tabs are excellent appendages on which to start and stop welding. Any defects in these starts and stops are located in areas that later will be discarded.

Selection of the material for runoff tabs is important. The composition of the tabs should not be allowed to adversely affect the properties of the weld metal. For example, for stainless steel which is intended for corrosion service, the runoff tabs should be of a compatible grade of stainless steel. Carbon steel tabs would be less costly, but fusion with the stainless steel filler metal would change the composition of the weld metal at the junction of the carbon steel tab and the stainless steel members of the joint. The weld metal at this location probably would not have adequate corrosion resistance.

PREHEATING

HEATING THE AREA to be welded before and during welding is required in order to achieve desired properties in the weld or the adjacent base metal, or both. Unnecessary preheat should be avoided as it takes time and energy. Excessive preheat temperatures are not cost effective and could degrade the properties and the quality of the joint. A

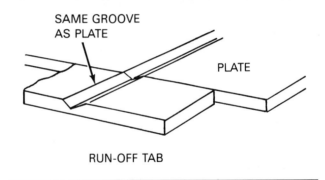

Figure 2.9—Runoff Tab at End of a Weld Joint

welder's discomfort increases with higher preheats and tends to reduce the quality of the work. Preheat temperatures employed would be based upon welding code requirements, competent technical evaluation, or the results of tests. In general, the temperature will depend on the material to be welded, the electrodes to be used, and the degree of restraint in the joint.

Hardenable steels, high-strength steels, and use of electrodes other than low hydrogen types normally require application of preheat.

Preheat is sometimes used when welding materials having high thermal conductivity, such as copper and aluminum alloys, to reduce the welding amperage required, improve penetration, and aid in fusing the weld metal to the base metal.

WELDING PROCEDURES

ELECTRODE DIAMETER

THE CORRECT ELECTRODE diameter is one that, when used with the proper amperage and travel speed, produces a weld of the required size in the least amount of time.

The electrode diameter selected for use depends largely on the thickness of the material to be welded, the position in which welding is to be performed, and the type of joint to be welded. In general, larger electrodes will be selected for applications involving thicker materials and for welding in the flat position in order to take advantage of their higher deposition rates.

For welding in the horizontal, vertical and overhead positions, the molten weld metal tends to flow out of the joint due to gravitational forces. This tendency can be controlled by using small electrodes to reduce the weld pool size. Electrode manipulation and increased travel speed along the joint also aid in controlling weld pool size.

Weld groove design must also be considered when electrode size is selected. The electrode used in the first few passes must be small enough for easy manipulation in the root of the joint. In V-grooves, small diameter electrodes are frequently used for the initial pass to control melt-through and bead shape. Larger electrodes can be used to complete the weld, taking advantage of their deeper penetration and higher deposition rates.

Finally, the experience of the welder often has a bearing on the electrode size. This is particularly true for out-of-position welding, since the welder's skill governs the size of the molten puddle that the welder can control.

The largest possible electrode that does not violate any pertinent heat input limitations or deposit too large a weld should be used. Welds that are larger than necessary are more costly and, in some instances, actually are harmful. Any sudden change in section size or in the contour of a weld, such as that caused by overwelding, creates stress concentrations. It is obvious that the correct electrode size is the one that, when used with the proper amperage and travel speed, produces a weld of the required size in the least amount of time.

WELDING CURRENT

SHIELDED METAL ARC welding can be accomplished with either alternating or direct current, when an appropriate electrode is used. The type of welding current, the polarity, and the constituents in the electrode covering influence the melting rate of all covered electrodes. For any given electrode, the melting rate is directly related to the electrical energy supplied to the arc. Part of this energy is used to melt a portion of the base metal and part is used to melt the electrode.

Direct Current

DIRECT CURRENT ALWAYS provides a steadier arc and smoother metal transfer than ac does. This is because the polarity of dc is not always changing as it is with ac. Most covered electrodes operate better on reverse polarity (electrode positive), although some are suitable for (and even are intended for) straight polarity (electrode negative). Reverse polarity produces deeper penetration, but straight polarity produces a higher electrode melting rate.

The dc arc produces good wetting action by the molten weld metal and uniform weld bead size even at low amperage. For this reason, dc is particularly suited to welding thin sections. Most combination ac-dc electrodes operate better on dc than on ac, even though they are designed to operate with either type of current.

Direct current is preferred for vertical and overhead welding and for welding with a short arc. The dc arc has less tendency to short out as globules of molten metal are transferred across it.

Arc blow may be a problem when magnetic metals (iron and nickel) are welded with dc. One way to overcome this problem is to change to ac.

Alternating Current

FOR SMAW, ALTERNATING current offers two advantages over dc. One is the absence of arc blow and the other is the cost of the power source.

Without arc blow, larger electrodes and higher welding currents can be used. Certain electrodes (specifically, those with iron powder in their coverings) are designed for operation at higher amperages with ac. The highest welding speeds for SMAW can be obtained in the drag technique with these electrodes on ac. Fixturing materials, fixture design, and workpiece connection location may not be as critical with ac.

An ac transformer costs less than an equivalent dc power source. The cost of the equipment alone should not be the sole criterion in the selection of the power source, however. All operating factors need to be considered.

Amperage

COVERED ELECTRODES OF a specific size and classification will operate satisfactorily at various amperages within some certain range. This range will vary somewhat with the thickness and formulation of the covering.

Deposition rates increase as the amperage increases. For a given size of electrode, the amperage ranges and the resulting deposition rates will vary from one electrode classification to another. This variation for several classifications of carbon steel electrodes of one size is shown in Figure 2.10.

With a specific type and size of electrode, the optimum amperage depends on several factors such as the position of welding and the type of joint. The amperage must be sufficient to obtain good fusion and penetration yet permit proper control of the molten weld pool. For vertical and overhead welding, the optimum amperages would likely be on the low end of the allowable range.

Amperage beyond the recommended range should not be used. It can overheat the electrode and cause excessive spatter, arc blow, undercut, and weld metal cracking. Figure 2.11(A), (B), and (C) show the effect of amperage on bead shape.

ARC LENGTH

THE ARC LENGTH is the distance from the molten tip of the electrode core wire to the surface of the molten weld pool. Proper arc length is important in obtaining a sound welded joint. Metal transfer from the tip of the electrode to the weld pool is not a smooth, uniform action. Instantaneous arc voltage varies as droplets of molten metal are transferred across the arc, even with constant arc length. However, any variation in voltage will be minimal when welding is done with the proper amperage and arc length. The latter requires constant and consistent electrode feed.

The correct arc length varies according to the electrode classification, diameter, and covering composition; it also varies with amperage and welding position. Arc length increases with increasing electrode diameter and amperage. As a general rule, the arc length should not exceed the diameter of the core wire of the electrode. The arc usually is shorter than this for electrodes with thick coverings, such as iron powder or "drag" electrodes.

Too short an arc will be erratic and may short circuit during metal transfer. Too long an arc will lack direction and intensity, which will tend to scatter the molten metal as it moves from the electrode to the weld. The spatter may be heavy and the deposition efficiency low. Also, the gas and flux generated by the electrode covering are not so effective in shielding the arc and weld metal. This can result in porosity and contamination of the weld metal by oxygen or nitrogen, or both.

Control of arc length is largely a matter of welder skill, involving the welder's knowledge, experience, visual perception, and manual dexterity. Although the arc length does change to some extent with changing conditions, certain fundamental principles can serve as a guide to the proper arc length for a given set of conditions.

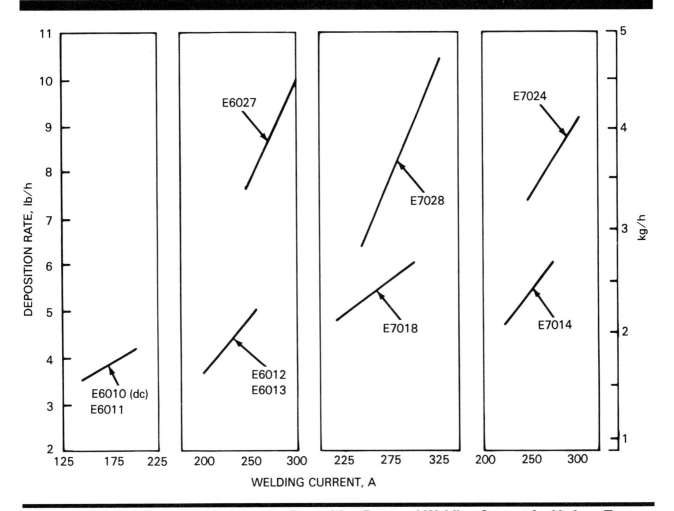

Figure 2.10–The Relationship Between Deposition Rate and Welding Current for Various Types of 3/16 in. (4.8 mm) Diameter Carbon Steel Electrodes

For downhand welding, particularly with heavy electrode coverings, the tip of the electrode can be dragged lightly along the joint. The arc length, in this case, is automatically determined by the coating thickness and the melting rate of the electrode. Moreover, the arc length is uniform. For vertical or overhead welding, the arc length is gaged by the welder. The proper arc length, in such cases, is the one that permits the welder to control the size and motion of the molten weld pool.

For fillet welds, the arc is crowded into the joint for highest deposition rate and best penetration. The same is true of the root passes in groove welds in pipe.

When arc blow is encountered, the arc length should be shortened as much as possible. The various classifications of electrodes have widely different operating characteristics, including arc length. It is important, therefore, for the welder to be familiar with the operating characteristics of the types of electrodes the welder uses in order to recognize the proper arc length and to know the effect of different arc lengths. The effect of a long and a short arc on bead appearance with a mild steel electrode is illustrated in Figures 2.11(D) and (E).

TRAVEL SPEED

TRAVEL SPEED IS the rate at which the electrode moves along the joint. The proper travel speed is the one which produces a weld bead of proper contour and appearance, as shown in Figure 2.11(A). Travel speed is influenced by several factors:

(1) Type of welding current, amperage, and polarity
(2) Position of welding
(3) Melting rate of the electrode
(4) Thickness of material
(5) Surface condition of the base metal

(6) Type of joint
(7) Joint fit-up
(8) Electrode manipulation

When welding, the travel speed should be adjusted so that the arc slightly leads the molten weld pool. Up to a point, increasing the travel speed will narrow the weld bead and increase penetration. Beyond this point, higher travel speeds can decrease penetration, cause the surface of the bead to deteriorate and produce undercutting at the edges of the weld, make slag removal difficult, and entrap gas (porosity) in the weld metal. The effect of high travel speed on bead appearance is shown in Figure 2.11 (G). With low travel speed, the weld bead will be wide and convex with shallow penetration, as illustrated in Figure 2.11 (F). The shallow penetration is caused by the arc dwelling on the molten weld pool instead of leading it and concentrating on the base metal. This, in turn, affects dilution. When dilution must be kept low (as in cladding), the travel speed, too, must be kept low.

Travel speed also influences heat input, and this affects the metallurgical structures of the weld metal and the heat-affected zone. Low travel speed increases heat input and this, in turn, increases the size of the heat-affected zone and reduces the cooling rate of the weld. Forward travel speed is necessarily reduced with a weave bead as opposed to the higher travel speed that can be attained with a stringer bead. Higher travel speed reduces the size of the heat-affected zone and increases the cooling rate of the weld. The increase in the cooling rate can increase the strength and hardness of a weld in a hardenable steel, unless preheat of a level sufficient to prevent hardening is used.

ELECTRODE ORIENTATION

ELECTRODE ORIENTATION, WITH respect to the work and the weld groove, is important to the quality of a weld. Improper orientation can result in slag entrapment, porosity, and undercutting. Proper orientation depends on the type and size of electrode, the position of welding, and the geometry of the joint. A skilled welder automatically takes these into account when the orientation to be used for a specific joint is determined. Travel angle and work angle are used to define electrode orientation.

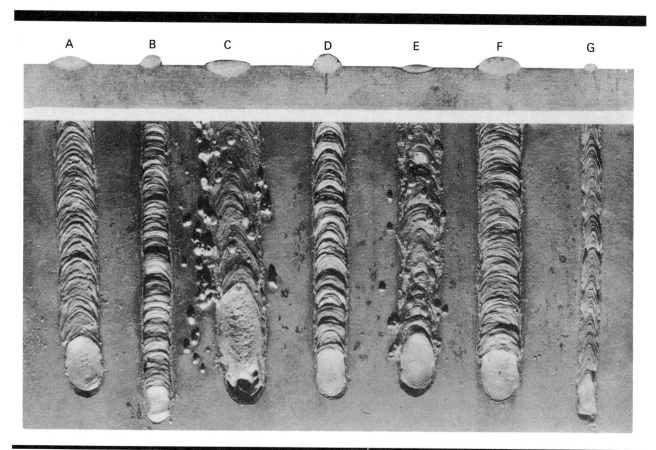

Figure 2.11–The Effect of Welding Amperage, Arc Length, and Travel Speed; (A) Proper Amperage, Arc Length, and Travel Speed; (B) Amperage Too Low; (C) Amperage Too High; (D) Arc Length Too Short; (E) Arc Length Too Long; (F) Travel Speed Too Slow; (G) Travel Speed Too Fast

Travel angle is the angle less than 90 degrees between the electrode axis and a line perpendicular to the weld axis, in a plane determined by the electrode axis and the weld axis. *Work angle* is the angle less that 90 degrees between a line perpendicular to the major workpiece surface and a plane determined by the electrode axis and the weld axis. When the electrode is pointed in the direction of welding, the *forehand technique* is being used. The travel angle, then, is known as the *push angle*. The *backhand technique* involves pointing the electrode in the direction opposite that of welding. The travel angle, then, is called the *drag angle*. These angles are shown in Figure 2.12.

Typical electrode orientation and welding technique for groove and fillet welds, with carbon steel electrodes, are listed in Table 2.3. These may be different for other materials. Correct orientation provides good control of the molten weld pool, the desired penetration, and complete fusion with the steel base.

A large travel angle may cause a convex, poorly shaped bead with inadequate penetration, whereas a small travel angle may cause slag entrapment. A large work angle can cause undercutting, while a small work angle can result in lack of fusion.

WELDING TECHNIQUE

THE FIRST STEP in SMAW is to assemble the proper equipment, materials, and tools for the job. Next, the type of welding current and the polarity, if dc, need to be determined and the power source set accordingly. The power source must also be set to give the proper volt-ampere characteristic (open circuit voltage) for the size and type of electrode to be used. After this, the work is positioned for welding and, if necessary, clamped in place.

The arc is struck by tapping the end of the electrode on the work near the point where welding is to begin, then quickly withdrawing it a small amount to produce an arc of proper length. Another technique for striking the arc is to use a scratching motion similar to that used in striking a match. When the electrode touches the work, there is a tendency for them to stick together. The purpose of the tapping and scratching motion is to prevent this. When the

electrode does stick, it needs to be quickly broken free. Otherwise, it will overheat, and attempts to remove it from the workpiece will only bend the hot electrode. Freeing it then will require a hammer and chisel.

The technique of restriking the arc once it has been broken varies somewhat with the type of electrode. Generally, the covering at the tip of the electrode becomes conductive when it is heated during welding. This assists in restriking the arc if it is restruck before the electrode cools. Arc striking and restriking are much easier for electrodes with large amounts of metal powders in their coverings. Such coverings are conductive when cold. When using heavily covered electrodes which do not have conductive coatings, such as E6020, low hydrogen, and stainless steel electrodes, it may be necessary to break off the projecting covering to expose the core wire at the tip for easy restriking.

Striking the arc with low hydrogen electrodes requires a special technique to avoid porosity in the weld at the point where the arc is started. This technique consists of striking the arc a few electrode diameters ahead of the place where welding is to begin. The arc is then quickly moved back, and welding is begun in the normal manner. Welding continues over the area where the arc originally was struck, refusing any small globules of weld metal that may have remained from striking the arc.

During welding, the welder maintains a normal arc length by uniformly moving the electrode toward the work as the electrode melts. At this same time, the electrode is moved uniformly along the joint in the direction of welding, to form the bead.

Any of a variety of techniques may be employed to break the arc. One of these is to rapidly shorten the arc, then quickly move the electrode sideways out of the crater. This technique is used when replacing a spent electrode, in which case welding will continue from the crater. Another technique is to stop the forward motion of the electrode and allow the crater to fill, then gradually withdraw the electrode to break the arc. When continuing a weld from a crater, the arc should be struck at the forward end of the crater. It should then quickly be moved to the back of the crater and slowly brought forward to continue the weld. In this manner, the crater is filled, and porosity

Table 2.3
Typical Shielded Metal Arc Electrode Orientation and Welding Technique
for Carbon Steel Electrodes

Type of Joint	Position of Welding	Work Angle, Deg	Travel Angle, Deg	Technique of Welding
Groove	Flat	90	5-10*	Backhand
Groove	Horizontal	80-100	5-10	Backhand
Groove	Vertical-Up	90	5-10	Forehand
Groove	Overhead	90	5-10	Backhand
Fillet	Horizontal	45	5-10*	Backhand
Fillet	Vertical-Up	35-55	5-10	Forehand
Fillet	Overhead	30-45	5-10	Backhand

* Travel angle may be 10° to 30° for electrodes with heavy iron powder coatings.

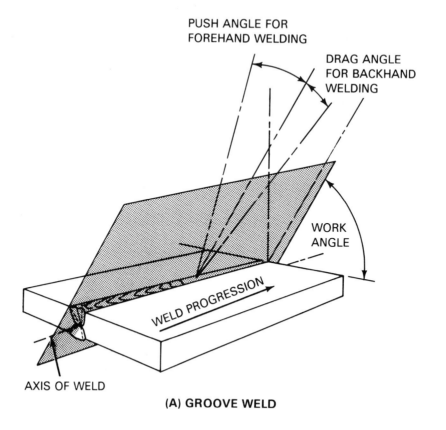

PUSH ANGLE FOR
FOREHAND WELDING

DRAG ANGLE
FOR BACKHAND
WELDING

WORK
ANGLE

WELD PROGRESSION

AXIS OF WELD

(A) GROOVE WELD

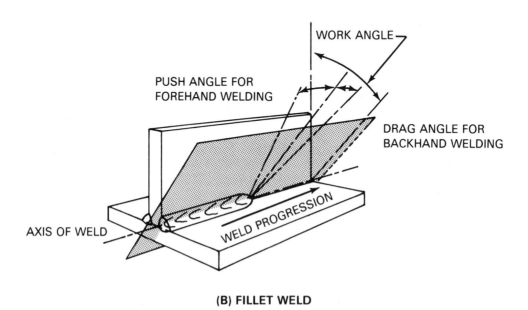

WORK ANGLE

PUSH ANGLE FOR
FOREHAND WELDING

DRAG ANGLE FOR
BACKHAND WELDING

WELD PROGRESSION

AXIS OF WELD

(B) FILLET WELD

Figure 2.12–Orientation of the Electrode

and entrapped slag are avoided. This technique is particularly important for low hydrogen electrodes.

SLAG REMOVAL

THE EXTENT TO which slag is removed from each weld bead before welding over the bead has a direct bearing on the quality of a multiple pass weld. Failure to thoroughly clean each bead increases the probability of trapping slag and, thus, producing a defective weld. Complete and efficient slag removal requires that each bead be properly contoured and that it blend smoothly into the adjacent bead or base metal.

Small beads cool more rapidly than large ones. This tends to make slag removal from small beads easier. Concave or flat beads that wash smoothly into the base metal or any adjoining beads minimize undercutting and avoid a sharp notch along the edge of the bead where slag could stick. Finally, it is most important that welders be able to recognize areas where slag entrapment is likely to occur. Skilled welders understand that complete removal of slag is necessary before continuing a weld.

WORKPIECE CONNECTION

PROPER CONNECTING OF the worklead is a necessary consideration in shielded metal arc welding. The location of the lead is especially important with dc welding. Improper location may promote arc blow, making it difficult to control the arc. Moreover, the method of attaching the lead is important. A poorly attached lead will not provide consistent electrical contact, and the connection will heat up. This can lead to an interruption of the circuit and a breaking of the arc. A copper contact shoe secured with a C-clamp is best. If copper pickup by this attachment to the base metal is detrimental, the copper shoe should be attached to a plate that is compatible with the work. The plate, in turn, is then secured to the work. For rotating work, contact should be made by shoes sliding on the work or through roller bearings on the spindle on which the work is mounted. If sliding shoes are used, at least two shoes should be employed. If loss of contact occurred with only a single shoe, the arc would be extinguished.

ARC STABILITY

A STABLE ARC is required if high quality welds are to be produced. Such defects as inconsistent fusion, entrapped slag, blowholes, and porosity can be the result of an unstable arc.

The following are important factors influencing arc stability:

(1) The open circuit voltage of the power source
(2) Transient voltage recovery characteristics of the power source
(3) Size of the molten drops of filler metal and slag in the arc
(4) Ionization of the arc path from the electrode to the work
(5) Manipulation of the electrode

The first two factors are related to the design and operating characteristics of the power source. The next two are functions of the welding electrode. The last one represents the skill of the welder.

The arc of a covered electrode is a transient arc, even when the welder maintains a fairly constant arc length. The welding machine must be able to respond rapidly when the arc tends to go out, or it is short circuited by large droplets of metal bridging the arc gap. In that case, a surge of current is needed to clear a short circuit. With ac, it is important that the voltage lead the current in going through zero. If the two were in phase, the arc would be very unstable. This phase shift must be designed into the welding machine.

Some electrode covering ingredients tend to stabilize the arc. These are necessary ingredients for an electrode to operate well on ac. A few of these ingredients are titanium dioxide, feldspar, and various potassium compounds (including the binder, potassium silicate). The inclusion of one or more of these arc stabilizing compounds in the covering provides a large number of readily ionized particles and thereby contributes to ionization of the arc stream. Thus, the electrode, the power source, and the welder all contribute to arc stability.

ARC BLOW

ARC BLOW, WHEN it occurs, is encountered principally with dc welding of magnetic materials (iron and nickel). It may be encountered with ac, under some conditions, but those cases are rare, and the intensity of the blow is always much less severe. Direct current, flowing through the electrode and the base metal, sets up magnetic fields around the electrode which tend to deflect the arc from its intended path. The arc may be deflected to the side at times, but usually it is deflected either forward or backward along the joint. Back blow is encountered when welding toward the workpiece connection near the end of a joint or into a corner. Forward blow is encountered when welding away from the lead at the start of the joint, as shown in Figure 2.13.

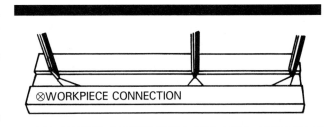

**Figure 2.13–The Effect of Workpiece
Connection Location on Magnetic Arc Blow**

Arc blow can result in incomplete fusion and excessive weld spatter and may be so severe that a satisfactory weld cannot be made. When welding with iron powder electrodes and electrodes which produce heavy slag, forward blow can be especially troublesome. It permits the molten slag, which normally is confined to the edge of the crater, to run forward under the arc.

The bending of the arc under these conditions is caused by the effects of an unbalanced magnetic field. When there is a greater concentration of magnetic flux on one side of the arc than on the other, the arc always bends away from the greater concentration. The source of the magnetic flux is indicated by the electrical rule which states that a conductor carrying an electric current produces a magnetic flux in circles around the conductor. These circles are in planes perpendicular to the conductor and are centered on the conductor.

In welding, this magnetic flux is superimposed on the steel and across the gap to be welded. The flux in the plate does not cause difficulty, but unequal concentration of flux across the gap or around the arc causes the arc to bend away from the heavier concentration. Since the flux passes through steel many times more readily than it does through air, the path of the flux tends to remain within the steel plates. For this reason, the flux around the electrode, when the electrode is near either end of the joint, is concentrated between the electrode and the end of the plate. This high concentration of flux on one side of the arc, at the start or the finish of the weld, deflects the arc away from the ends of the plates.

Forward blow exists for a short time at the start of a weld, then it diminishes. This is because the flux soon finds an easy path through the weld metal. Once the magnetic flux behind the arc is concentrated in the plate and the weld, the arc is influenced mainly by the flux in front of it as this flux crosses the root opening. At this point, back blow may be encountered. Back blow can occur right up to the end of the joint. As the weld approaches the end, the flux ahead of the arc becomes more crowded, increasing the back blow. Back blow can become extremely severe right at the very end of the joint.

The welding current passing through the work creates a magnetic field around it. The field is perpendicular to the path of current between the arc and the workpiece connection. The flux field around the arc is perpendicular to the one in the work. This concentrates the magnetic flux on the worklead side of the arc and tends to push the arc away. The two flux fields mentioned above are, in reality, one field. That field is perpendicular to the path of the current through the cable, the work, the arc, and the electrode.

Unless the arc blow is unusually severe, certain corrective steps may be taken to eliminate it or, at least, to reduce its severity. Some or all of the following steps may be necessary:

(1) Place worklead connections as far as possible from the joints to be welded.

(2) If back blow is the problem, place the workpiece connection at the start of welding, and weld toward a heavy tack weld.

(3) If forward blow causes trouble, place the workpiece connection at the end of the joint to be welded.

(4) Position the electrode so that the arc force counteracts the arc blow.

(5) Use the shortest possible arc consistent with good welding practice. This helps the arc force to counteract the arc blow.

(6) Reduce the welding current.

(7) Weld toward a heavy tack or runoff tab.

(8) Use the backstep sequence of welding.

(9) Change to ac, which may require a change in electrode classification.

(10) Wrap the worklead around the workpiece in a direction such that the magnetic field it sets up will counteract the magnetic field causing the arc blow.

QUALITY OF THE WELD

A WELDED JOINT must possess those qualities which are necessary to enable it to perform its expected function in service. To accomplish this, the joint needs to have the required physical and mechanical properties. It may need a certain microstructure and chemical composition to meet these properties. The size and shape of the weld also are involved, as is the soundness of the joint. Corrosion resistance may be required. All of these are influenced by the base materials, the welding materials, and the manner in which the weld is made.

Shielded metal arc welding is a manual welding process, and the quality of the joint depends on the skill of the welder who makes it. For this reason, the materials to be used must be selected with care, the welder must be proficient, and the procedure the welder uses must be correct.

Welded joints, by their nature, contain discontinuities of various types and sizes. Below some acceptable level, these are not considered harmful. Above that level, they are considered defects. The acceptance level can vary with the severity of the service to be encountered, or more commonly will be based on fabrication contract requirements or a specific code or specification.

The following discontinuities are sometimes encountered in welds made by the SMAW process:

(**1**) Porosity
(**2**) Slag inclusions
(**3**) Incomplete fusion
(**4**) Undercut
(**5**) Cracks

POROSITY

THIS TERM IS used to describe gas pockets or voids in the weld metal. These voids result from gas that forms from certain chemical reactions that take place during welding. They contain gas rather than solids, and, in this respect, they differ from slag inclusions.

Porosity usually can be prevented by using proper amperage and holding a proper arc length. Dry electrodes are also helpful in many cases. The deoxidizers which a covered electrode needs are easily lost during deposition when high amperage or a long arc is used. This leaves a supply which is insufficient for proper deoxidation of the molten metal.

SLAG INCLUSIONS

THIS TERM IS used to describe the oxides and nonmetallic solids that sometimes are entrapped in weld metal, between adjacent beads, or between the weld metal and the base metal. During deposition and subsequent solidification of the weld metal, many chemical reactions occur. Some of the products of these reactions are solid nonmetallic compounds which are insoluble in the molten metal. Because of their lower specific gravity, these compounds will rise to the surface of the molten metal unless they become entrapped within the weld metal.

Slag formed from the covering on shielded metal arc electrodes may be forced below the surface of the molten metal by the stirring action of the arc. Slag may also flow ahead of the arc if the welder is not careful. This can easily happen when welding over the crevasse between two parallel but convex beads or between one convex bead and a side wall of the groove. It can also happen when the welding is done downhill. In such cases, the molten metal may flow over the slag, entrapping the slag beneath the bead. Factors such as highly viscous or rapidly solidifying slag or insufficient welding current set the stage for this.

Most slag inclusions can be prevented by good welding practice and, in problem areas, by proper preparation of the groove before depositing the next bead of weld metal. In these cases, care must be taken to correct contours that are difficult to adequately penetrate with the arc.

INCOMPLETE FUSION

THIS TERM, AS it is used here, refers to the failure to fuse together adjacent beads of weld metal or weld metal and base metal. This condition may be localized or it may be extensive, and it can occur at any point in the welding groove. It may even occur at the root of the joint.

Incomplete fusion may be caused by failure to raise the base metal (or the previously deposited bead of weld metal) to the melting temperature. It may also be caused by failure to dissolve, because of improper fluxing, any oxides or other foreign material that might be present on the surfaces which must fuse with the weld metal.

Incomplete fusion can be avoided by making certain that the surfaces to be welded are property prepared and fitted and are smooth and clean. In the case of incomplete root fusion, the corrections are to make certain that the root face is not too large; the root opening is not too small; the electrode is not too large; the welding current is not too low; and the travel speed is not too high.

UNDERCUT

THIS TERM IS used to describe either of two situations. One is the melting away of the sidewall of a welding groove at the edge of the bead, thus forming a sharp recess in the sidewall in the area in which the next bead is to be deposited. The other is the reduction in thickness of the base metal at the line where the beads in the final layer of weld metal tie into the surface of the base metal (e.g., at the toe of the weld).

Both types of undercut usually are due to the specific welding technique used by the welder. High amperage and a long arc increase the tendency to undercut. Incorrect electrode position and travel speed also are causes, as is improper dwell time in a weave bead. Even the type of electrode used has an influence. The various classifications of electrodes show widely different characteristics in this respect. With some electrodes, even the most skilled welder may be unable to avoid undercutting completely in certain welding positions, particularly on joints with restricted access.

Undercut of the sidewalls of a welding groove will in no way affect the completed weld if the undercut is removed before the next bead is deposited at that location. A well-rounded chipping tool or grinding wheel will be required to remove the undercut. If the undercut is slight, however, an experienced welder who knows just how deep the arc will penetrate may not need to remove the undercut.

The amount of undercut permitted in a completed weld is usually dictated by the fabrication code being used, and the requirements specified should be followed because excessive undercut can materially reduce the strength of the joint. This is particularly true in applications subject to fatigue. Fortunately, this type of undercut can be detected by visual examination of the completed weld, and it can be corrected by blend grinding or depositing an additional bead.

CRACKS

CRACKING IN WELDED joints can be classified as either hot or cold cracking. Cracking can occur in the weld metal, base metal, or both. If cracking is observed during welding, the cracks should be removed prior to further welding, because weld metal deposited over a crack can result in continuation of that crack into the newly deposited weld metal.

Hot cracking is a function of chemical composition. The main cause of hot cracking is constituents in the weld metal which have a relatively low melting temperature and which accumulate at the grain boundaries during solidification. A typical example is iron sulfide in steel. The cracks are intergranular or interdendritic. They form as the weld metal cools. As solidification progresses in the cooling weld metal, the shrinkage stresses increase and eventually draw apart those grains which still have some liquid at their boundaries. Coarse-grained, single-phase structures have a marked propensity to this type of cracking. Solutions to cracking problems include:

(1) Changing the base metal (for instance, use a steel with manganese additions, or one produced to provide a fine grained structure)
(2) Changing filler metal (using filler metal with sufficient ferrite when welding austinetic stainless steel, for instance)

(3) Changing the welding technique/procedure by modifying the preheat and interpass temperatures and reducing the welding current

Cold cracking is the result of inadequate ductility or the presence of hydrogen in hardenable steels. This condition is caused by inadequate toughness in the presence of a mechanical or metallurgical notch and stresses of sufficient magnitude. These stresses do not have to be very high in some materials—large grained ferritic stainless steel, for instance.

To prevent cold cracking in hardenable steels, the use of dry low hydrogen electrodes and proper preheat is required. Preheat is also required for those materials which are naturally low in ductility or toughness. Materials which are subject to extreme grain growth (28 percent chromium steel, for instance) must be welded with low heat input and low interpass temperatures. Notches need to be avoided.

More information on the quality of welded joints can be found in the *Welding Handbook*, Chapter 5, Volume 1, 7th Edition and Chapter 6, Section 1, 6th Edition. *Welding Inspection*, published by AWS, also is a good reference.

SAFETY RECOMMENDATIONS

THE OPERATOR MUST protect eyes and skin from radiation from the arc. A welding helmet with a suitable filter lens should be used, as well as dark clothing, preferably wool, to protect the skin. Leather gloves and clothing should be worn to protect against burns from arc spatter.

Welding helmets are provided with filter plate windows, the standard size being 2 by 4-1/8 in. (51 by 130 mm). Larger openings are available. The filter plate should be capable of absorbing infrared rays, ultraviolet rays, and most of the visible rays emanating from the arc. Filter plates that are now available absorb 99 percent or more of the infrared and ultraviolet rays from the arc.

The shade of the filter plate suggested for use with electrodes up to 5/32 in. (4 mm) diameter is No. 10. For 3/16 to 1/4 in. (4.8 to 6.4 mm) electrodes, Shade No. 12 should be used. Shade No. 14 should be used for electrodes over 1/4 in. (6.4 mm).

The filter plate needs to be protected from molten spatter and from breakage. This is done by placing a plate of clear glass, or other suitable material, on each side of the filter plate. Those who are not welders but work near the arc also need to be protected. This protection usually is provided by either permanent or portable screens. Failure to use adequate protection can result in eye burn for the welder or for those working around the arc. Eye burn, which is similar to sunburn, is extremely painful for a period of 24 to 48 hours. Unprotected skin, exposed to the

arc, may also be burned. A physician should be consulted in the case of severe arc burn, regardless of whether it is of the skin or the eyes.

If welding is being performed in confined spaces with poor ventilation, auxiliary air should be supplied to the welder. This should be done through an attachment to the helmet.

The method used must not restrict the welder's manipulation of the helmet, interfere with the field of vision, or make welding difficult. Additional information on eye protection and ventilation is given in ANSI Z49.1, *Safety in Welding and Cutting*, published by the American Welding Society.

From time to time during welding, sparks or globules of molten metal are thrown out from the arc. This is always a point of concern, but it becomes more serious when welding is performed out of position or when extremely high welding currents are used. To ensure protection from burns under these conditions, the welder should wear flame-resistant gloves, a protective apron, and a jacket (see Figure 2.5). It may also be desirable to protect the welder's ankles and feet from slag and spatter. Cuffless pants and high work shoes or boots are recommended.

To avoid electric shock, the operator should not weld while standing on a wet surface. Equipment should be examined periodically to make sure there are no cracks or worn spots on electrode holder or cable insulation.

SUPPLEMENTARY READING LIST

American Society for Metals. "Welding, brazing, and Soldering." *Metals Handbook*, Vol. 6, 9th Ed., 75-95. Metals Park, Ohio: American Society for Metals, 1983.

Barbin, L. M. "The new moisture resistant electrodes." *Welding Journal* 56(7): 15-18; July 1977.

Chew, B. "Moisture loss and gain by some basic flux covered electrodes." *Welding Journal* 55(5): 127s-134s; May 1976.

Gregory, E. N. "Shielded metal arc welding of galvanized steel." *Welding Journal* 48(8): 631-638; August 1969.

Jackson, C. E. "Fluxes and slags in welding." Bulletin 190. New York: Welding Research Council, December 1973.

Lincoln Electric Company. *The procedure handbook of arc welding*, 12th Ed. Cleveland: Lincoln Electric Company, 1973.

Silva, E. A. and Hazlett, T. H. "Shielded metal arc welding underwater with iron powder electrodes." *Welding Journal* 50(6): 406s-415s; June 1971.

Stout, R. D., and Doty, W. D. *Weldability of Steels*, 2nd Ed., Ed. Epstein, S., and Somers, R. E. New York: Welding Research Council, 1971.

GAS TUNGSTEN ARC WELDING

PREPARED BY A COMMITTEE CONSISTING OF:

G. K. Hicken, Chairman
Sandia National Labs

R. D. Campbell
E. G. & G. Rocky Flats

G. J. Daumeyer, III
Allied Signal Aerospace

R. B. Madigan
Edison Welding Institute

S. J. Marburger
Sandia National Labs

B. Young
Westinghouse Savannah River Company

WELDING HANDBOOK COMMITTEE MEMBER:
M. J. Tomsic
Plastronic Incorporated

CHAPTER 3

GAS TUNGSTEN ARC WELDING

INTRODUCTION

GAS TUNGSTEN ARC welding (GTAW) is an arc welding process that uses an arc between a tungsten electrode (nonconsumable) and the weld pool. The process is used with shielding gas and without the application of pressure. The process may be used with or without the addition of filler metal. Figure 3.1 shows the gas tungsten arc welding process.

GTAW has become indispensable as a tool for many industries because of the high-quality welds produced and low equipment costs. The purpose of this chapter is to discuss the fundamentals of the GTAW process, the equipment and consumables used, the process procedures and variables, applications, and safety considerations.

The possibility of using helium to shield a welding arc and molten weld pool was first investigated in the 1920's.[1] However, nothing was done with this method until the beginning of World War II when a great need developed in the aircraft industry to replace riveting for joining reactive materials such as aluminum and magnesium. Using a tungsten electrode and direct current arc power with the electrode negative, a stable, efficient heat source was produced with which excellent welds could be made.

Helium was elected to provide the necessary shield because, at the time, it was the only readily available inert gas. Tungsten electrode inert gas torches typical of that period[2] are shown in Figure 3.2. The process has been called non-

1. H. M. Hobart U.S. Patent 1,746,081, 1926 and P. K. Devers U.S. patent 1,746,191, 1926.
2. R. Meredith, U.S. Patent 2,274,631

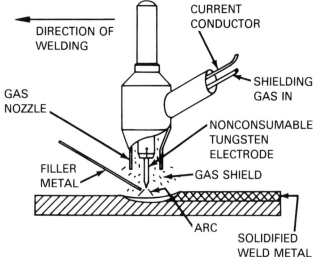

Figure 3.1–Gas Tungsten Arc Welding Operation

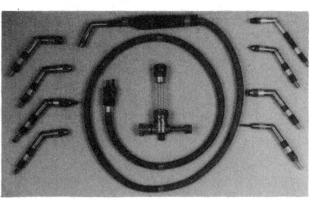

Figure 3.2–Early Gas Tungsten Arc Welding Heads, Circa 1943, With a Torch Body and an Early Flowmeter

consumable electrode welding and TIG (tungsten inert gas) welding. However, the AWS terminology for this process is gas tungsten arc welding (GTAW), because shielding gas mixtures which are not inert can be used for certain applications.

Since the early days of the invention, numerous improvements have been made to the process and equipment. Welding power sources have been developed specifically for the process. Some provide pulsed dc and variable polarity ac welding power. Water-cooled and gas-cooled torches were developed. The tungsten electrode has been alloyed with small amounts of active elements to increase its emissivity; this has improved arc starting, arc stability, and electrode life. Shielding gas mixtures have been identified for improved welding performance. Researchers are presently pursuing further improvements, in such areas as automatic controls, vision and penetration sensors, and arc length controls.

PRINCIPLES OF OPERATION

PROCESS DESCRIPTION

THE GAS TUNGSTEN arc welding process is illustrated in Figure 3.1. The process uses a nonconsumable tungsten (or tungsten alloy) electrode held in a torch. Shielding gas is fed through the torch to protect the electrode, molten weld pool, and solidifying weld metal from contamination by the atmosphere. The electric arc is produced by the passage of current through the conductive, ionized shielding gas. The arc is established between the tip of the electrode and the work. Heat generated by the arc melts the base metal. Once the arc and weld pool are established, the torch is moved along the joint and the arc progressively melts the faying surfaces. Filler wire, if used, is usually added to the leading edge of the weld pool to fill the joint.

Four basic components are common to all GTAW setups, as illustrated in Figures 3.1 and 3.3:

(**1**) Torch
(**2**) Electrode
(**3**) Welding power source
(**4**) Shielding gas

PROCESS ADVANTAGES

THE FOLLOWING ARE some advantages of the gas tungsten arc process:

(**1**) It produces superior quality welds, generally free of defects.
(**2**) It is free of the spatter which occurs with other arc welding processes.
(**3**) It can be used with or without filler metal as required for the specific application.
(**4**) It allows excellent control of root pass weld penetration.
(**5**) It can produce inexpensive autogenous welds at high speeds.
(**6**) It can use relatively inexpensive power supplies.
(**7**) It allows precise control of the welding variables.

(**8**) It can be used to weld almost all metals, including dissimilar metal joints.
(**9**) It allows the heat source and filler metal additions to be controlled independently.

PROCESS LIMITATIONS

THE FOLLOWING ARE some limitations of the gas tungsten arc process:

(**1**) Deposition rates are lower than the rates possible with consumable electrode arc welding processes.
(**2**) There is a need for slightly more dexterity and welder coordination than with gas metal arc welding or shielded metal arc welding for manual welding.
(**3**) It is less economical than the consumable electrode arc welding processes for thicker sections greater than 3/8 in. (10 mm).
(**4**) There is difficulty in shielding the weld zone properly in drafty environments.

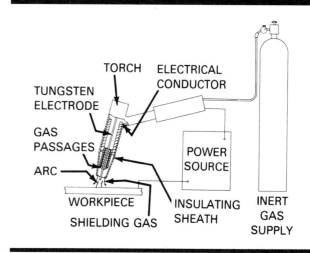

Figure 3.3–Gas Tungsten Arc Welding Equipment Arrangement

Potential problems with the process include:

(**1**) Tungsten inclusions can occur if the electrode is allowed to contact the weld pool.

(**2**) Contamination of the weld metal can occur if proper shielding of the filler metal by the gas stream is not maintained.

(**3**) There is low tolerance for contaminants on filler or base metals.

(**4**) Possible contamination or porosity is caused by coolant leakage from water-cooled torches.

(**5**) Arc blow or arc deflection, as with other processes.

PROCESS VARIABLES

THE PRIMARY VARIABLES in GTAW are arc voltage (arc length), welding current, travel speed, and shielding gas. The amount of energy produced by the arc is proportional to the current and voltage. The amount transferred per unit length of weld is inversely proportional to the travel speed. The arc in helium is more penetrating than that in argon. However, because all of these variables interact strongly, it is impossible to treat them as truly independent variables when establishing welding procedures for fabricating specific joints.

Arc Current

AS A GENERAL statement, arc current controls the weld penetration, the effect being directly proportional, if not somewhat exponential. Arc current also affects the voltage, with the voltage at a fixed arc length increasing in proportion to the current. For this reason, to keep a fixed arc length, it is necessary to change the voltage setting when the current is adjusted.

The process can be used with either direct or alternating current, the choice depending largely on the metal to be welded. Direct current with the electrode negative offers the advantages of deep penetration and fast welding speeds, especially when helium is used as the shield. Helium is the gas of choice for mechanized welding. Alternating current provides a cathodic cleaning (sputtering) which removes refractory oxides from the joint surfaces of aluminum and magnesium, allowing superior welds to be made. In this case, argon must be used for the shield because sputtering cannot be obtained with helium. Argon is the gas of choice for manual welding whether used with direct current or alternating current.

A third power option also is available, that of using direct current with the electrode positive. This polarity is used only rarely because it causes electrode overheating. These effects of polarity are explained in more detail in a following discussion of Direct Current.

Arc Voltage

THE VOLTAGE MEASURED between the tungsten electrode and the work is commonly referred to as the *arc voltage*.

Arc voltage is a strongly dependent variable, affected by the following:

(**1**) Arc current
(**2**) Shape of the tungsten electrode tip
(**3**) Distance between the tungsten electrode and the work
(**4**) Type of shielding gas

The arc voltage is changed by the effects of the other variables, and is used in describing welding procedures only because it is easy to measure. Since the other variables such as the shield gas, electrode, and current have been predetermined, arc voltage becomes a way to control the arc length, a critical variable that is difficult to monitor. Arc length is important with this process because it affects the width of the weld pool; pool width is proportional to arc length. Therefore, in most applications other than those involving sheet, the desired arc length is as short as possible.

Of course, recognition needs to be given to the possibility of short circuiting the electrode to the pool or filler wire if the arc is too short. However, with mechanized welding, using a helium shield, DCEN power, and a relatively high current, it is possible to submerge the electrode tip below the plate surface to produce deeply penetrating but narrow welds at high speeds. This technique has been called *buried arc*.

When arc voltage is being used to control arc length in critical applications, care must be taken to observe the other variables which affect arc voltage. Among them are electrode and shielding gas contaminants, improperly fed filler wire, temperature changes in the electrode, and electrode erosion. Should any of these change enough to affect the arc voltage during mechanized welding, the arc length must be adjusted to restore the desired voltage.

Travel Speed

TRAVEL SPEED AFFECTS both the width and penetration of a gas tungsten arc weld. However, its effect on width is more pronounced than that on penetration. Travel speed is important because of its effect on cost. In some applications, travel speed is defined as an objective, with the other variables selected to achieve the desired weld configuration at that speed. In other cases, travel might be a dependent variable, selected to obtain the weld quality and uniformity needed under the best conditions possible with the other combination of variables. Regardless of the objectives, travel speed generally is fixed in mechanized welding while other variables such as current or voltage are varied to maintain control of the weld.

Wire Feed

IN MANUAL WELDING, the way filler metal is added to the pool influences the number of passes required and the appearance of the finished weld.

In machine and automatic welding, wire feed speed determines the amount of filler deposited per unit length of weld. Decreasing wire feed speed will increase penetration and flatten the bead contour. Feeding the wire too slowly can lead to undercut, centerline cracking, and lack of joint fill. Increasing wire feed speed decreases weld penetration and produces a more convex weld bead.

EQUIPMENT

EQUIPMENT FOR GTAW includes torches, electrodes, and power supplies. Mechanized GTAW systems may incorporate arc voltage controls, arc oscillators, and wire feeders.

WELDING TORCHES

GTAW TORCHES HOLD the tungsten electrode which conducts welding current to the arc, and provide a means for conveying shielding gas to the arc zone.

Torches are rated in accordance with the maximum welding current that can be used without overheating. Typical current ranges are listed in Table 3.1. Most torches are designed to accommodate a range of electrode sizes and different types and sizes of nozzles.

The majority of torches for manual applications have a head angle (angle between the electrode and handle) of 120°. Torches are also available with adjustable angle heads, 90° heads, or straight-line (pencil type) heads. Manual GTAW torches often have auxiliary switches and valves built into their handles for controlling current and gas flow. Torches for machine or automatic GTAW are typically mounted on a device which centers the torch over the joint, may move the torch along the joint, and may change or maintain the torch-to-work distance.

Gas-Cooled Torches

THE HEAT GENERATED in the torch during welding is removed either by gas cooling or water cooling. Gas-cooled torches (sometimes called air-cooled) provide cooling by the flow of the relatively cool shielding gas through the torch, as shown in Figure 3.1. Gas cooled torches are limited to a maximum welding current of about 200 amperes.

Water-Cooled Torches

WATER-COOLED TORCHES ARE cooled by the continuous flow of water through passageways in the holder. As illustrated in Figure 3.4, cooling water enters the torch through the inlet hose, circulates through the torch, and exits through an outlet hose. The power cable from the power supply to the torch is typically enclosed within the cooling water outlet hose.

Water-cooled torches are designed for use at higher welding currents on a continuous duty cycle than similar sizes of gas-cooled torches. Typical welding currents of 300 to 500 amps can be used, although some torches have been built to handle welding currents up to 1000 amps. Most machine or automatic welding applications use water-cooled torches.

Water-cooled torches are typically cooled by tap water which flows through the torch and then down a drain. To conserve water, a closed system involving a reservoir, pump, and radiator or water chiller to disperse heat from the system can be used. The capacity of these systems ranges from one to fifty gallons. Automotive antifreeze can be added to the coolant to prevent freezing and corrosion and provide lubrication for the water pump.

Collets

ELECTRODES OF VARIOUS diameters are secured in the electrode holder by appropriately sized collets or chucks. Collets are typically made of a copper alloy. The electrode is gripped by the collet when the torch cap is tightened in place. Good contact between the electrode and the inside diameter of the collet is essential for proper current transfer and electrode cooling.

Nozzles

SHIELDING GAS IS directed to the weld zone by gas nozzles or cups which fit onto the head of the torch as illustrated in Figure 3.1. Also incorporated in the torch body are diffusers or carefully patterned jets which feed the shield gas to the nozzle. Their purpose is to assist in producing a laminar flow of the exiting gas shield. Gas nozzles are made of

Table 3.1
Typical Current Ratings for Gas- and Water-Cooled GTAW Torches

Torch Characteristic	Torch Size		
	Small	Medium	Large
Maximum current (continuous duty), A	200	200-300	500
Cooling method	Gas	Water	Water
Electrode diameters accommodated, in.	0.020 - 3/32	0.040 - 5/32	0.040 - 1/4
Gas cup diameters accommodated, in.	1/4 - 5/8	1/4 - 3/4	3/8 - 3/4

various heat-resistant materials in different shapes, diameters, and lengths. These nozzles are either threaded to the torch or held by friction fit.

Nozzle Materials. Nozzles are made of ceramic, metal, metal jacketed ceramic, fused quartz, or other materials. Ceramic nozzles are the least expensive and most popular, but are brittle and must be replaced often. Fused-quartz nozzles are transparent and allow better vision of the arc and electrode. However, contamination from metal vapors from the weld can cause them to become opaque, and they are also brittle. Water-cooled metal nozzles have longer life and are used mostly for machine and automatic welding applications where welding currents exceed 250 amps.

Sizes and Shapes of Nozzles. The gas nozzle or cup must be large enough to provide shielding gas coverage of the weld pool area and surrounding hot base metal. The nozzle diameter must be appropriate for the volume of shield gas needed to provide protection and the stiffness needed to sustain coverage in drafts. A delicate balance exists between the nozzle diameter and the flow rate. If the flow rate for a given diameter is excessive, the effectiveness of the shield is destroyed because of turbulence. High flow rates without turbulence require large diameters; these are essential conditions at high currents. Size selection depends on electrode size, type of weld joint, weld area to be effectively shielded, and access to the weld joint.

Suggested gas cup sizes for various electrode diameters are listed in Table 3.2. Use of the smallest nozzle listed permits welding in more restricted areas, and offers a better view of the weld. However, use of too small a nozzle may cause shielding gas turbulence and jetting, as well as melting of the lip of the nozzle. Larger nozzles provide better shielding gas coverage, especially for welding reactive metals such as titanium.

Nozzles are available in a variety of lengths to accommodate various joint geometries and the required clearance between the nozzle and the work. Longer nozzles generally produce stiffer, less turbulent gas shields.

The majority of gas nozzles are cylindrical in shape with either straight or tapered ends. To minimize shielding gas turbulence, nozzles with internal streamlining are available. Nozzles are also available with elongated trailing sections or flared ends which provide better shielding for welding metals such as titanium, which is highly susceptibility to contamination at elevated temperatures.

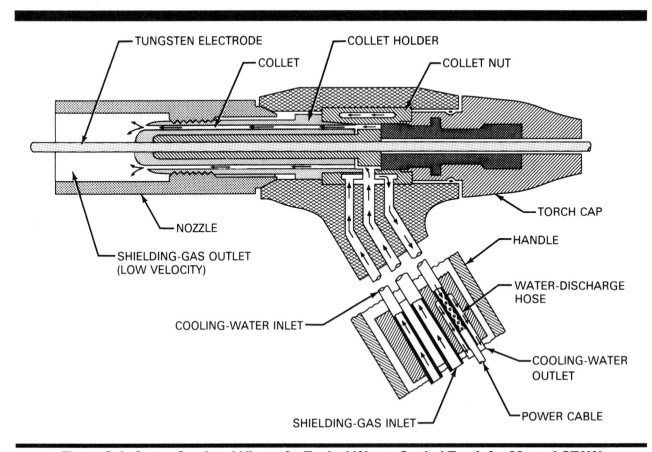

Figure 3.4–Cross-Sectional View of a Typical Water-Cooled Torch for Manual GTAW

Gas Lenses. One device used for assuring a laminar flow of shielding gas is an attachment called a *gas lens*. Gas lenses contain a porous barrier diffuser and are designed to fit around the electrode or collet. Gas lenses produce a longer, undisturbed flow of shielding gas. They enable operators to weld with the nozzle one in. (25.4 mm) or more from the work, improving their ability to see the weld pool and allowing them to reach places with limited access such as inside corners.

ELECTRODES

IN GTAW THE word *tungsten* refers to the pure element tungsten and its various alloys used as electrodes. Tungsten electrodes are nonconsumable if the process is properly used, because they do not melt or transfer to the weld. In other welding processes, such as SMAW, GMAW, and SAW, the electrode is the filler metal. The function of a tungsten electrode is to serve as one of the electrical termi-

nals of the arc which supplies the heat required for welding. Its melting point is 6170°F (3410°C). Approaching this high temperature, tungsten becomes thermionic; it is a ready source of electrons. It reaches this temperature by resistance heating and, were it not for the significant cooling effect of electrons boiling from its tip, resistance heating would cause the tip to melt. In fact, the electrode tip is much cooler than that part of the electrode between the tip and the externally-cooled collet.

Classification of Electrodes

TUNGSTEN ELECTRODES ARE classified on the basis of their chemical compositions, as specified in Table 3.3. Requirements for tungsten electrodes are given in the latest edition of ANSI–AWS A5.12, *Specification for Tungsten and Tungsten Alloy Electrodes for Arc Welding and Cutting*. The color code identification system for the various classes of tungsten electrodes is shown in Table 3.3.

Table 3.2
Recommended Tungsten Electrodes[a] and Gas Cups for Various Welding Currents

Electrode Diameter		Use Gas Cup I.D.	Direct Current, A		Alternating Current, A	
			Straight Polarity[b]	Reverse Polarity[b]	Unbalanced Wave[c]	Balanced Wave[c]
in.	mm	in.	DCEN	DCEP		
0.010	0.25	1/4	up to 15		up to 15	up to 15
0.020	0.50	1/4	5-20		5-15	10-20
0.040	1.00	3/8	15-80		10-60	20-30
1/16	1.6	3/8	70-150	10-20	50-100	30-80
3/32	2.4	1/2	150-250	15-30	100-160	60-130
1/8	3.2	1/2	250-400	25-40	150-210	100-180
5/32	4.0	1/2	400-500	40-55	200-275	160-240
3/16	4.8	5/8	500-750	55-80	250-350	190-300
1/4	6.4	3/4	750-1100	80-125	325-450	325-450

a. All values are based on the use of argon as the shielding gas.

b. Use EWTh-2 electrodes.

c. Use EWP electrodes.

Table 3.3
Color Code and Alloying Elements for Various Tungsten Electrode Alloys

AWS Classification	Color[a]	Alloying Element	Alloying Oxide	Nominal Weight of Alloying Oxide Percent
EWP	Green	—	—	—
EWCe-2	Orange	Cerium	CeO_2	2
EWLa-1	Black	Lanthanum	La_2O_3	1
EWTh-1	Yellow	Thorium	ThO_2	1
EWTh-2	Red	Thorium	ThO_2	2
EWZr-1	Brown	Zirconium	ZrO_2	.25
EWG	Gray	Not Specified[b]	—	—

a. Color may be applied in the form of bands, dots, etc., at any point on the surface of the electrode.

b. Manufacturer must identify the type and nominal content of the rare earth oxide addition.

Electrodes are produced with either a clean finish or ground finish. Electrodes with a clean finish have been chemically cleaned to remove surface impurities after the forming operation. Those with a ground finish have been centerless ground to remove surface imperfections.

Electrode Sizes and Current Capacities

TUNGSTEN AND THORIATED tungsten electrode sizes and current ranges are listed in Table 3.2, along with shield-gas cup diameters recommended for use with different types of welding power. It provides a useful guide for selecting the correct electrode for specific applications involving different current levels and power supplies.

Current levels in excess of those recommended for a given electrode size and tip configuration will cause the tungsten to erode or melt. Tungsten particles may fall into the weld pool and become defects in the weld joint. Current too low for a specific electrode diameter can cause arc instability.

Direct current with the electrode positive requires a much larger diameter to support a given level of current because the tip is not cooled by the evaporation of electrons but heated by their impact. In general, a given electrode diameter on DCEP would be expected to handle only 10 percent of the current possible with the electrode negative. With alternating current, the tip is cooled during the electrode negative cycle and heated when positive. Therefore, the current carrying capacity of an electrode on ac is between that of DCEN and DCEP. In general, it is about 50 percent less than that of DCEN.

EWP Electrode Classification. Pure tungsten electrodes (EWP) contain a minimum of 99.5 percent tungsten, with no intentional alloying elements. The current-carrying capacity of pure tungsten electrodes is lower than that of the alloyed electrodes. Pure tungsten electrodes are used mainly with ac for welding aluminum and magnesium alloys. The tip of the EWP electrode maintains a clean, balled end, which provides good arc stability. They may also be used with dc, but they do not provide the arc initiation and arc stability characteristics of thoriated, ceriated, or lanthanated electrodes.

EWTh Electrode Classifications. The thermionic emission of tungsten can be improved by alloying it with metal oxides that have very low work functions. As a result, the electrodes are able to handle higher welding currents without failing. Thorium oxide is one such additive. To prevent identification problems with these and other types of tungsten electrodes, they are color coded as shown in Table 3.3. Two types of thoriated tungsten electrodes are available. The EWTh-1 and EWTh-2 electrodes contain 1 percent and 2 percent thorium oxide (ThO_2) called *thoria*, respectively, evenly dispersed through their entire lengths.

Thoriated tungsten electrodes are superior to pure tungsten electrodes in several respects. The thoria provides

about 20 percent higher current-carrying capacity, generally longer life, and greater resistance to contamination of the weld. With these electrodes, arc starting is easier, and the arc is more stable than with pure tungsten or zirconiated tungsten electrodes.

The EWTh-1 and EWTh-2 electrodes were designed for DCEN applications. They maintain a sharpened tip configuration during welding, which is desirable for welding steel. They are not often used with ac because it is difficult to maintain the balled end, which is necessary with ac welding, without splitting the electrode.

Thorium is a very low-level radioactive material. The level of radiation has not been found to represent a health hazard. However, if welding is to be performed in confined spaces for prolonged periods of time, or if electrode grinding dust might be ingested, special precautions relative to ventilation should be considered. The user should consult the appropriate safety personnel.

A discontinued classification of tungsten electrodes is the EWTh-3 class. This tungsten electrode had a longitudinal or axial segment which contained 1.0 percent to 2.0 percent thoria. The average thoria content of the electrode was 0.35 percent to 0.55 percent. Advances in powder metallurgy and other processing developments have caused this electrode classification to be discontinued, and it is no longer commercially available.

EWCe Electrode Classification. Ceriated tungsten electrodes were first introduced into the United States market in the early 1980's. These electrodes were developed as possible replacements for thoriated electrodes because cerium, unlike thorium, is not a radioactive element. The EWCe-2 electrodes are tungsten electrodes containing 2 percent cerium oxide (CeO_2), referred to as *ceria*. Compared with pure tungsten, the ceriated electrodes exhibit a reduced rate of vaporization or burn-off. These advantages of ceria improve with increased ceria content. EWCe-2 electrodes will operate successfully with ac or dc.

EWLa Electrode Classification. The EWLa-1 electrodes were developed around the same time as the ceriated electrodes and for the same reason, that lanthanum is not radioactive. These electrodes contain 1 percent lanthanum oxide (La_2O_3), referred to as *lanthana*. The advantages and operating characteristics of these electrodes are very similar to the ceriated tungsten electrodes.

EWZr Electrode Classification. Zirconiated tungsten electrodes (EWZr) contain a small amount of zirconium oxide (ZrO_2), as listed in Table 3.3. Zirconiated tungsten electrodes have welding characteristics that generally fall between those of pure and thoriated tungsten. They are the electrode of choice for ac welding because they combine the desirable arc stability characteristics and balled end typical of pure tungsten with the current capacity and starting characteristics of thoriated tungsten. They have higher resistance to contamination than pure tungsten, and are preferred for radiographic-quality welding applications where tungsten contamination of the weld must be minimized.

EWG Electrode Classification. The EWG electrode classification was assigned for alloys not covered by the above classes. These electrodes contain an unspecified addition of an unspecified oxide or combination of oxides (rare earth or others). The purpose of the addition is to affect the nature or characteristics of the arc, as defined by the manufacturer. The manufacturer must identify the specific addition or additions and the nominal quantity or quantities added.

Several EWG electrodes are either commercially available or are being developed. These include additions of yttrium oxide or magnesium oxide. This classification also includes ceriated and lanthanated electrodes which contain these oxides in amounts other than as listed above, or in combination with other oxides.

Electrode Tip Configurations

THE SHAPE OF the tungsten electrode tip is an important process variable in GTAW. Tungsten electrodes may be used with a variety of tip preparations. With ac welding, pure or zirconiated tungsten electrodes form a hemispherical balled end. For dc welding, thoriated, ceriated, or lanthanated tungsten electrodes are usually used. For the latter, the end is typically ground to a specific included angle, often with a truncated end. As shown in Figure 3.5, various electrode tip geometries affect the weld bead shape and size. In general, as the included angle increases, the weld penetration increases and the width of the weld bead decreases. Although small diameter electrodes may be used with a square end preparation for DCEN welding, conical tips provide improved welding performance.

Regardless of the electrode tip geometry selected, it is important that a consistent electrode geometry be used once a welding procedure is established. Changes in electrode geometry can significantly influence the weld bead shape and size; therefore, electrode tip configuration is a welding variable that should be studied during the welding procedure development.

Tungsten tips are generally prepared by balling, grinding, or chemical sharpening. A tapered electrode tip is usually prepared on all but the smallest electrodes, even when the end later will be balled for ac welding.

Balling. With ac welding (usually performed with pure or zirconiated tungsten electrodes), a hemispherical tip is most desirable. Before use in welding, the electrode tip can be balled by striking an arc on a water-cooled copper block or other suitable material using ac or DCEP. Arc current is increased until the end of the electrode turns white hot and the tungsten begins to melt, causing a small ball to form at the end of the electrode. The current is down-

sloped and extinguished, leaving a hemispherical ball on the end of the tungsten electrode. The size of the hemisphere should not exceed 1-1/2 times the electrode diameter, otherwise it may fall off while it is molten.

Grinding. To produce optimum arc stability, grinding of tungsten electrodes should be done with the axis of the electrode perpendicular to the axis of the grinding wheel. The grinding wheel should be reserved for grinding only tungsten to eliminate possible contamination of the tungsten tip with foreign matter during the grinding operation. Exhaust hoods should be used when grinding thoriated electrodes to remove the grinding dust from the work area.

Thoriated, ceriated, and lanthanated tungsten electrodes do not ball as readily as pure or zirconiated tungsten electrodes. They maintain a ground tip shape much better. If used on ac these electrodes often split.

Chemical Sharpening. Chemical sharpening consists of submerging the red-hot end of a tungsten electrode into a container of sodium nitrate. The chemical reaction between the hot tungsten and the sodium nitrate will cause the tungsten to erode at a uniform rate all around the circumference and end of the electrode. Repeated heating and dipping of the tungsten into the sodium nitrate will form a tapered tip.

Electrode Contamination

CONTAMINATION OF THE tungsten electrode is most likely to occur when a welder accidentally dips the tungsten into the molten weld pool or touches the tungsten with the filler metal. The tungsten electrode may also become oxidized by an improper shielding gas or insufficient gas flow, during welding or after the arc has been extinguished. Other sources of contamination include: metal vapors from the welding arc, weld pool eruptions or spatter caused by gas entrapment, and evaporated surface impurities.

The contaminated end of the tungsten electrode will adversely affect the arc characteristics and may cause tungsten inclusions in the weld metal. If this occurs, the welding operation should be stopped and the contaminated portion of the electrode removed.

Contaminated tungsten electrodes must be properly dressed by breaking off the contaminated section and grinding to shape according to the manufacturer's suggested procedure.

WIRE FEEDERS

WIRE FEEDERS ARE used to add filler metal during automatic and machine welding. Either room temperature (cold) wire or preheated (hot) wire can be fed into the molten weld pool. Cold wire is fed into the leading edge and hot wire is fed into the trailing edge of the molten pool.

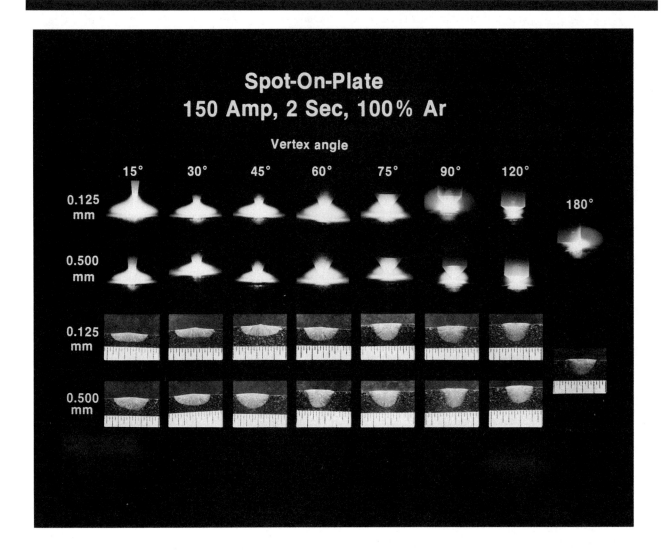

Figure 3.5–Arc Shape and Fusion Zone Profile as a Function of Electrode Tip Geometry in a Pure Argon Shield (150 A, 2.0 s Spot-On-Plate)

Cold Wire

THE SYSTEM FOR feeding of cold wire has three components:

(1) Wire drive mechanism
(2) Speed control
(3) Wire guide attachment to introduce the wire into the molten weld pool

The drive consists of a motor and gear train to power a set of drive rolls which push the wire. The control is essentially a constant-speed governor which can be either a mechanical or an electronic device. The wire is fed to the wire guide through a flexible conduit.

An adjustable wire guide is attached to the electrode holder. It maintains the position at which the wire enters the weld and the angle of approach relative to the electrode, work surface, and the joint. In heavy-duty applications, the wire guide is water cooled. Wires ranging from 0.015 to 3/32 in. (0.4 to 2.4 mm) in diameter are used. Special wire feeders are available to provide continuous, pulsed, or intermittent wire feed.

Hot Wire

THE PROCESS FOR hot wire addition is similar to that for cold wire, except that the wire is resistance heated to a temperature close to its melting point just before it contacts the molten weld pool. When using a preheated (hot) wire in machine and automatic gas tungsten arc welding in the flat position, the wire is fed mechanically to the weld pool through a holder from which inert gas flows to protect the hot wire from oxidation. This system is illustrated in Figure 3.6. Normally, a mixture of 75 percent helium-25 percent argon is used to shield the tungsten electrode and the molten weld pool.

Deposition rate is greater with hot wire than with cold wire, as shown in Figure 3.7. This rate is comparable to that in gas metal arc welding. The current flow is initiated when the wire contacts the weld surface. The wire is fed into the molten pool directly behind the arc at a 40 to 60 degree angle with respect to the tungsten electrode.

The wire is resistance heated by alternating current from a constant-voltage power source. Alternating current is used for heating the wire to avoid arc blow. When the heating current does not exceed 60 percent of the arc current, the arc oscillates 30 degrees in the longitudinal direction. The oscillation increases to 120 degrees when the heating and arc currents are equal. The amplitude of arc oscillation can be controlled by limiting the wire diameter to 0.045 in. (1.2 mm) and reducing the heating current below 60 percent of the arc current.

Preheated filler wire has been used successfully for joining carbon and low alloy steels, stainless steels, and alloys of copper and nickel. Preheating is not recommended for aluminum and copper because the low resistance of these filler wires requires high heating current, which results in excessive arc deflection and uneven melting.

POWER SUPPLIES

CONSTANT-CURRENT TYPE power sources are used for GTAW. Power required for both ac and dc GTAW can be supplied by transformer-rectifier power supplies or from rotating ac or dc generators. Advances in semiconductor electronics have made transformer-rectifier power sources popular for both shop and field GTAW, but rotating-type power sources continue to be widely used in the field.

GTAW power sources typically have either drooping or nearly true constant-current static output characteristics, such as those shown in Figure 3.8. The static output characteristic is a function of the type of welding current control used in the power source design.

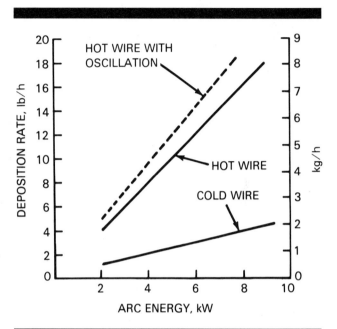

Figure 3.7–Deposition Rates for Gas Tungsten Arc Welding With Cold and Hot Steel Filler Wire

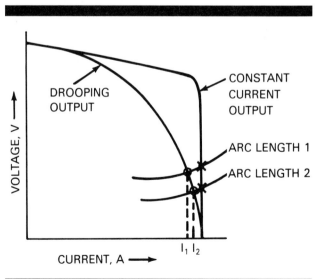

Figure 3.8–Static Volt-Ampere Characteristics for Drooping and Constant Current Power Supplies

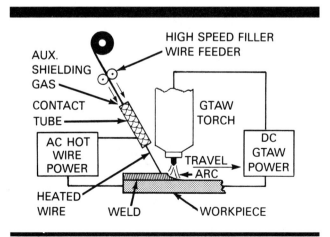

Figure 3.6–Gas Tungsten Arc Hot Wire System

A drooping volt-ampere characteristic is typical of magnetically controlled power source designs including the moving coil, moving shunt, moving core reactor, saturable reactor, or magnetic amplifier designs and also rotating power source designs. A truly constant-current output is available from electronically controlled power sources. The drooping characteristic is advantageous for manual welding where a remote foot pedal current control is not available at the site of welding. With a drooping characteristic, the welder may vary the current level slightly by changing the arc length. The degree of current control possible by changing arc length can be inferred from Figure 3.8.

In most of the magnetically controlled power sources, the current-level control is accomplished in the ac portion of the power source. As a result, these power sources are not typically used to provide pulsed current because of their slow dynamic response. The addition of a rectifier bridge allows these power sources to provide both ac and dc welding current. Those power sources which use a moving component for current control cannot readily be remotely controlled with a foot pedal, while the others typically can.

Most magnetically controlled power sources are considered to be open-loop controlled, in that the actual welding current for a given current setting depends on and may vary with the welding conditions. Single-phase power sources can provide both ac and dc current, while three-phase sources normally provide only dc. The dc current of a three-phase power source is typically smoother than that of a single-phase source because of reduced ripple-current amplitude.

The advantages of magnetically controlled power sources are that they are simple to operate, require little maintenance in adverse industrial environments, and are relatively inexpensive. The disadvantages are that they are large in size and weight and have a lower efficiency compared to electronically controlled power sources. Also, as mentioned, most magnetic-control techniques are open-loop which limits repeatability, accuracy, and response. An essentially constant-current volt-ampere characteristic can be provided by electronically controlled power sources, such as the series linear regulator, silicon controlled rectifier, secondary switcher, and inverter designs.

The essentially constant current characteristic is typically advantageous for machine and automatic welding, to provide sufficient accuracy and repeatability in current level from weld to weld. Most truly constant current power sources are closed-loop controlled, in which the actual current is measured and compared to the desired current setting. Adjustments are made electronically within the power source to maintain the desired current as welding conditions change.

Most electronically controlled power sources offer rapid dynamic response. As a result, these power sources can be used to provide pulsed welding current. Series linear regulator and switched secondary designs provide only dc welding current from single or three-phase input power.

Silicon controlled rectifier designs can provide ac and dc current from single-phase power and dc current from three-phase power. Depending on the design, inverters can provide ac and dc output from single or three phase input power. Inverter power sources are the most versatile, with many offering multi-process capabilities and variable welding current waveform output. Inverters are also lighter and more compact than other power source designs of equivalent current rating.

The advantages of electronically controlled power sources are that they offer rapid dynamic response, provide variable current waveform output, have excellent repeatability, and offer remote control. The disadvantages are that they are more complex to operate and maintain and are relatively expensive.

It is important to select a GTAW power source based on the type of welding current required for a particular application. The types of welding current include ac sine-wave, ac square-wave, dc, and pulsed dc. The next section of this chapter has more information on the types and effects of welding current. Many power sources are available with a variety of additional controls and functions such as water and shielding gas control, wire feeder and travel mechanism sequencing, current up-slope and down-slope, and multiple-current sequences. Refer to Chapter 1, "Arc Welding Power Sources," for more detailed information.

Direct Current

USING DIRECT CURRENT, the tungsten electrode may be connected to either the negative or the positive terminal of the power supply. In almost all cases, electrode negative (cathode) is chosen. With that polarity, electrons flow from the electrode to the work and positive ions are transferred from the work to the electrode, as shown for DCEN (straight polarity) in Figure 3.9. When the electrode is positive (anode), the directions of electron and positive ion flow are reversed, as shown for DCEP (reverse polarity) in Figure 3.9.

With DCEN and a thermionic electrode such as tungsten, approximately 70 percent of the heat is generated at the anode and 30 percent at the cathode. Since DCEN produces the greatest amount of heat at the workpiece, for a given welding current, DCEN will provide deeper weld penetration than DCEP (see Figure 3.9). DCEN is the most common configuration used in GTAW, and is used with argon, helium, or a mixture of the two to weld most metals.

When the tungsten electrode is connected to the positive terminal (DCEP), a cathodic cleaning action is created at the surface of the workpiece. This action occurs with most metals but is most important when welding aluminum and magnesium because it removes the refractory oxide surface that inhibits wetting of the weldment by the weld metal.

Unlike DCEN, in which the electrode tip is cooled by the evaporation of electrons, when the electrode is used as

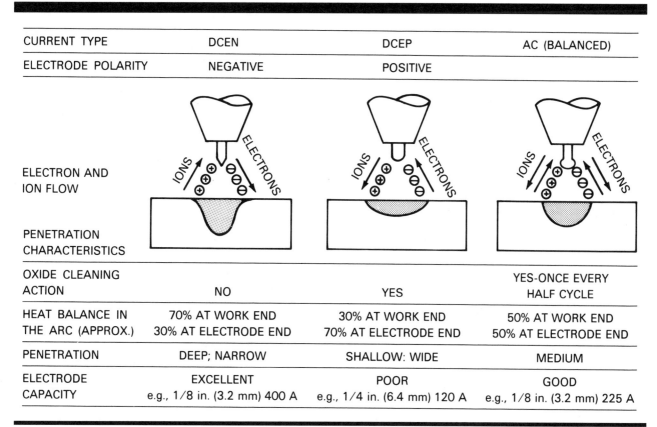

CURRENT TYPE	DCEN	DCEP	AC (BALANCED)
ELECTRODE POLARITY	NEGATIVE	POSITIVE	
ELECTRON AND ION FLOW PENETRATION CHARACTERISTICS			
OXIDE CLEANING ACTION	NO	YES	YES-ONCE EVERY HALF CYCLE
HEAT BALANCE IN THE ARC (APPROX.)	70% AT WORK END 30% AT ELECTRODE END	30% AT WORK END 70% AT ELECTRODE END	50% AT WORK END 50% AT ELECTRODE END
PENETRATION	DEEP; NARROW	SHALLOW: WIDE	MEDIUM
ELECTRODE CAPACITY	EXCELLENT e.g., 1/8 in. (3.2 mm) 400 A	POOR e.g., 1/4 in. (6.4 mm) 120 A	GOOD e.g., 1/8 in. (3.2 mm) 225 A

Figure 3.9–Characteristics of Current Types for Gas Tungsten Arc Welding

the positive pole, its tip is heated by the bombardment of electrons as well as by its resistance to their passage through the electrode. Therefore, to reduce resistance heating and increase thermal conduction into the electrode collet, a larger diameter electrode is required for a given welding current when reverse polarity is used. The current carrying capacity of an electrode connected to the positive terminal is approximately one-tenth that of an electrode connected to the negative terminal. DCEP is generally limited to welding sheet metal.

Pulsed Dc Welding. Pulsed dc involves the repetitive variation in arc current from a background (low) value to a peak (high) value. Pulsed dc power sources typically allow adjustments of the pulse current time, background current time, peak current level, and background current level, to provide a current output wave form suited to a particular application. Figure 3.10 shows a typical pulsed current waveform. Generally, the background and pulse duration times are adjustable so that the current can change levels anywhere from once every two seconds to 20 pulses per second. Pulsed dc is usually applied with the electrode negative (DCEN).

In pulsed dc welding, the pulse current level is typically set at 2 to 10 times the background current level. This combines the driving, forceful arc characteristics of high current with the low-heat input of low current. The pulse current achieves good fusion and penetration, while the background current maintains the arc and allows the weld area to cool.

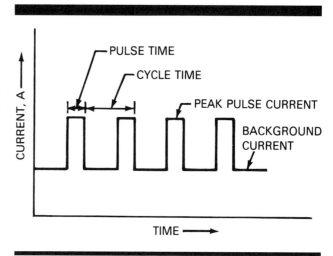

Figure 3.10–Pulsed DC Waveform

There are several advantages of pulsed current. For a given average current level, greater penetration can be obtained than with steady current, which is useful on metals sensitive to heat input and minimizes distortion. Because there is insufficient time for significant heat flow during the short duration of a pulse, metals of dissimilar thicknesses usually respond equally, and equal penetration can be achieved. For a similar reason, very thin metals can be joined with pulsed dc. In addition, one set of welding variables can be used on a joint in all positions, such as a circumferential weld in a horizontal pipe. Pulsed dc is also useful for bridging gaps in open root joints.

Although mostly used for machine and automatic GTAW, pulsing offers advantages for manual welding. Inexperienced welders find that they can improve their proficiency by counting the pulses (from 1/2 to 2 pulses per second) and using them to time the movement of the torch and the cold wire. Experienced welders are able to weld thinner materials, dissimilar alloys, and thicknesses with less difficulty.

High-Frequency Pulsed Welding. High-frequency switched dc involves the application of direct current which is switched from a low level to a high-current level at a rapid, fixed frequency of approximately 20 kHz, as shown in Figure 3.11.

The peak current "on" time is varied to change the average current level. The effect of high-frequency switching is to produce a "stiff" welding arc. Arc pressure is a measure of arc stiffness. As shown in Figure 3.12, as the switching frequency nears 10 kHz, arc pressure increases to nearly four times that of a steady dc arc. As arc pressure increases, lateral displacement of the arc, such as that produced by magnetic fields (arc blow) and shielding gas movement (wind), is reduced.

High-frequency switched dc is useful in precision machine and automatic applications where an arc with excep-

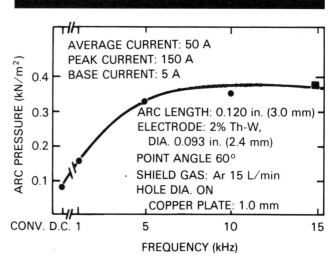

Figure 3.12—Relation Between Arc Pressure and Pulse Frequency

tional directional properties and stability is required. It is also used where a stable arc is needed at very low-average currents. The disadvantage of high-frequency switched dc is that the welding power sources are costly. Also, if the switching frequency is in the audible range, the arc sound can be very annoying.

Alternating Current

ALTERNATING CURRENT UNDERGOES periodic reversal in welding current polarity from electrode positive to electrode negative. Thus, ac can combine the work cleaning action of electrode positive (reverse polarity) with the deep penetration characteristic of electrode negative (straight polarity). AC welding is compared with DCEN and DCEP welding in Figure 3.9.

Conventional ac welding power sources produce a sinusoidal open circuit voltage output which is out-of-phase with the current by about 90°. The frequency of voltage reversal is typically fixed at the standard 60 Hz frequency of the primary power. The actual arc voltage is in phase with the welding current. The voltage measured is the sum of voltage drops in the electrode and the plasma and at the anode and cathode; all of which are the result of current flow.

When the current decays to zero, different effects will occur, depending on the polarity. When the thermionic tungsten electrode becomes negative, it supplies electrons immediately to reignite the arc. However, when the weld pool becomes negative, it cannot supply electrons until the voltage is raised sufficiently to initiate cold-cathode emission. Without this voltage, the arc becomes unstable. This is shown in Figure 3.13(A).

Some means of stabilizing the arc during voltage reversal is required with conventional sinusoidal welding power

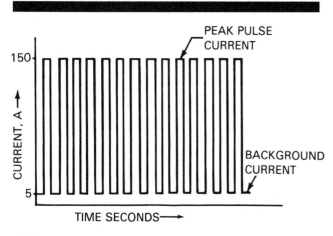

Figure 3.11—High-Frequency Switched DC Pulsed Welding Current Wave Form

sources. This has been done by using high open circuit power supplies; by discharging capacitors at the appropriate time in the cycle; by using high-voltage high-frequency sparks in parallel with the arc; and by using power supplies with a square wave output. The results of such stabilization are shown in Figure 3.13(B).

To improve arc stability, the open circuit voltage of the transformer can be increased. An open circuit voltage of about 100 V (rms) is needed with helium shielding. The necessary voltage can also be obtained by adding, in series with the transformer, a high-frequency voltage supply. The high-frequency voltage is generally on the order of several thousand volts, and its frequency can be as high as several megahertz. The current is very low. The high-frequency voltage may be applied continuously or periodically during welding. In the latter case, a burst of high-frequency voltage is set to occur during the time when the welding current passes through zero.

Square wave ac welding power sources can change the direction of the welding current in a short period of time. The presence of high voltage, coupled with high electrode and base metal temperature at current reversals, allows the arc to be reignited without the need for an arc stabilizer. Also, the lower "peak" current of the square wave form tends to increase the usable current range of the electrode.

Since it is easier to provide the electrons needed to sustain an arc when the electrode is negative, the voltage required also is less. The result is a higher welding current during the DCEN interval than during DCEP. In effect, the power supply produces both direct current and alternating current. Such rectification can cause damage to the power supply due to overheating or, with some machines, a decay in the output. Such rectification is eliminated by wave balancing as shown in Figure 3.13(C).

Early balanced-current power supplies involved either series-connected capacitors or a dc voltage source (such as a battery) in the welding circuit. Modern power supply circuits use electronic wave balancing. Balanced current flow is not essential for most manual welding operations. It is, however, desirable for high-speed machine or automatic welding. The advantages of balanced current flow are the following:

(1) Better oxide removal
(2) Smoother, better welding
(3) No requirement for reduction in output rating of a given size of conventional welding transformer (the unbalanced core magnetization that is produced by the dc component of an unbalanced current flow is minimized)

The following are disadvantages of balanced current flow:

(1) Larger tungsten electrodes are needed.
(2) Higher open circuit voltages generally associated with some wave balancing means may constitute a safety hazard.

(3) Balanced wave welding power sources are more costly.

Some square wave ac power sources adjust the current level during the electrode positive and electrode negative cycles at our standard 60 Hz frequency. More expensive power sources adjust the time of each polarity half cycle as well as the current level during that half cycle. Such variable wave forms will adjust the welding current to suit a particular application. The characteristics of variable square wave alternating current are shown in Figure 3.14.

ARC VOLTAGE CONTROL

ARC VOLTAGE CONTROLLERS (AVC) are used in machine and automatic GTAW to maintain arc length. In this case, the arc itself is a sensor, since it converts a measurement of length (arc gap) into an electrical signal (arc voltage).

The AVC compares the measured and desired arc voltages to determine which direction and at what speed the welding electrode should be moved. This determination, expressed as a voltage error signal, is amplified to drive motors in a slide that supports the torch. The changing voltage that results from the motion of the welding electrode is detected and the cycle repeats to maintain the desired arc voltage.

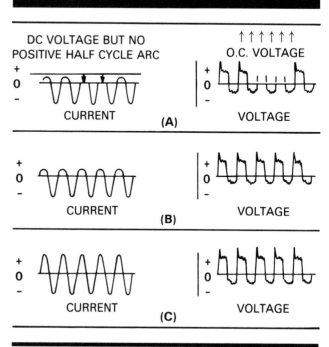

Figure 3.13–Voltage and Current Wave Forms for AC Welding: (A) Partial and Complete Rectification; (B) With Arc Stabilization; (C) With Current Balancing

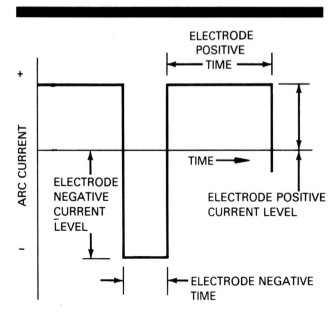

Figure 3.14–Characteristics of Variable Square Wave AC

ARC OSCILLATION

THE WIDTH OF gas tungsten arc welds can be increased by mechanical oscillation. Mechanical arc oscillation can be achieved by mounting a GTAW torch on a cross slide that provides movement of the torch transverse to the line of travel. Such equipment provides adjustable cross-feed speed, amplitude of oscillation, and dwell on each side of the oscillation cycle.

Better fusion of joint side-walls and a reduction of the disruptive effects of arc blow can be obtained by magnetic oscillation. Such oscillators deflect the arc longitudinally or laterally over the weld pool without moving the welding electrode. These oscillators consist of electromagnets, located close to the arc, that are powered by a variable-polarity, variable-amplitude power supply. Control features include adjustable oscillation frequency and amplitude, and separately adjustable dwell times.

SHIELDING GASES

SHIELDING GAS IS directed by the torch to the arc and weld pool to protect the electrode and the molten weld metal from atmospheric contamination. Backup purge gas can also be used to protect the underside of the weld and its adjacent base metal surfaces from oxidation during welding. Uniformity of root bead contour, freedom from undercutting, and the desired amount of root bead reinforcement are more likely to be achieved when using gas backup under controlled conditions. In some materials, gas backup reduces root cracking and porosity in the weld.

Types of Shielding Gases

ARGON AND HELIUM or mixtures of the two are the most common types of inert gas used for shielding. Argon-hydrogen mixtures are used for special applications.

Depending on the volume of usage, these gases may be supplied in cylinders or as a liquid in insulated tanks. The liquid is vaporized and piped to points within the plant, thus eliminating cylinder handling.

Argon.   Argon (Ar) is an inert monatomic gas with a molecular weight of 40. It is obtained from the atmosphere by the separation of liquified air.

Welding grade argon is refined to a minimum purity of 99.95 percent. This is acceptable for GTAW of most metals except the reactive and refractory metals, for which a minimum purity of 99.997 percent is required. Often, such metals are fabricated in chambers from which all traces of air have been purged prior to initiating the welding operation.

Argon is used more extensively than helium because of the following advantages:

(1) Smoother, quieter arc action
(2) Reduced penetration
(3) Cleaning action when welding materials such as aluminum and magnesium
(4) Lower cost and greater availability
(5) Lower flow rates for good shielding
(6) Better cross-draft resistance
(7) Easier arc starting

The reduced penetration of an argon shielded arc is particularly helpful when manual welding of thin material, because the tendency for excessive melt-through is lessened. This same characteristic is advantageous in vertical or overhead welding since the tendency for the base metal to sag or run is decreased.

Helium.   Helium (He) is an inert, very light monatomic gas, having an atomic weight of four. It is obtained by separation from natural gas. Welding grade helium is refined to a purity of at least 99.99 percent.

For given values of welding current and arc length, helium transfers more heat into the work than argon. The greater heating power of the helium arc can be advantageous for joining metals of high thermal conductivity and for high-speed mechanized applications. Also, helium is used more often than argon for welding heavy plate. Mixtures of argon and helium are useful when some balance between the characteristics of both is desired.

Characteristics Of Argon And Helium.   The chief factor influencing shielding effectiveness is the gas density. Argon is approximately one and one-third times as heavy as air and ten times heavier than helium. Argon, after leaving the torch nozzle, forms a blanket over the weld area.

Helium, because it is lighter, tends to rise around the nozzle. Experimental work has consistently shown that to produce equivalent shielding effectiveness, the flow of helium must be two to three times that of argon. The same general relationship is true for mixtures of argon and helium, particularly those high in helium content.

The important characteristics of these gases are the voltage-current relationships of the tungsten arc in argon and in helium that are illustrated in Figure 3.15. At all current levels, for equivalent arc lengths, the arc voltage obtained with helium is appreciably higher than that with argon. Since heat in the arc is roughly measured by the product of current and voltage (arc power), helium offers more available heat than argon. The higher available heat favors its selection when welding thick materials and metals having high thermal conductivity or relatively high melting temperatures.

However, it should be noted that at lower currents, the volt-ampere curves pass through a minimum voltage, at current levels approximately 90 amperes (A) apart, after which the voltage increases as the current decreases. For helium, this increase in voltage occurs in the range of 50 to 150 A where much of the welding of thin materials is done. Since the voltage increase for argon occurs below 50 A, the use of argon in the 50 to 150 A range provides the operator with more latitude in arc length to control the welding operation.

It is apparent that to obtain equal arc power, appreciably higher current must be used with argon than with helium. Since undercutting with either gas will occur at about equal currents, helium will produce satisfactory welds at much higher speeds.

The other influential characteristic is that of arc stability. Both gases provide excellent stability with direct current power. With alternating current power, which is used extensively for welding aluminum and magnesium, argon yields much better arc stability and the highly desirable cleaning action, which makes argon superior to helium in this respect.

Argon-Hydrogen Mixtures. Argon-hydrogen mixtures are employed in special cases, such as mechanized welding of light gage stainless steel tubing, where the hydrogen does not cause adverse metallurgical effects such as porosity and hydrogen-induced cracking. Increased welding speeds can be achieved in almost direct proportion to the amount of hydrogen added to argon because of the increased arc voltage. However, the amount of hydrogen that can be added varies with the metal thickness and type of joint for each particular application. Excessive hydrogen will cause porosity. Hydrogen concentrations up to 35 percent have been used on all thicknesses of stainless steel where a root opening of approximately 0.010 to 0.020 in. (0.25 to 0.5 mm) is used. Argon-hydrogen mixtures are limited to use on stainless steel, nickel-copper, and nickel-base alloys.

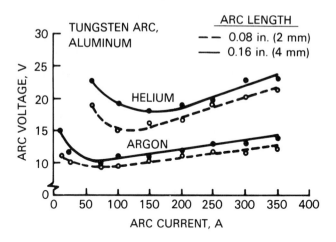

Figure 3.15–Voltage-Current Relationship With Argon and Helium Shielding

The most commonly used argon-hydrogen mixture contains 15 percent hydrogen. This mixture is used for mechanized welding of tight butt joints in stainless steel up to 0.062 in. (1.6 mm) thick at speeds comparable to helium (50 percent faster than argon). It is also used for welding stainless steel beer barrels, and tube to tubesheet joints in a variety of stainless steels and nickel alloys. For manual welding, a hydrogen content of five percent is sometimes preferred to obtain cleaner welds.

Selection of Shielding Gas. No set rule governs the choice of shielding gas for any particular application. Either argon, helium, or a mixture of argon and helium may be used successfully for most applications, with the possible exception of manual welding on extremely thin material, for which argon is essential. Argon generally provides an arc operates more smoothly and quietly, is handled more easily, and is less penetrating than an arc shielded by helium. In addition, the lower unit cost and the lower flow rate requirements of argon make argon preferable from an economic point of view. Argon is preferred for most applications, except where helium's higher heat penetration is required for welding thick sections of metals with high heat conductivity, such as aluminum and copper. A guide to the selection of gases is provided in Table 3.4.

Table 3.4
Recommended Types of Current, Tungsten Electrodes and Shielding Gases for Welding Different Metals

Type of Metal	Thickness	Type of Current	Electrode*	Shielding Gas
Aluminum	All	Alternating current	Pure or zirconium	Argon or argon-helium
	over 1/8 in.	DCEN	Thoriated	Argon-helium or argon
	under 1/8 in.	DCEP	Thoriated or zirconium	Argon
Copper, copper alloys	All	DCEN	Thoriated	Helium
	under 1/8 in.	Alternating current	Pure or zirconium	Argon
Magnesium alloys	All	Alternating current	Pure or zirconium	Argon
	under 1/8 in.	DCEP	Zirconium or thoriated	Argon
Nickel, nickel alloys	All	DCEN	Thoriated	Argon
Plain carbon, low-alloy steels	All	DCEN	Thoriated	Argon or argon-helium
	under 1/8 in.	Alternating current	Pure or zirconium	Argon
Stainless steel	All	DCEN	Thoriated	Argon or argon-helium
	under 1/8 in.	Alternating current	Pure or zirconium	Argon
Titanium	All	DCEN	Thoriated	Argon

* Where thoriated electrodes are recommended, ceriated or lanthanated electrodes may also be used.

Recommended Gas Flow Rates

SHIELDING GAS FLOW requirements are based on cup or nozzle size, weld pool size, and air movement. In general, the flow rate increases in proportion to the cross-sectional area at the nozzle (considering the obstruction caused by the collet). The nozzle diameter is selected to suit the size of the weld pool and the reactivity of the metal to be welded. The minimum flow rate is determined by the need for a stiff stream to overcome the heating effects of the arc and cross drafts. With the more commonly used torches, typical shielding gas flow rates are 15 to 35 cfh (7 to 16 L/min.) for argon and 30 to 50 cfh (14 to 24 L/min.) for helium. Excessive flow rates cause turbulence in the gas stream which may aspirate atmospheric contamination into the weld pool.

A cross wind or draft moving at five or more miles per hour can disrupt the shielding gas coverage. The stiffest, nonturbulent gas streams (with high-stream velocities) are obtained by incorporating gas lenses in the nozzle and by using helium as the shield gas. However, in the interest of cost, protective screens to block air flow are preferred to increasing shielding gas flow.

Backup Purge

WHEN MAKING THE root passes of welds, the air contained on the back side of the weldment can contaminate the weld. To avoid that problem, the air must be purged from this region. Argon and helium are satisfactory for the backup purge when welding all materials. Nitrogen may be used satisfactorily for backing up welds in austenitic stainless steel, copper, and copper alloys.

Gas flow requirements for the backup purge range from 1 to 90 CFH (0.5 to 42 L/min.), based on the volume to be purged. As a rule of thumb, a relatively inert atmosphere will be obtained by flushing with four times the volume to be purged. After purging is completed, the flow of backup gas during welding should be reduced until only a slight positive pressure exists in the purged area. After the root and first filler passes are completed, the backup purge may be discontinued.

Several devices are available to contain shielding gas on the back side of plate and piping weldments. One of these is shown in Figure 3.16. Refer to the latest edition of ANSI/AWS C5.5, *Recommended Practices for Gas Tungsten Arc Welding*, for more information.

When purging piping systems, provisions for an adequate vent or exhaust, as shown in Figure 3.17, are important to prevent excessive pressure buildup during welding. The area of vents through which the backup gas is exhausted to the atmosphere should be at least equal to the area of the opening through which the gas is admitted to the system. Extra care should be taken to ensure that the backup purge pressure is not excessive when welding the last inch or two on the root pass, to prevent weld pool blow-out or root concavity.

When using argon or nitrogen, the backup gas should preferably enter the system at a low point, to displace the atmosphere upwards, and be vented at points beyond the joint to be welded. Again refer to Figure 3.17. In piping systems which have several joints, all except the one being welded should be taped to prevent gas loss.

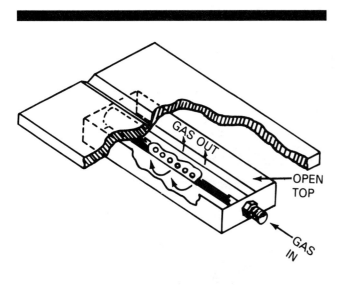

Figure 3.16–Backup Purge Gas Channel

Controlled Atmosphere Welding Chamber

MAXIMUM BENEFITS CAN be obtained when welding reactive metals if the entire object to be welded can be placed in a controlled atmosphere chamber. Such chambers, as shown in Figure 3.18, contain the pieces to be welded, the shielding gas, and welding equipment. After the parts have been put in the chamber, purging is started, and readings are taken on oxygen, nitrogen, and water vapor analyzing instruments to assure that welding is not started until contaminants are at a suitably low level, usually less than 50 PPM.

Trailing Shields

FOR SOME METALS, such as titanium, trailing shields are necessary if chambers or other shielding techniques are not available or practical. Use of a trailing shield ensures inert gas coverage over the weld area until the molten metal has cooled to the point that it will not react with the atmosphere. One type of trailing shield is shown in Figure 3.19.

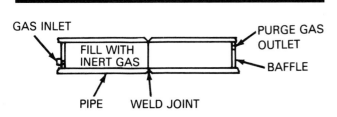

Figure 3.17–Backup Gas Purge Arrangement for Circumferential Pipe Joint (Note Baffles Used to Contain Purge Gas)

Fixed barriers, as illustrated in Figure 3.20, also aid in containing shielding gas within the area immediately surrounding the electrode.

ARC INITIATION METHODS

Scratch or Touch Start

WITH THE POWER supply energized, and the shielding gas flowing from the cup, the torch is lowered toward the workpiece until the tungsten electrode makes contact with the workpiece. The torch is quickly withdrawn a short distance to establish the arc.

The advantage of this method of arc initiation is its simplicity of operation for both manual and machine welding.

The disadvantage of touch starting is the tendency for the electrode to stick to the workpiece, causing electrode contamination and transfer of tungsten to the workpiece.

High-Frequency Start

HIGH-FREQUENCY STARTING CAN be used with dc or ac power sources for both manual and automatic applications. High-frequency generators usually have a spark-gap oscillator that superimposes a high-voltage ac output at radio frequencies in series with the welding circuit. The circuit is shown in Figure 3.21. The high voltage ionizes the gas between the electrode and the work, and the ionized gas will then conduct welding current that initiates the welding arc.

Because radiation from a high-frequency generator may disturb radio, electronic, and computer equipment, the use of this type of arc starting equipment is governed by regulations of the Federal Communications Commission. The user should follow the instructions of the manufacturer for the proper installation and use of high-frequency arc starting equipment.

Pulse Start

APPLICATION OF A high-voltage pulse between the tungsten electrode and the work will ionize the shielding gas and establish the welding arc. This method is generally used with dc power supplies in machine welding applications.

Pilot Arc Start

PILOT ARC STARTING may be used with dc welding power sources. The pilot arc is maintained between the welding electrode and the torch nozzle. The pilot arc supplies the ionized gas required to establish the main welding arc as shown in Figure 3.22. The pilot arc is powered by a small auxiliary power source and is started by high-frequency initiation.

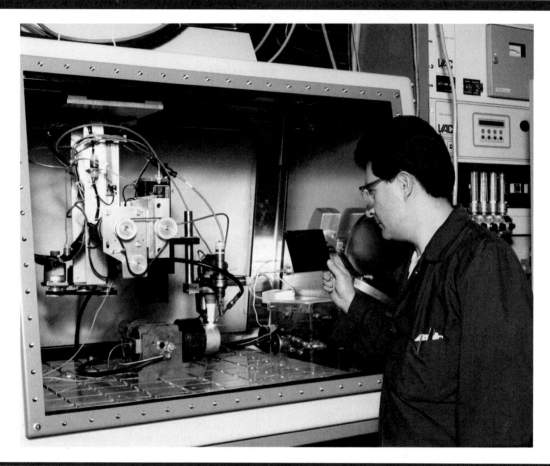

Figure 3.18–Controlled Atmosphere Chamber Used for Gas Tungsten Arc Welding of Reactive Metals (Note: Operator is Viewing the Arc Through A Plexiglass Window.)

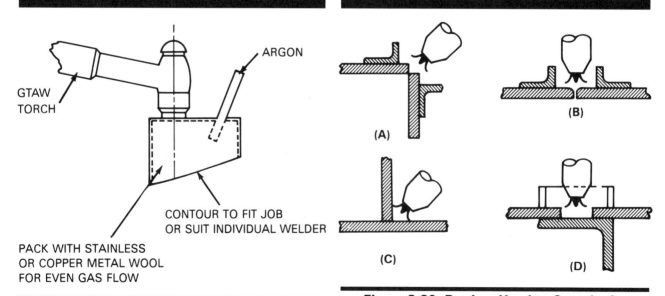

Figure 3.19–Trailing Shield for Manual Torch

Figure 3.20–Barriers Used to Contain the Shielding Gas Near the Joint to be Welded

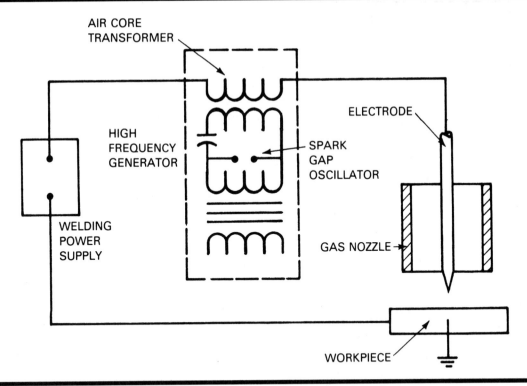

Figure 3.21–High Frequency Arc Starting

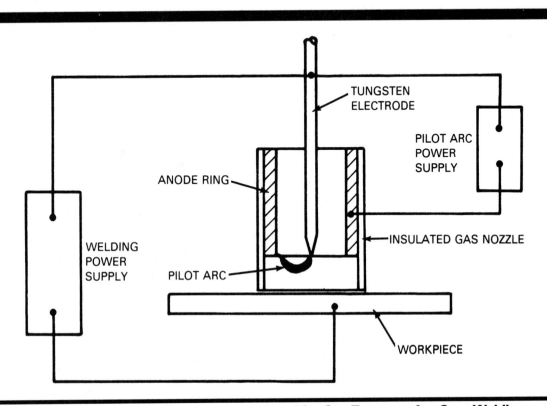

Figure 3.22–Pilot Arc Starting Circuit Used for Gas Tungsten Arc Spot Welding

GTAW TECHNIQUES

MANUAL WELDING

THE WORD "MANUAL" in the GTAW process implies that a person controls all the functions of the welding process. The functions include manipulation of the electrode holder, control of filler metal additions, welding current, travel speed, and arc length. The process is shown in Figure 3.23.

Manual Welding Equipment

IN ADDITION TO an appropriate power supply and a source of shielding gas, manual GTAW equipment includes the welding torch, hoses and electrical conductors, foot pedal (or switch on torch) for controlling welding current levels during the welding cycle, and gas flow controls.

Manual Welding Techniques

THE TECHNIQUE FOR manual welding is illustrated in Figure 3.24. Once the arc is started, the electrode is moved in a small circular motion until the desired weld pool is established. The torch is then held at an angle of 15° from the vertical as shown in the illustration and is moved along the joint to progressively melt the faying surfaces. Filler metal, if used, is added to the leading edge of the pool.

The electrode holder and welding rod must be moved progressively and smoothly so the weld pool, the hot welding rod end, and the hot solidified weld are not exposed to air that will contaminate the weld metal area or heat-affected zone. Generally, a large shielding gas envelope will prevent exposure to air.

The welding rod is usually held at an angle of about 15 degrees to the surface of the work and slowly fed into the molten pool. During welding, the hot end of the welding rod must not be removed from the protection of the inert gas shield.

MACHINE WELDING

MACHINE WELDING IS done with equipment that performs the welding operation under the constant observation and control of a welding operator. The equipment may or may not load and unload the workpieces.

Machine GTAW provides greater control over travel speed and heat input to the workpiece. The higher cost of equipment to provide these benefits must be justified by production and quality requirements.

Machine GTAW equipment, such as the orbital pipe welder shown in Figure 3.25, ranges from simple weld program sequencers and mechanical manipulators to orbital tube and pipe welding systems. Weld sequencers operate in an open loop control mode: variables are maintained at preset levels and no attempt is made to adjust them in response to changing weld quality. The sequencer automatically starts and completes the weld, stepping from one variable setting to other settings at predetermined times or locations along the weld joint. Part tolerances must be controlled closely and fixturing must be strong, since the sequencer cannot compensate for unwanted movement of the parts during welding. High precision parts and sturdy fixturing increase production costs, but welding sequencers usually cost less than more sophisticated automatic controllers.

SEMIAUTOMATIC WELDING

SEMIAUTOMATIC GTAW IS defined as welding with equipment which controls only the filler metal feed. Advance of the welding torch is controlled manually. Semiautomatic systems for GTAW were introduced about 1952 but have been used only for special applications.

AUTOMATIC WELDING

WELDING WITH EQUIPMENT that performs the welding operation without adjustment of the controls by a welding operator is called *automatic welding*. The equipment may or may not load and unload the workpieces. Figure 3.26 shows a typical automatic GTAW application in which the parts are automatically loaded and discharged.

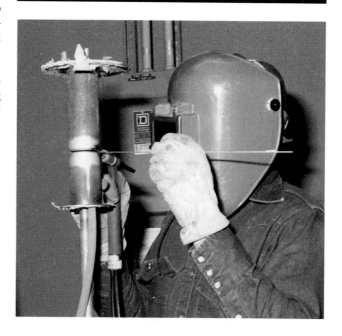

Figure 3.23–Manual Gas Tungsten Arc Welding of a Pipe Joint. Note Backup Purge Gas Hose

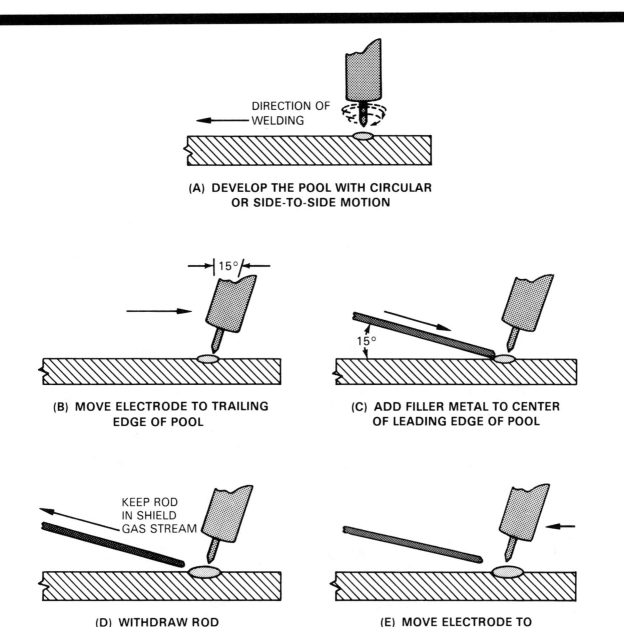

(A) DEVELOP THE POOL WITH CIRCULAR OR SIDE-TO-SIDE MOTION

(B) MOVE ELECTRODE TO TRAILING EDGE OF POOL

(C) ADD FILLER METAL TO CENTER OF LEADING EDGE OF POOL

(D) WITHDRAW ROD

(E) MOVE ELECTRODE TO LEADING EDGE OF POOL

Figure 3.24—Technique for Manual Gas Tungsten Arc Welding

Some modern automatic welding systems (frequently called *adaptive* or *feedback control*) make corrections to welding variables based on information gathered during welding. The objective is to maintain weld quality at a constant level in the presence of changing weld conditions. Automatic adjustment of individual weld variables, such as arc current or arc length, is made by monitoring a weld characteristic, such as pool width. Other feedback control systems are available to provide electrode guidance and constant joint fill.

ARC SPOT WELDING

GAS TUNGSTEN ARC spot welding is often done manually with a pistol-like holder that has a vented, water-cooled gas nozzle, a tungsten electrode that is concentrically positioned with respect to the gas nozzle, and a trigger switch for controlling the operation. Figure 3.27 illustrates such an arrangement. Gas tungsten arc spot welding electrode holders are also available for automatic applications.

The configuration of the nozzle is varied to fit the contour of the weldment. Edge locating devices can be used to prevent variations in the distance of spot weld locations from the workpiece edge. The nozzle is often used to press against the workpiece to assure tight fit-up of the faying surfaces. This technique also controls the electrode-to-work distance.

Spot welding may be done with either ac or DCEN. Automatic sequencing controls are generally used because of the relatively complex cycles involved. The controls automatically establish the preweld gas and water flow, start the arc, time the arc duration, and provide the required postweld gas and water flow.

Penetration is controlled by adjusting the current and the length of time it flows. In some applications, multiple pulses of current are preferred to one long sustained pulse. Variations in the shear strength, nugget diameter, and penetration of the spot weld can be minimized with accurate timers, current monitors, and tungsten electrodes that have precision ground tips.

A melted spot on the bottom of the lower workpiece is a positive indication of a good spot weld.

Figure 3.25—Machine Welding of Pipe Assembly Using Orbital Welder (NOTE: Baffle Taped on Nozzle to Contain Backup Gas)

Figure 3.26–Automatic GTAW Application in Which Workpieces Are Loaded and Unloaded Automatically

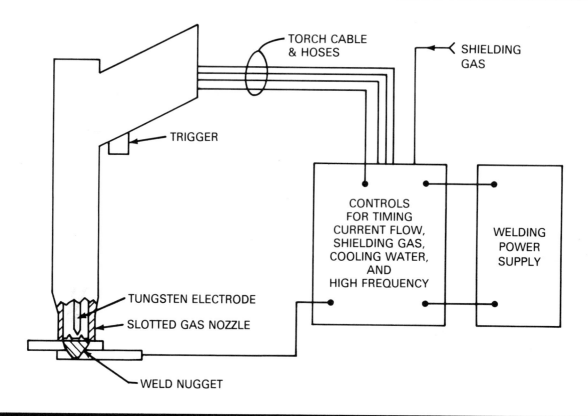

Figure 3.27—Schematic of Manual Gas Tungsten Arc Spot Welding

MATERIALS

THIS SECTION DESCRIBES the materials weldable with the GTAW process. Autogenous welds are made by melting only base metal. When filler metal is used, it can be in the form of welding wire or preplaced consumable inserts.

BASE METALS

MOST METALS CAN be welded by the GTAW process. Among these are grades of carbon, alloy, and stainless steels and other ferrous alloys; heat-resistant alloys of various types; aluminum alloys; magnesium alloys; copper and its alloys, such as copper-nickel, bronzes, and brasses; and nickel alloys. Certain metals must be welded with the GTAW process because it provides the greatest protection from contamination by the atmosphere. GTAW is especially useful for welding reactive and refractory metals and some nonferrous alloys. It is not used to weld metals such as cadmium, tin, or zinc, whose liquids have low-vapor pressures.

Details on the welding characteristics of specific metals and alloys can be found in the *Welding Handbook*, Vol. 4, 7th Ed.[3] Descriptions of the metallurgical responses of metals and alloys to the heat of welding can be found in the *Welding Handbook*, Vol. 1, 8th Ed., Chapter 4.

This section presents information on potential metallurgical problems unique to the gas tungsten arc welding process and special concerns when welding certain metals and alloys with this process. It includes suggested types of welding current, electrode compositions, and shielding gas compositions for optimum weld quality. A guide to the selection of electrodes and gases is provided in Table 3.4. In general, best welding results are obtained with DCEN for almost all metals, unless specified otherwise. The typical tungsten electrode composition is two percent thoriated, unless otherwise specified.

3. Volume 3 of the 8th Ed., when published, will also contain this information.

Carbon and Alloy Steels

THE QUALITY OF gas tungsten arc welds in carbon and alloy steels is more influenced by the base metal impurity content (e.g. sulfur, phosphorus, oxygen) than are welds made with SMAW or SAW. This is because fluxes are not present with GTAW to remove or tie-up these impurities.

High-strength low alloy (HSLA) steels are readily welded by the GTAW process. However, combined levels of phosphorus and sulfur in the base metal exceeding 0.03 percent can cause fusion zone and heat-affected zone cracking. Hydrogen embrittlement of these alloys is a problem if hydrocarbon or water-vapor contamination is present. Hydrogen induced cracking can be minimized by application of preheat or a postweld heat treatment, or, in some cases in high-humidity areas, with trailing gas shields.

Argon shielding gas is generally used for welding carbon and alloy steels up to 1/2 in. (12 mm) thick, because the molten weld pool is easier to control than with helium. When welding thicker sections, either argon or argon-helium mixtures can be employed depending on joint thickness.

Stainless Steels and Heat-Resistant Alloys

STAINLESS STEELS AND the iron-, nickel-, and cobalt-based heat-resistant superalloys are extensively welded with the GTAW process because they are protected from the atmosphere by the inert gas. Weld metal composition is essentially identical to base metal composition because the same alloys are used as filler metal, and because the filler enters the liquid weld pool without passing through the arc, where losses of volatile alloys might be expected.

Argon is recommended for manual welding of thicknesses up to approximately 1/2 in. (12 mm) because it provides better control of the molten weld pool. For thick sections, and for many machine and automatic applications, argon-helium mixtures or pure helium can be used to obtain increased weld penetration. Argon-hydrogen mixtures are used for some stainless steel welding applications to improve bead shape and wettability.

Alternating current can be used for automatic welding of the heat-resistant alloys when close control of arc length is possible.

Aluminum Alloys

GTAW IS IDEALLY suited for welding aluminum alloys. GTAW of aluminum can be performed on all thicknesses. Welding may be performed with or without filler metal.

Aluminum alloys form refractory surface oxides, which make joining more difficult. For this reason, most welding of aluminum is performed with alternating current (using high-frequency arc stabilization) because it provides the surface cleaning action of DCEP with the deeper penetration characteristics of DCEN. DCEP is sometimes used for welding thin aluminum sections. DCEN with helium shielding gas is used for high-current automatic welding of sections over 1/4-in. thick. Since DCEN produces no cleaning action, the aluminum parts must be thoroughly cleaned immediately prior to welding.

For welding with ac, pure tungsten, ceriated tungsten, and zirconiated tungsten electrodes are recommended. Thoriated tungsten electrodes are used on aluminum only with dc.

Argon shielding gas is generally used for welding aluminum with alternating current because it provides better arc starting, better cleaning action, and superior weld quality than does helium. When DCEN is used, helium gas provides faster travel speeds and deeper penetration. However, the poorer surface cleaning action of this combination may result in porosity.

Magnesium Alloys

MAGNESIUM ALLOYS FORM refractory surface oxides similar to aluminum alloys. Alternating current GTAW is typically used for welding of magnesium alloys because of the oxide cleaning action it provides. DCEP may be used for welding thicknesses less than 3/16 in. (5 mm) while alternating current provides better penetration for greater thicknesses. Argon provides the best quality welds, but helium or mixtures of the two are also used. Pure tungsten, ceriated, and zirconiated electrodes may be used.

Beryllium

BERYLLIUM IS A light metal and difficult to weld because of a tendency toward hot cracking and embrittlement. GTAW of beryllium is performed in an inert atmosphere chamber, generally using a shielding gas mixture of five parts helium to one part argon. Beryllium fumes are toxic.

Copper Alloys

GTAW IS WELL suited for copper and its alloys because of the intense heat generated by the arc, which can produce melting with minimum heating of the surrounding, highly conductive base metal. Most copper alloys are welded with DCEN and helium because of the high thermal conductivity. AC is sometimes used to weld beryllium coppers and aluminum bronzes because it helps break up the surface oxides which are present.

Nickel Alloys

NICKEL ALLOYS ARE often gas tungsten arc welded, typically with filler metal additions. DCEN is recommended for all welding, but ac with high-frequency stabilization may be used for machine welding. Argon, argon-helium, and helium are the most common shielding gases. Helium is preferred when no filler metal is to be added. Argon with small amounts of hydrogen (up to 5 percent) is sometimes used for single-pass welding.

High-purity nickel alloys can exhibit variable weld penetration caused by differences in surface active elements.

Refractory and Reactive Metals

GAS TUNGSTEN ARC is the most extensively used welding process for joining refractory and reactive metals. Refractory metals (notably tungsten, molybdenum, tantalum, niobium, and chromium) have extremely high melting temperatures and, like the reactive metals (such as titanium alloys, zirconium alloys, and hafnium), are readily oxidized at elevated temperatures unless protected by an inert gas cover. Absorption of impurities such as oxygen, nitrogen, hydrogen, and carbon will decrease toughness and ductility of the weld metal.

For these metals and alloys, GTAW provides a high concentration of heat and the greatest control over heat input, while providing the best inert gas shielding of any welding process. Welding these metals is typically performed in purged chambers containing high-purity inert gases. Occasionally GTAW is performed without special purge chambers, by providing the necessary inert gas atmosphere with torch, trailing, and backup shielding.

Argon is most frequently used for shielding, but helium and mixtures of the two gases can be used. Argon flow rates of 15 cfh or helium flow rates of 40 cfh are sufficient, even with the large diameter gas nozzles which are recommended.

Cast Irons

CAST IRON CAN be welded with the GTAW process because dilution of the base metal can be minimized with independent control of heat input and filler metal placement. A high level of operator skill is required to minimize dilution while maintaining acceptable penetration and fusion.

GTAW of cast irons is usually limited to repair of small parts. Nickel-based and austenitic stainless steel filler metals are recommended; they minimize cracking because of their ductility and their tolerance for hydrogen. Cracking can also be minimized by preheat and postweld heat treatment. DCEN is recommended, although ac may be used.

FILLER METALS

FILLER METALS FOR joining a wide variety of metals and alloys are available for use with gas tungsten arc welding. Filler metals, if used, should be similar, although not necessarily identical, to the metal that is being joined. When joining dissimilar metals, the filler metals will be different from one or both of the base metals.

Generally, the filler metal composition is adjusted to match the properties of the base metal in its welded (cast) condition. Such filler metals are produced with closer control on chemistry, purity, and quality than are base metals. Deoxidizers are frequently added to ensure weld soundness. Further modifications are made to some filler metal compositions to improve response to postweld heat treatments.

The choice of filler metal for any application is a compromise involving metallurgical compatibility, suitability for the intended service, and cost. The tensile and impact properties, corrosion resistance, and electrical or thermal conductivities that are required in a particular weldment also must be considered. Thus, the filler metal must suit both the alloy to be welded and the intended service.

Table 3.5 lists the AWS filler metal specifications which are applicable for gas tungsten arc welding. These specifications establish filler metal classifications based on the mechanical properties or chemical compositions, or both, of each filler metal. They also set forth the conditions under which the filler metals must be tested.

Appendices in the filler metal specifications provide useful background on the properties and uses of the filler metals within the various classifications. Manufacturers' catalogs provide useful information on the proper use of their

Table 3.5
AWS Specifications for Filler Metals Suitable for Gas Tungsten Arc Welding

Specification Number	Title
A 5.2	Iron and Steel Gas-Welding Rods
A 5.7	Copper and Copper Alloy Bare Welding Rods and Electrodes.
A 5.9	Corrosion-Resisting Chromium and Chromium-Nickel Steel Bare and Composite Metal Cored and Stranded Arc Welding Electrodes and Welding Rods
A 5.10	Aluminum and Aluminum Alloy Welding Rods and Bare Electrodes
A 5.13	Surfacing Welding Rods and Electrodes
A 5.14	Nickel and Nickel Alloy Bare Welding Rods and Electrodes
A 5.16	Titanium and Titanium Alloy Bare Welding Rods and Electrodes
A 5.18	Mild Steel Electrodes for Gas Metal Arc Welding
A 5.19	Magnesium-Alloy Welding Rods and Bare Electrodes
A 5.21	Composite Surfacing Welding Rods and Electrodes
A 5.24	Zirconium and Zirconium Alloy Bare Welding Rods and Electrodes

products. Brand name listings and addresses of vendors are shown in the latest edition of AWS *Filler Metal Comparison Charts*.

Filler metals for GTAW are available in most alloys in the form of straightened and cut lengths (rods), usually 36 in. (1 m) long, for manual welding, and spooled or coiled continuous wire for machine or automatic welding. The filler metal diameters range from about 0.020 in. (0.5 mm) for fine and delicate work to about 3/16 in. (5 mm) for high-current manual welding or surfacing.

Extra care must be exercised to keep the filler metals clean and free of all contamination while in storage, as well as in use. The hot end of the wire or rod should not be removed from the protection of the inert gas shield during the welding operation.

Consumable inserts provide filler metal additions for root pass welds in certain pipe and plate applications. Advantages include broader fit-up tolerances, lower operator-skill levels, more consistent weld bead fusion, and smooth, uniform underbeads.

JOINT DESIGN

DUE TO THE variety of base metals and their individual characteristics (such as surface tension, fluidity, melting temperature, etc.), joint geometries or designs that provide optimum welding conditions should be used. Factors affecting joint designs include metal composition and thickness, weld penetration requirements, joint restraint and joint efficiency requirements.

BASIC JOINT CONFIGURATION

THE FIVE BASIC joints (butt, lap, T, edge, and corner) shown in Figure 3.28 may be used for virtually all metals. Many variations are derived from these basic joints. In all instances the primary objective is to minimize welding cost while maintaining the desired weld quality and performance level for the design.

Factors affecting cost are preparation time, weld joint area to be filled, and setup time. While there are no fixed rules governing the use of a particular joint design for any one metal, certain designs were developed for specific purposes.

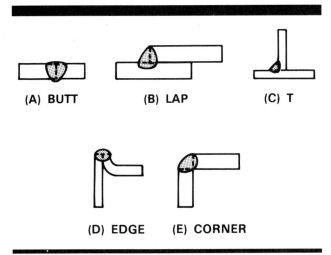

Figure 3.28—Five Basic Weld Joints

The primary variables of joint design are root opening, thickness of root face, and angle of bevel. All variables must be considered prior to joint preparation.

The amount of root opening and thickness of root face depend upon whether the GTAW process is to be manual or automatic, whether filler metal is to be added during the root pass, and if a consumable insert will be used. Backing strips are generally not used due to additional costs of material and fit-up, as well as difficulty in interpretation of radiographs.

The amount of bevel angle depends upon the thickness of the metal and the clearance needed for arc movement to assure adequate fusion on both sides of the joint. These variables are generally determined by welding sample joints which cover a variety of setups.

A major consideration in GTAW joint design is provision for proper accessibility. The groove angle must permit manipulation of the electrode holder to obtain adequate fusion of the groove face. Characteristics of the weld metal must also be considered. For example, the high nickel alloys are very sluggish when molten, and the weld metal does not wet the groove face well. Therefore, groove angles for high nickel alloys should be more open than those for carbon and alloy steel, to provide space for manipulation. However, opening the groove angle increases distortion, weld time, and cost, and should be limited as far as possible.

Specific information on joint designs may be found in the AWS *Welding Handbook*, Vol.1, 8th Ed. Chapter 5, or in the metals supplier's literature.

JOINT PREPARATION

AFTER A PARTICULAR joint design has been selected, the most important item for consideration is the method of joint preparation. There are many ways to remove metal to prepare a given joint configuration. However, many GTAW problems, or supposed problems, are a direct result of using improper methods to prepare the joint. Chief among these is the improper use of grinding wheels to prepare joints. Soft materials such as aluminum become impregnated with microsized abrasive particles which, unless subsequently removed, will result in excessive porosity. Grinding wheels should be cleaned and dedicated exclusively to the material being welded. The ideal joint preparation is obtained with cutting tools such as a lathe for round or cylindrical joints, or a milling cutter for longitudinal preparations. Care must be exercised in the choice of cutting fluid (if any) to be used. Cleaning after cutting or turning should be with safety-approved solvents that are free of residues.

Oxyfuel cutting and plasma arc cutting are also acceptable, provided any slag is removed by careful grinding.

Joint Tolerance

THE ALLOWABLE TOLERANCE of joint dimensions depends upon whether GTAW is to be done manually or by mechanized means. Manual welding applications can tolerate greater irregularities in joint fit-up than mechanized welding. The particular tolerance for a given application can be determined only by actual testing, and this tolerance should be specified for future work.

Cleaning

CLEANLINESS OF BOTH the weld joint areas and the filler metal is an important consideration when welding with the gas tungsten arc process. Oil, grease, shop dirt, paint, marking crayon, and rust or corrosion deposits all must be removed from the joint edges and metal surfaces, to a distance beyond the heat-affected zone. Their presence during welding may lead to arc instability and contaminated

welds. Depending upon the metallurgical response to these contaminants, welds can contain pores, cracks, and inclusions. Cleaning may be accomplished by mechanical means, by the use of vapor or liquid cleaners, or by a combination of these.

FIXTURING

FIXTURING MAY BE required if the parts to be welded cannot be self-supported during welding or if any resultant distortion cannot be tolerated or corrected by straightening. The fixturing should be massive enough to support the weight of the parts and weldment and to withstand welding stresses caused by thermal expansion and contraction. The fixtures must also handle the normal wear and tear that occurs during production.

The decision to use fixturing for the fabrication of a weldment is governed by economics and quality requirements. The proper use of fixturing, including heat sinks, can reduce welding time. The one-time fabrication of an assembly may not justify the use of fixturing; however, the fabrication of a large number of assemblies could justify even complex fixtures. Also, high-quality work may dictate that fixturing be used to maintain close tolerances required by the design, or for nondestructive examination requirements.

The following are primary functions of fixturing:

(1) Locate parts precisely within the assembly
(2) Maintain alignment during welding
(3) Minimize distortion in the weldment
(4) Control heat buildup

WELD QUALITY

DISCONTINUITIES AND DEFECTS

DISCONTINUITIES ARE INTERRUPTIONS in the typical structure of a weldment, and they may occur in the base metal, weld metal, and heat-affected zones. Discontinuities in a weldment that do not satisfy the requirements of an applicable fabrication code or specification are classified as defects, and are required to be removed because they could impair the performance of that weldment in service.

PROBLEMS AND CORRECTIONS

Tungsten Inclusions

ONE DISCONTINUITY FOUND only in gas tungsten arc welds is tungsten inclusions. Particles of tungsten from the electrode can be embedded in a weld when improper welding

procedures are used with the GTAW process. Typical causes are the following:

(1) Contact of electrode tip with molten weld pool
(2) Contact of filler metal with hot tip of electrode
(3) Contamination of the electrode tip by spatter from the weld pool
(4) Exceeding the current limit for a given electrode size or type
(5) Extension of electrodes beyond their normal distances from the collet (as with long nozzles) resulting in overheating of the electrode
(6) Inadequate tightening of the holding collet or electrode chuck
(7) Inadequate shielding gas flow rates or excessive wind drafts resulting in oxidation of the electrode tip
(8) Defects such as splits or cracks in the electrode

(9) Use of improper shielding gases such as argon-oxygen or argon-CO_2 mixtures that are used for gas metal arc welding

Corrective steps are obvious once the causes are recognized and the welder is adequately trained.

Lack of Shielding

DISCONTINUITIES RELATED TO the loss of inert gas shielding are the tungsten inclusions previously described, porosity, oxide films and inclusions, incomplete fusion, and cracking. The extent to which they occur is strongly related to the characteristics of the metal being welded. In addition, the mechanical properties of titanium, aluminum, nickel, and high-strength steel alloys can be seriously impaired with loss of inert gas shielding. Gas shielding effectiveness can often be evaluated prior to production welding by making a spot weld and continuing gas flow until the weld has cooled to a low temperature. A bright, silvery spot will be evident if shielding is effective.

WELDING PROBLEMS AND REMEDIES

NUMEROUS WELDING PROBLEMS may develop while setting up or operating a GTAW operation. Their solution will require careful evaluation of the material, the fixturing, the welding equipment, and the procedures. Some problems that may be encountered and possible remedies are listed in Table 3.6.

Table 3.6
Troubleshooting Guide for Gas Tungsten Arc Welding

Problem	Cause	Remedy
Excessive electrode consumption	1. Inadequate gas flow.	1. Increase gas flow.
	2. Operating on reverse polarity	2. Use larger electrode or change to straight polarity.
	3. Improper size electrode for current required.	3. Use larger electrode.
	4. Excessive heating in holder.	4. Check for proper collet contact.
	5. Contaminated electrode.	5. Remove contaminated portion. Erratic results will continue as long as contamination exists.
	6. Electrode oxidation during cooling.	6. Keep gas flowing after stopping arc for at least 10 to 15 seconds.
	7. Using gas containing oxygen or CO_2.	7. Change to proper gas.
Erratic arc	1. Base metal is dirty, greasy.	1. Use appropriate chemical cleaners, wire brush, or abrasives.
	2. Joint too narrow	2. Open joint groove; bring electrode closer to work; decrease voltage.
	3. Electrode is contaminated.	3. Remove contaminated portion of electrode.
	4. Arc too long	4. Bring holder closer to work to shorten arc.
Porosity	1. Entrapped gas impurities (hydrogen, nitrogen, air, water vapor).	1. Blow out air from all lines before striking arc; remove condensed moisture from lines; use welding grade (99.99%) inert gas.
	2. Defective gas hose or loose hose connections.	2. Check hose and connections for leaks.
	3. Oil film on base metal.	3. Clean with chemical cleaner not prone to break up in arc; DO NOT WELD WHILE BASE METAL IS WET.
Tungsten contamination of workpiece	1. Contact starting with electrode.	1. Use high frequency starter; use copper striker plate.
	2. Electrode melting and alloying with base metal.	2. Use less current or larger electrode; use thoriated or zirconium-tungsten electrode.
	3. Touching tungsten to molten pool.	3. Keep tungsten out of molten pool.

APPLICATIONS

THE GAS TUNGSTEN arc welding process offers advantages to many industries, ranging from the high quality required in the aerospace and nuclear industries and the high-speed autogenous welds required in tube and sheet metal manufacturing, to the ease and flexibility of GTAW which is so welcome in repair shops.

Gas tungsten arc welding provides precise control of heat input. For that reason it is preferred for joining thin gage metals and for making welds close to heat sensitive components. It is also used for small jobs and repair welding in many fabrication shops because of the ease of control of the process and the ability to add filler metal as necessary. Gas tungsten arc welding is used with or without filler metal to produce high-quality welds with smooth, uniform shapes. The GTAW process can also be used for spot welding in sheet metal applications.

The process can be used to weld almost all metals. It is especially useful for joining aluminum and magnesium, which form refractory oxides, and for reactive metals like titanium and zirconium, which can become embrittled if exposed to air while molten.

GTAW can be used to weld all types of joint geometries and overlays in plate, sheet, pipe, tubing, and other structural shapes. It is particularly appropriate for welding sections less than 3/8 in. (10 mm) thick. Welding of pipe is often accomplished using gas tungsten arc welding for the root pass and either SMAW or GMAW for the fill passes.

An aerospace application involving GTAW is shown in Figure 3.29. This pulsed current application shows the welding of a flanged joint between two stainless steel castings.

Figure 3.29–Autogenous Gas Tungsten Arc Weld on Two 17-4ph Stainless Steel Machined Castings (NOTE: Argon Shielding Gas is Used on This 7-in. (180 mm) Diameter Flanged Joint. Travel Speed is 16 in./min (40 cm/min) Using 41 Amps Average Pulsed Current.)

SAFE PRACTICES

THE GENERAL SUBJECT of safety and safe practices in welding, cutting, and allied processes is covered in ANSI Z49.1, *Safety in Welding and Cutting*.[4] All welding personnel should be familiar with the safe practices discussed in this document.

Using safe practices in welding and cutting will ensure the protection of persons from injury and illness and the protection of property from unwanted damage. The potential hazard areas in arc welding and cutting include, but are not limited to the handling of cylinders and regulators, gases, fumes, radiant energy, and electrical shock. The areas that are associated with GTAW are briefly discussed in this section. Safe practices should always be the foremost concern of the welder or welding operator.

SAFE HANDLING OF GAS CYLINDERS AND REGULATORS

COMPRESSED GAS CYLINDERS should be handled carefully. Knocks, falls, or rough handling may damage cylinders, valves, or safety devices and cause leakage or explosive rupture accidents. Valve protecting caps, when supplied, should be kept in place (hand-tight) except when cylinders are in use or connected for use. When in use, cylinders should be securely fastened to prevent accidental tipping. For further information, see CGA Pamphlet P-1, *Safe Handling of Compressed Gases in Containers*.[5]

GAS HAZARDS

THE MAJOR TOXIC gases associated with GTAW are ozone, nitrogen dioxide, and phosgene gas. Phosgene gas could be present as a result of thermal or ultraviolet decomposition of chlorinated hydrocarbon cleaning agents, such as trichlorethylene and perchlorethylene, located in the vicinity of welding operations. Degreasing or other cleaning operations involving chlorinated hydrocarbons should be performed where vapors from these operations are not exposed to radiation from the welding arc.

Ozone

THE ULTRAVIOLET LIGHT emitted by the welding arc acts on the oxygen in the surrounding atmosphere to produce ozone. The amount of ozone produced will depend upon the intensity of the ultraviolet energy, the humidity, the amount of screening afforded by the welding fume, and

other factors. Test results, based upon present sampling methods, indicate the average concentration of ozone generated in the GTAW process does not constitute a hazard under conditions of good ventilation and welding practice. (See ANSI Z49.1 for welding conditions requiring ventilation, particularly when welding is done in confined spaces.)

Nitrogen Dioxide

SOME TEST RESULTS show that high concentrations of nitrogen dioxide are found only within 6 in. (150 mm) of the arc. Natural ventilation quickly reduces these concentrations to safe levels in the welder's breathing zone. As long as the welder's head is kept out of the fumes, nitrogen dioxide is not thought to be a hazard in GTAW.

Inert Shielding Gases

PROVISION FOR ADEQUATE ventilation should be made when inert gas shielding and purging gases are used. Accumulation of these gases could cause suffocation of welding and inspection personnel.

Metal Fumes

THE WELDING FUMES generated by the GTAW process can be controlled by natural ventilation, general ventilation, local exhaust ventilation, or by respiratory protective equipment, as described in ANSI Z49.1. The method of ventilation required to keep the level of toxic substances in the welder's breathing zone within acceptable concentrations is directly dependent upon a number of factors, among which are the material being welded, the size of the work area, and the degree of confinement or obstruction to normal air movement where the welding is being done. Each operation should be evaluated on an individual basis in order to determine what will be required.

Acceptable levels of toxic substances associated with welding and designated as time-weighted average threshold limit values (TLVs) and ceiling values, have been published by the Occupational Safety and Health Administration. The OSHA standards for general industry are also known as the Code of Federal Regulation (29CFR 1910) and can be obtained through the U.S. Government Printing Office, Washington, D.C. 20402. Compliance with these acceptable levels can be checked by sampling the atmosphere inside the welder's helmet or in the immediate vicinity of the helper's breathing zone. Sampling should be in accordance with ANSI/AWS F1.1, *Method for Sampling Airborne Particulates Generated by Welding and Allied Processes*.

4. ANSI Z49.1 is available from the American Welding Society, 550 N.W. LeJeune Road, Miami, Florida 33126.
5. CGA P-1 is available from the Compressed Gas Association, Inc., 500 Fifth Avenue, New York, New York 10036.

RADIANT ENERGY

RADIANT ENERGY IS a hazard and may cause injury to the welder (or others exposed to the welding arc) in two areas; eyes and skin. The general subject of eye protection is covered in ANSI Z49.1, *Safety in Welding and Cutting*, and ANSI Z87.1, *Practice for Occupational and Educational Eye and Face Protection*. Any personnel within the immediate vicinity of a welding operation should have adequate protection from the radiation produced by the welding arc. Generally, the highest ultraviolet radiant energy intensities are produced when argon shielding gas is used and when aluminum or stainless steel is welded.

For the protection of eyes, filter glass or curtains should be used. The filter glass shades recommended for GTAW, as presented in ANSI Z49.1 as a guide, are shown in Table 3.7.

It is suggested that the welder use the darkest shade that is comfortable, but not lighter than recommended.

For the protection of skin, leather or dark wool clothing (to reduce reflection which could cause ultraviolet burns to the face and neck underneath the helmet) is recommended for GTAW. High-intensity ultraviolet radiation will cause rapid disintegration of cotton and some synthetic materials.

In addition, when an area or room is set aside for GTAW, the walls should be coated with pigments such as titanium dioxide or zinc oxide because these will reduce ultraviolet reflection. See ANSI Z49.1 for additional information on *Ultraviolet Reflection of Paint*, published by and available from the American Welding Society.

ELECTRICAL SHOCK

THE FOLLOWING IS from ANSI Z49.1 - 1983 section 11.4.9: "Avoidance of electrical shock is largely within the control of the welder; therefore, it is especially important that the welder be thoroughly instructed in detail how to avoid shock. Safe procedures shall be observed at all times when working with equipment having voltages necessary for arc welding. Even mild shocks can cause involuntary muscular contraction, leading to injurious falls from high places. Severity of shock is determined largely by the path, duration, and amount of current flowing through the body, which is dependent upon voltage and contact resistance of the area of skin involved. Clothing damp from perspiration or wet working conditions may reduce contact resistance and increase current to a value high enough to cause such violent muscular contraction that the welder cannot release contact with the live part."

WELDING EQUIPMENT SAFETY

ALL WELDING EQUIPMENT should be on an approved list from an NFPA recognized testing agency such as Factory Mutual or Underwriters Laboratory. Damaged equipment should be repaired properly before use. No welding should be done until all electrical connections, power supply, welding leads, welding machines, and work clamps are secure, and the welding power source frame is well grounded. The work clamp must be secure and the cable connecting it to the power supply must be in good condition. Any time the power supply is left unattended, it should be turned off. The line supply disconnect switch should also be placed in the "OFF" position.

Table 3.7
Recommended Lens Shades for Various Welding Current Ranges

Shade No.	Welding Current, A
8	Up to 75
10	75 to 200
12	200 to 400
14	Above 400

SUPPLEMENTAL READING LIST

Baeslack, W. A. III and Banas, C. M. "A comparative evaluation of laser and gas tungsten arc weldments in high temperature titanium alloys." *Welding Journal* 60(7): 121s-130s; July 1981.

Burgardt, P. and Heiple, C. R. "Interaction between impurities and welding variables in determining GTA weld shape." *Welding Journal* 65(6): 150s-156s; June 1986.

Correy, T. B. et al. "Radio frequency - free arc starting in gas tungsten arc welding." *Welding Journal* 64(2): 33-37; February 1986.

Geidt, W. H. et al. "GTA welding efficiency: calorimetric and temperature field measurements." *Welding Journal* 68(1): 28s-34s; January 1989.

Haberman, R. "GTAW torch performance relies on component materials." *Welding Journal* 66(12): 55-60; December 1987.

Heiple, C. R. et al. "Surface active elements effects on the shape of GTA laser, and electron beam welds." *Welding Journal* 62(3): 72s-77s; March 1983.

Kanne, W. R. "Remote reactor repair: GTA weld cracking caused by entrapped helium." *Welding Journal* 67(8): 33-38; August 1988.

Katoh, M. and Ken, H. W. "Investigation of heat-affected zone cracking of GTA welds of Al-Mg-Si alloys using the varestraint test." *Welding Journal* 66(12): 360s; December 1987.

Key, J. F. "Anode/cathode geometry and shielding gas interrelationships in GTAW." *Welding Journal* 59(12): 364s-370s; December 1980.

Kraus, H. G. "Experimental measurement of stationary SS 304, SS 316L and 8630 GTA weld pool surface temperatures." *Welding Journal* 68(7): 269s-279s; July 1989.

Kujanpaa, V. P. et al. "Role of shielding gases in flaw formation in GTAW of stainless steel strips." *Welding Journal* 63(5): 151s-155s; May 1984.

Lu, M. and Kou, S. "Power and current distributions in gas tungsten arcs." *Welding Journal* 67(2): 29s-36s; February 1988.

Malinowski-Brodnicka, M., et al. "Effect of electromagnetic stirring on GTA welds in austenitic stainless steel." *Welding Journal* 69(2): 52s; February 1990.

Metcalfe, J. C. and Quigley, M. C. B. "Arc and pool instability in GTA welding." *Welding Journal* 56(5): 133s-139s; May 1977.

Oomen, W. J. and P. A. Verbeek, P. A. "A real-time optical profile sensor for robot arc welding." *Robotic Welding*, Edited by J. D. Lane. United Kingdom: IFS Publications Ltd., 1987.

Patterson, R. A., et al. "Discontinuities formed in inconel GTA welds." *Welding Journal* 65(1): 19s-25s; January 1987.

Pearce, C. H. et al. "Development and applications of microprocessor controlled systems for mechanized TIG welding." *Computer Technology in Welding*. Edited by Lucas, W. The Welding Institute, June 1986.

Saede, H. R. and Unkel, W. "Arc and weld pool behavior for pulsed current GTAW." *Welding Journal* 67(11): 247s; November 1988.

Sicard, P. and Levine, M. D. "IEEE transactions on systems, man, and cybernetics." Volume 18, No. 2; March 1988.

Smith, J. S., et al. "A vision based seam tracker for TIG welding." *Computer Technology in Welding*. Edited by Lucas, W., The Welding Institute, June 1986.

Troyer, W. et al. "Investigation of pulsed wave shapes for gas tungsten arc welding." *Welding Journal* 56(1): 26-32; January 1977.

Voigt, R. C. and Loper, C. R. Jr. "Tungsten contamination during gas tungsten arc welding." *Welding Journal* 59(4): 99s-103s; April 1980.

Villafuerte, J. C. and Kerr, H. W. "Electromagnetic stirring and grain refinement in stainless steel GTA welds." *Welding Journal* 69(1): 1s; January 1990.

Walsh, D. W. and Savage, W. F. "Technical note: Autogenous GTA weldments-bead geometry variations due to minor elements." *Welding Journal* 64(2): 59s-62s; February 1985.

Walsh, D. W. and Savage, W. F. "Technical note: Bead shape variance in AISI 8630 steel GTAW weldments." *Welding Journal* 64(5): 137s-139s; May 1985.

Zacharia, T. et al. "Weld pool development during GTA and laser beam welding of type 3 or 4 stainless steel-part 1 and part 2." *Welding Journal* 68(12): 499s and 510s; December 1989.

CHAPTER 4

GAS METAL ARC WELDING

PREPARED BY A COMMITTEE CONSISTING OF:

D. B. Holliday, Chairman
Westinghouse Electric

S. R. Carter
Scott Paper Company

L. DeFreitas
College of San Mateo

D. A. Fink
Lincoln Electric Company

R. W. Folkening
FMC Corporation

D. D. Hodson
Tweco Products, Incorporated

R. H. Mann
Miller Electric Manufacturing Company

WELDING HANDBOOK COMMITTEE MEMBER:
P. I. Temple
Detroit Edison

CHAPTER 4

GAS METAL ARC WELDING

INTRODUCTION

DEFINITION AND GENERAL BACKGROUND

GAS METAL ARC welding (GMAW) is an arc welding process that uses an arc between a continuous filler metal electrode and the weld pool. The process is used with shielding from an externally supplied gas and without the application of pressure.

The basic concept of GMAW was introduced in the 1920's, but it was not until 1948 that it was made commercially available. At first it was considered to be, fundamentally, a high-current density, small diameter, bare metal electrode process using an inert gas for arc shielding. Its primary application was for welding aluminum. As a result, the term *MIG* (Metal Inert Gas) was used and is still a common reference for the process. Subsequent process developments included operation at low-current densities and pulsed direct current, application to a broader range of materials, and the use of reactive gases (particularly CO_2) and gas mixtures. This latter development has led to the formal acceptance of the term *gas metal arc welding* (GMAW) for the process because both inert and reactive gases are used.

A variation of the GMAW process uses a tubular electrode wherein metallic powders make up the bulk of the core materials (metal cored electrode). Such electrodes require a gas shield to protect the molten weld pool from atmospheric contamination.

Metal cored electrodes are considered a segment of GMAW by the American Welding Society. Foreign welding associations may group metal cored electrodes with flux cored electrodes.

GMAW may be operated in semiautomatic, machine, or automatic modes. All commercially important metals such as carbon steel, high-strength low alloy steel, stainless steel, aluminum, copper, titanium, and nickel alloys can be welded in all positions with this process by choosing the appropriate shielding gas, electrode, and welding variables.

USES AND ADVANTAGES

THE USES OF the process are, of course, dictated by its advantages, the most important of which are the following:

(**1**) It is the only consumable electrode process that can be used to weld all commercial metals and alloys.

(**2**) GMAW overcomes the restriction of limited electrode length encountered with shielded metal arc welding.

(**3**) Welding can be done in all positions, a feature not found in submerged arc welding.

(**4**) Deposition rates are significantly higher than those obtained with shielded metal arc welding.

(**5**) Welding speeds are higher than those with shielded metal arc welding because of the continuous electrode feed and higher filler metal deposition rates.

(**6**) Because the wire feed is continuous, long welds can be deposited without stops and starts.

(**7**) When spray transfer is used, deeper penetration is possible than with shielded metal arc welding, which may permit the use of smaller size fillet welds for equivalent strengths.

(**8**) Minimal postweld cleaning is required due to the absence of a heavy slag.

These advantages make the process particularly well suited to high production and automated welding applications. This has become increasingly evident with the advent of robotics, where GMAW has been the predominant process choice.

LIMITATIONS

AS WITH ANY welding process, there are certain limitations which restrict the use of gas metal arc welding. Some of these are the following:

(1) The welding equipment is more complex, more costly, and less portable than that for SMAW.

(2) GMAW is more difficult to use in hard-to-reach places because the welding gun is larger than a shielded metal arc welding holder, and the welding gun must be close to the joint, between 3/8 and 3/4 in. (10 and 19 mm), to ensure that the weld metal is properly shielded.

(3) The welding arc must be protected against air drafts that will disperse the shielding gas. This limits outdoor applications unless protective shields are placed around the welding area.

(4) Relatively high levels of radiated heat and arc intensity can result in operator resistance to the process.

FUNDAMENTALS OF THE PROCESS

PRINCIPLES OF OPERATION

THE GMAW PROCESS incorporates the automatic feeding of a continuous, consumable electrode that is shielded by an externally supplied gas. The process is illustrated in Figure 4.1. After initial settings by the operator, the equipment provides for automatic self-regulation of the electrical characteristics of the arc. Therefore, the only manual controls required by the welder for semiautomatic operation are the travel speed and direction, and gun positioning. Given proper equipment and settings, the arc length and the current (wire feed speed) are automatically maintained.

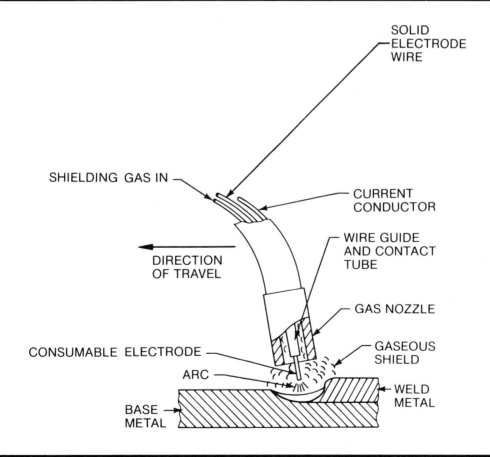

Figure 4.1–Gas Metal Arc Welding Process

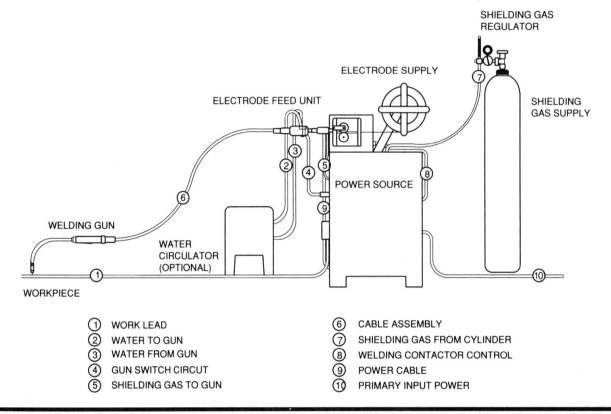

①	WORK LEAD	⑥	CABLE ASSEMBLY
②	WATER TO GUN	⑦	SHIELDING GAS FROM CYLINDER
③	WATER FROM GUN	⑧	WELDING CONTACTOR CONTROL
④	GUN SWITCH CIRCUT	⑨	POWER CABLE
⑤	SHIELDING GAS TO GUN	⑩	PRIMARY INPUT POWER

Figure 4.2–Diagram of Gas Metal Arc Welding Equipment

Equipment required for GMAW is shown in Figure 4.2. The basic equipment components are the welding gun and cable assembly, electrode feed unit, power supply, and source of shielding gas.

The gun guides the consumable electrode and conducts the electrical current and shielding gas to the work, thus providing the energy to establish and maintain the arc and melt the electrode as well as the needed protection from the ambient atmosphere. Two combinations of electrode feed units and power supplies are used to achieve the desirable self-regulation of arc length. Most commonly this regulation consists of a constant-potential (voltage) power supply (characteristically providing an essentially flat volt-ampere curve) in conjunction with a constant-speed electrode feed unit. Alternatively, a constant-current power supply provides a drooping volt-ampere curve, and the electrode feed unit is arc-voltage controlled.

With the constant potential/constant wire feed combination, changes in the torch position cause a change in the welding current that exactly matches the change in the electrode stick-out (electrode extension), thus the arc length remains fixed. For example, an increased stick-out produced by withdrawing the torch reduces the current output from the power supply, thereby maintaining the same resistance heating of the electrode.

In the alternative system, self-regulation results when arc voltage fluctuations readjust the control circuits of the feeder, which appropriately changes the wire feed speed. In some cases (welding aluminum, for example), it may be preferable to deviate from these standard combinations and couple a constant-current power source with a constant-speed electrode feed unit. This combination provides only a small degree of automatic self-regulation, and therefore requires more operator skill in semiautomatic welding. However, some users think this combination affords a range of control over the arc energy (current) that may be important in coping with the high thermal conductivity of aluminum base metals.

METAL TRANSFER MECHANISMS

THE CHARACTERISTICS OF the GMAW process are best described in terms of the three basic means by which metal is transferred from the electrode to the work:

(1) Short circuiting transfer
(2) Globular transfer
(3) Spray transfer

The type of transfer is determined by a number of factors, the most influential of which are the following:

(1) Magnitude and type of welding current
(2) Electrode diameter
(3) Electrode composition
(4) Electrode extension
(5) Shielding gas

Short Circuiting Transfer

SHORT CIRCUITING ENCOMPASSES the lowest range of welding currents and electrode diameters associated with GMAW. This type of transfer produces a small, fast-freezing weld pool that is generally suited for joining thin sections, for out-of-position welding, and for bridging large root openings. Metal is transferred from the electrode to the work only during a period when the electrode is in contact with the weld pool. No metal is transferred across the arc gap.

The electrode contacts the molten weld pool in a range of 20 to over 200 times per second. The sequence of events in the transfer of metal and the corresponding current and voltage are shown in Figure 4.3. As the wire touches the weld metal, the current increases [(A),(B), (C), (D) in Figure 4.3]. The molten metal at the wire tip pinches off at D and E, initiating an arc as shown in (E) and (F). The rate of current increase must be high enough to heat the electrode and promote metal transfer, yet low enough to minimize spatter caused by violent separation of the drop of metal. This rate of current increase is controlled by adjustment of the inductance in the power source.

The optimum inductance setting depends on both the electrical resistance of the welding circuit and the melting temperature of the electrode. When the arc is established, the wire melts at the tip as the wire is fed forward towards the next short circuit at (H), Figure 4.3. The open circuit voltage of the power source must be so low that the drop of molten metal at the wire tip cannot transfer until it touches the base metal. The energy for arc maintenance is partly provided by energy stored in the inductor during the period of short circuiting.

Even though metal transfer occurs only during short circuiting, shielding gas composition has a dramatic effect on the molten metal surface tension. Changes in shielding gas composition may dramatically affect the drop size and the duration of the short circuit. In addition, the type of gas influences the operating characteristics of the arc and the base metal penetration. Carbon dioxide generally produces high spatter levels compared to inert gases, but CO_2 also promotes deeper penetration. To achieve a good compromise between spatter and penetration, mixtures of CO_2 and argon are often used when welding carbon and low alloy steels. Additions of helium to argon increase penetration on nonferrous metals.

Globular Transfer

WITH A POSITIVE electrode (DCEP), globular transfer takes place when the current is relatively low, regardless of the type of shielding gas. However, with carbon dioxide and helium, this type of transfer takes place at all usable welding currents. Globular transfer is characterized by a drop size with a diameter greater than that of the electrode. The

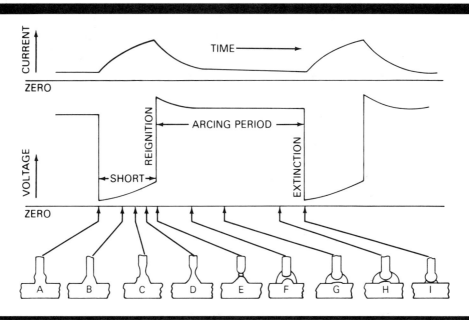

Figure 4.3—Schematic Representation of Short Circuiting Metal Transfer

large drop is easily acted on by gravity, generally limiting successful transfer to the flat position.

At average currents, only slightly higher than those used in short circuiting transfer, globular axially-directed transfer can be achieved in a substantially inert gas shield. If the arc length is too short (low voltage), the enlarging drop may short to the workpiece, become superheated, and disintegrate, producing considerable spatter. The arc must therefore be long enough to ensure detachment of the drop before it contacts the weld pool. However, a weld made using the higher voltage is likely to be unacceptable because of lack of fusion, insufficient penetration, and excessive reinforcement. This greatly limits use of the globular transfer mode in production applications.

Carbon dioxide shielding results in randomly directed globular transfer when the welding current and voltage are significantly above the range for short circuiting transfer. The departure from axial transfer motion is governed by electromagnetic forces, generated by the welding current acting upon the molten tip, as shown in Figure 4.4. The most important of these are the electromagnetic pinch force (P) and anode reaction force (R).

The magnitude of the pinch force is a direct function of welding current and wire diameter, and is usually responsible for drop detachment. With CO_2 shielding, the welding current is conducted through the molten drop and the electrode tip is not enveloped by the arc plasma. High-speed photography shows that the arc moves over the surface of the molten drop and workpiece, because force R tends to support the drop. The molten drop grows until it detaches by short circuiting (Figure 4.4B) or by gravity [Figure 4.4(A)], because R is never overcome by P alone. As shown in Figure 4.4(A), it is possible for the drop to become detached and transfer to the weld pool without disruption. The most likely situation is shown in Figure 4.4 (B), which shows the drop short circuiting the arc column and exploding. Spatter can therefore be severe, which limits its use of CO_2 shielding for many commercial applications.

Nevertheless, CO_2 remains the most commonly used gas for welding mild steels. The reason for this is that the spatter problem can be reduced significantly by "burying" the arc. In so doing, the arc atmosphere becomes a mixture of the gas and iron vapor, allowing the transfer to become almost spraylike. The arc forces are sufficient to maintain a depressed cavity which traps much of the spatter. This technique requires higher welding current and results in deep penetration. However, unless the travel speed is carefully controlled, poor wetting action may result in excessive weld reinforcement.

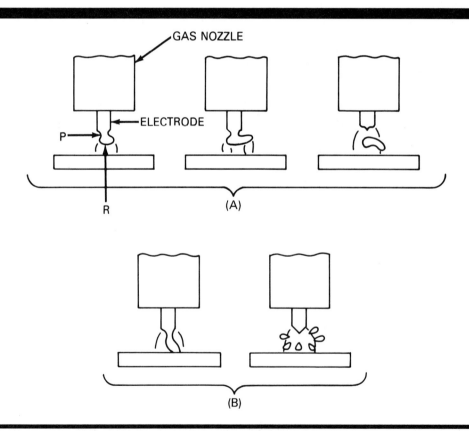

Figure 4.4–Nonaxial Globular Transfer

Spray Transfer

WITH ARGON-RICH SHIELDING it is possible to produce a very stable, spatter-free "axial spray" transfer mode as illustrated in Figure 4.5. This requires the use of direct current and a positive electrode (DCEP), and a current level above a critical value called the *transition current*. Below this current, transfer occurs in the globular mode described previously, at the rate of a few drops per second. Above the transition current, the transfer occurs in the form of very small drops that are formed and detached at the rate of hundreds per second. They are accelerated axially across the arc gap. The relationship between transfer rate and current is plotted in Figure 4.6.

The transition current, which is dependent on the liquid metal surface tension, is inversely proportional to the electrode diameter and, to a smaller degree, to the electrode extension. It varies with the filler metal melting temperature and the shielding gas composition. Typical transition currents for some of the more common metals are shown in Table 4.1.

The spray transfer mode results in a highly directed stream of discrete drops that are accelerated by arc forces to velocities which overcome the effects of gravity. Because of that, the process, under certain conditions, can be used in any position. Because the drops are smaller than the arc length, short circuits do not occur, and spatter is negligible if not totally eliminated.

Another characteristic of the spray mode of transfer is the "finger" penetration which it produces. Although the finger can be deep, it is affected by magnetic fields, which

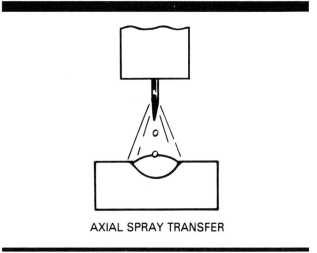

AXIAL SPRAY TRANSFER

Figure 4.5—Axial Spray Transfer

must be controlled to keep it located at the center of the weld penetration profile.

The spray-arc transfer mode can be used to weld almost any metal or alloy because of the inert characteristics of the argon shield. However, applying the process to thin sheets may be difficult because of the high currents needed to produce the spray arc. The resultant arc forces can cut through relatively thin sheets instead of welding them. Also, the characteristically high deposition rate may produce a weld pool too large to be supported by surface tension in the vertical or overhead position.

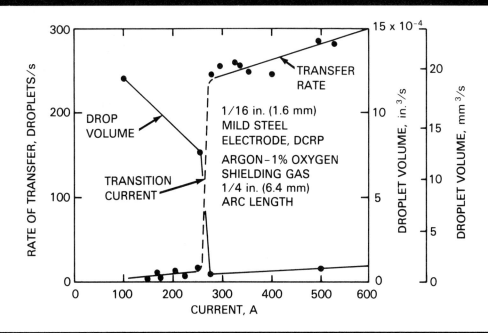

Figure 4.6—Variation in Volume and Transfer Rate of Drops With Welding Current (Steel Electrode)

Table 4.1
Globular-to-Spray Transition Currents for a Variety of Electrodes

Wire Electrode Type	Wire Electrode Diameter		Shielding gas	Minimum Spray Arc Current, A
	in.	mm		
Mild Steel	0.030	0.8	98% argon - 2% oxygen	150
Mild Steel	0.035	0.9	98% argon - 2% oxygen	165
Mild Steel	0.045	1.1	98% argon - 2% oxygen	220
Mild Steel	0.062	1.6	98% argon - 2% oxygen	275
Stainless Steel	0.035	0.9	98% argon - 2% oxygen	170
Stainless Steel	0.045	1.1	98% argon - 2% oxygen	225
Stainless Steel	0.062	1.6	98% argon - 2% oxygen	285
Aluminum	0.030	0.8	Argon	95
Aluminum	0.045	1.1	Argon	135
Aluminum	0.062	1.6	Argon	180
Deoxidized Copper	0.035	0.9	Argon	180
Deoxidized Copper	0.045	1.1	Argon	210
Deoxidized Copper	0.062	1.6	Argon	310
Silicon Bronze	0.035	0.9	Argon	165
Silicon Bronze	0.045	1.1	Argon	205
Silicon Bronze	0.062	1.6	Argon	270

The work thickness and welding position limitations of spray arc transfer have been largely overcome with specially designed power supplies. These machines produce carefully controlled wave forms and frequencies that "pulse" the welding current. As shown in Figure 4.7, they provide two levels of current; one a constant, low background current which sustains the arc without providing enough energy to cause drops to form on the wire tip; the other a superimposed pulsing current with amplitude greater than the transition current necessary for spray transfer. During this pulse, one or more drops are formed and transferred. The frequency and amplitude of the pulses control the energy level of the arc, and therefore the rate at which the wire melts. By reducing the average arc energy and the wire melting rate, pulsing makes the desirable features of spray transfer available for joining sheet metals and welding thick metals in all positions.

Many variations of such power sources are available. The simplest provide a single frequency of pulsing (60 or 120 pps) with independent control of the background and pulsing current levels. More sophisticated power sources, sometimes called synergic, automatically provide the optimum combination of background and pulse for any given setting of wire feed speed.

PROCESS VARIABLES

THE FOLLOWING ARE some of the variables that affect weld penetration, bead geometry and overall weld quality:

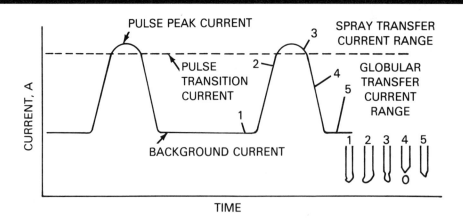

Figure 4.7–Pulsed-Spray Arc Welding Current Characteristic

(1) Welding current (electrode feed speed)
(2) Polarity
(3) Arc voltage (arc length)
(4) Travel speed
(5) Electrode extension
(6) Electrode orientation (trail or lead angle)
(7) Weld joint position
(8) Electrode diameter
(9) Shielding gas composition and flow rate

Knowledge and control of these variables is essential to consistently produce welds of satisfactory quality. These variables are not completely independent, and changing one generally requires changing one or more of the others to produce the desired results. Considerable skill and experience are needed to select optimum settings for each application. The optimum values are affected by (1) type of base metal, (2) electrode composition, (3) welding position, and (4) quality requirements. Thus, there is no single set of parameters that gives optimum results in every case.

Welding Current

WHEN ALL OTHER variables are held constant, the welding amperage varies with the electrode feed speed or melting rate in a nonlinear relation. As the electrode feed speed is varied, the welding amperage will vary in a like manner if a constant-voltage power source is used. This relationship of welding current to wire feed speed for carbon steel elec-

trodes is shown in Figure 4.8. At the low-current levels for each electrode size, the curve is nearly linear. However, at higher welding currents, particularly with small diameter electrodes, the curves become nonlinear, progressively increasing at a higher rate as welding amperage increases. This is attributed to resistance heating of the electrode extension beyond the contact tube. The curves can be approximately represented by the equation

$$WFS = aI + bLI^2 \tag{4.1}$$

where

$WFS =$ the electrode feed speed, in./min (mm/s)
$\quad a =$ a constant of proportionality for anode or cathode heating. Its magnitude is dependent upon polarity, composition, and other factors, in./(min. · A) [(mm/(s · A)]
$\quad b =$ constant of proportionality for electrical resistance heating, min^{-1} A^{-2} (s^{-1} A^{-2})
$\quad L =$ the electrode extension or stick out, in. (mm)
$\quad I =$ the welding current, A

As shown in Figures 4.8, 4.9, 4.10 and 4.11, when the diameter of the electrode is increased (while maintaining the same electrode feed speed), a higher welding current is required. The relationship between the electrode feed speed and the welding current is affected by the electrode chemical composition. This effect can be seen by comparing Figures 4.8, 4.9, 4.10 and 4.11, which are for carbon

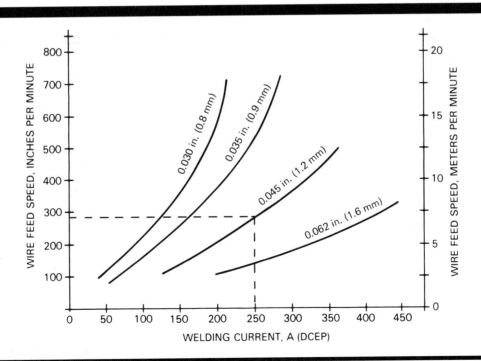

Figure 4.8–Typical Welding Currents Versus Wire Feed Speeds for Carbon Steel Electrodes

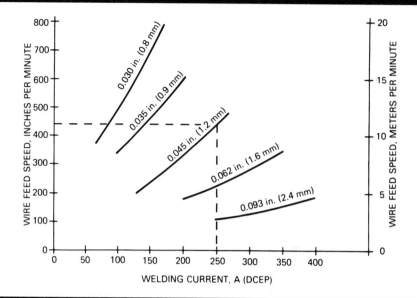

Figure 4.9–Welding Currents Versus Wire Feed Speed for ER4043 Aluminum Electrodes

steel, aluminum, stainless steel, and copper electrodes respectively. The different positions and slopes of the curves are due to differences in the melting temperatures and electrical resistivities of the metals. Electrode extension also affects the relationships.

With all other variables held constant, an increase in welding current (electrode feed speed) will result in the following:

(1) An increase in the depth and width of the weld penetration

(2) An increase in the deposition rate

(3) An increase in the size of the weld bead

Pulsed spray welding is a variation of the GMAW process in which the current is pulsed to obtain the advantages of the spray mode of metal transfer at average currents

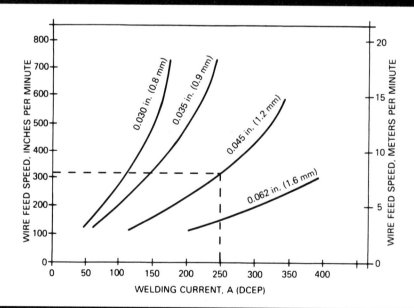

Figure 4.10–Typical Welding Currents Versus Wire Feed Speeds for 300 Series Stainless Steel Electrodes

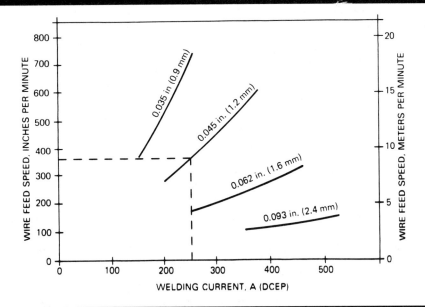

Figure 4.11–Welding Currents Versus Wire Feed Speed for ECu Copper Electrodes

equal to or less than the globular-to-spray transition current.

Since arc force and deposition rate are exponentially dependent on current, operation above the transition current often makes the arc forces uncontrollable in the vertical and overhead positions. By reducing the average current with pulsing, the arc forces and deposition rates can both be reduced, allowing welds to be made in all positions and in thin sections.

With solid wires, another advantage of pulsed power welding is that larger diameter wires [i.e., 1/16-in. (1.6 mm)] can be used. Although deposition rates are generally no greater than those with smaller diameter wires, the advantage is in the lower cost per unit of metal deposited. There is also an increase in deposition efficiency because of reduced spatter loss.

With metal cored wires, pulsed power produces an arc that is less sensitive to changes in electrode extension (stickout) and voltage compared to solid wires. Thus, the process is more tolerant of operator guidance fluctuations. Pulsed power also minimizes spatter from an operation already low in spatter generation.

Polarity

THE TERM *polarity* is used to describe the electrical connection of the welding gun with relation to the terminals of a direct current power source. When the gun power lead is connected to the positive terminal, the polarity is designated as direct current electrode positive (DCEP), arbitrarily called *reverse polarity*. When the gun is connected to the negative terminal, the polarity is designated as direct current electrode negative (DCEN), originally called

straight polarity. The vast majority of GMAW applications use direct current electrode positive (DCEP). This condition yields a stable arc, smooth metal transfer, relatively low spatter, good weld bead characteristics and greatest depth of penetration for a wide range of welding currents.

Direct current electrode negative (DCEN) is seldom used because axial spray transfer is not possible without modifications that have had little commercial acceptance. DCEN has a distinct advantage of high melting rates that cannot be exploited because the transfer is globular. With steels, the transfer can be improved by adding a minimum of 5 percent oxygen to the argon shield (requiring special alloys to compensate for oxidation losses) or by treating the wire to make it thermionic (adding to the cost of the filler metal). In both cases, the deposition rates drop, eliminating the only real advantage of changing polarity. However, because of the high deposition rate and reduced penetration, DCEN has found some use in surfacing applications.

Attempts to use alternating current with the GMAW process have generally been unsuccessful. The cyclic wave form creates arc instability due to the tendency of the arc to extinguish as the current passes through the zero point. Although special wire surface treatments have been developed to overcome this problem, the expense of applying them has made the technique uneconomical.

Arc Voltage (Arc Length)

Arc voltage and *arc length* are terms that are often used interchangeably. It should be pointed out, however, that they are different even though they are related. With GMAW, arc length is a critical variable that must be carefully controlled. For example, in the spray-arc mode with argon shielding, an arc that is too short experiences momentary short circuits. They cause pressure fluctuations which pump air into the arc stream, producing porosity or embrittlement due to absorbed nitrogen. Should the arc be too long, it tends to wander, affecting both the penetration and surface bead profiles. A long arc can also disrupt the gas shield. In the case of buried arcs with a carbon dioxide shield, a long arc results in excessive spatter as well as porosity; if the arc is too short, the electrode tip short circuits the weld pool, causing instability.

Arc length is the independent variable. Arc voltage depends on the arc length as well as many other variables, such as the electrode composition and dimensions, the shield gas, the welding technique and, since it often is measured at the power supply, even the length of the welding cable. Arc voltage is an approximate means of stating the physical arc length (see Figure 4.12) in electrical terms, although the arc voltage also includes the voltage drop in the electrode extension beyond the contact tube.

With all variables held constant, arc voltage is directly related to arc length. Even though the arc length is the variable of interest and the variable that should be controlled, the voltage is more easily monitored. Because of this, and the normal requirement that the arc voltage be specified in the welding procedure, it is the term that is more commonly used.

Arc voltage settings vary depending on the material, shielding gas, and transfer mode. Typical values are shown in Table 4.2. Trial runs are necessary to adjust the arc voltage to produce the most favorable arc characteristics and weld bead appearance. Trials are essential because the optimum arc voltage is dependent upon a variety of factors, including metal thickness, the type of joint, welding position, electrode size, shielding gas composition, and the type of weld. From any specific value of arc voltage, a voltage increase tends to flatten the weld bead and increase the width of the fusion zone. Excessively high voltage may cause porosity, spatter, and undercut. Reduction in voltage results in a narrower weld bead with a higher crown and deeper penetration. Excessively low voltage may cause stubbing of the electrode.

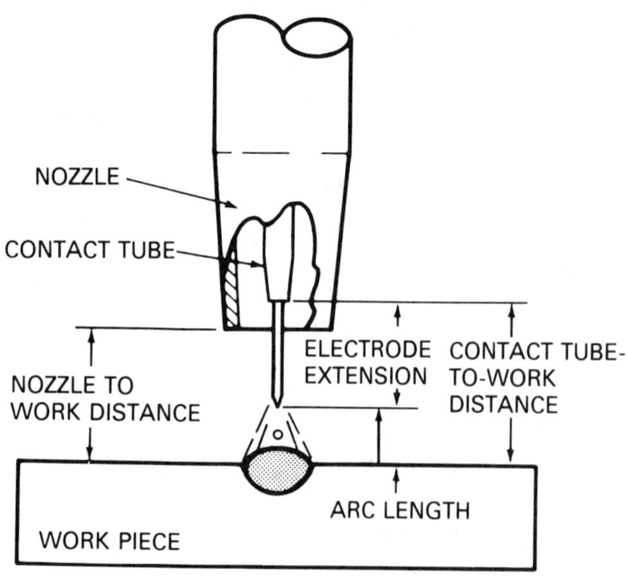

Travel Speed

TRAVEL SPEED IS the linear rate at which the arc is moved along the weld joint. With all other conditions held constant, weld penetration is a maximum at an intermediate travel speed.

When the travel speed is decreased, the filler metal deposition per unit length increases. At very slow speeds the welding arc impinges on the molten weld pool, rather than the base metal, thereby reducing the effective penetration. A wide weld bead is also a result.

As the travel speed is increased, the thermal energy per unit length of weld transmitted to the base metal from the arc is at first increased, because the arc acts more directly on the base metal. With further increases in travel speed, less thermal energy per unit length of weld is imparted to the base metal. Therefore, melting of the base metal first increases and then decreases with increasing travel speed. As travel speed is increased further, there is a tendency toward undercutting along the edges of the weld bead because there is insufficient deposition of filler metal to fill the path melted by the arc.

Electrode Extension

THE ELECTRODE EXTENSION is the distance between the end of the contact tube and the end of the electrode, as shown in Figure 4.12. An increase in the electrode extension results in an increase in its electrical resistance. Resistance heating in turn causes the electrode temperature to rise, and results in a small increase in electrode melting rate. Overall, the increased electrical resistance produces a greater voltage drop from the contact tube to the work. This is sensed by the power source, which compensates by decreasing the current. That immediately reduces the electrode melting rate, which then lets the electrode shorten the physical arc length. Thus, unless there is an increase in the voltage at the welding machine, the filler metal will be deposited as a narrow, high-crowned weld bead.

The desirable electrode extension is generally from 1/4 to 1/2 in. (6 to 13 mm) for short circuiting transfer and from 1/2 to 1 in. (13 to 25 mm) for other types of metal transfer.

NOZZLE

CONTACT TUBE

NOZZLE TO WORK DISTANCE

ELECTRODE EXTENSION

CONTACT TUBE-TO-WORK DISTANCE

ARC LENGTH

WORK PIECE

Figure 4.12–Gas Metal Arc Welding Terminology

Table 4.2
Typical Arc Voltages for Gas Metal Arc Welding of Various Metals[a]

Metal	Spray[b] Globular Transfer 1/16 in. (1.6 mm) Diameter Electrode					Short Circuiting Transfer Diameter Electrode			
	Argon	Helium	25% Ar-75% He	Ar-O$_2$ (1-5% O$_2$)	CO$_2$	Argon	Ar-O$_2$ (1-5% O$_2$)	75% Ar-25% CO$_2$	CO$_2$
Aluminum	25	30	29	—	—	19	—	—	—
Magnesium	26	—	28	—	—	16	—	—	—
Carbon steel	—	—	—	28	30	17	18	19	20
Low alloy steel	—	—	—	28	30	17	18	19	20
Stainless steel	24	—	—	26	—	18	19	21	—
Nickel	26	30	28	—	—	22	—	—	—
Nickel-copper alloy	26	30	28	—	—	22	—	—	—
Nickel-chromium-iron alloy	26	30	28	—	—	22	—	—	—
Copper	30	36	33	—	—	24	22	—	—
Copper-nickel alloy	28	32	30	—	—	23	—	—	—
Silicon bronze	28	32	30	28	—	23	—	—	—
Aluminum bronze	28	32	30	—	—	23	—	—	—
Phosphor bronze	28	32	30	23	—	23	—	—	—

a. Plus or minus approximately ten percent. The lower voltages are normally used on light material and at low amperage; the higher voltages are used on heavy material at high amperage.

b. For the pulsed variation of spray transfer the arc voltage would be from 18-28 volts depending on the amperage range used.

Electrode Orientation

As WITH ALL arc welding processes, the orientation of the welding electrode with respect to the weld joint affects the weld bead shape and penetration. Electrode orientation affects bead shape and penetration to a greater extent than arc voltage or travel speed. The electrode orientation is described in two ways: (1) by the relationship of the electrode axis with respect to the direction of travel (the travel angle), and (2) the angle between the electrode axis and the adjacent work surface (work angle). When the electrode points opposite from the direction of travel, the technique is called *backhand welding with a drag angle*. When the electrode points in the direction of travel, the technique is *forehand welding with a lead angle*. The electrode orientation and its effect on the width and penetration of the weld are illustrated in Figures 4.13 (A), (B), and (C).

When the electrode is changed from the perpendicular to a lead angle technique with all other conditions unchanged, the penetration decreases and the weld bead becomes wider and flatter. Maximum penetration is obtained in the flat position with the drag technique, at a drag angle of about 25 degrees from perpendicular. The drag technique also produces a more convex, narrower bead, a more stable arc, and less spatter on the workpiece. For all positions, the electrode travel angle normally used is a drag angle in the range of 5 to 15 degrees for good control and shielding of the molten weld pool.

For some materials, such as aluminum, a lead technique is preferred. This lead technique provides a "cleaning action" ahead of the molten weld metal, which promotes wetting and reduces base metal oxidation.

When producing fillet welds in the horizontal position, the electrode should be positioned about 45 degrees to the vertical member (work angle), as illustrated in Figure 4.14.

Weld Joint Position

MOST SPRAY TYPE GMAW is done in the flat or horizontal positions, while at low-energy levels, pulsed and short circuiting GMAW can be used in all positions. Fillet welds made in the flat position with spray transfer are usually more uniform, less likely to have unequal legs and convex profiles, and are less susceptible to undercutting than similar fillet welds made in the horizontal position.

To overcome the pull of gravity on the weld metal in the vertical and overhead positions of welding, small diameter electrodes are usually used, with either short circuiting metal transfer or spray transfer with pulsed direct current. Electrode diameters of 0.045 in. (1.1 mm) and smaller are best suited for out-of-position welding. The low-heat input allows the molten pool to freeze quickly. Downward welding progression is usually effective on sheet metal in the vertical position.

When welding is done in the "flat" position, the inclination of the weld axis with respect to the horizontal plane will influence the weld bead shape, penetration, and travel

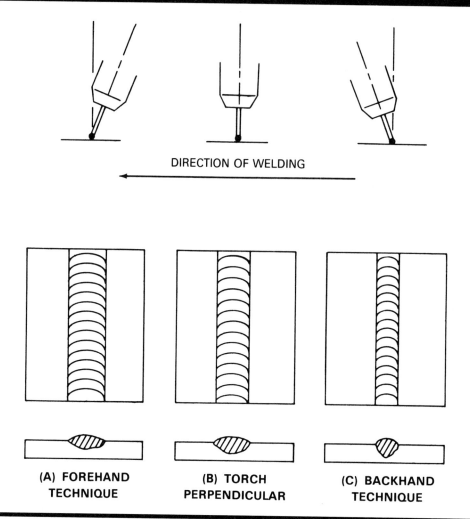

DIRECTION OF WELDING

(A) FOREHAND
TECHNIQUE

(B) TORCH
PERPENDICULAR

(C) BACKHAND
TECHNIQUE

Figure 4.13–Effect of Electrode Position and Welding Technique

speed. In flat position circumferential welding, the work rotates under the welding gun and inclination is obtained by moving the welding gun in either direction from top dead center.

By positioning linear joints with the weld axis at 15 degrees to the horizontal and welding downhill, weld reinforcement can be decreased under welding conditions that would produce excessive reinforcement when the work is in the flat position. Also, when traveling downhill, speeds can usually be increased. At the same time, penetration is lower, which is beneficial for welding sheet metal.

Downhill welding affects the weld contour and penetration, as shown in Figure 4.15(A). The weld puddle tends to flow toward the electrode and preheats the base metal, particularly at the surface. This produces an irregularly shaped fusion zone, called a *secondary wash*. As the angle of declination increases, the middle surface of the weld is depressed, penetration decreases, and the width of the

weld increases. For aluminum, this downhill technique is not recommended due to loss of cleaning action and inadequate shielding.

Uphill welding affects the fusion zone contour and the weld surface, as illustrated in Figure 4.15(B). The force of gravity causes the weld puddle to flow back and lag behind the electrode. The edges of the weld lose metal, which flows to the center. As the angle of inclination increases, reinforcement and penetration increase, and the width of the weld decreases. The effects are exactly the opposite of those produced by downhill welding. When higher welding currents are used, the maximum usable angle decreases.

Electrode Size

THE ELECTRODE SIZE (diameter) influences the weld bead configuration. A larger electrode requires higher minimum current than a smaller electrode for the same metal transfer

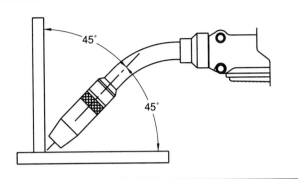

Figure 4.14—Normal Work Angle for Fillet Welds

characteristics. Higher currents in turn produce additional electrode melting and larger, more fluid weld deposits. Higher currents also result in higher deposition rates and greater penetration. However, vertical and overhead weld-

ing are usually done with smaller diameter electrodes and lower currents.

Shielding Gas

THE CHARACTERISTICS OF the various gases and their effect on weld quality and arc characteristics are discussed in detail in the consumables section of this chapter.

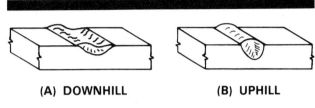

(A) DOWNHILL   **(B) UPHILL**

Figure 4.15—Effect of Work Inclination on Weld Bead Shape

EQUIPMENT

THE GMAW PROCESS can be used either semiautomatically or automatically. The basic equipment for any GMAW installation consists of the following:

(1) Welding gun (air or water cooled)
(2) Electrode feed unit
(3) Welding control
(4) Welding power supply
(5) A regulated supply of shielding gas
(6) A source of electrode
(7) Interconnecting cables and hoses
(8) Water circulation system (for water-cooled torches)

Typical semiautomatic and mechanized components are illustrated in Figures 4.2 and 4.16.

WELDING GUNS

DIFFERENT TYPES OF welding guns have been designed to provide maximum efficiency regardless of the application, ranging from heavy-duty guns for high current, high-production work, to lightweight guns for low current, out-of-position welding.

Water or air cooling and curved or straight nozzles are available for both heavy-duty and lightweight guns. An air-cooled gun is generally heavier than a water-cooled gun at the same rated amperage and duty cycle, because the air-cooled gun requires more mass to overcome its less efficient cooling.

The following are basic components of arc welding guns:

(1) Contact tube (or tip)
(2) Gas shield nozzle
(3) Electrode conduit and liner
(4) Gas hose
(5) Water hose
(6) Power cable
(7) Control switch

These components are illustrated in Figure 4.17.

The contact tube, usually made of copper or a copper alloy, transfers welding current to the electrode and directs the electrode towards the work. The contact tube is connected electrically to the welding power supply by the power cable. The inner surface of the contact tube should be smooth so the electrode will feed easily through this tube and also make good electrical contact. The instruction booklet supplied with every gun will list the correct size contact tube for each electrode size and material.

Generally, the hole in the contact tube should be 0.005 to 0.010 in. (0.13 to 0.25 mm) larger than the wire being used, although larger hole sizes may be required for aluminum. The contact tube must be held firmly in the torch and must be centered in the gas shielding nozzle. The positioning of the contact tube in relation to the end of the nozzle may be a variable depending on the mode of transfer being used. For short-circuiting transfer, the tube is usually flush or extended beyond the nozzle, while for

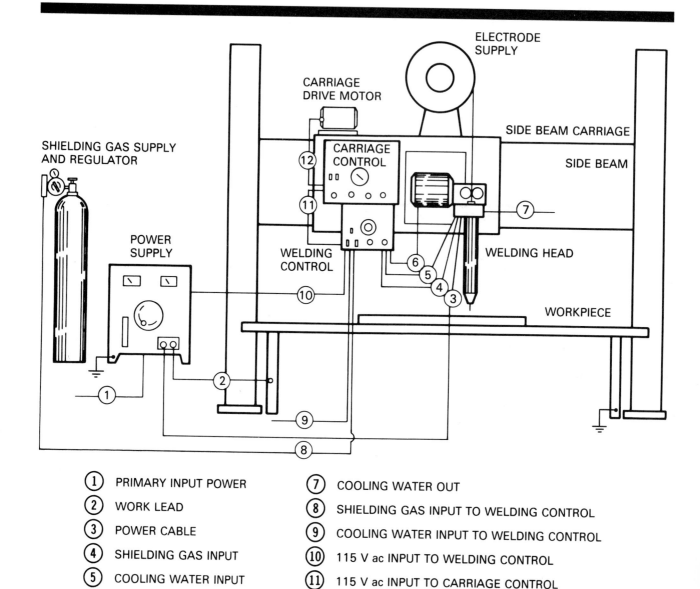

Figure 4.16—Mechanized Gas Metal Arc Welding Installation

1. PRIMARY INPUT POWER
2. WORK LEAD
3. POWER CABLE
4. SHIELDING GAS INPUT
5. COOLING WATER INPUT
6. ELECTRODE FEED UNIT INPUT
7. COOLING WATER OUT
8. SHIELDING GAS INPUT TO WELDING CONTROL
9. COOLING WATER INPUT TO WELDING CONTROL
10. 115 V ac INPUT TO WELDING CONTROL
11. 115 V ac INPUT TO CARRIAGE CONTROL
12. INPUT TO CARRIAGE DRIVE MOTOR

spray arc it is recessed approximately 1/8 in. During welding, it should be checked periodically and replaced if the hole has become elongated due to excessive wear or if it becomes clogged with spatter. Using a worn or clogged tip can result in poor electrical contact and erratic arc characteristics.

The nozzle directs an even-flowing column of shielding gas into the welding zone. An even flow is extremely important to assure adequate protection of the molten weld metal from atmospheric contamination. Different size nozzles are available and should be chosen according to

the application, i.e., larger nozzles for high-current work where the weld puddle is large, and smaller nozzles for low current and short circuiting welding. For spot welding applications the nozzles are made with ports that allow the gas to escape when the nozzle is pressed onto the workpiece.

The electrode conduit and its liner are connected to a bracket adjacent to the feed rolls on the electrode feed motor. The conduit supports, protects, and directs the electrode from the feed rolls to the gun and contact tube. Uninterrupted electrode feeding is necessary to insure

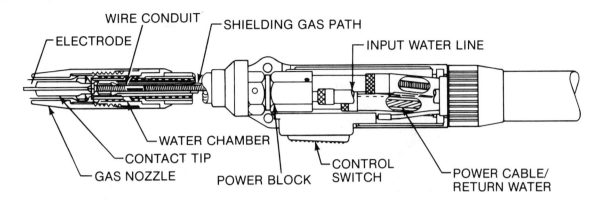

Figure 4.17–Cross-Sectional View of Typical Gas Metal Arc Welding Gun

good arc stability. Buckling or kinking of the electrode must be prevented. The electrode will tend to jam anywhere between the drive rolls and the contact tube if not properly supported.

The liner may be an integral part of the conduit or supplied separately. In either case, the liner material and inner diameter are important. Liners require periodic maintenance to assure they are clean and in good condition to assure consistant feeding of the wire.

A helical steel liner is recommended when using hard electrode materials such as steel and copper. Nylon liners should be used for soft electrode materials such as aluminum and magnesium.

Care must be taken not to crimp or excessively bend the conduit even though its outer surface is usually steel-supported. The instruction manual supplied with each unit will generally list the recommended conduits and liners for each electrode size and material.

The remaining accessories bring the shielding gas, cooling water, and welding power to the gun. These hoses and cables may be connected directly to the source of these facilities or to the welding control. Trailing gas shields are available and may be required to protect the weld pool during high-speed welding.

The basic gun, Figure 4.18, is connected to an electrode feed unit that pushes the electrode from a remote location through the conduit. Other designs are also available including a unit with a small electrode feed mechanism built into the gun, Figure 4.19. This gun will pull the electrode from the source, where an additional drive may also be located to simultaneously push the electrode into the conduit (i.e., a "push-pull" system). This type of gun is also useful for feeding small diameter or soft electrodes (e.g. aluminum), where pushing might cause the electrode to buckle. Another variation is the "spool-on-gun" type illustrated in Figure 4.20 in which the electrode feed mechanism and the electrode source are self-contained.

ELECTRODE FEED UNIT

THE ELECTRODE FEED unit (wire feeder) consists of an electric motor, drive rolls, and accessories for maintaining electrode alignment and pressure. These units can be integrated with the speed control or located remotely from it. The electrode feed motor is usually a direct current type. It pushes the electrode through the gun to the work. It should have a control circuit that varies the motor speed over a broad range.

Constant-speed wire feeders are normally used in combination with constant-potential power sources. They may

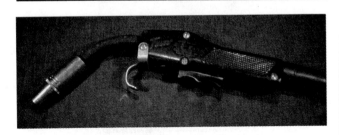

Figure 4.18–Commercially Available Gas Metal Arc Welding Gun

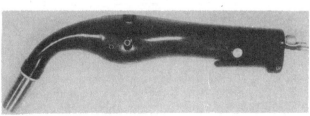

Figure 4.19–Pull-Type GMAW Gun

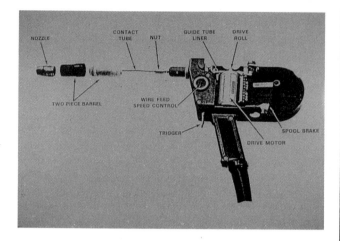

Figure 4.20–Exploded View of a Spool-on-Gun-Type Torch

Figure 4.21–Typical 4 Drive Roll Wire Feeding Unit

be used with constant-current power supplies if a slow electrode "run-in" circuit is added.

When a constant-current power source is used, an automatic voltage sensing control is necessary. This control detects changes in the arc voltage and adjusts the wire feed speed to maintain a constant arc length. This combination of variable speed wire feeder and constant-current power source is limited to larger diameter wires [over 1/16 in. (1.6 mm)], where the feed speeds are lower. At high wire feed speeds the adjustments to motor speed cannot normally be made quickly enough to maintain arc stability.

The feed motor is connected to a drive roll assembly. These drive rolls in turn transmit the force to the electrode, pulling it from the electrode source and pushing it through the welding gun. Wire feed units may use a two-roll or four-roll arrangement. A typical four-roll wire feeding unit is shown in Figure 4.21. The drive roll pressure adjustment allows for variable force to be applied to the wire, depending on its characteristics (e.g. solid or cored, hard or soft). The inlet and outlet guides provide for proper alignment of the wire to the drive rolls and support the wire to prevent buckling.

The type of feed rolls generally used with solid wires is shown in Figure 4.22A, where a grooved roll is combined with a flat backup roll. A V-groove is used for solid hard wires such as carbon and stainless steels, and a U-groove is used for soft wires such as aluminum.

Serrated or knurled feed rolls with a knurled back up roll, as shown in Figure 4.22B, are generally used with cored wires. The knurled design allows for maximum drive force to be transmitted to the wire with a minimum of drive roll pressure. These types of rolls are not recommended for softer wire, such as aluminum, because the rolls tend to cause a flaking of the wire which can eventually clog the gun or liner.

Welding Control

THE WELDING CONTROL and electrode feed motor for semi-automatic operation are available in one integrated package. The main function of the welding control is to regulate the speed of the electrode feed motor, usually through the use of an electronic governor. By increasing the wire feed speed the operator increases the welding current. Decreases in wire feed speed result in lower welding currents. The control also

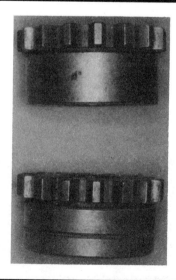

Figure 4.22A–Ground Feed Roll with Flat Backup Used to Feed Solid Wires

regulates the starting and stopping of the electrode feed through a signal received from the gun switch.

Also available are electrode feed control features that permit the use of a "touch-start" (the electrode feed is initiated when the electrode touches the work), or a "slow run-in" (the initial feed rate is reduced until the arc is initiated and then increases to that required for welding). These two features are employed primarily in conjunction with constant-current type power supplies, and are particularly useful for gas metal arc welding of aluminum.

Normally, shielding gas, cooling water, and welding power are also delivered to the gun through the control, requiring direct connection of the control to these facilities and the power supply. Gas and water flow are regulated to coincide with the weld start and stop by use of solenoid valves. The control may also sequence the starting and stopping of gas flow, and may energize the power source contactor. The control may allow some gas to flow before welding starts (prepurge) and after welding stops (postpurge) to protect the molten weld puddle. The control is usually independently powered by 115 V ac.

Power Source

THE WELDING POWER source delivers electrical power to the electrode and workpiece to produce the arc. For the vast majority of GMAW applications, direct current with electrode positive (DCEP) is used; therefore, the positive lead is connected to the gun and the negative lead to the workpiece. The major types of direct current power sources are engine-driven-generators (rotating) and transformer-rectifiers (static). Inverters are included in the static category. The transformer-rectifier type is usually preferred for in-shop fabrication where a source of either 230 V or 460 V is available. The transformer-rectifier type re-

Figure 4.22B—Knurled Feed Rolls Generally Used with Cored Wires

sponds faster than the engine-driven-generator type when the arc conditions change. The engine-driven generator is used when there is no other available source of electrical energy, e.g., remote locations.

Both types of power source can be designed and built to provide either constant current or constant potential. Early applications of the GMAW process used constant-current power sources (often referred to as a *droopers*). Droopers maintain a relatively fixed current level during welding, regardless of variations in arc length, as illustrated in Figure 4.23. These machines are characterized by high open circuit voltages and limited short circuit current levels. Since they supply a virtually constant current output, the arc will maintain a fixed length only if the contact-tube-to-work distance remains constant, with a constant electrode feed rate.

In practice, since this distance will vary, the arc will then tend to either "burn back" to the contact tube or "stub" into the workpiece. This can be avoided by using a voltage-controlled electrode feed system. When the voltage (arc length) increases or decreases, the motor speeds up or slows down to hold the arc length constant. The electrode feed rate is changed automatically by the control system. This type of power supply is generally used for spray transfer welding since the limited duration of the arc in short circuiting transfer makes control by voltage regulation impractical.

As GMAW applications increased, it was found that a constant-potential (CP) power source provided improved operation. Used in conjunction with a constant-speed wire feeder, it maintains a nearly constant voltage during the welding operation. The volt-ampere curve of this type power source is illustrated in Figure 4.24. The CP system compensates for variations in the contact-tip-to-work-piece distance, which occur during normal welding operations, by instantaneously increasing or decreasing the welding current to compensate for the changes in stickout due to the changes in gun-to-work distance.

The arc length is established by adjusting the welding voltage at the power source. Once this is set, no other changes during welding are required. The wire feed speed, which also becomes the current control, is preset by the welder or welding operator prior to welding. It can be adjusted over a considerable range before stubbing to the workpiece or burning back into the contact tube occurs. Welders and welding operators easily learn to adjust the wire feed and voltage controls with only minimum instruction.

The self-correction mechanism of a constant-potential power source is illustrated in Figure 4.25. As the contact tip-to-work distance increases, the arc voltage and arc length would tend to increase. However, the welding current decreases with this slight increase in voltage, thus compensating for the increase in stickout. Conversely, if the distance is shortened, the lower voltage would be accompanied by an increase in current to compensate for the shorter stickout.

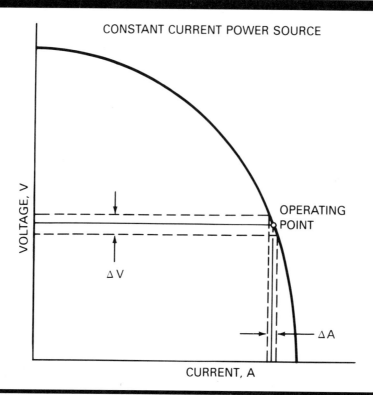

Figure 4.23–Volt-Amper Relationship for a Constant-Current (cc) Power Supply

The self-correcting arc length feature of the CP power source is important in producing stable welding conditions, but there are additional variables that contribute to optimum welding performance, particularly for short circuiting transfer.

In addition to the control of the output voltage, some degree of slope and inductance control may be desirable. The welder or welding operator should understand the effect of these variables on the welding arc and its stability.

Voltage. Arc voltage is the electrical potential between the electrode and the workpiece. Arc voltage is lower than the voltage measured directly at the power source because of voltage drops at connections and along the length of the welding cable. As previously mentioned, arc voltage is di-

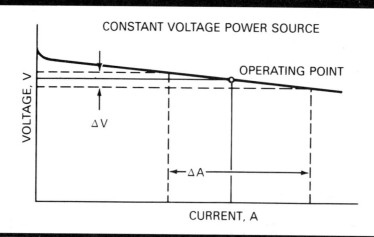

Figure 4.24–Volt-Ampere Relationship for a Constant-Potential (cp) Power Supply

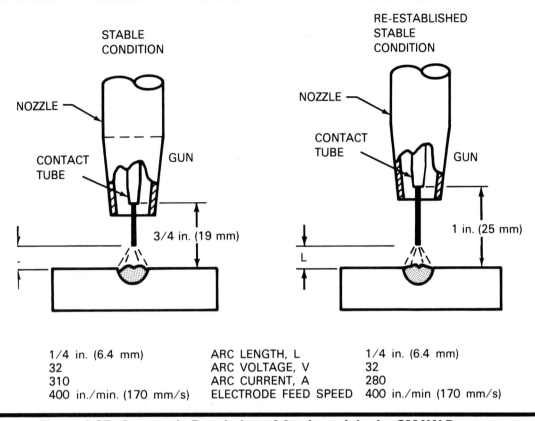

1/4 in. (6.4 mm)	ARC LENGTH, L	1/4 in. (6.4 mm)
32	ARC VOLTAGE, V	32
310	ARC CURRENT, A	280
400 in./min. (170 mm/s)	ELECTRODE FEED SPEED	400 in./min (170 mm/s)

Figure 4.25–Automatic Regulation of Arc Length in the GMAW Process

rectly related to arc length; therefore, an increase or a decrease in the output voltage at the power source will result in a like change in the arc length.

Slope. The static volt-ampere characteristics (static output) of a CP power source is illustrated in Figure 4.24. The slope of the output is the algebraic slope of the volt ampere curve and is customarily given as the voltage drop per 100 amperes of current rise.

The slope of the power source, as specified by the manufacturer, is measured at its output terminals and is not the total slope of the arc welding system. Anything that adds resistance to the welding system (i.e., power cables, poor connections, loose terminals, dirty contacts, etc.) adds to the slope. Therefore, slope is best measured at the arc in a given welding system. Two operating points are needed to calculate the slope of a constant-potential type welding system, as shown in Figure 4.26. It is not safe to use the open circuit voltage as one of the points, because of a sharp voltage drop with some machines at low currents. This is shown in Figure 4.24. Two stable arc conditions should be chosen at currents that envelope the range likely to be used.

Slope has a major function in the short-circuiting transfer mode of GMAW in that it controls the magnitude of the short circuit current, which is the amperage that flows when the electrode is shorted to the workpiece. In GMAW, the separation of molten drops of metal from the electrode is controlled by an electrical phenomenon called the *electromagnetic pinch effect*. Pinch is the magnetic "squeezing" force on a conductor produced by the current flowing through it. For short circuiting transfer, the effect is illustrated in Figure 4.27.

The short circuit current (and therefore the pinch effect force) is a function of the slope of the volt-ampere curve of the power source, as illustrated in Figure 4.28. The operating voltage and the amperage of the two power supplies are identical, but the short circuit current of curve A is less than that of curve B. Curve A has the steeper slope, or a greater voltage drop per 100 amperes, as compared to curve B thus, a lower short circuit current and a lower pinch effect.

In short circuiting transfer the amount of short circuit current is important since the resultant pinch effect determines the way a molten drop detaches from the electrode. This in turn affects the arc stability. When little or no slope is present in the power supply circuit, the short circuit current will rise rapidly to a high level. The pinch effect will also be high, and the molten drop will separate violently from the wire. The excessive pinch effect will abruptly

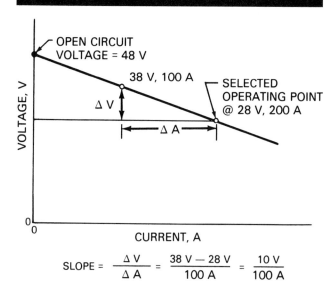

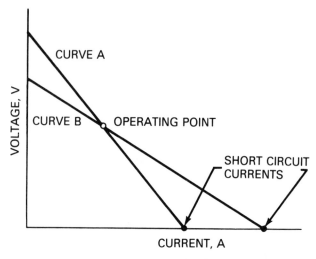

Figure 4.28–Effect of Changing Slope

$$\text{SLOPE} = \frac{\Delta V}{\Delta A} = \frac{38 V - 28 V}{100 A} = \frac{10 V}{100 A}$$

Figure 4.26–Calculation of the Slope for a Power Supply

squeeze the metal aside, clear the short circuit, and create excessive spatter.

When the short circuit current available from the power source is limited to a low value by a steep slope, the elec-

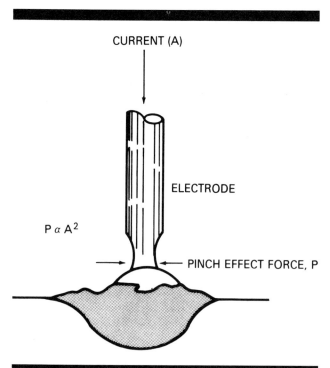

Figure 4.27–Illustration of Pinch Effect During Short Circuiting Transfer

trode will carry the full current, but the pinch effect may be too low to separate the drop and reestablish the arc. Under these conditions, the electrode will either pile up on the workpiece or freeze to the puddle. When the short circuit current is at an acceptable value, the parting of the molten drop from the electrode is smooth with very little spatter. Typical short circuit currents required for metal transfer with the best arc stability are shown in Table 4.3.

Many constant-potential power sources are equipped with a slope adjustment. They may be stepped or continuously adjustable to provide desirable levels of short circuit current for the application involved. Some have a fixed slope which has been preset for the most common welding conditions.

Inductance. When the electrode shorts to the work, the current increases rapidly to a higher level. The circuit characteristic affecting the time rate of this increase in current is inductance, usually measured in henrys. The effect of inductance is illustrated by the curves plotted in Figure 4.29. Curve A is an example of a current-time curve imme-

Table 4.3
Typical Short Circuit Currents Required for Metal Transfer in the Short Circuiting Mode

Electrode Material	Electrode Diameter		Short Circuit Current, Amperes (dcrp)
	in.	mm	
Carbon steel	0.030	0.8	300
Carbon steel	0.035	0.9	320
Aluminum	0.030	0.8	175
Aluminum	0.035	0.9	195

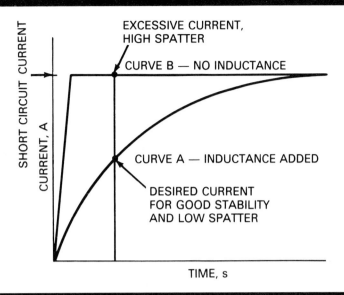

EXCESSIVE CURRENT,
HIGH SPATTER

CURVE B — NO INDUCTANCE

CURVE A — INDUCTANCE ADDED

DESIRED CURRENT
FOR GOOD STABILITY
AND LOW SPATTER

SHORT CIRCUIT CURRENT

CURRENT, A

TIME, s

Figure 4.29–Change in Rate of Current Rise Due to Added Inductance

diately after a short circuit when some inductance is in the circuit. Curve B illustrates the path the current would have taken if there were no inductance in the circuit.

The maximum amount of pinch effect is determined by the final short circuit current level. The instantaneous pinch effect is controlled by the instantaneous current, and therefore the shape of the current-time curve is significant. The inductance in the circuit controls the rate of current rise. Without inductance the pinch effect is applied rapidly and the molten drop will be violently "squeezed" off the electrode and cause excessive spatter. Higher inductance results in a decrease in the short circuits per second and an increase in the "arc-on" time. Increased arc-on time makes the puddle more fluid and results in a flatter, smoother weld bead.

In spray transfer, the addition of some inductance to the power source will produce a softer arc start without affecting the steady-state welding conditions. Power source adjustments required for minimum spatter conditions vary with the electrode material and diameter. As a general rule, higher short circuit currents and higher inductance are needed for larger diameter electrodes.

Power sources are available with fixed, stepped, or continuously adjustable inductance levels.

Shielding Gas Regulators

A SYSTEM IS required to provide constant shielding gas flow rate at atmospheric pressure during welding. A gas regulator reduces the source gas pressure to a constant working pressure regardless of variations at the source. Regulators may be single or dual stage and may have a built-in flowmeter. Dual stage regulators deliver gas at a more consistant pressure than single stage regulators when the source pressure varies.

The shielding gas source can be a high-pressure cylinder, a liquid-filled cylinder, or a bulk-liquid system. Gas mixtures are available in single cylinders. Mixing devices are used to obtain the correct proportions when two or more gas or liquid sources are used. The size and type of the gas storage source should be determined by the user, based on the volume of gas consumed per month.

Electrode Source

THE GMAW PROCESS uses a continuously fed electrode that is consumed at relatively high speeds. The electrode source must, therefore, provide a large volume of wire that can readily be fed to the gun to provide maximum process efficiency. This source usually takes the form of a spool or coil that holds approximately 10 to 60 pounds (.45 to 27 kg) of wire, wound to allow free feeding without kinks or tangles. Larger spools of up to 250 pounds (114 kg) are also available, and wire can be provided in drums or reels of 750 to 1000 pounds (340 to 450 kg). For spool-on-gun equipment small spools [1 to 2 pounds (.45 to .9 kg)] are used. The applicable AWS or military electrode specification defines standard packaging requirements. Normally, special requirements can be agreed to by the user and the supplier.

The electrode source may be located in close proximity to the wire feeder, or it can be positioned some distance away and fed through special dispensing equipment. Normally, the electrode source should be as close as possible to the gun to minimize feeding problems, yet far enough away to give flexibility and accessability to the welder.

CONSUMABLES

IN ADDITION TO equipment components, such as contact tips and conduit liners that wear out and have to be replaced, the process consumables in GMAW are electrodes and shielding gases. The chemical composition of the electrode, the base metal, and the shielding gas determine the weld metal chemical composition. This weld metal composition in turn largely determines the chemical and mechanical properties of the weldment. The following are factors that influence the selection of the shielding gas and the welding electrode:

(1) Base metal
(2) Required weld metal mechanical properties
(3) Base metal condition and cleanliness
(4) Type of service or applicable specification requirement
(5) Welding position
(6) Intended mode of metal transfer

ELECTRODES

THE ELECTRODES (FILLER metals) for gas metal arc welding are covered by various AWS filler metal specifications. Other standards writing societies also publish filler metal specifications for specific applications. For example, the Aerospace Materials Specifications are written by SAE, and are intended for Aerospace applications. The AWS specifications, designated as A5.XX standards, and a listing of GMAW electrode specifications are shown in Table 4.4. They define requirements for sizes and tolerances, packaging, chemical composition, and sometimes mechanical properties. The AWS also publishes *Filler Metal Comparison Charts* in which manufacturer's may show their trade name for each of the filler metal classifications.

Generally, for joining applications, the composition of the electrode (filler metal) is similar to that of the base metal. The filler metal composition may be altered slightly to compensate for losses that occur in the welding arc, or to provide for deoxidation of the weld pool. In some cases, this involves very little modification from the base metal composition. In certain applications, however, obtaining satisfactory welding characteristics and weld metal properties requires an electrode with a different chemical composition from that of the base metal. For example, the most satisfactory electrode for GMAW welding manganese bronze, a copper-zinc alloy, is either aluminum bronze or a copper-manganese-nickel-aluminum alloy electrode.

Electrodes that are most suitable for welding the higher strength aluminum and steel alloys are often different in composition from the base metals on which they are to be used. This is because aluminum alloy compositions such as 6061 are unsuitable as weld filler metals. Accordingly, electrode alloys are designed to produce the desired weld metal properties and to have acceptable operating characteristics.

Whatever other modifications are made in the composition of electrodes, deoxidizers or other scavenging elements are generally added. This is done to minimize porosity in the weld or to assure satisfactory weld metal mechanical properties. The addition of appropriate deoxidizers in the right quantity is essential to the production of sound welds. Deoxidizers most commonly used in steel electrodes are manganese, silicon, and aluminum. Titanium and silicon are the principal deoxidizers used in nickel alloy electrodes. Copper alloy electrodes may be deoxidized with titanium, silicon, or phosphorus.

The electrodes used for GMAW are quite small in diameter compared to those used for submerged arc or flux cored arc welding. Wire diameters of 0.035 to 0.062 in. (0.9 to 1.6 mm) are common. However, electrode diameters as small as 0.020 in. (0.5 mm) and as large as 1/8 in. (3.2 mm) may be used. Because the electrode sizes are small and the currents comparatively high, GMAW wire feed rates are high. The rates range from approximately 100 to 800 in./min. (40 to 340 mm/s) for most metals except magnesium, where rates up to 1400 in./min. (590 mm/s) may be required.

For such wire speeds, electrodes are provided as long, continuous strands of suitably tempered wire that can be fed smoothly and continuously through the welding equipment. The wires are normally wound on conveniently sized spools or in coils.

The electrodes have high surface-to-volume ratios because of their relatively small size. Any drawing compounds or lubricants worked into the surface of the electrode may adversely affect the weld metal properties. These foreign materials may result in weld metal porosity in aluminum and steel alloys, and weld metal or heat-affected zone cracking in high-strength steels. Consequently, the electrodes should be manufactured with a high-quality

Table 4.4
Specifications for Various GMAW Electrodes

Base Material Type	AWS Specification
Carbon Steel	A5.18
Low Alloy Steel	A5.28
Aluminum Alloys	A5.10
Copper Alloys	A5.7
Magnesium	A5.19
Nickel Alloys	A5.14
300 Series Stainless Steel	A5.9
400 Series Stainless Steel	A5.9
Titanium	A5.16

surface to preclude the collection of contaminants in seams or laps.

In addition to joining, the GMAW process is widely used for surfacing where an overlayed weld deposit may provide desirable wear or corrosion resistance or other properties. Overlays are normally applied to carbon or manganese steels and must be carefully engineered and evaluated to assure satisfactory results. During surfacing, the weld metal dilution with the base metal becomes an important consideration; it is a function of arc characteristics and technique. With GMAW, dilution rates from 10 to 50 percent can be expected depending on the transfer mode. Multiple layers are normally required, therefore, to obtain suitable deposit chemistry at the surface. Most weld metal overlays are deposited automatically to precisely control dilution, bead width, bead thickness, and overlaps by placing each bead against the preceding bead.

SHIELDING GASES

GENERAL

THE PRIMARY FUNCTION of the shielding gas is to exclude the atmosphere from contact with the molten weld metal. This is necessary because most metals, when heated to their melting point in air, exhibit a strong tendency to form oxides and, to a lesser extent, nitrides. Oxygen will also react with carbon in molten steel to form carbon monoxide and carbon dioxide. These varied reaction products may result in weld deficiencies, such as trapped slag, porosity, and weld metal embrittlement. Reaction products are easily formed in the atmosphere unless precautions are taken to exclude nitrogen and oxygen.

In addition to providing a protective environment, the shielding gas and flow rate also have a pronounced effect on the following:

(1) Arc characteristics
(2) Mode of metal transfer
(3) Penetration and weld bead profile
(4) Speed of welding
(5) Undercutting tendency
(6) Cleaning action
(7) Weld metal mechanical properties

The principal gases used in GMAW are shown in Table 4.5. Most of these are mixtures of inert gases which may also contain small quantities of oxygen or CO_2. The use of nitrogen in welding copper is an exception. Table 4.6 lists gases used for short circuiting transfer GMAW.

THE INERT SHIELDING GASES—ARGON AND HELIUM

ARGON AND HELIUM are inert gases. These gases and mixtures of the two are used to weld nonferrous metals and stainless, carbon, and low alloy steels. The physical differences between argon and helium are density, thermal conductivity, and arc characteristics.

Argon is approximately 1.4 times more dense than air, while the density of helium is approximately 0.14 times that of air. The heavier argon is most effective at shielding the arc and blanketing the weld area in the flat position. Helium requires approximately two to three times higher flow rates than argon to provide equal protection.

Helium has a higher thermal conductivity than argon and produces an arc plasma in which the arc energy is more uniformly distributed. The argon arc plasma, on the other hand, is characterized by a high-energy inner core and an outer zone of less energy. This difference strongly affects the weld bead profile. A welding arc shielded by helium produces a deep, broad, parabolic weld bead. An arc shielded by argon produces a bead profile characterized by a "finger" type penetration. Typical bead profiles for argon, helium, argon-helium mixtures and carbon dioxide are illustrated in Figure 4.30.

Helium has a higher ionization potential than argon, and consequently, a higher arc voltage when other variables are held constant. Helium can also present problems in arc initiation. Arcs shielded only by helium do not exhibit true axial spray transfer at any current level. The result is that helium-shielded arcs produce more spatter and have rougher bead surfaces than argon-shielded arcs. Argon shielding (including mixtures with as low as 80 percent argon) will produce axial spray transfer when the current is above the transition current.

MIXTURES OF ARGON AND HELIUM

PURE ARGON SHIELDING is used in many applications for welding nonferrous materials. The use of pure helium is generally restricted to more specialized areas because an arc in helium has limited arc stability. However, the desirable weld profile characteristics (deep, broad, and parabolic) obtained with the helium arc are quite often the objective in using an argon-helium shielding gas mixture. The result, illustrated in Figure 4.30, is an improved weld bead profile plus the desirable axial spray metal transfer characteristic of argon.

In short circuiting transfer, argon-helium mixtures of from 60 to 90 percent helium are used to obtain higher heat input into the base metal for better fusion characteristics. For some metals, such as the stainless and low alloy steels, helium additions are chosen instead of CO_2 addi-

Table 4.5
GMAW Shielding Gases for Spray Transfer

Metal	Shielding Gas	Thickness	Advantages
Aluminum	100% Argon	0 to 1 in. (0 to 25 mm)	Best metal transfer and arc stability; least spatter.
	35% argon -65% helium	1 to 3 in. (25 to 76 mm)	Higher heat input than straight argon; improved fusion characteristics with 5XXX series Al-Mg alloys.
	25% argon -75% helium	Over 3 in. (76 mm)	Highest heat input; minimizes porosity
Magnesium	100% Argon	—	Excellent cleaning action.
Steel carbon	95% Argon +3.5% oxygen	—	Improves arc stability; produces a more fluid and controllable weld puddle; good coalescence and bead contour; minimizes undercutting; permits higher speeds than pure argon.
	90% Argon +8/10% carbon dioxide	—	High-speed mechanized welding; low-cost manual welding.
Steel low-alloy	98% Argon -2% oxygen	—	Minimizes undercutting; provides good toughness.
Steel stainless	99% Argon -1% oxygen	—	Improves arc stability; produces a more fluid and controllable weld puddle, good coalescence and bead contour; minimizes undercutting on heavier stainless steels.
	98% Argon -2% oxygen	—	Provides better arc stability, coalescence, and welding speed than 1 percent oxygen mixture for thinner stainless steel materials.
Nickel, copper, and their alloys	100% Argon	Up to 1/8 in. (3.2 mm)	Provides good wetting; decreases fluidity of weld metal.
	Argon -helium	—	Higher heat inputs of 50 & 75 percent helium mixtures offset high heat dissipation of heavier gages.
Titanium	100% Argon	—	Good arc stability; minimum weld contamination; inert gas backing is required to prevent air contamination on back of weld area.

Table 4.6
GMAW Shielding Gases for Short Circuiting Transfer

Metal	Shielding Gas	Thickness	Advantages
Carbon steel	75% argon +25% CO_2	Less than 1/8 in. (3.2 mm)	High welding speeds without burn-through; minimum distrotion and spatter.
	75% argon +25% CO_2	More than 1/8 in. (3.2 mm)	Minimum spatter; clean weld appearance; good puddle control in vertical and overhead positions.
	Argon with 5-10% CO_2	—	Deeper penetration; faster welding speeds.
Stainless steel	90% helium + 7.5% argon + 2.5% CO_2	—	No effect on corrosion resistance; small heat-affected zone; no undercutting; minimum distortion.
Low alloy steel	60-70% helium +25-35% argon +4.5% CO_2	—	Minimum reactivity; excellent toughness; excellent arc stability, wetting characteristics, and bead contour; little spatter.
	75% argon +25% CO_2	—	Fair toughness; excellent arc stability, wetting characteristics, and bead contour; little spatter.
Aluminum, copper magnesium, nickel, and their alloys	Argon & argon + helium	Over 1/8 in. (3.2 mm)	Argon satisfactory on sheet metal; argon-helium preferred base material.

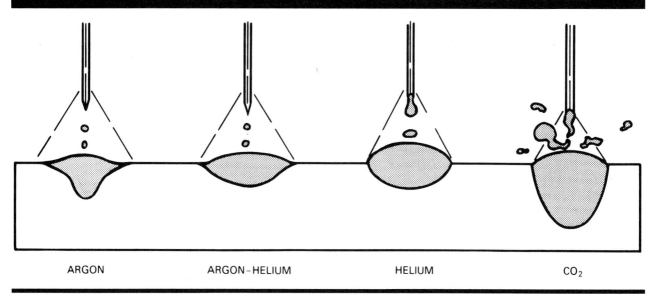

Figure 4.30–Bead Contour and Penetration Patterns for Various Shielding Gases

tions because CO_2 may adversely affect the mechanical properties of the deposit.

Mixtures of argon and 50 to 75 percent helium increase the arc voltage (for the same arc length) over that in pure argon. These gases are used for welding aluminum, magnesium, and copper because the higher heat input (from the higher voltage) reduces the effect of the high thermal conductivity of these base metals.

OXYGEN AND CO$_2$ ADDITIONS TO ARGON AND HELIUM

PURE ARGON AND, to a lesser extent, helium, produce excellent results in welding nonferrous metals. However, pure argon shielding on ferrous alloys causes an erratic arc and a tendency for undercut to occur. Additions to argon of from 1 to 5 percent oxygen or from 3 to 25 percent CO_2 produce a noticeable improvement in arc stability and freedom from undercut by eliminating the arc wander caused by cathode sputtering.

The optimum amount of oxygen or CO_2 to be added to the inert gas is a function of the work surface condition (presence of mill scale or oxides), the joint geometry, the welding position or technique, and the base metal composition. Generally, 2 percent oxygen or 8 to 10 percent CO_2 is considered a good compromise to cover a broad range of these variables.

Carbon dioxide additions to argon may also enhance the weld bead appearance by producing a more readily defined "pear-shaped" profile, as illustrated in Figure 4.31. Adding between 1 and 9 percent oxygen to the gas improves the

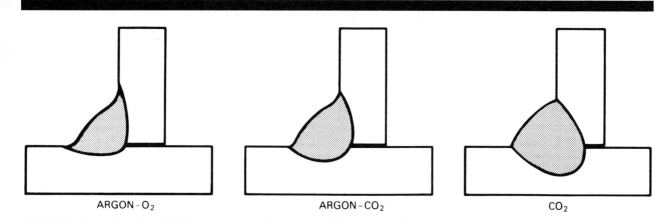

Figure 4.31–Relative Effect of Oxygen Versus Carbon Dioxide Additions to the Argon Shield

fluidity of the weld pool, penetration, and the arc stability. Oxygen also lowers the transition current. The tendency to undercut is reduced, but greater oxidation of the weld metal occurs, with a noticeable loss of silicon and manganese.

Argon-carbon dioxide mixtures are used on carbon and low alloy steels, and to a lesser extent on stainless steels. Additions of carbon dioxide up to 25 percent raise the minimum transition current, increase spatter loss, deepen penetration, and decrease arc stability. Argon - CO_2 mixtures are primarily used in short circuiting transfer applications, but are also usable in spray transfer and pulse arc welding.

A mixture of argon with 5 percent CO_2 has been used extensively for pulsed arc welding with solid carbon steel wires. Mixtures of argon, helium, and CO_2 are favored for pulsed arc welding with solid stainless steel wires.

MULTIPLE SHIELDING GAS MIXTURES

Argon-Oxygen-Carbon Dioxide

GAS MIXTURES OF argon with up to 20 percent carbon dioxide and 3 to 5 percent oxygen are versatile. They provide adequate shielding and desirable arc characteristics for spray, short circuiting, and pulse mode welding. Mixtures with 10 to 20 percent carbon dioxide are not in common use in the United States but are popular in Europe.

Argon-Helium-Carbon Dioxide

MIXTURES OF ARGON, helium, and carbon dioxide are used with short circuiting and pulse arc welding of carbon, low alloy, and stainless steels. Mixtures in which argon is the primary constituent are used for pulse arc welding, and those in which helium is the primary constituent are used for short circuiting arc welding.

Argon-Helium-Carbon Dioxide-Oxygen

THIS MIXTURE, commonly referred to as *quad mix*, is popular for high-deposition GMAW using the high-current-density metal transfer type arc. This mixture will give good mechanical properties and operability throughout a wide range of deposition rates. Its major application is welding low alloy, high-tensile base materials, but it has been used on mild steel for high-production welding. Weld econonmics are an important consideration in using this gas for mild steel welding.

CARBON DIOXIDE

CARBON DIOXIDE (CO_2) is a reactive gas widely used in its pure form for gas metal arc welding of carbon and low alloy steels. It is the only reactive gas suitable for use alone as a shield in the GMAW process. Higher welding speed, greater joint penetration, and lower cost are general characteristics which have encouraged extensive use of CO_2 shielding gas.

With a CO_2 shield, the metal transfer mode is either short circuiting or globular. Axial spray transfer requires an argon shield and cannot be achieved with a CO_2 shield. With globular transfer, the arc is quite harsh and produces a high level of spatter. This requires that CO_2 welding conditions be set to provide a very short "buried arc" (the tip of the electrode is actually below the surface of the work), in order to minimize spatter.

In overall comparison to the argon-rich shielded arc, the CO_2 shielded arc produces a weld bead of excellent penetration with a rougher surface profile and much less "washing" action at the sides of the weld bead, due to the buried arc. Very sound weld deposits are achieved, but mechanical properties may be adversely affected due to the oxidizing nature of the arc.

APPLICATIONS

GMAW CAN BE used on a wide variety of metals and configurations. Its successful application is dependent on proper selection of the following:

(1) Electrode - composition, diameter, and packaging
(2) Shielding gas and flow rate
(3) Process variables, including amperage, voltage, travel speed, and mode of transfer
(4) Joint design
(5) Equipment, including power source, gun, and wire feeder

ELECTRODE SELECTION

IN THE ENGINEERING of weldments, the objective is to select filler metals that will produce a weld deposit with two basic characteristics:

(1) A deposit that either closely matches the mechanical and physical properties of the base metal or provides some enhancement to the base material, such as corrosion or wear resistance
(2) A sound weld deposit, free from discontinuities

In the first case, a weld deposit, even one with composition nearly identical to the base metal, will possess unique metallurgical characteristics. This is dependent on factors such as the energy input and weld bead configuration. The second characteristic is generally achieved through use of a formulated filler metal electrode, e.g., one containing deoxidizers that produce a relatively defect-free deposit.

Composition

THE ELECTRODE MUST meet certain demands of the process regarding arc stability, metal transfer behavior, and solidification characteristics. It must also provide a weld deposit that is compatible with one or more of the following base metal characteristics:

(**1**) Chemistry
(**2**) Strength
(**3**) Ductility
(**4**) Toughness

Consideration should be given to other properties such as corrosion, heat-treatment response, wear resistance, and color match. All such considerations, however, are secondary to the metallurgical compatibility of the base metal and the filler metal.

American Welding Society specifications have been established for filler metals in common usage. Table 4.7 provides a basic guide to selecting appropriate filler metal types for the listed base metals, along with each applicable AWS filler metal specification.

Tubular Wires

BOTH SOLID AND tubular wires are used with GMAW. The tubular wires have a powdered metallic core which includes small amounts of arc stabilizing compounds. These wires have good arc stability and deposition efficiencies similar to a solid wire. This tubular approach permits the manufacture of low-slag, high-efficiency metallic electrodes in compositions which would be difficult to manufacture as a solid wire.

SHIELDING GAS SELECTION

AS NOTED IN earlier sections, the shielding gas used for the gas metal arc process can be inert (argon or helium), reactive (CO_2), or a mixture of the two types. Additions of oxygen and sometimes hydrogen can be made to achieve other desired arc characteristics and weld bead geometries.

The selection of the best shielding gas is based on consideration of the material to be welded and the type of metal transfer that will be used. For spray arc transfer, Table 4.5 lists the more commonly used shielding gases for various materials. Table 4.6 lists those gases used with the short circuiting mode of transfer. These tables do not list all the special gas combinations that are available.

SETTING PROCESS VARIABLES

THE SELECTION OF the process parameters (amperage, voltage, travel speed, gas flow rate, electrode extension, etc.) requires some trial and error to determine an acceptable set of conditions. This is made more difficult because of the interdependence of several of the variables. Typical ranges of seven variables have been established and are listed in Tables 4.8 through 4.13 for various base metals.

SELECTION OF JOINT DESIGN

TYPICAL WELD JOINT designs and dimensions for the GMAW process, as used in the welding of steel, are shown in Figure 4.32. The dimensions indicated will generally produce complete joint penetration and acceptable reinforcement with suitable welding procedures.

Similar joint configurations may be used on other metals, although the more thermally conductive types (e.g. aluminum and copper) should have larger groove angles to minimize problems with incomplete fusion.

The deep penetration characteristics of spray transfer GMAW may permit the use of smaller included angles. This reduces the amount of filler metal required and labor hours to fabricate weldments.

EQUIPMENT SELECTION

WHEN SELECTING EQUIPMENT, the buyer must consider application requirements, range of power output, static and dynamic characteristics, and wire feed speeds. If a major part of the production involves small diameter aluminum wire, for example, the fabricator should consider a push-pull type of wire feeder. If out-of-position welding is contemplated, the user should look into pulsed power welding machines. For the welding of thin gage stainless steel, a power supply with adjustable slope and inductance may be considered.

When new equipment is to be purchased, some consideration should be given to the versatility of the equipment and to standardization. Selection of equipment for single-purpose or high-volume production can generally be based upon the requirements of that particular application only. However, if a multitude of jobs will be performed (as in job shop operation), many of which may be unknown at the time of selection, versatility is very important.

Other equipment already in use at the facility should be considered. Standardizing certain components and complementing existing equipment will minimize inventory requirements and provide maximum efficiency of the overall operation. Details of equipment components are provided in earlier sections of this chapter.

Table 4.7
Recommended Electrodes for GMAW

Base Material		Electrode Classification	AWS Electrode Specification (Use latest edition)
Type	**Classification**		
Aluminum and aluminum alloys (ASTM Standards Volume 2.02)	1100	ER4043	A5.10
	3003, 3004	ER5356	
	5052, 5454	ER5554, ER5556, or ER5183	
	5083, 5086, 5456	ER5556 or ER5356	
	6061, 6063	ER4043 or ER5356	
Magnesium alloys (ASTM Standards Volume 2.02)	AZ10A	ERAZ61A, ERAZ92A	A5.19
	AZ31B, AZ61A, AZ80A	ERAZ61A, ERAZ92A	
	ZE10A	ERAZ61A, ERAZ92A	
	ZK21A	ERAZ92A	
	AZ63A, AZ81A, AZ91C	EREZ33A	
	AZ92A, AM100A	EREZ33A	
	HK31A, HM21A, HM31A	EREZ33A	
	LA141A	EREZ33A	
Copper and copper alloys (ASTM Standards Volue 2.01)	Commercially pure	ERCu	A5.7
	Brass	ERCuSi-A, ERCuSn-A	
	Cu-Ni alloys	ERCuNi	
	Manganese bronze	ERCuAl-A2	
	Aluminum bronze	ERCuAl-A2	
	Bronze	ERCuSn-A	
Nickel and nickel alloys (ASTM Standards Volume 2.04)	Commercially pure	ERNi	A5.14
	Ni-Cu alloys	ERNiCu-7	
	Ni-Cr-Fe alloys	ERNiCrFe-5	
Titanium and titanium alloys (ASTM Standards Volume 2.04)	Commercially pure	ERTi-1,-2,-3,-4	A5.16
	Ti-6 AL-4V	ERTi-6Al-4V	
	Ti-0.15Pd	ERTi-0.2Pd	
	Ti-5Al-2 5Sn	ERTi-5Al-2.5Sn	
	Ti-13V-11Cr-3AL	ERTi-13V-11Cr-3AL	
Austenitic stainless steels (ASTM Standards Volume 1.04)	Type 201	ER308	A5.9
	Types 301, 302, 304 & 308	ER308	
	Type 304L	ER308L	
	Type 310	ER310	
	Type 316	ER316	
	Type 321	ER321	
	Type 347	ER347	
Carbon steels	Hot and cold rolled plain carbon steels	E70S-3, or E70S-1	A5.18
		E70S-2, E70S-4	
		E70S-5, E70S-6	

Table 4.8
Typical Conditions for Gas Metal Arc Welding of Carbon and Low Alloy Steel in the Flat Position

Material Thickness			Wire Diameter		Current Voltage[1]		Wire Feed Speed			Gas Flow	
in.	mm	Type of Weld	in.	mm	amps	volts	IPM	mm/s	Shielding Gas[2]	CFH	LPM
.062	1.6	Butt[3]	.035	0.9	95	18	150	64	Ar 75%, CO_2 -25%	25	12
.125	3.2	Butt[3]	.035	0.9	140	20	250	106	Ar 75%, CO_2 25%	25	12
.187	4.7	Butt[3]	.035	0.9	150	20	265	112	Ar 75%, CO_2 25%	25	12
.250	6.4	Butt[3]	.035	0.9	150	21	265	112	Ar 75%, CO_2 25%	25	12
.250	6.4	Butt[4]	.045	1.1	200	22	250	106	Ar 75%, CO_2 -25%	25	12

1. Direct current electrode positive.
2. Welding grade CO_2 may also be used.
3. Root opening of .03 in. (0.8 mm).
4. Root opening of .062 in. (1.6 mm).

Table 4.9
Typical Conditions for Gas Metal Arc Welding of Aluminum in the Flat Position

Material Thickness			Wire Diameter		Current Voltage*		Wire Feed Speed		Shielding Gas	Gas Flow	
in.	mm	Type of Weld	in.	mm	amps	volts	IPM	mm/s		CFH	LPM
.062	1.6	Butt	.030	0.8	90	18	365	155	Argon	30	14
.125	3.2	Butt	.030	0.8	125	20	440	186	Argon	30	14
.187	4.8	Butt	.045	1.1	160	23	275	116	Argon	35	16
.250	6.4	Butt	.045	1.1	205	24	335	142	Argon	35	16
.375	9.5	Butt	.063	1.6	240	26	215	91	Argon	40	19

* Direct current electrode positive.

Table 4.10
Typical Conditions for Gas Metal Arc Welding of Austenitic Stainless Steel Using a Spray Arc in the Flat Position

Material Thickness			Wire Diameter		Current Voltage[1]		Wire Feed Speed		Shielding Gas	Gas Flow	
in.	mm	Type of Weld	in.	mm	amps	volts	IPM	mm/s		CFH	LPM
.125	3.2	Butt Joint with Backing	.062	1.6	225	24	130	55	Ar 98%, O_2 2%	30	14
.250(1)	6.4	V-Butt Joint 60ø Inc. Angle	.062	1.6	275	26	175	74	Ar 98%, O_2 2%	35	16
.375(1)	9.5	V-Butt Joint 60ø Inc. Angle	.062	1.6	300	28	240	102	Ar 98%, O_2 2%	35	16

1. Direct current electrode positive.
2. Two passes required.

Table 4.11
Typical Conditions for Gas Metal Arc Welding of Austenitic Stainless Steel Using a Short Circuiting Arc

Material Thickness		Type of Weld	Wire Diameter		Current	Voltage*	Wire Feed Speed		Shielding Gas	Gas Flow	
in.	mm		in.	mm	amps	volts	IPM	mm/s		CFH	LPM
.062	1.6	Butt Joint	.030	0.8	85	21	185	78	He 90%, Ar 7.5% CO_2 2.5%	30	14
.093	2.4	Butt Joint	.030	0.8	105	23	230	97	He 90%, Ar 7.5% CO_2 2.5%	30	14
.125	3.2	Butt Joint	.030	0.8	125	24	280	118	He 90%, Ar 7.5% CO_2 2.5%	30	14

* Direct current electrode positive.

Table 4.12
Typical Conditions for Gas Metal Arc Welding of Copper Alloys in the Flat Position

Material Thickness		Type of Weld	Wire Diameter		Current	Voltage*	Wire Feed Speed		Shielding Gas	Gas Flow	
in.	mm		in.	mm	amps	volts	IPM	mm/s		CFH	LPM
.125	3.2	Butt	.035		175	23	430	182	Argon	25	12
.187	4.8	Butt	.045		210	25	240	101	Argon	30	14
.250	6.4	Butt, Spaced	.062		365	26	240	101	Argon	35	16

* Direct current electrode positive.

Table 4.13
Typical Variable Settings for Gas Metal Arc Welding of Magnesium

Material Thickness		Type of Weld	Wire Diameter		Current	Voltage*	Wire Feed Speed		Argon Flow	
in.	mm		in.	mm	amps	volts	IPM	mm/s	CFH	LPM
.062	1.6	Square Groove or Fillet	.062	1.6	70	16	160	68	50	24
.090	2.3	Square Groove or Fillet	.062	1.6	105	17	245	104	50	24
.125	3.2	Square Groove or Fillet	.062	1.6	125	18	290	123	50	24
.250	6.4	Square Groove or Fillet	.062	1.6	265	25	600	254	60	28
.375	9.5	Square Groove or Fillet	.094	2.4	335	26	370	157	60	28

* Direct current electrode positive.

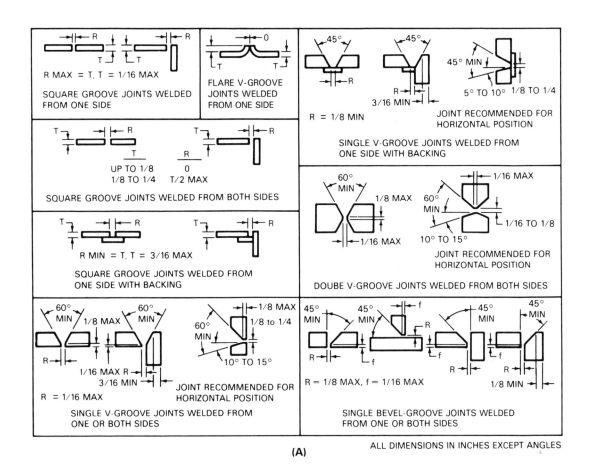

(A)

ALL DIMENSIONS IN INCHES EXCEPT ANGLES

Figure 4.32–Typical Weld Joint Designs and Dimensions for the Gas Metal Arc Welding Process

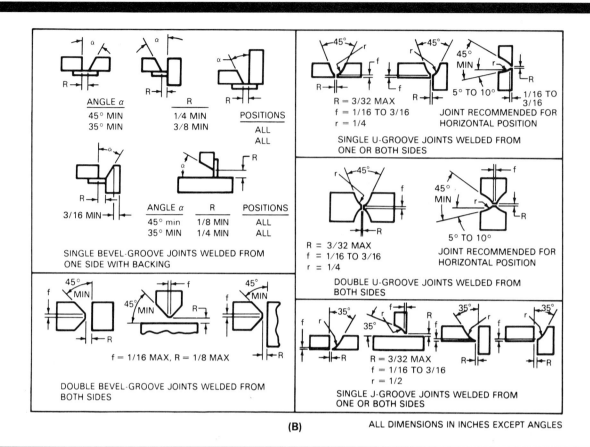

(B)   ALL DIMENSIONS IN INCHES EXCEPT ANGLES

Figure 4.32–Typical Weld Joint Designs and Dimensions for the Gas Metal Arc Welding Process

SPECIAL APPLICATIONS

SPOT WELDING

GAS METAL ARC spot welding is a variation of continuous GMAW wherein two pieces of sheet metal are fused together by penetrating entirely through one piece into the other. The process has been used for joining light-gage materials, up to approximately 3/16 in. (5 mm) thick, in the production of automobile bodies, appliances, and electrical enclosures. No joint preparation is required other than cleaning of the overlap areas. Heavier sections can also be spot welded with this technique by drilling or punching a hole in the upper piece, through which the arc is directed for joining to the underlying piece. This is called *plug welding*.

A comparison between a gas metal arc spot weld and a resistance spot weld is shown in Figure 4.33. Resistance spot welds are made through resistance heating and electrode pressure which melts the two components at their interface and fuses them together. In the gas metal arc spot weld, the arc penetrates through the top member and fuses the bottom component into its weld puddle. One big advantage of the gas metal arc spot weld is that access to only one side of the joint is necessary.

The spot weld variation does require some modifications to conventional GMAW equipment. Special nozzles are used which have ports to allow the shield gas to escape as the torch is pressed to the work. Timers and wire feed speed controls are also necessary, to provide regulation of the actual welding time and a current decay period to fill the weld crater, leaving a desirable reinforcement contour.

Joint Design

GAS METAL ARC spot welding may be used to weld lap joints in carbon steel, aluminum, magnesium, stainless steel, and copper-bearing alloys. Metals of the same or different thicknesses may be welded together, but the thinner sheet should always be the top member when different thickness are welded. Gas metal arc spot welding is normally restricted to the flat position. By modifying the nozzle design, it may be adapted to spotweld lap-fillet, fillet, and corner joints in the horizontal position.

Equipment Operation

THE SPOT WELDING GMAW gun is placed in position, pressing the workpieces together. The gun's trigger is de-

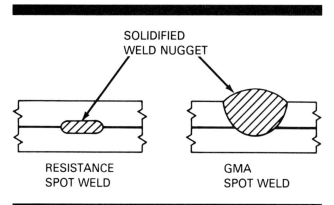

SOLIDIFIED
WELD NUGGET

RESISTANCE
SPOT WELD

GMA
SPOT WELD

**Figure 4.33–Comparison of Gas Metal Arc and
Resistance Spot Welds**

pressed to initiate the arc. The arc timer is started by a device that senses flow of welding current. The arc is maintained by the continuously-fed consumable electrode until it melts through the top sheet and fuses into the bottom sheet without gun travel. The time cycle is set to maintain an arc until the melt-through and fusing sequence is complete, i.e., until a spot weld has been formed. The electrode continues to feed during the arc cycle and should produce a reinforcement on the upper surface of the top sheet.

Effect of Process Variables on Weld Characteristics

THE WELD DIAMETER at the interface and the reinforcement are the two characteristics of a GMAW spot weld which determine whether the weld will satisfy the intended service. Three major process variables - weld current, voltage, and arc time - affect one or both of these characteristics.

Current.   Current has the greatest effect on penetration. Penetration is increased by using higher currents (with corresponding increase in wire feed speed). Increased penetration will generally result in a larger weld diameter at the interface.

Arc Voltage.   Arc voltage has the greatest effect on the spot weld shape. In general, with current being held constant, an increase in the arc voltage will increase the diameter of the fusion zone. However, it also causes a slight decrease in the reinforcement height and penetration. Welds made with arc voltages that are too low show a depression in the center of the reinforcement. Arc voltages that are too high create heavy spatter conditions.

Weld Time.   Welding conditions should be selected that produce a suitable weld within a time of 20 to 100 cycles of 60 Hz current (0.3 to 1.7 seconds) to join base metal up to 0.125 in. (3.2 mm) thick. Arc time up to 300 cycles (5 seconds) may be necessary on thicker materials to achieve adequate strength. The penetration, weld diameter, and reinforcement height generally increase with increased weld time.

As with conventional GMAW, the parameters for spot welding are very interdependent. Changing one usually requires changing one or more of the others. Some trial and error is needed to find a set or sets of conditions for a particular application. "Starting" parameters for gas metal arc spot welding of carbon steel are shown in Table 4.14.

NARROW GROOVE WELDING

NARROW GROOVE WELDING is a multipass technique for joining heavy section materials where the weld joint has a nearly square butt configuration with a minimal groove width [approximately 1/2 in. (13 mm)]. A typical narrow groove joint configuration is shown in Figure 4.34. The technique is used with many of the conventional welding processes, including GMAW, and is an efficient method of joining heavy section carbon and low alloy steels, with minimal distortion.

Using GMAW to weld joints in the narrow groove configuration requires special precautions to assure that the tip of the electrode is positioned accurately for proper fusion into the sidewalls. Numerous wire feeding methods for accomplishing this have been devised and successfully used in production environment. Examples of some of these are shown in Figure 4.35.

Two wires with controlled cast and two contact tubes are used in tandem, as shown in Figure 4.35(A). The arcs are directed toward each sidewall, producing a series of overlapping fillet welds.

The same effect can be achieved with one wire by means of a weaving technique, which involves oscillating the arc across the groove in the course of welding. This oscillation can be created mechanically by moving the contact tube across the groove (Figure 4.35(B)), but, because of the small contact tube-to-sidewall distance, this technique is not practical and is seldom used.

Another mechanical technique uses a contact tube bent to an angle of about 15 degrees - Figure 4.35(C). Along with a forward motion during welding, the contact tube twists to the right and left, which gives the arc a weaving motion.

A more sophisticated technique is illustrated in Figure 4.34(D). During feeding, this electrode is formed into a waved shape by the bending action of a "flapper plate" and feed rollers as they rotate. The wire is continuously deformed plastically into this waved shape, as the feed rollers press it against the bending plate. The electrode is almost straightened while going through the contact tube and tip, but recovers its waviness after passing through the tip. Continuous consumption of the waved electrode oscillates the arc from one side of the groove to the other. This technique produces an oscillating arc even in a very narrow groove, with the contact tube remaining centered in the joint.

The twist electrode technique, Figure 4.35(E), is another means that has been developed to improve sidewall penetration without moving the contact tube. The twist electrode consists of two intertwined wires which, when fed into the groove, generate arcs from the tips of the two wires. Due to the twist, the arcs describe a continuous rotational movement which increases penetration into the sidewall without any special weaving device.

Because these arc oscillation techniques often require special feeding equipment, an alternate method has been developed in which a larger diameter electrode [e.g. .093 to .125 in. (2.4 to 3.2 mm)] is fed directly into the center of the groove from a contact tip situated above the plate surface. With this technique, the wire placement is still critical, but there is less chance of arcing between the contact tube and the work, and standard welding equipment can be used. It does, however, have a more limited thickness potential and is normally restricted to the flat position.

The parameters for narrow groove welding are very similar to those used for conventional GMAW. A summary of some typical values is shown in Table 4.15. For the narrow groove application, however, the quality of the results is sensitive to slight changes in these parameters, voltage being particularly important. An excessive arc voltage (arc length) can cause undercut of the sidewall, resulting in oxide entrapment or lack of fusion in subsequent passes. High voltage may cause the arc to climb the sidewall and damage the contact tube. For this reason, pulsing power supplies have become widely used in this application. They can maintain a stable spray arc at low arc voltages.

Various shielding gases have been used with the narrow gap technique, as with conventional GMAW. A gas consisting of argon with 20 to 25 percent CO_2 has seen the widest application because it provides a good combination of arc characteristics, bead profile, and sidewall penetration. Delivering the shielding gas to the weld area is a challenge in the narrow groove configuration, and numerous nozzle designs have been developed.

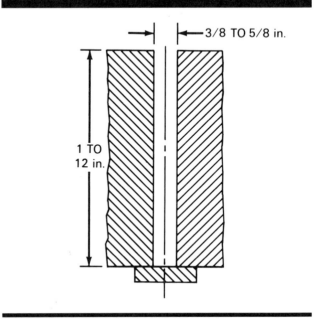

Figure 4.34–Typical Joint Configuration for Narrow Gap Welding

Table 4.14
Variable Settings for GMAW Spot Welding of Carbon Steel in the Flat Position (CO_2 Shielding Gas - 1/4 in. (6.4 mm) Diameter Nugget)

Electrode Size		Thickness			Arc Spot Time	Current Voltage*	
in.	mm	Gauge	in.	mm	s	A	V
0.030	0.8	24	0.022	0.56	1	90	24
		22	0.032	0.81	1.2	120	27
		20	0.037	0.94	1.2	120	27
0.035	0.9	18	0.039	0.99	1	190	27
		16	0.059	1.50	2	190	28
		14	0.072	1.83	5	190	28
0.045	1.2	14	0.072	1.83	1.5	300	30
		12	0.110	2.79	3.5	300	30
		11	0.124	3.15	4.2	300	30
0.063	1.6	11	1/8	3.15	1	490	32
			5/32	4.0	1.5	490	32

* Direct current electrode positive.

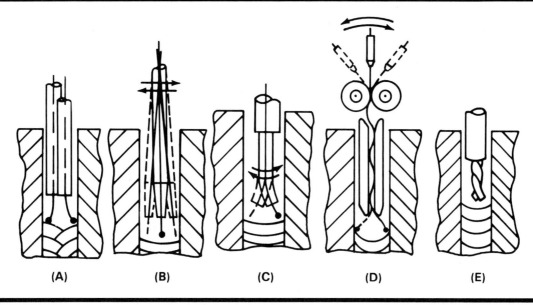

(A)   (B)   (C)   (D)   (E)

Figure 4-35–Typical Wire Feeding Techniques for Narrow Gap Gas Metal Arc Welding

Table 4.15
Typical Welding Conditions for the GMAW-NG Process

Technique, Weld Position	Groove Width, in.	mm	Current Amp	Voltage Volts[1]	Travel Speed, in./min.	mm/s	Gas Shield
NGW-I horiz.	0.375	9.5	260-270	25-26	40	17	Ar-CO_2
NGW-I horiz.	0.4-0.5	10-12	220-240	24-28[2]	13	6	Ar-CO_2
NGW-I flat	0.375	9.5	280-300	29[2]	9	4	Ar-CO_2
NGW-II flat	0.5	12.5	450	30-37.5	15	6	Ar-CO_2
NGW-II flat	0.47-0.55	12-14	450-550	38-42	20	8	Ar-CO_2

1. Direct current electrode positive.

2. Pulse power at 120 pulses per second.

INSPECTION AND WELD QUALITY

INTRODUCTION

WELD QUALITY CONTROL procedures for GMAW joints are quite similar to those used for other processes. Depending upon the applicable specifications, inspection procedures should provide for determining the adequacy of welder and welding operator performance, qualification of a satisfactory welding procedure, and making a complete examination of the final weld product.

Weld inspection on the assembled product is limited to nondestructive examination methods such as visual, liquid penetrant, magnetic particle, radiographic, and ultrasonic inspection. Destructive testing (tensile, shear, fatigue, impact, bend, fracture, peel, cross section, or hardness tests) is usually confined to engineering development, welding procedure qualification, and welder and welding operator performance qualification tests.

POTENTIAL PROBLEMS

Hydrogen Embrittlement

AN AWARENESS OF the potential problems of hydrogen embrittlement is important, even though it is less likely to occur with GMAW, since no hygroscopic flux or coating is used. However, other hydrogen sources must be considered. For example, shielding gas must be sufficiently low in moisture content. This should be well controlled by the gas supplier, but may need to be checked. Oil, grease, and drawing compounds on the electrode or the base metal may become potential sources for hydrogen pick-up in the weld metal. Electrode manufacturers are aware of the need for cleanliness and normally take special care to provide a clean electrode. Contaminants may be introduced during handling in the user's facility. Users who are aware of such possibilities take steps to avoid serious problems, particularly in welding hardenable steels. The same awareness is necessary in welding aluminum, except that the potential problem is porosity caused by the relatively low solubility of hydrogen in solidified aluminum, rather than hydrogen embrittlement.

Oxygen and Nitrogen Contamination

OXYGEN AND NITROGEN are potentially greater problems than hydrogen in the GMAW process. If the shielding gas is not completely inert or adequately protective, these elements may be readily absorbed from the atmosphere. Both oxides and nitrides can reduce weld metal notch toughness. Weld metal deposited by GMAW is not as tough as weld metal deposited by gas tungsten arc welding. It should be noted here, however, that oxygen in percentages of up to 5 percent and more can be added to the shielding gas without adversely affecting weld quality.

Cleanliness

BASE METAL CLEANLINESS when using GMAW is more critical than with SMAW or submerged arc welding (SAW). The fluxing compounds present in SMAW and SAW scavenge and cleanse the molten weld deposit of oxides and gas-forming compounds. Such fluxing slags are not present in GMAW. This places a premium on doing a thorough job of preweld and interpass cleaning. This is particularly true for aluminum, where elaborate procedures for chemical cleaning or mechanical removal of metallic oxides, or both, are applied.

Incomplete Fusion

THE REDUCED HEAT input common to the short circuiting mode of GMAW results in low penetration into the base metal. This is desirable on thin gauge materials and for out-of-position welding. However, an improper welding technique may result in incomplete fusion, especially in root areas or along groove faces.

WELD DISCONTINUITIES

SOME OF THE more common weld discontinuities that may occur with the GMAW process are listed in the following paragraphs.

Undercutting

THE FOLLOWING ARE possible causes of undercutting and their corrective actions (see Figure 4.36):

Possible Causes	Corrective Actions
(1) Travel speed too high	Use slower travel speed.
(2) Welding voltage too high	Reduce the voltage.
(3) Excessive welding current	Reduce wire feed speed.
(4) Insufficient dwell	Increase dwell at edge of molten weld puddle.
(5) Gun angle	Change gun angle so arc force can aid in metal placement.

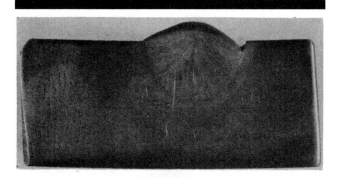

Figure 4.36–Undercutting at the Toe of the Weld

Porosity

THE FOLLOWING ARE possible causes of porosity and their corrective actions:

Figure 4.37–Porosity Resulting from Inadequate Shielding Gas Coverage.

Possible Causes	Corrective Actions
(1) Inadequate shielding gas coverage (see Figure 4.37)	Optimize the gas flow. Increase gas flow to displace all air from the weld zone. Decrease excessive gas flow to avoid turbulence and the entrapment of air in the weld zone. Eliminate any leaks in the gas line. Eliminate drafts (from fans, open doors, etc.) blowing into the welding arc. Eliminate frozen (clogged) regulators in CO_2 welding by using heaters. Reduce travel speed. Reduce nozzle-to-work distance. Hold gun at end of weld until molten metal solidifies.
(2) Gas contamination	Use welding grade shielding gas.
(3) Electrode contamination	Use only clean and dry electrode.
(4) Workpiece contamination	Remove all grease, oil, moisture, rust, paint, and dirt from work surface before welding. Use more highly deoxidizing electrode.
(5) Arc voltage too high	Reduce voltage.
(6) Excess contact tube-to-work distance	Reduce stick-out.

Incomplete Fusion

THE FOLLOWING ARE possible causes of incomplete fusion and their corrective actions:

Possible Causes	Corrective Actions
(1) Weld zone surfaces not free of film or excessive oxides	Clean all groove faces and weld zone surfaces of any mill scale impurities prior to welding.
(2) Insufficient heat input	Increase the wire feed speed and the arc voltage. Reduce electrode extension.
(3) Too large a weld puddle	Minimize excessive weaving to produce a more controllable weld puddle. Increase the travel speed.
(4) Improper weld technique	When using a weaving technique, dwell momentarily on the side walls of the groove. Provide improved access at root of joints. Keep electrode directed at the leading edge of the puddle.
(5) Improper joint design (see Figure 4.38)	Use angle groove large enough to allow access to bottom of the groove

and sidewalls with proper electrode extension, or use a "J" or "U" groove.

(6) Excessive travel speed — Reduce travel speed.

Incomplete Joint Penetration

THE FOLLOWING ARE possible causes of incomplete joint penetration and their corrective actions:

Possible Causes	Corrective Actions
(1) Improper joint preparation	Joint design must provide proper access to the bottom of the groove while maintaining proper electrode extension. Reduce excessively large root face. Increase the root gap in butt joints, and increase depth of back gouge.
(2) Improper weld technique	Maintain electrode angle normal to work surface to achieve maximum penetration. Keep arc on leading edge of the puddle.
(3) Inadequate welding current (see Figure 4.39)	Increase the wire feed speed (welding current).

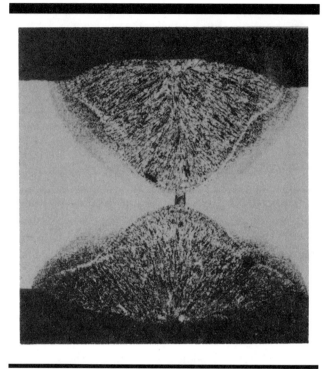

Figure 4.39–Incomplete Penetration

Excessive Melt-Through

THE FOLLOWING ARE possible causes of excessive melt-through and their corrective actions:

Possible Causes	Corrective Actions
(1) Excessive heat input	Reduce wire feed speed (welding current) and the voltage. Increase the travel speed.
(2) Improper joint penetration	Reduce root opening. Increase root face dimension.

Weld Metal Cracks

THE FOLLOWING ARE all possible causes of weld metal cracks and their corrective actions:

Possible Causes	Corrective Actions
(1) Improper joint design	Maintain proper groove dimensions to allow deposition of adequate filler metal or weld cross section to overcome restraint conditions.

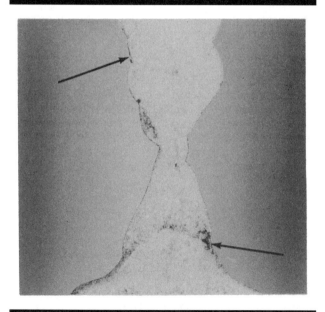

Figure 4.38–Incomplete Fusion Due to Narrow Groove Preparation

(2) Too high a weld depth-to-width ratio (see Figure 4.40)

Either increase arc voltage or decrease the current or both to widen the weld bead or decrease the penetration. Decrease travel speed to increase cross section of deposit.

(3) Too small a weld bead (particularly fillet and root beads)

(4) Heat input too high, causing excessive shrinkage and distortion

Reduce either current or voltage, or both. Increase travel speed.

(5) Hot-shortness

Use electrode with higher manganese content (use shorter arc length to minimize loss of manganese across the arc). Adjust the groove angle to allow adequate percentage of filler metal addition. Adjust pass sequence to reduce restraint on weld during cooling. Change to another filler metal providing desired characteristics.

(6) High restraint of the joint members

Use preheat to reduce magnitude of residual stresses. Adjust welding sequence to reduce restraint conditions.

(7) Rapid cooling in the crater at the end of the joint (see Figure 4.41)

Eliminate craters by backstepping technique.

Heat-Affected Zone Cracks

CRACKING IN THE heat-affected zone is almost always associated with hardenable steels.

Possible Causes

(1) Hardening in the heat-affected zone
(2) Residual stresses too high
(3) Hydrogen embrittlement

Corrective Actions

Preheat to retard cooling rate.
Use stress relief heat treatment.
Use clean electrode and dry shielding gas. Remove contaminants from the base metal. Hold weld at elevated temperatures for several hours before cooling (temperature and time required to diffuse hydrogen are dependent on base metal type).

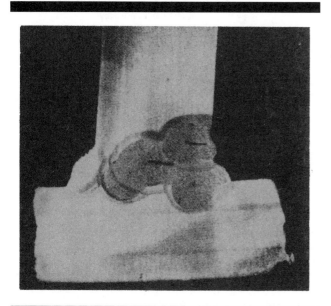

Figure 4.40–Too High a Weld Depth-to-Width Ratio

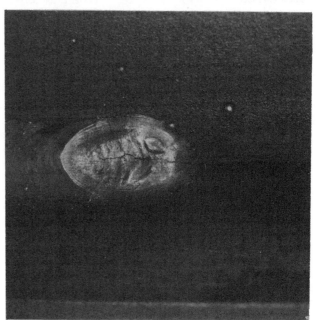

Figure 4.41–Weld Metal Cracking in Crater at the End of a Weld

TROUBLESHOOTING

TROUBLE SHOOTING OF any process requires a thorough knowledge of the equipment and the function of the various components, the materials involved, and the process itself. It is a more complicated task with gas metal arc than with manual processes such as SMAW and GTAW because of the complexity of the equipment, the number of variables and the inter-relationship of these variables.

For convenience, problems can be placed in one of the following three categories: electrical, mechanical, and process.

Tables 4.16 through 4.18 indicate some of the problems that are likely to be encountered, what the causes might be, and possible remedies. These are problems that occur during the welding operation or prevent the making of the weld as opposed to those that are discovered as a result of inspecting the final product. This latter type are covered in the "Inspection and Weld Quality" section of this chapter.

Table 4.16
Troubleshooting Electrical Problems Encountered in Gas Metal Arc Welding

Problem	Possible Cause	Remedy
Difficult arc starting	Wrong polarity	Check polarity; reverse leads if necessary.
	Poor work lead connection	Secure work lead connection
Irregular wire feed and burnback	Power circuit fluctuations	Check line voltage
	Polarity wrong	Check polarity; reverse leads if necessary
Welding cables overheating	Cables are too small or too long	Check current carrying requirements - replace or shorten if necessary
	Cable connections loose	Tighten
No wire feed speed control	Broken or loose wires in control circuit	Check and repair if necessary
	Bad P.C. board in governor	Replace P.C. board
Unstable arc	Cable connections are loose	Tighten connections
Electrode won't feed	Control circuit fuse blown	Replace fuse
	Fuse blown in power source	Replace fuse
	Defective gun trigger switch or broken wire leads	Check connections; replace switch
	Drive motor burned out	Check and replace
Wire feeds but no gas flows	Failure of gas valve solenoid	Replace
	Loose or broken wires to gas valve solenoid	Check and repair if necessary
Electrode wire feeds but is not energized (no arc)	Poor workpiece connection	Tighten if loose; clean work of paint, rust, etc.
	Loose cable connections	Tighten
	Primary contactor coil or points defective	Repair or replace
	Contactor control leads broken	Repair or replace
Porosity in weld	Loose or broken wires to gas solenoid valve	Repair or replace

Table 4.17
Troubleshooting Mechanical Problems Encountered in Gas Metal Arc Welding

Problem	Possible Cause	Remedy
Irregular wire feed and burnback	Insufficient drive roll pressure	Adjust
	Contact tube plugged or worn	Clean or replace
	Kinked electrode wire	Cut out, replace spool
	Coiled gun cable	Straighten cables, hang the wire feeder
	Conduit liner dirty or worn	Clean or replace
	Conduit too long	Shorten or use push-pull drive system
Electrode wire wraps around drive roll (•birdnesting•)	Excessive feed roll pressure	Adjust
	Incorrect conduit liner or contact tip	Match liner and contact tip to electrode size
	Misaligned drive rolls or wire guides	Check and align properly
	Restriction in gun or gun cable	Remove restriction
Heavily oxidized weld deposit	Air/water leaks in gun and cables	Check for leaks and repair or replace as necessary
	Restricted shield gas flow	Check and clean nozzle
Electrode wire stops feeding while welding	Excess or insufficient drive roll pressure	Adjust
	Wire drive rolls misaligned or worn	Realign and/or replace
	Liner or contact tube plugged	Clean or replace
Wire feeds but no gas flows	Gas cylinder is empty	Replace and purge lines before welding
	Gas cylinder valve closed	Open cylinder valve
	Flow meter not adjusted	Adjust to give flow specified in the procedure
	Restriction in gas line or nozzle	Check and clean
Porosity in the weld bead	Failed gas valve solenoid	Repair or replace
	Gas cylinder valve closed	Turn valve on
	Insufficient shielding gas flow	Check for restrictions in gas line or nozzle and correct
	Leaks in gas supply lines (including the gun)	Check for leaks (especially at connections) and correct
Wire feed motor operates but wire does not feed	Insufficient drive roll pressure	Adjust
	Incorrect wire feed rolls	Match feed rolls to wire size and type
	Excessive pressure on wire spool brake	Decrease brake pressure
	Restriction in the conduit liner or gun	Check liner and contact tip. Clean and/or replace
	Incorrect liner or contact tube	Check and replace with correct size
Welding gun overheats	Pinched or clogged coolant line	Check and correct
	Low coolant level in pump reservoir	Check and add coolant as necessary
	Water pump not functioning correctly	Check and repair or replace

Table 4.18
Troubleshooting Process Problems Encountered in Gas Metal Arc Welding

Problem	Possible Cause	Remedy
Unstable arc	Weld joint area dirty	Clean to remove scale, rust, etc.
Heavily oxidized weld deposit	Improper gun angle	Use approximately 15° lead or trail angle
	Excessive nozzle to work distance distance	Reduce. Should be approximately 1/2 to 3/4"
	Air drafts	Protect weld area from drafts
	Contact tube not centered in the gas nozzle distance	Center contact tube
Porosity in the weld bead	Dirty base material	Clean to remove scale, rust, etc.
	Excessive wire feed speed	Reduce.
	Moisture in the shielding gas	Replace gas cylinder
	Contaminated electrode	Keep wire protected while using. Clean wire before it enters feeder.
Electrode wire stubs into the workpiece	Excessive wire feed speed	Reduce speed
	Arc voltage too low	Increase voltage
	Excessive slope set on power source (for short circuiting transfer)	Reset to reduce slope
Excessive spatter	Excessive arc voltage	Reduce voltage
	Insufficient slope set on power source (for short circuiting transfer)	Increase slope setting
	Contact tube recessed too far in nozzle	Adjust or replace with longer one
	Excessive gas flow rate	Reduce flow
Welding gun overheats	Excessive amperage for gun	Reduce amperage or change to higher capacity gun

SAFE PRACTICES

INTRODUCTION

SAFETY IN WELDING, cutting, and allied processes is covered in ANSI Z49.1, *Safety in Welding and Cutting*,[1] and ANSI Z49.2, *Fire Prevention in the Use of Welding and Cutting Processes*,[2] and Chapter 16 Volume 1 of the *Welding Handbook*, 8th Edition.

Personnel should be familiar with the safe practices discussed in these documents.

In addition, there are other potential hazard areas in arc welding and cutting (including fumes, gases, radiant energy, noise, handling of cylinders and regulators, and electric shock) that warrant consideration. Those areas which may be associated with the GMAW process are briefly discussed in this chapter.

---

1. ANSI Z49.1 is available from the American Welding Society, 550 N.W. LeJeune Road, Miami, Florida 33135.
2. ANSI Z49.2 is available from the American National Standards Institute, 1430 Broadway, New York, NY 10018.

SAFE HANDLING OF GAS CYLINDERS AND REGULATORS

COMPRESSED GAS CYLINDERS should be handled carefully and should be adequately secured when stored or in use. Knocks, falls, or rough handling may damage cylinders, valves, and fuse plugs, and cause leakage or an accident. Valve protecting caps, when supplied, should be kept in place (hand tight) unless a regulator is attached to the cylinder. (See CGA Pamphlet P-1, *Safe Handling of Compressed Gas Cylinders*)[3].

The following should be observed when setting up and using cylinders of shielding gas:

(1) Properly secure the cylinder.
(2) Before connecting a regulator to the cylinder valve, the valve should momentarily be slightly opened and closed immediately ("cracking") to clear the valve of dust

---

3. CGA P-1 is available from the Compressed Gas Association, Inc., 500 Fifth Avenue, New York, NY 10036.

or dirt that otherwise might enter the regulator. The valve operator should stand to one side of the regulator gauges, never in front of them.

(3) After the regulator is attached, the pressure adjusting screw should be released by turning it counter-clockwise. The cylinder valve should then be opened slowly to prevent a rapid surge of high-pressure gas into the regulator. The adjusting screw should then be turned clockwise until the proper pressure is obtained.

(4) The source of the gas supply (i.e., the cylinder valve) should be shut off if it is to be left unattended, and the adjusting screw should be backed off.

GASES

THE MAJOR TOXIC gases associated with GMAW welding are ozone, nitrogen dioxide, and carbon monoxide. Phosgene gas could also be present as a result of thermal or ultraviolet decomposition of chlorinated hydrocarbon cleaning agents located in the vicinity of welding operations. Two such solvents are trichlorethylene and perchlorethylene. Degreasing or other cleaning operations involving chlorinated hydrocarbons should be located so that vapors from these operations cannot be reached by radiation from the welding arc.

Ozone

THE ULTRAVIOLET LIGHT emitted by the GMAW arc acts on the oxygen in the surrounding atmosphere to produce ozone, the amount of which will depend upon the intensity and the wave length of the ultraviolet energy, the humidity, the amount of screening afforded by any welding fumes, and other factors. The ozone concentration will generally increase with an increase in welding current, with the use of argon as the shielding gas, and when welding highly reflective metals. If the ozone cannot be reduced to a safe level by ventilation or process variations, it will be necessary to supply fresh air to the welder either with an air supplied respirator or by other means.

Nitrogen Dioxide

SOME TEST RESULTS show that high concentrations of nitrogen dioxide are found only within 6 in. (150 mm) of the arc. With normal natural ventilation, these concentrations are quickly reduced to safe levels in the welder's breathing zone, so long as the welder's head is kept out of the plume of fumes (and thus out of the plume of welding-generated gases). Nitrogen dioxide is not thought to be a hazard in GMAW.

Carbon Monoxide

CARBON DIOXIDE SHIELDING used with the GMAW process will be dissociated by the heat of the arc to form carbon monoxide. Only a small amount of carbon monoxide is created by the welding process, although relatively high concentrations are formed temporarily in the plume of fumes. However, the hot carbon monoxide oxidizes to carbon dioxide so that the concentrations of carbon monoxide become insignificant at distances of more than 3 or 4 in. (75 or 100 mm) from the welding plume.

Under normal welding conditions, there should be no hazard from this source. When welders must work over the welding arc, or with natural ventilation moving the plume of fumes towards their breathing zone, or where welding is performed in a confined space, ventilation adequate to deflect the plume or remove the fumes and gases should be provided (see ANSI Z49.1, *Safety in Welding and Cutting*).

METAL FUMES

THE WELDING FUMES generated by GMAW can be controlled by general ventilation, local exhaust ventilation, or by respiratory protective equipment as described in ANSI Z49.1. The method of ventilation required to keep the level of toxic substances within the welder's breathing zone below threshold concentrations is directly dependent upon a number of factors. Among these are the material being welded, the size of the work area, and the degree of confinement or obstruction to normal air movement where the welding is being done. Each operation should be evaluated on an individual basis in order to determine what will be required.

Acceptable exposure levels to substances associated with welding, and designated as time-weighted average threshold limit values (TLV) and ceiling values, have been established by the American Conference of Governmental Industrial Hygienists (ACGIH) and by the Occupational Safety and Health Administration (OSHA). Compliance with these acceptable levels of exposure can be checked by sampling the atmosphere under the welder's helmet or in the immediate vicinity of the welder's breathing zone. Sampling should be in accordance with ANSI/AWS F1.1, *Method for Sampling Airborne Particulates Generated by Welding and Allied Processes*.

RADIANT ENERGY

THE TOTAL RADIANT energy produced by the GMAW process can be higher than that produced by the SMAW process, because of its higher arc energy, significantly lower welding fume and the more exposed arc. Generally, the highest ultraviolet radiant energy intensities are produced when using an argon shielding gas and when welding on aluminum.

The suggested filter glass shades for GMAW, as presented in ANSI Z49.1 as a guide, are shown in Table 4.19. To select the best shade for an application, first select a very dark shade. If it is difficult to see the operation properly, select successively lighter shades until the operation is

Table 4.19
Suggested Filter Glass Shades for GMAW

Welding Current, A	Lowest Shade Number	Comfort Shade No.
Under 60	7	9
60-160	10	11
160-250	10	12
250-500	10	14

sufficiently visible for good control. However, do not go below the lowest recommended number, where given.

Dark leather or wool clothing (to reduce reflection which could cause ultraviolet burns to the face and neck underneath the helmet) is recommended for GMAW. The greater intensity of the ultraviolet radiation can cause rapid disintegration of cotton clothing.

NOISE—HEARING PROTECTION

PERSONNEL SHOULD BE protected against exposure to noise generated in welding and cutting processes in accordance with paragraph 1910.95 "Occupational Noise Exposure" of the Occupational Safety and Health Administration, U.S. Department of Labor.

ELECTRIC SHOCK

LINE VOLTAGES TO power supplies and auxilliary equipment used in GMAW range from 110 to 575 volts. Welders and service personnel should exercise caution not to come in contact with these voltages. See precautions listed in ANSI Z49.1, Safety in Welding and Cutting.

SUPPLEMENTARY READING LIST

Aldenhoff, B. J., Stearns, J. B., and Ramsey, P. W. "Constant potential power sources for multiple operation gas metal arc welding." *Welding Journal* 53(7): 425-429; July 1974.

Althouse, A. D., Turnquist, C. H. Bowditch, W. A. and Bowditch, K. E. *Modern welding.* South Holland, ILL: The Goodheart - Willcox Company, Inc., 1984.

American Welding Society. *Recommended safe practices for gas shielded arc welding,* AWS A6.1. American Welding Society Miami, Florida: 1966.

Baujet, V., and Charles, C. "Submarine hull construction using narrow-groove GMAW." *Welding Journal* 69(8): 31-36; August 1990.

Butler, C. A., Meister, R. P., And Randall, M. D. "Narrow gap welding—a process for all positions." *Welding Journal* 48(2): 102-108; February 1969.

Cary, Howard B. *Modern welding technology.* Englewood Cliffs, NJ: Prentice-Hall, Inc., 1979.

DeSaw, F. A. and Rodgers, J. E. "Automated welding in restricted areas using a flexible probe gas metal arc welding torch." *Welding Journal* 60(5): 17-22; May 1981.

Dillenbeck, V. R., Castagno, L. "The effects of various shielding gases and associated mixtures in GMA welding of mild steel." *Welding Journal* 66(9): 45-49; September 1987.

Hilton, D. E., Norrish, J. "Shielding gases for arc welding." *Welding and Metal Fabrication* 189-196; May-June 1988.

Kaiser Aluminum and Sales. *Welding kaiser aluminum,* 2nd Edition. Oakland, California: Kaiser Aluminum and Sales, Inc., 1978.

Kimura, S. et al. "Narrow-gap gas metal arc welding process in flat position." *Welding Journal* 58(7): 44-52; July 1979.

Kiyolara, M., et al. "On the stabilization of GMA welding of aluminum." *Welding Journal* 56(3): 20-28; March 1977.

Lesnewich, A. "MIG welding with pulsed power." Bulletin 170. New York; Welding Research Council, 1972.

———. "Control of melting rate and metal transfer in gas-shielded metal-arc welding. *Welding Journal* 37(8): 343-353; August 1958.

Lincoln Electric Company. *The procedure handbook of welding,* 12th Ed. Cleveland, Ohio: Lincoln Electric Company, 1973 .

Liu, S. and Siewart, T. A. "Metal transfer in gas metal arc welding: Droplet rate." *Welding Journal* 68(2): 52s; February 1989.

Lu, M. J. and Kou, S. "Power inputs in gas metal arc welding of aluminum," Part 1 and Part 2. *Welding Journal* 68 (9 and 11): 382s and 452s; September and November 1989.

Lyttle, K. A. "GMAW - A versatile process on the move." *Welding Journal.* 62(3): 15-23; March 1983.

———. "Reliable GMAW means understanding wire quality, equipment and process variables." *Welding Journal* 61(3): 43-48; March 1982.

Malin, V. Y. "The state-of-the-art of narrow gap welding," Part I. *Welding Journal* 62(4): 22-30; April 1983.

———. "The state-of-the-art of narrow gap welding," Part II. *Welding Journal* 62(6): 37-46; June 1983.

Manz, A. F. "Inductance vs. slope for control for gas metal arc power." *Welding Journal* 48(9): 707-712; September 1969.

————. *The welding power handbook*. American Welding Society, Miami, Florida, 1973.

Morris, R. W. "Application of multiple electrode gas metal arc welding to structural steel fabrication." *Welding Journal* 47(5): 379-385; May 1968.

Pan, J. L. et al. "Adaptive control GMA welding - a new technique for quality control." *Welding Journal* 68(3): 73; March 1989.

Pierre, Edward R. *Welding processes and power sources*, 3rd Edition. Minneapolis: Burgess Publishing Company, 1985.

Shackleton, D.N., and Lucas, W. "Shielding gas mixtures for high quality mechanized GMA welding of Q & T steels." *Welding Journal* 53(12): 537s-547s; December 1974.

Tekriwal, P. and Mazumder, J. "Finite element analysis of three-dimensiona l transient heat transfer in GMA welding." *Welding Journal* 67(7): 150s; July 1988.

Tsao, K. C. and Wir, C. S. "Fluid flow and heat transfer in GMA weld pools." *Welding Journal* 67(3): 70s; March 1988.

Union Carbide Corporation. *MIG welding handbook*. Danbury, Connecticut: Union Carbide Corporation, Linde Div., 1984.

Waszink, J. H. and Van Den Heurel, G. J. P. M. "Heat generation and heat flow in the filler metal in GMA welding." *Welding Journal* 61(8): 269s-282s; August 1982.

FLUX CORED ARC WELDING

PREPARED BY A COMMITTEE CONSISTING OF:

G. C. Barnes, Chairman
Alloy Rods

K. E. Banks
Teledyne McKay

J. E. Hinkle
Consultant

G. H. MacShane
Consultant

M. T. Merlo
Tri-Mark Incorporated

L. Soisson
Welding Consultants, Incorporated

WELDING HANDBOOK COMMITTEE MEMBER:
R. M. Walkosak
Westinghouse Electric Corporation

CHAPTER 5

FLUX CORED ARC WELDING

FUNDAMENTALS OF THE PROCESS

FLUX CORED ARC welding (FCAW) is an arc welding process that uses an arc between a continuous filler metal electrode and the weld pool. The process is used with shielding from a flux contained within the tubular electrode, with or without additional shielding from an externally supplied gas, and without the application of pressure.

The flux cored electrode is a composite tubular filler metal electrode consisting of a metal sheath and a core of various powdered materials. During welding an extensive slag cover is produced on the face of a weld bead.

The feature that distinguishes the FCAW process from other arc welding processes is the enclosure of fluxing ingredients within a continuously fed electrode. The remarkable operating characteristics of the process and the resulting weld properties are attributable to this electrode development.

Note that metal cored electrodes are not covered in this chapter, because their powdered core materials produce no more than slag islands on the face of a weld bead. Thus, they do not match the definition of flux cored electrodes. Metal cored electrodes are covered in Chapter 4 - Gas Metal Arc Welding.

FCAW offers two major process variations that differ in their method of shielding the arc and weld pool from atmospheric contamination (oxygen and nitrogen). One type, self-shielded FCAW, protects the molten metal through the decomposition and vaporization of the flux core by the heat of the arc. The other type, gas shielded FCAW, makes use of a protective gas flow in addition to the flux core action. With both methods, the electrode core material provides a substantial slag covering to protect the solidifying weld metal.

Flux cored electrodes are also used in electrogas welding (EGW). That process is a single pass, vertical-up welding process described in Chapter 7 - Electrogas Welding.

Flux cored arc welding is normally a semiautomatic process. The process is also used in machine and automatic welding.

HISTORY

GAS SHIELDED METAL arc welding processes have been in use since the early 1920's. Experiments at that time showed a significant improvement of weld metal properties when the arc and molten weld metal were protected from atmospheric contamination. However, the development of coated electrodes in the late 1920's reduced the interest in gas shielded methods.

Not until the early 1940's, with the introduction of the commercially-accepted gas tungsten arc welding process, did there become a renewed interest in these gas-shielded methods. Later in that same decade, the gas metal arc welding process was successfully commercialized. Argon and helium were the two primary shielding gases at that time.

Research work conducted on manual coated electrode welds dealt with an analysis of the gas produced in the disintegration of electrode coatings. Results of this analysis showed that the predominant gas given off by electrode coatings was CO_2. This discovery led quickly to the use of CO_2 for shielding of the gas metal arc process when used on carbon steels. Although early experiments with CO_2 as a shielding gas were unsuccessful, techniques were finally developed which permitted its use. Carbon dioxide shielded GMAW became commercially available in the mid-1950's.

About that same time, the CO_2 shielding was combined with a flux-containing tubular electrode which overcame many of the problems encountered previously. Operating characteristics were improved by the addition of the core materials and weld quality was improved by eliminating atmospheric contamination. The process was introduced publicly at the AWS Exposition held at Buffalo, New York, in May 1954. The electrodes and equipment were refined and introduced in essentially the present form in 1957.

The process is being continually improved. Power sources and wire feeders are now greatly simplified and more dependable than their predecessors. The new guns are lightweight and rugged. Electrodes are undergoing continuous improvement. Alloy electrodes and small diameter electrodes down to 0.035 in. (0.9 mm) are some of the later advances.

PRINCIPAL FEATURES

THE BENEFITS OF FCAW are achieved by combining three general features:

(1) The productivity of continuous wire welding
(2) The metallurgical benefits that can be derived from a flux
(3) A slag that supports and shapes the weld bead

FCAW combines characteristics of shielded metal arc welding (SMAW), gas metal arc welding (GMAW), and submerged arc welding (SAW).

The FCAW features, as well as those that distinguish the two major versions of the process, are shown in Figure 5.1, illustrating the gas-shielded version, and Figure 5.2, illustrating the self-shielded type. Both figures emphasize the melting and deposition of filler metal and flux, together with the formation of a slag covering the weld metal.

In the gas shielded method, shown in Figure 5.1, the shielding gas (usually carbon dioxide or a mixture of argon and carbon dioxide) protects the molten metal from the oxygen and nitrogen of the air by forming an envelope around the arc and over the weld pool. Little need exists for denitrification of the weld metal because air with its nitrogen is mostly excluded. However, some oxygen may be generated from dissociation of CO_2 to form carbon monoxide and oxygen. The compositions of the electrodes are formulated to provide deoxidizers to combine with small amounts of oxygen in the gas shield.

In the self-shielded method shown in Figure 5.2, shielding is obtained from vaporized flux ingredients which displace the air, and by slag compositions that cover the molten metal droplets, to protect the molten weld pool during welding. Production of CO_2 and introduction of deoxidizing and denitriding agents from flux ingredients right at the surface of the weld pool explain why self-shielded electrodes can tolerate stronger air currents than gas shielded electrodes. Thus self-shielded FCAW is the usual choice for field work such as that shown in Figure 5.3.

One characteristic of some self-shielded electrodes is the use of long electrode extensions. Electrode extension is the length of unmelted electrode extending beyond the end of the contact tube during welding. Self-shielded electrode extensions of 1/2 to 3-3/4 in. (19 to 95 mm) are generally used, depending on the application.

Increasing the electrode extension increases the resistance heating of the electrode. This preheats the electrode and lowers the voltage drop across the arc. At the same

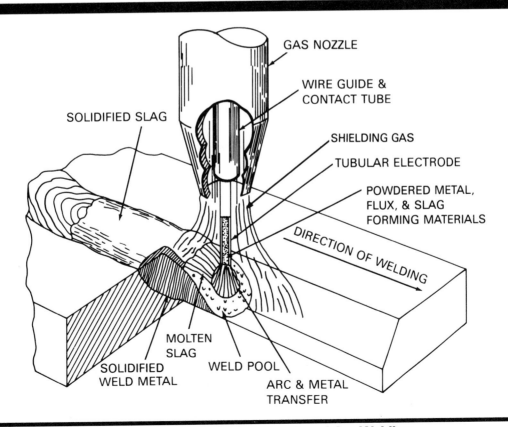

Figure 5.1–Gas Shielded Flux Cored Arc Welding

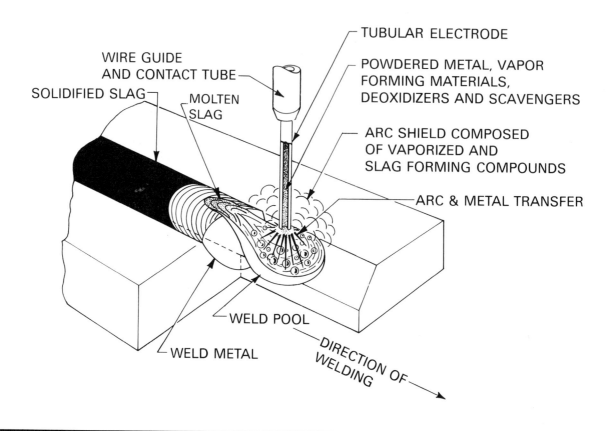

Figure 5.2–Self-Shielded Flux Cored Arc Welding

time, the welding current decreases, which lowers the heat available for melting the base metal. The resulting weld bead is narrow and shallow. This makes the process suitable for welding light gage material and for bridging gaps caused by poor fit-up. If the arc length (voltage) and welding current are maintained (by higher voltage settings at the power supply and higher electrode feed rates), longer electrode extension will increase the deposition rate.

On certain types of self-shielding flux cored electrodes, the polarity should be DCEN (straight polarity). This polarity results in less base metal penetration. As a result, small diameter electrodes such as 0.030 in. (0.8 mm), 0.035 in. (0.9 mm), and 0.045 in. (1.2 mm) have proven to be quite successful for work on thin gage materials. Some self-shielded electrodes have been developed specifically to weld the zinc-coated and aluminized steels which are now commonly used in automobile production.

In contrast, the gas-shielded method is suited to the production of narrow, deeply penetrating welds. Short electrode extensions and high welding currents are used for all wire diameters. For fillet welding, compared to SMAW, FCAW welds are narrower with larger throat lengths. The electrode extension principle cannot be equally applied to the gas shielded method because of adverse effects on the shielding.

PRINCIPAL APPLICATIONS

APPLICATION OF THE two methods of the FCAW process overlap. However, the specific characteristics of each method make each one suitable for different operating conditions. The process is used to weld carbon and low alloy steels, stainless steels, and cast irons. It is also used for arc spot welding of lap joints in sheet and plate, as well as for cladding and hardfacing.

The type of FCAW used depends on the type of electrodes available, the mechanical property requirements of the welded joints, and the joint designs and fit-up. Generally, the self-shielded method can often be used for applications that are normally done by shielded metal arc welding. The gas shielded method can be used for some applications that are welded by the gas metal arc welding process. The advantages and disadvantages of the FCAW process must be compared to those of other processes when it is evaluated for a specific application.

Higher productivity, compared to shielded metal arc welding, is the chief appeal of flux cored arc welding for many applications. This generally translates into lower overall costs per pound of metal deposited in joints that permit continuous welding and easy FCAW gun and equip-

Figure 5.3—Self-Shielded Flux Cored Arc Welding in a Field Application

ment accessibility. The advantages are higher deposition rates, higher operating factors, and higher deposition efficiency (no stub loss).

FCAW has found wide application in shop fabrication, maintenance, and field erection work. It has been used to produce weldments conforming to the *ASME Boiler and Pressure Vessel Code*, the rules of the American Bureau of Shipping, and ANSI/AWS D1.1, *Structural Welding Code - Steel*. FCAW enjoys prequalified status in ANSI/AWS D1.1.

Stainless steel, self-shielded, and gas shielded flux cored electrodes have been used in general fabrication, surfacing, joining dissimilar metals, and maintenance and repair.

The major disadvantages, compared to the SMAW process, are the higher cost of the equipment, the relative complexity of the equipment in setup and control, and the restriction on operating distance from the electrode wire feeder. FCAW may generate large volumes of welding fumes, which, except in field work, require suitable exhaust equipment. Compared to the slag-free GMAW process, the need for removing slag between passes is an added labor cost. This is especially true in making root pass welds.

EQUIPMENT

SEMIAUTOMATIC EQUIPMENT

As SHOWN IN Figure 5.4, the basic equipment for self-shielded and gas shielded flux cored arc welding is similar. The major difference is the provision for supplying and metering gas to the arc of the gas shielded electrode. The recommended power source is the dc constant-voltage type, similar to sources used for gas metal arc welding. The power supply should be capable of operating at the maximum current required for the specific application. Most semiautomatic applications use less than 500 A. The voltage control should be capable of adjustments in increments of one volt or less. Constant-current (dc) power sources of adequate capacity with appropriate controls and wire feeders are also used, but these applications are rare.

The purpose of the wire feed control is to supply the continuous electrode to the welding arc at a constant preset rate. The rate at which the electrode is fed into the arc determines the welding amperage that a constant-voltage power source will supply. If the electrode feed rate is changed, the welding machine automatically adjusts to maintain the preset arc voltage. Electrode feed rate may be controlled by mechanical or electronic means.

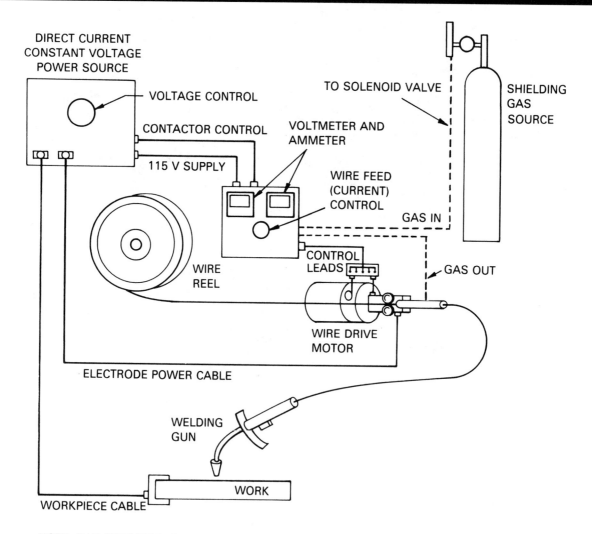

NOTE: GAS SHIELDING IS USED ONLY WITH FLUX CORED ELECTRODES THAT REQUIRE IT.

Figure 5.4–Typical for Semiautomatic Flux Cored Arc Welding Equipment

This process requires the use of drive rolls that will not flatten or otherwise distort the tubular electrode. Various grooved and knurled feed roll surfaces are used to advance the electrode. Some wire feeders have a single pair of drive rolls, while others have two pairs with at least one roll of each pair being driven. When all rolls are driven, the wire can be advanced with less pressure on the rolls.

Typical guns for semiautomatic welding are shown in Figures 5.5 and 5.6. They are designed for handling comfort, ease of manipulation, and durability. The guns provide internal contact with the electrode to conduct the welding current. Welding current and electrode feed are actuated by a switch mounted on the gun.

Welding guns may be either air cooled or water cooled. Air-cooled guns are favored because there is no requirement to deliver water. However, water-cooled guns are more compact, lighter in weight, and require less maintenance than air-cooled guns. Water-cooled guns generally have higher current ratings. Capacity ratings range up to 600 A, continuous duty. Guns may have either straight or curved nozzles. The curved nozzle can vary from 40° to 60°. In some applications, the curved nozzle enhances flexibility and ease of electrode manipulation.

Some self-shielded flux cored electrodes require a specific minimum electrode extension to develop proper shielding. Welding guns for these electrodes generally have guide tubes with an insulated extension guide to support the electrode and assure a minimum electrode extension. Details of a self-shielded electrode nozzle showing the insulated guide tube are illustrated in Figure 5.7.

AUTOMATIC EQUIPMENT

FIGURE 5.8 SHOWS the equipment layout for an automatic flux cored arc welding installation. For automatic operation, a dc constant-voltage power source designed for 100 percent duty cycle is recommended. The size of the power source is determined by the current required for the work

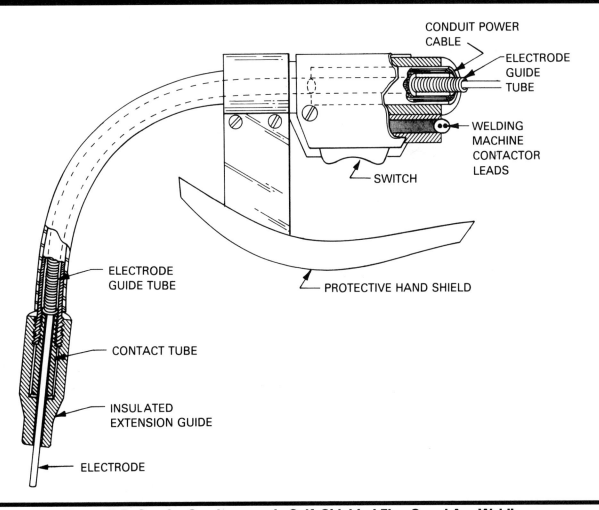

Figure 5.5—Gun for Semiautomatic Self-Shielded Flux Cored Arc Welding

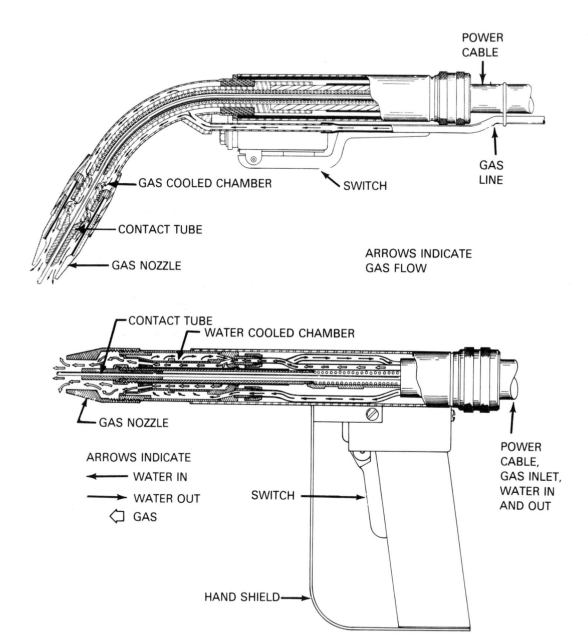

POWER
CABLE

GAS COOLED CHAMBER

SWITCH

CONTACT TUBE

GAS
LINE

GAS NOZZLE

ARROWS INDICATE
GAS FLOW

CONTACT TUBE

WATER COOLED CHAMBER

GAS NOZZLE

POWER
CABLE,
GAS INLET,
WATER IN
AND OUT

ARROWS INDICATE

WATER IN

WATER OUT

GAS

SWITCH

HAND SHIELD

Figure 5.6–Typical Guns for Gas-Shielded Flux Cored Arc Welding

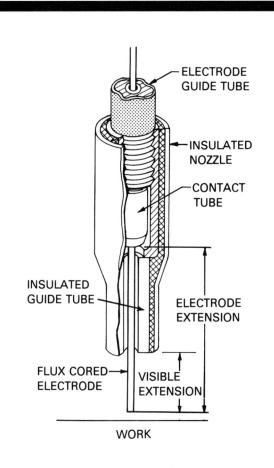

Figure 5.7–Self-Shielded Electrode Nozzle

Labels in figure:
ELECTRODE GUIDE TUBE
INSULATED NOZZLE
CONTACT TUBE
INSULATED GUIDE TUBE
ELECTRODE EXTENSION
FLUX CORED ELECTRODE
VISIBLE EXTENSION
WORK

equipment. Such installations may include a track-mounted manipulator supporting a multiple-electrode oscillating welding head with individual electrode feeders, and a track-mounted, power driven turning roll, in addition to power supply, electronic controls, and electrode supply system. Figure 5.11 illustrates the operating details of a six-electrode oscillating system for self-shielded surfacing of a vessel shell with stainless steel.

FUME EXTRACTORS

AS A RESULT of safety and health requirements for controlling air pollution, several manufacturers have introduced welding guns equipped with integral fume extractors. A fume extractor usually consists of an exhaust nozzle that encircles the gun nozzle. It can be adapted to gas-shielded and self-shielded guns. The nozzle is ducted to a filter canister and an exhaust pump. The aperture of the fume extracting nozzle is located at a sufficient distance behind the top of the gun nozzle to draw in the fumes rising from the arc without disturbing the shielding gas flow.

The chief advantage of this fume extraction system is that it is in close proximity to the fume source wherever the welding gun is used. In contrast, a portable fume exhaust may not generally be positioned so close to the fume source. It would also require repositioning of the exhaust hood for each significant change in welding location.

One disadvantage of the fume extractor system is that the added weight and bulk make semiautomatic welding more cumbersome for the welder. If not properly installed and maintained, fume extractors may cause welding problems by disturbing the gas shielding. In a well-ventilated welding area, a fume-extractor welding gun combination may not be necessary.

GAS SHIELDING APPARATUS

LIKE GMAW ELECTRODES, the gas shielded FCAW electrodes require gas shielding in addition to the internal flux. This involves a gas source, a pressure regulator, a flow metering device, and necessary hoses and connectors. Shielding gases are dispensed from cylinders, manifolded cylinder groups, or from bulk tanks which are piped to individual welding stations. Regulators and flowmeters are used to control pressure and flow rates. Since regulators can freeze during rapid withdrawal of CO_2 gas from storage tanks, heaters are available to prevent that complication. Welding grade gas purity is required because small amounts of moisture can result in porosity or hydrogen absorption in the weld metal. The dew point of shielding gases should be below -40°F (-40°C).

to be performed. Because large electrodes, high electrode feed rates, and long welding times may be required, electrode feeders necessarily have higher capacity drive motors and heavier duty components than similar equipment for semiautomatic operation.

Figure 5.9 shows two typical nozzle assemblies for automatic gas-shielded flux cored arc welding. Nozzle assemblies may be designed for side shielding or for concentric shielding of the electrode. Side shielding permits welding in narrow, deep grooves, and minimizes spatter buildup in the nozzle. Nozzle assemblies may be air or water cooled. In general, air-cooled nozzle assemblies are preferred for operation up to 600 A. Water-cooled nozzle assemblies are recommended for welding currents above 600 A. For higher deposition rates with gas shielded electrodes, tandem welding guns may be used, as shown in Figure 5.10.

For large-scale surfacing, increased productivity can be obtained from automatic multiple-electrode oscillating

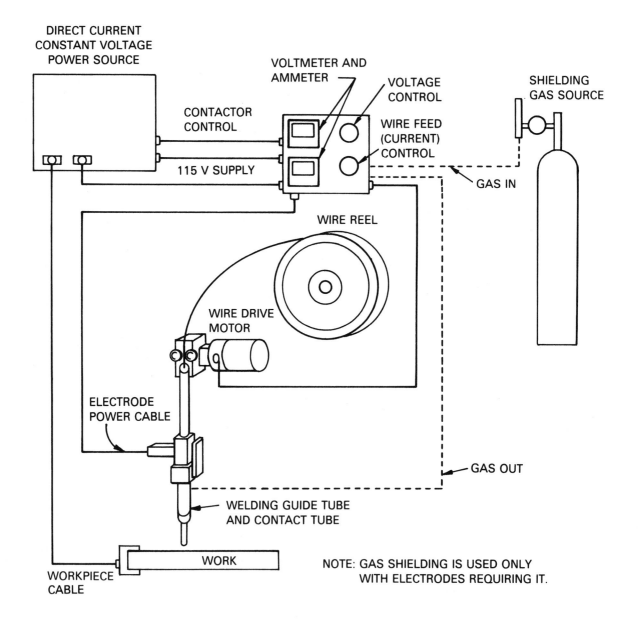

Figure 5.8–Typical Flux Cored Automatic Arc Welding Equipment

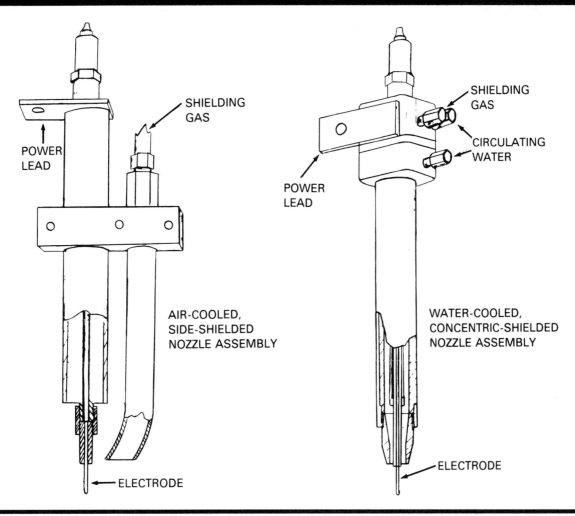

Figure 5.9–Typical Nozzle Assemblies for Automatic Gas-Shielded FCAW

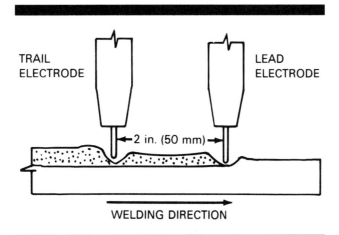

**Figure 5.10–Automatic Tandem Arc Welding
With Two Gas Shielded Flux Cored Electrodes**

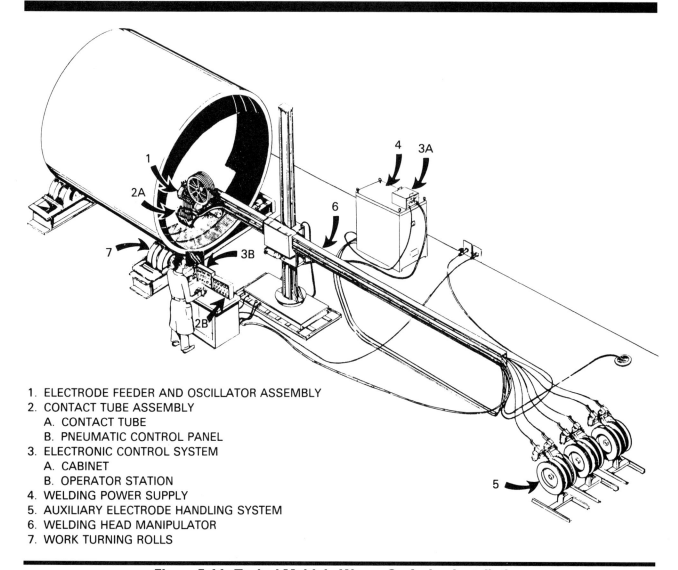

1. ELECTRODE FEEDER AND OSCILLATOR ASSEMBLY
2. CONTACT TUBE ASSEMBLY
   A. CONTACT TUBE
   B. PNEUMATIC CONTROL PANEL
3. ELECTRONIC CONTROL SYSTEM
   A. CABINET
   B. OPERATOR STATION
4. WELDING POWER SUPPLY
5. AUXILIARY ELECTRODE HANDLING SYSTEM
6. WELDING HEAD MANIPULATOR
7. WORK TURNING ROLLS

Figure 5.11–Typical Multiple Weave Surfacing Installation

MATERIALS

SHIELDING GASES

Carbon Dioxide

CARBON DIOXIDE (CO_2) is the most widely used shielding gas for flux cored arc welding. Two advantages of this gas are its low cost and deep weld penetration. Although it usually gives a globular metal transfer, some flux formulations produce a spray-like metal transfer in CO_2.

Carbon dioxide is relatively inactive at room temperature. When it is heated to high temperature by the welding arc, CO_2 dissociates to form carbon monoxide (CO) and oxygen (O), as indicated by the chemical equation

$$2CO_2 \rightarrow 2CO + O_2 \qquad (5.1)$$

Thus, the arc atmosphere contains a considerable amount of oxygen to react with elements in the molten metal. The oxidizing tendency of CO_2 shielding gas has been recognized in developing flux cored electrodes. Deoxidizing materials are added to the core of the electrode to compensate for the oxidizing effect of the CO_2.

In addition, molten iron reacts with CO_2, producing iron oxide and carbon monoxide in a reversible reaction:

$$Fe + CO_2 \rightleftarrows FeO + CO \qquad (5.2)$$

At red heat temperatures, some of the carbon monoxide dissociates to carbon and oxygen:

$$2CO \rightleftarrows 2C + O_2 \qquad (5.3)$$

The effect of CO_2 shielding on the carbon content of mild and low alloy steel weld metal is unique. Depending upon the original carbon contents of the base metal and the electrode, the CO_2 atmosphere can behave as either a carburizing or decarburizing medium. Whether the carbon content of the weld metal will be increased or decreased depends upon the carbon present in the electrode and the base metal. If the carbon content of the weld metal is below approximately 0.05 percent, the molten weld pool will tend to pick up carbon from the CO_2 shielding atmosphere. On the other hand, if the carbon content of the weld metal is greater than approximately 0.10 percent, the molten weld pool may lose carbon. The loss of carbon is attributed to the formation of carbon monoxide (CO), because of the oxidizing characteristics of CO_2 shielding gas at high temperatures.

When this reaction occurs, the carbon monoxide can be trapped in the weld metal as porosity. This tendency is minimized by providing an adequate level of deoxidizing elements in the core of the electrode. Oxygen will react with the deoxidizing elements rather than the carbon in the steel. That reaction results in formation of solid oxide compounds that float to the surface of the molten weld pool, where they form part of the slag covering.

Gas Mixtures

GAS MIXTURES USED in flux cored arc welding may combine the separate advantages of two or more gases. The higher the percentage of inert gas in mixtures with CO_2 or oxygen, the higher will be the transfer efficiencies of the deoxidizers contained in the core. Argon is capable of protecting the molten weld pool at all welding temperatures. Its presence in sufficient quantities in a shielding gas mixture results in less oxidation than occurs with 100 percent CO_2 shielding.

The mixture commonly used in gas shielded FCAW is 75 percent argon - 25 percent carbon dioxide. Weld metal deposited with this mixture generally has higher tensile and yield strengths than weld metal deposited with 100 percent CO_2 shielding. When welding with this mixture, spray transfer-type arc is achieved. The Ar-CO_2 mixture is primarily used for out-of-position welding; it has greater

operator appeal and better arc characteristics than 100 percent CO_2.

The use of shielding gas mixtures with high percentages of inert gas for electrodes designed for CO_2 shielding may cause an excessive buildup of manganese, silicon, and other deoxidizing elements in the weld metal. Such higher alloy content of the weld metal will change its mechanical properties. Therefore, electrode manufacturers should be consulted for the mechanical properties of weld metal obtained with specific shielding gas mixtures. If data are not available, tests should be made to determine the mechanical properties for the particular application.

Gas mixtures high in argon content, such as 95 percent argon - 5 percent oxygen, generally are not used with flux cored electrodes because the slag cover is lost.

BASE METALS WELDED

MOST STEELS THAT are weldable with the SMAW, GMAW, or SAW processes are readily welded using the FCAW process. Examples of these steels include the following:

(1) Mild steel, structural, and pressure vessel grades, such as ASTM A36, A515, and A516

(2) High-strength, low alloy structural grades, such as ASTM A440, A441, A572, and A588

(3) High-strength quenched and tempered alloy steels, such as ASTM A514, A517, and A533

(4) Chromium-molybdenum steels, such as 1-1/4 percent Cr-1/2 percent Mo and 2-1/4 percent Cr-1 percent Mo

(5) Corrosion-resistant wrought stainless steels, such as AISI Types 304, 309, 316, 347, 410, 430, and 502; also cast stainless steels such as ACI Types CF3 and CF8

(6) Nickel steels, such as ASTM A203

(7) Abrasion-resistant alloy steels when welded with filler metal having a yield strength less than that of the steel being welded

ELECTRODES

FLUX CORED ARC welding owes much of its versatility to the wide variety of ingredients that can be included in the core of a tubular electrode. The electrode usually consists of a low carbon steel or alloy steel sheath surrounding a core of fluxing and alloying materials. The composition of the flux core will vary according to the electrode classification and the particular manufacturer of the electrode.

Most flux cored electrodes are made by passing steel strip through rolls that form it into a U-shaped cross section. The formed strip is filled with a measured amount of granular core material (alloys and flux). The filled shape is then closed by closing rolls that round it and tightly compress the core material. The round tube is next pulled through drawing dies or rolls that reduce its diameter and further compress the core. The electrode is drawn to final size, and then wound on spools or in coils. Other methods of manufacture are also used.

Manufacturers generally consider the precise composition of their cored electrodes to be proprietary information. By proper selection of the ingredients in the core (in combination with the composition of the sheath), the following is possible:

(1) Produce welding characteristics ranging from high deposition rates in the flat position to proper fusion and bead shape in the overhead position.

(2) Produce electrodes for various gas shielding mixtures and for self shielding.

(3) Vary alloy content of the weld metal from mild steel for certain electrodes to high alloy stainless steel for others.

The primary functions of the flux core ingredients are to do the following:

(1) Provide the mechanical, metallurgical, and corrosion resistant properties of the weld metal by adjusting the chemical composition.

(2) Promote weld metal soundness by shielding the molten metal from oxygen and nitrogen in the air.

(3) Scavenge impurities from the molten metal by use of fluxing reactions.

(4) Produce a slag cover to protect the solidifying weld metal from the air, and to control the shape and appearance of the bead in the different welding positions for which the electrode is suited.

(5) Stabilize the arc by providing a smooth electrical path to reduce spatter and facilitate the deposition of uniformly smooth, properly sized beads.

Table 5.1 lists most of the elements commonly found in the flux core, their sources, and the purposes for which they are used.

In mild and low alloy steel electrodes, a proper balance of deoxidizers and denitrifiers (in the case of self-shielded electrodes) must be maintained to provide a sound weld deposit with adequate ductility and toughness. Deoxidizers, such as silicon and manganese, combine with oxygen to form stable oxides. This helps to control the loss of alloying elements through oxidation, and the formation of carbon monoxide which otherwise could cause porosity. The denitrifiers, such as aluminum, combine with nitrogen and tie it up as stable nitrides. This prevents nitrogen porosity and the formation of other nitrides which might be harmful.

CLASSIFICATIONS OF ELECTRODES

Mild Steel Electrodes

MOST MILD STEEL FCAW electrodes are classified according to the requirements of the latest edition of ANSI/AWS A5.20, *Specification for Carbon Steel Electrodes for Flux*

Table 5.1
Common Core Elements in Flux Cored Electrodes

Element	Usually Present As	Purpose in Weld
Aluminum	Metal powder	Deoxidize and denitrify
Calcium	Minerals such as fluorspar (CaF_2) and limestone ($CaCO_3$)	Provide shielding and form slag
Carbon	Element in ferroalloys such as ferromanganese	Increase hardness and strength
Chromium	Ferroalloy or metal powder	Alloying to improve creep resistance, hardness, strength and corrosion resistance
Iron	Ferroalloys and iron powder	Alloy matrix in iron base deposits, alloy in nickel base and other nonferrous deposits
Manganese	Ferroalloy such as ferromanganese or as metal powder	Deoxidize; prevent hot shortness by combining with sulfur to form MnS; increase hardness and strength; form slag
Molybdenum	Ferroalloy	Alloying to increase hardness strength, and in austenitic stainless steels to increase resistance to pitting-type corrosion
Nickel	Metal Powder	Alloying to improve hardness, strength, toughness and corrosion resistance
Potassium	Minerals such as potassium bearing feldspars and silicates and in frits	Stabilize the arc and form slag
Silicon	Ferroalloy such as ferrosilicon or silicomanganese; mineral silicates such as feldspar	Deoxidize and form slag
Sodium	Minerals such as sodium-bearing feldspars and silicates in frits	Stabilize the arc and form slag
Titanium	Ferroalloy such as ferrotitanium; in mineral, rutile	Deoxidize and denitrify; form slag; stabilize carbon in some stainless steels
Zirconium	Oxide or metal powder	Deoxidize and denitrify; form slag
Vanadium	Oxide or metal powder	Increase strength

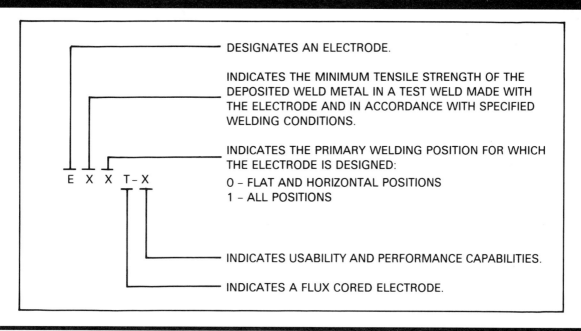

Figure 5.12–Identification System for Mild Steel FCAW Electrodes

Cored Arc Welding. The identification system follows the general pattern for electrode classification and is illustrated in Figure 5.12. It may be explained by considering a typical designation, E70T-1.

The prefix "E" indicates an electrode, as in other electrode classification systems. The first number refers to the minimum as-welded tensile strength in 10 000 psi units. In this example, the number "7" indicates that the electrode has a minimum tensile strength of 72 000 psi. The second number indicates the welding positions for which the electrode is designed. Here the "zero" means that the electrode is designed for flat groove and fillet welds, and horizontal groove and fillet welds.

However, some classifications may be suitable for vertical or overhead positions, or both. In those cases, a "1" would be used instead of the "0" to indicate all-position capability. The letter "T" indicates that the electrode is of tubular construction (a flux cored electrode). The suffix number (in this example "1") places the electrode in a particular grouping built around the chemical composition of deposited weld metal, method of shielding, and suitability of the electrode for single or multiple pass welds. Table 5.2 explains the significance of the last digit of the FCAW designations.

Mild steel FCAW electrodes are classified on the basis of whether they are self-shielded or whether carbon dioxide is required as a separate shielding gas, the type of current, and their usability for welding out of position. The classification also specifies whether the electrode is a single-pass or multiple-pass electrode, and the chemical composition and as-welded mechanical properties of deposited weld metal. Electrodes are designed to produce weld metals

having specified chemical compositions and mechanical properties when the welding and testing are done according to the specification requirements.

Electrodes are produced in standard sizes ranging from 0.045 to 5/32 in. (1.2 to 4.0 mm) diameter. Special sizes may also be available. Weld properties may vary appreciably depending on: electrode size, welding amperage, plate thickness, joint geometry, preheat and interpass temperatures, surface conditions, base metal composition and admixture with the deposited metal, and shielding gas (if re-

Table 5.2
Shielding and Polarity Requirements for Mild Steel FCAW Electrodes

AWS Classification	External Shielding Medium	Current and Polarity
EXTT-1 (Multiple-pass)	CO_2	dc, electrode positive
EXXT-2 (Single-pass)	CO_2	dc, electrode positive
EXXT-3 (Single-pass)	None	dc, electrode positive
EXXT-4 (Multiple-pass)	None	dc, electrode positive
EXXT-5 (Multiple-pass)	CO_2	dc, electrode positive
EXXT-6 (Multiple-pass)	None	dc, electrode positive
EXXT-7 (Multiple-pass)	None	dc, electrode negative
EXXT-8 (Multiple-pass)	None	dc, electrode negative
EXXT-10 (Single-pass)	None	dc, electrode negative
EXXT-11 (Multiple-pass)	None	dc, electrode negative
EXXT-G (Multiple-pass)	*	*
EXXT-GS (Single-pass)	*	*

* As agreed upon between supplier and user.

quired). Many electrodes are designed primarily for welding in the flat and horizontal positions. They may also be suitable for use in other positions if the proper choice of welding current and electrode diameter is made. Selected electrodes with diameters below 3/32 in. (2.4 mm) may be used for out-of-position welding at welding currents on the low side of the manufacturer's recommended range.

There are twelve different classifications of mild steel FCAW electrodes designated in ANSI/AWS A5.20. Their descriptions and intended uses are listed below.

EXXT-1. Electrodes of the T-1 group are designated for CO_2 shielding gas by this specification and are used with DCEP. However, gas mixtures of argon and CO_2 are also used to improve usability, especially for out-of-position applications. Decreasing amounts of CO_2 in the argon-CO_2 mixture will increase manganese and silicon in the deposit and may improve the impact properties. These electrodes are designed for single- or multiple-pass welding. The T-1 electrodes are characterized by a spray transfer, low spatter loss, flat to slightly convex bead configuration, and a moderate volume of slag which completely covers the weld bead.

EXXT-2. Electrodes of this classification are used with DCEP, are essentially T-1 electrodes with higher manganese or silicon or both, and are designed primarily for single-pass welding in the flat position and for horizontal fillets. The higher amounts of deoxidizers in these electrodes allow single-pass welding over scaled or rimmed steel. T-2 electrodes that use manganese as the principal deoxidizing element give good mechanical properties in both single- and multiple-pass applications. However, the manganese content and tensile strength will be high in multiple-pass applications. These electrodes can be used for welding material which has heavier mill scale, rust, or other foreign materials on its surface than can be tolerated by some electrodes of the T-1 classification, and will still produce welds of radiographic quality. The arc characteristics and deposition rates are similar to those of the T-1 electrodes.

EXXT-3. Electrodes of this classification are self-shielded, are used with DCEP, and have a spray-type transfer. The slag system is designed to give characteristics which make possible very high welding speeds. The electrodes are used to make single-pass welds in the flat, horizontal, and (up to 20°) downhill positions on sheet metal up to 3/16 in. (4.8 mm). They are not recommended for welding of materials greater than 3/16 in. (4.8 mm), nor for making multiple-pass welds.

EXXT-4. Electrodes of the T-4 classification are self-shielded, operate on DCEP, and have a globular-type transfer. The slag system is designed to give characteristics which permit high deposition rates while desulfurizing the weld metal to a low level, which makes the weld deposit

resistant to cracking. These electrodes are designed for low penetration, adapting them for use on joints with poor fit-up, and for single- and multiple-pass welding in the flat and horizontal positions.

EXXT-5. Electrodes of the T-5 group are designed to be used with CO_2 shielding gas (argon-CO_2 mixtures may be used, as with the T-1 types) for single- and multiple-pass welding in the flat position and for horizontal fillets. These electrodes are characterized by a globular transfer, slightly convex bead configurations, and a thin slag which may not completely cover the weld bead. Weld deposits produced by electrodes of this group have improved impact properties and crack resistance in comparison to the rutile types (EXXT-1 and EXXT-2).

EXXT-6. Electrodes of the T-6 classification are self-shielded, operate on DCEP, and have a spray-type transfer. The slag system is designed to give very good, low-temperature impact properties, deep penetration, and excellent deep groove slag removal. The electrodes are used for single- and multiple-pass welding in the flat and horizontal positions.

EXXT-7. Electrodes of the T-7 classification are self-shielded and operate on DCEN. The slag system is designed to give characteristics which allow the larger sized electrodes to be used at high deposition rates and the smaller sizes to be used for all-position welding. The slag system is also designed to desulfurize the weld metal thoroughly, which helps to make the weld deposit resistant to cracking. The electrodes are used for single- and multiple-pass welding.

EXXT-8. Electrodes of the T-8 classification are self-shielded and operate on DCEN. The slag system has characteristics which make it possible to use these electrodes for all-position welding. The slag system also achieves good, low-temperature impact properties in the weld metal and desulfurizes the weld metal to a low level, which helps resist weld cracking. The electrodes are used for both single- and multiple-pass applications.

EXXT-10. Electrodes of the T-10 classification are self-shielded and operate on DCEN. The slag system enables welds to be made at high travel speeds. The electrodes are used for making single-pass welds on material of any thickness in the flat, horizontal, and (up to 20°) downhill positions.

EXXT-11. Electrodes of the T-11 classification are self-shielded and operate on DCEN, and have a smooth spray-type arc. The slag system permits welding in all positions and at high travel speeds. These are general purpose electrodes for single- and multiple-pass welding in all positions.

EXXT-G. The EXXT-G classification is for new multiple-pass electrodes which are not covered under any of the presently defined classifications. The slag system, arc characteristics, weld appearance, and polarity are not defined.

EXXT-GS. The EXXT-GS classification is for new single-pass electrodes not covered by any other presently defined classification. The slag system, arc characteristics, weld appearance, and polarity are not defined.

Low Alloy Steel Electrodes

FLUX CORED ELECTRODES are commercially available for welding low alloy steels. They are described and classified in the latest edition of ANSI/AWS A5.29, *Specification for Low Alloy Steel Electrodes for Flux Cored Arc Welding*. The electrodes are designed to produce deposited weld metals having chemical compositions and mechanical properties similar to those produced by low alloy steel SMAW electrodes. They are generally used to weld low alloy steels of similar chemical composition. Some electrode classifications are designed for welding in all positions while others

are limited to flat and horizontal fillet positions only. Like the mild steel electrodes, there is an identification system used by AWS to describe the various classifications. Figure 5.13 illustrates the components of these designations.

ANSI/AWS A5.29 lists five different classifications of low alloy steel FCAW electrodes. Their descriptions and intended uses are summarized below.

EXXT1-X. Electrodes of the T1-X group are classified for use with CO_2 shielding gas. However, gas mixtures of argon and CO_2 may be used where recommended by the manufacturer to improve usability, especially for out-of-position applications. These electrodes are designed for single- and multiple-pass welding. The T1-X electrodes are characterized by a spray transfer, low spatter loss, flat to slightly convex bead configurations, and a moderate volume of slag, which completely covers the weld bead.

EXXT4-X. Electrodes of the T4-X classification are self-shielded, operate on DCEP, and have a globular type transfer. The slag system is designed to give characteristics that make possible high deposition rates and to desulfurize

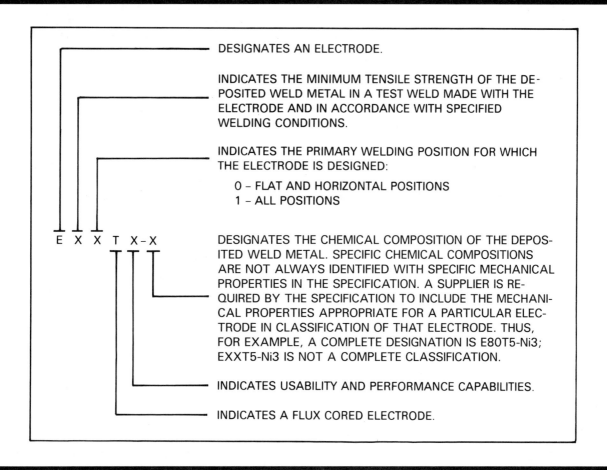

Figure 5.13–Identification System for Low Alloy Steel FCAW Electrodes

the weld metal to a low level, which helps make the weld deposit resistant to cracking. These electrodes are designed for low penetration, enabling them to be used on weld joints with poor fit-up and for single- and multiple-pass welding in the flat and horizontal positions.

EXXT5-X. Electrodes of the T5-X group are designed for use with DCEP and CO_2 shielding gas (argon-CO_2 mixtures may be used where recommended by the manufacturer, like the T1 types) for single- and multiple-pass welding in the flat position, and for horizontal fillets. Certain T5-X electrodes are designed to weld on DCEN with argon and CO_2 mixtures for out-of-position welding. These electrodes are characterized by a globular transfer, slightly convex bead configuration, and a thin slag, which may not completely cover the weld bead. Weld deposits produced by electrodes of this group have improved impact properties and crack-resistance in comparison with the T1-X types.

EXXT-8X Electrodes of the T8-x classification are self-shielded and operate on DCEN. The slag system has characteristics that make it possible to use these electrodes for all-position welding. The slag system also produces good, low-temperature impact properties in the weld metal and desulfurizes the weld metal thoroughly, which helps resist weld cracking. The electrodes are used for single- and multiple-pass welding.

EXXTX-G. EXXTX-G classification is for new multiple-pass electrodes that are not covered under any of the presently defined classifications. The slag system, arc characteristics, weld appearance, and polarity are not defined.

Most low alloy steel FCAW electrodes are designed for gas shielded welding using either a -T1-X or a -T5-X flux core formulation[1] and CO_2 shielding gas. However, the use of special formulations designed for shielding with 75 percent argon/25 percent CO_2 shielding gases is becoming increasingly common. They generally produce weld metals having Charpy V-notch impact strengths of 20 ft-lb (27 J) at 0°F (-18°C) or below. A few nickel steel electrodes with -T4-X or -T8-X formulations are available for self-shielded FCAW.

Charpy V-notch impact requirements for weld metal deposited with the -T4 formulation will generally meet 20 ft-lb (27 J) at 0°F (-18°C). Weld metal deposited with -T8 electrodes will generally meet 20 ft-lb (27 J) at -20°F (-29°C). A complete series of low alloy flux cored welding electrodes comparable to the variety of low alloy shielded metal arc electrodes described in ANSI/AWS A5.5, *Specification for Low Alloy Steel Covered Arc Welding Electrodes*, is described in the latest edition of ANSI/AWS Specification A5.29, *Specification for Low Alloy Steel Electrodes for Flux Cored Arc Welding*. As a result of the issuance of this specification, low alloy flux cored electrodes

have gained widespread acceptance in welding high-strength, low alloy steels.

Electrodes for Surfacing

FLUX CORED ELECTRODES are produced for certain types of surfacing applications, such as restoring usable service parts and hardfacing. Such electrodes possess many of the advantages of the electrodes used for joining, but there is less standardization of weld metal analysis and performance characteristics. Literature from various manufacturers should be consulted for details on flux cored surfacing electrodes.

Flux cored surfacing electrodes deposit iron base alloys which may be ferritic, martensitic, or austenitic. They may deposit weld metal that is high in carbides. The electrodes are variously designed to produce surfaces with corrosion resistance, wear resistance, toughness, or antigalling properties. They may be used to restore worn parts to original dimensions.

Stainless Steel Electrodes

THE CLASSIFICATION SYSTEM of ANSI/AWS A5.22, *Specification for Flux Cored Corrosion-Resisting Chromium and Chromium-Nickel Steel Electrodes*, prescribes requirements for flux cored corrosion resisting chromium and chromium-nickel steel electrodes. These electrodes are classified on the basis of the chemical composition of the deposited weld metal and the shielding medium to be employed during welding. Table 5.3 identifies the shielding designations used for classification, and indicates the respective current and polarity characteristics.

Electrodes classified EXXXT-1 that use CO_2 shielding suffer some minor loss of oxidizable elements and some increase in carbon content. Electrodes with the EXXXT-3 classifications, which are used without external shielding, suffer some loss of oxidizable elements and a pickup of nitrogen which may be significant. Low welding currents

Table 5.3
Shielding Designations and Welding Current Characteristics for Stainless Steel Flux Cored Electrodes

AWS Designations[a] (all classifications)	External Shielding Medium	Current and Polarity
EXXXT-1	CO_2	dcrp[b] (electrode positive)
EXXXT-2	Ar + 2% O	dcrp[b] (electrode positive)
EXXXT-3	None	dcrp[b] (electrode positive)
EXXXT-G	None Specified	Not Specified

a. The classifications are given in AWS A5.22, Specifications for Flux Cored Corrosion-Resisting Chromium and Chromium-Nickel Steel Electrodes. The letters "XXX" stand for the chemical composition (AISI Type) such as 308, 316, 410, and 502.

b. Direct current reverse polarity.

---

1. For an explanation of flux core designations, see the latest edition of ANSI/AWS A5.29, *Specification for Low Alloy Steel Electrodes for Flux Cored Arc Welding*, available from the American Welding Society.

coupled with long arc lengths (high arc voltages) increase the nitrogen pickup. Nitrogen stabilizes austenite and may therefore reduce the ferrite content of the weld metal.

The requirements of the EXXXT-3 classifications are different from those of the EXXXT-1 classifications because shielding with a flux system alone is not as effective as shielding with both a flux system and a separately applied external shielding gas. The EXXXT-3 deposits, therefore, usually have a higher nitrogen content than the EXXXT-1 deposits. This means that to control the ferrite content of the weld metal, the chemical compositions of the EXXXT-3 deposits must have different Cr/Ni ratios than those of the EXXXT-1 deposits. In contrast to self-shielded mild steel or low alloy steel electrodes, EXXXT-3 stainless steel electrodes generally do not contain strong denitriding elements such as aluminum.

The technology of the EXXXT-1 types has now been developed to the point that all-position stainless steel flux cored wires have become available. These wires have higher deposition rates than solid stainless wire when used out-of-position; they are easier to use than solid wire in the dip transfer mode; and they produce consistently sound welds with standard constant-potential power sources. These wires are available in sizes as small as 0.035 in. (0.9 mm) in diameter.

The mechanical properties of deposited weld metal are specified for each classification, including a minimum tensile strength and minimum ductility. Radiographic soundness requirements are also specified.

Although welds made with electrodes meeting AWS specifications are commonly used in corrosion or heat resisting applications, it is not practical to require electrode qualification tests for corrosion or scale resistance on welds or weld metal specimens. Special tests which are pertinent to an intended application should be established by agreement between the electrode manufacturer and the user.

Flux Cored Nickel Base Electrodes

AT THE TIME of this writing, a new AWS specification, A5.34, is being drafted to classify flux cored nickel base electrodes. Such electrodes have appeared commercially for a few nickel-base alloys. Their slag systems and operating characteristics have much in common with the stainless steel electrodes classified by ANSI/AWS A5.22. Consult AWS A5.34 as soon as it is published, for helpful, additional information.

Protection from Moisture

PROTECTION FROM MOISTURE pickup is essential with most flux cored electrodes. Moisture pickup can result in "worm tracks," or porosity in the weld bead. A return to the original package is recommended for overnight storage.

Reconditioning of exposed wire by baking at 300 to 600°F (150 to 315°C) is recommended by certain manufacturers. This assumes that the wire is spooled or coiled on a metal device.

PROCESS CONTROL

WELDING CURRENT

WELDING CURRENT IS proportional to electrode feed rate for a specific electrode diameter, composition, and electrode extension. The relationship between electrode feed rate and welding current for typical mild steel gas shielded electrodes, self-shielded mild steel electrodes, and self-shielded stainless steel electrodes are presented in Figures 5.14, 5.15, and 5.16 respectively. A constant voltage power source of the proper size is used to melt the electrode at a rate that maintains the preset output voltage (arc length). If the other welding variables are held constant for a given diameter of electrode, changing the welding current will have the following major effects:

(1) Increasing current increases electrode deposition rate.

(2) Increasing current increases penetration.

(3) Excessive current produces convex weld beads with poor appearance.

(4) Insufficient current produces large droplet transfer and excessive spatter.

(5) Insufficient current can result in pickup of excessive nitrogen and also porosity in the weld metal when welding with self-shielded flux cored electrodes.

As welding current is increased or decreased by changing electrode feed rate, power supply output voltage should be changed to maintain the optimum relationship of arc voltage to current. For a given electrode feed rate, measured welding current varies with the electrode extension. As the electrode extension increases, welding current will decrease, and vice versa.

ARC VOLTAGE

ARC VOLTAGE AND arc length are closely related. The voltage shown on the meter of the welding power supply is the sum of the voltage drops throughout the welding circuit. This includes the drop through the welding cable, the elec-

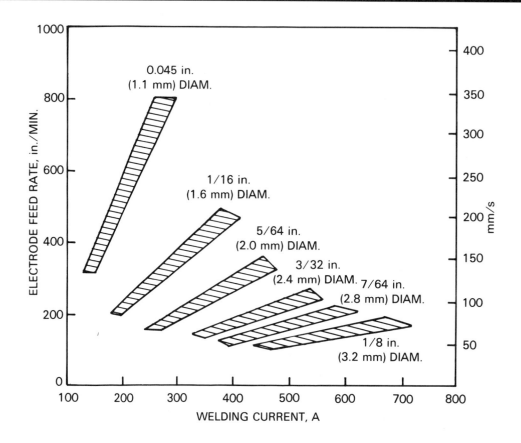

Figure 5.14—Electrode Feed Rate Versus Welding Current Range for E70-1 Steel Electrodes with CO₂ Shielding

trode extension, the arc, the workpiece and the worklead cable. Therefore, arc voltage will be proportional to the meter, reading provided all other circuit elements (including their temperatures) remain constant.

The appearance, soundness, and properties of welds made with flux cored electrodes can be affected by the arc voltage. Too high an arc voltage (too long an arc) can result in excessive spatter and wide, irregularly shaped weld beads. With self-shielded electrodes, too high an arc voltage will result in excessive nitrogen pickup. With mild steel electrodes, this may cause porosity. With stainless steel electrodes, it will reduce the ferrite content of the weld metal, and this in turn may result in cracking. Too low an arc voltage (too short an arc) will result in narrow convex beads with excessive spatter and reduced penetration.

ELECTRODE EXTENSION

THE UNMELTED ELECTRODE that extends beyond the contact tube during welding (electrode extension) is resistance heated in proportion to its length, assuming other variables remain constant. As explained earlier, electrode tempera-

ture affects arc energy, electrode deposition rate, and weld penetration. It also can affect weld soundness and arc stability.

The effect of electrode extension as an operating factor in FCAW introduces a new variable that must be held in balance with the shielding conditions and the related welding variables. For example, the melting and activation of the core ingredients must be consistent with that of the containment tube, as well as with arc characteristics. Other things being equal, too long an extension produces an unstable arc with excessive spatter. Too short an extension may cause excessive arc length at a particular voltage setting. With gas shielded electrodes, it may result in excessive spatter buildup in the nozzle that can interfere with the gas flow. Poor shielding gas coverage may cause weld metal porosity and excessive oxidation.

Most manufacturers recommend an extension of 3/4 to 1-1/2 in., (19 to 38 mm) for gas shielded electrodes and from approximately 3/4 to 3-3/4 in. (19 to 95 mm) for self-shielded types, depending on the application. For optimum settings in these ranges, the electrode manufacturer should be consulted.

TRAVEL SPEED

TRAVEL SPEED INFLUENCES weld bead penetration and contour. Other factors remaining constant, penetration at low travel speeds is greater than that at high travel speeds. Low travel speeds at high currents can result in overheating of the weld metal. This will cause a rough appearing weld with the possibility of mechanically trapping slag, or melting through the base metal. High travel speeds tend to result in an irregular, ropy bead.

SHIELDING GAS FLOW

FOR GAS SHIELDED electrodes, the gas flow rate is a variable affecting weld quality. Inadequate flow will result in poor shielding of the molten pool, resulting in weld porosity and oxidation. Excessive gas flow can result in turbulence and mixing with air. The effect on the weld quality will be the same as inadequate flow. Either extreme will increase weld metal impurities. Correct gas flow will depend on the type and the diameter of the gun nozzle, distance of the

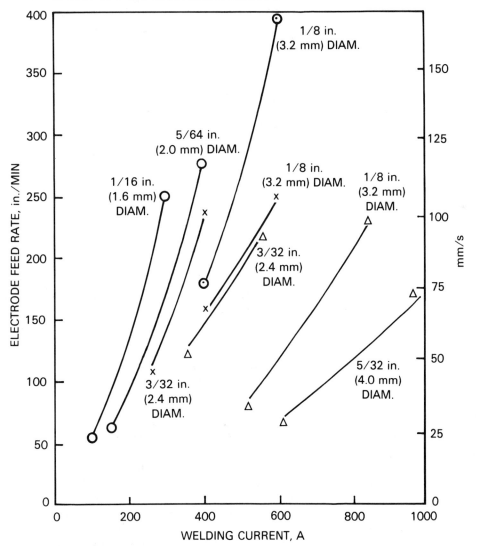

O E71T-7, 1 in. (25 mm) EXTENSION
x E70T-4, 2-3/4 in. (70 mm) EXTENSION
⊙ E70T-4, 3-3/4 in. (95 mm) EXTENSION
△ E70T-G, 1-1/4 in. (32 mm) EXTENSION

Figure 5.15—Electrode Feed Rate Versus Welding Current for Self-Shielded Mild Steel Electrodes

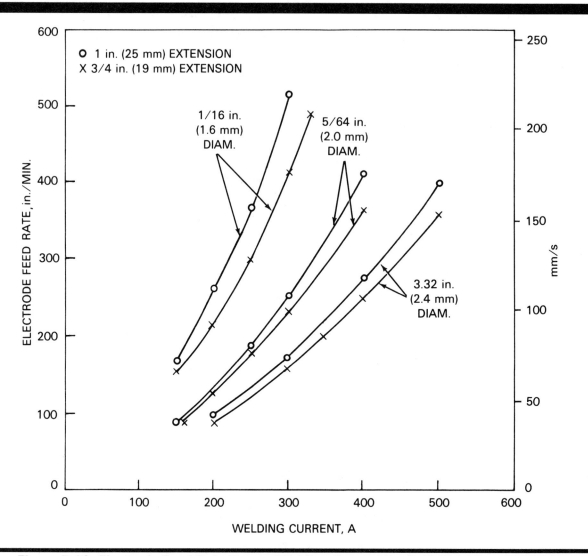

Figure 5.16–Electrode Feed Rate Versus Welding Current for Self-Shielding E308T-3

nozzle from the work, and air movements in the immediate region of the welding operation.

DEPOSITION RATE AND EFFICIENCY

DEPOSITION RATE IN any welding process is the weight of material deposited per unit of time. Deposition rate is dependent on welding variables such as electrode diameter, electrode composition, electrode extension, and welding current. Deposition rates versus welding current for various diameters of gas shielded and self-shielded mild steel electrodes and self-shielded stainless steel electrodes are presented in Figures 5.17, 5.18, and 5.19 respectively.

Deposition efficiencies of FCAW electrodes will range from 80 to 90 percent for those used with gas shielding,

and from 78 to 87 percent for self-shielded electrodes. Deposition efficiency is the ratio of weight of metal deposited to the weight of electrode consumed.

ELECTRODE ANGLE

THE ANGLE AT which the electrode is held during welding determines the direction in which the arc force is applied to the molten weld pool. When welding variables are properly adjusted for the application involved, the arc force can be used to oppose the effects of gravity. In the FCAW and SMAW processes, the arc force is used not only to help shape the desired weld bead, but also to prevent the slag from running ahead of and becoming entrapped in the weld metal.

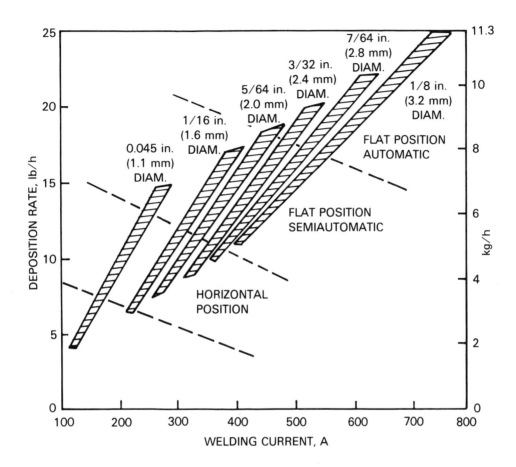

Figure 5.17–Deposition Rate Versus Welding Current for E70T-1 Mild Steel Electrodes with CO2 Shielding

When making groove and fillet welds in the flat position, gravity tends to cause the molten weld pool to run ahead of the weld. To counteract this, the electrode is held at an angle to the vertical with the electrode tip pointing backwards toward the weld, i.e., away from the direction of travel. This travel angle, defined as the *drag angle*, is measured from a vertical line in the plane of the weld axis, as shown in Figure 5.20(A).

The proper drag angle depends on the FCAW method used, the base metal thickness, and the position of welding. For the self-shielded method, drag angles should be about the same as those used with shielded metal arc welding electrodes. For flat and horizontal positions, drag angles will vary from approximately 20 to 45 degrees. Larger angles are used for thin sections. As material thickness increases, the drag angle is decreased to increase penetration.

For vertical-up welding, the drag angle should be 5 to 10 degrees.

With the gas shielded method, the drag angle should be small, usually 2 to 15 degrees, but not more than 25 degrees. If the drag angle is too large, the effectiveness of the shielding gas will be lost.

When fillet welds are made in the horizontal position, the weld pool tends to flow both in the direction of travel and at right angles to it. To counteract the side-flow, the electrode should point at the bottom plate close to the corner of the joint. In addition to its drag angle, the electrode should have a work angle of 40° to 50° from the vertical member. Figure 5.20(B) shows the electrode offset and the work angle used for horizontal fillets.

For vertical-up welding, a small leading electrode angle can be used.

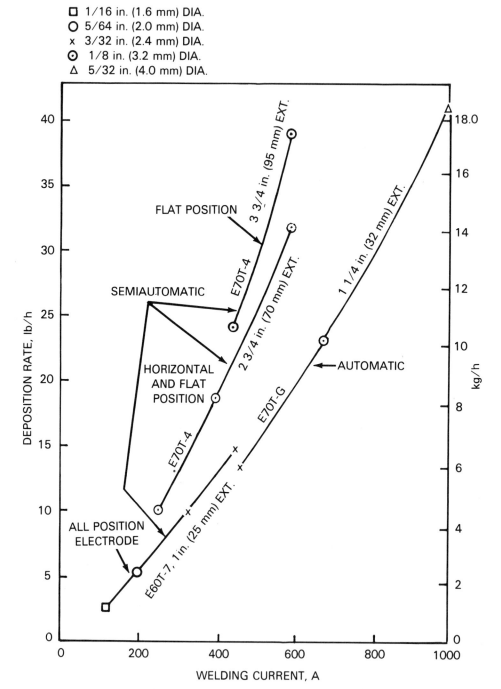

Figure 5.18–Deposition Rate Versus Welding Current for Self-Shielded Mild Steel Electrodes

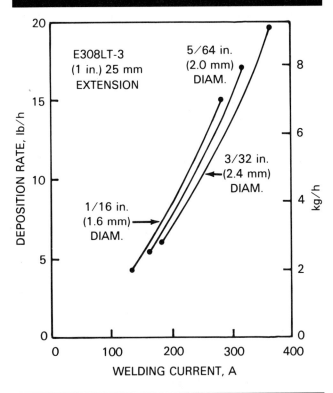

Figure 5.19–Deposition Rate Versus Welding Current for Self-Shielded E308LT-3 Stainless Steel Electrodes

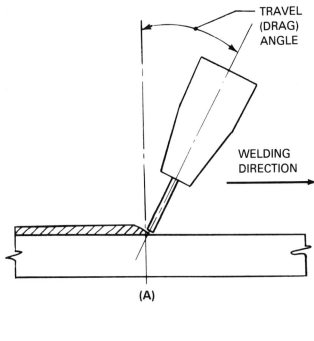

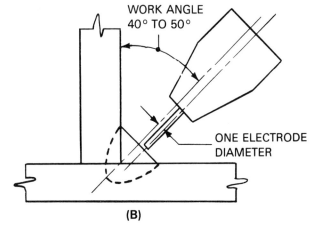

Figure 5.20–Welding Electrode Positions

JOINT DESIGNS AND WELDING PROCEDURES

THE JOINT DESIGNS and welding procedures appropriate for flux cored arc welding will depend on whether the gas shielded or self-shielded method is used. However, all the basic joint types can be welded by either method. All the basic weld groove shapes to which shielded metal arc welding is commonly applied can be welded by both FCAW methods.

There may be some differences in specific groove dimensions for a specific joint between the two methods of FCAW and between the FCAW and SMAW processes. Because FCAW electrode formulations, purpose and operating characteristics differ between classifications, the values of their welding procedure variables may also differ.

GAS SHIELDED ELECTRODES

IN GENERAL, JOINTS can be designed to take advantage of the penetration achieved by high-current densities. Narrower grooves with smaller groove angles, narrower root openings, and larger root faces than are practical with SMAW can be used with the gas shielded method.

For the basic butt joint designs, the following points should be considered:

(1) The joint should be designed so that a constant electrode extension can be maintained when welding succeeding passes in the joint.

(2) The joint should be designed so that the root is accessible, and any necessary electrode manipulation during welding is easily done.

Groove angles for various metal thicknesses are properly designed when they provide welding accessibility for the appropriate gas nozzle and electrode extension. Side shielding nozzles for automatic welding permit better accessibility into narrower joints and also permit smaller groove angles than do concentric nozzles. Sound welds can be obtained with the proper welding procedure. Typical joint designs and welding procedures for gas shielded FCAW of carbon steel are given in Table 5.4. Welding current and voltage may vary depending on the electrode source.

Adequate gas shielding is required to obtain sound welds. Flow rates required depend on nozzle size, draft conditions, and electrode extension. Welding in still air requires flow rates in the range of 30 to 40 ft^3/h (14 to 19 liters/min.). When welding in moving air or when electrode extension is longer than normal, flow rates of up to 55 ft^3/h (26 liters/min.) may be needed. Flow rates for side shielded nozzles are generally the same or slightly higher than those for concentric nozzles. Nozzle openings must be maintained free from excessive adhering spatter.

If there is brisk air movement in the welding area, such as when welding outdoors, curtains should be used to screen the weld zone and avoid loss of gas shielding.

SELF-SHIELDED MILD STEEL ELECTRODES

THE BASIC JOINT types that are suitable for the gas shielded FCAW and SMAW processes are also suitable for self-shielded FCAW. Although the general shapes of weld grooves are similar to those used for shielded metal arc welding, specific groove dimensions may differ. The differences are largely required by the higher deposition rates and shallower penetration available with self-shielded FCAW.

Electrode extension introduces another welding procedure variable that may influence joint design. When using a long electrode extension for making flat position groove welds without backing, adequate root penetration must be planned. Welding of the first pass in the groove may best be accomplished with the SMAW process for better control of fusion and penetration. Similarly, in grooves with backing, the root opening must be sufficient to permit complete fusion by globular metal transfer.

Typical joint designs and procedures for welding carbon steel when using self-shielded electrodes are given in Table 5.5. The welding current and voltage may vary for a specific electrode size and classification from different manufacturers. Depending on the joint spacing and technique used for root pass welding, back gouging and welding may be required when no backing strip is used.

When welding in the flat position, techniques similar to those used with low hydrogen covered electrodes are followed. When making vertical welds on plates 3/4 in. (19 mm) and thicker, the root pass may be deposited vertically down for joints without backing and vertical-up for joints with backing. With some self-shielded electrodes, root passes may be deposited in any position without backing. Subsequent passes are deposited vertical-up, using a technique similar to that used for low-hydrogen covered electrodes. Because self-shielded mild steel and low alloy steel electrodes (but not self-shielded stainless steel electrodes) contain considerable denitriders which may have undesirable metallurgical effects when diluted into gas shielded deposits, it may not be advisable to use self-shielded electrodes for root passes followed by gas-shielded electrodes for fill passes. Before attempting such a procedure, the electrode manufacturer should be contacted for recommendations.

SELF-SHIELDED STAINLESS STEEL ELECTRODES

TYPICAL JOINT DESIGNS and welding procedures for self-shielded stainless steel electrodes are given in Table 5.6. These electrodes, in their present state of development, are limited to butt welding in the flat position, fillet welding in the flat and horizontal positions, and surfacing in the flat and horizontal positions. If welding on stainless steels in other positions is required, EXXXT-1 electrodes with all-position capability are now available. In general, joint geometry for butt welds should be approximately the same as that used for shielded metal arc welding. When applying a surfacing weld on carbon or low alloy steels, special precautions must be taken to control dilution during the initial surfacing passes.

EDGE PREPARATION AND FIT-UP TOLERANCES

EDGE PREPARATION FOR welding with flux cored electrodes can be done by oxyfuel gas cutting, plasma arc cutting, air carbon arc gouging, or machining, depending on the type of base metal and joint design required. For best radiographic quality, dross from cutting or gouging and any machining lubricants should be completely removed

Table 5.4
Typical Gas Shielded Flux Cored Arc Welding Procedures for Carbon and Low Alloy Steel Electrodes (EXXT-1 Types)

Joint Design	Thickness, T (in.)	(mm)	Root Opening, R (in.)	(mm)	Total Passes	Electrode Diameter (in.)	(mm)	Welding Power, dcrp (ep) (V)	(A)	Wire Feed Speed (in./min)	(mm/s)
	Flat position groove welds (semiautomatic)										
	1/4	6	1/8	3	1	5/64	2.0	30	425	275	116
	1/2	13	1/4	6	2	3/32	2.4	32	450	195	80
	1/2	13	0	0	2	3/32	2.4	30	480	225	95
	1	25	0	0	6	3/32	2.4	32	480	225	95
	5/8	16	3/16	5	3	3/32	2.4	32	480	225	95
	1	25	3/16	5	6	3/32	2.4	32	480	225	95
	1	25	0	0	6	3/32	2.4	32	450	195	80
	2	51	0	0	14	3/32	2.4	32	450	195	80
	1	25	0	0	4	3/32	2.4	32	450	195	80
	2	51	0	0	10	3/32	2.4	32	450	195	80
	Horizontal position groove weld (semiautomatic)										
	1/2	13	1/8	3	6	5/64	2.0	28	350	175	75
	1	25	1/8	3	18	5/64	2.0	28	350	175	75
	Vertical position groove welds (semiautomatic)										
	3/8	10	0	0	2	1/16	1.6	23	220	165	70
	1/2	13	0	0	3	1/16	1.6	23	220	165	70
	Flat position groove welds (automatic)										
	1/4	6	1/8	3	1	3/32	2.4	30	450	195	83
	1/2	13	1/4	6	1	3/32	2.4	30	450	195	83

Table 5.5
Typical Self-Shielded Flux Cored Arc Welding Procedures for Carbon and Low Alloy Steel Electrodes

Joint Design	Plate Thickness, T		Root Opening		Total Passes	Electrode Diameter		Welding Power, dc		Wire Feed Speed		Electrode Extension	
	in.	mm	in.	mm		in.	mm	A	V(P)[a]	in./min	mm/s	in.	mm
					Flat position groove welds (semiautomatic)								
	0.14	3.4	5/32	4	1	3/32[c]	2.4	300	29+	150	65	2-3/4	70
	3/8	10	3/8	10	2	1/8[c]	3.2	500	33+	200	85	2-3/4	70
	1/2	13	3/8	10	3	1/8[c]	3.2	500	32+	200	85	2-3/4	70
	1	25	3/8	10	6	1/8[c]	3.2	550	36+	300	125	3-3/4	95
	1/2	13	3/32	2	2	3/32[c]	2.4	350	29+	190	80	2-3/4	70
Weld by SMAW	3	76	3/32	2	26	1/8[c]	3.2	550	36+	300	125	3-3/4	95
	3/8	10	3/8	10	2	1/8[c]	3.2	500	32+	200	85	2-3/4	70
	1-1/4	32	3/8	10	7	1/8[c]	3.2	550	36+	300	125	3-3/4	95
					Vertical position groove welds (semiautomatic)								
Vertical down	0.105	2.7	1/8	3	1	5/64[d]	2.0	250	20-	110	55	1	25
	1/4	6	7/32	5	3	5/64[d]	2.0	350	25-	230	100	1	25
					Vertical position groove welds (semiautomatic)								
SMAW (1/16 in.) Vertical Up	5/16	8	3/32	2	1	1/16[d]	1.6	150	18-	90	40	1	25
	1	25	3/32	2	1	1/16[d]	1.6	195	21-	120	50	1	25
	3/8	10	3/16	5	2	1/16[d]	1.6	170	19-	105	45	1	25
Vertical up	1	25	3/16	5	6	5/64[d]	2.0	190	19-	110	45	1	25
	3/8	10	1/4	6	1	1/16[d]	1.6	170	19-	105	45	1	25
Vertical up	1-1/2	38	1/4	6	4	5/64[d]	2.0	190	19-	110	45	1	25

Table 5.5
(Continued)

Joint Design	Plate Thickness, T in.	mm	Root Opening in.	mm	Total Passes	Electrode Diameter in.	mm	Welding Power, dc A	V(P)[a]	Wire Feed Speed in./min	mm/s	Electrode Extension in.	mm
Vertical up	1/4	6	0	0	1	1/16[d]	1.6	130	18	80	35	1	25
	5/8	16	0	0	1	1/16[d]	1.6	185	21	108	45	1	25
	1-1/2	38	0	0	4	1/16[d]	1.6	190	21	110	45	1	25

Vertical position fillet weld (semiautomatic)

Joint Design	Plate Thickness, T in.	mm	Root Opening in.	mm	Total Passes	Electrode Diameter in.	mm	Welding Power, dc A	V(P)[a]	Wire Feed Speed in./min	mm/s	Electrode Extension in.	mm
	5/16	8	3/16	5	3	3/32[c]	2.4	300	28+	150	65	2-3/4	70
	1-1/4	32	3/16	5	16	1/8[c]	3.2	400	29+	160	70	2-3/4	70

Horizontal position groove welds (semiautomatic)

Joint Design	Plate Thickness, T in.	mm	Root Opening in.	mm	Total Passes	Electrode Diameter in.	mm	Welding Power, dc A	V(P)[a]	Wire Feed Speed in./min	mm/s	Electrode Extension in.	mm
	3/4	19	3/32	2	6	3/32[c]	2.4	300	28+	140	60	2-3/4	70
	1-1/2	38	3/32	2	12	1/8[c]	3.2	400	29+	160	70	2-3/4	70

Horizonal position groove welds (semiautomatic)

Joint Design	Plate Thickness, T in.	mm	Root Opening in.	mm	Total Passes	Electrode Diameter in.	mm	Welding Power, dc A	V(P)[a]	Wire Feed Speed in./min	mm/s	Electrode Extension in.	mm
	0.105	2.7	0	0	1	1/16[d]	1.6	150	18−	100	40	1	25
	3/4	19	0	0	6	1/16[d]	1.6	180	19−	115	50	1	25

Overhead position groove and fillet welds (semiautomatic)

Joint Design	Plate Thickness, T in.	mm	Root Opening in.	mm	Total Passes	Electrode Diameter in.	mm	Welding Power, dc A	V(P)[a]	Wire Feed Speed in./min	mm/s	Electrode Extension in.	mm
Weld by SMAW	5/16	8	1/16	1.6	2	1/16[d]	1.6	150	18−	90	40	1	25
	1	25	1/16	1.6	8	1/16[d]	1.6	170	19−	105	45	1	25

Joint Design	Plate Thickness, T in.	mm	Root Opening in.	mm	Total Passes	Electrode Diameter in.	mm	Welding Power, dc A	V(P)[a]	Wire Feed Speed in./min	mm/s	Electrode Extension in.	mm
	0.105	2.7	0	0	1	5/64[d]	2.0	235	20−	105	45	1	25
	3/16	5	0	0	1	3/32[d]	2.4	335	21−	110	45	1	25

Horizontal position fillet welds (semiautomatic)

Joint Design	Plate Thickness, T in.	mm	Root Opening in.	mm	Total Passes	Electrode Diameter in.	mm	Welding Power, dc A	V(P)[a]	Wire Feed Speed in./min	mm/s	Electrode Extension in.	mm
	1/4	6	0	0	1	3/32[c]	2.4	325	29+	150	65	1	25
	1	25	0	0	5	1/8[c]	3.2	450	29+	175	75	2-3/4	70

Joint Design	Plate Thickness, T in.	mm	Root Opening in.	mm	Total Passes	Electrode Diameter in.	mm	Welding Power, dc A	V(P)[a]	Wire Feed Speed in./min	mm/s	Electrode Extension in.	mm
	5/16	8	0	0	1	3/32[c]	2.4	350	30+	190	80	2-3/4	70
	1	25	0	0	4	1/8[c]	3.2	580	27+	330	140	3-3/4	95

Flat position welds (semiautomatic)

Table 5.5
(Continued)

Joint Design	Plate Thickness, T		Root Opening		Total Passes	Electrode Diameter		Welding Power, dc		Wire Feed Speed		Electrode Extension	
	in.	mm	in.	mm		in.	mm	A	V(P)[a]	in./min	mm/s	in.	mm
	\multicolumn{13}{Flat position groove welds (automatic)}												
	0.05	1.2	0	0	1	3/32[c]	2.4	425	26+	160	70	1	25
	3/16	5	0	0	1	5/32[c]	4.0	950	27+	150	65	1-1/4	32
	\multicolumn{13}{Horizontal position fillet welds (automatic)}												
	0.05	1.2	0	0	1	3/32[c]	2.4	475	26+	170	70	1	25
	3/16	5	0	0	1	5/32[c]	4.0	900	26+	140	60	1-1/4	32
	0.06	1.5	0	0	1	3/32[c]	2.4	425	26+	160	70	1	25
	3/16	5	0	0	1	5/32[c]	4.0	875	27+	130	55	1-1/4	32

a. (p)—Polarity: + electrode positive; − electrode negative
b. Production rate at 100 percent operator factor
c. E70T-4 electrode
d. E60T-7 electrode
e. E70T-G electrode

Table 5.6
Typical Self-Shielded Flux Cored Arc Welding Procedures for Stainless Steels Using Stainless Steel Electrodes

Joint Design	Weld Size, T		Root Opening R		Total Passes	Electrode Diameter		Welding Power, dcrp (ep)		Wire Feed Speed		Electrode Extension	
	in.	mm	in.	mm		in.	mm	A	V	in./min	mm/s	in.	mm
	\multicolumn{13}{Flat position groove welds}												
	1/4	6	1/8	3	1	3/32	2.4	300	27.5	190	70	1	25
	3/8	10	1/8	3	2	3/32	2.4	300	27.5	170	70	1	25
	1/2	13	3/16	5	2	3/32	2.4	300	27.5	170	70	1	25
	3/4	19	3/16	5	4	3/32	2.4	300	27.5	170	70	1	25
	7/8	22	3/8	10	6	3/32	2.4	300	27.5	170	70	1	25
	1-1/4	32	3/8	10	8	3/32	2.4	300	27.5	170	70	1 to 1-1/4	25-32
	1/2	13	1/8	3	2	3/32	2.4	300	27.5	170	70	1	25
	3	76	1/8	3	25	3/32	2.4	300	27.5	170	70	1 to 1-1/4	25-32

Table 5.6
(Continued)

Joint Design	Weld Size, T		Root Opening R		Total Passes	Electrode Diameter		Welding Power, dcrp (ep)		Wire Feed Speed		Electrode Extension	
	in.	mm	in.	mm		in.	mm	A	V	in./min	mm/s	in.	mm
	3/8	10	3/8	10	3	3/32	2.4	300	27.5	170	70	1	25
	1-1/4	32	3/8	10	8	3/32	2.4	300	27.5	170	70	1 to 1-1/4	25-32
					Flat position fillet weld								
	3/8	10	0	0	1	3/32	2.4	300	27.5	170	70	1	25
	3/4	19	0	0	3	3/32	2.4	300	27.5	170	70	1	25
					Horizontal position fillet weld								
	1/8	3	0	0	1	1/16	1.6	185	24	265	110	1/2	13
	3/8	10	0	0	1	3/32	2.4	300	27	170	70	1	25

before welding. Fit-up tolerances for welding will depend on the following:

(1) Overall tolerance of completed assembly
(2) Level of quality required for the joint
(3) Method of welding (gas shielded or self-shielded; automatic or semi-automatic)
(4) Thickness of the base metal welded
(5) Type and size of electrode
(6) Position of welding

In general, mechanized and automatic flux cored arc welding preparations require close tolerances in joint fit-up. Welds made with semiautomatic equipment can accept somewhat wider tolerances.

WELD QUALITY

THE QUALITY OF welds that can be produced with the FCAW process depends on the type of electrode used, the method (gas shielded or self-shielded), condition of the base metal, weld joint design, and welding conditions. Particular attention must be given to each of these factors to produce sound welds with the best mechanical properties.

The impact properties of mild steel weld metal may be influenced by the welding method. Some self-shielded electrodes are highly deoxidized types that may produce weld metal with relatively low notch toughness. Other self-shielded electrodes have excellent impact properties. Gas-shielded and self-shielded electrodes are available that meet the Charpy V-notch impact requirements of specific classifications of the AWS filler metal specifications. Notch toughness requirements should be considered before selecting the method and the specific electrode for an application.

A few FCAW mild steel electrodes are designed to tolerate a certain amount of mill scale and rust on the base metals. Some deterioration of weld quality should be expected when welding dirty materials. When these electrodes are used for multipass welding, cracking may occur in the weld metal caused by accumulated deoxidizing agents.

In general, sound FCAW welds can be produced in mild and low alloy steels that will meet the requirements of several construction codes. Particular attention given to all factors affecting weld quality should assure meeting code requirements.

When less stringent requirements are imposed, advantages of higher welding speeds and currents can be enjoyed. Minor discontinuities that are not objectionable from the design and service standpoints may be permitted in such welds.

Flux cored arc welds can be produced in stainless steels with qualities equivalent to those of gas metal arc welds. Welding position and arc length are significant factors when self-shielded electrodes are used. Out-of-position welding procedures should be carefully evaluated with respect to weld quality. Excessive arc length generally causes an increased nitrogen pickup in the weld metal. Because nitrogen is an austenite stabilizer, absorption of excess nitrogen into the weld may prevent the formation of sufficient ferrite and thus may increase the susceptibility to microfissuring.

Low alloy steels may be welded with the gas shielded method using TX-1 or TX-5 electrode core formulations when good, low-temperature toughness is required. The combination of gas shielding and proper flux formulation generally produces sound welds with good mechanical properties and notch toughness. Self-shielded electrodes containing nickel for good strength and impact properties as well as aluminum as a denitrider, are also available. In general, the composition of the electrode should be similar to that of the base metal.

TROUBLESHOOTING

SEVERAL TYPES OF discontinuities can result from bad procedures or practices. Although many of the discontinuities are innocuous, they adversely affect the weld appearance, and therefore adversely affect the reputation of FCAW. These problems and discontinuities along with their causes and remedies are shown in Table 5.7.

ADVANTAGES OF FCAW

FLUX CORED ARC welding has many advantages over the manual SMAW process. It also provides certain advantages over the SAW and GMAW processes. In many applications, the FCAW process provides high-quality weld metal at lower cost with less effort on the part of the welder than SMAW. It is more forgiving than GMAW, and is more flexible and adaptable than SAW. These advantages can be listed as follows:

(1) High-quality weld metal deposit
(2) Excellent weld appearance — smooth, uniform welds
(3) Excellent contour of horizontal fillet welds
(4) Many steels weldable over a wide thickness range
(5) High operating factor — easily mechanized
(6) High deposition rate — high current density
(7) Relatively high electrode deposit efficiency
(8) Economical engineering joint designs
(9) Visible arc — easy to use
(10) Less precleaning required than GMAW
(11) Reduced distortion over SMAW
(12) Up to 4 times greater deposition rate than SMAW
(13) Use of self-shielded electrodes eliminates need for flux handling or gas apparatus, and is more tolerant to windy conditions present in outdoor construction (see disadvantage "6" below for gas shields)
(14) Higher tolerance for contaminants that may cause weld cracking
(15) Resistant to underbead cracking

LIMITATIONS OF FCAW

THE FOLLOWING ARE some of the limitations of this process:

(1) FCAW is presently limited to welding ferrous metals and nickel base alloys.
(2) The process produces a slag covering which must be removed.
(3) FCAW electrode wire is more expensive on a weight basis than solid electrode wires, except for some high alloy steels.
(4) The equipment is more expensive and complex than that required for SMAW; however, increased productivity usually compensates for this.
(5) The wire feeder and power source must be fairly close to the point of welding.
(6) For the gas shielded version, the external shield may be adversely affected by breezes and drafts. Except in very high winds this is not a problem with self-shielded electrodes because the shield is generated at the end of the electrode, which is exactly where it is required.
(7) Equipment is more complex than that for SMAW, so more maintenance is required.
(8) More smoke and fumes are generated (compared to GMAW and SAW).

Table 5.7
Flux Cored Arc Welding Troubleshooting

Problem	Possible Cause	Corrective Action
Porosity	Low gas flow	Increase gas flowmeter setting clean spatter clogged nozzle.
	High gas flow	Decrease to eliminate turbulence
	Excessive wind drafts	Shield weld zone from draft/wind
	Contaminated gas	Check gas source Check for leak in hoses/fittings
	Contaminated base metal	Clean weld joint faces
	Contaminated filler wire	Remove drawing compound on wire Clean oil from rollers Avoid shop dirt Rebake filler wire
	Insufficient flux in core	Change electrode
	Excessive voltage	Reset voltage
	Excess electrode stick out	Reset stickout & balance current
	Insufficient electrode stickout (self-shielded electrodes)	Reset stickout & balance current
	Excessive travel speed	Adjust speed
Incomplete Fusion or Penetration	Improper manipulation	Direct electrode to the joint root
	Improper parameters	Increase current Reduce travel speed Decrease stickout Reduce wire size Increase travel speed (self-shielded electrodes)
	Improper joint design	Increase root opening Reduce root face
Cracking	Excessive joint restraint	Reduce restraint Preheat Use more ductile weld metal Employ peening
	Improper electrode Insufficient deoxidizers or inconsistent flux fill in core	Check formulation and content of flux
Electrode feeding	Excessive contact tip wear	Reduce drive roll pressure
	Melted or stuck contact tip	Reduce voltage Adjust backburn control Replace worn liner
	Dirty wire conduit in cable	Change conduit liner. Clean out with compressed air

SAFETY

WELDING SHOULD BE done in a manner that provides maximum safety for the welder as well as those in the immediate vicinity of the welding area. From the standpoint of electrical safety and eye protection, flux cored arc welding requires the same precautions as GMAW. Welding fumes, however, are generated.

Flux cored arc welding electrodes generate welding fumes at a rate per pound of deposited metal comparable to SMAW. Since FCAW deposition rates are several times those of SMAW, the fume generation rate, in grams per minute, is much higher than that with SMAW. It is important to make sure that the welding fume concentration remains below the permissible exposure limit (PEL), specified at 5 mg/cubic meter by the Occupational Safety and Health Administration of the U.S. Department of Labor (OSHA). Note that local requirements may be more stringent.

Special precautions must be taken to protect the operator from breathing manganese-containing fumes when welding Hadfield manganese products. Also, welding of stainless steels and surfacing with chromium alloys presents the problem of chromium-containing fumes.

Safety requires being aware that shielding gases present a hazard of their own when welding is done in confined spaces. These shielding gases are not poisonous, but are asphyxiants, and will displace oxygen. Welding with shielding gases high in argon will generate substantial ultraviolet radiation, which will react with oxygen in the vicinity of the arc to form ozone.

Safe use of the flux cored welding process calls for careful evaluation of these factors and instituting appropriate corrective measures before welding.

SUPPLEMENTARY READING LIST

American Society for Metals. *Metals Handbook*, Vol. 6, 9th Ed. 96-113, Metals Park, Ohio: American Society for Metals, 1983.

American Welding Society. *Method for sampling airborne particulates generated by welding and Allied Processes*, AWS F1.1. Miami: American Welding Society, 1976.

————. *Safety in welding and cutting*, ANSI Z49.1. Miami: American Welding Society, 1973.

————. *Specification for carbon steel electrodes for flux cored arc welding*, ANSI/AWS A5.20. Miami: American Welding Society, 1979.

————. *Specification for flux cored corrosion-resisting chromium and chromium-nickel steel electrodes*, ANSI/AWS A5.22. Miami: American Welding Society, 1974.

Barnes, G. "Comparison of AWS filler metal specification for low alloy covered electrodes and low alloy flux Ccred electrodes." *Welding Journal* 61(8): 57-61; August 1982.

Bishel, R. A. "Flux-cored electrode for cast iron welding." *Welding Journal* 52(6): 372-381; June 1973.

Cary, Howard "Match wire to the job." *Welding Engineer* 55(4): 44-46; April 1970.

Hinkel, J. L. "Long stickout welding - a practical way to increase deposition rates." *Welding Journal* 47(3): 869-874; March 1968.

Hoitomt, M., and Lee, R. K. "All-position production welding with flux-cored gas-shielded electrodes." *Welding Journal* 51(11): 765-768' November 1972.

The Lincoln Electric Company. *The procedure handbook of arc welding*, 12th Ed. Cleveland, Ohio: The Lincoln Electric Company, 1973.

Wick, W. C., and Lee, R. K. "Welding low-alloy steel castings with the flux-cored process." *Welding Journal* 47 (5): 394-397; May 1968.

Zvanut, A. J., and Farmer, H. N., Jr. "Self-shielded stainless steel flux cored electrodes." *Welding Journal* 51(11): 775-780; November 1972.

SUBMERGED ARC WELDING

PREPARED BY A COMMITTEE CONSISTING OF:

R. M. Nugent, Chairman
Houston Lighting and Power Company

R. J. Dybas
General Electric Company

J. F. Hunt
Babcock & Wilcox Company

D. W. Meyer
L-Tec Welding and Cutting Systems

WELDING HANDBOOK COMMITTEE MEMBER:
D. R. Amos
Westinghouse Turbine Generator Plant

CHAPTER 6

SUBMERGED ARC WELDING

PROCESS FUNDAMENTALS

DESCRIPTION

SUBMERGED ARC WELDING (SAW) produces coalescence of metals by heating them with an arc between a bare metal electrode and the work. The arc and molten metal are "submerged" in a blanket of granular fusible flux on the work. Pressure is not used, and filler metal is obtained from the electrode and sometimes from a supplemental source such as welding rod or metal granules.

In submerged arc welding, the arc is covered by a flux. This flux plays a main role in that (1) the stability of the arc is dependent on the flux, (2) mechanical and chemical properties of the final weld deposit can be controlled by flux, and (3) the quality of the weld may be affected by the care and handling of the flux.

Submerged arc welding is a versatile production welding process capable of making welds with currents up to 2000 amperes, ac or dc, using single or multiple wires or strips of filler metal. Both ac and dc power sources may be used on the same weld at the same time.

PRINCIPLES OF OPERATION

IN SUBMERGED ARC welding, the end of a continuous bare wire electrode is inserted into a mound of flux that covers the area or joint to be welded. An arc is initiated using one of six arc-starting methods, described later in this chapter. A wire-feeding mechanism then begins to feed the electrode wire towards the joint at a controlled rate, and the feeder is moved manually or automatically along the weld seam. For machine or automatic welding, the work may be moved beneath a stationary wire feeder.

Additional flux is continually fed in front of and around the electrode, and continuously distributed over the joint. Heat evolved by the electric arc progressively melts some of the flux, the end of the wire, and the adjacent edges of the base metal, creating a pool of molten metal beneath a layer of liquid slag. The melted bath near the arc is in a highly turbulent state. Gas bubbles are quickly swept to the surface of the pool. The flux floats on the molten metal and completely shields the welding zone from the atmosphere.

The liquid flux may conduct some electric current between the wire and base metal, but an electric arc is the predominant heat source. The flux blanket on the top surface of the weld pool prevents atmospheric gases from contaminating the weld metal, and dissolves impurities in the base metal and electrode and floats them to the surface. The flux can also add or remove certain alloying elements to or from the weld metal.

As the welding zone progresses along the seam, the weld metal and then the liquid flux cool and solidify, forming a weld bead and a protective slag shield over it.

It is important that the slag is completely removed before making another weld pass. The submerged arc process is illustrated in Figure 6.1.

Factors that determine whether to use submerged arc welding include:

(1) The chemical composition and mechanical properties required of the final deposit
(2) Thickness of base metal to be welded
(3) Joint accessibility
(4) Position in which the weld is to be made
(5) Frequency or volume of welding to be performed

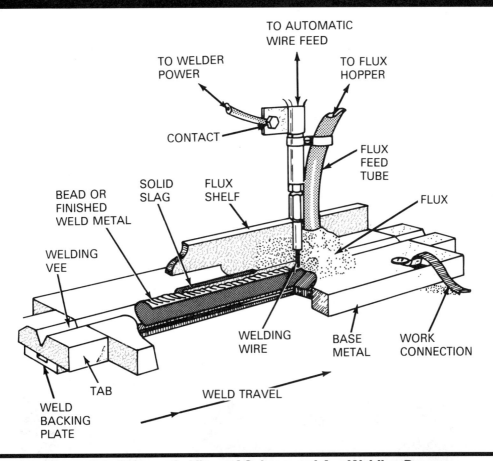

TO AUTOMATIC WIRE FEED

TO WELDER POWER

TO FLUX HOPPER

CONTACT

FLUX FEED TUBE

SOLID SLAG

FLUX SHELF

FLUX

BEAD OR FINISHED WELD METAL

WELDING VEE

WELDING WIRE

BASE METAL

WORK CONNECTION

TAB

WELD TRAVEL

WELD BACKING PLATE

Figure 6.1–Schematic View of Submerged Arc Welding Process

GENERAL METHODS

SUBMERGED ARC WELDING can be applied in three different modes: semiautomatic, automatic, and machine. Each method requires that the work be positioned so that the flux and the molten weld pool will remain in place until they have solidified. Many types of fixtures and positioning equipment are available or can be built to satisfy this requirement.

Semiautomatic Welding

SEMIAUTOMATIC WELDING IS done with a hand-held welding gun, which delivers both flux and the electrode. The electrode is driven by a wire feeder. Flux may be supplied by a gravity hopper mounted on the gun or pressure fed through a hose. This method features manual guidance using relatively small diameter electrodes and moderate travel speeds. The travel may be manual or driven by a small gun-mounted driving motor. See Figure 6.2.

Figure 6.2–Hand-Held Submerged Arc Welding Gun

Automatic Welding

AUTOMATIC WELDING IS done with equipment that performs the welding operation without requiring a welding operator to continually monitor and adjust the controls. Expensive self-regulating equipment can be justified in order to achieve high-production rates. Automatic SAW of piping is shown in Figure 6.3.

Machine Welding

MACHINE WELDING EMPLOYS equipment that performs the complete welding operation. However, it must be monitored by a welding operator to position the work, start and stop welding, adjust the controls, and set the speed of each weld. A typical machine welding operation is shown in Figure 6.4.

PROCESS VARIATIONS

SUBMERGED ARC WELDING lends itself to a wide variety of wire and flux combinations, single and multiple electrode arrangements, and use of ac or dc welding power sources.

The process has been adapted to a wide range of materials and thicknesses. Various multiple arc configurations may be used to control the weld profile and increase the deposition rates over single arc operation. Weld deposits may range from wide beads with shallow penetration for surfacing, to narrow beads with deep penetration for thick joints. Part of this versatility is derived from the use of ac arcs.

The principles which favor the use of ac to minimize arc blow in single arc welding are often applied in multiple arc welding to create a favorable arc deflection. The current flowing in adjacent electrodes sets up interacting magnetic fields that can either reinforce or diminish each other. In the space between the arcs, these magnetic fields are used to produce forces that will deflect the arcs (and thus distribute the heat) in directions beneficial to the intended welding application.

Various types of power sources and related equipment are designed and manufactured especially for multiple arc welding. These relatively sophisticated machines are intended for high production on long runs of repetitive type applications.

Figure 6.3–Automatic Submerged Arc Welding of Piping Using Five Wire Feed Heads

Figure 6.4—Mechanized Submerged Arc Weld Made on a Vessel Head

EQUIPMENT

THE EQUIPMENT REQUIRED for submerged arc welding consists of (1) a power supply, (2) an electrode delivery system, (3) a flux distribution system, (4) a travel arrangement, and (5) a process control system. Optional equipment includes flux recovery systems and positioning or manipulating equipment.

POWER SOURCES

THE POWER SOURCE chosen for a submerged arc welding system plays a major operating role.

Several types of power supply are suitable for submerged arc welding. A dc power supply may be a transformer-rectifier or a motor or engine generator, which will provide a constant voltage (CV), constant current (CC), or a selectable CV/CC output. AC power supplies are generally transformer types, and may provide either a CC output or a CV square wave output. Because SAW is generally a high-current process with high-duty cycle, a power supply capable of providing high amperage at 100 percent duty cycle is recommended.

Dc Constant-Voltage Power Sources

DC CONSTANT-VOLTAGE POWER supplies are available in both transformer-rectifier and motor-generator models. They range in size from 400 A to 1500 A models. The smaller supplies may also be used for GMAW and FCAW. These power sources are used for semiautomatic SAW at currents ranging from about 300 to 600 A with 1/16, 5/64, and 3/32 in. (1.6, 2.0, and 2.4 mm) diameter electrodes. Automatic welding is done at currents ranging from 300 to over 1000 A, with wire diameters generally ranging from 3/32 to 1/4 in. (2.4 to 6.4 mm). However, applications for dc welding at over 1000 A are limited because severe arc blow may occur at such high current. A typical dc constant-voltage power supply is shown in Figure 6.5.

With some older CV supplies, the minimum useful current density is about 40 000 A/in² (62 A/mm²), based on electrode diameter. Below this current density, the arc becomes unstable. However, this problem has been overcome by more recent power supplies and a stable arc can be maintained at current densities as low as 15 000 A/in² (23 A/mm²).

A constant-voltage power supply is self-regulating, so it can be used with a constant-speed wire feeder. No voltage or current sensing is required to maintain a stable arc, so very simple wire feed speed controls may be used. The wire feed speed and wire diameter control the arc current, and the power supply controls the arc voltage.

Constant-voltage dc power supplies are the most commonly used supplies for submerged arc welding. They work well for most applications where the arc current does not exceed 1000 A, and may work without a problem at higher currents. The CV dc power supply is the best choice for high-speed welding of thin steel.

Dc Constant-Current Power Sources

CONSTANT-CURRENT DC POWER sources are available in both transformer-rectifier and motor-generator models, with rated outputs up to 1500 A. Some CC dc power sources may also be used for GTAW, SMAW, and air carbon arc cutting. With the exception of high-speed welding of thin steel, CC dc sources can be used for the same range of applications as CV dc supplies.

Constant-current sources are not self-regulating, so they must be used with a voltage-sensing variable wire feed speed control. This type of control adjusts the wire feed speed in response to changes in arc voltage. The voltage is monitored to maintain a constant arc length. With this sys-

Figure 6.5–Typical DC Constant Voltage Power Supply for Submerged Arc Welding

tem, the arc voltage is dependent upon the wire feed speed and the wire diameter. The power source controls the arc current. Because voltage-sensing variable wire feed speed controls are more complex, they are also more expensive than the simple, constant wire feed speed controls that may be used with CV systems.

CV/CC Combination Power Sources

POWER SOURCES THAT can be switched between CV and CC modes are also available. Sources rated at up to 1500 A are available, but machines rated at 650 A or less are much more common. The value of these power sources lies in their versatility, since they can be used for SMAW, GMAW, GTAW, FCAW, air carbon arc cutting, and stud welding, in addition to submerged arc welding.

Alternating Current Power Sources

POWER SOURCES FOR ac welding are most commonly transformers. Sources rated for 800 to 1500 A at 100 percent duty cycle are available. If higher amperages are required, these machines can be connected in parallel.

Conventional ac power sources are the constant-current type. The output voltage of these machines approximates a square wave, and the output current approximates a sine wave as shown in Figure 6.6(A). The output of these machines drops to zero with each polarity reversal, so a high, open circuit voltage (greater than 80 V) is required to ensure reignition of the arc. Even at that high open circuit voltage, arc reignition problems are sometimes encountered with certain fluxes. Because these power supplies are the constant-current type, the speed controls must be voltage sensing, variable wire feed type.

The constant-voltage square wave ac power source is a relatively new type. Both the output current and the output voltage from these supplies approximate square waves. Because polarity reversals are instantaneous with square wave supplies, as is shown in Figure 6.6(B), arc reignition problems are not as severe as those encountered with conventional ac supplies. Hence, some fluxes that do not work with conventional ac sources will work with square wave ac supplies. Relatively simple, constant wire feed speed controls can be used with square wave supplies, since they supply constant voltage.

The most common uses of ac power for SAW are high-current applications, multiwire applications, narrow-gap welding, and applications where arc blow is a problem.

CONTROLS

THE CONTROL SYSTEMS used for semiautomatic submerged arc welding are simple wire feed speed controls. Controls used with constant-voltage power supplies maintain a constant wire feed speed. Controls used with constant-current

power supplies monitor the arc voltage and adjust the wire feed speed to maintain a constant voltage.

The simplest wire feeders have one-knob analog controls that maintain constant wire feed speed. A typical unit can be seen in Figure 6.7.

The state-of-the-art wire feeders used for automatic SAW, such as the one shown in Figure 6.8, have microprocessor-based digital controls. These controls have feed-

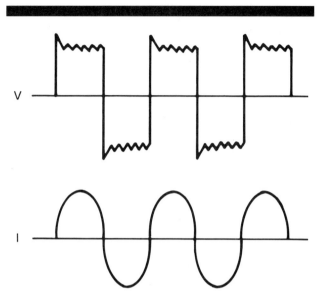

(A) CONVENTIONAL AC WAVEFORMS

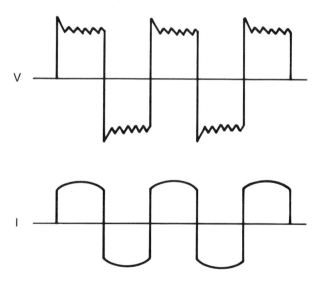

(B) SQUARE WAVE AC WAVEFORMS

Figure 6.6–Difference Between Conventional and Square Wave AC Waveforms

back loops interfaced with the power supply and wire feed motor, to maintain the welding voltage and wire speed at preset values. The great advantage of digital controls is their precise control of the welding process. The disadvantages are that the controls are not compatible with some power supplies, and they are slightly less rugged than most analog controls.

Digital controls are currently available only for use with constant voltage power supplies. These controls provide for wire feed speed adjustment (current control), power supply adjustment (voltage control), weld start-stop, automatic and manual travel on-off, cold wire feed up-down, run-in and crater fill control, burnback, and flux feed on-off. Digital current, voltage, and wire feed speed meters are standard on digital controls.

Analog controls are available for use with both constant-voltage and constant-current power supplies. Basic controls consist of a wire feed speed control (adjusts current in CV systems; controls voltage in CC systems), a power supply control (adjusts voltage in CV systems; adjusts current in CC systems), a weld start-stop switch, automatic or manual travel on-off, and cold wire feed up-down. These controls have the same advantages as analog controls for semiautomatic SAW, but they are prone to drift and do not allow precise process control.

Figure 6.8–Digital Control for Two-Wire Submerged Arc Welder

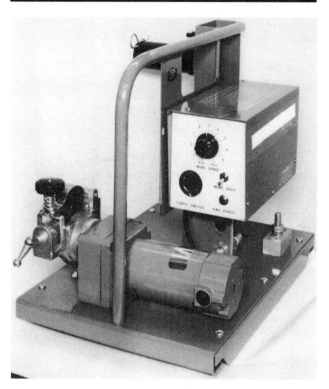

Figure 6.7–Example of a Wire Feeder Showing Control, Wire Feed Motor and Wire Drive Assembly

WELD HEADS AND TORCHES

A SUBMERGED ARC welding head comprises the wire feed motor and feed roll assembly, the torch assembly and contact tip, and accessories for mounting and positioning the head. A flux nozzle is usually mounted on the weld head, to deposit the flux either slightly ahead of or concentric with the welding wire.

Wire feed motors are typically heavy duty, permanent magnet-type motors with an integral reducing gearbox, feeding wire at speeds in the range of 20 to 550 in./min. (8 to 235 mm/sec).

The feed roll assembly may have one drive and one idler roll, two drive rolls, or four drive rolls. Four-roll drive assemblies are reported to provide positive feeding with the least wire slippage. Feed rolls may be knurled-V or smooth-V type; knurled-V rolls are the most common. In some cases, where the wire is being pushed through a conduit, smoother feeding will result if smooth V- groove rolls are used.

Torch assembly designs are numerous, but their purpose is always the same. The torch assembly guides the wire through the contact tip to the weld zone, and also delivers welding power to the wire at the contact tip.

Special equipment is needed for standard submerged arc welding, narrow groove (SAW-NG), and strip electrode SAW. Parallel wire SAW uses special feed roll and torch assemblies that provide positive feeding of two wires through one torch body. Strip electrode SAW also requires a special feed roll and torch assembly. Torches that feed strip are generally adjustable to accommodate several sizes of strip, typically 1.2, 1.8, 2.4, 3.5 in. wide, and up to 0.04 in. thick (30, 45, 60, 90 mm wide; up to 1 mm thick). The assemblies for parallel wire and strip electrode SAW are generally designed for mounting on standard welding heads with little or no modification.

The special SAW-NG equipment has long narrow torch assemblies and long narrow flux nozzles to deliver the flux and wire to the bottom of deep narrow grooves. These systems may also have some means to bend the wire to assure good side wall fusion in the narrow groove. Simple SAW-NG adaptors can be mounted directly on standard weld heads; more complex systems are available as complete weld head assemblies.

For semiautomatic SAW, the weld head may be a GMAW-type wire feeder that pushes the electrode through a conduit to the torch assembly. Such wire feeders accept any of the drive roll systems discussed above and are generally capable of feeding wire up to 3/32 in. (2.4 mm) in diameter at wire feed speeds over 550 in./min. (235 mm/s). The torch-conduit assembly allows for welding at up to 15 ft (4.6 m) from the wire feeder. Flux feed is provided either by a small 4 lb (1.8 kg) gravity feed flux hopper mounted on the torch, or from a remote flux tank that uses compressed air to push the flux to the weld zone. In both cases, the flux is delivered through the torch surrounding the welding wire. A typical semiautomatic SAW system is shown in Figure 6.2.

ACCESSORY EQUIPMENT

ACCESSORY EQUIPMENT COMMONLY used with SAW include travel equipment, flux recovery units, fixturing equipment, and positioning equipment.

Travel Equipment

WELD HEAD TRAVEL in SAW is generally provided by a tractor-type carriage, a side beam carriage, or a manipulator. Basic information about this equipment can be found in Chapter 10 of the *Welding Handbook*, Volume 1 (8th Edition).

A tractor-type carriage, as shown in Figure 6.9, provides travel along straight or gently curved weld joints by riding on tracks set up along the joint, or by riding on the workpiece itself. Trackless units use guide wheels or some other type of mechanical joint-tracking device. The weld head, control, wire supply, and flux hopper are generally mounted on the tractor. Maximum travel speeds possible with tractors are about 100 in./min. (45 mm/s). Tractors find the most use in field welding where their relative portability is necessary because the workpiece cannot be moved.

Side beam carriages provide linear travel only, and are capable of travel speeds in excess of 200 in./min. (85 mm/s). Because side beam systems are generally fixed and the workpiece must be brought to the weld station, their greatest use is for shop welding. The weld head, wire, flux hopper, and sometimes the control are mounted on the carriage. Figure 6.10 shows two weld heads mounted on a single carriage for a cladding operation.

Manipulators are similar to side beams, in that they are fixed and the workpiece must be brought to the welder. Manipulators are more versatile than side beams in that they are capable of linear motion in three axes. The weld head, wire, flux hopper, and often the control and operator ride on the manipulator. See Figures 6.11 and 6.12.

Figure 6.9–Submerged Arc Welding Head, Control, Wire Supply and Flux Hopper Mounted on a Tractor Type Carriage

Figure 6.10–Two Submerged Arc Welding Heads Mounted on a Side Beam Carriage for a Cladding Operation

Flux Recovery Units

FLUX RECOVERY UNITS are frequently used to maximize flux utilization and minimize manual clean-up. Flux recovery units may do any combination of the following:

(1) Remove unfused flux and fused slag behind the weld head.
(2) Screen out fused slag and other oversized material.
(3) Remove magnetic particles.
(4) Remove fines.
(5) Recirculate flux back to a hopper for reuse.
(6) Heat flux in a hopper to keep it dry.

Pneumatic flux feeding is commonly used in semiautomatic SAW and frequently in automatic SAW.

Positioners and Fixtures

BECAUSE SAW IS limited to flat position welding, positioners and related fixturing equipment find widespread use. Commonly used positioners include:

(1) Head-tailstock units, turning rolls, or both, to rotate cylindrical parts under the weld head (Figure 6.13)
(2) Tilting-rotating positioners, to bring the area to be welded on irregular parts into the flat position (Figure 6.14)

Custom fixturing often includes positioners to aid in setting up, positioning, and holding the workpiece. Turnkey systems are available.

Figure 6.11—Multiple Pass Submerged Arc Weld on a Heavy Wall Fixture Used to Test Blow-Out Preventers (A Positioner Rotates the Fixture Beneath a Stationary Head)

Figure 6.12—Hardfacing Material Being Applied to the Inside of a Ball Mill Shell Using the Submerged Arc Process (The Welding Head is Riding on a Beam-Mounted Travel Carriage)

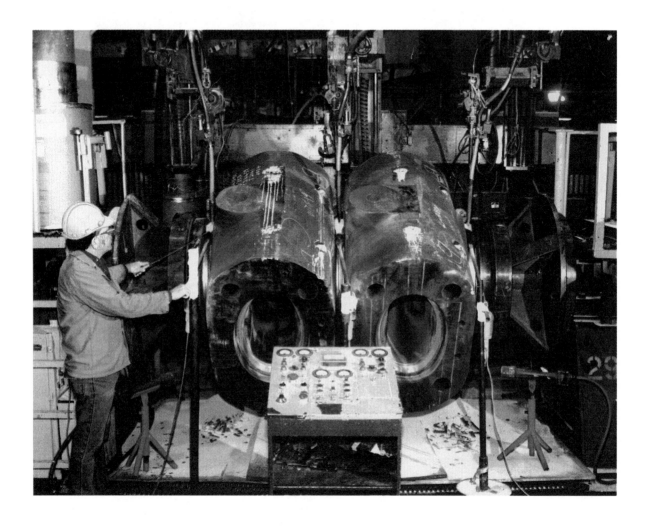

Figure 6.13–Three Circumferential Submerged Arc Welds Made Simultaneously on Heavy Wall Forgings to Fabricate a Blow-Out Preventer for an Oil Well (This Blow-Out Preventer, Low-Temperature Qualified to -70°F, is Designed for Arctic Service)

Figure 6.14–Multiple Pass Deep Groove Submerged Arc Weld Being Made on a Steam Chest and Throttle Valve for a Steam Turbine [A Positioner Rotates the Workpiece Beneath the Welding Head. The Base Metal is 2¹/₄ Cr, 1 Mo Steel and is 8-in (203 mm) Thick at the Weldment.]

MATERIALS

SUBMERGED ARC WELDING is used to fabricate most materials in use today, from "plain" carbon steels to exotic nickel-base alloys. Most steels and alloys are readily weldable with commercially available wires and fluxes. However, some metals require special heats of electrode wire manufactured to precise chemistries, and special fluxes designed to obtain specific weld joint properties.

BASE METALS

THE FOLLOWING ARE general classes of base metals welded:

(1) Carbon steels up to 0.29 percent carbon
(2) Low alloy steels [up to 100 ksi (690 MPa) yield strength]
(3) Chromium - molybdenum steels (1/2 to 9 percent Cr and 1/2 to 1 percent Mo)
(4) Stainless steels
(5) Nickel-base alloys

The alloy compositions that may be submerged arc welded have expanded with the availability of suitable electrodes and fluxes. Electrode-flux combinations are usually classified by code specifications. Data on special wire-flux combinations for base metals which are less widely used can be obtained from flux manufacturers.

ELECTRODES

SUBMERGED ARC ELECTRODES produce weld deposits matching carbon steel, low alloy steel, high carbon steels, special alloy steels, stainless steels, nickel alloys, and special alloys for surfacing applications. These electrodes are supplied as bare solid wire and as composite metal-cored electrodes (similar to flux-cored arc welding electrodes).

Electrode manufacturers prepare composite electrodes that duplicate complex alloys by enclosing required alloying elements in a tube of more available composition (stainless steel or other metals).

Electrodes are normally packaged as coils or drums ranging in weight from 25 to 1000 lb (11 to 454 kg). Large electrode packages are economical. They increase operating efficiency and eliminate end-of-coil waste.

Steel electrodes are usually copper coated, except those for welding corrosion resisting materials or for certain nuclear applications. The copper coating provides good shelf life, decreases contact tube wear, and improves electrical conductivity. Electrodes are packaged to ensure long shelf life when stored indoors under normal conditions.

Submerged arc welding electrodes vary in size from 1/16 to 1/4 in. (1.6 to 6.4 mm) in diameter. General guidelines for amperage range selection are presented in Table 6.1. The wide amperage ranges are typical of submerged arc welding.

FLUXES

FLUXES SHIELD THE molten weld pool from the atmosphere by covering the metal with molten slag (fused flux). Fluxes clean the molten weld pool, modify the chemical composition of the weld metal, and influence the shape of the weld bead and its mechanical properties. Fluxes are granular mineral compounds mixed according to various formulations. Based on the choice of several manufacturing methods, the different types of fluxes are fused, bonded (also known as agglomerated), and mechanically mixed.

Fused Fluxes

TO MANUFACTURE A fused flux, the raw materials are dry mixed and melted in an electric furnace. After melting and any final additions, the furnace charge is poured and cooled. Cooling may be accomplished by shooting the melt through a stream of water or by pouring it onto large chill blocks. The result is a product with a glassy appearance which is then crushed, screened for size, and packaged.

Fused fluxes have the following advantages:

Table 6.1
Submerged Arc Wires -Diameters vs. Current Range

Wire Diameter		Current Range
in.	mm	(Amperes)
5/64	2.3	200 - 500
3/32	2.4	300 - 600
1/8	3.2	300 - 800
5/32	4.0	400 - 900
3/16	4.8	500 - 1200
7/32	5.6	600 - 1300
1/4	6.4	600 - 1600

(1) Good chemical homogeneity

(2) Easy removal of the fines without affecting the flux composition

(3) Not hygroscopic normally, which simplifies handling, storage, and welding problems

(4) Readily recycled through feeding and recovery systems without significant change in particle size or composition

Their main disadvantage is the difficulty of adding deoxidizers and ferro-alloys to them during manufacture without segregation or extremely high losses. The high temperatures needed to melt the raw ingredients limit the range of flux compositions.

Bonded Fluxes

TO MANUFACTURE A bonded flux, the raw materials are powdered, dry mixed, and bonded with either potassium silicate, sodium silicate, or a mixture of the two. After bonding, the wet mix is pelletized and baked at a temperature lower than that used for fused fluxes. The pellets are then broken up, screened to size, and packaged.

The advantages of bonded fluxes include the following:

(1) Easy addition of deoxidizers and alloying elements; alloying elements are added as ferro-alloys or as elemental metals to produce alloys not readily available as electrodes, or to adjust weld metal compositions.

(2) Usable with thicker layer of flux when welding

(3) Color identification

The disadvantages are the following:

(1) Tendency for some fluxes to absorb moisture in a manner similar to coatings on some shielded metal arc electrodes

(2) Possible gas evolution from the molten slag

(3) Possible change in flux composition due to segregation or removal of fine mesh particles

Mechanically Mixed Fluxes

TO PRODUCE A mechanically mixed flux, two or more fused or bonded fluxes are mixed in any ratio necessary to yield the desired results.

The advantage of mechanically mixed fluxes is that several commercial fluxes may be mixed for highly critical or proprietary welding operations.

The following are disadvantages of mechanically mixed fluxes:

(1) Segregation of the combined fluxes during shipment, storage, and handling

(2) Segregation occuring in the feeding and recovery systems during the welding operation

(3) Inconsistency in the combined flux from mix to mix

Particle Size and Distribution

FLUX PARTICLE SIZES and their uniform distribution within the bulk flux are important because that influences feeding and recovery, amperage level, and weld bead smoothness and shape. As amperage increases, the average particle size for fused fluxes should be decreased and the percentage of small particles should be increased. If the amperage is too high with a given particle size, the arc may be unstable and leave ragged, uneven bead edges. When rusty steel is welded, coarse particle fluxes are preferable, because they allow gases to escape more easily.

Some flux manufacturers may mark their packages with sizing information presented in the form of two mesh numbers. The numbers represent the largest and smallest particle sizes present when standardized screens are used to measure them. The first number identifies the mesh (screen) through which essentially all of the particles will pass, and the second number identifies the mesh through which most or all of the particles will not pass.

Those particle sizes do not provide all the information that may be needed. For instance, designating a flux as having a mesh size 20 x 200 does not indicate whether the flux is coarse with some fines, or is fine with some coarse particles. All that is known is the range.

Some flux manufacturers offer each of their fluxes in only one particle size range, customized for a general area of flux application.

Flux Usage

IF THE FLUX is too fine, it will pack and not feed properly. If a fine flux or a flux with small amounts of fine particles is recovered by a vacuum system, the fine particles may be trapped by the system. Only the coarser particles will be returned to the feeding system for reuse, which may cause welding problems.

In applications where low hydrogen considerations are important, fluxes must be kept dry. Fused fluxes do not contain chemically bonded H_2O, but the particles hold surface moisture. Bonded fluxes contain chemically bonded H_2O, and may hold surface moisture as well. Bonded fluxes need to be protected in the same manner as low-hydrogen shielded metal arc electrodes. The user should follow the directions of the flux manufacturer for specific baking procedures.

When alloy-bearing fluxes are used, it is necessary to maintain a fixed ratio between the quantities of flux and electrode melted, to obtain a consistent weld metal composition. This ratio is actually determined by the variables of the welding procedure. For example, deviation from an established volt-ampere relation will change the alloy content of the weld metal by changing the flux-electrode melting ratio.

Fluxes are also identified as chemically basic, chemically acid, or chemically neutral. The basicity or acidity of a flux is related to the ease with which the component oxides of the flux ingredients dissociate into a metallic cation and an oxygen anion. Chemically basic fluxes are normally high in MgO or CaO, while chemically acid fluxes are normally high in SiO_2.

The basicity or acidity of a flux is often referred to as the ratio of CaO or MgO to SiO_2. Fluxes having ratios greater than one are called *chemically basic*. Ratios near unity are *chemically neutral*. Those less than unity are *chemically acidic*.

Basic fluxes have recently become the prime fluxes for welding in critical applications where close controls on deposit properties and chemistry are required. Most of the basic group are formulated for specific wire deposits, i.e., fluxes that stabilize chromium or carbon loss. They limit transfers of silicon/manganese/oxygen from the slag to the weld metal. Basic fluxes are available to suit any material weldable by submerged arc.

WELDING OF CARBON STEEL MATERIALS

CARBON STEEL MATERIALS are usually welded with electrode-and- flux combinations classified under AWS standard A5.17, *Carbon Steel Electrodes and Fluxes for Submerged Arc Welding*. Typical steels that are welded with these consumables are listed in ANSI/AWS D1.1, *Structural Welding Code-Steel*, as Group I and II classifications. These steels include ASTM A106 Grade B, A36, A516 Grades 55 to 70, A537 Class 1, A570 Grades 30 to 50, API 5LX Grades X42 to X52, and ABS Grades A to EH36. These steels are usually supplied in the as-rolled or the normalized condition.

Table 6.2 lists minimum mechanical properties for various wire/flux combinations. When selecting SAW welding consumables, it is required that both the minimum tensile and minimum yield strengths as well as the notch toughness properties (when required) of the weld metal be matched with the base metal. AWS *Filler Metal Comparison charts* shows the commercial products that meet the AWS wire-flux classifications listed in Table 6.2. In special applications, particularly carbon steel weldments subject to long term postweld heat treatment, low alloy submerged arc welding consumables covered by ANSI/AWS A5.23, *Specifications for Low Alloy Steel Electrodes and Fluxes*, may be required to meet tensile properties of the base metal. ANSI/AWS A5.23 lists welding consumables used with carbon steel base materials to meet special notch toughness requirements.

Preheat and Maximum Interpass Temperature Requirements

MINIMUM PREHEAT TEMPERATURES for SAW of structural steels are stated in Table 4.3 of ANSI/AWS D1.1-90. Alternate methods of determining minimum preheat and interpass temperatures for similar steels are shown in Appendix XI of the same document. Although these minimum

Table 6.2
Minimum Mechanical Properties with Carbon Steel Consumables Covered by AWS A5.17

AWS Classification	Welding Condition	Tensile Strength		Yield Strength		% Elongation in 2 inches	Charpy Impact Values		
		KSI	MPA	KSI	MPA		(Ft-Lbs)	(Joules)	Test Temp.
F6A2-EL12	AW	60	414	48	331	22	20	27	-20°F (-29°C)
F6A6-EL12	AW	60	414	48	331	22	20	27	-60°F (-51°C)
F7A2-EL12	AW	70	483	58	400	22	20	27	-20°F (-29°C)
F6P4-EM12K	SR	60	414	48	331	22	20	27	-40°F (-40°C)
F7A2-EM12K	AW	70	483	58	400	22	20	27	-20°F (-29°C)
F7A6-EM12K	AW	70	483	58	400	22	20	27	-60°F (-51°C)
F7A2-EH14	AW	70	483	58	400	22	20	27	-20°F (-29°C)

1. Actual mechanical properties obtained may significantly exceed minimum values shown.
2. Type of welding flux (manufacture) greatly influences CVN impact properties of the weld metal.
3. Caution should be used when these weld deposits are stress relieved, they may fall below base metal strengths.
4. Test data on 1 inch thick plate (ASTM A36 plate)
5. Stress relieved condition [1150°F (621°C)] for 1 hour.

preheat temperatures are generally acceptable, it must be recognized that certain joint designs and higher strength welding consumables result in high residual stress when used on carbon steel base materials, which may require higher preheats to avoid both liquation cracking and delayed (hydrogen) cracking. Other codes or engineering specifications may require higher preheats than specified or calculated using ANSI/AWS D1.1.

The maximum interpass temperature for welding carbon steels is typically limited to 500°F (260°C). Studies have shown that as preheat and maximum interpass temperature increase, the size of the heat-affected zone increases, tensile strength decreases, and notch toughness properties of both the deposited weld metal and the heat-affected zone decrease. Lower maximum interpass temperatures may be specified when special mechanical properties must be obtained, such as minimum notch toughness properties at low temperatures.

Special Service Conditions

SOME CARBON STEEL components that are submerged arc welded will be used in special service conditions where the hardness of the weld metal, heat-affected zone, and plate must not exceed a specified maximum level. This is usually required in the oil industry, where the component will be exposed to wet hydrogen sulfide gas. It has been found that if the hardness is kept below a prescribed level, depending on the type of material and the service conditions, stress corrosion cracking will generally not occur.

Hardness is controlled by selecting welding consumables that will produce low weld metal hardness, less than 200 Brinell for carbon steel, or by preheating the work or

raising the welding heat input. Increasing preheat and heat input slows the weld cooling rate thus producing softer weld metal and heat-affected zone microstructures.

Postweld heat treatment may also reduce the hardness of the weld metal, the heat-affected zone, and the base metal.

CARBON STEEL ELECTRODES AND FLUXES

AWS SPECIFICATION A5.17 prescribes the requirements for electrodes and fluxes for submerged arc welding of carbon steels. Solid electrodes are classified on the basis of chemical composition (as manufactured), while composite electrodes are based on deposit chemistry. Fluxes are classified on the basis of weld metal properties obtained when used with specific electrodes. Table 6.3 shows the classification system for flux-electrode combinations.

Carbon steels are defined as those steels having additions of carbon up to 0.29 percent, manganese up to 1.65 percent, silicon up to 0.60 percent, and copper up to 0.60 percent, with no specified range of other alloying elements.

Fluxes are classified on the basis of the chemical composition and mechanical properties of the deposited weld metal with some particular classification of electrode. Selection of SAW consumables will depend on the chemical and mechanical properties required for the component being fabricated, the welding position (1G, 2G, 2F), and any required surface preparation of the steel to be welded.

SAW consumable manufacturers produce electrode/flux combinations which are formulated to meet specific chemical and mechanical property requirements and weldability conditions. Purchasing consumables always from the same

Table 6.3
Classification System for Flux-Electrode Combination
See AWS Publication A5.17 (Latest Edition) for Additional Information

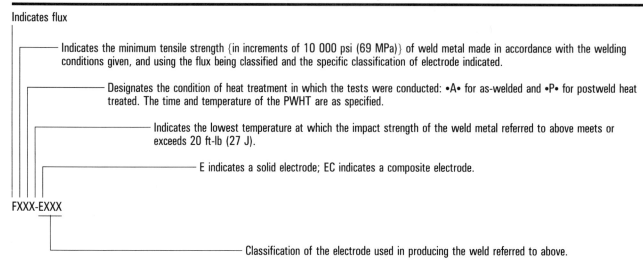

Indicates flux

Indicates the minimum tensile strength {in increments of 10 000 psi (69 MPa)} of weld metal made in accordance with the welding conditions given, and using the flux being classified and the specific classification of electrode indicated.

Designates the condition of heat treatment in which the tests were conducted: •A• for as-welded and •P• for postweld heat treated. The time and temperature of the PWHT are as specified.

Indicates the lowest temperature at which the impact strength of the weld metal referred to above meets or exceeds 20 ft-lb (27 J).

E indicates a solid electrode; EC indicates a composite electrode.

FXXX-EXXX

Classification of the electrode used in producing the weld referred to above.

Examples

F7A6-EM12K is a complete designation. It refers to a flux that will produce weld metal which, in the as-welded condition, will have a tensile strength no lower than 70 000 psi (480 MPa) and Charpy V-notch impact strength of a least 20 ft-lb (27 J) at -60°F (-51°C) when produced with an EM12K electrode under the conditions called for in this specification.

F7A4-EC1 is a complete designation for a flux when the trade name of the electrode used in classification is indicated as well. It refers to a flux that will produce weld metal with that electrode, which in the as-welded condition, will have a tensile strength no lower than 70 000 psi (480 MPa) and Charpy V-notch energy of at least 20 ft-lb (27 J) at -40°F (-40°C) under the conditions called for in this specification.

---

manufacturer is recommended when composite electrodes are used, whereas with solid electrodes the fabricator can pick and choose among available fluxes to be used with a given AWS electrode classification. It should be noted that the electrode chosen has the greatest influence on the resulting deposited weld metal chemistry, while the flux chosen has a great effect on the Charpy V-notch (CVN) impact properties and the overall weldability of the electrode/flux combination. The following items must be considered when selecting SAW consumables:

(1) Whether to choose a "neutral" or "active" flux. A neutral flux adds little or no alloying elements to the weld deposit, whereas an active flux adds alloying elements to the deposited weld metal. Active fluxes are usually chosen for single pass welding operations; their application for multiple pass use may be limited by engineering specifications because of concern that excessive alloy build-up will take place in the deposited weld metal.

(2) Whether the fluxes being considered are correctly balanced in chemical composition, for use with a given electrode classification.

(3) The mechanical property requirements required. This includes CVN impact properties as well as the strength and ductility of the resulting deposit.

(4) Usability of a given electrode/flux combination, including wetting of side walls without undercut or cold lap, ability to weld over rust and scale, and ease of slag removal.

WELDING OF LOW ALLOY STEEL MATERIALS

LOW ALLOY STEEL materials are welded with electrode and flux combinations classified under ANSI/AWS A5.23. Low alloy steels are divided into many subgroups, using the chemical composition and tensile strength of the steels as the factor which determines which SAW consumables should be chosen to weld the joints. In many cases, it is essential that the weld metal composition be similar to the base metal composition in terms of alloy elements, in order to meet service requirements. This is particularly true for components that will be used at temperatures greater than 650°F (345°C), where oxidation resistance and elevated tensile properties are important. Major groupings of low alloy steels are described below.

Higher Strength Low Alloy Steels

HIGHER STRENGTH LOW alloy steels are steels with relatively low chemical additions, usually less than 1 percent of the chemical composition, of Cr, Cu, Ni, Cb, and V. These steels are usually supplied in the as-rolled, normalized, or quenched and tempered condition from the manufacturer, depending on the material specification requirements. Steels that are welded with SAW consumables covered by ANSI/AWS A5.23 include ASTM specifications A242, A537 Class 1 and 2, A572 Grades 42-65, A588, and A633 Grades A-E. It must be noted that some of these steels can also be welded with consumables specified in ANSI/AWS A5.17, the choice depends on the mechanical property requirements.

ASTM A242 and A588 steels are resistant to rusting. Some applications require welding these materials with electrode/flux combinations that will result in the same appearance and oxidation resistance as the base material.

Some of the steels listed above can be produced in "microalloyed" versions generally available from Japan or Europe. The microalloy additions are amounts less than 0.1 percent of boron, columbium, and vanadium. Many of the weldable microalloyed steels have tensile strengths up to approximately 80 ksi (552 MPa).

High Yield Strength, Quenched and Tempered Low Alloy Steel

HIGH YIELD STRENGTH, quenched and tempered low alloy steels are similar to the higher strength low alloys steels mentioned above, except that they have higher amounts of alloy additions, up to approximately 2 percent each of Cr, Cu, Ni, Cb, and V. These steels are always supplied in the heat-treated condition. Typical steels that are welded with SAW consumables in ANSI/AWS A5.23 include ASTM A514 and A517. There are many different grades of these steels with varying chemical composition. When welding these steels, the manufacturer of the steel should be consulted to determine which consumables are recommended. Selection of the electrode/flux combination depends on the thickness of the steels and the mechanical properties required, including notch toughness properties.

Carbon-Molybdenum Steels

CARBON-MOLYBDENUM STEELS ARE similar to carbon steels except they have an addition of approximately 0.5 percent of molybdenum. These steels are used in pressure vessels or pipe lines operating at elevated temperature. They include ASTM A204 Grades A, B, C, and ASTM A182 Grade F1 forgings.

Carbon-molybdenum steels are produced in the as-rolled or normalized conditions. After completion of welding, the weldments should usually be postweld heat-treated.

Chromium-Molybdenum Steels

CHROMIUM-MOLYBDENUM STEELS ARE steels containing varying amounts of chromium, up to a nominal 9 percent chemical composition, and molybdenum up to a nominal 1 percent chemical composition. These steels usually come from the steel manufacturer in the annealed, normalized and tempered, or quenched and tempered condition. These steels are also used in pressure vessels and pipe lines operating at elevated temperature. They include ASTM A387 grades 2, 5, 7, 9, 11, 12, 21, and 22, and ASTM A182 forging grades F2, F5, F7, F9, F11, F12, F21, and F22.

Other Alloy Steels

A LARGE NUMBER of nickel, nickel-molybdenum, nickel-chromium-molybdenum and other steels may be welded using consumables listed in ANSI/AWS A5.23.

LOW ALLOY STEEL ELECTRODES AND FLUXES

LOW ALLOY STEELS usually have less than 10 percent of any one alloying element. Low alloy steel weld metal may be deposited by solid alloy steel electrodes, fluxes containing the alloying elements, and composite electrodes where the core contains the alloying elements. Alloy steel electrodes and composite electrodes are normally welded under a neutral flux. Alloy-bearing fluxes are generally used with carbon steel electrodes to deposit alloyed weld metal. Many electrode-flux combinations are available.

ANSI/AWS A5.23 prescribes requirements for solid and composite electrodes and fluxes for welding low alloy steels. The fluxes are classified according to the weld metal properties obtained when used with specific electrodes. The required chemical composition for the electrodes or deposits, or both, and other information, are detailed in the latest revision of the specification.

STAINLESS STEELS

STAINLESS STEELS ARE capable of meeting a wide range of final needs such as corrosion resistance, strength at elevated temperatures, and toughness at cryogenic temperatures, and are selected for a broad range of applications.

The stainless steels most widely used for welded industrial applications are classified as follows:

(**1**) Martensitic
(**2**) Ferritic
(**3**) Austenitic
(**4**) Precipitation hardening
(**5**) Duplex or ferritic-austenitic

Filler metals for fabricating these steels are specified in ANSI/AWS A5.9, *Specification for Corrosion-Resisting*

Chromium and Chromium-Nickel Steel Bare and Composite Metal Cored and Stranded Welding Electrodes and Welding Rods.

Not all stainless steels are readily weldable by the submerged arc process, and some require that special considerations be followed. In stainless steels and nickel base alloys the main advantage of submerged arc welding, its high deposition rates, sometimes becomes a disadvantage. As deposition rates increase, so does heat input, and in stainless alloys high heat inputs may cause deleterious microstructural changes. Brief comments about each class of stainless steels and pertinent welding considerations are presented in the following sections. Metallurgical summaries will be found in Volume 4 of the Welding Handbook, 7th Edition.

MARTENSITIC STAINLESS STEELS

THE MARTENSITIC AISI "400" series stainless steels have 11.5 to 18 percent Cr as the major alloying element. AISI type 410, which may also be produced as a casting known as CA-15, has 11.5 to 13.5 percent Chromium. Now a relatively new low carbon 13 Cr-4.5 Ni-0.5 Mo casting grade (CA-6NM) provides higher strength, increased toughness, better weldability, and greater corrosion resistance than the traditional CA-15 casting grade. Both alloys are martensitic stainless steels.

The AISI 500 series (e.g., type 502 with 5 Cr-0.5 Mo, and type 505 with 9 Cr-1 Mo) are heat-resisting steels, although not classed as stainless because their chromium content is well under the required 11 percent minimum. They are nevertheless martensitic.

Martensitic stainless steels when rapidly cooled have a brittle structure. Preheating the base metal retards the rate of cooling, reducing shrinkage stresses and allowing dissolved hydrogen to escape.

FERRITIC STAINLESS STEELS

THE AISI 400 series also covers the ferritic stainless steels. As chromium increases beyond 18 percent in steels, the predominant metallurgical structure is ferrite even at elevated temperatures, if the carbon content is low. Ferritic stainless steels are relatively nonhardening. Because they are quite magnetic, they are subject to arc blow during welding. If the heat of welding causes carbon and nitrogen to combine with chromium to form carbides and nitrides at grain boundaries, the result may be intergranular corrosion, although not to the extent experienced with austenitic stainless steels. Carbide precipitation is discussed in detail in Volume 4 of the *Welding Handbook*, 7th Edition.

Because of embrittlement problems, the ferritic stainless steels are not considered readily weldable, and standard filler metals are not readily available. Flux cored stainless wire can be purchased on special order from several manufacturers. Austenitic filler metals 309, 310, and 312 are often used where the application can reconcile the different corrosion resistant characteristics and the greater coefficients of linear expansion of an austenitic weld metal. Where postweld annealing at 1450°F (790°C) is specified, the austenitic filler metal should be a stabilized grade or a low carbon grade to avoid carbide precipitation.

AUSTENITIC STAINLESS STEELS

AUSTENITIC STAINLESS STEELS are essentially chromium-nickel alloys. They are covered by AISI 300 classifications. Corresponding AWS ER3XX filler metals or over-matching 300 series filler metals are used in over 90 percent of stainless steel welding applications. Such welds have good corrosion resistance and excellent strength at both low and high temperatures, characterized by a high degree of toughness even in the as-welded condition. Being austenitic, these weldments are virtually nonmagnetic and not subject to arc blow in welding.

Although the 300 series austenitic stainless steels are welded with greater ease than the 400 series, there are several factors peculiar to the 300 series stainless steels that must be considered to ensure the production of satisfactory weldments. When compared with plain carbon, low alloy and 400 series stainless steels, the austenitic stainless steels have lower melting points, higher electrical resistance, one-third the thermal conductivity, and as much as 50 percent greater coefficients of expansion. For these reasons less heat input (less current) is required for fusion, and the heat concentrates in a small zone adjacent to the weld. Greater thermal expansion may result in warping or distortion, especially in thin sections, suggesting greater need for jigging to maintain dimensional control.

The structure of austenitic stainless steel weld metals varies from fully austenitic in type 310 to dual phase austenitic-ferritic in types 308, 309, 312, and 316. Some ferrite is desired in such weldments for crack resistance. Techniques for measuring the ferrite number (FN) of stainless steel weld metals are described in ANSI/AWS A4.2, *Standard Procedures for Calibrating Magnetic Instruments to Measure the Delta Ferrite Content of Austentic Stainless Steel Weld Metal.* The ferrite number (FN) should exceed 7 to ensure adequate crack resistance and good low temperature impact strength, but it should not exceed 10 FN to prevent formation of brittle sigma phase when the weldment is exposed to a temperature range of 900 to 1700°F.

PRECIPITATION HARDENING STAINLESS STEELS

THE PRECIPITATION HARDENING (PH) stainless steels are a family of Fe-Cr-Ni alloys with additives such as copper, molybdenum, columbium, titanium, and aluminum. Precipitation hardening results when a supercooled solid solution (solution annealed) changes its metallurgical structure on aging. The advantage of a PH steel is that components can be fabricated in the annealed condition and subsequently hardened (strengthened) by treatment at 900 to

1100°F (480 to 590°C), minimizing the problems associated with high-temperature quenching. Strength levels up to 260 ksi (1800 MPa) can be achieved (exceeding the strength of martensitic stainless steels), with corrosion resistance similar to that of Type 304.

PH steels fall within three general groups: martensitic PH, semiaustenitic PH, and austenitic PH stainless steels. No AWS "PH" specifications have been issued. Proprietary electrodes are available.

DUPLEX STAINLESS STEELS

DUPLEX STAINLESS STEELS have a microstructure that is part austenitic and part ferritic. Many austenitic stainless steel grades are "dual-phase", with ferrite levels from about 7FN in type 308 to over 28FN in type 312. The *duplex* grades have approximately 50 percent ferrite. They are low in carbon, and usually contain about 0.10 to 0.20 percent Nitrogen. Weldments in duplex stainless steels which do not contain nitrogen may have wholly ferritic heat-affected zones and may contain the brittle sigma phase. Type 312 is not a duplex grade because of its high carbon content.

Duplex stainless steels combine some of the better features of austenitic stainless steels (which are vulnerable to stress- corrosion cracking [SCC] in chloride environments) and ferritic stainless steels (which are brittle). Compared with lower ferrite austenitic grades, duplex austenite-ferrite filler metal grades exhibit high strength (more than twice the yield strength), dramatically better resistance to SCC in chloride solutions, but lower ductility and lower toughness.

Proprietary duplex stainless steel filler metals have been successfully used in many applications. Such filler metals should be intentionally rich in austenitizers, mainly nickel, but also nitrogen. There are presently no AWS specifications for them.

STAINLESS STEEL ELECTRODES AND FLUXES

ANSI/AWS A5.9, COVERS filler metals for welding corrosion or heat resisting chromium and chromium-nickel steels. This specification includes steels in which chromium exceeds 4 percent and nickel does not exceed 50 percent of the composition. Solid wire electrodes are classified on the basis of their chemical composition, as manufactured, and composite electrodes on the basis of the chemical analysis of a fused sample. The American Iron and Steel Institute numbering system is used for these alloys.

Fluxes for stainless steel SAW are proprietary. Manufacturers of fluxes should be consulted for recommendations. Submerged arc fluxes are available in fused and bonded types for welding stainless alloys. Some bonded fluxes contain chromium, nickel, molybdenum, or columbium to replace elements lost across the arc. The newer chemically basic fluxes have shown more consistant element recovery than earlier less basic or acid types. Performance of fluxes for stainless steel weldments may depend on the user's care in flux handling and reuse. Over-recycled fluxes will become depleted in compensating elements. Refer to the manufacturer's recommendation for handling and recycling of flux.

Nickel Alloy Steels

NICKEL ALLOY STEELS contain between 1 and 3.5 percent nickel. These steels are used in low temperature applications [below -50°F (-46°C)] because they have good notch toughness properties at low temperatures. These steels include ASTM A203 Grades A, B, C, D, E and F and ASTM A350 Grades LF3, LF5, and LF9.

When electrode/flux consumables are selected, the important factors to match are both the minimum tensile strength and the minimum notch toughness of the base metal. It is usually difficult to produce impact values exceeding 20 ft-lbs at -100°F in SAW weldments.

Nickel and Nickel Alloy Electrodes and Fluxes

NICKEL AND NICKEL alloy electrodes in wire form are available for submerged arc welding. ANSI/AWS A5.14, *Specification for Nickel and Nickel Alloy Bare Welding Electrodes and Rods*, covers nickel and nickel alloy filler metals. The electrodes are classified according to their chemical compositions, as manufactured. For specific information on these electrodes, consult the latest revision of the specification. Fluxes for nickel and nickel alloy SAW welding have proprietary compositions. The flux manufacturers should be consulted for recommendations.

GENERAL PROCESS APPLICATIONS

SAW IS USED in a wide range of industrial applications. High weld quality, high deposition rates, deep penetration, and adaptability to automatic operation make the process suitable for fabrication of large weldments. It is used extensively in pressure vessel fabrication, ship and barge building, railroad car fabrication, pipe manufacturing, and the fabrication of structural members where long welds are required. Automatic SAW installations manufacture mass produced assemblies joined with repetitive short welds.

The process is used to weld materials ranging from 0.06 in. (1.5 mm) sheet to thick, heavy weldments. Submerged arc welding is not suitable for all metals and alloys. It is widely used on carbon steels, low alloy structural steels, and stainless steels. It joins some high-strength structural steels, high-carbon steels, and nickel alloys. However, better joint properties are obtained with these metals by using a process with lower heat input to the base metal, such as gas metal arc welding.

Submerged arc welding is used to weld butt joints in the flat position, fillet welds in the flat and horizontal positions, and for surfacing in the flat position. With special tooling and fixturing, lap and butt joints can be welded in the horizontal position.

OPERATING VARIABLES

CONTROL OF THE operating variables in submerged arc welding is essential if high production rates and welds of good quality are to be obtained. These variables, in their approximate order of importance, are the following:

(1) Welding amperage
(2) Type of flux and particle distribution
(3) Welding voltage
(4) Welding speed
(5) Electrode size
(6) Electrode extension
(7) Type of electrode
(8) Width and depth of the layer of flux

The operator must know how the variables affect the welding action and what changes should be made to them.

WELDING AMPERAGE

WELDING CURRENT IS the most influential variable because it controls the rate at which the electrode is melted and therefore the deposition rate, the depth of penetration, and the amount of base metal melted. If the current is too high at a given travel speed, the depth of fusion or penetration will be too great. The resulting weld may tend to melt through the metal being joined. High current also leads to waste of electrodes in the form of excessive reinforcement. This overwelding increases weld shrinkage and causes greater distortion.

If the current is too low, inadequate penetration or incomplete fusion may result. The effect of current variation is shown in Figure 6.15.

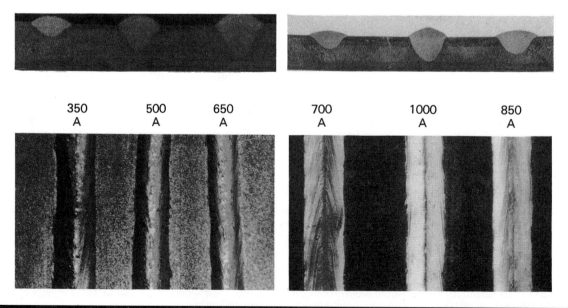

SEMIAUTOMATIC WELDING
3/32 in. (2.4 mm) WIRE, 35 V,
24 in./MIN (10 mm/s)

FULLY AUTOMATIC WELDING
7/32 in. (5.6 mm) WIRE, 34 V,
30 in./MIN (13 mm/s)

| 350 A | 500 A | 650 A | 700 A | 1000 A | 850 A |

Figure 6.15–Effect of Amperage Variation on Weld Bead Shape and Penetration

Following are three rules concerning welding current:

(**1**) Increasing current increases penetration and melting rate.
(**2**) Excessively high current produces a digging arc and undercut, or a high, narrow bead.
(**3**) Too low welding current produces an unstable arc.

WELDING VOLTAGE

WELDING VOLTAGE ADJUSTMENT varies the length of the arc between the electrode and the molten weld metal. If the overall voltage is increased, the arc length increases; if the voltage decreased, the arc length decreases.

Voltage has little effect on the electrode deposition rate, which is determined by welding current. The voltage principally determines the shape of the weld bead cross section and its external appearance. Figure 6.16 illustrates this effect.

Increasing the welding voltage with constant current and travel speed will:

(**1**) Produce a flatter and wider bead.
(**2**) Increase flux consumption.
(**3**) Tend to reduce porosity caused by rust or scale on steel.
(**4**) Help bridge an excessive root opening when fit-up is poor.

(**5**) Increase pickup of alloying elements from an alloy flux.

Excessively high-arc voltage will:

(**1**) Produce a wide bead shape that is subject to cracking.
(**2**) Make slag removal difficult in groove welds.
(**3**) Produce a concave shaped weld that may be subject to cracking.
(**4**) Increase undercut along the edge(s) of fillet welds.

Lowering the voltage produces a "stiffer" arc, which improves penetration in a deep weld groove and resists arc blow. An excessively low voltage produces a high, narrow bead and causes difficult slag removal along the bead edges.

TRAVEL SPEED

WITH ANY COMBINATION of welding current and voltage, the effects of changing the travel speed conform to a general pattern. If the travel speed is increased, (1) power or heat input per unit length of weld is decreased, and (2) less filler metal is applied per unit length of weld, resulting in less weld reinforcement. Thus, the weld bead becomes smaller, as shown in Figure 6.17.

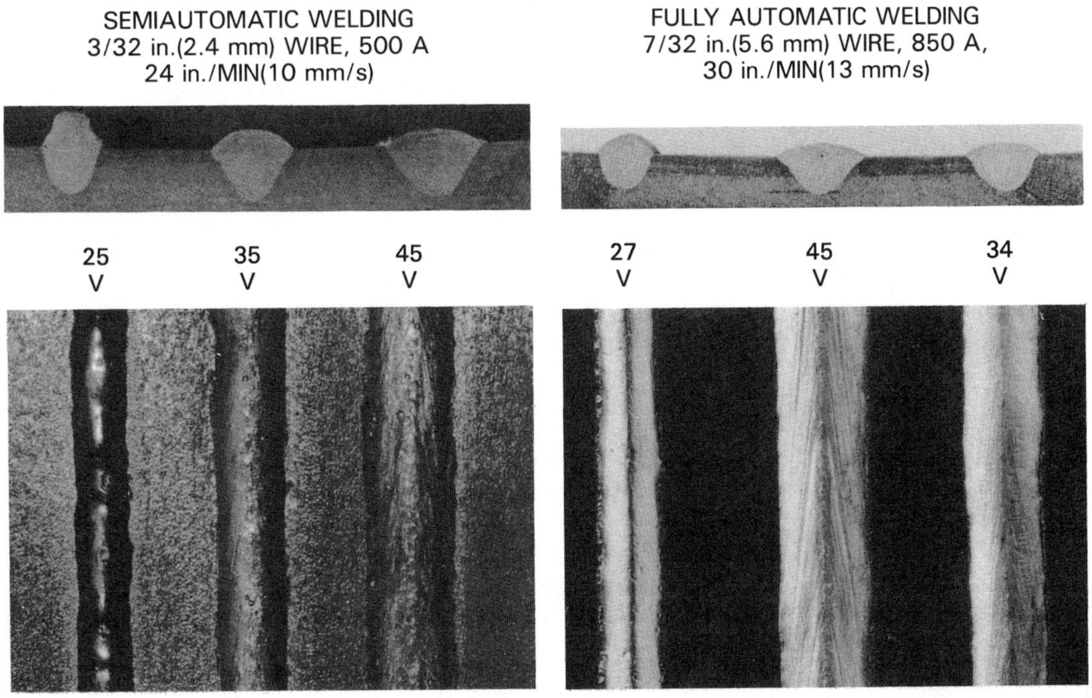

SEMIAUTOMATIC WELDING
3/32 in.(2.4 mm) WIRE, 500 A
24 in./MIN(10 mm/s)

25 V 35 V 45 V

FULLY AUTOMATIC WELDING
7/32 in.(5.6 mm) WIRE, 850 A,
30 in./MIN(13 mm/s)

27 V 45 V 34 V

Figure 6.16–Effect of Arc Voltage Variations on Weld Bead Shape and Penetration

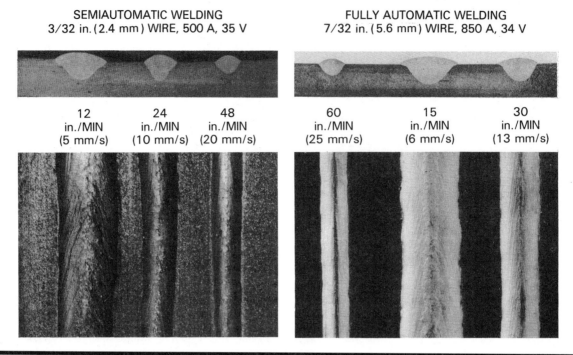

SEMIAUTOMATIC WELDING
3/32 in. (2.4 mm) WIRE, 500 A, 35 V

FULLY AUTOMATIC WELDING
7/32 in. (5.6 mm) WIRE, 850 A, 34 V

| 12 in./MIN (5 mm/s) | 24 in./MIN (10 mm/s) | 48 in./MIN (20 mm/s) | 60 in./MIN (25 mm/s) | 15 in./MIN (6 mm/s) | 30 in./MIN (13 mm/s) |

Figure 6.17—Effect of Travel Speed Variation on Weld Bead Shape and Penetration

Weld penetration is affected more by travel speed than by any variable other than current. This is true except for excessively slow speeds when the molten weld pool is beneath the welding electrode. Then the penetrating force of the arc is cushioned by the molten pool. Excessive speed may cause undercutting.

Within limits, travel speed can be adjusted to control weld size and penetration. In these respects, it is related to current and the type of flux. Excessively high travel speeds promotes undercut, arc blow, porosity, and uneven bead shape. Relatively slow travel speeds provide time for gases to escape from the molten metal thus reducing porosity. Excessively slow speeds produce (1) a convex bead shape that is subject to cracking, (2) excessive arc exposure, which is uncomfortable for the operator, and (3) a large molten pool that flows around the arc, resulting in a rough bead and slag inclusions.

ELECTRODE SIZE

ELECTRODE SIZE AFFECTS the weld bead shape and the depth of penetration at a fixed current, as shown in Figure 6.18. Small diameter electrodes are used with semiautomatic equipment to provide flexibility of movement. They are also used for multiple electrode, parallel power equipment. Where poor fit-up is encountered, a larger diameter electrode is better than small ones for bridging large root openings.

Electrode size also influences the deposition rate. At any given current, a small diameter electrode will have a higher current density and a higher deposition rate than a larger electrode. However, a larger diameter electrode can carry more current than a smaller electrode, and produce a higher deposition rate at higher amperage. If a desired electrode feed rate is higher (or lower) than the feed motor can maintain, changing to a larger (or smaller) size electrode will permit the desired deposition rate.

ELECTRODE EXTENSION

AT CURRENT DENSITIES above 80 000 A/in^2 (125 A/mm^2), electrode extension becomes an important variable. At high-current densities, resistance heating of the electrode between the contact tube and the arc increases the electrode melting rate. The longer the extension, the greater is the amount of heating and the higher the melting rate. This resistance heating is commonly referred to as I^{2}R heating.

In developing a procedure, an electrode extension of approximately eight times the electrode diameter is a good starting point. As the procedure is developed, the length is modified to achieve the optimum electrode melting rate with fixed amperage.

Increased electrode extension adds a resistance element in the welding circuit and consumes some of the energy previously supplied to the arc. With lower voltage across the arc, bead width and penetration decrease (see Figure

FULLY AUTOMATIC WELDING
600 A, 30 V, 30 in./MIN (13 mm/s)

ELECTRODE SIZE:

1/8 in.	5.32 in.	7.32 in.
(3.2 mm)	(4.0 mm)	(5.6 mm)

Figure 6.18–Effect of Electrode Size on Weld Bead Shape and Penetration

6.16). Because lower arc voltage increases the convexity of the bead, the bead shape will be different from one made with a normal electrode extension. Therefore, when the electrode extension is increased to take advantage of the higher melting rate, the voltage setting on the machine should be increased to maintain proper arc length.

The condition of the contact tube affects the effective electrode extension. Contact tubes should be replaced at predetermined intervals to insure consistent welding conditions.

Deposition rates can be increased from 25 percent to 50 percent by using long electrode extensions with no change in welding amperage. With single electrode automatic SAW, the deposition rate may approach that of the two-wire method with two power sources.

An increase in deposition rate is accompanied by a decrease in penetration, therefore changing to a long electrode extension is not recommended when deep penetration is needed. When melt-through is a problem, as may be encountered when welding thin gage material, increasing the electrode extension may be beneficial. However, as the electrode extension increases, it is more difficult to maintain the electrode tip in the correct position with respect to the joint.

The following are suggested maximum electrode extensions for solid steel electrodes for SAW:

(1) For 5/64, 3/32, and 1/8 in. (2.0, 2.4, and 3.2 mm) electrodes, 3 in. (75 mm)

(2) For 5/32, 3/16, and 7/32 in. (4.0, 4.8, and 5.6 mm) electrodes, 5 in. (125 mm)

WIDTH AND DEPTH OF FLUX

THE WIDTH AND depth of the layer of granular flux influence the appearance and soundness of the finished weld as well as the welding action. If the granular layer is too deep, the arc is too confined and a rough ropelike appearing weld will result. The gases generated during welding cannot readily escape, and the surface of the molten weld metal becomes irregularly distorted. If the granular layer is too shallow, the arc will not be entirely submerged in flux. Flashing and spattering will occur. The weld will have a poor appearance, and it may be porous. Figure 6.19 shows the effects on weld bead surface appearance of proper and shallow depths of flux.

An optimum depth of flux exists for any set of welding conditions. This depth can be established by slowly increasing the flow of flux until the welding arc is submerged and flashing no longer occurs. The gases will then puff up quietly around the electrode, sometimes igniting.

During welding, the unfused granular flux can be removed a short distance behind the welding zone after the fused flux has solidified. However, it may be best not to disturb the flux until the heat from welding has been evenly distributed throughout the section thickness.

Fused flux should not be forcibly loosened while the weld metal is at a high temperature [above 1100°F (600°C)]. Allowed to cool, the fused material will readily detach itself. Then it can be brushed away with little effort.

CORRECT DEPTH	TOO SHALLOW
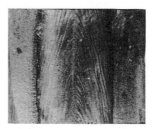	
3/4 in. (19 mm) DEPTH 5/8 in. (16 mm) PLATE RESULT: SMOOTH TOP, 　　　　 SOUND WELD 　　　　 STRUCTURE	1/4 in. (6 mm) DEPTH 5/8 in. (16 mm) PLATE RESULT: GAS POCKETS 　　　　 IN WELD; OPEN 　　　　 ARCING 　　　　 OCCURRED

Figure 6.19–Effect of Proper and Improper Depth of Flux on Weld Appearance

Sometimes a small section may be forcibly removed for quick inspection of the weld surface appearance.

It is important that no foreign material be picked up when reclaiming the flux. To prevent this, a space approximately 12 in. (300 mm) wide should be cleaned on both sides of the weld joint before the flux is laid down. If the recovered flux contains fused pieces, it should be passed through a screen with openings no larger than 1/8 in. (3.2 mm) to remove the coarse particles.

The flux is thoroughly dry when packaged by the manufacturer. After exposure to high humidity, it should be dried by baking it before it is used. Moisture in the flux will cause porosity in the weld. The manufacturer's recommendation should be followed.

TYPES OF WELDS

SUBMERGED ARC WELDING is used for making groove, fillet, plug, and surfacing welds. Groove welds are usually made in the flat position and fillet welds are usually made in the flat and horizontal positions. This is because the molten weld pool and the flux are most easily contained in these positions. However, simple techniques are available for producing groove welds in the horizontal welding position. Good submerged arc welds can be made downhill at angles up to 15 degrees from the horizontal. Surfacing and plug welding are done in the flat position.

Welds made by this process may be classified with respect to: the following:

(1) Type of joint
(2) Type of groove
(3) Welding method (semiautomatic or machine)
(4) Welding position (flat or horizontal)
(5) Single or multiple pass deposition
(6) Single or multiple electrode operation
(7) Single or multiple power supply (series, parallel, or separate connections).

GROOVE WELDS

GROOVE WELDS ARE commonly made in butt joints ranging from 0.05 (1.2 mm) sheet metal to thick plate. The greater penetration inherent in submerged arc welding permits square groove joints 1/2 in. (13 mm) or more in thickness to be completely welded from one side, provided some form of backing is used to support the molten metal (see Tables 6.4 and 6.5 for typical data). Single pass welds up to 5/16 in. (7.9 mm) thickness and two pass welds up to 5/8 in (15.9 mm) are made in steel with a square groove butt joint, no root opening, and a weld backing.

Table 6.4
Typical Welding Conditions for Single Electrode, Machine Submerged Arc Welding of Steel Plate Using One Pass (Square Groove)

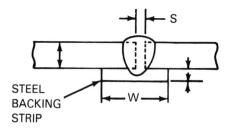

Plate Thickness T		Root Opening S		Current	DCEP Voltage	Travel Speed		Electrode Diameter		Electrode Consumption		t, min		W, min	
in.	mm	in.	mm	A	V	in./min	mm/s	in.	mm	lb/ft	kg/m	in.	mm	in.	mm
10 ga.	3.6	0-1/16	1.6	650	28	48	20	1/8	3.2	0.070	0.104	1/8	3.2	5/8	15.9
3/16	4.8	1/16	1.6	850	32	36	15	3/16	4.8	0.13	0.194	3/16	4.8	3/4	19.0
1/4	6.4	1/8	3.2	900	33	26	11	3/16	4.8	0.20	0.248	1/4	6.4	1	25.4
3/8	9.5	1/8	3.2	950	33	24	10	7/32	5.6	0.24	0.357	1/4	6.4	1	25.4
1/2	12.7	3/16	4.8	1100	34	18	8	7/32	5.6	0.46	0.685	3/8	9.5	1	25.4

Table 6.5
Typical Welding Conditions for Single Electrode, Two Pass Submerged Arc Welding of Steel Plate (Square Groove)

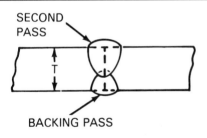

		Second Pass					Backing Pass						Electrode	
Plate Thickness T		DCEP Current	DCEP Voltage	Travel Speed		Electrode Diameter	DCEP Current	Voltage	Travel Speed		Electrode Diameter		Electrode Consumption	
in.	mm	A	V	in./min	mm/s	in.	mm	A	V	in./min	mm/s	in.	mm	lb/ft	kg/m

Semiautomatic Welding

Plate Thickness T in.	mm	DCEP Current A	DCEP Voltage V	Travel Speed in./min	mm/s	Electrode Diameter in.	mm	DCEP Current A	Voltage V	Travel Speed in./min	mm/s	Electrode Diameter in.	mm	Electrode Consumption lb/ft	kg/m
10 ga.	3.6	325	27	50	21	1/16	1.6	250	25	50	21	1/16	1.6	0.070	0.104
3/16	4.8	350	32	46	19	1/16	1.6	300	29	46	19	1/16	1.6	0.088	0.131
1/4	6.4	375	33	42	18	1/16	1.6	325	34	42	18	1/16	1.6	0.106	0.158
3/8	9.5	475	35	28	12	5/64	2.0	425	33	28	12	5/64	2.0	0.18	0.268
1/2	12.7	500	36	21	9	5/64	2.0	475	34	21	9	5/64	2.0	0.28	0.417
5/8	15.9	500	37	16	7	5/64	2.0	500	35	16	7	5/64	2.0	0.43	0.640

Machine Welding

Plate Thickness T in.	mm	DCEP Current A	DCEP Voltage V	Travel Speed in./min	mm/s	Electrode Diameter in.	mm	DCEP Current A	Voltage V	Travel Speed in./min	mm/s	Electrode Diameter in.	mm	Electrode Consumption lb/ft	kg/m
1/4	6.4	575	32	48	20	5/32	4.0	475	29	48	20	5/32	4.0	0.11	0.164
3/8	9.5	850	35	32	14	5/32	4.0	500	33	32	14	5/32	4.0	0.23	0.343
1/2	12.7	950	36	27	11	3/16	4.8	700	35	27	11	3/16	4.8	0.34	0.506
5/8	15.9	950	36	22	9	3/16	4.8	900	36	22	9	3/16	4.8	0.50	0.745

With multiple pass welding, using single or multiple electrodes, any plate thickness can be welded. Welds in thick material, when deposited from both sides, can use V- or U-grooves on one or both sides of the plate.

Groove welding of butt joints in the horizontal position is known as three o'clock welding. This type of weld may be made from both sides of the joint simultaneously, if desired. In most cases, the electrodes are positioned at a 10 to 30 degree angle above the horizontal. To support the flux and molten metal, some form of sliding support (or a proprietary moving belt) is used.

FILLET WELDS

USING A SINGLE electrode, fillet welds up to 3/8 in. (9.5 mm) throat size can be made in the horizontal position with one pass. Larger single-pass, horizontal position fillet welds may be made with multiple electrodes. However, welds larger than 5/16 in. (7.9 mm) are usually made in the flat position or by multiple passes in the horizontal. Fillet welds made by the submerged arc welding process can have greater penetration than those made by shielded metal arc welding, thereby exhibiting higher shear strength for the same size weld.

PLUG WELDS

SUBMERGED ARC WELDING is used to make high-quality plug welds. The electrode is positioned in the center of the hole and remains in this position until the weld is complete. The time required is dependent on welding amperage and hole size. Because of the deep penetration obtained with this process, it is essential to have adequate thickness in the weld backing.

SURFACING WELDS

BOTH SINGLE AND multiple electrode SAW methods are used to provide a base metal with special surface properties. The purpose may be to repair or reclaim worn equipment that is otherwise serviceable, or to impart desired properties to surfaces of original equipment. The high deposition rates achieved by submerged arc welding are well suited to large area surfacing applications.

WELDING PROCEDURES

TO FULLY REALIZE the high production benefits of SAW, consideration must be given to preweld operations. Joint design, edge preparation, fit-up, and fixturing the workpiece must be considered.

JOINT DESIGN AND EDGE PREPARATION

TYPES OF JOINTS used in submerged arc welding include chiefly butt, T-, and lap joints, although edge and corner joints can also be welded. The principles of joint design and methods of edge preparation are similar to other arc welding processes. Typical welds include fillet, square groove, single and double V-groove, and single and double U-groove welds.

Joint designs, especially for plate welding, often call for a root opening of 1/32 to 1/16 in. (0.8 to 1.6 mm) to prevent angular distortion or cracking due to shrinkage stresses. However, a root opening that is larger than that required for proper welding will increase welding time and costs. This is true for both groove and fillet welding.

Edge preparation may be done by any of the thermal cutting methods or by machining. The accuracy of edge preparation is important, especially for machine or automatic welding. For example, if a joint designed with a 1/4 in. (6.4 mm) root face were actually produced with a root face that tapered from 5/16 to 1/8 in. (7.9 to 3.2 mm) along the length of the joint, the weld might be unacceptable because of lack of penetration at the start and excessive melt-through at the end. In such a case, the capability of the cutting equipment, as well as the skill of the operator, should be checked and corrected.

JOINT FIT-UP

JOINT FIT-UP IS an important part of the assembly or subassembly operations, and it can materially affect the quality, strength, and appearance of the finished weld. When welding plate thicknesses, the deeply penetrating characteristics of the submerged arc process emphasize the need for close control of fit-up. Uniformity of joint alignment and of the root opening must be maintained.

WELD BACKING

SUBMERGED ARC WELDING creates a large volume of molten weld metal which remains fluid for an appreciable period of time. This molten metal must be supported and contained until it has solidified.

There are several methods commonly used to support molten weld metal when complete joint penetration is required:

(1) Backing strips

(2) Backing welds
(3) Copper backing bars
(4) Flux backing
(5) Backing tapes

In the first two methods, the backing may become a part of the completed joint. The other three methods employ temporary backing which is removed after the weld is completed. Methods 1 and 2 may require removing the backing, depending on the design requirements of the joint.

In many joints, the root face is designed to be thick enough to support the first pass of the weld. This method may be used for butt welds (partial joint penetration), for fillet welds, and for plug or slot welds. Supplementary backing or chilling is sometimes used. It is most important that the root faces of groove welds be tightly butted at the point of maximum penetration of the weld.

Backing Strip

IN THIS METHOD, the weld penetrates into and fuses with a backing strip which temporarily or permanently becomes an integral part of the assembly.

Backing strips must be compatible with the metal being welded. When the design permits, the joint is located so that a part of the structure forms the backing. It is important that the contact surfaces be clean and close together; otherwise porosity and leakage of molten weld metal may occur.

Backing Weld

IN A JOINT backed by weld metal, the backing pass is usually made with some other process, such as FCAW, GMAW, or SMAW. This backing pass forms a support for subsequent SAW passes made from the opposite side. Manual or semiautomatic welds are used as backing for submerged arc welds when alternate backing methods are not convenient because of inaccessibility, poor joint penetration or fit-up, or difficulty in turning the weldment.

The weld backing may remain as a part of the completed joint or may be removed by oxygen or arc gouging, by chipping, or by machining after the submerged arc weld has been made. It is then replaced by a permanent submerged arc surfacing bead.

Copper Backing

WITH SOME JOINTS, a copper backing bar is used to support the molten weld pool but does not become a part of the weld. Copper is used because of its high thermal conductivity, which prevents the weld metal from fusing to the backing bar. Where it is desirable to reinforce the underside of the weld, the backing bar may be grooved to the desired shape of the reinforcement. The backing bar must have enough mass to prevent it from melting beneath the arc, which would contaminate the weld with copper. Caution must be used to prevent copper pickup in the weld caused by harsh arc starts. Sometimes water is passed through the interior of the copper backing bar to keep it cool, particularly for high-production welding applications. Care must be taken to prevent water condensation from forming on the copper backing bar.

Copper backing is sometimes designed to slide so a relatively short length can be used in the vicinity of the arc and the molten weld pool. In still other applications, the copper backing is a rotating wheel.

Flux Backing

FLUX, UNDER MODERATE pressure, is used as backing material for submerged arc welds. Loose granular flux is placed in a trough on a thin piece of flexible sheet material. Beneath the flexible sheet, there is an inflatable rubberized canvas fire hose. The hose is inflated to no more than 5 to 10 psi (35 to 70 kPa) to develop moderate flux pressure on the backside of the weld. This concept is illustrated in Figure 6.20.

FIXTURING

THE MAIN PURPOSE of fixturing is to hold a workpiece assembly in proper alignment during handling and welding. Some assemblies may require stiffening fixtures to maintain their shape. In addition, some type of clamping or fixturing may be required to hold the joint alignment for welding and to prevent warpage and buckling from the heat of welding.

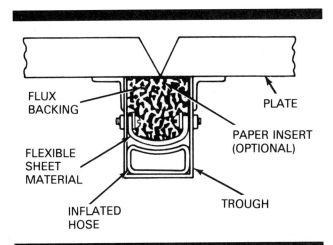

Figure 6.20–A Method of Supporting Flux Backing for Submerged Arc Welding

For assemblies that are inherently rigid, tack welding alone may suffice. Heavy section thicknesses in themselves offer considerable restraint against buckling and warpage. In intermediate cases, a combination of tack welding, fixturing, and weld sequencing may be required. For joints of low restraint in light gage materials, clamping is needed. Clamping bars maintain alignment and remove heat to reduce or prevent warpage. Tack welds are usually necessary.

Fixtures also include the jigs and tooling used to facilitate the welding operation. Weld seam trackers and travel carriages are used to guide machine or automatic welding heads. Turning rolls are used to rotate cylindrical workpieces during fit-up and welding. Rotating turntables with angular adjustment are used to position weld joints in the most favorable position for welding. Manipulators with movable booms are used to position the welding head and sometimes the weld operator, for hard-to-reach locations.

INCLINATION OF WORK

THE INCLINATION OF the work during welding can affect the weld bead shape, as shown in Figure 6.21. Most submerged arc welding is done in the flat position. However, it is sometimes necessary or desirable to weld with the work slightly inclined so that the weld progresses downhill or uphill. For example, in high-speed welding of 0.050 in. (1.3 mm) steel sheet, a better weld results when the work is inclined 15 to 18 degrees, and the welding is done downhill. Penetration is less than when the sheet is in a horizontal plane. The angle of inclination should be decreased as plate thickness increases to increase penetration.

Downhill welding affects the weld, as shown in Figure 6.21B. the weld pool tends to flow under the arc and preheat the base metal, particularly at the surface. This produces an irregularly shaped fusion zone. As the angle of inclination increases, the middle surface of the weld is depressed, penetration decreases, and the width of the weld increases.

Uphill welding affects the fusion zone contour and the weld surface, as illustrated in Figure 6.21C. The force of gravity causes the weld pool to flow back and lag behind the welding electrode. The edges of the base metal melt and flow to the middle. As the angle of inclination increases, reinforcement and penetration increase, and the width of the weld decreases. Also, the larger the weld pool, the greater the penetration and center buildup. These effects are exactly the opposite of those produced by downhill welding. The limiting angle of inclination when welding uphill with currents up to 800 amperes is about 6 degrees, or a slope of approximately one in ten. When higher welding currents are used, the maximum workable angle decreases. Greater inclination than approximately 6 degrees makes the weld uncontrollable. Lateral inclination of the workpiece produces the effects shown in Figure 6.21D. The limit for a lateral slope is approximately 3 de-

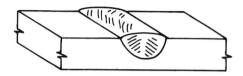

(A) FLAT POSITION WELD

(B) DOWNHILL WELD (1/8 SLOPE)

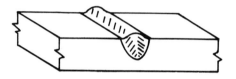

(C) UPHILL WELD (1/8 SLOPE)

(D) LATERAL WELD (1/19 SLOPE)

Figure 6.21–Effect of Work Inclination on Weld Bead Shape

grees or 1 in 20. Permissible lateral slope varies somewhat, depending on the size of the weld puddle.

WORKLEAD CONNECTIONS

THE ELECTRIC WELDING circuit consists of a lead from the power supply to the welding head and a return lead from the workpiece to the power supply. The latter is called the *worklead connection*. The worklead connection is sometimes called the ground lead, which is an incorrect term. The work may also be grounded to the building or earth, but for best results in submerged arc welding, a worklead connection directly to the power source is required.

The method of attachment and the location of the worklead connection are important considerations in submerged arc welding since they can affect the arc action, the quality of the weld, and the speed of welding. A poor worklead location can cause or increase arc blow, resulting in porosity, lack of penetration, and poor bead shape. Testing may be necessary since it is often difficult to predict the effect of the worklead location. Generally, the best direction of welding is away from the work connection.

There may be a tendency for the welding current to change slowly in welding long seams. This can happen because the path and electrical characteristics of the circuit change as the weld progresses. A more uniform weld can frequently be obtained by attaching workleads to both ends of the object being welded.

When the longitudinal seam of a light gage cylinder is welded in a clamping fixture with copper backing, it is usually best to connect the worklead on the bottom of the cylinder at the start end. If this is not possible, then the worklead should be attached to the fixture at the start end. It is undesirable to connect the worklead to a copper backing bar, because the welding current will enter or leave the work at the point of best electrical contact, not necessarily beneath the arc. If the current sets up a magnetic field around some length of the backing bar, that may cause arc blow.

When current return is through a sliding shoe, two or more shoes should always be used. This will prevent interruptions of current.

ARC STARTING METHODS

THE METHOD USED to start the arc in a particular application will depend on such factors as the time required for starting relative to the total setup and welding time, the number of pieces to be welded, and the importance of starting the weld at a particular place on the joint. There are six methods of starting:

Steel Wool Ball Start

A TIGHTLY ROLLED ball of steel wool about 3/8 in. (10 mm) in diameter is positioned in the joint directly beneath the welding electrode. The welding electrode is lowered onto the steel wool until the ball is compressed to approximately one-half its original height. The flux is then applied and welding is started. The steel wool ball creates a current path to the work, but it melts rapidly while creating an arc.

Sharp Wire Start

THE WELDING ELECTRODE, protruding from the contact tube, is snipped with wire cutters. This forms a sharp, chisel-like configuration at the end of the wire. The electrode is then lowered until the end just contacts the workpiece. The flux is applied and welding is commenced. The chisel point melts away rapidly to start the arc.

Scratch Start

THE WELDING ELECTRODE is lowered until it is in light contact with the work, and the flux is applied. Next, the carriage is started and the welding current is immediately applied. The motion of the carriage prevents the welding wire from fusing to the workpiece.

Molten Flux Start

WHENEVER THERE IS a molten puddle of flux, an arc may be started by simply inserting the electrode into the puddle and applying the welding current. This method is regularly used in multiple-electrode welding. When two or more welding electrodes are separately fed into one weld pool, it is only necessary to start one electrode to establish the weld pool. Then the other electrodes will arc when they are fed into the molten pool.

Wire Retract Start

RETRACT ARC STARTING is one of the most positive methods, but the welding equipment must be designed for it. It is cost effective when frequent starts have to be made and when starting location is important.

Normal practice is to move the electrode down until it just contacts the workpiece. Then the end of the electrode is covered with flux, and the welding current is turned on. The low voltage between the electrode and the work signals the wire feeder to withdraw the tip of the electrode from the surface of the workpiece. An arc is initiated as this action takes place. As the arc voltage builds up, the wire feed motor quickly reverses direction to feed the welding electrode toward the surface of the workpiece. Electrode feed speeds up until the electrode melting rate and arc voltage stabilize at the preset value.

If the workpiece is light gage metal, the electrode should make only light contact, consistent with good electrical contact. The welding head should be rigidly mounted. The end of the electrode must be clean and free of fused slag. Wire cutters are used to snip off the tip of the electrode (preferably to a point) before each weld is made. The electrode size should be chosen to permit operation with high-current densities since they enable easier starting.

High-Frequency Start

THIS METHOD REQUIRES special equipment but requires no manipulation by the operator other than closing a starting switch. It is particularly useful as a starting method for intermittent welding, or for welding at high-production rates where many starts are required.

When the welding electrode approaches to within approximately 1/16 in. (1.6 mm) above the workpiece, a high-frequency, high-voltage generator in the welding circuit causes a spark to jump from the electrode to the workpiece. This spark produces an ionized path through which the welding current can flow, and the welding action begins. This is a commonly used arc starting technique.

ARC TERMINATION

IN SOME ELECTRICAL systems, the travel and the electrode feed will stop at the same time when the "stop weld" button is pushed. Other systems will stop the travel, but the electrode will continue to feed for a controlled length of time. A third type of system reverses the direction of travel for a controlled length of time while welding continues. The two latter systems fill the weld crater.

ELECTRODE POSITION

IN DETERMINING THE proper position of the welding electrode, three factors must be considered:

(1) The alignment of the welding electrode in relation to the joint
(2) The angle of tilt in the lateral direction, that is, the tilt in a plane perpendicular to the joint (work angle)
(3) The forward or backward direction in which the welding electrode points (travel angle)

Forward is the direction of travel. Hence, a forward pointing electrode is one that makes an acute angle with the finished weld. A backward pointing electrode makes an obtuse angle with the finished weld.

Most submerged arc welds are made with the electrode axis in a vertical position. Pointing the electrode forward or backward becomes important when multiple arcs are being used, when surfacing, and when the workpiece cannot be inclined. Pointing the electrode forward results in a weld configuration similar to downhill welding; pointing the electrode backward results in a weld similar to uphill welding. Pointing the electrode forward or backward does not affect the weld configuration as much as uphill or downhill positioning of the workpieces.

When butt joints are welded between plates of equal thicknesses, the electrode should be aligned with the joint center line, as shown in Figure 6.22(A). Improper alignment may cause lack of joint penetration as shown in Figure 6.22B. When unequal thicknesses are butt welded, the electrode must be located over the thick section to melt it at the same rate as the thin section. Figure 6.22(C) shows this requirement.

When welding horizontal fillets, the centerline of the electrode should be aligned below the root of the joint and toward the horizontal piece at a distance equal to one-fourth to one-half the electrode diameter. The greater distance is used when making larger sizes of fillet welds. Careless or inaccurate alignment may cause undercut in the vertical member or produce a weld with unequal legs.

When horizontal fillet welds are made, the electrode is tilted between 20 and 45 degrees from the vertical (work angle). The exact angle is determined by either or both of the following factors:

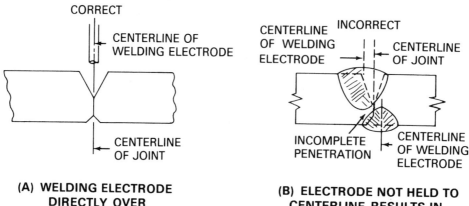

(A) **WELDING ELECTRODE DIRECTLY OVER JOINT CENTERLINE**

(B) **ELECTRODE NOT HELD TO CENTERLINE RESULTS IN INCOMPLETE PENETRATION**

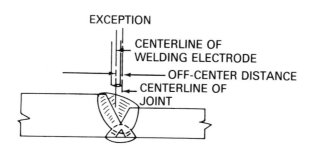

(C) **OFF-CENTER ALIGNMENT IS SOMETIMES REQUIRED WHEN BUTT-WELDING PLATES OF DIFFERENT THICKNESS**

Figure 6.22–Effect of Electrode Location With Respect to the Weld Groove

(1) Clearance for the welding torch, especially when structural sections are being welded to plate

(2) The relative thicknesses of the members forming the joint (If the possibility of melting through one of the members exists, it is necessary to direct the electrode toward the thicker member.)

Normally, horizontal fillet welding should be done with the welding electrode positioned perpendicular to the axis of the weld.

When fillet welding is done in the flat position, the electrode axis is normally in a vertical position and bisects the angle between the workpieces, as shown in Figure 6.23(A). When making positioned fillet welds where greater than normal penetration is desired, the electrode and the workpieces are positioned as shown in Figure 6.23(B). The electrode is positioned so that its centerline intersects the joint near its center. The electrode may be tilted to avoid undercutting.

RUN-ON AND RUN-OFF TABS

WHEN A WELD starts and finishes at the abrupt end of a workpiece, it is necessary to provide a means of supporting the weld metal, flux, and molten slag so that spillage does not occur. Tabs are the method most commonly used. An arc is started on a run-on tab that is tack welded to the start end of the weld, and it is stopped on a run-off tab at the finish end of the weld. The tabs are large enough so that the weld metal on the work itself is properly shaped at the ends of the joint. When the tabs are prepared, the groove should be similar to the one being welded, and the tabs must be wide enough to support the flux.

A variation of the tab is a copper dam that holds the flux, which in turn supports the weld metal at the ends of the joint.

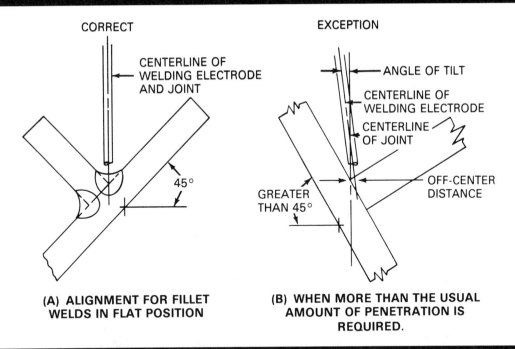

(A) ALIGNMENT FOR FILLET WELDS IN FLAT POSITION

(B) WHEN MORE THAN THE USUAL AMOUNT OF PENETRATION IS REQUIRED.

Figure 6.23–Electrode Positions for Fillet Welds in the Flat Position

CIRCUMFERENTIAL WELDS

CIRCUMFERENTIAL WELDS DIFFER from those made in the flat position because of the tendency for the molten flux and weld metal to flow away from the arc. To prevent spillage or distortion of the bead shape, welds must solidify as they pass the 6 or 12 o'clock positions. Figure 6.24 illustrates the bead shapes that result from various electrode positions with respect to the 6 and 12 o'clock positions.

Too little arc displacement on an outside weld or too much arc displacement on an inside weld produces deep penetration and a narrow, very convex bead shape. Undercutting may also take place. Too much displacement on an outside weld or too little displacement on an inside weld produces a shallow, concave bead.

The flux is granular and will spill off small diameter work if it is not contained. If spilling of flux occurs, the arc is uncovered and poor quality welds result. One method of

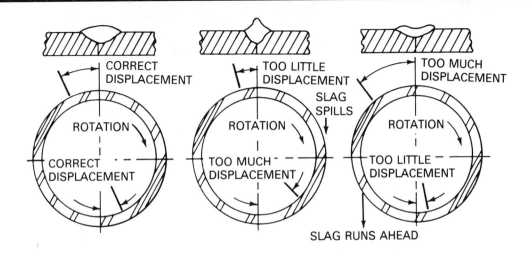

Figure 6.24–Effect of Electrode Positions on Weld Bead Shape When Circumferential Welding

overcoming this is to use a nozzle assembly that pours the flux concentric with the arc; this gives it less chance to spill. A wire brush or some other flexible heat-resisting material can be attached so that it rides the work ahead of the arc and contains the flux. See Figure 6.25.

Regardless of electrode position, if the molten pool is too big for the diameter of work, the molten weld metal will spill simply because it cannot freeze fast enough. Bead size, as measured by the volume of deposited metal per unit length of weld, depends on the amperage and travel speed used. Lower amperages and higher travel speeds will reduce the size of the bead.

SLAG REMOVAL

ON MULTIPLE PASS welds, slag removal becomes important because no subsequent passes should be made where slag is present. The factors that are particularly important in dealing with slag removal are bead size and bead shape. Smaller beads tend to cool more quickly and slag adherence is reduced. Flat to slightly convex beads that blend evenly with the base metal make slag removal much easier than very concave or undercut beads. For this reason, a decrease in voltage will improve slag removal in narrow grooves. On the first pass of two-pass welds, a concave bead that blends smoothly to the top edges of the joint is much easier to clean than a convex bead that does not blend well.

Figure 6.25–Circumferential Submerged Arc Weld. A Wire Brush is Used to Contain the Flux Around the Electrode

PROCESS VARIATIONS

SUBMERGED ARC WELDING lends itself to a wide variety of wire and flux combinations, single and multiple electrode arrangements, and use of ac or dc welding power sources. The process has been adapted to a wide range of materials and thicknesses. Various multiple arc configurations may be used to control the weld profile and increase the deposition rates over single arc operation. Weld deposits may range from wide beads with shallow penetration for surfacing to narrow beads with deep penetration for thick joints. Part of this versatility is derived from the use of ac arcs.

The principles which favor the use of ac to minimize arc blow in single arc welding are often applied in multiple arc welding to create a favorable arc deflection. The current flowing in adjacent electrodes sets up interacting magnetic fields that can either reinforce or diminish each other. In the space between the arcs, these magnetic fields are used to produce forces that will deflect the arcs (and thus distribute the heat) in directions beneficial to the intended welding application.

Various types of power sources and related equipment are designed and manufactured especially for multiple arc welding. These relatively sophisticated machines are intended for high production on long runs of repetitive-type applications.

The following are typical SAW process configurations used in production welding today. They may be used for welding both carbon steel and low alloy steels within the limitations noted previously.

SINGLE ELECTRODE WELDING

SINGLE ELECTRODE WELDING is the most common of all SAW process configurations, using only one electrode and one power source. It is normally used with direct current electrode positive (DCEP) polarity, but may also be used with direct current electrode negative (DCEN) polarity when less penetration into the base metal is required. The process may be used in the semiautomatic mode where the welder manipulates the electrode, or in the machine mode.

A single electrode is frequently used with special welding equipment for completing horizontal groove welds in large storage tanks and process vessels. The unit rides on the top of each ring as it is constructed and welds the circumferential joint below it. A special flux belt or other equipment is used to hold the flux in place against the shell ring. In addition, both sides of the joint (inside and outside) are usually welded simultaneously to reduce fabrication time.

NARROW GROOVE WELDING

A NARROW GROOVE configuration is often adopted for welding material 2 in. (50 mm) thick and greater, with a root opening between 1/2 and 1 in. (13 and 25 mm) wide at the bottom of the groove and a total included groove angle between 0 and 8 degrees. This process variation usually powers a single electrode with either DCEP or alternating current, depending on the type of electrode and flux being used. It is essential to use welding fluxes that have been developed for narrow groove welding because of the difficulty in removing slag. These fluxes have special characteristics for easier removal from the narrow groove.

MULTIPLE WIRE WELDING

MULTIPLE WIRE SYSTEMS combine two or more welding wires feeding into the same puddle. The wires may be current-carrying electrodes or cold fillers. They may be supplied from single or multiple power sources. The power sources may be dc or ac or both.

Multiwire welding systems not only increase weld metal deposition rates, but also improve operating flexibility and provide more efficient use of available weld metal. This increased control of metal deposition can also achieve higher welding speeds, up to five times those obtainable with a single wire.

Twin Electrode SAW Process

THIS WELDING CONFIGURATION uses two electrodes feeding into the same weld pool. The two electrodes are connected to a single power source and wire feeder, and are normally used with DCEP. Because two electrodes are melted, this mode offers increased deposition rates compared to single electrode submerged arc welding. The process is used in the machine or automatic welding mode and can be used for flat groove welds and horizontal fillet welding.

Tandem Arc SAW Process

THERE ARE TWO variations of two-electrode tandem arc SAW. One configuration uses a DCEP lead electrode and an alternating current trail electrode. The electrodes are separated 0.75 in. (19 mm) but are active in the same weld puddle. This process offers higher deposition rates compared to the single electrode SAW process, up to 40 lbs per hour when using larger diameter electrodes. This configuration is used in the machine or automatic modes for welding thicker materials, 1 in. (25.4 mm) and greater, in the flat welding position (Figure 6.26). It should be noted that additional ac trailing electrodes may be added to the configuration to increase deposition rates even more.

The second configuration uses two ac power sources electrically connected as shown in Figure 6.27. This configuration is called a *Scott connection*, and the interaction of the magnetic fields of the two arcs result in a forward deflection of the trail arc. The forward deflection allows for greater welding speeds without undercutting the base metal.

Triple-Arc Tandem Arc SAW

THERE ARE TWO popular variations of triple arc SAW. In the one variation all three electrodes are connected to ac transformers. The transformers are connected to the three phase primary as shown in Figure 6.28. The first electrodes in this system are Scott connected the same as shown in Figure

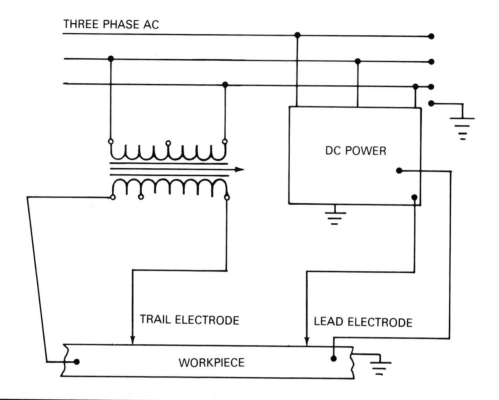

Figure 6.26–DC-AC Multi-Power Welding

6.27, and the trail electrode is in phase with the lead electrode. This connection results in a powerful forward deflection of the trail arc and promotes high travel speeds. This variation is used in many pipe mills. It is also popular in shipbuilding in one-side welding applications.

The second variation of triple arc SAW uses a dcep lead arc and two ac trail arcs connected in Scott as shown in Figure 6.27.

COLD WIRE ADDITION

COLD WIRE ADDITIONS have proven feasible using both solid and flux cored wires without deterioration in weld properties. The technique has not gained widespread industrial use. The equipment required is the same for any multiwire application, but one wire is not connected to a power supply. Increases in deposition rates up to 73 percent are possible; rates 35 to 40 percent greater are consistently achievable. The higher deposition at fixed heat input results in lower penetration.

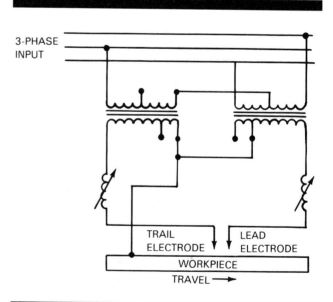

Figure 6.27—AC—Arc Welding Transformers — Scott Connections

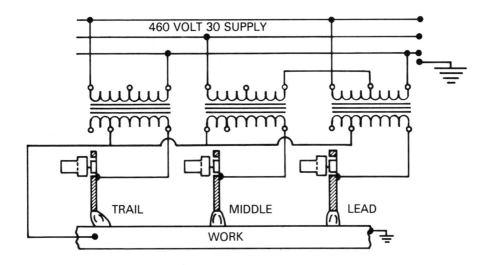

Figure 6.28–Three Wire AC-AC-AC Welding

HOT WIRE ADDITION

HOT WIRE ADDITIONS are much more efficient than cold wires or adding an additional arc, because the current introduced is used entirely to heat the filler wire and not to melt base material or flux. Deposition can be increased 50 to 100 percent without impairing the weld metal properties. The process does require additional welding equipment, additional control of variables, considerable setup time, and closer operator attention.

METAL POWDER ADDITION

METAL POWDER ADDITIONS may increase deposition rates up to 70 percent. The technique gives smooth fusion, improved bead appearance, and reduced penetration and dilution. Metal powders can also modify the chemical composition of the final weld deposit. These powders can be added ahead of the weld pool or directly into the pool, either by gravity feed or using the magnetic field surrounding the wire to transport the powder. See Figure 6.29.

Testing of metal powder additions has confirmed that the increase in deposition rate does not require additional arc energy, does not deteriorate weld metal toughness, nor increase risks of cracking. These tests also indicate that weld properties may be enhanced by control of resulting grain structures due to the lower heat input and restoration of diluted weld metal chemistries.

WELD SURFACING

THE TERM *surfacing*, as used with SAW, refers to the application by welding of a layer of material to a surface to obtain desired properties or dimensions, as opposed to making a joint.

The SAW process is often used to surface carbon steel with stainless steel as an economical way to obtain a corrosion resistant layer on a steel workpiece. To end up with an overlay of specified composition, the filler metal must be enriched sufficiently to compensate for dilution. For any given filler metal composition, changes in weld procedure can cause variations in dilution and consequently undesirable overlay compositions. Therefore, to ensure consistently satisfactory results, the welding procedure must be carefully controlled.

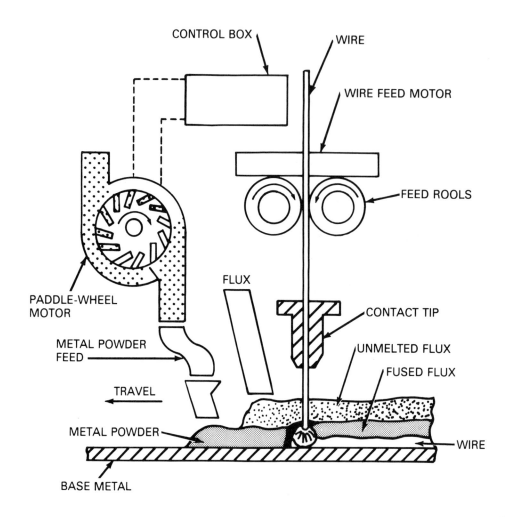

Figure 6.29–Typical Metal Powder Addition Diagram

Methods of calculating weld dilution and the approximate weld metal content of any element are shown in Figure 6.30.

Welding variables and procedure deviations may result in variation in dilution, with consequent variation in composition. Procedure qualification records should include chemical analysis of the weld deposit. The welding procedure specification should contain the acceptable limits of chemical composition of the deposit.

CLAD STEELS

WHEN WELDING CLAD steels, the dilution effects discussed above are equally important. Stainless clad carbon or low alloy steel plates are sometimes welded with stainless filler metal throughout the whole plate thickness, but usually carbon or low alloy steel filler metal is used on the unclad side, followed by removal of a portion of the cladding and completion of the joint with stainless filler metal.

Inexperienced fabricators should consult the manufacturer of the clad steel for recommendations of detailed welding procedures and subsequent postweld heat treatments. Joining clad steel to unclad steel sections normally requires making the butt weld and restoring the clad section in a fashion similar to joining two clad plates.

A typical welding procedure for stainless clad steel is shown in Figure 6.31. Note that alternate Step No. 6 shows a center "annealing" bead technique for use when 400 series stainless cladding is involved.

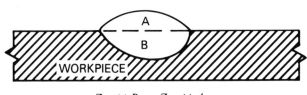

$$Z_{WM} = \frac{Z_{BM} \times B}{A + B} + \frac{Z_{FM} \times A}{A + B}$$

where

Z_{WM} = weld metal content of element Z
Z_{BM} = base metal content of element Z
Z_{FM} = filler metal content of element Z

$$\% \text{ DILUTION} = \frac{B}{A \& B} \times 100$$

Figure 6.30–A Method of Calculating Weld Dilution

STEP NO. 1—EDGE PREPARATION.

STEP NO. 2—PLATE FIT-UP BEFORE WELDING. THE LIP OF STEEL ABOVE CLADDING PROTECTS STEEL WELD FROM HIGH ALLOY PICK-UP.

STEP NO. 3—STEEL SIDE WELDED USING STEEL ELEC-TRODE. NOTE THAT STEEL WELD HAS NOT PENETRATED INTO THE CLADDING.

STEP NO. 4—CLAD SIDE PRE-PARED FOR WELDING BY GOUG-ING, CHIPPING OR GRINDING.

STEP NO. 5—WELDING THE CLAD SIDE WITH AN APPRO-PRIATE STAINLESS STEEL FILLER METAL.

STEP NO. 6—COMPLETING THE JOINT WITH STAINLESS WELD METAL.

STEP NO. 6—ALTERNATE. NOTE "ANNEALING" BEAD NO. 4 FOR 400 SERIES CLAD.

Figure 6.31–Typical Welding Procedure. Note That the Steel Side is Welded First.

Careful consideration of filler metals is required. The unclad side requires a low carbon or low alloy filler metal of appropriate composition (XX18 for manual shielded electrodes). For the clad side, type 309L could be used for 405, 410, 430, 304, or 304L; type 309Cb for 321 or 347; 310 for 310; 309Mo for 316. Some applications may require an essentially matching composition in the case of type 400 clad surfaces. The user should consider the manufacturers recommendations in chosing filler metals.

SPECIAL SERVICE CONDITIONS

SOME LOW ALLOY steel components that are submerged arc welded will be used in special service conditions where weld metal, heat-affected zone, and plate hardness must not exceed maximum engineering specification hardness requirements, similar to those previously discussed for carbon steel components. This is usually required for oil industry components that will be exposed to wet hydrogen sulfide gas. It has been found that if hardness is kept below a prescribed level, depending on the type of material and the service condition, cracking due to exposure to hydrogen will generally not occur. Typical maximum hardness requirements for carbon-molybdenum steels and chromium-molybdenum steels as established by API and NACE are shown in Table 6.6.

Hardness may be reduced in several ways: by selecting consumables that will produce low weld metal hardness, by raising the preheat temperature or welding heat input to produce weld metals and heat-affected zones with softer microstructures, and by postweld heat treatment.

Table 6.6
Typical Maximum Hardness Requirements for C-Mo and Cr-Mo Steels That Will be Exposed to Wet Hydrogen Sulphide gas

	Maximum Hardness
Carbon - 1/2% Molybdenum Steel	225 BHN
1-1/4% Cr - 1/2% Mo Steels	225 BHN
2-1/4% Cr - 1% Mo Steels	225 BHN
5% Cr - 1/2% Mo Steels	235 BHN
9% Cr - 1% Mo Steels	241 BHN

(BHN = Brinell Hardness Number)

WELD QUALITY

POROSITY PROBLEMS

SUBMERGED ARC DEPOSITED weld metal is usually clean and free of injurious porosity because of the excellent protection afforded by the blanket of molten slag. When porosity does occur, it may be found on the weld bead surface or beneath a sound surface. Various factors that may cause porosity are the following:

(1) Contaminants in the joint
(2) Electrode contamination
(3) Insufficient flux coverage
(4) Contaminants in the flux
(5) Entrapped flux at the bottom of the joint
(6) Segregation of constituents in the weld metal
(7) Excessive travel speed
(8) Slag residue from tack welds made with covered electrodes

As with other welding processes, the base metal and electrode must be clean and dry. High travel speeds and associated fast weld metal solidification do not provide time for gas to escape from the molten weld metal. The travel speed can be reduced, but other solutions should be investigated first to avoid higher welding costs. Porosity from covered electrode tack welds can be avoided by using electrodes that will not leave a porosity-causing residue. Recommended tack weld electrodes are E6010, E6011, E7016, and E7018.

CRACKING PROBLEMS

CRACKING OF WELDS in steel is usually associated with liquid metal cracking (center bead cracking). This cause may be traced to the joint geometry, welding variables, or stresses at the point where the weld metal is solidifying. This problem can occur in both butt welds and in fillet welds, including grooves and fillet welds simultaneously welded from two sides.

One solution to this problem is to keep the depth of the weld bead less than or equal to the width of the face of the weld. Weld bead dimensions may best be measured by sectioning and etching a sample weld. To correct the problem, the welding variables or the joint geometry must be changed. To decrease the depth of penetration compared to the width of the face of the joint, the welding travel speed as well as the welding current can be reduced.

Cracking in the weld metal or the heat-affected zone may be caused by diffusible hydrogen in the weld metal. The hydrogen may enter the molten weld pool from the following sources: flux, grease or dirt on the electrode or base metal, and hydrogen in the electrode or base metal. Cracking due to diffusible hydrogen in the weld metal is usually associated with low alloy steels and with increasing tensile and yield strengths. It sometimes can occur in carbon steels. There is always some hydrogen present in deposited weld metal, but it must be limited to relatively small amounts. As tensile strength increases, the amount of diffusible hydrogen that can be tolerated in the deposited weld decreases.

Cracking due to excessive hydrogen in the weld is called *delayed cracking*; it usually occurs several hours, up to approximately 72 hours, after the weld has cooled to ambient temperature. Hydrogen will diffuse out of the base metal at elevated temperatures [above approximately 200°F (93°C)] without resulting in cracking. It is at ambient temperatures that hydrogen accumulated at small defects in the weld metal or base metal results in cracking.

To keep the hydrogen content of the weld metal low:

(1) Remove moisture from the flux by baking in an oven (follow the manufacturer's recommendations).
(2) Remove oil, grease, or dirt from the electrode and base material.
(3) Increase the work temperature to allow more hydrogen to escape during the welding operation. This may be done by continuing the "preheat" until the seam is completely welded, or by postheating the weld joint for several hours before letting it cool to ambient temperature.

SAFETY RECOMMENDATIONS

FOR DETAILED SAFETY information, refer to the equipment manufacturer's instructions and the latest editions of ANSI Z49.1, *Safety in Welding and Cutting*. For mandatory federal safety regulations established by the U.S. Labor Department's Occupational Safety and Health Administration, refer to the latest edition of OSHA Standards, Code of Federal Regulations, Title 29 Part 1910, available from the Superintendent of Documents, U.S. Printing Office, Washington, DC 20402.

Operators should always wear eye protection to guard against weld spatter, arc glare exposure, and flying slag particles.

Power supplies and accessory equipment such as wire feeders should be properly grounded. Welding cables should be kept in good condition.

Certain elements, when vaporized, can be potentially dangerous. Alloy steels, stainless steels, and nickel alloys contain such elements as chromium, cobalt, manganese, nickel, and vanadium. Material safety data sheets should be obtained from the manufacturers to determine the content of the potentially dangerous elements and their threshold limit values. For many of these elements the limit is 1.0 milligram per cubic meter or less.

The submerged arc process greatly limits exposure of operators to air contaminants because few welding fumes escape from the flux overburden. Adequate ventilation will generally keep the welding area clear of hazards. The type of fan, exhaust, or other air movement system will be dependent on the work area to be cleared. The various manufacturers of such equipment should be consulted for a particular application.

SUPPLEMENTARY READING LIST

Allen, L. J. et al. "The formation of chevron cracks in submerged arc weld metal." *Welding Journal* 61(7): 212s-221s; July 1981.

American Society for Metals. *Metals handbook*, Vol. 6, 9th Ed, 114-152. Metals Park, Ohio: American Society for Metals, 1983.

Bailey, N. and Jones, S. B. "The solidification mechanics of ferritic steel during submerged arc welding." *Welding Journal* 57(8): 217s-231s; August 1978.

Butler, C. A., and Jackson, C. E. "Submerged arc welding characteristics of the CaO-TiO$_2$-SiO$_2$ system." *Welding Journal* 46(10): 448s-456s; October 1967.

Chandel, R. S. "Mathematical modeling of melting rates for submerged arc welding." *Welding Journal* 66(5): 135s-139s; May 1987.

Dallam, C.B., Liu, S., and Olson, D.L. "Flux composition dependence of microstructure and toughness of submerged arc HSLA weldments." *Welding Journal* 64(5): 1405-1515; May 1985.

Eager, T. W. "Sources of weld metal oxygen contamination during submerged arc welding." *Welding Journal* 57 (3): 76s-80s; March 1978.

Ebert, H. W. and Winsor, F. J. "Carbon steel submerged arc welds - Tensile strength vs. corrosion resistance." *Welding Journal* 59(7): 193s-198s; July 1980.

Fled, N. A. et al. "The role of filler metal wire and flux composition in submerged arc weld metal transformation kinetics." *Welding Journal* 65(5): 113s; May 1986.

Gowrisankar, I. et al. "Effect of the number of passes on the structure and properties of submerged arc welds of AISI type 316L stainless steel." *Welding Journal* 66(5): 147s-151s; May 1987.

Hantsch, H. et al. "Submerged arc narrow-gap welding of thick walled components." *Welding Journal* 61(7): 27-34; July 1982.

Hinkel, J. E., and Forsthoefel, F. W. "High current density submerged arc welding with twin electrodes." *Welding Journal* 55(3): 175-180; March 1976.

Indacochea, J. E. et al. "Submerged arc welding: Evidence for electrochemical effects on the weld pool." *Welding Journal* 68(3): 77s-81s; March 1989.

Jackson, C. E. "Fluxes and slags in welding." Bulletin 190. New York: Welding Research Council, December 1973.

Keith, R. H. "Weld backing comes of age." *Welding Journal* 54(6): 422-430; June 1975.

Konkol, P. J. and Koons, G. F. "Optimization of parameters for two-wire ac-ac submerged arc welding." *Welding Journal* 57(12): 367s-374s; December 1978.

Kubli, R. A., and Sharav, W. B. "Advancements in submerged arc welding of high impact steels. *Welding Journal* 40(11): 497s-502s; November 1961.

Lau, T. et al. "Gas/metal/slag reactions in submerged arc welding using CaO-Al$_2$O$_3$ based fluxes." *Welding Journal* 65(2): 31s; February 1986.

———. "The sources of oxygen and nitrogen contamination in submerged arc welding using CaO-Al$_2$O$_3$ based fluxes." *Welding Journal* 64(12): 343s-348s; December 1985.

Lau, T. et al, Dallan, C.B., et al. "Flux composition dependence of microstructure and toughness of submerged arc HSLA weldments." *Welding Journal* 64(5): 1045-1515; May 1985.

Lewis, W. J., Faulkner, G. E., and Rieppel, P. J. "Flux and filler wire developments for submerged arc welding HY-80 steel. *Welding Journal* 40(8): 337s-345s; August 1961.

Lincoln Electric Company. *The procedure handbook of arc welding*, 12th Ed. Cleveland, Ohio: Lincoln Electric Co., 1973.

Majetich, J.C. "Optimization of conventional SAW for severe abrasion - wear hardfacing application." *Welding Journal* 64(11): 3145-3215; November 1985.

Mallya, V.D. and Srinivas, H.S. "Bead characteristic in submerged arc strip cladding." *Welding Journal* 68(12): 30; December 1989.

McKeighan, J. S. "Automatic hard facing with mild steel electrodes and agglomerated alloy fluxes." *Welding Journal* 34(4): 301-308; April 1955.

North, T.H. et al. "Slag/metal interaction, oxygen and toughness in submerged arc welding." *Welding Journal* 57(3): 635-755; March 1978.

Patchett, B. M. "Some influences of slag composition on heat transfer and arc stability." *Welding Journal* 53(5): 203s-210s; May 1974.

Polar, A., et al "Electrochemically generated oxygen contamination in submerged arc welding." *Welding Journal* 69(2): 68s-??s; February 1990.

Renwick, B. G., and Patchett, B. M. "Operating characteristics of the submerged arc process." *Welding Journal* 55 (3): 69s-76s; March 1976.

Smith, N.J. et al. "Microstructure/mechanical property relationships of submerged arc welds in HSLA 80 steel." *Welding Journal* 68(3): 1125; March 1989.

Troyer, W., and Mikurak, J. "High deposition submerged arc welding with iron powder joint-fill." *Welding Journal* 53(8): 494-504; August 1974.

Union Carbide Corporation. *Submerged arc welding handbook*. New York: Union Carbide Corp., Linde Div., 1974.

Uttrachi, G. D., and Messina, J. E. "Three-wire submerged arc welding of line pipe." *Welding Journal* 47(6): 475-481; June 1968.

Wittstock, G. G. "Selecting submerged arc fluxes for carbon and low alloy steels." *Welding Journal* 55(9): 733-741; September 1976.

Wilson, R. A. "A selection guide for methods of submerged arc welding." *Welding Journal* 35(6): 549-555; June 1956.

CHAPTER 7

ELECTROGAS WELDING

PREPARED BY A COMMITTEE CONSISTING OF:

B. L. Shultz, Chairman
Taylor-Winfield Corp.

G. A. Gix
Neyer, Tiseo and Hindo, Ltd.

J. Gonzalez
The Lincoln Electric Co.

J. R. Hannahs
Midmark Corporation

D. Metzler
Owen Steel Co.

O. W. Seth
Chicago Bridge and Iron Co.

WELDING HANDBOOK COMMITTEE MEMBER:
J. R. Hannahs
Midmark Corporation

CHAPTER **7**

ELECTROGAS WELDING

INTRODUCTION

THE FIRST AVAILABLE thick-plate single-pass vertical welding method was electroslag welding. Demand arose immediately for equipment that would apply the process to thinner sections. Almost all vertical joints were then being welded with the manual shielded metal arc (SMAW) process or by semiautomatic gas metal arc welding (GMAW). Then in 1961, laboratory studies with an electroslag welding machine adapted to feed auxiliary gas shielding around a flux cored electrode demonstrated that plate as thin as 1/2 in. (13 mm) could be satisfactorily welded in the vertical position in a single pass. This technique is called *electrogas welding* (EGW).

EGW has enjoyed steady growth in the United States in the past decade. This growth has occurred because the process uses simple base metal edge preparations and exhibits high deposition rates and efficiencies. Electrogas welds normally exhibit excellent weld metal soundness. Thus, the process offers economic benefits and quality enhancements.

The mechanical aspects of electrogas welding are strikingly similar to those of the electroslag process from which it was developed. There are two variations of the process commonly used in the United States. Based on the GMAW process, it can feed a solid electrode into the joint; based on

the flux cored arc welding process (FCAW), it can incorporate a flux within a tubular electrode. Both variations use retaining shoes (dams) to confine the molten weld metal, which permits welding in the vertical position. Gas shielding, when needed, is provided through inlet ports in the dams or a gas cup around the electrode, or both. When using a self-shielded FCAW electrode, no gas is added.

Applications in welding plain carbon steels, structural steels, and pressure vessel steels will be discussed in detail later. Applications involving aluminum alloys and stainless steel will not be discussed, although successful examples have been reported.

Some of the advantages associated with EGW have resulted in considerable cost savings, particularly in joining thicker materials. Savings have been achieved where components can be joined in the vertical position with a continuous vertical weld. In thicker materials, EGW is often less expensive than the more conventional joining methods such as submerged arc welding and flux cored arc welding. Even in some applications involving thinner base materials, EGW may result in cost savings because of its efficiency and simple joint preparation.

FUNDAMENTALS OF THE PROCESS

PROCESS DESCRIPTION

ELECTROGAS WELDING IS an arc welding process that uses an arc between a continuous filler metal electrode and the weld pool, employing vertical position welding with backing to confine the molten weld metal. The process is used with or without an externally supplied shielding gas and without the application of pressure.

A square-groove or single-V groove joint is positioned so that the axis or length of the weld is vertical. There is no repositioning of the joint once welding has started; welding continues to completion, so that the weld is made in one pass. Electrogas welding is a machine welding process. The nature of the melting and solidification during welding results in a high quality weld deposit. There is little or no angular distortion of the base metal with single-pass welds. The welding action is quiet, with little spatter.

PRINCIPLES OF OPERATION

THE CONSUMABLE ELECTRODE, either solid or flux cored, is fed downward into a cavity formed by the base metals to be welded and the retaining shoes. A sump (starting tab) is used at the beginning of the weld to allow the process to stabilize before the molten weld metal reaches the work. An arc is initiated between the electrode and the sump.

Heat from the arc melts the continuously fed electrode and the groove faces. Melted filler metal and base metal collect in a pool beneath the arc and solidify to form the weld. The electrode may be oscillated horizontally through

the joint for uniform distribution of heat and weld metal. As the cavity fills, one or both shoes may move upward. Although the weld travel is vertical, the weld metal is actually deposited in the flat position at the bottom of the cavity.

ELECTRODE VARIATIONS

Solid Electrode

A SCHEMATIC VIEW of a typical electrogas welding installation using a solid electrode is shown in Figure 7.1. The electrode is fed through a welding gun, called a *nonconsumable guide*. The electrode may be oscillated horizontally to weld thicker materials. Gas shielding, normally carbon dioxide (CO_2) or an argon-carbon dioxide (Ar-CO_2)

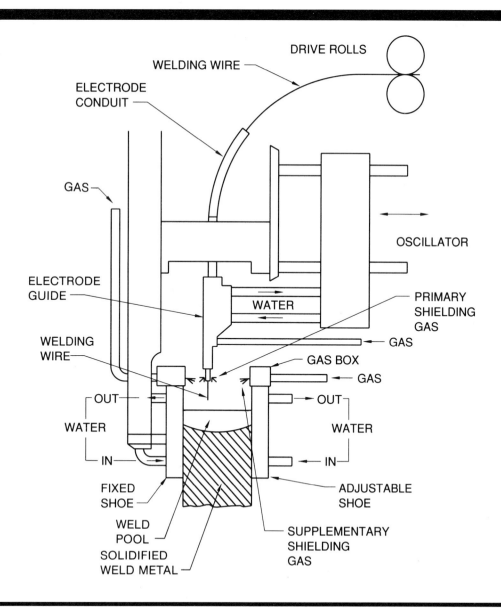

Figure 7.1–Electrogas Welding With a Solid Electrode

mixture, is provided to the weld cavity through gas ports, boxes, or nozzles.

Water-cooled copper retaining shoes are normally used on each side of the joint to retain the molten weld metal. The shoes are usually attached to the welding machine and move vertically as the machine moves. Vertical movement of the welding machine must be consistent with the deposition rate. Vertical movement may be automatic or controlled by the welding operator.

Electrogas welding with solid electrodes can be used to weld base metals ranging in thickness from approximately 3/8 in. (10 mm) to 4 in. (100 mm). Base metal thicknesses most commonly welded are between 1/2 in. (13 mm) and 3 in. (76 mm). Electrode diameters most commonly used are 1/16, 5/64, 3/32, and 1/8 in. (1.6, 2.0, 2.4, and 3.2 mm).

Flux Cored Electrode

FIGURE 7.2 ILLUSTRATES electrogas welding using a self-shielded flux cored electrode. The principles of operation and characteristics are identical to the solid electrode variation, except that no separate gas shielding is needed. The flux cored electrode creates a thin layer of slag between the weld metal and copper shoes to provide a smooth weld surface.

Electrogas welding with a flux cored electrode may be done with an external gas shield or a self-shielding elec-trode. Self-shielded electrodes operate at higher current levels and deposition rates than gas shielded types.

Diameters of flux cored electrodes commonly vary from 1/16 in. to 1/8 in. (1.6 mm to 3.2 mm). The wire (electrode) feeder must be capable of smooth continuous feeding of small diameter wires at high speeds and larger diameter wires at slower speeds.

CONSUMABLE GUIDE PROCESS

EGW WITH A consumable guide is similar to consumable guide electroslag welding. This variation of EGW is primarily used for short weldments in shipbuilding, and in column and beam fabrication. Consumable guide EGW uses relatively simple equipment, as shown in Figure 7.3. The principle difference is that none of the equipment moves vertically during consumable guide welding. Instead, the electrode is fed through a hollow consumable guide tube which extends to about 1 in. (25 mm) from the bottom of the joint. As the weld progresses vertically, the electrode melts back to the guide tube. Initially, the wire electrode penetrates about an inch beyond the end of the guide tube. Then a steady-state relationship develops between melting of the end of guide tube and the electrode wire, as shown in Figure 7.4. The relationship indicated in Figure 7.4 remains as the weld is completed.

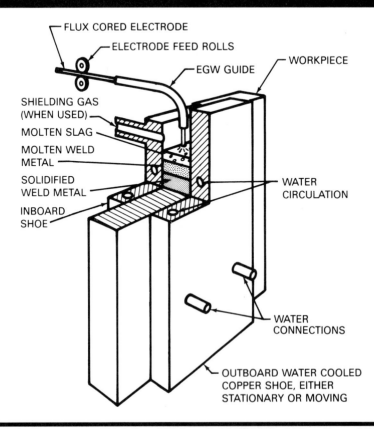

Figure 7.2–Electrogas Welding With a Self Shielded Flux Cored Electrode

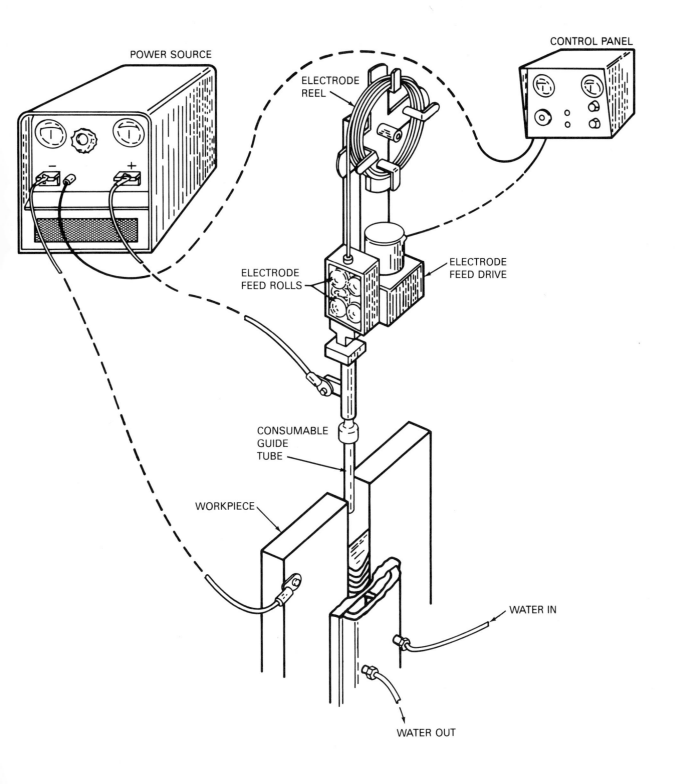

POWER SOURCE

CONTROL PANEL

ELECTRODE REEL

ELECTRODE FEED DRIVE

ELECTRODE FEED ROLLS

CONSUMABLE GUIDE TUBE

WORKPIECE

WATER IN

WATER OUT

Figure 7.3–Consumable Guide Electrogas Welding Equipment

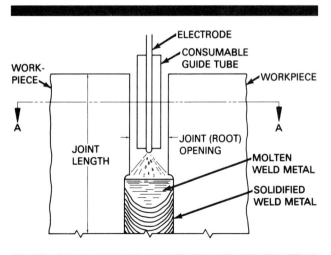

Figure 7.4–Schematic Diagram of Consumable Guide Electrogas Welding Process

The consumable guide provides approximately 5 to 10 percent of the deposited metal, with the balance supplied from the wire electrode. Arc initiation is begun in the same general fashion as in the conventional process.

A collar or clamp device, usually made of copper, holds the guide tube in position in the consumable guide process. The consumable guide is selected so that its chemical composition is compatible with the base material. Guide tubes are usually 1/2 to 5/8 in. (13 to 16 mm) O.D. and 1/8 to 3/16 in. (3 to 5 mm) I.D. Guide tubes are available in various lengths and for some applications may be welded together to accommodate longer joints. Oscillation with a consumable guide tube over 3 ft (1 m) long becomes increasingly more difficult as weld length increases because the bottom end of the tube starts a "whipping" motion and may distort due to heating. This tendency can be eliminated by adding one or more additional electrodes and thereby reducing or eliminating the oscillation. Circular insulators may be used approximately every 12 to 18 in. (300 to 450 mm) along the length of the tube to prevent shorting of the tube to the wall of the groove.

In consumable guide EGW, the retaining shoes are independent of the welding equipment, as indicated in Figure 7.3. The shoes may be held in place with wedges placed between the shoes and strong backs welded to the workpiece. On short welds, shoes may be the same length as the joint. Several sets of shoes may be stacked one on top of the other for longer joints.

One complication in consumable guide EGW is the buildup of slag on top of the molten weld pool. Excessive slag buildup can lead to poor penetration or nonfusion. Slag buildup can be overcome by either providing weep holes in the retaining shoes or by shimming the retaining shoes intermittently along the length of the weld. Either method will permit the excess slag to escape from the weld groove.

Consumable guide tubes may sometimes be used for complex shapes. The guides may be preformed to fit along the length of a curved joint.

EQUIPMENT

THE BASIC MECHANICAL equipment for electrogas welding consists of a direct current power supply, a device for feeding the electrode, shoes for retaining molten metal, an electrode guide, a mechanism for oscillating the electrode guide, and equipment needed for supplying shielding gas, when used. In a typical electrogas welding system, the essential components (with the exception of the power supply) are incorporated in an assembly that moves vertically as welding progresses.

Power Supply

DIRECT CURRENT ELECTRODE positive (reverse polarity) is normally used for EGW, with the power supply being either constant voltage or constant current. The power source should be capable of delivering the required current without interruption during the welding of a seam that may be of considerable length. Power sources used for electrogas welding usually have capacities of 750 to 1000 amperes at 30 to 55 volts and 100 percent duty cycle. Direct current is usually supplied by transformer-rectifier power sources, although motor-driven and engine-driven generators may be used.

Electrode Feeder

THE WIRE FEEDER for the electrode is of the push type, such as used with automatic GMAW or FCAW. The wire feeder is normally mounted as an integral part of the vertical-moving welding machine. Wire feed speeds may vary up to 550 in./min. (230 mm/s).

The wire feed system may include a wire straightener to eliminate the cast and helix set in the electrode and thus minimize electrode wander at the joint. Since the electrode extension in EGW is relatively long, 1-1/2 in. (40 mm) or more, a straight length of electrode projecting from the guide is necessary to assure positive arc location in the joint.

Electrode Guide

ELECTRODE GUIDES IN many respects are similar to the welding guns used for semiautomatic GMAW or for FCAW. The guide may have a shielding gas outlet to deliver gas around the protruding electrode (usually supplementary to the shielding gas supplied by ports in the shoes), and it may or may not be water cooled. The electrode guide (or at least part of it, in the case of a curved guide) rides inside the gap directly above the molten weld pool. It must be narrow enough to be contained in the gap with clearance to permit horizontal oscillation, if used, between the two shoes. For this reason, the width of an electrode guide is often limited to 3/8 in. (10 mm), so as to accommodate the 11/16 in. (18 mm) minimum gaps customarily used.

Electrode Guide Oscillators

WITH BASE METALS 1-1/4 in. to 4 in. (30 mm to 100 mm) thick, it is necessary to move the arc back and forth between the shoes and over the molten weld pool to achieve uniform metal deposition and ensure fusion to both workpieces. Horizontal oscillation of the arc is usually not needed for joints in base metal less than 1-1/4 in. (30 mm) thick, but is sometimes used with thinner base metals to minimize penetration into the work and to improve weld properties. This horizontal motion is accomplished by a system that oscillates the electrode guide and provides adjustable dwell times at either end of the oscillation. The arc is normally oscillated to about 1/4 in. from the retaining shoe on each side of the joint. Dwell time may be adjusted to produce good fusion at the base metal surface.

Retaining Shoes

RETAINING SHOES, ALSO called *dams*, are pressed against each side of the gap between the base metals to be welded to retain (dam) the molten weld metal in the groove. One or both shoes may move upward as welding progresses. In some weldments, a steel backing bar, which fuses to the weld, may replace the outboard shoe. Nonfusing ceramic backups have also been used. Retaining shoes are usually water cooled; moving shoes must be water cooled. The shoes are usually grooved to develop the desired weld reinforcement. Sliding shoes may or may not contain gas ports for supplying shielding gas directly into the cavity formed by the shoes and the weld groove. When gas ports are not used in the shoes, a "gas box" arrangement may be mounted on the shoes so as to surround the electrode and the welding arc with shielding gas. Gas ports in the shoes and "gas boxes" are not required when using self-shielded flux cored electrodes.

Controls

WITH THE EXCEPTION of the vertical travel control, EGW controls are primarily adaptations of the devices used with GMAW and FCAW. Vertical travel controls, either electrical, optical, or manual, maintain a given electrode extension, with the top of the movable shoe a specific distance above the molten weld pool.

On equipment using a self-shielded flux cored electrode, starting procedures must be carefully controlled to minimize porosity in the starting sump area. Welds are usually started at lower wire feed speeds and high voltages. These lower start settings must be maintained until the arc stabilizes and the starting area is heated. After a preset time, the equipment automatically changes the feed speed and voltage to the operating welding speed and voltage. Automatic rather than manual changing reduces the chance that operator error will cause starting porosity.

CONSUMABLES

ELECTRODES

BOTH FLUX CORED and solid electrodes are used with the electrogas welding process.

Flux cored EGW electrodes contain a lower percentage of slagging compounds than typical electrodes used in the FCAW process. These special electrodes allow a thin slag layer to form between the shoes and the weld to provide a smooth weld surface. Flux cored electrodes are supplied in sizes from 1/16 to 1/8 in. (1.6 to 3.2 mm). Only flux cored electrodes specifically designed for EGW should be used.

Solid electrodes are generally identical to those used for GMAW. They are supplied in sizes from 0.030 to 1/8 in. (0.8 to 3.2 mm).

Both flux cored and solid electrodes are available in various chemical compositions designed to introduce the necessary alloying elements to achieve strength, impact properties, or appropriate combinations of these and other properties in the deposited weld metal. For steel welding, these additional elements include manganese, silicon, and nickel. The required properties are normally obtained in the as-welded condition.

The American Welding Society publishes ANSI/AWS A5.26, *Specification for Consumables Used for Electrogas Welding of Carbon and High Strength Low Alloy Steels*, which prescribes requirements for solid and flux cored electrodes for electrogas welding.

This specification classifies solid electrodes on the basis of as-manufactured chemical composition and weld metal mechanical properties in the as-welded condition. It classifies flux cored electrodes on the basis of whether an external shielding gas is required, on the weld metal chemical composition, and as-deposited weld metal mechanical properties. The general classification system of electrodes for EGW is illustrated in Figure 7.5.

The classification requirements for the chemical composition of as-manufactured solid wires and as-deposited flux cored electrogas weld metals are shown in Tables 7.1 and 7.2 respectively. The required as-deposited weld metal mechanical properties are shown in Table 7.3. Note that there are two strength classifications, and that different base metals are used in the standard test weldment for each classification. Note also that either flux cored or solid electrodes can be classified in any of the mechanical properties classifications. The use of specific base metals for electrode classification tests accommodates the high dilu-

tion rate characteristic of single-pass electrogas welding. For each level of strength, electrode classifications are provided that define three levels of minimum toughness as determined by the Charpy V-notch impact test.

SHIELDING

SELF-SHIELDED FLUX CORED electrodes for EGW contain core materials that shield the molten weld metal. Other flux cored electrodes require an additional external shielding gas (normally carbon dioxide, but mixtures of argon and carbon dioxide are also used). Recommended gas flow rates range from 30 to 140 CFH (14 to 66 L/min). Gas flow rates depend upon the equipment design, and manufacturer recommendations should be followed. Mixtures of argon and carbon dioxide are normally used for welding steel with solid electrodes and may be used with flux cored electrodes.

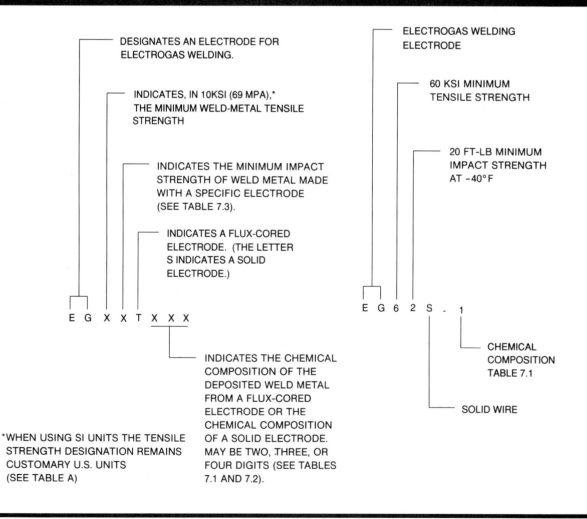

Figure 7.5–Electrogas Welding Electrode Classification System

Table 7.1
Chemical Composition Requirements For Solid Electrodes

AWS Classification[b]	C	Mn	S	P	Si	Wt. Percent[a] Ni	Mo	Cu[c]	Ti	Zr	Al
EGXXS-1	0.07 to 0.19	0.90 to 1.40	0.035	0.025	0.30 to 0.50	-	-	0.35	-	-	-
EGXXS-2	0.07	0.90 to 1.40	0.035	0.025	0.40 to 0.70	-	-	0.35	0.05 to 0.15	0.02 to 0.12	0.05 to 0.15
EGXXS-3	0.06 to 0.15	0.90 to 1.40	0.035	0.025	0.45 to 0.70	-	-	0.35	-	-	-
EGXXS-5	0.07 to 0.19	0.90 to 1.40	0.035	0.025	0.30 to 0.60	-	-	0.35	-	-	0.50 to 0.90
EGXXS-6	0.07 to 0.15	1.40 to 1.85	0.035	0.025	0.80 to 1.15	-	-	0.35	-	-	-
EGXXS-D2[d]	0.07 to 0.12	1.6 to 2.1	0.035	0.025	0.50 to 0.80	0.15	0.40 to 0.60	0.35	-	-	-
EGXXS-G	-	Not Specified[e]									

a. Single values are maximums.

b. The letters "XX" refer to either 6Z, 60, 62, 7Z, 70, 72, and 8Z, 80, and 82 mechanical property designations of Table 7.3.

c. The copper limit includes copper that may be applied as a coating on the electrode.

d. Formerly EGXXS-1B.

e. Composition shall be reported; the requirements are those agreed to by the purchaser and the supplier.

Table 7.2
Chemical Composition Requirements for Weld Metal from Composite Flux Cored and Metal Cored Electrodes

AWS Classification[b]	Shielding Gas	C	Mn	P	S	Weight Percent[a] Si	Ni	Cr	Mo	Cu	V	Other Elements Total
EGXXT-1	None	(c)	1.7	0.03	0.03	0.50	0.30	0.20	0.35	0.35	0.08	0.50
EGXXT-2	CO_2	(c)	2.0	0.03	0.03	0.90	0.30	0.20	0.35	0.35	0.08	0.50
EGXXT-Ni1 (Formerly EGXXT-3)	CO_2	0.10	1.0 to 1.8	0.03	0.03	0.50	0.7 to 1.1	-	0.30	0.35	-	0.50
EGXXT-NM1 (Formerly EGXXT-4)	Ar/CO_2 or CO_2	0.12	1.0 to 2.0	0.02	0.03	0.15 to 0.50	1.5 to 2.0	0.20	0.40 to 0.65	0.35	0.05	0.50
EGXXT-NM2 (Formerly EGXXT-6)	CO_2	0.12	1.10 to 2.10	0.03	0.03	0.20 to 0.60	1.1 to 2.0	0.20	0.10 to 0.35	0.35	0.05	0.50
EGXXT-W (Formerly EGXXT-5)	CO_2	0.12	0.50 to 1.3	0.03	0.03	0.30 to 0.80	0.40 to 0.80	0.45 to 0.70	-	0.30 to 0.75	-	0.50
EGXXT-G				Not specified[d]								

a. Single values are maximums.

b. The letter "XX", in this table refer to either the 6Z, 60, 62, 7Z, 70, 72, 8Z, 80 or 82 mechanical property designations of Table 7.3

c. Composition range for carbon not specified for these classifications, but the amount shall be determined and reported.

d. Composition shall be reported; the requirements are those agreed to between the purchaser and supplier.

Table 7.3
As-Welded Mechanical Property Requirements For Deposited Weld Metal

AWS Classification[a]	Tensile Strength		Yield Strength at 0.2% Offset (min)		Elongation in 2 in. (50mm)	Charpy V-Notch Impact Strength, min[b]	
	ksi	MPa	ksi	MPa	min, percent	ft-lb.	J
(When using ASTM A36 plate)[c]							
EG6ZXXX	60	420				Not req'd.	Not specified
EG60XXX	to	to	36	250	24	20 at 0°F[d]	27 at −18°C[d]
EG62XXX	80	550				20 at −20°F[d]	27 at −29°C[d]
(When using ASTM structural steels, A242, A441, A572 Grade 50, or A588)[c]							
EG7ZXXX	70	490				Not req'd.	Not specified
EG70XXX	to	to	50	350	22	20 at 0°F[d]	27 at −18°C[d]
EG72XXX	95	650				20 at −20°F[d]	27 at −29°C[d]
EG8ZXXX	80	550				Not req'd.	Not specified
EG80XXX	to	to	60	420	20	20 at 0°F[d]	27 at −18°C[d]
EG82XXX	100	690				20 at −20°F[d]	27 at −29°C[d]

a. The letters XXX in this table refer to classifications T1, T2, T3, S-1, S-2, S-3, and so forth. Shielding gas for flux cored electrodes is as noted in Table 7.2. Shielding gas is CO_2 for solid electrodes. Other gases may be used if agreed upon between supplier and purchaser.

b. The lowest value obtained, together with the highest value obtained, may be disregarded for this test. Two of the three remaining values shall be greater than the specified 20 ft-lb (27 J) energy level. The computed average value of the three values shall be equal to or greater than the 20 ft-lb (27 J) energy level.

c. Restriction of plate materials for tensile/yield strengths listed is necessary due to inherent characteristics of the electrogas process.

d. Note that if a specific electrode meets the requirements of a given classification, this classification also meets the requirements of all lower numbered classifications in the series. For instance, an electrode meeting the requirements of the EG62 classification, also meets the requirements of EG6Z. This applies to the EG7X series also.

OPERATING VARIABLES

BASE METALS

THE METALS MOST commonly joined by electrogas welding are plain carbon, structural, and pressure vessel steels. Some examples of these steel grades are shown in Table 7.4. In addition to the steels shown in Table 7.4, EGW has been used to join aluminum and some grades of stainless steels.

Table 7.4
Examples of Steel Grades That are Commonly Joined by Electrogas Welding

Application	Grades
Plain Carbon Steels	AISI 1010, 1018, 1020
Structural Steels	ASTM A36, A131, A242, A283, A441, A572, A573, A588
Pressure Vessel Steels	ASTM A36, A285, A515, A516, A537

Electrogas welding of quenched and tempered or normalized steels may not be permitted by some codes. For example, ANSI/AWS D1.1, *Structural Welding Code—Steel*, does not permit electrogas welding of quenched-and-tempered steels.

JOINT DESIGN

SQUARE BUTT JOINTS ARE commonly used with approximately 7/8 in. (22 mm) spacing. This joint normally uses two moving shoes as shown in Figure 7.6(A). Single-V-groove welds are also used with one moving and one stationary shoe, Figure 7.6(B). The root opening of single-V-groove welds is usually about 5/32 in. (4 mm), and the face opening is normally 7/8 in. (22 mm). The moving shoe molds the metal at the weld face, and the stationary shoe molds the root face.

Examples of other joint configurations that have been used successfully are shown in Figure 7.7. Variations of

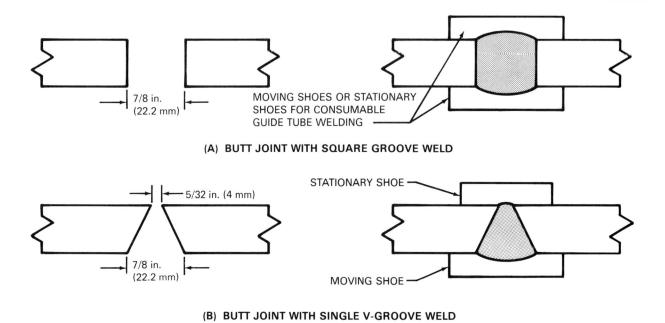

(A) BUTT JOINT WITH SQUARE GROOVE WELD

MOVING SHOES OR STATIONARY
SHOES FOR CONSUMABLE
GUIDE TUBE WELDING

7/8 in.
(22.2 mm)

5/32 in. (4 mm)

7/8 in.
(22.2 mm)

STATIONARY SHOE

MOVING SHOE

(B) BUTT JOINT WITH SINGLE V-GROOVE WELD

Figure 7.6–Typical Electrogas Welding Joint Designs

these joint configurations are also acceptable, and the welding engineer should conduct experiments to determine the range of joint dimensions that is best suited to the engineer's application.

FIT-UP AND ASSEMBLY

Misalignment

BASE METAL SURFACE misalignment should be less than 1/8 in. (3.2 mm) on both front and back surfaces. Grinding may be required to stay within this tolerance. Greater misalignment creates a variety of problems, including leakage of weld metal, irregular bead shape, poor tie-in, overlap, undercut, and loss of shielding. In addition, when using steel backing, excessive misalignment can result in melt-through.

When welding plates of different thicknesses, the thicker plate should be tapered, and the retaining shoe should be tapered to match it. For structural welding applications a 1 : 2–1/2 taper is satisfactory.

Strongbacks

SUITABLE HOLDING CLAMPS, brackets, or strongbacks are required to hold the base materials in alignment during welding. U-shaped brackets shown in Figure 7.8 allow clearance for the moving shoe or backing. The brackets should hold the base material in alignment, but should not be so rigid as to cause excessive restraint. Generally,

smaller brackets can be used for fixed backup applications than are needed for clearance of a moving shoe.

Starting Sump

A STARTING SUMP permits striking the arc below the actual joint so starting discontinuities will be contained in discard materials. The sump should be deep enough to allow the process to switch from starting to welding procedures and for the arc and travel to stabilize before the molten weld pool rises into the joint being welded. The sump is cut away after welding.

The sump thickness and gap configuration should be the same as those of the base materials being welded. The sump is attached to the base materials as shown in Figure 7.9. The top of the sump, where it fits against the bottom of the base materials, should be clean and dry. The starting sump should be seal welded at the joint, but should not be sealed all around. An opening should be left so that gases expanding between the sump and the work as a result of the welding heat are driven away from the molten weld metal.

Runoff Tabs

RUNOFF TABS ARE required because the rapid cooling of the large molten weld pool at the end of the weld creates a shrinkage crater with a tendency to trap slag or gas. Typical runoff tabs are shown in Figure 7.9. Fixed backup bars must reach to the top of the runoff area, finishing 1 to 2 in. (25 to 50 mm) above the top of the workpiece. Runoff tabs should be removed after the weld is completed.

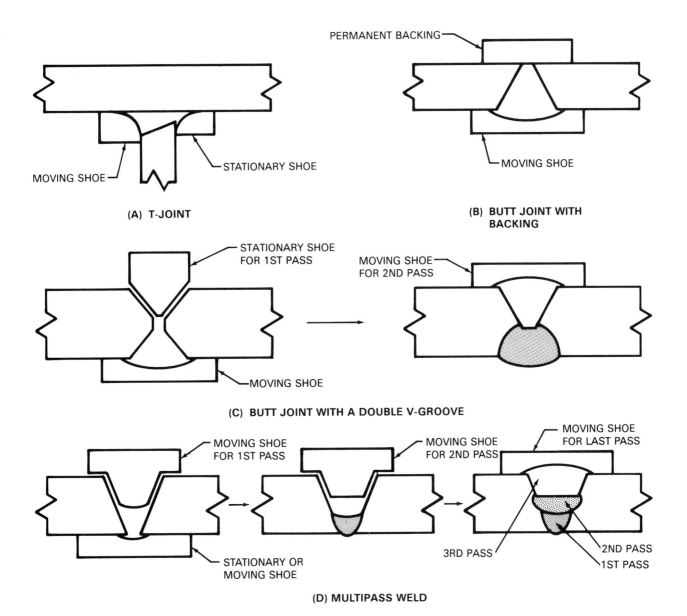

Figure 7.7–Alternate Electrogas Welding Joint Designs

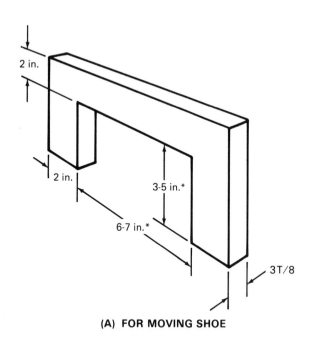

(A) FOR MOVING SHOE

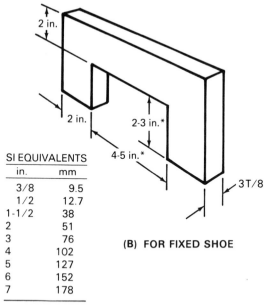

SI EQUIVALENTS

in.	mm
3/8	9.5
1/2	12.7
1-1/2	38
2	51
3	76
4	102
5	127
6	152
7	178

(B) FOR FIXED SHOE

*ACTUAL DIMENSIONS TO FIT DAM OR SHOE

T = BASE METAL THICKNESS

Figure 7.8–Typical Dimensions for Electrogas Welding Strongbacks

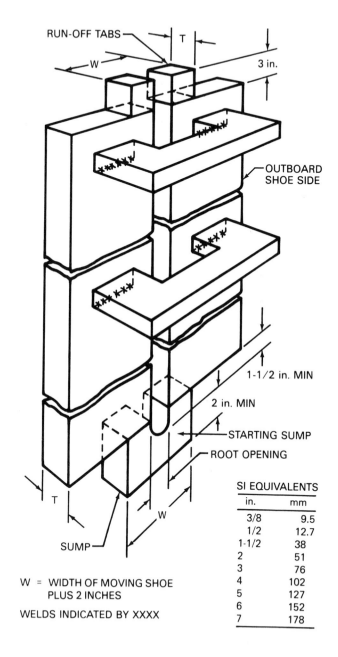

W = WIDTH OF MOVING SHOE PLUS 2 INCHES

WELDS INDICATED BY XXXX

SI EQUIVALENTS

in.	mm
3/8	9.5
1/2	12.7
1-1/2	38
2	51
3	76
4	102
5	127
6	152
7	178

Figure 7.9–Typical Arrangement of Sump, Run-Off Tabs and Strongbacks for Electrogas Welding Joints

APPLICATIONS

APPLICATIONS OF ELECTROGAS welding include storage tanks, ship hulls, structural members, and pressure vessels. EGW should be considered for any joint to be welded in a vertical position in materials ranging in thickness from 3/8 to 4 in. (10 to 100 mm). Some typical EGW applications are illustrated in Figures 7.10 to 7.14.

METALLURGICAL CONSIDERATIONS

A DETAILED DESCRIPTION of the metallurgical effects of weld thermal cycles may be found in Volume 1 of the *Welding Handbook*, 8th Edition, Chapter 4.

Electrogas Thermal Cycle

THE THERMAL CYCLE of EGW is prolonged because of the relatively slow travel speeds [1–1/2 to 8 in./min. (0.2 to 3 mm/s)]. Therefore, the electrogas weld metal structure contains large grains with a marked tendency to columnar growth. The heat-affected zone is wider than in conventional arc welding, and also contains a wider grain-coarsened region. However, the heat-affected zone and the grain-coarsened regions of electrogas welds are narrower than those of electroslag welds of comparable size. The prolonged thermal cycle results in a relatively slow weld cooling rate. The slow cooling rates produce heat-affected zones without undesirable hard structures that often occur in conventional arc welds in carbon and low alloy steels.

The hardness profiles of an electrogas and a shielded metal arc weld are shown in Figures 7.15 and 7.16 respectively. The hardness profile of the electrogas weld is relatively uniform, whereas the shielded metal arc weld exhibits a significant increase in hardness in the heat-affected zone near the weld interface. The high heat-affected-zone hardness results from the metallurgical reactions that occur when the shielded metal arc weld cools at a rapid rate from the welding temperature.

In some cases, this hardness can approach the maximum quench hardness of the material. The heat-affected-zone pattern shown in Figure 7.16 is typical for all commonly used multipass welding processes, such as shielded metal arc welding, submerged arc welding, flux cored arc welding, and gas metal arc welding.

In addition to slowing the cooling rate, the protracted thermal cycle of electrogas welding allows more time at high temperatures where grain growth occurs. Thus, the weld metal and the heat-affected zone of electrogas welds exhibit larger grains and larger coarse grain regions.

Weld Structure

THE DOMINANT FEATURE of weld microstructure in electrogas welds is the large columnar grains that result from the weld solidification. Weld solidification is covered in detail in Chapter 4 of the *Welding Handbook*, Volume 1, 8th Edition. The contour of the solid-liquid interface can be estimated from a vertical cross section of a completed weld. Columnar grain growth occurs perpendicular to the solid-liquid interface, and an etched vertical macrosection of an electrogas weld would be identical to the electroslag weld cross section shown in Figure 7.17. Thus, during welding, the approximate boundary between the liquid and solid metal is essentially perpendicular to the columnar grains as shown in Figure 7.17.

Figure 7.10—Electrogas Welding of a Ship Hull

Figure 7.11–Electrogas Welding of a Field Storage Tank

Figure 7.12–A Portable Electrogas Welding Unit Housed in a Special Elevator Cage

NOTE: PANELS SHIELD THE ELECTROGAS UNIT FROM AIR CURRENTS WHICH WOULD DISPERSE THE SHIELDING GAS

Figure 7.13—A Self-Contained Electrogas Welding Unit Used in Shipbuilding

Figure 7.14–Electrogas Welding of Surge Tanks

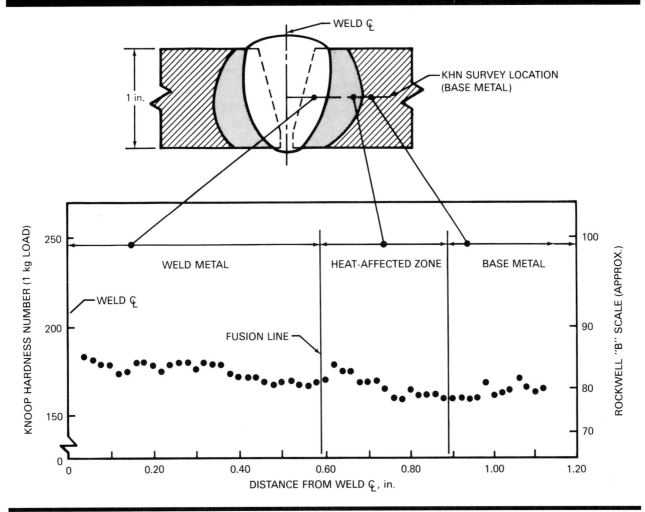

Figure 7.15—Knoop Microhardness Survey Across an Electrogas Weld in A-283 Grade C Base Metal

Form factor, the ratio of the width of the weld pool to its depth, is the term most often used to describe the shape of the weld pool in electrogas welding. Generally, a high form factor is desirable because the freezing progression is vertical and any impurities, segregates, and low melting point constituents are maintained in the weld pool, float upwards, and freeze harmlessly in the runoff tab outside of the production weld metal. By contrast, welds with low form factors may trap the low melting point constituents and impurities along the centerline of the weld and result in a plane of weakness. This condition increases the tendency for weld cracking which usually occurs at high temperatures in conjunction with or immediately after solidification.

The weld structure resulting from electrogas welding consists of large columnar grains; the resultant heat-affected zone is wide with considerable grain coarsening near the fusion line. Welds with large columnar grains generally exhibit lower notch toughness and a higher transi-

tion temperature than welds with fine equiaxed or dendritic grains. Properly made electrogas welds in the as-welded condition generally meet the minimum impact properties specified for hot-rolled carbon and low alloy steels. Often, the weld metal and heat-affected-zone impact properties are also adequate for normalized steel grades. Postweld heat treatment may enhance the toughness to allow the use of electrogas welding for lower service temperature applications.

Preheat

PREHEATING THE BASE metal is usually unnecessary when electrogas welding low and medium carbon steels. Most of the heat generated in electrogas welding is conducted into the workpieces. The conducted heat serves to preheat the base metal due to the relatively slow advance of the welding arc.

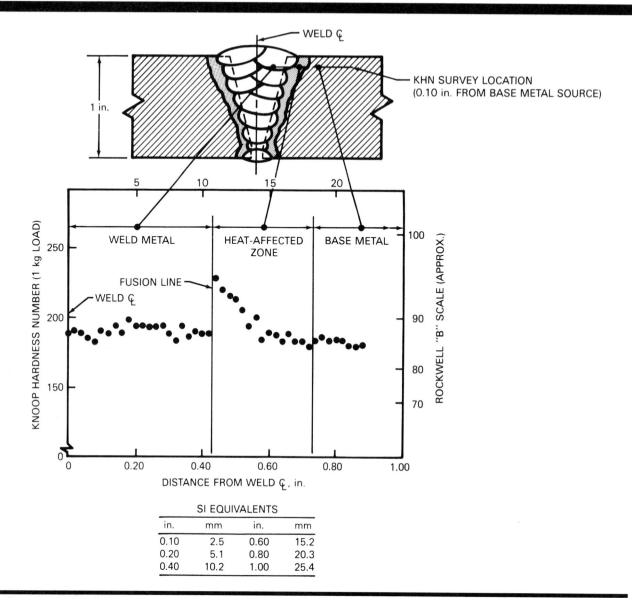

Figure 7.16–Knoop Microhardness Survey Across a Shielded Metal Arc Weld (Vertical Up) in A-283 Grade C Base Metal

Although the heat of welding provides adequate preheat for most applications, preheat requirements should be determined by the user for the specific conditions of each application. Preheat may increase resistance to cracking when welding:

(1) High-strength steels
(2) Steels with high carbon or alloy content
(3) Thick base metals [over 3 in. (75 mm)]
(4) Highly restrained base metals
(5) Base metals below 32°F (0°C)

Preheating the start area of the weld also improves edge wetting. However, excessive preheat slows the weld cooling rate, increases the joint penetration, and may contribute to weld metal leakage and melt-through.

Sometimes, because of an equipment malfunction or other causes, welding may be terminated prior to completion, and the weld must be restarted in the joint. In this case, an insert or wedge needs to be jammed into the open joint immediately after welding stops. The insert will prevent squeezing the head assembly in the weld groove as the base metals contract with the removal of the welding heat. Preheating the restart area reverses the shrinkage forces.

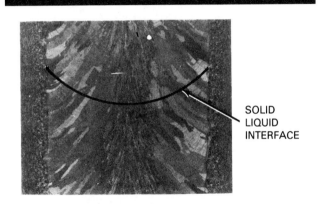

SOLID
LIQUID
INTERFACE

Figure 7.17–Typical Grain Growth Pattern of Vertical Section of an Electrogas Weld

This will allow the welding head assembly to track up the joint. After the arc has been restarted, the weld groove joint opening should expand enough to remove the wedge. It is important to note that certain defects may occur as a result of the restart. The weld should be carefully examined in these regions.

Postweld Heat Treatment

MOST APPLICATIONS OF electrogas welding, particularly the welding of field erected structures, require no postweld heat treatments. Stress-relief heat treatment of electrogas weldments typically results in a slight drop in the yield and tensile strength and a slight improvement in weld and heat-affected-zone notch toughness. In weldments requiring optimum notch toughness, a postweld heat treatment such as normalizing may be required. Electrogas welded quenched-and-tempered steels will exhibit lower strength in the weld metal and the heat-affected zone. Matching strength may be restored by a full heat treatment (austenitize, quench, and temper), but this procedure is often impractical.

Residual Stresses and Distortion

SINCE SOLIDIFICATION OF an electrogas weld begins at the shoes, the outer surfaces of the weld are under compressive residual stresses, and the center of the weld is in tension. This unique residual stress pattern is just the reverse of the residual stresses that are likely to occur with multipass arc welding processes.

Angular distortion in the horizontal plane is virtually nonexistent in single-pass electrogas welds. This is due to the symmetry of most electrogas joint designs about the mid-thickness of the base material, which results in uniform shrinkage in the thickness direction. As welding progresses up the joint, the parts are drawn together by the weld shrinkage. Therefore, the root opening at the top of the joint should be approximately 3/32 to 1/8 in. (2.4 to 3.2 mm) more than at the bottom to allow for this shrinkage. Factors influencing shrinkage allowance include material type, joint thickness, joint length, and degree of restraint of the parts being joined.

Angular distortion can occur during multiple-pass applications of electrogas welding. It may be compensated for by presetting the parts. Experience will suggest the proper amount of preset for each application.

Mechanical Properties

EGW INVOLVES HIGH base-metal penetration, and therefore, melted base metal may contribute up to 35% of the total weld metal. Thus, weld composition and mechanical properties can vary substantially with different types of steel and even with different heats of the same steel grade.

Mechanical properties also vary with welding conditions such as current and voltage settings, joint design, type of backup, and cooling rate. Electrogas electrodes are designed to accommodate as much variation as practical, and consistent mechanical properties can be achieved when the welding conditions are controlled within the normal limitation of variables found in fabrication standards.

AWS B2.1, *Standard for Welding Procedure and Performance Qualification*, describes the welding procedure qualification variables and their limits that should be controlled for electrogas and other welding processes. Similar procedure qualification requirements directed at specific applications are described in some codes.

ANSI/AWS D1.1, *Structural Welding Code—Steel*, for example, includes qualification requirements for structural welding. Thus, the results of welding procedure qualification tests are representative of mechanical properties to be expected in production welds. Typical weld metal and heat-affected zone mechanical properties for flux cored electrogas welds in several carbon steel grades are listed in Table 7.5A. Table 7.5B shows similar results in SI Units.

One problem encountered in electrogas welding has been low notch toughness properties of welds and heat-affected zones in the as-welded condition. By using suitably alloyed electrodes and controlled welding conditions, the notch toughness properties of the weld metal can be equal to or better than those of the base metal for most structural and pressure vessel steels. In most cases, these notch toughness properties can be achieved in the as-welded condition. For many structural and pressure vessel steel grades, the heat-affected zone notch toughness is more dependent on the base-metal notch toughness than the welding conditions. Base metal, heat-affected zone, and weld metal toughness of an A283 steel electrogas weldment are shown in Table 7.6.

Table 7.5A
Typical As-Welded Mechanical Properties of Electrogas Welds Fabricated with Flux Cored Electrodes (US Customary Units)

Materials (ASTM)	Type Application	Thickness in.	Electrode Class (A5.26)	Shielding Gas[2]	Tensile Properties			Charpy V-Notch Properties		
					Yield Strength ksi	Tensile Strength ksi	Elongation in 2 in., Percent	Test Temp. °F	Energy Absorption ft-lb WM	HAZ
1020[1]	Structural	1/2	EG72T1	-	72	80	28	-20	32	-
1020[1]	Structural	1	EG72T1	-	64	79	27	-20	24	-
A36	Structural	1	EG72T1	-	65	79	27	-20	30	-
A36	Structural	1	EG72T4	CO_2	70	92	26	-22	26	-
A36	Structural	3	EG72T4	CO_2	67	89	26	-22	23	-
A36	Structural	3	EG72T4	Ar/CO_2	-	87	23	0	39	-
A131-C	Shipbuilding	1-1/2	EG72T3	CO_2	-	71	30	-30	33	22
A441	Structural	3/4	EG72T1	-	74	82	23	-20	30	-
A441	Structural	1	EG72T1	-	66	81	23	-20	27	-
A441	Structural	2	EG72T1	-	65	81	24	-20	22	-
A441	Structural	2	EG72T4	CO_2	68	92	24	0	22	-
A572-50	Structural	1	EG72T4	Ar/CO_2	-	-	-	+14	45	25
A572-50	Structural	1-1/2	EG72T1	-	57	77	23	0	10	-
A588	Structural	3	EG72T4	CO_2	-	95	23	0	41	-
A203	Pressure Vessel	1-5/8	EG72T4	Ar/CO_2	53	72	32	-40	21	62
A516	Pressure Vessel	1	EG72T1	-	58	74	26	-20	21	-
A516	Pressure Vessel	1-1/2	EG72T4	Ar/CO_2	78	90	29	-20	30	48
A537-1	Pressure Vessel	3/4	EG72T1	-	70	86	24	-4	32	-
A537-1	Pressure Vessel	1	EG72T4	CO_2	62	83	29	-20	25	29
A537-1	Pressure Vessel	1-1/8	EG72T4	Ar/CO_2	74	100	29	-22	34	-
A633	Pressure Vessel	1	EG72T1	-	74	88	25	0	46	-

1. American Iron and Steel Institute

2. Ar/CO_2 is 80% argon and 20% CO_2.

Table 7.5B
Typical As-Welded Mechanical Properties of Electrogas Welds Fabricated with Flux Cored Electrodes (SI Units)

Materials (ASTM)	Type Application	Thickness mm	Electrode Class (A5.26)	Shielding Gas[2]	Tensile Properties			Charpy V-Notch Properties			
					Yield Strength MPa	Tensile Strength MPa	Elongation in 2 in., Percent	Test Temp. °C	Energy Absorption ft-lb		
									WM	HAZ	
1020[1]	Structural	12	EG72T1	-	496	552	28	−29	43	-
1020[1]	Structural	25	EG72T1	-	441	545	27	−29	33	-
A36	Structural	25	EG72T1	-	448	545	27	−29	41	-
A36	Structural	25	EG72T4	CO_2	483	634	26	−30	35	-
A36	Structural	75	EG72T4	CO_2	462	614	26	−30	31	-
A36	Structural	75	EG72T4	Ar/CO_2	-	600	23	−18	53	-
A131-C	Shipbuilding	38	EG72T3	CO_2	-	490	30	−34	45	30
A441	Structural	19	EG72T1	-	510	565	23	−29	41	-
A441	Structural	25	EG72T1	-	455	558	23	−29	36	-
A441	Structural	50	EG72T1	-	448	558	24	−29	30	-
A441	Structural	50	EG72T4	CO_2	468	634	24	−18	30	-
A572	Structural	25	EG72T4	Ar/CO_2	393	531	-	−10	61	34
A572-50	Structural	38	EG72T1	-	-	655	23	−18	14	-
A588	Structural	76	EG72T4	CO_2	-	-	23	−18	56	-
A203	Pressure Vessel	41	EG72T4	Ar/CO_2	365	496	32	−40	23	-
A516	Pressure Vessel	25	EG72T1	-	400	510	26	−29	28	-
A516	Pressure Vessel	38	EG72T4	Ar/CO_2	538	621	29	−29	41	64
A537-1	Pressure Vessel	19	EG72T1	-	483	593	24	−20	43	-
A537-1	Pressure Vessel	25	EG72T4	CO_2	427	572	29	−29	34	39
A537-1	Pressure Vessel	28	EG72T4	Ar/CO_2	510	689	26	−30	46	-
A633	Pressure Vessel	25	EG72T1	-	510	614	25	−18	62	-

1. American Iron and Steel Institute

2. Ar/CO_2 is 80% argon and 20% CO_2.

Table 7.6
As-Welded Notch Toughness Properties of an Electrogas Weld in 1 in. (25 mm) Thick A283 Steel

| | Charpy V-Notch* ft-lb (J) | | |
	+70°F (21°C)	+20°F (−7°C)	0°F (−18°C)
Unaffected plate	25, 18, 9 (34, 24, 12)	5, 5, 6 (7, 7, 8)	4, 3, 5 (5, 4, 7)
Heat-affected zone	35, 28, 12 (47, 38, 16)	5, 5, 5 (7, 7, 7)	6, 6, 5 (8, 8, 7)
Center of weld	35, 32, 18 (47, 43, 24)	22, 23, 11 (30, 31, 15)	12, 10, 8 (16, 14, 11)

* Notch perpendicular to the plate surface.

The electrogas process should be used with caution and be thoroughly tested for applications that require low temperature notch toughness or that involve a reversal of stress. The electrogas process may degrade the heat-affected zone toughness to the extent that use of the process should only be considered in connection with a full postweld heat treatment.

PROCESS VARIABLES

PROCESS VARIABLES FOR electrogas welding include arc voltage, current, electrode feed speed, electrode extension, electrode oscillation and dwell, and joint opening. The proper selection of these variables affects the operation and economics of the process and the resultant weld quality. The effects of each variable must be fully understood, since they differ significantly from gas metal arc and flux cored arc welding. As an example, weld penetration and depth of fusion in conventional arc welding processes (GMAW and FCAW) are in line with the axis of the electrode. Penetration and depth of fusion are generally increased by increasing the welding current and decreasing the voltage. However, in electrogas welding, the joint faces of the base metal are parallel to the axis of the electrode. Increasing the welding current or decreasing the voltage in electrogas welding results in a deeper weld pool and a narrower depth of fusion (sidewall penetration).

Welding Voltage

THE WELDING VOLTAGE is the major variable affecting weld width and base metal melting in electrogas welding. Welding voltages of 30 to 35 volts are normally used. Increasing the voltage increases the sidewall penetration and width of the weld as illustrated in Figure 7.18. For thicker base metals or for higher deposition rates the voltage should be increased. However, excessively high voltage may cause the electrode to arc to the joint sidewalls above the molten weld pool, causing unstable operation.

Welding Current—Electrode Feed Speed

WELDING CURRENT AND electrode feed speed are proportional for a given electrode diameter and extension. Increasing the electrode feed speed increases the deposition rate, welding current, and travel speed (fill rate).

For a given set of conditions, increasing the current decreases the sidewall penetration and width of the weld, as shown in Figure 7.19. Low current levels result in slow travel speeds and wide welds. Excessive welding current may cause a severe reduction in weld width and sidewall penetration. Excessive welding current also causes a low form factor which contributes to centerline weld crack susceptibility. Welding currents of 300 to 400 amperes for 1/16 in. (1.6 mm), 400 to 800 amperes for 3/32 in. (2.4 mm), and 500 to 1000 amperes for 1/8 in (3.2mm) diameter electrodes are commonly used. The power source and welding cables must be sufficiently rated for the high currents and extended arc times typical of this welding process.

Form Factor

THE RESISTANCE OF the weld to centerline cracking is greatly influenced by the manner in which it solidifies.

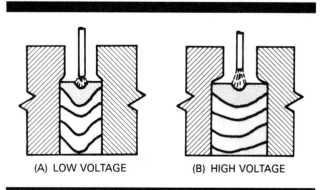

(A) LOW VOLTAGE   (B) HIGH VOLTAGE

Figure 7.18–Effect of Welding Voltage on the Shape of the Molten Weld Pool

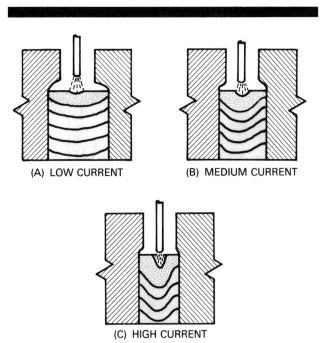

(A) LOW CURRENT   (B) MEDIUM CURRENT

(C) HIGH CURRENT

Figure 7.19–Effect of Welding Current on the Shape of the Molten Weld Pool

Heat is removed from the molten weld metal by the base metal and retaining shoes. Solidification begins at these cooler areas and progresses toward the center of the weld. Since filler metal is continuously added, there is a progressive solidification from the bottom of the joint, and there is always molten weld metal above the solidifying weld metal. The resulting solidification pattern is shown by the sketch of a vertical section of an electrogas weld in Figure 7.20.

The molten weld pool solidification pattern can be expressed by the term *form factor*. The form factor, as previously noted, is the ratio of the maximum molten pool width (gap opening plus sidewall penetration) to its maximum depth. The form factor is measured on a vertical sec-

tion of the electrogas weld, usually taken at the mid-thickness of the base metal. Welds having high (1.5) form factors (wide width and shallow depth) have maximum resistance to centerline cracking. Welds having low (0.5) form factors (narrow width and deep molten weld pool) tend to have low resistance to centerline cracking. Thus, the form factor is an empirical number that provides a relative measure of resistance to centerline cracking and describes the shape of the weld metal solidification pattern. However, form factor alone does not determine cracking propensity. The chemical composition of the filler metal and the base metal (especially carbon content) and joint restraint also contribute to centerline crack susceptibility.

The shape of the molten weld pool and its resultant form factor are controlled by the welding variables. In general, increasing root opening or the voltage increases the form factor, while increasing the current or decreasing the root opening decreases the form factor. Since the welding variables are usually specified in a qualified welding procedure, the form factor is rarely measured or recorded.

Electrode Extension

IN EXTERNALLY GAS shielded electrogas welding, an electrode extension of about 1-1/2 in. (40 mm) is recommended for solid and flux cored electrodes. For most self-shielded flux cored electrogas welding, an extension of 2 to 3 in. (50 to 75 mm) is recommended.

The power supply modifies the effect of electrode extension. When using a constant-potential power source, increasing the electrode extension by jogging the welding head up will reduce the arc voltage. The width of the resultant weld will be reduced, as indicated in Figure 7.21. Increasing the electrode extension by increasing the electrode feed speed will increase the deposition rate and decrease the sidewall penetration and weld width.

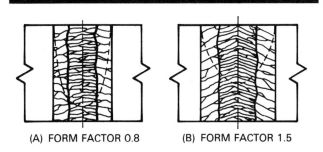

(A) FORM FACTOR 0.8   (B) FORM FACTOR 1.5

Figure 7.20–Sketches of Weld Macrostructure of a Vertical Section of an Electrogas Weld

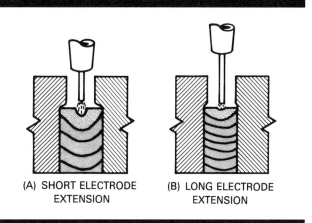

(A) SHORT ELECTRODE EXTENSION   (B) LONG ELECTRODE EXTENSION

Figure 7.21–Effect of Electrode Extension (With CP Power Supply) on the Shape of the Molten Weld Pool

For a constant-current (or drooping) power source, increasing the electrode extension, by jogging the welding head up, will cause an increase in the voltage. This will result in a wider weld. Increasing the electrode extension by increasing the wire feed speed will decrease the arc voltage and result in a higher deposition rate (and fill rate) with a narrower weld width.

Electrode Oscillation

THE NECESSITY FOR electrode oscillation depends upon the welding conditions including the current, voltage, and electrode diameter. Therefore, the following comments are general, and users should develop procedures that best suit their operation.

Base metals up to 3/4 in. (19 mm) thick are commonly welded with a stationary electrode, and base metals between 3/4 and 1-1/4 in. (32 mm) may be welded with stationary electrodes, but oscillation is commonly employed for greater thicknesses. Base metals with a thickness between 1-1/4 and 1-1/2 in. (40 mm) have been welded with a stationary electrode, but high voltages and special techniques are required to prevent lack of edge fusion.

Electrode oscillation should be controlled so that the arc stops at a distance of approximately 3/8 in. (10 mm) from each shoe. Oscillation speed is commonly 20 to 50 in./min. (13 to 16 mm/s). A dwell period at each end of the oscillation travel is often needed to assure complete fusion at the joint corners. The dwell time may vary from 1/2 to 3 seconds depending on joint thickness.

Joint Opening

A MINIMUM JOINT opening is needed to give sufficient clearance for the electrode guide, which usually reaches inside the joint. Increasing the joint opening increases the weld width. Excessive joint openings will increase the welding time and consume extra filler metals and gases, which will increase welding costs. Excessive joint openings may also cause lack of edge fusion if the voltage is not increased. Joint openings for the square-groove weld design are generally 11/16 in. (17 mm) to 1-1/4 in. (32 mm). Joint openings for the single-V-groove weld design are generally 11/16 to 1-1/4 in. (17 to 32 mm) at the face and 5/32 to 3/8 in. (4 to 10 mm) at the root.

Worklead

THE WORKLEAD is usually connected to the starting sump, but it can be split and attached to each side of the joint at the bottom. Although this worklead location has proved satisfactory in most applications, severe arc blow can cause excessive starting porosity, lack of fusion on one side of the joint, or both. The final worklead location must be determined by the user to suit the specific conditions of each application.

Commonly used settings for electrogas welds are shown in Table 7.7. Users should optimize these settings to meet their specific requirements.

Methods of Inspection

CONVENTIONAL NONDESTRUCTIVE EXAMINATION (NDE) methods can be used to inspect electrogas welded joints.

Nondestructive Examination techniques normally used with EGW include:

(1) VT—visual
(2) RT—radiography (isotope or x-ray)
(3) MT—magnetic particle
(4) PT—liquid penetrant
(5) UT—ultrasonic (the large grain size may require special techniques)

Table 7.7
Typical Conditions for Electrogas Welds Using .120 in. Diameter AWS Class EG72T1 Electrode With Moving Shoes

Plate Thickness in.	Current Amps	DCEP Volts[1]	Electrode Feed in./min.	Travel Speed in./min	Electrode Extension in.	Oscillation Distance in.	Joint Opening in.	Start Sequence sec
1/2	450-500	35-37	300	6.0	2+1/8	N.A.	1/2	2[2]
5/8	475-525	36-38	340	4.4	2+1/8	N.A.	5/8	3[2]
3/4	525-575A	37-39	380	3.9	2+1/8	N.A.	3/4	3[2]
1	625-675	40-42	350	3.4	3+1/8	N.A.	3/4	5[3]
1 1/4	625-675	40-42	350	2.6	3+1/8	N.A.	3/4	8[3]
1 1/2	625-675	40-42	350	1.7	3+1/8	3/4	3/4	10[3]

1. Measured at power supply terminals.

2. Starting conditions: 150 in./min electrode feed at 29 volts.

3. Starting conditions: 250 in./min electrode feed at 37 volts.

All NDE should be performed in accordance with qualified procedures by qualified technicians. In addition to the usual training of NDE personnel, some training in interpretation of weld discontinuities found in EGW is necessary. Typical discontinuities found in EGW will be discussed later in the chapter.

The inspection method used will depend not only upon the requirements of the governing code but also the owner's or purchaser's contract specifications.

Every joint should have thorough visual examination (VT) to reveal incomplete edge fusion, underfill, and, in some cases, cracks. When the sump and runoff tabs are removed, defects such as cracking and porosity may be detected. However, to find internal defects (such as porosity, inclusions, cracking, and, in rare instances, lack of internal fusion), RT and UT testing are the most effective means of examination. Magnetic particle (MT) testing may also be used to detect cracks and lack of fusion, but it is limited to the surface and immediate subsurface.

Acceptance

MOST WELDING ACCEPTANCE criteria are established by the customer or some regulatory agency, or both. EGW applications are almost always welded to various code requirements. Some standards presently permitting use of the EGW process (sometimes with additional restrictions) are listed below:

ANSI/AWS D1.1 *Structural Welding Code—Steel*
ANSI/ASME *Boiler and Pressure Vessel Code*
API Standards 620 and 650
CSA Standards W59
ABS *Rules for Building and Classing Steel Vessels*
Lloyds *Register of Shipping Rules and Regulations for Classification of Ships*

Occasionally, customer specifications may add to the requirements of the regulatory agency.

Quality Control

A WRITTEN PROCEDURE that identifies the essential variables to be used in EGW should be prepared. This procedure should be reviewed to determine its adequacy by comparison with applicable code requirements.

Only approved welding procedures should be issued to the welding foreman or welding operator. Some completed typical EGW procedure specification forms are shown in Tables 7.8 and 7.9. Any additional instructions considered necessary should be made part of the procedure specification.

A checklist should be used by the welding operator to assure that the equipment is properly set up and that all of the required operating adjustments are made. A typical EGW checklist is shown in Table 7.10.

Proper quality control procedures should be used to assure that the proper filler metal is being used and that the qualified welding procedure is being followed.

Rework

IN GENERAL, REPAIRS of defective joints can be minimized by a good preventive maintenance program on the equipment, by using qualified and trained welding operators, and by implementing a sound welding procedure.

A defect such as underfill can often be repaired by building up with the SMAW process without gouging or grinding. Defects such as lack of fusion (at the joint surface), overlap, copper pickup on the weld face, and metal spillage are visible defects that can be repaired by superficial gouging or grinding to sound metal and rewelding with the SMAW process. Defects such as porosity, cracking, and lack of fusion (internal) are usually detected by radiographic or ultrasonic testing and can be repaired by gouging deeply to sound metal and rewelding with the SMAW process. Some graphic examples of discontinuties that are peculiar to EGW are shown in Figure 7.22.

Economic considerations should dictate whether EGW or some other process will be used for repair. Generally, repairs will be made with the SMAW process using electrodes appropriate for the base metal and a qualified welding procedure.

Restarting of electrogas welds should be avoided if at all possible. However, when it becomes necessary to restart a weld, the starting defect can be confined to the near surface area by employing the technique illustrated in Figure 7.23. As shown in Figure 7.23, the starting cavity is sloped by air carbon arc gouging. The start area is preheated to a minimum of 300°F (135°C), and the arc initiated near the front shoe. As the sloped crater fills, the arc is moved toward the center of the groove until the normal running position is achieved. For welds in thick plates where oscillation is required, the oscillation distance is expanded following the same principle. This technique should produce a shallow starting discontinuity on the near side that can be easily removed. The cavity can be filled with sound metal deposited by conventional arc welding processes.

TRAINING OF WELDING OPERATORS

PROPERLY TRAINED OPERATORS are vital to the successful application of the EGW process. EGW is generally a single-pass welding process, but with special techniques multiple pass welds can be made. The operator should be trained to set up and properly operate the equipment.

Table 7.8
Typical EGW Procedure Specification with Moving Shoe(s)*

Welding Process: Welding shall be done by the electrogas welding (EGW) process, using one electrode wire. [If two (2) electrodes are used, the electrode spacing should be specified.]

Base Material: The base metal shall conform to the specification for ASTM A588   .

Base Material Thickness: This procedure will cover the welding of the base material from   2   in. (50   mm) thick to   2   in. (50   mm) thick.

Filler Metal: The electrode shall conform to AWS specification A5.26 for classification EG   70S-1B . The diameter of the electrode shall be   0.062 in. (1.6   mm).

Shielding Gas: Electrode is used with externally supplied shield gas.
Shielding gas, if used: Composition   Ar (80%) CO_2 (20%)
Flow Rate   130 ft^3/h (4 m^3/h)

Position: Welding shall be done in the vertical position.

Shoes: The welding shall be done using a water-cooled copper inboard shoe. The outboard shoe shall be water-cooled copper .

Preheating: None required by this procedure specification. However, no welding shall be performed when the temperature of the base metal at the point of welding is below 32°F (0°C).

Postweld Heat Treatment: None required by this procedure specification.

Base Material Preparation: The edges or surfaces of the parts to be joined shall be prepared by oxygen cutting and shall be cleaned of oil, grease, moisture, scale, rust, or other foreign material. The surfaces upon which the copper shoes must slide shall be flat and smooth.

Welding Current: The welding current shall be direct current, electrode positive (reverse polarity).

Welding Technique: The welding shall be done in a single pass, vertical up. Starting and run-off tabs shall be used. An electrode extension of 1-1/2 in. (38   mm) shall be maintained during welding. The welding power supply shall have a constant current characteristic.

Welding Conditions: All welding shall be performed using the conditions given below:

Plate thickness
[in.(mm)]   2 in. (50 mm)

Root opening
[in.(mm)]   11/16 — 13/16 (17.5 — 20.6)

Welding current (amps)   350

Electrode
WFS [in/min(mm/s)]   controlled by current

Welding voltage(volts)   35

Oscillation:
Distance[in.(mm)]   1-1/2   (38)

Period(sec)   4.2

Dwell(sec)   2

Traverse time(sec)   0.1

Travel Speed: Speed of vertical travel is a function of deposition rate and need not be specified.

Procedure Qualification: Procedure Qualification tests have been made in accordance with ANSI/AWS D1.1

Joint Design: The joint design shall be as detailed above:

* A typical EGW Procedure Specification should include but not necessarily be limited to the items included herein.

Table 7.9
Typical EGW Procedure Specification* Using Consumable Guide Tubes and Fixed Copper Shoes

Welding Process: Welding shall be done by the electrogas welding (EGW) process, using a consumable guide tube.

Base Metal: The base metal shall conform to the specification for ASTM A36   .

Base Material Thickness: This procedure will cover the welding of the base material from __4__ in. (__100__ mm) thick to __4__ in. (__100__ mm) thick.

Filler Metal: The electrode shall conform to AWS specification A5.26 for classification EG __72T1__ . The diameter of the electrode shall be __0.120__ in. (__3.2__ mm).

Consumable Guide Tube: 1/2 in. O.D. x 5/32 in. I.D. C1008/C1020

Shielding Gas: Electrode is used without externally supplied shield gas.
Shielding gas, if used: Composition __NONE__
Flow rate __NONE__

Shoes: The fixed copper shoes shall be water cooled/air-cooled and shall be one piece/stacked.

Preheating: None required by this procedure specification. However, no welding shall be performed when the temperature of the base metal at the point of welding is below 32°F (0°C). Start area of weldment may be preheated to __100__ °F (__38__ °C) to improve bead wetting at start of weld, especially on heavier plates.

Postweld Heat Treatment: None required by this procedure specification.

Base Material Preparation: The edges or surfaces of the parts to be joined shall be prepared by __oxygen cutting__ and shall be cleaned of oil, grease, moisture, scale, rust, or other foreign material. The surfaces against which the copper shoes must fit shall be flat and smooth.

Welding Current: The welding current shall be direct current, electrode positive (reverse polarity).

Welding Technique: The welding shall be done in a single pass, vertical up. Starting and run-off tabs shall be used. The welding power supply shall have a *constant potential* characteristic.

Welding Conditions: All welding shall be performed using the conditions given below:

Plate thickness [in.(mm)] __4 (100)__

Root opening [in.(mm)] __1 to 1-1/8 (25-30)__

Electrode WFS [in/min(mm/s)] __400 (170)__

Welding current (amps) __controlled by wire feed speed__

Welding voltage (volts) __45-47__

Oscillation:

Distance [in.(mm)] __3 (75)__

Period __(sec.) 18__

Dwell __(sec.) 3-4__

Traverse time (sec.) __5 1/2__

Travel Speed: Speed of vertical travel is a function of deposition rate and root opening and need not be specified.

Procedure Qualification: Procedure Qualification tests have been made in accordance with __ANSI/AWS D1.1__

Joint Design: The joint design shall be as detailed above:

BACK (UNSLOTTED) SHOE
FRONT (SLOTTED) SHOE
4 in.
3 in.
1/4 in.
1 in. – 1-1/8 in.

* A typical EGW Procedure Specification should include but not necessarily be limited to the items included herein.

Table 7.10
Typical Electrogas Check List with Moving Shoe(s)

Electrogas welding is designed to continuously weld the entire joint, usually in a single pass. Before starting to weld, it is vital that the operator check all elements of the operation.

CHECK EACH ITEM AS JOB PROGRESSES

1. Check material preparation. Make sure that the weld edge and adjacent surface have been prepared properly along entire length.

2. Fit starting sump at bottom of joint and run-off tabs at top. Minimum distance from starting surface in the sump to bottom edge of joint shall be 1-1/2 in. (38 mm).

3. Fit seam per procedure. Maintain required tolerance.

4. Attach and tighten work lead cables.

5. Make sure the machine is properly fitted to the joint.

6. Check vertical rise. Make sure there are no obstructions that would prevent or stop the rise of the machine and that equipment leads are of adequate length.

7. Check quantity of electrode wire and gas (if used). Make sure there is enough to finish the weld.

8. Check water and gas connections. Make sure both are in operating condition.

9. Check the face of the shoe(s). Make sure there are no worn or irregular areas that would cause a "hang up" during welding.

10. Check wire guide tip. Do not start with a worn tip; it could cause a shutdown or, unstable welding conditions.

11. Check head and electrode line-up. Line up the head and electrode control. Specific line-up settings and tolerances are listed in each welding procedure.

    11a. Electrode angle to joint is always 90°.

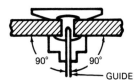

    11b. Adjust the wire straightener so that the electrode is straight or has a maximum underbend of 3/8 in. (9 mm).

Table 7.10
Continued

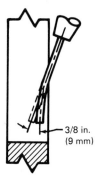

    11c. Adjust electrode extension. Do *not* change the electrode extension when making other adjustments.

    11d. Adjust drag angle. (See figure 11c.)

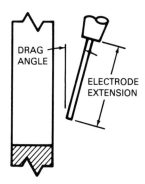

    11e. Set the electrode location by measuring from the front face of the plate after the electrode is in the seam. Note: The electrode location changes as the guide tip wears. Replace the guide tip when it becomes worn or the end is fused or deformed.

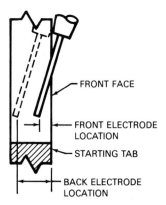

    11f. Set sufficient spacing between the guide tip and top of the movable shoe. When oscillation of the electrode in the joint is required, set the front location (for the electrode), oscillation distance, oscillation time, and dwell time(s) as specified.

Table 7.10
Continued

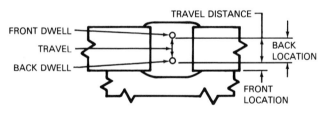

11g. Set 1/2 in. (13 mm) spacing between the end of the electrode and the sump.

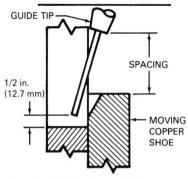

12. Check moving shoe(s). Clamp with the correct operating pressure.

13. Apply glass tape or putty to any opening between the shoe face and work caused by the weld reinforcement groove in the shoe. A run-out may occur at start-up if this is not done.

14. Set water and gas flow rates.

15. Have a few tools (e.g., pliers, screwdriver, and open end wrench) handy.

16. Check operation of ventilating equipment if used.

17. Have eye shield ready.

18. Set start-up volts and amperes.

19. Energize contactor.

20. Recheck and adjust welding conditions.

21. Turn on oscillator, if used.

EGW is unlike manual and semiautomatic processes where welding can be temporarily stopped or delayed and easily restarted. EGW is designed to be continuous from beginning to end. If the weld is not sound, the discontinuities are often continuous, and therefore, the entire weld may have to be removed and the joint rewelded. Depending upon the type of application and setup, at least some portion of welding must be completed before the welding operator can visually inspect, evaluate, and take any necessary corrective action. In some applications, the entire weld is completed before visual inspection can be per-

formed. For these reasons, electrogas operators should understand the process and follow the procedure in every detail. Operators should be patient, conscientious, alert, and sufficiently experienced to recognize when unsatisfactory welds are being deposited.

QUALIFICATIONS

THE ELECTROGAS PROCESS has been proven successful on many applications, including those previously discussed. Successful users of EGW maintain their equipment, use properly trained and qualified operators, and follow welding procedures that meet the requirements of the relevant code or specifications. Successful users are familiar with the recommendations of the equipment manufacturer. An operator checklist, similar to that shown in Table 7.10 should be developed for each new application.

TROUBLESHOOTING GUIDE

Porosity at Start

SEVERAL FACTORS CONTRIBUTE to starting porosity in electrogas welds. Among them are the following:

(1) Low electrode feed speed, high voltage, short electrode extension, insufficient time in sump
(2) Contaminants in sump or between sump and base metals
(3) Condensation on shoes or between shoes and base metals
(4) Fast cooling rates caused by thick plates, or large copper starting sumps, and low ambient temperatures
(5) Sump too shallow
(6) Arc location too close to shoes
(7) Water leakage in sump
(8) Poor shoe or sump fit
(9) Insufficient gas coverage or contaminated gas
(10) Erratic welding arc

Porosity in Weld

WELD METAL POROSITY, as illustrated in Figure 7.22(A), is generally caused by expanding gases that are insoluble in the freezing metal, but gas-like voids can also be caused by mechanical problems. Some of the causes are the following:

(1) Starting porosity that extends into the production weld
(2) Excessive voltage
(3) Low electrode feed speed
(4) Electrode extension too short
(5) A "cold" weld (low electrode feed speed and voltage)
(6) Contaminants in weld area

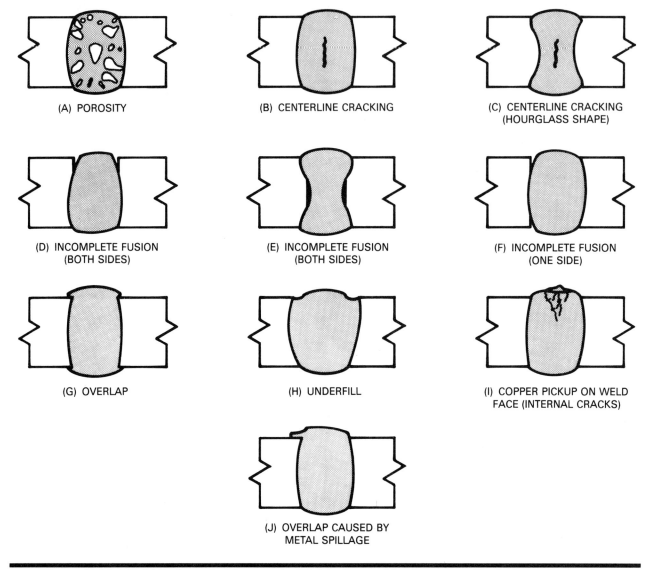

(A) POROSITY

(B) CENTERLINE CRACKING

(C) CENTERLINE CRACKING
(HOURGLASS SHAPE)

(D) INCOMPLETE FUSION
(BOTH SIDES)

(E) INCOMPLETE FUSION
(BOTH SIDES)

(F) INCOMPLETE FUSION
(ONE SIDE)

(G) OVERLAP

(H) UNDERFILL

(I) COPPER PICKUP ON WELD
FACE (INTERNAL CRACKS)

(J) OVERLAP CAUSED BY
METAL SPILLAGE

Figure 7.22–Weld Defects in Electrogas Welds Caused by Improper Technique, Defective Equipment, or Both

(7) Loose fit of shoe allowing air to bleed into weld
(8) Insufficient gas coverage or contaminated gas
(9) Water leakage from shoe
(10) Erratic welding arc

Porosity at Weld Termination

POROSITY AT THE end of the weld can be caused by the same factors as porosity within the weld. In addition, the following factors can cause porosity:

(1) Runoff tabs or stationary shoe too short
(2) Slag leaking due to improperly attached runoff tabs

(3) Arc blow caused by improper location of worklead

Centerline Weld Cracking

CENTERLINE CRACKING, AS illustrated in Figures 7.22(B) and 7.22(C), may be related to procedure variables, thermal conditions, and weld restraint. Some steel grades are more susceptible to centerline cracking than others. For the more susceptible steels, each of the other factors becomes more critical.

As previously discussed, the procedure variables that contribute to a low form factor contribute to centerline cracking. Among these conditions are the following:

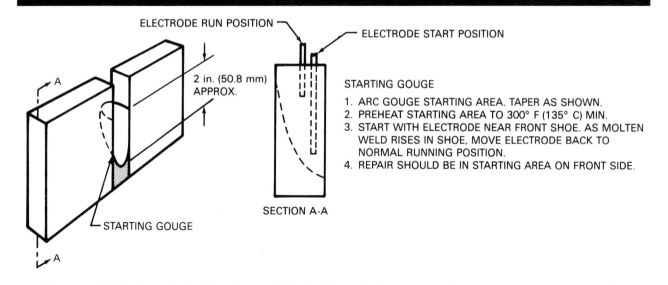

ELECTRODE RUN POSITION

ELECTRODE START POSITION

2 in. (50.8 mm) APPROX.

STARTING GOUGE

STARTING GOUGE
1. ARC GOUGE STARTING AREA. TAPER AS SHOWN.
2. PREHEAT STARTING AREA TO 300° F (135° C) MIN.
3. START WITH ELECTRODE NEAR FRONT SHOE. AS MOLTEN WELD RISES IN SHOE, MOVE ELECTRODE BACK TO NORMAL RUNNING POSITION.
4. REPAIR SHOULD BE IN STARTING AREA ON FRONT SIDE.

SECTION A-A

Figure 7.23–Procedure for Restarting an Electrogas Weld

(1) Electrode feed speed too high (excessive current)
(2) Arc voltage too low
(3) Gap too narrow
(4) Dwell time too long

Rapid cooling rates also contribute to centerline cracking. Some of the factors that contribute to rapid cooling rates are large shoes with excessive water flow and lack of preheat for thick plates at low abient temperatures.

Finally, excessive restraint is a contributing factor in all occurrences of cracking.

Incomplete Fusion to Both Sidewalls

INCOMPLETE FUSION AT both sidewalls, as illustrated in Figures 7.22(D) and 7.22(E), is caused by unsatisfactory thermal conditions that prevent the sidewalls from melting. Poor heat distribution as well as insufficient heat can result in this defect. The following are among the contributing factors:

(1) Cold weld (low voltage or slow electrode feed speed and low voltage)
(2) Excessive electrode feed speed (fast fill rate)
(3) Gap too narrow (fast fill rate)
(4) Oscillation speed too fast
(5) Excessive slag on top of molten weld pool

An excessive slag burden can be avoided by better shoe design. By machining a deeper and wider groove in the shoe, more slag is permitted to coat the weld reinforcement. This reduces the burden. If the weld pool is maintained higher in the shoe, then more slag escapes as spatter. This also will reduce the slag burden.

Incomplete Fusion to One Sidewall

INCOMPLETE FUSION TO one sidewall; as illustrated in Figure 7.22(F), is caused by asymmetric thermal conditions. These can result from the following:

(1) Arc location off-center
(2) Electrode angled toward one sidewall
(3) Arc blow caused by improper location of worklead

Overlap

OVERLAP IS CAUSED by weld metal flow out of the joint without melting the base metal. The condition is illustrated in Figures 7.22(G) and 7.22(J). The condition shown in Figure 7.22(J) is often caused by a poor fit of the base metal. It can result if the sliding shoe is lifted from contact with the base metal by foreign material such as spatter. This allows weld metal to leak and freeze against the plate.

(1) Overlap on front face
   (a) Arc location too far back: improper settings of wire straightener, drag angle, or guide location; worn tip guide
   (b) Bevel angle too large
   (c) Cold weld (low voltage or low electrode feed and voltage)
(2) Overlap on back face
   (a) Arc location too far forward: improper settings of wire straightener, drag angle, or guide location; worn guide tip
   (b) Cold weld (low voltage or low electrode feed and voltage)
(3) Overlap on both faces

(a) Cold weld (low voltage or low electrode feed and voltage)

(b) Groove in copper shoe(s) too wide

(c) Excessive quenching from shoe(s)—improper shoe design or excessive water flow

(d) Excessive travel speed

(e) Joint opening too narrow

(f) Arc blow

(g) Incorrect oscillation cycle

Underfill

UNDERFILL UNDER CERTAIN conditions may be acceptable, but at best it represents marginal workmanship. Moreover, it is fairly easy to prevent. Underfill, illustrated in Figure 7.22(H), can be caused by excessive melting of base metal beyond the shoe or too narrow a groove in the shoe.

Melt-Through in Starting Sump

BECAUSE MELT-THROUGH IN the starting sump occurs outside the production weld, it is not a weld defect. It does, however, prevent production of the workpiece. It is easily prevented by using adequate thickness of material on the bottom of the sump, or by attaching a backup plate to the sump. Poorly fitted backing shoes can result in the same problem.

Hot Cracking

HOT CRACKING CAN be caused by the partial dissolution of the copper molding shoes. The cracks are generally at or near the surface. This type of cracking is shown in Figure 7.22(I). The solution of copper in the weld metal can be caused by arcing on shoe(s) or melting of shoes due to poor shoe cooling.

Electrogas welds of consistent quality can be fabricated by maintaining the equipment and following the weld procedure exactly. Operators should inspect their equipment before each weld is started to verify that the contact tip is not worn and the welding electrode feeds freely and straight. Operators should also check the workpiece to verify that the joint is correct, and that the base metal condition will allow free passage of the welding head to the top of the joint. Lastly, prior to starting the weld, the electrode should be set and oscillated within the joint to assure that it is free to oscillate and that the oscillation speed and dwell are correctly adjusted. When the weld is stabilized, the arc current and voltage should be determined to be within the limits prescribed by the welding procedure specification.

SAFETY

SPECIFIC INSTRUCTIONS FOR safe operation of electrogas welding equipment should be found in the manufacturer's literature. General safety instructions for all welding and cutting can be found in ANSI/ASC Z49.1, *Safety in Welding and Cutting*, published by AWS. Mandatory Federal safety regulations are established by the U. S. Labor Department's Occupational Safety and Health Administration. The latest edition of OSHA Standards, *Code of Federal Regulations*, Title 29 Part 1910, is available from the Superintendent of Documents, U. S. Printing Office, Washington, D. C. 20402.

RADIANT ENERGY

EYE PROTECTION WITH the proper shade of filter glass for the current being used is required for EGW. The total radiant energy produced by the EGW process can be higher than that produced by the SMAW process because EGW has a more exposed arc. Generally, the highest ultraviolet radiant energy intensities are produced when using an argon shielding gas and when welding on aluminum. The suggested filter glass shades for EGW are shown in Table 7.11.

The choice of filter shade may be made on the basis of visual acuity and may, therefore, vary widely from one individual to another, using different current densities, materials, and welding processes. However, the degree of protection from radiant energy afforded by the filter plate or lens when chosen to allow for visual acuity will still remain in excess of the needs of eye filter protection. Filter plate shields as low as shade eight have proven suitably radiation-absorbent for protection from arc welding processes.

Suitable clothing should be worn to protect skin from being exposed to arc radiation. Leather or wool clothing that is dark in color (to reduce reflection, which could cause

**Table 7.11
Recommended Shade of Filter Glass for EGW of Several Metals**

Base Metal to Be Welded	Shade*
Aluminum	13
Ferrous Metals	12
Nonferrous Metals (except aluminum)	11

* A #2 filter lens is recommended for flash goggles.

ultraviolet burns to the face and neck underneath the helmet) is recommended. The intensity of ultra-violet radiation tends to cause rapid disintegration of cotton clothing.

GENERAL SAFETY NOTES

THE WELDING MACHINE should be turned off except when welding is actually in progress or at service check for open circuit voltage (when electrode cannot feed).

The main power switch should be turned off before opening the control cabinet.

Accessory equipment, such as wire feeders, travel mechanisms, oscillators, etc., should be grounded. If they are not grounded, insulation breakdown might cause these units to become electrically "hot" with respect to ground.

Adequate ventilation should be provided, especially when welding with self-shielded flux cored electrodes.

If chlorinated solvents have been used to degrease or clean the workpiece, a check should be made to ensure that the solvent has been removed before welding. Welding should not be permitted near degreasing tanks.

The primary power for the welding power source should be turned off, and any shielding gas supply should be shut off at the supply source when leaving the work or stopping the work for any appreciable time, or when moving the machine.

NOISE AND HEARING PROTECTION

PERSONNEL SHOULD BE protected against exposure to noise generated in welding and cutting processes in accordance with Paragraph 1910.95, Occupational Noise Exposure, of the Occupational Safety and Health Standards, Occupational Safety and Health Administration, U. S. Department of Labor. In addition, *Arc Welding and Cutting Noise* is available from the American Welding Society, 550 N.W. LeJeune Road, P.O. Box 351040, Miami, Florida 33135.

SUPPLEMENTARY READING LIST

American Society for Metals. "Welding and brazing." *Metals Handbook*, Vol. 6, 9th Ed. Metals Park, Ohio: American Society for Metals.

American Welding Society. "Welding process—arc and gas welding and cutting, brazing, and soldering." *Welding Handbook*, Vol. 2, 7th Ed., 225-260. Miami, Florida: American Welding Society, 1978.

Arnold, P. C. and Bertossa, D. C. "Multiple-pass automatic vertical welding." *Welding Journal* 45(8): 651-660; August 1966.

Campbell, H. C. "Electroslag, electrogas, and related welding processes." Bulletin No. 154. New York: Welding Research Council, 1970.

Franz, R. J. and Wooding, W. H. "Automatic vertical welding and its industrial applications." *Welding Journal* 42(6): 489-494; June 1963.

Irving, R. R. "Vertical welding goes into orbit." *Iron Age* 50; October 26, 1972.

Normando, N. J., Wilcox, D. V., and Ashton, R. F. "Electrogas vertical welding of aluminum." *Welding Journal* 52(7): 440-448; July 1973.

Schwartz, N. B. "New way to look at welded joints." *Iron Age* 54-55; August 20, 1970.

Warner, Basil. "Welding offshore drillings rigs." *American Machinist*, October 14, 1974.

ELECTROSLAG WELDING

PREPARED BY A COMMITTEE CONSISTING OF:

J. R. Hannahs, Chairman
Midmark Corporation

D. R. Amos
Westinghouse Electric Corporation

R. J. Christoffel
General Electric Company

R. H. Frost
Colorado School of Mines

S. A. LaClair
Foster Wheeler Corporation

S. Liu
Colorado School of Mines

J. E. Sims
CBI-NA-CON

R. B. Smith
L-TEC Welding and Cutting Systems

W. E. Wood
Oregon Graduate Center

WELDING HANDBOOK COMMITTEE MEMBER:

J. R. Hannahs
Midmark Corporation

CHAPTER 8

ELECTROSLAG WELDING

INTRODUCTION

HISTORY

SINGLE-PASS WELDING OF heavy plates has long been desired as a means of avoiding multipass welding techniques. Before 1900, graphite molds were placed on each side of a space between vertical plates to contain molten metal created by graphite electrodes, fusing the edges to form the weld. Graphite molds were replaced by copper or ceramic molds, and conventional welding arcs, gas torches and thermit mixtures were devised for generating the molten metal with a degree of superheat sufficient to obtain uniform coalescence.

In the early 1950's, Russian scientists from the Paton Institute of Electric Welding in Kiev announced the development of machines that employed the principle of an electrically conductive slag to make single-pass vertical welds. Subsequent work at the Bratislava Institute of Welding in Czechoslovakia was made available to engineers in Belgium in 1958, and through them to the rest of the Western world. An electroslag unit was introduced in the United States in 1959. Since then, many refinements and modifications have been made, resulting in production machines capable of meeting the standards of our industry.

SCOPE

THIS CHAPTER DESCRIBES the electroslag welding (ESW) process including material on fundamentals, equipment, safety, consumables, applications, quality control, qualifications, training, troubleshooting and definitions associated with the process.

USES AND ADVANTAGES

ESW IS MOST often employed to join metals in the vertical or near vertical position, usually in a single pass. However, it has been shown that ESW can be used at angles of 45° or greater from vertical. Some of the advantages associated with ESW have resulted in considerable cost savings, particularly in joining thicker materials. Savings have been achieved where components are joined to make larger units instead of initially producing massive castings or forgings. ESW is often less expensive than more conventional joining methods such as submerged arc welding in thicker section weldments. Even in some applications involving thinner base materials ESW has resulted in cost savings because of its efficiency and simple joint preparation.

The ESW process offers many opportunities for reducing welding costs on specific types of joints. Process advantages are the following:

(1) Extremely high metal deposition rates; ESW has a deposition rate of 35 to 45 lbs per hour per electrode.

(2) Ability to weld very thick materials in one pass; there is one equipment setup and no interpass cleaning since there is only one pass.

(3) Preheating is normally not required, even on materials of high hardenability.

(4) High-quality weld deposit; the weld metal stays molten for an appreciable time, allowing gases to escape and slag to float to the top of the weld.

(5) Minimum joint preparation and fit-up requirements. Mill edges and flame-cut squares edges are normally employed.

(6) High-duty cycle; the process is automatic and once started continues to completion; there is little operator fatigue.

(7) Minimum materials handling; the work needs to be positioned only to place the axis of the weld is vertical or near-vertical; there is no manipulation of the parts once welding has started.

(8) Elimination of weld spatter, which results in 100 percent filler metal deposition efficiency.

(9) Low flux consumption; approximately 1 pound of flux is used for each 20 pounds of weld metal.

(10) Minimum distortion; there is no angular distortion in the horizontal plane; distortion is minimum in the vertical plane, but this is easily compensated for.

(11) Minimum welding time; ESW is the fastest welding process for large, thick material.

LIMITATIONS

THE LIMITATIONS OF the ESW process are the following:

(1) The ESW process welds only carbon and low alloy steels, and some stainless steels.

(2) Joints must be positioned in the vertical or near vertical position.

(3) Once welding has started, it must be carried to completion or a defective area is likely to result.

(4) ESW cannot be used on materials thinner than about 3/4 in. (19 mm).

(5) Complex material shapes may be difficult or impossible to weld using ESW.

FUNDAMENTALS

PRINCIPLES OF OPERATION

ELECTROSLAG WELDING (ESW) is a welding process producing coalescence of metals with molten slag that melts the filler metal and the surfaces of the workpieces to be welded. The weld pool is shielded by this slag, which moves along the full cross section of the joint as welding progresses. The process is initiated by an arc that heats a granulated flux and melts it to form the slag. The arc is then extinguished by the conductive slag which is kept molten by its resistance to electric current passing between the electrode and the workpieces.

Usually a square groove joint is positioned so that the axis or length of the weld is vertical or nearly vertical. Except for circumferential welds, there is no manipulation of the work once welding has started. Electroslag welding is a machine welding process, and once started, it continues to completion. Since no arc exists, the welding action is quiet and spatter-free. Extremely high metal deposition rates allow the welding of very thick sections in one pass. A high-quality weld deposit results from the nature of the melting and solidification during welding. There is no angular distortion of the welded plates.

The process is initiated by starting an electric arc between the electrode and the joint bottom. Granulated welding flux is then added and melted by the heat of the arc. As soon as a sufficiently thick layer of molten slag (flux) is formed, all arc action stops, and the welding current passes from the electrode through the slag by electrical conduction. Welding is started in a sump or on a starting tab to allow the process to stabilize before the welding action reaches the work.

Heat generated by the resistance of the molten slag to passage of the welding current is sufficient to fuse the welding electrode and the edges of the workpiece. The interior temperature of the bath is in the vicinity of 3500°F

(1925°C). The surface temperature is approximately 3000°F (1650°C). The melted electrode and base metals collect in a pool beneath the molten slag bath and slowly solidify to form the weld. There is progressive solidification from the bottom upward, and there is always molten metal above the solidifying weld metal.

Run-off tabs are required to allow the molten slag and some weld metal to extend beyond the top of the joint. Both starting and run-off tabs are usually removed flush with the ends of the joint.

PROCESS VARIATIONS

THERE ARE TWO variations of electroslag welding that are in general use. One variation uses a wire electrode with a nonconsumable guide (contact) tube to direct the electrode into the molten slag bath. This variation will be referred to as the "conventional method." The other variation is similar to the first, except that a consumable guide extends down the length of the joint. This variation will be called the "consumable guide method." With the conventional method, the welding head moves progressively upward as the weld is deposited. With the consumable guide method, the welding head remains stationary at the top of the joint, and both the guide tube and the electrode are progressively melted by the molten slag.

Conventional Method

THE CONVENTIONAL METHOD of electroslag welding is illustrated in Figure 8.1. One or more electrodes are fed into the joint, depending on the thickness of the material being welded. The electrodes are fed through nonconsumable wire guides which are maintained 2 to 3 in. (50 to 75 mm) above the molten slag. Horizontal oscillation of the electrodes may be used to weld very thick materials.

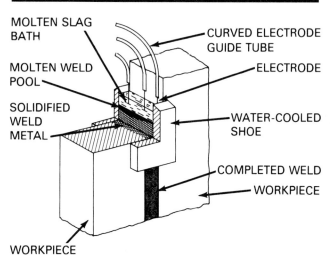

MOLTEN SLAG BATH

CURVED ELECTRODE GUIDE TUBE

ELECTRODE

MOLTEN WELD POOL

SOLIDIFIED WELD METAL

WATER-COOLED SHOE

COMPLETED WELD

WORKPIECE

WORKPIECE

Figure 8.1–Nonconsumable Guide Method of Electroslag Welding (Three Electrodes)

Water-cooled copper shoes (dams) are normally used on both sides of the joint to contain the molten weld metal and slag bath. The shoes are attached to the welding machine and move vertically with the machine. Vertical movement of the welding machine is consistent with the electrode deposition rate. Movement may be either automatic or controlled by the welding operator.

Vertical movement of the shoes exposes the weld surfaces. There is normally a slight reinforcement on the weld, which is shaped by a groove in the shoe. The weld surfaces are covered with a thin layer of slag. This slag consumption must be compensated for during welding by the addition of small amounts of flux to the molten slag bath. Fresh flux is normally added manually. Flux-cored wires may be used to supply flux to the bath.

The conventional method of electroslag welding can be used to weld plates ranging in thickness from approximately 1/2 to 20 in. (13 to 500 mm). Thicknesses from 3/4 to 18 in. (19 to 460 mm) are most commonly welded. One oscillating electrode will successfully weld up to 5 in. (120 mm) thickness, two electrodes up to 9 in. (230 mm), and three electrodes up to 20 in. (500 mm). With each electrode, the process will deposit from 25 to 45 lb (11 to 20 kg) of filler metal per hour. The diameter of the electrode used is generally 1/8 in. (3.2 mm). Electrode metal transfer efficiency is almost 100 percent. The normally large weld made by electroslag welding will consume approximately 5 lb (2.3 kg) of flux for each 100 lb (45 kg) of deposited weld metal.

Consumable Guide Method

THE CONSUMABLE GUIDE method of ESW is shown in Figure 8.2. In this method, filler metal is supplied by both an elec-

trode and its guiding member. The electrode wire is directed to the bottom of the joint by a guide tube extending the entire joint length (height). Welding current is carried by the guide tube, and it melts off just above the surface of the slag bath. Thus, the welding machine does not move vertically, and stationary or nonsliding shoes are used. On short welds, shoes may be the same length as the joint. Several sets of shoes may be required for longer joints. As the metal solidifies, a set of shoes is removed from below the weld pool and placed above the top shoes. This "leap frog" pattern is repeated until the weld is complete.

As welding proceeds and the slag bath rises, the consumable guide melts and becomes part of the weld metal. The consumable guide provides approximately 5 to 15 percent of the filler metal.

As with the conventional method, one or more electrodes may be used, and they may oscillate horizontally in the joint. Since the guide tube carries electrical current, it may be necessary to insulate it from the joint side walls (base plate) and shoes.

A flux coating may be provided on the outside of the consumable guide to insulate it and to help replenish the slag bath. Other forms of insulation include doughnut shaped insulators, fiberglass sleeves, and tape.

The consumable guide method can be used to weld sections of virtually unlimited thickness. When using stationary electrodes, each electrode will weld approximately 2.5 in. (63 mm) of plate thickness. One oscillating electrode will successfully weld up to 5 in. (130 mm), two oscillating electrodes up to 12 in. (300 mm), and three oscillating electrodes up to 18 in. (450 mm). Weld lengths up to 30 ft (9 m) have been routinely accomplished with a single stationary electrode. Oscillation control problems may occur on long weld lengths. Therefore, if the required amount of oscillation cannot be properly controlled, additional electrodes may be added and oscillation reduced or eliminated.

Wing-type guide tubes can be used for certain applications where a round guide tube cannot properly heat the entire cross-sectional area, or for weldments that are irregular in shape. The most common use of a wing-type guide is for joints that would ordinarily be welded using one oscillating guide tube, but where oscillation equipment is unavailable. Also, wing-type guides are used when the joint is large enough to require two guides, but only one wire feeder is available. This application is limited to 4 in. (102 mm) joints. Joints that exceed 4-1/2 in. (114 mm) in thickness are marginal when welding with one guide.

Wing-type guide tubes are made by tack welding light carbon steel bar stock to the sides of a round guide tube. This is illustrated in Figure 8.3. Normally, wing-type guides extend to within 1/4 in. (6 mm) of the edge of the joint. The additional cross-section passes current into the slag bath, which keeps the bath temperature high enough remote from the filler wire to melt the plate edges. Because of this extra amount of current, the power demand equals that of a multiple electrode setup.

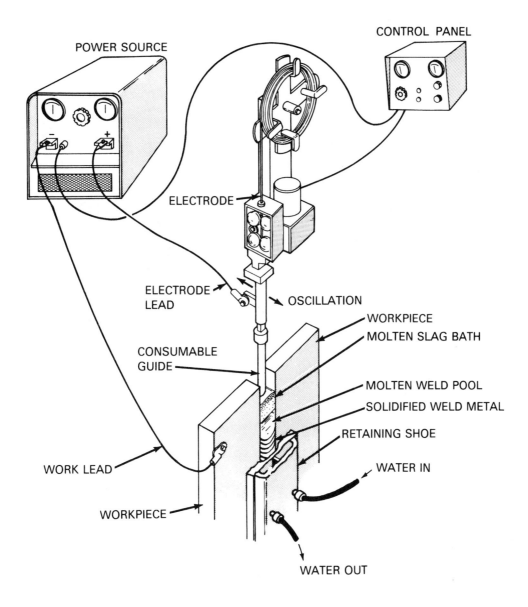

Figure 8.2–Consumable Guide Method of Electroslag Welding

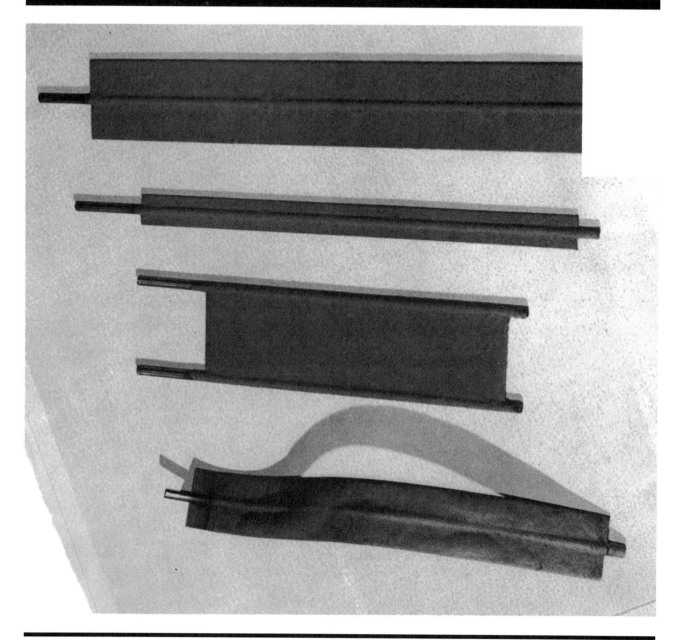

Figure 8.3—Wing-Type Guide Tubes

EQUIPMENT

THE EQUIPMENT FOR both ESW process methods is the same except for the design of the electrode guide tubes and the requirements for vertical travel. The following are major components of electroslag welding equipment:

(1) Power supply

(2) Wire feeder and oscillator
(3) Electrode guide tube
(4) Welding controls
(5) Welding head
(6) Retaining shoes (dams)

POWER SUPPLY

POWER SOURCES ARE typically of the constant voltage transformer-rectifier type with ratings of 750 to 1000 A, dc, at 100 percent duty cycle. The sources are similar to those used for submerged arc welding. Load voltages generally range from 30 to 55 V; therefore, the minimum open circuit voltage of the power source should be 60 V. Constant-voltage ac power sources of similar ratings are used for some applications. A separate power source is required for each electrode.

The power source is generally equipped with a contactor, a means for remote control of output voltage, a means for balancing multiple electrode installations, a main power switch, a range control, an ammeter, and a voltmeter.

WIRE FEEDER AND OSCILLATOR

THE FUNCTION OF a wire feed device is to deliver the wire electrode at a constant speed from the wire supply through the guide tube to the molten slag bath. The wire feeder is usually mounted on the welding head.

In general, each wire electrode is driven by its own drive motor and feed rolls. A dual gearbox to drive two electrodes from one motor may be used, but it does not provide redundancy in the event of a feeding problem. In the case of multiple electrode welding, the failure of one wire drive unit need not shut down the welding operation if a corrective measure can be accomplished quickly. It should be stressed, however, that for successful electroslag welding, it is vital to avoid a shut down because weld repair at the restart can be costly. At times, fifty or more continuous operating hours are demanded of these wire drives for heavy, long weldments.

The motor-driven wire feeders are similar in design and operation to those used for other continuous electrode welding processes, such as gas metal arc and submerged arc welding. A typical wire feed unit is shown in Figure 8.4. The feed rolls normally consist of a geared pair, the driving force thus being applied by both rolls. Configuration of the roll groove may vary, depending on whether a solid or cored electrode is used. With solid wire, care must be taken that the wire feeds without slipping but is not squeezed so tightly that it becomes knurled. Knurled wire can create the effect of a file and abrade the components between the feed rolls and the weld. Rolls with an oval groove configuration have been found to perform best for both types of wire without danger of crushing metal cored electrodes.

A wire straightener, either of the simple three-roll design or the more complex revolving design should be used to remove the cast in the wire electrode. Cast will cause the electrode to wander as it emerges from the guide. This, in turn, can cause changes in the position of the molten weld pool which may cause defects, such as lack of fusion. Cast

in the electrode causes a more severe problem on large, heavy weldments.

Electrode speed depends on the current required for the desired deposition rate, and also the diameter and type of electrode being used. Generally, a speed range of 40 to 350 in./min (17 to 150 mm/s) is entirely suitable for use with 3/32 in. (2.4 mm) or 1/8 in. (3.2 mm) diameter metal cored or solid electrodes.

Electrodes should be packaged for uniform, uninterrupted feeding. The wire supply should allow feeding with minimal driving torque requirements so that binding or wire stoppage will not take place. The wire package must be of adequate size to complete the entire weldment without stopping.

Electrode oscillation devices are required when the joint thickness exceeds approximately 2-1/4 in. (57 mm) per electrode. Oscillation of the electrode guide tube(s) can be provided by motor-operated mechanical drive mechanisms, such as a lead screw or a rack and pinion. The drive must be adjustable for travel distance, travel speed, and

Figure 8.4—Typical Wire Feed Unit (Consumable Guide Method)

variable delay at the end of each stroke. Control of oscillation movement is generally done using electronic circuitry.

ELECTRODE GUIDE TUBE

Conventional Method

IN CONVENTIONAL ELECTROSLAG welding, the nonconsumable wire guide tube (the "snorkel") guides the electrode from the wire feed rolls into the molten slag bath. It also functions as an electrical contact to energize the electrode. The exit end of the tube is positioned close to the molten slag, and it will deteriorate with time.

Guide tubes are generally made of beryllium copper alloy and supported by two narrow, rectangular bars brazed to them. Beryllium copper is used because it retains reasonable strength at elevated temperatures. The tubes are wrapped with insulating tape to prevent short circuiting to the work.

To feed the electrodes vertically into the molten slag, the guides must be curved and narrow enough to fit into the joint root opening. They are generally less than 1/2 in. (13 mm) in diameter. To overcome cast in the electrode, integral wire straightening can be designed into the tubes.

Consumable Guide Method

THE CONSUMABLE GUIDE tube is made of a steel that is compatible with the base metal and slightly longer than the joint to be welded. It is commonly 1/2 to 5/8 in. (12 to 16 mm) in outside diameter and 1/8 to 3/16 in. (3.2 to 4.8 mm) inside diameter. Smaller diameters are necessary for welding sections less than 3/4 in. (19 mm) thickness.

The guide tube is attached to a copper alloy support tube, which is mounted on the welding head. Welding current is transmitted from the copper tube to the steel tube, and then to the electrode.

For welds more than 2 to 3 ft (600 to 900 mm) long, it is necessary to insulate the guide tubes to prevent short circuiting with the work. The entire tube length may be coated with flux; or insulator rings, spaced 12 to 18 in. (300 to 450 mm) apart, are slipped over the tube and held in place with small weld buttons on the tube. The flux covering or insulator rings melt and help replenish the slag bath as the guide tube is consumed.

WELDING CONTROLS

ELECTROSLAG WELDING CONTROLS consist of a console mounted near the welding head which contains the following component groups:

(1) Alarm systems for equipment or system malfunctions.
(2) Ammeters, voltmeters, and remote contactor controls for each of the power sources, and remote voltage

control in the form of either a manual rheostat or a motorized rheostat which is triggered by a reversing switch.
(3) A speed control for each of the wire drive motors. This control will also jog and reverse the wire drive. On some consoles, the power contactor and wire drive can be activated by a common switch.
(4) Oscillator controls to move the guides back and forth in the weld joint. Adjustable limit switches mounted on the welding head control the established length of stroke; timers control dwell duration at each end of the stroke.
(5) Control for the vertical rise of the welding head (conventional method only). The type of control depends on whether rise is activated manually or automatically.

An electric eye sensor may be used with the rise device for automatic control. The sensor is aimed at a point below the top of the containment shoe and adjusted to detect the top of the molten slag bath. When the bath rises above the aim point, the rise drive is activated and moves the welding head and shoe up until the bath is no longer detectable. A continuous incremental rise is obtained automatically in this manner.

In a given joint, if the plate thickness, root opening, number of electrodes, and electrode feed speed are known, then the vertical rate of rise may be computed approximately. This speed may be set on a variable-speed motor by the welding operator. As welding progresses, minor adjustments may be made to keep the molten slag bath and liquid metal pool within the containment shoes.

WELDING HEAD

IN ELECTROSLAG WELDING, the weldments are relatively large and heavy. Therefore, it is convenient to establish a single location for the power source and then use long cables, a remote control, and a portable welding head at the weld joint location. The control boxes are generally lightweight and contain a minimum of components. The controls interconnect between the power source and wire feeders on the welding head.

The welding head will include the wire feeder, electrode supply, wire guide tube(s), electrical connections to the guide tubes, and a means of attaching it to the work. It may also contain provisions for multiple electrode operation and an electrode oscillation drive unit. Where portability is required, the wire feeder and electrode supply may be located a short distance away from the welding head, as in semiautomatic gas metal arc welding.

An electroslag plate crawler is shown in Figure 8.5. A single electrode is fed into the joint through a guide tube. Once welding is initiated, the crawler assembly is propelled upward by a serrated drive wheel that tracks the weld joint. The vertical speed of the plate crawler is set by remote control. Two water-cooled copper shoes slide along the joint to contain the molten slag and weld metal. The

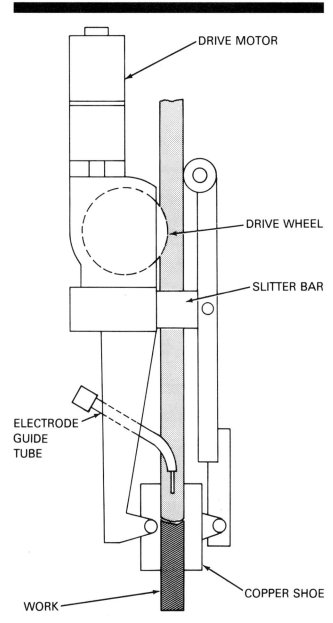

DRIVE MOTOR

DRIVE WHEEL

SLITTER BAR

ELECTRODE
GUIDE
TUBE

COPPER SHOE

WORK

**Figure 8.5—Conventional Method of Electroslag
Welding Using a Plate Crawler**

copper shoes are held tight against the plate by spring tension between the front and rear sections of the plate crawler.

The plate crawler may be used to weld plates ranging in thickness from 1/2 in. (13 mm) to 2 in. (51 mm). Normally, either a square groove or single V-groove joint is used. Vertical travel speeds of up to 7 in./min (3 mm/s) are possible.

RETAINING SHOES (DAMS)

THE RETAINING SHOES and the associated water circulation system are included in this category. The function of the shoes is to maintain the molten metal and slag bath within the weld cavity. The shoes are fabricated of copper and generally include water passages at critical heat build-up points to prevent overheating or melting. Each shoe, shown in Figure 8.2, generally has a cavity machined in the side toward the weld to provide for a slight reinforcement of weld metal.

The shoe may be cooled by a water circulation system or by tap water. Water circulators must have a 30 000 to 40 000 Btu/h (32 to 42 kJ/h) heat removal capacity. A recirculating system does not normally cause condensation on the shoes. Tap water frequently is at a lower temperature than ambient air, causing condensation on the shoes. If the condensation runs down the shoes and collects in the starting tabs prior to starting the weld, weld porosity will likely occur. Condensation on the inside of the shoes during welding will evaporate ahead of the advancing slag bath. Thus, it is best to turn on the tap water just before starting the weld.

With conventional welding, water-cooled shoes are mounted on the welding head and travel upward as welding progresses. With the consumable guide method, the shoes do not move. However, they can be repositioned, leap frog fashion, as welding progresses upward. Sometimes the shoes are not water-cooled, but they must be massive to avoid melting. The shoes are clamped in place, usually by wedges against U-shaped bridges (strongbacks) across the joint, or by large C-clamps on short welds made by the consumable guide method.

SAFETY

AS IN ANY type of welding, reasonable care must be exercised in the set-up, welding, and post-welding procedures for ESW. A number of hazards exist, some minor and others serious, but all can be eliminated. Failure to use safety protection equipment or follow safe practices can result in

damages to production parts, equipment, and facilities, and physical danger to personnel.

Adequate and suitable clothing must be worn to protect against arc radiation from other welding operations, radiant heat from hot parts, and the hazards of hot molten slag. Gloves should be worn at all times.

Safety glasses with side shields are recommended because of hot slag or "spatter balls" which may fly from the joint during welding. Shade No. 12 is recommended if the arc must be observed before the slag bath is established. Shade No. 4 is recommended for observing the slag bath.

Considerable caution should be used when removing the cooling shoes. Solidified slag is essentially glass and will break up and splinter erratically.

The hazard of electrical shock is possible with any electrical equipment. However, with ESW, the operator does not touch the equipment except for occasional adjustments. The electrode wire, and in effect, everything coming in contact with it, is "electrically hot". The electrode wire and other "live" parts of the welding system should not be touched by the operator or other personnel in the area. As a normal procedure, the operator should make a preliminary check of the equipment to spot potential problems. Electrical hazards such as loose or worn connections, frayed insulation, or cables in or around water should be repaired or the hazard should be removed. Care should be taken to prevent the drainage of water from cooling shoes or circulators around ESW equipment. The manufacturer's recommendations for cable size, installation, use, and maintenance of the equipment should always be followed.

Parts to be electroslag welded are generally large and because they are welded vertically, are sometimes tall and awkwardly positioned. Caution should be used in positioning these components, and qualified welders should be used to weld needed braces, clamps, strong-backs, and other restraints or fixtures. The same care should be exercised when removing the fixtures by cutting or air carbon arc gouging. When removing sumps and run-off blocks, ensure that they fall in a proper place, not onto cutting-gas lines or electric cables.

Fumes are generated by ESW, as with other welding processes, but it is not necessary for the ESW operator to be constantly in or around them. The process is automatic by nature, although the operator does occasionally monitor the process. If welding is being performed in a poorly ventilated area, exhaust fans or smoke extractors should be used.

Hot slag may spill out the top of the joint, or leak around the cooling shoe. Poor fit-up of the cooling shoe or misalignment of sumps and runoffs also may open gaps where leaks can occur. Slag leakage can be especially dangerous because of the large volume of slag bath associated with the ESW process. For this same reason, special attention should be given to shoe fit-up and alignment, and the cooling shoes should be left in place after completion of the weld until the slag bath has solidified.

Water-cooled shoes are much safer to handle than solid shoes, but both types can cause severe burns.

Do not weld around combustible materials. Make sure that solvents used to clean the weld joint area are completely removed. Do not place aerosol cans near the weld joint, as they may explode.

For detailed safety information, refer to the manufacturer's instructions and the latest editions of the following publications: ANSI Z49.1, *Safety in Welding and Cutting*, ANSI Z87.1, *Practice for Occupational and Educational Eye and Face Protection*, AWS *Welding Handbook* (Eighth Edition), "Safe Practices in Welding and Cutting."

For mandatory Federal Safety Regulations established by the U.S. Labor Department's Occupational Safety and Health Administration, refer to the latest edition of OSHA Standards, *Code of Federal Regulations*, Title 29 Part 1910, available from the Superintendent of Documents, U.S. Printing Office, Washington, D.C., 20402, Accident Prevention Manual for Industrial Operations (Seventh Edition) National Safety Council.

CONSUMABLES

THE COMPOSITION OF electroslag weld metal is determined by base metal and filler metal compositions and their relative dilution. The filler metal and consumables used in electroslag welding include the electrode, the flux, and, in the case of consumable guide welding, a comsumable guide and its insulation. The welding consumables can effectively control the final chemical composition and mechanical properties of the weld metal.

ELECTRODES

THERE ARE TWO types of electrodes used with the electroslag welding process: solid and metal cored electrodes. The solid electrodes are more widely used. Various chemical compostions are available with each type of electrode to produce desired weld metal mechanical properties.

Metal cored electrodes permit adjustment of filler metal composition for welding alloy steels through alloy additions (ferro alloys) in the core. They also provide a means for replenishing flux in the molten bath. The metal tube is low carbon steel. The use of metal cored electrodes may result in excessive buildup of slag in the bath when the core is composed entirely of flux (flux cored).

In the electroslag welding of carbon steels and high strength, low alloy steels, the electrodes usually contain less carbon than the base metal. Weld metal strength and toughness are achieved by alloying with a variety of elements. This approach reduces the tendency towards weld metal cracking in steels containing up to 0.35 percent carbon.

The electrode wire compositions used for welding higher alloy steels normally match the base metal compositions. Higher alloy steels generally develop their mechanical properties by a combination of chemical composition and heat treatment. Usually it will be necessary to heat-treat an electroslag weld in a higher alloy steel to develop the desired weld metal and heat-affected zone properties. Thus, the best approach is to select a weld metal composition and a base material composition designed so that both will respond to heat treatment to approximately the same degree.

When the electrode wire for electroslag welding is selected, its dilution with the base metal must be considered. In a typical electroslag weld, the dilution generally runs from 30 percent to 50 percent base metal. The amount of dilution from base metal melting is dependent upon the welding procedure. The filler metal and melted base metal thoroughly mix to provide a weld with almost uniform chemical composition throughout.

The most popular electrode sizes are 3/32 in. (2.4 mm) and 1/8 in. (3.2 mm) diameter. However, 1/16 to 5/32 in. (1.6 to 4.0 mm) diameter electrodes have been successfully used. Smaller diameter electrodes provide a higher deposition rate than do larger electrodes at the same welding amperage. From practical applications, it has been found that either 3/32 in. (2.4 mm) or 1/8 in. (3.2 mm) diameter electrodes provide the optimum combination of deposition rate, feedability, welding amperage ranges, and ability to be straightened.

Electrodes for electroslag welding are supplied in the forms of 60 lb (27 kg) coils, large spools typically 600 lb (270 kg), or large drums. Since it is essential that enough electrode be available to complete the entire weld joint, it has been found most practical and economical to use spools or drums up to 750 lb (340kg).

FLUX

THE FLUX IS a major part of the successful operation of the electroslag welding process. Flux composition is of utmost importance since its characteristics determine how well the electroslag process operates. During the welding operation, the flux is melted into slag that transforms the electrical energy into thermal energy to melt the filler metal and base metal. Also, the slag (flux) must conduct the welding current, protect the molten weld metal from the atmosphere, and provide for a stable operation.

An electroslag welding flux must have several important characteristics. When molten, it must be electrically conductive and yet have adequate electrical resistance to generate sufficient heat for welding. However, if its resistance is too low, arcing will occur between the electrode and the slag bath surface. The molten slag viscosity must be fluid enough for good circulation to ensure even distribution of heat in the joint. A slag that is too viscous will cause slag inclusions in the weld metal, and one that is too fluid will leak out of small openings between the work and the retaining shoes.

The melting point of the flux must be well below that of the metal being welded, and its boiling point must be well above the operating temperature to avoid losses that could change operating characteristics. Generally, the molten slag is chosen to be relatively inert to reactions with the metal being welding, and it should be stable over a wide range of welding conditions and slag bath sizes. However, it is sometimes advantageous to select a flux which produces a reactive slag. Reactive slags may be used to refine the weld metal or to adjust the level of impurities such as oxygen.

Solidified slag on the weld surfaces should be easy to remove, although some commercial fluxes may not have this characteristic.

Electroslag flux in original unopened packages should be protected against moisture pickup by the packaging under normal conditions. Flux reconditioning may be necessary if the flux has been exposed to high humidity.

Fluxes for electroslag welding are usually combinations of complex oxides of silicon, manganese, titanium, calcium, magnesium, and aluminum, with some calcium fluoride always present to adjust the electrical characteristics. Special characteristics are achieved by variations in the composition of the flux.

Only a relatively small amount of flux is used during electroslag welding. An initial quantity of flux is required to establish the process. Flux solidifies as slag in a thin layer on the cold surfaces of the retaining shoes and on both weld faces. It is necessary to add flux to the molten bath during welding to maintain the required depth. Neglecting losses by leakage, the total flux used is approximately 1 lb (0.5 kg) for each 20 lb (9 kg) of deposited metal. However, as the plate thickness or weld length increases, the flux consumption approaches 1 lb (0.5 kg) for each 80 lb (36 kg) of deposited metal.

CONSUMABLE GUIDE TUBE

THE PRIMARY FUNCTION of the consumable guide tube is to provide support to the electrode wire from the welding head to the molten slag bath, and to act as the primary current path. The consumable guide melts periodically just at the top of the rising molten slag bath. Its use permits the welding head to be fixed in position at the top of the vertical seam. The electrode cable is attached to the guide tube. Welding current is conducted to the electrode as it passes through the end of the guide tube, and then into the molten slag bath.

Most consumable guide tubing is 1/2 in. (13 mm) or 5/8 in. (16 mm) outside diameter. The inside diameter of the tubing is normally determined by the size of the electrode wire being used. The amount of metal contributed to the weld by the melted guide tube is generally small, except when welding thin sections. Since the guide tube becomes part of the weld, its composition should be compatible with the desired weld metal composition.

When making short welds, a bare consumable guide tube can be used. However, for long welds, the tube must be insulated to prevent electrical contact with the base metal. A flux coating can be used to provide electrical insulation, and, at the same time, to add flux to the slag bath. Other forms of insulation include doughnut shaped insulators, fiberglass sleeves, and tape. Since the insulation becomes part of the slag pool, it should be selected so as not to affect either the deposited weld metal or the operating characteristics of the flux.

SPECIFICATION FOR CONSUMABLES

ANSI/AWS A5.25, SPECIFICATION for Consumables Used for Electroslag Welding of Carbon and High Strength Low Alloy Steels, classifies electrodes and fluxes for ESW. Metal cored electrodes are classified on the basis of chemical analysis of weld metal taken from an undiluted ingot. The solid electrodes are classified on the basis of their chemical composition as manufactured. Since the consumable guide

tube usually contributes only a small amount of filler metal to the joint, it does not change the flux-electrode classification. However, the guide tubes must conform to AISI Specifications for 1008 to 1020 carbon steel tubing.

Metal cored electrodes deposit weld metals of low carbon steel and low alloy steels. The low alloy steel deposits contain small amounts of nickel and chromium, and either copper or molybdenum. Carbon is less than 0.15 percent. Solid electrodes are divided into three classes: medium manganese (approximately 1 percent), high manganese (approximately 2 percent), and special classes.

In the flux-electrode classification system, both metal cored and solid electrodes are used in any combination with six fluxes. The fluxes are classified on the basis of the mechanical properties of a weld deposit made with a particular electrode and a specified base metal. The compositions of the fluxes are left to the discretion of the manufacturer. Two levels of tensile strength for flux-electrode combinations are specified: 60 to 80 ksi (415 to 550 MPa), and 70 to 95 ksi (485 to 655 MPa). For each level of strength, two of the three flux-electrode classifications must meet minimum toughness requirements as determined by the Charpy V-notch impact test.

APPLICATIONS

BASE METALS

MANY TYPES OF carbon steels can be electroslag welded in production, such as AISI 1020, AISI 1045, ASTM A36, ASTM A441, and ASTM A515. They generally can be welded without postweld heat treatment.

In addition to carbon steels, other steels are successfully electroslag welded. They include AISI 4130, AISI 8620, ASTM A302, HY80, austenitic stainless steels, ASTM A514, ingot iron, and ASTM A387. Most of these steels require special electrodes and a grain refining postweld heat treatment to develop required weld or weld heat-affected zone properties.

JOINT DESIGN

THERE IS ONE basic type of joint, which is the square groove butt joint. Square edge plate preparations can be used to produce other types of joints such as corner, T-, and edge joints. It is also possible to make transition joints, fillet welds, cross-shaped joints, overlays, and weld pads with the ESW process. Typical ESW joint designs and the outlines of the final welds are shown in Figure 8.6. Specially designed retaining shoes are needed for joints other than butt, corner, and T-joints.

PROCESS VARIABLES

WELDING VARIABLES ARE those factors which affect the operation of the process, weld quality, and economics of the process. A smooth running process and a quality weld deposit result when all of the variables are in proper balance. In electroslag welding, it is essential that the effects of each variable be fully understood since they differ from those with conventional arc welding processes.

Form Factor

THE SLOW COOLING rate and solidification patterns of an electroslag weld are similar to metal cooling in a mold. In ESW, heat is removed from the molten weld metal by the cool base metal and the water-cooled retaining shoes. Solidification begins at these cooler areas and progresses toward the center of the weld as shown in Figure 8.7(A) and (B). However, since filler metal is added continuously and the joint fills during welding, solidification progresses from the bottom of the joint, as indicated by the solidified grain structure shown in Figure 8.7(B).

The angle at which the grains meet in the center is determined by the shape of the molten weld pool. Weld pool shape can be expressed by the term form factor. The form factor is the ratio of the weld pool width to its maximum depth. Width is the root opening plus the total penetration

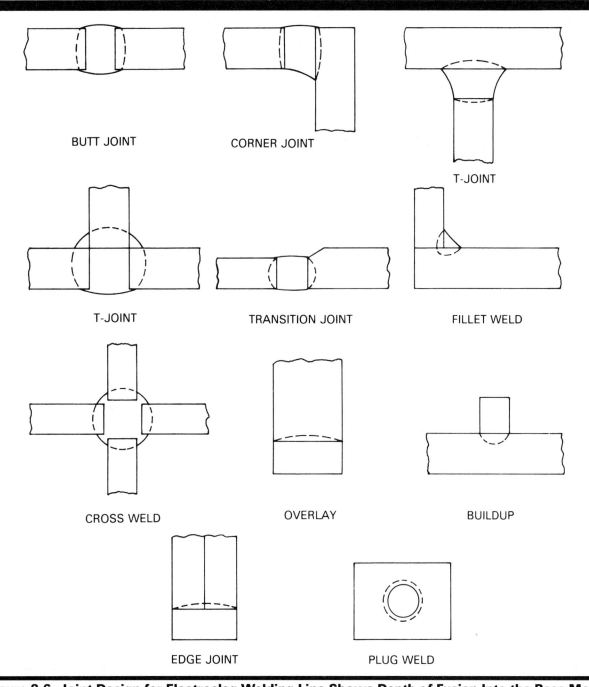

Figure 8.6—Joint Design for Electroslag Welding Line Shows Depth of Fusion Into the Base Metal

into the base metal. Depth is the distance from the top of the molten weld pool to the lowest level of the liquid-solid interface. The dendrites grow competitively into the weld pool at an angle of approximately 90° to the solid liquid interface.

A line drawn perpendicular to the dendritic grains will approximate the shape of the solid-liquid interface. This is illustrated in Figure 8.7(B). This boundary will define the width and depth of the molten pool, and the form factor can be easily determined. Welds having a high form factor (wide width and shallow weld pool) tend to solidify with the grains meeting at an acute angle. Welds having a low form factor (narrow width and deep weld pool) tend to solidify with the grains meeting at an obtuse angle. Thus, the form factor indicates how the grains from opposite sides of the weld meet at the center.

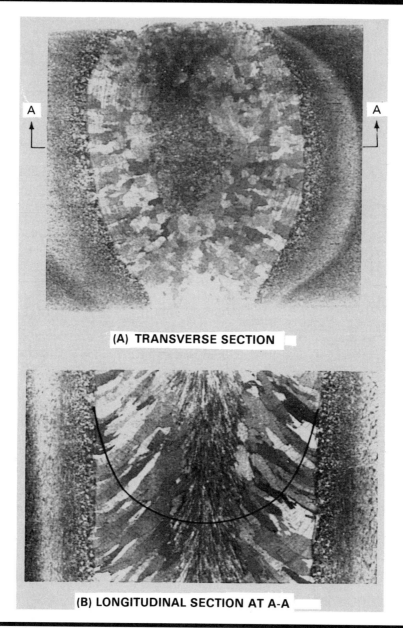

(A) TRANSVERSE SECTION

(B) LONGITUDINAL SECTION AT A-A

Figure 8.7–Transverse and Longitudinal Section Through a 4 in. (100 mm) Thick Electroslag Weld

The angle at which the grains meet in the center determines whether the weld will have a high or low resistance to hot center line cracking. If the dendritic grains meet head-on at an obtuse (large) included angle, the cracking resistance will be low. However, if the angle is acute (small), the cracking resistance will be high. Therefore, maximum resistance to cracking is obtained with a high form factor.

The shape of the molten weld pool and the resultant form factor is controlled by the welding variables. However, form factor alone does not control cracking. The base metal composition (especially the carbon content), the filler metal composition, and joint restraint also have a significant effect on cracking propensity. Some studies have indicated that manganese content, or the manganese/silicon ratio, is important in cracking.

Welding Amperage

WELDING AMPERAGE AND electrode feed rate are directly proportional and can be treated as one variable. Increasing the electrode feed speed increases the welding amperage and the deposition rate when a constant-voltage power source is used.

As the welding amperage is increased, so is the depth of the molten weld pool. When welding with a 1/8 in. (3.2 mm) diameter electrode below approximately 400 A, an increase in amperage also increases weld width. The net result is a slight decrease in form factor. However, when operating a 1/8 in. (3.2 mm) diameter electrode above 400 A, an increase in amperage reduces weld width. Thus, the net effect of increasing the welding amperage is to decrease the form factor and thus lower the resistance to cracking.

Welding amperages of 500 to 700 A are commonly used with 1/8 in. (3.2 mm) electrodes. Metals or conditions prone to cracking may require a high form factor associated with welding amperages below 500 amperes.

Welding Voltage

WELDING VOLTAGE IS an extremely important variable. It has a major effect on the depth of fusion into the base metal and on the stable operation of the process. Welding voltage is the primary means for controlling depth of fusion. Increasing the voltage increases both depth of fusion and width of the weld. Depth of fusion must be somewhat greater in the center of the weld than at the edges to assure complete fusion at the outside edges, where the chilling effect of the water-cooled shoes must be overcome.

Since an increase in welding voltage increases the weld width, it also increases the form factor and thereby increases cracking resistance.

The voltage must also be maintained within limits to assure stable operation of the process. If the voltage is low, short circuiting or arcing to the weld pool will occur. Too high a voltage may produce unstable operation because of slag spatter and arcing on the top of the slag bath. Welding voltages of 32 to 55 volts per electrode are used. Higher voltages are used with thicker sections.

Electrode Extension

THE DISTANCE BETWEEN the slag bath surface and the end of the guide (contact) tube, when the conventional welding method is used, is referred to as the *dry electrode extension*. There is generally no dry extension in the consumable guide method, since the guide tube melts primarily by heat conduction from the molten slag. However, at high-heat input ranges, radiation heat transfer from the slag bath may be sufficient to melt the guide above the slag pool. Using constant-voltage power and constant-electrode feed speed and increasing the dry electrode extension will increase the resistance. This causes the power supply to reduce its current output, which slightly increases the form factor.

Electrode extensions of 2 to 3 in. (50 to 75 mm) are generally used. Extensions of less than 2 in. (50 mm) usually cause overheating of the guide tube. Those greater than 3 in. (75 mm) cause overheating of the electrode because of the increased electrical resistance. Hence, at long extensions, the electrode will melt at the slag bath surface instead of in the bath. This will result in instability and improper slag bath heating.

Electrode Oscillation

PLATES UP TO 3 in. (75 mm) thick can be welded with a stationary electrode and high voltage. However, the electrode is usually oscillated horizontally across the plate thickness when the material exceeds 2 in. (50 mm). The oscillation pattern distributes the heat and helps to obtain better edge fusion. Oscillation speeds vary from 20 to 100 in./min (8 to 40 mm/s), with the speed increasing to match the plate thickness. Generally, oscillation speeds are based on a traverse time of 3 to 5 seconds. Increasing the oscillation speed reduces the weld width and hence the form factor. Thus, the oscillation speed must be balanced with the other variables. A dwell period is used at each end of the oscillation travel to obtain complete fusion with the base metal and to overcome the chilling effect of the retaining shoes. The dwell time may vary from 2 to 7 seconds.

Slag Bath Depth

A MINIMUM SLAG bath depth is necessary so that the electrode will enter into the bath and melt beneath the surface. Too shallow a bath will cause slag spitting and arcing on the surface. Excessive bath depth provides excessive area for heat transfer into the retaining shoes and base metal. This reduces the overall temperature of the slag bath which reduces the weld width and hence the form factor. Slag bath circulation is poor with excessive depth, and the cooler slag may tend to solidify on the surface of the base metal so that slag inclusions may result. A bath depth of 1-1/2 in. (38 mm) is optimum, but it can be as low as 1 in. (25 mm) or as high as 2 in. (51 mm) without significant effect.

Number of Electrodes and Spacing

AS THE METAL thickness per electrode increases, the weld width decreases slightly; but the weld pool depth decreases greatly. Thus, the form factor improves as the thickness of material increases for a given number of electrodes. However, a point is reached where the weld width at the cool retaining shoes is less than the root opening, and lack of edge fusion results. At this point, the number of electrodes must be increased. In general, one oscillating electrode can be utilized for sections up to 5 in. (130 mm) thick, and two oscillating electrodes up to 12 in. (300 mm) thick. Each additional oscillating electrode will accommodate approximately 6 in. (150 mm) of additional thickness. This applies to both the wire and consumable guide methods. If nonoscillating electrodes are used, each electrode will handle approximately 2-1/2 in. (65 mm) of plate thickness.

Root Opening

A MINIMUM ROOT opening is needed for sufficient slag bath size, good slag circulation, and, in the case of the consumable guide method, clearance for the guide tube and its insulation. Increasing the root opening does not affect the weld pool depth. However, it does increase the weld width and hence the form factor. Excessive root openings will require extra amounts of filler metal, which may not be economical. Also, excessive root openings may cause a lack of edge fusion. Root openings are generally in the range of 3/4 to 1-1/2 in. (20 to 40 mm), depending on the base metal thickness, the number of electrodes, and the use of electrode oscillation.

WELDING PROCEDURES

Joint Preparation

ONE OF THE major advantages of the electroslag process is the relatively simple joint preparation. It is basically a square groove joint, so the only preparation required is a flat, straight edge on each groove face, which can be produced by thermal cutting, machining, etc. If sliding retaining shoes are used, the plate surfaces on either side of the groove must be reasonably smooth to prevent slag leakage and jamming of the shoes.

The joint should be free of any oil, heavy mill scale, or moisture, which is true for any welding process. However, the joint need not be as clean as would normally be required for other welding processes. Oxygen-cut surfaces should be free of adhering slag, but a slightly oxidized surface is not detrimental.

Care should be exercised to protect the joint prior to initiating the weld. Moisture bearing materials packed around the shoes, commonly known as mud, will cause porosity. It should be completely dry before welding. Also, leaking water-cooled shoes can cause porosity or defects at the weld face.

Joint Fit-Up

PRIOR TO WELDING, components should be set up with the proper joint alignment and root opening. Rigid fixturing or strongbacks that bridge the joint should be used. Strongbacks are bridge-shaped plates that are welded to each component along the joint so that alignment during welding is maintained. They are designed with clearance to span fixed-position and movable retaining shoes. After welding is completed, they are removed.

For the consumable guide method, imperfect joint alignment can be accommodated to some degree. Plates with large misalignments may be welded by special retaining shoes that are adaptable to the fit-up. Alternatively, the space between the shoes and the work is packed with refractory material or steel strips (the steel must be similar in composition to the base metal). Afterwards, the steel strips may be removed in such a way as to blend the weld faces smoothly with the adjacent base metal.

Experience will dictate the proper root opening for each application. As welding progresses up the joint, the parts are drawn together by weld shrinkage. Therefore for long welds, the root opening at the top of the joint should be approximately 1/8 to 1/4 in. (3 to 6 mm) more than at the bottom to allow for this shrinkage. Factors influencing shrinkage allowance include material type, joint thickness, and joint length.

The use of proper root opening and shrinkage allowance is important for maintaining the dimensions of the weldment. However, if the root opening proves incorrect, the welding conditions may be varied to compensate for it, within limits. For example, if the initial root opening is too small, the wire feed speed may be lowered to reduce deposition rate and increase penetration. If the root opening is too large, the wire feed speed may be increased within good operating limits. Regardless of the type of base metal, the voltage should be increased to account for a wider root opening.

If the root opening is greater than the capability of the normal number of electrodes used, an additional electrode may be added if there is space available. In some cases, excessive root opening may be compensated for by oscillation of the electrode(s).

When a root opening is too small, the joint may fill too fast, causing weld cracks or lack of edge fusion. It is also possible for small root openings to close up from weld shrinkage and stop an oscillating guide tube from traversing.

Inclination of Work

THE AXIS OF the weld joint is generally in the vertical or near vertical position. If it deviates by more than 10 to 15 degrees from vertical, special welding procedures must be used. With greater deviations, it becomes increasingly difficult to weld without slag inclusions and lack of edge fusion. However, acceptable welds have been produced at 45 degrees and more from the vertical.

Alignment of the guide tube can be a problem with inclined welds using the consumable guide method. Large insulators (flux rings) or flux-coated guide tubes with spring clips are often required to keep the tube aligned in the joint.

Electrical Connection to the Workpiece

A GOOD ELECTRICAL return (work connection) is important because of the relatively high welding currents used for the

ESW process. Normally, two 4/0 welding cables are sufficient for each electrode. It is best to attach the worklead directly under the sump, that is, below the electrode. In that location, the effect of any strong magnetic field in the weldment on filler metal transfer will be minimized.

Spring-type ground clamps are not recommended because they tend to overheat. A more positive connection, such as a "C" type clamp, is best.

Run-On and Run-Off Tabs and Starting Plate

WHERE FULL PENETRATION is required for the full length of the joint, run-on and run-off tabs will be required. The run-on tabs, frequently referred to as *starting tabs*, are located at the bottom of the joint. They are used in conjunction with a starting plate to initiate the welding process. Generally, the tabs and starting plate are made from metal the same as or similar to the base metal. The starting tabs and plate form a sump in which the weld is started. In this case, the sump is removed and discarded after welding. The tabs and starting plate are the same thickness as the base plates.

Where the run-on tabs and starting plate are disposable, a tab is welded to the bottom of each base plate. The starting plate is welded across the bottom of the tabs to form the sump. The faces of the sump are flush with the base plate surfaces.

Copper sumps may be used, in which case water cooling is usually necessary. The arc is not started on the copper sump because it would melt through the water jacket. Normally, one or two small blocks of base metal are placed in the bottom of the copper sump, and the arc is started on them.

Disposable run-off tabs should also be of the same metal or a metal similar to the base metal. Copper tabs may be used, but they must be water cooled. The run-off tabs should be the same thickness as the base metal and securely attached to both plates at the end of the joint. The weld is completed in the cavity that they form above the base plates.

Electrode Position

THE POSITION OF the electrode will determine where the greatest amount of heat is generated. The electrode should normally be centered in the joint. However, if the electrode is cast toward one side, then the wire guide may be displaced in the opposite direction to compensate for the cast. In the welding of corner and T-joints, or any joint where a fillet is to be formed, the electrode may need to be offset to produce the required weld metal geometry.

Initiation and Termination of Welding

UNLESS MOLTEN SLAG is poured into the joint, the normal starting method is to strike an arc between the electrode and the starting plate. This may be done by two methods: (1) a steel wool ball is inserted between the electrode and the base metal, and the power is turned on; or (2) the electrode is advanced towards the starting plate with the power on. The latter method requires a chisel point on the end of the electrode. Once an arc is struck, flux is slowly added until the arc is extinguished. The process is then in the electroslag mode.

It is extremely important that the welding operation proceed without interruption. The equipment should be checked, and the electrode and flux supplies determined adequate before the weld is started. The weld must not be interrupted to replenish the electrode supply. The welding equipment must be capable of operating continuously until the weld is finished.

When the weld has reached the run-off tabs, termination should follow a procedure which fills the crater, or crater cracking may occur. Usually the electrode feed is gradually reduced when the slag reaches the top of the run-off tabs and the crater fills. The welding amperage decreases simultaneously. When the electrode feed stops, the power source is turned off. The run-on and run-off tabs and the excess weld metal are then cut off flush with the top and bottom edges of the weldment.

Slag Removal

A CHIPPING GUN will be effective in slag removal, although a slag hammer or pick will also do the job. Slag will also adhere to the copper shoes and copper sump, if used. They must be cleaned before another weld is made. Eye protection should be worn during slag removal operations.

Circumferential Welds

A SQUARE GROOVE butt joint is used to join two cylindrical components end to end. The weld is made by rotating the parts and allowing the slag and molten weld metal to remain at the 3 o'clock position with clockwise rotation. The welding head is stationary until the finish of the weld. Special starting and run-off techniques are required to complete the seam. Small diameter weldments may not be economically feasible. However, as wall thickness and diameter increase, the cost savings may make the process attractive.

METALLURGICAL CONSIDERATIONS

DURING ELECTROSLAG WELDING, heat is generated by the resistance to current flow as it passes from the electrode(s) through the molten slag into the weld pool. The molten slag, being conductive, is electromagnetically stirred in a vigorous motor action. Heat diffuses throughout the entire cross section being welded. Temperatures attained in the electroslag welding process are considerably lower than those in arc welding processes. However, the molten slag bath temperature must be higher than the melting range of the base metal for satisfactory welding. The molten zone, in both slag and weld metals, advances relatively slowly, usually approximately 1/2 to 1-1/2 in./min (13 to 38 mm/min), and the weld is generally completed in one pass. There are several differences between electroslag welds and consumable electrode arc welds. The following are some differences:

(1) In the as-deposited condition, a generally favorable residual-stress pattern is developed in electroslag welded joints. The weld surfaces and heat-affected zones are normally in compression, and the center of the weld is in tension.

(2) Because of the symmetry of most vertical ESW butt welds (square groove joints welded in a single pass), there is no angular distortion in the weldment. There is slight distortion in the vertical plane caused by weld metal contraction, but compensation can be made for this during joint fit-up.

(3) The weld metal stays molten long enough to permit some slag refining action, as in electroslag remelting. The progressive solidification allows gases in the weld metal to escape and nonmetallic inclusions to float up and mix with the slag bath. High quality, sound weld deposits are generally produced by ESW.

(4) The circulating slag bath washes the groove faces and melts them in the lower portion of the bath. The weld deposit contains up to 50 percent admixed base metal, depending on welding conditions. Therefore, the composition of steel being welded and the amount melted significantly affect both the chemical composition of the weld metal and the resultant weld joint mechanical properties.

(5) The prolonged thermal cycle results in a weld metal structure that consists of large prior austenite grains that generally follow a columnar solidification pattern. The grains are oriented horizontally at the weld metal edges and turn to a vertical orientation at the center of the weld, as shown in Figure 8.7(B).

The microstructure of electroslag welds in low carbon steels generally consists of acicular ferrite and pearlite grains with proeutectoid ferrite outlining the prior austenite grains. It is very common to observe coarse, prior austenite grains at the periphery of the weld and a much finer grained region near the center of the weld. This fine-grained region appears equiaxed in a transverse cross section; however, longitudinal sections reveal its columnar nature, as shown in Figure 8.7(B). Changes in weld metal composition, and to a lesser extent welding procedure, can markedly change the relative proportions of the coarse- and fine-grained regions, to the extent that only one may be present.

(6) The relatively long time at high temperatures and the slow cooling rate after welding result in wide heat-affected zones with a relatively coarse grain structure. The cooling rate is slow enough to form only relatively soft, high-temperature transformation products. For most steels, this is an advantage, particularly if stress-corrosion cracking may be a problem.

Preheat and Postheat

PREHEATING IS NOT required or generally used in electroslag welding. This is a significant advantage over arc welding for many types of steel. By its nature, the process is self-preheating in that a significant amount of heat is conducted into the workpieces and preheats them ahead of the weld. Also, because of the very slow cooling rate after welding, postheating is usually unnecessary.

Postweld Heat Treatment

MOST APPLICATIONS OF electroslag welding, particularly the welding of structural steel, require no postweld heat treatment. As discussed previously, as-deposited electroslag welds have a favorable residual stress pattern which is negated by a postweld heat treatment. Subcritical postweld heat treatments (stress relief) can be either detrimental or beneficial to mechanical properties, particularly notch toughness. They are generally not employed after electroslag welding.

The properties of carbon and low alloy steel welds can be greatly altered by heat treatment. Normalizing removes nearly all traces of the cast structure of the weld and almost equalizes the properties of the weld metal and base metal. This may improve the resistance to brittle fracture initiation and propagation above certain temperatures as measured by the Charpy V-notch impact test.

Quenched and tempered steels are not usually joined by ESW. They must be heat-treated after welding to obtain adequate mechanical strength properties in the weld and heat-affected zones. Such a heat treatment is very difficult to apply to large, thick structures.

WELDING SCHEDULES

TYPICAL WELDING CONDITIONS which will produce sound butt welds in normal situations are listed in Table 8.1. However, the typical welding conditions shown are not necessarily the only conditions that can be used for a particular weld. It is possible that as the particular requirements of a repetitive weld become better known, the settings can be adjusted to obtain optimum welding results. Qualification tests often required for code work may provide conditions different from those listed in the table.

The data shown in Table 8.1 are based on butt welds in carbon steels using water-cooled shoes. Adjustments may need to be made depending upon the flux, electrode diameter, and joint design used. Welding voltages shown are measured directly across the slag bath, not at the power supply. There is a voltage drop in the welding cables and consumable guide tube(s).

Table 8.1
Typical Electroslag Welding Conditions

(A) One Electrode - Non-Oscillating

Plate Thickness		Joint Opening		Welding Current	Welding Voltage
in.	mm	in.	mm	Amperes	Volts
3/4	19	1	25	500	35
1	25	1	25	600	38
2	51	1	25	700	39
3	76	1	25	700	52

(B) One Electrode - Oscillating

Plate Thickness		Joint Opening		Oscillation Distance		Oscillation Speed*		Welding Current	Welding Voltage
in.	mm	in.	mm	in.	mm	ipm	mm/s	Amperes	Volts
2	51	1 1/4	32	1 1/4	32	25	11	700	39
3	76	1 1/4	32	2 1/4	57	45	19	700	40
4	102	1 1/4	32	3 1/4	83	65	27	700	43
5	127	1 1/4	32	4 1/4	108	85	36	700	46

*A dwell time of 2 second is used at each shoe.

(C) Two Electrodes - Non-Oscillating

Plate Thickness		Joint Opening		Electrode Spacing		Welding Current	Welding Voltage
in.	mm	in.	mm	in.	mm	Amperes	Volts
3	76	1	25	2 1/2	64	425/wire	40
4	102	1	25	2 1/2	64	425/wire	43
5	127	1	25	2 1/2	64	425/wire	46

(D) Two Electrodes - Oscillating[1]

Plate Thickness		Joint Opening		Oscillation Distance		Oscillation Speed[2]		Welding Current	Welding Voltage
in.	mm	in.	mm	in.	mm	ipm	mm/s	Amperes	Volts
5	127	1 1/4	32	1	25	20	8	700/wire	41
6	152	1 1/4	32	2	51	40	17	700/wire	42
8	203	1 1/4	32	4	102	80	34	700/wire	45
10	254	1 1/4	32	6	152	120	51	700/wire	48
12	305	1 1/4	32	8	203	120	51	700/wire	51

1. Electrode Spacing is 3 1/4 in. (83 mm) for two (2) oscillating electrodes.
2. A dwell time of 2 seconds is used at each shoe.

(E) Four Electrodes[1] - Oscillating

Plate Thickness		Joint Opening		Electrode Spacing		Oscillation Distance		Oscillation Speed[2]	
in.	mm	in.	mm	in.	mm	in.	mm	ipm	mm/s
24	609	1 1/2	38	5	127	3	76	75	32
25	635	1 1/2	38	5 1/4	133	3	76	82.5	35
26	660	1 1/2	38	5 1/2	140	3	76	90	38
27	686	1 1/2	38	5 3/4	146	3	76	97.5	41
28	711	1 1/2	38	6	152	3	76	105	44
29	737	1 1/2	38	6	152	4	102	112.5	48
30	762	1 1/2	38	6 1/4	159	4	102	120	51

1. Initially use 600 amperes and 55 volts on each electrode. After experience is gained, adjust welding current and voltage in accordance with Section on Welding Variables.
2. A dwell time of 4 seconds is used at each shoe.

Table 8.1
(continued)

(F) Six Electrodes[1] - Oscillating

Plate Thickness		Joint Opening		Electrode Spacing		Oscillation Distance		Oscillation Speed[2]	
in.	mm	in.	mm	in.	mm	in.	mm	ipm	mm/s
30	762	1 1/2	38	5 1/4	133	2 3/4	70	82.5	35
31	787	1 1/2	38	5 1/2	140	2 1/2	64	90	38
32	813	1 1/2	38	5 5/8	143	2 7/8	73	94	40
33	838	1 1/2	38	5 3/4	146	3 1/4	83	99.5	42
34	864	1 1/2	38	6	152	3	76	105	44
35	889	1 1/2	38	6 1/8	156	3 3/8	86	109	46
36	914	1 1/2	38	6 1/4	159	3 3/4	95	112.5	48

1. Initially use 600 amperes and 55 volts on each electrode. After experience is gained, adjust welding current and voltage in accordance with Section on Welding Variables.
2. A dwell time of 4 seconds is used at each shoe.

(G) Three Electrodes - Non-Oscillating

Plate Thickness		Joint Opening		Electrode Spacing		Welding Current Amperes	Welding Voltage Volts
in.	mm	in.	mm	in.	mm		
6	152	1	25	2 1/2	64	500/wire	41
7	178	1	25	2 1/2	64	550/wire	45
8	203	1	25	2 3/4	70	600/wire	49
9	229	1	25	3	76	625/wire	53

(H) Three Electrodes[1] - Oscillating

Plate Thickness		Joint Opening		Electrode Spacing		Oscillation Distance		Oscillation Speed[2]	
in.	mm	in.	mm	in.	mm	in.	mm	ipm	mm/s
12	305	1 1/2	38	4 1/2	114	2	51	60	25
13	330	1 1/2	38	5	127	2 1/2	64	75	32
14	356	1 1/2	38	5 1/4	133	2 3/4	70	82.5	35
15	381	1 1/2	38	5 1/2	140	3	76	90	38
16	406	1 1/2	38	6	152	3 1/2	89	105	44
17	432	1 1/2	38	6 1/4	159	3 3/4	95	112.5	48
18	457	1 1/2	38	6 1/2	165	4	102	120	51

1. Initially use 600 amperes and 55 volts on each electrode. After experience is gained, adjust welding current and voltage in accordance with Section on Welding Variables.
2. A dwell time of 4 seconds is used at each shoe.

(I) Four Electrodes[1] - Oscillating

Plate Thickness		Joint Opening		Electrode Spacing		Oscillation Distance		Oscillation Speed[2]	
in.	mm	in.	mm	in.	mm	in.	mm	ipm	mm/s
18	457	1 1/2	38	5	127	2	51	75	32
19	483	1 1/2	38	5 1/4	133	2 1/4	57	82.5	35
20	508	1 1/2	38	5 1/2	140	2 1/2	64	90	38
21	533	1 1/2	38	5 3/4	146	2 3/4	70	97.5	41
22	559	1 1/2	38	6	152	3	76	105	44
23	584	1 1/2	38	6 1/4	159	3 1/4	83	112.5	48
24	610	1 1/2	38	6 1/2	165	3 1/2	89	120	51

1. Initially use 600 amperes and 55 volts on each electrode. After experience is gained, adjust welding current and voltage in accordance with Section on Welding Variables.
2. A dwell time of 4 seconds is used at each shoe.

APPLICATIONS AND OTHER TYPICAL USES

ELECTROSLAG WELDING IS recognized by all of the important national codes. Several of the codes have requirements that may differ from other welding processes. For example, in ANSI/AWS D1.1, *Structural Welding Code - Steel*, electroslag welding is not permitted as a prequalified welding process. This means that a contractor must prepare a welding procedure qualification test plate and destructively test the joint. That test must demonstrate that the contractor is capable of successfully using the process. Electroslag welds in pressure vessels that are fabricated in accordance with the *ASME Boiler and Pressure Vessel Code* are required to be normalized after welding. As indicated previously, this refines the weld metal and heat-affected zone grain structure.

Other fabrication and construction codes such ANSI/API 1104, *Standard for Welding Pipelines and Related Facilities* and *ABS Rules for Building and Classing of Steel Vessels*, do not require any special testing or heat treating of electroslag welds. However purchase contracts may permit owners, owners representatives, and regulatory agencies to require special tests prior to approval of weld procedures, and these perogatives may be exercised prior to approving electroslag welding.

Structural

PROBABLY THE WIDEST use of the ESW process is for structural applications. The electroslag process has many unique advantages that make it a highly desirable welding process. The high weld metal deposition rates, the low percentage of weld defects, and the fact that it is an automatic process are strong reasons for the use of electroslag welding. For thick sections, electroslag welding is a low-cost process if the weldments meet the design requirements and service conditions. However, if the welding process is stopped during the welding of a joint for any reason, the restart area must be carefully inspected for discontinuities. Those which are considered unacceptable for the application must be repaired by a more acceptable welding process.

A common structural application of the electroslag process is the transition joint between different flange thicknesses, which is a type of butt joint. By using copper shoes designed for this type of joint configuration, the varying thicknesses present no problem.

Electroslag welding is often used to weld stiffeners in box columns and wide flanges. In all cases, the stiffener weld would be a T-joint. This application is illustrated in Figure 8.8.

Machinery

MANUFACTURERS OF LARGE presses and machine tools work with large, heavy plates. Quite often the design requires plates that are larger than the mill can produce in one piece. Electroslag welding is used to splice two or more plates together.

Other machinery applications include kilns, gear blanks, motor frames, press frames, turbine rings, shrink rings, crusher bodies, rebuilding metal mill rolls, and rims for road rollers. These parts are formed from plate and welded along a longitudinal seam.

Pressure Vessels

PRESSURE VESSELS FOR the chemical, petroleum, marine, and power generating industries are made in all shapes and sizes, with wall thicknesses from less than 1/2 in. (13 mm) to greater than 16 in. (400 mm). In current practice, plate may be rolled to form the shell of the vessel, and the longitudinal seam welded. In very large or thick-walled vessels, the shell may be fabricated from two or more curved plates and joined by several longitudinal electroslag seam welds.

Steels used in pressure vessel construction are generally heat-treated. Consequently, when welding these steels with a high-heat input, such as with electroslag welding, the weld heat-affected zone does not have adequate mechanical properties. To improve the mechanical properties, weldments are given the required heat treatment.

Ships

ELECTROSLAG WELDING IS used in the ship building industry for both in-shop and on-ship applications. Main hull section joining is done with the conventional method. Vertical welding of the side shell, from the bilge area to, but not including, the sheer strake may be done with an electroslag plate crawler. Plate thickness of 1/2 to 1 1/4 in. (13 to 32 mm) are commonly found in marine side shells. The weld length may be 40 to 70 ft (12 to 21 m), depending on the size of the ship.

Castings

ELECTROSLAG WELDING IS often used to fabricate cast components. The metallurgical characteristics of a casting and an electroslag weld are similar, and both respond to postweld heat treatment in a similar way. Many large difficult-to-cast components may be produced in smaller, higher quality units, and then be electroslag welded together. Costs are reduced and the quality is usually improved. Compatible weld metal produces a homogeneous structure. Color match, machinability, and other desirable properties are thus produced.

MECHANICAL PROPERTIES

THE MECHANICAL PROPERTIES of electroslag welds will depend on the type and thickness of base metal, electrode composition, electrode-flux combination, and the welding conditions. All of these will influence the chemical composition, metallurgical structure, and mechanical properties of the weldment. In general, electroslag welds are used in structures that will be loaded under static or fluctuating

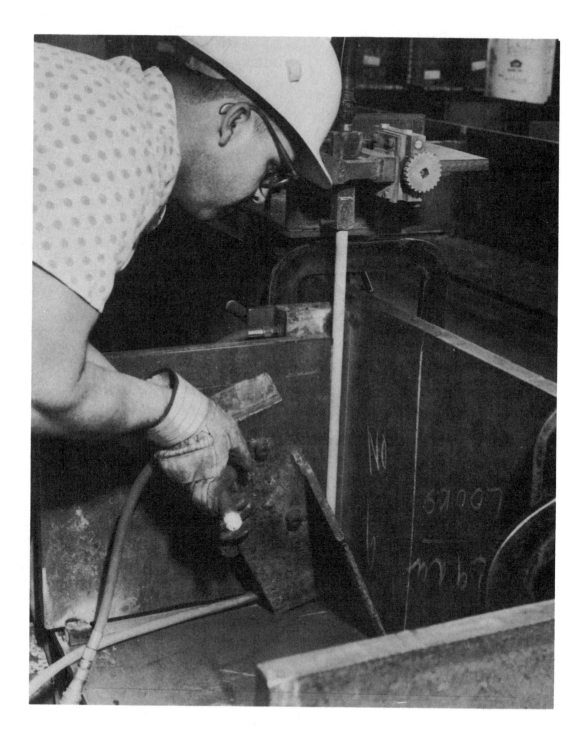

Figure 8.8—Electroslag Welding a Stiffener Plate in a Building Column

load conditions. One major concern is the notch toughness of the weld metal and heat-affected zones under service conditions, particularly at low temperatures. This must be carefully evaluated for each particular application, so that design requirements will be met and the weldments will perform satisfactorily.

Typical mechanical properties of deposited weld metal in selected structural steels are shown in Table 8.2, and for some carbon and low alloy steels in Table 8.3. The number and type of electrodes used for welding are reported in the table. Some variables that could affect the mechanical properties are unknown and, therefore, the data should not be used for design purposes.

ECONOMICS

TO APPRECIATE THE true overall economy of electroslag welding, the first consideration is the cost of joint preparation. A square oxygen-cut joint is suitable preparation for the process in carbon steels. No elaborate joint matching or close fit-up is required. In welds 3 in. (75mm) or more in thickness, electroslag welding requires much less weld metal, and as much as 90 percent less flux, than a comparable submerged arc weld.

Next, compare welding times. Once the pieces are in place for welding, the electroslag weld is completed without stopping if the process remains under control. Downtime or nonproductive time encountered in most arc welding processes can range from 30 percent to 75 percent. Nevertheless, it must be clearly recognized that stoppage of a heavy electroslag weld in process can be very costly. Moreover, to restart without producing a defect is difficult and may be impossible.

The deposition rate for ESW is approximately 35 to 45 lb/h (16 to 20 kg/h) per electrode. In very heavy plates, using three electrodes, 105 to 135 lb/h (47 to 61 kg/h) of weld metal can be deposited. Using a joint spacing of 1-1/8 in. (29 mm), the rate of welding is shown in Figure 8.9. Heavy plates ranging from 3 to 12 in. (76 to 305 mm)

in thickness are welded at speeds between 2 to 4 ft/h (610 to 1220 mm/h).

Another significant saving is achieved by the elimination of angular distortion and subsequent rework. Angular distortion can become a major factor in heavy multiple-pass welding, whether it is done from one side or from both sides.

Electroslag welding normally produces a high percentage of defect-free weldments, thereby minimizing repair costs. Slag entrapment, porosity, and lack of fusion can be avoided in most cases.

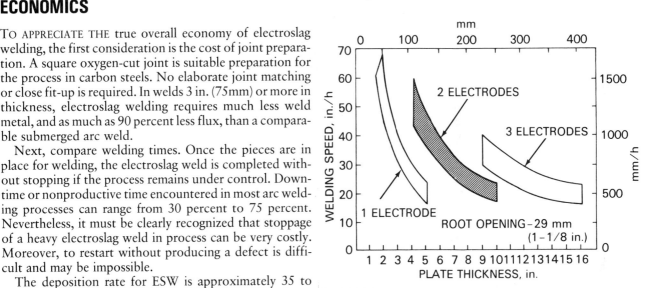

Figure 8.9–Electroslag Welding Speeds for Various Plate Thickness

Table 8.2
Typical Mechanical Properties of Weld Metal from Electroslag Welds in Structural Steels, As-Welded (Consumable Guide Method)

Base Metal, ASTM	Thickness		Electrode		Yield Strength		Tensile Strength		Elong. % in 51 mm (2 in.)	Reduction of area, %	Impact Strength*	
	in.	mm	Type (AWS)	No.	ksi	MPa	ksi	MPa			ft-lb	J
A441	1	25	EM13K-EW	1	49.9	344	75.8	523	28.0	59.0	17	23
A441	2-1/2	64	EM13K-EW	1	45.8	316	73.3	505	26.5	66.0	27	37
A36	6	152	EM13K-EW	2	46.0	317	79.5	548	28.5	52.8	—	—
A36	12	305	EM13K-EW	2	37.0	255	67.3	464	33.5	71.0	—	—
A572 Gr. 42	8	203	EM13K-EW	2	58.2	400	84.8	585	25.0	67.6	28	38
A572 Gr. 60	2-1/4	57	EH10Mo-EW	1	61.5	423	98.5	680	18.0	35.6	—	—

* Impacts at - 17.8°C (0°F)

Table 8.3
Typical Mechanical Properties of Weld Metal from Electroslag Welds in Carbon and Alloy Steels

Base Metal, ASTM	Thickness		Electrode		Heat Treatment	Yield Strength		Tensile Strength		Elong., % in 51 mm (2 in.)	Charpy Impact Tests, -12.2° C (10° F) Notch Location					
											WMFG[a]		WMCG[b]		BM HAZ[c]	
	in.	mm	No.	Type		ksi	MPa	ksi	MPa		ft-lb	J	ft-lb	J	ft-lb	J
A204-A	3-1/2	89	1	Mn-Mo	NT[d]	55.5	382	81.5	562	27	34	46	29	39	13	18
A515 GR70	1-1/2	38	1	Mn-Mo	SR[e]	52.0	358	84.1	579	26	45	61	26	35	7	10
A515 GR70	2	51	1	Mn-Mo	SR	68.1	469	85.2	587	23	46	63	21	29	5	7
A515 GR70	3-3/8	86	2	Mn-Mo	NT	57.5	396	78.0	537	29	34	46	33	45	22	30
A515 GR70	6-3/4	171	2	Mn-Mo	NT	45.5	313	74.3	512	31	24	33	21	29	12	16
A302-B	3	76	1	Mn-Mo-Ni	NT	57.0	393	82.0	565	28	53	72	52	71	63	86
A387-C	3	76	2	1-1/4 Cr-1/2 Mo	NT	46.5	320	73.0	503	29	70	95	76	103	57	78
A387-D	3-1/4	83	1	2-1/4 Cr-1 Mo	SR	57.5	396	82.0	565	25	46	63	50	68	48	65
A387-D	7-1/2	191	2	2-1/4 Cr-1 Mo	SR	80.0	551	95.5	658	20	62	84[f]	75	102[f]	83	113[f]

a. WMFG - Weld metal, fine grain
b. WMCG - Weld metal, coarse grain
c. BM HAZ - Base metal, heat-affected zone

d. NT - Normalized and tempered
e. SR - Stress relieved
f. Impacts at 10°C (50°F)

INSPECTION AND QUALITY CONTROL

BECAUSE OF THE nature of the ESW process, the weld defects that may occur do not always resemble those of a multi-pass arc weld. Weld metal defects that may be encountered in the electroslag process are discussed in the section on Troubleshooting.

Conventional nondestructive examination can be used to determine the soundness of the ESW joint. NDE techniques normally used with ESW include the following:

(1) VT-Visual examination
(2) PT-Liquid penetrant inspection
(3) MT-Magnetic particle inspection
(4) RT-Radiography (isotope or x-ray)
(5) UT-Ultrasonic testing (the large grain size may require special techniques)

All NDE should be performed in accordance with qualified procedures by qualified technicians.

METHODS OF INSPECTION

THE INSPECTION METHOD used will depend not only upon the requirements of the code or standard pertinent to the weldment but also the owner's or purchaser's contract specifications.

Every joint should have thorough visual examination (VT), which should reveal lack of edge fusion, undercutting, and surface cracks. When the sump and run-off tabs are removed, defects such as inclusions, cracking, and porosity may be seen. However, for internal defects (such as porosity, inclusions, cracking, and, in rare instances, lack of internal fusion), radiographic (RT) and ultrasonic (UT) testing are the most effective means of examination. Magnetic particle (MT) testing may also be used in searching for cracks and lack of fusion, but it is limited to the surface and immediate subsurface.

ACCEPTANCE

MOST WELDING ACCEPTANCE criteria are established by the customer or some regulatory agency or both. ESW applications are almost always welded to various code requirements. Occasionally, customer requirements will supplement or alter these requirements.

QUALITY CONTROL

A WRITTEN PROCEDURE that identifies the essential variables to be used in ESW should be prepared. This procedure should be reviewed to determine its adequacy on the basis of comparison with applicable code requirements.

Only an approved welding procedure should be issued to the welding foreman or welding operator, together with any additional instructions considered necessary. A check-off list may be used by the welding operator to assure that the equipment is properly set up and that all of the required operating adjustments are made. In welding joints not readily visible from both sides, the operator may need a helper. Proper quality control procedures should be used to provide assurance that the qualified welding procedure is being followed.

REWORK

IN GENERAL, REPAIRS of defective joints can be minimized by a good preventive maintenance program on the equipment, by using well-qualified and properly trained welding operators, and by implementing a sound welding procedure.

A defect such as undercutting can often be repaired by rewelding with the SMAW or GMAW process without gouging or grinding. Defects such as lack of fusion (at the joint surface), overlap, copper pickup on the weld face, and metal spillage around the shoes are visible defects that can be repaired by gouging or grinding to sound metal and rewelding. Defects such as porosity, cracking, and lack of fusion (internal) are usually detected by radiographic or ultrasonic testing and can be repaired by gouging to sound metal and rewelding. Economic considerations should dictate whether ESW or some other process will be used for repair.

Restarting of ESW welds should be avoided if possible. Where such restarts are made in a partially completed ESW weld, some repairs in the starting area are almost inevitable. One method of preparing the plate for a restart using an arc-gouged start area is illustrated in Figure 8.10. The restart area should be closely inspected for defects.

TESTING OF WELDS

ALL CODES AND specifications have definite rules for testing qualification welds to determine compliance with requirements. Most frequently required for groove welds are mechanical tests, such as tensile and bend tests, cut from specific locations in the welds. Fillet welds do not readily lend themselves to mechanical bend tests. In such cases, fillet weld break tests or macroetch tests or both may be required. Test procedures and methods of determining the mechanical properties are detailed in AWS B4.0, *Standard Methods for Mechanical Testing of Welds*.

Radiographic testing is also sometimes allowed as an alternative to mechanical testing when qualifying operators.

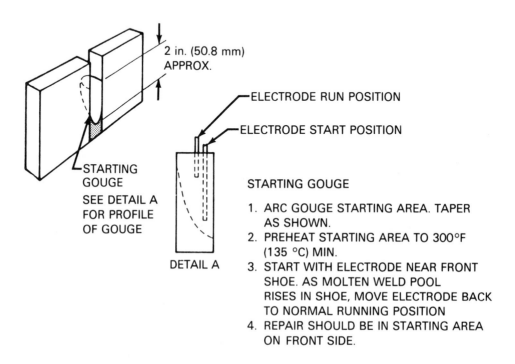

2 in. (50.8 mm) APPROX.

ELECTRODE RUN POSITION

ELECTRODE START POSITION

STARTING GOUGE
SEE DETAIL A
FOR PROFILE
OF GOUGE

STARTING GOUGE

1. ARC GOUGE STARTING AREA. TAPER AS SHOWN.
2. PREHEAT STARTING AREA TO 300°F (135 °C) MIN.
3. START WITH ELECTRODE NEAR FRONT SHOE. AS MOLTEN WELD POOL RISES IN SHOE, MOVE ELECTRODE BACK TO NORMAL RUNNING POSITION
4. REPAIR SHOULD BE IN STARTING AREA ON FRONT SIDE.

DETAIL A

Figure 8.10–Typical ESW Restart Procedure

TROUBLESHOOTING

WELDS MADE BY the ESW process under proper operating conditions are high quality and free from harmful discontinuities. In any welding process, however, abnormal conditions may occur during welding and cause discontinuities in the weld. Some of these discontinuities, their possible causes, and remedies are shown in Table 8.4. The information is primarily applicable to electroslag welded joints in carbon and low alloy steels.

Table 8.4
Electroslag Weld Discontinuities, Their Causes and Remedies

Location	Discontinuity	Causes	Remedies
Weld	1. Porosity	1. Insufficient slag depth 2. Moisture, oil, or rust 3. Contaminated or wet flux	1. Increase flux additions 2. Dry or clean workpiece 3. Dry or replace flux
	2. Cracking	1. Excessive welding speed 2. Poor form factor 3. Excessive center-to-center distance between electrodes or guide tubes	1. Slow electrode feed rate 2. Reduce current; raise voltage; decrease oscillation speed 3. Decrease spacing between electrodes or guide tubes
	3. Nonmetallic inclusions	1. Rough plate surface 2. Unfused nonmetallics from plate laminations	1. Grind plate surfaces 2. Use better quality plate
Fusion line	1. Lack of fusion	1. Low voltage 2. Excessive welding speed 3. Excessive slag depth 4. Misaligned electrodes or guide tubes 5. Inadequate dwell time 6. Excessive oscillation speed 7. Excessive electrode to shoe distance 8. Excessive center-to-center distance between electrodes	1. Increase voltage 2. Decrease electrode feed rate 3. Decrease flux additions; allow slag to overflow 4. Realign electrodes or guide tubes 5. Increase dwell time 6. Slow oscillation speed 7. Increase oscillation width or add another electrode 8. Decrease spacing between electrodes
	2. Undercut	1. Too slow welding speed 2. Excessive voltage 3. Excessive dwell time 4. Inadequate cooling of shoes 5. Poor shoe design 6. Poor shoe fit-up	1. Increase electrode feed rate 2. Decrease voltage 3. Decrease dwell time 4. Increase cooling water flow to shoes or use larger shoe 5. Redesign groove in shoe 6. Improve fit-up; seal gap with refractory cement dam
Heat-affected zone	1. Cracking	1. High restraint 2. Crack-sensitive material 3. Excessive inclusions in plate	1. Modify fixturing 2. Determine cause of cracking 3. Use better quality plate

SUPPLEMENTARY READING LIST

Brosholen, A. Skaug, E, and Visser, J. J. "Electroslag welding of large castings for ship construction." *Welding Journal* 56(8): 26-30; August 1977.

Campbell, H. C. "Electoslag, electrogas, and related welding processes." Bulletin No. 154, New York: Welding Research Council, 1970.

des Ramos, J. B., Pense, A. W., and Stout, R. D. "Fracture toughness of electroslag welded A537G steel." *Welding Journal* 55(12): 389s-399s; December 1976.

Dilawari, A. H., Eager, T. W., and Szekely, J. "An analysis of heat and fluid flow phenomena in electroslag welding." *Welding Journal* 57(1): 24s-30s; January 1978

Dorschu, K. E., Norcross, J. E., and Gage, C. C. "Unusual electroslag welding applications." *Welding Journal* 52 (11): 710-716; November 1973.

Eichhorn, E., Remmel, J., and Wubbels, B. "High speed electroslag welding." *Welding Journal* 63(1): 37-41; January 1984.

Forsber, S. G. "Resistance electroslag (RES) surfacing." *Welding Journal* 63(1): 37-41; January 1984.

Frost, R. H., Edwards, G. R., and Rheinlander, M. D. "A constitutive equation for the critical energy input during electroslag welding." *Welding Journal* 60(1): 1s-6s; January 1981.

Frost, R. H., Olson, D. L., and Edwards, G. R. In: Modelling of Casting and Welding Processes II. Proceedings 1983 Engineering Foundation Conference. Henniker, New Hampshire: 31 July-5 Aug. 1983. Ed: Dantzig, J. A. and Berry, J. T. Warrendale, PA: The Metallurgical Society of AIME, 1984.

Hannahs, J. R. and Daniel, L. "Where to consider electroslag welding." *Metal Progress* 98(5): 62-64; May 1970.

Kenyon, N., Redfern, G. A., and Richardson, R. R. "Electroslag welding of high nickel alloys." *Welding Journal* 54(7): 235s-239s; July 1977.

Konkol, P. J. "Effects of electrode composition, flux basicity, and slag depth on grain-boundary cracking in electroslag weld metals." *Welding Journal* 62(3): 63s-71s; March 1983.

Lawrence, B. D. "Electroslag welding curved and tapered cross-sections." *Welding Journal* 52(4): 240-246; April 1973.

Liu, S., and Su, C. T. "Grain refinement in electroslag weldments by metal powder addition." *Welding Journal* 68 (4): 132s; April 1989.

Malin, V. Y. "Electroslag welding of titanium and its alloys." *Welding Journal* 64(2): 42-49; February 1985.

Myers, R. D. "Electroslag welding eliminates costly field machining on large mining shovel." *Welding Journal* 59 (4): 17-22; April 1980.

Noruk, J. S. "Electroslag welding used to fabricate world's largest crawler driven dragline." *Welding Journal* 61(8): 15-19; August 1982.

Oh, Y. K., Devletian, J. H., and Chen, S. J. "Low-dilution electroslag cladding for shipbuilding." *Welding Journal* 69(8): 27-44; August 1990.

Okumura, M., et al. "Electroslag welding of heavy section 2 1/4 Cr-1Mo steel." *Welding Journal* 55(12): 389s-399s; December 1976.

Parrott, R. S., Ward, S. W., and Uttrachi, G. D. "Electroslag welding speeds shipbuilding." *Welding Journal* 53 (4): 218-222; April 1974.

Patchett, B. M. and Milner, D. R. "Slag-metal reactions in the electroslag process." *Welding Journal* 51(10): 491s-505s; October 1972.

Paton, B. E. "Electroslag welding of very thick material." *Welding Journal* 41(12): 1115-1123; December 1962.

Paton, B. E., ed. *Electroslag welding*, 2nd ed. Miami, Florida: American Welding Society, 1962.

Pense, A. et al. "Recent experiences with electroslag welded bridges." *Welding Journal* 60(12): 33-42; December 1981.

Ricci, W. S. and Eagar, T. W. "A parametric study of the electroslag welding process." *Welding Journal* 61(12): 397s-400s; December 1982.

Ritter, J. C., Dixon, B. F., and Phillips, R. H. "Electroslag Welding of ship propeller support frames." *Welding Journal* 66(10): 29-39; October 1987.

Schilling, L. G. and Klippstein, K. H. "Tests of electroslag-welded bridge girders." *Welding Journal* 60(12): 23-30; December 1981.

Scholl, M. R., Turpin, R. B., Devletian, J. H., and Wood, W. E. "Consumable guide tube electroslag welding of high carbon steel of irregular cross-section." ASM Paper 8201-072. Metals Park, OH: American Society for Metals, 1982.

Shackleton, D. N. "Fabricating steel safely using the electroslag welding process." Part 1. *Welding Journal* 60(12): 244s-251s; December 1981.

———. "Fabricating steel safely using the electroslag welding process." Part 2. *Welding Journal* 61(1): 23s-32s; January 1982.

Solari, M. and Biloni, H. "Effect of wire feed speed on the structure in electroslag welding of low carbon steel." *Welding Journal* 56(9): 274s-280s; September 1977.

Tribau, R. and Balo, S. R. "Influence of electroslag weld metal composition on hydrogen cracking." *Welding Journal* 62(4): 97s-104s; April 1983.

Yu, D., Ann, H. S., Devletian, J. H., and Wood, W. E. "Solidification study of narrow-gap electroslag welding." In: Welding Research: The State of the Art. Proceedings: Joining Division Council, University Research Symposium, Toronto, Canada, 15-17 Oct. 1985. Eds Nippes, E. F. and Ball, D. J. Metals Park, OH: American Society for Metals, 1986.

STUD WELDING

PREPARED BY A COMMITTEE CONSISTING OF:

D. E. Kuehn, Chairman
Nelson Stud Welding

J. C. Jenkins, Co-Chairman
TRW Nelson Stud Welding Div.

R. W. Folkening
FMC Corp.

R. McClellan
Ingalls Shipbuilding

C. C. Pease
Erico Fastening Systems, Inc.

WELDING HANDBOOK COMMITTEE MEMBER:
J. C. Papritan
Ohio State University

CHAPTER 9

STUD WELDING

INTRODUCTION

STUD WELDING IS a general term for joining a metal stud or similar part to a workpiece. Welding can be done by a number of welding processes including arc, resistance, friction, and percussion. Of these processes, the one that utilizes equipment and techniques unique to stud welding is arc welding. This process, known as stud arc welding (SW), will be covered in this chapter. The other processes use conventionally designed equipment with special tooling for stud welding. Those processes are covered in other chapters in this Volume.[1]

In stud arc welding, the base (end) of the stud is joined to the other work part by heating the stud and the work with an arc drawn between the two. When the surfaces to be joined are properly heated, they are brought together under low pressure. Stud welding guns are used to hold the studs and move them in proper sequence during welding. There are two basic power supplies used to create the arc for welding studs. One type uses dc power sources similar to those used for shielded metal arc welding. The other type uses a capacitor storage bank to supply the arc power. The stud arc welding processes using these two types of power sources are commonly known as arc stud welding and capacitor discharge stud welding, respectively.

ARC STUD WELDING

ARC STUD WELDING, the more widely used of the two major stud welding processes, is similar in many respects to manual shielded metal arc welding. The heat necessary for welding of studs is developed by a dc arc between the stud (electrode) and the plate (work) to which the stud is to be welded. The welding current is supplied by either a dc motor-generator or a dc transformer-rectifier power source, similar to those used for shielded metal arc welding. Welding time and the plunging of the stud into the molten weld pool to complete the weld are controlled automatically. The stud, which is held in a stud welding gun, is positioned by the operator, who then actuates the unit by pressing a switch. The weld is completed quickly, usually in less than one second. This process generally uses a ceramic arc shield, called a *ferrule*. It surrounds the stud to contain the molten metal and shield the arc. A ferrule is not used with some special welding techniques, nor with some nonferrous metals.

CAPACITOR DISCHARGE STUD WELDING

CAPACITOR DISCHARGE stud welding derives its heat from an arc produced by the rapid discharge of electrical energy stored in a bank of capacitors. During or immediately following the electrical discharge, pressure is applied to the stud, plunging its base into the molten pool of the workpiece. The arc may be established either by rapid resistance heating, and vaporization of a projection on the stud weld base, or by drawing an arc as the stud is lifted away from the workpiece. In the first type, arc times are about three to six milliseconds; in the second type, they range from six to fifteen milliseconds. The capacitor discharge process does not require a shielding ceramic ferrule because of the short arc duration and small amount of molten metal expelled from the joint. It is suited for applications requiring small to medium sized studs.

For either process, a wide range of stud styles is available. They include such types as threaded fasteners, plain or slotted pins, internally threaded fasteners, flat fasteners with rectangular cross section, and headed pins with various upsets. Studs may be used as holddowns, standoffs, heat transfer members, insulation supports, and in other fastening applications. Most stud styles can be rapidly applied with portable equipment.

---

1. See Chapter 17, "Spot, Seam, and Projection Welding", Chapter 18, "Flash, Upset and Percussion Welding", and Chapter 23, "Friction Welding".

PROCESS CAPABILITIES AND LIMITATIONS

CAPABILITIES

BECAUSE STUD ARC welding time cycles are very short, heat input to the base metal is very small compared to conventional arc welding. Consequently, the weld metal and heat-affected zones are very narrow. Distortion of the base metal at stud locations is minimal. The local heat input may be harmful when studs are welded onto medium and high carbon steels. The unheated portion of the stud and base metal will cool the weld and heat-affected zones very rapidly, causing these areas to harden. The resulting lack of weld joint ductility may be detrimental under certain types of loading, such as cyclic loads. On the other hand, when stud welding precipitation hardened aluminum alloys, a short weld cycle minimizes overaging and softening of the adjacent base metal. Metallurgical compatibility between stud material and the base metal must also be considered.

Studs can be welded at the appropriate time during construction or fabrication without access to the back side of the base member. Drilling, tapping, or riveting for installation is not required.

Using this process, designers need not specify thicker materials nor provide heavy bosses and flanges to obtain required tap depths for threaded fasteners. With stud welded designs of lighter weight, not only can material be saved, but the amount of welding and machining needed to join parts can be reduced.

Small studs can be welded to thin sections by the capacitor discharge method. Studs have been welded to sheet as thin as 0.03 in. (0.75 mm) without melt-through. They have been joined to certain materials (stainless steel, for example) in thicknesses down to 0.01 in. (0.25 mm). Because the depth of melting is very shallow, capacitor discharge welds can be made without damage to a prefinished opposite side. No subsequent cleaning or finishing is required.

Capacitor discharge power permits the welding of more dissimilar metals and alloys than arc stud welding. While both can join steel to stainless steel, only the capacitor discharge welding system can join brass to steel, copper to steel, brass to copper, aluminum to die-cast zinc, and similar combinations.

LIMITATIONS

ONLY ONE END of a stud can be welded to the workpiece. If a stud is required on both sides of a member, a second stud must be welded to the other side. Stud shape and size are limited because the stud design must permit chucking of the stud for welding. The stud base size is limited for thin base metal thicknesses.

Studs applied by arc stud welding usually require a disposable ceramic ferrule around the base. It is also necessary to provide flux in the stud base or a protective gas shield to obtain a sound weld.

Most studs applied by capacitor discharge power require a close tolerance projection on the weld base to initiate the arc. Stud diameters that can be attached by this method generally range from 1/8 to 3/8 in. (3.2 to 9.5 mm). Above this size, arc stud welding is more economical.

A welding power source located convenient to the work area is required for stud welding. For arc stud welding, 230 or 460 V ac power is required to operate the dc welding power source. For most capacitor discharge welding, a single phase 110 V ac main supply will serve, but high production units require three phase ac, 230 or 460 V, for operation.

ARC STUD WELDING

PRINCIPLES OF OPERATION

THE ARC STUD welding process involves the same basic principles as any of the other arc welding processes. Application of the process consists of two steps.

(1) Welding heat is developed with an arc between the stud and the plate (work).
(2) The two pieces are brought into intimate contact when the proper temperature is reached.

The most basic equipment arrangement consists of the stud gun, a control unit (timing device), studs and ferrules, and an available source of dc welding current. Typical equipment is illustrated in Figure 9.1A. Equipment is now available in which the power source and gun timing device are integrated into one unit as shown in Figure 9.1B. The mechanics of the process are illustrated in Figure 9.2. The stud is loaded into the chuck, the ferrule (also known as an arc shield) is placed in position over the end of the stud, and the gun is properly positioned for welding [Figure 9.2 (A)]. The trigger is then depressed, starting the automatic welding cycle.

A solenoid coil within the body of the gun is energized. This lifts the stud off the work and, at the same time, creates an arc [Figure 9.2(B)]. The end of the stud and the

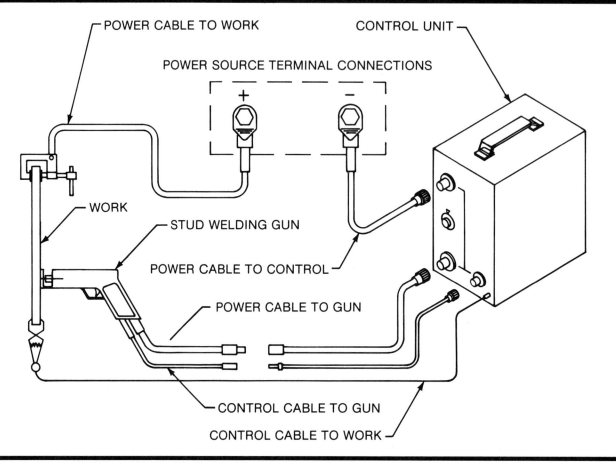

POWER CABLE TO WORK

CONTROL UNIT

POWER SOURCE TERMINAL CONNECTIONS

+ —

WORK

STUD WELDING GUN

POWER CABLE TO CONTROL

POWER CABLE TO GUN

CONTROL CABLE TO GUN

CONTROL CABLE TO WORK

Figure 9.1A–Basic Equipment Setup for Arc Stud Welding of Steel

workpiece are melted by the arc. When the preset arc period is completed, the welding current is automatically shut off and the solenoid is de-energized by the control unit. The mainspring of the gun plunges the stud into the molten pool on the work to complete the weld [Figure 9.2 (C)]. The gun is then lifted from the stud, and the ferrule is broken off [Figure 9.2(D)].

The time required to complete a weld varies with the cross-sectional area of the stud. For example, typical weld time is about 0.13 seconds for a 10 gage (0.135 in. or 3.4 mm diameter) stud, and 0.92 seconds for a 7/8 in. (22 mm) diameter stud. An average rate is approximately 6 studs per minute, although a rate of 15 studs per minute can be achieved for some applications.

The equipment involved in stud welding compares with that of manual shielded metal arc welding with regard to portability and ease of operation. The initial cost of such equipment varies with the size of the studs to be welded.

The gun and the control unit are connected to a dc power source. The control unit connections shown in

Figure 9.1A are for power sources designed for secondary current interruption, as is the case with motor-generator sets, battery units, and most rectifier type welding machines.

DESIGNING FOR ARC STUD WELDING

WHEN A DESIGN calls for stud type fasteners or supports, arc stud welding should be considered as a means for attaching them. Compared to threaded studs, the base (work) material thickness required to obtain full strength is less for arc stud welding. The use of arc welded studs may reduce the thickness of bosses at attachment points or may eliminate them. Cover plate flanges may be thinner than those required for threaded fasteners. Thus, there is potential weight savings when the process is used.

The weld base diameters of steel studs range from 1/8 to 1-1/4 in. (3.2 to 32 mm). For aluminum, the range is 1/8 to 1/2 in. (3.2 to 13 mm), and for stainless steels, it is 1/8 to 1 in. (3.2 to 25 mm). For design purposes, the smallest cross-sectional area of the stud should be used for load

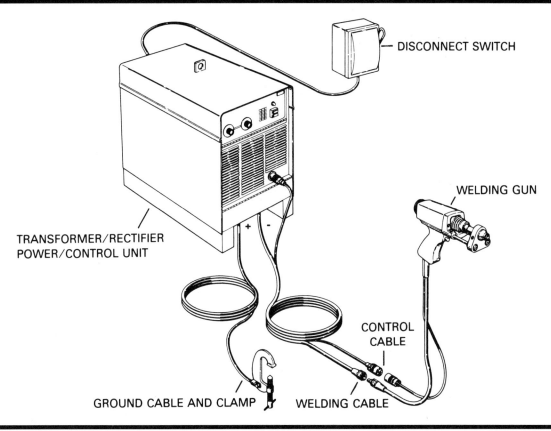

Figure 9.1B—Stud Welding Equipment With Timing Control Integrated Into Power Supply

determination, and adequate safety factors should be considered.

To develop full fastener strength, the plate (work) thickness should be a minimum of approximately one third the weld base diameter. A minimum plate thickness is required for each stud size to permit arc stud welding without melt-through or excessive distortion, as shown in Table 9.1. For steel, a 1:5 minimum ratio of plate thickness to stud weld base diameter is the general rule.

Fasteners can be stud welded with smaller edge distances than those required for threaded fasteners. However, loading and deflection requirements must be considered at stud locations.

STUDS

Stud Materials

THE MOST COMMON stud materials welded with the arc stud welding process are low carbon steel, stainless steel, and aluminum. Other materials are used for studs on a special application basis. Typical low carbon steel studs have a chemical composition as follows (all values are maximum): 0.23 percent carbon, 0.90 percent manga-

nese, 0.040 percent phosphorus, and 0.050 percent sulfur. They have a minimum tensile strength of 55 000 psi (380 MPa) and a minimum yield strength of 50 000 psi (345 MPa). The typical tensile strength for stainless steel studs is 85 000 psi.

High-strength studs, meeting the SAE steel fastener Grade 5 tensile strength of 120 000 psi (825 MPa) minimum, are also available. These studs are basically carbon steels that are heat treated to meet the tensile strength requirement.

Low carbon and stainless steel studs require a quantity of welding flux within or permanently affixed to the end of the stud. The main purposes of the flux are to deoxidize the weld metal and to stabilize the arc. Figure 9.3 shows the methods for securing the flux to the base of the stud.

Aluminum studs do not use flux on the weld end. Argon or helium shielding is required to prevent oxidation of the weld metal and stabilize the arc. The studs usually have a small tip on the weld end to aid arc initiation.

Stud Designs

MOST STUD WELD bases are round. However, there are many applications which use a square or rectangular

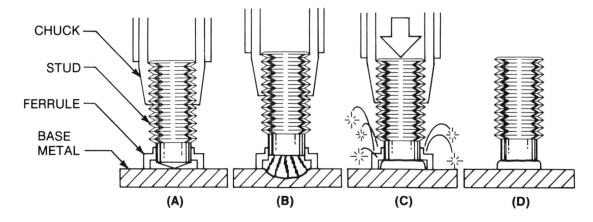

(A) Gun is Properly Positioned, (B) Trigger is Depressed and Stud is Lifted, Creating an Arc, (C) Arcing Period is Completed and Stud is Plunged Into Molten Pool of Metal on Base Metal, (D) Gun is Withdrawn From the Welded Stud and Ferrule is Removed

Figure 9.2—Steps in Arc Stud Welding

shaped stud. With rectangular studs, the width-to-thickness ratio at the weld base should not exceed five to obtain satisfactory weld results. Figure 9.4 shows a wide variety of sizes, shapes, and types of stud weld fasteners. In addition to conventional straight threaded studs, they include eyebolts, J-bolts, and punched, slotted, grooved, and pointed studs.

Stud designs are limited in that (1) welds can be made on only one end of a stud; (2) the shape must be such that a ferrule (arc shield) that fits the weld base can be produced; (3) the cross section of the stud weld base must be within the range that can be stud welded with available equipment; and (4) the stud size and shape must permit chucking or holding for welding. A number of standard stud designs are produced commercially. The stud manufacturers can

provide information on both standard and special designs for various applications.

One important consideration in designing or selecting a stud is to recognize that some of its length will be lost due to welding, since the stud and the base metal melt. The molten metal is then expelled from the joint. The stud length reductions shown in Table 9.2 are typical, but they may vary to some degree depending upon the materials, geometries, and welding variables involved.

Part of the material from the length reduction appears as flash in the form of a fillet around the stud base. This flash must not be confused with a conventional fillet weld because it is formed in a different manner. When properly formed and contained, the flash indicates complete fusion over the full cross section of the stud base. It also suggests

Table 9.1
Recommended Minimum Steel and Aluminum Plate Thicknesses for Arc Stud Welding

Stud Base Diameter		Steel Without Backup		Aluminum			
				Without Backup		With Backup*	
in.	mm	in.	mm	in.	mm	in.	mm
3/16	4.8	0.04	1.0	0.13	3.3	0.13	3.3
1/4	6.4	0.05	1.3	0.13	3.3	0.13	3.3
5/16	7.9	0.06	1.5	0.19	4.8	0.13	3.3
3/8	9.5	0.08	2.0	0.19	4.8	0.19	4.8
7/16	11.1	0.09	2.3	0.25	6.4	0.19	4.8
1/2	12.7	0.12	3.0	0.25	6.4	0.25	6.4
5/8	15.9	0.15	3.8	--	-	--	-
3/4	19.1	0.19	4.8	--	-	--	-
7/8	22.2	0.25	6.4	--	-	--	-
1	25.4	0.38	9.5	--	-	--	-

* A metal backup to prevent melt-through of the plate.

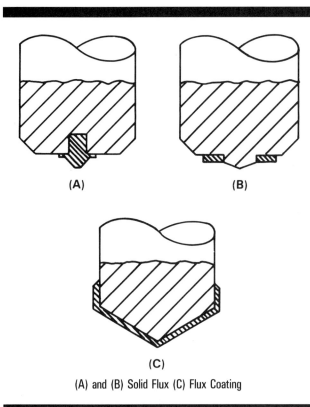

(A) and (B) Solid Flux (C) Flux Coating

**Figure 9.3—Methods of Containing Flux on the
End of a Welding Stud**

that the weld is free of contaminants and porosity. The stud weld flash may not be fused along its vertical and horizontal legs. This lack of fusion is not considered detrimental to the stud weld joint quality.

The dimensions of the flash are closely controlled by the design of the ferrule, where one is required. Since the diameter of the flash is generally larger than the diameter of the stud, some consideration is required in the design of mating parts. Counterbore and countersink dimensions commonly used to provide clearance for the flash of round studs are shown in Table 9.3. Flash size and shape will vary with stud material and ferrule clearance. Therefore, test welds should be made and checked. Three other methods of accommodating flash are shown in Figure 9.5.

FERRULES

FERRULES ARE REQUIRED for most arc stud welding applications. One of them is placed over the stud at the weld end where it is held in position by a grip or holder on the stud welding gun. The ferrule performs the following important functions during welding:

(1) Concentrating the heat of the arc in the weld area
(2) Restricting the flow of air into the area, which helps to control oxidation of the molten weld metal

**Table 9.2
Typical Length Reductions of Studs
in Arc Stud Welding**

Stud Diameters		Length Reductions	
in.	mm	in.	mm
3/16 thru 1/2	5 thru 13	1/8	3
5/8 thru 7/8	16 thru 22	3/16	5
1 and over	25 and over	3/16 to 1/4	5 to 6

(3) Confining the molten metal to the weld area
(4) Preventing the charring of adjacent non-metallic materials

The ferrule also shields the operator from the arc. However, safety glasses with No. 3 filter lenses are recommended for eye protection.

Ferrules are made of a ceramic material and are easily removed by breaking them. Since ferrules are designed to be used only once. their size is minimized for economy, and their dimensions are optimized for the application. A standard ferrule is generally cylindrical in shape and flat across the bottom for welding to flat surfaces. The base of the ferrule is serrated to vent gases expelled from the weld area. Its internal shape is designed to form the expelled molten metal into a cylindrical flash around the base of the stud. Special ferrule designs are used for special applications such as welding at angles to the work and welding to contoured surfaces. Ferrules for such applications are designed so that their bottom faces match the required surface contours.

SPECIAL PROCESS TECHNIQUES

THERE ARE SEVERAL special process techniques that employ the basic arc stud welding process, but each is limited to very specific types of applications.

One special process technique, referred to as *gas-arc*, uses an inert gas for shielding the arc and molten metal from the atmosphere. A ferrule is not used. This technique is suitable for both steel and aluminum stud welding applications, but its primary use is with aluminum. It is usually limited to production type applications because a fixed setup must be maintained, and also the welding variables fall into a very narrow range. Without a ferrule, there is greater susceptibility to arc blow and poorer control of the fillet around the base.

Another special process technique, which again does not use a ferrule, is called *short cycle* welding. It uses a relatively high weld current for a very short time to minimize oxidation and nitrification of the molten metal. Short cycle welding is generally limited to small studs, 0.25 in. (6.4 mm) diameter and under, where the amount of metal melted is minimal. One application is the welding of studs to thin base materials where shallow penetration is required and backside marking is not a consideration.

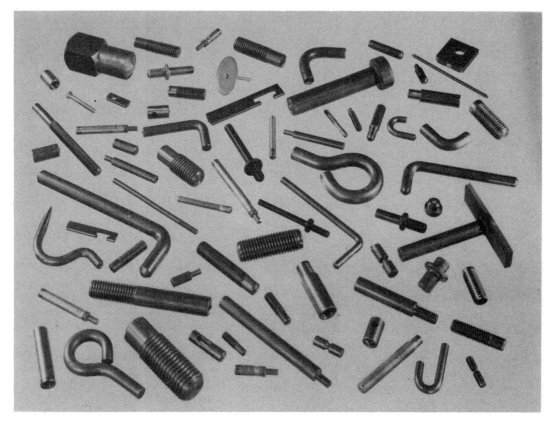

NOTE: Stud Stock May be Round, Square, or Rectangular in Cross Section

Figure 9.4–Studs and Fastening Devices Commonly Used for Arc Stud Welding

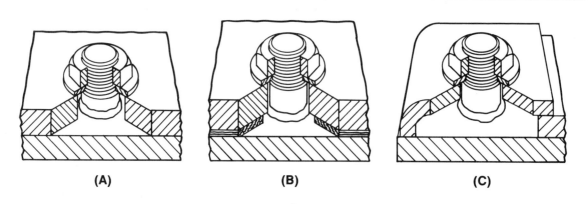

(A)   (B)   (C)

Figure 9.5–Methods of Accomodating Flash

Table 9.3
Weld Fillet Clearances for Arc Stud Welds

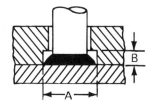

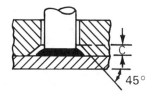

Stud Base Diameter		Counterbore				90 Degree Countersink	
		A		B		C	
in.	mm	in.	mm	in.	mm	in.	mm
1/4	6.4	0.437	11.1	0.125	3.2	0.125	3.2
5/16	7.9	0.500	12.7	0.125	3.2	0.125	3.2
3/8	9.5	0.593	15.1	0.125	3.2	0.125	3.2
7/16	11.1	0.656	16.7	0.187	4.7	0.125	3.2
1/2	12.7	0.750	19.1	0.187	4.7	0.187	4.7
5/8	15.9	0.875	22.2	0.218	5.5	0.187	4.7
3/4	19.1	1.125	28.6	0.312	7.9	0.187	4.7

ARC STUD WELDING EQUIPMENT

THE NECESSARY EQUIPMENT for stud welding consists of a stud welding gun, an integrated power-control unit or a power source and a control unit to control the time of the current flow, and proper connecting cables and accessories (Figure 9.1).

Types of Guns

THERE ARE TWO types of stud welding guns, portable hand-held and fixed production types, shown in Figure 9.6. The principle of operation is the same for both types.

The portable stud welding gun resembles a pistol. It is made of a tough plastic material and weighs between 4.5 and 9 lb. (2 and 4 kg), depending upon the type and size of gun. A small gun is used for studs from 1/8 to 5/8 in. (3.2 to 16 mm) in diameter; a larger heavy-duty gun is used for studs up through 1-1/4 in. (32 mm) diameter. The large gun can be used for the entire stud range. However, in applications where only small diameter studs are used, it is advantageous to use a small, lighter weight gun.

A gun consists basically of the body, a lifting mechanism, a chuck holder, an adjustable support for the ferrule holder, and the connecting weld and control cables (Figure 9.6). The portable gun body is usually made of a high impact strength plastic. The stud lifting mechanism consists of a solenoid, a clutch, and a mainspring. The mechanism is actuated by the solenoid to provide positive control of the lift. The lift will be consistent over a range of 0.03 to 0.125 in. (0.8 to 3.2 mm), and will be constant regardless of the length of stud protrusion within limits of the gun. An added feature of some guns

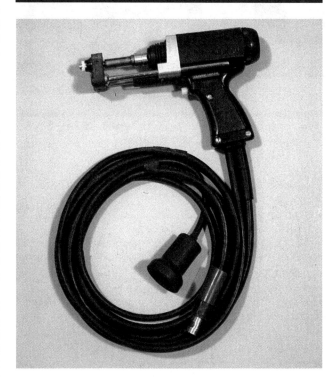

(A) Portable Hand Operated Type

Figure 9.6–Two Types of Arc Stud Welding Guns

(B) Fixed Production Type

Figure 9.6 (Continued)–Two Types of Arc Stud Welding Guns

is a cushioning arrangement to control the plunging action of the stud to complete the weld. Controlled plunge eliminates the excessive spatter normally associated with the welding of large diameter studs with no control.

The fixed or production gun is mounted on an automatic positioning device which is usually air operated and electrically controlled. The workpiece is positioned under the gun with suitable locating fixtures. Tolerances of ± 0.005 in. (0.13 mm) on location and 0.010 in. (± 0.26 mm) in height may be obtained when a production gun is used. A production unit may contain a number of guns, depending upon the nature of the job and the production rate required.

Control Unit

THE CONTROL UNIT consists fundamentally of a contactor suitable for conducting and interrupting the welding cur-

rent and a weld timing device with associated electrical controls. The adjustable weld timer is graduated in either cycles or number settings. Once set, the control unit maintains the proper time interval for the size of stud being welded. The time interval may vary from 0.05 to 2 seconds depending upon the diameter of the stud.

The control unit has two connectors for the welding cables. One is for the cable from the terminal of the dc welding power source, and the other is for the cable to the stud welding gun. Most control units also have a work cable for connection to the workpiece. As with stud welding guns, control units are of two sizes. For welding studs up to 5/8 in. (16 mm) diameter, a small control unit can be used. A large control unit must be used for large diameter studs.

Power Sources

A DC TYPE power source is used for arc stud welding. Alternating current is not suitable for stud welding. There are three basic types of dc power sources that can be used: (1) transformer-rectifier, (2) motor-generator (motor or engine driven), and (3) battery.

The following are general characteristics desired in a stud welding power source:

(1) High open-circuit voltage, in the range of 70 to 100 V.
(2) A drooping output volt-ampere characteristic.
(3) A rapid output current rise to the set value.
(4) High current output for a relatively short time. The current requirements are higher, and the duty cycle is much lower for stud welding than for other types of arc welding.

There are many standard dc arc welding power sources available which meet these requirements and are entirely satisfactory for stud welding. However, dc welding power sources with a constant voltage characteristic are not suitable for stud welding. With this type power source, weld current control can be difficult, and it may not be possible to obtain the proper weld current range for the application.

Many stud welding current range requirements extend beyond those used for SMAW. Because of the maximum current output limitation, standard dc power sources are generally used for welding only 0.5 in. (13 mm) diameter and smaller studs. For large studs, two standard dc type power sources wired in parallel, or a single unit designed specifically for arc stud welding must be used.

When the applications require high welding currents (sometimes over 2500 A) and short weld times, special stud welding power sources are recommended. These special power sources yield higher efficiency not only from the standpoint of weld current output relative to their size and weight, but also from the fact that they cost less than two or more standard arc welding machines.

Duty Cycle

THE BASIS FOR rating special stud welding power sources is different from that of conventional arc welding machines. Because stud welding requires a high current for a relatively short time, the current output requirements of a stud welding power source are higher, but the duty cycle is much lower than those for other types of arc welding. Also, the load voltage is normally higher for stud welding. Cable voltage drop is greater with stud welding than arc welding because of the higher current requirements.

The duration of a stud weld cycle is generally less than one second. Therefore, load ratings and duty cycle ratings are made on the basic of one second. The rated output of a machine is its average current output at 50 V for a period of one second. Thus, a rating of 1000 A at 50 V means that during a period of one second the current output will average 1000 A and the terminal voltage will average 50 V.

Oscillographic traces show that the current output of a motor-generator stud welding power source is higher at the start of welding than at the end. Thus, it is necessary to use the average current for rating purposes.

The duty cycle for stud arc welding machines is based on the formula

% Duty cycle = 1.7 x no. of 1 s loads per minute

where the one second load is the rated output.

Thus, if a machine can be operated six times per minute at rated load without causing its components to exceed their maximum allowable temperatures, then the machine would have a 10 percent duty cycle rating.

Power Control Units

TRANSFORMER-RECTIFIER TYPE POWER sources developed specifically for stud welding are of two basic types: (1) those that require a separate stud welding control unit, and (2) those that incorporate the stud welding gun control and timing circuits as an integral part of the power source. This latter type is generally referred to as a power-control unit. Power-control type units use silicon controlled rectifiers for initiating and interrupting the weld current, and solid-state components for gun control and timing circuitry. Power-control type units are available for both three-phase and single-phase power input. The three-phase units are preferred for stud welding larger diameter studs because they provide a balanced load on the incoming power line. Single-phase units are low cost, portable types for welding studs 1/2 in. (12.7 mm) diameter and under.

The trend in power-control units has been towards regulated current machines. This has been made possible by the development of high power solid state components. Controls are designed to include current feedback circuitry that monitors and maintains or regulates the unit output to desired current irrespective of changes in primary voltage, cable length, or the cable resistance due to heat buildup. This type unit is recommended when the ultimate in current and time control is necessary. Figure 9.1B shows a typical equipment setup for arc stud welding of steel with an integrated power-control unit.

Nonregulated power sources may be severely hampered by the use of either very small diameter or very long cables. This factor is often overlooked when the problem of inadequate welding power arises. When examining cable length, the total cable in the welding circuit must be taken into account. For any given length of cable, the welding current can be increased approximately 10 percent by using a cable of the next larger diameter.

Other major factors to be considered in connection with power sources for stud welding are the incoming power and the cable size and length (primary power and welding cables). Both motor-generator and rectifier power sources normally operate on 230 or 460 V, ac, three phase power. Because of the high currents required for stud welding, line voltage regulation sometimes becomes a problem. Satisfactory operation of either type of equipment can only be assured if the power line voltage regulation will remain within prescribed limits while a weld is in progress.

Welding Cable

THE WELDING CABLE length, including both the gun and ground cables, and the cable size are very important in stud welding. Many times, there is significant power loss in the welding circuit caused by the use of either too small or too long welding cables. The current available for welding at a given machine setting may vary as much as 50 percent, depending upon the size and length of welding cables used.

Figure 9.7 illustrates the effect of cable size and cable length on welding current. The tests made to determine these curves were run with a 2000 A motor generator power supply at maximum setting. Only the cable length and cable size were changed. In this case, the maximum welding current was 2360 A with 30 ft (9 m) of AWG #1 cable. When the same size of cable was lengthened to 180 ft (55 m), the available current decreased 38 percent to 1450 A. On the other hand, when 180 ft (55 m) of #4/0 cable was used, the current was 2050 A, a decrease of only 13 percent. While regulated machines maintain a constant output as cable is added, they will only do so until the maximum output of the unit is reached. Larger cables will extend this limit. Thus, when the distance from the power source to the welding gun increases significantly, larger welding cable should be used.

Automatic Feed Systems

STUD WELDING SYSTEMS with automatic stud feed are available for both portable and fixed welding guns. The studs are automatically oriented in a parts feeder, transferred to the gun (usually through a flexible feed tube), and loaded

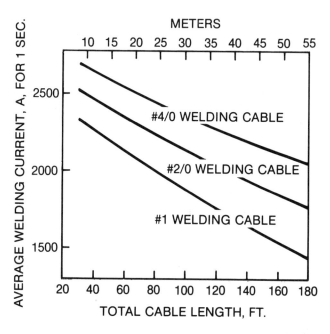

Figure 9.7–Effect of Cable Size and Length on Available Welding Current From a 2000 A Power Source

into the welding gun chuck. Generally, a ferrule is hand loaded for each weld. However, automatic ferrule feed is available with fixed gun production type systems. Figure 9.8 illustrates both types of equipment. Automated portable equipment, using solid state controls and no ferrule, is available for welding 3/8 in. (9.5 mm) diameter studs and smaller.

STUD LOCATING TECHNIQUES

THE METHOD OF locating studs depends upon the intended use of the studs and the accuracy of location required. For applications where extreme accuracy is required, special locating fixtures and fixed (production type) stud welding equipment is recommended. The extent of tooling will be a function of the required production rate as well as total production.

Several methods and procedures are used for positioning studs with a portable stud welding gun. The simplest and most common procedure is to either lay out the work or employ a template for center punch locations. A stud is then located by placing the point of the stud in the punch mark. Although operator skill is always a factor in accuracy, location tolerances of ± 0.050 in. (1.2 mm) can be obtained. Cover plates that have been punched or drilled can be used as templates. When a number of pieces are to be stud welded, common practice is to weld directly through holes in a template without preliminary marking, as shown in Figure 9.9. A simple template positions the stud by locating the ferrule. Because of manufacturing tol-

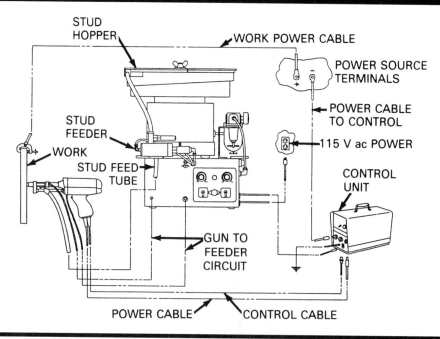

Figure 9.8–Conventional Portable Arc Stud Welding Equipment with an Automatic Stud Feed System

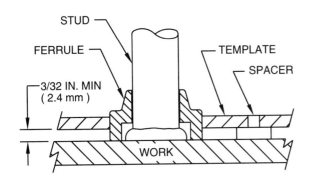

Figure 9.9–Simple Template Used to Locate Studs Within ± 1/32 in. (0.8 mm)

erances on ferrules, the tolerance on stud location with this method is usually ± 0.030 in. (0.8 mm).

When accurate stud location and alignment are required, a tube-type template is used. The stud is centered indirectly by inserting a tube adaptor on the gun in a locating bushing in the template. Figure 9.10 illustrates this type of template. The template uses a hardened and ground bushing with a closely machined tube adaptor. Because standard ferrule grips are used with this adaptor, standardization of templates is possible. It is only necessary to change ferrule grips to weld studs of different diameters. With this type of template, a tolerance of ± 0.015 in. (0.4 mm) can be held on stud location. This method also maintains perpendicular alignment of the stud.

WELDING CURRENT—TIME RELATIONSHIPS

THE CURRENT AND time required for a proper arc stud weld are dependent on the cross-sectional area of the stud.

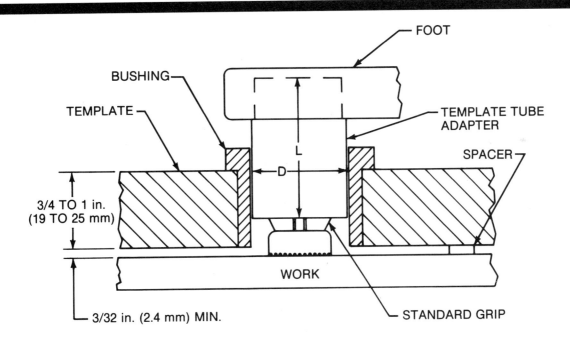

STUD DIAMETER		D		L	
in.	mm	in.	mm	in.	mm
1/2 AND UNDER	13 AND UNDER	1-1/4	32	2	51
5/8 TO 3/4	16 TO 19	1-9/16	40	2-1/2	64
7/8 AND OVER	22 AND OVER	2-1/8	54	2-1/2	64

Figure 9.10–Template With Hardened and Ground Bushing and Welding Gun Adapter Used to Locate Studs Within ± 0.015 in. (0.4 mm)

The total energy input (joules or watt-seconds) is a function of welding current, arc voltage, and arc time. Arc voltage is determined by the lift distance set in the stud gun. Proper lift distance is usually recommended by the stud manufacturer. Therefore, arc energy is basically a function of the welding current and weld time settings with a set or constant lift distance.

The same energy input can be obtained by using a range of current and time settings. It is possible, within certain limitations, to compensate for low or high welding current by changing the weld time. There is a fairly broad range of combinations for each stud size. Under some conditions, such as welding studs to a vertical member or to thin gage material, the allowable range is much smaller.

Although energy input is a major criterion for satisfactory welds, it is not the only factor involved. Other factors such as arc blow, plate surface conditions (rust, scale, moisture, paint), and operator technique can cause poor welds, even though the proper weld energy input was used.

METALLURGICAL CONSIDERATIONS

THE METALLURGICAL STRUCTURES encountered in arc stud welds are generally the same as those found in any arc weld where the heat of an electric arc is used to melt both a portion of the base metal and the electrode (stud) in the course of welding. Acceptable mechanical properties are obtained when the stud and base material are metallurgically compatible. Properly executed stud welds are usually characterized by the absence of inclusions, porosity, cracks, and other defects.

A typical stud weld macrosection, Figure 9.11, shows that molten weld metal is pushed to the perimeter of the stud to form a flash. The amount of weld metal (cast structure) in the joint is minimal. Because of the short welding cycle, the heat-affected zones common to arc welding are present, but they are small. Chapter 4, "Welding Metallurgy," *Welding Handbook*, Volume 1, 8th Edition, contains helpful information on welding metallurgy that is applicable to arc stud welding of various materials.

MATERIALS WELDED

Steels

Low Carbon Steel. Low carbon (mild) steels can be stud welded with no major metallurgical problems. The upper carbon limit for steel to be arc stud welded without preheat is usually 0.30 percent. If work sections are relatively thin for the stud diameters being welded (below those in Table 9.1), the carbon limit may be somewhat higher because of the decreased cooling effect of the work. The most important factor regarding work section thickness is that the material must be heavy enough to permit the welding of the studs without melt-through.

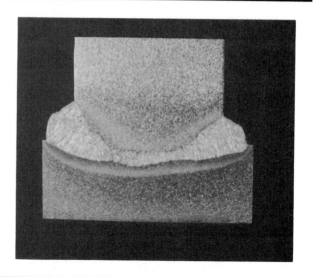

Figure 9.11—Macrostructure of a Typical Arc Stud Weld

Medium And High Carbon Steel. If medium and high carbon steels are to be stud welded, it is imperative that preheat be used to prevent cracking in the heat-affected zones. In some instances, a combination of preheating and postheating may be used to obtain satisfactory results. In cases where the welded assemblies are to be heat-treated for hardening after the welding operation, the preheating or postheating operation may be eliminated if the parts are handled in a manner that prevents damage to the studs.

Low Alloy Steel. Generally, the high strength low alloy steels are satisfactorily stud welded when the carbon content is 0.15 percent or lower. If carbon content exceeds 0.15 percent, it may be necessary to use a low preheat temperature to obtain desired toughness in the weld area.

When the hardness of the heat-affected zones and fillet do not exceed 30 Rockwell C, studs can be expected to perform well under almost any type of severe service. Although good results have been obtained when the hardness ranges up to 35 Rockwell C, it is best to avoid extremely high working stresses and fatigue loading. In special cases where microstructures are important, the weld should be evaluated and qualified for the specific application. Since alloy steels vary in toughness and ductility at high hardness levels, weld hardness should not be used as the sole criterion for weld evaluation.

Heat-Treated Structural Steel. Many structural steels used in shipbuilding and in other construction are heat-treated at the mill. Heat-treated steels require that attention be given to the metallurgical characteristics of the heat-affected zone. Some of these steels are sufficiently hardenable that the heat-affected zones will be martensitic. This structure will be quite sensitive to underbead cracking, and it will have insufficient ductility to carry impact loads. Therefore, for maximum toughness in these steels, a preheat of 700°F (370°C) is recommended. Consideration of the application and end use of the stud will further influence the welding procedures to be followed.

Stainless Steels. Most classes of stainless steel may be arc stud welded. The exceptions are the free machining grades. However, only the austenitic stainless steels (3XX grades) are recommended for general application. The other types are subject to air hardening, and they tend to be brittle in the weld area unless heat-treated after welding. The weldable stainless steel grades include AISI Types 304, 305, 308, 309, 310, 316, 321, and 347. Types 302 HQ, 304, and 305 are most commonly used for stud welding.

Stainless steel studs may be welded to stainless steel or to mild steel as the application may require. The welding setup used is the same as that recommended for low carbon steel except for an increase of approximately 10 percent in power requirement. Where stainless steel studs are welded to mild steel, it is essential that the carbon content of the base metal not exceed 0.20 percent. When welding stainless steel studs to mild steel with 0.20 to 0.28 percent carbon, or to low carbon hardenable steels, Type 308, 309, or 310 studs are recommended. Because of the composition of the weld metal when chromium-nickel alloy studs are welded to mild steel, the weld zone may be quite hard. The hardness will depend on the carbon content in the base metal and whether the molten metal is predominantly austenitic. It is possible to overcome this by using studs with high alloy content such as Type 309 or 310. It is also suggested when welding stainless steel studs to mild steel that a fully annealed stainless steel stud be used.

Nonferrous Metals

Aluminum. The basic approach to aluminum stud welding is similar to that used for mild steel stud welding. The power sources, stud welding equipment, and controls are the same. The stud welding gun is modified slightly by the addition of a dampening device to control the plunging rate of the stud at the completion of the weld time. Also, a special gas adaptor foot ferrule holder is used to contain the high purity inert shielding gas during the weld cycle. Argon is generally used, but helium may be useful with large studs to take advantage of the higher arc energy.

Reverse polarity is used with the stud (electrode) positive and the work negative. An aluminum stud differs from a steel stud in that no flux is used on the weld end. A cylindrical or cone shaped projection is used on the base of the stud. The projection dimensions on the welding end are designed for each size stud to give the best arc action. The projection serves to initiate the long arc used for aluminum stud welding.

Studs range in weld base diameters from 1/4 to 1/2 in. (6.4 to 13 mm). Their sizes and shapes are similar to steel studs.

Aluminum studs are commonly made of aluminum-magnesium alloys, including 5086 and 5356, that have a typical tensile strength of 40 000 psi (275 MPa). These alloys have high strength and good ductility. They are metallurgically compatible with the majority of aluminum alloys used in industry.

In general, all plate alloys of the 1100, 3000, and 5000 series are considered excellent for stud welding; alloys of the 4000 and 6000 series are considered fair; and the 2000 and 7000 series are considered poor. The minimum aluminum plate thickness, with and without backup, to which aluminum studs of 3/16 to 1/2 in. (4.8 to 13 mm) base diameter may be welded are given in Table 9.1.

Figure 9.12 illustrates a cross section of a typical aluminum alloy stud weld. Table 9.4 gives typical conditions for aluminum arc stud welding.

Magnesium. The gas shielded arc stud welding process used for aluminum also produces high strength welds in magnesium alloys. A ceramic ferrule is not needed. Helium shielding gas and dc reverse polarity (electrode positive) should be used. A gun with plunge dampening will avoid spattering and base metal undercutting.

Breaking loads up to 1500 lb. (6.7 kN) for 1/4 in. (6.4 mm) diameter studs and up to 4500 lb. (20 kN) for 0.50 in. (13 mm) diameter studs have been obtained with AZ31B alloy studs welded to 1/4 in. (6.4 mm) thick AZ31B or ZE10A base metal.

Minimum base metal thicknesses to which 1/4 and 1/2 in. (6.4 and 13 mm) diameter studs may be attached,

Figure 9.12–Macrostructure of a 3/8 in. (9.5 mm) Diameter Type 5356 Aluminum Alloy Stud Welded to a 1/4 in. (6.4 mm) Type 5053 Aluminum Alloy Plate

Table 9.4
Typical Conditions for Arc Stud Welding of Aluminum Alloys

Stud Weld Base Diameter		Weld Time, Seconds	Welding Current (a) A	Shielding Gas Flow (b)	
in.	mm			ft³/h	liter/min
1/4	6.4	0.33	250	15	7.1
5/16	7.9	0.50	325	15	7.1
3/8	9.5	0.67	400	20	9.4
7/16	11.1	0.83	430	20	9.4
1/2	12.7	0.92	475	20	9.4

a. The currents shown are actual welding current and do not necessarily correspond to power source dial settings.

b. Shielding Gas - 99.95% pure argon.

without melt-through or great loss in strength, are 1/8 and 1/4 in. (3.2 and 6.4 mm), respectively. If strength is not a consideration, 1/2 in. (13 mm) diameter studs can be welded to 3/16 in. (4.8 mm) thick plate without melt-through.

Other Materials. On a moderate scale, arc stud welding is being done in industry on various brass, bronze, nickel-copper, and nickel-chromium-iron alloys. The applications are usually very special ones requiring careful evaluation to determine suitability of design.

Nickel, nickel-copper, nickel-chromium-iron, and nickel-chromium-molybdenum alloys are best stud welded with dc using reverse polarity (electrode positive). Nickel, nickel-copper, and nickel-chromium-iron alloy stud welds tend to contain porosity and crevices. The mechanical strengths, however, are usually high enough to meet most requirements. The weld itself should not be exposed to corrosive media.

QUALITY CONTROL AND INSPECTION

WELD QUALITY ASSURANCE requires the proper materials, equipment, setup, and operating procedures, and also a trained operator. Proper setup includes such things as gun retraction (lift), stud extension beyond the ferrule (plunge), and proper welding current and time.

Weld quality is maintained by close attention to the factors that may produce variations in the weld. To maintain weld quality and consistency, the following is necessary:

(1) Have sufficient welding power for the size and type of stud being welded.

(2) Use dc straight polarity for steels and dc reverse polarity for aluminum and magnesium.

(3) Ensure a good work connection.

(4) Have welding cables of sufficient size with good connections.

(5) Use correct accessories and ferrules.

(6) Clean the work surface where the stud is to be welded.

(7) Adjust the gun so that the stud extends the recommended distance beyond the ferrule and also retracts it the proper distance for good arc characteristics. Stud extension will be about equal to the length reductions in Table 9.2.

(8) Hold the gun steady at the proper angle to the work. Generally it is perpendicular. Accidental movement of the gun during the weld cycle may cause a defective weld.

(9) Keep stud welding equipment properly cleaned and maintained.

(10) Make test welds before starting and at selected intervals during the job.

Steel Studs

THE LATEST EDITION of ANSI/AWS D1.1, *Structural Welding Code-Steel*, contains provisions for the installation and inspection of steel studs welded to steel components. Quality control and inspection requirements for stud welding are also included. ANSI/AWS C5.4, *Recommended Practices for Stud Welding*, latest edition, briefly covers inspection and testing of both steel and aluminum stud welds.

Welded studs may be inspected visually for weld appearance and consistency, and also mechanically. Production studs can be proof tested by applying a specified load (force) on them. If they do not fail, the studs are considered acceptable. Production studs should not be bent or twisted for proof testing.

Visual Inspection. The weld flash around the stud base is inspected for consistency and uniformity. Lack of flash may indicate a faulty weld. Figure 9.13(A) indicates a satisfactory stud weld with a good weld flash formation. In contrast, Figure 9.13(B) shows a stud weld in which the plunge was too short. Prior to welding, the stud should always project the proper length beyond the bottom of the ferrule. (This type of defect may also be caused by arc blow.) Figure 9.13(C) illustrates "hang up." The stud did not plunge into the weld pool. This condition may be corrected by realigning the accessories to insure completely

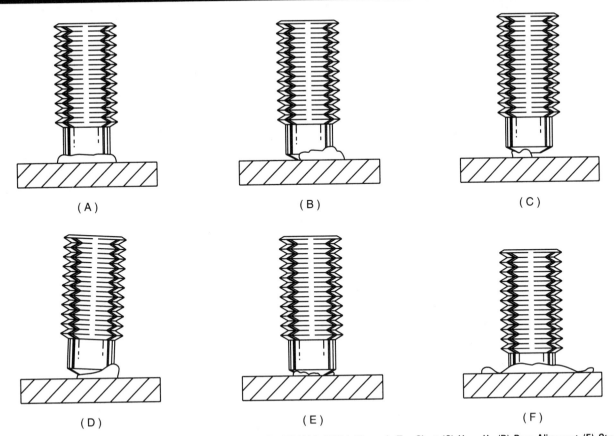

(A) Satisfactory Stud Weld With a Good Fillet Formation (B) Stud Weld in Which Plunge is Too Short (C) Hang-Up (D) Poor Alignment (E) Stud Weld Made With Low Current (F) Stud Weld Made With High Current

Figure 9.13—Satisfactory and Unsatisfactory Arc Stud Welds

free movement of the stud during lift and plunge. Arc length may also require adjustment.

Figure 9.13(D) shows poor alignment, which may be corrected by positioning the stud gun perpendicular to the work. Figure 9.13(E) shows the results of low weld current. To correct this problem, the worklead and all connections should be checked. Also, the current setting, the time setting, or both, should be increased. It may also be necessary to adjust the arc length. The effect of too much weld current is shown in Figure 9.13(F). Decreasing the current setting or the welding time, or both, will lower the weld power.

Mechanical Testing. Mechanical tests should be made as part of procedure and performance qualification before initiation of production welding to insure that the welding schedule is satisfactory. They may also be made during the production run or at the beginning of a shift to insure that the welding conditions have not changed. Arc stud welds are tested by bending the stud or by applying a proof tensile load.

Bending may be done by striking the stud with a hammer or by using a bending tool such as a length of a tube or pipe, as shown in Figure 9.14. The angle through which the stud will bend without weld failure will depend on the stud and base metal compositions and conditions (cold worked, heat-treated) and stud design. Acceptable bending should be determined when the welding procedure specification is established or from the applicable welding code. Bend testing may damage the stud and should preferably be done on qualification samples only. For some applications, however, studs can remain in the bent condition.

The method used to apply tensile load on an arc welded stud will depend on the stud design. Special tooling may be required to grip the stud properly without damage, and a special loading device may be needed. A simple method that can be used for straight threaded studs is shown in Figure 9.15. A steel sleeve of appropriate size is placed over the stud. A nut of the same material as the stud is tightened against a washer bearing on the sleeve with a torque

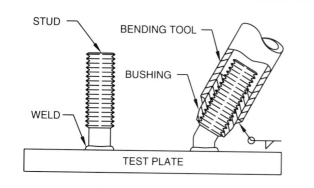

Figure 9.14—Bend Test for Welded Studs to Determine Acceptable Welding Procedures

wrench This applies a tensile load (and some shear) on the stud.

The relationship between nut torque, T, and tensile load, F, can be estimated using the equation

$$T = kFd \qquad (9.1)$$

where

d = the nominal thread diameter

k = a constant related to such factors as thread angle, helix angle, thread diameters, and coefficients of friction between the nut and thread, and the nut and washer.

For mild steel, k is approximately 0.2 for all thread sizes and for both coarse and fine thread. However, the many factors that influence friction will influence the value of k. These include the stud, nut, and washer materials, surface

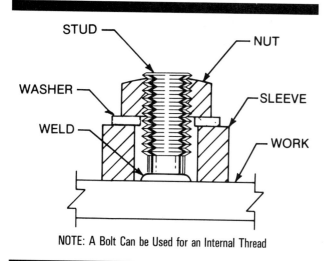

NOTE: A Bolt Can be Used for an Internal Thread

Figure 9.15—Method of Applying a Tensile Load to a Welded Stud Using Torque

finishes, and lubrication. For other materials, k may have some other value because of the differences in friction between the parts.

Aluminum Studs

VISUAL INSPECTION OF aluminum stud welds for acceptance is limited because the appearance of the weld fillet does not necessarily indicate quality. Therefore, visual inspection of aluminum stud welds is recommended only to determine complete fusion and absence of undercut around the periphery of the weld.

Aluminum studs can be tested to establish acceptable welding procedures using the bend test shown in Figure 9.14. If the stud bends about 15 degrees or more from the original axis without breaking the stud or weld, the welding procedures should be considered satisfactory. Production studs should not be bent and then straightened because of possible damage to them. In this case, the torque test or separate qualification test plates may be substituted.

Torque testing of threaded aluminum studs is done in the same manner as that used for steel studs. Torque is applied to a predetermined value or until the stud fails. Typical torque tests gave the failure loads shown in Table 9.5. For a particular application, the acceptable proof load should be established by suitable laboratory tests relating applied torque to tensile loading.

APPLICATIONS

ARC STUD WELDING has been widely accepted by all the metalworking industries. Specifically, stud welding is being used extensively in the following fields: automotive, boiler and building and bridge construction, farm and industrial equipment manufacture, railroads, and shipbuilding. Defense industry applications include missile containers, armored vehicles, and tanks.

Some typical applications are attaching wood floors to steel decks or framework; fastening linings or insulation in tanks, boxcars, and other containers; securing inspection covers; mounting machine accessories; securing tubing and wire harnesses; and welding shear connectors and concrete anchors to structures.

Table 9.5
Typical Nut Torques Causing Failure of Aluminum Alloy Studs

Thread Size	Failure Load	
	lbf · in.	N · m
1/4-20	60	7
5/16-18	115	13
3/8-16	195	22
7/16-14	290	33
1/2-13	435	49

CAPACITOR DISCHARGE STUD WELDING

DEFINITION AND GENERAL DESCRIPTION

CAPACITOR DISCHARGE STUD welding is a stud arc welding process where dc arc power is produced by a rapid discharge of stored electrical energy with pressure applied during or immediately following the electrical discharge. The process uses an electrostatic storage system as a power source in which the weld energy is stored in capacitors of high capacitance. No ferrule or fluxing is required.

PRINCIPLES OF OPERATION

THERE ARE THREE different types of capacitor discharge stud welding: initial contact, initial gap, and drawn arc. They differ primarily in the manner of arc initiation. Initial contact and initial gap capacitor discharge stud welding studs have a small, specially designed projection (tip) on the weld end of the stud. Drawn arc stud welding creates a pilot arc as the stud is lifted off the workpiece by the stud gun. That version is similar to arc stud welding.

Initial Contact Method

IN INITIAL CONTACT stud welding, the stud is first placed against the work as shown in Figure 9.16(A). The stored energy is then discharged through the projection on the base of the stud. The small projection presents a high resistance to the stored energy, and it rapidly disintegrates from the high current density as shown in Figure 9.16(B). This creates an arc that melts the surfaces to be joined. During arcing, Figure 9.16(C), the pieces to be joined are being brought together by action of a spring, weight, or an air cylinder. When the two surfaces come in contact, Figure 9.16(D), fusion takes place, and the weld is completed.

Initial Gap Method

THE SEQUENCE OF events in initial gap stud welding is shown in Figure 9.17. Initially, the stud is positioned off the work, leaving a gap between it and the work, as shown in Figure 9.17(A). The stud is released and continuously moves toward the work under gravity or spring loading, Figure 9.17(B). At the same time, open-circuit voltage is applied between the stud and the work. When the stud contacts the work, Figure 9.17(C), high current flashes off the tip and initiates an arc. The arc melts the surfaces of the stud and work as the stud continues to move forward, Figure 9.17(D). Finally, the stud plunges into the work, and the weld is completed, Figure 9.17(E).

With proper design of the electrical characteristics of the circuit and size of the projection, it is possible to produce a high current arc of such short duration (about 0.006 seconds) that its effect upon the stud and workpiece is purely superficial. A surface layer only a few thousandths of an inch (hundredths of a millimeter) in thickness on each surface reaches the molten state.

Drawn Arc Method

IN THE DRAWN arc method, arc initiation is accomplished in a manner similar to that of arc stud welding. The stud does not require a tip on the weld face. An electronic control is used to sequence the operation. Weld time is controlled by an electronic circuit in the unit. The welding gun is similar to that used for arc stud welding.

The operating sequence is shown in Figure 9.18. In sequence, the stud is positioned against the work as shown in Figure 9.18(A). The trigger switch on the stud welding gun is actuated, energizing the welding circuit and a solenoid coil in the gun body. The coil motion lifts the stud from the work, Figure 9.18(B), drawing a low amperage pilot arc between them. When the lifting coil is de-energized, the stud starts to return to the work. The welding capacitors are then discharged across the arc. The high amperage from the capacitors melts the end of the stud and the adjacent work surface, Figure 9.18(C). The spring action of the welding gun plunges the stud into the molten metal, Figure 9.18(D), to complete the weld.

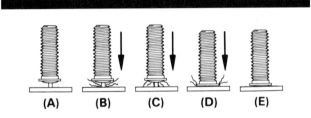

(A)   (B)   (C)   (D)   (E)

Figure 9.16–Steps in Initial Contact Capacitor Discharge Stud Welding

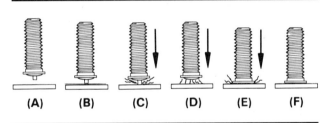

(A)   (B)   (C)   (D)   (E)   (F)

Figure 9.17–Steps in Initial Gap Capacitor Discharge Stud Welding

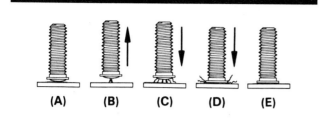

Figure 9.18–Steps in Drawn Arc Capacitor Discharge Stud Welding

DESIGNING FOR CAPACITOR DISCHARGE STUD WELDING

THE ABILITY OF the capacitor discharge method to weld studs to thin sections in an important design feature. Material as thin as 0.030 in. (0.75 mm) can be welded without melt-through. Studs have been successfully welded to some materials (stainless steel, for example) in thicknesses as low as 0.010 in. (1/4 mm).

Another design feature of this system of stud welding is its ability to weld studs to dissimilar metals. The penetration into the work from the arc is so shallow that there is very little mixing of the stud metal and work metal. Steel to stainless steel, brass to steel, copper to steel, brass to copper, and aluminum to die cast zinc are a few of the combinations that may be used. Many other unusual metal combinations, not normally considered weldable by fusion processes, are possible with this process.

Another feature is the elimination of post-weld cleaning or finishing operations on the side of the base metal opposite to the stud attachment (face surface). The process can be used on parts that have had the face surface painted, plated, polished, or coated with ceramic or plastic.

STUDS

Stud Materials

THE MATERIALS THAT are commonly capacitor discharge stud welded are low carbon steel, stainless steel, aluminum, and brass. Low carbon steel and stainless steel studs are generally the same compositions as those used for arc stud welding. For aluminum, 1100 and 5000 series alloys are generally used. Copper alloy studs are mostly No. 260 and No. 268 compositions (brasses).

Stud Designs

STUD DESIGNS FOR capacitor discharge stud welding range from standard shapes to complex forms for special applications. Usually, the weld base of the fastener is round. The shank may be almost any shape or configuration. These include threaded, plain, round, square, rectangular, tapered, grooved, and bent configurations, or flat stampings.

The size range is 1/16 to 1/2 in. (1.6 to 12.7 mm) diameter, with the great bulk of attachments falling in the 1/8 to 3/8 in. (3.2 to 9.5 mm) diameter range. Figure 9.19 shows some common stud designs.

Initial contact and initial gap capacitor discharge studs are designed with a tip or projection on the weld end. The size and shape of this tip is important because it is one of the variables involved in the achievement of good quality welds. The standard tip is cylindrical in shape. For special applications, a conically shaped tip is used. The detailed weld base design is determined by the stud material, the base diameter, and sometimes by the particular application. The weld base is tapered slightly to facilitate the expulsion of the expanding gases that develop during the welding cycle. Usually, the weld base diameter is larger than that of the stud shank. The weld area is larger than the stud cross section to provide a joint strength equal to or higher than that of the stud.

Drawn arc capacitor discharge studs are designed without a tip or projection on the weld end. However, the weld end is tapered or slightly spherical so that the arc will initiate at the center of the base. As with the other capacitor discharge methods, these studs are generally designed with a large base in the form of a flange.

Stud melt-off or reduction in length due to melting is almost negligible when compared to the arc stud welding method. Stud melt-off is generally in the range of 0.008 to 0.015 in. (0.2 to 0.4 mm).

WELDING EQUIPMENT

CAPACITOR DISCHARGE STUD welding requires a stud gun and a combination power-control unit with associated interconnecting cables. Both portable and stationary production units are available.

Solid state control circuitry provides signals for automatic sequencing of several events during the welding cycle. The events include one or more of the following:

(1) Energize the gun solenoid or air cylinder for initial gap and drawn arc methods.
(2) Initiate the pilot arc in the drawn arc method.
(3) Discharge the welding current from the capacitor bank at the proper time in the welding sequence.
(4) De-energize the solenoid or air cylinder of the gun.
(5) Control the changing voltage of the capacitor bank.

Portable Units

THE HAND-HELD STUD gun is usually made of high impact strength plastic. The gun holds and positions the stud for welding. A trigger initiates the welding cycle through a control cable to the power source control unit. By changing the chuck that holds the stud, various diameters and shapes of studs can be accommodated.

The power source control unit provides the welding current, and it contains the necessary circuitry for charging

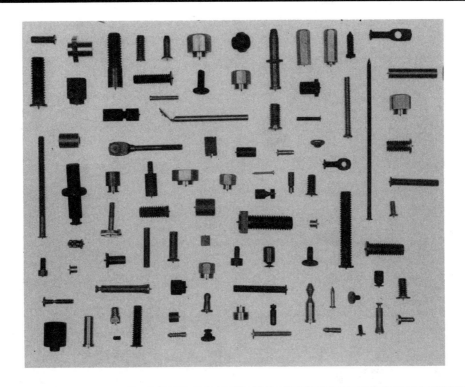

Figure 9.19–Commonly Used Studs for Capacitor Discharge Stud Welding

the capacitors. Variable discharge currents are obtained by varying the charge voltage on the capacitors. Control of the charging and discharging currents is done automatically by the welding machine. The units generally operate on 115 V, 60 Hz power.

Typical portable capacitor discharge equipment is illustrated in Figure 9.20. The stored energy of such a unit would be in the neighborhood of 70 000 μF charged to 170 V, and it would be capable of welding 1/4 in. (6.4 mm) diameter studs at a rate of eight to ten per minute.

Stationary Production Equipment

THIS TYPE OF equipment consists of either an air actuated, an electrically actuated, or a gravity drop stud gun (or guns) mounted above a work surface. The electrical controls for the air systems and for charging the capacitors are usually located under the work table.

The power-control units are generally designed for a specific application because automatic sequencing of clamping, indexing, and unloading devices may be incorporated. The capacitance of production units ranges from about 20 000 to 200 000 μF. The capacitor charge voltage does not exceed 200 V, and it is isolated from the stud chuck until welding is initiated. Power input is 230 or 460 V, single or three phase.

Depending upon the amount of automation in the fixturing and in the feeding of studs and parts to be welded, high production rates can be obtained with this equipment. Up to 45 welds per minute have been made with a single gun.

Automatic Stud Feed Systems

CAPACITOR DISCHARGE STUD welding is well suited for high speed automatic stud feed applications because no ceramic ferrules are required. Portable drawn arc type capacitor discharge equipment with automatic stud feed is available for studs ranging from #6 through 1/4 in. (3.5 mm through 6.4 mm) diameter. Using this system, weld rates of approximately 42 studs per minute can be achieved. Such a unit is shown in Figure 9.21.

STUD LOCATION

THE METHOD OF locating studs depends on several factors: the accuracy and consistency of positioning required, the type of welding equipment to be used (portable or fixed), the required rate of production, and to some extent the geometry or shape of the workpiece. In general, the fixed production type welding unit affords greater precision in stud location than does the portable hand-held unit.

Accuracy of location with a portable gun is usually dependent upon the care used in laying out the location(s) on the workpiece. However, with the application of various types of spacers, bushings, and template, the accuracy range can be within a tolerance of ± 0.020 in. (0.5 mm).

Standard production type units will provide tolerance limits of ± 0.005 in. (0.12 mm). Precision location requires not only accurate and well maintained welding equipment and tooling, but also exceptionally precise, high-quality studs.

WELD ENERGY REQUIREMENTS

IN CAPACITOR DISCHARGE stud welding, arc power is obtained by discharging a capacitor bank through the stud to the work. Arc times are significantly shorter and welding currents are much higher than those used for arc stud welding. It is the very short weld time that accounts for the shallow weld penetration into the work and also the small stud melt-off length.

Depending upon stud size and type of equipment used, the peak welding current can vary from about 600 to

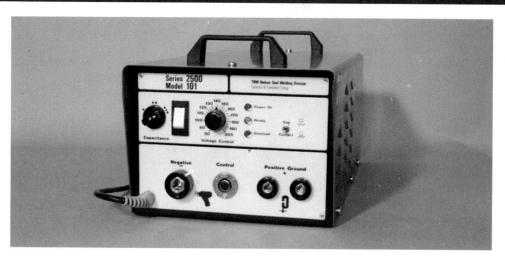

(A) Control-Power Source Unit

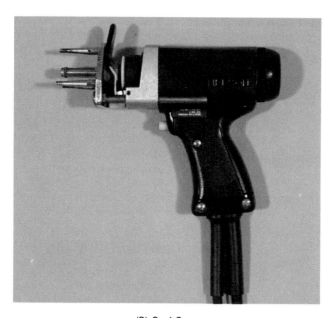

(B) Stud Gun

Figure 9.20–Portable Capacitor Discharge Stud Welding Equipment

Figure 9.21–Capacitor Discharge Production Stud Welding Machine With Automatic Stud Feed System

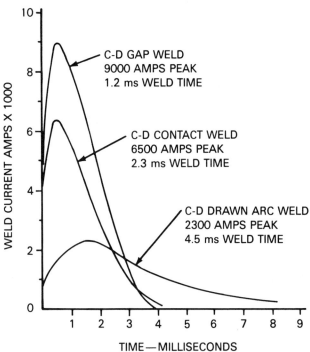

Figure 9.22–Typical Current Versus Time Curves for the Three Capacitor Discharge Stud Welding Methods

20 000 A. The total time to make a weld depends on the welding method used. For the drawn arc method, weld time is in the range of 4 to 6 milliseconds. Figure 9.22 illustrates typical current-time relationships for the three welding methods. Note that the arc current for the initial contact or initial gap method is much higher than for the drawn arc method.

MATERIAL WELDED

IN GENERAL, THE same metal combinations that can be joined by the arc stud welding method also can be joined by the capacitor discharge method. These include carbon steel, stainless steel, and aluminum alloys.

In addition, some dissimilar metal combinations that present metallurgical problems with arc stud welding can be successfully capacitor discharge stud welded. The reason for this is the small volume of metal melted in the very short capacitor discharge time. The small volume and its expulsion, when the stud plunges into the plate, result in a very thin layer of weld metal in the joint. If the weld metal is sound and strong, the stud will carry its design load. Weld metal ductility is not a significant factor.

Weldable stud and base metal combinations of commonly used alloys are listed in Table 9.6. The applications are not limited to these materials. The relative electrical conductivities or melting temperatures of the materials are not of great significance unless there are great differences between them.

Typical macrostructures of capacitor discharge stud welds are shown in Figure 9.23. Note the very narrow weld line between the brass and steel sheet in Figure 9.23(B).

Because of the very short weld times, flux or shielding is not normally required to prevent weld metal contamina-

tion from air. One exception is that argon shielding be used when stud welding aluminum and some other metals with the drawn arc method; arc time is long enough for harmful oxidation to occur. The stud gun should then be equipped with a gas adaptor foot. Welding grade argon (99.95 percent pure) should be used at the flow rate recommended by the manufacturer.

QUALITY CONTROL AND INSPECTION

QUALITY CONTROL OF a capacitor discharge stud weld is more difficult than that of an arc stud weld because of the absence of a steady welding arc and a weld flash. The operator does not hear and see the welding arc, nor can the operator use the characteristics of a weld flash to evaluate weld quality. However, there should be some flash at the weld joint.

The best method of quality control for capacitor discharge stud welding is to destructively test studs that have been welded to base metal similar to that to be used in the actual production. The destructive test should be a bend, torque, or tensile test. Once a satisfactory welding schedule is established, the production run can begin. It is best to check weld quality at regular intervals during production and especially after maintenance to ascertain that the welding conditions have not changed.

**Table 9.6
Typical Combinations of Base Metal and Stud Metal for Capacitor Discharge Stud Welding**

Base Metal	Stud Metal
Low carbon steel, AISI 1006 to 1022	Low carbon steel, AISI 1006 to 1010, stainless steel, series 300*, copper alloy 260 and 268 (brass)
Stainless steel, series 300* and 400	Low carbon steel, AISI 1006 to 1010, stainless steel, series 300*
Aluminum alloys, 1100, 3000 series, 5000 series, 6061 and 6063	Aluminum alloy 1100, 5086, 6063
ETP copper, lead free brass, and rolled copper	Low carbon steel, AISI 1006 to 1010, stainless steel, series 300*, copper alloys 260 and 268 (brass)
Zinc alloys (die cast)	Aluminum alloys 1100 and 5086

* Except for the free-machining Type 303 stainless steel.

**(A) STEEL STUD, 3/16 in. (5 mm) DIAMETER,
TO 0.024 in. (0.6 mm) THICK STEEL SHEET**

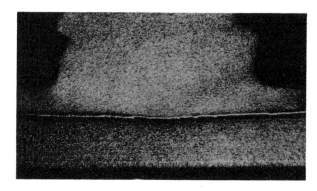

**(B) BRASS STUD, 1/4 in. (6.4 mm) DIAMETER,
TO 1/16 in. (1.6 mm) THICK MILD STEEL**

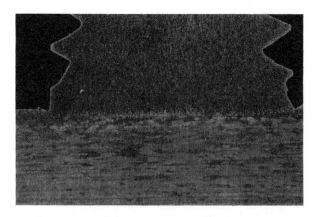

**(C) 6061 T6 ALUMINUM STUD, 3/8 in. (9.5 mm) DIAMETER,
TO 1/8 in. (3.2 mm) ALUMINUM SHEET OF THE SAME ALLOY**

Figure 9.23–Macrostructures of Three Capacitor Discharge Stud Welds

The following are some points to consider for producing and maintaining good capacitor discharge stud welds:

(1) Power source of sufficient size for stud size being welded.

(2) Properly maintained and operating equipment.

(3) Tight cable connections.

(4) Proper handling of studs and stud gun during the welding process.

(5) Welding surface cleanliness. The surface should be free from excessive oils, grease, and other lubricants and from rust, mill scale, and other oxides. These conditions contribute to high electrical resistance in areas of welding and grounding.

(6) Welding surface imperfections, such as extreme roughness, which can prevent complete fusion in the weld area.

(7) Perpendicularity of stud axis to the work surface. This is important for complete fusion.

(8) Proper weld end design on the stud. The tip size, face angle, and weld base diameter must be correct for the application.

Capacitor discharge stud welds may be inspected both visually and mechanically. The success of visual inspection methods depends on the interpretation of the appearance of the weld. Figure 9.24 illustrates good and bad capacitor discharge stud welds.

If a questionable weld is evident after the welds have been visually inspected, the weld should be mechanically tested.

MECHANICAL TESTING

MECHANICAL TESTING OF capacitor discharge stud welds should be done by using the same methods described for arc stud welding. These include bend testing and proof tensile loading. The tests are used to establish welding conditions and also to qualify production studs.

Maximum nut torque values for proof testing studs by this method are given in Table 9.7 for various stud materials and sizes. The torque values listed will produce tensile

(A) GOOD WELD   **(B) WELD POWER TO HIGH**

(C) WELD POWER TOO LOW

Figure 9.24–Examples of Satisfactory and Unsatisfactory Capacitor Discharge Stud Welds

stresses in those studs that are slightly below the material yield strengths.

The table also gives the tensile loads that will develop approximately the nominal tensile strength of the stud material in the different diameters. The maximum shear load that the studs can carry is also listed for information. For proof testing or stud selection, appropriate safety factors should be applied by the user.

APPLICATIONS

SOME INDUSTRIAL APPLICATIONS of capacitor discharge stud welding are aircraft and aerospace, appliances, building construction, maritime construction, metal furniture, stainless steel equipment, and transportation. Capacitor discharge welded studs are also used, where appropriate, as fasteners or supports.

Table 9.7
Torque, Tensile, and Shear Loads for Capacitor Discharge Welded Studs
of Various Materials and Sizes

Stud Material	Stud Size	Maximum Fastening Torque (a)		Maximum Tensile Load (b)		Maximum Shear Load	
		lbf · in.	N · m	lb	kN	lb	kN
Low-carbon, Copper-Flashed Steel	6-32	6	0.7	500	2.2	375	1.7
	8-32	12	1.4	765	3.4	575	2.6
	10-24	14	1.6	960	4.3	720	3.2
	1/4-20	43	4.9	1750	7.8	1300	5.8
	5/16-18	72	8.1	2900	13	2200	9.8
	3/8-16	106	12	4300	19	3250	14
Stainless Steel 304 or 305	6-32	10	1.1	790	3.5	590	2.6
	8-32	20	2.3	1260	5.6	940	4.2
	10-24	23	2.6	1530	6.8	1150	5.1
	1/4-20	75	8.5	2880	13	2160	9.6
	5/16-18	126	14	3750	17	3100	14
	3/8-16	186	21	4850	22	4550	20
Aluminum Alloy 1100	6-32	2.5	0.3	200	0.9	125	0.6
	8-32	5	0.6	295	1.3	185	0.8
	10-24	6.5	0.7	380	1.7	235	1.0
	1/4-20	21.5	2.4	670	3.0	415	1.9
	5/16-18	36	4.1	1125	5.0	695	3.1
	3/8-16	53	6.0	1660	7.4	1000	4.4
Aluminum Alloy 5086	6-32	3.5	0.4	375	1.7	235	1.0
	8-32	7.5	0.8	585	2.6	365	1.6
	10-24	10	1.1	735	3.3	460	2.0
	1/4-20	32.5	3.7	1360	6.1	850	3.8
	5/16-18	54.5	6.2	2300	10	1400	6.2
	3/8-16	81	9.2	3400	15	2100	9.4
Copper Alloy (Brass) 260 and 268	6-32	8	0.9	600	2.7	390	1.7
	8-32	16	1.8	860	3.8	560	2.5
	10-24	18.5	2.1	1040	4.6	680	3.0
	1/4-20	61	6.4	1950	8.7	1275	5.7
	5/16-18	102	12	3280	15	2140	9.5
	3/8-16	150	16	4800	21	3160	14

a. These values should develop stud tensile stresses to slightly below the yield strengths of the materials.

b. These values should develop the nominal tensile strengths of the materials.

PROCESS SELECTION AND APPLICATION

THERE ARE SOME types of applications for which the capabilities of the arc stud welding process and the capacitor discharge stud welding process overlap, but generally the selection between these two basic processes is well defined. A process selection chart is shown in Table 9.8. The area in which selection is usually more difficult is the method of capacitor discharge stud welding that should be used, i.e., contact, gap, or drawn arc. The main criteria for selecting which basic type of stud welding process should be used are fastener size, base metal thickness, and base metal composition. Using these criteria, it is almost always possible to select the best method.

FASTENER SIZE

FOR STUDS OVER 5/16 in. (8.0 mm) diameter, the arc stud welding process must be used for portable applications. The capacitor discharge stud welding process is limited to 5/16 in. (8.0 mm) diameter with hand-held guns and to 3/8 in. (9.5 mm) diameter studs with fixed or production type equipment. Applications suitable for the capacitor

Table 9.8
Stud Welding Process Selection Chart

Factors to be Considered	Arc Stud Welding	Capacitor Discharge Stud Welding	
		Initial Gap and Initial Contact	Drawn Arc
Stud Shape			
Round	A	A	A
Square	A	A	A
Retangular	A	A	A
Irregular.......................	A	A	A
Stud Diameter or Area			
1/16 to 1/8 in. (1.6 to 3.2 mm) diam	D	A	A
1/8 to 1/4 in. (3.2 to 6.4 mm) diam	C	A	A
1/4 to 1/2 in. (6.4 to 12.7 mm) diam	A	B	B
1/2 to 1 in. (12.7 to 25.4 mm) diam	A	D	D
up to 0.05 in.^{2} (32.3 mm^2)	C	A	A
over 0.05 in.^{2} (32.3 mm^2)	A	D	D
Stud Metal			
Carbon Steel	A	A	A
Stainless Steel.....................	A	A	A
Alloy Steel	B	C	C
Aluminum.........................	B	A	B
Brass	C	A	A
Base Metal			
Carbon Steel	A	A	A
Stainless Steel.....................	A	A	A
Alloy Steel	B	A	C
Aluminum.........................	B	A	B
Brass	C	A	A
Base Metal Thickness			
under 0.015 in. (0.4 mm)	D	A	B
0.015 to 0.062 in. (0.4 to 1.6 mm)........	C	A	A
0.062 to 0.125 in. (1.6 to 3.2 mm)........	B	A	A
over 0.125 in. (3.2 mm)	A	A	A
Strength Criteria			
Heat Effect on Exposed Surfaces	B	A	A
Weld Fillet Clearance	B	A	A
Strength of Stud Governs	A	A	A
Strength of Base Metal Governs	A	A	A

Legend

A -- Applicable without special procedures, equipment, etc.

B -- Applicable with special techniques or on specific applications which justify preliminary trials or testing to develop welding procedure and technique.

C -- Limited application.

D -- Not recommended.

discharge type process with studs in the 5/16 to 3/8 in. (8.0 to 9.5 mm) diameter range generally involve thin base materials where avoidance of reverse side marking is the foremost requirement.

BASE METAL THICKNESS

FOR BASE THICKNESSES under 1/16 in. (1.6 mm), the capacitor discharge stud welding process should be used. Using this process, the base metal can be as thin as 0.020 in.

(0.5 mm) without melt-through occurring. On such thin material, the sheet will tear when the stud is loaded excessively. Reverse side marking is the principal effect involved in appearance.

Using the arc stud welding process, the base metal thickness should be at least one-third the weld base diameter of the stud to assure maximum weld strength. Where strength is not the foremost requirement, the base metal thickness may be a minimum of one-fifth the weld base diameter.

BASE METAL COMPOSITION

FOR MILD STEEL, austenitic stainless steel, and various aluminum alloys, either process can be used. For copper, brass, and galvanized steel sheet, the capacitor discharge process is best suited.

CAPACITOR DISCHARGE METHOD

USING THE ABOVE criteria, if the capacitor discharge process is chosen as the best one for the application, then the methods within this process must be evaluated. Since there is considerable overlap in the stud welding capabilities of the three methods, there are many applications where more than one of them can be used. On the other hand, there are many instances where one method is best suited for the application. Setting up specific guidelines for selection of the best capacitor discharge method is rather difficult. However, usage of the three different methods is generally as follows:

Initial Contact Method

THE INITIAL CONTACT method is used only with portable equipment, principally for welding mild steel studs. Equipment simplicity makes it ideal for welding mild steel insulation pins to galvanized duct work.

Initial Gap Method

THIS METHOD IS used with portable and fixed equipment for welding mild steel, stainless steel, and aluminum. Generally, it is superior to both the drawn arc and initial contact methods for welding dissimilar metals and aluminum. Inert gas is not needed for aluminum welding.

Drawn Arc Method

THE TYPES OF equipment and materials welded are the same as those of the initial gap method. The stud does not require a special tip. The method is ideally suited for high-speed production applications involving automatic feed systems with either portable equipment or fixed production type equipment. Inert gas is required for aluminum welding.

APPLICATION CONSIDERATIONS

STUDS CAN BE welded with the work in any position, i.e., flat, vertical, and overhead. The use of the gravity drop head principle, of course, is limited to the flat position. Stud welding in the vertical position is presently limited to studs with a maximum diameter of 3/4 in. (19 mm).

Studs can be welded to curved or angled surfaces. However, using the arc stud process which melts considerably more metal, the ceramic ferrule must be designed to fit the contour of the work surface.

The arc stud welding process is much more tolerant of work surface contaminants, such as light coatings of paint, scale, rust, or oil, than is the capacitor discharge stud welding process. The long arc duration with the arc stud welding process tends to burn the contaminants away. Also, the

molten metal expulsion tends to wash any residue out of the joint.

On the other hand, the percussive nature of the capacitor discharge arc tends to expel metallic coatings, such as those applied by electroplating and galvanizing, out of the joint. This makes the process suitable for welding small diameter studs to thin gage galvanized sheet metal. The arc stud welding process is suitable for welding through thick galvanized coatings using special welding procedures, provided the base material is thick enough to withstand the long arc time.

Arc stud welding and capacitor discharge stud welding have been widely accepted by all the metalworking industries.

SAFETY PRECAUTIONS

PERSONNEL OPERATING STUD welding equipment should be provided with face and skin protection to guard against burns from spatter produced during welds. Eye protection in the form of goggles or a face shield with a No. 3 filter lens should be worn to protect against arc radiation.

Before repairs to equipment are attempted, electrical power should be turned off and electric switch boxes locked out. Capacitors used in capacitor discharge equipment should be completely drained of electrical charge before attempting repairs.

SUPPLEMENTARY READING LIST

American Society of Metals. "Welding and brazing." *Metals Handbook,* Vol. 6, 9th Ed., 729–738. Metals Park, Ohio: American Society Metals, 1983.

American Welding Society. *Recommended practices for stud welding,* ANSI/AWS C5.4. Miami: American Welding Society, 1974.

————. *Structural Welding Code—Steel*, ANSI/AWS D1.1-90. Miami: American Welding Society, 1990.

"Automated system welds heat transfer studs." *Welding Journal* 53(1): 29–30; January 1974.

Baeslack, W. A., Fayer, G., Ream, S., and Jackson, C. E. "Quality control in arc stud welding." *Welding Journal* 54(11): 789–798; November 1975.

Hahu, O. and Schmitt, K. G. "Microcomputerized quality control of capacitor discharge stud welding." *Proceedings: 4th International JWS Symposium,* OsaKa, November 24–26, 1982, Vol. 2, 633–637. Japan, 1982.

Lockwood, L. F. "Gas shielded steel welding of magnesium." *Welding Journal* 46(4): 168s–174s; April 1967.

Masubuchi, K. et. al. "An initial study of remotely manipulated stud welding for space applications." *Welding Journal* 67(4); 25–34, April 1988.

Pease, C. C. "Capability studies of capacitor discharge stud welding on aluminum alloy." *Welding Journal* 48(6): 253s–257s; June 1969.

Pease, C. C., and Preston, F. J. "Stud welding through heavy galvanized decking." *Welding Journal* 51(4): 241–244; April 1972.

Pease, C. C., Preston, F. J., and Taranto, J. "Stud welding on 5083 aluminum and 9% nickel steel for cryogenic use." *Welding Journal* 52(4): 232–237; April 1973.

Shoup, T. E. "Stud welding." Bulletin 214. New York: Welding Research Council, April 1976.

PLASMA ARC WELDING

PREPARED BY A COMMITTEE CONSISTING OF:

S. E. Barhorst, Chairman
Hobart Brothers Co.

E. H. Daggett
Consultant

S. A. Hilton
Pratt & Whitney

J. T. Perozek
Hobart Brothers

E. Spitzer
Merrick Engineering Corp.

WELDING HANDBOOK COMMITTEE MEMBER:
J. R. Condra
E. I. DuPont de Nemours & Co.

PLASMA ARC WELDING

INTRODUCTION

PLASMA ARC WELDING (PAW) is an arc welding process that produces coalescence of metals by heating them with a constricted arc between an electrode and the workpiece (transferred arc) or between the electrode and the constricting nozzle (nontransferred arc). Shielding is generally obtained from the hot, ionized gas issuing from the torch. This plasma gas is usually supplemented by an auxiliary source of shielding gas. Shielding gas may be a single inert gas or a mixture of inert gases. Pressure is not applied, and filler metal may or may not be added.

Plasma arc welding, like gas tungsten arc welding (GTAW), uses a nonconsumable electrode. The PAW torch has a nozzle that creates a gas chamber surrounding the electrode. The arc heats the gas fed into the chamber to a temperature where it becomes ionized and conducts electricity. This ionized gas is defined as *plasma*. Plasma issues from the nozzle orifice at a temperature of about 30 000°F (16 700°C).

Plasma arc welding can be applied to join most metals in all positions. It provides better directional control of the arc and smaller heat-affected zones as compared to GTAW. The narrow, constricted arc pattern is also more tolerant to variations in torch standoff distance.

The major disadvantage of plasma arc welding is the relatively high expense of the equipment. Also, compared to GTAW, there are more process control variables, requiring more complex welding procedures and more extensive operator training.

HISTORY OF PLASMA ARCS

ONE OF THE earliest plasma arc systems was a gas vortex stabilized device introduced by Schonherr in 1909.[1] In this unit, gas was blown tangentially into a tube through which an arc was struck. The centrifugal force of the gas stabilized the arc along the axis of the tube by creating a low pressure axial core. Arcs up to several meters in length were produced, and the system proved useful for arc studies.

Gerdien and Lotz[2] built a water vortex arc-stabilizing device in 1922. In this device, water injected tangentially into the center of a tube was swirled around the inner surface and ejected at the ends. When an arc struck between carbon electrodes was passed through the tube, the water concentrated the arc along its axis, producing higher current densities and temperatures than were otherwise available. The Gerdien and Lotz invention had no practical metal-working applications because of the rapid consumption of its carbon electrodes and the presence of water vapor in the plasma jets.

While working on the arc melting of refractory metals in 1953, Gage[3] observed the similarity in appearance between a long electric arc and an ordinary gas flame. Efforts to control the heat intensity and velocity of the arc led to the development of the modern plasma arc torch.

The first practical plasma arc metal-working tool was a cutting torch introduced in 1955. This device was similar to a gas tungsten arc welding torch in that it used a tungsten electrode and a "plasma" gas. However, the electrode was recessed in the torch, and the arc was constricted by passing it through an orifice in the torch nozzle. The usual circuitry for gas tungsten arc welding was supplemented in the plasma arc cutting torch with a pilot arc circuit for arc initiation.

Commercial equipment for plasma arc surfacing emerged in 1961, and plasma arc welding was introduced in 1963.

---

1. *Encyclopedia of physics*, XXII, 300. Springer-Verlog, Berlin, 1956.
2. Loc cit.
3. Gage, R. M., U.S. Patent No. 2,806,124.

PRINCIPLES OF OPERATION

PLASMA ARC WELDING is basically an extension of the gas tungsten arc welding (GTAW) process. However, it has a much higher arc energy density and higher plasma gas velocity by virtue of the arc plasma being forced through a constricting nozzle, as shown in Figure 10.1.

The orifice gas is the gas directed through the torch to surround the electrode. It becomes ionized in the arc to form the plasma, and issues from the orifice in the torch nozzle as the plasma jet. For most operations, auxiliary shielding gas is provided through an outer gas cup, similar to gas tungsten arc welding. The purpose of the auxiliary shielding gas is to blanket the area of arc plasma impingement on the workpiece to avoid contamination of the weld pool.

The arc constricting nozzle through which the arc plasma passes has two main dimensions: orifice diameter and throat length. The orifice may be cylindrical or have a converging or diverging taper.

The distance that the electrode is recessed within the torch is the electrode setback. The dimension from the outer face of the torch nozzle to the workpiece is known as the *torch standoff distance.*

The plenum or plenum chamber is the space between the inside wall of the constricting nozzle and the electrode. The orifice gas is directed into this chamber and then through the orifice toward the work. A tangential vector may be imparted to the gas flow to form a swirl through the orifice.

The basic arrangement of both GTAW and PAW systems is shown in Figure 10.2. The electrode in the GTAW torch extends beyond the end of the shielding gas nozzle. The gas tungsten arc is not constricted and assumes an approximately conical shape, producing a relatively wide heat pattern on the workpiece. For a given welding current, the area of impingement of the cone-shaped arc on the workpiece varies with the electrode-to-work distance. Thus, a small change in arc length produces a relatively large change in heat input per unit area.

By contrast, the electrode in the plasma arc torch is recessed within the constricting nozzle. The arc is collimated and focused by the constricting nozzle on a relatively small area of the workpiece. Because the shape of the arc is essentially cylindrical, there is very little change in the area of contact on the workpiece as torch standoff varies. Thus, the PAW process is less sensitive to variations in torch-to-work distance than the GTAW process.

Since the electrode of the plasma arc torch is recessed inside the arc-constricting nozzle, it is not possible for the electrode to touch the workpiece. This feature greatly reduces the possibility of contaminating the weld with electrode metal.

As the orifice gas passes through the plenum chamber of the plasma arc torch, it is heated by the arc, expands, and exits through the constricting orifice at high velocity. Since too powerful a gas jet can cause turbulence in the weld puddle, orifice gas flow rates are generally held to within 0.5 to 10 ft^3/h (0.25 to 5 L/min). The orifice gas alone is not generally adequate to shield the weld pool from atmospheric contamination. Therefore, auxiliary shielding gas must be provided through an outer gas nozzle. Typical shielding gas flow rates are in the range of 20 to 60 ft^3/h (10 to 30 L/min).

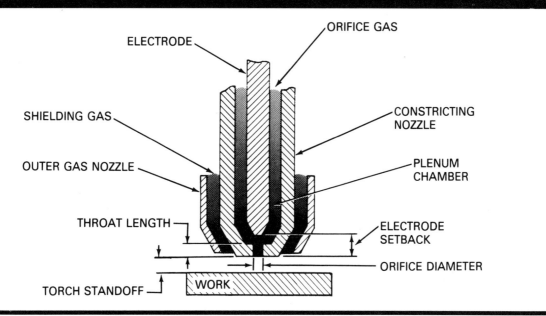

Figure 10.1–Plasma Arc Torch Terminology

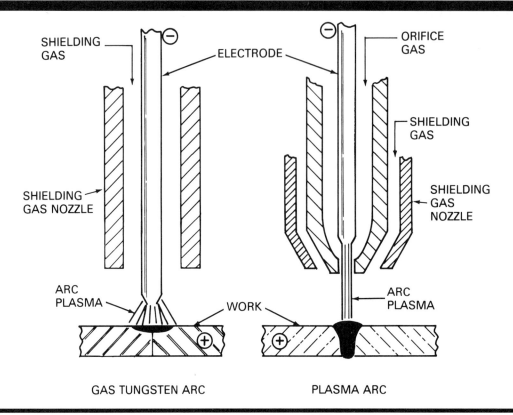

Figure 10.2–Comparison of Gas Tungsten Arc and Plasma Arc Welding Processes

Purposes of Arc Constriction

SEVERAL IMPROVEMENTS IN performance over open arc operation (GTAW) can be obtained by passing the plasma arc through a small orifice. The most noticeable improvement is the directional stability of the plasma jet. A conventional gas tungsten arc is attracted to the nearest work connection back to the power supply, and it is deflected by low-strength magnetic fields. On the other hand, a plasma jet is comparatively stiff, it tends to go in the direction in which it is pointed, and it is less affected by magnetic fields.

High current densities and high energy concentration can be produced by arc constriction. The higher current densities result in higher temperatures in the plasma arc.

The higher temperatures and electrical changes brought about by constricting an arc are compared in Figure 10.3. The left side of this figure represents a normal unconstricted tungsten arc operating at 200 A, DCEN, in argon at a flow rate of 40 ft^3/h (19 L/min). The right side illustrates an arc, with the same current and gas flow, that is constricted by passing it through a 3/16 in. (4.8 mm) diameter orifice. Under these conditions, the constricted arc shows a 100 percent increase in arc power and a 30 percent increase in temperature over the open arc. The spectroscopic methods used to measure the temperatures of arcs are based on the analysis and interpretation of emission spectra.

The increased temperature of the constricted arc is not its chief advantage, since the temperature in the gas tungsten arc far exceeds the melting points of the metals generally welded by the process. The main advantages of the plasma arc are its directional stability and focusing effect brought about by arc constriction, and its relative insensitivity to variations in torch standoff distance.

The plasma arc efficiently uses the supplied arc energy. The degree of arc collimation, arc force, energy density on the workpiece, and other characteristics are primarily functions of the following:

(1) Plasma current
(2) Orifice diameter and shape
(3) Type of orifice gas
(4) Flow rate of orifice gas
(5) Type of shield gas

The fundamental differences among the plasma arc metal working processes arise from the relationship of these five factors. They can be adjusted to provide very high or very low thermal energies. For example, the high-energy concentration and high jet velocity necessary for plasma arc cutting dictate a high arc current, a small diameter orifice, high orifice gas flow rate, and a gas having high thermal conductivity. For welding, on the other hand, a

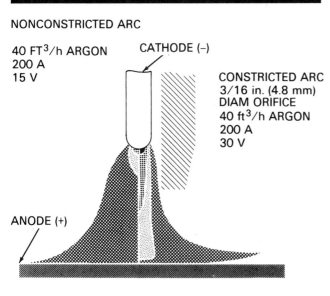

NONCONSTRICTED ARC

40 FT³/h ARGON
200 A
15 V

CATHODE (–)

CONSTRICTED ARC
3/16 in. (4.8 mm)
DIAM ORIFICE
40 ft³/h ARGON
200 A
30 V

ANODE (+)

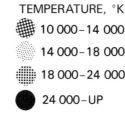

TEMPERATURE, °K

10 000 – 14 000

14 000 – 18 000

18 000 – 24 000

24 000 – UP

Figure 10.3–Effect of Arc Constriction on Temperature and Voltage

low plasma jet velocity is necessary to prevent weld metal expulsion from the workpiece. This calls for larger orifices, considerably lower gas flow rates, and lower arc currents.

The constricted arc is much more effective than an open arc for heating the gas to be used for a particular operation. When the gas passes directly through a constricted arc, it is exposed to higher energy concentrations than when it passes around a conventional gas tungsten arc, as shown in Figure 10.3.

Arc Modes

TWO ARC MODES are used in plasma arc welding: transferred arc and nontransferred arc. Figure 10.4 illustrates the two modes. With a transferred arc, the arc "transfers" from the electrode to the workpiece. The workpiece is part of the electrical circuit, and heat is obtained from the anode spot on the workpiece as well as from the plasma jet.

With a nontransferred arc, the arc is established and maintained between the electrode and the constricting orifice. The arc plasma is forced through the orifice by the plasma gas. The workpiece is not in the arc circuit. Useful heat is obtained from the plasma jet only.

Transferred arcs have the advantage of greater energy transfer to the work, and this is the mode generally used for welding. Nontransferred arcs are useful for cutting and joining nonconductive workpieces or for applications where relatively low energy concentration is desirable.

If there is insufficient orifice gas flow or excessive arc current for a given nozzle geometry, or if the nozzle is touched to the work, the nozzle may be damaged by a phenomenon known as *double arcing*. In double arcing, the metallic torch nozzle forms part of the current path from the electrode back to the power supply. In essence, two arcs are formed, as shown in Figure 10.5. The first arc is from the electrode to the nozzle, and the second is from the nozzle to the work. Heat generated at the cathode and anode spots, formed where the two arcs attach to the nozzle, invariably causes damage to this part.

Types of Welding Current

DIRECT CURRENT, ELECTRODE negative (DCEN) power is used for most plasma arc welding applications. A pure tungsten or thoriated tungsten electrode and a transferred arc are used for most applications. The current range for DCEN plasma welding is from approximately 0.1 to 500 amps. Current pulsation is often used. Steel alloys, stainless steels, nickel alloys, and titanium are commonly welded. Direct current, electrode positive (DCEP) is used to a limited extent for welding aluminum. Excessive electrode heating is the primary limitation using electrode positive. Its maximum current is usually less than 100 amps.

Sine wave alternating current with continuous high frequency stabilization can be used to weld aluminum and magnesium alloys. The current range is generally limited between 10 and 100 amps. Higher amperages generally cannot be used due to excessive electrode deterioration during the electrode positive current cycle. The primary reason for using alternating current when welding magnesium and aluminum alloys is oxide removal. During the electrode positive half-cycle of alternating current, positive ions are released from the electrode which bombard the oxides on the surface of the workpiece. This bombardment, called *cathodic etching*, removes the oxides and exposes clean aluminum alloy to be welded. Square wave ac has largely replaced sine wave ac for welding aluminum and magnesium alloys.

Square wave alternating current, with unbalanced electrode positive and negative current half-cycles (variable polarity plasma arc) is highly efficient for welding magnesium and aluminum alloys. High frequency stabilization is not required. Unbalanced square wave retains the desired property of cathodic etching for oxide removal, with most of the energy anodic for maximum heat transfer. See Chapter 1 of this volume for a discussion of sine wave and square wave alternating current waveforms. Variable polarity plasma arc (VPPA) welding is described further under the section on Equipment.

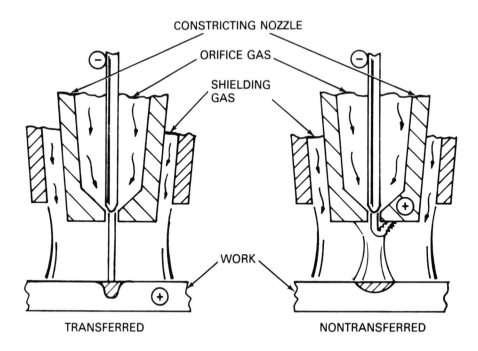

Figure 10.4–Transferred and Nontransferred Plasma Arc Modes

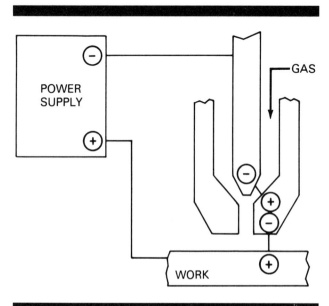

Figure 10.5–Schematic View of Double Arcing

Arc Length

THE COLUMNAR NATURE of the constricted arc makes the plasma arc process less sensitive to variations in arc length than the gas tungsten arc process. Since the unconstricted gas tungsten arc has a conical shape, the area of heat input to the workpiece varies as the square of the arc length. A small change in arc length, therefore, causes a relatively large change in the unit area heat transfer rate. With the essentially cylindrical plasma jet, however, as the arc length is varied within normal limits, the area of heat input and the intensity of the arc are virtually constant.

The collimated plasma jet also permits the use of a much longer torch-to-work distance (torch standoff) than is possible with the GTAW process, and reduces the operator skill required to manipulate the torch. Typical arc lengths used to weld thin gage material at approximately 10 amperes are shown in Figure 10.6. The plasma arc is approximately 1/4 in. (6.4 mm) long as compared to the .06 in. (1.5 mm) long gas tungsten arc.

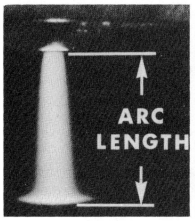 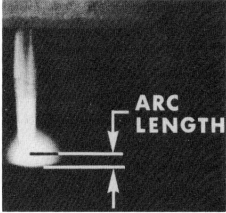

PLASMA ARC   GAS TUNGSTEN ARC

Figure 10.6–Comparison of Plasma Arc and Gas Tungsten Arc Lengths Commonly Used for Welding Very Thin Metal Sections at 10A

PROCESS TECHNIQUES

THE PLASMA ARC welding process is well accepted in the fabrication, aerospace, and nuclear industries. It offers process fabrication latitude and economy, while maintaining quality and weld joint reliability. All metals weldable with the GTAW process can be satisfactorily welded with the PAW process; therefore, few exceptions are required in the establishment of weldment acceptance specifications.

The advantage of the low-current plasma arc process, operating in the range of 0.1 to 50 amps, is the very stable and controllable arc available at low currents for welding thin materials. Through use of a pilot arc, transferred arc starting is more reliable than GTAW starting at these current levels. Because of the collimated arc, edge joints result in uniform bead contours when welded manually or automatically. Other applications include turbine blades, seal edges, bellows, pacemakers, and diaphragms. For these applications, plasma arc welding has often been the economical choice over laser welding.

High-current welding procedures, in the range of 50 to 400 amps, often use the melt-in mode, which produces a weld similar to that obtained with conventional gas tungsten arc welding. The melt-in mode is generally preferred over the gas tungsten arc process in mechanized applications for consistent control of weld quality. Again, due to arc stability and stiffness, arc penetration into the weld joint is more controlled. It can also reduce welding time. Applications include welding lamination stacks, tube mill welding, cladding, and cover passes on keyhole welding.

ADVANTAGES

THE LOW-CURRENT and high-current (melt-in) modes have the following advantages over gas tungsten arc welding:

(1) Energy concentration is greater, hence
    (a) Welding speeds are higher in some applications.
    (b) Lower current is needed to produce a given weld and results in less shrinkage. Distortion may be reduced by as much as 50 percent.
    (c) Penetration can be controlled by varying welding variables.
(2) Arc stability is improved.
(3) Arc column has greater directional stability.
(4) Narrower beads (higher depth-to-width ratio) for a given penetration, resulting in less distortion.
(5) Need for fixturing is less for some applications.
(6) Where the addition of filler metal is desirable, this operation is much easier since torch standoff distance is generous and the electrode cannot touch the filler or puddle. This also results in less downtime for tungsten repointing and eliminates tungsten contamination of the weld.
(7) Reasonable variations in torch standoff distance have little effect on bead width or heat concentration at the work; this makes out-of-position welding much easier.

LIMITATIONS

Some of the limitations associated with low-current and high-current (melt-in) plasma arc welding include:

(1) Due to the narrow constricted arc, the process has little tolerance for joint misalignment.

(2) Manual plasma welding torches are generally more difficult to manipulate than a comparable GTAW torch.

(3) For consistent weld quality, the constricting nozzle must be well maintained and regularly inspected for signs of deterioration.

KEYHOLE WELDING TECHNIQUE

IN PLASMA ARC welding of certain ranges of metal thicknesses, appropriate combinations of plasma gas flow, arc current, and weld travel speed will produce a relatively small weld pool with a hole penetrating completely through the base metal. It is called a *keyhole* and is shown in Figure 10.7. The keyhole technique is generally performed in the downhand position on material thicknesses ranging from 1/16 to 3/8 in. (1.6 to 9.5 mm). However, using appropriate welding conditions on certain metal thicknesses, keyhole welding can be done in any position. The plasma arc process is the only gas shielded welding process typically operated with this unusual characteristic.

In the keyhole operation, molten metal is displaced to the top bead surface by the plasma stream (as it penetrates the weld joint) to form the keyhole. As the plasma arc torch then is moved along the weld joint, metal melted at the front side of the keyhole by the arc flows around the plasma stream to the rear where the weld pool progressively solidifies. The principal advantage of keyhole welding is making welds in a single pass.

An open keyhole also forms an escape path through the thin molten lining for impurities to flow to the surface and gases to be expelled before solidification. The maximum weld pool volume and the resultant underbead root surface profile are largely determined by the force balance between the surface tension of the molten weld metal, the plasma arc current, and the velocity of the ionized gas exiting the orifice.

The high-current keyhole technique of welding operates just below conditions that would actually cut rather than weld. For cutting, slightly higher orifice gas velocity blows the molten metal away. In welding, the lower gas velocity lets surface tension hold the molten metal in the joint. Consequently, orifice gas flow rates for welding are critical and must be closely controlled. Flow no more than 0.25 ft³/h (0.12 L/min) is recommended. This rate is quite small.

A summary of the advantages of plasma welding with the keyhole technique over GTAW include:

(1) The plasma gas flushing through the open keyhole helps remove gases that would, under other circumstances, be trapped as porosity in molten metal.

(2) The symmetrical fusion zone of the keyhole weld reduces the tendency for transverse distortion.

(3) The greater joint penetration permits a reduction in the number of passes required for a given joint. Many welds can be completed in a single pass.

(4) Square butt joints are generally used, thereby reducing joint preparation and machining costs.

Limitations of the plasma welding keyhole technique include:

(1) Welding procedures involve more process variables and can have narrow operating windows.

(2) Increased operator skill is required for manual operation, particularly on thicker materials.

(3) Except for aluminum alloys, most keyhole technique plasma welding is restricted to the 1G position (it can be used in all positions, however)

(4) The plasma torch must be well maintained for consistent operation.

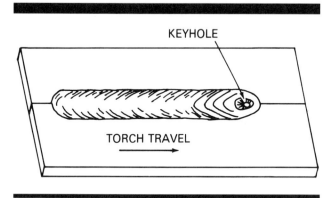

Figure 10.7–Pictorial Representation of the Keyhole in Plasma Arc Welding

EQUIPMENT

THE BASIC EQUIPMENT for plasma arc welding is shown in Figure 10.8. Plasma arc welding is done with both manual and mechanized equipment.

A complete system for manual plasma arc welding consists of a torch, control console, power source, orifice and shielding gas supplies, source of torch coolant, and acces-

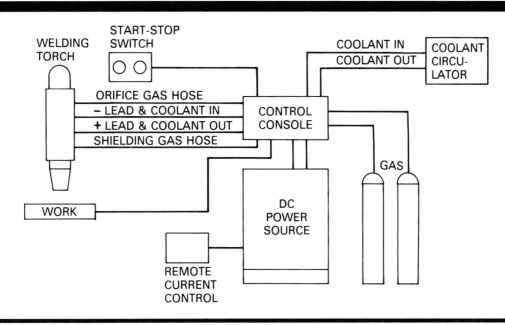

Figure 10.8–Typical Equipment for Plasma Arc Welding

sories such as an on-off switch, gas flow timers, and remote current control. Equipment is presently available for operation in the current range of 0.1 to 225 A, DCEN.

Mechanized equipment must be used to achieve the high welding speeds and deep penetration advantages associated with high-current plasma arc welding. A typical mechanized installation consists of a power source, control unit, machine welding torch, torch stand or travel carriage, coolant source, high frequency power generator, and supplies of shielding gases. Accessory units such as an arc voltage control and filler wire feed system may be used as required. Machine welding torches are available for welding with currents up to 500 A, DCEN.

ARC INITIATION

THE PLASMA ARC cannot be started with the normal techniques used with gas tungsten arc welding. Since the electrode is recessed in the constricting nozzle, it cannot be touch-started against the workpiece. It is first necessary to ignite a low-current pilot arc between the electrode and the constricting nozzle. Pilot arc power is normally provided either by a separate power supply within the control console or from the welding power supply itself. The pilot arc is generally initiated through use of high-frequency ac power or by a high-voltage dc pulse superimposed on the welding circuit. These methods help to break down the arc gap and ionize the orifice gas so that it will conduct the pilot arc current.

The basic circuitry for a plasma arc welding system with a high-frequency generator is shown in Figure 10.9. The constricting nozzle is connected to the power source posi-

tive terminal through a current limiting resistor. A low-current pilot arc is initiated between the electrode and the nozzle by the high-frequency generator. The electrical circuit is completed through the resistor. The ionized gas from the pilot arc forms a low resistance path between the electrode and work. When the power source is energized, the main arc is initiated between the electrode and the work. The pilot arc is used only to assist in starting the main arc. After the main arc starts, the pilot arc may be extinguished.

POWER SUPPLY—NONPULSED

STEADY-CURRENT DC power sources for plasma arc welding come in various amperage capacities from 0.1 amp up to several hundred amperes. Conventional type power sources with a drooping volt-amp characteristic are used for DCEN plasma arc welding. These are typically the same types of power sources that are used for GTAW and are available in amperage ranges from 0.1 amp up to 400 amps with 60 percent through 100 percent duty cycles. Rectifier type units are preferred over motor generator units because of electrical output characteristics.

A rectifier with an open circuit voltage in the range of 65 to 80 volts is satisfactory for plasma arc welding with argon or with argon-hydrogen gas mixtures containing up to 7 percent hydrogen. However, if helium or an argon-hydrogen gas mixture containing more than 7 percent hydrogen is used, additional open circuit voltage is required for reliable arc ignition. This may be obtained by connecting two power sources in series. If erratic arc ignition is experienced, another approach is to strike the arc in pure argon

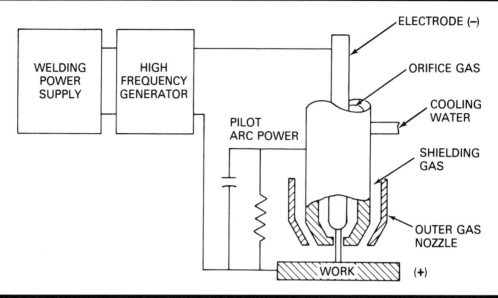

Figure 10.9—Plasma Arc Welding System With High Frequency Pilot Arc Initiation

and then switch over to the desired argon-hydrogen mixture or to helium for the welding operation. Constant-current power sources are available with several options such as a programmed upslope of current, a programmed taper or decay of weld current, and a programmed downslope of weld current. These special features of the power source are used for various applications, primarily automatic welding.

POWER SUPPLY—PULSED CURRENT

FOR SOME PLASMA arc welding applications, it is essential that pulsed current be used. Pulsed-current power sources similar to those used with gas tungsten arc welding are used for plasma arc welding. A pulsed-current power source is a conventional drooping volt-amp characteristic power source which has the capability of pulsing to a high level referred to as a *peak current*. Pulsed-current power sources used for plasma arc welding have variable pulse frequencies and pulse width ratios. See Figure 10.10 for definition of terms for pulsed current.

Transistorized, inverter, and SCR type power sources are available with pulsed-current capabilities built into them. For conventional steady-current power sources, add-on packages are available which will provide pulsed current within a limited range of pulse frequencies. Upslope and downslope of weld current can be obtained on pulsed-current power sources. For a more detailed discus-

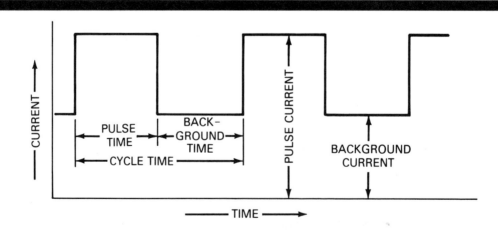

Figure 10.10—Pulsed Current Terminology

sion of the technical aspects of this type equipment, see Chapter 1.

VARIABLE POLARITY PLASMA ARC WELDING

KEYHOLE PLASMA ARC welding of aluminum has been done using square wave ac with variable polarity (VPPA). The variable polarity waveform is shown in Figure 10.11. This type of waveform, where the duration and magnitude of the DCEN and DCEP current excursions can be controlled separately, is obtainable using solid-state technology. A description of power supplies using silicon controlled rectifiers and transistors is provided in Chapter 1.

Cleanliness of the aluminum workpiece surface is of the utmost importance in avoiding porosity in the weld. Usual cleaning procedures involve mild alkaline solutions or vapor degreasing. Welding should occur soon after cleaning. With the variable polarity process, oxide removal before welding is not required for most aluminum alloys. However, the 5000 series alloys with their extremely tenacious surface oxides require removal of the surface oxides by scraping prior to welding.

Although a dc pilot arc is maintained during VPPA to stabilize the welding arc, continuous high frequency is not used.

Duration of Current Excursions

THE MOST IMPORTANT variable in keyhole plasma arc welding of aluminum was found to be DCEN and DCEP duration times. The proper ratio of DCEN and DCEP times was determined empirically. Best results were obtained with DCEN current flowing from 15 to 20 milliseconds and DCEP current flowing from 2 to 5 milliseconds. Refer to Figure 10.11. With DCEP durations shorter than 2 milliseconds, the weld was porous. When the DCEP time

exceeded 6 milliseconds, tungsten deterioration and double-arcing tendencies became apparent.

Note that the DCEP current amplitude shown in Figure 10.11 is greater than the DCEN current amplitude. The added DCEP current provides an additional spike of cleaning action to break up surface oxides on the workpiece, yet produces minimal heat input on the electrode and torch orifice. Appropriate cleaning of the weld face and root face can be accomplished by increasing the DCEP current an additional 30 to 80 A..

Applications

TYPICAL KEYHOLE VPPA welding conditions for 1/4-in. thick aluminum in the flat, horizontal and overhead positions are shown in Table 10.1.

The process has been used to make single pass, full penetration welds on space shuttle fuel tanks, hulls for hydrofoil boats, and aluminum tanks.

PLASMA CONTROL CONSOLE

THE PLASMA CONSOLE is a device that controls the main plasma arc functions. A typical plasma console will contain controls for setting plasma gas flow, shielding gas flow, provide a junction box for gas and water hoses, may or may not provide a high frequency circuit for the starting of the pilot arc, and may have a small power source that will provide the current to the pilot arc. Other features that might be provided are a high-low option so that plasma gas flow rate can be easily switched between melt-in mode and keyhole mode, and an arc pressure gage that would measure the plasma gas back-pressure at the orifice.

The plasma console might have built-in capability for programmed upslope and downslope of the plasma gas to start and close a keyhole, and the plasma console might also have the water circulator built into the same package. Figure 10.12 illustrates a plasma control console.

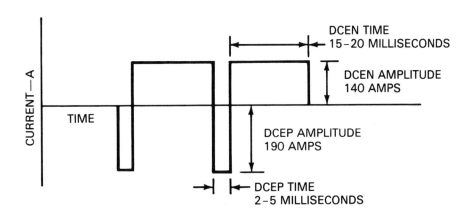

Figure 10.11–Typical Variable Polarity Current Waveform

Table 10.1
VPPA Welding Conditions for 0.250 in. Thick Aluminum in the Flat,
Horizontal and Overhead Positions

Position	Flat		Horizontal		Overhead	
Material thickness, in. (mm)	1/4	(6.4)	1/4	(6.4)	1/4	(6.4)
Type Aluminum	2219		3003		1100	
Filler Metal Diameter, in. (mm)	1/16	(1.6)	1/16	(1.6)	1/16	(1.6)
Filler Metal Grade	2319		4043		4043	
DCEN welding current, A	140		140		170	
DCEN welding time, ms	19		19		19	
Additional DCEP current, A	50		60		80	
DCEP current time, ms	3		4		4	
Plasma gas cfh (L/min) (start)	Ar 2	(0.9)	Ar 2.5	(1.2)	Ar 2.5	(1.2)
Plasma gas cfh (L/min) (run)	Ar 5	(2.4)	Ar 4.5	(2.1)	Ar 5	(2.4)
Shielding gas flow cfh (L/min)	Ar 30	(14)	Ar 40	(19)	Ar 45	(21)
Electrode size, in. (mm)	1/8	(3.2)	1/8	(3.2)	1/8	(3.2)
Travel speed, ipm (mm/s)	8	(3.4)	8	(3.4)	7.5	(3.2)

WELDING TORCHES

PLASMA ARC WELDING torches are more complex than those used for gas tungsten arc welding. A series of passages are necessary to supply the torch with orifice gas, shielding gas, and liquid coolant to chill the constricting nozzle assembly.

In most instances, two dual-function cables provide both electrical energy and circulating coolant. One cable supplies current for the pilot-arc, while the other supplies the welding current. Two additional hoses provide the orifice (plasma) and shielding gases. Cooling water is necessary to dissipate the heat generated in the constricting nozzle by the pilot arc and main welding arc. The electrode holder assembly for a plasma arc welding torch is made of copper, and most are designed to center the electrode automatically within the central section of the torch nozzle. Any misalignment of the electrode in this section could cause melting of the copper nozzle near the orifice, possible weld contamination, and undercutting.

The orifice gas supplied to the torch has a low flow rate and thus does not provide sufficient gas to protect the molten weld pool from atmospheric contamination. In addition, turbulence caused by the high velocity of the plasma stream while keyhole welding further reduces effectiveness of plasma gas coverage. Necessary shielding gas is supplied through the shielding gas nozzle that surrounds the orifice tip portion of the torch. In some applications, additional shielding gas trailers are required for further protection.

Manual Torches

A CROSS SECTION of a typical torch design for manual plasma arc welding is shown in Figure 10.13. The torch generally is lightweight and has a handle, some device for securing the tungsten electrode in position and conducting current to it, separate passages for orifice and shielding gases, a water-cooled constricting copper nozzle with inlet and outlet passages, and a shielding gas nozzle (usually made of a ceramic material).

Manual plasma arc welding torches are available with 70 degree and 90 degree head angles. They are available for operation on DCEN at currents up to 225 A, and also for operation on DCEP up to about 70 A. DCEP is used to a limited extent with tungsten or water-cooled copper electrodes for welding aluminum.

Controls for welding current and gas are usually separate from the torch and are operated either by foot control or automatically. Holders to mount the torch for mechanized application are available.

Machine Torches

TORCHES FOR AUTOMATED plasma arc welding are similar to manual torches, except they are designed with straight-line or offset configurations.

Mechanized plasma arc welding torches are available for operation on either DCEN, DCEP, or square wave ac, with current ratings generally ranging from 50-500 amps. DCEN is used with a tungsten electrode for most welding applications, with an optional mode of pulsed current welding. In the pulsing mode, the current is allowed to fluctuate between two set currents, a higher and a lower level of amperage, which allows the molten weld puddle to solidify at the lower level. The pulsing mode can help alleviate problems of distortion by reducing the total heat input along the weld joint. Also available is variable polarity (square wave ac) plasma arc welding for welding aluminum.

Arc Constricting Nozzles

NOZZLES IN WIDE variety have been designed for use in PAW. These include single orifice and multiple orifice noz-

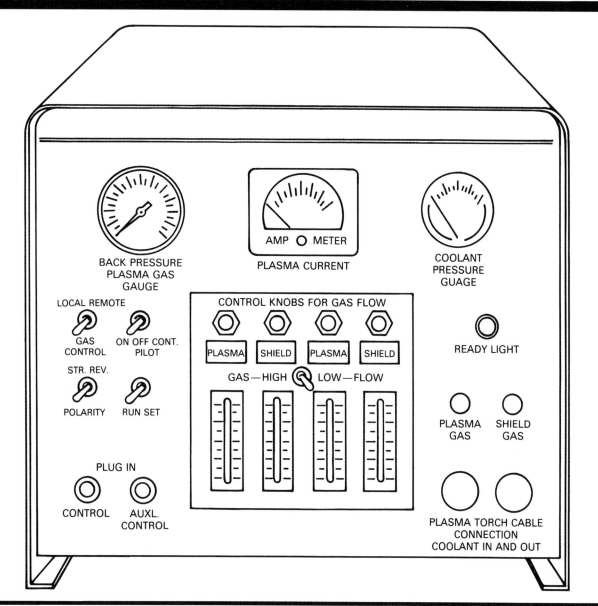

Figure 10.12—Plasma Welding Control Console

zles, with holes arranged in circles, rows, and other geometric patterns. Single orifice nozzles are the most widely used. Among the multiple port nozzles, the design most widely used has the center orifice bracketed by two smaller auxiliary gas orifices, all with a common centerline. These two types are shown in Figure 10.14.

With a single orifice nozzle, the arc and all of the plasma gas pass through the single orifice. With the multiport nozzle, the arc and some of the orifice gas pass through the larger center orifice, while the remainder of the gas moves through the smaller auxiliary orifices.

Multiport nozzles can be used to an advantage on several types of weld joints. When the multiport nozzle is aligned to place the common centerline of all three ports perpendicular to the weld groove, the plasma jet is concentrated on the joint by these adjacent gas streams. The results are a narrowed weld bead and higher welding speeds.

Each given orifice size and orifice gas flow rate has a maximum current rating. For example, an orifice with 0.081 in. (2.1 mm) diameter might be rated at 75 A with 1.9 ft.^{3}/hr. (0.9 L/min.) argon flow rate. If the flow rate of the orifice gas dropped below 1.9 ft.^{3}/hr. (0.9 L/min.), the maximum current rating of the orifice would also decrease.

The electrode in the plasma arc torch is recessed in the arc-constricting nozzle. As the arc passes through the nozzle, it is collimated so that the arc heat is concentrated

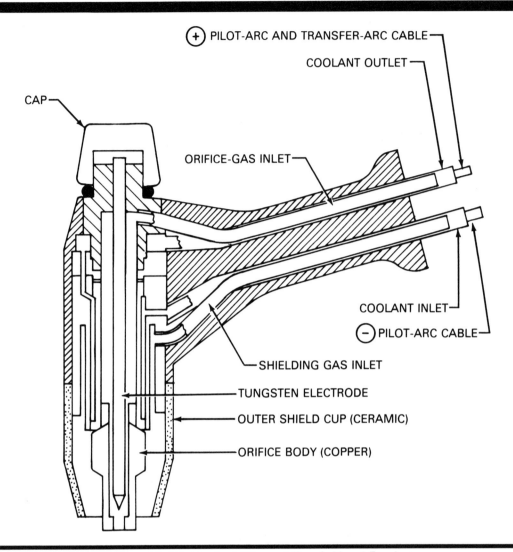

Figure 10.13–Typical Manual Torch Head Design

onto a relatively small area of the workpiece. The increased heat concentration, along with the characteristically more forceful plasma stream, produces a narrower weld fusion zone while increasing weld penetration and preventing arc wander.

During normal operation, the arc column within the torch nozzle is surrounded by a layer of nonionized gas. This layer of relatively cool nonconductive gas at the nozzle wall provides thermal and electrical insulation that protects the inside surface of the nozzle. The most commonly used nozzle material is copper. When water-cooled, copper can be used to constrict an arc plasma with temperature in excess of 30 000°F (16 600°C). If the protective layer of gas is disturbed, as when there is insufficient orifice gas flow, or if there is excessive arc current

for a particular nozzle geometry, double arcing may occur and damage the nozzle. This phenomenon was described earlier.

ACCESSORY EQUIPMENT

Wire Feeders

AS WITH THE GTAW process, conventional filler wire feed systems can be used with the PAW process. The filler metal is added to the leading edge of the weld pool or the key-hole at a predetermined feed rate. A wire feed system may alleviate occurrences of undercut or underfill when welding thicker materials.

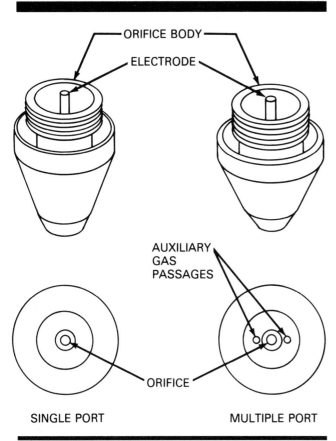

**Figure 10.14—Single and Multiple Port
Constricting Nozzles**

Hot wire feed systems may also be used and should be fed into the trailing edge of the weld pool. Initiation and termination of wire feed may be controlled and programmed with automatic welding equipment.

A popular technique when pulse welding is to pulse filler material into a weld joint in synchronism with the pulsing of the plasma arc current. Variations of this technique are used in many edge build-up applications in the automatic modes of welding.

Arc Voltage Control

SINCE THE PLASMA arc process is relatively insensitive to arc length variations, arc voltage control equipment is not necessary for many applications. Arc voltage control can nevertheless be used when plasma arc welding uneven or contoured joint geometries. The control must be de-activated or "locked-out" when the current or plasma gas flow rate is sloped during weld starts or crater filling because changing these variables will also cause a change in the arc voltage.

Positioning Equipment

POSITIONING EQUIPMENT FOR PAW is similar to that used for GTAW. Depending on the application, either the workpiece may be manipulated or the torch motion can be controlled. Workpiece manipulation generally involves a rotary positioner with the capability of tilt control. Moving the torch while the workpiece remains stationary requires a carriage on tracks or a side beam carriage for following linear joints. Combining the movement of the torch and workpiece as a system would require the use of computer programming for coordinating the operations.

MATERIALS

BASE METALS

THE PLASMA ARC welding process can be used to join all metals weldable by the GTAW process. Most material thicknesses from 0.01 to 0.25 in. (0.3 to 6.4 mm) can be welded in one pass with a transferred arc. All metals except aluminum and magnesium and their alloys are welded with DCEN. Square wave ac is used to effectively remove refractory oxides when welding aluminum and magnesium. AC welding reduces the current capacity of the electrode unless the power source is capable of minimizing the duration of the positive electrode cycle. One pass keyhole welds can be made in aluminum alloys up to 1/2 in. (12.7 mm) thick.

Metallurgical effects of the heat from the plasma and gas tungsten arc welding processes are similar, except the smaller diameter plasma arc will usually melt less base metal, resulting in narrower and deeper penetration. Preheat, postheat, and gas shielding procedures are similar for both processes. Each base material has its requirements that maximize weld quality.

CONSUMABLES

Filler Metals

FILLER METALS USED to weld the work base materials are the same as those used with the GTAW and GMAW processes. They are added in rod form for manual welding or wire form for mechanized welding. Table 10.2 lists the AWS specifications for appropriate filler metals.

Table 10.2
AWS Specifications for Filler Metals Used for Plasma Arc Welding

AWS Specification	Filler Metals
A5.7	Copper and copper alloy welding rods
A5.9	Corrosion resistant chromium and chromium nickel steel bare electrodes
A5.10	Aluminum and aluminum alloy welding rods and bare electrodes
A5.14	Nickel and nickel alloy bare welding rods and electrodes
A5.16	Titanium and titanium alloy bare welding rods and electrodes
A5.18	Mild steel electrodes for gas metal arc welding
A5.19	Magnesium alloy welding rods and bare electrodes
A5.24	Zirconium and zirconium alloy bare welding rods and electrodes

Electrodes

THE ELECTRODE IS the same as used for gas tungsten arc welding. Pure tungsten rods and tungsten with small additions of thoria, zirconia, or ceria may be used for DCEN welding. Electrodes are made to ANSI/AWS A5.12, *Specification for Tungsten Arc Welding Electrodes*. Pure tungsten electrodes are generally selected for ac welding.

Direct current electrode positive welding is not widely practiced with the plasma arc welding process because of its intense electrode heating and reduced current capacity. The arc end of the electrode is ground to a cone with an angle of 20 to 60 degrees, as specified by the torch maker.

It is essential to have a smooth concentric shape. The electrode collet must hold the electrode in the exact center of the nozzle orifice. A gage is usually specified to set the electrode axial position. If these precautions are not observed, the consistency of the welds will suffer, and excessive deterioration of the constricting nozzle will occur.

For square wave ac welding, the electrode is usually prepared with a balled or flat end. The recommended shapes help prevent electrode overheating and provide greater current carrying capability.

GASES

THE CHOICE OF gases to be used for plasma arc welding depends on the metal to be welded. For many PAW applications, the shielding gas is often the same as the orifice gas. Typical gases used to weld various metals are shown in Table 10.3.

The orifice gas must be inert with respect to the tungsten electrode to avoid rapid deterioration of the electrode. The shielding gases are generally inert. Active shielding gas can be used if it does not adversely affect the weld properties.

Argon is the preferred orifice gas for low current plasma arc welding because its low ionization potential assures a dependable pilot arc and reliable arc starting. Since the pilot arc is used only to maintain ionization in the plenum chamber, pilot arc current is not critical; it can remain fixed for a wide variety of operating conditions. The recommended orifice gas-flow rates are usually less than $1 \text{ ft}^3/\text{h}$ (.5 L/min.), and the pilot arc current may be fixed at five amperes.

Table 10.3
Gas Selection Guide for High Current Plasma Arc Welding[a]

Metal		Thickness		Welding Technique	
		in.	mm	Keyhole	Melt-In
Carbon steel	under	1/8	3.2	Ar	Ar
(aluminum killed)	over	1/8	3.2	Ar	75% He-25% Ar
Low alloy steel	under	1/8	3.2	Ar	Ar
	over	1/8	3.2	Ar	75% He-25% Ar
Stainless steel	under	1/8	3.2	Ar, 92.5% Ar-7.5% H_2	Ar
	over	1/8	3.2	Ar, 95% Ar-5% H_2	75% He-25% Ar
Copper	under	3/32	2.8	Ar	75% He-25% Ar
	over	3/32	2.8	Not recommended[b]	He
Nickel alloys	under	1/8	3.2	Ar, 92.5% Ar-7.5% H_2	Ar
	over	1/8	3.2	Ar, 95% Ar-5% H_2	75% He-25% Ar
Reactive metals	under	1/4	1/4	Ar	Ar
	over	1/4	6.4	Ar-He (50 to 75% He)	75% He-25% Ar

a. Gas selections are for both orifice and shielding gases.

b. The underbead will not form correctly. The technique can be used for copper-zinc alloys only.

Typical shielding gases for low current welding are given in Table 10.4. Argon is used for welding carbon steel, high-strength steels, and reactive metals such as titanium, tantalum, and zirconium alloys.

Argon-hydrogen mixtures are often used as the orifice and shielding gases for making keyhole welds in stainless steel, nickel base alloys, and copper-nickel alloys. Permissible hydrogen percentages vary, from the 5 percent used on 1/4 in. (6.4 mm) thick stainless steel to the 15 percent used for the highest welding speeds on 0.150 in. (3.8 mm) and thinner wall stainless tubing in tube mills. In general, the thinner the workpiece, the higher the permissible percentage of hydrogen in the gas mixture, up to 15 percent maximum. However, when argon-hydrogen mixtures are used as an orifice gas, the rating of the orifice diameter for a given welding current is usually reduced because of the higher arc temperature.

Additions of hydrogen to argon produce a hotter arc and more efficient heat transfer to the workpiece. In this way, higher welding speeds are obtained with a given arc current. The amount of hydrogen that can be used in the mixture is limited because excessive hydrogen additions tend to cause porosity or cracking in the weld bead. With the plasma keyhole technique, a given metal thickness can be welded with higher percentages of hydrogen than are possible with the gas tungsten arc welding process. The ability to use higher percentages of hydrogen without inducing porosity may be associated with the keyhole effect and the different solidification pattern it produces.

Helium additions to argon produce a hotter arc for a given arc current. The mixture must contain at least 40 percent helium before a significant change in heat can be detected; mixtures containing over 75 percent helium behave about the same as pure helium. Argon-helium mixtures containing between 50 and 75 percent helium are generally used for making keyhole welds in heavier titanium and aluminum sections, and for filler passes on all metals when the additional heat and wider heat pattern are desirable.

The shielding gas provided through the gas shielding nozzle and around the nozzle can be argon, an argon-hydrogen mixture, or an argon-helium mixture, depending on the welding application. Shielding gas flow rates are usually in the range of 20 to 30 ft³/h (10 to 15 L/min.) for low current applications; for high current welding, flow rates of 30 to 60 ft³/h (15 to 30 L/min.) are used.

Use of helium as an orifice gas increases the heat load on the torch nozzle and reduces its service life and current capacity. Because of the lower mass of helium, it is difficult, at reasonable flow rates, to obtain a keyhole condition with this gas. Therefore, helium is used only for making melt-in welds.

Since the shielding gas does not contact the tungsten electrode, reactive gases such as CO_2 can sometimes be used. Flow rates for CO_2 are in the range of 20 to 30 ft³/h (10 to 15 L/min.). Plasma welding of lamination stacks often uses 75% Ar - 25% CO_2 as the shielding gas.

When the gas flow and current must be varied during the weld, or at the start and end of a keyhole weld, an electronic programmable gas control system is used.

Back Purge and Trailing Gases

WHEN WELDING REACTIVE metals such as titanium, zirconium, or tantalum, it is essential to shield the hot metals from atmospheric contamination until they have cooled below the reaction point. Auxiliary shielding provided by backup and trailing shields are necessary. A trailing shield device can be attached to the rear of a plasma torch. The

Table 10.4
Shielding Gas Selection Guide for Low Current Plasma Arc Welding*

Metal		Thickness		Welding Technique	
		in.	mm	Keyhole	Melt-In
Aluminum	under	1/16	1.6	Not recommended	Ar, He
	over	1/16	1.6	He	He
Carbon steel	under	1/16	1.6	Not recommended	Ar, 25% He-75% Ar
(aluminum killed)	over	1/16	1.6	Ar, 75% He-25% Ar	Ar, 75% He-25% Ar
Low alloy steel	under	1/16	1.6	Not recommended	Ar, He, Ar-H₂ (1-5% H₂)
	over	1/16	1.6	75% He-25% Ar, Ar-H₂ (1-5% H₂)	Ar, He, Ar-H₂ (1-5% H₂)
Stainless steel		All		75% He-25% Ar, Ar-H₂ (1-5% H₂)	Ar, He, Ar-H₂ (1-5% H₂)
Copper	under	1/16	1.6	Not recommended	25% He-75% Ar, 75% He-25% Ar, He
	over	1/16	1.6	75% He-25% Ar, He	He
Nickel alloys		All		Ar, 75% He-25% Ar, Ar-H₂ (1-5% H₂)	Ar, He, Ar-H₂ (1-5% H₂)
Reactive metals	under	1/16	1.6	Ar, 75% He-25% Ar, He	Ar
	over	1/16	1.6	Ar, 75% He-25% Ar, He	Ar, 75% He-25% Ar

* Gas selections are for shielding gas only. Argon is the orifice gas in all cases.

trailing shield provides increased gas coverage and time for cooling.

PAW of highly reactive materials can be done in a welding chamber or a glove box similar to GTAW. For such metals as stainless steel where the root of the weld is exposed to the atmosphere, a back purge of a nonreactive gas such as argon, helium, or nitrogen is used. See Figure 10.15.

FIT-UP AND FIXTURING

FIT-UP AND FIXTURING for PAW should be the same as used for GTAW. Joint edges for butt fusion welds should be in intimate contact and clamp down pressure should be applied to the weld joint during welding. Burrs, notches, and gaps along the weld joint can cause separation of the weld joint because of melt back and should be eliminated.

Fit-up, fixturing, and cleanliness become critical when welding very thin materials such as 0.004 in. stainless steel. Joint fit-up must be precise and consideration should be given to chill bars to remove heat from the joint.

Figure 10.16 shows a graph that gives a general idea of fixturing and clamping for welding of 304 stainless steel.

Weld Backing

WELD BACKING IS used to locate parts, take heat away from the weld, and provide gas backup to shield the underside of the weld area.

Chill clamping and gas backup also help reduce the heat-affected zone of the weld. The type of material being welded will largely determine if gas backup is needed. Metals such as titanium, zirconium, and niobium need inert gas backup. Some weld joints may not allow gas backup because of design constraints. Gas backup inside of parts can also help cool and protect items such as electronic components that might be inside a part during welding.

Torch Positions

PAW CAN BE done in all positions similar to GTAW welding. Users must remember that the plasma arc is very directional, and when welding it is more difficult to keep the arc exactly on the weld joint. The arc does not bend to follow the centerline of the joint. Side to side variations of the torch must be extremely limited.

Fortunately the plasma arc is forgiving of variations of standoff distance. Thus, torch to work changes are not so critical as in GTAW welding.

Filler Metal

FILLER METAL CAN be used with plasma welding in the same manner as with GTAW welding.

WELDING PROCEDURES

WELDING CONDITIONS FOR a variety of materials are shown in Tables 10.5 through 10.8.

Manual Welding

USUALLY, MANUAL WELDING with plasma is best done in the low current range from 0.1 amp to 50.0 amps. When manually welding above 50.0 amps, the GTAW process is often easier for the operator to manipulate and more economical. Manual PAW is often used for welding wire mesh when each wire has to be joined or edge fusion welded. Because standoff distance in PAW is not so critical as in GTAW, PAW is superior for this application. Usually the standoff distance can be raised or lowered as much as 0.035 in. (.9 mm) without affecting weld quality.

Mechanized Melt-In Welding

MECHANIZED MELT-IN PAW is very popular, especially for welding small intricate components such as medical equipment, lighting components, instrumentation, batteries, wires, and bellows.

In many applications, plasma is used with microprocessor controls which control initial current, upslope, pulsing, downslope, and final current.

Because high frequency is used only to start the pilot arc, there is no high frequency burst when the arc is transferred for welding. This allows plasma to be used around electronic test equipment, robots, microprocessor controls, and programmable controls without the need to isolate or shield components that would need protection in automated systems.

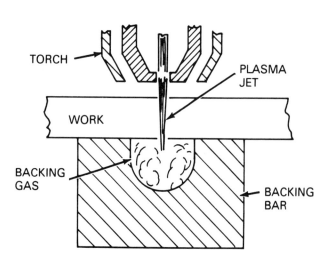

Figure 10.15–Typical Backing Bar for Keyhole Plasma Arc Welding

Table 10.5
Typical Plasma Arc Welding Conditions for Butt Joints in Stainless Steel

Thickness		Travel Speed		Current (DCSP) A	Arc Voltage V	Nozzle Type[a]	Gas Flow[b]				Remarks[d]
in.	mm	in./min	mm/s				Orifice[c]		Shield[c]		
							ft³/H	L/min	ft³/H	L/min	
0.092	2.4	24	10	115	30	111M	6	3	35	17	Keyhole, square-groove weld
0.125	3.2	30	13	145	32	111M	10	5	35	17	Keyhole, square-groove weld
0.187	4.8	16	7	165	36	136M	13	6	45	21	Keyhole, square-groove weld
0.250	6.4	14	6	240	38	136M	18	8	50	24	Keyhole, square-groove weld

a. Nozzle type: number designates orifice diameter in thousandths of an inch; "M" designates design.

b. Gas underbead shielding is required for all welds.

c. Gas used: 95% Ar-5% H.

d. Torch standoff: 3/16 in. (4.8 mm.)

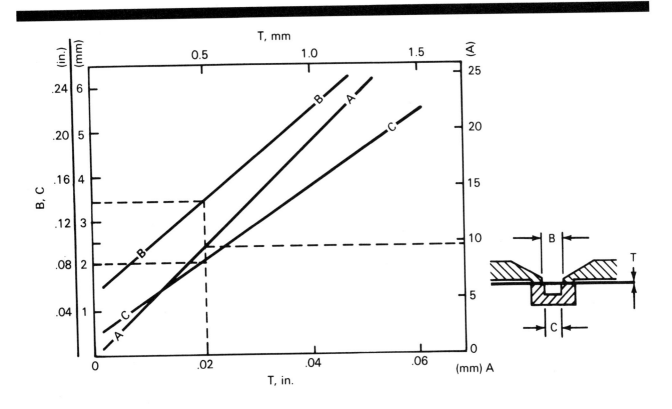

T	=	MATERIAL THICKNESS	0.02 in.	0.5 mm
B	=	UPPER CLAMP SPACING	0.14 in.	3.5 mm
C	=	BACK-UP BAR SPACING	0.08 in.	2.0 mm
A	=	WELD CURRENT	9 AMPS	9 AMPS

Figure 10.16–Graph Showing Welding Data for Low Current Butt Welding of Stainless Steel (On the Right, the Jig Geometry is Schematically Represented)

Table 10.6
Typical Plasma Arc Welding Conditions for Butt Joints in Carbon and Low Alloy Steels

Metal	Thickness in.	Thickness mm	Travel Speed in./min	Travel Speed mm/s	Current (DCSP) A	Arc Voltage V	Nozzle Type[a]	Gas Flow[b] Orifice[c] ft³/H	Gas Flow[b] Orifice[c] L/min	Shield[c] ft³/H	Shield[c] L/min	Remarks[d]
Mild Steel	0.125	3.2	12	5	185	28	111M	13	6	60	28	Keyhole, square-groove weld
4130 steel	0.170	4.3	10	4	200	29	136M	12	6	60	28	Keyhole, square-groove weld, 1.2mm (3/64 in.) diam filler wire added at 30 in./min (13mm/s)
D6AC steel	0.250	6.4	14	6	275	33	136M	15	7	60	28	Keyhole, square-groove weld, 600°F (315°C) preheat

a. Nozzle type: number designates orifice diameter in thousandths of an inch; "M" designates design.

b. Gas underbead shielding is required for all welds.

c. Gas used: argon.

d. Torch standoff: 3/64 in. (1.2 mm) for all welds.

Mechanized Keyhole Welding

IT IS RECOMMENDED that plasma keyhole welding be done in an automated mode. Keyhole welding requires accurate control of travel speed, plasma gas flow, and wire feed speed. The development of good mass flow controllers has led to more accurate control of the plasma gas during welding.

WELD QUALITY CONTROL

PLASMA ARC WELD discontinuities include both surface and subsurface types as listed in Table 10.9.

Surface discontinuities such as reinforcement, underfill, undercut, and mismatch, which are associated with weld bead contour and joint alignment, are easily detected visually or dimensionally. Lack of penetration is also detected visually through the absence of a root bead. Weld cracks which are open to the surface are usually detected using dye penetrants. Finally, surface contamination, which results from insufficient shielding gas coverage, is usually revealed by severe discoloration of the weld bead or adjacent heat-affected zones.

Subsurface weld discontinuities are generally more prevalent in manual PAW than in mechanized PAW. In either case, subsurface discontinuities are detected primarily using radiography or ultrasonic testing.

Table 10.7
Typical Plasma Arc Welding Conditions for Butt Joints in Titanium

Thickness in.	Thickness mm	Travel Speed in./min	Travel Speed mm/s	Current (DCSP) A	Arc Voltage V	Nozzle Type[a]	Gas Flow[b] Orifice ft³/H	Gas Flow[b] Orifice L/min	Shield ft³/H	Shield L/min	Remarks[c]
0.125	3.2	20	8.5	185	21	111M	8[d]	3.8	60[d]	28	Keyhole, square-groove weld
0.187	4.8	13	5.5	175	25	136M	18[d]	9	60[d]	28	Keyhole, square-groove weld
0.390	9.9	10	4.2	225	38	136M	32[e]	15	60[e]	28	Keyhole, square-groove weld
0.500	12.7	10	4.2	270	36	136M	27[f]	13	60[f]	28	Keyhole, square-groove weld

a. Nozzle type: number designates orifice diameter in thousandths of an inch; "M" designates design.

b. Gas underbead shielding is required for all welds.

c. Torch standoff: 3/16 in. (4.8 mm).

d. Gas used: argon.

e. Gas used: 75% He - 25% Ar.

f. Gas used: 50% He - 50% Ar.

Table 10.8
Typical Plasma Arc Welding Conditions for Welding Stainless Steels—Low Amperage

| Thickness | | | Travel Speed | | Current (DCSP) | Orifice Diam. | | Gas Flow Orifice[a,b,c] | | Torch Standoff | | Electrode Diam. | | |
in.	mm	Type of Weld	in./min	mm/s	A	in.	mm	ft³/h	L/min	in.	mm	in.	mm	Remarks
0.030	0.76	Square-groove weld, butt joint	5.0	2	11	0.030	0.76	0.6	0.3	1/4	6.4	0.040	1.0	Mechanized
0.060	1.5	Square-groove weld, butt joint	5.5	2	28	0.047	1.2	0.8	0.4	1/4	6.4	0.060	1.5	Mechanized
0.030	0.76	Fillet weld, tee joint	--	--	8	0.030	0.76	0.6	0.3	1/4	6.4	0.040	1.0	Manual, filler metal[d]
0.060	1.5	Fillet weld, tee joint	--	--	22	0.047	1.2	0.8	0.4	1/4	6.4	0.060	1.5	Manual, filler metal[d]
0.030	0.76	Fillet weld, lap joint	--	--	9	0.030	0.76	0.3	0.6	3/8	9.5	0.040	1.0	Manual, filler metal[d]
0.060	1.5	Fillet weld, lap joint	--	--	22	0.047	1.2	0.8	0.4	3/8	9.5	0.060	1.5	Manual, filler metal[e]

a. Orifice gas: argon.

b. Shielding gas: 95% Ar-5% H at 20 ft 3/h (10 L/min).

c. Gas underbead shielding: argon at 10 ft 3/h (5 L/min).

d. Filler wire: 0.045 in. (1.1 mm) diameter 310 stainless steel.

e. Filler wire: 0.055 in. (1.4 mm) diameter 310 stainless steel.

Porosity is the most common subsurface discontinuity encountered.

Tunneling is a severe void which runs along the axis of the joint. This discontinuity may result from a combination of torch misalignment and improper welding variables, particularly travel speed.

Lack of fusion discontinuities are most prevalent in repair areas, either single or multipass. The discontinuities result from insufficient heat input to obtain complete fusion.

Subsurface contamination in PAW can result when copper from the torch nozzle is expelled into the weld. This condition usually results when the torch nozzle gets too close to the weld, overheats, and copper melts into the puddle. The resulting contamination, which may be detrimental, may be undetectable welding by conventional NDT procedures. The best way to avoid copper contamination is by properly training the operator and by developing good torch manipulation techniques.

Table 10.9
Plasma Arc Weld Discontinuities

Surface Discontinuities	Internal Discontinuities
Reinforcement	Porosity
Underfill	Tunneling (voids)
Undercut	Lack of fusion
Mismatch	Contamination
Lack of penetration	Cracks
Cracks	
Contamination	

SAFETY RECOMMENDATIONS

FOR DETAILED SAFETY information, refer to the manufacturer's instructions and the latest edition of ANSI Z49.1, *Safety in Welding and Cutting.* For mandatory federal safety regulations established by the U.S. Labor Department's Occupational Safety and Health Administration, refer to the latest edition of OSHA Standards, Code of Federal Regulations, Title 29 Part 1910, available from the Superintendent of Documents, U.S. Printing Office, Washington, D.C. 20402.

When welding with a transferred arc at currents up to 5 A, spectacles with side shields or other types of eye protection with a No. 6 filter lens are recommended. Although face protection is not normally required for this current range, its use depends on personal preference. When welding with transferred arc currents between 5 and 15 A, a full face plastic shield is recommended in addition to eye protection with a No. 6 filter lens. At current levels over 15 A, a standard welder's helmet with the proper shade of filter plate for the current being used is required.

When a pilot arc is operated continuously, normal precautions should be used for protection against arc flash and heat burns. Suitable clothing must be worn to protect exposed skin from arc radiation. Welding power should be turned off before electrodes are adjusted or replaced. Adequate eye protection should be used when observation of a high frequency discharge is required to center the electrode.

Accessory equipment, such as wire feeders, arc voltage controls, and oscillators should be properly grounded. If they are not grounded, insulation breakdown might cause these units to become electrically "hot" with respect to ground.

Adequate ventilation should be used, particularly when welding metals with high copper, lead, zinc, or beryllium contents.

SUPPLEMENTARY READING LIST

American Welding Society. *Recommended practices for plasma-arc welding*, C5.1. Miami: American Welding Society, 1973.

Ashauer, R. C., and Goodman, S. "Automatic plasma arc welding of square butt pipe joints." *Welding Journal* 46 (5): 405-415; May 1967.

Filipski, S. P. "Plasma arc welding." *Welding Journal* 43 (11): 937-943; November 1964.

Garrabrant, E. C., and Zuchowski, R. S. "Plasma arc-hot wire surfacing—A new high deposition process." *Welding Journal* 48(5): 385-395; May 1969.

Gorman, E. F. "New developments and applications in manual plasma arc welding." *Welding Journal* 48(7): 547-556; July 1969.

Gorman, E. F., Skinner, G. M., and Tenni, D. M. "Plasma needle arc for very low current work." *Welding Journal* 45(11): 899-908; November 1966.

Holko, K. H. "Plasma arc welding 2-1/4 Cr - 1 Mo Tubing." *Welding Journal* 57(5): 23-31; May 1978.

Keanini, R. G. and Rubinsky, B. "Plasma arc welding under normal and zero gravity." *Welding Journal* 69(6): 41; June 1990.

Kyselica, S. "High-frequency reversing arc switch for plasma arc welding of aluminum." *Welding Journal* 66 (1): 31-35; January 1987.

Langford, G. J. "Plasma arc welding of structural titanium joints. *Welding Journal* 47(2): 102-113; February 1968.

Metcalfe, J. C., and Quigley, M. B. C. "Heat transfer in plasma-arc welding." *Welding Journal* 54(3): 99-103; March 1975.

———. "Keyhole stability in plasma arc welding." *Welding Journal* 54(11): 401-404; November 1975.

Miller, H. R., and Filipski, S. P. "Automated plasma arc welding for aerospace and cryogenic fabrications." *Welding Journal* 45(6): 493-501; June 1966.

Nunes, A. C. et. al. "Variable polarity plasma arc welding on the space shuttle external tank." *Welding Journal* 27-35; September 1984.

O'Brien, R. L. "Arc plasmas for joining, cutting, and surfacing." Bulletin No. 131. New York: Welding Research Council, July 1968.

Ruprecht, W. J., and Lundin, C. D. "Pulsed current plasma arc welding." *Welding Journal* 53(1): 11-19; January 1974.

Steffans, H. D., and Kayser, H. "Automatic control for plasma arc welding." *Welding Journal* 51(6): 408-418; June 1972.

Tomsic, M., and Barhorst, S. "Keyhole plasma arc welding of aluminum with variable polarity power." *Welding Journal* 63(2): 25-32; February 1984.

OXYFUEL GAS WELDING

PREPARED BY A COMMITTEE CONSISTING OF:

G. R. Meyer, Chairman
Victor Equipment Company

J. D. Compton
College of the Canyons

R. D. Green
Airco-Mapp

J. F. Leny
Harnischfeger Corporation

C. R. McGowan
Consultant

WELDING HANDBOOK COMMITTEE MEMBER:
B. R. Somers
Consultant

CHAPTER 11

OXYFUEL GAS WELDING

FUNDAMENTALS OF THE PROCESS

OXYFUEL GAS WELDING (OFW) includes any welding operation that uses combustion with oxygen as a heating medium. The process involves melting the base metal and usually a filler metal, using a flame produced at the tip of a welding torch. Fuel gas and oxygen are combined in the proper proportions inside a mixing chamber which may be part of the welding tip assembly. Molten metal from the plate edges, and filler metal, if used, intermix in a common molten pool and coalesce upon cooling.

One advantage of this welding process is the control a welder can exercise over heat input and temperature, independent of the addition of filler metal. Weld bead size, shape, and weld puddle viscosity are also controlled in the welding process. OFW is ideally suited for repair welding and for welding thin sheet, tubes, and small diameter pipe. Thick section welds, except for repair work, are not economical when compared to the many available arc welding processes.

The equipment used in oxyfuel gas welding is low in cost, usually portable, and versatile enough to be used for a variety of related operations, such as bending and straightening, preheating, postheating, surfacing, brazing, and braze welding.

Cutting attachments, multiflame heating nozzles, and a variety of special application accessories add greatly to the overall versatility of the basic OFW equipment. With relatively simple equipment changes, manual and mechanized oxygen cutting operations can be performed. Metals normally welded include carbon and low alloy steels, and most nonferrous metals, but generally not refractory or reactive metals.

Commercial fuel gases have one common property - they all require oxygen to support combustion. To be suitable for welding operations, a fuel gas, when burned with oxygen, must have the following:

(**1**) High flame temperature
(**2**) High rate of flame propagation
(**3**) Adequate heat content
(**4**) Minimum chemical reaction of the flame with base and filler metals

Among commercially available fuel gases, acetylene most closely meets all these requirements. Other fuel gases, such as methylacetylene-propadiene products, propylene, propane, natural gas, and proprietary gases based on these, offer sufficiently high flame temperatures but exhibit lower flame propagation rates. These latter gas flames are excessively oxidizing at oxygen-to-fuel gas ratios that are high enough to produce usable heat transfer rates. Therefore, flame holding devices, such as counterbores on the tips, are necessary even at the higher ratios for stable operation and good heat transfer. Commercial fuel gases, however, are frequently used for oxygen cutting. They are also used for torch brazing, soldering, and other operations where demands upon the flame characteristics and heat transfer rates are not the same as those for welding.

CHARACTERISTICS OF FUEL GASES

GENERAL CHARACTERISTICS

TABLE 11.1. LISTS SOME of the pertinent characteristics of commercial gases. In order to appreciate the significance of the information in this table, it is necessary to understand some terms and concepts involved in burning fuel gases.

Specific Gravity

THE SPECIFIC GRAVITY of a fuel gas with respect to air indicates how the gas may accumulate in the event of a leak. For example, gases with a specific gravity less than air tend to rise, and can collect in corners of rooms lofts, and ceiling spaces. Those gases with a specific gravity greater than air tend to accumulate in low, still areas.

Volume-to-Weight Ratio

A SPECIFIC QUANTITY of gas at standard temperature and pressure can be described by volume or weight. The values shown in Table 11.1 give the volume per unit weight at 60°F (15.6°C) and under atmospheric pressure. Multiplying these figures by the known weight will give the volume. If the volume is known, multiplying the reciprocal of the figures shown by the volume will give the weight.

Combustion Ratio

TABLE 11.1 INDICATES the volume of oxygen theoretically required for complete combustion of each of the fuel gases shown. These oxygen-to-fuel gas ratios (called *stoichiometric mixtures*) are obtained from the balanced chemical equations given in Table 11.2. The values shown for complete combustion are useful in calculations. They do not represent the oxygen-to-fuel gas ratios actually delivered by an operating torch, because, as will be explained later, the complete combustion is partly supported by oxygen in the surrounding air.

Heat of Combustion

THE TOTAL HEAT of combustion (heat value) of a hydrocarbon fuel gas is the sum of the heat generated in the primary and secondary reactions that take place in the overall flame. This is shown in Table 11.1. The combustion of hydrogen takes place in a single reaction. The theoretical basis for these chemical reactions and their heat effects is discussed in the *Welding Handbook*, Volume 1, 8th Edition.

Typically, the heat content of the primary reaction is generated in an inner, or primary, flame. This is where combustion is supported by oxygen supplied by the torch. The secondary reaction takes place in an outer, or secondary, flame envelope where combustion of the primary reaction products is supported by oxygen from the air.

Although the heat of the secondary flame is important in most applications, the more concentrated heat of the primary flame is a major contribution to the welding capability of an oxyfuel gas system. The primary flame is said to be neutral when the chemical equation for the primary reaction is exactly balanced, yielding only carbon monoxide and hydrogen. Under these conditions the primary flame atmosphere is neither carburizing nor oxidizing.

Since the secondary reaction is necessarily dependent upon the primary reaction end-products, the term *neutral* serves as a convenient reference point for (1) describing combustion ratios, and (2) comparing the various heat characteristics of different fuel gases.

Flame Temperature

THE FLAME TEMPERATURE of a fuel gas will vary according to the oxygen-to-fuel ratio. Although the flame temperature gives an indication of the heating ability of the fuel gas, it is only one of the many physical properties to consider in making an overall evaluation. Flame temperatures are usually calculated since there is, at present, no simple method of physically measuring these values.

The flame temperatures listed in Table 11.1 are for the so-called neutral flame, i.e., the primary flame that is neither oxidizing nor carburizing in character. Flame temperatures higher than those listed may be achieved, but, in every case, that flame will be oxidizing, an undesirable condition in the welding of many metals.

Combustion Velocity

A CHARACTERISTIC PROPERTY of a fuel gas, its combustion velocity (flame propagation rate), is an important factor in the heat output of the oxyfuel gas flame. This is the velocity at which a flame front travels through the adjacent unburned gas. It influences the size and temperature of the primary flame. Combustion velocity also affects the velocity at which gases may flow from the torch tip without causing a flame standoff or backfire. A flame standoff occurs when combustion takes place some distance away from the torch tip rather than right at the torch tip. Backfire is the momentary recession of the flame into the welding tip, followed by the reappearance or complete extinction of the flame.

Table 11.1
Characteristics of the Common Fuel Gases

Fuel Gas	Formula	Specific Gravity[a] (Air = 1)	Volume to Weight Ratio[a]		Oxygen-to-Fuel Gas Combustion Ratio[b]	Flame Temperature for Oxygen[c]		Heat of Combustion					
								Primary		Secondary		Total	
			ft³/lb	m³/kg		°F	°C	Btu/ft³	MJ/m³	Btu/ft³	MJ/m³	Btu/ft³	MJ/m³
Acetylene	C_2H_2	0.906	14.6	0.91	2.5	5589	3087	507	19	963	36	1470	55
Propane	C_2H_3	1.52	8.7	0.54	5.0	4579	2526	255	10	2243	94	2498	104
Methylacetylene-propadiene (MPS) (d)	C_3H_4	1.48	8.9	0.55	4.0	5301	2927	571	21	1889	70	2460	91
Propylene	C_3H_6	1.48	8.9	0.55	4.5	5250	2900	438	16	1962	73	2400	89
Natural gas (methane)	CH_4	0.62	23.6	1.44	2.0	4600	2538	11	0.4	989	37	1000	37
Hydrogen	H_2	0.07	188.7	11.77	0.5	4820	2660					325	12

a. At 60°F (15.6°C).

b. The volume units of oxygen required to completely burn a unit volume of fuel gas according to the formulae shown in Table 11.2. A portion of the oxygen is obtained from the atmosphere.

c. The temperature of the neutral flame.

d. May contain significant amounts of saturated hydrocarbons.

Table 11.2
Chemical Equations for the Complete Combustion of the Common Fuel Gases

Fuel Gas	Reaction With Oxygen
Acetylene	$C_2H_2 + 2.5O_2 \rightarrow 2CO_2 + H_2O$
Methylacetylene-propadiene (MPS)	$C_3H_4 + 4O_2 \rightarrow 3CO_2 + 2H_2O$
Propylene	$C_3H_6 + 4.5O_2 \rightarrow 3CO_2 + 3H_2O$
Propane	$C_3H_8 + 5O_2 \rightarrow 3CO_2 + 4H_2O$
Natural gas (methane)	$CH_4 + 2O_2 \rightarrow CO_2 + 2H_2O$
Hydrogen	$H_2 + 0.5O_2 \rightarrow H_2O$

As shown in Figure 11.1, the combustion velocity of a fuel gas varies in a characteristic manner according to the proportions of oxygen and fuel in the mixture.

Combustion Intensity

FLAME TEMPERATURES AND heating values of fuels have been used almost exclusively as the criteria for evaluating fuel gases. These two factors alone, however, do not provide sufficient information for a complete appraisal of fuel gases for heating purposes. A concept known as *combustion intensity* or "specific flame output" is used to evaluate different oxygen-fuel gas combinations. Combustion intensity takes into account the burning velocity of the flame, the heating value of the mixture of oxygen and fuel gas, and the area of the flame cone issuing from the tip.

Combustion intensity may be expressed as the following:

$$C_i = C_v \times C_h \qquad (11.1)$$

where:

C_i = combustion intensity in $Btu/ft^2 \cdot s \, (J/m^2 \cdot s)$
C_v = normal combustion velocity of flame in ft/s (m/s)
C_h = heating value of the gas mixture under consideration in $Btu/ft^3 \, (J/m^3)$

Combustion intensity (C_i), therefore, is maximum when the product of the normal burning velocity of the flame (C_v) and the heating value of the gas mixture (C_h) is maximum.

Like the heat of combustion, the combustion intensity of a gas can be expressed as the sum of the combustion intensities of the primary and secondary reactions. However, the combustion intensity of the primary flame, located near the torch tip where it can be concentrated on the workpiece, is of major importance in welding. The secondary combustion intensity influences the thermal gradient in the vicinity of the weld.

Figures 11.2 and 11.3 show the typical rise and fall of the primary and secondary combustion intensities of several fuels with varying proportions of oxygen and fuel gas. Figure 11.4 gives the total combustion intensities for the same gases. These curves show that, for the gases plotted, acetylene produces the highest combustion intensities.

ACETYLENE

ACETYLENE IS THE fuel gas of choice for of welding because of its high-combustion intensity. While the other fuel gases are rarely, if ever, used for welding, their characteristics are described below.

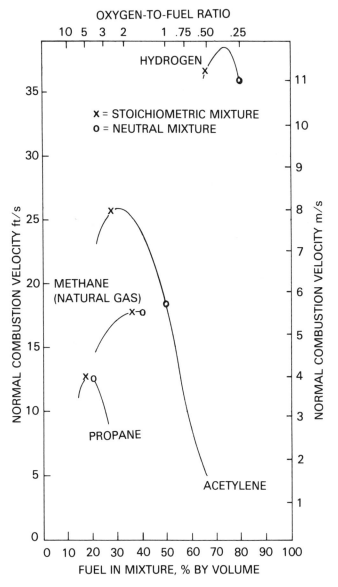

Figure 11.1–Normal Combustion Velocity (Flame Propagation Rate) of Various Fuel Gas-Oxygen Mixtures

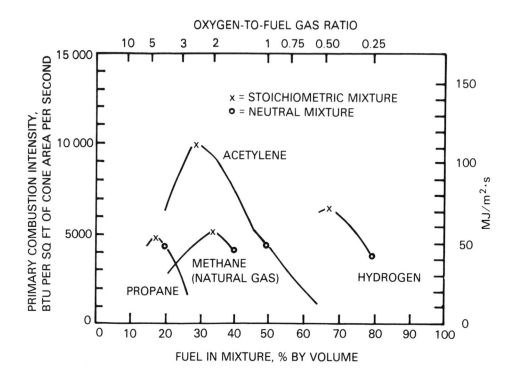

Figure 11.2–Primary Combustion Intensity of Various Fuel Gas-Oxygen Mixtures

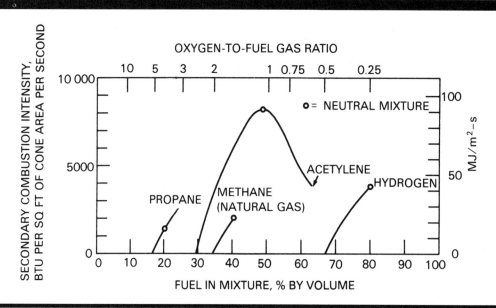

Figure 11.3–Secondary Combustion Intensity of Various Fuel Gas-Oxygen Mixtures

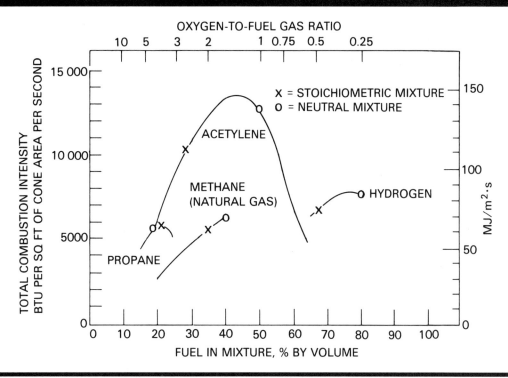

Figure 11.4–Total Combustion Intensity of Various Fuel Gas-Oxygen Mixtures

Acetylene is a hydrocarbon compound, C_2H_2, which contains a larger percentage of carbon by weight than any of the other hydrocarbon fuel gases. Colorless and lighter than air, it has a distinctive odor resembling garlic. Acetylene contained in cylinders is dissolved in acetone and therefore has a slightly different odor from that of pure acetylene.

At temperatures above 1435°F (780°C) or at pressures above 30 psig (207 kPa), gaseous acetylene is unstable and decomposition may result even in the absence of oxygen. This characteristic has been taken into consideration in the preparation of a code of safe practices for the generation, distribution, and use of acetylene gas. The accepted safe practice is never to use acetylene at pressures exceeding 15 psig (103 kPa), in generators, pipelines, or hoses.

THE OXYACETYLENE FLAME

THEORETICALLY, THE COMPLETE combustion of acetylene is represented by the chemical equation

$$C_2H_2 + 2.5\ O_2 \rightarrow 2CO_2 + H_2O \qquad (11.2)$$

This equation indicates that one volume of acetylene (C_2H_2) and 2.5 volumes of oxygen (O_2) react to produce two volumes of carbon dioxide (CO_2) and one volume of water vapor (H_2O). The volumetric ratio of oxygen to acetylene is 2.5 to one.

As noted earlier, the reaction of equation (11.2) does not proceed directly to the end products shown. Combustion takes place in two stages. The primary reaction takes place in the inner zone of the flame (called the *inner cone*) and is represented by the chemical equation:

$$C_2H_2 + O_2 \rightarrow 2CO + H_2 \qquad (11.3)$$

Here, one volume of acetylene and one volume of oxygen react to form two volumes of carbon monoxide and one volume of hydrogen. The heat content and high temperature (Table 11.1) of this reaction result from the decomposition of the acetylene and the partial oxidation of the carbon resulting from that decomposition.

When the gases issuing from the torch tip are in the one-to-one ratio indicated in equation (11.3), the reaction produces the typical brilliant blue inner cone. This relatively small flame creates the combustion intensity needed for welding steel. The flame is termed neutral because there is no excess carbon or oxygen to carburize or oxidize the metal. The end products are actually in a reducing status, a benefit when welding steel.

In the outer envelope of the flame, the carbon monoxide and hydrogen produced by the primary reaction burn with oxygen from the surrounding air. This forms carbon dioxide and water vapor respectively, as shown in the following secondary reaction:

$$2CO + H_2 + 1.5O_2 \rightarrow 2CO_2 + H_2O \qquad (11.4)$$

Although the heat of combustion of this outer flame is greater than that of the inner, its combustion intensity and temperature are lower because of its large cross-sectional area. The final end products are produced in the outer flame because they cannot exist in the high temperature of the inner cone.

The oxyacetylene flame is easily controlled by valves on the welding torch. By a slight change in the proportions of oxygen and acetylene flowing through the torch, the chemical characteristics in the inner zone of the flame and the resulting action of the inner cone on the molten metal can be varied over a wide range. Thus, by adjusting the torch valves, it is possible to produce a neutral, oxidizing, or carburizing flame.

PRODUCTION

ACETYLENE IS THE product of a chemical reaction between calcium carbide (CaC_2) and water. In that reaction, the carbon in the calcium carbide combines with the hydrogen in the water, forming gaseous acetylene. At the same time the calcium combines with oxygen and hydrogen to form a calcium hydroxide residue. The chemical equation is

$$CaC_2 + 2H_2O \rightarrow C_2H_2 + Ca\,(OH)_2 \qquad (11.5)$$

The carbide used in this process is produced by smelting lime and coke in an electric furnace. When removed from the furnace and cooled, the carbide is crushed, screened, and packed in airtight containers. The most common of these holds 100 lb (45 kg) of the hard, grayish solid. Ten cubic ft (0.28 m^3) of acetylene can be generated from approximately 2.2 lb (1 kg) of calcium carbide.

Acetylene is also frequently produced in petrochemical plants, and may be used for a variety of processes other than oxyfuel gas welding and cutting.

ACETYLENE GENERATORS

THE TWO PRINCIPAL methods currently employed to generate acetylene are carbide-to-water and water-to-carbide. In the United States, the carbide-to-water method is used almost exclusively. The construction of the generator used for this method allows small lumps of carbide to be discharged from a hopper into a relatively large body of water. This type of generator is shown schematically in Figure 11.5. The details of its construction vary with different manufacturers. All carbide-to-water generators can be classified as low-pressure or medium-pressure types. The former operate at about 1 psig or less, while the latter produce acetylene at 1 to 15 psig.

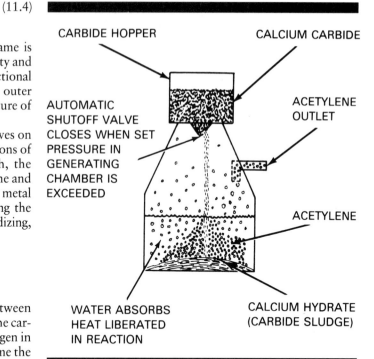

Figure 11.5—An Acetylene Generator in which Calcium Carbide is Added to Water

The water-to-carbide type of acetylene generator is rarely used in this country, but is popular in Europe. Fundamentally, the operating principal is the same as that of the carbide-to-water type, but the method differs. Water from a tank is allowed to drip onto a bed of carbide, and the gas evolved is piped from the generator. The carbide is usually in the form of bricks or cakes, to limit the surface area exposed to the water.

The generation of acetylene evolves a considerable amount of heat, which must be dissipated because of the instability of acetylene at elevated temperatures. The relatively large volume of water employed in the carbide-to-water generator makes this type highly efficient in dissipating the heat. The water-to-carbide type, on the other hand, uses the minimum amount of water, and its heat dissipation is slow.

Acetylene generators are available in both stationary and portable units, in a large range of sizes and gas production rates. Generating capacities of these units range from 12 ft^3/h (0.34 m^3/h) for small portable units to about 6000 ft^3/h (170 m^3/h) for the large stationary units at industrial installations. Most modern generators operate automatically after the operating pressure is initially set.

ACETYLENE CYLINDERS

BECAUSE FREE ACETYLENE, under certain pressure and temperature conditions, may dissociate explosively into its hydrogen and carbon constituents, cylinders to be filled with acetylene are initially packed with a porous filler. Acetone, a solvent capable of absorbing 25 times its own volume of acetylene per atmosphere of pressure, is added to the filler. By so dissolving the acetylene and dividing the interior of the cylinder into small, partly separated cells within the porous filler, a safe acetylene-filled container is produced.

Acetylene cylinders are available in sizes containing from 10 to 420 ft^3 (0.28 to 12 m^3) of the gas. The cylinders are equipped with fusible safety plugs made of a metal that melts at about 212°F (100°C). This allows the gas to escape if the cylinder should be subjected to excessive heat, resulting in a relatively controlled burn rather than rupturing the cylinder.

MPS FUEL GAS TYPES

SEVERAL COMMERCIALLY PREPARED fuel-gas mixtures are available for welding, but they are not generally used for this purpose. They are more extensively used for cutting, torch brazing, and other heating operations. One group of mixed fuel gases has compositions approximating methylacetylene-propadiene (MPS) and containing mixtures of propadiene, propane, butane, butadiene, and methylacetylene. One characteristic of these mixed fuel gases is that the heat distribution within the flame is more even than with acetylene, thus requiring less manipulation of the gas torch for controlling the heat input. The flame temperatures of these gases are lower than that of acetylene with neutral gas mixtures. Temperatures can be increased by making the flames oxidizing. The gases are popular because they may be less costly than acetylene, and the cylinders contain a greater volume of fuel for a given size and weight.

PROPYLENE

A SINGLE COMPONENT fuel gas, propylene (C_3H_6), is an oil refinery product with performance characteristics similar to the MPS type gases. Although not suitable for welding, propylene is used for oxygen cutting, brazing, flame spraying, and flame hardening. The gas makes use of equipment similar in design to that used with the MPS type gases.

PROPANE

PROPANE (C_3H_8) IS used primarily for preheating in oxygen cutting, and for heating operations. The main source of this gas is the crude-oil and gas mixtures obtained from active oil and natural gas wells. Propane is also produced in certain oil refining processes, and in the recycling of natural gas. It is sold and transported in steel cylinders containing up to 100 lb (45 kg) of the liquefied gas. Large consumers are supplied by tank car and bulk delivery. Small self-contained propane torch sets are available for home workshop use as well as for incidental heating operations.

NATURAL GAS (METHANE)

NATURAL GAS IS obtained from wells and distributed by pipelines. Its chemical composition varies widely, depending upon the locality from which it is obtained. The principal constituents of most natural gases are methane (CH_4) and ethane (C_2H_6). The volumetric requirement of natural gas is, as a rule, about 1-1/2 times that of acetylene to provide an equivalent amount of heat. Natural gas finds its principal use in the welding industry as a fuel gas for oxygen cutting and heating operations.

HYDROGEN

THE RELATIVELY LOW heat content of the oxy-hydrogen flame restricts the use of hydrogen to certain torch brazing operations and to the welding of aluminum, magnesium, lead, and similar metals. Other welding processes, however, are largely supplanting all forms of oxyfuel gas welding for many of these materials.

Hydrogen is available in seamless, drawn, steel cylinders charged to a pressure of about 2000 psig (14 MPa) at a temperature of 70°F (21°C). It may also be supplied as a liquid, either in individual cylinders or in bulk. At the point of use, the liquid hydrogen is vaporized into gas.

OXYGEN

OXYGEN IN THE gaseous state is colorless, odorless, and tasteless. It occurs abundantly in nature. A chief source of oxygen is our atmosphere, which contains approximately 21 percent oxygen by volume. Although there is sufficient oxygen in air to support fuel gas combustion, the use of pure oxygen speeds up burning reactions and increases flame temperatures.

Most oxygen used in the welding industry is extracted from the atmosphere by liquefaction techniques. In the extraction process, air may be compressed to approximately 3000 psig (20 MPa), although some types of equipment operate at much lower pressure. The carbon dioxide and any impurities in the air are removed; the air passes through coils, and is allowed to expand to a rather low pressure. The air becomes substantially cooled during the expansion, and then it is passed back over the coils, further cooling the incoming air, until liquefaction occurs. The liquid air is sprayed on a series of evaporating trays or plates in a rectifying tower. Nitrogen and other gases boil at lower temperatures than the oxygen and, as these gases escape from the top of the tower, high-purity liquid oxygen collects in a receiving chamber at the base. Some plants are designed to produce bulk liquid oxygen; in other plants, gaseous oxygen is withdraw for compression into cylinders.

OXYFUEL GAS WELDING EQUIPMENT

BASIC WELDING EQUIPMENT

THE MINIMUM BASIC equipment needed to perform oxyfuel gas welding is shown schematically in Figure 11.6. This equipment setup is completely self-sufficient and relatively inexpensive. It consists of fuel gas and oxygen cylinders, each with a gas regulator for reducing cylinder pressure, hoses for conveying the gases to the torch, and a torch and tip combination for adjusting the gas mixtures and producing the desired flame.

Each of these units plays an essential part in the control and application of the heat necessary for welding. The same basic equipment is used for torch brazing and for many heating operations. By a simple substitution of the proper torch and tip combination, the equipment is readily converted to manual or carriage-controlled oxygen cutting. Since the use of this equipment is controlled by the operator, he or she must be thoroughly familiar with the capabilities and limitations of the equipment and the rules of safe operation.

A variety of equipment is obtainable for most welding operations. Some of this equipment is designed for general use and some is produced for specific operations. The most suitable equipment should be selected for each particular operation.

WELDING TORCHES

A TYPICAL WELDING torch consists of a torch handle, mixer and tip assembly. It provides a means of independently controlling the flow of each gas, a method of attaching a variety of welding tips or other apparatus, and a convenient handle for controlling the movement and direction of the flame. Figure 11.7 is a simplified schematic drawing of the basic elements of a welding torch.

The gases pass through the control valves, through separate passages in the torch handle, and to the torch head. They then pass into a mixer assembly where the oxygen and fuel gas are mixed, and finally pass out through an orifice at the end of the tip. The tip is shown as a simple tube, narrowed at the front end to produce a suitable welding cone. Sealing rings or surfaces are provided in the torch head or on the mixer seats to facilitate leak-tight assembly.

Types of Torch Handles

WELDING TORCH HANDLES are manufactured in a variety of styles and sizes, from the small size for extremely light (low gas flow) work to the extra heavy (high gas flow) handles generally used for localized heating operations.

A typical small welding torch used for sheet metal welding will pass acetylene at volumetric rates ranging from about 0.25 to 35 ft^3/h (0.007 to 1.0 m^3/h)). Medium sized torches are designed to provide acetylene flows from about 1 to 100 ft^3/h (0.028 to 2.8 m^3/h). Heavy-duty heating torches may permit acetylene flows as high as 400 ft^3/h (11 m^3/h). Fuel gases other than acetylene may be used with even larger torches that have fuel-gas flow rates as high as 600 ft^3/h (17 m^3/h).

Torch handles can be used with a variety of mixer and tip designs, as well as special purpose nozzles, cutting attachments, and heating nozzles. See Figure 11.8.

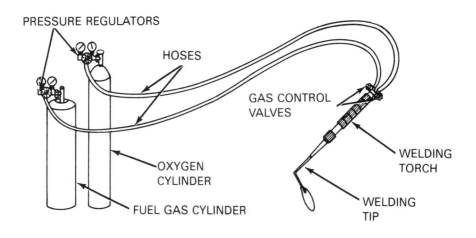

Figure 11.6–Basic Oxyfuel Gas Welding Equipment

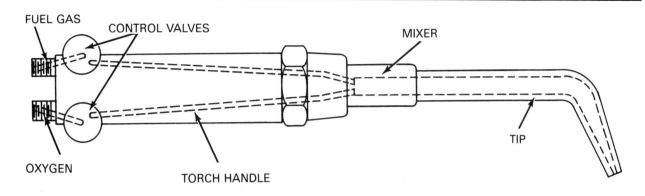

Figure 11.7–Basic Elements of an Oxyfuel Gas Welding Torch

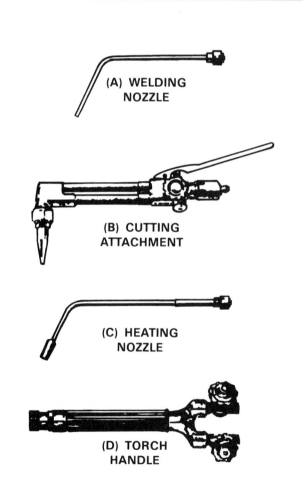

Figure 11.8–Typical Torch Handle to be Used With Welding Nozzles, Cutting Attachments, and Heating Nozzles

Types of Mixers

TWO GENERAL TYPES of oxygen fuel-gas mixers are the *positive- pressure* (also called *equal* or *medium-pressure*) type, and the *injector* or *low-pressure type*.

The positive-pressure-type mixer requires that the gases be delivered to the torch at pressures above 2 psig (14kPa). In the case of acetylene, the pressure should be between 2 and 15 psig (14 and 103 kPa). Oxygen generally is supplied at approximately the same pressure. There is, however, no restrictive limit on the oxygen pressure. It can, and sometimes does, range up to 25 psig (172kPa) with the larger sized tips.

The purpose of the injector-type mixer is to increase the effective use of fuel gases supplied at pressures of 2 psig (14kPa) or lower. In this torch, oxygen is is supplied at pressures ranging from 10 to 40 psig (70 to 275 kPa), the pressure increasing to match the tip size. The relatively high velocity of the oxygen flow is used to aspirate or draw in more fuel gas than would normally flow at the low-supply pressures.

Gas mixers come in various styles and sizes, according to the manufacturer's design. The chief function of these units is to mix the fuel gas and oxygen thoroughly to assure smooth combustion. Because of their construction, mixers also serve as a heat sink to help prevent the flame from flashing back into the mixer or torch.

A flashback is the recession of flame into or back of the mixing chamber of the oxyfuel gas torch. In some cases, the flashback travels back through the hose to the gas regulator, causing a fire at the gas cylinder.

A typical mixer for a positive-pressure torch is shown in Figure 11.9(A). The oxygen enters through a center duct, and the fuel gas enters through several angled ducts to effect the mixing. Mixing turbulence decreases to a laminar flow as the gas passes through the tip.

Gas mixers designed for injector-type torches employ the principle of the venturi tube to increase the fuel gas flow. In this case (Figure 11.9B), the high-pressure oxygen passes through the small central duct creating a high-velocity jet. The oxygen jet crosses the openings of the angled fuel gas ducts at the point where the venturi tube is restricted. This action produces a pressure drop at the fuel gas openings, causing the low-pressure fuel-gas flow to increase as the mixing gases pass into the enlarged portion of the venturi.

Care of Torches

WELDING TORCH HANDLES, mixers, and tips are designed to withstand the tough operating conditions to which they are exposed. In order to provide safe and efficient usage, they must be properly maintained in good working condition at all times. They should be used only with the proper fuel gas and for the purpose for which they were designed.

It is important to carefully follow the manufacturer's operating instructions and recommendations for safe use.

If torches require service, this work should be done only by qualified repair technicians.

Welding Tips

THE WELDING TIP is that portion of the torch through which the gases pass just prior to ignition and burning. The tip enables the welder to guide the flame and direct it to the work with maximum ease and efficiency.

Tips are generally made of a nonferrous metal, such as a copper alloy, with high thermal conductivity to reduce the danger of overheating. Tips are generally manufactured by

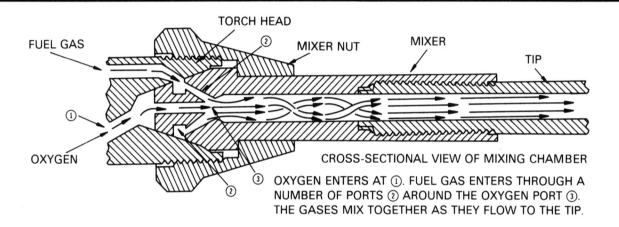

(A) POSITIVE-PRESSURE TYPE

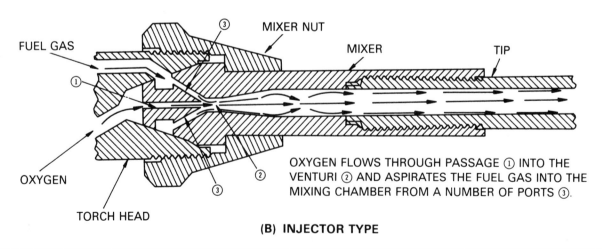

(B) INJECTOR TYPE

Figure 11.9–Typical Design Details of Gas Mixers for Positive Pressure and Injector-Type Welding Torches

drilling bar stock to the proper orifice size or by swaging tubing to the proper diameter over a mandrel. The bore in both types must be smooth in order to produce the required flame cone. The front end of the tip should also be shaped to permit easy use and provide a clear view of the welding operation being performed.

Welding tips are available in a great variety of sizes, shapes, and constructions. Two methods of combining tips and mixers are employed. A special tip may be used for each size of mixer, or one or more mixers can cover the entire range of tip sizes. In the latter method, the tip unscrews from its mixer, and each size of mixer has a particular thread size to prevent improper coupling of a tip and a mixer.

A single mixer is used for some classes of welding. It has a "gooseneck" into which the various sizes of tips may be threaded.

Since tips generally are made of a relatively soft copper alloy, care must be taken to guard against damaging them. The following precautions should be observed:

(1) Tips should be cleaned using tip cleaners designed for this purpose.

(2) Tips should never be used for moving or holding the work.

(3) Tip and mixer threads and all sealing surfaces must be kept clean and in good condition. A poor seal can result in leaks, and a backfire or backflash may result.

When performing a welding operation, care should be taken to obtain the correct flame adjustment with the proper size of torch, mixer, and tip. The proper methods for obtaining the desired flame characteristics are given elsewhere in this chapter.

When a series of welding tips is selected for a variety of metal thicknesses, the thickness range covered by one tip should slightly overlap that covered by the next tip. Since there is no single standard for tip size designations, the manufacturer's recommendations should be followed.

Volumetric Rate of Flow. The most important factor in determining the usefulness of a torch tip is the action of the flame on the metal. If it is too violent, it may blow the metal out of the molten pool. Under such conditions, the volumetric flow rates of acetylene and oxygen should be reduced to a velocity at which the metal can be welded. This condition represents the maximum volumetric flow rate that can be handled by a given size of welding tip. As a general rule, the larger the volumetric rate of gas that can be handled by a specific size of tip, the greater is the heat.

A flame may also be too "soft" for easy welding. When the flame is too soft, the volumetric flow rates must be increased.

Tips having a hooded or cup-shaped end are available for gases with low combustion velocities, such as propane. These tips are usually used for heating, brazing, and soldering.

Flame Cones. The purpose of a welding flame is to raise the temperature of metal to the point of fusion. This can best be accomplished when the welding flame (or cone) permits the heat to be directed easily. Consequently, the cone characteristics become important. Laminar or streamlined gas flow throughout the length of the tip becomes of paramount importance, especially during passage through the front portion.

A high-velocity flame cone presents striking proof of the velocity gradient extending across a circular orifice when the existing flow is laminar (Figure 11.10). Since the greatest velocity exists at the center of the stream, the flame at the center is the longest. Similarly, since the velocity of the gas stream is lowest at the wall of the tip (bore) where the flowing friction is greatest, that portion of the flame bordering the wall is the shortest. From the analysis of the principles that underlie the formation of a flame cone, it is possible to understand the flow conditions that exist along the last portion of the gas passageway in a tip. The shape of the flame cone will depend upon a number of factors, such as the smoothness of the bore, the ratio of lead-in to final run diameter, and the sharpness of neck-down.

Generally speaking, the cone produced by a small tip will vary from a pointed to a semipointed shape. Cones from medium-sized tips will vary from a semipointed to a medium shape, and cones from a large-sized tip will vary from a semiblunt to a blunt shape (Figure 11.11).

Hoses

HOSES USED IN oxyfuel gas welding and allied operations are manufactured specifically to meet the utility and safety requirements for this service. For mobility and ease of manipulation in welding, the hoses must be flexible. They must also be capable of withstanding high line pressures at moderate temperatures.

Each hose should have a check valve at the regulator and another at the torch. The purpose of check valves is to prevent flashbacks within the hose and regulator.

For rapid identification, all fuel-gas hoses are colored red. As a further precaution, the swivel nuts used for making fuel-gas hose connections are identified by a groove cut into the outside of the nut. The nuts also have left-hand threads to match the fuel gas regulator outlet and the fuel gas inlet fitting on the torch.

Oxygen hoses are colored green, and the connections each have a plain nut with right-hand threads matching the oxygen regulator outlet and the oxygen inlet fitting on the torch.

The standard means of specifying hose is by inside diameter and applications. Nominal inside diameters most commonly used are 1/8, 3/16, 1/4, 15/16, 3/8, and 1/2 in. (3.2, 4.8, 6.4, 7.9, 9.5, and 12.7mm), although larger sizes are available. Standard industrial welding hoses and fittings have a maximum working pressure of 200 psig.

Wherever possible, hoses should be supported in an elevated position to avoid damage by falling objects, truck

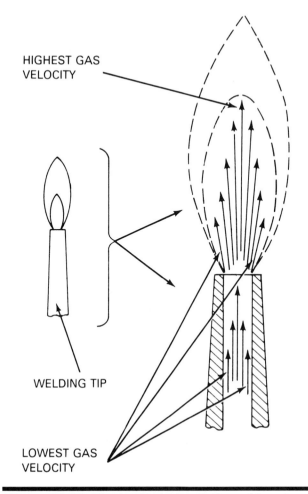

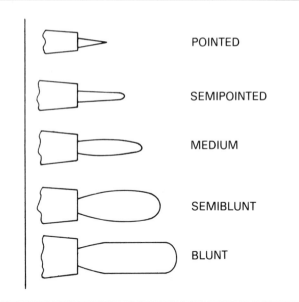

Figure 11.11–Representative Flame Cone Shapes Produced by Welding Tips

Figure 11.10–Vector Representation of Laminar Flow Velocity in a Welding Tip and in the Formation of a Uniform Flame Cone

wheels, or hot metal. Damaged hose should be replaced or repaired with the proper fittings intended for this purpose.

Lengths of hose over 25 ft (8m) and small diameter hose, may restrict the flow of gas to the torch. In some cases, this restriction may be overcome by using a higher regulator pressure, but usually a larger diameter hose is recommended. It is recommended that hoses be as short as practical.

Regulators

A REGULATOR CAN be described as a mechanical device for maintaining the delivery of a gas at some substantially constant reduced pressure even though the pressure at the source may change. Regulators used in oxyfuel gas welding and allied applications are adjustable pressure reducers, designed to operate automatically after an initial setting. Except for minor differences, all of these regulators operate on the same basic principle. They fall into different application categories according to their designed capabilities for handling specific gases, different pressure ranges, and different volumetric flow rates.

Regulators are generally classed as single stage or two stage, depending on whether the pressure is reduced in one or two steps.

The output pressure of the single-stage type exhibits a characteristic known as *rise* or *drift*. This is a slight rise or drop in the delivery pressure that occurs as the cylinder pressure is depleted. This characteristic is usually detrimental only when a large quantity of the gas is withdrawn from a high-pressure cylinder at a single usage. Periodic readjustment of regulator pressure will correct any detrimental effects.

Two-stage regulators are essentially two single-stage regulators operating in series within one housing. They provide constant delivery pressure as cylinder pressure is depleted.

Operating Principle. The components of a pressure reducing regulator are shown schematically in Figure 11.12. The following are principal operating elements:

(1) An adjusting screw that controls the thrust of a bonnet spring
(2) A bonnet spring that transmits this thrust to a diaphragm
(3) A diaphragm that contacts a stem on a movable valve seat
(4) A valve consisting of a nozzle and the movable valve seat
(5) A small spring located under the moveable valve seat

The bonnet spring force tends to hold the seat open while the forces on the underside of the diaphragm tend to cause the seat to close. When gas is withdrawn at the outlet, the pressure under the diaphragm is reduced, thus further opening the seat and admitting more gas until the forces on either side of the diaphragm are equal.

A given set of conditions, such as constant inlet pressure, constant volumetric flow, and constant outlet pressure, will produce a balanced condition such that the nozzle and its mating seat member will maintain a fixed relationship. As noted earlier, the inlet pressure from cylinders drops as gas is used, causing a gradual drift in regulator outlet pressure. The factors affecting the extent of this drift depend on the type of single-stage regulator.

Single-Stage Types. There are two basic types of pressure reducing regulators:

(1) The stem type, closed by inlet pressure (sometimes referred to as the inverse or negative type), illustrated in Figure 11.12A
(2) The nozzle type, opened by inlet pressure (sometimes referred to as the direct acting or positive type), illustrated in Figure 11.12B

In the stem-type regulator, inlet pressure tends to close the seat member (pressure closing) against the nozzle. The outlet pressure of this type of regulator has a tendency to

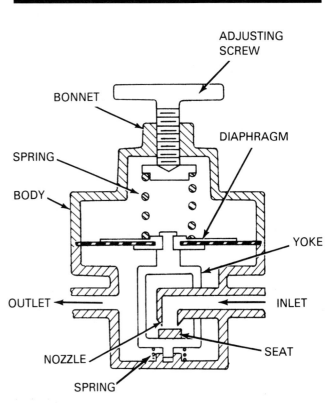

Figure 11.12B–Cross Section Showing the Major Components of a Typical Single-Stage-Nozzle-Type Regulator

increase somewhat as the inlet pressure decreases. This increase is caused by a decrease in the force produced by the inlet gas pressure against the seating area as the inlet pressure decreases.

The gas outlet pressure for any particular setting of the adjusting screw is regulated by a balance of forces between the bonnet spring thrust and the opposing forces created by

(1) The gas pressure against the underside of the diaphragm
(2) The force created by the inlet pressure against the valve seat
(3) The force of the small spring located under the valve seat

When the inlet pressure decreases, its force against the seat member decreases, allowing the bonnet spring force to move the seat member away from the nozzle. Thus, more gas pressure is allowed to build up against the diaphragm to re-establish the balance condition.

In the nozzle-type regulator, the inlet pressure tends to move the seat member away (pressure opening) from the nozzle, thus opening the valve. The outlet pressure of this

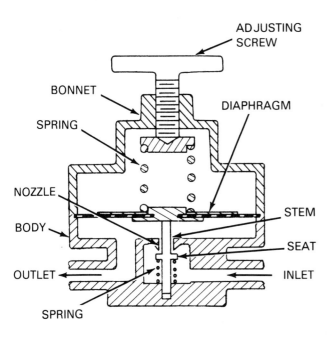

Figure 11.12A–Cross Section Showing the Major Components of a Typical Single-Stage-Stem-Type Regulator

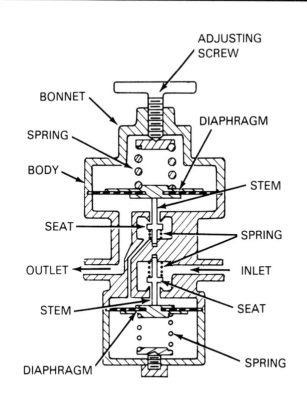

Figure 11.12C–Cross Section Showing the Major Components of a Typical Two-Stage-Type Regulator

type of regulator decreases somewhat as the inlet pressure decreases, because the force tending to move the seat member away from the nozzle is reduced as the inlet pressure decreases. A small outlet pressure on the underside of the diaphragm is then required to close the seat member against the nozzle.

Two-Stage Regulators. The two-stage regulator provides more precise regulation over a wide range of varying inlet pressures. A two-stage regulator, as illustrated in Figure 11.12C, is actually two single-stage regulators in series incorporated as one unit.

The outlet pressure from the first stage is usually preset to deliver a specified inlet pressure to the second stage. In this way, a practically constant delivery pressure may be obtained from the outlet of the regulator as the supply pressure decreases.

The combinations employed to make a two stage regulator are as follows:

(1) Nozzle type first-stage and stem type second-stage
(2) Stem type first-stage and nozzle type second-stage
(3) Two stem types, as illustrated in Figure 11.12C
(4) Two nozzle types

Regardless of the combination used, the increase or decrease in outlet pressure is usually so slight (and apparent only at very low inlet supply pressures), that for all practical purposes the variation in delivery pressure is disregarded in welding and cutting operations. Two-stage regulators are suggested for precise work, such as continuous machine cutting, in order to maintain a constant working pressure and a controlled volumetric flow at the welding or cutting torch.

Applications of Regulators. Regulators are produced with different capacities for pressure and volumetric flow, depending on the application and the source of supply. They should, therefore, be used only for the purpose intended. In oxyfuel gas welding, the requirements for cylinder regulators are considerably different from those of station regulators.

In the commonly used one-torch setup shown in Figure 11.6, oxygen and acetylene are supplied from single cylinders; each is connected to a cylinder regulator, which may be either the single- stage or the two-stage type. Each regulator is equipped with two pressure gages, one indicating the inlet or cylinder pressure, the other indicating the outlet or torch working pressure. Cylinder regulators and cylinder pressure gages are built to withstand high pressures with a safe overload margin. Working pressure gages are built and graduated to accommodate the intended service application.

Pipeline pressures for oxygen seldom exceed 200 psig; for acetylene, the pressure must not exceed 15 psig. Station regulators are, therefore, built for low-pressure operation, although they may have high volumetric flow capacity. Station regulator requirements are adequately met by single-stage types equipped only with a working pressure gage. Due to their capacity limitations, station regulators should never be substituted for cylinder regulators because of the possibility of a serious accident.

Cylinder regulators should not be used on station outlets because they may not have adequate flow capacity, due to the low inlet pressures.

Regulator Inlet and Outlet Connections. Cylinder outlet connections are of different sizes and shapes to preclude the possibility of connecting a regulator to the wrong cylinder. Regulators must, therefore, be made with different inlet connections to fit the various gas cylinders. The Compressed Gas Association (CGA) has standardized noninterchangeable cylinder valve outlet connections. These specifications are published by the American National Standards Institute as ANSI B57.1, *Compressed Gas Cylinder Valve Outlet and Inlet Connections*, latest edition, which should be consulted for information.

Regulator outlet fittings also differ in size and thread, depending upon the gas and regulator capacity. Oxygen outlet fittings have right-hand threads; fuel outlet fittings have left-hand threads, with grooved nuts.

Regulator Safety

THE FOLLOWING SAFETY precautions should always be observed to prevent accidents when using regulators:

(1) The operator should be trained in the proper use of regulators or be under competent supervision. It is important to follow carefully the manufacturer's recommended operating procedures.

(2) The regulator should always be clean and in good working condition.

(3) Cylinders should be secured to a wall, post, or cart so they will not tip or fall.

(4) Cylinder valves should be inspected for damaged threads, dirt, dust, oil, or grease. Remove dust and dirt with a clean cloth. DO NOT ATTACH THE REGULATOR TO A GAS CYLINDER IF OIL, GREASE, OR DAMAGE IS PRESENT! Inform your gas supplier of this condition. Regulators should only be repaired or serviced by qualified technicians.

(5) Crack open the cleaned cylinder valve for an instant, and then close it quickly. This will blow out any foreign matter that may be inside the valve port.
CAUTION: If the cylinder valve is opened too much, the cylinder may tip over due to the force of escaping gas. Do not stand in front of the valve port.

(6) Attach the regulator to the cylinder valve and tighten securely with a wrench.

(7) Before opening the cylinder valve, turn the regulator adjusting screw counter-clockwise until the adjusting spring pressure is released.

(8) Stand to the side of the regulator when opening the cylinder valve. NEVER stand in front of or behind a regulator. CAREFULLY and SLOWLY open the cylinder valve until the cylinder pressure is indicated on the high pressure gauge.
NOTE: Never open acetylene cylinder valves more than one (1) complete turn. All other cylinder valves should be opened completely to seal the valve packing.

(9) Turn the regulator adjusting screw clockwise to attain the desired delivery pressure for the apparatus you are using.

(10) The system should be tested for leaks by using the methods recommended by the manufacturer of the regulator.

(11) Keep the cylinder valves closed at all times, except when the cylinder is in use. When you are finished using the apparatus, close both cylinder valves. To bleed the system, open the fuel gas valve on the torch, and allow the gas to escape at a safe location. Then close the torch fuel valve and turn the regulator pressure adjusting screw counterclockwise until the screw turns freely. Repeat this operation on the oxygen system. It is important not to bleed both systems at the same time. A reverse flow or mixing of gases may result, which may be dangerous.

Accessories

IN ADDITION TO the equipment and materials described above, a wide variety of auxiliary equipment may be used in the process of gas welding. Only a brief description of such items is included here.

Two of the most universally required articles are the friction lighter, which should always be used to ignite the gas, and check valves at both ends of the hoses for safety.

Other accessories, such as tip cleaners, cylinder trucks, clamps, and holding jigs and fixtures are also important auxiliary aids for gas welding.

Welders should at all times use goggles or eyeshields as a protection against sparks and the intense glare and heat radiated from the flame and molten metal. Suitable gloves, leather aprons, sleeves, and leggings should also be worn. In some situations, a forced ventilation or supplemental breathing system may be required.

STORAGE AND DISTRIBUTION

THE DISTRIBUTION OF gases to the work facility is dependent upon location, size, consumption requirements, and application of the various oxyfuel processes. Methods of delivery can be by single cylinders, portable or stationary manifolds, bulk supply systems, and pipelines.

Individual cylinders of gaseous oxygen and acetylene provide an adequate supply of gas for welding and cutting torches that consume limited quantities of gas. Cylinder trucks are used extensively to provide a convenient, safe support for a cylinder of oxygen and a cylinder of acetylene. Gases are transported readily by such means.

Oxygen may be brought to the user in individual cylinders as a compressed gas or as a liquid; there are also several bulk distribution methods. Gaseous oxygen in cylinders is usually under a pressure of approximately 2200 psig (15 170 kPa). Cylinders of various capacities are used, holding approximately 70, 80, 122, 244, and 300 ft^3 (2, 2.3, 3.5, 6.9, and 8.5 m^3) of oxygen. Liquid oxygen cylinders contain the equivalent of approximately 3000 ft^3 (85 m^3) of gaseous oxygen. These cylinders are used for applications which do not warrant a bulk oxygen supply system but which are too large to be supplied conveniently by gaseous oxygen in cylinders. The liquid oxygen cylinders are equipped with liquid-to-gas converters, or may use external gas vaporizors.

Care must be taken not to exceed a certain withdrawal rate from a given size acetylene cylinder. If the volumetric demand is too high, acetone may be drawn from the cylinder with the acetylene. It has, therefore, become standard practice to limit the withdrawal of acetylene from a single cylinder to an hourly rate not exceeding one seventh of the cylinder's volumetric contents. Two or more acetylene cylinders may be manifolded together to provide high flow rates.

Cylinders for liquefied fuel gas contain no filler material. These welded steel or aluminum cylinders hold the liquefied fuel gas under pressure. The pressure in the cylinder is a function of the temperature. Liquefied fuel gas cylinders have relief valves which are set at predetermined pressures to prevent overpressurization should the cylinder be exposed to temperatures approaching 200°F (93°C).

Should these temperatures be reached, the rapid discharge of fuel gas through the relief valve causes the cylinder to cool down, the cylinder pressure drops, and the relief valve closes. In a fire, the cylinder relief valve opens and shuts intermittently until all the fuel in the cylinder has been discharged or the source of the extreme heat has been removed.

The withdrawal rate of liquefied fuel gases from cylinders is a function of the following:

(1) The temperature
(2) The amount of fuel in the cylinder
(3) The desired operating pressure

Pertinent information should be obtained from the gas supplier.

Manifolded Cylinders

INDIVIDUAL CYLINDERS CANNOT supply high rates of gas flow, particularly for continuous operation over long periods of time. Manifolding of cylinders is one answer to this problem. A reasonably large volume of fuel gas is provided by this means, and it can be discharged at a moderately rapid rate.

Manifolds are of two types, portable and stationary. Portable manifolds (see Figure 11.13) may be installed with a minimum of effort and are useful where moderate volumes of gas are required for jobs of a nonrepetitive nature, either in the shop or the field.

Stationary manifolds (see Figure 11.14) are installed in shops where larger volumes of gas are required. Such a manifold feeds a pipeline system distributing the gas to various stations throughout the plant. This arrangement enables many operators to work from a common pipeline system without interruption. Alternatively, it may supply large automatic torch brazing or oxygen cutting operations.

An important protective device for the stationary fuel system is its hydraulic seal or hydraulic flashback arrestor. This device will stop a flashback originating at the station from passing further into the system. It consists of a small pressure vessel partly filled with water, through which the fuel supply flows. The gas continues through the space above the water level and through the vessel head to the station regulator. A flashback or high-pressure backup will cause a relief valve in the vessel head to vent the pressure to the atmosphere outside. A check valve prevents the water from backing up into the line.

Bulk Systems

TO SATISFY THE large consumption of some industries, gaseous oxygen may be transported from the producing plant to the user in multiple cylinder portable banks or in long, high-pressure tubes, mounted on truck trailers. The trailers may hold as many as 30 000 to 50 000 ft³ (850 to 1420 m³) in the larger units and 10 000 ft³ (285 m³) in the smaller units.

Bulk oxygen may also be distributed as a liquid in large insulated containers mounted on truck trailers or railroad cars. The liquid oxygen is transferred to an insulated storage tank on the consumer's property. The oxygen is withdrawn, converted to gas, and passed into distribution pipelines, as needed, by means of automatic regulating equipment.

Liquefied fuel gases may be distributed from onsite bulk tanks of 500 to 12 000 gal (2000 to 45 000 l) capacity. The tanks are filled periodically by truck delivery.

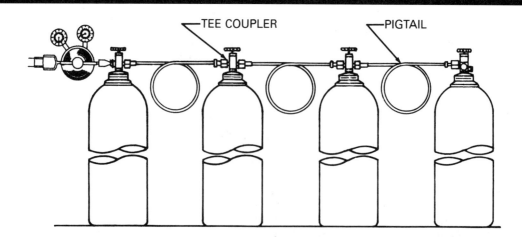

Figure 11.13–Typical Arrangement for a Portable Oxygen Manifold

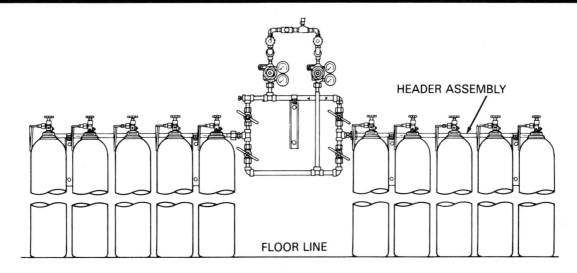

Figure 11.14–Typical Arrangement for a Stationary Gas Manifold

Other Systems

WHEN STATIONARY MANIFOLDS, truck trailers, or bulk supply systems are used as sources of oxygen, and large acetylene generators are used to produce acetylene in a consumer's plant, the gases are distributed by pipelines to the points of use. The pipelines should be designed properly to handle and distribute the gases in sufficient volume without undue pressure drop. They should also incorporate all necessary safety devices.

SAFE PRACTICES FOR MANIFOLDS AND PIPELINES

THE RULES AND regulations set forth in the current issue of the *Standard for the Installation and Operation of Oxygen-Fuel Gas Systems for Welding and Cutting*, as Recom- mended by the National Fire Protection Association (NFPA No. 51)[1], govern the installation of oxygen and fuel gas manifolds and pipelines. Local regulations or ordinances should be consulted also to be sure of compliance with them. Manifolds should, in all cases, be obtained from reliable manufacturers. They should be installed by personnel familiar with proper construction and installation of oxygen and acetylene manifolds and pipelines.

It is important to note that copper tubing, while satisfactory for use as oxygen pipelines, should never be used for acetylene piplines. The reason is that acetylene, in proximity to copper, may form copper acetylide, which may explode spontaneously.

1. Obtainable from the National Fire Protection Association, Battery Park, Quincy, Massachusetts 02269

APPLICATIONS OF OXYFUEL GAS WELDING

METALS WELDED

OXYFUEL WELDING CAN be used on a wide range of commercial ferrous and nonferrous metals and alloys. As in any welding process, however, physical dimensions and chemical composition may limit the weldability of certain materials and pieces.

During welding, the metal is taken through a temperature range almost the same as that of the original casting procedure. The base metal in the weld area loses those properties that were given to it by prior heat treatment or cold working. The ability to weld such materials as high carbon and high alloy steels is limited by the equipment available for heat treating after welding. These metals are successfully welded when the size or nature of the piece permits postheat treating operations.

The welding procedure for plain carbon steels is straightforward and offers little difficulty to the welder. Sound welds are produced in other materials by variations in preheating, technique, heat treating, and fluxing.

The oxyfuel welding process can be used for repair welding metal of considerable thickness and for the usual assemblies encountered in maintenance and repair. Very thick cast iron machinery frames have been repaired by braze welding or by welding with a cast iron filler rod.

Steels and Cast Iron

LOW CARBON, LOW alloy, and cast steels are the materials most easily welded by the oxyacetylene process. Fluxes are usually required when welding these materials.

In oxyfuel welding, steels having more than 0.35 percent carbon are considered high carbon steels and require special care to maintain their particular properties. Alloy steels of the air-hardening type require extra precautions to maintain their properties, even though the carbon content may be 0.35 percent or less. The joint area usually is preheated to retard the cooling of the weld by conduction of heat into surrounding base metal. Slow cooling prevents the hardness and brittleness associated with rapid cooling. A full furnace anneal or heat treatment may be required immediately after welding air-hardening steels.

The welder should use a neutral or slightly carburizing flame for welding and should be careful not to overheat and decarburize the base metal. The preheating temperature required depends upon the composition of the steel to be welded. Temperatures ranging from 300° to 1000°F (150° to 540°C) have been used.

In addition to finding the proper preheat temperature, it is important that it be maintained uniformly during welding. A uniform temperature can be maintained by protecting the part with a heat retaining covering. Other means of shielding can also be used to retain the temperature in the part. Generally, the interpass temperature should be maintained within 150°F (65°C) of the preheat temperature. Lower interpass temperatures cause excessive shrinking forces that can result in either distortion or cracking at the weld or other sections. This type of cracking frequently occurs in the welding of circular structures made of brittle metals, such as cast iron.

Modifications in procedures are required for stainless and similar steels. Because of their high chromium-nickel content, these steels have relatively low thermal conductivity, and a smaller flame than that used for equal thicknesses of plain carbon steel is recommended. Because chromium oxidizes easily, a neutral flame is employed to minimize oxidation. A flux is used to dissolve oxides and protect the weld metal. Filler metal of high chromium or nickel-chromium steel is used. Even with these precautions, whenever a weld of high caliber is required, a process other than oxyfuel welding is generally recommended. Table 11.3 summarizes the basic information for welding ferrous metals.

Cast iron, malleable iron, and galvanized iron all present particular problems in welding by any method. The gray cast iron structure in cast iron can be maintained through the weld area by the use of preheat, a flux, and an appropriate cast iron welding rod.

Nodular iron requires materials in the welding filler metal that will assist in promoting agglomeration of the free graphite, in order to maintain ductility and shock resistance in the heat-affected area. The filler metal manufacturer should be consulted to obtain information on preheat and interpass temperature control for the filler metal being used.

There are, of course, instances in which cast irons are welded without preheat, particularly in salvage work. In most applications, however, preheat from 400° to 600°F (200° to 320°C), with control of interpass temperature

Table 11.3
General Conditions for Oxyacetylene Welding of Various Ferrous Metals

Metal	Flame Adjustment	Flux	Welding Rod
Steel, cast	Neutral	No	Steel
Steel pipe	Neutral	No	Steel
Steel plate	Neutral	No	Steel
Steel sheet	Neutral	No	Steel
	Slightly oxidizing	Yes	Bronze
High carbon steel	Slightly carburizing	No	Steel
Wrought iron	Neutral	No	Steel
Galvanized iron	Neutral	No	Steel
	Slightly oxidizing	Yes	Bronze
Cast iron, gray	Neutral	Yes	Cast iron
	Slightly oxidizing	Yes	Bronze
Cast iron, malleable	Slightly oxidizing	Yes	Bronze
Cast iron pipe, gray	Neutral	Yes	Cast iron
	Slightly oxidizing	Yes	Bronze
Cast iron pipe	Neutral	Yes	Cast iron or base metal composition
Chromium-nickel steel castings	Neutral	Yes	Base metal composition or 25-12 chromium-nickel steel
Chromium-nickel steel (18-8 and 25-12)	Neutral	Yes	Columbium stainless steel or base metal composition
Chromium steel	Neutral	Yes	Columbium stainless steel or base metal composition
Chromium iron	Neutral	Yes	Columbium stainless steel or base metal composition

and provision for slow cooling, will assure more consistent results. Protection such as a heat resistant covering can be used to assure slow and uniform cooling. Care should be taken that localized cooling is not allowed to occur. It should also be stressed that in the salvage of cast iron, removal of all foundry sand and slag is necessary for consistent repair results.

Nonferrous Metals

THE PARTICULAR PROPERTIES of each nonferrous alloy should be considered when selecting the most suitable welding technique. When the necessary precautions are taken, acceptable results can be obtained.

Aluminum, for instance, gives no warning by changing color prior to melting, but it appears to collapse suddenly at the melting point. Consequently, practice in welding is required to learn to control the rate of heat input. Aluminum and its alloys suffer from hot shortness, and welds should be supported adequately in all areas during welding. Finally, any exposed aluminum surface is always covered with a layer of oxide that, when combined with the flux, forms a fusible slag which floats on top of the molten metal.

When copper is welded, allowances are necessary for the chilling of the welds because of the very high thermal conductivity of the metal. Preheating is often required. Considerable distortion can be expected in copper because the thermal expansion is higher than in other commercial metals. These characteristics obviously pose difficulties that must be overcome for satisfactory welding.

Parts to be welded should be fixtured or tack welded securely in place. The section least subject to distortion should be welded first so that it forms a rigid structure for the balance of the welding.

When the design of the structure permits welding from both sides, distortion can be minimized by welding alternately on each side of the joint. Strongbacks or braces can be applied to sections most likely to distort. The welds may be peened to reduce distortion. This method, if performed properly, can overcome severe warpage. A backstep sequence of welding may be used to control distortion. This method consists primarily of making short weld increments in the direction opposite to the progress of welding the joint. The weldment should be designed so that distortion during welding will be minimized.

Metallurgical Effects

THE TEMPERATURE OF the base metal varies during welding, from that of the molten weld puddle down to room temperature in areas most remote from the weld. When steels are involved, the weld and the adjacent heat-affected zones are heated considerably above the transformation temperature of the steel. This results in a coarse grain structure in the weld and adjacent base metal. The coarse grain structure can be refined by a normalizing heat treatment, such as heating to the austenitizing temperature range (approximately 1650°F or 900°C) and cooling in air after welding.

Hardening of the base metal heat-affected zone heated to above the transformation temperature of the steel can occur if the steel contains sufficient carbon and the cooling rate is high enough. Hardening can be avoided in most hardenable steels by using the torch to keep heat on the weld for a short time after the weld is completed. If air-hardening steels are welded, the best heat treatment is full furnace anneal of the weldment.

The oxyfuel flame allows a degree of control to be maintained over the carbon content of the deposited metal and over the portion of the base metal that is heated to its melting temperature. When an oxidizing flame is used, a rapid reaction results between the oxygen and the carbon in the metal. Some of the carbon is lost in the form of carbon monoxide, and the steel and the other constituents are also oxidized. When the torch is used with an excess acetylene flame, carbon is introduced into the molten weld metal.

When heated to a temperature range between 800° and 1600°F (430° and 870°C), carbide precipitation occurs in unstabilized austenitic stainless steel. Chromium carbides gather at the grain boundaries and lower the corrosion resistance of the heat-affected zone. If this occurs, a heat treatment after welding is required, unless the steel is an alloy stabilized by the addition of columbium or titanium and welded with the aid of a columbium-bearing stainless steel welding rod. The columbium combines with the carbon and so minimizes the formation of chromium carbide. All the chromium is left dissolved in the austenitic matrix, the form in which it can best resist corrosion.

Another factor to be considered in welding is the possible tendency toward hot shortness of the metal (a marked loss in strength at high temperatures). Some of the copper base alloys have this tendency to a high degree. If the base metal has this tendency, it should be welded with care to prevent hot cracking in the weld zone. Allowances should be made in the welding technique used with these metals, and jigging or clamping should be done with caution. Proper welding sequence and multiple layer welding with narrow stringer beads help to reduce hot cracking.

Oxidation and Reduction

CERTAIN METALS HAVE such a high affinity for oxygen that oxides form on the surface almost as rapidly as they are removed. In oxyacetylene operations, these oxides are usually removed by means of fluxes. This affinity for oxygen can be a useful characteristic in certain welding operations. The manganese and silicon contained in plain carbon steel, for example, are important in oxyfuel welding because they deoxidize the weld puddle. The correct manganese and silicon content of steel welding rods is, therefore, important.

The type of flame used in welding various metals plays an important part in securing the most desirable weld metal deposit. The proper type of flame with the correct welding technique can be used as a shielding medium which will reduce the effect of oxygen and nitrogen (in the atmosphere) on the molten metal. Such a flame also has the effect of stabilizing the molten weld metal and preventing the loss of carbon, manganese, and other alloying elements.

The proper type of oxyfuel flame for any application is determined by the type of base and filler metals involved and the thickness of the base metal. For most metals a neutral flame is used. An exception is the welding of aluminum and high carbon steel, where a slightly carburizing (reducing) flame is used.

FILLER METALS

THE PROPERTIES OF the weld metal should closely match those of the base metal. Because of this requirement, welding rods of various chemical compositions are available for welding many ferrous and nonferrous materials. Obviously, it is important that the correct welding filler metal be selected.

The welding process itself influences the filler metal composition, since certain elements are lost during welding. Filler metals are available for joining almost all common base materials. The standard diameters of rods vary from 1/16 to 3/8 in. (1.6 to 10mm), and the standard lengths for rods are 24 and 36 in. (610 and 914 mm).

The chemical composition of a filler metal must be within the limits specified for that particular material. There are many proprietary filler metals on the market recommended for specific applications. Filler metal should be free from porosity, pipes, nonmetallic inclusions, and any other foreign matter. The metal should deposit smoothly.

Allowances for changes taking place during welding are made in the production of welding rods so that the deposited metal will have the correct chemical composition. Deposits should be made with free-flowing filler metal that unites readily with the base metal to produce sound, clean welds.

In maintenance and repair work it is not always necessary that the composition of the welding rod match that of the base metal. A steel welding rod of nominal strength can be used to repair parts made of alloy steels broken by overloading or accident. Every effort should be made, however, to match the filler metal and base metal. Where it is necessary to heat-treat a steel part after welding, carbon can be added to a deposit of mild steel by the judicious use of the carburizing flame. It is preferable, however, to use a welding rod of low alloy steel.

The AWS Committee on Filler Metal has prepared a number of specifications. Many of the oxyfuel gas welding filler metals meet these specifications. Information on the welding and selection of filler materials will be found in the chapters on the various metals and alloys in the *Welding Handbook*, Volume 4, 7th Edition.

Welding rods for steel are listed in ANSI/AWS A5.2, *Specification for Carbon and Low Alloy Steel Rods for Oxyfuel Gas Welding*. The rods are classified on the basis of strength. The most commonly used is RG60 (60 ksi or 414 MPa minimum tensile strength), which has properties compatible with most low carbon steels.

For the oxyfuel and braze welding of cast iron, both cast iron and copper base welding rods are used. See ANSI/AWS A5.15, *Specification for Welding Electrodes and Rods for Cast Iron*. These filler metals are classified on the basis of chemical composition.

FLUXES

ONE OF THE most important ways to control weld quality is to remove oxides and other impurities from the surface of the metal to be welded. Unless the oxides are removed, fusion may be difficult, the joint may lack strength, and inclusions may be present. The oxides will not flow from the weld zone but will remain to become entrapped in the solidfying metal, interfering with the addition of filler metal. These conditions may occur when the oxides have a higher melting point than the base metal, and a means must be found to remove those oxides. Fluxes are applied for this purpose.

Steel and its oxides, and slags which form during welding, do not fall into the above category, and they need no fluxing. Aluminum, however, forms an oxide with a very high melting point, and the oxide must be removed from the welding zone before satisfactory results can be obtained. Certain substances will react chemically with the oxides of most metals, forming fusible slags at welding temperature. These substances, either singly or in combination, make efficient fluxes. GTAW or GMAW is generally used to weld aluminum to avoid these slag problems.

A good flux should assist in removing the oxides during welding by forming fusible slags that will float to the top of the molten puddle and not interfere with the deposition and fusion of filler metal. A flux should protect the molten puddle from the atmosphere and also prevent the puddle from absorbing or reacting with gases in the flame. This must be done without obscuring the welder's vision or hampering manipulation of the molten puddle.

During the preheating and welding periods, the flux should clean and protect the surfaces of the base metal and, in some cases, the welding rod. Flux should not be used as a substitute for base metal cleaning during joint preparation. Fluxes are excellent metal cleaners, but if their activity is used for cleaning dirty metal, that will interfere with their primary functions.

Flux may be prepared as a dry powder, a paste or thick solution, or a preplaced coating on the welding rod. Some fluxes operate much more favorably if they are used dry. Braze welding fluxes and fluxes for use with cast iron are usually in this class. These fluxes are applied by heating the end of the welding rod and dipping it into the powdered flux. Enough will adhere to the rod for adequate fluxing. Dipping the hot rod into the flux again will coat another portion.

Dropping some of the dry powder on the base metal ahead of the welding zone will sometimes help, especially in the repair of "dirty" castings.

Fluxes in paste form are usually painted on the base metal with a brush, and the welding rod can either be painted or dipped.

Commercial precoated rods can be used without further preparation, and, when required, additional flux can be placed on the base metal. Sometimes a precoated rod will need to be dipped in powdered flux, if the flux melts off too far from the end of the rod during welding.

The common metals and welding rods requiring fluxes are bronze, cast iron, brass, silicon bronze, stainless steel, and aluminum.

TYPICAL APPLICATIONS

THE WIDE FIELD of possible applications, as well as the convenience and the economy of oxyacetylene welding, are recognized in most metal working industries. The process is used for fabricating sheet metal, tubing, pipe, and other metal shapes in industries such as industrial piping and automotive. It is also used in welding pipelines up to 2 in. (51mm) in diameter.

The process is almost universally used and accepted in the field of maintenance and repair, where its flexibility and mobility result in great savings of time and labor. The typical self-contained unit, consisting of a welding torch, an oxygen cylinder, and an acetylene cylinder on a two-wheeled cart, can be readily moved about in a plant. It can be carried easily into the field on a small truck to service a breakdown wherever it may have occurred. Oxyacetylene welding is also well suited for use in machine and automobile repair shops. It is equally useful in shops devoted entirely to welding, where the repair of large and small industrial, agricultural, and household equipment may be the main business.

The oxyfuel process is used for many surfacing operations, some of which are not possible with arc welding processes. For instance, the application of materials high in zinc content, such as admiralty metal, can be accomplished with the gas welding torch. Automatic procedures are used for this type of surfacing application, such as on tube sheets or heat exchangers.

The same gases and equipment can be used in cutting, brazing, soldering, heat treating, surfacing, and welding. This makes the oxyacetylene process particularly attractive from the viewpoint of initial investment.

WELDING PROCEDURES

OPERATING PRINCIPLES

THE OXYFUEL GAS welding torch serves as the implement for mixing the combustible and combustion-supporting gases and provides the means for applying the flame at the desired location. A range of tip sizes is provided for obtaining the required volume or size of welding flame. Tip flames may vary from a short, small diameter needle flame to a flame 3/16 in. (4.8 mm) or more in diameter and 2 in. (50 mm) or more in length.

The inner cone or vivid blue flame of the burning mixture of gases issuing from the tip is called the *working flame*. The closer the end of the inner cone is brought to the surface of the metal being heated or welded, the more effective is the heat transfer from flame to metal. The flame can be made soft or harsh by varying the gas flow. Too low a gas flow for a given tip size will result in a soft, ineffective flame sensitive to backfiring. Too high a gas flow will result in a harsh, high velocity flame that is hard to handle and will blow the molten metal from the puddle.

The chemical action of the flame on a molten pool of metal can be altered by changing the ratio of the volume of oxygen to the volume of acetylene issuing from the tip.

Most oxyacetylene welding is done with a neutral flame having approximately a 1:1 gas ratio. An oxidizing action can be obtained by increasing the oxygen flow, and a reducing action will result from increasing the acetylene flow. Both adjustments are valuable aids in welding.

Flame Adjustment

WELDING TORCHES SHOULD be lighted with a friction lighter or a pilot flame. The instructions of the equipment manufacturer should be observed when adjusting operating pressures at the gas regulators and torch valves, before the gases issuing from the tip are ignited. Three types of oxyacetylene flame adjustment are shown in Figure 11.15.

The neutral flame is obtained most easily by adjustment from an excess acetylene flame, which is recognized by the feather extension of the inner cone. The feather will diminish as the acetylene flow is decreased or the flow of oxygen is increased. The flame is neutral just at the point of disappearance of the "feather" extension of the inner cone. This neutral flame, while neither carburizing or oxidizing within itself, may have a reducing effect on the metal being welded.

(A) PURE ACETYLENE FLAME

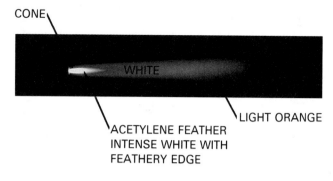

CONE

WHITE

ACETYLENE FEATHER
INTENSE WHITE WITH
FEATHERY EDGE

LIGHT ORANGE

(B) CARBURIZING FLAME

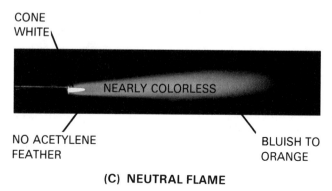

CONE
WHITE

NEARLY COLORLESS

NO ACETYLENE
FEATHER

BLUISH TO
ORANGE

(C) NEUTRAL FLAME

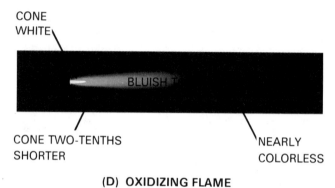

CONE
WHITE

BLUISH

CONE TWO-TENTHS
SHORTER

NEARLY
COLORLESS

(D) OXIDIZING FLAME

Figure 11.15–Oxyacetylene Flame Adjustments

A practical method for determining the amount of excess acetylene in a reducing flame is to compare the length of the feather with the length of the inner cone, measuring both from the torch tip. An excess acetylene flame with twice the acetylene required for a neutral flame has an acetylene feather that is twice the length of the inner cone. Starting with a neutral flame adjustment, the welder can produce the desired acetylene feather by increasing the acetylene flow.

The oxidizing flame adjustment is sometimes given as the amount by which the length of a neutral inner cone should be reduced — for example, one tenth. Starting with the neutral flame, the welder can increase the oxygen until the length of the inner cone is decreased the desired amount.

Forehand and Backhand Welding

OXYACETYLENE WELDING MAY be performed with the torch tip pointed forward in the direction the weld progresses. This method is called the *forehand technique*. Welding can also be done in the opposite direction, with the torch pointing toward the completed weld; this method is known as *backhand welding*. Each method has its advantages, depending upon the application, and each method imposes some variations in deposition technique.

In general, the forehand method is recommended for welding material up to 1/8 in. (3.2 mm) thick, because it provides good control of a small weld puddle, resulting in a smoother weld at both top and bottom. The puddle of molten metal is small and easily controlled. A great deal of pipe welding is done using the forehand technique, even in 3/8 in. (9.5 mm) wall thicknesses.

Increased speeds and better control of the puddle are possible with the backhand technique when metal 1/8 in. (3.2 mm) and thicker is welded. This recommendation is based on careful study of the speeds normally achieved with this technique, and on the greater ease of obtaining fusion at the root of the weld. Backhand welding may be used with a slightly carburizing flame (slight acetylene feather) when it is desirable to melt a minimum amount of steel in making a joint. The increased carbon content obtained from this flame lowers the melting point of a thin layer of steel, and increases welding speed. This technique greatly increases the speed of making pipe joints where the wall thickness is 1/4 to 5/16 in. (6.4 to 7.9 mm) and the groove angle is less than normal. Backhand welding is sometimes used in surfacing operations.

Base Metal Preparation

CLEANLINESS ALONG THE joint and on the sides of the base metal is of the utmost importance. Dirt, oil, and oxides can cause incomplete fusion, slag inclusions, and porosity in the weld.

The spacing between the parts to be joined should be considered carefully. The root opening for a given thickness of metal should permit the gap to be bridged without difficulty, yet it should be large enough to permit full penetration. Specifications for root openings should be followed exactly.

The thickness of the base metal at the joint determines the type of edge preparation for welding. Thin sheet metal is easily melted completely by the flame. Thus, edges with square face can be butted together and welded. This type of joint is limited to material under 3/16 in. (4.8 mm) in thickness. For thicknesses of 3/16 to 1/4 in. (4.8 to 6.4 mm), a slight root opening or groove is necessary for complete penetration, but filler metal must be added to compensate for the opening.

Joint edges 1/4 in. (6.4 mm) and greater in thickness should be beveled. Beveled edges at the joint provide a groove for better penetration and fusion at the sides. The angle of bevel for oxyacetylene welding varies from 35 to 45 degrees, which is equivalent to a variation of the included angle of the joint from 70 to 90 degrees, depending upon the application. A root face 1/16 in. (1.6 mm) wide is normal, but feather edges are sometimes used. Plate thicknesses 3/4 in. (19 mm) and above are double beveled when welding can be done from both sides. The root face can vary from 0 to 1/8 in. (0 to 3.2 mm). Beveling both sides reduces by approximately one-half the amount of filler metal required. Gas consumption per unit length of weld is also reduced.

A square groove edge preparation is the easiest to obtain. This edge can be machined, chipped, ground, or oxygen cut. The thin oxide coating on an oxygen-cut steel surface does not need to be removed because it is not detrimental to the welding operation or to the quality of the joint. A bevel angle can be oxygen cut.

MULTIPLE LAYER WELDING

MULTIPLE LAYER WELDING is used when maximum ductility of a steel weld in the as-welded or stress-relieved condition is desired, or when several layers are required to weld thick metal. Multiple layer welding is accomplished by depositing filler metal in successive passes along the joint until it is filled. Since the area covered with each pass is small, the weld puddle is reduced in size. This procedure enables the welder to obtain complete joint penetration without overheating while the first few passes are being deposited. The smaller puddle is more easily controlled, and the welder can thus avoid oxides, slag inclusions, and incomplete fusion with the base metal.

An increase in ductility in the deposited steel results from grain refinement in the underlying passes when they are reheated. The final layer will not possess this refinement, unless an extra pass is added and removed, or unless the torch is passed over the joint to bring the last deposit up to normalizing temperature.

Oxyacetylene welding is not recommended for high-strength heat treatable steels, especially when they are being fabricated in the heat treated condition. In welding quenched and tempered steels, the slow rate of heat input with oxyacetylene welding may cause metallurgical changes in the heat-affected area and so may destroy the heat treated base metal properties. This type of metal should be welded with one of the arc welding processes.

WELD QUALITY

THE APPEARANCE OF a weld does not necessarily indicate its quality. If discontinuities exist in a weld, they can be grouped into two broad classifications: those that are apparent to visual inspection and those that are not. Visual examination of the underside of a weld will determine whether there is complete penetration and whether there are excessive globules of metal. Inadequate joint penetration may be due to insufficient beveling of the edges, too thick a root face, too high a welding speed, or poor torch and welding rod manipulation.

Oversized and undersized welds can be observed readily. Weld gages are available to determine whether a weld has excessive or insufficient reinforcement. Undercut or overlap at the sides of the welds can usually be detected by visual examination.

Although other discontinuities, such as incomplete fusion, porosity, and cracking, may or may not be externally apparent, excessive grain growth and the presence of hard spots cannot be determined visually. Incomplete fusion may be caused by insufficient heating of the base metal, too rapid travel, or gas or dirt inclusions. Porosity is a result of entrapped gases, usually carbon monoxide, which may be avoided by more careful flame manipulation and adequate fluxing where needed. Hard spots and cracking are a result of metallurgical characteristics of the weldment. For further details, see Chapter 4, Welding Metallurgy, the *Welding Handbook*, Volume 1, 8th Edition.

INSPECTION

THE TERM *inspection* usually implies a formal inspection, prescribed by a code or by the requirements of a purchaser, that is given to welds and welded structures. The minimum requirements of welding codes are inflexible, and must be met.

WELDING WITH OTHER FUEL GASES

PRINCIPLES OF OPERATION

GASES SUCH AS propane, natural gas, butane, MPS, propylene, and other similar gases are not suitable for welding ferrous materials due to their oxidizing characteristics. Many nonferrous and ferrous metals can be braze welded with these gases, taking care of the adjustment of flame and the use of flux. It is important to use tips designed for the fuel gas being employed. These gases are extensively used for both manual and mechanized brazing and soldering operations.

These fuel gases have relatively low flame propagation rates. When standard welding tips are used, the maximum flame velocity is so low that it interferes seriously with heat transfer from the flame to the work. The highest flame temperatures of the gases are obtained at high oxygen-to-fuel gas ratios. These ratios produce highly oxidizing flames which prevent the satisfactory welding of most metals.

Tips having flame-holding devices, such as skirts, counterbores, and holder flames, should be used to permit higher gas velocities before they leave the tip. This makes it possible to use these fuel gases for many heating applications with excellent heat transfer efficiency.

Air contains approximately 80 percent nitrogen by volume. Since nitrogen does not support combustion, fuel gases burned with air produce lower flame temperatures than those burned with oxygen. The total heat content is also lower. The air-fuel gas flame is suitable only for welding light sections of lead and for light brazing and soldering operations.

EQUIPMENT

STANDARD OXYACETYLENE EQUIPMENT, with the exception of torch tips and regulators, can be used to distribute and burn these gases. Special regulators may be obtained, and specialized heating and cutting tips are available. Natural gas is supplied by pipelines; other fuel gases are stored in cylinders or delivered in liquid form to storage tanks on the user's property. Delivery is made through the pipeline equipment to the points of use.

The torches for use with air-fuel gas generally are designed to aspirate the proper quantity of air from the atmosphere to support combustion. The fuel gas flows through the torch at a supply pressure of 2 to 40 psig and aspirates the needed air. For light work, the fuel gas usually is supplied from a small cylinder that is easily transportable.

The plumbing, refrigeration, and electrical trades use propane in small cylinders for many heating and soldering applications. The propane flows through the torch at a supply pressure from 3 to 60 psig, which serves to aspirate the air. The torches are used for soldering electrical connections, the joints in copper pipelines, and light brazing jobs.

Safety precautions listed earlier in this chapter should be observed when using these fuel gases. Storage and distribution systems should be installed according to the applicable national, state, or local codes.

APPLICATIONS

AIR-FUEL GAS IS used for welding lead up to approximately 1/4 in. (6.4 mm) in thickness. Perhaps its greatest field of application, however, is in the plumbing and electrical industry. There it is used extensively for soldering copper tubing.

SAFE PRACTICES

NO ONE SHOULD attempt to operate any oxyfuel apparatus until trained in its proper use or working under competent supervision. It is most important that the manufacturer's recommendations for safe use and operating instructions be followed closely.

Oxygen by itself does not burn or explode, but it does support combustion. Oxygen under high pressure may react violently with oil, grease, or other combustible material. Cylinders, fittings, and all equipment to be used with oxygen should be kept away from oil, grease, and other contaminents at all times. Oxygen cylinders should never be stored near highly combustible materials. Oxygen should never be used to operate pneumatic tools, to start internal combustion engines, to blow out pipelines, to dust clothing, or for any other potentially unsafe use.

Acetylene is a fuel gas and will burn readily. It must, therefore, be kept away from open flames. Acetylene cylinder and manifold pressures must always be reduced through pressure-reducing regulators. Cylinders should always be protected against excessive temperature rises and should be stored in well ventilated, clean, dry locations, free from other combustibles. They should be stored and used with the valve end up.

All liquefied fuel gas cylinders must always be used in an upright position. If these cylinders are laid on their side, it is possible to withdraw liquid, rather than vapor, from the cylinder. This could cause damage to the apparatus and produce a large, uncontrollable flame. Loose carbide should not be scattered about or allowed to remain on floors because it will absorb moisture from the air and generate acetylene.

Acetylene in contact with copper, mercury, or silver may form acetylides, especially if impurities are present. These compounds are violently explosive and can be detonated by a slight shock or by the application of heat. Alloys containing more than 67 percent copper, other than tips and nozzles, should not be used in any acetylene system.

When gas is taken from an acetylene cylinder that is lying on its side, acetone may be withdrawn along with the acetylene, as noted above. This can cause damage to the apparatus or contaminate the flame, resulting in welds of inferior quality.

Cylinders of other fuel gases are also pressurized and should be handled with care. These cylinders should also be stored in clean, dry, well-ventilated locations.

Refrigerated oxygen cylinders are of double wall construction like a thermos bottle, with a vacuum between the inner and outer shell. They should be handled with extreme care to prevent damage to the internal piping and the loss of vacuum. Such cylinders should always be transported and used in an upright position.

Cylinders can become a hazard if tipped over and care should be exercised to avoid this possibility. If the cylinder valve is ruptured as the result of a fall, the escaping gas can cause the cylinder to become a dangerous projectile. It is standard practice to fasten the cylinders on a cylinder truck or to secure them against a rigid support.

SUPPLEMENTARY READING LIST

Ballis, W. L., et al. "Training of oxyacetylene welding to weld mild steel pipe." *Welding Journal* 56(4): 15-19; April 1977.

Fay, R. H. "Heat transfer from fuel gas flames." *Welding Journal* 46(8): 380s-383s; August 1967.

International Acetylene Association. *Oxyacetylene welding and its applications.* New York: International Acetylene Association, 1958. (Obtained from the Compressed Gas Association.)

Koziarski, J. "Hydrogen vs acetylene vs inert gas in welding aluminum alloys." *Welding Journal* 36(2): 141-148; February 1957.

Kugler, A. N. *Oxyacetylene Welding and Oxygen Cutting Instruction Course.* New York: Airco, Inc., revised 1966.

Lewis, B., and Von Elbe, G. *Combustion Flames and Explosions of Gases.* New York: Academic Press, Inc., 1961.

Moen, W. B., and Campbell, J. "Evaluation of fuels and oxidants for welding and associated processes." *Welding Journal* 34(9): 870-876; September 1955.

National Fire Protection Association. *Gas systems for welding and cutting,* NFPA No. 51. Quincy, Massachusetts: National Fire Protection Association.

Postman, B. F. "Safety in installation and use of welding equipment." *Welding Journal* 34(4): 337-344; April 1955.

Sosnin, H. A. "Efficiency and economy of the oxyacetylene process." *Welding Journal* 61(10): 46-48; October 1982.

The National Training Fund for the Sheet Metal and Air Conditioning Industry. *Welding book I,* 1st Ed. Alexandria, Virginia, The National Training Fund for the Sheet Metal and Air Conditioning Industry, 1979.

Union Carbide Corporation. *The oxyacetylene handbook,* 2nd Ed. New York: Union Carbide Corporation, Linde Div., 1960.

BRAZING

PREPARED BY A COMMITTEE CONSISTING OF:

M. J. Lucas, Chairman
General Electric Corporation

R. L. Peaslee
Wall Colmonoy Corporation

WELDING HANDBOOK COMMITTEE MEMBER:
M. J. Tomsic
Plastronic, Inc.

BRAZING

INTRODUCTION

DEFINITION AND GENERAL DESCRIPTION

BRAZING JOINS MATERIALS by heating them in the presence of a filler metal having a liquidus above 840°F (450°C) but below the solidus of the base metals. Heating may be provided by a variety of processes. The filler metal distributes itself between the closely fitted surfaces of the joint by capillary action. Brazing differs from soldering, in that soldering filler metals have a liquidus below 840°F (450°C).

Brazing with silver alloy filler metals is sometimes called *silver soldering*, a nonpreferred term. Silver brazing filler metals are not solders; they have liquidus temperatures above 840°F (450°C).

Brazing does not include the process known as braze welding. Braze welding is a method of welding with a brazing filler metal. In braze welding, the filler metal is melted and deposited in grooves and fillets exactly at the points where it is to be used. Capillary action is not a factor in distribution of the brazing filler metal. Indeed, limited base metal fusion may occur in braze welding. Braze welding is described in greater detail beginning on page 414.

Brazing must meet each of three criteria:

(1) The parts must be joined without melting the base metals.
(2) The filler metal must have a liquidus temperature above 840°F (450°C).
(3) The filler metal must wet the base metal surfaces and be drawn into or held in the joint by capillary action.

To achieve a good joint using any of the various brazing processes described in this chapter, the parts must be properly cleaned and must be protected by either flux or atmosphere during the heating process to prevent excessive oxidation. The parts must be designed to afford a capillary for the filler metal when properly aligned, and a heating process must be selected that will provide the proper brazing temperature and heat distribution.

APPLICATIONS

THE BRAZING PROCESS is used to join together various materials for numerous reasons. By using the proper joint design, the resulting braze can function better than the base metals being joined. In many instances it is desireable to join different materials to obtain the maximum benefit of both materials and have the most cost- or weight-effective joint. Applications of brazing cover the entire manufacturing arena from inexpensive toys to highest quality aircraft engines and aerospace vehicles. Brazing is used because it can produce results which are not always available with other joining processes. Advantages of brazing to join components include:

(1) Economical for complex assemblies
(2) Simple way to join large joint areas
(3) Excellent stress and heat distribution
(4) Ability to preserve coatings and claddings
(5) Ability to join dissimilar materials
(6) Ability to join nonmetals to metals
(7) Ability to join widely different thicknesses
(8) Capability of joining precision parts
(9) Joints require little or no finishing
(10) Can do many parts at one time (batch processing)

Throughout this chapter, examples of brazing illustrate when to select brazing and how to design the joint and select braze materials best suited for the individual application.

PROCESS ADVANTAGES AND DISADVANTAGES

LIKE ANY JOINING process, brazing has both advantages and disadvantages. The advantages vary with the heating method employed, but in general, brazing will be very economical when done in large batches. A major benefit of brazing is the ability to take brazed joints apart at a later time. It can also join dissimilar metals without melting the base metals as required by other joining methods. In many instances, several hundred parts with many feet of braze joints can be brazed at one time. When protective atmosphere brazing is used, parts are kept clean and the heat treatment cycle may be employed as part of the brazing cycle.

Since the brazing process uses a molten metal to flow between the materials to be joined, there is the possibility of liquid metal interactions which are unfavorable. Depending on the material combinations involved and the thickness of the base sheets, base metal erosion may occur. In many cases, the erosion may be of little consequence, but when brazing heavily loaded or thin materials, the erosion can weaken the joint and make it unsatisfactory for its intended application. Also, the formation of brittle intermetallics or other phases can make the resulting joint too brittle to be acceptable.

A disadvantage with some of the manual brazing processes is that highly skilled technicians are required to perform the operation. This is especially true for gas torch brazing using a high melting point brazing filler metal.

Nevertheless, with the proper joint design, brazing filler metal, and process selection, a satisfactory brazing technique can be developed for most joining applications where it is not feasible to join the materials with a fusion welding process because of strength or economic considerations.

PRINCIPLES OF OPERATION

CAPILLARY FLOW IS the dominant physical principle that assures good brazements whenever both faying surfaces to be joined are wet by the molten filler metal. The joint must be spaced to permit efficient capillary action and resulting coalescence. More specifically, capillarity is a result of surface tension between base metal(s) and filler metal, protected by a flux or atmosphere, and promoted by the contact angle between base metal and filler metal. In actual practice, brazing filler metal flow is influenced by dynamic considerations involving fluidity, viscosity, vapor pressure, gravity, and especially the effects of metallurgical reactions between filler metal and base metal.

The typical brazed joint has a relatively large area and very small gap. In the simplest brazing application, the surfaces to be joined are cleaned to remove contaminants and oxides. Next, they are coated with flux. A flux is a material which is capable of dissolving solid metal oxides and also preventing new oxidation. The joint area is then heated until the flux melts and cleans the base metals, which are protected against further oxidation by the layer of liquid flux.

Brazing filler metal is then melted at some point on the surface of the joint area. Capillary attraction between the base metal and the filler metal is much higher than that between the base metal and the flux. Accordingly, the flux is displaced by the filler metal. The joint, upon cooling to room temperature, will be filled with solid filler metal, and the solid flux will be found on the joint periphery.

Joints to be brazed are usually made with clearances of 0.001 to 0.010 in. (0.025 to 0.25 mm). The fluidity of the filler metal, therefore, is an important factor. High fluidity is a desirable characteristic of brazing filler metal since capillary action may be insufficient to draw a viscous filler metal into closely fitted joints.

Brazing is sometimes done under an active gas, such as hydrogen, or in an inert gas or vacuum. Atmosphere brazing eliminates the necessity for post cleaning and insures absence of corrosive mineral flux residue. Carbon steels, stainless steels, and superalloy components are widely processed in atmospheres of reacted gases, dry hydrogen, dissociated ammonia, argon, or vacuum. Large vacuum furnaces are used to braze zirconium, titanium, stainless steels, and the refractory metals. With good processing procedures, aluminum alloys can also be vacuum furnace brazed with excellent results.

Brazing is economically attractive for the production of high strength metallurgical bonds while preserving desired base metal properties.

BRAZING PROCESSES

BRAZING PROCESSES ARE customarily designated according to the sources or methods of heating. Industrial methods currently significant are the following:

(1) Torch brazing
(2) Furnace brazing
(3) Induction brazing
(4) Resistance brazing
(5) Dip brazing
(6) Infrared brazing

Whatever the process used, the filler metal has a melting point above 840°F (450°C), but below that of the base metal, and it spreads within the joint by capillary action.

TORCH BRAZING

TORCH BRAZING IS accomplished by heating with one or more gas torches.[1] Depending upon the temperature and

---

1. Chapter 11 contains information on gas torches used for welding and brazing.

the amount of heat required, the fuel gas (acetylene, propane, city gas, etc.) may be burned with air, compressed air, or oxygen. Manual torch brazing is shown in Figure 12.1.

Air-natural gas torches provide the lowest flame temperature as well as the least heat. Acetylene under pressure is used in the air-acetylene torch with air at atmospheric pressure. Both air-natural gas and air-acetylene torches can be used to advantage on small parts and thin sections.

Torches which employ oxygen with natural gas, or other cylinder gases (propane, butane) have higher flame temperatures. When properly applied as a neutral or slightly reducing flame, excellent results are obtainable with many brazing applications.

Oxyhydrogen torches are often used for brazing aluminum and nonferrous alloys. The lower temperature reduces the possibility of overheating the assembly during brazing. An excess of hydrogen provides the joint with additional cleaning and protection.

Specially designed torches having multiple tips or multiple flames can be used to an advantage to increase the rate of heat input. Care must be exercised to avoid local overheating by constantly moving the torch with respect to the work.

For manual torch brazing, the torch may be equipped with a single tip, either single- or multiple-flame. Manual torch brazing is particularly useful on assemblies involving sections of unequal mass. Machine operations can be set up, where the rate of production warrants, using one or more torches equipped with single or multiple-flame tips. The machine may be designed to move either the work or the torches, or both. For premixed city gas and air flames, a refractory type burner is used.

Torch heating for brazing is limited in use to filler metals supplied with flux or self-fluxing. The list includes aluminum-silicon, silver, copper-phosphorus, copper-zinc, and nickel. With the exception of the copper-phosphorus filler metals, they all require fluxes. For certain applications even the self-fluxing copper-phosphorus filler metals require added flux, as shown in Table 12.1.

The filler metal can be preplaced on the joint and fluxed before heating, or it may be face-fed. Heat is applied to the joint, first melting the flux, then continuing until the brazing filler metal melts and flows into the joint. Overheating of the base metal and brazing filler metal should be avoided because rapid diffusion and "drop through" of the metal may result. Natural gas is well suited for torch brazing because its relatively low flame temperature reduces the danger of overheating.

Brazing filler metal may be preplaced at the joint in the forms of rings, washers, strips, slugs, or powder, or it may be fed from hand-held filler metal, usually in the form of wire or rod. In any case, proper cleaning and fluxing are essential.

Torch brazing techniques differ from those used for oxyfuel gas welding. Operators experienced only in welding techniques may require instruction in brazing techniques. It is good practice, for example, to prevent the inner cone of the flame from coming in contact with the joint except during preheating, since melting of the base metal and dilution with the filler metal may increase its liquidus temperature and make the flow more sluggish. In addition, the flux may be overheated and thus lose its ability to promote capillary flow, and low melting constituents of the filler metal may evaporate.

FURNACE BRAZING

FURNACE BRAZING, AS illustrated in Figure 12.2, is used extensively when (1) the parts to be brazed can be preassembled or jigged to hold them in the correct position, (2) the brazing filler metal can be placed in contact with the joint, (3) multiple brazed joints are to be formed simultaneously on a completed assembly, (4) many similar assemblies are to be joined, and (5) complex parts must be heated uniformly to prevent the distortion that would result from local heating of the joint area.

Electric, gas, or oil heated furnaces with automatic temperature control capable of holding the temperature within $\pm 10\,°F$ ($\pm 6\,°C$) should be used for furnace brazing. Fluxes or specially controlled atmospheres that perform fluxing functions must be provided.

Parts to be brazed should be assembled with the filler metal and flux, if used, located in or around the joints. The

Figure 12.1—Manual Torch Brazing

Table 12.1
Classification of Brazing Fluxes with Brazing or Braze Welding Filler Metals

Classification*	Form	Filler Metal Type	Activity Temperature Range	
			°F	°C
FB1-A	Powder	BA1Si	1080-1140	580-615
FB1-B	Powder	BA1Si	1040-1140	560-615
FB1-C	Powder	BA1Si	1000-1140	540-615
FB2-A	Powder	BMg	900-1150	480-620
FB3-A	Paste	BAg and BCuP	1050-1600	565-870
FB3-C	Paste	BAg and BCuP	1050-1700	565-925
FB3-D	Paste	BAg, BCu, BNi, BAu and RBCuZn	1400-2200	760-1205
FB3-E	Liquid	BAg and BCup	1050-1600	565-870
FB3F	Powder	BAg and BCuP	1200-1600	650-870
FB3G	Slurry	BAg and BCuP	1050-1600	565-870
FB3-H	Slurry	BAg	1050-1700	565-925
FB3-I	Slurry	BAg, BCu, BNi, BAu and RBCuZn	1400-2200	760-1205
FB3-J	Powder	BAg, BCu, BNi, BAu and RBCuZn	1400-2200	760-1205
FB3-K	Liquid	BAg and RBCuZn	1400-2200	760-1205
FB4-A	Paste	BAg and BCuP	1100-1600	595-870

* Flux 3B shown in the Brazing Manual, 3rd Edition, 1976 has been discontinued. Type 3B has been divided into types FB3C and FB3D.

Note: The selection of a flux designation for a specific type of work may be based on the form, the filler metal type, and the description above, but the information here is generally not adequate for flux selection. Refer to the latest issue of the Brazing Manual for further assistance.

preplaced filler metal may be in the form of wire, foil, filings, slugs, powder, paste, or tape. The assembly is heated in the furnace until the parts reach brazing temperature and brazing takes place. The assembly is then removed. These steps are shown in Figure 12.2. A laboratory setup for induction brazing in vacuum is shown in Figure 12.3. Many commercial fluxes are available for both general and specific brazing operations. Satisfactory results are obtained if dry powdered flux is sprinkled along the joint. Flux paste is satisfactory in most cases, but in some cases it retards the flow of brazing alloy. Flux pastes containing water can be dried by heating the assembly at 350 to 400°F (175 to 200°C) for 5 to 15 minutes in drying ovens or circulating air furnaces.

Brazing time will depend somewhat on the thickness of the parts and the amount of fixturing necessary to position them. The brazing time should be restricted to that necessary for the filler metal to flow through the joint to avoid excessive interaction between the filler metal and base metal. Normally, one or two minutes at the brazing temperature is sufficient to make the braze. A longer time at the brazing temperature will be beneficial where the filler metal remelt temperature is to be increased and where diffusion will improve joint ductility and strength. Times of 30 to 60 minutes at the brazing temperature are often used to increase the braze remelt temperature.

Furnaces used for brazing are classified as (1) batch type with either air or controlled atmosphere, (2) continuous type with either air or controlled atmosphere, (3) retort type with controlled atmosphere, or (4) vacuum. A high temperature, high vacuum brazing furnace with control panel and charging carriage is shown in Figure 12.3. Most brazing furnaces have a temperature control of the potentiometer type connected to thermocouples and gas control valves or contactors. The majority of furnaces are heated by electrical resistance using silicon-carbide, nickel-chromium, or refractory metal (Mo, Ta, W) heating elements. When a gas or oil flame is used for heating, the flame must not impinge directly on the parts.

With controlled atmosphere furnaces, a continuous flow of the atmosphere gas is maintained in the work zone to avoid contamination from outgassing of the metal parts and dissociation of oxides. If the controlled atmosphere is flammable or toxic, adequate venting of the work area and protection against explosion are necessary.

Batch type furnaces heat each workload separately. They may be top loading (pit type), side loading, or bottom loading. When a furnace is lowered over the work, it is called a *bell furnace*. Gas or oil fired batch type furnaces without retorts require that flux be used on the parts for brazing. Electrically heated batch type furnaces are often equipped for controlled atmosphere brazing, since the heating elements can usually be operated in the controlled atmosphere.

Continuous furnaces receive a steady flow of incoming assemblies. The heat source may be gas or oil flames, or

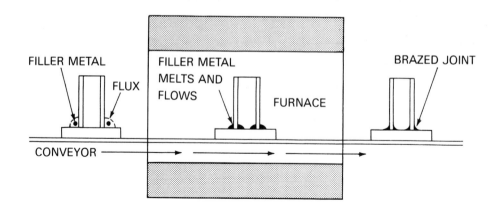

Figure 12.2—Illustration of Furnace Brazing Operation

electrical heating elements. The parts move through the furnace either singly or in trays or baskets. Conveyor types (mesh belts or roller hearth), shaker hearth, pusher, or slot type continuous furnaces are commonly used for high production brazing. Continuous furnaces usually contain a preheat or purging area which the parts enter first. In this area, the parts are slowly brought to a temperature below the brazing temperature. If brazing atmosphere gas is used in the brazing zone it also flows over and around the parts in the preheat zone, under positive pressure. The gas flow removes any entrapped air and starts the reduction of surface oxides. Atmosphere gas trails the parts into the cooling zone.

Retort type furnaces are batch furnaces in which the assemblies are placed in a sealed retort for brazing. The air in the retort is purged by controlled atmosphere gas and the retort is placed in the furnace. After the parts have been brazed, the retort is removed from the furnace, cooled, and its controlled atmosphere is purged. The retort is opened, and the brazed assemblies are removed. A protective atmosphere is sometimes used within a high temperature furnace to reduce external scaling of the retort.

Vacuum furnace brazing is widely used in the aerospace and nuclear fields, where reactive metals are joined or where entrapped fluxes would be intolerable. If the vacuum atmosphere is maintained by continuous pumping, it will remove volatile constituents liberated during brazing.

Vacuum brazing equipment is currently used to a large extent to braze stainless steels, superalloys, aluminum alloys, titanium alloys, and metals containing refractory or reactive elements. Vacuum is a relatively economical "atmosphere" which prevents oxidation by removing air from around the assembly. Surface cleanliness is nevertheless re-

quired for good wetting and flow. Base metals containing chromium and silicon can be vacuum brazed. Base metals that can generally be brazed only in vacuum are those containing more than a few percent of aluminum, titanium, zirconium, or other elements with particularly stable oxides. However, a nickel plated barrier is still preferred to obtain optimum quality.

Vacuum brazing furnaces are of three types:

(1) *Hot retort, or single pumped retort furnace.* This is a sealed retort, usually of fairly thick metal. The retort with work loaded inside is sealed, evacuated, and heated from the outside by a furnace. Most brazing work requires vacuum pumping continuously throughout the heat cycle to remove gases being given off by the workload. The furnaces are gas fired or electrical. The retort size and its maximum operating temperature are limited by the ability of the retort to withstand the collapsing force of atmospheric pressure at brazing temperature. Top temperature for vacuum brazing furnaces of this type is about 2100°F (1150°C).

Argon, nitrogen, or other gas is often introduced into the retort to accelerate cooling after brazing.

(2) *Double pumped or double wall hot retort vacuum furnace.* The typical furnace of this type has an inner retort containing the work, within an outer wall or vacuum chamber. Also within the outer wall are the thermal insulation and electrical heating elements. A moderately reduced pressure, typically 1.0 to 0.1 torr (133 to 13.3 Pa), is maintained within the outer wall, and a much lower pressure, below 10^{-2} torr (1.3 Pa), within the inner retort. Again most brazing requires continuous vacuum pumping of the inner retort throughout the heat cycle to remove gases given off by the workload.

Figure 12.3–A High Temperature, High Vacuum Brazing Furnace with Control Panel and Charging Dolly

In this type of furnace, the heating elements and the thermal insulation are not subjected to the high vacuum. Heating elements are typically of nickel-chromium alloy, graphite, stainless steel, or silicon carbide materials. Thermal insulation is usually silica or alumina brick, or castable or fiber materials.

(3) *Cold wall vacuum furnace.* A typical cold wall vacuum furnace has a single vacuum chamber, with thermal insulation and electrical heating elements located inside the chamber. The vacuum chamber is usually water cooled. The maximum operating temperature is determined by the materials used for the thermal insulation (the

heat shield) and the heating elements, which are subjected to the high vacuum as well as the operating temperature of the furnace.

Heating elements for cold wall furnaces are usually made of high temperature, low vapor pressure materials, such as molybdenum, tungsten, graphite, or tantalum. Heat shields are typically made of multiple layers of molybdenum, tantalum, nickel, or stainless steel. Thermal insulation may be high purity alumina brick, graphite, or alumina fibers sheathed in stainless steel. The maximum operating temperature and vacuum obtainable with cold wall vacuum furnaces depends on the heating element material and the thermal insulation or heat shields. Temperatures up to 4000°F (2200°C) and pressures as low as 10^{-6} torr (1.33 x 10^{-4} Pa) are obtainable.

Configurations for all three types of furnaces include side loading (horizontal), bottom loading, and top loading (pit type). Work zones are usually rectangular for side loading furnaces, and circular for bottom and top loading types.

Vacuum pumps for brazing furnaces may be oil sealed mechanical types for pressures from 0.1 to 10 torr (13 to 1300 Pa). Brazing of base metals containing chromium, silicon, or other rather strong oxide formers usually requires pressures of 10^{-2} to 10^{-3} torr (1.3 to 0.13 Pa), which are best obtained with a high-speed, dry Roots, or turbo-mechanical type pump. Vacuum pumps of this type are not capable of exhausting directly to atmosphere and require a roughing vacuum pump.

Brazing of base materials containing more than a few percent of aluminum, titanium, zirconium, which form very stable oxides, requires vacuum of 10^{-3} torr (0.13 Pa) or lower. Vacuum furnaces for such brazing usually require a diffusion pump that will obtain pressures of 10^{-2} to 10^{-6} torr (1.3 to 0.0001 Pa). The diffusion pump is backed by a mechanical vacuum pump or by both a Roots-type pump and a mechanical pump.

INDUCTION BRAZING

THE HEAT FOR brazing with this process is obtained from an electric current induced in the parts to be brazed, hence the name *induction brazing*. For induction brazing, the parts are placed in or near a water-cooled coil carrying alternating current. They do not form a part of the electrical circuit. Parts to be heated act as the short circuited secondary of a transformer where the work coil, which is connected to the power source, is the primary. On both magnetic and nonmagnetic parts, heating is obtained from the resistance of the parts to currents induced in them by the transformer action. See Figure 12.4.

The brazing filler metal is preplaced. Careful design of the joint and the coil setup are necessary to assure that the surfaces of all members of the joint reach the brazing temperature at the same time. Flux is employed except when an atmosphere is specifically introduced to perform the same function.

Frequencies for induction brazing generally vary from 10 KHz to 450 khz. The lower frequencies are obtained with solid-state generators and the higher frequencies with vacuum tube oscillators. Induction generators are manufactured in sizes from one kilowatt to several hundred kilowatts output. Various induction brazing coil designs are illustrated in Figure 12.5. One generator may be used to energize several individual workstations in sequence, using a transfer switch, or assemblies in holding fixtures may be indexed or continuously processed through a conveyor-type coil for heating to brazing temperature.

Induction brazing is used when very rapid heating is required. Time for processing is usually in the range of

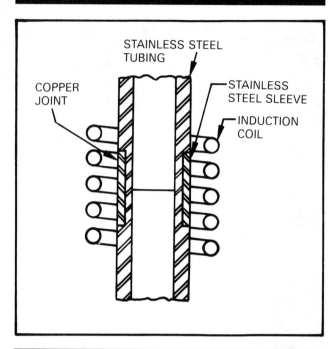

Figure 12.4—Joint in Stainless Steel Tubing Induction Brazed in a Controlled Atmosphere. Note Placement of Joint in Induction Coil.

seconds when large numbers of parts are handled automatically. Induction brazing has been used extensively to produce consumer and industrial products; structural assemblies; electrical and electronic products; mining, machine, and hand tools; military and ordnance equipment; and aerospace assemblies. An aerospace application of vacuum induction brazing is shown in Figure 12.6.

Assemblies may be induction brazed in a controlled atmosphere by placing the components and coil in a nonmetallic chamber, or by placing the chamber and work inside the coil. The chamber can be quartz Vycor or tempered glass. A dual station bell jar fixture of this type is shown in Figure 12.7.

RESISTANCE BRAZING

THE HEAT NECESSARY for resistance brazing is obtained from the flow of an electric current through the electrodes and the joint to be brazed. The parts comprising the joint become part of the electric circuit. The brazing filler metal, in some convenient form, is preplaced or face-fed. Fluxing is done with due attention to the conductivity of the fluxes. (Most fluxes are insulators when dry.) Flux is employed except when an atmosphere is specifically introduced to perform the same function. The parts to be brazed are held between two electrodes, and proper pressure and current are applied. The pressure

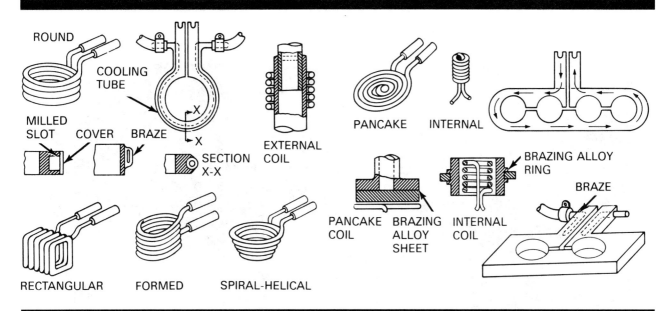

Figure 12.5–Typical Induction Brazing Coils and Plates

should be maintained until the joint has solidified. In some cases, both electrodes may be located on the same side of the joint with a suitable backing to maintain the required pressure.

Brazing filler metal is used in the form of preplaced wire, shims, washers, rings, powder, or paste. In a few instances, face feeding is possible. For copper and copper alloys, the copper-phosphorus filler metals are most satisfactory since they are self-fluxing. Silver base filler metals may be used, but a flux or atmosphere is necessary. A wet flux is usually applied as a very thin mixture just before the assembly is placed in the brazing fixture. Dry fluxes are not used because they are insulators and will not permit sufficient current to flow.

The parts to be brazed must be clean. The parts, brazing filler metal, and flux are assembled and placed in the fixture and pressure applied. As current flows, the electrodes become heated, frequently to incandescence, and the flux and filler metal melt and flow. The current should be adjusted to obtain uniform rapid heating in the parts. Overheating risks oxidizing or melting the work, and the electrodes will deteriorate. Too little current lengthens the time of brazing. Experimenting with electrode compositions, geometry, and voltage will give the best combination of rapid heating with reasonable electrode life.

Quenching the parts from an elevated temperature will help flux removal. The assembly first must cool sufficiently to permit the braze to hold the parts together. When brazing insulated conductors it may be advisable to quench the parts rapidly while they are still in the electrodes to prevent overheating of the adjacent insulation. Water-cooled clamps prevent damage to the insulation.

Resistance brazing is most applicable to joints which have a relatively simple configuration. It is difficult to obtain uniform current distribution, and therefore uniform heating, if the area to be brazed is large or discontinuous or is much longer in one dimension. Parts to be resistance brazed should be so designed that pressure may be applied to them without causing distortion at brazing temperature. Wherever possible, the parts should be designed to be self-nesting, which eliminates the need for dimensional features in the fixtures. Parts should also be free to move as the filler metal melts and flows in the joint.

The equipment consists of tongs or clamps with the electrodes attached at the end of each arm. The tongs should preferably be water cooled to avoid overheating. The arms are current-carrying conductors attached by leads to a transformer.

One common source of current for resistance brazing is a stepdown transformer whose secondary circuit can furnish sufficient current at low voltage (2-25 V). The current will range from about 50 A for small, delicate jobs to many thousands of amperes for larger jobs. Commercial equipment is available for resistance brazing.

Electrodes for resistance brazing are made of high resistance electrical conductors, such as carbon or graphite blocks, tungsten or molybdenum rods, or even steel in some instances. The heat for brazing is mainly generated in the electrodes and flows into the work by conduction. It is generally unsatisfactory to attempt to use the resistance of the workpieces alone as a source of heat.

The pressure applied by a spot welding machine, clamps, pliers, or other means must be sufficient to maintain good electrical contact and to hold the pieces firmly together as

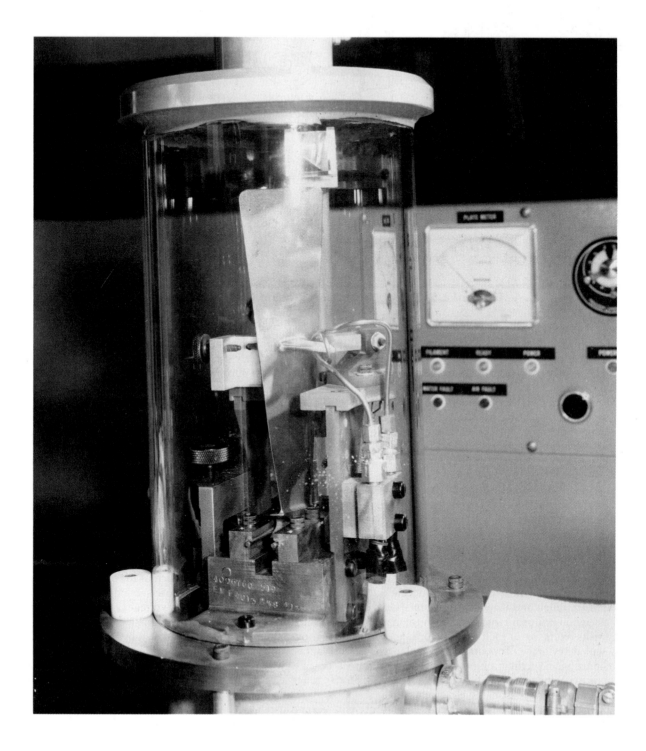

Figure 12.6–Example of Vacuum Induction Brazing. A Tungsten Carbide Wear Pad is Being Brazing to A Titanium Compressor Blade

Figure 12.7–Production Arrangement for Induction Brazing in Controlled Atmosphere or Vacuum Showing Dual Station Bell Jar Fixture, Induction Generator, Movable Stand, Generator and Gas Controls, and Supports to Facilitate Up and Down Movement of Bell Jar

the filler metal melts. The pressure must be maintained during the time of current flow and after the current is shut off until the joint solidifies. The time of current flow will vary from about one second for small, delicate work to several minutes for larger work. This time is usually controlled manually by the operator, who determines when brazing has occurred by the temperature and the extent of filler metal flow.

DIP BRAZING

TWO METHODS OF dip brazing are molten metal bath dip brazing and molten chemical (flux) bath dip brazing.

Molten Metal Bath Method

THIS METHOD IS usually limited to the brazing of small assemblies, such as wire connections or metal strips. A crucible, usually made of graphite, is heated externally to the required temperature to maintain the brazing filler metal in fluid form. A cover of flux is maintained over the molten filler metal. The size of the molten bath (crucible) and the heating method must be such that the immersion of parts in the bath will not lower the bath temperature below brazing temperature. Parts should be clean and protected with flux prior to their introduction into the bath. The ends of the wires or parts must be held firmly together when they are removed from the bath until the brazing filler metal has fully solidfied.

Molten Chemical (Flux) Bath Method

THIS BRAZING METHOD requires either a metal or ceramic container for the flux and a method of heating the flux to the brazing temperature. Heat may be applied externally with a torch or internally with an electrical resistance heating unit. A third method involves electrical resistance heating of the flux itself; in that case, the flux must be initially melted by external heating. Suitable controls are provided to maintain the flux within the brazing temperature range. The size of the bath must be such that immersion of parts for brazing will not cool the flux below the brazing temperature. See Figure 12.8.

Parts should be cleaned, assembled, and preferably held in jigs prior to immersion into the bath. Brazing filler metal is preplaced as rings, washers, slugs, paste, or as a cladding on the base metal. Preheat may be necessary to assure dryness of parts and to prevent the freezing of flux on parts which may cause selective melting of flux and brazing filler metal. Preheat temperatures are usually close to the melt-

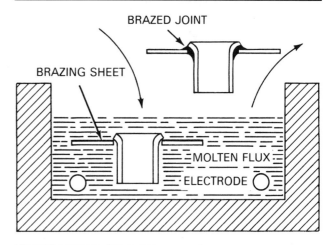

Figure 12.8–Illustration of Chemical Bath Dip Brazing

ing temperature of the flux. A certain amount of flux adheres to the assembly after brazing. Molten flux must be drained off while the parts are hot. Flux remaining on cold parts must be removed by water or by chemical means.

INFRARED BRAZING

INFRARED BRAZING MAY be considered a form of furnace brazing with heat supplied by long-wave light radiation. Heating is by invisible radiation from high intensity quartz lamps capable of delivering up to 5000 watts of radiant energy. Heat input varies inversely as the square of the distance from the source, but the lamps are not usually shaped to follow the contour of the part to be heated. Concentrating reflectors focus the radiation on the parts.

For vacuum brazing or inert-gas protection, the assembly and the lamps are placed in a bell jar or retort that can be evacuated or filled with inert gas. The assembly is then heated to a controlled temperature, as indicated by thermocouples. Figure 12.9 shows an infrared brazing arrangement. The part is moved to the cooling platens after brazing.

SPECIAL PROCESSES

Blanket Brazing

BLANKET BRAZING USES a blanket that is resistance heated; the heat is transferred to the parts by conduction and radiation, but mostly by radiation.

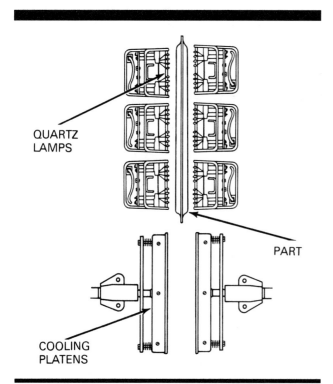

Figure 12.9–Infrared Brazing Apparatus

EXOTHERMIC BRAZING

EXOTHERMIC BRAZING IS a special process which heats a commercial filler metal by a solid-state exothermic chemical reaction. An exothermic chemical reaction generates heat released as the free energy of the reactants. Nature has provided countless numbers of such reactions; those solid-state or nearly solid-state metal-metal oxide reactions are suitable for use in exothermic brazing units.

Exothermic brazing uses simplified tooling and equipment. The reaction heat brings adjoining metal interfaces to a temperature at which preplaced brazing filler metal melts and wets the base metal interface surfaces. Several commercially available brazing filler metals have a suitable flow temperature. The process is limited only by the thickness of the base metal and the effect of brazing heat, or any previous heat treatment, on the metal properties.

BRAZING AUTOMATION

THE IMPORTANT VARIABLES involved in brazing are the temperature, time at temperature, filler metal, and brazing atmosphere. Other variables are joint fit-up, amount of filler metal, and rate and mode of heating. All of these features may be automated.

Heating by welding torches may be automated. So may furnace brazing (e.g., vacuum and atmosphere), resistance brazing, induction brazing, dip brazing, and infrared brazing. Generally, the amount of heat supplied to the joint is automated by controlling temperature and time at temperature.

Brazing filler metal may be preplaced at the joints during assembly of components, or automatically fed into the joints while at brazing temperature. So also may fluxing be provided.

Further automation may include in-line inspection and cleaning (flux removal), simultaneous brazing of multiple joints in an assembly, and continuous brazing operations.

Generally, the more automated a process becomes, the more rigorous must be its economic justification. Usually the increased cost of automation is justified by increased productivity. In the case of brazing, further justification may well be found in the energy saved with efficient joint heating.

Basically, the major advantages of automatic brazing are these:

(1) High production rates
(2) High productivity per worker
(3) Filler metal savings
(4) Consistency of results
(5) Energy savings
(6) Adaptability and flexibility

Manual torch brazing, totally unautomated, represents the simplest brazing technique, but it has economic justification. First, the braze joint is visible to the operator, who adjusts the process based on observation. Second, heat is directed only to the joint area. Whenever energy costs represent a large fraction of the cost of a braze joint, this is an important consideration.

Nevertheless, torch brazing is labor intensive and low in productivity. A continuous belt furnace increases production but loses in-line inspection and lowers energy efficiency because the entire assembly is heated.

Automatic brazing machines improve torch brazing. Typically, heat is directed just to the joint area by one or more torches. Similar effects can be obtained by induction heating. A typical machine has provisions for assembly and fixturing, automatic fluxing, preheating (if needed), brazing, air or water quenching, part removal, and inspection.

BRAZING FILLER METALS

CHARACTERISTICS

BRAZING FILLER METALS must have the following properties:

(1) Ability to form brazed joints with mechanical and physical properties suitable for the intended service application
(2) Melting point or melting range compatible with the base metals being joined, and sufficient fluidity at brazing temperature to flow and distribute themselves into properly prepared joints by capillary action
(3) Composition of sufficient homogeneity and stability to minimize separation of constituents (liquation) during brazing
(4) Ability to wet surfaces of base metals and form a strong, sound bond
(5) Depending on requirements, ability to produce or avoid filler-metal interactions with base metals

MELTING AND FLUIDITY

PURE METALS MELT at a constant temperature and are generally very fluid. Binary compositions (two metals) have differing characteristics, depending upon the relative contents of the two metals. Figure 12.10 is the equilibrium diagram for the silver-copper binary system. The solidus line, ADCEB, traces the start-of-melting temperature of the alloys, while the liquidus line, ACB, shows the temperatures at which the alloys become completely liquid. At point C the two lines meet (72 percent silver-28 percent copper), indicating that that particular alloy melts at that fixed temperature (the eutectic temperature). This alloy is the eutectic composition; it is as fluid as a pure metal, while the other alloy combinations are mushy between their solidus and liquidus temperatures. The wider that temperature spread, the more sluggish are the alloys with respect to flow in a capillary joint.

The α region is a solid solution of copper in silver, the β region is a solid solution of silver in copper. The central solid zone consists of an intimate mixture of α and β solid solutions. Above the liquidus line, the silver and copper atoms are thoroughly interspersed as a liquid solution.

LIQUATION

BECAUSE THE SOLID and liquid alloy phases of a brazing filler metal generally differ, the composition of the melt will gradually change as the temperature increases from the solidus to the liquidus. If the portion that melts first is allowed to flow out, the remaining solid may not melt and so may remain behind as a residue or "skull." Filler metals with narrow melting ranges do not tend to separate, so they flow quite freely into joints with extremely narrow clearance. Filler metals with wide melting ranges need rapid heating or delayed application to the joint until the base metal reaches brazing temperature, to minimize separation, which is called *liquation*. Filler metals subject to liquation have a sluggish flow, require wide joint clearances, and form large fillets at joint extremities.

WETTING AND BONDING

TO BE EFFECTIVE, a brazing filler metal must alloy with the surface of the base metal without (1) undesirable diffusion into the base metal, (2) dilution with the base metal, (3) base metal erosion, and (4) formation of brittle com-

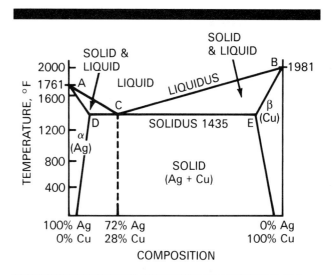

Figure 12.10–Silver-Copper Constitutional Diagram

pounds. Effects (1), (2), and (3) depend upon the mutual solubility between the brazing filler metal and the base metal, the amount of brazing filler metal present, and the temperature and time duration of the brazing cycle.

Some filler metals diffuse excessively, changing the base metal properties. To control diffusion, select a suitable filler metal, apply the minimum quantity of filler metal, and follow the appropriate brazing cycle. If the filler metal wets the base metal, capillary flow is enhanced. In long capillaries between the metal parts, mutual solubility can change the filler metal composition by alloying. This will usually raise its liquidus temperature and cause it to solidify before completely filling the joint.

Base metal erosion (3) occurs if the base metal and the brazing filler metal are mutually soluble. Sometimes such alloying produces brittle intermetallic compounds (4) that reduce the joint ductility.

Compositions of brazing filler metals are adjusted to control the above factors and to provide desirable characteristics, such as corrosion resistance in specific media, favorable brazing temperatures, or material economies. Thus, to overcome the limited alloying ability (wettability) of silver-copper alloys used to braze iron and steel, those filler metals contain zinc or cadmium, or both, to lower the liquidus and solidus temperatures. Tin is added in place of zinc or cadmium when constituents with high vapor pressure would be objectionable.

Similarly, silicon is used to lower the liquidus and solidus temperatures of aluminum and nickel-base brazing filler metals. Other brazing filler metals contain elements such as lithium, phosphorus, or boron, which reduce surface oxides on base metal and form compounds with melting temperatures below the brazing temperature. Those molten oxides then flow out of the joint, leaving a clean metal surface for brazing. These filler metals are essentially self-fluxing.

FILLER METAL SELECTION

FOUR FACTORS SHOULD be considered when selecting a brazing filler metal:

(1) Compatibility with base metal and joint design
(2) Service requirements for the brazed assembly
Compositions should be selected to suit operating requirements, such as service temperature (high versus cryogenic), thermal cycling, life expectancy, stress loading, corrosive conditions, radiation stability, and vacuum operation.
(3) Brazing temperature required
Low brazing temperatures are usually preferred to economize on heat energy, minimize heat effects on base metal (annealing, grain growth, warpage), minimize base metal-filler metal interaction, and increase the life of fixtures and other tools.

High brazing temperatures are used in order to take advantage of a higher melting, but more economical, brazing

filler metal; to combine annealing, stress relief, or heat treatment of the base metal with brazing; to permit subsequent processing at elevated temperatures; to promote base metal-filler metal interactions to increase the joint remelt temperature; or to promote removal of certain refractory oxides by vacuum or an atmosphere.
(4) Method of heating
Filler metals with narrow melting ranges—less than 50°F (28°C) between solidus and liquidus—can be used with any heating method, and the brazing filler metal may be preplaced in the joint area in the form of rings, washers, formed wires, shims, powder, or paste.

Alternatively, such alloys may be manually or automatically face-fed into the joint after the base metal is heated. Filler metals that tend to liquate should be used with heating methods that bring the joint to brazing temperature quickly, or the brazing filler metal should be introduced after the base metal reaches the brazing temperature.

To simplify filler metal selection, ANSI/AWS A5.8, *Specification for Brazing Filler Metal*, divides filler metals into seven categories and various classifications within each category. The specification lists products which are common, commercially available filler metals. Suggested base metal-filler metal combinations are given in Table 12.2. Other brazing filler metals not currently covered by the specification are available for special applications.

ALUMINUM-SILICON FILLER METALS

THIS GROUP IS used for joining aluminum grades 1060, 1100, 1350, 3003, 3004, 3005, 5005, 5050, 6053, 6061, 6951, and cast alloys A712.0 and C711.0. All types are suited for furnace and dip brazing, while some types are also suited for torch brazing, using lap joints rather than butt joints.

Brazing sheet or tubing is a convenient source of aluminum filler metal. It consists of a core of aluminum alloy and a coating of lower melting filler metal. The coatings are aluminum-silicon alloys, applied to one or both sides of the sheet. Brazing sheet is frequently used as one member of an assembly, with the mating piece made of an unclad brazeable alloy. The coating on the brazing sheet or tubing melts at brazing temperature and flows by capillary attraction and gravity to fill the joints.

MAGNESIUM FILLER METALS

MAGNESIUM FILLER METAL (BMg-1) is used to join AZ10A, K1A, and M1A magnesium alloys by torch, dip, or furnace brazing processes. Heating must be closely controlled to prevent melting of the base metal. Joint clearances of 0.004 to 0.010 in (0.10 to 0.25 mm) are best for most applications. Corrosion resistance is good if the flux is completely removed after brazing. Brazed assemblies are generally suited for continuous service up to 250°F (120°C) or intermittent service to 300°F (150°C), subject to the usual limitations of the actual operating environment.

Table 12.2
Base Metal-Filler Metal Combinations

	Al & Al Alloys	Mg & Mg Alloys	Cu & Cu Alloys	Carbon & Low Alloy Steels	Cast Iron	Stainless Steel	Ni & Ni Alloys	Ti & Ti Alloys	Be, Zr, & Alloys (Reactive Metals)	W, Mo, Ta, Cb & Alloys (Refractory Metals)	Tool Steels
Al & Al alloys	BAlSi										
Mg & Mg alloys	X	BMg									
Cu & Cu alloys	X	X	BAg, BAu, BCuP, RBCuZn	BNi							
Carbon & low alloy steels	BAlSi	X	BAg, BAu, RBCuZN, BNi	BAg, BAu, BCu, RBCuZn, BNi							
Cast iron	X	X	BAg, BAu, RBCuZN, BNi	BAg, RBCuZn, BNi	BAg, BAu, BCu, BNi						
Stainless steel	BAlSi	X	BAg, BAu	BAg, BAu, BCu, BNi	BAg, BAu, BCu, BNi	BAg, BAu, BCu, BNi					
Ni & Ni alloys	X	X	BAg, BAu, RBCuZn, BNi	BAg, BAu, BCu, RBCuZn, BNi	BAg, BCu, RBCuZn	BAg, BAu, BCu, BNi	BAg, BAu, BCu, BNi				
Ti & Ti alloys	BAlSi	X	BAg	BAg	BAg	BAg	BAg	Y			
Be, Zi & alloys (reactive metals)	X BAlSi(Be)	X	BAg	BAg, BNi*	BAg, BNi*	BAg, BNi*	BAg, BNi*	Y	Y		
W, Mo, Ta, Cb & alloys (refractory metals)	X	X	BAg, BNi	BAg, BCu, BNi*	BAg, BCu, BNi*	BAg, BCu, BNi*	BAg, BCu, BNi*	Y	Y	Y	
Tool steels	X	X	BAg, BAu, RBCuZn, BNi	BAg, BAu, BCu, RBCuZn, BNi	BAg, BAu, RBCuZn, BNi	BAg, BAu, BCu, BNi	BAg, BAu, BCu, RBCuZn, BNi	X	X	X	BAg, BAu, BCu, RBCuZn, BNi

Note: Refer to AWS Specification A5.8 for information on the specific compositions within each classification.
X—Not recommended; however, special techniques may be practicable for certain dissimilar metal combinations.
Y—Generalizations on these combinations cannot be made. Refer to the Brazing Handbook for usable filler metals.
*—Special brazing filler metals are available and are used successfully for specific metal combinations.

Filler Metals:
BAlsi—Aluminum   BCuP—Copper phosphorus
BAg—Silver base   RBCuZn—Copper zinc
BAu—Gold base   BMg—Magnesium base
BCu—Copper   BNi—Nickel base

COPPER AND COPPER-ZINC FILLER METALS

THESE BRAZING FILLER metals are used to join ferrous metals and nonferrous metals. The corrosion resistance of the copper-zinc alloy filler metals is generally inadequate for joining copper, silicon bronze, copper-nickel alloys, or stainless steel.

The essentially pure copper brazing filler metals are used to join ferrous metals, nickel-base alloys, and copper-nickel alloys. They are free flowing and often used in furnace brazing with a combusted gas, hydrogen, or dissociated ammonia atmosphere without flux. Copper filler metals are available in wrought and powder forms.

One copper filler metal is a copper oxide to be suspended in an organic vehicle.

Copper-zinc filler metals are used on steel, copper, copper alloys, nickel and nickel-base alloys, and stainless steel where corrosion resistance is not a requirement. They are used with the torch, furnace, and induction brazing processes. Fluxing is required, and a borax-boric acid flux is commonly used.

COPPER-PHOSPHORUS FILLER METALS

THESE FILLER METALS are primarily used to join copper and copper alloys. They have some limited use for joining silver, tungsten, and molybdenum. They should not be used on ferrous or nickel-base alloys, nor on copper-nickel alloys with more than 10 percent nickel. These filler metals are suited for all brazing processes and have self-fluxing properties when used on copper. They tend to liquate if heated slowly.

SILVER FILLER METALS

THESE FILLER METALS are used to join most ferrous and nonferrous metals, except aluminum and magnesium, with all methods of heating. They may be preplaced in the joint or fed into the joint area after heating.

Silver-copper alloys high in silver do not wet steel well when brazing is done in air with a flux. Copper forms alloys with cobalt and nickel much more readily than silver does. Thus, copper wets many of these metals and their alloys satisfactorily, where silver does not. When brazing in certain protective atmospheres without flux, silver-copper alloys will wet and flow freely on most steels at the proper temperature.

Zinc is commonly used to lower the melting and flow temperatures of silver-copper alloys. It is by far the most helpful wetting agent when joining alloys based on iron, cobalt, or nickel. Alone, or in combination with cadmium or tin, zinc produces alloys that wet the iron group metals but do not alloy with them to any appreciable depth.

Cadmium is incorporated in some silver-copper-zinc filler metals alloys to further lower the melting and flow temperatures, and to increase the fluidity and wetting action on a variety of base metals. Since cadmium oxide fumes are a health hazard, cadmium-bearing filler metals should be used with caution.

Tin has a low vapor pressure at normal brazing temperatures. It is present in silver brazing filler metals in place of zinc or cadmium when volatile constituents are objectionable, such as when brazing is done without flux in atmosphere or vacuum furnaces, or when the brazed assemblies will be used in high vacuum at elevated temperatures. Silver-copper filler metals with tin additions have wide melting ranges. Fillers containing zinc wet ferrous metals more effectively than those containing tin, and where zinc is tolerable, they are preferred to fillers with tin.

Stellites, cemented carbides, and other molybdenum- and tungsten-rich refractory alloys are brazed with filler metals with added manganese, nickel, and, infrequently, cobalt to increase wettability.

When stainless steels and alloys that form refractory oxides are brazed in reducing or inert atmospheres without flux, silver brazing filler metals containing lithium as the wetting agent are quite effective. The heat of formation of Li_2O is high, so lithium metal reduces adherent oxides on the base metal. The resultant lithium oxide is readily displaced by the brazing filler metal.

GOLD FILLER METALS

GOLD FILLER METALS are used to join parts in electron tube assemblies where volatile components are undesirable. They are used to braze iron, nickel, and cobalt-base metals where resistance to oxidation or corrosion is required. They are commonly used on thin sections because of their low rate of interaction with the base metal.

NICKEL FILLER METALS

NICKEL BRAZING FILLER metals are generally used on 300 and 400 series stainless steels, nickel and cobalt-base alloys, even carbon steel, low alloy steels, and copper when specific properties are desired. They exhibit good corrosion and heat resistance properties. They are normally applied as powders, pastes, rod, foil, or in the form of sheet or rope with plastic binders.

Nickel filler metals have the very low vapor pressure needed in vacuum systems and vacuum tube applications at elevated temperatures.

The phosphorus-containing filler metals suffer from low ductility because they form nickel phosphides. The boron-containing filler metals must be carefully controlled when used to braze thin sections, to prevent erosion.

COBALT FILLER METAL

THIS FILLER METAL is used for its high temperature properties and its compatibility with cobalt-base metals. Brazing in a high quality atmosphere or diffusion brazing gives optimum results. Special high temperature fluxes are available for torch brazing.

FILLER METALS FOR REFRACTORY METALS

BRAZING IS EXCELLENT for fabricating assemblies of refractory metals, in particular those involving thin sections. However, only a few filler metals have been specifically designed for both high temperature and high corrosion applications.

Those filler metals and pure metals used to braze refractory metals are given in Table 12.3. Low melting filler metals, such as silver-copper-zinc, copper-phosphorus, and copper, are used to join tungsten for electrical contact applications, but these filler metals cannot operate at high temperatures. The use of higher melting rare metals, such as tantalum and columbium, is warranted in those cases.

Nickel-base and precious-metal-base filler metals may also be used to join tungsten.

Various brazing filler metals will join molybdenum. The effect of brazing temperature on base metal recrystallization must be considered. When brazing above the recrystallization temperature, brazing time must be kept short. If high temperature service is not required, copper and silver-base filler metals may be used.

Columbium and tantalum are brazed with a number of refractory or reactive-metal-base filler metals. The metal systems Ti-Zr-Be and Zr-Cb-Be are typical, also platinum, palladium, platinum-iridium, platinum-rhodium, titanium, and nickel-base filler metals (such as nickel-chromium-silicon alloys). Copper-gold alloys containing gold in amounts between 46 and 90 percent form age hardening compounds which are brittle. Silver-base filler metals are not recommended because they may embrittle the base metals.

Table 12.3
Brazing Filler Metals for Refractory Metals[a]

Brazing Filler Metal	Liquidus Temperature		Brazing Filler Metal	Liquidus Temperature	
	°F	°C		°F	°C
Cb	4380	2416	Mn-Ni-Co	1870	1021
Ta	5425	2997			
Ag	1760	960	Co-Cr-Si-Ni	3450	1899
Cu	1980	1082	Co-Cr-W-Ni	2600	1427
Ni	2650	1454	Mo-Ru	3450	1899
Ti	3300	1816	Mo-B	3450	1899
Pd-Mo	2860	1571	Cu-Mn	1600	871
Pt-Mo	3225	1774	Cb-Ni	2175	1190
Pt-30W	4170	2299			
Pt-50Rh	3720	2049	Pd-Ag-Mo	2400	1306
			Pd-Al	2150	1177
Ag-Cu-Zn-Cd-Mo	1145-1295	619-701	Pd-Ni	2200	1205
Ag-Cu-Zn-Mo	1324-1450	718-788	Pd-Cu	2200	1205
Ag-Cu-Mo	1435	780	Pd-Ag	2400	1306
Ag-Mn	1780	971	Pd-Fe	2400	1306
			Au-Cu	1625	885
Ni-Cr-B	1950	1066	Au-Ni	1740	949
Ni-Cr-Fe-Si-C	1950	1066	Au-Ni-Cr	1900	1038
Ni-Cr-Mo-Mn-Si	2100	1149	Ta-Ti-Zr	3800	2094
Ni-Ti	2350	1288			
Ni-Cr-Mo-Fe-W	2380	1305	Ti-V-Cr-Al	3000	1649
Ni-Cu	2460	1349	Ti-Cr	2700	1481
Ni-Cr-Fe	2600	1427	Ti-Si	2600	1427
Ni-Cr-Si	2050	1121	Ti-Zr-Be[b]	1830	999
			Zr-Cb-Be[b]	1920	1049
			Ti-V-Be[b]	2280	1249
			Ta-V-Cb[b]	3300-3500	1816-1927
			Ta-V-Ti[b]	3200-3350	1760-1843

a. Not all the filler metals listed are commercially available.

b. Depends on the specific composition.

FLUXES AND ATMOSPHERES

METALS AND ALLOYS may react with the atmosphere to which they are exposed, more so as the temperature is raised. The common reaction is oxidation, but nitrides and carbides are sometimes formed.

Fluxes, gas atmospheres, and vacuum are used to prevent undesirable reactions during brazing. Some fluxes and atmospheres may also reduce oxides already present.

Titanium, zirconium, columbium (niobium), and tantalum become permanently embrittled when brazed in any atmosphere containing hydrogen, oxygen, or nitrogen. Hydrogen will embrittle copper that has not been thoroughly deoxidized.

The use of flux or atmosphere does not eliminate the need to clean parts prior to brazing. Recommended cleaning procedures are contained in Chapter 7 of the *AWS Brazing Manual*, 3rd edition, 1976. The functions of individual fluxing ingredients are discussed in Chapter 4 of that Manual.

Since the purpose of a braze filler is to flow over the base material and into capillaries, it also may flow over portions of the piece being joined. This may be undesirable from a cosmetic viewpoint or there may be holes or features on the part that must not be filled or plugged for the device to function properly. When extraneous flow must be prevented, the brazer applies a "stopoff" material to retard the flow of the filler material. Great care must be exercised to prevent the stopoff material from getting into the actual braze joint because this would produce an unbonded condition. Stopoff materials are generally oxides applied by brush, tape, spray, or a hypodermic needle system. The common stopoffs are oxides of titanium, calcium, aluminum, or magnesium.

Stopoffs retard braze flow by intentionally putting oxides on the surface of the materials being joined. This works quite well when furnace brazing without flux. However, when flux is used, the cleaning action of the flux may counteract the stopoff effect. After brazing, the stopoff material can be removed by washing with hot water or by chemical or mechanical stripping.

APPLICATIONS

SELECTION OF BASE METALS

THE EFFECT OF brazing on the mechanical properties of the metal in a brazement and the final joint strength must be considered. Base metals strengthened by cold working will be annealed by brazing process temperatures and times in the annealing range of the base metal being processed. "When brazed, "hot-cold worked", heat-resistant base metals will also exhibit only the annealed physical properties. The brazing cycle by its very nature will usually anneal cold worked base metal unless the brazing temperature is very low and the time at temperature is very short.

It is not practical to cold work the base metal after the brazing operation.

When a brazement must have strength after brazing that will be above the annealed properties of the base metal, a heat treatable base metal should be selected. The base metal can be an oil-quench type, an air-quench type that can be brazed and hardened in the same or a separate operation, or a precipitation-hardening type that can be brazed and solution treated in a combined cycle. Parts already hardened may be brazed with a low temperature filler metal using short times at temperature to maintain the mechanical properties.

ALUMINUM AND ALUMINUM ALLOYS

THE NONHEAT TREATABLE wrought aluminum alloys that are brazed most successfully are the ASTM 1XXX and 3XXX series, and low magnesium alloys of the ASTM 5XXX series. Available filler metals melt below the solidus temperatures of all commercial wrought, nonheat treatable alloys.

The heat treatable wrought alloys most commonly brazed are the ASTM 6XXX series. The ASTM 2XXX and 7XXX series of aluminum alloys are low melting and, therefore, not normally brazeable, with the exception of 7072 and 7005 alloys.

Aluminum sand and permanent mold casting alloys most commonly brazed are ASTM 443.0, 356.0, and 712.0 alloys. Aluminum die castings are generally not brazed because of blistering from their high gas content.

Table 12.4 lists the common aluminum base metals that can be brazed.

Most aluminum brazing is done by torch, dip, or furnace processes. Furnace brazing may be done in air or controlled atmosphere, including vacuum.

Additional information on brazing aluminum and aluminum alloys is contained in Chapter 12, *Brazing Manual*, 3rd Edition.

MAGNESIUM AND MAGNESIUM ALLOYS

BRAZING TECHNIQUES SIMILAR to those used for aluminum are used for magnesium alloys. Furnace, torch, and dip brazing can be employed, although the latter process is the most widely used.

Magnesium alloys that are considered brazeable are given in Table 12.5. Furnace and torch brazing experience

Table 12.4
Nominal Composition and Melting Range of Common Brazeable Aluminum Alloys

Commercial Designation	ASTM Alloy	Brazeability Rating[b]	Nominal Composition[a]						Approximate Melting Range	
			Cu	Si	Mn	Mg	Zn	Cr	°F	°C
EC	EC	A		Al 99.45% min					1195-1215	646-657
1100	1100	A		Al 99% min					1190-1215	643-657
3003	3003	A	--	--	1.2	--	--	--	1190-1210	643-654
3004	3004	B	--	--	1.2	1.0	--	--	1165-1205	629-651
3005	3005	A	0.3	0.6	1.2	0.4	0.25	0.1	1180-1215	638-657
5005	5005	B	--	--	--	0.8	--	--	1170-1210	632-654
5050	5050	B	--	--	--	1.2	--	--	1090-1200	588-649
5052	5052	C	--	--	--	2.5	--	--	1100-1200	593-649
6151	6151	C	--	1.0	--	0.6	--	0.25	1190-1200	643-649
6951	6951	A	0.25	0.35	--	0.65	--	--	1140-1210	615-654
6053	6053	A	--	0.7	--	1.3	--	--	1105-1205	596-651
6061	6061	A	0.25	0.6	--	1.0	--	0.25	1100-1205	593-651
6063	6063	A	--	0.4	--	0.7	--	--	1140-1205	615-651
7005	7005	B	0.1	0.35	0.45	1.4	4.5	0.13	1125-1195	607-646
7072	7072	A	--	--	--	--	1.0	--	1125-1195	607-646
Cast 43	Cast 443.0	A	--	5.0	--	--	--	--	1065-1170	629-632
Cast 356	Cast 356.0	C	--	7.0	--	0.3	--	--	1035-1135	557-613
Cast 406	Cast 406	A		Al 99% min					1190-1215	643-657
Cast A612	Cast A712.0	B	--	--	--	0.7	6.5	--	1105-1195	596-646
Cast C612	Cast C712.0	A	--	--	--	0.35	6.5	--	1120-1190	604-643

a. Percent of alloying elements: aluminum and normal impurities constitute remainder.

b. Brazeability ratings: A = Alloys readily brazed by all commercial methods and procedures.
                         B = Alloys that can be brazed by all techniques with a little care.
                         C = Alloys that require special care to braze.

is limited to M1A alloy. Dip brazing can be used for AZ10A, AZ31B, AZ61A, K1A, M1A, ZE10A, ZK21A, and ZK60A alloys.

The filler metals used for brazing magnesium are also summarized in Table 12.5. BMg-1 brazing filler metal is suitable for the torch, dip, or furnace brazing process. The BMg-2a alloy is usually preferred in most brazing applications because of its lower melting range. A zinc base filler metal known as GA432 is an even lower melting composition suitable only for dip brazing use.

BERYLLIUM

BRAZING IS THE preferred method for metallurgically joining beryllium.[2] Suitable brazing filler metal systems and their temperature ranges include:

(1) Zinc: 800-850°F (427-454°C)
(2) Aluminum-silicon: 1050-1250°F (566-677°C)
(3) Silver-copper: 1200-1660°F (649-904°C)
(4) Silver: 1620-1750°F (882-954°C)

2. Beryllium and its compounds are toxic. Proper handling and identification of beryllium metal is required by federal regulations.

Zinc melts below 840°F, the temperature defined by AWS for brazing filler metal. Nevertheless, it is generally accepted as the lowest melting filler metal for brazing beryllium.

Aluminum-silicon filler metals can be used in high-strength, wrought beryllium assemblies because the brazing temperature is well below the base metal recrystallization temperature. BA1Si-4 type filler metal brazes well with fluxes. Fluxless brazing requires stringent control. Aluminum-base filler metals have less metallurgical interaction with the base metal than silver-base fillers. This is a significant advantage when thin beryllium sections or foils are to be joined.

Silver and silver-base brazing filler metals find use in structures exposed to elevated temperatures. Atmosphere brazing with these alloy systems is straight forward and may be performed in purified atmospheres or vacuum.

COPPER AND COPPER ALLOYS

THE COPPER ALLOY base metals include copper-zinc alloys (brass), copper-silicon alloys (silicon bronze), copper-aluminum alloys (aluminum bronze), copper-tin alloys (phosphor bronze), copper-nickel alloys, and several others. The brazing of copper and copper alloys and appropriate filler

Table 12.5
Brazeable Magnesium Alloys and Filler Metals

AWS A5.8 Classification	ASTM Alloy Designation	Avail. Forms	Solidus		Liquidus		Brazing Range		Suitable Filler	
			°F	°C	°F	°C	°F	°C	BMg-1	BMG-2a
					Base Metal					
—	AZ10A	E	1170	632	1190	643	1080-1140	582-616	X	X
—	AZ31B	E. S	1050	566	1160	627	1080-1100	582-593		X
—	K1A	C	1200	649	1202	650	1080-1140	582-616	X	X
—	M1A	E. S	1198	648	1202	650	1080-1140	582-616	X	X
—	ZE10A	S	1100	593	1195	646	1080-1100	582-593		X
—	ZK21A	E	1159	626	1187	642	1080-1140	582-616	X	X
					Filler Metal					
BMg-1	AZ92A	W. R. ST. P	830	443	1110	599	1120-1140	604-616	—	—

E = Extruded shapes and structural sections
S = Sheet and plate
C = Castings
W = Wire
R = Rod
ST = Strip
P = Powder

metals are discussed in detail in Chapter 14, *Brazing Manual*, 3rd Edition.

LOW CARBON AND LOW ALLOY STEELS

LOW CARBON AND low alloy steels are brazed without difficulty. They are frequently brazed at temperatures above 1980°F (1080°C) with copper filler metal in a controlled atmosphere, or at lower temperatures with silver base filler metals.

For alloy steels, the filler metal should have a solidus well above any heat-treating temperature to avoid damage to joints that will be heat-treated after brazing. In some cases, air hardening steels can be brazed and then hardened by quenching from the brazing temperature.

A filler metal with brazing temperature lower than the critical temperature of the steel can be used when no change in the metallurgical properties of the base metal is wanted.

HIGH-CARBON AND HIGH-SPEED TOOL STEELS

HIGH-CARBON STEELS contain more than 0.45 percent carbon. High-carbon tool steels usually contain 0.60 to 1.40 percent carbon.

Brazing of high-carbon steels is best accomplished prior to or during the hardening operation. Hardening temperatures for carbon steels range from 1400 to 1500°F (760 to 820°C). Filler metals having brazing temperatures above 1500°F (820°C) should be used. When brazing and hardening are done in one operation, the filler metal should have a solidus at or below the austenitizing temperature.

Tempering and brazing can be combined for high-speed tool steels and high-carbon, high-chromium alloy tool steels which have tempering temperatures in the range of 1000 to 1200°F (540°C to 650°C). Filler metals with brazing temperatures in that range are used. The part is removed from the tempering furnace, brazed by localized heating methods, and then returned to the furnace for completion of the tempering cycle.

CAST IRONS

CAST IRONS GENERALLY require special brazing considerations. The types of cast iron include white, gray, malleable, and ductile. White cast iron is seldom brazed.

Prior to brazing, faying surfaces generally are cleaned electrochemically or chemically, seared with an oxidizing flame, or grit blasted. When low-melting silver brazing filler metals are used, wetting by the brazing filler metal is easiest. Ductile and malleable cast irons should be brazed below 1400°F (760°C).

When high carbon cast iron is brazed with copper, the brazing temperature should be low to avoid melting of localized areas of the cast iron, particularly in light sections.

STAINLESS STEELS

ALL OF THE stainless steel alloys are difficult to braze because of their high chromium content. Brazing of these alloys is best accomplished in purified (dry) hydrogen or in a vacuum. Dew points below -60°F (-51°C) must be maintained because wetting becomes difficult following the formation of chromium oxide. Torch brazing requires fluxing to reduce any chromium oxides present.

Most of the silver alloy, copper, and copper-zinc filler metals are used for brazing stainless steels. Silver alloys containing nickel are generally best for corrosion resistance. Filler metals containing phosphorus should not be used on highly stressed parts because brittle nickel and iron phosphides may be formed at the joint interface.

Boron-containing nickel filler metals are generally best for stainless steels containing titanium or aluminum, or both, because boron has a mild fluxing action which aids in wetting these base metals. Diffusion brazing produces joints with improved physical properties.

Brazing of the austenitic chromium-nickel stainless steels is discussed further in Chapter 18, *Brazing Manual*, 3rd Edition.

Chromium Irons and Steels

THE MARTENSITIC STAINLESS steels (403, 410, 414, 416, 420, and 431) air harden upon cooling from brazing, which occurs above their austenitizing temperature range. Therefore, they must be annealed after brazing or during the brazing operation. These steels are also subject to stress cracking with certain brazing filler metals.

The ferritic stainless steels (405, 406, and 430) cannot be hardened and their grain structure cannot be refined by heat treatment. These alloys degrade in properties when brazed at temperatures above 1800°F (980°C), because of excessive grain growth. They lose ductility after long heating times between 650 and 1100°F (340 and 600°C). However, some of the ductility can be recovered by heating the brazement to approximately 1450°F (790°C) for a suitable time.

Precipitation-Hardening Stainless Steels

THESE STEELS ARE basically stainless steels with additions of one or more of the elements copper, molybdenum, aluminum, and titanium. Such alloying additions make it possible to strengthen the alloys by precipitation hardening heat treatments. When alloys of this type are brazed, the brazing cycle and temperature must match the heat treatment cycle of the alloy. Manufacturers of these alloys have developed recommended brazing procedures for their particular steels.

NICKEL AND HIGH-NICKEL ALLOYS

NICKEL AND THE high nickel alloys are embrittled by sulfur and low-melting metals present in brazing alloys, such as zinc, lead, bismuth, and antimony. Base metal surfaces must be thoroughly cleaned prior to brazing to remove any substances that may contain these elements. Sulfur and sulfur compounds must also be excluded from the brazing atmosphere.

Nickel and its alloys are subject to stress cracking in the presence of molten brazing filler metals. Parts should be annealed prior to brazing to remove residual stresses, or carefully stress relieved during the braze cycle.

Silver brazing filler metals are commonly used. In corrosive environments, high silver brazing alloys are preferred. Cadmium-free brazing filler metals are chosen to avoid stress corrosion cracking.

Nickel-base brazing filler metals offer the greatest corrosion and oxidation resistance and elevated temperature strength.

Brazing is a preferred method for joining dispersion-strengthened nickel alloys that must function at elevated temperatures. High strength brazements have been made with special nickel-base brazing filler metals and then tested up to 2400°F (1300°C).

HEAT-RESISTANT ALLOYS

HEAT-RESISTANT ALLOYS are generally brazed in a hydrogen atmosphere or high temperature vacuum furnaces using nickel-base or special filler metals.

The cobalt-base alloys are the easiest of the super alloys to braze because most of them do not contain titanium or aluminum. Alloys that are high in titanium or aluminum are difficult to braze in dry hydrogen because titanium and aluminum oxides are not reduced at brazing temperatures.

TITANIUM AND ZIRCONIUM

TITANIUM AND ZIRCONIUM combine readily with oxygen, and react to form brittle intermetallic compounds with many metals and with hydrogen and nitrogen. Parts must be cleaned before brazing and brazed immediately after cleaning.

Silver and silver-based filler metals were used in early brazing of titanium, but brittle intermetallics were formed and crevice corrosion resulted. Type 3003 aluminum foil will join thin, lightweight structures, such as complex honeycomb sandwich panels. Electroplating various elements on the base metal faying surfaces will let them react in situ with the titanium during brazing to form a titanium alloy eutectic. That transient liquid phase flows well and forms fillets, then solidifies due to interdiffusion.

Other brazing filler metals with high service capability and corrosion resistance include Ti-Zr-Ni-Be, Ti-Zr-Ni-Cu, and Ti-Ni-Cu alloys. The best braze processing is obtained in high vacuum furnaces using closely controlled temperatures in the range of 1650 to 1750° (900 to 955°C).

CARBIDES AND CERMETS

CARBIDES OF THE refractory metals tungsten, titanium, and tantalum that are bonded with cobalt are used for cutting tools and dies. Closely related materials called *cermets* are ceramic particles bonded with various metals.

Brazing carbides and cermets is more difficult than brazing metals. Torch, induction, or furnace brazing is used, often with a sandwich brazing technique: a layer of weak, ductile metal (pure nickel or pure copper) is interposed between the carbide or cermet and a hard metal support. The cooling stresses cause the soft metal to deform instead of cracking the ceramic.

Silver-base brazing alloys, copper-zinc alloys, and copper are often used on carbide tools. Silver alloys containing nickel are preferred for their better wettability. The nickel base alloys containing boron and a 60% Pd - 40% Ni alloy may be satisfactory for brazing nickel- and cobalt-bonded cermets of tungsten carbide, titanium carbide, and columbium carbide.

CERAMICS

ALUMINA, ZIRCONIA, MAGNESIA, forsterite (Mg_2SiO_4), beryllia, and thoria are ceramic materials which can be joined by brazing. They are inherently difficult to wet with conventional filler metals. Differences in thermal expansion, heat conduction, and ductility result in cracking and crack propagation at relatively low stresses.

If the ceramic is premetallized to facilitate wetting, copper, silver-copper, and gold-nickel filler metals are used. Titanium or zirconium hydride can be decomposed at the ceramic-metal interface to form an intimate bond.

Nonmetallized ceramics are brazed with silver-copper-clad or nickel-clad titanium wires. Useful titanium and zirconium alloys are Ti-Zr-Be, Ti-V-Zr, Zr-V-Cb, Ti-V-Be, and Ti-V-Cr.

PRECIOUS METALS

THE PRECIOUS METALS silver, gold, platinum, and palladium present few brazing difficulties. Their thin oxide films are readily removed by fluxes and reducing atmospheres.

Resistance or furnace brazing is common for electrical contacts. Silver (BAg) and precious metal (BAu) filler metals braze contacts to holders.

REFRACTORY METALS

TUNGSTEN, MOLYBDENUM, TANTALUM, and columbium brazing is still in the developmental stages.

Tungsten

TUNGSTEN CAN BE brazed to itself and to other metals and nonmetals with nickel-base filler metals, but interaction between tungsten and nickel will recrystallize the base metal. The tungsten should be stress relieved by heat treat-ment prior to brazing, and the brazing cycle should be short to limit interaction with the filler metal.

Molybdenum

MOLYBDENUM AND ITS alloys are brazed with palladium-base filler metals and molybdenum-base metals (Mo-0.5Ti) with high recrystallization temperatures. Chromium plating, as a barrier layer, prevents formation of intermetallic compounds. Most high-temperature brazing filler metals are suitable for oxidation resistant service for coating applications.

Tantalum and Columbium

TANTALUM AND COLUMBIUM require special techniques to be satisfactorily brazed. All reactive gases must be removed from the brazing atmosphere. These include oxygen, nitrogen, carbon monoxide, ammonia, and hydrogen. Tantalum forms oxides, nitrides, carbides, and hydrides very readily, leading to a loss of ductility. For oxidation protection at high temperatures, tantalum and columbium are often electroplated with copper or nickel. The brazing filler metal must be compatible with any plating used.

DISSIMILAR METAL COMBINATIONS

MANY DISSIMILAR METAL combinations may be brazed, even those with metallurgical incompatibility that precludes welding.

Important criteria to be considered start with differences in thermal expansion. If a metal with high thermal expansion surrounds a low expansion metal, clearances at room temperature which are satisfactory for capillary flow will be too great at brazing temperature. Conversely, if a low expansion metal surrounds a high expansion metal, no clearance may exist at brazing temperature. For example, when brazing a molybdenum plug in a copper block, the parts must be a press fit at room temperature; if a copper plug is to be brazed in a molybdenum block, a properly centered loose fit at room temperature is required.

In brazing tube-and-socket type joints between dissimilar base metals, the tube should be the low expansion metal and the socket the high expansion metal. At brazing temperature, the clearance will be maximum and the capillary will fill with brazing alloy. When the joint cools to room temperature, the brazed joint and the tube will be in compression.

A tongue-in-groove joint should place the groove in the low expansion material. The fit at room temperature should be designed to give capillary joint clearances on both sides of the tongue at brazing temperature. Longitudinal shear stresses in the braze metal are limited by making overlap distances small.

"Sandwich brazing" is commonly used to manufacture carbide-tipped metal cutting tools. A relatively ductile metal is coated on both sides with brazing filler metal, and the composite is used in the joint. This places a third material in that joint which will deform during cooling and reduce the stresses caused by differential contraction of the parts brazed together.

The filler metal used to braze dissimilar metals must be compatible with both base metals. It should have corrosion or oxidation resistance at least equal to the poorer of the two metals being brazed. It should not form galvanic couples which could promote crevice corrosion in the braze area. Brazing filler metals form low melting phases with many base metals, requiring adaption of the brazing cycle, quantity and placement of filler metal, and joint design.

Metallurgical reactions between the brazing filler metal and dissimilar base metals may be objectionable. One example is the brazing of aluminum to copper. Copper reacts with aluminum to form a low melting brittle compound. Such problems can be overcome by coating one of the base metals with a metal which is compatible with the brazing filler metal. To braze aluminum to copper, the copper is plated with silver, or a high silver alloy. The joint is then brazed at 1500°F (816°C) with a standard aluminum brazing filler metal. Nickel plating would also form a suitable diffusion barrier.

JOINT DESIGN

BASICALLY TWO TYPES of joints are used in brazing: the lap joint and, the butt joint. These joints are shown in Figure 12.11.

The lap joint may be made as strong as the weaker member, even when using a low strength filler metal or in the presence of small defects in the joint, by using an overlap at least three times the thickness of the thinner member. Lap joints feature high joint efficiency and ease of fabrication; they have the disadvantage that the increased metal thickness at the joint creates a stress concentration at those abrupt changes in cross section.

Butt joints are used where the lap joint thickness would be objectionable, and where the strength of a brazed butt joint will satisfactorily meet service requirements. The joint strength depends only partly on the filler metal strength.

The scarf joint is a variation of the butt joint. As shown in Figure 12.12, the cross-sectional area of this joint is increased without an increase in metal thickness. Two disadvantages which limit its use: the sections are difficult to align, and the joint is difficult to prepare, particularly in thin members. Since the joint is at an angle to the axis of tensile loading, the load-carrying capacity is that of a lap joint.

JOINT CLEARANCE

JOINT CLEARANCE HAS a major effect on the mechanical performance of a brazed joint. This applies to all types of loading, such as static, fatigue, and impact, and to all joint designs. Several effects of joint clearance on mechanical performance are (1) the purely mechanical effect of restraint to plastic flow of the filler metal by a higher strength base metal, (2) the possibility of slag entrapment, (3) the possibility of voids, (4) the relationship between joint clearance and capillary force which accounts for filler metal distribution, and (5) the amount of filler metal that must be diffused with the base metal when diffusion brazing.

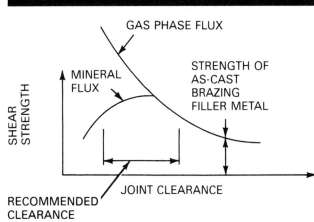

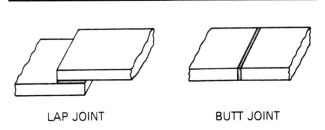

Figure 12.11—Basic Lap and Butt Joints for Brazing

Figure 12.12—Relationship Between Joint Clearance and Shear Strength for two Fluxing Methods

If the brazed joint is free of defects (no flux inclusions, voids, unbrazed areas, pores, or porosity), its strength in shear depends upon the joint thickness, as illustrated in Figure 12.13. This figure indicates the change in joint shear strength with joint clearance. Table 12.6 may be used as a guide for clearances at brazing temperature when designing brazed joints for maximum strength.

Some specific clearance versus strength data for silver brazed butt joints in steel are shown in Figures 12.14 and 12.15.[3] Figure 12.14 shows the optimum shear values obtained with joints in 0.5 in. (12.7 mm) round drill rod using pure silver. The rods were butt brazed by induction heating in a dry 10 percent hydrogen-90 percent nitrogen atmosphere. Figure 12.15 relates tensile strength to joint thickness for butt brazed joints of the same size. Note how the strength decreased at extremely small clearances.

Preplaced filler is brazing filler metal placed in the joint, such as foil placed between two plates. In this application, the clearances noted in Table 12.6 generally do not apply. In applications using preplaced filler metal, the members being joined should be preloaded so that the joint clearance will decrease during the brazing operation. That forces the filler metal into voids created by the normal roughness of the faying surfaces. In some applications, additional filler metal is made available by extending the filler metal shim out beyond the joint edges.

The type of fluxing will have an important bearing on the joint clearance to used to accomplish a given brazement.

_____

3. The data in Figures 12.14 and 12.15 were obtained with nonstandard test specimens.

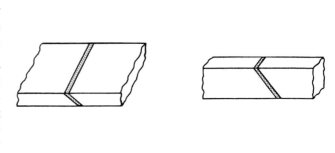

Figure 12.13–Typical Scarf Joint Designs

A mineral flux must melt at a temperature below the melting range of the brazing filler metal, and it must flow into the joint ahead of the filler metal. When the joint clearance is too small, the mineral flux may be held in the joint and not be displaced by the molten filler metal. This will produce joint defects. When the clearance is too large, the molten filler metal will flow around pockets of flux, causing excessive flux inclusions.

The joint clearance at the brazing temperature of a joint between dissimilar base metals must be calculated from thermal expansion data. Figure 12.16 shows thermal expansion data for some materials. Figure 12.17 can be used to find the diametral clearance at brazing temperature between dissimilar metals.

Table 12.6
Recommended Joint Clearance at Brazing Temperature

Filler Metal AWS Classification[a]	in.	mm	Joint Clearance[b]
BA1Si Group	0.006-0.010	0.15-0.25	For length at lap less than 1/4 in. (6.35 mm)
	0.010-0.025	0.25-0.61	For length at lap greater than 1/4 in. (6.35 mm)
BCuP Group	0.001-0.005	0.03-0.12	
BAg Group	0.002-0.005	0.05-0.12	Flux brazing (mineral fluxes)
	0.001-0.002[c]	0.03-0.05	Atmosphere brazing (gas phase fluxes)
BAu Group	0.002-0.005	0.05-0.12	Flux brazing (mineral fluxes)
	0.000-0.002[c]	0.00-0.05	Atmosphere brazing (gas phase fluxes)
BCu Group	0.000-0.002[c]	0.00-0.05	Atmosphere brazing (gas phase fluxes)
BCuZn Group	0.002-0.005	0.05-0.12	Flux brazing (mineral fluxes)
BMg Group	0.004-0.010	0.10-0.25	Flux brazing (mineral fluxes)
BNi Group	0.002-0.005	0.05-0.12	General applications (flux or atmosphere)
	0.000-0.002	0.00-0.05	Free flowing types, atmosphere brazing

a. See Table 12.2 for an explanation of filler metals.

b. Clearance on the radius when rings, plugs, or tubular members are involved. On some applications it may be necessary to use the recommended clearance on the diameter to assure not having excessive clearance when all the clearance is on one side. An excessive clearance will produce voids. This is particularly true when brazing is accomplished in a high quality atmosphere (gas phase fluxing).

c. For maximum strength, a press fit of 0.001 mm/mm or in./in. of diameter should be used.

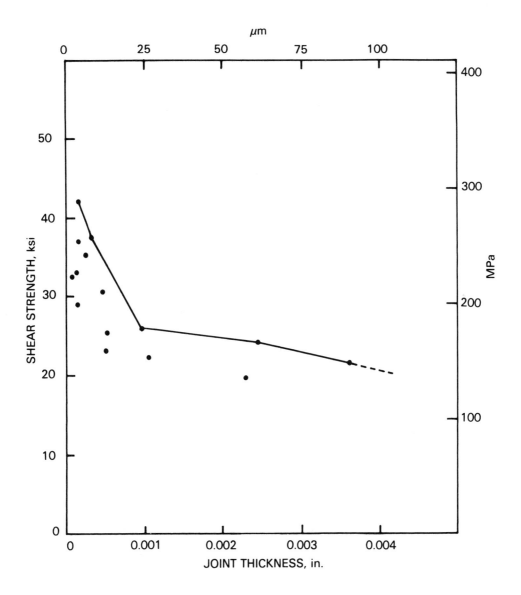

Figure 12.14–Relationship of Shear Strength to Brazed Joint Thickness for Pure Silver Joints in 0.5 in. (12.7 mm) Diameter Steel Drill Rod

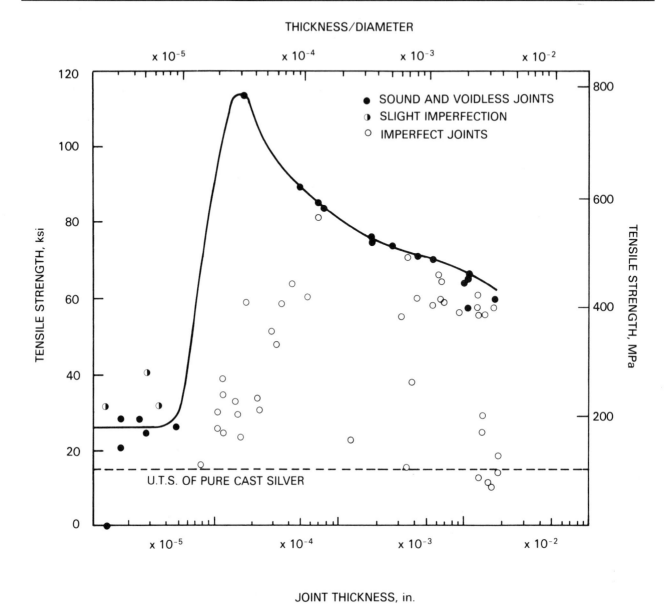

THICKNESS/DIAMETER

JOINT THICKNESS, in.

Figure 12.15—Relationship of Tensile Strength to Brazed Joint Thickness of 0.5 in. (12.7 mm) Diameter Silver Brazed Butt Joints in 4340 Steel

To withstand high differential thermal expansion of two metals being brazed, the brazing filler metal must be strong enough to resist fracture and the base metal must yield during cooling. Some residual stress will remain in the final brazement. Thermal cycling of such a brazement during its service life will repeatedly stress the joint area, which may shorten the service life. Dissimilar metal brazements should be designed so that residual stresses do not add to the stress imposed during service.

STRESS DISTRIBUTION

HIGH-STRENGTH BRAZEMENTS ARE designed to fail in the base metal. In brazements where joints will be lightly loaded, it is economical to use simplified joint designs which may break in the brazed joint if overstressed in testing or in service.

A good brazement design will incorporate joints that avoid high-stress concentration at the edges of the braze

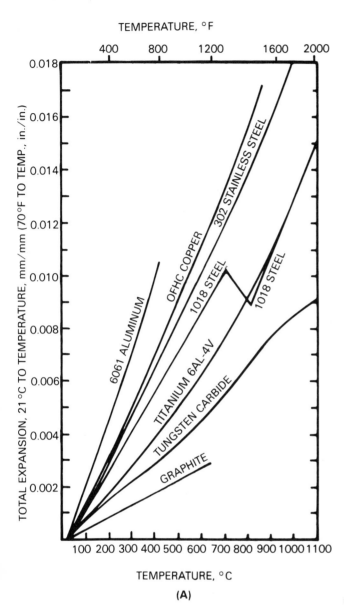

(A)

**Figure 12.16—Thermal Expansion Curves for
Some Common Materials**

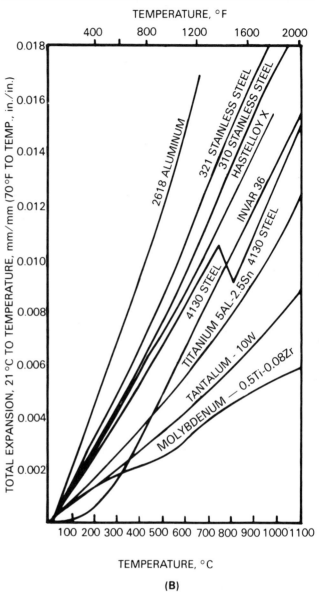

(B)

**Figure 12.16 (Continued)—Thermal Expansion
Curves for Some Common Materials**

and will distribute the stresses uniformly into the base metal. Typical designs are shown in Figures 12.18 through 12.21.

A fillet of brazing filler metal is not good brazing design. It is seldom possible to make the brazing filler metal consistently form a desired fillet size and contour. When the fillets become too large, shrinkage or piping porosity will act as a stress concentration.

ELECTRICAL CONDUCTIVITY

BRAZING FILLER METALS in general have low electrical conductivity compared to copper. However, a braze joint will not add appreciable resistance to the circuit when properly designed.

With butt joints, the brazed joint thickness (resistance) is very small compared to the length-wise resistance of the conductor, even though the unit resistivity of the filler metal is much higher than that of the base metal. Never-

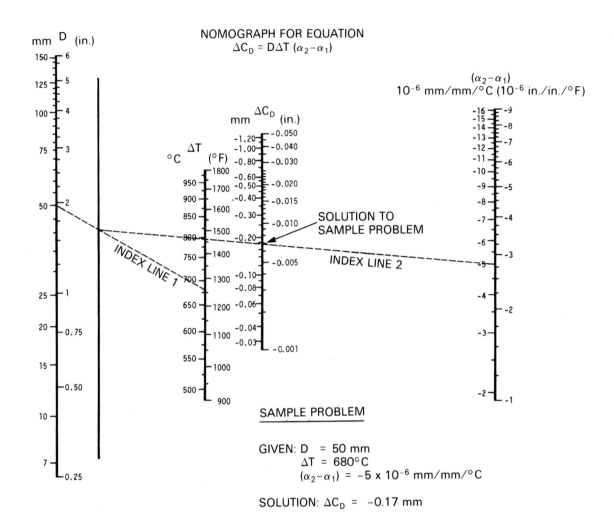

NOMOGRAPH FOR EQUATION
$$\Delta C_D = D \Delta T (\alpha_2 - \alpha_1)$$

SAMPLE PROBLEM

GIVEN: D = 50 mm
ΔT = 680°C
$(\alpha_2 - \alpha_1)$ = −5 × 10⁻⁶ mm/mm/°C

SOLUTION: ΔC_D = −0.17 mm

NOTES:
1. This nomograph gives change in diameter caused by heating. Clearance to promote brazing filler metal flow must be provided at brazing termperature.

2. D = nominal diameter of joint, mm (in.)
ΔC_D = change in clearance, mm (in.)
ΔT = brazing temperature minus room temperature, °C (°F)
$\alpha 1$ = mean coefficient of thermal expansion, male member, mm/mm/°C (in./in./°F)
$\alpha 2$ = mean coefficient of thermal expansion, female member, mm/mm/°C (in./in./°F)

3. This nomograph assumes a case where α_1 exceeds α_2 so that scale value for $(\alpha_1 - \alpha_2)$ is negative. Resultant values for ΔC_D are therefore also negative, signifying that the joint gap reduces upon heating. Where $(\alpha_2 - \alpha_1)$ is positive, values of ΔC_D are read as positive, signifying enlargement of the joint gap upon heating.

Figure 12.17–Nomograph for Finding the Change in Diametral Clearance in Dissimilar Metal Joints

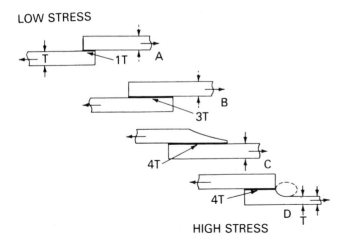

LOW STRESS

HIGH STRESS

Figure 12.18–Brazed Lap Joint Designs for use at Low and High Stresses—Flexure of Right Member in C and D will Distribute the Load Through the Base Metal

theless, a filler metal with low resistivity should be used, provided it will meet all other requirements of the project.

Since voids in the brazed joint will reduce the effective area of the electrical path, lap joints are recommended. A lap length at least 1-1/2 times the thickness of the thinner member will have a joint resistance approximately equal to the same length in solid copper.

TESTING OF BRAZED JOINTS

STANDARDIZING TESTING TO evaluate the strength of brazed joints must be adopted. Different designs of test specimens yield different results. Note in Figure 12.22 that

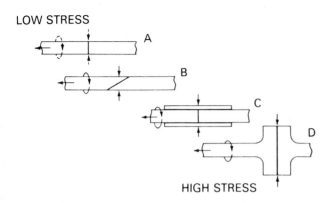

LOW STRESS

HIGH STRESS

Figure 12.19–Brazed Butt Joint Designs to Increase Capacity of Joint for High Stress and Dynamic Loading

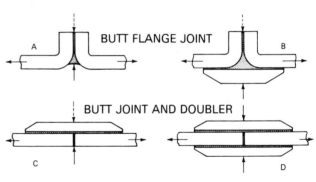

BUTT FLANGE JOINT

BUTT JOINT AND DOUBLER

Figure 12.20–Butt Joing Designs for Sheet Metal Brazements—The Loading in Joint A cannot be Symmetrical

the "apparent joint strength" measured for a low overlap distance is high in comparison to the long overlap strength. Two laboratories that each test only one overlap distance may be testing at opposite ends of the curve, with widely different conclusions. The entire usable overlap range of the curve must be sampled to obtain adequate data.

The load-carrying capacity of the joint is best revealed in the right-hand portion of the base metal curve. The brazement should be designed to fail in the base metal without an excessive overlap.

For further information, refer to the latest edition of AWS C3.2, *Standard Method for Evaluating the Strength of Brazed Joints in Shear.*

BRAZING METALLURGY

BRAZING TEMPERATURES ARE below the solidus of the metal(s) being joined. Metallurgical changes that accom-

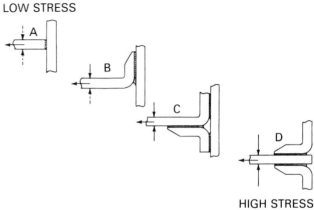

LOW STRESS

HIGH STRESS

Figure 12.21–T-joint Designs for Sheet Metal Brazements

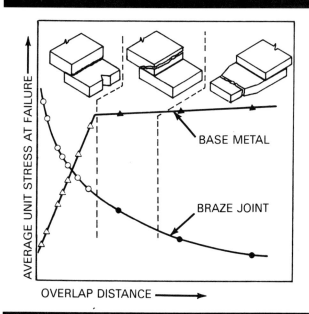

Figure 12.22–Average Unit Shear Stress in the Brazed Lap Joint and Average Unit Tensile Strength in the Base Metal as Functions of Overlap Distance—(Open Symbols Represent Failures in the Filler Metal; Filled Symbols Represent Failures in the Base Metal)

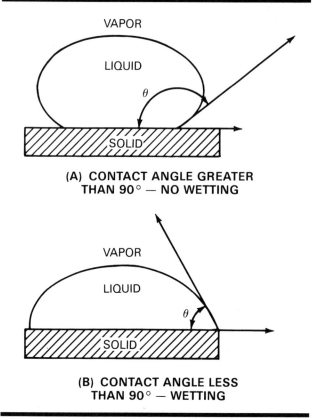

Figure 12.23–Wetting Angles of Brazing Filler Metals

pany brazing are restricted to solid-state reactions in the base metal, solidification and interface reactions between the brazing filler metal and base metal, reactions within the solid filler metal.

The capillary flow of brazing metal depends upon its surface tension, wetting characteristics, and physical and metallurgical reactions with the base material, flux or atmosphere, and oxides on the base metal surface. The flow is further controlled by hydrostatic pressure within the joint. Figure 12.23 is an idealized presentation of the wetting concept.

A contact angle less than 90 degrees measured between the solid and liquid usually identifies a positive wetting characteristic. Contact angles greater than 90 degrees indicate no wetting (dewetting).

In some brazing processes, wetting and spreading are assisted by the addition of flux. In vacuum brazing, flow and wetting depend entirely upon surface interactions between the liquid metal and base metal. Most oxides are readily displaced or removed by flux. Oxides of chromium, aluminum, titanium, and manganese require special treatments.

At peak temperature in the brazing cycle, when liquid filler metal is present in the joint, erosion can occur in the base metal. The rate of dissolution of the base metal by the filler metal depends on the mutual solubility limits, the quantity of brazing filler metal available to the joint, the brazing temperature, and the potential formation of lower temperature eutectics.

Sometimes an interlayer of intermetallic compound may form between the filler metal and the base metal during the joining operation. Phase diagrams are used to predict intermetallic compound formation.

Once the filler metal has solidified to form the joint, subsequent effects may be controlled by diffusion phenomena. When joining super alloys with a nickel-base filler metal containing boron, subsequent thermal cycles diffuse the boron into the base metal. This method of metallurgical joining is called *liquid-activated diffusion welding*, but actually, it is an extension of the joining mechanism in brazing.

Liquid filler metal penetration between base metal grain boundaries may occur. Base metals in a stressed state are particularly susceptible to liquid metal penetration. Copper-based filler metals used on high iron-nickel alloys under stress fail rapidly. Alloying elements diffuse more rapidly into grain boundaries than into a crystal lattice.

If a eutectic is formed, being low-melting it may fill any grain-boundary crack as it separates; then little damage may be done. This is known as an *intrusion*.

The dynamic characteristics of the brazing process are receiving increasing recognition, and careful consideration is being given to the subsequent diffusion and metallurgical changes that can occur in service. At elevated temperatures, changes may occur in the solid-state as a direct result of diffusion, oxidation, or corrosion. This means that the metallurgical and mechanical properties of these joints may change in service and must be evaluated as part of the joint qualification procedure.

BRAZING PROCEDURES

PRECLEANING AND SURFACE PREPARATION

CLEAN, OXIDE-FREE SURFACES are essential to ensure sound brazed joints of uniform quality. Grease, oil, dirt, and oxides prevent the uniform flow and bonding of the brazing filler metal, and they impair fluxing action resulting in voids and inclusions. With the refractory oxides or critical atmosphere brazing applications, precleaning must be more thorough and the cleaned components must be preserved and protected from contamination.

The length of time that cleaning remains effective depends upon the metals involved, the atmospheric conditions, the amount of handling the parts may receive, the manner of storage, and similar factors. It is recommended that brazing be done as soon as possible after the parts have been cleaned.

Degreasing is generally done first. The following degreasing methods are commonly used, and their action may be enhanced by mechanical agitation or by applying ultrasonic vibrations to the bath:

(1) Solvent cleaning: petroleum solvents or chlorinated hydrocarbons
(2) Vapor degreasing: stabilized trichloroethylene or stabilized perchloroethylene
(3) Alkaline cleaning: commercial mixtures of silicates, phosphates, carbonates, detergents, soaps, wetting agents and, in some cases, hydroxides
(4) Emulsion cleaning: mixtures of hydrocarbons, fatty acids, wetting agents, and surface activators
(5) Electrolytic cleaning: both anodic and cathodic

Scale and oxide removal can be accomplished mechanically or chemically. Prior degreasing allows intimate contact of the pickling solution with the parts, and vibration aids in descaling with any of the following solutions:

(1) Acid cleaning: phosphate type acid cleaners
(2) Acid pickling: sulfuric, nitric, and hydrochloric acid
(3) Salt bath pickling: electrolytic and nonelectrolytic

The selection of chemical cleaning agent will depend on the nature of the contaminant, the base metal, the surface condition, and the joint design. For example, base metals containing copper and silver should not be pickled with nitric acid. In all cases, the chemical residue must be removed by thorough rinsing to prevent formation of other equally undesirable films on the joint surfaces, or subsequent chemical attack of the base metal.

Mechanical cleaning removes oxide and scale and also roughens the mating surfaces to enhance capillary flow and wetting by the brazing filler metal. Grinding, filing, machining, and wire brushing can be used. Grit blasting can be done with clean blasting material such as silica sand, alumina, and other nonmetallics. They must not leave any deposit on the surfaces that would impair brazing.

FLUXING AND STOPOFF

WHEN A FLUX is selected for use, it must be applied as an even coating, completely covering the joint surfaces of the parts. Fluxes are most commonly applied in the form of pastes or liquids. Dry powdered flux may be sprinkled on the joint or applied by dipping the heated end of the filler metal rod into the flux container. The particles should be small and thoroughly mixed to improve metal coverage and fluxing action. The areas surrounding the joints may be kept free from discoloration and oxidation by applying flux to a wide area on each side of the joint.

The paste and liquid flux should adhere to clean metal surfaces. If the metal surfaces are not clean, the flux will ball up and leave bare spots. Thick paste fluxes can be applied by brushing. Less viscous consistencies can be applied by dipping, hand squirting, or automatic dispensing. The proper consistency depends upon the types of oxides present, as well as the heating cycle. For example, ferrous oxides formed during fast heating of the base metal are soft and easy to remove, and only limited fluxing action is required. However, when joining copper or stainless steel or when the heating cycle is long, a concentrated flux is required. Flux reacts with oxygen, and once it becomes saturated, it loses all its effectiveness. The viscosity of the flux may be reduced without dilution by heating it to 120 to 140°F (50 to 60°C), preferably in a ceramic-lined flux or glue pot with a thermostat control. Warm flux has low surface tension and adheres to the metal more readily.

When filler metal flow must be restricted to definite areas, "stopoffs" are employed to outline the areas that are not to be brazed. Some commercial stopoff preparations are a slurry in water or an organic binder of oxides of aluminum, chromium, titanium, or magnesium. Others are called *parting compounds* and *surface reaction stopoffs*.

BRAZING FILLER METAL PLACEMENT

WHEN DESIGNING A brazed joint, the brazing process to be used and the manner in which the filler metal will be placed in the joint should be established. In most manually brazed joints, the filler metal is simply fed from the face side of the joint. For furnace brazing and high production brazing, the filler metal is preplaced at the joint. Automatic dispensing equipment may perform this operation.

Brazing filler metal is available in the form of wire, shims, strip, powder, and paste. Figures 12.24 and 12.25 illustrate methods of preplacing brazing filler metal in wire and sheet forms. When the base metal is grooved to accept preplaced filler metal, the groove should be cut in the heavier section. When computing the strength of the intended joint, the groove area should be subtracted from the joint area, since the brazing filler metal will flow out of the groove and into the joint interfaces, as shown in Figure 12.26.

Powdered filler metal can be applied in any of the locations indicated in Figure 12.24. It can be applied dry to the joint area and then wet down with binder, or it can be premixed with the binder and applied to the joint. The density of powder is usually only 50 to 70 percent of a solid metal, so the groove volume must be larger for powder.

Where preplaced shims are used, the sections being brazed should be free to move together when the shims melt. Some type of loading may be necessary to move them

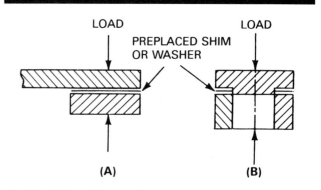

Figure 12.25–Preplacement of Brazing Filler Shims

together and force excess filler metal and flux out of the joint.

ASSEMBLY

THE PARTS TO be brazed should be assembled immediately after fluxing, before the flux has time to dry and flake off. Assemblies designed to be self-locating and self-supporting are the most economical.

When fixtures are needed to maintain alignment or dimensions, the mass of a fixture should be minimized. It should have pinpoint or knife-edge contact with the parts, away from the joint area. Sharp contacts minimize heat loss through conduction to the fixture. The fixture material must have adequate strength at brazing temperature to support the brazement. It must not readily alloy at elevated temperatures with the work at the points of contact. In torch brazing, extra clearance will be needed to access the joint with the torch flame as well as the brazing filler metal. In induction brazing, fixtures are generally made of ceramic materials to avoid putting extraneous metal in the field of the induction coil. Ceramic fixtures may be designed to serve as a heat shield or a heat absorber.

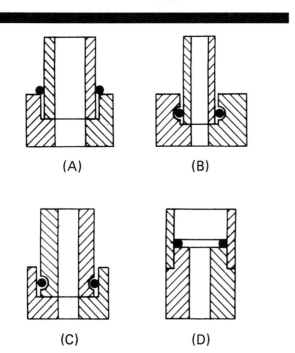

(A)   (B)

(C)   (D)

Figure 12.24–Methods of Preplacing Brazing Filler Wire

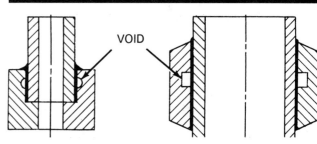

Figure 12.26–Brazed Joints with Grooves for Preplacement of Filler Metal; After the Brazing Cycle the Grooves are Void of Filler Metal

Flux Removal

FOR ALL PROCESSES, all traces of flux should be removed from the brazement. Flux residues usually may be removed by rinsing with hot water. Oxide-saturated flux is glass-like and more difficult to remove. If the metal and joint design can withstand quenching, saturated flux can be removed by quenching the brazement from an elevated temperature. This treatment cracks off the flux coating. In stubborn cases, it may be necessary to use a warm acid solution, such as 10 percent sulfuric acid, or one of the proprietary cleaning compounds which are available commercially. Nitric acid should not be used on alloys containing copper or silver.

Fluxes used for brazing aluminum are not readily soluble in cold water. They are usually rinsed in very hot water, above 180°F (82°C), with a subsequent immersion in nitric acid, hydrofluoric acid, or a combination of those acids. A thorough water after-rinse is then necessary.

Oxidized areas adjacent to the joint may be restored by chemical cleaning or by mechanical methods, such as wire brushing or blast cleaning.

Stopoff Removal

STOPOFF MATERIALS OF the "parting-compound" type can be easily removed mechanically by wire brushing, air blasting, or water flushing. The "surface-reaction" type used on corrosion and heat resistant base metals can best be removed by pickling in hot nitric acid-hydrofluoric acid, except in assemblies containing copper and silver. Sodium hydroxide (caustic soda) or ammonium bifluoride solutions can be used in all applications, including copper and silver, because they will not attack base metals or filler metals. A few stop off materials can readily be removed by dipping in 5 to 10 percent nitric or hydrochloric acid.

INSPECTION

INSPECTION OF BRAZEMENTS should always be required to protect the ultimate user, but it's often specified by regulatory codes and by the fabricator. Inspection of brazed joints may be conducted on test specimens or by tests of the finished brazed assembly. The tests may be nondestructive or destructive.

Generally, brazing discontinuities are of three general classes:

(1) Those associated with drawing or dimensional requirements
(2) Those associated with structural discontinuities in the brazed joint
(3) Those associated with the braze metal or the brazed joint

NONDESTRUCTIVE TESTING METHODS

THE OBJECTIVES OF nondestructive inspection of brazed joints should be (1) to seek out discontinuities defined in quality standards or codes, and (2) to obtain clues to the causes of irregularities in the fabricating process.

Visual Inspection

EVERY BRAZED JOINT should be examined visually. It is a convenient preliminary test when other test methods are to be used.

The joint should be free from foreign materials: grease, paint, oil, oxide film, flux, and stopoff. Visual examination should reveal flaws due to damage, misalignment and poor fit-up of parts, dimensional inaccuracies, inadequate flow of brazing filler metals, exposed voids in the joint, surface flaws such as cracks or porosity, and heat damage to base metal.

Visual inspection will not detect internal flaws, such as flux entrapment in the joint or incomplete filler metal flow between the faying surfaces.

Proof Testing

PROOF TESTING IS a method of inspection that subjects the completed joint to loads slightly in excess of those that will be experienced during its subsequent service life. These loads can be applied by hydrostatic methods, by tensile loading, by spin testing, or by numerous other methods. Occasionally, it is not possible to assure a serviceable part by any of the other nondestructive methods of inspection, and proof testing then becomes the most satisfactory method.

Leak Testing

PRESSURE TESTING DETERMINES the gas or liquid tightness of a closed vessel. It may be used as a screening method to find gross leaks before adopting sensitive test methods. A low pressure air or gas test may be done by one of three methods (sometimes used in conjunction with a pneumatic proof test): (1) submerging the pressurized vessel in water and noting any signs of leakage by rising air bubbles; (2) pressurizing the assembly, closing the air or gas inlet source, and then noting any change in internal pressure over a period of time (corrections for temperature may be necessary); or (3) pressurizing the assembly and checking for leaks by brushing the joint area with a soap solution or a commercially available liquid and noting any bubbles and their source.

A method sometimes used in conjunction with a hydrostatic proof test is to examine the brazed joints visually for indications of the hydrostatic fluid escaping through the joint.

The leak testing of brazed assemblies with freon is extremely sensitive. The part under test is pressurized using either pure freon gas, or a gas such as nitrogen containing a tracer, usually Freon 12. Areas are sniffed or probed with a sampling device which is sensitive to the halide ion. The detection of a leak is indicated by a meter or an audible alarm. A leak may be measured quantitatively by this method. Precaution must be taken to avoid contaminating the surrounding air with freon which will decrease the sensitivity of the method.

A less sensitive method is to probe for leaks of the tracer gas with a butane gas torch flame. The presence of Freon 12 is indicated by a change in the flame color. Flame testing must not be used near combustible material.

The mass spectrometer leak test is the most sensitive and accurate way of detecting extremely small leaks. A tracer gas, such as helium or hydrogen, is used in conjunction with a mass spectrometer in one of two ways: (1) Evacuate the brazed assembly and surround the area to be tested with the tracer gas—the mass spectrometer is coupled to the interior; or (2) Pressurize the brazed assembly with the tracer gas and sniff the exterior with the mass spectrometer probe. A sensitive-sensing device detects the tracer gas and converts it to an electrical signal.

Liquid Penetrant Inspection

THIS NDT METHOD finds cracks, porosity, incomplete flow, and similar surface flaws in a brazed joint. Commercially colored or fluorescent penetrants penetrate surface openings by capillary action. After the surface penetrant has been removed, any penetrant in a flaw will be drawn out by a white developer that is applied to the surface. Colored penetrant is visible under ordinary light. Fluorescent penetrant flaw indications will glow under an ultraviolet (black) light source. Since penetration of minute openings is involved, interpretation is sometimes difficult because of the irregularities in braze fillets and residues of flux deposits. Inpection by another method must be used to differentiate surface irregularities from joint discontinuities.

Radiographic Inspection

RADIOGRAPHIC INSPECTION OF brazements detects lack of bond or incomplete flow of filler metal. The joints should be uniform in thickness and the exposure made straight through the joint. The sensitivity of the method is generally limited to two percent of the joint thickness. X-ray absorption by certain filler metals, such as gold and silver, is greater than absorption by most base metals. Therefore, areas in the joint that are void of braze metal show much darker than the brazed area on the film or viewing screen.

Ultrasonic Inspection

THE ULTRASONIC TESTING method using low energy, high frequency mechanical vibration (sound waves) readily detects, locates, or identifies discontinuities in brazed joints. The applicability to brazements of this method depends largely on the design of the joint, surface condition, material grain size, and the configuration of adjacent areas.

Thermal Heat Transfer Inspection

INSPECTION BY HEAT transfer will detect lack of bond in such brazed assemblies as honeycomb and covered skin panel surfaces. With one technique, the surfaces are coated with a developer which is a low melting point powder. The developer melts and migrates to cool areas upon the application of heat from an infrared lamp. The bonded areas act as heat sinks, resulting in a thermal gradient to which the developer will react. Sophisticated techniques use phosphors, liquid crystals, and temperature-sensitive materials.

Infrared-sensitive electronic devices with some form of readout are available to monitor temperature differences less than $2°F$ ($1°C$) which indicate variations in braze quality.

DESTRUCTIVE TESTING METHODS

DESTRUCTIVE METHODS OF inspection clearly show whether a brazement design will meet the requirements of intended service conditions. Destructive methods must be restricted to partial sampling. It is used to verify the nondestructive methods of inspection, by sampling production material at suitable intervals.

Metallographic Inspection

THIS METHOD REQUIRES the removal of sections from the brazed joints and preparing them for macroscopic or microscopic examination. This method detects flaws (especially porosity), poor flow of brazing filler metal, excessive base metal erosion, the diffusion of brazing filler metal, improper fit-up of the joint, and it will reveal the microstructure of the brazed joint.

Peel Tests

PEEL TESTS ARE frequently employed to evaluate lap type joints. One member of the brazed specimen is clamped rigidly in a vise, and the free member is peeled away from the joint. The broken parts reveal the general quality of the bond and the presence of voids and flux inclusions in the joint. The permissible number, size, and distribution of these discontinuities should be defined in the job contract, specification, or code.

Tension and Shear Tests

THESE TESTS DETERMINE quantitatively the strength of the brazed joint, or verify the relative strengths of the joint and base metal. This method is widely used when developing a brazing procedure. Random sampling of brazed joints is used for quality control and verification of brazing performance.

Torsion Tests

THE TORSION TEST evaluates brazed joints with a stud, screw, or tubular member brazed to a base member. The base member is clamped rigidly and the stud, screw, or tube is rotated to failure which will occur in either the base metal or the brazing alloy.

COMMON IMPERFECTIONS IN BRAZED JOINTS

NONDESTRUCTIVE AND DESTRUCTIVE inspections identify the following types of brazing imperfections. The limits of acceptability should be specifically defined.

Lack of Fill (Voids, Porosity)

LACK OF FILL can be the result of improper cleaning, excessive clearances, insufficient filler metal, entrapped gas, and movement of the mating parts caused by improper fixturing. The filler metal is vulnerable when in the liquid or partially liquid state. Lack of fill reduces the strength of the joint by reducing the load-carrying area, and it may provide a path for leakage.

Flux Entrapment

ENTRAPPED FLUX MAY be found in any brazing operation where a flux is added to prevent and remove oxidation during the heating cycle. Flux trapped in the joint prevents flow of the filler into that area, thus reducing the joint strength. It may also falsify leak- and proof-test indications. Entrapped corrosive flux may reduce service life.

Noncontinuous Fillets

MISSING FILLETS ARE usually noted during visual inspection. Whether their omissions can be waived depends upon the job contract.

Base Metal Erosion

EROSION RESULTS WHEN the brazing filler metal alloys with the base metal. It may result in undercuts or the disappearance of the mating surface. Erosion reduces the strength of the joint by changing the composition of the materials and by reducing the base metal cross-sectional area.

Unsatisfactory Surface Appearance

UNSATISFACTORY BRAZING FILLER metal appearance, including excessive spreading and roughness, is objectionable for more than aesthetic reasons. Appearance defects may act as stress concentrations, corrosion sites, or may interfere with inspection of the brazement.

Cracks

CRACKS REDUCE BOTH strength and service life. They act as stress raisers, lowering the mechanical strength of the brazement and causing premature fatigue failure.

TROUBLESHOOTING

POOR BRAZING IS usually the result of the following failures:

(1) No wetting - no capillary flow, which leaves voids
(2) Excessive wetting - too much filler metal where it is not desired, e.g., in holes, or on machined surfaces
(3) Erosion - attack on the base metal by the brazing filler metal, which reduces the thickness of parent metal areas

If the basic cause of each of these failures can be identified, the solution of the brazing problem will be at hand. Table 12.7 lists items to consider for each of these failure problems.

Table 12.7
Solutions to Typical Brazing Problems

PROBLEM —No Flow, No Wetting

CAUSES:
—Braze filler—different lot or wrong one
—Low temp—poor technique, thermocouple/controller error
—Time—too short
—Dirty parts—not cleaned properly
—Poor atmosphere—too little flux, wrong flux, bad gas or vacuum
—No Ni-plate—allowing oxidation of base metal
—Gap too large—poor fitup control

PROBLEM —Excess Flow or Wetting—Causes Hole Plugging, Brazing Wrong Joints

CAUSES:
—Temperature too high—poor technique, furnace error
—Time—too long
—Too much filler metal—poor technique, different gap size
—Braze filler—different lot or wrong one
—No stopoff used

PROBLEM Erosion—Braze Filler Metal Eats Away Parent Metal

CAUSES:
—Temperature too high—poor technique, furnace error
—Time at temperature too long—poor technique, controller error
—Excessive braze filler metal—poor technique, change in gap, parts in different attitude
—Cold worked parts—highly susceptible—change in part manufacturer—not stress relieved
—Braze filler metals are too high above liquidus or high concentration of melting point depressants.

BRAZE WELDING

INTRODUCTION

BRAZE WELDING IS accomplished using a brazing filler metal having a liquidus above 840°F (450°C) but below the solidus of the base metals to be welded. As noted on the first page of this chapter, braze welding differs from brazing in that the filler metal is not distributed in the joint by capillary attraction. The filler metal is added to the joint as welding rod or is deposited from an arc welding electrode.[4] The base metals are not melted, only the filler metal melts. Bonding takes place between the deposited filler metal and the hot unmelted base metals in the same manner as conventional brazing, but without intentional capillary flow. Joint designs for braze welding are similar to those used for oxyacetylene welding.

Braze welding was originally developed to repair cracked or broken cast iron parts. Fusion welding of cast iron requires extensive preheating and slow cooling, to minimize the development of cracks and the formation of hard cementite. With braze welding, cracks and cementite are easier to avoid, and fewer expansion and contraction problems are encountered.

Most braze welding is done with an oxyfuel gas welding torch, a copper alloy brazing rod, and a suitable flux. Braze welding also is done with carbon arc, gas tungsten arc, and plasma arc torches, without flux. The carbon arc torch is used to weld galvanized sheet steel. The GTAW and PAW torches, which use inert gas shielding, braze weld with filler metals that have relatively high melting temperatures.

Braze welding has the following advantages over conventional fusion welding processes:

(1) Less heat is required to accomplish bonding, which permits faster joining and lower fuel consumption. The process produces little distortion from thermal expansion and contraction.
(2) The deposited filler metal is relatively soft and ductile, readily machinable, and under low residual stress.
(3) Welds have strength adequate for many applications.

---

4. Braze welding of cast iron is sometimes done by the shielded metal arc welding process. See Chapter 2.

(4) The equipment is simple and easy to use.

(5) Metals that are brittle, such as gray cast iron, can be braze welded without extensive preheat.

(6) The process provides a convenient way to join dissimilar metals, for example copper to steel and cast iron, and nickel-copper alloys to cast iron and steel.

Braze welding does have these disadvantages:

(1) Weld strength is limited to that of the filler metal.

(2) Permissible performance temperatures of the product are lower than those of fusion welds because of the lower melting temperature of the filler metal. With copper alloy filler metal, service is limited to 500°F (260°C) or lower.

(3) The braze welded joint may be subject to galvanic corrosion and differential chemical attack.

(4) The brazing filler metal color may not match the base metal color.

EQUIPMENT

CONVENTIONAL BRAZE WELDING is done using an oxyfuel gas welding torch and the associated equipment described in Chapter 11. In some applications, an oxyfuel preheating torch may be needed. Special applications use carbon arc, gas tungsten arc, or plasma arc welding equipment described in other chapters of the Handbook.

Clamping and fixturing equipment may also be needed to hold the parts in place and align the joint.

MATERIALS

Base Metals

BRAZE WELDING IS generally used to join cast iron and steel. It can also be used to join copper, nickel, and nickel alloys. Other metals can be braze welded with suitable filler metals that wet and form a strong metallurgical bond with them.

Dissimilar metal weldments between many of the above metals are possible with braze welding if suitable filler metals are used.

Filler Metals

COMMERCIAL BRAZE WELDING filler metals are the brasses containing approximately 60 percent copper and 40 percent zinc. Brazing alloys with small additions of tin, iron, manganese, and silicon have improved flow characteristics, decreased volatilization of the zinc, and they scavenge oxygen and increase the weld strength and hardness. Filler metal with added nickel (10 percent) has a whiter color and higher weld metal strength.

Chemical compositions and properties of three standard copper-zinc welding rods used for braze welding are given in Table 12.8. The minimum joint tensile strength will be approximately 40 to 60 ksi (275 to 413 MPa). The joint strength decreases rapidly when the weldment is above 500°F (260°C).

Because a braze weld is a bimetal joint, corrosion must be considered in its application. The completed joint will be subject to galvanic corrosion in certain environments, and the filler metal may be less resistant to certain chemical solutions than the base metal.

Fluxes

FLUXES FOR BRAZE welding are proprietary compounds developed for braze welding of stated base metals with brass filler metal rods. They are designed for use at temperatures higher than met in brazing operations, and so they remain active for longer times at temperature than similar fluxes used for capillary brazing. The following types of flux are in general use for braze welding of iron and steels:

(1) A basic flux that cleans the base metal and weld beads and assists in the precoating (tinning) of the base metal. It is used for steel and malleable iron.

(2) A flux that performs the same functions as the basic flux and also suppresses the formation of zinc oxide fumes.

(3) A flux that is formulated specifically for braze welding of gray or malleable cast iron. It contains iron oxide or manganese dioxide to combine with free carbon on the cast iron surface and so remove it.

Flux may be applied by one of the following four methods:

Table 12.8
Copper-Zinc Welding Rods for Braze Welding

AWS Classification*	Approximate Chemical Composition, %					Min Tensile Strength		Liquidus Temperature	
	Copper	Zinc	Tin	Iron	Nickel	ksi	MPa	°F	°C
RBCuZn-A	60	39	1			40	275	1650	900
RBCuZn-C	60	38	1	1		50	344	1630	890
RBCuZn-D	50	40			10	60	413	1714	935

* See AWS Specifications A5.7 and A5.8 for additional information.

(1) The heated filler rod may be dipped into the flux and transferred to the joint during braze welding.

(2) The flux may be brushed on the joint prior to brazing.

(3) The filler rod may be precoated with flux.

(4) The flux may be introduced through the oxyfuel gas flame.

METALLURGICAL CONSIDERATIONS

THE BOND BETWEEN filler metal and base metal in braze welding is the same bonding that occurs with conventional brazing. The clean base metal is heated to a temperature at which its surface is wet by the molten filler metal, producing a metallurgical bond between them. Cleanliness is prerequisite. The presence of dirt, oil, grease, oxide film, or carbon will inhibit wetting.

Following wetting, atomic diffusion takes place between the brazing filler metal and the base metal in a narrow zone at the interface. Indeed, with some base metals the brazing filler metal may slightly penetrate the grain boundaries of the base metal, further contributing to bond strength.

Braze welding filler materials are alloys that have sufficient ductility as-cast to let them flow plastically during solidification and subsequent cooling. The alloys thereby accommodate shrinkage stresses. Two-phase alloys that have a low-melting grain boundary constituent are not useable—those boundaries crack open during solidification and cooling.

GENERAL PROCESS APPLICATIONS

THE GREATEST USE of braze welding is the repair of broken or defective steel and cast iron parts. Since large components can be repaired in place, significant cost savings result. Braze welding also rapidly joins thin-gage mild steel sheet and tubing where fusion welding would be difficult.

Galvanized steel duct work is braze welded using a carbon arc heat source. The brazing temperature is held to below the vaporization temperature of the zinc. This minimizes the loss of the protective zinc coating from the steel surfaces, but it exposes the welder to a significant amount of zinc fumes, requiring exhaust ventilation.

The thicknesses of metals that can be braze welded range from thin gage sheet to very thick cast iron sections. Fillet and groove welds are used to make butt, corner, lap, and T-joints.

BRAZE WELDING PROCEDURE

Fixturing

ADEQUATE FIXTURING IS usually required to hold parts in their proper location and alignment for braze welding. In repairing cracks and defects in cast iron parts, fixturing may not be necessary unless the part is broken apart.

Joint Preparation

JOINT DESIGNS FOR braze welds are similar to those for oxyacetylene welding. For thicknesses over 3/32 in. (2 mm), single- or double-V-grooves are prepared with 90 to 120 degrees included angle, to provide large bond areas between base metal and filler metal. Square grooves may be used for thickness less than 3/32 in. (2 mm).

The prepared joint faces and adjacent surfaces of the base metal must be cleaned to remove all oxide, dirt, grease, oil, and other foreign material. On cast iron, the joint faces must also be free of graphite smears caused by prior machining. Graphite smears can be removed by quickly heating the cast iron to a dull red color and then wire brushing it after it cools to black heat. If the casting has been heavily soaked with oil, it should be heated in the range of 600 to 1200°F (320 to 650°C) to burn off the oil. The surfaces should be wire brushed to remove any residue.

In production braze welding of cast iron components, the surfaces to be joined are usually cleaned by immersion in an electrolytic molten salt bath.

Preheating

PREHEATING MAY BE required to prevent cracking from thermally induced stresses in large cast iron parts. Preheating copper reduces the amount of heat required from the brazing torch and the time required to complete the joint.

Preheating may be local or general. The temperature should be 800 to 900°F (425 to 480°C) for cast iron. Higher temperatures can be used for copper. When braze welding is completed on cast iron parts, they should be thermally insulated for slow cooling to room temperature, to minimize the development of thermally induced stresses.

Technique

THE JOINT TO be oxyfuel gas braze welded must be aligned and fixtured in position. Braze welding flux, when required, is applied to preheated filler rod (unless precoated) and also sprinkled on thick joints during heating with the torch. The base metal is heated until the filler metal melts, wets the base metal, and flows onto the joint faces (precoating). The braze welding operation then progresses along the joint, precoating the faces, then filling the groove with one or more passes using operating techniques similar to oxyfuel gas welding. With an oxyacetylene flame, the inner cone should not be directed on copper-zinc alloy filler metals nor on iron or steel base metal.

With electric arc torches the technique is similar to oxyfuel gas braze welding, except that flux is not generally used.

TYPES OF WELDS

GROOVE, FILLET, AND edge welds are used to braze weld assemblies made from sheet and plate, pipe and tubing, rods and bars, castings, and forgings. To obtain good joint strength, adequate bond area between the brazing filler metal and the base metal is required. Weld groove geome-

try should provide adequate groove face area so that the joint will not fail along the interfaces.

SAFE PRACTICES IN BRAZING

HAZARDS ENCOUNTERED WITH brazing operations are similar to those associated with welding and cutting. At brazing temperatures some elements vaporize, producing toxic gases. Personnel and property need protection against hot materials, gases, fumes, electrical shock, radiation, and chemicals.

Minimum brazing safety requirements are specified in the American National Standard Z49.1, *Safety in Welding and Cutting*,[5] published by the American Welding Society, Miami, Florida. This standard applies to brazing, braze welding, and soldering, as well as other welding and cutting processes.

GENERAL AREA SAFE PRACTICE

BRAZING EQUIPMENT, MACHINES, cables, and other apparatus should be placed so that they present no hazard to personnel in work areas, in passageways, on ladders, or on stairways. Good housekeeping should be maintained.

Precautionary signs conforming to the requirements of ANSI Z535.2, *Environment and Facility Safety Signs*, should be posted designating the applicable hazard(s) and safety requirements.

PERSONNEL PROTECTION

Ventilation

IT IS ESSENTIAL that adequate ventilation be provided so that personnel will not inhale gases and fumes generated while brazing. Some filler metals and base metals contain toxic materials such as cadmium, beryllium, zinc, mercury, or lead, which are vaporized during brazing. Fluxes contain chemical compounds of fluorine, chlorine, and boron, which are harmful if they are inhaled or contact the eyes or skin.

Solvents such as chlorinated hydrocarbons and cleaning compounds, such as acids and alkalies, may be toxic or flammable or cause chemical burn when present in the brazing environment.

To avoid suffocation, care must be taken with atmosphere furnaces to insure that the furnace is purged with air before personnel enter it.

Eye and Face Protection

EYE AND FACE protection shall comply with ANSI Z87.1, *Practices for Occupational and Educational Eye and Face Protection*. Goggles or spectacles with shade number four or five filter lenses should be worn by operators and helpers for torch brazing. Operators of resistance, induction, or salt bath dip brazing equipment and their helpers should use face shields, spectacles, or goggles as appropriate, to protect their faces and eyes.

Protective Clothing

APPROPRIATE PROTECTIVE CLOTHING for brazing should provide sufficient coverage and be made of suitable materials to minimize skin burns caused by spatter or radiation. Heavier material such as woolen or heavy cotton clothing are preferable to lighter materials because they are more difficult to ignite. All clothing shall be free from oil, grease, and combustible solvents. Brazers shall wear protective heat-resistant gloves made of leather or other suitable materials.

Respiratory Protective Equipment

WHEN CONTROLS SUCH AS ventilation fail to reduce air contaminants to allowable levels and where the implementation of such controls is not feasible, respiratory protective equipment should be used to protect personnel from hazardous concentrations of airborne contaminants. Only approved respiratory protection equipment should be used. Approvals of respiratory equipment are issued by the National Institute of Occupational Safety and Health (NIOSH) and the Mine Safety and Health Administration (MSHA). Selection of the proper equipment should be in accordance with ANSI Z88.2.

PRECAUTIONARY LABELING AND MATERIAL SAFETY DATA SHEETS

BRAZING OPERATIONS POSE potential hazard from fumes, gases, electric shock, heat, and radiation. Personnel should be warned against these hazards, where applicable, by use of adequate precautionary labeling as defined in ANSI/ASC Z49.1. Examples of labeling are shown in Figures 12.27 through 12.30.

---

5. American National Standards Institute (ANSI) 1430 Broadway, New York, NY 10018.

WARNING: PROTECT yourself and others. Read and understand this label.

FUMES AND GASES can be dangerous to your health.
ARC RAYS can injure eyes and burn skin.
ELECTRIC SHOCK can KILL.

- Before use, read and understand the manufacturer's instructions, Material Safety Data Sheets (MSDSs), and your employer's safety practices.

- Keep your head out of the fumes.

- Use enough ventilation, exhaust at the arc, or both, to keep fumes and gases from your breathing zone and the general area.

- Wear correct eye, ear, and body protection.

- Do not touch live electrical parts.

- See American National Standard Z49.1, *Safety in Welding and Cutting*, published by the American Welding Society, 550 N.W. LeJeune Rd., P.O. Box 351040, Miami, Florida 33135; OSHA Safety and Health Standards, 29 CFR 1910, available from U.S. Government Printing Office, Washington, DC 20402.

DO NOT REMOVE THIS LABEL.

Figure 12.27–Warning Label for Arc Welding Processes and Equipment

WARNING: PROTECT yourself and others. Read and understand this label.

FUMES AND GASES can be dangerous to your health.
HEAT RAYS (INFRARED RADIATION from flame or hot metal) can injure eyes.

- Before use, read and understand the manufacturer's instructions, Material Safety Data Sheets (MSDSs), and your employer's safety practices.

- Keep your head out of the fumes.

- Use enough ventilation, exhaust at the flame, or both, to keep fumes and gases from your breathing zone and the general area.

- Wear correct eye, ear, and body protection.

- See American National Standard Z49.1, *Safety in Welding and Cutting*, published by the American Welding Society, 550 N.W. LeJeune Rd., P.O. Box 351040, Miami, Florida 33135; OSHA Safety and Health Standards, 29 CFR 1910, available from U.S. Government Printing Office, Washington, DC 20402.

DO NOT REMOVE THIS LABEL.

Figure 12.28–Warning Label for Oxyfuel Gas Processes

DANGER: CONTAINS CADMIUM, Protect yourself and others. Read and understand this label.

FUMES ARE POISONOUS AND CAN KILL.

- Before use, read and understand the manufacturer's instructions, Material Safety Data Sheets (MSDSs), and your employer's safety practices.

- Do not breathe fumes. Even brief exposure to high concentrations should be avoided.

- Use enough ventilation, exhaust at the work, or both, to keep fumes and gases from your breathing zone and the general area. If this cannot be done, use air supplied respirators.

- Keep children away when using.

- See American National Standard Z49.1, *Safety in Welding and Cutting*, published by the American Welding Society, 550 N.W. LeJeune Rd., P.O. Box 351040, Miami, Florida 33135; OSHA Safety and Health Standards, 29 CFR 1910, available from U.S. Government Printing Office, Washington, DC 20402.

If chest pain, shortness of breath, cough, or fever develop after use, obtain medical help immediately.

DO NOT REMOVE THIS LABEL.

Figure 12.29–Warning Label for Brazing Filler Metals Containing Cadmium

Resistance and Induction Brazing Processes

As a MINIMUM, the information shown in Figure 12.27, or its equivalent, shall be placed on stock containers of consumable materials and on major equipment such as power supplies, wire feeders, and controls used in electrical resistance or induction brazing processes. The information shall be readily visible to the worker and may be on a label, tag, or other printed form as defined in ANSI Z535.2 and ANSI Z535.4, *Product Safety Signs and Labels*.

Oxyfuel Gas, Furnace, Dip Brazing Processes

As a MINIMUM, the information shown in Figure 12.28, or its equivalent, should be placed on stock containers of consumable materials and on major equipment used in oxyfuel gas, furnace (except vacuum), and dip brazing processes. The information should be readily visible to the worker and may be on a label, tag, or other printed form as defined in ANSI Z535.2 and ANSI Z535.4.

Filler Metals Containing Cadmium

As a MINIMUM, brazing filler metals containing more cadmium than 0.1 percent by weight should carry the information shown in Figure 12.29, or its equivalent on tags, boxes, or other containers, and on any coils or wire or strip not supplied to the user in a labeled container. Label requirements should also conform to ANSI Z535.4.

Brazing Fluxes Containing Fluorides

As a MINIMUM, brazing fluxes and aluminum salt bath dip brazing salts containing fluorine compounds should have precautionary information as shown in Figure 12.30, or its equivalent, on tags, boxes, jars, or other containers. Labels for other fluxes should conform to the requirements of ANSI Z129.1, *Precautionary Labeling for Hazardous Industrial Chemicals*.

Material Safety Data Sheets (MSDSs)

THE SUPPLIERS OF brazing materials shall provide Material Safety Data Sheets, or equivalent, which identify the hazardous materials, if any, present in their products. The MSDS shall be prepared and distributed to users in accordance with OSHA 29CFR 1910.1200, *Hazard Communications Standard*.

A number of potentially hazardous materials may be present in fluxes, filler metals, coatings, and atmospheres used in brazing processes. When the fumes or gases from a product contain a component whose individual limiting

WARNING: CONTAINS FLUORIDES. Protect yourself and others. Read and understand this label.

FUMES AND GASES CAN BE DANGEROUS TO YOUR HEALTH. BURNS EYES AND SKIN ON CONTACT. CAN BE FATAL IF SWALLOWED.

- Before use, read and understand the manufacturer's instructions, Material Safety Data Sheets (MSDSs), and your employer's safety practices.

- Keep your head out of the fumes.

- Use enough ventilation, exhaust at the work, or both, to keep fumes and gases from your breathing zone and the general area.

- Avoid contact of flux with eyes and skin.

- Do not take internally.

- Keep out of reach of children.

- See American National Standard Z49.1, *Safety in Welding and Cutting*, published by the American Welding Society, 550 N.W. LeJeune Rd., P.O. Box 351040, Miami, Florida 33135; OSHA Safety and Health Standards, 29 CFR 1910, available from U.S. Government Printing Office, Washington, DC 20402.

First Aid: If flux comes in contact with eyes, flush immediately with clean water for at least 15 minutes. If swallowed, induce vomiting. Never give anything by mouth to an unconscious person. Call a physician.

DO NOT REMOVE THIS LABEL.

Figure 12.30—Warning Label for Brazing and Gas Welding Fluxes Containing Fluorides

value will be exceeded before the general brazing fume limit of 5 mg/m^3 is reached, the component shall be identified on the MSDS. These include, but are not limited to, the low PEL materials listed earlier.

FIRE PREVENTION AND PROTECTION

FOR DETAILED INFORMATION on fire prevention and protection in brazing processes, NFPA 51B, *Fire Protection in Use of Cutting and Welding Processes*, should be consulted.

Brazing should preferably be done in specially designated areas which have been designed and constructed to minimize fire risk. No brazing shall be done unless the atmosphere is either nonflammable or unless gases (such as hydrogen) which can become flammable when mixed with air are confined and prevented from being released into the atmosphere.

Sufficient fire extinguishing equipment shall be ready for use where brazing work is being done. The fire extinguishing equipment may be pails of water or a water hose, buckets of sand, hose, portable extinguishers, or an automatic sprinkler system, depending upon the nature

and quantity of combustible material in the adjacent area.

Before brazing is begun in a location not specifically designated for such purposes, inspection and authorization by a responsible person shall be required.

When repairing containers that have held flammable or other hazardous materials, there is the possibility of explosions, fires, and the release of toxic vapors. Brazers must be fully familiar with American Welding Society ANSI/AWS F4.1, *Recommended Safe Practices for the Preparation for Welding and Cutting of Containers and Piping That Have Held Hazardous Substances*".

For more information, consult the applicable state, local, and Federal specifications, as well as the AWS *Brazing Manual*.

Brazing Atmospheres

FLAMMABLE GASES ARE sometimes used as atmospheres for furnace brazing operations. These include combusted fuel gas, hydrogen, dissociated ammonia, and nitrogen-hydrogen mixtures. Prior to introducing such atmospheres, the furnace or retort must be purged of air by safe procedures recommended by the furnace manufacturer.

Adequate area ventilation must be provided that will exhaust and discharge to a safe place explosive or toxic gases that may emanate from furnace purging and brazing operations. Local environmental regulations should be consulted when designing the exhaust system.

Steam Hazard From Moist Materials

IN DIP BRAZING and in dip soldering, the parts to be immersed in the bath must be completely dry. Moisture on the parts will cause an instantaneous generation of steam that may expel the contents of the dip pot explosively.

Predrying the parts prevents this danger. If supplementary flux must be added, it must be dried to remove both surface moisture and also water of hydration.

ELECTRICAL HAZARDS

ALL ELECTRICAL EQUIPMENT used for brazing should conform to ANSI/NFPA 70, *National Electric Code* (latest edition). The equipment should be installed by qualified personnel under the direction of a competent technical supervisor. Prior to production use, the equipment should be inspected by competent safety personnel to ensure that it is safe to operate.

SUPPLEMENTARY READING LIST

American Society for Metals. *Metals handbook*, Vol. 6, 9th Ed. Metals Park, Ohio: American Society for Metals, 1983.

American Welding Society. *Brazing Manual*, 3rd Ed. Miami: American Welding Society, 1976.

————. *Recommended practices for design, manufacture and inspection of critical components*, C3.3-80. Miami: American Welding Society, 1980.

————. *Safety in welding and cutting*, ANSI Z49.1. (published by) Miami: American Welding Society, 1988.

————. *Specification for filler metals for brazing*, ANSI/AWS A5.8-89. Miami: American Welding Society, 1989.

————. *Standard method for evaluating the strength of brazed joints in shear*, C3.2-82. Miami: American Welding Society, 1982.

Cole, N. C., Gunkel, R. W., and Koger, J. W. "Development of corrosion resistant filler metals for brazing molybdenum." *Welding Journal* 52(10): 446s–473s; October 1973.

Gilliland, R. G., and Slaughter, G. M. "The development of brazing filler metals for high temperature service." *Welding Journal* 48(10), 463s–469s; October 1969.

Hammond, J. P., et al. "Brazing ceramic oxides to metals at low temperature." *Welding Journal* 67(10): 227s; October 1988.

Helgesson, C. I. "Ceramic-to-metal bonding." Cambridge, Mass.: Boston Technical Publishers, Inc., 1968.

Jones, T. A., and Albright, C. E. "Laser beam brazing of small diameter copper wires to laminated copper circuit boards." *Welding Journal* 63(12): 34–37; December 1984.

Kawakatsu, I. "Corrosion of BAg brazed joints in stainless steel." *Welding Journal* 52(6): 223s–239s; June 1973.

Lugscheider, E., and Cosack, T. "High temperature brazing of stainless steel with low-phosphorus nickel-based filler metal." *Welding Journal* 67(11): 215s–219s; October 1988.

Lugscheider, E., and Krappitz, H. "The influence of brazing conditions on the impact strength of high-temperature brazed joints." *Welding Journal* 65(10): 261s–; October 1986.

Lugscheider, E., et al. "Thermal and metallurgical influences on AISI316 and Inconel 625 by high temperature brazing with nickel base filler metals." *Welding Journal* 61(10): 329s–333s; October 1982.

Lugscheider, E., et al. "Surface reactions and welding mechanisms of titanium- and aluminum-containing nickel-base and iron-base alloys during brazing under vacuum." *Welding Journal* 62(10): 295s–300s; October 1983.

————. "Metallurgical aspects of additive-aided wide-clearance brazing with nickel-based filler metals." *Welding Journal* 68(1): 9s–13s; January 1989.

McDonald, M. M., et al. "Wettability of brazing filler metals on molybdenums and TMZ." *Welding Journal* 389s–393s; October 1989.

Mizuhara, H., and Mally, K. "Ceramic-to-metal joining with active brazing filler metal." *Welding Journal* 27–32; October 1985.

Moorhead, A. J., and Becher, P. F. "Development of a test for determining fracture toughness of brazed joints in ceramic materials." *Welding Journal* 66(1): 26s–31s; January 1987.

Patrick, E. P. "Vacuum brazing of aluminum." *Welding Journal* 54(6): 159–163; March 1975.

Pattee, H. E. "Joining ceramics to metals and other materials." Bulletin 178. New York: Welding Research Council, November 1972.

————. "High-temperature brazing." Bulletin 187. New York: Welding Research Council, September 1973.

Rugal, V., Lehka, N., and Malik, J. K. "Oxidation resistance of brazed joints in stainless steel." *Metal Construction and British Welding Journal.* 183–176; June 1974.

Sakamoto, A., et al. "Optimizing processing variables in high-temperature brazing with nickel-based filler metals." *Welding Journal* 68(3): 63–67; March 1989.

Schmatz, D. J. "Grain boundary penetration during brazing of aluminum." *Welding Journal* 62(10): 267s–271s; October 1983.

Schultze, W., and Schoer, H. "Fluxless brazing of aluminum using protective gas." *Welding Journal* 52(10): 644–651; October 1973.

Schwartz, M. M. "The fabrication of dissimilar metal joints containing reactive and refractory metals. Bulletin 210. New York: Welding Research Council, October 1975.

Schwartz, M. M. "Brazed honeycomb structures." Bulletin 182. New York: Welding Research Council, April 1973.

Schwartz, M. M. *Modern metal joining techniques.* John Wiley & Sons, September 1969.

Swaney, O. D., Trace, D. E., and Winterbottom, W. L. "Brazing aluminum automotive heat exchangers in vacuum." *Welding Journal* 49–57; May 1986.

Terrill, J. R., et al., "Understanding the mechanisms of aluminum brazing." *Welding Journal* 50(12): 833–839; December 1971.

The Aluminum Association. *Aluminum brazing handbook.* New York: The Aluminum Association, 1971.

Winterbottom, W. L. "Process control criteria for brazing under vacuum." *Welding Journal* 63(10): 33–39; October 1984.

Witherell, C. E., and Ramos, T. J. "Laser brazing." *Welding Journal* 59(10): 267s–277s; October 1980.

CHAPTER 13

SOLDERING

PREPARED BY A COMMITTEE CONSISTING OF:

R. E. Beal, Chairman
Amalgamated Technologies

W. G. Bader
Consultant

WELDING HANDBOOK COMMITTEE MEMBER:
C. W. Case
Inco Alloys International

CHAPTER 13

SOLDERING

HISTORY

SOLDERING IS A technology that has been in continuous development from ancient times. Many artifacts discovered in archeological excavations were joined by soldering. The technology seems to have existed for several thousand years with changes as metallurgical knowledge and new metals were discovered.

Copper and lead alloys were the first to be joined. Early metallurgists learned to identify eutectics in binary systems. The use of eutectic alloys permitted soldering to join simple shapes into complex items of jewelry and utensils. The industrial revolution promoted widespread use of soldered joints. Advancements in alloy joining, processing techniques, and applications continue today. Soldering is now used in industrial applications, satellite communications, computers, and the space program.

FUNDAMENTALS

DEFINITION

SOLDERING IS DEFINED as a group of joining processes that produce coalescence of materials by heating them to the soldering temperature and by using a filler metal (solder) having a liquidus not exceeding 840°F (450°C) and below the solidus of the base metals. The solder is distributed between closely fitted faying surfaces of the joint by capillary action.

PRINCIPLES AND PRACTICES IN SOLDERING

THE SOLDERED JOINT is generally considered to be a metallurgical bond between the solder filler metal and the base metals being joined. Strength of the joint can be enhanced by mechanical configuration of the joint. Some solder joints do not have a metallurgical bond, but are held together by adhesion properties of the interfaces.

The metallurgical solder joint is produced by reaction of the base metals and the filler metal. The solder alloy is applied as a liquid metal that wets and spreads in the joint, and generally forms a layer of an intermetallic compound with a small amount of the base metal. Upon solidification, the joint is held together by the same attraction between adjacent atoms that holds a piece of solid metal together.

A sound soldered joint is achieved by the selection and use of proper materials and processes. There are many soldering filler metals, processes, methods, procedures, and equipments, and many metal alloys that are joined. Specific applications require consideration of all these factors to obtain the optimum manufacturing and service results. This chapter is essentially a summary of soldering technology. It covers soldering principles and practices in some detail. Filler metal selection, joint design, metal cleaning, heating methods, fluxes, and joint properties are covered. Temperature ranges of commonly used soldering alloys are compared with base metal melting points in Figure 13.1.

Soldering is an attractive metal joining process. A major factor in its popularity is that such a low temperature process has minimum effect on base-metal properties. The low temperature used for joining requires little energy input and allows precise control of the process. A wide range of heating methods can be adopted, giving flexibility in design and manufacturing procedures. Modern automation produces large numbers of joints in electrical and electronic circuits. High joint reliability can be obtained with carefully controlled procedures. The occasional defective soldered joint can be easily repaired. Soldering technology is an essential technique in modern industry.

PHYSICAL AND CHEMICAL PROPERTIES IN SOLDERING

MANY PHYSICAL SCIENCE principles are involved in the preparation for and execution of the soldered joint. Chem-

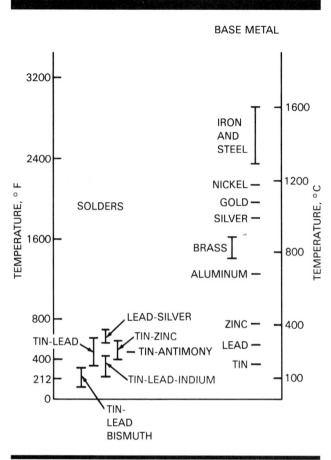

Figure 13.1–Solder Temperature Ranges Compared with Base Metal Melting Points

quires the presence of a flux. The flux cleans the metal to be joined and lowers the surface tension between the molten metal and the solid substrate. This flux behavior improves the wetting and spreading of the solder metal.

Wetting is said to take place when the solder leaves a continuous permanent film on the base metal surface. Alloying depends on the solubility of the base metal in molten solder metal. A high level of alloying between the base metal and solder metal can retard spreading. Therefore, good solder filler metals usually dissolve only a moderate amount of metal. Intermetallic compounds may form depending upon the metal systems involved.

Many solder joints are designed with gaps that require capillarity between the solder and base metal. Capillary attraction is improved by lowering surface tension, narrowing the gap in the joint, and using a highly compatible displacement type flux.

Principles at work during soldering require that the surfaces of the materials to be joined are cleaned of dirt, oxides, or other contaminants for good soldering to take place. One function of a flux is to provide final cleaning by chemical reaction with the metal surface. This attack should be slight but effective. Covering the surface with flux is no substitute for prior cleaning procedures.

When heated, the flux is activated, cleans contacted surfaces, and protects the cleaned areas from oxidation during soldering. The solder filler metal is applied when the joint has been heated to the soldering temperature. The surfaces are protected by the activated flux during soldering action. When soldered joints have been cooled, some residual flux may be present that needs to be removed to prevent early joint deterioration.

Physical problems affecting wetting, spreading, and capillar action can result in unsatisfactory joints. Unsatisfactory joints generally result from a poor surface condition or improper flux. Some metals, for example chromium, cannot be readily wet by most known solder filler metals. De-wetting is the retraction of solder on an already wetted surface which leaves areas of incomplete coverage. Inadequate cleaning, poor flux selection, and wrong solder composition are the main causes of de-wetting.

istry, physics, and metallurgy are the main disciplines involved in solder joining.

Wetting and spreading of solder filler metals on metallic surfaces are conditioned by the surface tension properties of the materials involved and the degree of alloying taking place during the soldering action. Soldering normally re-

BASIC STEPS IN GOOD SOLDERING

BASE METAL SELECTION

BASE METALS ARE usually selected for specific properties that are needed for the component or part design. These include strength, ductility, electrical conductivity, weight, and corrosion resistance. When soldering is required, the solderability of the base materials must also be considered. The selection of flux and surface preparation will be affected by the solderability of the base materials to be joined.

SOLDER SELECTION

THE SOLDER IS selected to provide good flow, penetration, and wettability in the soldering operation, and the desired joint properties in the finished product.

FLUX SELECTION

FLUX IS INTENDED to enhance the wetting of base materials by the solder by removing tarnish films from precleaned

surfaces, and by preventing oxidation during the soldering operation. The selection of the type of flux usually depends on the ease with which a material can be soldered. Rosin fluxes are used with base metals in electrical and electronic applications or with metals that are precoated with a solderable finish. Inorganic fluxes are often used in industrial soldering such as plumbing and vehicle radiators. The flux requirements for soldering a number of alloys and metals are indicated in Table 13.1.

JOINT DESIGN

JOINTS SHOULD BE designed to fulfill the requirements of the finished assembly and to permit application of the flux and solder by the soldering process that will be used. Joints should be designed so that proper clearance is maintained during heating. Special fixtures may be necessary, or the components can be crimped, clinched, wrapped, or otherwise held together.

Table 13.1
Flux Requirements for Metals, Alloys, and Coatings

Base Metal, Alloy, or Applied Finish	Rosin	Organic	Inorganic	Special Flux and/or Solder	Soldering Not Recommended*
Aluminum	-	-	-	X	-
Aluminum-bronze	-	-	-	X	-
Beryllium	-	-	-	-	X
Beryllium-copper	-	X	X	-	-
Brass	X	X	X	-	-
Cadmium	X	X	X	-	-
Cast iron	-	-	-	X	-
Chromium	-	-	-	-	X
Copper	X	X	X	-	-
Copper-chromium	-	-	X	-	-
Copper-nickel	X	X	X	-	-
Copper-silicon	-	-	X	-	-
Gold	X	X	X	-	-
Inconel	-	-	-	X	-
Lead	X	X	X	-	-
Magnesium	-	-	-	-	X
Manganese-bronze (high tensile)	-	-	-	-	X
Monel	-	X	X	-	-
Nickel	-	X	X	-	-
Nickel-iron	-	X	X	-	-
Nichrome	-	-	-	X	-
Palladium	X	X	X	-	-
Platinum	X	X	X	-	-
Rhodium	-	-	X	-	-
Silver	X	X	X	-	-
Stainless steel	-	-	X	-	-
Steel	-	-	X	-	-
Tin	X	X	X	-	-
Tin-bronze	X	X	X	-	-
Tin-lead	X	X	X	-	-
Tin-nickel	-	X	X	-	-
Tin-zinc	X	X	X	-	-
Titanium	-	-	-	-	X
Zinc	-	X	X	-	-
Zinc die castings	-	-	-	-	X

* With proper procedures, such as precoating, most metals can be soldered.

PRECLEANING

ALL METAL SURFACES to be soldered should be cleaned before assembly to facilitate wetting of the base metal by the solder. Flux should not be considered as a substitute for precleaning. Precoating may be necessary for base materials that are difficult to solder.

SOLDERING PROCESS

THE SOLDERING PROCESS should be selected to provide the proper soldering temperature, heat distribution, and rate of heating and cooling required for the product being assembled.

Application of the solder and flux will be dictated by the selection of the soldering process.

FLUX RESIDUE TREATMENT

FLUX RESIDUES SHOULD be removed after soldering unless the flux is specifically designed to be consumed during the process.

SOLDERS

SOLDERS HAVE MELTING points or melting ranges generally below 800°F (425°C). There is a wide range of commercially available solder filler metals designed to work with most industrial metals and alloys. These generally flow satisfactorily with the appropriate fluxes, and produce good surface wetting and result in joints with satisfactory properties. Tin-lead alloys are the most widely used solder filler metals. These alloys and other common filler metals are discussed in the following sections.

TIN-LEAD SOLDERS

TIN-LEAD-ALLOY SOLDERS ARE those most widely used in the joining of metals. Most commercial fluxes, cleaning methods, and soldering processes may be used with tin-lead solders.

In describing solders, it is customary to identify the tin content first. As an example, 40/60 solder is 40 percent tin and 60 percent lead.

The behavior of the various tin-lead alloys can best be illustrated by their constitutional diagram. This diagram is shown in Figure 13.2. The following terms are used to describe this diagram:

(1) *Solidus temperature* is the highest temperature at which a metal or alloy is completely solid. This is curve ACEDB of Figure 13.2.
(2) *Liquidus temperature* is the lowest temperature at which a metal or alloy is completely liquid. This is curve AEB of Figure 13.2.
(3) *Eutectic alloy* is an alloy that melts at one temperature and not over a range. The eutectic temperature is the solidus temperature at curve CED of Figure 13.2. The eutectic alloy is the composition noted at point E, Figure 13.2. This alloy is approximately 63 percent tin by weight.
(4) *Melting range* is the temperature between the solidus ACEDB and the liquidus AEB, where the solder is partially melted. As shown in Figure 13.2, pure lead melts at 621°F

(327°C) (point A), and pure tin melts at 450°F (232°C) (point B). Solders containing 19.5 percent tin (point C) up to 97.5 percent tin (point D) have the same solidus temperature, namely the eutectic temperature, which is 361°F (183°C). The eutectic composition is completely liquid above 361°F (183°C). Any other composition will contain some solid metal in equilibrium with the liquid. These compositions do not melt completely until above the liquidus temperature. For example, 50/50 solder has a solidus temperature of 361°F (183°C) and a liquidus temperature of 417°F (214°C), which is a melting range covering 56°F (31°C). The melting range is the temperature difference between the solidus and liquidus.

Melting characteristics of specific tin-lead solders are shown in Table 13.2.

The 5/95 tin-lead solder is a relatively high melting temperature solder with a small melting range. Wetting and flow characteristics of 5/95 tin-lead solder are less attractive compared to solders with higher tin contents. Proper wetting and flow of 5/95 tin-lead solder requires extra care in surface preparation. High lead solders have better creep properties at 300°F (149°C) than solders containing more tin. The high soldering temperature limits the use of organic base fluxes, such as rosin or those of the intermediate type. This solder is particularly adaptable to torch, dip, induction, or oven soldering. It is used for sealing precoated containers, for coating and joining metals, and for automotive radiators and other moderately elevated temperature uses.

The 10/90, 15/85, and 20/80 tin-lead solders have lower liquidus and solidus temperatures, but they have wider melting ranges than 5/95 tin-lead solder. Their wetting and flow characteristics are also better with most fluxes. However, to prevent hot tearing, extreme care must be taken to avoid movement of these solders during solidification. These solders are used for sealing cellular automobile heater cores, some radiators, and for the coating and joining of metals.

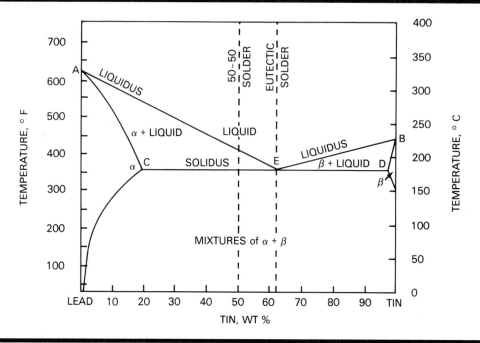

Figure 13.2–Constitutional Diagram for the Tin-Lead Alloy System

The 25/75 and 30/70 tin-lead solders have lower liquidus temperatures than all previously mentioned alloys but have the same solidus temperature as the 20/80 tin-lead solder. Therefore, their melting ranges are narrower than that of the 20/80 solder. All standard cleaning, fluxing, and soldering techniques can be used with these solders. Torch soldering and dipping are widely used. These alloys are used in radiators, radiator repair, and a variety of other industrial applications.

The 35/65, 40/60, and 50/50 tin-lead solders have low liquidus temperatures. The solidus temperature is the same as the 20 percent to 30 percent tin solders, and the melting ranges are narrower. Solders of this group have very good wetting properties, better strength, and economy for electronic applications. Extensive use of these solders is also found in sheet-metal work. The 50/50 combination is used for non-potable water plumbing and piping. These solders are also used as rosin-cored wires and in wave soldering industrial applications.

Table 13.2
Tin-Lead Solder Melting Properties

ASTM Solder Classification*	Composition, wt. %		Solidus		Liquidus		Melting Range	
	Tin	Lead	°F	°C	°F	°C	°F	°C
5	5	95	572	300	596	314	24	14
10	10	90	514	268	573	301	59	33
15	15	85	437	225	553	290	116	65
20	20	80	361	183	535	280	174	97
25	25	75	361	183	511	267	150	84
30	30	70	361	183	491	255	130	72
35	35	65	361	183	477	247	116	64
40	40	60	361	183	455	235	94	52
45	45	55	361	183	441	228	80	65
50	50	50	361	183	421	217	60	34
60	60	40	361	183	374	190	13	7
70	70	30	361	183	378	192	17	9

* See ASTM Specification B32, Standard Specification for Solder Metal

The 60/40 alloy and the 63/37 eutectic tin-lead solder are used when the joint cannot be exposed to high temperatures. Such applications are delicate instruments and electronic assemblies. The 60/40 composition is close enough to that of the eutectic tin-lead alloy to have an extremely narrow melting range. All methods of cleaning, fluxing, and heating may be used with this solder. These solders are widely used for electronic wave soldering, vapor phase processes, and incorporation in solder pastes.

The 70/30 tin-lead solder is a special purpose solder used where a high tin content is required. All soldering techniques are applicable.

IMPURITIES IN TIN-LEAD SOLDERS

IMPURITIES IN TIN-LEAD solders can occur during manufacture of the alloys or may result from contamination during usage. Specifications for solder alloys usually limit the maximum total impurity content, with specific limitations for certain metals. Solder metals covered by ASTM B32, *Standard Specifications for Solder Metals*, are shown in Table 13.3. Individual users sometimes need additional restrictions on impurities for particular applications.

Impurities can cause a reduction in wetting properties, sluggishness of flow within solder joints, an increase in oxidation rate, and changes in melting temperature ranges. Strength properties of joints can be adversely affected, with increased tendencies to cracking of the solder or difficulties with adhesion to the base materials.

Impurity elements affect the appearance and quality of a molten solder. Combined effects can be disastrous in soldered joints. Manufacturing specifications recognize the problem and also take into account the quality of solder needed for different applications. Care should be taken in purchasing solder materials to obtain the appropriate alloy and grade.

Aluminum

ALUMINUM CAN CAUSE grittiness at levels of more than 0.005 percent in the solder. A noticeable deterioration in oxidation of a solder bath surface can be an indication of aluminum contamination.

Antimony

ANTIMONY IS OFTEN used in solders as a desirable addition. This metal tends to reduce wetting and spreading qualities and can cause adhesion problems when present at levels higher than required.

Arsenic

ARSENIC IS A cause of de-wetting with as little as 0.005 percent as present. The problem becomes more severe at higher levels, and therefore less than 0.002 percent in the solder metal is desirable.

Bismuth

LOW LEVELS OF bismuth can be tolerated although it can change metallurgical characteristics of the joints.

Cadmium

THIS METAL AS a contaminant increases the surface tension of solders and can cause such deleterious effects as bridging and icicle formations on printed circuit boards.

Copper

THE AMOUNT OF copper that can be present in a solder without causing problems depends largely upon the application. ASTM specifications limit copper content of tin-lead solders to 0.08 percent. However, copper can be present up to 0.3 percent without any observable reduction in soldering properties.

Iron and Nickel

IRON AND NICKEL are not normally present in solder alloys. Specifications usually limit the iron and nickel content to 0.02 percent maximum. Severe reductions in wetting properties have been observed with higher levels.

Phosphorus and Sulfur

PHOSPHORUS AND SULFUR should be kept at the absolute minimal level to prevent oxidation and grittiness problems.

Zinc

ZINC AFFECTS WETTING and surface tension properties of molten solder. Thus tin-lead solders should contain less than 0.005 percent zinc. De-wetting on copper surfaces has been ascribed to zinc at only 0.01 percent in the alloy.

TIN-ANTIMONY SOLDER

THE 95 PERCENT tin, 5 percent antimony solder has the melting characteristics shown in Table 13.4. It provides a narrow melting range at a temperature higher than the tin-lead eutectic. This solder is used in many plumbing, refrigeration, and air conditioning applications because it has good creep properties.

Table 13.3
Solder Compositions and Melting Properties

Alloy Grade	Composition, % (a, b)											Melting Range (c)			
												Solidus		Liquidus	
	Sn	Pb	Sb	Ag	Cu	Cd	Al	Bi	As	Fe	Zn	°F	°C	°F	°C
Sn96	Rem	0.10	0.12 max	3.4-3.8	0.08	0.005	0.005	0.15	0.01 max	0.02	0.005	430	221	430	221
Sn95	Rem	0.10	0.12	4.4-4.8	0.08	0.005	0.005	0.15	0.01	0.02	0.005	430	221	473	245
Sn94	Rem	0.10	0.12	5.4-5.8	0.08	0.005	0.005	0.15	0.01	0.02	0.005	430	221	536	280
Sn70	69.5-71.5	Rem	0.50	0.015	0.08	0.001	0.005	0.25	0.03	0.02	0.005	361	183	377	193
Sn63	62.5-63.5	Rem	0.50	0.015	0.08	0.001	0.005	0.25	0.03	0.02	0.005	361	183	361	183
Sn62	61.5-62.5	Rem	0.50	1.75-2.25	0.08	0.001	0.005	0.25	0.03	0.02	0.005	354	179	372	189
Sn60	59.5-61.5	Rem	0.50	0.015	0.08	0.001	0.005	0.25	0.03	0.02	0.005	361	183	374	190
Sn50	49.5-51.5	Rem	0.50	0.015	0.08	0.001	0.005	0.25	0.025	0.02	0.005	361	183	421	216
Sn45	44.5-46.5	Rem	0.50	0.015	0.08	0.001	0.005	0.25	0.025	0.02	0.005	361	183	441	227
Sn40A	39.5-41.5	Rem	0.50	0.015	0.08	0.001	0.005	0.25	0.02	0.02	0.005	361	183	460	238
Sn40B	39.5-41.5	Rem	0.50	0.015	0.08	0.001	0.005	0.25	0.02	0.02	0.005	365	185	448	231
Sn35A	34.5-36.5	Rem	1.8-2.4	0.015	0.08	0.001	0.005	0.25	0.02	0.02	0.005	361	183	477	247
Sn35B	34.5-36.5	Rem	0.50	0.015	0.08	0.001	0.005	0.25	0.02	0.02	0.005	365	185	470	243
Sn30A	29.5-31.5	Rem	1.6-2.0	0.015	0.08	0.001	0.005	0.25	0.02	0.02	0.005	361	183	491	255
Sn30B	29.5-31.5	Rem	0.50	0.015	0.08	0.001	0.005	0.25	0.02	0.02	0.005	365	185	482	250
Sn25A	24.5-26.5	Rem	1.1-1.5	0.015	0.08	0.001	0.005	0.25	0.02	0.02	0.005	361	183	511	266
Sn25B	24.5-26.5	Rem	0.50	0.015	0.08	0.001	0.005	0.25	0.02	0.02	0.005	365	185	504	263
Sn20A	19.5-21.5	Rem	0.8-1.2	0.015	0.08	0.001	0.005	0.25	0.02	0.02	0.005	361	183	531	277
Sn20B	19.5-21.5	Rem	0.50	0.015	0.08	0.001	0.005	0.25	0.02	0.02	0.005	363	184	517	270
Sn15	14.5-16.5	Rem	0.50	0.015	0.08	0.001	0.005	0.25	0.02	0.02	0.005	437	225	554	290
Sn10A	9.0-11.0	Rem	0.20	1.7-2.4	0.08	0.001	0.005	0.25	0.02	0.02	0.005	514	268	576	302
Sn10B	9.0-11.0	Rem	0.50	0.015	0.08	0.001	0.005	0.03	0.02	0.02	0.005	514	268	570	299
Sn5	4.5-5.5	Rem	0.50	0.015	0.08	0.001	0.005	0.25	0.02	0.02	0.005	586	308	594	312
Sn2	1.5-2.5	Rem	4.5-5.5	0.015	0.08	0.001	0.005	0.25	0.02	0.02	0.005	601	316	611	322
Sb5	94.0 min	0.20	0.40	0.015	0.08	0.03	0.005	0.15	0.05	0.04	0.005	450	233	464	240
Ag1.5	0.75-1.25	Rem	0.40	1.3-1.7	0.30	0.001	0.005	0.25	0.02	0.02	0.005	588	309	588	309
Ag2.5	0.25	Rem	0.40	2.3-2.7	0.30	0.001	0.005	0.25	0.02	0.02	0.005	580	304	580	304
Ag5.5	0.25	Rem	0.40	5.0-6.0	0.30	0.001	0.005	0.25	0.02	0.02	0.005	580	304	716	380

a. Limits are % max unless shown as a range or stated otherwise.

b. For purposes of determining conformance to these limits, an observed value or calculated value obtained from analysis shall be rounded to the nearest unit in the last right-hand place of figures used in expressing the specified limit, in accordance with rounding method of Recommended Practice E 29.

c. Temperatures given are approximations and for information only.

Table 13.4
Tin-Antimony Solder Melting Properties

Composition, Weight%		Solidus		Liquidus		Melting Range	
Tin	Antimony	°F	°C	°F	°C	°F	°C
95	5	450	232	464	240	14	8

TIN-ANTIMONY-LEAD SOLDERS

ANTIMONY MAY BE added to a tin-lead solder as a substitute for some of the tin. The addition of antimony increases the mechanical properties of the joint with only slight impairment to the soldering characteristics. All standard methods of cleaning and heating may be used. Specialized fluxes are needed for best results with these alloys.

TIN-SILVER, TIN-COPPER-SILVER, AND TIN-LEAD-SILVER SOLDERS

SOLDERS CONTAINING SILVER with their melting characteristics are listed in Table 13.5. The 96 percent tin-4 percent silver solder is free of lead and is often used to join stainless steel for food handling equipment. It has good shear and creep strengths and excellent flow characteristics.

The tin-silver and tin-copper-silver solders are the standard alloys used with copper pipe and tubes in potable water systems. Lead is eliminated for health reasons.

The 62 percent tin-36 percent lead-2 percent silver solder is used when soldering to silver-coated surfaces in electronic applications. The silver addition retards the dissolution of the silver coating during the soldering operation. The addition of silver also increases creep resistance.

High-lead solders containing tin and silver provide higher temperature solders for many applications including automobile radiators. They exhibit good tensile, shear, and creep strengths, and they are recommended for cryogenic applications. Inorganic fluxes are generally recommended for use with these solders.

TIN-ZINC SOLDERS

A LARGE NUMBER of tin-zinc solders, some of which are listed in Table 13.6, have come into use for joining aluminum. Galvanic corrosion of soldered joints in aluminum is minimized if the solder and the base metal are close to each other in the electrochemical series. Alloys containing 70 to 80 percent tin with the balance zinc are recommended for soldering aluminum. The addition of 1 to 2 percent aluminum, or an increase of the zinc content to as high as 40 percent, improves corrosion resistance. However, the liquidus temperature rises correspondingly, and those solders are therefore more difficult to apply. The 91/9 and 60/40 tin-zinc solders may be used at temperatures above 300°F (140°C). The 80/20 and 70/30 tin-zinc solders are more widely used for coating parts before the soldering operation.

CADMIUM-SILVER SOLDER

THE 95 PERCENT-CADMIUM 5 percent silver solder has the melting characteristics shown in Table 13.7. Its primary use is in applications where service temperatures will be higher than permissible with lower melting solders. Butt joints in copper can be made to produce room temperature tensile strengths of 25 000 psi (170 MPa). At 425°F (219°C), the tensile strength is 2600 psi (18 MPa).

Joining aluminum to itself or to other metals is possible with 95 percent cadmium-silver solder. Improper use of solders containing cadmium may lead to health hazards. Therefore, care should be taken in their application, particularly with respect to fume inhalation.

Table 13.5
Tin-Silver and Tin-Lead-Silver Solder Melting Properties

Composition, Weight%			Solidus		Liquidus		Melting Range	
Tin	Lead	Silver	°F	°C	°F	°C	°F	°C
96	-	4	430	221	430	221	0	0
62	36	2	354	180	372	190	18	10
5	94.5	0.5	561	294	574	301	13	7
2.5	97	0.5	577	303	590	310	13	7
1	97.5	1.5	588	309	588	309	0	0

Table 13.6
Tin-Zinc Solder Melting Properties

Composition, Weight%		Solidus		Liquidus		Melting Range	
Tin	Zinc	°F	°C	°F	°C	°F	°C
91	9	390	199	390	199	0	0
80	20	390	199	518	269	128	70
70	30	390	199	592	311	202	112
60	40	390	199	645	340	255	141
30	70	390	199	708	375	318	176

CADMIUM-ZINC SOLDERS

CD-ZN SOLDERS ARE useful for soldering aluminum. Their melting characteristics are given in Table 13.8. The cadmium-zinc solders develop joints with intermediate strength and corrosion resistance, when used with the proper flux. The 40 percent cadmium-60 percent zinc solder has found considerable use for soldering aluminum lamp bases. Improper use of this solder may lead to health hazards, particularly with respect to fume inhalation.

ZINC BASED SOLDERS

ZINC ALUMINUM SOLDER, shown in Table 13.9, is specifically for use on aluminum. It develops joints with high strength and good corrosion resistance. The solidus temperature is high, which limits its use to applications where soldering temperatures in excess of 700°F (371°C) can be tolerated. A major application is in dip soldering the return bends of aluminum air conditioner coils. These coils are also made by flame brazing with fluxes. Ultrasonic solder pots that do not require the use of flux are also employed. In manual soldering operations, the heated aluminum surface is rubbed with the solder stick to promote wetting without a flux.

Table 13.7
Cadmium-Silver Solder Melting Properties

Composition, Weight%		Solidus		Liquidus		Melting Range	
Cadmium	Silver	°F	°C	°F	°C	°F	°C
95	5	640	338	740	393	100	55

Table 13.8
Cadmium-Zinc Solder Melting Properties

Composition, Weight%		Solidus		Liquidus		Melting Range	
Cadmium	Zinc	°F	°C	°F	°C	°F	°C
82.5	17.5	509	265	509	265	0	0
40	60	509	265	635	335	126	70
10	90	509	265	750	399	241	134

Table 13.9
Zinc-Aluminum Solder Melting Properties

Composition, Weight%		Solidus		Liquidus		Melting Range	
Zinc	Aluminum	°F	°C	°F	°C	°F	°C
95	5	720	382	720	382	0	0

Zinc solders with 95 percent zinc and other additions to restrict copper dissolution, and to improve wetting, strength, and corrosion resistance have been developed specifically for automobile radiator application. Melting temperatures in the range of 800°F (425°C) are involved. The solders are used with all heating processes. A series of inorganic fluxes is available for use with these solders.

FUSIBLE ALLOYS

FUSIBLE ALLOYS MAKE substantial use of bismuth and are used in soldering operations where soldering temperatures below 361°F (183°C) are required. The melting characteristics and compositions of a representative group of fusible alloys are shown in Table 13.10.

The following are where low-melting temperature solders apply:

(1) Heat-treated base metals are soldered and higher soldering temperatures would result in softening the part.

(2) Materials adjacent to soldered joints are sensitive to temperature and would deteriorate at higher soldering temperatures.

(3) Step soldering operations are used to avoid destroying a nearby joint that has been made with a higher melting temperature solder.

(4) Temperature-sensing devices, such as fire sprinkler systems, are activated when the fusible alloy melts at relatively low temperature.

Many of these solders, particularly those containing a high percentage of bismuth, are very difficult to use successfully in high-speed soldering operations. Particular attention must be paid to the cleanliness of metal surfaces. Strong, potentially corrosive fluxes must be used to make satisfactory joints on uncoated surfaces of metals, such as copper or steel. If the surface can be plated for soldering with such metals as tin or tin-lead, noncorrosive rosin fluxes may be satisfactory. However, these are not effective below 350°F (177°C).

INDIUM SOLDERS

INDIUM SOLDERS HAVE properties that make them valuable for many electronic and special applications. Melting characterics and compositions of a representative group of these solders are shown in Table 13.11.

A 50 percent tin-50 percent indium alloy adheres to glass readily and may be used for glass-to-metal and glass-to-glass soldering. The low vapor pressure of this alloy makes it useful for seals in vacuum systems.

High fatigue resistance, especially to thermal cycling, has resulted in increased use of indium alloys, particularly indium-lead and indium-lead-silver solders in electronic systems.

Indium solders do not require special handling techniques. All of the soldering methods, fluxes, and processes used with the tin-lead solders are applicable to indium solders. They are, however, sensitive to corrosion in the presence of chlorides. Joints should be cleaned after soldering. They perform best when covered by conformal coatings or in hermetically sealed conditions.

Table 13.10
Typical Fusible Alloys Melting Properties

| Composition, Weight% | | | | Solidus | | Liquidus | | Melting Range | |
Lead	Bismuth	Tin	Other	°F	°C	°F	°C	°F	°C
26.7	50	13.3	10 Cd	158	70	158	70	0	0
25	50	12.5	12.5 Cd	158	70	165	74	7	4
40	52	-	8 Cd	197	91	197	91	0	0
32	52.5	15.5	- -	203	95	203	95	0	0
28	50	22	- -	204	96	225	107	25	11
28.5	48	14.5	9 Sb	217	102	440	227	223	125
44.5	55.5	-	- -	255	124	255	124	0	0

Table 13.11
Typical Indium Solder Melting Properties

| Composition, Weight% | | | Solidus | | Liquidus | | Melting Range | |
Tin	Indium	Lead	°F	°C	°F	°C	°F	°C
50	50	-	243	117	257	125	14	8
37.5	25	37.5	230	138	230	138	0	0
-	50	50	356	180	408	209	52	29

SOLDER SPECIFICATIONS

SPECIFICATIONS FOR SOLDERS are published by the ASTM (ASTM B32, *Standard Specification for Solder Metal*; ASTM B284, *Standard Specification for Rosin Flux-Cored Solder*; and ASTM B486, *Standard Specification for Paste Solder*) and by the United States Government (Federal Specification QQ-S-571, Solders), in addition to various military specifications.

Solders are commercially available in various forms and products which can be grouped into about a dozen classifications. The major groups of solder product forms are listed in Table 13.12. This listing is by no means complete, inasmuch as any desired size, weight, or shape of any form is available on special order.

Table 13.12
Commercial Solder Product Forms

Pig	Available in 50 and 100 lb (25 and 45 kg) pigs
Ingots	Rectangular or circular in shape, weighing 3, 5, and 10 lb (1.4, 2.3, and 4.5 kg)
Bars	Available in numerous cross sections, weights, and lengths
Paste or cream	Available as a mixture of powdered solder and flux
Foil, sheet, or ribbon	Available in various thicknesses and widths
Segment or drop	Triangular bar or wire cut into any desired number of pieces or lengths
Wire, solid	Diameters of 0.010 to 0.250 in. (0.25 to 6.35 mm) on spools
Wire, flux cored	Solder cored with rosin, organic, or inorganic fluxes. Diameters of 0.010 to 0.250 in. (0.25 to 6.35 mm)
Preforms	Unlimited range of sizes and shapes to meet special requirements

FLUXES

THE PURPOSE OF a flux in soldering is to activate a cleaned metal surface, protect that cleaned surface during heating processes, and be available to protect the molten solder at the proper processing temperature. The flux must have sufficient staying power to continue these functions until the joint has been completely soldered.

A soldering flux may be liquid, solid, or gaseous material which, when heated, promotes or accelerates the wetting of metals by solder. A soldering flux should remove and exclude small amounts of oxides and other surface compounds from surfaces being soldered. Anything that interferes with the attainment of uniform contact between the surface of the base metal and the molten solder will prevent the formation of a sound joint. An efficient flux prevents reoxidation of the surfaces during the soldering process and is readily displaced by the molten solder.

A functional method of classifying fluxes is based on their ability to remove metal tarnishes (activity). Fluxes may be classified in three groups: inorganic fluxes (most active), organic fluxes (moderately active), and rosin fluxes (least active).

A generalized metal solderability chart and flux selector guide covering different base materials and flux types is shown in Table 13.13.

INORGANIC FLUXES

THE INORGANIC CLASS of flux as includes inorganic acids and salts. These fluxes are used to best advantage where conditions require rapid and highly active fluxing action. They can be applied as solutions, pastes, or dry salts. They function equally well with torch, oven, resistance, or induction soldering methods, since they do not char or burn. These fluxes can be formulated to provide stability over a wide range of soldering temperatures.

The chloride-based inorganic fluxes have one distinct disadvantage in that the residue remains chemically active after soldering. This residue, if not removed, may cause severe corrosion at the joint. Adjoining areas may also be attacked by residues from the spraying of flux and from flux vapors.

The bromide family of inorganic fluxes is used widely by the automotive radiator industry with and without washing facilities. Certain compositions of these fluxes can be used without washing, and those residues will not cause corrosion of the parts that have been soldered.

The following are typical inorganic flux constituents:

(1) Zinc chloride
(2) Ammonium chloride
(3) Tin chloride
(4) Hydrochloric acid
(5) Phosphoric acid
(6) Other metal chlorides

Table 13.13
Metal Solderability Chart and Flux Selector Guide

Metals	Solderability	Rosin Fluxes			Organic Fluxes (Water Soluble)	Inorganic Fluxes (Water Soluble)	Special Flux and/or Solder
		Non-Activated	Mildly Activated	Activated			
Platinum, gold, copper, silver, cadmium plate, tin (hot dipped), tin plate, solder plate	Easy to solder	Suitable	Suitable	Suitable	Suitable	Not recommended for electrical soldering	. . .
Lead, nickel plate, brass, bronze, rhodium, beryllium copper	Less easy to solder	Not suitable	Not suitable	Not suitable	Suitable	Suitable	. . .
Galvanized iron, tin-nickel, nickel-iron, low-carbon steel	Difficult to solder	Not suitable	Not suitable	Not suitable	Suitable	Suitable	. . .
Chromium, nickel-chromium, nickel-copper, stainless steel	Very difficult to solder	Not suitable	Not suitable	Not suitable	Not suitable	Suitable	. . .
Aluminum, aluminum-bronze	Most difficult to solder	Not suitable	Not suitable	Not suitable	Not suitable	. . .	Suitable
Beryllium, titanium	Not solderable	. . .	. . .	. . .	. . .	. . .	. . .

ORGANIC FLUXES

ORGANIC FLUXES, WHILE less active than the inorganic materials, are effective at soldering temperatures from 200 to 600°F (90 to 320°C). They consist of organic acids and bases and often certain of their derivatives, such as hydrohalides. They are active at soldering temperatures, but the period of activity is short because of their susceptibility to thermal decomposition. Their tendency to volatilize, char, or burn when heated limits their use with torch or flame heating. When properly used, residues from these fluxes are relatively inert and can be removed with water.

Organic fluxes are particularly useful in applications where controlled quantities of flux can be applied and where sufficient heat can be used to fully decompose or volatilize the corrosive constituents. Precaution are necessary to prevent undecomposed flux from spreading onto insulating sleeving. Care must also be taken when soldering in closed systems where corrosive fumes may condense on critical parts of the assembly.

The following are typical organic flux constituents:

(1) Abietic acid
(2) Ethylene diamine
(3) Glutamic acid
(4) Hydrazine hydrobromide
(5) Oleic acid
(6) Stearic acid
(7) A wide range of other acid-based or acid-forming organic chemicals.

ROSIN FLUXES

Nonactive Rosin

WATER-WHITE ROSIN DISSOLVED in a suitable organic solvent is the closest approach to a noncorrosive flux. Rosin fluxes possess important physical and chemical properties which make them particularly suitable for use in the electrical industry. The active constituent, abietic acid, becomes mildly active at soldering temperatures between 350°F (177°C) and 600°F (316°C). The residue is hard, nonhygroscopic, electrically nonconductive, and not corrosive.

Mildly Activated Rosin

BECAUSE OF THE low activity of rosin, mildly activated rosin fluxes have been developed to increase their fluxing action without significantly altering the noncorrosive nature of the residue. These are the preferred fluxes for military, telephone, and other high reliability electronic products.

Activated Rosin

A THIRD AND still more active type of rosin-base flux is called *activated rosin*. These fluxes are widely used in commercial electronics and in high reliability applications where the residue needs to be completely removable after soldering. The activating material can be an organic that reacts to release chlorides, or other halides or a low level of organic acid.

SPECIAL FLUXES

REACTION FLUXES ARE a special group of fluxes that are useful when soldering aluminum. These fluxes are also finding uses with other metals. In practice, the decomposition of the flux cleans and displaces oxides and deposits a metallic film on the metal surface, which allows for wetting and spreading.

FLUXING ACTIONS

A GENERAL FLUX selection guide for various soldering applications including all the above classifications of materials is shown in Table 13.14 to help direct the user towards the most appropriate flux materials.

There are many fluxes available for soldering. These fluxes are designed specifically for the applications involved. Fluxes are available for electronics, plumbing, radiators, different metals, and a wide variation of industrial products. Determination of the appropriate flux is important to assure a successful soldering operation.

Desirable general properties of fluxes include the ability to remove oxides, protect metal surfaces, and melt below the soldering temperature. Desirable post soldering properties include being electrically nonconductive and being corrosion resistant.

Each flux is designed for a heating process and has an optimum processing temperature range for best results.

There is no universal test that can identify all the necessary properties of a flux for an specific application. Therefore, a number of tests have been developed that relate to flux characteristics and value in the fabrication of particular components. Manufacturers should perform a thorough review before selecting a flux. They should not rely entirely on commercially available data that may not be relevant to a particular application.

FLUX FORMS

FLUX IS AVAILABLE as single or multiple cores in wire solder, and in liquids, pastes, and dry powder forms. Not all fluxes are available in each form.

Table 13.14
Typical Fluxing Agents

Type	Composition	Carrier	Uses	Temperature Stability	Ability to Remove Tarnish	Corrosiveness	Recommended Cleaning After Soldering
INORGANIC							
Acids	Hydrochloric, hydrofluoric, orthophosphoric	Water, petrolatum paste	Structural	Good	Very good	High	Hot water rinse and neutralize; organic solvents
Salts	Zinc chloride, ammonium chloride, tin chloride	Water, petrolatum paste, polyethylene glycol	Sturctural	Excellent	Very good	High	Hot water rinse and neutralize; 2% HCl solution; hot water rinse and neutralize; organic solvents
ORGANIC							
Acids	Lactic, oleic, stearic glutamic, phthalic	Water, organic solvents, petrolatum paste, polyethylene glycol	Structural, electrical	Fairly good	Fairly good	Moderate	Hot water rinse and neutralize; organic solvents
Halogens	Aniline hydrochloride, glutamic hydrochloride, bromide derivatives of palmitic acid, hydrazine hydrochloride (or hydrobromide)	Same as organic acids	Sturctural, electrical	Fairly good	Fairly good	Moderate	Same as organic acids
Amines and amides	Urea, ethylene diamine	Water, organic solvents, petrolatum paste, polyethylene glycol	Structural, electrical	Fair	Fair	Noncorrosive normally	Hot water rinse and neutralize; organic solvents
Activated rosin	Water-white rosin	Isopropyl alcohol, organic solvents, polyethylene glycol	Electrical	Poor	FairFair	Noncorrosive normally	Water-based detergents; isopropyl alcohol; organic solvents
Water-white rosin	Rosin only	Same as activated	Electrical	Poor	Poor	None	Same as activated water white rosin but does not normally require post-cleaning

JOINT DESIGN

THE SELECTION OF a joint design for a specific application will depend largely on the service requirements of the assembly. It may also depend on such factors as the heating method to be used, the fabrication techniques prior to soldering, the number of items to be soldered, and the method of applying the solder.

When service requirements of a joint are severe, it is generally necessary to design the joint so that it does not limit the function of the assembly. Solders have low strength compared to the metals that are usually soldered; therefore, the soldered joint should be designed to avoid dependence on the strength of the solder. The necessary strength can be provided by shaping the parts to be joined so that they engage or interlock, requiring the solder only to seal and stiffen the assembly.

There are industrial uses of soldering where the solder joint itself must carry the joint load. A typical example is pipe joints in plumbing systems, where lap joints are used with no additional mechanical support. In these cases, the properties of the solder alloy and the joint as manufactured are important to the service operation involved.

There are two basic types of joint design used for soldering: the lap joint and the lock seam joint. Typical joint designs frequently used in soldering are illustrated in Figure 13.3. Butt joints are not often used.

The lap or lock seam type of joint should be employed whenever possible, since they offer the best possibility of obtaining joints with maximum strength.

An important factor in joint design is the manner in which the solder will be applied to the joint. The designer must consider the number of joints per assembly and the number of assemblies to be manufactured. For limited production, using a manual soldering process, the solder may be face-fed into the joint with little or no problems. However, for large numbers of assemblies containing multiple joints, an automated process such as wave soldering may be advantageous. In this case, the design must provide accessible joints suitable for automated fluxing, soldering, and cleaning.

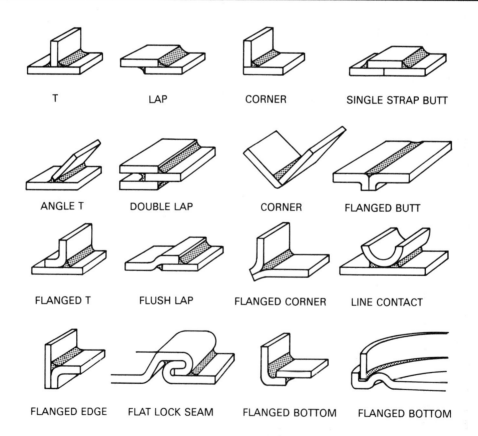

Figure 13.3–Joint Designs Frequently Used in Soldering

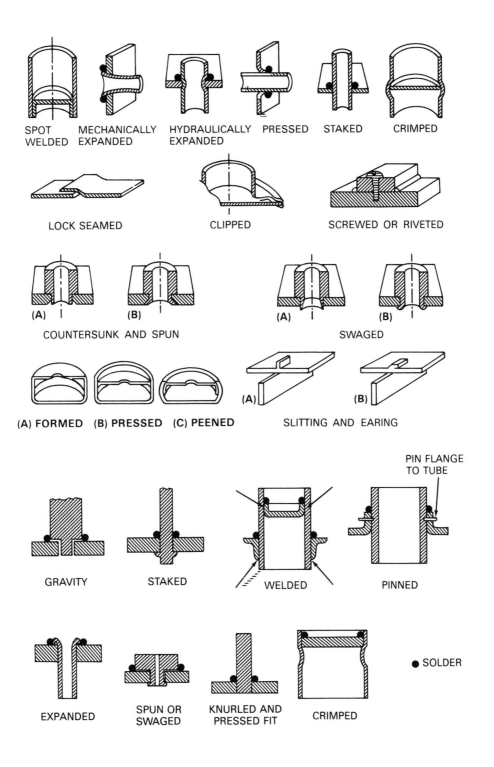

Figure 13.4–Self-Jigging Solder Joint Designs

Clearance between the parts to be joined should allow the solder to be drawn into the space between them by capillary action, but not so wide that the solder cannot fill the gap. Joint clearances up to 0.003 in. (0.075 mm) are preferred for optimum strength, but variations are allowable in specific instances. For example, when soldering precoated metals, a clearance as low as 0.001 in. (0.025 mm) is possible.

Twenty-one designs for self-jigging solder joints are illustrated in Figure 13.4. Various means of improving the strength of joints in printed circuits are shown in Figure 13.5.

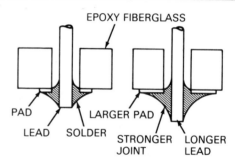

(A) USING LARGER PADS AND LONGER LEADS

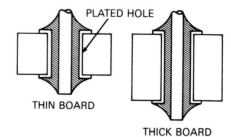

(B) THICK BOARDS PROVIDE LARGER SHEAR SURFACES.

Figure 13.5—Methods of Improving Joint Strength

PRECLEANING AND SURFACE PREPARATION

AN UNCLEAN SURFACE prevents the solder from flowing and makes soldering difficult or impossible. Materials such as oil, grease, paint, pencil markings, drawing and cutting lubricants, general atmospheric dirt, oxides, and rust films should be removed before soldering. To insure sound soldered joints, the importance of cleanliness cannot be overemphasized.

DEGREASING

SOLVENT OR ALKALINE degreasing is recommended for cleaning oily or greasy surfaces. Of the solvent degreasing methods, the vapor condensation type solvents leave the least residual film on the surface. In the absence of vapor degreasing apparatus, immersion in liquid solvents or in detergent solutions is a suitable procedure. Hot alkali detergents are widely used for degreasing. All cleaning solutions must be thoroughly removed before soldering. Residues from hard-water rinses may later interfere with soldering.

Cleaning methods are often designed for a specific soldering operation, hence their suitability for a critical application should be investigated thoroughly.

PICKLING

THE PURPOSE OF pickling or acid cleaning is to remove rust, scale, and oxides or sulfides from the metal to provide a clean surface for soldering. The inorganic acids (hydrochloric, sulfuric, phosphoric, nitric, and hydrofluoric), singly or mixed, all fulfill this function, although hydrochloric and sulfuric are the most widely used. The pieces should be washed thoroughly in hot water after pickling and dried as quickly as possible.

MECHANICAL CLEANING

MECHANICAL CLEANING INCLUDES the following methods:

(1) Mechanical sanding or grinding
(2) Hand filing or sanding
(3) Cleaning with steel wool
(4) Wire brushing or scraping
(5) Grit or shotblasting

Soft metals such as copper are best cleaned by gentle wire brushing or sanding, or with steel wool on plumbing materials. Mechanical cleaning is best avoided for electronic components. Aluminum can be soldered better after oxides are removed by mechanical means; wire brushing or scraping is best. Steel or stainless steel can be brushed or blasted. Shot blasting is preferable to sanding because it avoids embedding silica particles. Stainless steel shot should be used for stainless surfaces. For best results, cleaning should extend beyond the joint area.

PRECOATING

COATING OF BASE metal surfaces with a more solderable metal or alloy prior to the soldering operation is sometimes desirable to facilitate soldering. Coatings of tin, copper, silver, cadmium, iron, nickel, and alloys of tin-lead, tin-zinc, tin-copper, and tin-nickel are used for this purpose. The advantages of precoating are twofold: (1) soldering becomes more rapid and uniform, and (2) strong acidic fluxes can be avoided during soldering. The precoating of metals that have tenacious oxide films, such as aluminum, aluminum bronzes, highly alloyed steels, and cast iron, is almost mandatory. Precoating of steel, brass, and copper can sometimes be useful.

Precoating of the metal surfaces may be accomplished by a number of different methods. Solder or tin may be applied with a soldering iron or an abrasive wheel, by ultrasonic soldering, by immersion in molten metal, by electrodeposition, or by chemical displacement.

Hot dipping may be accomplished by fluxing and dipping the parts in molten tin or solder. Small parts are often placed in wire baskets, cleaned, fluxed, dipped in the molten metal, and centrifuged to remove excess metal. Coating by hot dipping is applicable to carbon steel, alloy steel, cast iron, copper, and certain copper alloys. Prolonged immersion in molten tin or solder should be avoided to prevent excessive formation of intermetallic compounds at the interface between the coating and the base metal.

Precoating by electrodeposition may be done in stationary tanks, in conveyorized plating units, or in barrels. These methods are applicable to all steels, copper alloys, and nickel alloys. The coating metals are not limited to tin and solder. Copper, cadmium, silver, precious metals, nickel, iron, and alloy platings such as tin-copper, tin-zinc, and tin-nickel are also in common use.

Certain combinations of electrodeposited metals (duplex coatings), where one metal is plated over another, are becoming more popular as an aid to soldering. A coating of 0.0002 in (0.005 mm) of copper plus 0.0003 in. (0.008 mm) of tin is particularly useful for brass. The solderability of aluminum is assisted by a coating of 0.0005 in. (0.013 mm) of nickel followed by 0.0003 in. (0.008 mm) of tin, or by a combination of zincate (zinc), 0.0002 in. (0.005 mm) copper, and tin. An iron plating followed by tin plating is extremely useful over a cast iron surface.

Immersion coatings or chemical displacement coatings of tin, silver, or nickel may be applied to some of the common base metals. These coatings are usually very thin and generally have a poor shelf life.

The shelf life of a coating is defined as the ability of the coating to withstand storage conditions without impairment of solderability. Hot tinned and flow brightened electrotinned coatings have an excellent shelf life; inadequate thicknesses of electrotinned or immersion tinned coatings have a limited shelf life. Coating thicknesses of 0.0001 in. (0.003 mm) to 0.0003 in. (0.008 mm) of tin or solder are recommended to assure maximum solderability after prolonged storage.

SOLDERING PROCESS CONSIDERATIONS

SOLDERING IS CARRIED out by many methods covered in the next section. There are common conditions that occur in all soldering practices that should be carefully considered in deciding which process or method is best for a particular job.

The basic stages in soldering are joint preparation, cleaning, fluxing, preheating, soldering, and final cleaning. Soldering is a low-temperature joining method so that fluxes used need to have good activation and reaction at these low temperatures. Each choice of a solder and flux has a process that provides the best results.

Solderability tests are widely used on materials. These tests give important information but do not cover such effects as future storage, variation in materials, or the ability to clean already prepared components.

Soldering often must be done in close proximity to other heat sensitive materials or metals that have been given a specific heat treatment. Coldworked metals can become softened or relaxed during the soldering process. This should be taken into account in designing a finished part. Soldering requires the maintenance of close tolerances to ensure quality joints. It is often advisable to make sample parts with the intended process and subject them to destructive testing to be sure that production parts will be satisfactory. Soldering processes can be highly automated when all the material and processing variables have been evaluated and are carefully controlled. In contrast, soldering can also be carried out successfully and efficiently with individual parts or small lots using hand-held soldering torches.

SOLDERING METHODS AND EQUIPMENT

PROPER APPLICATION OF heat is of paramount importance in any soldering operation. The heat should be applied in such a manner that the solder melts while the surface is heated to permit the molten solder to wet and flow over the surface. A number of tools and methods are available as heat sources.

SOLDERING IRONS

THE TRADITIONAL SOLDERING tool is the soldering iron with a copper tip which may be heated electrically or by oil, coke, or gas burners. To lengthen the usable life of a copper tip, a coating of solder-wettable metal, such as iron, with or without additional coatings, is applied to the surface of the copper. The rate of dissolution of the iron coating in molten solder is substantially less than the rate for copper. The iron coating also shows less wear, oxidation, and pitting than uncoated copper.

The selection of soldering irons can be simplified by classifying them into four groups: (1) soldering irons for servicemen, (2) transformer type, low-voltage pencil irons; (3) special quick-heating and plier-type irons, and (4) heavy-duty industrial irons. A selection of the more common types of soldering irons is shown in Table 13.15.

Regardless of the heating method, the tip performs the following functions:

(1) Stores and conducts heat from the heat source to the parts being soldered
(2) Stores molten solder
(3) Conveys molten solder
(4) Withdraws surplus molten solder

The performance of electrical industrial irons cannot be measured solely by the wattage rating of the heating element. The materials used and the design of the iron affect the heat reserve and temperature recovery of the copper tip.

The angle at which the copper tip is applied to the work is important in delivering the maximum heat to the work. The flat side of the tip should be applied to the work to obtain the maximum area of contact. Flux cored solders should not be melted on the soldering tip because this destroys the effectiveness of the flux. The cored solder should be touched to the soldering tip to initiate good heat transfer, and then the solder should be melted on the work parts to complete the solder joint.

Modern hand-soldering irons are manufactured with closely controlled temperatures in the tip, in a wide range of tip sizes especially designed to work with certain solder wire diameters and to maintain the required soldering temperatures.

TORCH SOLDERING

THE SELECTION OF a gas torch for soldering should match the size, mass, and configuration of the assembly to be soldered. Flame temperature is controlled by the nature of the gas or gases used. Fuel gas, when burned with oxygen, will provide higher flame temperatures than when burned with air. The highest flame temperatures are attained with acetylene and lower temperatures with propane, butane, natural gas, and manufactured (city) gas, roughly in the order given. The flame of a fuel gas burned with oxygen will be sharply defined; with air, the flame will be bushy and flared.

Multiple flame tips, or burners, of shapes suitable to the work are frequently used. They may be designed to operate on oxygen and fuel gas, compressed air and fuel gas, or bunsen-type torches.

In adjusting tips or torches, care should be taken to avoid adjustment which results in a "sooty" flame; the carbon deposited on the work will prevent the flow of solder.

Complex automated torch systems with many flames are used in some industrial applications.

Table 13.15
Selection of Soldering Irons

Work to be Done	Tip Diameter Range		Power Range, Watts
	in.	mm	
Miniature printed circuits, thin substrates, temperature-sensitive components	1/32 - 1/8	1-3	10-20
Intermittent light assembly work, printed circuits, instruments, jewelry	1/8 - 3/16	3-5	20-35
Repetitive assembly work, telephone and applicances, art glass	3/16 - 1/4	5-6	40-60
High speed production soldering, light tinware, general duty, medium electrical, light plumbing	1/4 - 1/2	6-13	70-150
Medium tinware, light roofing, shipboard repair, heavy electrical, heavy plumbing	1/2 - 1-1/2	13-38	170-350
Heavy tinware, roofing, radiators, armatures, transformer cans	1-1/2 - 2	38-53	350-1250

DIP SOLDERING

THIS SOLDERING METHOD uses a molten bath of solder to supply both the heat and the solder necessary to join the workpieces.

Two techniques of dip soldering are illustrated in Figure 13.6(A). When conducted properly, this method is useful and economical in that an entire unit comprising any number of joints can be soldered in one operation. Fixtures are usually required to hold the parts and maintain joint clearances during solidification of the solder.

The soldering pot should be large enough to maintain the rate of production. Parts being dipped should not appreciably lower the temperature of the solder bath. Pots of adequate size can be held at lower operating temperatures and still supply sufficient heat to solder the dipped joints.

WAVE SOLDERING

IN WAVE SOLDERING, as shown in Figure 13.6(B), the solder is pumped out of a narrow slot above the solder pot to produce a wave or series of waves. The work conveyor can pass over the waves at a small angle to the horizontal, to assist in draining the solder, and double waves or special wave forms may also be used for this purpose. Wave-solder systems are excellent for oxide-free solder surfaces.

A alternate technique of wave soldering is cascade soldering, as illustrated in Figure 13.6(C). The solder flows down a trough by gravity and is returned by pump to the upper reservoir.

Integrated wave-soldering systems for printed circuit assemblies provide units that can apply the flux, dry and preheat the board, solder components, and clean the completed assembly. Some of these systems have special features where the flux is applied by passing through a wave, by spraying, by rolling, or by dipping. Several systems employ oil mixed with the solder to aid in the elimination of icicles (also called *bridging*) between conductor paths.

Another system features dual waves with the solder alloy flowing in the direction opposite to the board travel.

VAPOR PHASE SOLDERING (CONDENSATION)

THIS METHOD USES the latent heat of vaporization of a condensing saturated liquid to provide the heat required for soldering work with preplaced flux and solder. A reservoir of saturated vapor over a boiling liquid provides a constant controlled temperature with rapid heat transfer that is useful for soldering large assemblies as well as temperature-sensitive parts. Commercial equipment uses conveyors to provide an in-line continuous process for electronics manufacturing. Condensing fluids are fluorinated organic compounds with boiling points between 420 and 490°F (215 and 253°C).

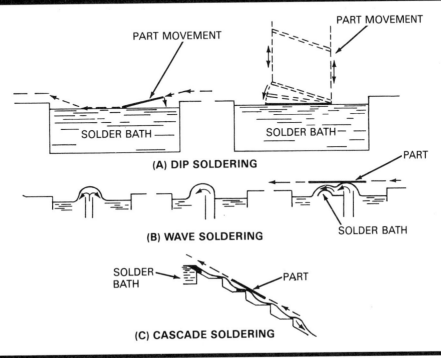

Figure 13.6–Several Soldering Techniques Used for Large Production Runs

OVEN OR FURNACE SOLDERING

THERE ARE MANY applications, especially in high-volume soldering, where furnace soldering produces consistent and satisfactory results.

Oven heating should be considered under the following circumstances:

(1) When entire assemblies can be brought to the soldering temperature without damage to any of the components

(2) When production is sufficiently great to allow expenditure for jigs and fixtures to hold the parts during soldering

(3) When the assembly is complicated, making other heating methods impractical

Proper clamping fixtures are important during oven or furnace soldering. Movement of the joint during solidification of the solder may result in a poor joint.

Oven or furnace soldering is usually carried out with inorganic fluxes because of the temperature and time requirements. The use of a reducing atmosphere in the oven allows joints to be made with less aggressive types of fluxes, depending on the metal and solder combination. The use of inert atmospheres will prevent further oxidation of the parts but still requires adequate and appropriate fluxing.

It is often advantageous to accelerate the cooling of the parts on their removal from the oven. An air blast has been found satisfactory.

Furnaces should be equipped with adequate temperature controls since the flow of solder has an optimum temperature range, depending upon the flux used. The optimum heating condition exists when the heating capacity of the oven is sufficient to heat the parts rapidly under controlled flux application.

RESISTANCE SOLDERING

RESISTANCE SOLDERING REQUIRES the work to be placed either between a ground and a movable electrode or between two movable electrodes to complete an electrical circuit. Heat is applied to the joint both by the electrical resistance of the metal being soldered and by conduction from the moving electrode, which is usually carbon.

Production assemblies may use multiple electrodes, rolling electrodes, or special electrodes, whichever will be advantageous with regard to soldering speed, localized heating, and power consumption.

Resistance soldering electrode tips cannot be tinned, and the solder must be fed into the joint or supplied by preforms or solder coatings on the parts.

INDUCTION SOLDERING

THE MATERIAL THAT is to be induction soldered must be an electrical conductor. The rate of heating is dependent upon the induced current flow, while the distribution of heat obtained with induction heating is a function of the induced wave frequency. Higher frequencies concentrate the heat at the surface. Types of equipment available for induction heating include the vacuum tube oscillator, the resonant-spark gap system, motor-generator units, and solid-state electrical supplies.

Induction soldering is generally chosen for the following:

(1) Large scale production
(2) Application of heat to a localized area
(3) Minimum oxidation of surface adjacent to the joint
(4) Good appearance and consistently high joint quality
(5) Simple joint design, which lends itself to mechanization

The induction technique requires that parts being joined have clean surfaces and accurate joint clearances. High-grade solders spread rapidly and produce good capillary flow. Preforms often are the best means of supplying the correct amount of solder and flux to the joint.

When soldering dissimilar metals by induction, particularly joints composed of both magnetic and nonmagnetic components, attention must be given to the design of the induction coil in order to bring both parts to approximately the same temperature.

INFRARED SOLDERING

OPTICAL SOLDERING SYSTEMS are available based on focusing infrared light (radiant energy) on the joint by means of a lens. Lamps ranging from 45 to 1500 watts can be used for different applications. The devices can be programmed through a solid-state controlled power supply with an internal timer.

HOT GAS SOLDERING

HOT GAS SOLDERING uses a fine jet of inert gas, heated to above the liquidus of the solder. The gas acts as a heat transfer medium and as a shield to reduce access of air to the joint.

ULTRASONIC SOLDERING

EQUIPMENT IS AVAILABLE for ultrasonic dip soldering and hand-soldering operations. An ultrasonic transducer produces high-frequency vibrations which break up tenacious oxide films on base metals. The freshly exposed base metal is readily wet action without the use of flux, or with a less aggressive flux. Ultrasonic units are useful in soldering return bends to the sockets of aluminum air conditioner coils. Ultrasonic soldering is also used to apply solderable coatings on difficult-to-solder metals.

SPRAY GUN SOLDERING

THIS METHOD IS generally selected when the contour of the part is difficult to handle with more conventional techniques.

Gas fired or electrically heated guns are available, each designed to spray molten or semimolten solder on the work from a continuously fed solid solder wire.

Gas fired guns use propane with oxygen, or natural gas with air, to heat and spray a continuously-fed solid solder wire, approximately 1/8 in. (3.2 mm) in diameter. About 90 percent of the solder wire is melted by the flame of the gun. The solder strikes the workpiece in a semiliquid form. The workpiece, heated also by the flame, then supplies the balance of the heat required to melt and flow the solder. Adjustments can be made within the spray gun to control the solder spray.

Electrically heated guns are similar to the gas fired guns except that they use a heating element to melt the solder. Compressed air is then used to spray the molten solder on the workpiece.

FLUX RESIDUE TREATMENT

AFTER THE JOINT is soldered, flux residues which may corrode the base metal or otherwise prove harmful to the effectiveness of the joint must be removed. The removal of flux residues is especially important where joints are subjected to humid environments.

Zinc chloride based fluxes leave a fused residue that will absorb water from the atmosphere. Removal is best accomplished by thorough washing in hot water containing two percent of concentrated hydrochloric acid, followed by a hot water rinse. The acidified water removes the white crust of zinc oxy-chloride, which is insoluble in water alone. Complete removal can also be accomplished by further washing in hot water which contains some washing soda (sodium carbonate), followed by a clear water rinse. Occasionally some mechanical scrubbing may also be required.

The inorganic type flux residues containing inorganic salts and acids should be removed completely. Residues from the organic type fluxes that are composed of very mild organic acids, such as stearic acid, oleic acid, and ordinary tallow, or the highly corrosive combinations of urea plus various organic hydrochlorides, should also be removed.

To determine whether all of the salts have been removed, the joint should be washed with warm water containing a few drops of silver nitrate. If any chloride salts are present the wash will turn milky with the precipitation of silver chloride.

Residues from the organic fluxes are usually quite soluble in hot water. Double rinsing in warm water is always advisable.

Generally, rosin flux residues may be left on the joint unless appearance is the prime factor, or if the joint area is to be painted or subsequently coated. Activated rosin fluxes may be treated in the same manner, but they should be removed for critical electronic applications.

If rosin residues must be removed, alcohol or chlorinated hydrocarbons may be used. Certain rosin activators are insoluble in water but soluble in organic solvents. These flux residues require removal by organic solvents, followed by a water rinse.

The residues from reaction type fluxes used on aluminum are usually removed with a rinse in warm water. If this does not remove all traces of residue, the joint may be scrubbed with a brush and then immersed in two percent sulfuric acid, followed by immersion in one percent nitric acid. A final warm-water rinse is then required.

Soldering pastes for plumbing systems are usually emulsions of petroleum jelly and a water solution of zinc ammonium chloride. Because of the corrosive nature of the acid salts contained in the flux, residues must be removed to prevent corrosion of the soldered joints and the copper pipes. Oily or greasy flux paste residues are generally removed with an organic solvent.

INSPECTION AND TESTING

VISUAL INSPECTION

VISUAL INSPECTION IS normally adequate for soldered joints. Soldered joints should be smooth and free of obvious voids, holes, or porosity. The profile between the soldered joint and the material being joined should show a smooth transition with a relatively low angle of contact between the solder and the base metal. Examination for any areas that have not been successfully wetted should be made. Non-wetting can be seen where the metal retains its original color. De-wetting occurs where solder has originally flowed across the joining surfaces and then pulled back into globules, leaving a discolored, dirty-looking surface. These defects are usually related to poor surface precleaning or use of an inappropriate flux.

Solder joints can be readily overheated or underheated. Overheated joints can be detected by the presence of burned fluxes and oxides on the solder joint. Underheated joints generally show poor flow characteristics, and the solder lumps appear stuck to the surface. These features indicate that no metallurgical bonding has occurred.

Soldered printed circuit boards produce a set of defects peculiar to that product. Bridging of solder may occur between electrical connections that are closely spaced, and should be insulated from each other. Bridging can be caused by the alloy composition or the processing conditions. Another defect unique to circuit boards is called *icicling*, which produces spikes of solder beneath the board. This may cause electrical interference in the finished product. Icicling is promoted by impurities such as cadmium or zinc and by lack of flux activity.

Some types of porosity can be caused by the design or the material of the printed circuit board.

All of these defects can be found by visual inspection.

OTHER INSPECTION METHODS

OTHER METHODS OF nondestructive testing are used to inspect some soldered products. Pressure-vacuum fluid-seal testing and leakage-rate testing can be used on closed systems. Examples are plumbing systems checked by water pressure tests, vehicle radiators checked by air pressure tests, food cans checked by vacuum tests, and gas-filled systems checked by halogen-leak testing.

Radiography can be used for pipe joints or other applications where large surface areas of lead solder joints are present.

Laser inspection techniques are finding use in electronic fabrications. Heat generated by the laser provides an indication of solder joint quality. Surface dimensions may also be checked.

Acoustic emission testing is useful, but this process may affect the joint quality.

Normal destructive testing techniques, including mechanical tests, corrosion evaluation, and metallurgical analysis, are applied to soldered joints in all areas of their application.

PROPERTIES OF SOLDERS AND SOLDER JOINTS

THE PHYSICAL AND mechanical properties of solders are usually provided by the supplier and are used in specifications to ensure the consistent quality of filler metals. Typical properties reported by filler metal manufacturers may not apply to commercial products and applications. Therefore, users should conduct tests on their manufactured products to determine the suitability of the filler metal and the solder process. The reported properties of these alloys serve only to provide a basis to choose among several available solder filler metals.

Soldered joints are used mainly in shear as lap joints, or in peel as lock seam or material-supported joints. The test method must be appropriate to the product for mechanical property evaluation. Short-time tensile tests are good for manufacturing quality control and for comparisons. Most solder joints are subject to some stress in service, and therefore results of creep, stress-rupture, and fatigue tests are important indicators of product performance. Ultimately, the total soldered product must be tested to closely simulate actual service, otherwise serious deficiencies can occur by premature joint failures. Mechanical properties of soldered products are very much dependent upon the product design, alloy selection, manufacturing process, and service conditions. Each individual product should be studied for all these factors so that an optimum balance can be obtained between costs and utility. Additional information can be found in the Supplementary Reading List at the end of the chapter.

SAFETY PRACTICES IN SOLDERING

SOLDERING OPERATIONS SHOULD be carried out under safe conditions. Care must be taken to read all labels on the solder filler wires and fluxes supplied to ensure freedom from handling problems, to recognize any potential for toxic metals or chemicals, and to use these materials only for the purposes intended. All handsoldering operations should be carried out in a ventilated area with working surfaces kept clean of solder droplets, particles, and residual fluxes. Workers using solders and fluxes should always wash exposed skin areas before consuming food.

Industrial soldering operations often require electrical supplies at relatively high levels of power. All soldering irons and equipment should be properly grounded. Where electrical heaters are used for dip soldering operations, current leakage safety devices should be used for worker protection.

Overheated solder pots can give off toxic metal vapors and fumes. Ventilation systems should be installed to eliminate these fumes.

Employees should be kept aware of all factors involved in soldering that could have an influence on their health and safety.

SUPPLEMENTARY READING LIST

Aluminum Company of America (Alcoa). *Soldering alcoa aluminum*. Pittsburgh: Aluminum Company of America, 1972.

American Society for Metals. *Metals handbook*, Vol. 6, 9th Ed. Metals Park, Ohio: American Society for Metals, 1983.

American Society for Testing and Materials. Papers on Soldering. ASTM Special Publication No. 319. Philadelphia: American Society for Testing and Materials, 1962.

——. *Symposium on solder*, ASTM Special Publication No. 189. Philadelphia: American Society for Testing and Materials, 1956.

American Welding Society. *Soldering manual*. Miami: American Welding Society, 1978.

Bannos, T. S. "Lead free solder to meet new safe drinking water regulations." *Welding Journal* 67(10): 23–27; October 1988.

Beal, R. E. "Flux technology of inorganic materials for soldering." *Welding Journal* 58(2); 27–33; February 1979.

Beeferman, D. C. "Soldering Creams for electronic surface mounted devices." *Welding Journal* 65(1): 37–41; January 1986.

C.D.A. Auto Radiator Seminar, Copper Development Assoc., 1983.

Coombs, C. F., Jr., Editor. *Printed circuits handbook*. New York: McGraw-Hill, 1967.

Klein Wassink, R. J. *Soldering in electronics*. Ayr, Scotland: Electrochemical Publications Limited, 1984.

Manko, H. H. *Solders and soldering*. New York: McGraw-Hill, 1979. Thwaits, C. J. *Soft soldering handbook*, Publication 533. Columbus, OH: International Tin Research Institute, 1977.

OXYGEN CUTTING

PREPARED BY A COMMITTEE CONSISTING OF:

G. R. Meyer, Chairman
Victor Equipment Company

R. D. Green
Airco-Mapp

J. F. Leny
Harnischfeger Corporation

C. R. McGowen
Consultant

WELDING HANDBOOK COMMITTEE MEMBER:
B. R. Somers
Consultant

CHAPTER 14

OXYGEN CUTTING

INTRODUCTION

OXYGEN CUTTING (OC) describes a group of cutting processes used to sever or remove metals by high-temperature exothermic reaction of oxygen with the base metal. With some oxidation-resistant metals, the reaction can be aided by the use of a chemical flux or metal powder. Typical oxygen cutting processes are oxyfuel gas, oxygen arc, oxygen lance, chemical flux, and metal powder cutting.

OXYFUEL GAS CUTTING

FUNDAMENTALS OF THE PROCESS

Definition and General Description

OXYFUEL GAS CUTTING (OFC) processes sever or remove metal by the chemical reaction of oxygen with the metal at elevated temperatures. The necessary temperature is maintained by a flame of fuel gas burning in oxygen. In the case of oxidation resistant metals, the reaction is aided by adding chemical fluxes or metal powders to the cutting oxygen stream.

The process has been called various other names, such as as burning, flame cutting, and flame machining. The actual cutting operation is performed by the oxygen stream. The oxygen-fuel gas flame is the mechanism used to raise the base metal to an acceptable preheat temperature range and to maintain the cutting operation.

The OFC torch is a versatile tool that can be readily taken to the work site. It is used to cut plates up to 7 ft (2 m) thick. Because the cutting oxygen jet has a 360° "cutting edge", it provides a rapid means of cutting both straight edges and curved shapes to required dimensions without expensive handling equipment. Cutting direction can be continuously changed during operation.

Principles of Operation

THE OXYFUEL GAS cutting process employs a torch with a tip (nozzle). The functions of the torch are to produce preheat flames by mixing the gas and the oxygen in the correct proportions and to supply a concentrated stream of high-purity oxygen to the reaction zone. The oxygen oxidizes the hot metal and also blows the molten reaction products from the joint. Features of cutting torches are shown in Figures 14.1 and 14.2. The cutting torch mixes the fuel and oxygen for the preheating flames and aims the oxygen jet into the cut. The torch cutting tip contains a number of preheat flame ports and a center passage for the cutting oxygen.

The preheat flames are used to heat the metal to a temperature where the metal will react with the cutting oxygen. The oxygen jet rapidly oxidizes most of the metal in a narrow section to make the cut. Metal oxides and molten metal are expelled from the cut by the kinetic energy of the oxygen stream. Moving the torch across the workpiece at a proper rate produces a continuous cutting action. The torch may be moved manually or by a mechanized carriage.

The accuracy of a manual operation depends largely on the skill of the operator. Mechanized operation generally improves the accuracy and speed of the cut and the finish of the cut surfaces.

Kerf. When a piece is cut by an OC process, a narrow width of metal is progressively removed. The width of the cut is called a *kerf*, as shown in Figure 14.3. Control of the kerf is important in cutting operations where dimensional accuracy of the part and squareness of the cut edges are significant factors in quality control. With the OFC process, kerf width is a function of the size of oxygen port, type of tip used, speed of cutting, and flow rates of cutting oxygen and preheating gases. As material thickness in-

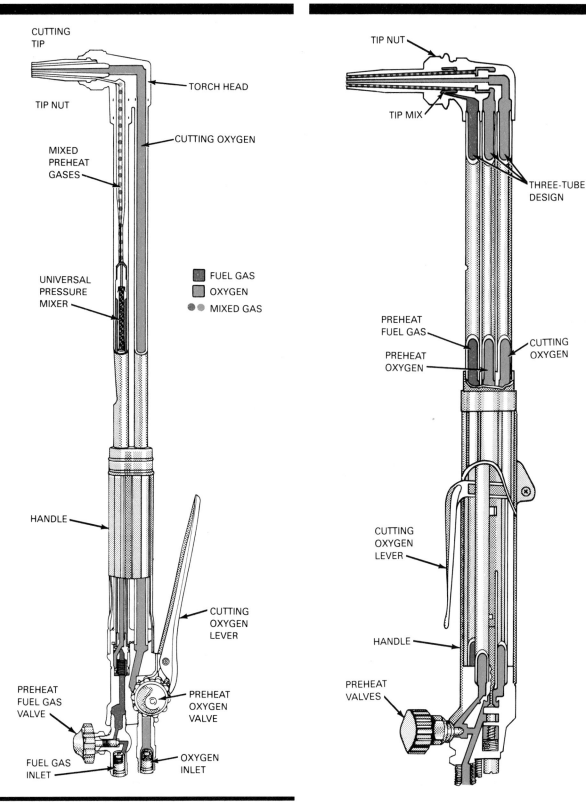

CUTTING
TIP

TORCH HEAD

TIP NUT

CUTTING OXYGEN

MIXED
PREHEAT
GASES

UNIVERSAL
PRESSURE
MIXER

FUEL GAS
OXYGEN
MIXED GAS

HANDLE

CUTTING
OXYGEN
LEVER

PREHEAT
FUEL GAS
VALVE

PREHEAT
OXYGEN
VALVE

FUEL GAS
INLET

OXYGEN
INLET

TIP NUT

TIP MIX

THREE-TUBE
DESIGN

PREHEAT
FUEL GAS

CUTTING
OXYGEN

PREHEAT
OXYGEN

CUTTING
OXYGEN
LEVER

HANDLE

PREHEAT
VALVES

Figure 14.1–Typical Premixing-Type Cutting Torch

Figure 14.2–Typical Tip Mix Cutting Torch

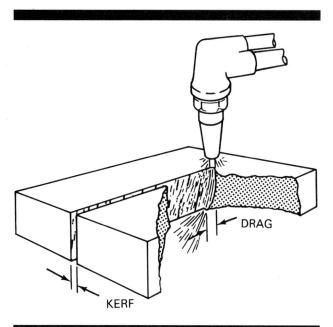

Figure 14.3—Kerf and Drag in Oxyfuel Gas Cutting

creases, oxygen flow rates must usually be increased. Cutting tips with larger cutting oxygen ports are required to handle the higher flow rates. Consequently, the width of the kerf increases as the material thickness being cut increases.

Kerf width is especially important in shape cutting. Compensation must be made for kerf width in the layout of the work, or the design of the template. Generally, on materials up to 2 in. (50 mm) thick, kerf width can be maintained within +1/64 in. (+0.4 mm).

Drag. When the speed of the cutting torch is adjusted so that the oxygen stream enters the top of the kerf and exits from the bottom of the kerf along the axis of the tip, the cut will have zero drag. If the speed of cutting is increased, or if the oxygen flow is decreased, the oxygen available in the lower regions of the cut decreases. With less oxygen available, the oxidation reaction rate decreases, and also the oxygen jet has less energy to carry the reaction products out of the kerf. As a result, the most distant part of the cutting stream lags behind the portion nearest to the torch tip. The length of this lag, measured along the line of cut, is referred to as the *drag*. This is shown in Figure 14.3.

Drag may also be expressed as a percentage of the cut thickness. A ten percent drag means that the far side of the cut lags the near side of the cut by a distance equal to ten percent of the material thickness.

An increase in cutting speed with no increase in oxygen flow usually results in a larger drag. This may cause a decrease in cut quality. There is also a strong possibility of loss of cut at excessive speeds. Reverse drag may occur when the cutting oxygen flow is too high or the travel speed is too low. Under these conditions, poor-quality cuts usually result. Cutting stream lag caused by incorrect torch alignment is not considered to be drag.

Cutting speeds below those recommended for best quality cuts usually result in irregularities in the kerf. The oxygen stream inconsistently oxidizes and washes away additional material from each side of the cut. Excessive preheat flame results in undesirable melting and widening of the kerf at the top.

Chemistry of Oxygen Cutting

THE PROCESS OF oxygen cutting is based on the ability of high-purity oxygen to combine rapidly with iron when it is heated to its ignition temperature, above 1600°F (870°C). The iron is rapidly oxidized by the high-purity oxygen and heat is liberated by several reactions.

The balanced chemical equations for these reactions are the following:

(1) $Fe + O \rightarrow FeO$ + heat (267 kJ), first reaction
(2) $3Fe + 2O_2 \rightarrow Fe_3O_4$ + heat (1120 kJ) second reaction
(3) $2Fe + 1.5O_2 \rightarrow Fe_2O_3$ + heat (825 kJ), third reaction

The tremendous heat release of the second reaction predominates over that of the first reaction, which is supplementary in most cutting applications. The third reaction occurs to some extent in heavier cutting applications. Stoichiometrically, 104 ft^3 (0.29 m^3) of oxygen will oxidize 2.2 lb (1 kg) of iron to Fe$_3$O$_4$.

In actual operations, the consumption of cutting oxygen per unit mass of iron varies with the thickness of the metal. Oxygen consumption per unit mass is higher than the ideal stoichiometric reaction for thicknesses less than approximately 1-1/2 in. (40 mm), and it is lower for greater thicknesses. For thicker sections, the oxygen consumption is lower than the ideal stoichiometric reaction because only part of the iron is completely oxidized to Fe$_3$O$_4$. Some unoxidized or partly oxidized iron is removed by the kinetic energy of the rapidly moving oxygen stream.

Chemical analysis has shown that, in some instances, over 30 percent of the slag is unoxidized metal. The heat generated by the rapid oxidation of iron melts some of the iron adjacent to the reaction surface. This molten iron is swept away with the iron oxide by the motion of the oxygen stream. The concurrent oxidizing reaction heats the layer of iron at the active cutting front.

The heat generated by the iron-oxygen reaction at the focal point of the cutting reaction (the hot spot) must be sufficient to continuously preheat the material to the ignition temperature. Allowing for the loss of heat by radiation and conduction, there is ample heat to sustain the reaction. In actual practice, the top surface of the material is frequently covered by mill scale or rust. That layer must be melted away by the preheating flames to expose a clean metal surface to the oxygen stream. Preheating flames help to sustain the cutting reaction by providing heat to the surface. They also shield the oxygen stream from turbulent interaction with air.

The alloying elements normally found in carbon steels are oxidized or dissolved in the slag without markedly interfering with the cutting process. When alloying elements are present in steel in appreciable amounts, their effect on the cutting process must be considered. Steels containing minor additions of oxidation resistant elements, such as nickel and chromium, can still be oxygen cut. However, when oxidation resistant elements are present in large quantities, modifications to the cutting technique are required to sustain the cutting action. This is true for stainless steels.

OXYGEN

OXYGEN USED FOR cutting operations should have a purity of 99.5 percent or higher. Lower purity reduces the efficiency of the cutting operation. A one percent decrease in oxygen purity to 98.5 percent will result in a decrease in cutting speed of approximately 15 percent, and an increase of about 25 percent in consumption of cutting oxygen. The quality of the cut will be impaired, and the amount and tenacity of the adhering slag will increase. With oxygen purities below 95 percent, the familiar cutting action disappears, and it becomes a melt-and-wash action that is usually unacceptable.

PREHEATING FUELS

FUNCTIONS OF THE preheat flames in the cutting operation are the following:

(1) Raise the temperature of the steel to the ignition point
(2) Add heat energy to the work to maintain the cutting reaction
(3) Provide a protective shield between the cutting oxygen stream and the atmosphere
(4) Dislodge from the upper surface of the steel any rust, scale, paint, or other foreign substance that would stop or retard the normal forward progress of the cutting action

A preheat intensity that raises the steel to the ignition temperature rapidly will usually be adequate to maintain cutting action at high travel speeds. However, the quality of the cut will not be the best. High-quality cutting can be carried out at considerably lower preheat intensities that those normally required for rapid heating. On most larger cutting machines, dual range gas controls are provided that limit high-intensity preheating to the starting operation. Then the preheat flames are reduced to lower intensity during the cutting operation, to save fuel and oxygen and provide a better cut surface.

A number of commercially available fuel gases are used with oxygen to provide the preheating flames. Some have proprietary compositions. Fuel gases are generally selected because of availability and cost. Properties of some commonly used fuel gases are listed in Table 14.1. To understand the significance of the information in this table, it is necessary to understand some of the terms and concepts involved in the burning of fuel gas. These terms and concepts are discussed in Chapter 11. Combustion intensity or specific flame output for various fuel gases is also covered in that chapter. This property is an important consideration in fuel gas selection.

Fuel Selection

THE FOLLOWING ARE some general factors for consideration when selecting a preheat fuel:

(1) Time required for preheating when starting cuts on square edges and rounded corners, and also when piercing holes for cut starts
(2) Effect on cutting speeds for straight line, shape, and bevel cutting
(3) Effect of the above factors on work output
(4) Cost and availability of the fuel in cylinder, bulk, and pipeline volumes
(5) Cost of the preheat oxygen required to burn the fuel gas efficiently
(6) Ability to use the fuel efficiently for other operations, such as welding, heating, and brazing, if required
(7) Safety in transporting and handling the fuel gas containers

For best performance and safety, the torches and tips should be designed for the particular fuel selected.

Acetylene

ACETYLENE IS WIDELY used as a fuel gas for oxygen cutting and also for welding. Its chief advantages are availability, high flame temperature, and widespread familiarity of users with its flame characteristics.

Combustion of acetylene with oxygen produces a hot, short flame with a bright inner cone at each preheat port. The hottest point is at the end of this inner cone. Combustion is completed in the long outer flame.

The sharp distinction between the two flames helps to adjust the oxygen-to-acetylene ratio for the desired flame characteristics.

Depending on this ratio, the flame may to be adjusted to reducing (carburizing), neutral, or oxidizing, as shown in Figure 14.4. The neutral flame, obtained with a ratio of approximately one part oxygen to one part acetylene, is used for manual cutting. As the oxygen flow is decreased, a bright streamer begins to appear. This indicates a reducing flame, which is sometimes used to rough-cut cast iron.

Table 14.1
Properties of Common Fuel Gases

	Acetylene	Propane	Propylene	Methyl-acetylene-propadiene (MPS)	Natural Gas
Chemical Formula	C_2H_2	C_8H_8	C_3H_6	C_3H_4 (Methylacetylene, propadiene)	CH_4 (Methane)
Neutral flame temperature					
°F	5600	4580	5200	5200	4600
°C	3100	2520	2870	2870	2540
Primary flame heat emission					
btu/ft^3	507	255	433	517	11
MJ/m^3	19	10	16	20	0.4
Secondary flame heat emission					
btu/ft^3	963	2243	1938	1889	989
MJ/m^3	36	94	72	70	37
Total heat value (after vaporization)					
btu/ft^3	1470	2498	2371	2406	1000
MJ/m^3	55	104	88	90	37
Total heat value (after vaporization)					
btu/lb	21 500	21 800	21 100	21 100	23 900
kJ/kg	50 000	51 000	49 000	49 000	56 000
Total oxygen required (neutral flame)					
vol. O_2/vol. fuel	2.5	5.0	4.5	4.0	2.0
Oxygen supplied through torch (neutral flame)					
vol. O_2/vol. fuel	1.1	3.5	2.6	2.5	1.5
ft^3 oxygen/lb fuel (60°F)	16.0	30.3	23.0	22.1	35.4
m^3 oxygen/kg (15.6°C)	1.0	1.9	1.4	1.4	2.2
Maximum allowable regulator pressure					
psi	15	150	150	150	Line
kPa	103	1030	1030	1030	
Explosive limits in air: percent	2.5-80	2.3-9.5	2.0-10	3.4-10.8	5.3-14
Volume-to-weight ratio					
ft^3/lb (60°F)	14.6	8.66	8.9	8.85	23.6
m^3/kg (15.6°C)	0.91	0.54	0.55	0.55	1.4
Specific gravity of gas (60°F, 15.6°C) Air = 1	0.906	1.52	1.48	1.48	0.62

When excess oxygen is supplied, the inner flame cone shortens and becomes more intense. The flame temperature increases to a maximum at an oxygen-to-acetylene ratio of about 1.5 to 1. An oxidizing flame is used for short preheating times and for cutting very thick sections.

The high flame temperature and heat transfer characteristics of the oxyacetylene flame are particularly important for bevel cutting. They are also an advantage for operations in which the preheat time is an appreciable fraction of the total time for cutting, such as short cuts.

(A) CARBURIZING FLAME

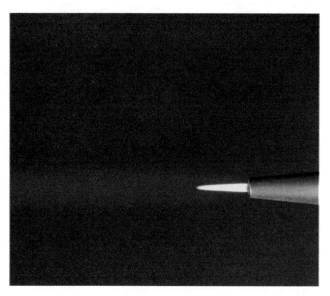

(B) NEUTRAL FLAME

Figure 14.4–Types of Oxyacetylene Flames

Acetylene in the free state should not be used at pressures higher than 15 psi (103 kPa) gage, or 30 psi (207 kPa) absolute pressure. At higher pressures, it may decompose with explosive force when exposed to heat or shock.

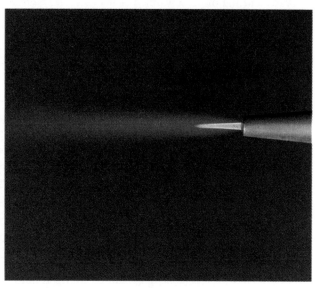

(C) OXIDIZING FLAME

Figure 14.4–Types of Oxyacetylene Flames

Chapter 11 contains additional information on acetylene, its production and storage, and on the oxyacetylene flame.

Methylacetylene-Propadiene Stabilized (MPS)

MPS IS A liquefied, stabilized acetylenelike fuel that can be stored and handled similarly to liquid propane. MPS is a mixture of several hydrocarbons, including propadiene (allene), propane, butane, butadiene, and methylacetylene. Methylacetylene, like acetylene, is an unstable, high-energy, triple-bond compound. The other compounds in MPS dilute the methylacetylene sufficiently to make the mixture safe for handling. The mixture burns hotter than either propane or natural gas. It also affords a high release of energy in the primary flame cone, another characteristic similar to acetylene. The outer flame gives relatively high heat release, like propane and propylene. The overall heat distribution in the flame is the most even of any of the gases.

A neutral flame is achieved at a ratio of 2.5 parts of torch-supplied oxygen to 1 part MPS. Its maximum flame temperature is reached at a ratio of 3.5 parts of oxygen to 1 part of MPS. These ratios are used for the same applications as the acetylene flame.

Although MPS gas is similar in many characteristics to acetylene, it requires about twice the volume of oxygen per volume of fuel for a neutral preheat flame. Thus, oxygen cost will be higher when MPS gas is used in place of acetylene for a specific job. To be competitive, the cost of MPS gas must be lower than acetylene for the job.

MPS gas does have an advantage over acetylene for underwater cutting in deep water. Because acetylene outlet pressure is limited to 30 psi (207 kPa) absolute, it usually is not applicable at depths below 20 ft (6m) of water. MPS can be used there and at greater depths, as can hydrogen. For a particular underwater application, MPS, acetylene, and hydrogen should be evaluated for preheat fuel.

Natural Gas

THE COMPOSITION OF natural gas varies depending on its source. Its main component is methane (CH_4). The ratio of torch supplied oxygen to natural gas is 1.5 to 1 for a neutral flame. The flame temperature with natural gas is lower than with acetylene. It is also more diffused and less intense. The characteristics of the flame for carburizing, neutral, or oxidizing conditions are not as distinct as with the oxyacetylene flame.

Because of the lower flame temperature and the resulting lower heating efficiency, significantly greater quantities of natural gas and oxygen are required to produce heating rates equivalent to those of oxygen and acetylene. To compete with acetylene, the cost and availability of natural gas and oxygen, their higher gas consumptions, and their longer preheat times must be considered. The use of tips designed to provide a heavy preheat flame, or cutting machines that allow a high-low preheat setting, may compensate for deficiencies in the lower heat output of natural gas.

The torch and tip designs for natural gas are different from those for acetylene. The delivery pressure for natural gas is generally low and the combustion ratios are different (see Table 14.1)

Propane

PROPANE IS USED regularly for oxygen cutting in a number of plants because of its availability and its much higher total heat value (MJ/m^3) than natural gas (see Table 14.1). For proper combustion during cutting, propane requires 4 to 4 1/2 times its volume of preheat oxygen. This requirement is offset somewhat by its higher heat value. It is stored in liquid form and is easily transported to the work site.

Propylene

PROPYLENE, UNDER MANY different brand names, is used as fuel gas for oxygen cutting. One volume of propylene requires 2.6 volumes of torch supplied oxygen for a neutral flame and 3.6 volumes for maximum flame temperature. Cutting tips are similar to those used for MPS.

ADVANTAGES AND DISADVANTAGES

OXYFUEL GAS CUTTING has a number of advantages and disadvantages compared to other metal cutting operations, such as sawing, milling, and arc cutting.

Advantages

SEVERAL ADVANTAGES OF OFC are as follows:

(1) Steels can generally be cut faster by OFC than by mechanical chip removal processes.
(2) Section shapes and thicknesses that are difficult to produce by mechanical means can be severed economically by OFC.
(3) Basic manual OFC equipment costs are low compared to machine tools.
(4) Manual OFC equipment is very portable and can be used in the field.
(5) Cutting direction can be changed rapidly on a small radius during operation.
(6) Large plates can be cut rapidly in place by moving the OFC torch rather than the plate.
(7) OFC is an economical method of plate edge preparation for bevel and groove weld joint designs.

Disadvantages

THERE ARE A number of disadvantages with oxyfuel gas cutting of metals. Several important ones are as follows:

(1) Dimensional tolerances are significantly poorer than machine tool capabilities.
(2) The process is essentially limited commercially to cutting steels and cast iron, although other readily oxidized metals, such as titanium, can be cut.
(3) The preheat flames and expelled red hot slag present fire and burn hazards to plant and personnel.
(4) Fuel combustion and oxidation of the metal require proper fume control and adequate ventilation.
(5) Hardenable steels may require preheat, postheat, or both to control their metallurgical structures and mechanical properties adjacent to the cut edges.
(6) Special process modifications are needed for OFC of high alloy steels and cast irons.

EQUIPMENT

THERE ARE TWO basic types of OFC equipment: manual and machine. The manual equipment is used primarily for maintenance, for scrap cutting, cutting risers off castings, and other operations that do not require a high degree of accuracy or a high quality cut surface. Machine cutting equipment is used for accurate, high quality work, and for large volume cutting, such as in steel fabricating shops. Both types of equipment operate on the same principle.

No one should attempt to operate any oxyfuel apparatus until trained in its proper use or under competent supervision. It is important to follow closely the manufacturer's recommendations and operating instructions for safe use. See Volume I, Chapter 16, for more information on safe practices.

Manual Equipment

A SETUP FOR manual OFC requires the following:

(1) One or more cutting torches suitable for the preheat fuel gas to be used and the range of material thicknesses to be cut

(2) Required torch cutting tips to cut a range of material thicknesses

(3) Oxygen and fuel gas hoses

(4) Oxygen and fuel gas pressure regulators

(5) Sources of oxygen and fuel gases to be used

(6) Flame strikers, eye protection, flame and heat resistant gloves and clothing, and safety devices

(7) Equipment operating instructions from the manufacturer

Torches. The functions of an OFC torch are as follows:

(1) To control the flow and mixture of fuel gas and preheat oxygen

(2) To control the flow of cutting oxygen

(3) To discharge the gases through the cutting tip at the proper velocities and volumetric flow rates for preheating and cutting

These functions are partially controlled by the operator, by the pressures of incoming gases, and by the design of the torch and cutting tips.

For manual cutting, a torch that can be readily manipulated by the operator is preferred. Manual oxygen cutting torches are available in various sizes. Torch and tip selection generally depend on the thickness range of the steel to be cut. Tips used in manual cutting equipment have varied designs, depending on the fuel gas and type of work to be done. For example, for cutting rusty or scaly steel, a tip furnishing a great amount of preheat should be selected.

There are two basic types of OFC torches: (1) the tip mixing type, in which the fuel and oxygen for preheating flames are mixed in the tip; and (2) the premixing type, in which the mixing takes place within the torch. Premixing-type torches are further designated as equal (positive) pressure design, or injector (low pressure) design. The positive-pressure-type torches are used where there is sufficient fuel gas pressure to supply the torch mixer with the required volume of gas. The injector-type torches are used where the fuel gas pressure (usually natural gas at less than 2 psig) is such that the fuel gas must be drawn into the torch by the venturi action of the injector mixer. The two types of torches are shown in Figures 14.1 and 14.2 respectively. Some manufactures offer a mixer design that will operate effectively at both low and high fuel pressures. This design is referred to as a universal pressure mixer.

Manual Cutting Tips. Cutting tips are precision machined copper-alloy parts of various designs and sizes. They are held in the cutting torch by a tip nut. All oxygen cutting tips have preheat flame ports, usually arranged in a circle around a central cutting oxygen orifice. The preheat flame ports and the cutting oxygen orifice are sized for the thickness range of metal that the tip is designed to cut. Cutting tips are designated as standard or high speed. Standard tips have a straight bore oxygen port, and they are usually used with oxygen pressures from 30 to 60 psi (205 to 415 kPa). High-speed tips differ from standard tips in that the exit end of the oxygen orifice flares out or diverges. The divergence allows the use of higher oxygen pressures, typically 60 to 100 psi (415 to 690 kPa), while maintaining a uniform oxygen jet at supersonic velocities. High-speed tips are ordinarily used for machine cutting only. They usually permit cutting at speeds approximately 20 percent greater than speeds available with standard tips. Both types of tips are shown in Figure 14.5.

Cutting oxygen orifice size and design are not usually affected by the type of fuel used. However, preheat flame port design does depend on the fuel. Various fuel gases require different volumes of oxygen and fuel, and they burn at different velocities. Therefore, the preheat flame port size and number are designed to provide both a stable flame and adequate preheat for applications with the particular fuel gas being used. Acetylene tips are usually one piece with drilled or swaged flame ports. They are flat on the flame end. Tips for use with other fuel gases are either one piece, similar to acetylene tips, or two pieces with

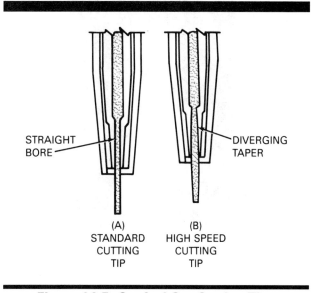

STRAIGHT BORE

DIVERGING TAPER

(A) STANDARD CUTTING TIP

(B) HIGH SPEED CUTTING TIP

Figure 14.5–Oxyfuel Gas Cutting Tips

milled splines on the inner member, as illustrated in Figure 14.6.

Tips for MPS have a flat surface on the flame end. Most propylene tips have a slight recess, and natural gas and propane tips usually have a deeper recess or cupped end.

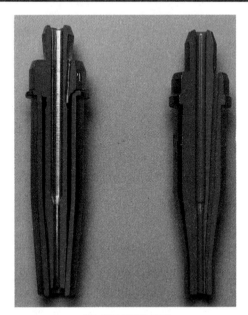

(A) ONE PIECE TIPS

(B) TWO PIECE TIPS

Figure 14.6–Cross Section View of One Piece and Two Piece Tips Used for Fuel Gases Other Than Acetylene

Cutting tips, although considered consumable items, are precision tools. The tip is considered to have the greatest influence on cutting performance. Proper maintenance of tips can greatly extend their useful life and provide continued high-quality performance.

The accumulation of slag in and around the preheat and cutting oxygen passages disturbs the preheat flame and oxygen stream characteristics. This can result in an obvious reduction in performance and quality of cut. When it happens, the tip should be taken out of service and either restored to a good working condition or replaced.

Gas Pressure Regulators.   The ability to make a successful cut requires not only the proper choice of cutting torch and tip for the fuel gas selected, but also a means of precisely regulating the proper gas pressures and volumes. Regulators are pressure control devices used to reduce high source pressures to required working pressures by manually adjusted pressure valves. They vary in design, performance, and convenience features. Gas pressure regulators are designed for use with specific types of gases and for definite pressure ranges.

Gas pressure regulators used for OFC are generally similar in design to those used for oxyfuel gas welding (OFW),which are discussed in Chapter 11. Regulators for most other fuel gases are similar in design to acetylene regulators. For OFC, regulators with higher capacities and delivery pressure ranges than those used for OFW may be required for multitorch operations and heavy cutting.

Hoses.   Oxygen and fuel gas hoses used for OFC are the same as those used for OFW. They are discussed in Chapter 11.

Other Equipment.   Tinted goggles or other appropriate eye protection devices are available in a number of different shades. Tip cleaners, wrenches, strikers, and all appropriate safety devices including protective clothing should be used.

Mechanized Equipment

MECHANIZED OFC WILL require additional facilities depending on the application:

(1) A machine to move one or more torches in the required cutting pattern
(2) Torch mounting and adjusting arrangements on the machine
(3) A cutting table to support the work
(4) Means for loading and unloading the cutting table
(5) Automatic preheat ignition devices for multiple torch machines

Mechanized OFC equipment can vary in complexity from simple hand-guided machines to very sophisticated numerically-controlled units. The mechanized equipment is analogous to the manual equipment in principle, but differs in design to accommodate higher fuel pressures, faster cutting speeds, and means for starting the cut. Many machines are designed for special purposes, such as those for making vertical cuts, edge preparation for welding, and pipe cutting and beveling. Many variations of mechanized cutting systems are commercially available.

Machine Torches. A typical machine cutting torch consists of a barrel, similar to a manual torch but with heavier construction, and a cutting tip. See Figure 14.7. The torch body and barrel encase the oxygen and fuel gas tubes, which carry the gases to the end where the cutting tip is secured by a tip nut. The body of the torch may have a rack for indexing the tip to a desired position from the work surface. A machine torch will have either two or three gas (hose) inlets. Torches with two gas inlet fittings have a fuel-line connection and one oxygen connection with two valves. Torches with three inlet fittings have separate connections for fuel gas, preheat oxygen, and cutting oxygen. Three inlet torches permit separate regulation of preheat and cutting oxygen. They are generally recommended when remote control operation is desired.

Machine Cutting Tips. Machine cutting tips are designed to operate at higher oxygen and fuel pressures than those normally used for manual cutting. The two-piece divergent tip is one type used for operation at high cutting speeds [see Figure 14.5(B)]. Divergent cutting tips are based on the principles of gas flow through a venturi. High velocities are reached as the gas emerges from the venturi nozzle. Divergent cutting tips are precision machined to minimize any distortion of the gases when they exit from the nozzle. They are used for the majority of machine cutting applications because of their superior cutting characterics for materials up to 6 in. (150 mm) thick. They are not recommended for cutting materials over 10 in. (250 mm) thick.

Regulators. When natural gas or propane is used as a preheat fuel in machine cutting, fuel and oxygen can be conserved by using combination high-low pressure regulating systems. Because these fuels burn at lower heat transfer intensities than acetylene, high flow rates of fuel and preheat oxygen are required to heat the metal to ignition temperature in a reasonable time. Once the cut is started, less heat is needed to maintain cutting action with an appropriate savings in gas costs.

High-low pressure regulating systems permit the starting gas flow rates to be reduced to a predetermined level when the flow of cutting oxygen is initiated. This reduction may be done manually or automatically, depending on the regulator and control system design.

Cutting Machines. Oxyfuel gas cutting machines are either portable or stationary. Portable machines are usu-

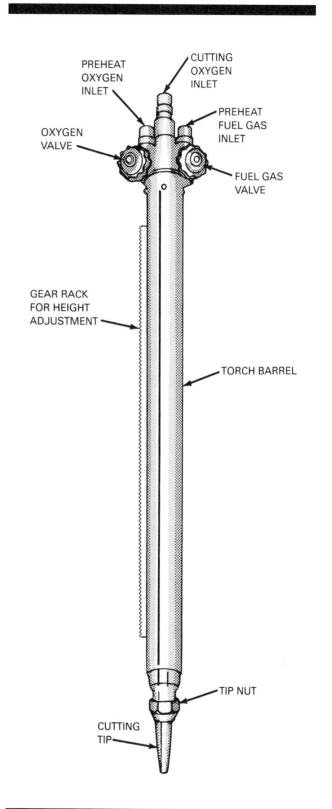

Figure 14.7–Three Hose Machine Cutting Torch

ally moved to the work. Stationary machines are fixed in location and the work is moved to the machine.

Portable Machines. Portable cutting machines are primarily used for straight line cutting, although they can be adapted to cut circles and shapes. Portable machines usually consist of a motor driven carriage with an adjustable mounting for the cutting torch. See Figure 14.8. In most cases, the machine travels on a track, which performs the function of guiding the torch. The carriage speed is adjustable over a wide range. The degree of cutting precision depends upon both the accuracy of the track, or guide, and the fit between the track and the driving wheels of the carriage. Portable machines are of various weights and sizes, depending on the type of work to be done. The smallest machines weigh only a few pounds. They are limited to carrying light-duty torches for cutting thin materials. Large, portable cutting machines are heavy and rugged. They can carry one or more heavy-duty torches and the necessary auxiliary equipment for cutting thick sections.

Generally, the operator must follow the carriage to make adjustments, as required, to produce good quality cuts. The operator ignites the torch, positions it at the starting point, and initiates the cutting oxygen flow and carriage travel. The operator adjusts torch height to main-

Figure 14.8–Machine Cutting Torch Mounted on a Portable Carriage

tain the preheat flames at the correct distance from the work surface. At the completion of the cut, the operator shuts off the cutting torch and carriage.

Stationary Machines. Stationary machines are designed to remain in a single location. The raw material is moved to the machine, and the cut shapes are transported away. The work station is composed of the machine, a system to supply the oxygen and preheat fuel to the machine, and a material handling system.

The torch support carriage runs on tracks. The structure either spans the work with a gantry-type bridge across the tracks or it is cantilevered off to one side of the tracks. These types of equipment are shown in Figures 14.9 and 14.10 respectively. They are usually classified according to the width of plate that can be cut (transverse motion). The length that can be cut is the travel distance on the tracks. The maximum cutting length is dictated by physical limitations of gas and electric power supply lines. An operator station with consolidated controls for gas flow, torch movement, and machine travel is generally a part of the machine.

A number of torches can be mounted on a shape cutting machine, depending on the size of the machine. The machine can cut shapes of nearly any complexity and size. In multiple torch operations, several identical shapes can be cut simultaneously. The number depends on the part size, plate size, and the number of available torches.

A rectilinear or coordinate drive-type machine often has a sine-cosine potentiometer to coordinate separate drive motors for longitudinal and transverse motion of the torch. The carriage and the cross arm, each with its own driving motor, are driven in the proper directions, and the linear speed of the torch remains at a constant preselected value. This type of construction permits the design and manufacture of cutting machines with sufficient rigidity to carry all modern control equipment.

It is possible to feed information to the electric drive motors of the carriage and cross arm from any suitable control. One method uses a photoelectric cell tracer that can follow line drawings or silhouettes. Numerical control machines use profile programs placed on punched or magnetic tapes or computer disks. These storage devices, in turn, control the shape cutting by appropriate signals to the cutting machine drive motors.

GENERAL PROCESS APPLICATIONS

MANUAL OFC IS widely used for the severing of steel and some other iron alloys. Portability permits taking the equipment to the job site. Structural shapes, pipe, rod, and similar materials can be cut to length for construction and maintenance, or cut up in scrap and salvage operations. In a steel mill or foundry, extraneous projections, such as caps, gates, and risers, are quickly severed from billets and castings. Mechanical fastenings, such as bolts, rivets, and

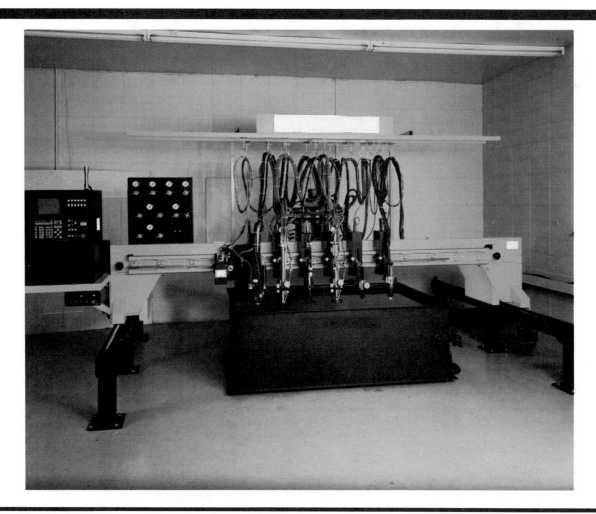

Figure 14.9–Gantry Type Shape Cutting Machine with Computer Numerical Control Drive

pins, are rapidly severed for disassembly using OFC. Holes can be made rapidly in steel components by piercing and cutting.

Machine OFC is used in many industries and steel warehouses to cut steel plate to size, to cut various shapes from plate, and to prepare plate edges for welding. Many machine parts such as gears, clevises, frames, and tools are made by oxygen-cutting procedures.

Machines capable of cutting to tolerances of 1/32 to 1/16 in. (0.8 to 1.6 mm) are used to produce parts that can be assembled into final product form without intermediate machining. They are also used for rapid material removal prior to machining to close tolerances.

Oxyfuel gas cutting is used to cut a wide range of steel thicknesses from approximately 1/8 to 84 in. (3 to 2100 mm). Thicknesses over approximately 20 in. (500 mm) are not generally cut except in steel mill operations, where the pieces are cut while still at high temperatures.

OPERATING PROCEDURES

IN THE OPERATION of OFC equipment, the recommendations of the equipment manufacturer in assembling and using the equipment should always be closely followed. This will prevent damage to the equipment and also insure its proper and safe use.

Regulators

THE OXYGEN AND fuel gas regulators must be clean and in good working condition. If there is oil, grease, or foreign material on a regulator or other equipment, or if the equipment is damaged, it must not be used prior to being properly cleaned or serviced by a qualified repair technician. Hoses must be in good condition and of appropriate size to provide adequate volume and pressure of both oxygen and fuel gas to the cutting torch.

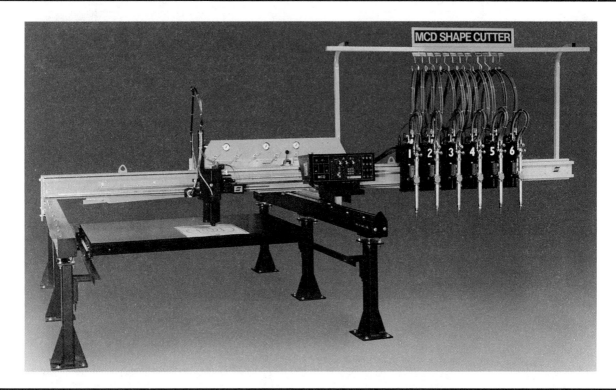

Figure 14.10–Cantilever Type Mechanized Shape Cutting Machine Equipped with Photocell Tracer and Six Oxyfuel Gas Cutting Torches

Flashback and Back Fire

A FLASHBACK IS the burning of the flame in or behind the torch mixing chamber. It is a serious condition, and corrective action must be taken to extinguish it. The torch oxygen valve should be turned off immediately and then the fuel gas valve. One cause of flashback is failure to purge the hose lines before lighting the torch; another cause is the overheating of the torch tip.

A backfire is the momentary recession of the flame into the torch tip followed by immediate reappearance or complete extinguishing of the flame. After this condition, the torch is still workable. If backfiring continues, the torch or tips, or both, should be removed from service for cleaning and possible repair.

Operating the Torch

THE MANUFACTURER'S RECOMMENDATIONS for lighting, testing, and using the equipment should always be followed. Only a spark lighter or other recommended lighting device should be used. Shaded or tinted eye protection and other appropriate clothes must be worn.

The most widely accepted manner to light the torch is to open the fuel-gas valve slightly, and light the gas with a spark lighter. Adjust the fuel gas until a stable flame is maintained at the end of the tip. Open the oxygen preheat valve slowly and increase the flow until the desired flame is attained. The intensity of the flame may be adjusted by slightly increasing or decreasing the volumes of both gases.

Flame Adjustment

FLAME ADJUSTMENT IS a critical factor in attaining satisfactory torch operation. The amount of heat produced by the flame depends on the intensity and type of flame used. Three types of flames can be set by properly adjusting the torch valves. See Figure 14.4.

A carburizing flame with acetylene, MPS, or propylene is indicated by trailing feathers on the primary flame cone or by long yellow-orange streamers in the secondary flame envelope. Propylene-based fuels, propane, and natural gas have a long, rounded primary flame cone. A carburizing flame is often used for the best finish and for stack cutting of thin material.

A neutral flame with acetylene, MPS, or propylene is indicated by a sharply defined, dark primary flame cone and a pale blue secondary flame envelope. Propane and propylene base fuels and natural gas have a short and sharply defined cone. This flame is obtained by adding oxygen to a carburizing flame. It is the flame most frequently used for cutting.

An oxidizing flame for acetylene or MPS has a light color primary cone and a smaller secondary flame shroud. It also generally burns with a harsh whistling sound. With propane and propylene base fuels and natural gas, the primary flame cones are longer, less sharply defined, and have a lighter color. This flame is obtained by adding some oxygen to the neutral flame. This type of flame is frequently used for fast, low-quality cutting, and selectively in piercing and quality beveling.

CUTTING PROCEDURES

Manual Cutting

SEVERAL METHODS CAN be used to start a cut on an edge. The most common method is to place the preheat flames halfway over the edge, holding the end of the flame cones 1/16 to 1/8 in. (1.5 to 3 mm) above the surface of the material to be cut. The tip axis should be aligned with the plate edge. When the top corner reaches a reddish yellow color, the cutting oxygen valve is opened and the cutting process starts. Torch movement is started after the cutting action reaches the far side of the edge.

Another starting method is to hold the torch halfway over the edge, with the cutting oxygen turned on, but not touching the edge of the material. When the metal reaches a reddish yellow color, the torch is moved onto the material and cutting starts. This method wastes oxygen, and starting is more difficult than with the first method. It should only be used for cutting thin material where preheat times are very short.

A third method is to put the tip entirely over the material to be cut. The preheat flame is held there until the metal reaches its kindling temperature. The tip is then moved to the edge of the plate so the oxygen stream will just clear the metal. With the cutting oxygen on, the cut is initiated. This method has the advantage of producing sharper corners at the beginning of the cut.

Once the cut has been started, the torch is moved along the line of cut with a smooth, steady motion. The operator should maintain as constant a tip-to-work distance as possible. The torch should be moved at a speed that produces a light ripping sound and a smooth spark stream.

For plate thicknesses of 1/2 in. (13 mm) or more, the cutting tip should be held perpendicular to the plate. For the thin plate, the tip can be tilted in the direction of the cut. Tilting increases the cutting speed and helps prevent slag from freezing across the kerf. When cutting material in a vertical position, start on the lower edge of the material and cut upward.

It is often necessary to start a cut at some point other than on the edge of a piece of metal. This technique is known as *piercing*. Piercing usually requires a somewhat larger preheat flame than the one used for an edge start. In addition, the flame should be adjusted to slightly oxidizing to increase the heat energy. The area where the pierce cut is to begin should be located in a scrap area. Hold the torch tip in one spot until the steel surface turns a yellowish red and a few sparks appear from the surface of the metal. The tip should be angled and lifted up as the cutting oxygen valve is opened. The torch is held stationary until the cutting jet pierces through the plate.

Torch motion is then initiated along the cut line. If the cutting oxygen is turned on too quickly and the torch is not lifted, slag may be blown into the tip and may plug the gas ports.

Machine Cutting

OPERATING CONDITIONS FOR mechanized oxygen cutting will vary depending on the fuel gas and the style of cutting torch being used. Tip size designations, tip design, and operating data can be obtained from the torch manufacturer.

Start up and shutdown procedures for machine OFC are essentially the same as those previously given for hand torch operation. However, proper adjustment of operating conditions is more critical if high-speed, high-quality cuts are to be obtained. The manufacturer's or supplier's cutting chart should be used to select the proper tip size for the material thickness to be cut. In addition to the tip size, initial fuel and oxygen pressure settings and travel speeds should be selected from the chart. Frequently the chart will also list gas flow rates, drill size of the oxygen orifice, preheat cone lengths, and kerf width. Operating conditions should then be adjusted to give the desired cut quality.

Proper tip size and cutting oxygen pressure are important in making a quality machine cut. If the proper tip size is not used, maximum cutting speed and the best quality of cut will not be achieved. The cutting oxygen pressure setting is an essential condition; deviations from the recommended setting will greatly affect cut quality. For this reason, some manufacturers specify setting the pressure at the regulator and operating with a given length of hose. When longer or shorter hoses are used, an adjustment in pressure should be made. An alternative is to measure oxygen pressure at the torch inlet. Pressure settings for cutting oxygen are then adjusted to obtain the recommended pressure at the torch inlet, rather than at the regulator outlet.

Other adjustments, such as the preheat fuel and oxygen pressure settings and the travel speed, are also important. Once the regulators have been adjusted, the torch valves are used to throttle gas flows to give the desired preheat flame. If sufficient flow rates are not obtained, pressure settings at the regulator can be increased to compensate. Cleanliness of the nozzle, type of base metal, purity of cutting oxygen, and other factors have a direct effect on performance.

Manufacturers differ in their recommended travel speeds. Some give a range of speeds for specific thicknesses, while others list a single speed. In either case, the settings are intended only as a guide. In determining the proper speed for an application, begin the cut at a slower speed than that recommended. Gradually increase the speed until cut quality falls below the required level. Then reduce the speed until the cut quality is restored, and continue to operate at that speed.

Typical data for cutting low carbon steel, using commonly available fuel gases, are shown in Table 14.2. The gas flow rates and cutting speeds are to be considered only as guides for determining more precise settings for a particular job. When a new material is being cut, a few trial cuts

should be made to obtain the most efficient operating conditions.

Heavy Cutting

HEAVY CUTTING IS considered the cutting of steel over approximately 12 in. (300 mm) thick. The basic reactions that permit oxygen cutting of thick steel are the same as

Table 14.2
Data for Manual and Machine Cutting of Clean Low Carbon Steel Without Preheat

U.S. Customary Units

Thickness of Steel in.	Diameter of cutting Orifice, in.	Cutting Speed in./min.	Gas Flow, ft³/h				
			Cutting Oxygen	Acetylene	MPS	Natural Gas	Propane
1/8	0.020-0.040	16-32	15-45	3-9	2-10	9-25	3-10
1/4	0.030-0.060	16-26	30-55	3-9	4-10	9-25	5-12
3/8	0.030-0.060	15-24	40-70	6-12	4-10	10-25	5-15
1/2	0.040-0.060	12-23	55-85	6-12	6-10	15-30	5-15
3/4	0.045-0.060	12-21	100-150	7-14	8-15	15-30	6-18
1	0.045-0.060	9-18	110-160	7-14	8-15	18-35	6-18
1-1/2	0.060-0.080	6-14	110-175	8-16	8-15	18-35	8-20
2	0.060-0.080	6-13	130-190	8-16	8-20	20-40	8-20
3	0.065-0.085	4-11	190-300	9-20	8-20	20-40	9-22
4	0.080-0.090	4-10	240-360	9-20	10-20	20-40	9-24
5	0.080-0.095	4-8	270-360	10-25	10-20	25-50	10-25
6	0.095-0.105	3-7	260-500	10-25	20-40	25-50	10-30
8	0.095-0.110	3-5	460-620	15-30	20-40	30-55	15-32
10	0.095-0.110	2-4	580-700	15-35	30-60	35-70	15-35
12	0.110-0.130	2-4	720-850	20-40	30-60	45-95	20-45

SI Units

Thickness of Steel mm	Diameter of cutting Orifice, mm	Cutting Speed mm/s	Gas Flow, L/min				
			Cutting Oxygen	Acetylene	MPS	Natural Gas	Propane
3.2	0.51-1.02	6.8 -13.5	7.2- 21.2	2- 4	2- 4	4-12	2- 5
6.4	0.76-1.52	6.8 -11.0	14.2- 26.0	2- 4	2- 5	4-12	2- 6
9.5	0.76-1.52	6.4 -10.1	18.9- 33.0	3- 5	2- 5	5-12	3- 7
13	1.02-1.52	5.1 - 9.7	26.0- 40.0	3- 5	2- 5	7-14	3- 8
19	1.14-1.52	5.1 - 8.9	47.2- 70.9	3- 6	3- 5	7-14	3- 9
25	1.14-1.52	3.8 - 7.6	51.9- 75.5	4- 7	4- 7	8-17	4- 9
38	1.52-2.03	2.5 - 5.9	51.9- 82.6	4- 8	4- 8	9-17	4-10
51	1.52-2.03	2.5 - 5.5	61.4- 89.6	4- 8	4- 8	9-19	4-10
76	1.65-2.16	1.7 - 4.7	89.6-142	4- 9	4-10	10-19	5-11
102	2.03-2.29	1.7 - 4.2	113 -170	5-10	4-10	10-19	5-11
127	2.03-2.41	1.7 - 3.4	127 -170	5-10	5-10	12-24	5-12
152	2.41-2.67	1.3 - 3.0	123 -236	5-12	5-12	12-24	6-19
203	2.41-2.79	1.3 - 2.1	217 -293	7-14	10-19	14-30	7-15
254	2.41-2.79	0.85 - 1.7	274 -331	7-17	10-19	16-33	7-15
305	2.79-3.30	0.85 - 1.7	340 -401	9-19	15-29	20-75	10-22

Notes:

1. Preheat oxygen consumptions: Preheat oxygen for acetylene = 1.1 to 1.25 x acetylene flow ft³/h; preheat oxygen for natural gas = 1.5 to 2.5 x natural gas flow ft³/h; preheat oxygen for propane = 3.5 to 5 x propane flow ft³/h.

2. Operating notes: Higher gas flows and lower speeds are generally associated with manual cutting, whereas lower gas flows and higher speeds apply to machine cutting. When cutting heavily scaled or rusted plate, use high gas flow and low speeds. Maximum indicated speeds apply to straight line cutting; for intricate shape cutting and best quality, lower speeds will be required.

those for the cutting of thinner sections. Thicknesses ranging from 12 to 60 in. (300 to 1525 mm) may be cut using heavy-duty torches. Preheat and cutting oxygen flows increase, and cutting speed decreases, as thickness increases.

For heavy cutting, the most important factor is oxygen flow. Tip size and operating pressure must provide the necessary cutting oxygen flow required for the thickness being cut. Oxygen cutting pressures in the range of 10 to 55 psi (70 to 380 kPa), measured at the cutting torch, have been found adequate for the heaviest cutting using the proper tip size and equipment. The oxygen flow at the torch entry is of paramount importance when comparing results of different cutting operations. By relating performance to oxygen flow rate rather than pressure, heavy cutting data can be plotted as a continuous curve.

In terms of flow, it is possible to arrive at an approximate demand constant that will be useful as a guide in selecting equipment suitable for a given job. These demand constants may vary, but in terms of thickness, they usually fall within the approximate range of 80 to 125 ft^3 of oxygen per in. (89 to 139L of oxygen per mm) of thickness. Table 14.3 gives the range of operating conditions that cover normal heavy cutting operations.

Heavy cutting covers a wide variety of operations, such as ingot cropping, scrap cutting, and riser cutting. The data in Table 14.3 may not be entirely suitable for all heavy cutting operations, although the values given have been used successfully. They may be used as a guide in selecting the correct equipment and operating conditions. The actual values for most efficient operation of a specific cutting application are always best found by trial cuts.

When heavy cutting is performed with the torch in a horizontal position, the cutting oxygen pressure may need to be increased to aid in removing slag from the kerf.

Recommended travel speeds are not included in Table 14.3, but speeds from 2 to 6 in./min (0.85 to 2.5 mm/s) are used in the range of thicknesses covered. A speed of 3 in./min (1.3 mm/s) is possible for thicknesses up to at least 36 in. (910 mm). The correct speed is obtained by observing the operating conditions carefully and making suitable adjustments while actual cutting is in progress.

Because heavy pieces usually have a scale covered surface, techniques of starting the cut differ from those used with clean, thin material. The start is made more slowly on the rougher edges. Figure 14.11 indicates correct and incorrect starting procedures. Figure 14.11(A) shows the desirable starting position with the preheat flames on the top corner and extending down the face of the material. The cutting reaction starts at the top corner. It proceeds down the face of the material to the bottom as the torch moves forward. Figures 14.11(B), (C), (D), (E), and (F) show problems occurring from incorrect procedures.

When the cut proceeds properly with correct oxygen flow and forward speed, the reaction will proceed to the end of the cut without leaving a skipped corner. Figure 14.12 illustrates various correct and incorrect terminating conditions and also proper drag conditions. Conditions producing a drop cut are depicted in Figure 14.12(A).

In general, the following conditions are required for successful heavy cutting on a production basis:

(1) Adequate gas supply sufficient to complete the cut; this is necessary because a lost cut on heavy materials is extremely difficult, if not impossible, to restart.

(2) Equipment of sufficient size structurally to maintain rigidity and to carry the equipment needed, and of sufficient capacity to handle the range of speeds and gas flows required.

(3) Skilled personnel that are trained in proper heavy cutting techniques.

Stack Cutting

IF DATA ON machine OFC speeds and gas requirements are plotted against the material thickness, the requirements are not directly proportional to material thickness, "t". Gas consumption per unit of thickness, "t", decreases as the thickness, "t", increases. Consequently, cutting costs

Table 14.3
Data for Oxyfuel Gas Cutting of Thick Low Carbon Steel

Material Thickness		Cutting Oxygen					
		Orifice Diameter		Flow Rate		Pressure at Torch	
in.	mm	in.	mm	ft^3/h	L/min	psi	kPa
12	305	0.147-0.221	3.74- 5.61	1000-1500	472- 708	56-33	386-228
16	406	0.170-0.290	4.32- 7.36	1300-2000	614- 944	54-25	372-172
20	508	0.194-0.332	4.93- 8.44	1700-2500	803-1180	52-22	359-152
24	610	0.221-0.332	5.61- 8.44	2000-3000	944-1416	48-29	331-200
28	711	0.250-0.375	6.35- 9.53	2300-3500	1087-1652	41-26	283-179
32	813	0.250-0.375	6.35- 9.53	2700-4000	1274-1888	51-30	352-207
36	914	0.290-0.422	7.37-10.72	3000-4500	1416-2120	40-26	276-179
40	1016	0.290-0.422	7.37-10.72	3400-5000	1605-2360	46-30	317-207
44	1118	0.290-0.468	7.37-11.90	3800-5500	1792-2600	51-26	352-179
48	1219	0.332-0.468	8.44-11.90	4000-6000	1888-2830	40-28	276-193

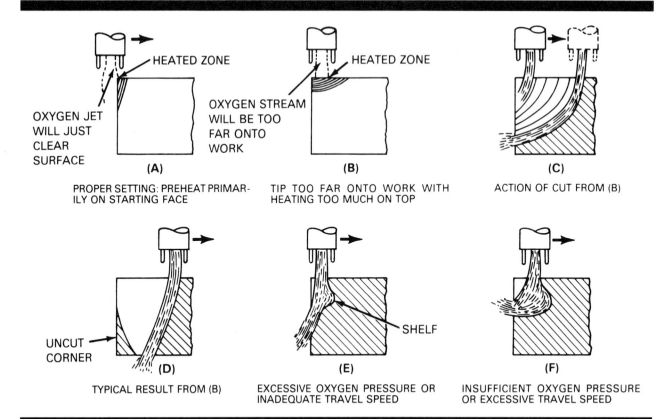

Figure 14.11–Starting Procedures for Heavy Cutting

per "t" may decrease as "t" increases when "t" is below a specific value, depending on the material being cut. Stacking of material for cutting can be more economical than cutting individual pieces, particularly when the material thickness is under 1/4 in. (6 mm). Stack cutting is limited to sheet and plate up to 1/2 in. (13 mm) thick because of the difficulty in clamping heavier material in a tight stack. A stack cutting operation is shown in Figure 14.13.

Stack cutting is also a means of cutting sheet material that is too thin for ordinary OFC methods. Sheet thicknesses of 20 gage (0.9 mm) and over are the most practical. Stack cutting is used in place of shearing or stamping, particularly where volume does not justify expensive dies. The flame cut sheet edges are square with no burrs.

Successful stack cutting requires clean, flat sheet or plate. Dirt, mill scale, rust, and paint may interrupt the cut and reduce cut quality. The stack must be securely clamped, particularly at the cut location, with the edges aligned at the point where the cut is to start.

Piercing of stacks with the OF torch to start a cut is impractical. Holes must be drilled though the stacks to start an interior cut.

The total thickness of the stack is determined by the cutting tolerance requirement and the thickness of the top piece. With a cutting tolerance of 1/32 in. (0.8 mm), stack height should not exceed 2 in. (50 mm); with a 1/16 in. (1.6 mm) tolerance, the thickness may be up to 4 in. (100 mm). The maximum practical limit of thickness is about 6 in. (150 mm).

When stack cutting material less than 3/16 in. (5 mm) thick, a waster plate 1/4 in. (6 mm) thick is used on top. It insures better starting, a sharper edge on the top production piece, and no buckling of the top sheet.

Starting the cut must be done with extreme care so that it will extend through the stack. One method of starting is to align the sheet edges exactly in a vertical line. A vertical strip along the aligned face is preheated with a hand torch to ignition temperature. The machine torch is quickly positioned at the starting point and cutting initiated. Another procedure is to position each sheet so that its edge projects slightly over the edge below. This is advantageous for sheared sheet stacked with the burr down. Cutting is initiated on the top plate (waster plate) and progresses from one sheet to the other through the stack. A third method is to run a vertical weld bead down the stack to form a continuous strip of metal. The cut is started through the weld bead and progresses into the stack.

Even when extreme care has been exercised, there is always the possibility of an interruption of cutting with possible loss of the entire stack. The application of flux cut-

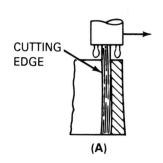

CUTTING EDGE

(A)

NO DRAG PERMITS STREAM TO BREAK THROUGH FACE UNIFORMLY AT ALL POINTS. TYPICAL OF BAL-ANCED CONDITIONS

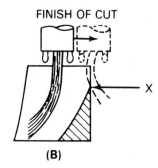

FINISH OF CUT

X

(B)

DRAG CAUSES ACTION TO CARRY THROUGH AT X AND AND TO PASS BEYOND MATERIAL, LEAVING UNCUT CORNER. TYPICAL OF INSUFFICIENT OXYGEN OR EXCESSIVE SPEED

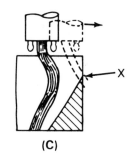

X

(C)

FORWARD DRAG CAUSES STREAM TO BREAK THROUGH AT X AND BECOME DEFLECTED, LEAVING UN-CUT CORNER. TYPICAL OF HIGH CUT-TING OXYGEN PRESSURE OR TOO LITTLE SPEED

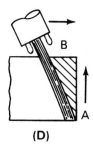

(D)

IF CUTTING FACE IS SUCH THAT BREAKTHROUGH AT BOTTOM, A, LIES AHEAD OF ENTRY POINT, B, AND AT NO POINT DOES FACE EXTEND BEYOND A, ACTION WILL SEVER FROM A, UPWARD

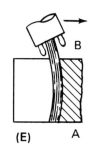

(E)

ANGULAR TIP DISPOSITION SIMILAR TO (D) SHOWING LIMIT OF EFFEC-TIVENESS. SIMILAR TO (A)

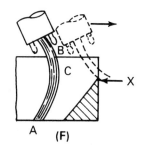

X

(F)

IF CONDITIONS ARE SUCH THAT A AND B ARE IN LINE OR OTHERWISE DISPOSED, BUT C LIES AHEAD OF A, STREAM WILL BREAK AT X, LEAVING UNCUT CORNER, SIMILAR TO (C)

Figure 14.12–Terminating Conditions for Heavy Cutting — A, B, and C With Torch Vertical; D, E, and F With Torch Angled in Direction of Cutting

ting and powder cutting processes greatly minimizes this hazard. These methods assist in propagating the oxidation reaction within the cut. Appreciable air gaps that otherwise might inhibit cutting can be tolerated between plates. The use of divergent tips with high velocity cutting jets also appears to aid this transfer action.

Regardless of the procedure employed, the economy of a stack cutting operation must be carefully compared with the total costs involved, including such items as material preparation, stack makeup, clamping devices, and increased skill and care requirements.

Plate Edge Preparation

BEVEL, V-, AND U-groove joint joint designs are used for welding steel components together. The preparation of the edges to be welded together can be done by oxygen cutting or gouging. Single and double bevels are produced using standard cutting tips and torches, usually mechanized, for straight-line beveling. Oxygen gouging is done by using specially designed cutting tips to produce U-groove joints.

Plate Beveling. The beveling of plate edges before welding is necessary in many applications to insure proper dimensions and fit, and also to accommodate standard welding techniques. Beveling may be done by using a single torch or multiple torches operating simultaneously. Although single beveling can be done manually, beveling is best done by machine for accurate control of the cutting variables. When cutting bevels with two or three torches, plate riding devices should be used to insure constant tip position above the plate, as shown in Figure 14.14.

In single-torch beveling, the amount and type of torch preheat is a dominant factor. With bevel angles of less than 15°, the loss of preheat efficiency is small. When the bevel

Figure 14.13–Typical Stack Cutting Operation With the Plates Clamped by Vertical Welds

angle is above 15°, the heat transferred from the preheat flames to the plate decreases rapidly as the bevel angle increases. Considerably greater preheat input is required, particularly for thicknesses up to 1 in. (25 mm). Best results are obtained by positioning the tip very close to the work and using high oxygen to fuel ratios. For bevels greater than 30°, or on heavy plate, special bevel tips will provide the additional preheat capacity required.

An auxiliary torch (with only preheat flames burning) mounted perpendicular to the work or an auxiliary adapter, which divides the preheat and applies a portion of it at right angles to the work, may be used to obtain faster beveling speeds. Either method actually consumes less total preheat gas than a single angled tip.

The best quality of cut face is usually not obtained at the highest cutting speed. The cut face finish can usually be improved by operating at lower speeds. When speed is reduced to obtain improved surface finish, the preheat flames should be decreased to prevent excessive meltdown of the top edge of the faces.

Figures 14.15, 14.16, and 14.17 illustrate the torch positions to cut the three basic beveled edges. In each case, torch position spacings A and B are governed by plate thickness, tip size, and speed of cutting. The cutting torches are positioned at spacings that are practical without interrupting the cutting action of any of the three cutting oxygen streams. When the lengths of A or B or both are too great, the cutting action of the trailing torch does not span the kerf of the leading torch. This causes the oxygen stream to be deflected into the kerf of the leading torch, and it gouges the cut face. This produces a rough surface and usually a light slag adhering to the underside of the prepared edge.

The positioning of the torches in a lateral direction for multibevel cutting is usually accomplished by trial and error. However, this can be costly and result in lengthy reworking or possible scrap. A simple machined template, which is typical of the desired edge geometry, is quite useful for torch alignment. A kerf-centering device is attached to each cutting tip, as shown in Figure 14.18. The torches

Figure 14.14–Mechanized Cutting Arrangement for Beveling a Plate Edge

are then properly angled and adjusted to the edge template. The multiple torch cutting head is now ready to duplicate the template profile.

To obtain close dimensional tolerance when preparing plate edges, precise torch-conveying equipment is necessary. For reproducibility, accuracy, and maximum efficiency, large gantry and rail-type cutting machines are used. Such apparatus may be classified in the same category as a machine tool. A plate is placed on a flat cutting table between the rails of a three-gantry type cutting machine, as shown in Figure 14.19. The machine can prepare all four edges of the plate without repositioning it. It can also cut the plate into smaller segments at the same time.

Gouging. Gouging of steel plate using the OFC processes is usually limited to steel plate thicknesses up to 1 in. (25 mm). The process is frequently used on the underside of a welded joint to remove defects that are in the original root pass. OFC gouging is also frequently used to remove defective weld joints or cracks when repairing previously fabricated metal.

The gouging process usually requires a special gouging tip with extra-heavy preheat capacity and a central oxygen orifice that causes a high level of turbulence in the oxygen stream. This turbulence causes a wide flow of oxygen that can be controlled by the operator to achieve the desired width and depth of gouge. Other factors used to determine the shape of the gouge are speed, tip angle, pressure, amount of preheat, and tip size. One of the significant advantages of oxyfuel gouging is that no additional equipment other than that already used in the OFC process is required.

UNDERWATER CUTTING

UNDERWATER CUTTING IS used for salvage work and for cutting below the water line on piers, dry docks, and ships. The two methods most widely used are oxyfuel gas cutting and oxygen arc cutting.

The technique for underwater cutting with OFC is not materially different from that used in cutting steel in open air. An underwater OFC torch embodies the same features as a standard OFC torch with the additional feature of supplying its own ambient atmosphere. In the underwater cutting torch, fuel and oxygen are mixed together and and burned to produce the preheat flame. Cutting oxygen is provided through the tip to sever the steel. In addition, the torch provides an air bubble around the cutting tip. The air bubble is maintained by a flow of compressed air around the tip, as shown in Figure 14.20. The air shield stabilizes the preheat flame and at the same time displaces the water from the cutting area.

The underwater cutting torch has connections for three hoses to supply compressed air, fuel gas, and oxygen. A combination shield and spacer device is attached at the cutting end of the torch. The adjustable shield controls the formation of the air bubble. The shield is adjusted so that the preheat flame is positioned at the correct distance from the work. The feature is essential for underwater work because of poor visibility and reduced operator mobility caused by cumbersome diving suits. Slots in the shield allow the burned gases to escape. A short torch is used to reduce the reaction force produced by the compressed air and cutting oxygen pushing against the surrounding water.

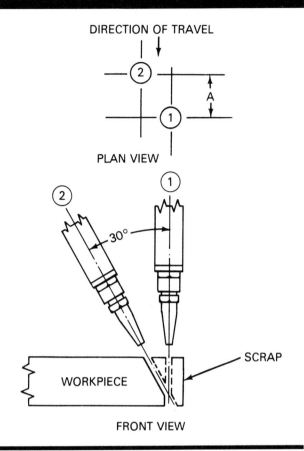

Figure 14.15–Cutting a Single Bevel Edge Preparation With a Root Face

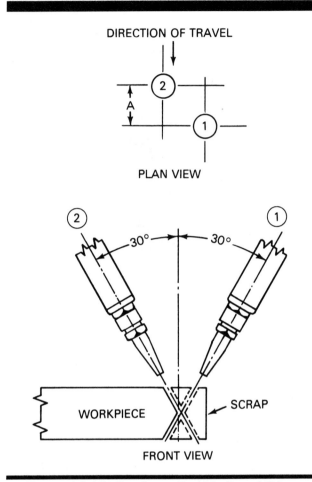

Figure 14.16–Cutting a Double Bevel Edge Preparation With No Root Face

As the depth at which the cutting is being done increases, the gas pressures must be increased to overcome both the added water pressure and the frictional losses in the longer hoses. Approximately 1/2 psi (3.5 kPa) for each 12 in. (300 mm) of depth must be added to the basic gas pressure requirements used in air for the thickness being cut.

MPS, propylene, and hydrogen are the best all-purpose preheat gases, because they can be used at any depths to which divers can descend and perform satisfactorily. Acetylene must not be used at depths greater than approximately 20 ft (6 m), because its maximum safe operating pressure is 15 psi (100 kPa) gage.

The oxyfuel gas cutting torch experiences no great difficulty underwater in severing steel plate in thicknesses from 1/2 in. (13 mm) to approximately 4 in. (101 mm). Under 1/2 in. (13 mm) thickness, the constant quenching effect of the surrounding water lowers the efficiency of preheating. This requires much larger preheating flames and preheat gas flows. Cutting oxygen orifice size is considerably larger for underwater cutting than for cutting in air. A spe-

cial apparatus for lighting the preheat flames under water is also needed.

Some manufacturers have developed a spacing sleeve to be used for underwater cutting with a standard cutting torch. This device clamps over the cutting tip and provides a guide for the proper tip-to-work distance. A source of compressed air is not required for this unit.

The recommendations of the manufacturer should be followed for setting up and operating underwater OFC equipment.

QUALITY OF CUTTING

ACCEPTABLE QUALITY OF OFC depends on the job requirements. Salvage operations and severing members for scrap do not require high-quality cutting. Oxygen cutting is used to rapidly complete the operations with little regard to the quality of the cut surfaces.

When the cut materials are used in fabrications with no other processing of the cut surfaces, the quality of the sur-

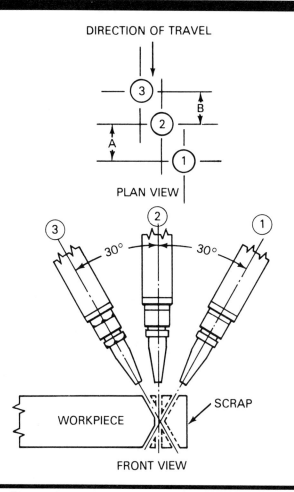

DIRECTION OF TRAVEL

PLAN VIEW

30°   30°

WORKPIECE   SCRAP

FRONT VIEW

Figure 14.17—Cutting a Double Bevel Edge Preparation With a Root Face

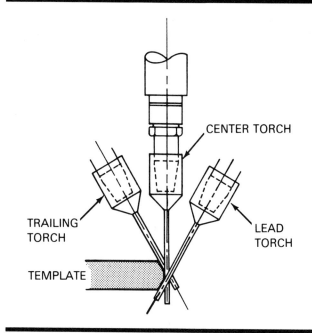

CENTER TORCH

TRAILING TORCH

LEAD TORCH

TEMPLATE

Figure 14.18—Kerf Centering and Bevel Angle Setting Method

faces may be significant. Cutting quality may include such things as:

(1) Proper angle of the cut surface with adjacent surfaces
(2) Flatness of the surface
(3) Sharpness of the cut preheat edge
(4) Dimensional tolerances of the cut shape
(5) Adherence of tenacious slag
(6) Cut surface defects, such as cracks and pockets

Close control of these items is generally confined to machine OFC. Good control of torch position, initiation of the cut, travel speed, and template stability are required for high-quality cutting. Also, consistent maintenance and cleanliness of the equipment is needed.

With the proper equipment in good condition, a well-trained operator, and reasonably clean and well-supported work, shapes can be cut to tolerances of 1/32 to 1/16 in.

(0.8 to 1.6 mm) from material not more than 2 in. (51 mm) thick. Correct cutting tip, preheat flame adjustment, cutting oxygen pressure and flow, and travel speed must be used.

Regardless of operating conditions, drag lines are inherent to oxygen cutting. They are the lines that appear on the cut surface, shown in Figure 14.21, resulting from the way that the iron oxidizes in the kerf. Light drag lines on the cut surface are not considered detrimental. The amount of drag is important. If it is too great, the corner at the end of the cut may not be completely severed and the part will not drop.

Cut surface quality is dependent on many variables, the most significant being the following:

(1) Type of steel
(2) Thickness of the material
(3) Quality of steel (freedom from segregations, inclusions, etc.)
(4) Condition of the steel surface
(5) Intensity of the preheat flames and the preheat oxy-fuel gas ratio
(6) Size and shape of the cutting oxygen orifice
(7) Purity of the cutting oxygen
(8) Cutting oxygen flow rate
(9) Cleanliness and flatness of the exit end of the nozzle
(10) Cutting speed

For any given cut, the variables listed should be evaluated so that the required quality of cut may be obtained

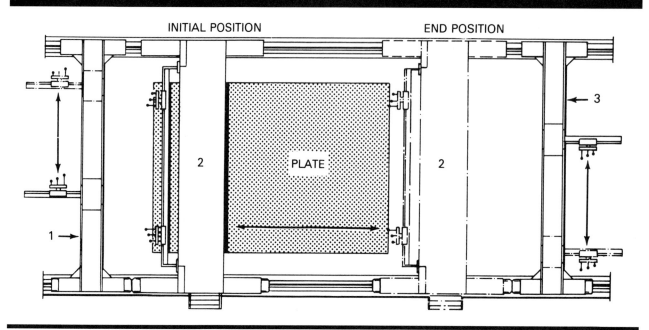

Figure 14.19—Plan View of a Three-Gantry Cutting Machine

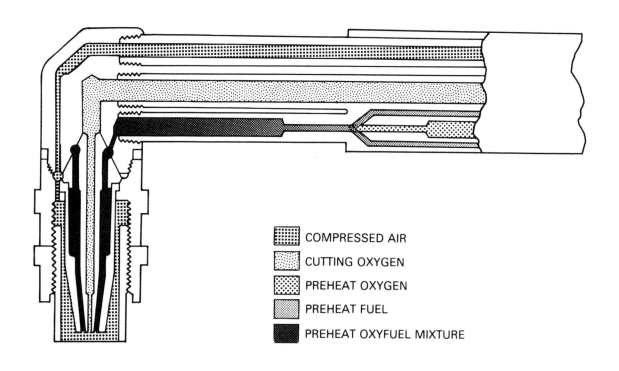

COMPRESSED AIR

CUTTING OXYGEN

PREHEAT OXYGEN

PREHEAT FUEL

PREHEAT OXYFUEL MIXTURE

Figure 14.20—Basic Design of an Underwater Oxyfuel Gas Cutting Torch

Figure 14.21–Drag Lines on the Kerf Wall Resulting From Oxygen Cutting

with the minimum aggregate cost in oxygen, fuel gas, labor, and overhead. Figures 14.22 and 14.23 show typical edge conditions resulting from variations in the cutting procedure for material of uniform type and thickness.

Dimensional tolerance and surface roughness must be considered together when judging the quality of a cut, because they are somewhat dependent on each other. Most specifications include dimensional tolerances. These include straightness of edge, squareness of edge, and permissible variation in plate width. All of these are primarily a function of the cutting equipment and its mechanical operation. When the torch is held rigidly and advanced at a constant speed, as in machine OFC, dimensional tolerances can be maintained within reasonable limits. The degree of longitudinal precision of a machine cut depends primarily on such factors as the condition of the equipment, trueness of guide rails, clearances in the operating mechanism, and the uniformity of speed control of the drive unit. In addition to equipment, dimensional accuracy is dependent on the control of thermal expansion of the material being cut. Lack of dimensional tolerance may re-

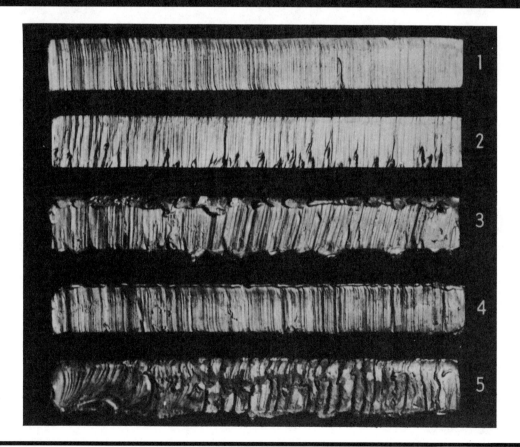

Figure 14.22–Typical Edge Conditions Resulting From Oxyfuel Gas Cutting Operations: (1) Good Cut in 1 in. (25 mm) Plate - the Edge is Square, and the Drag Lines are Essentially Vertical and Not Too Pronounced; (2) Preheat Flames Were Too Small for This Cut, and the Cutting Speed Was Too Slow, Causing Bad Gouging at the Bottom; (3) Preheating Flames Were Too Long, With the Result That the Top Surface Melted Over, the Cut Edge is Irregular, and There is an Excessive Amount of Adhering Slag; (4) Oxygen Pressure Was Too Low, With the Result That the Top Edge Melted Over Because of the Slow Cutting Speed; (5) Oxygen Pressure Was Too High and the Nozzle Size Too Small, With the Result That Control of the Cut Was Lost

Figure 14.23–Typical Edge Conditions Resulting From Oxyfuel Gas Cutting Operations: (6) Cutting Speed Was Too Slow, With the Result That the Irregularities of the Drag Lines are Emphasized; (7) Cutting Speed Was Too Fast, With the Result That There is a Pronounced Break in the Dragline, and the Cut Edge is Irregular; (8) Torch Travel Was Unsteady, With the Result That the Cut Edge is Wavy and Irregular; (9) Cut was Lost and Not Carefully Restarted, Causing Bad Gouges at the Restarting Point

sult from buckling of the material (thin plate or sheet), warpage resulting from the heat being applied to one edge, or shifting of the material while it is being cut.

The OFC operation should be planned carefully to minimize the effect of the variables on dimensional accuracy. For instance, when trimming opposite edges of a plate, warpage will be minimized if both cuts are made simultaneously in the same direction. Distortion can often be controlled when cutting irregular shapes from plates by inserting wedges in the kerf following the cutting torch, to limit movement of the metal from thermal expansion and contraction. In cutting openings in the middle of a plate, distortion may be limited by making a series of unconnected cuts. The section is left attached to the plate in a number of places until cutting is almost completed, then connecting locations are finally cut though. The intermittent cutting will reduce cut quality somewhat.

Thin material is often stack cut to eliminate warping and buckling. Another technique is to cut the thin plate while it is partially submerged in water to remove the heat.

MATERIALS CUT

FOR MOST STEEL cutting, standard oxygen cutting equipment is satisfactory. For high alloy and stainless steel cutting, it may be necessary to use a special OFC process, such as flux injection or powder cutting, or one of the arc cutting processes. The cutting process and type of operation (manual or mechanized) selected depend on the material that is being cut, production requirements, and the ultimate use of the product.

CARBON AND LOW ALLOY STEELS

CARBON STEELS ARE readily cut by the OFC process. Low carbon steels are cut without difficulty using standard procedures. Typical data for cutting low carbon steel, using commonly available fuel gases, are shown in Table 14.2. The gas flow rates and cutting speeds listed are to be considered as guides for determining more precise settings for a particular job. When a new material is being cut, a few trial cuts should be made to obtain the most efficient operating conditions.

It should be noted that the tables end at 12 in. (300 mm), which is the maximum thickness normally encountered for shape cutting in production shops. The division has been made arbitrarily. The cutting of steel plate over approximately 12 in. (300 mm) thick is considered heavy cutting. The characteristics of heavy cutting are discussed later.

Effects Of Alloying Elements

ALLOYING ELEMENTS HAVE two possible effects on the oxygen cutting of steel. They may make the steel more difficult to cut, or they may give rise to hardened or heat-checked cut surfaces, or both. The effects of alloying elements are roughly evaluated in Table 14.4.

A large quantity of heat energy is liberated in the kerf when steel is cut with an oxygen jet. Much of this energy is transferred to the sides of the kerf, where it raises the temperature of the steel adjacent to the kerf above its critical temperature. Since the torch is moving forward, the source of heat quickly moves on. The mass of cold metal near the kerf acts as a quenching medium, rapidly cooling the hot steel. This quenching action may harden the cut surfaces of high carbon and alloy steels.

The depth of the heat-affected zone depends on the carbon and alloy contents, on the thickness of the base metal, and the cutting speed employed. Hardening of the heat-affected zones of steels containing up to 0.25 percent carbon is not critical in the thicknesses usually cut. Higher carbon steels and some alloy steels are hardened to a degree that the thickness may become critical.

Typical depths of the heat-affected zones in oxygen cut steel are shown in Table 14.5. For most applications of oxygen cutting, the affected metal need not be removed. However, if it is removed, removal should be by mechanical means.

Preheating and Postheating

THE MATERIAL BEING cut may be preheated to provide desired mechanical and metallurgical characteristics or to improve the cutting operation.

Preheating the work can accomplish several useful purposes:

(1) It can increase the efficiency of the cutting operation by permitting higher travel speed. Higher travel speed will reduce the total amount of oxygen and fuel gas required to make the cut.

(2) It will reduce the temperature gradient in the steel during the cutting operation. This in turn, will reduce or give more favorable distribution to thermally induced stresses and prevent the formation of quenching or cooling cracks. Distortion will also be reduced.

Table 14.4
Effect of Alloying Elements on Resistance of Steel to Oxygen Cutting

Element	Effect of Element on Oxygen Cutting
Carbon	Steels up to 0.25% carbon can be cut without difficulty. Higher carbon steels should be preheated to prevent hardening and cracking. Graphite and cementite (Fe_3C) are detrimental, but cast irons containing 4% carbon can be cut by special techniques.
Manganese	Steels with about 14% manganese and 1.5% carbon are difficult to cut and should be preheated for best results.
Silicon	Silicon, in amounts usually present, has no effect. Transformer irons containing as much as 4% silicon are being cut. Silicon steel containing large amounts of carbon and manganese must be carefully preheated and postannealed to avoid air hardening and possible surface fissures.
Chromium	Steels with up to 5% chromium are cut without much difficulty when the surface is clean. Higher chromium steels, such as 10% chromium steels, require special techniques (see the section Oxidation Resistant Steels), and the cuts are rough when the usual oxyacetylene cutting process is used. In general, carburizing preheat flames are desirable when cutting this type of steel. The flux injection and iron powder cutting processes enable cuts to be readily made in the common straight chromium irons and steels as well as in stainless steel.
Nickel	Steels containing up to 3% nickel may be cut by the normal oxygen cutting processes; up to about 7% nickel content, cuts are very satisfactory. Cuts of excellent quality may be made in the common stainless steels (18-8 to about 35-15 as the upper limit) by the flux injection or iron powder cutting processes.
Molybdenum	This element affects cutting about the same as chromium. Aircraft quality chrome-molybdenum steel offers no difficulties. High molybdenum-tungsten steels, however, may be cut only by special techniques.
Tungsten	The usual alloys with up to 14% tungsten may be cut very readily, but cutting is difficult with a higher percentage of tungsten. The limit seems to be about 20% tungsten.
Copper	In amounts up to about 2%, copper has no effect.
Aluminum	Unless present in large amounts (on the order of 10%), the effect of aluminum is not appreciable.
Phosphorus	This element has no effect in amounts usually tolerated in steel.
Sulfur	Small amounts, such as are present in steels, have no effect. With higher percentages of sulfur, the rate of cutting is reduced and sulfur dioxide fumes are noticeable.
Vanadium	In the amounts usually found in steels, this alloy may improve rather than interfere with cutting.

Table 14.5
Approximate Depths of Heat-Affected Zones in Oxygen Cut Steels*

Thickness		Depth			
		Low Carbon Steels		High Carbon Steels	
in.	mm	in.	mm	in.	mm
Under 1/2	Under 13	Under 1/32	Under 0.8	1/32	0.8
1/2	13	1/32	0.8	1/32 to 1/16	0.8 to 1.6
6	152	1/8	3.2	1/8 to 1/4	3.2 to 6.4

* The depth of the fully hardened zone is considerably less than the depth of the heat-affected zone.

(3) It may prevent hardening the cut surface by reducing the cooling rate.

(4) It will decrease migration of carbon toward the cut face by lowering the temperature gradient in the metal adjacent to the cut.

The temperatures used for preheating generally range from 200 to 1300°F (90 to 700°C) depending upon the part size and the type of steel to be cut. The majority of carbon and alloy steels can be cut with the steel heated to the 400 to 600°F (200 to 315°C) temperature range. The higher the preheat temperature, the more rapid is the reaction of the oxygen with the iron. This permits higher cutting speeds.

It is essential that the preheat temperature be fairly uniform through the section in the areas to be cut. If the metal near the surfaces is at a lower temperature than the interior metal, the oxidation reaction will proceed faster in the interior. Large pockets may form in the interior and either produce unsatisfactory cut surfaces or cause slag entrapment that may interrupt the cutting action. If the material is preheated in a furnace, cutting should be started as soon as possible after the material is removed from the furnace, to take advantage of the heat in the plate.

If furnace capacity is not available for preheating the entire piece, local preheating in the vicinity of the cut will be of some benefit. For light cutting, preheating may be accomplished by passing the cutting torch preheating flames slowly over the line of the cut until the desired preheat temperature is reached. Another method which may give better results is to preheat with a multiflame heating torch mounted ahead of the cutting torch.

To reduce thermally induced internal stresses in the cut parts, they may be annealed, normalized, or stress relieved. Using a proper postheat treatment, most metallurgical changes caused by the cutting heat can be eliminated. If a furnace of the required size is not available for postheat treatment, the cut surface may be reheated to the proper temperature by the use of multiple flame heating torches.

CAST IRON

THE HIGH CARBON content of cast iron resists the ordinary OFC techniques used for cutting low carbon steels. Cast irons contain some of the carbon in the form of graphite flakes or nodules, and some in the form of iron carbide (Fe_3C). Both of these constituents hinder the oxidation of the iron. High-quality production cuts typical of steels cannot be obtained with cast iron. Most cutting is done to remove risers, gates, or defects, to repair or alter castings, or for scrapping.

Cast iron can usually be manually cut by using an oscillating motion of the cutting torch, as shown in Figure 14.24. The degree of motion depends on the section thickness and carbon content. Torch oscillation helps the oxygen jet to blow the slag and molten metal out of the kerf. The kerf is normally wide and rough.

A larger cutting tip and higher gas flow than those used for steel are required for cutting the same thickness of cast iron. A hot carburizing flame is used, with the streamer extending to the far side of the cast iron section. The excess fuel gas helps to maintain preheat in the kerf as it burns.

Cast iron is also sometimes cut by using the special techniques for cutting oxidation resistant steels. These are waster plate cutting, metal powder cutting (POC), and chemical flux cutting (FOC), which are described later in this chapter. Cast iron is readily cut using the air carbon arc cutting (CAC-A) and plasma arc cutting (PAC) processes, and these are frequently preferred over the OFC processes.

OXIDATION RESISTANT STEELS

THE ABSENCE OF alloying materials in pure iron permits the oxidation reaction to proceed rapidly. As the quantity and number of alloying elements in iron increase, the oxidation rate decreases from that of pure iron. Cutting becomes more difficult.

Oxidation of the iron in any alloy steel liberates a considerable amount of heat. The iron oxides produced have melting points near the melting point of iron. However, the oxides of many of the alloying elements in steels, such as aluminum and chromium, have melting points higher than those of iron oxides. These high-melting oxides, which are refractory in nature, may shield the material in the kerf so that fresh iron is not continuously exposed to the cutting oxygen stream. Thus, the speed of cutting de-

MOVEMENT WHEN CUTTING THIN CAST IRON

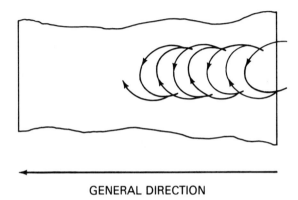

MOVEMENT WHEN CUTTING HEAVY CAST IRON

GENERAL DIRECTION

Figure 14.24—Typical Cutting Torch Manipulation for Cutting Cast Iron

creases as the amount of refractory oxide-forming elements in the iron increases.

For ferrous metals with high alloy content, such as stainless steel, the use of plasma arc cutting (PAC) and in some cases air carbon are cutting (CAC-A) should be considered. If these options are not available or practical, then variations of OFC must be used.

There are several variations for oxygen cutting of oxidation resistant steels, which are also applicable to cast irons. The important ones are the following:

(**1**) Torch oscillation
(**2**) Waster plate
(**3**) Wire feed
(**4**) Powder cutting
(**5**) Flux cutting

When the above methods are used to cut oxidation resistant metals, the quality of the cut surface is somewhat impaired. Scale and slag may adhere to the cut faces. Pickup of carbon or iron, or both, usually appears on the cut surfaces of stainless steels and nickel alloy steels. This may affect the corrosion resistance and magnetic properties of the metal. If the corrosion resistance or magnetic properties of the material are important, approximately 1/8 in. (3 mm) of metal should be machined from the cut edges.

Torch Oscillation

THIS TECHNIQUE IS the one described previously for cast iron cutting. Low alloy content stainless steels up to 4 in. (100 mm) thick can sometimes be severed with a standard cutting torch and oscillation. The entire thickness of the starting edge must be preheated to a bright red color before the cut is started. This technique should be combined with some of the other cutting methods listed.

Waster Plate

ONE METHOD OF cutting oxidation resistant steels is to clamp a low carbon steel "waster" plate on the upper surface of the material to be cut. The cut is started in the low carbon steel material. The heat liberated by the oxidation of the low carbon steel provides additional heat at the cutting face to sustain the oxidation reaction. The iron oxide from the low carbon steel helps to wash away the refractory oxides from stainless steel. The thickness of the waster plate must be in proportion to the thickness of the material being cut. Several undesirable features of this method are the cost of the waster plate material, the additional setup time, the slow cutting speeds, and the rough quality of the cut.

Wire Feed

WITH THE APPROPRIATE equipment, a small diameter low carbon steel wire is fed continuously into the torch preheat flames, ahead of the cut. The end of the wire should melt rapidly into the surface of the alloy steel plate. The effect of the wire addition on the cutting action is the same as that of the waster plate. The deposition rate of the low carbon steel wire must be adequate to maintain the oxygen cutting action. It should be determined by trial cuts. The thickness of the alloy plate and the cutting speed are also factors that must be considered in the process. A motor-driven wire feeder and wire guide, mounted on the cutting torch, are needed as accessory equipment.

Metal Powder Cutting

THE METAL POWDER cutting process (POC) is a technique for supplying an OFC torch with a stream of iron-rich powdered material. The powdered material accelerates and propagates the oxidation reaction and also the melting and spalling action of hard-to-cut materials. The powder is directed into the kerf through either the cutting tip or single or multiple jets external to the tip. When the first method is used, gas-conveyed powder is introduced into the kerf by special orifices in the cutting tip. When the powder is introduced externally, the gas conveying the powder imparts sufficient velocity to the powder particles to carry them through the preheat envelope into the cutting oxygen stream. Their short time in the preheat envelope is sufficient to produce the desired reaction in the cutting zone.

Some of the powders react chemically with the refractory oxides produced in the kerf and increase their fluidity. The resultant molten slags are washed out of the reaction zone by the oxygen jet. Fresh metal surfaces are continuously exposed to the oxygen jet and powder. Iron powder and mixtures of metallic powders, such as iron and aluminum, are used.

Cutting of oxidation resistant steels by the powder method can be done at approximately the same speeds as oxygen cutting of carbon steel of equivalent thickness. The cutting oxygen flow must be slightly higher with the powder process.

Powder Cutting Equipment

DISPENSERS OF POWDER for the POC process are of two general types. One type of dispenser is a vibratory device in which the quantity of powder dispensed from the hopper is governed by a vibrator. Desired amounts of powder can be obtained by adjusting the amplitude of vibration. The vibratory-type dispenser is generally used where uniform and accurate powder flow is required.

The other type of dispenser is a pneumatic device. In the bottom of a low pressure vessel there is an ejector or fluidizing unit. The powder-conveying gas is brought into the dispenser in a manner that fluidizes the powder. The powder flows uniformly into an ejector unit where it is picked up by a gas stream that serves as the transporting medium to the torch.

In addition to the fuel and oxygen hoses, another hose is used to convey the powder to the torch. A special, manual powder cutting torch mixes the oxygen and fuel gas and then discharges this mixture through a multiplicity of orifices in the cutting tip. The powder valve is an integral part of the torch. The cutting oxygen lever on the torch also opens the powder valve in proper sequence. The powder carried by the conveying gas is brought through a separate tube into a chamber forward of the preheat gas chamber in the torch head. The powder then enters a separate group of passages in a two-piece cutting tip. From there, it discharges at the mouth of the tip in a conical pattern. The powder emerges with sufficient velocity to pass through the burning preheat gas and surrounds the central cutting oxygen stream.

Flux Cutting

THIS PROCESS IS primarily intended for cutting stainless steels. The flux is designed to react with oxides of alloying elements, such as chromium and nickel, to produce compounds with melting points near those of iron oxides. A special apparatus is required to introduce the flux into the kerf. With a flux addition, stainless steels can be cut at a uniform linear speed without torch oscillation. Cutting speeds approaching those for equivalent thicknesses of carbon steel can be attained. The tip sizes will be larger, and the cutting oxygen flow will be somewhat greater than for the carbon steels.

Flux Cutting Equipment

TO USE THE flux process, a flux feed unit is required. The cutting oxygen passes through the feed unit, and so transports the flux to the torch. The flux is held in a dispenser designed to operate at normal cutting oxygen pressures. The flux is transported through a hose from a dispenser to a conventional three-hose cutting torch. A mixture of oxygen and flux flows from the cutting oxygen orifice of the torch tip. Special operating procedures are used to prevent buildup of flux in the cutting oxygen hose and the cutting torch.

OXYGEN LANCE CUTTING

DEFINITION AND DESCRIPTION

OXYGEN LANCE CUTTING (LOC) is an oxygen cutting process that uses oxygen supplied through a consumable steel pipe or lance. The preheat required to start the cutting is obtained by other means.

The earliest version of LOC used a plain black iron pipe as a lance, with oxygen flowing through it. An oxyfuel gas cutting or welding torch is used to heat the cutting end of the lance to a cherry red, and then the oxygen flow is started. The iron pipe burns in a self-sustaining, exothermic reaction, and the heating torch is removed. When the burning end of the lance is brought close to the workpiece, the work is melted by the heat of the flame.

The oxygen lancing operation is shown schematically in Figure 14.25.

An improved version of the lance involves a number of low carbon steel wires packed into the steel tube. This increases the cutting life and capability of the lance. Commercially available tubes are typically 10 1/2 ft (3.2 m) long and 0.625 in. (16 mm) in diameter.

LOC can be used to pierce virtually all materials. It has been used successfully on aluminum, cast iron, steel, and reinforced concrete.

Oxygen lancing of a 40 in. (1 m) diameter cast iron roll used in a paper mill is shown in Figure 14.26. Cutting oxygen was supplied at 80 to 120 psi (550 to 870 kPa). Holes pierced in the roll are shown in Figure 14.27. The variable

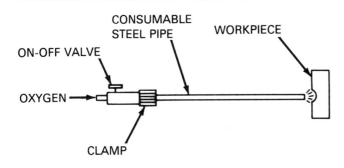

Figure 14.25–Schematic View of Oxygen Lance Cutting

angle bracket shown in Figure 14.27 was found to be helpful in guiding the lance.

A 2-1/2 in. diameter hole can be made in 24 in. of reinforced concrete at a rate of about 4 in./min. (100 mm/min.). This operation would use about 60 cu. ft (1.7 m^3) of oxygen.

The process has been used to open furnace tap holes and to remove solidified material from vessels, ladles, and

Figure 14.27–Holes Pierced in a Cast Iron Roll Using an Oxygen Lance

molds. It can be used to cut refractory brick, mortar, and slag.

The LOC process can be used underwater. The lance must be lighted before it is placed underwater, but then piercing proceeds essentially as in air. The violent bubbling action produced restricts visibility.

ARC-STARTED OXYGEN LANCING

A VARIATION OF the oxygen lancing process uses an arc to start the iron-oxygen reaction. This equipment uses tubes typically 18 in. (45 cm) long and either 0.25 or 0.375 in. (6.4 or 9.5 mm) in diameter. A 12-volt battery can be used as a power source, with the cutting tube connected to one battery terminal and a copper striker plate connected to the other. To start the burning operation, the operator starts the oxygen flow and draws the steel tube across the copper plate at a 45° angle. Sparking at the copper plate will ignite the tube. The burning rod can then be used for cutting, piercing, or beveling steel. It can also be used to remove pins, rivets, and bolts.

Figure 14.26–Severing a 40 in. (1 m) Diameter Cast Iron Roll by Multiple Hole Piercing Using an Oxygen Lance

SAFE PRACTICES

SAFE PRACTICES FOR the installation and operation of oxy-fuel gas systems for welding and cutting are given in American National Standard Z49.1, latest edition. These practices and those recommended by the equipment manufacturer should always be followed by the person operating the equipment.

Fumes are a potential health hazard. When the process is used in an enclosed or semi-enclosed area, exhaust ventilation should be provided and the operator should be equipped with a respirator. Noise from the operation may exceed safe levels in some circumstances. When necessary, ear protection should be provided for the operator. Fire is a potential hazard and combustible materials should be cleared away from the cutting area for a distance of at least 35 ft (11 m).

Appropriate protective clothing and equipment for any cutting operation will vary with the nature and location of the work to be performed. Some or all of the following may be required:

(1) Tinted goggles or face shields with filter lens; the recommended filter lenses for various cutting operations are

   (a) Light cutting, up to 1 in. (25 mm) — shade 3 or 4
   (b) Medium cutting, 1 to 6 in. (25 to 150 mm) — shade 4 or 5
   (c) Heavy cutting, over 6 in. (150 mm) — shade 5 or 6
(2) Flame resistant gloves
(3) Safety glasses
(4) Flame resistant jackets, coats, hoods, aprons, etc.
   (a) Woolen clothing preferably, not cotton or synthetic materials
   (b) Sleeves, collars, and pockets kept buttoned
   (c) Cuffs eliminated
(5) Hard hats
(6) Leggings and spats
(7) Safety shoes
(8) Flame extinguishing protective equipment
(9) Supplemental breathing equipment
(10) Other safety equipment

SUPPLEMENTARY READING LIST

Broco, Inc. "Underwater cutting process surfaces for new application." *Welding Journal* 68(6): July 1989.

Canonico, D. A. "Depth of heat-affected zone in thick pressure vessel plate due to flame cutting (technical note)." *Welding Journal* 47(9): 410s-419s; September 1968.

Couch, M. F. "Economic evaluation of fuel gases for oxy-fuel gas cutting in steel fabrication." *Welding Journal* 46 (10): 825-832; October 1967.

Fay, R. H. "Heat transfer from fuel gas flames." *Welding Journal* 46(8): 380s-383s; August 1967.

Hembree, J. D., Belfit, R. W., Reeves, H. A., and Baughman, J. P. "A new fuel gas - stabilized methylacetylene-propadiene." *Welding Journal* 42(5): 395-404; May 1963.

Ho, N. J., Lawrence, F. V. Jr., and Altstetter. "The fatigue resistance of plasma and oxygen cut steel." *Welding Journal* 60(11): 231s-236s; November 1981.

Jolly, W. D. et al. "Control factors for automation of oxy-fuel gas cutting." *Welding Journal* 64(7): 19-25; July 1985.

Kandel, C. "Underwater cutting and welding." *Welding Journal* 25(3): 209-212; March 1946.

Khuong-Huu, D., White, S. S., and Adams, C. M., Jr. "Combustion of liquid hydrocarbon fuels for oxygen cutting." *Welding Journal* 37(3): 101s-106s; March 1958.

Manhart, D. C. "CIM oxyfuel gas cutting." *Welding Journal* 66(1): 33; January 1987.

Moss, C. E. and Murray, W. E. "Gas welding, torch brazing, and oxygen cutting." *Welding Journal* 58(9): 37-46; September 1979.

Phelps, H. C. "Iron powder/oxypropane cutting of stainless steel." *Welding Journal* 56(4): 38-39; April 1977.

Slottman, G. V., and Roper, E. H. *Oxygen cutting*. New York: McGraw-Hill, 1951.

Worthington, J. C. "Analytical study of natural-gas oxygen cutting, theory and application." *Welding Journal* 39(3): 229-235; March 1960.

CHAPTER 15

ARC CUTTING AND GOUGING

PREPARED BY A COMMITTEE CONSISTING OF:

D. O'Hara, Co-Chairman
Thermal Dynamics

L. R. Soisson, Co-Chairman
Welding Consultants, Inc.

D. G. Anderson
L-Tec

R. P. Sullivan
L-Tec

P. I. Temple
Detroit Edison

WELDING HANDBOOK COMMITTEE MEMBER:
P. I. Temple
Detroit Edison

CHAPTER 15

ARC CUTTING AND GOUGING

INTRODUCTION

ARC CUTTING (AC) covers a group of thermal cutting processes that sever or remove metal by melting it with the heat of an arc between an electrode and the workpiece.

Thermal gouging is a thermal cutting process variation that removes metal by melting or burning the entire removed portion, to form a bevel or groove.

This definition covers a number of processes that are or have been used for cutting or gouging metals. This includes:

Plasma Arc Cutting	PAC
Air Carbon Arc Cutting	CAC-A
Shielded Metal Arc Cutting	SMAC

Gas Metal Arc Cutting	GMAC
Gas Tungsten Arc Cutting	GTAC
Oxygen Arc Cutting	AOC
Carbon Arc Cutting	CAC

Each of these processes offers the user certain advantages and disadvantages. When selecting a process, consideration must be given to costs relating to the volume of cutting, equipment requirements, and operator skill requirements. Plasma arc and air carbon arc cutting are addressed separately in this chapter due to their broad usage. The others are discussed in the final section of the chapter.

PLASMA ARC CUTTING

DESCRIPTION

THE PLASMA ARC cutting (PAC) process severs metal by using a constricted arc to melt a localized area of a workpiece, removing the molten material with a high-velocity jet of ionized gas issuing from the constricting orifice. The ionized gas is a plasma, hence the name of the process. Plasma arcs operate typically at temperatures of 18 000°-25 000°F (10 000°-14 000°C).

PAC was invented in the mid 1950's and became commercially successful shortly after its introduction to industry. The ability of the process to sever any electrically conductive material made it especially attractive for cutting nonferrous metals that could not be cut by the oxyfuel cutting (OFC) process. It was initially used for cutting stainless steel and aluminum. As the cutting process was developed, it was found that it had advantages over other cutting processes for cutting carbon steel as well as nonferrous metals. These advantages are summarized below.

When compared to mechanical cutting processes, the amount of force required to hold the workpiece in place and move the torch (or vice versa) is much lower with the "non-contact" plasma arc cutting process. Compared to OFC, the plasma cutting process operates at a much higher energy level, resulting in faster cutting speeds. In addition to its higher speed, PAC has the advantage of instant start-up without requiring preheat. Instantaneous starting is particularly advantageous for applications involving interrupted cutting, such as severing mesh.

There are notable limitations to PAC. When compared to most mechanical cutting means, PAC introduces hazards such as fire, electric shock, intense light, fumes and gases, and noise levels that may not be present with mechanical processes. It is also difficult to control PAC as precisely as some mechanical processes for close tolerance work. When compared to OFC, the PAC equipment tends to be more expensive, requires a fairly large amount of electric power, and introduces electrical shock hazards.

An arc plasma is a gas which has been heated by an arc to at least a partially ionized condition, enabling it to conduct an electric current. A plasma exists in any electric arc, but the term *plasma arc* is associated with torches which utilize a constricted arc. The principle feature which distinguishes plasma arc torches from other arc torches is that, for a given current and gas flow rate, the arc voltage is higher in the constricted arc torch.

The arc is constricted by passing it through an orifice downstream of the electrode. The basic terminology and the arrangement of the parts of a plasma cutting torch are shown in Figure 15.1. As plasma gas passes through the arc, it is heated rapidly to a high temperature, expands, and is accelerated as it passes through the constricting orifice toward the workpiece. The intensity and velocity of the plasma is determined by several variables including the type of gas, its pressure, the flow pattern, the electric current, the size and shape of the orifice, and the distance to the workpiece.

PAC circuitry is shown in Figure 15.2. The process operates on direct current, straight polarity. The orifice directs the super-heated plasma stream from the electrode toward the workpiece. When the arc melts the workpiece, the high-velocity jet blows away the molten metal to form the kerf or cut. The cutting arc attaches to or "transfers" to the workpiece, and is referred to as a *transferred arc*.

The different gases used for plasma arc cutting include nitrogen, argon, air, oxygen, and mixtures of nitrogen/hydrogen and argon/hydrogen.

PAC torches are available in various current ranges, generally categorized as low power [those operating at 30 amperes (A) or less], medium power level [30-100 (A)], and high power [from 100-1000 (A)]. Different power levels are ap-

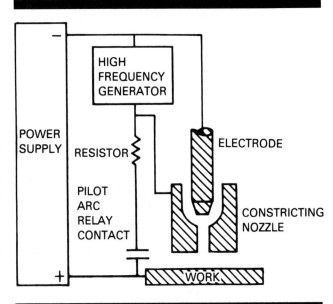

Figure 15.2–Basic Plasma Arc Cutting Circuitry

propriate for different applications, with the higher power levels being used for cutting thicker metal at higher speeds.

One of two starting methods is used to initiate the cutting arc: pilot arc starting or electrode (or tip) retract starting.

A pilot arc is an arc between the electrode and the torch tip. This arc is sometimes referred to as a *nontransferred arc* because it does not transfer or attach to the workpiece, as compared to the transferred arc which does. A pilot arc provides an electrically conductive path between the electrode in the torch and the workpiece so that the main cutting arc can be initiated.

The most common pilot arc starting technique is to strike a high-frequency spark between the electrode and the torch tip. A pilot arc is established across the resulting ionized path. When the torch is close enough to the workpiece so the plume or flame of the pilot arc touches the workpiece, an electrically conductive path from the electrode to the workpiece is established. The cutting arc will follow this path to the workpiece.

Retract starting torches have a moveable tip or electrode so that the tip and electrode can be momentarily shorted together and then separated or "retracted" to establish the cutting arc.

EQUIPMENT

Torches

THE PLASMA CUTTING process is used with either a hand-held torch or a mechanically-mounted torch. There are several types and sizes of each, depending on the thickness

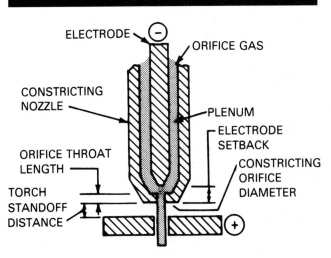

Figure 15.1–Plasma Arc Torch Terminology

of metal to be cut. Some torches can be dragged along in direct contact with the workpiece, while others require that a standoff be maintained between the tip of the torch and workpiece.

Mechanized torches can be mounted either on a tractor or on a computer-controlled cutting machine or robot. Usually a standoff is maintained between the torch tip and workpiece for best cut quality. The standoff distance must be maintained within fairly close tolerances to achieve uniform results. Some mechanized torches are equipped with an automatic standoff controlling device to maintain a fixed distance between the torch and workpiece. In other cases mechanical followers are used to accomplish this.

PAC torches operate at extremely high temperatures, and various parts of the torch must be considered to be consumable. The tip and electrode are the most vulnerable to wear during cutting, and cutting performance usually deteriorates as they wear. The timely replacement of consumable parts is required to achieve good quality cuts.

Modern plasma torches have self-aligning and self-adjusting consumable parts. As long as they are assembled in accordance with the manufacturer's instructions, the torch should require no further adjustment for proper operation.

Other torch parts such as shield cups, insulators, seals, etc. may also require periodic inspection and replacement if they are worn or damaged. Again, the manufacturer's instructions should be followed.

Power Supplies

PAC REQUIRES A constant-current or drooping volt-ampere characteristic, relatively high-voltage direct-current power supply. To achieve satisfactory arc starting performance, the open circuit voltage of the power supply is generally about twice the operating voltage of the torch. Operating voltages will range from 50 or 60 volts (V) to over 200 volts (V), so PAC power supplies will have open circuit voltages ranging from about 150 to over 400 volts.

There are several types of PAC power supplies, the simplest being the fixed output type which consists of a transformer and rectifier. The transformer of such a machine is wound with a "drooping" characteristic, so that the output voltage drops as the cutting current increases.

In some cases, several outputs are available from a single power supply through a switching arrangement. This switching arrangement can select between taps provided on the transformer or reactor of the power supply.

Variable output power supplies are also available. The most widely used units utilize a saturable reactor and current feedback circuit so that the output can be stabilized at the desired current level.

Other types of controls are available on plasma cutting power supplies, including electronic phase control and various types of "switch mode" power supplies. The switch mode power supplies utilize high-speed, high-current semiconductors to control the output. They can either regulate the output of a standard DC power supply, the so-called "chopper" power supply, or they can be incorporated in an inverter-type power supply. As new types of semiconductors become commercially available, it can be expected that improved versions of this type of power supply will appear. Switch mode supplies have the advantage of higher efficiency and smaller size, and are attractive for applications where portability and efficiency are important considerations.

Cutting Controls

PAC CONTROLS ARE relatively simple. Most manual torches are controlled by a trigger switch. This switch is pressed to start the cutting arc and released to stop the cut.

For mechanized cutting, starting and stopping the cutting arc can be manual by pushbutton or automatic by the motion controls of the system. Cutting controls can also sequence the entire operation, including varying the gas flow and power level if necessary.

Several interlocks are normally used with PAC systems. If the plasma torch is run without an adequate supply of gas, the torch may be damaged by internal arcing. For this reason, a gas-pressure switch is usually included in the circuit to ensure that adequate gas pressure is present before the torch can operate. This interlock will also shut down the torch in the event of a gas supply failure during cutting.

High-current torches are liquid cooled, and in this case an additional interlock is included in the coolant system. The interlock prevents operation of the torch without coolant flow and will shut the power supply off to prevent damage if coolant flow is interrupted during operation.

Motion Equipment

A VARIETY OF motion equipment is available for use with plasma cutting torches. This can range from straight-line tractors to numerically-controlled or direct computer-controlled machines with parts nesting capabilities, etc. Plasma cutting equipment can also be adapted to robotic actuators for cutting other than flat plates.

Environmental Controls

THE PLASMA CUTTING process is inherently a noisy and fume generating process. Several different devices and techniques are available to control and contain the hazards. One commonly used approach to reduce noise and fume emissions is to cut over a water table and surround the arc with a water shroud. This method requires a cutting table filled with water up to the work-supporting surface, a water shroud attachment to go around the torch, and a recirculating pump to draw water from the cutting table and pump it through the shroud. In this case, a relatively high (15 to 20 gpm [55 to 75 L/min]) water flow is used.

Another method, underwater plasma cutting, is also in common use. With this method, the working end of the torch and the plate to be cut are submerged under approximately 3 in. (75 mm) of water. While the torch is underwater but not cutting, a constant flow of compressed air is maintained through the torch to keep water out.

The primary requirements in water-table design are adequate strength for supporting the work, sufficient scrap capacity to hold the dross or slag resulting from cutting, procedure for removing the slag, and ability to maintain the water level in contact with the work. When the table is used for underwater cutting, it is necessary to provide a means of rapidly raising and lowering the water level. This can be accomplished by pumping the water in and out of a holding tank, or by displacing it with air in an enclosure under the surface of the water.

A cutting table for mechanized or hand plasma cutting is usually equipped with a down-draft exhaust system. This is vented to the outdoors in some cases, although fume removal or filtering devices may be required to meet air pollution regulations.

APPLICATIONS

THE FIRST COMMERCIAL application of plasma arc cutting was the mechanized cutting of manway holes on aluminum railroad tank cars. The process has since been used on a wide variety of aluminum applications. Table 15.1 shows typical conditions for mechanized cutting of aluminum plate.

Typical conditions for mechanized cutting of stainless steel plate are shown in Table 15.2.

Manual plasma arc cutting is widely used in automobile body repair for cutting high-strength low alloy steel. Instant starting and high travel speeds reduce heat input to the HSLA steel and help maintain its strength.

The chief application of mechanized plasma arc cutting to carbon steel is for thicknesses up to 1/2 in. (13 mm). The higher cost of plasma arc equipment compared to OFC equipment can be justified by its higher cutting speeds. Conditions for mechanized plasma arc cutting of carbon steel plate are given in Table 15.3.

The plasma process has been used for stack cutting of carbon steel, stainless steel, and aluminum. The plates to be stack cut should preferably be clamped together, but PAC can tolerate wider gaps between plates than OFC.

Table 15.1
Typical Conditions for Plasma Arc Cutting of Aluminum Alloys

Thickness		Speed		Orifice Diam*		Current	Power kW
in.	mm	in./min	mm/s	in.	mm	(dcsp), A	
1/4	6	300	127	1/8	3.2	300	60
1/2	13	200	86	1/8	3.2	250	50
1	25	90	38	5/32	4.0	400	80
2	51	20	9	5/32	4.0	400	80
3	76	15	6	3/16	4.8	450	90
4	102	12	5	3/16	4.8	450	90
6	152	8	3	1/4	6.4	750	170

* Plasma gas flow rates vary with orifice diameter and gas used from about 100 ft^3/h (47 L/min.) for a 1/8 in. (3.2 mm) orifice to about 250 ft^3/h (120 L/min.) for a 1/4 in. (6.4 mm) orifice. The gases used are nitrogen and argon with hydrogen additions from 0 to 35%. The equipment manufacturer should be consulted for each application.

Table 15.2
Typical Conditions for Plasma Arc Cutting of Stainless Steels

Thickness		Speed		Orifice Diam*		Current	Power kW
in.	mm	in./min	mm/s	in.	mm	(dcsp), A	
1/4	6	200	86	1/8	3.2	300	60
1/2	13	100	42	1/8	3.2	300	60
1	25	50	21	5/32	4.0	400	80
2	51	20	9	3/16	4.8	500	100
3	76	16	7	3/16	4.8	500	100
4	102	8	3	3/16	4.8	500	100

* Plasma gas flow rates vary with orifice diameter and gas used from about 100 ft^3/h (47 L/min.) for a 1/8 in. (3.2 mm) orifice to about 200 ft^3/h (94 L/min.) for a 3/16 in. (4.8 mm) orifice. The gases used are nitrogen and argon with hydrogen additions from 0 to 35%. The equipment manufacturer should be consulted for each application.

Table 15.3
Typical Conditions for Plasma Arc Cutting of Carbon Steel

Thickness		Speed		Orifice Diam*		Current	Power kW
in.	mm	in./min	mm/s	in.	mm	(dcsp), A	
1/4	6	200	86	1/8	3.2	275	55
1/2	13	100	42	1/8	3.2	275	55
1	25	50	21	5/32	4.0	425	85
2	51	25	11	3/16	4.8	550	110

* Plasma gas flow rates vary with orifice diameter and gas used from about 200 ft^3/h (94 L/min.) for a 1/8 in. (3.2 mm) orifice to about 300 ft^3/h (104 L/min.) for a 3/16 in. (4.8 mm) orifice. The gases used are usually compressed air, nitrogen with up to 10% hydrogen additions, or nitrogen with oxygen added downstream from the electrode (dual flow). The equipment manufacturer should be consulted for each application.

Plate and pipe edge beveling is done by using techniques similar to those for OFC. One to three PAC torches are used depending on the joint preparation required.

CUT QUALITY

FACTORS TO CONSIDER in evaluating the quality of a cut include surface smoothness, kerf width, kerf angle, dross adherence, and sharpness of the top edge. These factors are affected by the type of material being cut, the equipment being used, and the cutting conditions.

Plasma cuts in plates up to approximately 3 in. (75 mm) thick may have a surface smoothness very similar to that produced by oxyfuel gas cutting. Surface oxidation is almost nonexistent with mechanized equipment that uses water injection or water shielding. On thicker plates, low travel speeds produce a rougher surface and discoloration. On very thick stainless steel, 5 to 7 in. (125 to 180 mm) in thickness, the plasma arc process has little advantage over oxyfuel gas powder cutting.

Kerf widths of plasma arc cuts are 1-1/2 to 2 times the width of oxyfuel gas cuts in plates up to 2 in (50 mm) thick. For example, a typical kerf width in 1 in. (25 mm) stainless steel is approximately 3/16 in. (5 mm). Kerf width increases with plate thickness. A plasma cut in 7 in. (180 mm) stainless steel made at approximately 4 in./min. (3mm/s) has a kerf width of 1-1/8 in. (28 mm).

The plasma jet tends to remove more metal from the upper part of the kerf than from the lower part. This results in beveled cuts wider at the top than at the bottom. A typical included angle of a cut in 1 in. (25 mm) steel is four to six degrees. This bevel occurs on one side of the cut when orifice gas swirl is used. The bevel angle on both sides of the cut tends to increase with cutting speed.

Dross is the material that melts during cutting and adheres to the bottom edge of the cut face. With present mechanized equipment, dross-free cuts can be produced on aluminum and stainless steel up to approximately 3 in. (75 mm) thickness and on carbon steel up to approximately 1-1/2 in. (40 mm) thickness. With carbon steel,

selection of speed and current are more critical. Dross is usually present on thick materials.

Top edge rounding will result when excessive power is used to cut a given plate thickness or when the torch standoff distance is too large. It may also occur in high-speed cutting of materials less than 1/4 in. (6 mm) thick.

METALLURGICAL EFFECTS

DURING PAC, THE material at the cut surface is heated to its melting temperature and ejected by the force of the plasma jet. This produces a heat-affected zone along the cut surface, as with fusion welding operations. The heat not only alters the structure of the metal in this zone but also introduces internal tensile stresses from the rapid expansion, upsetting, and contraction of the metal at the cut surface.

The depth to which the arc heat will penetrate the workpiece is inversely proportional to cutting speed. The heat-affected zone on the cut face of a 1 in. (25 mm) thick stainless steel plate severed at 50 in./min. (21 mm/s) is 0.003 to 0.005 in. (0.08 to 0.13 mm) deep. This measurement was determined from microscopic examination of the grain structure at the cut edge of a plate.

Because of the high cutting speed on stainless steel and the quenching effect of the base plate, the cut face passes through the critical 1200°F (650°C) temperature very rapidly. Thus, there is virtually no chance for chromium carbide to precipitate along the grain boundaries, so corrosion resistance is maintained. Measurements of the magnetic properties of Type 304 stainless steel made on base metal and on plasma arc cut samples indicate that magnetic permeability is unaffected by arc cutting.

Metallographic examination of cuts in aluminum plates indicates that the heat-affected zones in aluminum are deeper than those in stainless steel plate of the same thickness. This results from the higher thermal conductivity of aluminum. Microhardness surveys indicate that the heat effect penetrates about 3/16 in. (5 mm) into a 1 in. (25 mm) thick plate. Age hardenable aluminum alloys of the 2000 and 7000 series are crack-sensitive at the cut surface. Cracking appears to result when a grain boundary eutectic film melts and separates under stress. Machining to remove the cracks may be necessary on edges that will not be welded.

Hardening will occur in the heat-affected zone of a plasma arc cut in high carbon steel if the cooling rate is very high. The degree of hardening can be reduced by preheating the workpiece to reduce the cooling rate at the cut face.

Various metallurgical effects may occur when long, narrow, or tapered parts, or outside corners are cut. The heat generated during a preceding cut may reach and adversely affect the quality of a following cut.

PLASMA ARC GOUGING

Process Description

PLASMA ARC GOUGING is an adaptation of the plasma cutting process. For gouging, arc constriction is reduced, resulting in a lower arc stream velocity. The temperature of the arc and the velocity of the gas stream are used to melt and expel metal in a similar manner to other gouging processes. A major difference compared to other gouging processes is that the gouge is bright and clean, particularly on nonferrous material such as aluminum and stainless steel. Virtually no post-cleaning is required when the plasma gouged surface is to be welded. A plasma arc gouging operation on stainless steel plate is shown in Figure 15.3.

Equipment

THE BASIC EQUIPMENT for plasma gouging is the same as for plasma cutting. Most plasma cutting equipment can be used for plasma gouging providing that the volt-ampere output curve of the power source is steep enough and the

Figure 15.3–Plasma Arc Gouging of Stainless Steel Plate

voltage high enough to sustain the long arc used for plasma gouging.

The torch utilizes a gouging tip which is designed to give a softer, wider arc and proper stream velocity. The torch used is the same as a plasma cutting torch and may be either single- or dual-gas flow and air or water cooled.

Gases

THE RECOMMENDED PLASMA gas for all gouging is argon plus 35-40 percent hydrogen. The gas can be supplied from cylinders or prepared using a gas-mixing device. Helium may be substituted for the argon-hydrogen mixture, but the resulting gouge will be shallower. The secondary or cooling gas, when used, is argon, nitrogen, or air. Selection is based on brightness of gouge desired, fume generation, and cost.

Air is sometimes used for the plasma gas on air operating systems but is generally limited to carbon steel gouging. Most manual air cutting systems are limited to 100 A output and this restricts the size and speed of plasma gouging.

Operating Procedure

THE TECHNIQUE FOR plasma gouging is essentially the same as for other gouging methods. The torch is angled approximately 30 degrees from the horizontal. Gouge depth is determined by speed of travel. It is important not to attempt removal of too much metal in a single pass.

Applications

PLASMA GOUGING CAN be used on all metals. It is particularly effective on aluminum or stainless steel, where the gouges produced are clean and devoid of any carbon contamination.

SAFETY

THE POTENTIAL HAZARDS of plasma arc cutting and gouging are similar to those of most arc welding and cutting processes. The obvious hazards such as fires, burns, etc. that are related to the heat of the arc are discussed at the end of this section. Emphasis here is placed on the less obvious hazard categories of electrical shock, fume and gas generation, noise, and radiation.

The equipment should not be operated until the manufacturer's instructions have been read and understood. In addition, other potential physical hazards such as those due to the high-pressure gas and water systems must be considered.

Some cutting gas mixtures contain hydrogen. Inadvertent release of such gases can result in explosion and fire hazards. Do not operate equipment when gas leaks are suspected. The manufacturer should be contacted if there is a question about the equipment operation with certain gases.

Electrical

VOLTAGES USED IN plasma cutting equipment range from 150 to 400 V direct current. Electric shock can be fatal. The equipment must be properly grounded and connected as recommended by the manufacturer.

Emergency first aid should be available. Prompt, trained emergency response may reduce the extent of injury due to accidental electrical shock. Only trained personnel should be permitted to operate or maintain the equipment. In addition to the manufacturer's instructions, the following may be of assistance:

(1) ANSI C-2, the National Electrical Safety Code
(2) ANSI Z49.1, Safety in Welding and Cutting
(3) 29CFR1910, OSHA General Industry Standards and NFPA Standard 51B, Fire Prevention in the Use of Cutting and Welding Processes

Some additional safety items are listed below:

(1) Keep all electrical circuits dry. Moisture may provide an unexpected path for current flow. Equipment cabinets that contain water and gas lines as well as electrical circuits should be checked periodically for leaks.
(2) All electrical connections should be kept mechanically tight. Poor electrical connections can generate heat and start fires.
(3) High-voltage cable should be used. Make sure cables and wires are kept in good repair. Consult the manufacturer's instructions for proper cable and wire sizes.
(4) Do not touch live circuits. Keep equipment access doors closed.
(5) The risk of electrical shock is probably the greatest when replacing used torch parts. Operators must make sure that the primary power to the power supplies and the power to the control circuitry is disconnected when replacing torch parts.
(6) Operators and maintenance personnel should be aware that plasma arc cutting equipment, due to the higher voltages, presents a greater hazard than conventional welding equipment.

Fumes And Gases

PAC PRODUCES FUMES and gases which can harm your health. The composition and rate of generation of fumes and gases depend on many factors including arc current, cutting speed, material being cut, and gases used. The fume and gas by-products will usually consist of the oxides of the metal being cut, ozone, and oxides of nitrogen.

These fumes must be removed from the work area or eliminated at the source by using an exhaust system. Codes may require that the exhaust be filtered before being vented to the atmosphere.

Several alternative fume removal systems are available for mechanized cutting. One system consists of two parts, a cutting table which maintains a bed of water that contacts the bottom surface of the workpiece, and an annular nozzle which generates a water shroud around the arc.

Another system also uses a water bed, but instead of having the level of the water contact only the bottom surface of the workpiece, the water totally submerges the workpiece. This system is referred to as *underwater cutting* and does not require the use of a water-shroud nozzle. It does require that the level of the water be periodically lowered for loading and unloading the plate, positioning of the torch and plate, etc. Since the operator cannot see the plate during cutting with this system, it is intended for use with numerically controlled systems.

There is a possibility of hydrogen detonation beneath the workpiece when cutting aluminum or magnesium plate on a water table. The actual cause of such detonations is not fully understood, but they are believed to be due to hydrogen released by the interaction of molten aluminum or magnesium and water. The hydrogen can accumulate in pockets under the workpiece and ignite when the cutting arc is near the pocket. Before cutting aluminum or magnesium on a water table, the equipment manufacturer should be contacted for recommended practices.

Noise

THE AMOUNT OF noise generated by a PAC torch operated in the open depends primarily on the cutting current. A torch operating at 400 A typically generates approximately 100 dBA measured at about six feet. At 750 A the noise level is about 110 dBA. Much of the noise is in the frequency range of 5000—20 000 HZ. Such noise levels can damage your hearing. Hearing protection should be worn when the noise level exceeds specified limits. These values may vary locally and are specified by OSHA for most industrial environments.

The water-shroud technique described earlier is commonly used to reduce noise in mechanized cutting applications. The water effectively acts as a sound absorbing enclosure around the torch nozzle. The water directly below the plate keeps noise from coming through the kerf opening. Noise reduction is typically about 20 dBA. This reduction will usually be sufficient to bring the operation within OSHA limits.

The water-shroud technique should not be confused with water injection or water shielding, since neither of those process variations use sufficient water to significantly reduce noise.

Underwater PAC provides greater noise reduction than the water shroud because the nozzle end of the torch and the arc are totally submerged.

Radiation

THE PLASMA ARC emits intense visible and invisible (ultraviolet and infrared) radiation. In addition to potential harm to the eyes and skin, this radiation may produce ozone, oxides of nitrogen, or other toxic fumes in the surrounding atmosphere.

It is necessary to wear eye and skin protection when exposure to radiation is unavoidable. The recommended eye protection is shown in Table 15.4. The likelihood of radiation exposure may be reduced by the use of mechanical barriers such as walls and welding curtains. The water shroud will also act as a light-absorbing shield, especially when dye is added to the water in the table. When the use of dye is contemplated, contact the equipment manufacturer for information on the type and concentration to use. It is advisable to provide operator eye protection, even when using these dyes, because of the possibility of unexpected interruption of water flow through the water shroud.

Table 15.4
Recommended Eye Protection for Plasma Arc Cutting (Source: ANSI/AWS C5.2-83, Recommended Practices for Plasma Arc Cutting)

Cutting Current Amperes	Lens Shade Number
Up to 300	9
300–400	12
400–800	14

Underwater plasma cutting reduces the amount of radiation because of the greater depth of the water. Additional dye is not generally required.

AIR CARBON ARC CUTTING

DESCRIPTION

AIR CARBON ARC cutting (CAC-A) is a carbon arc cutting process variation that removes molten metal with a jet of air. In the air carbon arc cutting process, the intense heat of the arc between a carbon-graphite electrode and the workpiece melts a portion of the workpiece. Simultaneously, a jet of air of sufficient volume and velocity is passed through the arc to blow away the molten material. The exposed solid metal is then melted by the heat of arc, and the sequence continues. The process is used for severing and gouging.

Air carbon arc cutting does not depend on oxidation to maintain the cut, so it is capable of cutting metals that OFC will not cut. The process is used successfully on carbon steel, stainless steel, many copper alloys, and cast irons. The melting rate is a function of current. The metal removal rate is dependent upon the melting rate and the efficiency of the air jet in removing the molten metal. The air must be capable of blowing the molten metal out and clear of the arc region before it can resolidify. The process is shown schematically in Figure 15.4.

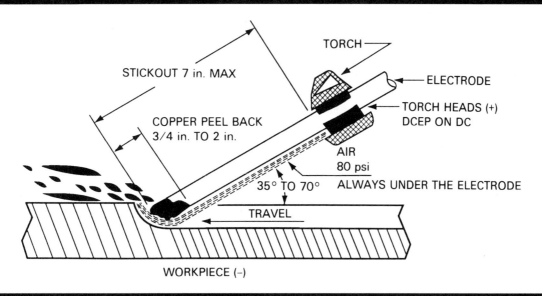

Figure 15.4–Typical Operating Procedures for Air Carbon Arc Gouging

Air carbon arc cutting was developed in the 1940's as an extension of carbon arc cutting (CAC). CAC must be done in the vertical or overhead position to permit gravity to remove the melted metal. The air CAC version enables the operator to remove metal in all positions.

First attempts at achieving an air-blast version of CAC involved two operators. The first held a CAC torch to melt the metal and the second directed a nozzle with an air jet at the molten pool. A single torch combining the air blast with the carbon electrode holder evolved shortly as the forerunner to today's improved CAC-A torches. The first commercial CAC-A torch was introduced in 1948.

EQUIPMENT AND CONSUMABLES

THE PROCESS REQUIRES an electrode holder, cutting electrodes, a power source, and an air supply. For mechanized cutting, a control and carriage are also required. Figure 15.5 shows a typical arrangement for CAC-A equipment.

Cutting Torches

MANUAL ELECTRODE HOLDERS for CAC-A are similar to conventional heavy-duty shielded metal arc welding holders, as shown in Figure 15.6. The electrode is held in a rotatable head which contains one or more air orifices, so that, regardless of the angle at which the electrode is set

with respect to the cutting torch, the air jet remains in alignment with the electrode. A valve is provided for turning the air on and off. A cross section diagram of a manual CAC-A torch is shown in Figure 15.7.

Torches available range from light-duty farm and hobby shop sizes to extra heavy-duty foundry torches. Following is a guide for torch usage:

Light Duty. These are recommended for small shops, farms, and maintenance operations with limited air supply. Maximum current is about 450 amps dc.

General Purpose. These torches are for general purpose applications in shipyards, fabrication shops, and general maintenance. Limited to a maximum of 1000 amps.

Heavy Duty. These torches are for general foundry work, padwashing, and cutoff. High amperage work in shipyards and fabrication shops. Limited to 1600 amps with air-cooled cables and 2000 amps with water-cooled cables.

Mechanized. Mechanized electrode holders are used for edge preparation and high production applications. They are used with 5/16 through 3/4 in. (8 through 19 mm) jointed carbons. Typical automatic CAC-A equipment is shown in Figure 15.8.

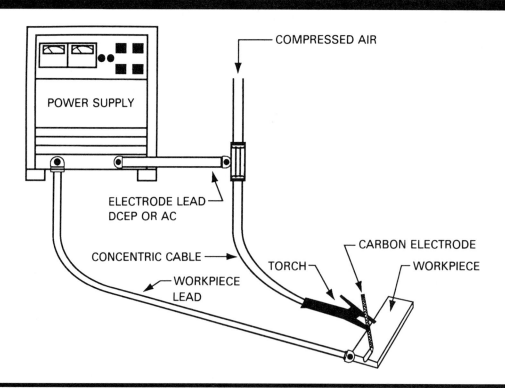

COMPRESSED AIR

POWER SUPPLY

ELECTRODE LEAD DCEP OR AC

CONCENTRIC CABLE

WORKPIECE LEAD

TORCH

CARBON ELECTRODE

WORKPIECE

Figure 15.5–Typical Air Carbon Arc Gouging Equipment

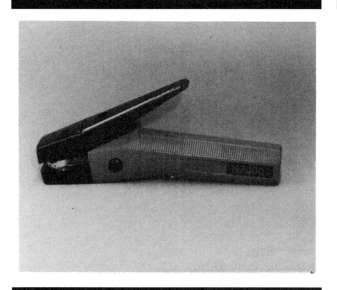

15.6–Typical 400 Ampere Manual Air Carbon Arc Gouging Electrode Holder

15.8–Typical Automatic CAC-A Equipment

Controls. There are three types of controls for mechanized CAC-A. All systems are capable of making grooves of consistent depth to a tolerance of ± 0.025 in. (0.6 mm). These units are used where high-quality, high-production gouges are desired. They are as follows:

(1) An amperage-controlled type which maintains the arc current by amperage signals through solid-state controls. This type of system controls the electrode feed speed, which maintains the preset amperage and can be operated with constant-voltage power sources only.

(2) A voltage-controlled type which maintains arc length by voltage signals through solid-state electronic

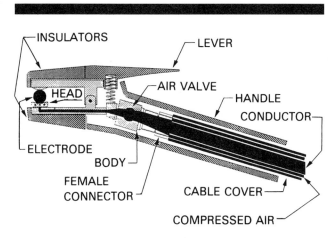

15.7–Schematic Cross Section of an Air Carbon Arc Gouging Torch

controls. This type controls the arc length determined by the preset voltage, and can be used with constant-current power supplies only.

(3) A dual system which can be adjusted for either amperage control or voltage control by means of a selector switch located in the control.

Electrodes

THREE TYPES OF electrodes are used for CAC-A: dc copper coated, dc plain, and ac copper coated. Round electrodes are the shape most frequently used. Flat and half-round electrodes are available to produce rectangular grooves.

DC Copper Coated Electrodes. This type is most widely used because of its comparatively long electrode life, stable arc characteristics, and groove uniformity. These electrodes are made from a special mixture of carbon and graphite with a suitable binder. The mixture is extruded and baked to produce dense, homogeneous graphite electrodes of low electrical resistance. The electrodes are then coated with a controlled thickness of copper. These electrodes are available in diameters ranging from 1/8 to 3/4 in. (3.2 to 19.1mm).

Jointed electrodes are available for operation without stub loss. They are furnished with a female socket and a matching male tenon, and are available in diameters ranging from 5/16 to 1 in. (8 to 25.4 mm).

DC Plain Electrodes. Of limited use, these electrodes have no copper coating. During cutting they are consumed more rapidly than the coated electrodes. Plain electrodes are available in sizes ranging from 1/8 to 1 in. (3.2 to

25.4 mm) diameter, but their principal use is in diameters less than 3/8 in. (9.5 mm).

AC Coated Electrodes. These electrodes are made from a mixture of carbon and graphite with rare-earth materials added to provide arc stabilization for cutting with an alternating current. These electrodes, coated with a controlled thickness of copper, are available in diameters ranging from 3/16 to 1/2 in. (4.8 to 12.7 mm).

Power Sources

MOST STANDARD WELDING power sources can be used for the air carbon arc cutting process. The open circuit voltage should be sufficiently higher than the required arc voltage to allow for the voltage drop in the circuit. The arc voltages used in air carbon arc gouging and cutting range from 35 to 55 V. An open circuit voltage of at least 60 V is adequate. The actual arc voltage in air carbon arc gouging and cutting is governed to a large extent by the size of the electrode and the application. Recommended power sources are given in Table 15.5.

The power supply manufacturer should be consulted concerning its use for CAC-A because some types of power supplies that are satisfactory for use with welding cannot be used for CAC-A.

Electrical leads in the cutting circuit should be standard welding cables recommended for arc welding. Cable size is determined by the maximum cutting current that will be used.

Air Supply

COMPRESSED AIR WITH pressure ranging from 80 to 100 psi (560 to 700 kPa) is normally required for air carbon arc gouging. Light-duty electrode holders allow for gouging with as little as 40 psi (280 kPa) at 3 ft^3/min (8.5 liter/min). Compressed nitrogen or inert gas may be used where compressed air is not available. Oxygen should not be used in a CAC-A electrode holder.

The air stream must be of sufficient volume and velocity to properly remove the melted slag from the kerf. The orifices in air carbon arc torches are designed to provide an adequate air stream for gouging. However, poor-quality gouging may result if the air pressure falls below the minimum specified by the torch manufacturer, or if the volume of air is restricted by hoses or fittings that are too small.

While gouges or cuts made with insufficient air may not always look particularly bad, they may be loaded with slag and carbon deposits. For this reason, it is important that the air pressure be at or above the minimum specified for the type of torch being used. The inside diameter of all hoses and fittings must be large enough to allow the intended volume of air to reach the electrode holder.

Hoses and fittings with an inside diameter of 1/4 in. (6.4 mm) are sufficient for light-duty holders. A minimum I.D. of 3/8 in. (9.5 mm) is required for general purpose and heavy-duty electrode holders. Automatic gouging holders should be equipped with hoses and fittings with a minimum inside diameter of 1/2 in. (12.7 mm).

APPLICATIONS

THE AIR CARBON arc cutting process can be used to sever and gouge carbon, low alloy, and stainless steels; cast iron; and alloys of aluminum, magnesium, copper, and nickel. Gouging may be used to prepare plate and pipe edges for welding. Two edges may be butted together and a U-groove gouged along the joint, as shown in Figure 15.9. The root of a weld may be gouged out to sound metal before completing the weld on the second side. Similarly, defective weld metal may be gouged out for repair. Another application is the removal of old surfacing material before a part is resurfaced.

OPERATING PROCEDURES

AIR CARBON ARC cutting electrodes are designed to operate with ac or dc, or both, depending on the material being cut. Table 15.6 gives the recommended electrodes and types of current for cutting several common alloys.

Table 15.5
Power Sources for Air Carbon Arc Cutting and Gouging

Type of Current	Type of Power Source	Remarks
dc	Constant current motorgenerator, rectifier, or resistor grid unit	Recommended for all electrode sizes.
dc	Constant potential motorgenerator or rectifier	Recommended for 6.4mm (1/4 in.) and larger diameter electrodes only. May cause carbon deposit with small electrodes. Not suitable for automatic torches with voltage control.
ac	Constant current transformer	Recommended for ac electrodes only.
ac or dc	Constant current	DC supplied from three phase transformer-rectifier supplies is satisfactory, but dc from single phase sources gives unsatisfactory arc characrictics. AC output from ac/dc units is satisfactory provided ac electrodes are used.

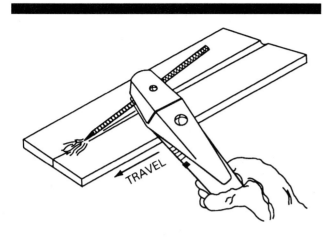

15.9–Manual Air Carbon Arc Gouging Operation in the Flat Position

Current ranges for commonly used CAC-A electrodes are shown in Table 15.7. The actual current used for a given electrode size will depend on the operating conditions such as the material being cut, type of cut, cutting speed, cutting position, and required cut quality. Recommendations of the manufacturer should be followed for the operation and maintenance of the equipment and consumables.

Gouging

THE ELECTRODE SHOULD be gripped, as shown in Figure 15.4, so that a maximum of 7 in. (178 mm) extends from the cutting torch. For nonferrous materials, this extension should be reduced to 3 in. (76.5 mm).

The air jet should be turned on before striking the arc, and the cutting torch should be held as shown in Figure 15.9. The electrode slopes back from the direction of travel with the air jet behind the electrode. Under proper operating conditions, the air jet will sweep beneath the electrode end and remove all molten metal. The arc may be struck by lightly touching the electrode to the workpiece. The electrode should not be drawn back once the arc is struck. The gouging technique is different from that of arc welding because metal is removed instead of deposited. A short arc should be maintained by progressing in the direction of the cut fast enough to keep up with metal removal. The steadiness of progression controls the smoothness of the resulting cut surface.

When using jointed carbon electrodes, it is important to strike the arc with the open or blunt end of the electrode. The reason for this becomes apparent when the electrode has been almost completely consumed and is approaching the jointed section. If the arc had been struck on the tapered end of the electrode, the jointed section would consist of a tapered end surrounded by a loose red-hot sleeve of carbon. This hot sleeve tends to be ejected violently from the gouging arc and, similar to weld spatter, can cause burns or set fire to combustibles.

When the arc is struck with the open end of the electrode, and the electrode is consumed to the jointed section, the sleeve is part of the incoming electrode and is restrained from violent ejection.

When gouging a workpiece in the vertical position, gouging should be done downhill, to let gravity assist in removing the molten metal. Gouging in the horizontal position may be done either to the right or to the left, but always in the forehand direction.

In gouging to the left, the cutting torch should be held as shown in Figure 15.9. In gouging to the right, the cutting torch will be reversed to locate the air jet behind the electrode. When gouging overhead, the electrode and torch should be held at an angle that will prevent molten metal from falling on the operator.

Table 15.6
Electrode and Current Recommendations for Air Carbon Arc Cutting of Several Alloys

Alloy	Electrode Type	Current Type	Remarks
Carbon, low alloy, and stainless steels	dc	dcrp	
	ac	ac	Only 50% as efficient as dcrp
Cast irons	ac	dcsp	At middle of electrode current range
	ac	ac	
	dc	dcrp	At maximum current only
Copper alloys:			
copper 60% or less	dc	dcrp	At maximum current
copper over 60%	ac	ac	
Nickel alloys	ac	ac	
	ac	dcsp	
Magnesium alloys	dc	dcrp	Before welding, surface must be cleaned.
	dc	dcrp	Electrode extension should not exceed 4 in. (100 mm).
Aluminum alloys			Before welding, surface must be cleaned.

Table 15.7
Suggested Current Ranges for the Commonly Used CAC-A Electrode Types and Sizes

Electrode Diameter		DC Electrode with DCEP, A		AC Electrode with ac, A		AC Electrode with DCEN, A	
in.	mm	min	max	min	max	min	max
5/32	4.0	90	150	--	--	--	--
3/16	4.8	150	200	150	200	150	180
1/4	6.4	200	400	200	300	200	250
5/16	7.9	250	450	--	--	--	--
3/8	9.5	350	600	300	500	300	400
1/2	12.7	600	1000	400	600	400	500
5/8	15.9	800	1200	--	--	--	--
3/4	19.1	1200	1600	--	--	--	--
1	25.4	1800	2200	--	--	--	--

The depth of the groove produced is controlled by the travel speed. Slow travel speeds will produce a deep groove, while fast speeds will produce a shallow groove. Grooves up to 1 in. (25 mm) deep may be made. However, the deeper the groove, the more experience required on the part of the operator.

The width of the groove is determined by the size of the electrode used and is usually about 1/8 in. (3.2 mm) wider than the electrode diameter. A wider groove may be made by oscillating the electrode with a circular or weaving motion.

When gouging, a push angle of 35 degrees from the surface of the workpiece is used for most applications. A steady rest is recommended in gouging to ensure a smoothly gouged surface. This is particularly advantageous for use in the overhead position. Proper travel speed depends on the size of the electrode, type of base metal, cutting amperage, and air pressure. An indication of proper speed and good gouge quality is a smooth hissing sound in the arc.

Severing

IN GENERAL, THE technique for severing is the same as for gouging, except that the electrode is held at a steeper angle—between 70 and 80 degrees to the surface of the workpiece.

For cutting thick nonferrous metals, the electrode should be held perpendicular to the workpiece surface, with the air jet in front of the electrode in the direction of travel. With the electrode in this position, the metal may then be severed by moving the arc up and down through the metal with a sawing motion.

Washing

IN USING THE air carbon arc cutting process for removing metal from large areas, such as the removal of surfacing metal or riser pads on castings, the proper position of the electrode is shown in Figure 15.10. The electrode should be oscillated from side-to-side while pushing forward at the depth desired. In pad washing operations, an angle of 15 to 70 degrees to the workpiece surface is used. The 15 degree angle is used for light finishing passes, while the steeper angles allow deeper rough cutting to be done with greater ease.

Cutting torches with fixed angle heads that hold the electrode at the correct angle are particularly well-suited for this application. With other types of torches, care should be taken to keep the air behind the electrode. The steadiness of the cutter's hand determines the smoothness of the surface produced.

METALLURGICAL EFFECTS

TO AVOID DIFFICULTIES with carburized metal, users of the air carbon arc cutting process should be aware of the metallurgical events that occur during gouging and cutting. When the carbon electrode is positive (reverse polarity), the current flow carries ionized carbon atoms from the electrode to the base metal. The free carbon particles are rapidly absorbed by the melted base metal. Since this absorption cannot be avoided, it is important that all carburized molten metal be removed from the kerf, preferably by the air jet.

When the air carbon arc cutting process is used under improper conditions, the carburized molten metal left on the surface can usually be recognized by its dull, gray-black color. This is in contrast to the bright blue color of the properly made groove. Inadequate air flow may leave small pools of carburized metal in the bottom of the groove. Irregular electrode travel, particularly in a manual operation, will produce ripples in the groove wall that tend to trap the carburized metal. Finally, an improper electrode angle may cause small beads of carburized metal to remain along the edge of the groove.

The effect of carburized metal that remains in the kerf or groove through a subsequent welding operation depends on

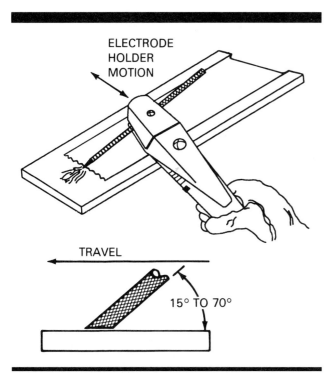

15.10–Pad Washing Techique with Air Carbon Arc Electrode Holder. Electrode to Work Angle is 15° to 70° with the Steeper Angle Being Used on Cast Iron

many factors including the amount of carburized metal present, the welding process to be employed, the kind of base metal, and the weld quality required. Although it may seem that filler metal deposited during welding would dissolve small pools or beads of carburized metal, experience with steel base metals shows that traces of metal containing approximately 1 percent carbon may remain along the weld bond line. Carbon pickup in the weld metal becomes significant with demands for increasing weld strength and toughness. Increased carbon content can decrease weld toughness, especially in quenched and tempered steels.

There is no evidence that the copper from copper-coated electrodes is transferred to the cut surface in the base metal.

Carburized metal on the cut surface may be removed by grinding, but it is much more efficient to conduct air carbon arc gouging and cutting properly within prescribed conditions, which will completely avoid the retention of undesirable metal.

Studies have been conducted on stainless steel to determine whether air carbon arc gouging, carried out in the prescribed manner, would adversely affect corrosion resistance. Corrosion rates typical for Type 304L stainless steel were obtained, and the studies showed no significant difference in the corrosion rates of welds prepared by CAC-A and those prepared by grinding. Had any appreciable carbon absorption occurred, the corrosion rates for welds

prepared by CAC-A would have been significantly higher. However, surfaces prepared using CAC-A may be more susceptible to stress corrosion cracking depending on the service environment. If a question exists, the surfaces should be mechanically dressed following CAC-A.

Compared to oxyfuel gas cutting, CAC-A is a lower heat input process. For that reason, a workpiece gouged or cut by CAC-A will have less distortion than one prepared by OFC.

SAFE PRACTICES

THE GENERAL SUBJECTS of safety and safe practices in welding and thermal cutting processes, such as air carbon arc, are covered in ANSI Z49.1, *Safety in Welding and Cutting*, and NFPA 51B, *Fire Prevention in Use of Welding and Cutting Processes*. Air carbon arc cutters and their supervisors should be familiar with the practices discussed in these documents.

Furthermore, there are other potential hazard areas in arc cutting and gouging. Fumes, gases, noise, and radiant energy warrant additional consideration. Those areas associated with air carbon arc cutting and and gouging are discussed in this section.

Gases

THE MAJOR TOXIC gases which may be produced during arc cutting are ozone, nitrogen dioxide, and carbon monoxide. Phosgene gas could be present as a result of thermal or ultraviolet decomposition of chlorinated hydrocarbon cleaning agents or suspension agents used in some aerosol anti-spatter agents or paints. Degreasing or other operations involving chlorinated hydrocarbons should be located so that vapors from these operations are not exposed to radiation from the arc.

Ozone

THE ULTRAVIOLET LIGHT emitted from the arc acts on the oxygen in the surrounding atmosphere to produce ozone. The amount of ozone produced will depend upon the intensity of the ultraviolet energy, the humidity, the amount of screening afforded by the fume, and other factors. The ozone concentration will generally be increased with an increase in current and when aluminum is gouged. The concentration can be controlled by natural ventilation, local exhaust ventilation or by respiratory protective equipment described in ANSI Z49.1.

Nitrogen Dioxide

TESTS HAVE SHOWN that high concentrations of nitrogen dioxide are found only close to the arc. Natural ventilation reduces these concentrations quickly to safe levels in the cutter's breathing zone, so long as the cutter keeps his or her head out of the cutting fumes.

Metal Fumes

THE METAL FUMES generated by the CAC-A process can be controlled by natural ventilation, local exhaust ventilation, or by respiratory protective equipment described in ANSI Z49.1. The method of ventilation required to keep the level of toxic substances in the cutters breathing zone within acceptable concentrations is directly dependent upon a number of factors, among which are the metal being cut, the size of the work area and the degree of confinement or obstruction to normal air movement where the cutting is taking place. Each operation should be evaluated on an individual basis in order to determine what will be required.

Acceptable levels of toxic substances associated with cutting and designated as time-weighted average threshold limit values (TLVs) and ceiling values have been established by the American Conference of Governmental Hygienists and by the Occupational Safety and Health Administration. Compliance with these levels can be tested by sampling the atmosphere under the cutter's helmet or in the immediate vicinity of the cutter's breathing zone. Sampling should be in accordance with ANSI/AWS F1.1 *Method for Sampling Airborne Particulates Generated by Welding and Allied Processes.*

Fire Prevention

CAC-A REQUIRES SPECIAL fire prevention precautions because of the metal removal process. All combustibles within 35 ft (11 m) of the work area should be removed. Protection such as metal screens should be placed in the line of hot metal ejected by the compressed air stream if ample room for dissipation is not available.

Noise

NOISE FROM CAC-A gouging may exceed safe levels. When necessary, ear protection should be provided for the operator.

Radiant Energy

ANY PERSON WITHIN the immediate vicinity of the cutting arc should have adequate eye and skin protection from radiation produced by the cutting arc. The filter shade recommended for CAC-A is shade twelve. Leather or wool clothing that is dark in color is recommended to reduce reflection which could cause ultraviolet burns to the neck and face inside the helmet.

OTHER ARC CUTTING PROCESSES

THIS CONCLUDING SECTION of the chapter provides a brief explanation of five remaining processes. In general, these are not widely used because of economic considerations. However, the reader should be aware of them because they can be used when other processes are not available. Consult the supplementary reading list for more information.

SHIELDED METAL ARC CUTTING

Principles of Operation

SHIELDED METAL ARC cutting (SMAC) is an arc cutting process that uses a covered electrode. A constant-current power source operating on direct current straight polarity (dcen) is preferred. The principal function of the electrode covering in cutting is to serve as electrical insulation, permitting insertion of the electrode into the gap of the cut without short-circuiting the sides of the electrode, and to act as an arc stabilizer, thereby concentrating and intensifying the arc action. Effectiveness of this procedure in cutting heavy thicknesses is a function of manipulation of the

electrode. E6010, E6012, and E6020 electrode types are usually employed, but cutting can be achieved using virtually any SMAW electrode. Electrodes with coverings specially made for cutting are also available.

Equipment

ALTHOUGH A DC constant welding machine is preferred for SMAC, an ac constant-current power source may also be used. For shielded metal arc cutting in air, heavy-duty electrode holders should be used with 3/16 in. diameter and larger electrodes. For SMAC under water, specially constructed, fully insulated electrode holders are mandatory. A straight polarity power source must be used to protect the holder and the metal parts of the diver's outfit from electrolytic corrosion.

Applications

SMAC HAS BEEN used for cutting risers and gates in nonferrous foundries and to cut nonferrous scrap for remelting. The workpiece should be positioned so that gravity assists in removing the molten metal. Generally, the process does not provide satisfactory edge preparation for welding without considerable cleanup by chipping or grinding.

OXYGEN ARC CUTTING

Principles of Operation

OXYGEN ARC CUTTING (AOC) is an oxygen cutting process that uses an arc between the workpiece and a consumable tubular electrode through which oxygen is directed to the workpiece. Mild steel is cut by using the arc to raise the temperature of the material to its kindling point in the presence of oxygen. The combustion reaction that occurs is self-sustaining, liberating sufficient heat to maintain the kindling temperature on all sides of the cut. The necessary preheat at the start of cutting is provided by the electric arc. A schematic illustration of the process is shown in Figure 15.11.

For oxidation-resistant metals, the cutting mechanism is more of a melting action. In these instances, the covering on the electrode provides a flux that helps the molten metal flow from the cut.

Metallurgical Effects

THE OXYGEN ARC method of cutting produces metallurgical effects in the heat-affected zone comparable to those that occur in shielded metal arc welding. The power input approaches that of shielded metal arc welding, but the heat penetration is generally not as deep in AOC because of the faster speed of travel. This produces a somewhat more pronounced quench effect. Metals that do not require a postheat treatment after welding may be severed by this process without detrimental effect. Grades of austenitic stainless steels that are sensitive to corrosion attack when subjected to shielded metal arc welding will be sensitized along the cut when severed by this process.

Oxygen arc cuts in cast iron and medium carbon, low alloy steels are apt to develop cracks on the face of the cut. The extent and frequency of cracking depend on the composition and hardenability of the steel.

Equipment

OXYGEN ARC CUTTING may be performed using either constant current ac or dc power sources of sufficient capacity. Direct current electrode negative is preferred for rapid cutting. The electrode holder used for oxygen-arc cutting is of special design; it must convey not only electric current to the electrode but also oxygen to the cut. This is accomplished by bringing oxygen to the electrode holder and passing it through the bore of the electrode into the arc.

For cutting in air, a fully insulated electrode holder is desirable. When used for underwater cutting, a fully insulated holder equipped with a suitable flash-back arrester is required.

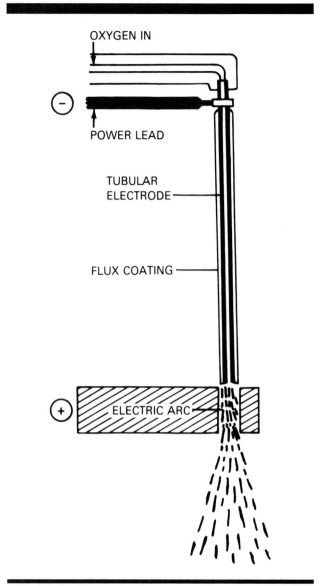

15.11–Schematic of Oxygen–Arc Electrode in Operation

Tubular steel electrodes are available in 3/16 and 5/16 in. (5 to 8 mm) diameter sizes, 18 in. (46 cm) long, with bore diameter approximately 1/16 in. (1.6 mm). The extruded covering is comparable to a mild steel electrode of AWS classification E6013. Underwater electrodes are steel tubes with a waterproof coating.

Cutting Techniques

IN THE OXYGEN arc method of cutting, piercing, and gouging, the coating is kept in contact with the base metal at all times. The coating insulates the core from the work and automatically maintains the proper length of arc.

The start of cutting and piercing operations is the same. The tip of the electrode is tapped on the work at the desired location as though striking an arc for welding, and the arc is maintained for a moment while the oxygen valve is opened. Piercing action begins immediately and the electrode is pushed through the plate as the hole is formed. The coating insulates the electrode core from shorting against the sides of the hole.

In cutting, the electrode is dragged along the plate surface at the speed of travel dictated by the progress of the cut. The inclination of the electrode and the speed of motion are adjusted to give the most efficient and highest quality cut.

Template-guided cutting is common. The electrode is pressed against the template; it is insulated from it by the coating. For straight-line cuts, any straight edge may be clamped along the line to be cut. The cut is accomplished by holding the electrode against the guide and the plate at the same time. Circular openings in tanks have been cut by using the circumference of a suitably sized pipe as the guiding template.

When cutting in air (up to 3 in. of mild steel or 1/2 in. of certain nonferrous alloys), the technique is to drag the electrode along the intended line of cut while applying slight pressure. In underwater cutting, regardless of the thickness of the metal being cut, positive pressure must be maintained against the metal being cut.

Gouging is performed by striking the arc, releasing the oxygen stream, and inclining the rod until it is almost parallel to the plate surface and pointed away from the operator along the line of the prospective gouge. The arc and oxygen melt the plate surface, and the molten metal is blown away by the force of the oxygen jet.

Applications

OXYGEN ARC CUTTING electrodes were developed primarily for use in underwater cutting and were later applied to cutting in air. In either application, oxygen arc electrodes can cut ferrous and nonferrous metals in any position.

Oxygen arc cutting has been used effectively by foundries and scrap yards for cutting mild and low alloy steels, stainless steel, cast iron, and nonferrous metals in any position. The usefulness of the process varies with the thickness and composition of the material being cut.

The edges of metal cut by the oxygen arc torch are somewhat uneven and usually require a light surface preparation to make them suitable for welding.

GAS TUNGSTEN ARC CUTTING

Principles of Operation

GAS TUNGSTEN ARC cutting can be used to sever nonferrous metals and stainless steel in thicknesses up to 1/2 in. using standard gas tungsten arc welding equipment. Metals

cut include aluminum, magnesium, copper, silicon-bronze, nickel, copper-nickel, and various types of stainless steels. This cutting process can be used either manually or mechanized. The same electric circuit is used for cutting as for welding. Higher current is required to cut a given thickness of plate than to weld it. An increased gas flow is also required to melt through and sever the plate.

In practice, a 5/32 in. (4 mm) diameter 2 percent thoriated tungsten electrode is extended approximately 1/4 in. (6.4 mm) beyond the end of a 3/8 in. (9.5 mm) diameter metallic or ceramic gas cup. A mixture of approximately 65 percent argon and 35 percent hydrogen is delivered to the torch at a flow rate of 60 cfh. Nitrogen can also be used, but the quality of the cut is not as good as that obtained with an argon-hydrogen mixture. Best cutting results are obtained using dcsp, but alternating current with superimposed high frequency has produced satisfactory cuts on material up to 1/4 in. (6.4 mm) thick.

Arc starting can be accomplished with either a high-frequency spark or by scratching the electrode on the workpiece. An electrode-to-work distance of 1/16 to 1/8 in. (1.6 to 3.2 mm) is used, but this is not a critical factor. As the torch is moved over the plate, a small section of the plate is melted by the heat of the arc and the molten metal is blown away by the gas stream to form the kerf. At the end of the cut, the torch is raised from the workpiece to break the arc.

One face of the cut is usually dross-free, with dross adhering to the side of the workpiece away from the worklead. The cut quality on the dross-free side is usually acceptable while the other requires considerable cleanup.

Equipment

STANDARD GAS TUNGSTEN arc welding torches can be used for cutting. As shown in Table 15.8, cutting currents up to 600 A are used. Welding torches can be used for cutting at currents up to 175 percent of their nominal ratings because there is little reflected heat from the cutting operation. For example, a 300 A torch can be used for cutting with 500 A for short periods.

A constant-current dc power supply, either rectifier or motor generator, with a minimum open circuit voltage of 70 V is recommended for cutting. Cuts made with ac power have a plate thickness limitation of 1/4 in. (6.4 mm). The major difficulty encountered when using ac power is the loss of tungsten from the electrode at the high currents required.

GAS METAL ARC CUTTING

Principles of Operation

GAS METAL ARC cutting (GMAC) is an arc cutting process that uses a continuous consumable electrode and a shielding gas. GMAC was developed soon after the commercial introduction of the gas metal arc welding process. GMAC first occurred accidentally during a welding operation, when it was found that if the electrode feed rate was set too high, it would penetrate through the plate. When the torch was moved, a cut was made.

The chief drawbacks to use of GMAC are the high consumption of welding electrode and high cutting currents (up to 2000 amps) required.

Applications

GMAC HAS BEEN used to cut shapes in stainless steel and aluminum. Using normal welding equipment and a 3/32 in. (2.4 mm) diameter carbon steel electrode, stainless steel up to 1-1/2 in. thick and aluminum up to 3 in. thick can be cut.

CARBON ARC CUTTING

CARBON ARC CUTTING is the oldest arc cutting process and is rarely used today. The process utilized an arc between a carbon (graphite) electrode and the base metal to melt the surface of the workpiece. Since the process depends on gravity to remove the molten metal, it can only be used in the vertical and overhead positions.

One variation used the arc force to assist in pushing metal out of the kerf by using higher amperages. The cuts produced required extensive cleanup of dross and slag. Prior to welding, the cut edges required grinding to remove the melted area remaining on the metal, which was high in carbon picked up from the carbon electrode.

Table 15.8
Conditions for Gas Tungsten Arc Cutting

Material	Thickness in.	Travel Speed, ipm	Current dcsp amps	Type of Gas
Stainless Steel	1/8	20	350	80% A + 20% H(2)
Stainless Steel	1/4	20	500	65% A + 35% H(2)
Stainless Steel	1/2	15	600	65% A + 35% H(2)
Aluminum	1/8	30	200	80% A + 20% H(2)
Aluminum	1/4	20	300	65% A + 35% H(2)
Aluminum	1/2	20	450	65% A + 35% H(2)

SUPPLEMENTARY READING LIST

Plasma Arc Cutting and Gouging

Alban, J. F. "Revival of a lost art: plasma arc gouging of aluminum." *Welding Journal* 64(5): 954-959; November 1976.

Couch, R. W., Jr. and Dean, D. C., Jr. "High quality water arc cutting." *Welding Journal* 50(4): 233-237; April 1971.

Frappier, M. B. "Plasma arc cutting supplies explained." *Welding Journal* 67(2): 48; February 1988.

Hebble, C. M., Jr. "Cutting with low current broadens application of plasma process." *Welding Journal* 52(9): 587-589; September 1973.

Heflin, R. L. "Plasma arc gouging of aluminum." *Welding Journal* 64(5): 16-19; May 1985.

McGough. M. S. et al. "Underwater plasma arc cutting in Three Mile Island's reactor. *Welding Journal* 68(7): 22-26; July 1989.

Na, S. et al. "A microprocessor-based shape and velocity control system for plasma arc cutting." *Welding Journal* 67(2): 27-33; February 1988.

O'Brien, R. L. "Arc plasmas for joining, cutting, and surfacing." Bulletin No. 131. New York: Welding Research Council, July 1968.

O'Brien, R. L., Wickham, R. J., and Keane, W. P. "Advances in plasma arc cutting." *Welding Journal* 43(12): 1015-1021; December 1964.

Shamblin, J. E., and Armstead, B. H. "Plasma arc cutting." *Welding Journal* 43(10): 470s-472s; October 1964.

Skinner, G. M., and Wickham, R. J. "High quality plasma arc cutting and piercing." *Welding Journal* 46(8): 657-664; August 1967.

Spies, G. R., Jr. "Comparison of plasma and oxyfuel gas cutting." *Welding Journal* 44(10): 815-828; October 1965.

Wodtke, C. H., Plunkett, W. A., and Firzzell, D. R. "Development of underwater plasma arc cutting." *Welding Journal* 55(1): 15-24; January 1976.

Air Carbon Arc Cutting

American Welding Society. *Recommended practices for air carbon arc gouging and cutting*, C5.3-82. Miami, Florida: American Welding, Society 1982.

Coughlin, W. J. and Fayer, G. IV. "Growth of the air carbon arc gouging process." *Welding Journal* 60(6): 26-31; June 1981.

Marshall, W. J. et al. "Optical radiation levels produced by air carbon arc cutting processes." *Welding Journal* 59(3): 43-46; March 1980.

Panter, D. "Air carbon arc gouging." *Welding Journal* 56 (5): 32-37; May 1977.

Shielded Metal Arc Cutting

Thielsch, H. and Quass, J. "Shielded-metal-arc cutting and grooving." *Welding Journal* 33(5): 438-446; 1954.

U.S. Government Printing Office. *Underwater cutting and welding manual*, NAVSHIPS 250-692-9. Washington, D.C. Supt. of Documents, U. S. Govt. Printing Office.

Gas Metal Arc Cutting

Babcock, R. S. "Inert-gas metal arc-cutting." *Welding Journal* 34(4): 309-315; 1955.

Blackman, P. R., et al. "Electric arc cutting." U.S. Patent 3,115,568, December 24, 1963.

Hull, W. G. "Use of gas-shielded arc processes for cutting non ferrous metals." *Welding and Metal Fabrication*, May 1954.

Gas Tungsten Arc Cutting

Conner, G. A. "Tungsten arc cutting of stainless steel." *Welding Journal* 39(3): 215-222; March 1960.

Wait, J. D., and Resh, S. H. "Tungsten arc cutting of stainless steel shapes in steel warehousing operations." *Welding Journal* 38(6): 576-581; June 1959.

"Tungsten-arc welding torch cuts light-gage metal." *Iron Age* 186 (152) November 17, 1960.

Oxygen Arc Cutting

Campbell, H. C. "The theory of oxyarc cutting." *Welding Journal* 26(10): 889-903; 1947.

"A New Combination Oxygen-Arc Cutting Process," *Industry and Welding* 20(1): 48; 1947.

Clauser, H. R. "New oxygen-arc process for cutting ferrous and non-ferrous alloys." *Materials and Methods* 25 (1): 78; 1947.

Hughey, Howard G. "Stainless steel cutting." *Welding Journal* 26(5): 393-400; 1947.

Kandel, Charles "Underwater cutting and welding." *Welding Journal* 25(3): 209-212; 1946.

"Machine makes smooth cuts in honeycomb materials." *Iron Age* 141-3, November 17, 1960.

Sibley, C. R. "Electric arc cutting." U.S. Patent 2,906,853, September 29, 1959.

Warren, W. G. "Electric arc-cutting of aluminum." *Welding and Metal Fabrication*, March 1953.

LASER BEAM AND WATER JET CUTTING

PREPARED BY A COMMITTEE CONSISTING OF:

J. C. Chennat, Chairman
Ford Motor Co.

C. E. Albright
Ohio State University

C. O. Brown
United Technologies Industrial Lasers

R. Chellevold
*Ingersoll-Rand
Waterjet Cutting Systems*

D. L. Havrilla
Rofin-Sinar Lasers

T. A. Johnson
Ferranti-Sciaky, Inc.

D. Kautz
Lawrence Livermore National Labs

L. W. Lamb
Flow Systems, Inc.

F. Mason
American Machinist & Automated Mfg.

L. R. Migliore
Amada Laser Systems and Service, Inc.

G. White
Coherent General

WELDING HANDBOOK COMMITTEE MEMBER:
G. N. Fischer
Fischer Engineering Company

CHAPTER 16

LASER BEAM AND WATER JET CUTTING

LASER BEAM CUTTING

INTRODUCTION

LASER BEAM CUTTING is a thermal cutting process that severs material by locally melting or vaporizing with the heat from a laser beam. The process is used with or without assist gas to aid the removal of molten and vaporized material.

Drilling with a laser is a pulsed operation involving higher power densities and shorter dwell times than laser cutting. Holes are produced by single or multiple pulses. Laser drilling is a cost-effective alternative to mechanical drilling, electro-chemical machining, and electrical-discharge machining for making holes of relatively shallow depths.

A laser is a heat source with some unique characteristics. See Chapter 22 for a description of the equipment used to produce a laser beam. Relatively modest amounts of laser energy can be focused to very small spot sizes, resulting in high power densities. In cutting and drilling, these power densities are in the range of 6.5×10^6 to 6.5×10^8 W/in.^{2} (10^4 to 10^6 W/mm^2). Such high concentrations of energy cause melting and vaporization of the workpiece material, and material removal is enhanced by a jet of gas. Depending on the material, a jet of reactive gas such as oxygen can be applied coaxially with the beam, improving process speed and cut edge quality. The physical mechanisms involved in the material removal process are quite complex, involving material properties and several process variables.

Among laser material processing applications, cutting is the most common process, enjoying excellent growth rate worldwide. The first laser material processing application was drilling diamonds for wire drawing dies. Today, laser cutting and the related processes of drilling, trimming, and scribing account for more than 50 percent of the international industrial laser installations.

A high-power CO_2 laser can cut up to 1 in. (25 mm) thick carbon steel. However, good quality cuts on steel are typically made on metal thinner than 0.375 in. (9.5 mm), because of the limited depth of focus of the laser beam. CO_2 lasers in the range of 400 to 1500 W dominate the cutting area. Neodymium-doped, yttrium aluminum garnet (Nd:YAG) lasers are also used.

Laser cutting has the advantages of high speeds, narrow kerf widths, high-quality edges, low-heat input, and minimal workpiece distortion. It is an easily automated process that can cut most materials. The cut geometry can be changed without the major rework required with mechanical tools; there is no tool wear involved, and finishing operations are not usually required. Within its thickness range, it is an alternative to punching or blanking, and to oxyfuel gas and plasma arc cutting. Laser cutting is especially advantageous for prototyping studies and for short production runs. Compared to most conventional processes, noise, vibration, and fume levels involved in laser cutting are quite low.

Laser cutting results are highly reproducible, and laser systems have achieved operating uptimes greater than 95 percent. Relative movement between the beam and the workpiece can be easily programmed using CNC workstations currently available. High precision and good edge quality are quite common even in three-dimensional laser cutting. Lasers also have the flexibility for power and time sharing so that cost effectiveness of full-time beam operation can be maximized.

LASER DRILLING

HOLE DIAMETERS PRODUCED by laser beam drilling typically range from about 0.0001 to 0.060 in. (0.0025 to 1.5 mm). Depths achieved are usually less than 1 in. (25 mm) because of beam focusing limitations. Examples of laser drilling on jet engine blades and a rotor component are shown in Figure 16.1.

The process produces clean holes with very small recast layers. When large holes are required, a trepanning technique is used where the beam cuts a circle with the required diameter.

Laser drilling shares most of the advantages found in laser cutting. It is especially advantageous when the required hole diameters are less than 0.020 in. (0.5 mm) and when holes are to be made in areas inaccessible to conventional tools. Beam-entry angles can be very close to zero, a situation where mechanical tools are susceptible to breakage. The high-intensity pulsed outputs from solid-state lasers with shorter wavelengths such as Nd:YAG, Nd:glass, and ruby, are more suitable for drilling. The industrial laser drilling area is dominated by Nd:YAG lasers. The elements of a Nd:YAG laser are shown schematically in Figure 16.2. CO_2 lasers are usually used for drilling nonmetals like ceramics, composites, plastics, and rubber.

The two most discouraging aspects of laser material processing have been its high equipment cost and the intimidation workers feel about a "high technology" process, requiring high operator skills and good knowledge of laser-material interaction. More laser and system manufacturers are entering the field worldwide, and more reliable products with lower prices and new features are being produced. The industrial laser market is presently enjoying a 20 percent annual growth rate, and the prices of laser cut-

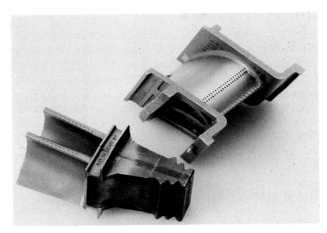

Figure 16.1–Jet Engine Blades and a Rotor Component Showing Laser Drilled Holes

ting systems are expected to drop by a small percentage every year for the next several years.

New software packages and easy-to-learn programming are making laser cutting more welcome in low operator-skill areas. Fully integrated laser-robot systems and easy interfacing with personal computers are offering better control of laser systems and operating variables. CO_2 lasers with reduced size and weight, multi-kilowatt pulsable CO_2 lasers with better beam quality, single mode YAG lasers, and YAG laser outputs up to 1.5 KW are some of the improvements being made. Improved system designs are leading to higher accuracy and repeatability of the process.

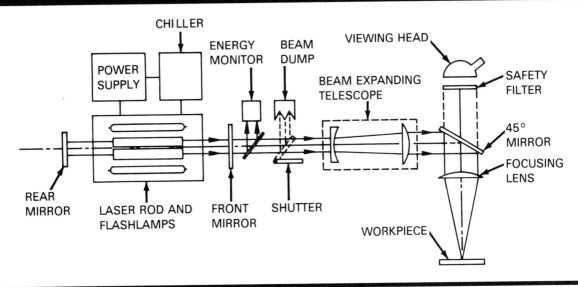

Figure 16.2–Schematic Representation of the Elements of A Nd: YAG Laser

PROCESS PRINCIPLES AND CHARACTERISTICS

LASER BEAM CUTTING (LBC) and laser beam drilling (LBD) are two completely different processes for removing material. These processes have been widely investigated both experimentally and theoretically to understand the mechanisms which take place in the material removal process.

Both processes can use either pulsed or continuous lasers as the primary energy source. Many factors, as listed in Table 16.1, are involved in laser cutting and drilling. The engineering disciplines involved include lasers themselves, optics, fluid dynamics, and materials.

Simplistic Model

Cutting. The laser beam cutting process can be described quite simply. It requires simultaneous action of a focused laser beam with a power density greater than 6.5×10^6 W/in.^{2} (10^4 W/mm^2), and an assist gas jet, which together produce a kerf in the workpiece. The laser beam acts as a line heat source which produces a keyhole after the initial transient conditions have come to steady-state. The assist gas jet ejects the molten material within the keyhole through the root of the kerf. In certain cases, an active gas can be used to improve the cutting efficiency by an exothermic chemical reaction. Commonly used assist gases are listed in Table 16.2.

The advantages of laser cutting over other processes include: (1) narrow kerf width, (2) narrow heat-affected zone, (3) high cutting speeds, (4) good cut quality, (5) adaptability to automation, and (6) no mechanical contact between the cutting device and the workpiece.

Drilling. The process of laser drilling requires merely a pulsed laser with beam focused to power densities of

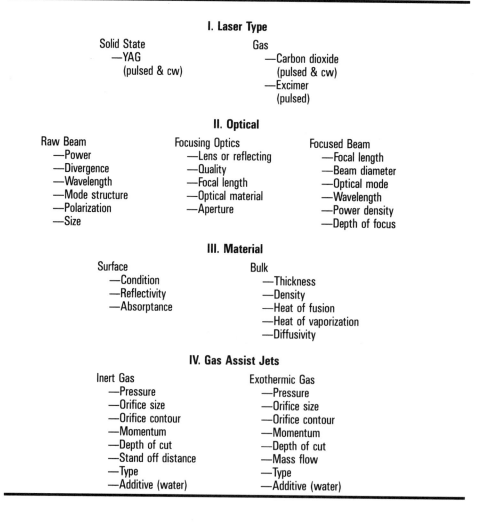

Table 16.1
Factors Influencing Laser Drilling and Cutting Processes

I. Laser Type

Solid State
—YAG
 (pulsed & cw)

Gas
—Carbon dioxide
 (pulsed & cw)
—Excimer
 (pulsed)

II. Optical

Raw Beam
—Power
—Divergence
—Wavelength
—Mode structure
—Polarization
—Size

Focusing Optics
—Lens or reflecting
—Quality
—Focal length
—Optical material
—Aperture

Focused Beam
—Focal length
—Beam diameter
—Optical mode
—Wavelength
—Power density
—Depth of focus

III. Material

Surface
—Condition
—Reflectivity
—Absorptance

Bulk
—Thickness
—Density
—Heat of fusion
—Heat of vaporization
—Diffusivity

IV. Gas Assist Jets

Inert Gas
—Pressure
—Orifice size
—Orifice contour
—Momentum
—Depth of cut
—Stand off distance
—Type
—Additive (water)

Exothermic Gas
—Pressure
—Orifice size
—Orifice contour
—Momentum
—Depth of cut
—Mass flow
—Type
—Additive (water)

Table 16.2
Assist Gases Used for Laser Beam Cutting of Various Materials

Assist Gas	Material	Comments
Air	Aluminum	Good result up to 0.060 in. (1.5 mm)
	Plastic	
	Wood	
	Composites	
	Alumina	All gases react similarly; air is the least expensive
	Glass	
	Quartz	
Oxygen	Carbon Steel	Good finish, high speed; oxide layer on surface
	Stainless Steel	Heavy oxide on surface
	Copper	Good surface up to 1/8 in. (3 mm)
Nitrogen	Stainless Steel	
	Aluminum	Clean, oxide-free edges to 1/8" (3 mm)
	Nickel Alloys	
Argon	Titanium	Inert assist gas required to produce good cutting of various materials

greater than 6.5×10^7 W/in.^{2} (10^5 W/mm^2). When the focused beam strikes a surface, material is melted and volatilized, and the molten and vaporized material is violently ejected, forming a hole. Depths normally achieved are approximately six times the diameter of the hole. Thus, multiple pulses may be required to completely penetrate the thickness of the material. Materials up to 1 in. (25 mm) thick have been drilled in the pulsed mode to date.

The advantages of laser drilling include (1) short drilling times, (2) adaptability to automation, (3) ability to penetrate difficult-to-drill materials, and (4) no mechanical contact.

Laser Types

LASER BEAM CUTTING and drilling require a precisely focused, coherent laser beam. Two primary laser sources, the pulsed YAG laser operating at a wavelength of 1.06 microns, and the CO_2 laser operating either pulsed or continuously at a wavelength of 10.6 microns, are predominately used for these applications. Table 16.3 describes the basic cutting mechanisms and the laser used in each case. For the pulsed YAG laser, its interaction with material results in evaporation and removal at very high power densities. The continuous and pulsed CO_2 laser removes most material by first melting it, which then must be blown from the kerf with an inert gas assist. If the inert gas is replaced by a reactive gas such as oxygen, the process becomes exothermic and additional energy is supplied by oxidation of the material.

More recently, the excimer laser operating at a wavelength of 248 nanometers has been added as a laser source for drilling. The process of material removal using this laser is thought to be by photo ablation when used on polymers having bond energies below the excimer photon energies.

In either of the processes (cutting or drilling, Figure 16.1), it is necessary to achieve power densities from 6.5×10^6 to 6.5×10^8 W/in.^{2} (10^4 to 10^6 W/mm^2). This is accomplished by focusing the beam with either lenses or reflective optics, depending upon the laser type and wavelength. In either case, the beam spot size is defined the same way and is given by the relationship,

$$d_s = 2.44 \, K \, F/D \qquad (16.1)$$

where

d_s = focused spot diameter μ in. (μ mm)
K = a constant dependent upon optical beam mode (See Table 16.4)
F = focal length of lens or mirror in. (mm)
D = aperture diameter of beam on focusing mirror in. (mm)
λ = laser optical wavelength μ in. (μ mm)

Table 16.4 gives the values of K, which are dependent upon the optical beam mode structure and its divergence, for three of the most common continuous laser beams.

In the case of drilling, short focal-length lenses are used to focus the high-peak power optical beams from the pulsed lasers to spot sizes on the order of 0.024 in. (0.6 mm) diameter to achieve power density levels exceeding 6.5×10^7 W/in.^{2} (10^5 W/mm^2). Under these conditions, the material is volatilized and ejected from the workpiece, leaving a partially drilled hole. Multiple pulses are used to achieve full penetration.

In most cutting applications the laser is of the continuous type operating at power levels between 400 and 1500 W, somewhat lower than the peak powers of the pulsed lasers described for the drilling applications. As a result, the required power densities are lower and generally in the range of 6.5×10^6 to 6.5×10^7 W/in.^{2} (10^4 to

---

Table 16.3
Basic Cutting Mechanisms

A. Solid-State Laser—YAG

   Evaporation
      —Material removal by volatilization at $>6.5 \times 10^7$ W/in.^{2} ($>10^5$ W/mm^2)
      —Pulsed only.

B. Gas Laser—CO_2

   Fusion
      —Most material removed in liquid state by means of inert gas assist with beam intensities
      of 6.5×10^6 W/in.^{2}(10^4 W/mm^2).

   Exothermic
      —Most material removed in liquid state at beam intensities of 6.5×10^6 W/in.^{2}
      (10^4 W/mm^2).
      —Additional energy supplied by oxygen gas assist.

C. Excimer Laser

   Photo Ablation
      —Material removed by photo ablation when used on polymers having bond energies below
      the eximer photon energy level.

10^5 W/mm^2). This requires spot sizes on the order of 0.04 in. (1 mm) at the CO_2 laser wavelength in order to achieve the required power densities.

Laser-Material Interactions

A FUNDAMENTAL AND important feature of LBC and LBD is the interaction of the laser beam with the material surface. Figure 16.3 shows the relationship between the optical beam, the focusing system, the assist gas jet, and the workpiece to be cut. The optical lens or mirror focuses the input laser beam to a spot size given by equation 16.1. The location (generally aimed within the thickness of the workpiece) of the focal plane relative to the workpiece surface depends upon several factors, all governed by the relationships listed above. In practice, the exact location is determined experimentally, to suit the application.

Because most metals are highly reflective at the laser wavelengths under consideration, the coupling of the beam and workpiece is very inefficient and absorption is low. However, the absorption coefficient is a function of the temperature of the material, which changes during the transient phase of the process. This relationship is shown in Figure 16.4.

The initial weak absorption at the surface of the workpiece begins to increase the workpiece temperature directly under the optical beam, and that decreases the reflectivity quite rapidly. Temperature and absorption

---

Table 16.4
Effect of Beam Mode on Focusability

Type of Laser Beam	K
1. Uniform Wavefront	1.0
2. Gaussian Beam	0.86
3. Unstable Resonator*	
a. M=2**	4.0
b. M=4	3.5

* Magnifications 'M' most used

** M, Magnification ratio of an annular beam, $= \dfrac{\text{Beam OD}}{\text{Beam ID}}$

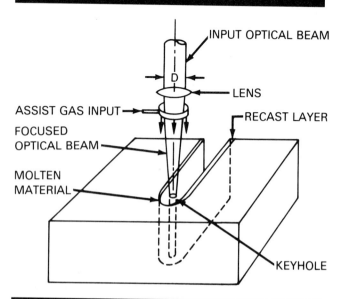

Figure 16.3–Schematic View of Laser Cutting Operation

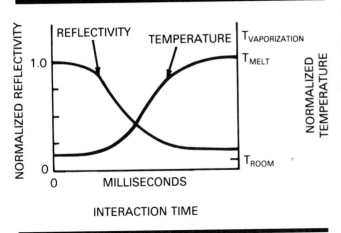

Figure 16.4–Reflectivity and Temperature Transient Time for Typical Metals

ejected through the root of the workpiece to produce a cut.

For the keyhole cutting process to be initiated, it is essential that the power density be high enough to overcome the reflection barrier. The depth of the cut is then controlled by the melting and vaporization relationships depicted in Figure 16.5. At power density levels below 3.25 x 10^6 W/in.^{2} (5 x 10^3 W/mm^2), only surface melting is achieved. To develop a keyhole, power densities in the range of 6.5 x 10^6 to 6.5 x 10^7 W/in.^{2} (10^4 to 10^5 W/mm^2) are required. Within the keyhole range, both melting and vaporization occur. Complete vaporization required for drilling is achieved above this range.

Gas Jet Assist

THE LIQUID COLUMN formed by the laser during welding is supported against gravity by both surface tension and capillary action. An assist gas jet, as shown in Figure 16.3, is used to remove the molten metal before resolidification can occur. This action prevents the formation of a weld. The momentum of the gas from the jet ejects a large percentage of the molten material from the root of the kerf. A very thin recast layer is left along the sidewall of the kerf. A beam delivery system for laser cutting with gas assist is shown in Figure 16.6.

increase until melting and vaporization temperatures are reached. That permits a keyhole or radiation trap to form. The laser beam then acts as an energetic line heat source within the material and forms a molten pool. When the pool is exposed to a high-pressure gas jet, molten metal is

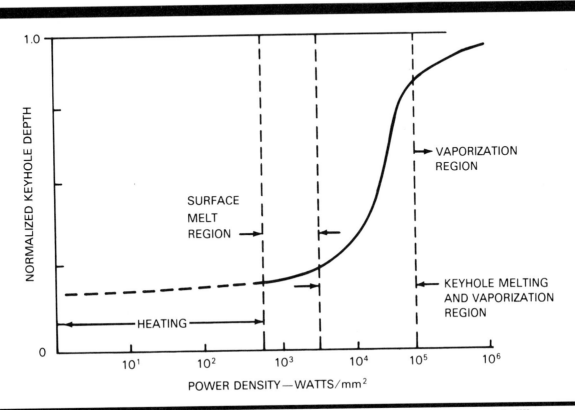

Figure 16.5–Power Density Requirements for Laser Keyhole Cutting and Drilling

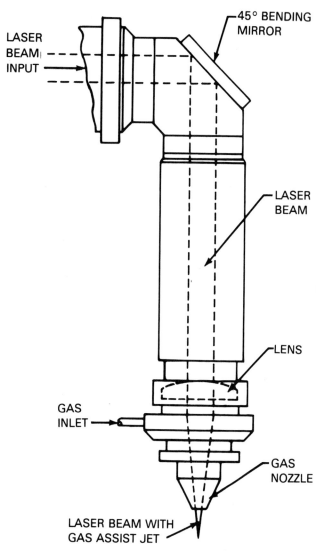

Figure 16.6—Beam Delivery System for Laser Cutting With Gas Assist

THEORY AND PHYSICAL MECHANISMS

IN RECENT YEARS, the theory and physical mechanisms of laser cutting have been studied in detail. These studies should be continued so that the technical limits of the process can be evaluated and extended into regions not presently possible. It has been the purpose of these studies not only to increase the cutting and drilling depths, but to improve the quality of the cut surfaces as well. This section is a brief overview of the mechanisms; greater detail is provided in the referenced articles.

The primary factors which influence the laser-cutting process (see Table 16.1) are the power level, mode, polarization, and such optical variables as the focal length, aperture diameter, depth of focus, and location of the focal plane relative to the workpiece.

The energy balance of the laser-cutting process is shown in Figure 16.7. The energy sources are the laser and the reactive gas. The primary losses are the heat of conduction, reflection from the erosion front, heat of vaporization, convection, radiation, and the energy contained in material ejected at the root of the cut.

The most critical process taking place is the absorption of the incident radiation on the erosion front. Without absorption of the incoming beam, cutting would not be possible. It is also very important that this process be efficient. Absorption efficiency is dependent upon several factors, including the cut width, the instantaneous slope of the erosion front, polarization of the input optical beam, and the optical beam intensity distribution in the longitudinal and radial directions.

A factor which limits thick-section cutting at the power levels available is the narrow kerf width. Narrow kerfs are desirable from an applications standpoint; however, they are a detriment to the gas-assist approach because of the small jet nozzle diameters required to deliver gas into a small kerf. The coherent length of an overexpanded free jet is typically on the order of a few jet orifice diameters. This leads to an overexpansion of the gas jet within the keyhole cavity and limits the effective length of the jet. This limits the depth and surface smoothness of the cut. As a result, edges toward the bottom of a laser cut in thick material are generally rougher than those produced by other cutting methods.

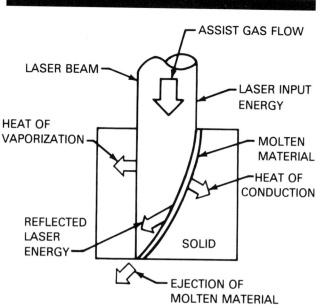

Figure 16.7—Schematic View of Laser Cutting Showing Energy Balance

The erosion front, merely sketched in Figure 16.7, is the interface between the incoming beam with its gas assist, and the molten layer of material. This front is the equilibrium surface along which the ejected material moves out of the root of the base metal due to the momentum of the assist gas. Its shape almost always has a lagging angle relative to the vertical, depending upon the forward speed of the incoming beam across the workpiece. The angle of the incident optical beam relative to the molten layer along the erosion front determines the efficiency of absorption. This depends on the polarization of the optical beam. The beam can be either linearly polarized, in which case the quality of the cut depends on the cutting direction, or circularly polarized, in which case the cut is good in two dimensions, assuring uniformly contoured cuts in a plane.

Cutting with a circularly polarized beam is generally less efficient than with a linearly polarized beam. As an example, up to 80 percent of a linearly polarized beam is absorbed at an incident angle of 85 degrees. With a circularly polarized beam at its optimum ample, the peak absorptance drops to about 40 percent. As the angle of incidence varies above or below the optimum angles, the absorption begins to fall. Thus, the energy absorption by the erosion front is dependant upon the shape of the erosion front, the spatial intensity of the input beam, and its polarization.

EQUIPMENT

CO₂ LASERS

THE CARBON DIOXIDE (CO_2) laser is the standard beam source for contour cutting applications. This is because it is the most powerful and reliable type of laser in general use.

The CO_2 laser is a gas-discharge device: it operates by sending an electric current through a gas. In industrial lasers, high efficiency is obtained by using a mixture of helium, nitrogen, and carbon dioxide. Electrical energy is coupled to the gases by establishing a glow discharge in the nitrogen. The nitrogen transmits this energy to the CO_2 molecules by collisions which put a large percentage of them in an elevated state. Laser emission at 10.6 microns in the infrared zone is produced when these molecules drop to an intermediate state. Collisions with the helium bring the CO_2 back to energy ground level, where the process can begin again. The gas is typically passed through a heat exchanger where it is cooled before being recycled.

Temporal Characteristics

CO_2 LASERS CAN operate by continuous wave (CW) or in a variety of pulsed modes. The pulse frequency may be as high as 10 kHz. The most common types of pulsing are termed *gated* and *enhanced*. In the gated mode, the laser operates at a peak power level that is within its normal CW range. The output is modulated to generate a reduced duty cycle. Gated pulses can be any length that is compatible with the chosen repetition rate. Lasers that can produce enhanced pulses have peak powers that are several times their CW rating. Enhanced pulses are usually about 100 microseconds long regardless of the repetition rate.

Spatial Characteristics

THE LOW-DENSITY AND high-thermal diffusivity of a gaseous laser medium reduce its tendency to distort the light that goes through it. This allows even high-power CO_2 lasers to have good optical quality. Beams from many lasers with outputs of up to 1500 watts are close approximations of the fundamental Gaussian mode TEM₀₀. Such beams may be focused to the limit set by the diffraction of light. A spot size of .004 in. (0.1 mm) is easily achieved by normal focusing lenses for CO_2 lasers.

Another property of TEM₀₀ beams is *low divergence*, a term describing the angle at which the laser beam spreads out as it propagates. Typical values are in the range of 1 milliradian, which allows great flexibility in machine design since the laser can be distant from the focusing lens.

Slow-Flow Lasers

THE EARLIEST INDUSTRIAL CO_2 lasers consisted of glass tubes with mirrors on both ends. The laser gas flowed through the tube while electricity was applied near each mirror. These devices very simple and reliable, but are limited to about 50 watts per meter of discharge length, because there is no way to cool the gas. Such slow-flow lasers become unwieldy if more than 400 watts is required. They are in use today because they can produce stable high-quality outputs, and because the large volume of active medium allows for massive pulse enhancement.

Transverse-Flow Lasers

THE TRANSVERSE-FLOW LASER was developed to produce high power in a small package. It does this by circulating the laser gas through the discharge region at high speed and then cooling it with a heat exchanger so that it can be reused. Transverse-flow lasers tend to have asymmetrical modes because of the gain characteristics of the discharge currents. Despite these limitations, transverse-flow machines have been highly successful as cutting lasers.

The newest laser design in use today is the fast axial-flow type. This is a modification of the slow-flow laser, using a Rootes pump to circulate the gas. Fast axial lasers are small, powerful, and inexpensive to build. While they have appeared in many laser systems, most models have severe problems with instability of the output beam. This results in roughness on the surface of laser-cut parts.

YAG LASERS

THE YAG (MORE correctly, neodymium-doped, yttrium aluminum garnet) is the standard drilling laser in industry. Some contour cutting is also appropriate for its characteristics.

Principle of Operation

A YAG LASER contains a crystalline rod surrounded by xenon or krypton lamps. The crystal is an yttrium aluminum garnet (YAG) which has been doped with neodymium. Light from the lamps "pump" the neodymium atoms to an excited state, where they emit light at a wavelength of 1.06μm. Water flowing around the rod cools the atoms to the ground state.

Temporal Characteristics

INDUSTRIAL YAG LASERS generally operate in the pulsed mode for cutting or drilling. The repetition rate is generally below 200 Hz. Control of the power going into the lamps allows tailoring of the shape and duration of the laser pulse. The solid laser medium has a high concentration of light-emitting atoms, so the peak power can be very large. High-energy pulses of short duration remove the material being cut or drilled.

Spatial

LASER RODS GENERATE heat in the center and are cooled on the outside. Whenever substantial power is produced, a temperature gradient develops across the rod's diameter. That gradient induces changes in the rod's refractive index, which degrades the optical performance of the laser. High power YAG lasers have multimode outputs with high divergence, which limits the ability of the system to focus the beam to a small spot and requires the laser head to be near the work area.

OTHER TYPES

Glass

GLASS LASERS ARE very similar to YAG lasers. The laser rod is made of neodymium-doped glass rather than garnet. When glass rather than YAG is used as a matrix, a higher concentration of neodymium atoms can be incorporated in the laser rod. This allows glass lasers to produce stronger pulses than YAG lasers, which makes them more appropriate for deep drilling. The disadvantage of glass is that its poor thermal conductivity limits the pulse repetition rate to about 1 pps, making it useless for contour cutting.

Ruby

RUBY WAS THE first material in which laser emission was observed. The ruby laser is a flashlamp-pumped, solid-state device like the YAG and glass lasers, but emits visible light. Although largely replaced by other types, it is still suitable for drilling, with characteristics similar to Nd-glass lasers.

Excimer Lasers

EXIMER LASERS ARE pulsed high-pressure gas lasers which emit at wavelengths in the ultraviolet band. The term *excimer* is a contraction of the words "excited dimer". A dimer is basically a molecule that exists only in the excited state, such as krypton fluoride (KrF). Such molecules are formed when the appropriate gas mixture (typically a noble gas and a halogen) is excited in a pulsed electrical discharge. Lasing occurs when the excited molecule relaxes to the lower state.

SYSTEMS

IN ORDER TO cut, a laser must be integrated with a mechanism to deliver the beam and with means to handle the workpiece. Today, laser contour cutters are controlled by some sort of computer. The most common type of control is one which reads numerical data and transforms it into axis commands. Such devices are called *computer numerical controls*, or CNC's. The cutting head, consisting of a focusing lens and provision for assist gas, must be kept at a certain distance from the part to be cut. These components are enclosed, to provide safety for personnel, in a package termed a *laser cutting system*.

There is considerable variety in the design of these systems. Standard machines are available for work such as contour cutting of sheet metal or drilling of turbine blades, while special units can be obtained for tasks such as slitting of sheet materials on production lines.

For optimum cut quality, the optics should be held motionless, since any vibration or misalignment in the beam delivery system results in poor cut quality or inaccuracy. Fixed optics, however, require that the workpiece move, which becomes more complicated with large sheets. The minimum floor area for a fixed beam system is four times the maximum sheet size, which again is a problem with large workpieces. Automatic sheet feeding and part removal are difficult, as is accurate contouring with widely varying loads.

Under these conditions, moving the optics simplifies the laser system. With a "moving beam" system, sheets move only when they are being loaded onto or removed from the cutting table. The drive system always handles the same load, allowing servo response to be optimized. There are, however, several problems with moving the optical system:

(1) Beam divergence—Laser beams do not propagate unchanged through space. Beam diameters and other properties vary as a function of distance from their source. Since, with a moving beam system, the focusing lens intercepts the laser beam at different locations, the focal-point location and spot size will vary. The net result is that cutting conditions will vary at different locations on the table.

(2) Alignment—With fixed optics, it is only necessary to get the laser beam through the delivery system without it being clipped by any apertures. For a moving beam to function properly, the beam must be allowed to travel across the entire workpiece without a change in alignment.

(3) Rigidity—A fixed cutting head can be made rigid by using a massive support structure, and there is little penalty for doing this. When the head is moving, however, vibration and deflection are more difficult to suppress. This results in surface roughness or deviations from the programmed path, especially when the machine is making sharp corners.

(4) Beam path cleanliness—All high-power laser systems are sensitive to dirt on their optical elements. Dust particles that settle on lenses and mirrors are heated by the beam, causing damage to the components. As a result, all industrial laser systems must seal the beam path against the contaminants that exist in shop environments. This is, again, simple to do for a fixed beam but more complex when the elements are moving. In many moving beam systems, the laser optics share enclosures with gears, motors, and other sources of contaminants, shortening the life of the optics.

The ultimate extension of a moving beam system is the 6-axis gantry robot. Engineering difficulties involved in making a gantry are similar to those for a moving beam, only more severe. The ability of gantries to cut complex contours, however, makes the effort worthwhile.

A cantilevered robot with a moving beam delivery system is shown in Figure 16.8.

FOCUSING HEADS

THERE ARE FEW fixed-beam systems in use because of production limitations. Most systems move the lens to focus it, this motion ranging from a few thousandths of an inch to many feet. The relation between the surface of the workpiece and the focal point of the lens is one of the most important variables in laser cutting, so control of it is essential to maintaining process consistency. Depth of focus for CO_2 systems is on the order of .010 in. (0.25 mm). In many cases, variation in part thickness is greater than this. It is important, then, for the system to provide some way to hold focus when cutting uneven materials.

Machines that cut flat sheets often have heads that ride on the surface of the work with ball bearings. This approach works well but can mar the work finish on some parts, and it is unsuitable for contoured material. A more sophisticated approach is to attach a drive motor to the focusing mechanism and control it with a sensor. Capacitive probes work well in all orientations and do not protrude from the cutting head but are restricted to conductive workpieces. Contact probes, consisting of a fork or cup around the cutting nozzle, work on any material but function only in the vertical direction.

CONSUMABLES

THE PRIMARY COSTS associated with operating laser systems are those for electricity, optics, flashlamps (solid-state only), and gases. Gases are used for two purposes: for generating CO_2 light and to assist in cutting.

Operating Cost for a Co2 Laser Cutting System

TYPICAL CONSUMABLE COSTS for a CO_2 laser system operating at 1500 watts are shown in Table 16.5. Gas costs are based on averages from different parts of the country. Depending on the material being cut, hourly operating costs range from $9.89 to $25.04 per hour.

Operating Cost for a Yag Laser Drilling System

THE PRIMARY COST in most YAG processes is flashlamp replacement. Lamp life is in the range of 1 to 10 million pulses, depending on the power used. A cost analysis made for a YAG laser drilling at 20 pps is shown in Table 16.6.

Table 16.5
Typical Laser Beam Cutting Costs for a CO2
System Operating at 1500 Watts

Consumable	Hourly Cost
Electricity	2.10
Internal laser optics*	2.06
Laser gas	1.03
Focusing lens*	1.10
Assist gas:	
O_2 for 10 ga. carbon steel	3.60
N_2 for .060 in. stainless steel	3.60
Ar for .060 in. titanium	18.75

* Cost based on manufacturer's estimate of operating life for these parts.

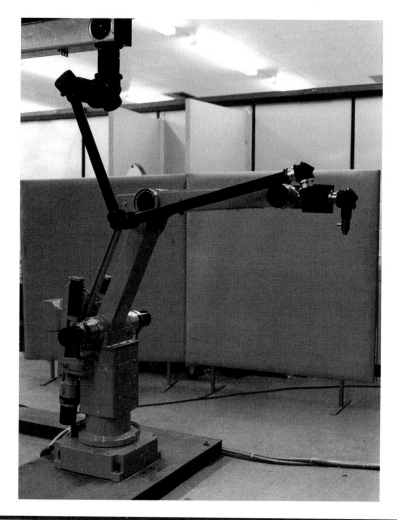

Figure 16.8–Laser Contour Cutter Using a Robot and a Beam Delivery System

Table 16.6
Typical Operating Costs for a YAG Laser Drilling System

Consumable	Hourly Cost
Electricity	.75
Laser optics*	1.00
Flashlamps*	2.00
Assist gas	4.00
Total	7.75

* Cost based on manufacturer's estimate of operating life for these parts.

MATERIALS

LASER CUTTING IS a thermal process: materials are cut because the laser beam heats them until they melt, decompose, or vaporize. It is therefore useful to examine the thermal properties of materials to determine their response to laser radiation. Equally important are a substance's optical properties, since energy is transferred in the form of light. In many cases, reactive or inert gases are used to assist cutting, so the chemical behavior of the material is important.

Compared to other classes of materials, metals have high thermal diffusivities and optical reflectivities. They also melt without decomposing and have very high boiling points. Laser metal cutting, then, requires high power densities to put energy into the material faster than it is conducted away, along with an assist gas to remove the liquid metal from the kerf. Within this broad characterization, there are significant variations among metals in their suitability for laser cutting. Typical laser cutting conditions for various materials are shown in Table 16.7.

CARBON STEEL

CARBON STEEL IS one of the easiest metals to cut with a laser. An examination of the energy balance during cutting, when oxygen is used as the assist gas, shows that most of the heat comes from the exothermic reaction of iron and oxygen, with the laser beam serving as a pilot or preheat energy source. The metal heated by the laser burns in the oxygen stream, leaving the surrounding material unaffected. The cut edge can be extremely smooth, with finishes better than 32 micro-inches achievable in 0.06 in. (1.5 mm) thick sheet.

ALLOY STEEL

THE TERM *alloy steel* covers a wide range of metals. Low alloy steels, such as AISI 4140 and 8620, cut much like carbon steel. The generally lower impurity levels found in low alloy steels result in improved cut quality compared to commercial cold rolled carbon steel. Increasing quantities of alloying elements change the steel's behavior. Tool steels with high tungsten additions cut slowly and with some slag adherence. Chromium additions reduce the steel's reactivity with oxygen and produce adherent scale on the cut edge.

STAINLESS STEEL

STAINLESS STEELS ARE a subset of alloy steels with two primary classifications: austenitic (300 series) and ferritic/martensitic (400 series). Stainless steels have relatively low thermal conductivity, which should make them easy to cut with a thermal process. However, the alloying elements that give stainless steels their corrosion resistant properties make them resistant to oxidation. This makes such materials react quite differently to laser energy than carbon steels.

The 400 series stainless steels, which have chromium as their primary alloying element, cut cleanly with oxygen assist but have a tenacious chrome oxide layer on the cut edge.

The austenitic materials, which have nickel and chromium additions, tend to have tenacious slag on the bottom of the kerf edges in addition to the oxide layer. This slag and the oxide are serious problems in production, since

Table 16.7
Typical Laser Cutting Variables

Material	Thickness Inches	Travel Speed IPM	Power Watts	Assist Gas
Carbon steel	.060	150	400	O_2
	.125	120	800	O_2
	.250	80	1200	O_2
	.375	50	1500	O_2
Stainless steel	.060	150	1500	N_2
	.125	40	1500	N_2
	.250	40	650	O_2
	.375	30	800	O_2
Titanium Ti6Al4V	.060	150	1500	argon
Kevlar-epoxy	.125	250	400	air
	.250	250	1500	air
G10 glass-polyester	.060	600	1000	air
Boron-Aluminum	.040	300	1500	air
Silicon Carbide	.030	25	150	argon

they require additional operations to produce a finished part. Slag can be removed by a grinder. The oxide must be removed prior to welding.

Use of an inert assist gas has proved successful in making cuts in stainless steel without oxide or slag adherence. The edges produced in this manner can be welded with no additional operations.

ALUMINUM

SEVERING ALUMINUM WAS a problem in the early days of laser cutting. Because of aluminum's very high diffusivity and reflectivity, it required large amounts of laser power to melt it. If the focus was incorrect, the aluminum reflected the beam back into the laser, often destroying the laser. As lasers with one kilowatt or more of power became available, along with accurate focusing methods, cutting problems diminished, but cut quality remained poor. Laser-cut aluminum had heavy slag on the bottom edges of the kerf sides. The cost involved in removing this slag usually made laser cutting uncompetitive compared to other methods. The recent development of inert gas cutting has made it possible to produce high-quality aluminum cuts with a CO_2 laser.

COPPER

COPPER, WITH DIFFUSIVITY and reflectance both higher than aluminum, is very difficult to cut with low-power lasers. However, copper is easily cut with kilowatt-class CO_2 lasers, as long as they have good TEM modes and the system keeps the beam focused on the work. YAG lasers, with their high-pulse power and shorter wavelength, cut copper with no problems.

COPPER ALLOYS

CUTTING RESULTS ON copper alloys such as brass are similar to those obtained on aluminum.

NICKEL BASE ALLOYS

MOST NICKEL-BASE ALLOYS are intended for some form of severe service, like high temperatures or corrosive environments. While these metals are easily laser cut, it is usually necessary to examine the part for such metallurgical defects as microcracking and grain growth to ensure that the part will perform properly. Recent tests with inert gas laser cutting show higher quality than with oxygen-assist cutting.

TITANIUM

TITANIUM AND ITS alloys react with oxygen and nitrogen to form brittle compounds at the cut edge, generally an unacceptable condition. Therefore, it is necessary to use argon as the assist gas for titanium cutting. Argon ionizes easily under laser cutting conditions, which can lead to plasma formation above the workpiece. When this happens, the laser output must be revised to obtain consistent results.

NONMETALS

ONE OF THE laser's attributes is that it can cut an extremely wide range of materials without regard to their hardness or electrical conductivity. It is convenient to divide materials into the categories of metals and nonmetals, and to subdivide the nonmetals into organic and inorganic categories.

Inorganic Materials

NONMETALLIC, INORGANIC MATERIALS, as a class, have low vapor pressures and poor thermal conductivities. These characteristics, combined with their generally high absorption of 10.6 μm wavelength light, should make them good candidates for laser cutting. Unfortunately, many of the common varieties have very high melting points and poor thermal shock resistance. This tends to make them harder to process than metals.

Alumina. Alumina (Al_2O_3) is often cut or scribed by lasers. Cutting is performed using high-power pulses to vaporize the material, since recast melted material is a problem. The high melting point of alumina, coupled with the low average power of lasers operating in the enhanced pulse mode, results in low cutting speeds.

The process of scribing is the standard method of preparing alumina substrates for hybrid microcircuits. Scribing is performed by drilling rows of holes partially though the material. These perforations make it possible to snap the ceramic apart along the lines. For typical 0.025 in. (0.64 mm) thick alumina substrates, holes are drilled 0.008 in. (0.2 mm) deep and 0.007 in. (0.18 mm) apart. For such conditions a laser pulsing at 1000 Hz can scribe at 7 in./s (175 mm/s).

Quartz. Quartz can be processed much like metal because it has a high thermal shock resistance. Continuous CO_2 radiation is used, since quartz is quite transparent to the 1.06 μm light emitted by YAG lasers. Strains caused by thermal stresses must often be relieved by annealing the parts after cutting.

Glass. The laser cutting of glass is limited by the poor thermal shock resistance of most compositions. This causes complex glass parts to crack apart after cutting. Glass also tends to form recast material on the cut edge, because it does not have a well-defined melting point.

Organic Materials

ORGANIC MATERIALS ARE generally decomposed by laser light. The energy required to do this is usually much lower than that required to melt inorganic substances, so cutting can often be done at high speeds or with lower power lasers. The large volume of decomposition products causes some problems: gases in the kerf have trouble escaping, limiting process speeds and degrading edge quality. In addition, many organic materials evolve toxic compounds during laser cutting. These effluents must be handled in a manner to eliminate hazards to operators and to the environment.

Cloth. Since cloth is so thin, it presents few problems for laser cutting. Most of the difficulties are related to the construction of systems capable of moving fast enough to fully utilize laser cutting capability.

Plastics. A wide variety of polymers are cut with lasers. The beam causes melting, vaporization, and decomposition of the material. Thermoplastics such as polypropylene and polystyrene are cut by shearing of molten material, while thermosets such as as phenolics or epoxies are cut by decomposition. Materials which decompose in the beam leave a carbon residue on the cut edge. This must often be removed by some operation such as bead blasting before the parts may be used. Decomposition products of laser cut polymers have been found to be quite hazardous, and precautions must be taken to protect operating personnel.

Composites

COMPOSITES ARE MATERIAL consisting of two or more distinct constituents. Usually, one component is fibrous, while the other forms a surrounding matrix. By selecting appropriate matrices and reinforcing elements, the material can be engineered to have properties optimized for a specific use. From the standpoint of laser cutting, the main differences between composites are whether the matrix, the fibers, or both are organic.

Organic. If organic fibers are set in an organic matrix, the laser has little difficulty cutting. Kevlar (aramid) fibers in an epoxy matrix, a common high-performance composite, is readily laser-cut in thicknesses up to 1/4 in. Thicker sections exhibit considerable charring of the cut edge.

Organic-Inorganic Materials. The presence of inorganic materials changes the response of composites to laser heating. To cut fiberglass-epoxy, the laser must melt the glass. This takes much more energy than decomposing the epoxy, and so controls the processing rate. Graphite-epoxy is extremely difficult to cut because graphite must be heated to 6500 F to vaporize it. Since graphite has fairly good thermal conductivity, the epoxy near the cutting zone is exposed to high temperatures which decompose it for a significant distance from the cut edge. Laser cutting of graphite-epoxy is thus limited to relatively thin [1/16 in (1.6 mm) or less] sections.

Inorganic Materials. Some of the highest-performance materials available today are metal-matrix composites. The addition of refractory fibers to a superalloy matrix produces tremendous strength at high temperatures, combined with high toughness. Unfortunately, these characteristics also make it difficult to machine such materials. Lasers have successfully cut several types of metal-matrix composites, and should see increasing use for this application. One effect that must be controlled is the melting back of the matrix from the cut edge, leaving exposed fibers. The use of high-energy pulses, as produced by YAG lasers, minimizes that problem.

LASER CUTTING VARIABLES

A GREAT NUMBER of variables affect the results of laser cutting. They can be divided into material-related, laser-related, and process-related variables.

MATERIAL-RELATED VARIABLES

THE PRIMARY VARIABLES that make materials behave differently have been discussed above. Any specific material, however, can behave differently depending on its condition.

Thickness

THICKNESS IS THE most important variable affecting how a given material can be cut with a laser. In general, cutting speed is inversely proportional to thickness.

Surface Finish

FOR HIGHLY REFLECTIVE materials, such as pure aluminum or copper, the surface finish can affect the initial coupling of laser energy. Sheets with extremely shiny surfaces may not cut consistently.

Carbon steel often has rust and scale on it. These oxides interfere with the oxygen-assisted cutting process and cause poor edge quality. One of the factors that makes hot-rolled plate less amenable to laser cutting than cold-rolled is the generally poorer surface of hot-rolled material.

Highly finished sheets of stainless steel and aluminum for decorative applications often have coatings of paper or plastic to protect the surface from scratches during manufacturing. While these layers do not have much direct effect on the cutting process, they can cause problems when the assist gas gets under them and lifts them. This action can foul mechanical parts such as focusing heads.

LASER-RELATED VARIABLES

MANY CUTTING VARIABLES are related to the laser itself. The suitability of a laser for cutting is a function of all of these variables.

Power Limitations

MOST LASERS ARE characterized by their maximum continuous power output. While this is a useful quantity, it does not describe the machine completely. Lasers that operate only in the continuous mode are fairly well defined by a power specification. Others can be pulsed to high peak powers but produce low average powers. YAG lasers, which typically operate in the pulsed mode, will deliver their rated average powers only under specific pulse conditions.

Some materials, such as thick steel, require high continuous power, while others such as alumina must be cut with short, high-energy pulses. As power is increased, cutting speed for a specific material also increases.

The ability to vary laser power using CNC control is important when cutting intricate shapes, because the motion system often cannot maintain constant speed for all features of a part.

Mode

THE IDEAL LASER output for cutting is the fundamental Gaussian mode TEM_{00}. This can be focused on the smallest spot and has the greatest depth of focus (the least change in power density with distance) of all possible modes. CO_2 lasers with power not over 2000 watts can be made to produce beams which closely approximate the Gaussian profile.

The method of evaluating the beam is important: lasers which appear to have a good beam as indicated by the traditional acrylic-mode burn can, in fact, be unstable on a millisecond time scale. Recent work has demonstrated that these short-time variations in the mode are common and have significant effects on cut quality.

Beam stability and focusing ability are especially important in the contour cutting of thin [0.125 in. (3.2 mm) or less] carbon steel sheet. A stable, low-order beam is required to produce surface finishes of 32 μin. or better. Since this is a significant market, considerable attention has been devoted towards achieving such results. When the material thickness exceeds the depth of focus of the laser system, the focusing quality of the beam has less effect on the edge quality.

Duty Cycle

THE HIGHEST SPEED (and often the highest quality) is achieved using a beam that is on all the time. Many situations exist, however, that make it necessary to pulse the beam.

When cutting intricate steel parts with a CO_2 laser, a motion system may not be able to maintain the linear speeds that are appropriate for good cut quality. Reducing the continuous wave (CW) power with travel speed is useful, but is ineffective at speeds below about 20 inches per minute because of bulk heating of the workpiece. The solution to this difficulty is to maintain CW power and pulse the beam to reduce the percent of time that it is on. A typical schedule is to have the beam on 25 percent of the time with a repetition rate of 500 Hz. The actual repetition rate is dependent on the ability of the laser to generate clean pulses. There should be no laser emission during the off part of the cycle, because that heats the material and reduces the benefit of pulsing. The pulses themselves should be uniform in duration and power.

Certain electronically pulsable lasers have high "simmer" levels. Simmer current is applied to ensure uniform response to the pulse current, and can result in significant CW output. Cutting of thin materials with this type of beam gives poor results. Mechanical choppers (used with nonpulsable lasers) actually have an advantage here, since they produce highly uniform pulses and reduce output to absolutely nothing between them. The main disadvantages of mechanical pulsing are limited repetition rate and slow response to commands to change the cycle.

Another type of pulse which is used in cutting is termed *enhanced pulsing* or *superpulsing*. This involves circuitry designed to trigger a pulse whose duration and power are preset. The pulse is usually repeated at a frequency of 10 to 200 Hz for YAG lasers and 100-5000 Hz for CO_2. YAG lasers generally operate this way, and many CO_2 lasers can be set up to do it. Slow-flow CO_2 lasers can produce several times their CW outputs when operated in enhanced pulse mode. Fast-flow lasers, because of their small volume of active medium, cannot deliver the same degree of enhancement and lose effectiveness at high repetition rates.

Operations such as ceramic scribing and cutting of refractory materials are usually done with enhanced pulsing. Short, high-intensity pulses vaporize substances before they have time to conduct heat away. This reduces the volume of molten material, minimizing recast. The same technique produces good results in metal-matrix composites.

Beam Propagation

THE FOCUS DISTANCE and spot size that result when a lens focuses a laser beam are well defined functions of the distance between the lens and the laser. Because of high divergence (the beam spreads out rapidly as it leaves the laser head), YAG lasers are usually set within 4 ft (1.2 m) of the focusing lens and maintained at that distance. The small size of a YAG laser head allows it to be set on a moving axis, so that there are no significant in-process variations in the focus.

CO$_2$ laser heads are large and best kept stationary, and their low divergence allows the beam to propagate 30 ft (10 m) or more. A potential problem arises when the process variables are set correctly for a specific laser-to-lens distance, and then the distance changes, as with a moving-beam system. Large changes in distance will change the focal point, with possible loss of process quality.

PROCESS-RELATED VARIABLES

ONCE A LASER system is built, many of the above variables are fixed. There are, however, a large number of variables that must be controlled to get reliable cutting.

Focusing Lens

THE FOCUSING LENS controls the spot size and depth of focus. For a CO$_2$ laser with a raw beam diameter of 0.8 in. (20 mm), a 5 in. (125 mm) focal length produces a spot 0.01 in. (0.25 mm) in diameter and has a depth of focus of 0.020 in. (0.5 mm). This works well for metals from 0.010 to 0.38 in. (0.25 to 10 mm) thick, and so it is the most popular focal length for such laser systems.

For thin material, a focal length of 2.5 in. (64 mm) gives better results because its spot size is half that of the 5 in. (125 mm) lens. The smaller spot allows higher travel speeds, produces a smoother surface, and leaves a narrower kerf. The depth of focus, however, is only a quarter that of the 5 in. (125 mm) lens and limits the utility of the 2.5 in. (64 mm) lens to materials 1/8 in. (3.2 mm) thick or less.

For thick metal or organic materials, a 7.5 or 10 in. (190 or 250 mm) lens is sometimes used. The long depth of focus provided by such lenses results in straighter kerfs than those made by shorter lenses.

Assist Gas Variables

ALMOST ALL LASER cutting is gas-assisted. Gas-related variables have a significant effect on cutting results. Oxygen reacts with most metals and many nonmetals. Carbon steel is usually cut with oxygen to get the best surface and process rates. Acrylic plastic may be cut with oxygen to achieve very high cutting speeds.

Air is used for cutting aluminum and alumina. Since it is the cheapest assist gas available, it is commonly used for nonmetals, where the gas composition does not make much difference.

Nitrogen gives good results with aluminum, stainless steel, and nickel base alloys. It is reactive with respect to titanium, and should not be used on that metal. Argon, which is inert, must be used to get clean edges on titanium.

Assist Gas Pressure. Material is removed from the cut by gas pressure. This pressure varies from near zero for acrylic to 120 psi (830 kPa) for inert gas cuts. Generally speaking, as the pressure increases, the effectiveness of the gas-sweeping action improves.

For certain applications, however, the assist gas pressure cannot exceed specific limits. For example, in oxygen-assisted cutting of carbon steel, excess pressure causes uncontrolled burning of the material. Thick plate is usually cut at pressures of 10 to 20 psi (70 to 140 kPa) measured in the cutting head.

In thick organic materials, high assist gas pressure results in incandescent decomposition products in the kerf. These radiate energy and widen the kerf in the middle of the cut face.

Assist Gas Nozzle. The gas pressure in the laser head is transmitted to the workpiece through a nozzle which is coaxial with the laser beam. For laminar flow, a nozzle must have a high-aspect (length to diameter) ratio. Such a design isn't compatible with beam focusing optics, so compromises must be made.

Cutting nozzle diameters vary from 0.030 to 0.125 in. (0.75 to 3.2 mm). The smaller sizes are used with thin materials. Cutting 1/4 in. (6.4 mm) steel with a nozzle smaller than 0.06 in (1.5 mm) diameter gives poor results because the pressure profile of a small nozzle doesn't extend far enough from the beam centerline to clean up the bottom of the kerf. A nozzle too large for a given material uses excessive amounts of assist gas.

Nozzle damage has serious effects on cut quality. Asymmetry in the opening causes changes in performance as the direction of cutting varies. It is not possible to get good results in metal cutting with a dented or burned nozzle.

Nozzle Standoff. The distance between the nozzle and work controls the pressure in the kerf. The relationship is not linear because most laser cutting is done at supersonic flow velocities, and the resulting shock waves produce complex pressure patterns. The pressure at the workpiece can, in fact, decrease as the nozzle is brought closer to it. Typical standoff distances are of the order of the nozzle diameter. It is often more critical to control the nozzle standoff than it is to maintain beam focus.

Travel Speed. One of the reasons that laser cutting is used is that the process rates are high. In contour cutting, process rate is the same as travel speed. For a given material, thickness, and laser power, there is a range of speeds that gives satisfactory results. Above the maximum speed, the cut doesn't go through or has excessive slag. Below the minimum, the heat from the cutting process destroys the edge of the work.

For most materials, cutting speed at constant laser power is more or less inversely proportional to thickness. There is a characteristic maximum thickness, above which no cutting will occur at any travel speed, and there are dynamic effects which reduce process efficiency at very high speeds.

It is often impossible to maintain the linear speed that gives the best results. For example, 16 gage or 0.060 in. (1.5 mm) cold-rolled steel should be cut at about 150 in./min (64 mm/sec) with 500 watts of CW power from a CO_2 laser. Typical laser-cut parts, however, are too intricate for most motion systems to trace out at this speed. Corners, for example, require that one axis decelerate to zero and the other accelerate to the cutting speed. If motion accelerates at 0.1 g (3.2 ft/sec^2), the table must travel 0.080 in. (2.0 mm) before it reaches 150 in./min (64 mm/sec). The reduced speed in the corner can cause burning of the part.

Laser systems incorporate several ways of dealing with this. One way is to vary CW power as a function of speed. This is very effective when the right relationship is used and the laser responds quickly to power commands. Another method is to change to pulsed operation and cut at low speed. While simpler to implement than power control, pulsing has the obvious disadvantage of increasing processing time.

Controlling the duty cycle as a function of speed has the potential of maximizing speed and quality: at full speed, the laser runs CW. As speed drops in corners or small radii, the laser is pulsed at a high repetition rate. The percent of time that the beam is on is varied to suit the instantaneous speed. The range of travel speeds accomodated by a variable duty cycle is much greater than the range that varying CW power can handle. With a suitable schedule of duty cycle vs. speed, optimum quality can be achieved on any geometry.

Characteristics of Cuts

LASERS ARE USED for cutting because of the high quality of the cuts produced. The attributes of laser cutting are narrow kerf width, smooth surface finish, clean edges, and good dimensional accuracy.

Kerf Width. Kerf widths produced by CO_2 lasers range from .004 to .040 in. (0.1 to 1.0 mm). The usual goal is to generate the narrowest kerf possible, since that minimizes the amount of material that is removed. This has two ad-

vantages: The heat input is reduced and accuracy is increased. Short focal length lenses, which have small focused spot sizes, are used to produce narrow kerfs. As material thickness increases, the kerf width tends to widen. Narrow kerfs in thick material make it difficult for the cut material to be ejected. Carbon steel has a tendency to start burning back from the cut line, further widening the kerf.

Roughness. One gage of cut quality is the degree of surface roughness. The ability to produce finished parts can depend on maintaining acceptable smoothness. It is possible to cut 20 gauge or 0.036 in. (0.92 mm) carbon steel sheet with an average roughness (R_a) less than 32 μin. This type of finish is adequate for most purposes. Laser stability, motion-system smoothness, and beam-delivery rigidity must all be optimized to achieve such results. As steel gets thicker, the roughness of the edge increases. The best finish achievable on 3/8 in. (9.5 mm) plate is on the order of 250 μin. R_a. Inert gas cutting, used on many metals to obtain weld-ready edges, uses high pressure to cut. The turbulence created by this pressure increases surface roughness to about 63 μin. on 0.063 in. (1.6 mm) thick material.

Other materials have different characteristics. Acrylic plastic, which vaporizes during cutting, can have an 8 μin. finish on a 1 in. (25 mm) section if the assist gas flow is low enough to avoid turbulence. Plastics such as polycarbonate, which decompose in the beam, are much rougher. It is hard to produce finishes better than 250 μin. on polycarbonate.

Dross. Gas-assisted laser cutting of metal works by pushing molten material out of the narrow channel created by a focused laser beam. Under some circumstances a portion of this material adheres to the bottom of the cut edge. This slag or dross is always undesirable and often unacceptable. With carbon steel, dross appears when the focus is incorrect, the gas pressure is too low, or the travel speed too high. Cuts in stainless steel and aluminum are very likely to have slag adherence; extremely high-assist gas pressures are often needed to eliminate it, even in thin sections. Anti-spatter coatings such as graphite can be used to reduce the adhesion of recast material to the bottom of a laser-cut sheet.

Dimensional Accuracy

THE ACCURACY ATTAINED by laser cut parts is a function of the following:

(1) Table accuracy
(2) Ability of the CNC to contour to the programmed path
(3) Stability of the laser beam

(4) Distortion induced in the workpiece by the cutting process

Machines which produce close-tolerance parts must limit their travel speeds to keep motion errors to a minimum. Once a table and its control are able to follow a programmed path accurately, a beam delivery system must be constructed that will inhibit vibration and deflection during cutting. In addition, changes in focal position or focused spot size will change the effective cutting size, which will alter the dimensions of the part. The workpiece itself is the last source of dimensional error. If the workpiece moves because of thermal expansion during cutting, the parts cut out of it will not match the tool path. As laser cutters approach accuracies of 0.0001 in. (0.0025 mm), thermal effects will be more apparent. The only way to deal with them at present is to distort the part program in the opposite direction.

Setting up for CO$_2$ Cutting

AS INDICATED ABOVE, several areas must be considered before consistent quality cutting can be done.

Alignment. The beam coming from the laser goes through several optical elements before it hits the work. Correct alignment of the beam-delivery system is essential for proper operation.

It is relatively easy to align a fixed beam system. As long as the beam does not clip (hit something opaque like the side of a mirror housing) and goes through the middle of the focusing lens and gas nozzle, the system is aligned. The stationary elements of a fixed beam delivery also tend to stay aligned because they aren't subject to shaking or vibration.

A moving-beam system is aligned if there is no change in the beam location when the axes are run through their range of motion. This is usually checked for each axis at both extremes of its travel, and mirrors are adjusted until the beam stays in place. Moving-beam systems have a tendency to become misaligned because they have many mirrors, long beam paths, and moving parts.

Gantry-type systems are aligned much like moving beams. Rotational axes add some difficulty because they require that the beam be parallel to the axis within 0.2 milliradian to maintain nozzle alignment when the axis rotates.

Beam Focus. Laser-cutting quality depends on the focusing of the beam. The relation of the focal point to the surface of the work is one of the most important variables in the process.

Finding the Focal Point. Since there is significant variation between different lenses of the same nominal fo-

cal length, it is necessary to test each one under power. There are several tests for focal point. One method is to make a flat position weld along a sloping plate and measure from the nozzle to the narrowest part of the weld bead. Another is to make a series of cuts in thin metal while changing focus and find the thinnest kerf. Whatever method is used, it is important to be consistent so that process data have continuity.

Setting the Focus Point Position. Focus for most metal cutting is at or slightly below the surface of the work. With inert gas cutting, slag is minimized by locating the focus deeper into the material. The focus can be set with calipers, feeler gauges, or through CNC commands.

Maintaining Focus. It is important to keep the focus in the same place throughout the cutting process. It is easy to do this with flat sheets, but most material has some warping. The cutting system must have some form of focus control to accommodate out-of-flat sheets.

The focusing lens is part of the pressurized head and is limited in the pressure that it can stand. A standard 1.1 in. (28 mm) diameter x 0.10 in. (2.5 mm) thick zinc selenide lens will take up to 80 psi (550 kPa). Higher pressures, such as are used in inert gas cutting of metals, require thicker lenses.

At high pressures, the cost of operating the system increases, from increased gas consumption. In addition, there is more chance of leakage and seal damage.

Assist Gas. Table 16.2 shows commonly used combinations of assist gases and the materials on which they are used.

Concentricity. The focused laser beam must go through the center of the assist gas nozzle to get uniform cutting performance in all directions. All laser systems provide some means of adjusting for concentricity, and there are several ways to check it. One of the most accurate ways is to pierce a hole in thin [0.030 in. (0.75 mm)] steel while observing the material to see the direction that metal is ejected. The lens or nozzle is then adjusted to make the ejected metal form a uniform starburst around the nozzle. This will occur when the beam and nozzle are concentric to within 0.002 in. (50 μmm), which is the order of accuracy needed.

TROUBLE SHOOTING

CONSIDERING THE FACT that carbon dioxide lasers are now being used to process a wide variety of metallic and nonmetallic materials, it can often be difficult to identify causes of poor-quality cuts. A deterioration in cutting performance will usually be attributable to one of the following conditions:

Incorrect Cutting Speed

THE EFFECT OF cutting speed on cut quality for individual materials has been discussed in previous sections. Often the speed which gives the best quality is somewhat slower than the maximum speed. But slowing down beyond a certain point will also reduce the quality. Consistent results will be obtained when the optimum speed is determined empirically.

Relatively small changes in chemical composition of ferrous metals can produce significant changes in optimum cutting speed when cutting with oxygen as the assist gas.

Generally, cutting speed is directly related to laser power and the power density at the workpiece. If it becomes necessary to reduce the cutting speed from a previously determined optimum, then a fault involving loss of power or power density should be suspected. Loss of power from the laser itself will usually be indicated by a lower reading on a power meter internal to the laser. Loss of power could also occur along the path of the beam between the laser and the focusing lens, if any of the reflecting mirrors become dirty. If laser power has not changed, and the material being cut has not changed, then the need to reduce cutting speed will likely have resulted from reduced power density, caused by a larger focused spot at the work surface. The larger spot usually produces a wider cut than was previously obtained.

Other potential causes of reduced power density include a distorted laser output coupler, and organic or other absorbing vapors in the beam path. Freon, trichloroethylene, paint solvents, and polymer plasticizing agents are some such absorbing vapors. A small, positive flow of clean, dry air or nitrogen into one end of the beam path between the laser and the focussing lens is usually sufficient to keep such vapors out.

Incorrect Cutting Gas or Cutting Gas Pressure

WHEN CHANGING FROM cutting one type of material to another, it may be necessary to change the type of gas used. Attempting to cut flammable materials with pure oxygen is a potential fire hazard. Attempting to cut most metals with air or inert gas would give the appearance of cutting with insufficient power.

A deterioration of cut quality can also be noted as the pressure of assist gas varies from its optimum level. One example of this occurs as a gas cylinder empties. The effect noted would be a greater accumulation of oxide slag when cutting metal.

Incorrect Nozzle Height

IN METAL CUTTING, the nozzle should be relatively close to the surface [0.02 to 0.08 in. (0.5 to 2 mm)], to ensure maximum removal of molten slag. When cutting materials where there is no molten cutting product to be removed, the spacing is less critical. In the case of plastics that are softened by heat, such as acrylics, there can be a frosting effect on the cutting edge produced by the gas flow from the nozzle. This effect can be minimized by increasing the nozzle to workpiece distance and by using minimum gas flow.

A height-control probe can be used to maintain a constant nozzle-to-workpiece distance. Both contact and noncontact sensors are available to detect workpiece undulations. Noncontact devices, such as capacitive sensors, are best suited to metals.

Incorrect Lens Focal Length or Beam-Focus Setting

THIS SITUATION IS most likely to occur after changing a lens.

If the point of focus is considerably above or below the nozzle tip, the nozzle will intercept part of the beam and hence become very hot. There will be less power reaching the workpiece, resulting in a reduction of cutting performance. There may be reflections off the bore of the nozzle which can cause burn marks at the side of the cut; this can be particularly noticeable in thermally sensitive materials, such as paper and plastics.

If the focus is inside or just above the nozzle tip, the beam may pass safely through the nozzle orifice, but the beam will be diverging when it reaches the surface of the workpiece; this will result in a wider kerf than normal, and because of the loss of power density, a lower cutting speed may result.

Defective or Dirty Lens

IF THE LENS becomes defective or dirty, then the position of the focal point will change during cutting operations due to thermal lensing. If this happens during a cutting operation, its effect would be as described above for incorrect beam focus setting.

It should be noted that a reduction in focal length can also occur due to a thermally focussing laser output coupler.

Incorrect Alignment of the Beam in the Cutting Head

IF THE LASER beam, as it exits from the nozzle, is not concentric with the gas jet, an asymmetric cutting action can take place. If the misalignment is such that the beam clips the nozzle, overheating of the nozzle will occur.

The effect of asymmetric metal cutting is to induce a burn-out action preferentially on one side of the cut, or to produce a cut with asymmetric dross adherence on the bottom surface. When preferential burn-out occurs, it is due to the beam being offset in the nozzle towards the side where burn out occurs.

Damaged Nozzle Tip

THIS CAN OCCUR as a result of molten oxide blown on the nozzle when piercing metal or when attempting to cut metal too fast.

The effect is the same as a misalignment of the beam in the nozzle, because the gas jet profile will be permanently asymmetric through damage.

Effect of Polarization

POLARIZATION OF THE laser beam is particularly important when cutting ferrous and other reactive metals with oxygen. Laser light may be polarized in several different ways: linearly, elliptically, circularly, or randomly, depending on the design of the laser. The best results in oxygen-assisted cutting of metals are obtained by using circular polarization. Linear and elliptical polarizations will not cut the same in all directions of travel, and they tend to produce a slanted cut edge in some directions. Random polarization will produce an acceptable cut only if it remains consistently random. A laser which produces linearly polarized light can be made to cut well by inserting optical devices (known as *phase shifters* or *circular polarizers*) into the beam path which convert the linear polarization to circular polarization.

INSPECTION AND QUALITY CONTROL

INSPECTION

INSPECTION CRITERIA FOR laser cuts are largely dependent on the material to be cut. Three areas of concern when inspecting laser cut materials are physical appearance, dimensional accuracy, and thermal alterations.

Visual inspection is the first and often the only inspection method in laser cutting. A laser-cut surface is visually inspected for dross (resolidified metal attached at the bottom of the cut), which is usually unacceptable. Surface roughness is viewed qualitatively to determine if the cut is similar to previous acceptable cuts produced in the same metal. Color of the cut metal edge is also a consideration.

Some metals, such as titanium, stainless steels, and nickel based alloys are usually cut with inert gas to produce oxide free cuts with a bright silver appearance. Oxide-free cuts are advantageous when the cut component will subsequently be welded, or when the cut surface is exposed in the end product. The angle of the striations in the laser cut is viewed because of its relationship to the cut speed. If cut rates are near the maximum speed, the vertical striations deflect at the root of the cut. Slower cutting speeds will yield striations that are completely vertical.

Nonmetals such as plastics, ceramics, wood, and composites are often cut with lasers. The appearance of the cut surface for these materials varies greatly. Cuts made with proper conditions produce a fire polished edge on thermoplastics. Thermoset plastics are cut with the objective of minimizing charring or discoloration. Ceramics are visually inspected for cracks due to their low ductility and toughness.

Dimensional accuracy is another factor in cut quality. Components can be inspected with traditional measuring devices and accuracies of ± 0.001 in. (25 μmm) are commonly achieved. A controlling factor in dimensional accuracy is the surface finish of the cut.

The surface roughness on laser cut metals varies through the thickness of the cut. Typically, the top surface is smoother than the bottom surface. Therefore, the surface roughness measurements should always be taken in the same location.

Taper or parallelism is another dimensional value on which laser cuts are evaluated. The minimum value for parallelism is dependent on the material cut. Parallelism in metal cutting can be held within 5 to 25 angular minutes for sheet metal.

Thermal alterations to the substrate can have dramatic effects on the service life of the laser-cut component. Inspection for thermal alterations is usually accomplished destructively.

Metals that are cut with lasers are inspected for the size of the heat-affected zone (HAZ), the amount of resolidified metal on the cut surface (recast), and the length and number of microcracks penetrating into the recast, HAZ, and base metal.

The HAZ in laser-cut metals varies with composition and thickness. The width of the HAZ is usually between 0.001 and 0.010 in. (0.025 to 0.25 mm). The HAZ is uniform along the face of the cut. Dross on the bottom of the cut can increase the HAZ at the root of the cut.

Laser cutting of metals produces a liquid phase in the metal, which is removed with a coaxial gas jet. Some of the molten phase clings to the base metal and resolidifies on the walls of the cut surface. This resolidified metal is known as *recast* or *remelt*. The depth of the recast is usually only a few thousandths of an inch in laser cutting.

Microcracks can result from the thermal input of laser cutting. The laser cutting process can produce high thermal stresses at the cut edge which may result in the nucleation of microcracks. These small cracks can be detrimental to the service life of the laser-cut component if the material has poor toughness. Some metals will microcrack easier than others. For instance, heat treatable aluminum alloys lose ductility at elevated temperatures, a phenomenon known as hot shortness. These metals are particularly sensitive to the formation of microcracks.

Microcracks are quantified by metallographic cross-section to determine either maximum crack length, average crack length, or the total number of cracks. The location of the microcracks is also pertinent. Microcracks in the recast layer may be acceptable, but microcracks extending into the HAZ or parent metal may not be acceptable. Acceptability of the size, number, and location of microcracks is dependent on the toughness of the metal, the intended service for the laser-cut component, and industry specifications.

Thermal alterations to nonmetals could be advantageous or detrimental. A laser cut in a fibrous material in a thermoplastic will seal the edge, while mechanical cuts will leave a frayed edge. Delamination caused by laser cutting in other composites can lead to premature failure.

QUALITY

HIGH QUALITY LASER cuts can be produced when the proper procedures are followed. The high-energy density achievable with this process allows materials to be separated with minimal-heat input and minimum alteration of the cut surface.

A key factor in obtaining good quality with minimum-heat input to the material is the laser mode. The mode governs the energy distribution across the laser beam. The optimum laser mode has a gaussian distribution. This gaussian distribution in laser modes is referred to as TEM_{oo}. A gaussian mode allows the laser beam to be focused to the smallest spot size for a given focal-length lens. The smallest spot size will yield minimum heat input and maximum feed rates.

The focal length of the lens also affects quality. Usually, as the material thickness increases, the focal length should also be increased for a given beam diameter. The longer focal length lens will have greater depth of field, which will maintain the proper power density to cut the material and minimize taper.

Focal position in the material is important to maintain consistant results. Often this is the only variable controlled in real time using autofocus techniques. The two most common autofocus methods are mechanical and capacitive sensor. The mechanical method operates on a spring-loaded mechanism that rides on the material being cut to maintain proper focus. This method is primarily used when cutting flat sheet. The capacitive sensor method is used on conductive materials.

The proper combination of the above variables will produce excellent quality cuts in a wide variety of materials.

LASER CUTTING SAFETY

THE AREAS OF safety concern for laser cutting may be divided into the following categories:

(1) General safety
(2) High-voltage power supplies
(3) Exposure to direct or reflected light
(4) Fumes from materials being cut

Each of these areas is discussed separately in the following sections. The section on general safety applies to the other sections since it includes definitions and terms used throughout this guide.

Laser safety guidelines should be stressed and understood by all persons who operate or work in the vicinity of lasers.

GENERAL SAFETY

THE STANDARD USED in the United States to design a laser facility is ANSI Z136.1 (latest edition), *Safe Use of Lasers*. This specification details the minimum criteria to be met for facility construction and defines the common terminology for laser safety. Although new facilities should have no difficulty in meeting these requirements, it should be remembered that modifications to existing facilities should also meet these requirements.

ANSI Z136.1 also defines the hazard classifications for lasers. Four classes are defined, of which only Class IV lasers ("high power") are typically used for cutting. However, some laser cutting systems do use a "low-power," visible-light, helium-neon laser (He-Ne) for beam-alignment purposes. Proper warning signs or signals should be posted around areas that are exposed to laser beams. Some form of light-tight enclosure must surround areas in which these beams are exposed to the atmosphere. It should be remembered that a high-power collimated or unfocused beam is more dangerous over large distances than focused beams, which diverge much more rapidly.

Some lasers may be extremely noisy, especially if used in enclosed areas. A hearing protection specialist should be consulted to recommend the proper methods to guard against excessive noise.

HIGH-VOLTAGE POWER SUPPLIES

SINCE HIGH VOLTAGES as well as large capacitive storage devices are associated with lasers, the possibility for lethal electric shock is ever present. All reported laser-related deaths have been associated with the high voltage present in the laser system.

All electrical components should comply with NEMA Standards and ANSI/NFPA 70 (latest edition). All personnel working around the high-voltage components of a laser should be trained in the proper safety techniques for electrical systems. Appropriate grounding and interlocking devices should be employed around any high-voltage components. There should be provisions for discharging capacitors before human access to areas containing electrically charged components.

EXPOSURE TO DIRECT OR REFLECTED LIGHT

BEAM EXPOSURE IS the most common safety hazard associated with laser cutting. Lasers that can cut engineering materials can also cause great damage to the human body. Laser beam exposure can cause eye damage, including burning of the cornea or the retina, or both. Lasers may also cause severe skin and tissue damage on unprotected areas of the body.

Two main references are available for eye protection around lasers: ANSI Z87.1 (latest edition) on eye and face protection, and the Laser Institute of America's *Guide for the Selection of Laser Eye Protection*, provide guidelines for proper eye protection. The main concern when choosing eye protection for lasers is blocking the wavelength of light being used for welding or cutting. At high-beam power, lasers tend to produce plasma plumes of extreme brilliance. Shaded glasses must be worn to protect against these brilliant light sources. Frequent eye examinations should also be part of the eye protection program to ensure that eye protection is adequate.

FUMES FROM MATERIALS BEING CUT

MANY MATERIALS THAT are cut with lasers emit toxic vapors, dusts, or fumes. Studies have shown that laser cutting of polymethyl metharcrylate, polyvinyl chloride, and Kelvar produces byproducts containing toxicants and carcinogenic compounds. Precautions should be taken so that proper ventilation is supplied in the area of laser operation. Before cutting any material, Material Safety Data Sheets should be consulted to determine associated health hazards and prevention techniques. Fire extinguishers should also be available in case a fire is started by the laser cutting process.

WATER JET CUTTING

INTRODUCTION

WATER JET MACHINING, also called *hydrodynamic machining*, cuts a wide variety of materials, both nonmetals and metals, using a high-velocity water jet. The jet is formed by forcing water through a 0.004 to 0.024 in. (0.1 to 0.6 mm) diameter orifice in a man-made sapphire under high pressure (30,000 to 60,000 psi [207 to 414 MPa]). Jet velocities range from 1700 to 3000 ft/s (520 to 914 m/s). At these speeds and pressures, the water erodes many materials rapidly, acting like a saw blade. The water stream, with a flow rate of 0.1 to 5 gallons/min. (0.4 to 19 L/min) is usually manipulated by a robot or gantry system, but small workpieces may be guided past a stationary water jet by hand. A typical range of nozzle to work distances is 0.010 to 1.0 in. (0.25 to 25 mm), with distances under 1/4 in. (6.4 mm) being preferred.

Metals and other hard materials are cut by adding an abrasive in powder form to the water stream. With this method, called hydroabrasive machining or abrasive-jet machining, the abrasive particles (often garnet) are accelerated by the water and accomplish most of the cutting.

Higher flow rates of water are required to accelerate the abrasive particles.

Materials are cut cleanly, without ragged edges (unless the traverse speed is too high), without heat, and generally faster than on a bandsaw. A narrow (0.030 to 0.100 in. [0.8 to 2.5 mm]), smooth kerf is produced. There are no thermal, delamination, or deformation problems when properly applied. Dust is nonexistent.

HISTORY

THE ANCIENT EGYPTIANS used sand combined with water for mining and cleaning. Sandblasters in this century used a pressurized stream [500 psi (3400 kPa)] for cleaning and paint removal. In 1968, Franze patented a concept for a very high-pressure water jet cutting system. His patent for producing a coherent cutting stream involved the addition of a long-chain liquid polymer to the water stream to prevent it from breaking up as it left the exit orifice of the pressurized chamber.

Prior to its application as a cutting tool in industry, high-pressure water was used for cutting in both forestry and mining. In the 1970's, high pressure (30 000 to 55 000 psi [207 to 379 MPa]) water jet cutting technology was developed to cut nonmetals. The first commercial water jet cutting system was sold in 1971, to cut furniture shapes from laminated paper stock that bandsaws, reciprocating saws, and routers couldn't handle well. In 1983, the process was modified by the addition of abrasives such as silica and garnet particles to the stream to cut metals, composites, and other hard materials.

SCOPE

WATER JET AND abrasive water jet systems compete with such processes as bandsaws, the reciprocating knife, flame cutting, plasma, and laser cutting. They can handle materials that suffer heat damage from thermal processes or gum up mechanical cutting tools. In some cases, they can cost-effectively replace three operations: rough-cutting, milling, and deburring of contoured shapes.

The extremely wide range of materials which may be cut can be seen from Table 16.8. Water and abrasive jet machining are often thought of as sheet-material processing systems, but this need not be the case. Examples of cuts made to test the limits of the process are 7.5 in. (190 mm) thick carbon steel, 3 in. (75 mm) thick 7075 T-6 aluminum, 2.5 in. (64 mm) thick graphite/epoxy with 470 plies, and 10 in. (250 mm) thick titanium.

USES AND ADVANTAGES

THE WIDE APPLICATION range and lack of heat are the major advantages of water jet cutting. The versatility of the process is demonstrated by the simultaneous cuts through carbon steel, brass, copper, aluminum, and stainless steel shown in Figure 16.9. An abrasive jet is particularly good for cutting laminates of different materials, including sandwiches of metals and nonmetals. Since the abrasive jet can penetrate most materials, no predrilling is required to start, and cutting may be omnidirectional. Multiple shapes can be nested and cut, depending on the limits of the control system and the workpiece size. Tapering of the kerf is gen-

erally not a problem unless the cutting speed is too high, the workpieces are too thick, or worn nozzles are involved. Minimal or no deburring is required. The process is easily adapted to robotic control.

There are no tools to wear out, other than the orifice and the nozzle; perhaps there will be some wear on the robot mechanism. Minimal lateral forces are generated, simplifying fixturing.

Tolerances depend on the equipment and the workpiece material and thickness, but can be as close as ± 0.004 in. (0.1 mm) on dimensions and ± 0.002 (50 μm) on positioning. Laser cutting achieves closer tolerances.

Finishes vary widely. Abrasive water jet finishes on aerospace components have been reported in the 63 to 250 μin. Ra range.

In simple water jet cutting, the kerf width is usually 0.005 in. (0.13 mm) or wider; in abrasive water jet cutting it is usually 0.032 in. (0.8 mm) or larger. The water jet tends to spread as it leaves the nozzle, so the kerf is wider at the bottom than at the top. Kerf tapering may be reduced by adding long chain polymers, such as polyethelyne oxide, to the water, or by reducing cutting speed.

With the exception of sophisticated systems for aerospace applications, most abrasive and water jet CNC systems are relatively easy to program.

LIMITATIONS

RELATIVELY LOW CUTTING speeds are the chief limitation of the water jet cutting system. Typical cutting speeds are shown in Table 16.9. Another limitation is that a device must be provided to collect the exhaust liquid from the cutting stream. Initial capital costs are high because of the pumps and pressure chamber required to propel and direct the water jet.

The material to be cut must be softer than the abrasive used. Very thin ductile metals tend to suffer bending stress from an abrasive jet and show exit burrs. Ceramics cut with a water jet show a decrease in as-fired strength.

Nozzles must be replaced every two to four hours (sometimes even more frequently) in abrasive water jet sys-

Table 16.8
Water Jet Cutting Speeds on Various Materials

Material	Thickness		Travel Speed	
	in.	mm	in./min.	mm/s
ABS Plastic	0.080	2.0	80	34
Cardboard	0.055	1.4	240	102
Corrugated cardboard	0.250	6.4	120	51
Circuit board	0.103	2.6	100	423
Leather	0.063	1.6	3800	1600
Plexiglass	0.118	3.0	35	15
Rubber	0.050	1.3	3600	1500
Rubber-backed carpet	0.375	9.5	6000	2500
Wood	0.125	3.2	40	17

Figure 16.9—Abrasive Jet Stack Cutting of Various Metals

tems. The abrasive grit wears the carbide nozzles to an out-of-round condition, and the jet loses its symmetry, causing cut quality to deteriorate.

The water supply should optimally be deionized water filtered to 0.5 micron particle size to reduce maintenance, but other water-treatment options are possible. Many systems operate successfully with simple line filters on the incoming municipal water supply, if the water is relatively soft. Waste water and slurry from the cutting operation must be disposed of properly.

The fatigue life of abrasive water jet cut edges in critical aerospace structures can be lower than for a raw-sheared

edge if the abrasive particle used is a coarse 60 grit. Decreasing the particle size to 150 grit increases fatigue life 50 percent or more, but at a corresponding decrease in cutting speed.

FUNDAMENTALS

INCOMING WATER FIRST passes through a booster pump to pressurize it to about 190 psi (1300 kPa) and to filter it. Then an intensifier pump (a hydraulically driven double acting reciprocating type pump) creates a water pressure of 30 000 to 60 000 psi (207 to 414 MPa) with a flow rate of up to 3.5 gallons/min. (13.3 L/min.). Forced through a sapphire orifice, the stream forms a water jet. The jet velocity depends on the water pressure.

For abrasive cutting, dry abrasives may be fed from a hopper into a mixing chamber. There the water accelerates the particles to supersonic velocities. The high-speed slurry is focused and then exits the nozzle in a stream 0.020 to 0.090 in. (0.5 to 2.3 mm) in diameter. Water jets can be made with jet diameters down to 0.003 in. (80 μmm) in diameter, suitable for cutting paper. Abrasive jets are generally not made smaller than 0.009 in. (0.23 mm) in diameter.

Depending on the properties of the target material, the actual cutting is a result of erosion, shearing, or failure under rapidly changing localized stress fields. The process does not produce thermal or mechanical distortions. There is a slight work hardening of metals at the cut surface. Downstream of the kerf, the water or water-abrasive stream is collected in a tank or catcher.

PROCESS VARIATIONS

CUTTING DEPTH AND surface characteristics of the cut vary with the following variables: (1) waterjet pressure and diameter; (2) the size, type, and flow rate of the abrasive material; (3) traverse speed; (4) angle of cutting; and (5) number of passes.

Increasing the pressure, increasing the jet diameter, and lowering the traverse speed all increase the thickness and density of workpieces that can be cut with a water jet. Increasing the flow rate of the water, the abrasive, or both and increasing the abrasive size will increase the cutting speed of an abrasive jet. Use of smaller abrasive particle sizes and slower cutting speeds will improve the edge quality of both cuts.

Increasing the water pressure in abrasive jet cutting increases the plate thickness cutting capability because of increased particle velocities. The optimum pressure tends to remain in the 30 000 to 45 000 psi (207 to 310 MPa) range, because higher pressures result in increased equipment maintenance costs with only slight process advantages.

Fine abrasive particles, below 150 mesh, are relatively ineffective; the most effective general purpose size for cutting metals is 60 or 80 mesh. For very hard ceramics, boron

Table 16.9
Cutting Speeds on Various Materials With Abrasive Water Jet

Material	Thickness		Travel Speed	
	in.	mm	in./min.	mm/s
Aluminum	0.125	3.2	40	17
Aluminum	0.50	12.7	18	8
Aluminum	0.75	19.0	5	2
Brass	0.125	3.2	20	8.5
Brass	0.425	10.8	5	2
Bronze	1.0	25.4	1	0.5
Copper	0.063	1.6	35	15
Copper	0.625	15.9	8	3
Lead	2.0	50.8	8	3
Carbon steel	0.75	19.1	8	3
Cast iron	1.5	38.1	1	0.5
Stainless steel	0.1	2.5	25	25
Stainless steel (304)	1.0	25.4	4	2
Stainless steel(304)	4.0	101.6	1	0.5
Armor plate	0.75	19.1	10	4
Inconel	0.625	15.9	8	3
Inconel 718	1.25	31.8	1	0.5
Titanium	0.025	0.6	60	25
Titanium	0.500	12.7	12	5
Tool steel	0.250	6.4	10	4
Ceramic (99.6% aluminum)	0.025	0.6	6	2.5
Fiberglass	0.100	2.5	200	85
Fiberglass	0.250	6.4	100	42
Glass	0.250	6.4	100	42
Glass	0.75	19.1	40	17
Graphite/epoxy	0.250	6.4	80	34
Graphite/epoxy	1.0	25.4	15	6
Kevlar	0.375	9.5	40	17
Kevlar	1.0	25.4	3	1.3
Lexan	0.5	12.7	12	5
Metal-matrix composite	0.125	3.2	30	13
Pheonolic	0.5	12.7	10	4
Plexiglass	0.175	4.4	50	21
Rubber belting	0.300	7.6	200	85

carbide abrasive is sometimes used.

High-abrasive flow rates result in high cutting costs: a nominal 2 lb/min flow rate at $0.12/lb will result in an hourly cost of $14.40, not including cleaning and disposal costs. This represents a large portion of the total hourly cost. These high flow rates also result in rapid wear of the mixing nozzles.

While many operations are completed in a single pass, the optimum cutting of thick metals may require multiple passes at an optimum traverse rate. This will increase the standoff distance at each pass, requiring a lowered traverse rate.

EQUIPMENT

THE KEY PIECES of equipment for a water jet or an abrasive water jet system are (1) the special high-pressure pump or intensifier used to provide the stream of water, (2) the plumbing and tank or catcher unit to handle the water, (3) the gantry, robotic, or other delivery system to traverse and guide the water jet, and (4) the nozzle assembly unit, which forms the jet. In the case of abrasive water jets, there is an abrasive delivery system including a hopper, a metering valve, and a mixing unit, which mixes the abrasive par-

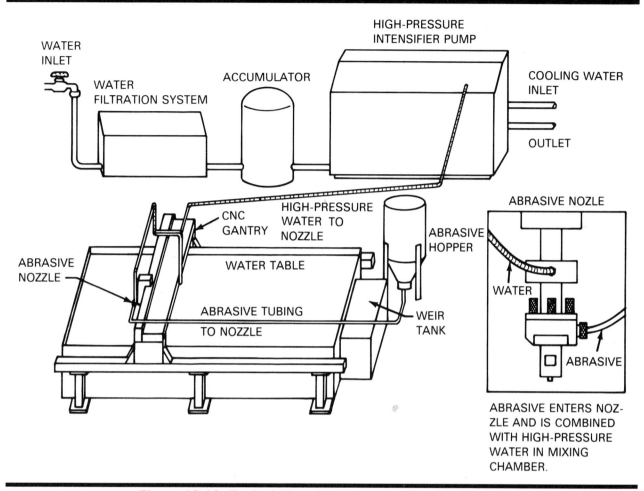

WATER INLET

WATER FILTRATION SYSTEM

ACCUMULATOR

HIGH-PRESSURE INTENSIFIER PUMP

COOLING WATER INLET

OUTLET

CNC GANTRY

HIGH-PRESSURE WATER TO NOZZLE

ABRASIVE HOPPER

ABRASIVE NOZZLE

WATER TABLE

WATER

ABRASIVE NOZZLE

ABRASIVE TUBING TO NOZZLE

WEIR TANK

ABRASIVE

ABRASIVE ENTERS NOZZLE AND IS COMBINED WITH HIGH-PRESSURE WATER IN MIXING CHAMBER.

Figure 16.10–Typical Abrasive Water Jet Cutting System

ticles into the water stream. See Figure 16.10.

Equipment is available in a range from individual components to finished machine tools. More complex systems, such as 5-axis robotic systems, tend to be custom-built. There are several instances of flame cutting machines converted to water jet cutting.

Consumables

THE MAIN WEAR item on the equipment is the sapphire orifice and, on abrasive systems, the carbide abrasive nozzle. On pure water jet systems, a man-made sapphire may last up to 200 hrs. In abrasive systems, the carbide abrasive nozzles last only two to four hours. Other consumables are water, abrasive, and electricity. Abrasive particles are used at the rate of 0.25 lb. to 3.0 lbs. (0.1 to 1.4 kg) per minute.

Accessory Equipment

AUXILIARY EQUIPMENT FOR loading and unloading workpieces, such as cranes, gantry robots, or pedestal mounted robots, may used. This work handling equipment is generally distinct from the system, robotic or other, which drives the water jet cutting head.

For contour cutting in five axes it may be necessary to have a special catcher device to stop the water jet and dissipate its energy.

Hard water may require a water-treatment system.

Periodic cleaning of the water table to remove abrasive grit and metal particles generated during cutting is a necessary operation.

APPLICATIONS

THERE ARE NOW hundreds of factory applications in place in dozens of countries, including over 100 water-jet-equipped robots. Industries which use the technology include automotive, aerospace and defense, building supplies, circuit boards, fabrication shops, foundries, food, glass, job shops, mining, oil and gas well equipment, packaging, paper, rubber, shipyards, and steel service centers. A steel, circular saw-blade cut using hydroabrasive machining is shown in Figure 16.11.

Aerospace applications include the abrasive jet cutting of advanced composite structures; titanium, nickel, and cobalt super-alloys; and stack cutting of metals and fiberglass. Abrasive water jet is particularly useful for cutting composites because of the absence of both delamination and heat damage.

Automotive companies and their suppliers use water jets and abrasive jets for trimming carpeting, composite panels and bumpers, door panel linings, and glass.

Foundries use abrasive jets to remove exterior burned-in sand from iron castings and mono-shell ceramic coatings from investment castings. Degating and definning are also common applications.

ECONOMICS

THE TOTAL HOURLY cost of operating a $200 000 (capital equipment cost) abrasive water jet system has been estimated at $27 per hour. This includes maintenance, electricity, abrasive additive, and nozzle wear. Labor cost would be extra.

SAFETY CONSIDERATIONS

SINCE THE WATER jet or abrasive jet would easily cut flesh or bone, operator protection is required. Noise generated during cutting is typically in the range of 80-95 decibels, but may reach 120 dB. Safety enclosures provided to protect the operator from the cutting operation are designed to deaden sound, but the operator should use ear protection.

Maintenance personnel need to be trained to handle the high-pressure equipment and water lines. Each cutting installation should be designed to provide shielding to prevent a discharge of high-pressure water if the high pressure should rupture any of the tubing. Pressure sensors are used to shut down the system in case of tubing failure.

Figure 16.11–Steel Saw-Blade Cut Using Hydroabrasive Machining

SUPPLEMENTARY READING LIST

ASM "Nontraditional machining." Conference Proceedings. (published by) Metal Park, Ohio: ASM, December 1985.

American Society of Mechanical Engineers. Proceedings of the Fourth U.S. Water Jet Conference, August 1987, Berkeley, CA (published by) New York, NY: American Society of Mechanical Engineers, 1987

Behringer-Ploskonka, C. A., "Waterjet cutting—a technology afloat on a sea of potential." *Manufacturing Engineering*, November 1987.

Firestone, R. F. "Lasers and other nonabrasive machining methods for ceramics." Advanced Ceramics Conference, February 1987, Cincinnati, OH. Hubbard Woods, IL: Metals Science Co., 1987.

Hashih, M. "Abrasive water jet cutting studies." Kent Washington: Flow Industries Inc., 1984.

Holland, C. L. "Implemeting abrasive water jet cutting." Fabtech Conference, Chicago, IL, SME Tech paper #MF85-875, Chula Vista, CA: Rohr Industries, Inc., September 1985.

Jones, E. P. "Water jet and abrasive water jet and their application in the automotive industry." Presented at the Tracking Robotic Applications in Automotive Manufacturing Conference, Detroit, MI, September 1986. Kent Washington: Flow Systems, 1986.

Martin, J. M., Assistant Editor. "Using water as a cutting tool." *American Machinist*, April 1980.

Schwartz, B. L. "Principles and applications of water and abrasive jet cutting." Conference paper.

Slattery, T. J. "Abrasive water jet carves out metalworking niche." *Machine & Tool Blue Book*, August 1987.

Sprow, E. E., Special Projects Editor "Cutting composites: three choices for any budget." *Tooling and Production*, December 1987.

Steinhauser, J. "Abrasive water jets: on the cutting edge of technology." Presented at Fabtech Conference, Chicago, IL, September 1985. Kent Washington: Flow Systems, 1985.

Wightman, D. F. "Water jets on the cutting edge of machining." Delivered at the FMS Conference, Chicago, IL, SME Tech Paper MS86-171, March 1986. Elmhurst, IL: Ingersoll-Rand Water Jet Cutting Systems, 1986.

Wightman, D. F. "Hydroabrasive near-net shaping of titanium parts and forgings." Delivered at the March 1988 Westec '88 Conference, Los Angeles, CA. SME Tech paper MR88-141.

SPOT, SEAM, AND PROJECTION WELDING

PREPARED BY A COMMITTEE CONSISTING OF:

P. Dent, Chairman
Grumman Aerospace Corporation

J. C. Bohr
General Motors

R. G. Gasser
Ferranti/Sciaky, Incorporated

J. M. Gerken
Lincoln Electric Corporation

D. L. Hallum
Bethlehem Steel Corporation

J. W. Lee
Textron Lycoming

R. B. McCauley
McCauley Associates

D. H. Orts
Armco, Incorporated

G. W. Oyler
Welding Research Council

W. T. Shieh
General Electric Company

K. C. Wu
Pertron/Square D

WELDING HANDBOOK COMMITTEE MEMBER:
A. F. Manz
A. F. Manz Associates

SPOT, SEAM, AND PROJECTION WELDING

FUNDAMENTALS OF THE PROCESSES

DEFINITION AND GENERAL DESCRIPTION

SPOT, SEAM, AND projection welding are three resistance welding processes in which coalescence of metals is produced at the faying surfaces by the heat generated by the resistance of the work to the passage of electric current. Force is always applied before, during, and after the application of current to confine the weld contact area at the faying surfaces and, in some applications, to forge the weld metal during postheating. Figure 17.1 illustrates the three processes.

In spot welding, a nugget of weld metal is produced at the electrode site, but two or more nuggets may be made simultaneously using multiple sets of electrodes. Projection welding is similar except that nugget location is determined by a projection or embossment on one faying surface, or by the intersection of parts in the case of wires or rods (cross-wire welding). Two or more projection welds can be made simultaneously with one set of electrodes.

Seam welding is a variation of spot welding in which a series of overlapping nuggets is produced to obtain a continuous, leak tight seam. One or both electrodes are generally wheels that rotate as the work passes between them. A

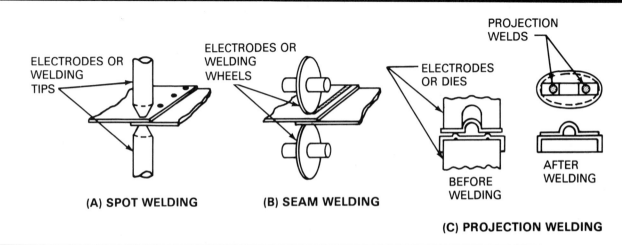

(A) SPOT WELDING   **(B) SEAM WELDING**   **(C) PROJECTION WELDING**

Figure 17.1–Simplified Diagrams Depicting the Basic Process of

seam weld can be produced with spot welding equipment but the operation will be much slower.

A series of separate spot welds may be made with a seam welding machine and wheel electrodes by suitably adjusting the travel speed and the time between welds. Movement of the work may or may not be stopped during the spot weld cycle. This procedure is known as *roll spot welding*.

PRINCIPLES OF OPERATION

SPOT, SEAM, AND projection welding operations involve a coordinated application of electric current and mechanical pressure of the proper magnitudes and durations. The welding current must pass from the electrodes through the work. Its continuity is assured by forces applied to the electrodes, or by projections which are shaped to provide the necessary current density and pressure. The sequence of operation must first develop sufficient heat to raise a confined volume of metal to the molten state. This metal is then allowed to cool while under pressure until it has adequate strength to hold the parts together. The current density and pressure must be such that a nugget is formed, but not so high that molten metal is expelled from the weld zone. The duration of weld current must be sufficiently short to prevent excessive heating of the electrode faces. Such heating may bond the electrodes to the work and greatly reduce their life.

The heat required for these resistance welding processes is produced by the resistance of the workpieces to an electric current passing through the material. Because of the short electric current path in the work and limited weld time, relatively high welding currents are required to develop the necessary welding heat.

Heat Generation

IN AN ELECTRICAL conductor, the amount of heat generated depends upon three factors: (1) the amperage, (2) the resistance of the conductor (including interface resistance), and (3) the duration of current. These three factors affect the heat generated as expressed in the formula

$$Q = I^2 Rt \qquad (17.1)$$

where:

Q = heat generated, joules
I = current, amperes
R = resistance of the work, ohms
t = duration of current, seconds

The heat generated is proportional to the square of the welding current and directly proportional to the resistance and the time. Part of the heat generated is used to make the weld and part is lost to the surrounding metal.

The welding current required to produce a given weld is approximately inversely proportional to the square root of the time. Thus, if the time is extremely short, the current required will be very high. A combination of high current and insufficiently short time may produce an undesirable distribution of heat in the weld zone, resulting in severe surface melting and rapid electrode deterioration.

The secondary circuit of a resistance welding machine and the work being welded constitute a series of resistances. The total resistance of the current path affects the current magnitude. The current will be the same in all parts of the circuit regardless of the instantaneous resistance at any location in the circuit, but the heat generated at any location in the circuit will be directly proportional to the resistance at that point.

An important characteristic of resistance welding is the rapidity with which welding heat can be produced. The temperature distribution in the work and electrodes, in the case of spot, seam, and projection welding, is illustrated in Figure 17.2. There are, in effect, at least seven resistances connected in series in a weld that account for the temperature distribution. For a two thickness joint, these are the following:

(**1**) 1 and 7, the electrical resistance of the electrode material.

(**2**) 2 and 6, the contact resistance between the electrode and the base metal. The magnitude of this resistance depends upon the surface condition of the base metal and the electrode, the size and contour of the electrode face, and the electrode force. (Resistance is roughly inversely proportional to the contacting force.) This is a point of high heat generation, but the surface of the base metal does not reach its fusion temperature during the current passage, due to the high thermal conductivity of the electrodes(1 and 7) and the fact that they are usually water cooled.

(**3**) 3 and 5, the total resistance of the base metal itself, which is directly proportional to its resistivity and thickness, and inversely proportional to the cross-sectional area of the current path.

(**4**) 4, the base metal interface resistance at the location where the weld is to be formed. This is the point of highest resistance and, therefore, the point of greatest heat generation. Since heat is also generated at points 2 and 6, the heat generated at interface 4 is not readily lost to the electrodes.

Heat is generated at all of these locations, not at the base metal interface alone. The flow of heat to or from the base metal interface is governed by the temperature gradient established by the resistance heating of the various components in the circuit. This in turn assists or retards the creation of the proper localized welding heat.

Heat will be generated in each of the seven locations in Figure 17.2 in proportion to the resistance of each. Welding heat, however, is required only at the base metal interface, and the heat generated at all other locations should be minimized. Since the greatest resistance is located at 4, heat is

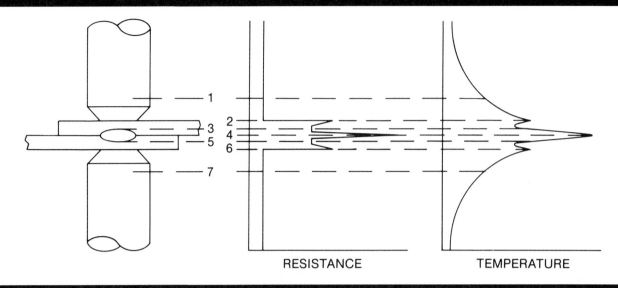

RESISTANCE   TEMPERATURE

Figure 17.2–Graphs of Resistance and Temperature as a Function of Location in the Diagrammed Circuit

most rapidly developed at that location. Points of next lower resistance are 2 and 6. The temperature rises rapidly at these points also, but not as fast as at 4. After about 20 percent of the weld time, the heat gradient may conform to the profile shown in Figure 17.2. Heat generated at 2 and 6 is rapidly dissipated into the adjacent water-cooled electrodes 1 and 7. The heat at 4 is dissipated much more slowly into the base metal. Therefore, while the welding current continues, the rate of temperature rise at plane 4 will be much more rapid than at 2 and 6. The welding temperature is indicated on the chart at the right of Figure 17.2, by the number of dots within the drawing leading to the matching curve.

In a well-controlled weld, the welding temperature will first be reached at numerous point contacts at the interface that melt and with time quickly grow into a nugget.

Factors that affect the amount of heat generated in the weld joint by a given current for a unit of weld time are (1) the electrical resistances within the metal being welded and the electrodes, (2) the contact resistances between the workpieces and between the electrodes and the workpieces, and (3) the heat lost to the workpieces and the electrodes.

Effect of Welding Current. In the formula, $Q = I^2Rt$, current has a greater effect on the generation of heat than either resistance or time. Therefore, it is an important variable to be controlled. Two factors that cause variation in welding current are fluctuations in power line voltage and variations in the impedance of the secondary circuit with AC machines. Impedance variations are caused by changes in circuit geometry or by the introduction of varying masses of magnetic metals into the secondary loop of the machine. Direct current machines are not significantly affected by magnetic metals in the secondary loop and are little affected by circuit geometry.

In addition to variations in welding current magnitude, current density may vary at the weld interface. This can result from shunting of current through preceding welds and contact points other than those at the weld. An increase in electrode face area, or projection size in the case of projection welding, will decrease current density and welding heat. This may cause a significant decrease in weld strength.

A minimum current density for a finite time is required to produce fusion at the interface. Sufficient heat must be generated to overcome the losses to the adjacent base metal and the electrodes.

Weld nugget size and strength increase rapidly with increasing current density. Excessive current density will cause molten metal expulsion (resulting in internal voids), weld cracking, and lower mechanical strength properties. Typical variations in shear strength of spot welds as a function of current magnitude are shown in Figure 17.3. In the case of spot and seam welding, excessive current will overheat the base metal and result in deep indentations in the parts and, it will cause overheating and rapid deterioration of the electrodes.

Effect of Weld Time. The rate of heat generation must be such that welds with adequate strength will be produced without excessive electrode heating and rapid deterioration. The total heat developed is proportional to weld time. Essentially, heat is lost by conduction into the surrounding base metal and the electrodes; a very small amount is lost by radiation. These losses increase with increases in weld time and in metal temperature, but they are essentially uncontrollable.

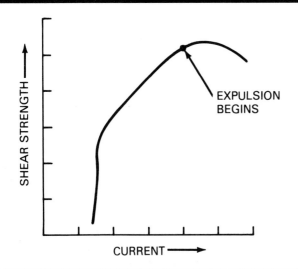

Figure 17.3–Effect of Welding Current on Spot Weld Shear Strength

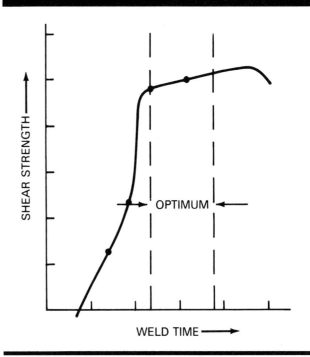

Figure 17.4–Tensile-Shear Strength as a Function of Weld Time

During a spot welding operation, some minimum time is required to reach melting temperature at some suitable current density. If current is continued, the temperature at plane 4 in the weld nugget will far exceed the melting temperature, and the internal pressure may expel molten metal from the joint. Generated gases or metal vapor may be expelled together with minute metal particles. If the work surfaces are scaly or pitted, gases and particles may also be expelled at planes 2 and 6.

Excessively long weld time will have the same effect as excessive amperage on the base metal and electrodes. Furthermore, the weld heat-affected zone will extend farther into the base metal.

In most cases, the heat losses at some point during an extended welding interval will equal the heat input; temperatures will stabilize. An example of the relationship between weld time and spot weld shear strength is shown in Figure 17.4, assuming all other conditions remain constant.

To a certain extent, weld time and amperage may be complementary. The total heat may be changed by adjusting either the amperage or the weld time. Heat transfer is a function of time and the development of the proper nugget size requires a minimum length of time, regardless of amperage.

When spot welding heavy plates, welding current is commonly applied in several relatively short impulses without removal of electrode force. The purpose of pulsing the current is to gradually build up the heat at the interface between the workpieces. The amperage needed to accomplish welding can rapidly melt the metal if the heat pulse time is too long, resulting in expulsion.

Effect of Welding Pressure. The resistance R in the heat formula is influenced by welding pressure through its effect on contact resistance at the interface between the workpieces. Welding pressure is produced by the force exerted on the joint by the electrodes. Electrode force is considered to be the net dynamic force of the electrodes upon the work, and it is the resultant pressure produced by this force that affects the contact resistance.

Pieces to be spot, seam, or projection welded must be clamped tightly together at the weld location to enable the passage of the current. Everything else being equal, as the electrode force or welding pressure is increased, the amperage will also increase up to some limiting value. The effect on the total heat generated, however, may be the reverse. As the pressure is increased, the contact resistance and the heat generated at the interface will decrease. To increase the heat to the previous level, amperage or weld time must be increased to compensate for the reduced resistance.

The surfaces of metal components, on a microscopic scale, are a series of peaks and valleys. When they are subjected to light pressure, the actual metal-to-metal contact will be only at the contacting peaks, a small percentage of the area. Contact resistance will be high. As the pressure is increased, the high spots are depressed and the actual metal-to-metal contact area is increased, thus decreasing the contact resistance. In most applications, the electrode material is softer than the workpieces; consequently, the application of a suitable electrode force will produce better contact at the electrode-to-work interfaces than at the interface between the workpieces.

Influence of Electrodes. Electrodes play a vital role in the generation of heat because they conduct the welding current to the work. In the case of spot and seam welding, the electrode contact area largely controls the welding current density and the resulting weld size. Electrodes must have good electrical conductivity, but they must also have adequate strength and hardness to resist deformation caused by repeated applications of high electrode force. Deformation or "mushrooming" of the electrode face increases the contact area and decreases both current density and welding pressure. Weld quality will deteriorate as tip deformation proceeds; consequently, the electrodes must be reshaped or replaced at intervals to maintain adequate heat generation for acceptable weld properties.

When the electrodes are slow in following a sudden decrease in total work thickness, a momentary reduction in pressure will occur. If this happens while the welding current is on, interface contact resistance at locations 2, 4, and 6 and the rate of heat generation will increase. An excessive heating rate at the three contacting surfaces tends to cause overheating and violent expulsion of molten metal. Molten metal is retained at each interface by a ring of unfused metal surrounding the weld nugget. A momentary reduction in electrode force permits the internal metal pressure to rupture this surrounding ring of unfused metal. Internal voids or excessive electrode indentation may result. Weld properties may fall below acceptable levels, and electrode wear will be greater than normal.

Influence of Surface Condition. The surface condition of the parts influences heat generation because contact resistance is affected by oxides, dirt, oil, and other foreign matter on the surfaces. The most uniform weld properties are obtained when the surfaces are clean.

The welding of parts with a nonuniform coating of oxides, scale, or other foreign matter on the surface causes variations in contact resistance. This produces inconsistencies in heat generation. Heavy scale on the work surfaces may also become embedded in the electrode faces, causing rapid electrode deterioration. Oil and grease will pick up dirt which also will contribute to electrode deterioration.

Influence of Metal Composition. The electrical resistivity of a metal directly influences resistance heating during welding. In high-conductivity metals such as silver and copper, little heat is developed even under high-current densities. The small amount of heat generated is rapidly transmitted into the surrounding work and the electrodes.

The composition of a metal determines its specific heat, melting temperature, latent heat of fusion, and thermal conductivity. These properties govern the amount of heat required to melt the metal and produce a weld. However, the amounts of heat necessary to raise unit masses of most commercial metals to their fusion temperatures are very nearly the same. For example, stainless steel and aluminum require the same Btu's per pound (joules per gram) to reach fusion temperature, even though they differ widely in spot welding characteristics. As a result, the electrical and thermal conductivities become dominant. The conductivities of aluminum are about ten times greater than those of stainless steel. Consequently, the heat lost into the electrodes and surrounding metal is greater with aluminum. Accordingly the welding current for aluminum must be considerably greater than that for stainless steel.

Heat Balance

HEAT BALANCE OCCURS when the depths of fusion (penetration) in the two workpieces are approximately the same. The majority of spot and seam welding applications are confined to the welding of equal thicknesses of the same metal, with electrodes of the same alloy, shape, and size. Heat balance in these cases is automatic; however, in many applications, the heat generated in the parts is unbalanced.

Heat balance may be affected by the following:

(1) Relative electrical and thermal conductivities of the metals to be joined
(2) Relative geometry of the parts at the joint
(3) Thermal and electrical conductivities of the electrodes
(4) Geometry of the electrodes

Heating will be unbalanced when pieces to be welded have significantly different compositions, different thicknesses, or both. The unbalance can be minimized in many cases by part design, electrode material and design, or projection location (in the case of projection welding). Heat balance can also be improved by using the shortest weld time and lowest current that will produce acceptable welds.

Heat Dissipation

DURING WELDING, HEAT is lost by conduction into the adjacent base metal and the electrodes, as shown in Figure 17.5. This heat dissipation continues at varying rates during current application and afterward, until the weld has cooled to room temperature. It may be divided into two phases: (1) during the time of current application, and (2) after the cessation of current. The extent of the first phase depends upon the composition and mass of the workpieces, the welding time, and the external cooling means. The composition and mass of the workpieces are determined by the design. External cooling depends upon the welding setup and the welding cycle.

The heat generated by a given amperage is inversely proportional to the electrical conductivity of the base metal. The thermal conductivity and temperature of the base metal determine the rate at which heat is dissipated or conducted from the weld zone.[1] In most cases, the thermal

---

1. Heat flow in welding is discussed in the *Welding Handbook*, Vol. 1, 8th Ed.

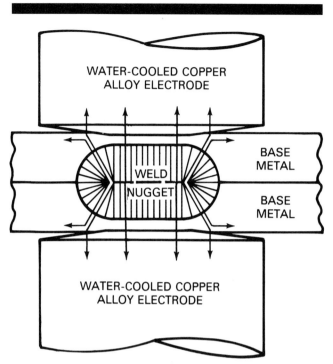

WATER-COOLED COPPER ALLOY ELECTRODE

BASE METAL

WELD NUGGET

BASE METAL

WATER-COOLED COPPER ALLOY ELECTRODE

Figure 17.5—Heat Dissipation During Resistance Welding into the Surrounding Base Metal and Electrodes

and electrical conductivities of a metal are similar. In a high-conductivity metal, such as copper or silver, high amperage is needed to produce a weld and compensate for the heat that is dissipated rapidly into the adjacent base metal and the electrodes. Spot, seam, and projection welding of these metals is very difficult.

If the electrodes remain in contact with the work after weld current ceases, they rapidly cool the weld nugget. The rate of heat dissipation into the surrounding base metal decreases with longer welding times because a larger volume of base metal will have been heated. This reduces the temperature gradient between the base metal and the weld nugget. For thick sheets of metal where long welding times are generally required, the cooling rates will be slower than with thin sheets and short weld times.

If the electrodes are removed from the weld too quickly after the welding current is turned off, problems may result. With thin sheets, this procedure may cause excessive warpage. With thick sheets, adequate time is needed to cool and solidify the large weld nugget while under pressure. It is usually best, therefore, to have the electrodes in contact with the work until the weld cools to a temperature where it is strong enough to sustain any loading imposed when the pressure is released.

The cooling time for a seam weld nugget is short when the electrodes are rotated continuously. Therefore, welding is commonly done with water flowing over the workpieces to remove the heat as rapidly as possible.

It is not always good practice to cool the weld zone rapidly. With quench-hardenable alloy steels, it is usually best to retract the electrodes as quickly as possible to minimize heat dissipation to the electrodes, and thus to retard the cooling rate of the weld.

WELDING CYCLE

THE WELDING CYCLE for spot, seam, and projection welding consists basically of four phases:

(1) Squeeze time - the time interval between timer initiation and the first application of current; the time interval is to assure that the electrodes contact the work and establish the full electrode force before welding current is applied.
(2) Weld time - the time that welding current is applied to the work in making a weld in single-impulse welding.
(3) Hold time - the time during which force is maintained to the work after the last impulse of current ends; during this time, the weld nugget solidifies and is cooled until it has adequate strength.
(4) Off time - the time during which the electrodes are off the work and the work is moved to the next weld location; the term is generally applied where the welding cycle is repetitive.

Figure 17.6 shows a basic welding cycle. One or more of the following features may be added to this basic cycle to improve the physical and mechanical properties of the weld zone:

(1) Precompression force to seat the electrodes and workpieces together
(2) Preheat to reduce the thermal gradient in the metal at the start of weld time
(3) Forging force to consolidate the weld nugget
(4) Quench and temper times to produce the desired weld strength properties in hardenable alloy steels
(5) Postheat to refine the weld grain size in steels
(6) Current decay to retard cooling on aluminum

In some applications, the welding current is supplied intermittently during a weld interval time; it is on during heat time and ceases during cool time. Figure 17.7 shows the sequence of operations in a more complex welding cycle.

WELDING CURRENT

BOTH ALTERNATING CURRENT (ac) and direct current (dc) are used to produce spot, seam, and projection welds. The welding machine transforms line power to low voltage, high amperage welding power. Some applications use single phase ac of the same frequency as the power line, usually 60 Hz. Direct current is used for applications that require high amperage because the load can be balanced on a 3-phase power line. Its use also reduces the power losses in

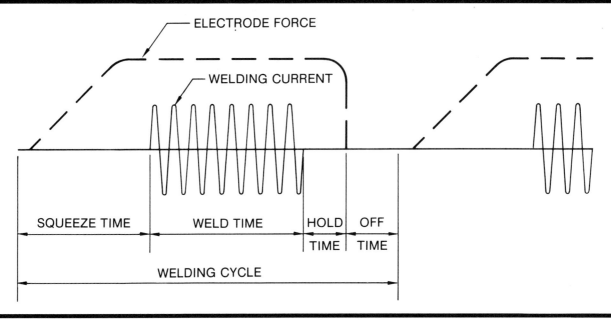

Figure 17.6–Basic Single Impulse Welding Cycle for Spot and Projection Welding

the secondary circuit. Direct current may be essentially constant for a timed period or in the form of a high-peaked pulse. The latter is normally produced from stored electrical energy.

Current Programming

WITH DIRECT ENERGY machines, the rate of current rise and fall can be programmed. The current rise period is commonly called *upslope time* and the current fall period is called *downslope time* (see Figure 17.7). These features are available on machines equipped with electronic control systems.

Upslope is generally used to avoid overheating and expulsion of metal at the beginning of weld time, when the base metal interface resistance is high. Downslope is used to control weld nugget solidification, to avoid cracking in metals that are quench-hardenable or subject to hot tearing.

Prior to welding, the base metal can be preheated using a low current. Following the formation of the weld nugget, the current can be reduced to some lower value for postheating of the weld zone. This may be part of the weld interval, as shown in Figure 17.7, or a separate application of current following a quench time period.

WELD TIME

THE TIME OF current application, or weld time for other than stored energy power, is controlled by electronic, mechanical, manual, or pneumatic means. Times commonly

range from one half cycle (1/120 sec.) for very thin sheets to several seconds for thick plates. For the capacitor or magnetic type of stored energy machines, the weld time is determined by the electrical constant of the system.

Single Impulse Welding

THE USE OF one continuous application of current to make an individual weld is called *single impulse welding* (see Figure 17.6). Up or down current slope may be included in the time period.

Multiple Impulse Welding

MULTIPLE IMPULSE WELDING consists of two or more pulses of current separated by a preset cool time (see Figure 17.7). This sequence is used to control the rate of heating at the interface while spot welding relatively thick steel sheet.

ELECTRODE FORCE

COMPLETION OF THE electrical circuit through the electrodes and the work is assured by the application of electrode force. This force is produced by hydraulic, pneumatic, magnetic, or mechanical devices. The pressure developed at the interfaces depends upon the area of the electrode faces in contact with the workpieces. The functions of this force or pressure are to (1) bring the various interfaces into intimate contact, (2) reduce initial contact resistance at the interfaces, (3) suppress the expulsion of

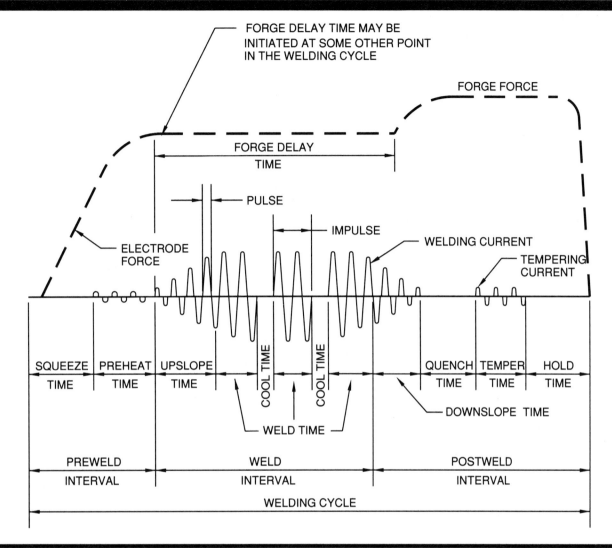

Figure 17.7–Enhanced Welding Cycle which Includes: Preheat Time, Upslope Time, Downslope Time, Quench Time, Temper Time, and Forging Force

molten weld metal from the joint, and (4) consolidate the weld nugget.

Forces may be applied during the welding cycle as follows:

(1) A constant weld force
(2) Precompression and weld forces - a high initial level to reduce initial contact resistance and bring the parts into intimate contact, followed by a lower level for welding

(3) Precompression, weld, and forging forces - the first two levels as described in (2), followed by a forging force near the end of the weld time; forging is used to reduce porosity and hot cracking in the weld nugget
(4) Weld and forging forces

EQUIPMENT

SPOT, SEAM, AND projection welding equipment consists of three basic elements: an electrical circuit, the control equipment, and a mechanical system.[2]

ELECTRICAL CIRCUIT

THIS CIRCUIT COMPRISES a welding transformer, a primary contactor, and a secondary circuit. The secondary circuit includes the electrodes that conduct the welding current to the work, and the work itself. In some cases, a means of storing electrical energy is also included in the circuit. Both alternating current and direct current are used for resistance welding. The welding machine converts 60 Hz line power to low voltage, high amperage power in the secondary circuit of the welding machine.

Alternating Current

SOME RESISTANCE WELDING machines produce single-phase alternating current (ac) of the same frequency as the power line, usually 60 Hz. These machines contain a single-phase transformer that provides the high welding currents required at low voltage. Depending upon the thickness and type of material to be welded, currents may range from 1000 to 100 000 amperes. A typical electrical circuit designed for this type of machine is shown in Figure 17.8.

---

2. Resistance welding equipment is covered in Chapter 19.

Direct Current

WELDING MACHINES MAY produce direct current of continuous polarity, pulses of current of alternating polarity, or high-peaked pulses of current. The latter type is produced by stored electrical energy.

Rectifier Type Machines. These machines are direct energy types, in that ac power from the plant distribution system passes through a welding transformer and is then rectified to dc power. Silicon diode rectifiers are widely used in secondary circuits because of their inherent reliability and efficiency. The system can be single phase; however, one of the advantages of direct current systems is the ability to use a three-phase transformer to feed the rectifier system in the secondary circuit. This makes it possible to use balanced three-phase line power.

Frequency Converter Machines. This type of machine has a special welding transformer with a three-phase primary and a single-phase secondary. The primary current is controlled by ignitron tubes or silicon-controlled rectifiers (SCRs). Half cycles of three-phase power, either positive or negative, are conducted to the transformer for a timed period that depends upon the transformer design. The transformer output is a pulse of direct current. By switching the polarity of the primary half cycles, the polar-

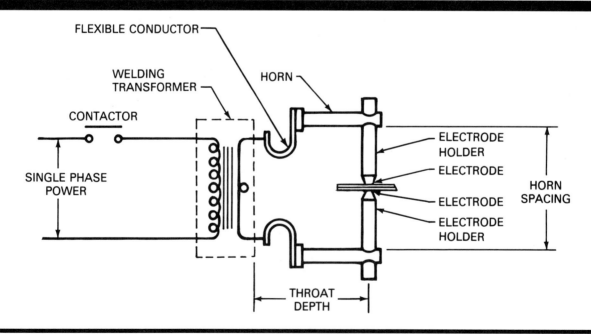

Figure 17.8–Typical Single-Phase Spot Welding Circuit

ity of the secondary current is reversed. A weld may be made with one or more dc pulses.

Stored Energy Machines. Stored energy machines are of electrostatic design. They draw power from a single-phase system, store it, and then discharge it in a very short pulse period to make the weld. These machines draw power from the supply line over a relatively long time between welds, accumulating power to deliver to the electrodes during a short weld time.

The equipment for electrostatic stored energy welding consists primarily of a bank of capacitors, a circuit for charging these capacitors to a predetermined voltage, and a system for discharging the capacitors through a suitable welding transformer. High- voltage capacitors are generally used, the most common varying from 1500 to 3000 volts.

ELECTRODES

RESISTANCE WELDING ELECTRODES[3] perform four functions:

(1) Conduct the welding current to the work, and for spot and seam welding, fix the current density in the weld zone. In projection welding, the current density is determined by the size, shape, and number of projections.
(2) Transmit a force to the workpieces.
(3) Dissipate part of the heat from the weld zone.
(4) Maintain relative alignment and position of the workpieces in projection welding.

If the application of pressure were not involved, electrode material selection could be made almost entirely on the basis of electrical and thermal conductivity. Since the electrodes are subjected to forces that are often of considerable magnitude, they must be capable of withstanding the imposed stresses at elevated temperatures without excessive deformation. Proper electrode shape is important because the current must be confined to a fixed area to achieve needed current density.

When only one spot or seam weld is to be made at a time, only one pair of electrodes is required. In this case, the force and current are applied to each weld by shaped electrodes. Several closely spaced projection welds can be made with one pair of welding dies (electrodes).

Electrodes of several copper alloys with satisfactory physical and mechanical properties are available commercially. Generally speaking, the harder the alloy, the lower are its electrical and thermal conductivities. The choice of a suitable alloy for any application is based on a compromise of its electrical and thermal properties with its mechanical qualities. Electrodes selected for aluminum welding, for instance, should have high conductivity at the

expense of high compressive strength, to minimize sticking of electrodes to the work. Electrodes for welding stainless steel, on the other hand, should sacrifice high conductivity to obtain good compressive strength, to withstand the required electrode force.

Resistance to deformation or mushrooming depends upon the proportional limit and hardness of the electrode alloy. The proportional limit is largely established by heat treatment. The temperature of the electrode face is the governing factor, because this is where softening takes place.

The sizes and shapes of electrodes are usually determined by the sheet thickness and the metal to be welded.

CONTROL EQUIPMENT

WELDING CONTROLS MAY provide one or more of the following principal functions:

(1) Initiate and terminate current to the welding transformer
(2) Control the magnitude of the current.
(3) Actuate and release the electrode force mechanism at the proper times

They may be divided into three groups based on their purposes: welding contactors, timing and sequencing controls, other current controls and regulators.

A welding contactor connects and disconnects the primary power and the welding transformer. Electronic contactors use silicon-controlled rectifiers (SCRs), ignitron tubes, or thyratron tubes to interrupt the primary current.

The timing and sequence control establishes the welding sequence and the duration of each function of the sequence. This includes application of electrode force and current, as well as the time intervals following each function.

The welding current output of a machine is controlled by transformer taps, or an electronic heat control, or both. An electronic heat control is used in conjunction with ignitron tubes or SCRs. It controls current by delaying the firing of the ignitron tubes or SCRs during each half cycle (1/120 sec.). Varying the firing delay time can be used to gradually increase or decrease the primary (rms) amperage. This provides upslope and downslope control of welding current.

Transformer taps are used to change the number of primary turns connected across the ac power line. This changes the turns ratio of the transformer, with an increase or decrease in open circuit secondary voltage. Decreasing the turns ratio will increase the open circuit secondary voltage, the primary current, and the welding current.

MECHANICAL SYSTEMS

SPOT, SEAM, AND projection welding machines have essentially the same types of mechanical operation. The electrodes approach and retract from the work at controlled

---

3. Resistance welding electrodes are discussed more fully in Chapter 19.

times and rates. Electrode force is applied by hydraulic, pneumatic, magnetic, or mechanical means. The rate of electrode approach must be rapid, but controlled, so that the electrode faces are not deformed from repeated blows. The locally heated weld metal expands and contracts rapidly during the welding cycle and the electrodes must follow this movement to maintain welding pressure and electrical contact. The ability of the machine to follow motion is influenced by the mass of the moving parts, or their inertia, and by friction between the moving parts and the machine frame.

If the pressure between the electrodes and work drops rapidly during weld time, the contact surfaces of the electrodes and workpieces may overheat and result in burning or pitting of the electrode faces. The electrodes may stick to the work and, in some cases, the surfaces of the parts that are being welded may vaporize from the very high energy.

The electrode force used during the melting of the weld nugget may not be adequate to consolidate the weld metal and to prevent internal porosity or cracking. Multilevel force machines may be employed to provide a high forging pressure during weld solidification. The magnitude of this pressure should suit the composition and thickness of the metal and the geometry of the parts. The forging pressure is often two to three times the welding pressure. Since the weld cools from the periphery inward, the forging pressure must be applied at or close to the current termination time.

SURFACE PREPARATION

FOR ALL TYPES of resistance welding, the condition of the surfaces of the parts to be welded largely controls how consistent the weld quality will be. The contact resistance of the faying surfaces has a significant influence on the amount of welding heat generated; hence, the electrical resistance of these surfaces must be highly uniform for consistent results. They must be free of high-resistance materials such as paint, scale, thick oxides, and heavy oil and grease. If it is necessary to use a primer paint on the faying surfaces prior to welding, as is sometimes the case, the welding operation must be performed immediately after applying the primer, or special conducting primers must be used. For best results, the primer should be as thin as possible so that the electrode force will displace it and give metal-to-metal contact.

Paint should never be applied to outside base metal surfaces before welding, because it will reduce electrode life and produce poor surface appearance. Heavy scale should be removed by mechanical or chemical methods. Light oil on steel is not harmful unless it has picked up dust or grit. Drawing compounds containing mineral fillers should be removed before welding.

The methods used for preparing surfaces for resistance welding differ for various metals and alloys. A brief description of surface conditions and methods of cleaning follows.

ALUMINUM

THE CHEMICAL AFFINITY of aluminum for oxygen causes it to become coated with a thin film of oxide whenever it is exposed to air. The thin oxide film that forms on a freshly cleaned aluminum surface does not cause sufficient resistance to be troublesome for resistance welding. The permissible holding period, or elapsed time between cleaning and welding, may vary from 8 to 48 hours or more, depending upon the cleaning process used, cleanliness of the shop, the particular alloy, and the application.

An aluminum surface may be mechanically cleaned for resistance welding with a fine grade of abrasive cloth, fine steel wool, or a fine wire brush. Clad aluminum may also be cleaned by mechanical means, but care must be taken not to damage the cladding. Numerous commercial chemical cleaners are available for aluminum. Chemical cleaning is usually preferred in large volume production for reasons of economy as well as uniformity and control.

MAGNESIUM

CLEANING MAGNESIUM ALLOYS is particularly important since they readily alloy with copper at elevated temperatures. The contact resistance between the electrode and the work must be kept as low as possible. Magnesium alloys are supplied coated with an oil or are chrome-pickled, to protect the metal from oxidation during shipment and storage. To obtain sound and consistent welds, the protective coating must be removed to facilitate the removal of residual magnesium oxide.

COPPER

CLEANING OF COPPER alloys is important. The beryllium-coppers and aluminum-bronzes are particularly difficult to clean by chemical means. Mechanical means are preferred. In some instances, a flash coating of tin is applied to produce a uniformly higher surface resistance than pure copper would have.

NICKEL

NICKEL AND ITS alloys demand high standards of material cleanliness for successful resistance welding. The presence of grease, dirt, oil, and paint increases the probability of sulfur embrittlement during welding, and will result in defective welds. Oxide removal is necessary if heavy oxides are present from prior thermal treatments. Machining, grinding, blasting, or pickling may be employed. Wire brushing is not satisfactory.

TITANIUM

BEFORE WELDING, THE surfaces of titanium parts should be scrupulously clean. Materials such as oil, grease, dirt, oxides and paint can adversely affect both weld consistency and chemical composition. Titanium and titanium alloys react with many elements and compounds at welding temperatures. Contamination by oxygen, hydrogen, nitrogen, and carbon, which enter the microstructure interstitially, can significantly reduce weld ductility and toughness. Scale-free surfaces may be welded either after degreasing or after degreasing plus acid pickling. The surfaces may be degreased with acetone, methylethylketone, or a dilute solution of sodium hydroxide. Chlorinated solvents should not be used. Titanium and its alloys are susceptibile to stress corrosion.

Pickling may be used to remove light oxide scale before welding. Pickling is usually performed with HF-HNO3 solutions containing two to five percent HF and 30 to 40 percent HNO3 by volume, balance water.

STEELS

PLAIN CARBON AND low alloy steels have relatively low resistance to corrosion in ordinary atmosphere; hence, these metals are usually protected by a thin oil film during shipment, storage, and processing. This oil film has no harmful effects on the weld, provided the oily surfaces are not contaminated with shop dirt or other poorly conductive or dielectric materials.

Steels are supplied with various surface finishes. Some of the more common are (1) hot-rolled, unpickled; (2) hot-rolled, pickled, and oiled; and (3) cold-rolled with or without an anneal. Unpickled hot-rolled steel must be pickled or mechanically cleaned prior to welding. Hot-rolled pickled steel is weldable in the as-received condition, except for possible wiping to remove loose dirt. Cold-rolled steel presents the best welding surface and, if properly protected by oil, requires no cleaning prior to welding other than wiping to remove loose dirt.

High alloy steels and stainless steels are noncorrosive and usually require no involved cleaning before resistance welding. When exposed to elevated temperatures, stainless steels will acquire an oxide film; the thickness depends upon the temperature and time of exposure. The scale is an oxide of chromium which is effectively removed by pickling. Oil and grease should be removed with solvents or by vapor degreasing prior to welding.

COATED STEELS

THE COATINGS AND platings applied to carbon steel to provide corrosion resistance or for decoration lend themselves satisfactorily to resistance welding with few exceptions. In general, good results may be obtained without special cleaning processes.

Welding of aluminized steel results in less expulsion and pickup if the surfaces are wire brushed.

Phosphate coatings increase the electrical resistance of the surfaces to a degree that welding current cannot pass through the sheets with low welding pressures. Higher pressures will produce welds, but slight variations in coating thickness may prevent welding.

SURFACE PREPARATION CONTROL

SURFACE PREPARATION CONTROL can be maintained by periodically measuring the room temperature contact resistance of the workpieces immediately following cleaning. The measurement is most readily taken tip to tip between two RW electrodes, through two or more thicknesses of metal. Unit surface resistance varies inversely with pressure, temperature, and area of contact. The test conditions must be specified, to make the measurements significant in control of surface cleanliness.

RESISTANCE SPOT WELDING

APPLICATIONS

RESISTANCE SPOT WELDING (RSW) is used to fabricate sheet metal assemblies up to about 0.125 in. (3.2 mm) thickness, when the design permits the use of lap joints and leak tight seams will not be required. Occasionally the process is used to join steel plates 1/4 in. (6.35 mm) thick or thicker; however, loading of such joints is limited and the joint overlap adds weight and cost to the assembly when compared to the cost of an arc welded butt joint.

The process is used in preference to mechanical fastening, such as riveting or screwing, when disassembly for maintenance is not required. It is much faster and more economical because separate fasteners are not needed for assembly.

Spot welding is used extensively for joining low carbon steel sheet metal components for automobiles, cabinets, furniture, and similar products. Stainless steel, aluminum, and copper alloys are commonly spot welded commercially.

ADVANTAGES AND LIMITATIONS

THE MAJOR ADVANTAGES of resistance spot welding are its high speed and adaptability for automation in high-rate production of sheet metal assemblies. Spot welding is also economical in many job shop operations, because it is faster than arc welding or brazing and requires less skill to perform.

The process also has some limitations:

(1) Disassembly for maintenance or repair is very difficult.

(2) A lap joint adds weight and material cost to the product, when compared to a butt joint.

(3) The equipment costs are generally higher than the costs of most arc welding equipment.

(4) The short time, high-current power requirement produces unfavorable line power demands, particularly with single phase machines.

(5) Spot welds have low tensile and fatigue strengths, because of the notch around the periphery of the nugget between the sheets.

(6) The full strength of the sheet cannot prevail across a spot welded joint, because fusion is intermittent and loading is eccentric due to the overlap.

PROCESS VARIATIONS

VARIATIONS OF THE resistance spot welding process differ in the application of welding current and pressure, and the arrangement of the secondary circuit.

Direct and Indirect Welding

DIRECT WELDING IS a resistance welding secondary circuit variation in which welding current and electrode force are applied to the workpieces by directly opposing electrodes.

Indirect welding is a variation in which the welding current paths are through the workpieces away from, as well as at, the spot weld locations. Typical arrangements for direct and indirect spot welding are shown schematically in Figure 17.9. Figures 17.9 (A), (B), and (C) apply to direct resistance welding. In Figure 17.9 (A), direct welding is done using electrodes with similar geometries. A larger electrode against one workpiece in Figure 17.9 (B) provides an increased contact surface area for applications requiring better welding heat balance, or in order to reduce the marking on the lower sheet by the electrodes. The larger electrode surface area will conduct heat away from the weld joint more rapidly. Figure 17.9 (C) is a schematic of two or more electrodes connected in series to a single transformer. Electrode arrangements such as this can make spot welds in rapid succession with one set of electrodes in contact with the work at a time. This arrangement is economical with respect to equipment costs.

Figures 17.9 (D) through (G) represent indirect resistance welding arrangements. A backing plate arrangement in Figure 17.9 (D) provides a current path and pressure when the backing plate is made of a conducting material. If the backing plate is non-conductive, the plate only provides welding pressure, and the current path will be from the top electrode, through the faying surface location, along the lower workpiece to the return connection further down the joint. Figure (E) is similar to (D) with the exception that the non-pressure electrode is away from the lap joint. Current may be through a conducting backing plate between the electrodes, or through the base material between the electrodes. Figures 17.9 (F) and (G) are similar to the joints in (E) and (D), but are for spot welding high-resistance materials which require higher voltages. The two secondary circuits are in series, and are connected to two transformers. The primary circuits may be connected either in series or in parallel. The two secondary circuits provide the sum of their respective voltages at the spot welds.

Parallel and Series Welding

PARALLEL AND SERIES resistance welding are secondary circuit variations used for multiple spot welding applications. Examples of parallel and series welding arrangements are shown schematically in Figure 17.10.

Parallel welding arrangements divide and conduct the secondary current through the workpieces and electrodes in parallel electrical paths, simultaneously forming spot welds. Figure 17.10 (A) and (B) represent parallel welding arrangements. In Figure (A), the welding current is from a single transformer with multiple electrodes in the secondary circuit in parallel. Figure (B) represents a parallel welding system operating with a three-phase primary. This particular system is limited to three work stations.

In series welding (Figures 17.10 C and D), the secondary circuit current is conducted through the workpieces and the electrodes in a series electrical path, simultaneously forming multiple spot welds at the electrode locations. A series welding arrangement requires equal resistance values at the faying surfaces in order to obtain uniform heating at each spot weld. When spot welding with two electrodes in series, a portion of the current will travel through the adjacent workpiece from one electrode to the other, bypassing the faying surfaces. This shunted current does not contribute to the spot weld, and must be taken into account when developing a series spot welding procedure.

HEAT BALANCE

HEAT BALANCE IN a spot weld occurs when the depths of fusion in the workpieces are approximately the same. Problems with heat balance arise when joints are made using metals of different thicknesses, different electrical conductivities, or a combination of both. Electrode configurations and compositions can be used to overcome unbalanced heating to some extent, as shown in Figure 17.11. This sketch illustrates general methods for over-

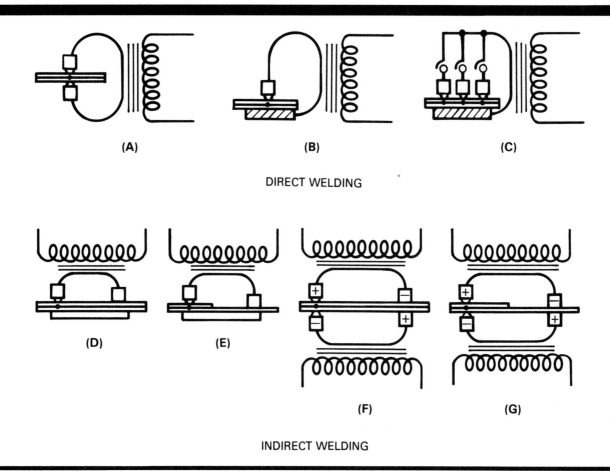

DIRECT WELDING

INDIRECT WELDING

Figure 17.9–Typical Secondary Circuit Arrangements for Direct and Indirect Resistance Spot Welding

coming improper joint heating involving base metals with different electrical conductivities.

In Figure 17.11 (A), an electrode with a smaller face area is used on the metal having the higher conductivity. The smaller contact area will increase the current density in the higher conducting metal. Less heat is conducted away from the joint by the base metal and the electrode. More heat is generated in the workpiece, and the fusion area will shift from the lower conducting metal towards the higher conducting metal. An alternative would be to apply an electrode with higher resistance against the higher-conducting metal, in order to limit the heat loss through that electrode [Figure 17.11 (B)]. Figure 17.11 (C) presents the combination of a higher-resistance electrode and a smaller electrode face area applied to the more conductive metal. Better heat balance can also be obtained by increasing the thickness of the more conductive metal, as seen in Figure 17.11 (D), resulting in an increase in the effective resistance of that sheet.

Spot welding metals with similar electrical characteristics but differing thicknesses will also result in uneven joint heating. The thicker workpiece will exhibit a higher resis-

tance (lower conductivity) than the thinner sheet, resulting in deeper penetration into the thicker sheet. The heat balance can be improved by decreasing the current density in the thicker sheet, by decreasing the heat loss from the thinner sheet, or by a combination of both. Applying a large diameter electrode on the thicker sheet will concentrate the current density into the thinner metal, shifting the nugget penetration deeper into the thinner sheet.

In order to effectively spot weld two or more dissimilar thicknesses of the same metal, a maximum section-thickness ratio of the outer sheets is suggested. For carbon steels the suggested maximum section-thickness ratio is 4 to 1. Joining three different thicknesses of carbon steel using pointed electrodes, an outer sheet thickness ratio of 2.5 to 1 is suggested. To accommodate joints with higher thickness ratios, altering the electrode face diameter and the electrode composition are important methods of balancing the heat produced in each member of the joint.

In multiple layers of dissimilar thicknesses, a long weld time permits more uniform distribution of heat in the asymmetrical resistance path between the electrodes. Cor-

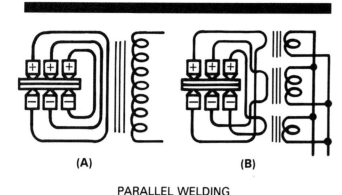

PARALLEL WELDING

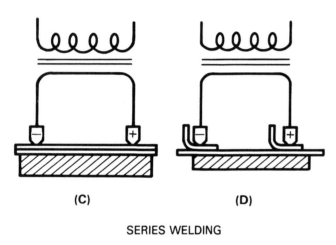

SERIES WELDING

Figure 17.10—Typical Secondary Arrangements for Direct Multiple Spot Welding

rect heat balance may be obtained by using multiple impulse (pulsation) welding, or a single impulse of continuous current for an equivalent time.

JOINT DESIGN

THE JOINT DESIGN in all cases of spot welding consists of a lap joint.

One or more of the welded members may be part flanges, or formed sections such as angles and channels. The use of standard resistance welding machines, portable welding guns, and special-purpose machines must be considered when designing the lap joint configuration. The joint design for indirect welding must allow access to both sides of the joint by the electrodes or backup dies.

Factors that should be considered when designing for spot welding include:

(1) Edge distance
(2) Joint overlap
(3) Fit-up
(4) Weld spacing
(5) Joint accessibility
(6) Surface marking
(7) Weld strength

Edge Distance

THE EDGE DISTANCE is the distance from the center of the weld nugget to the edge of the sheet. Enough base metal must be available to resist the expulsion of molten metal from the joint. Spot welds made too close to the edge of one or both members will cause the base metal at the edge of the member to overheat and upset outward. See Figure 17.12. The restraint by the base metal at the edge on the molten nugget is reduced and expulsion of molten metal may occur due to the high internal pressure in the molten nugget. The weld nugget may be unsound, the electrode indentation excessive, and the weld strength low. The required minimum edge distance is a function of base metal composition and strength, section thicknesses, electrode face contour, and the welding cycle.

Joint Overlap

THE MINIMUM PERMISSIBLE joint overlap is twice the minimum edge distance. The overlap must include the base metal requirement for avoiding edge overheating and expulsion for both sheet members. Other factors such as electrode clearance, however, may require a larger overlap. If the overlap is too small, the edge distance will automatically be insufficient, as sketched in Figure 17.12.

Fit-Up

THE MATING PARTS should fit together along the joint with very little or no gap between them. Any force required to overcome gaps in the joint reduces the effective welding force. The force required to close the joint may vary as welding progresses, and consequently may change the actual welding force. The ultimate result may be significant variations in the strengths of the individual welds.

Weld Spacing

WHEN NUMEROUS SPOT welds are made successively along a joint, a portion of the secondary current shunts through the adjacent welds. The shunting of the current must be considered when establishing the distance between adja-

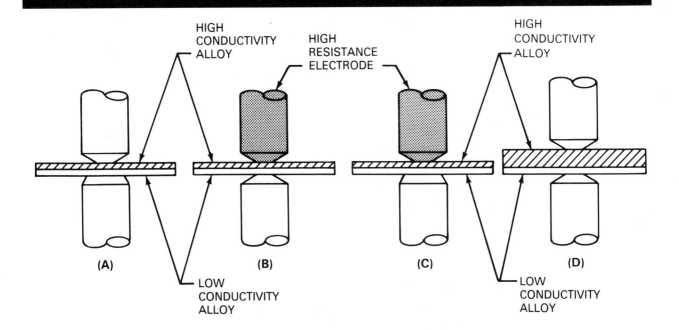

(A) ELECTRODE WITH SMALLER FACE AREA AGAINST HIGH-CONDUCTIVITY ALLOY
(B) HIGH-ELECTRICAL RESISTANCE ELECTRODE AGAINST HIGH-CONDUCTIVITY ALLOY
(C) SAME AS B. WITH ADDITION OF LARGER ELECTRODE FACE AGAINST LOW-CONDUCTIVITY MATERIAL
(D) INCREASE THICKNESS OF HIGH-CONDUCTIVITY WORKPIECE

Figure 17.11–Typical Techniques for Improving Joint Heat Balance When Spot Welding Metals with Different Electrical Conductivities

cent spot welds and when establishing the welding machine settings.

The division of current will depend primarily upon the ratio of the resistances of the two paths, one through the adjacent welds and the other across the interface between the sheets. If the path length through the adjacent weld is long compared to the joint thickness, that resistance will be high compared to the resistance of the joint, and the shunting effect will be negligible. The minimum spacing between spot welds is related to the conductivity and thickness of the base metal, the diameter of the weld nugget, and the cleanliness of the faying surfaces. For example, metals with higher conductivities or thicker sections will require greater spacing between spot welds. The suggested minimum spacing between adjacent spot welds increases when joining three or more sheets. The spot spacing for a weld joining three thicknesses is generally 30 percent greater than the spacing required for welding two sections

of the thicker outer sheet. Current levels may be increased in order to provide more current to the weld and so offset the shunting effects; however, the higher heat inputs may cause expulsion if applied to the first spot weld, which is unshunted. An auxiliary weld timer or current control may be provided to produce the first spot weld using lower heat input.

Joint Accessibility

THE JOINT DESIGN should consider the size and shape of commercially available electrodes and electrode holders, as well as the type of spot welding equipment on which the welding will be done. Each side of the weld joint should be accessible to the electrodes mounted on the welding machine or to backup dies in the case of indirect welding. See Chapter 19 for information on electrodes and electrode holder designs.

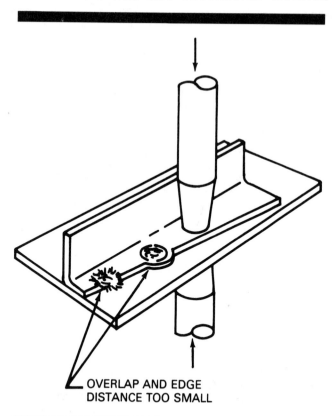

OVERLAP AND EDGE
DISTANCE TOO SMALL

Figure 17.12–Effect of Improper Overlap and Edge Distance

Surface Marking

SURFACE MARKING RESULTS from workpiece shrinkage, caused by a combination of the heat of welding and electrode penetration into the surface of the workpiece.

When the welding current is on, the work is resistance heated locally and tries to expand in all directions. Because of the pressure exerted by the electrodes, expansion transverse to the plane of the sheets is restricted. As the weld cools, contraction takes place almost entirely in the transverse direction and produces concave surfaces or marks at the electrode locations. See Figure 17.13. This weld

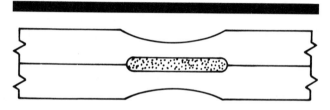

Figure 17.13–Surface Irregularity Produced by Spot Welding

shrinkage is not to be confused with excessive electrode indentation into the work caused by improper welding procedures. The contraction shrinkage seldom exceeds a few thousandths of an inch.

A circular ridge around the spot weld concavity occurs when the expanding workpiece upsets in the plane of the sheets around the electrode face. See Figure 17.13. This ridge is caused by the relatively high electrode force and will occur to some extent with all shaped electrodes.

After some finishing operations, such as painting, the marks may be very conspicuous. It is difficult to eliminate the marks completely, but they can be reduced materially by modifying the welding procedure. For example, the depth of fusion into the sheet can be minimized by welding in the shortest practical time.

Various techniques are used to minimize these markings. The common method is to use a large flat-faced electrode against the *show side* of the joint. (The show side of the joint is the side that is visable when the assembly is in use.) This electrode should be made of a hard copper alloy to minimize wear. Another technique is to use indirect welding arrangements such as those shown in Figures 17.9 and 17.14.

Surface marking may also occur when an electrode or its holder accidentally contacts the workpiece adjacent to where the spot weld is to be made. Resultant arcing may produce a small pit in the work which is undesirable in some applications. If localized melting occurs as a result of the contact, cracks may result in some materials.

Electrode misalignment, skidding, or deflection of the supporting machine component under load may also result in undesirable surface marking. Localized overheating and electrode deflection will not be a problem if the proper joint design, electrodes, and equipment are used.

Weld Strength

THE STRENGTH OF a single spot weld in shear is determined by the cross-sectional area of the nugget in the plane of the faying surfaces. Strength tests for spot welds are discussed in Chapter 12, Volume 1 of the 8th Edition of the *Welding Handbook*. For additional information on spot weld test procedures, see AWS C1.1, *Recommended Practices for Resistance Welding*.

Lap joints tested with the weld in shear will experience an eccentricity of loading resulting in rotation of the joint at the weld as the test load increases. Resistance to joint rotation increases with increasing sheet thicknesses. The joint may fail either by shear through the nugget, or by tearing of the base metal adjacent to the weld nugget. See Figure 17.15. Normally, low weld strengths are associated with weld nugget shear failure, and high strengths are associated with base metal tearing. A minimum nugget diameter is required in order to obtain failure by base metal tear-

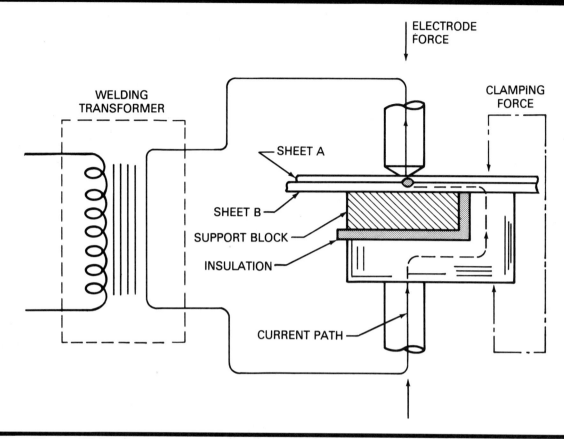

Figure 17.14–Application of Indirect Welding to Minimize Marking on One Side

ing. The minimum nugget diameter is unique to the type of base material, surface condition, and, if applicable, coating type.

Increasing the nugget diameter above this minimum value may provide some increase in the weld strength. Figure 17.16 shows the slight increase in strength values for low carbon steel as the nugget size increases.

Spot welds have relatively low strengths when stressed in tension by loading transverse to the plane of the sheets. This is due to the sharp notch between the sheets at the periphery of the weld nugget; consequently, spot welded joints should not be loaded in this manner.

The strength of multiple spot welded joints is dependent upon material thickness, spot weld spacing, and weld pattern. The spacing between adjacent spot welds may alter the joint weld strength due to current shunting through previous welds. As the spacing between adjacent spot welds decreases, the joint shear strength may decrease.

Figure 17.17 shows the effect of shunt distance (spot spacing) on tensile shear strength of spot welds. Data were taken from welds made on 1/4 in. (6.3 mm) thick by 3 in. (76 mm) wide strips of mild steel. All welds were made with one shunt circuit. Average shear strength of twenty-four welds was 17 570 lbs (8000 kg).

To obtain a desired joint strength, the number of required welds must satisfy minimum spacing requirements in order to minimize current shunting effects. A staggered weld pattern of multiple rows of welds rather than a rectangular pattern will provide better strengths, by distributing the load more efficiently among the spot welds.

To summarize the relationship between resistance spot welding variables and joint strength, Table 17.1 lists suggested resistance spot welding variables for welding uncoated low carbon steel, showing the resultant minimum shear strengths and the nugget "button" diameters.

ELECTRODE MAINTENANCE

MAINTENANCE OF ELECTRODES is necessary for the production of consistent welds. An abnormal increase in the size of the electrode faces contacting the work is detrimental to strength and quality. For example, if a 1/4 in.

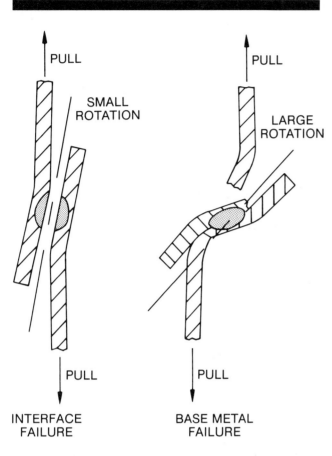

Figure 17.15—Failure Types in Tensile Shear Test as a Function of Rotation

(6.35 mm) diameter electrode face is allowed to increase to 5/16 in. (7.94 mm) diameter by mushrooming, the contact area will increase 50 per cent, with a corresponding decrease in current density and pressure. Depending somewhat upon the weld schedule, the result may be weak or defective welds. A danger sign is the production of poorly shaped spots, which may be caused by the following:

(1) Noncircular electrode faces
(2) Too large a flat face on the electrode
(3) Concavity or convexity of the electrode face
(4) Misalignment of the electrodes with respect to the work

Correct electrode alignment is relatively easy to maintain with stationary welding machines and proper supporting fixtures; however, misalignment is common with portable gun type machines. The seriousness of this condition is dependent upon the ease with which the equipment can be manipulated and correctly positioned for welding. It is likely that the electrodes will have longer life between dressings on positioned work(stationary machines) than on nonpositioned work(portable welding guns).

WELDBONDING

WELDBONDING IS A combination of resistance spot welding and adhesive bonding. Paste or film adhesive is placed between the members to be joined, and resistance welds are then made through the adhesive layer. The adhesive is allowed to cure either at ambient temperature or heated in an oven, as required by the adhesive manufacturer. The spot welds principally hold the joint together during curing; they are fewer in number than otherwise required and so do not contribute greatly to the strength of the joint. Common structures joined by weldbonding are found in the aerospace and transportation industries. Weldbonding is used to attach beaded panels to aircraft skins, and aircraft or truck skins to channels, angles, and other types of reinforcement.

The adhesive, whether paste or film type, can be applied to one or both joint surfaces. The electrode force during welding squeezes out the adhesive at the spot weld locations, creating a current path through the sheets. The adhesive must have good wetting and flow characteristics, in order to bond the faying surfaces securely. Premature curing of the adhesive, during or prior to spot welding, may hamper proper adhesive movement and result in high resistance between the faying surfaces. High resistance may impede weld current, or result in excessive heating and subsequent metal expulsion. Application of a precompression electrode force prior to the welding cycle may help displace the adhesive at each weld site.

Weldbonding improves the fatigue life and durability of the joint over that obtained with spot welding alone. The process may also improve stress distribution, joint rigidity, and buckling resistance in thin sheets. The adhesive in the joint dampens vibration and noise, and provides some corrosion resistance. In some aircraft components, greater cost effectiveness is obtainable with weldbonding than with mechanical fastening or adhesive bonding alone.

Disadvantages of weldbonding, in most applications, include the additional costs of the adhesive, the added curing operation, and the time and labor costs for cleaning the components, treating the surfaces, and applying the adhesive. In addition, operating temperatures for the component are limited to the effective service temperature of the adhesive.

The presence of adhesive in the joint makes welding more difficult, and may contribute to significant variations in weld quality. Regardless of welding conditions, not all of the adhesive will be displaced from between the sheets, and therefore the contact resistance will be higher than with clean sheets.

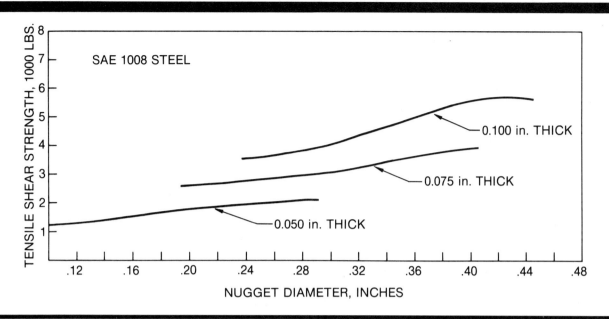

Figure 17.16—Effect of Nugget Size and Sheet Thickness on Tensile Shear Strength, Failure Occuring by Base Metal Tear Out

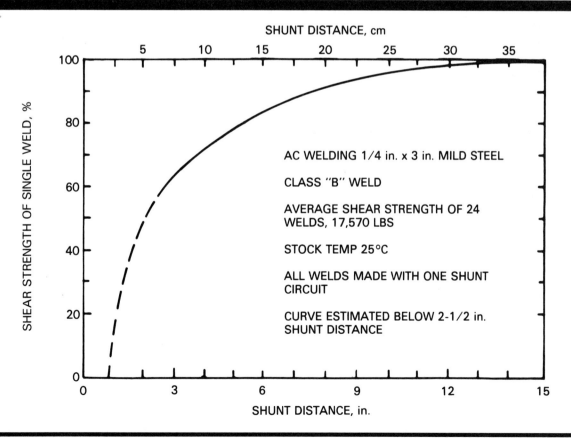

Figure 17.17—The Effect of Shunt Distance (Spot Spacing) on Tensile Shear Strength Loss

Table 17.1
Suggested Schedules for Spot Welding Uncoated Low Carbon Steel Sheet

| Thickness in. | Electrode | | | Force, lb | Weld Time, (60Hz) cy | Welding Current, (Approx.) A | Minimum Contact Overlap, in. | Minimum Weld Spacing | | Minimum Shear Strength, lb | Button Dia., in. |
	Face Dia., in.	Shape *	Bevel Angle, Degrees**					2 stack, in.	3 stack, in.		
0.020	0.188	E,A,B	45	400	7	8,500	0.44	0.38	0.62	320	0.10
0.025	0.188	E,A,B	45	450	8	9,500	0.47	0.62	0.88	450	0.12
0.030	0.250	E,A,B	45	500	9	10,500	0.47	0.62	0.88	575	0.14
0.035	0.250	E,A,B	45	600	9	11,500	0.53	0.75	1.06	750	0.16
0.040	0.250	E,A,B	45	700	10	12,500	0.53	0.75	1.06	925	0.18
0.045	0.250	E,A,B	45	750	11	13,000	0.59	0.94	1.18	1150	0.19
0.050	0.312	E,A,B	30	800	12	13,500	0.59	0.94	1.18	1350	0.20
0.055	0.312	E,A,B	30	900	13	14,000	0.63	1.06	1.31	1680	0.21
0.060	0.312	E,A,B	30	1000	14	15,000	0.63	1.06	1.31	1850	0.23
0.070	0.312	E,A,B	30	1200	16	16,000	0.66	1.18	1.50	2300	0.25
0.080	0.312	E,A,B	30	1400	18	17,000	0.72	1.38	1.60	2700	0.26
0.090	0.375	E,A,B	30	1600	20	18,000	0.78	1.56	1.88	3450	0.27
0.105	0.375	E,A,B	30	1800	23	19,500	0.84	1.68	2.00	4150	0.28
0.120	0.375	E,A,B	30	2100	26	21,000	0.88	1.81	2.50	5000	0.30

* Shape Definitions: E = Truncated Cone
        A = •A• Nose Pointed
        B = 3 in. Radius

** Applies to truncated cone electrodes only and is measured from the plane of the electrode face.

Notes:
1. For intermediate thicknesses, force and weld time may be interpolated.
2. Minimum weld spacing is measured from centerline to centerline.
3. The data within this table were supplied by the AWS D8 Committee and represent an average of typical variables used by the automotive industry.

RESISTANCE SEAM WELDING

APPLICATIONS

A RESISTANCE SEAM weld (RSEW) is made on overlapping workpieces and is a continuous weld formed by overlapping weld nuggets, by a continuous weld nugget, or by forging the joint as it is heated to the welding temperature by its resistance to the welding current.

Seam welds are typically used to produce continuous gas- or liquid-tight joints in sheet assemblies, such as automotive gasoline tanks. The process is also used to weld longitudinal seams in structural tubular sections that do not require leak-tight seams. In most applications, two wheel electrodes, or one translating wheel and a stationary mandrel, are used to provide the current and pressure for resistance seam welding (Figure 17.18). Seam welds can also be produced using spot welding electrodes; this requires the purposeful overlapping of the spot welds in order to obtain a leak-tight seam weld. Overlapping spot welding requires an increase in power after the first spot weld, to offset the shunting effect in order to obtain adequate nugget formation as welding progresses.

ADVANTAGES AND LIMITATIONS

RESISTANCE SEAM WELDING has the same advantages and limitations as resistance spot welding. An additional advantage is the ability to produce a continuous leak-tight weld.

Seam welds must be made in a straight or uniformly curved path. Abrupt changes in welding direction or in joint contour along the path cannot be welded leaktight. This limits the design of the assembly.

Strength properties of seam welded lap joints are generally lower than those of fusion welded butt joints, due to the eccentricity of loading on lap joints and the built-in notch along the nugget at the sheet interface.

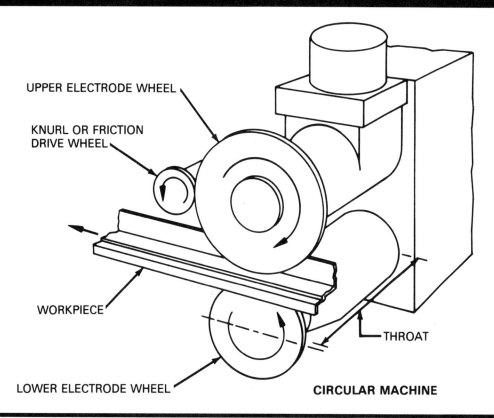

UPPER ELECTRODE WHEEL

KNURL OR FRICTION
DRIVE WHEEL

WORKPIECE

LOWER ELECTRODE WHEEL

THROAT

CIRCULAR MACHINE

Figure 17.18—Position of Electrode Wheels on Resistance Seam Welder

PROCESS VARIATIONS

RESISTANCE SEAM WELDING variations are shown in Figure 17.19.

Lap Seam Welding

LAP JOINTS CAN be seam welded using two wheel electrodes [Figure 17.19 (A)] or with one wheel and a mandrel. The minimum joint overlap is the same as for spot welding, namely twice the minimum edge distance (distance from the center of the weld nugget to the edge of the sheet).

Mash Seam Welding

MASH SEAM WELDING is a resistance welding variation that makes a lap joint primarily by high-temperature plastic forming and diffusion, as opposed to melting and solidification. The joint thickness after welding is less than the original assembled thickness.

Mash seam welding [Figure 17.19 (B)] requires considerably less over lap than the conventional lap joint. The overlap is about 1 to 1.5 times the sheet thickness, with proper weld-

ing procedures. Wide, flat-faced wheel electrodes, which completely cover the overlap, are used. Mash seam welding requires high electrode force, continuous welding current, and accurate control of force, current, welding speed, overlap, and joint thickness in order to obtain consistent welding characteristics. Overlap is maintained at close tolerances, usually by rigidly clamping or tack welding the pieces.

Typically, the exposed or show side of the welded component is placed against a mandrel which acts as an electrode and supports the members to be joined. A welding wheel electrode is applied to the side of the joint that does not show. The show surface of the joint must be mashed as nearly flat as possible so that it will present a good appearance. Proper positioning of the wheel with respect to the joint is required to obtain a smooth weld face. Some polishing of the weld area may be required before painting or coating, when the appearance of the finished product is important.

Mash seam welding produces continuous seams, which have good appearance and are free of crevices. Crevice-free joints are necessary in applications having strict contamination or cleanliness requirements, such as joints in food containers or refrigerator liners.

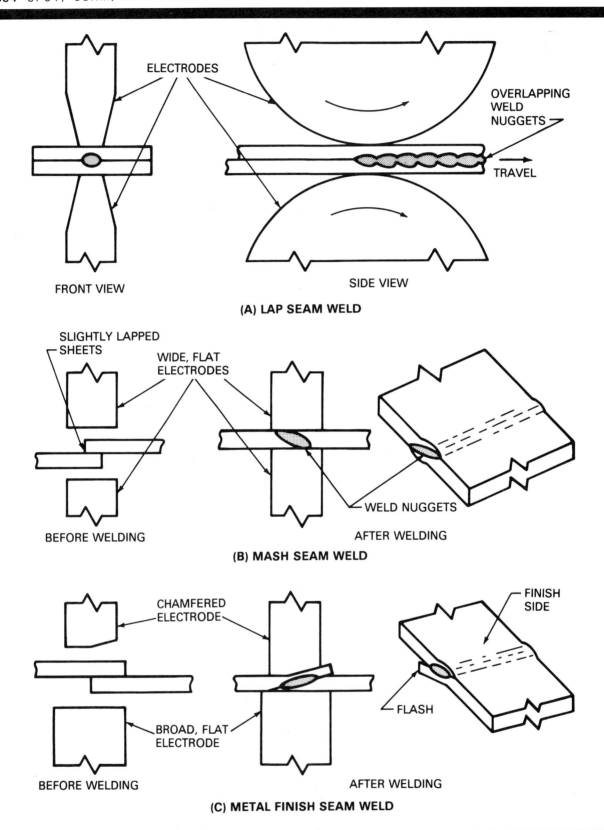

Figure 17.19–Resistance Seam Welding Variations

Disadvantages of the mash seam welding process include the following:

(1) Offset at the joint, due to the inability of the process to completely flatten the seam

(2) Distortion: the inherent lateral flow of metal as it is welded is restrained by fixturing or by tack welds

(3) Very rigid fixturing, required to resist weld distortion

In order to obtain acceptable welds, the materials to be joined by mash seam welding must have wide plastic temperature ranges. Low carbon steel and stainless steel can be mash seam welded for certain applications.

Metal Finish Seam Welding

LAP AND MASH seam welds differ with respect to the amount of forging, or, as the name implies, mash down. The lap weld has practically no mash down, while the thickness of a mash seam weld approaches that of one sheet thickness. In metal finish seam welding, mash down occurs on only one side of the joint (Figure 17.19C), and is a compromise between lap and mash seam welding.

The amount of deformation, or mash, is affected by the geometry of one electrode wheel face and the position of the joint with respect to that face. The wheel face is beveled on one side of the midpoint (Figure 17.20). This varies the amount of deformation across the joint. Good surface finish can be produced on the side of the joint against the flat wheel by using proper welding procedures.

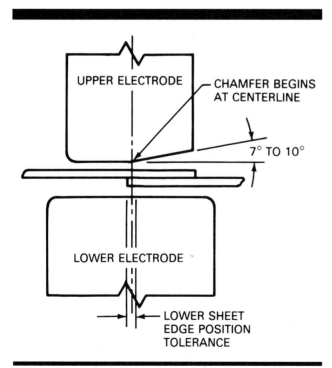

Figure 17.20—Electrode Face Contour and Joint Position for Metal Finish Seam Welding

The location of the edge of the sheet contacting the flat-faced electrode, relative to the bevel on the other electrode, must be held within close tolerance (Figure 17.20). With 0.031 in.(0.8 mm) thick low carbon steel sheet, for example, the edge must be within 0.016 in. (0.4 mm) of center. The overlap distance is not critical.

Higher amperage and electrode force are required than those for mash seam welding, because of the greater overlap distance. Materials that are easily mash seam welded (those with wide plastic temperature ranges) are also easily welded using the metal finish seam welding variation.

Electrode Wire Seam Welding

THE ELECTRODE WIRE seam welding process uses an intermediate wire electrode between each wheel electrode and the workpiece (Figure 17.21). The electrode wire seam welding process is used almost exclusively for seam welding of tin mill products to fabricate cans. The copper wire travels around the wheel electrodes at the welding speed. The copper wire electrode provides a continuously renewed surface, but it is not consumed in the welding operation. The tin build-up which would occur on a copper wheel electrode is avoided. The copper wire electrode may have a circular or flat cross section.

The process requires specially designed welding systems. The seam welds may be made with two wheel electrodes, or one wheel electrode and a mandrel electrode.

The temperature range of an electrode wire seam weld should provide good solid phase bonding and not exceed the melting point of the base material. If the seam weld reaches temperatures greater than the melting point of the base metal, spikes or splashes of molten metal are expelled from the seam weld. Splashes of material can lead to corrosion of the welded component, and are, therefore, undesirable. Electrode wire seam welding has small tolerance for temperature variation. Variations in welding temperature resulting from fluctuations in electrical power or electrode pressure, and changes in overlap distances are usually acceptable.

Butt Joint Seam Welding

BUTT JOINT SEAM welding is done with the edges of the sheets forming a butt joint. A thin, narrow strip of metal, fed between the workpieces and the wheel electrode, is welded to one or both sides of the joint. The metal strip bridges the gap between the workpieces, distributes the welding current to both sheet edges, offers added electrical resistance, and contains the molten weld nugget as the nugget forms. The strip serves as a filler metal and produces a flush or slightly reinforced weld joint. See Figure 17.22.

Strip electrode configurations can be circular, triangular, or flat. The metal strip must be guided accurately and centered on the joint to secure even current distribution to both sheet edges. The strip may be roll spot welded to the

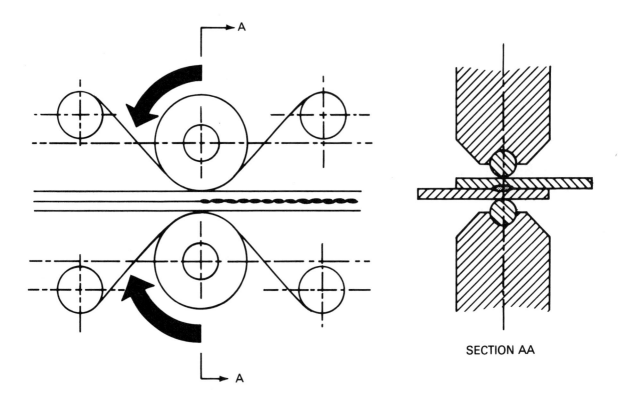

SECTION AA

Figure 17.21–Electrode Wire Seam Weld

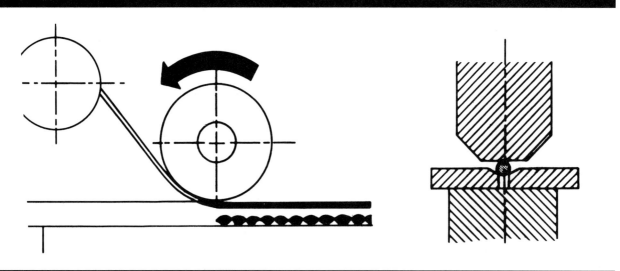

Figure 17.22–Butt Joint Seam Weld

sheets at low power before the joint is seam welded. Very little forging of the joint during welding is required; therefore, there is lower distortion of the joint compared to lap seam welding. The travel speed for butt joint seam welding of low-carbon steel is comparable to that for lap seam welding. Welding procedures must ensure welding of the strips over their entire width, in order to avoid reduced corrosion resistance.

Other Types of Seam Welding

AS WITH SPOT welding, two seam welds can be made in series, using two weld heads. The two heads may be mounted side by side or in tandem. Two seams can be welded with the same welding current, and power demand will be only slightly greater than for a single weld.

A tandem wheel arrangement can reduce welding time by 50 percent, since both halves of a joint can be welded simultaneously. Thus, for a joint 72 in. (182 cm) long, two welding heads can be placed 36 in. (91 cm) apart, with the welding current path through the work from one wheel electrode to the other. A third continuous electrode is used on the other side of the joint. The full length of the joint can be welded with only 36 in. (91 cm) of travel.

HEAT BALANCE

THE SEAM WELDING of dissimilar metals or unequal thicknesses presents the same heat balance problems as with spot welding. The techniques for improving joint heat balance are similar. On workpieces requiring a lower current density or faster cooling, the contact area between the work and the electrode can be enlarged by increasing the diameter or width of the wheel electrode. As an alternative, one of the electrode wheels or mandrels may be made of an alloy of higher thermal conductivity in order to facilitate heat conduction from the workpiece via the electrode.

WELDING CYCLE

RESISTANCE SEAM WELDS typically require higher welding currents than resistance spot welds, due to the shunting of the welding current through previously made welds.

The resistance welding current is normally supplied in timed pulses (heating times) which are separated by periods of cool times (hold times). A weld nugget is produced during each pulse of current. For a given welding speed and heating-cooling cycle, the welding current determines the depth of weld penetration. The welding-current schedule and weld speed control the weld nugget overlap. As the welding speed increases, the ratio of heating time to cool time must be increased in order to maintain weld nugget overlapping. The heat time controls the size of the weld nugget. Figure 17.23 shows the effect that cool time has on weld nugget penetration (macrosection perpendicular to

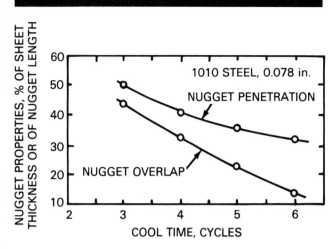

NOTE: WELDING CONDITIONS: HEAT TIME, 6 CYCLES; ELECTRODE FORCE, 1500 LB; WELDING CURRENT, 18 950 A; WELDING SPEED, 55 in./MIN. (SOURCE: RWMA BULLETIN 23)

Figure 17.23–Effects of Cool Time on Nugget Penetration and Nugget Overlap in Seam Welding

weld centerline) and weld overlap (macrosection parallel to weld centerline).

Welding currents higher than those required to obtain proper joint properties may result in excessive indentation, or burning of the welded workpieces. Short heating times, or fast weld speeds, require higher current levels for proper weld nugget formation, but may produce greater electrode wear.

Either pulsed (intermittent) current or continuous alternating current is used for resistance seam welding. As an example of the lap seam welding process, Table 17.2 provides suggested resistance seam welding conditions for welding uncoated low carbon steel sheet.

Pulsed Current

PULSED CURRENT IS usually desirable for most seam welding operations for the following reasons:

(1) Good control of the heat is obtained.
(2) Each weld nugget in the seam is allowed to cool under pressure.
(3) Distortion of the work parts is minimized.
(4) It is easy to control expulsion or burning.
(5) Sound welds with better surface appearance are possible.

To produce a leak-tight seam, the nuggets should overlap 15 to 20 percent of the nugget diameter. For maximum

Table 17.2
Suggested Schedules for Seam Welding Uncoated Low Carbon Steel Sheet

Thickness	W, Min.	E, Max.	Force (lb.)	On Time (cycles) (60 per second)	Off Time (cycles)	Weld Speed (in./min.)	Welds Per Inch	Current (Amp.)	(in.)
0.010	0.380	0.18	400	2	1	80	15.0	8000	0.38
0.021	0.380	0.19	550	2	2	75	12.0	11000	0.44
0.031	0.500	0.25	900	3	2	72	10.0	13000	0.50
0.040	0.500	0.25	980	3	3	67	9.0	15000	0.50
0.050	0.500	0.31	1050	4	3	65	8.0	16500	0.56
0.062	0.500	0.31	1200	4	4	63	7.0	17500	0.62
0.078	0.062	0.38	1500	6	5	55	6.0	19000	0.69
0.094	0.062	0.44	1700	7	6	50	5.5	20000	0.75
0.109	0.750	0.50	1950	9	6	48	5.0	21000	0.81
0.125	0.750	0.50	2200	11	7	45	4.5	22000	0.88

Notes:
1. Type of steel - SAE 1010
2. Material should be free from scale, oxides, paint, grease and oil.
3. Welding conditions determined by thickness outside piece "T".
4. Data for total thickness of pile-up not exceeding 4 "T". Maximum ratio between thicknesses; 3 to 1.

5. Electrode material.   Class 2
   Minimum conductivity - 75% of Copper
   Minimum hardness - 75 Rockwell "B"
6. For large assemblies minimum contacting overlap indicated should be increased 30 percent.

strength, the overlap should be 40 to 50 percent. The size of the nugget will depend upon the heat time for a given welding speed and current. The amount of overlap will depend upon the cool time.

For a particular metal and sheet thickness, the number of welds (nuggets) per inch that can be produced economically will fall within a range. In general, as the sheet thickness decreases, the number of welds per unit of length must increase to obtain a strong, leak-tight seam weld. The ratio of welds per inch to welds per minute will establish the welding speed in inches per minute. The number of welds per minute is the number of cycles of ac per minute, divided by the sum of the heat and cool times (in cycles) for a single weld.

To obtain the minimum number of welds per inch that will produce the required seam at a given welding speed, the heat time and welding current should be adjusted to give the required weld nugget geometry. The cool time should then be set to give the necessary nugget overlap. Since decreasing cool time may increase the heat buildup, nugget penetration may increase.

Roll Spot Welding

ROLL SPOT WELDING consists of making a series of spaced spot welds in a row with a seam welding machine without retracting the electrode or removing the electrode force between welds. Electrode wheel rotation may or may not be stopped during the welding cycle. The radius of the wheel electrode, the contour of its face, and the weld time influence the shape of the nugget. The nugget is usually oval-shaped.

The weld spacing is obtained by adjustment of cool time with the wheel electrodes continuously rotating at a set speed. Hold time is effectively zero. Roll spot welding may also be done with interrupted electrode rotation when a hold time period is needed to consolidate the weld nugget as it cools.

When continuously moving electrodes are employed, as is commonly the case, weld time is usually shorter and welding amperage higher than those used for conventional spot welding. The higher amperage employed may sometimes require the use of a higher electrode force. Otherwise, recommended practices for spot welding apply.

Continuous Current

WITH LOW CARBON steel, welding current can be applied continuously along the length of the seam with high travel speeds, if the current wave form available will produce the proper nugget size and spacing. In this case weld quality is secondary to high-production requirements. Continuous current can be used for sheet up to and including 0.040 in. (1 mm) thickness. Above this thickness, surface condition has a significant effect on welding, and electrode life is short. Continuous current welds in a particular thickness can be made over a wide range of speeds. For example, two thicknesses of 0.040-in.(1 mm.) steel stock can be welded at speeds ranging from 105 to 310 in./min. (44 to 131 mm/s). The required amperage increases with speed.

A problem that may arise when using continuous alternating current operations is arcing between the wheel electrode and a localized region of the weld assembly on the exit side of the electrode. The arcing may produce superficial melting of the sheet surface and the electrode. In steels, the rapid cooling of molten metal that results from workpiece-to-electrode arcing, can result in brittle martensite formation. The localized martensitic microstructure may provide initiation sites for crack formation.

WELDING SPEED

THE SPEED OF welding depends upon the metal being welded, stock thickness, and the weld strength and quality requirements. In general, permissible welding speeds are much lower with stainless steels and nonferrous metals, because of restrictions on heating rate to avoid weld metal expulsion.

In some applications, it is necessary to stop the movement of the electrodes and work as each weld nugget is made. This is usually the case for sections over 0.188 in. (4.78 mm) thick, and for metals that require postheating or forging cycles to produce the desired weld properties. Interrupted motion significantly reduces welding speed because of the relatively long time required for each weld.

With continuous motion, the welding current must be increased and heat time decreased as welding speed is increased to maintain weld quality and joint strength. There is a speed beyond which the required welding current may cause undesirable surface burning and electrode pickup. This will accelerate electrode wear.

ELECTRODES

SEAM WELDING ELECTRODES are normally wheels with diameters ranging from 2 to 24 in. (50 to 600 mm). Common sizes have diameters of 7 to 12 in. (175 to 300 mm) and widths of 0.375 to 0.75 in. (10 to 19 mm).

The width of the weld cross section at the interface of the two workpieces should range from 1.5 to 3 times the thickness of the thinner member. The ratio of weld width to sheet thickness normally decreases as the thickness increases. The weld width is always slightly less than the electrode face width when commercial welding schedules are used.

For more information on seam welding electrodes, see Chapter 19, "Resistance Welding Equipment".

EXTERNAL COOLING

FLOOD COOLING, IMMERSION, or mist cooling is commonly used with seam welding. This is generally in addition to any internal cooling of the components in the secondary circuit of the welding machine. When external cooling is not used, electrode wear and distortion of the work may be excessive. For welding nonferrous metals and stainless steel, clean tap water is satisfactory. For ordinary steels, a 5 percent borax solution is commonly used to minimize corrosion.

JOINT DESIGN

THE VARIOUS REQUIREMENTS that must be met in designing spot welded joints apply to seam welded joints. With seam welding, electrode design together with mounting and leak-tightness requirements place some limitations on design.

The wheel-type electrodes are relatively large and require unobstructed access to the joint. Since the electrodes rotate during welding, they cannot be inserted into small recesses or internal corners. External flanges must change direction over large radii in order to produce a strong, leaktight seam weld. Joint designs that incorporate corners having small radii may result in welding problems when resistance seam welding. Decreased welding speeds are sometimes required in order to maintain weld quality.

Figure 17.24 presents some common designs of seam welded joints. They are similar to those used for spot welding applications. The lap joint (Figure 17.24A) is the most common design. The workpiece edges must overlap sufficiently to prevent expulsion of the weld metal from the edges of the workpiece. Excessive overlapping however, may entrap dirt or moisture within the joint, and may cause subsequent manufacturing or service problems. Lap seam welds are used for the longitudinal seams in cans, buckets, water tanks, mufflers, and large diameter, thin-walled pipes.

Flange joints are forms of lap joints. The design in Figure 17.24 (B), in which one of the pieces is straight, is commonly used to weld flanged ends to containers of various types. In Figure 17.24 (C), both pieces are flanged. This design is used to join the two sections of automotive gasoline tanks. Often the flanged pieces are dished to obtain added strength, in which case it is necessary to mount one or both wheels at an angle to clear the work, as shown in Figure 17.24 (D). A practical limit is 6 degrees because greater angles cause excessive bearing thrust.

Specialized workpiece designs may require the adjustment of the shape and contour of the wheel electrode. Workpieces that contain regularly spaced contours may be welded with notched or segmented wheel electrodes (Figure 17.25).

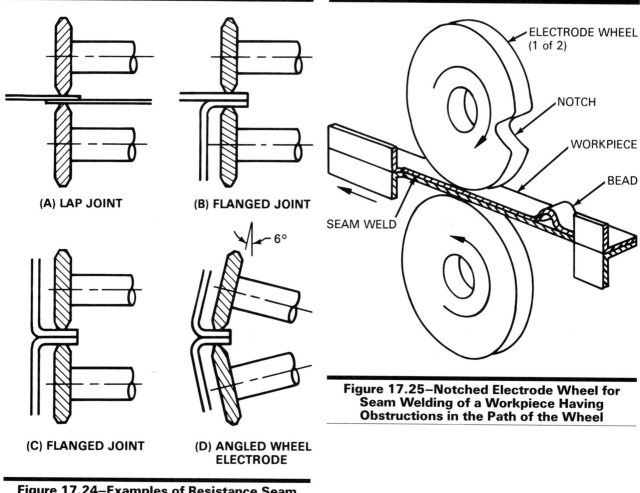

(A) LAP JOINT

(B) FLANGED JOINT

(C) FLANGED JOINT

(D) ANGLED WHEEL ELECTRODE

Figure 17.24–Examples of Resistance Seam Welded Joints

Figure 17.25–Notched Electrode Wheel for Seam Welding of a Workpiece Having Obstructions in the Path of the Wheel

PROJECTION WELDING

APPLICATIONS

PROJECTION WELDING IS primarily used to join a stamped, forged, or machined part to another part. One or more projections are produced on the parts during the forming operations. Fasteners or mounting devices, such as bolts, nuts, pins, brackets, and handles, can be projection welded to a sheet metal part. Projection welding is especially useful for producing several weld nuggets simultaneously between two parts. Marking of one part can be minimized by placing the projections on the other part.

The process is generally used for section thicknesses ranging from 0.02 to 0.125 in. (0.5 to 3.2 mm) thick. Thinner sections require special welding machines capable of following the rapid collapse of the projections. Various carbon and alloy steels and some nickel alloys can be projection welded.

ADVANTAGES AND LIMITATIONS

IN GENERAL, PROJECTION welding can be used instead of spot welding to join small parts to each other and to larger parts. Selection of one method over another depends upon the ecomonics, advantages, and limitations of the two processes. The chief advantages of projection welding include the following:

(1) A number of welds can be made simultaneously in one welding cycle of the machine. The limitation on the

number of welds is the ability to apply uniform electrode force and welding current to each projection.

(**2**) Less overlap and closer weld spacings are possible, because the current is concentrated by the projection, and shunting through adjacent welds is not a problem.

(**3**) Thickness ratios of at least 6 to 1 are possible, because of the flexibility in projection size and location. The projections are normally placed on the thicker section.

(**4**) Projection welds can be located with greater accuracy and consistency than spot welds, and the welds are generally more consistent because of the uniformity of the projections. As a result, projection welds can be smaller in size than spot welds.

(**5**) Projection welding generally results in better appearance, on the side without the projection, than spot welding can produce. The most deformation and greatest temperature rise occur in the part with the projection, leaving the other part relatively cool and free of distortion, particularly on the exposed surface.

(**6**) Large, flat-faced electrodes are used; consequently, electrode wear is much less than that with spot welding and this reduces maintenance costs. In some cases, the fixturing or part locators are combined with the welding dies or electrodes when joining small parts together.

(**7**) Oil, rust, scale, and coatings are less of a problem than with spot welding, because the tip of the projection tends to break through the foreign material early in the welding cycle; however, weld quality will be better with clean surfaces.

The most important limitations of projection welding are the following:

(**1**) The forming of projections may require an additional operation unless the parts are press-formed to design shape.

(**2**) With multiple welds, accurate control of projection height and precise alignment of the welding dies are necessary to equalize the electrode force and welding current.

(**3**) With sheet metal, the process is limited to thicknesses in which projections with acceptable characteristics (see Projection Designs - Sheet Metal) can be formed, and for which suitable welding equipment is available.

(**4**) Multiple welds must be made simultaneously, which requires higher capacity equipment than does spot welding. This also limits the practical size of the component that contains the projections.

TYPES OF JOINTS

AS WITH SPOT and seam welding, projection welding can be used to produce lap joints. The number and shape of the projections depend upon the requirements for joint strength.

Circular or annular ring projections can be used to weld parts requiring either gas-tight or water-tight seals, or to obtain a larger area weld than button-type projections can provide.

PROJECTION DESIGNS

THE MEANS OF producing projections depends upon the material in which they are to be produced. Projections in sheet metal parts are generally made by embossing, as opposed to projections formed in solid metal pieces which are made by either machining or forging. In the case of stamped parts, projections are generally located on the edge of the stamping.

The purpose of a projection is to localize the heat and pressure at a specific location on the joint. The projection design determines the current density. Various types of projection designs are shown in Figure 17.26.

Sheet Metal

A PROJECTION DESIGN for sheet metal should meet the following requirements:

(**1**) Be sufficiently rigid to support the initial electrode force before welding current is applied.

(**2**) Have adequate mass to heat a spot on the other surface to welding temperature. If too small, the projection will collapse before the other surface is adequately heated.

(**3**) Collapse without metal expulsion between the sheets or sheet separation after welding.

(**4**) Be easy to form and not be partially sheared from the sheet during the forming operation. Such projections may be weak and the resulting welds may be easily torn from the sheet on loading.

(**5**) Cause little distortion of the part during forming or welding.

The general design of a projection suitable for steel sheet is shown in Figure 17.27. This design avoids the tendency for the forming operation to shear the sheet or to significantly thin the projection wall. The designs of the punch and die that form this projection shape are illustrated in Figure 17.28. The projection sizes recommended for various sheet thicknesses, and the punch and die dimensions to produce the projections, are given in Table 17.3.

Projections may be elongated to increase nugget size, and thus the strength of the weld. In this case, the contact between the projection and the mating section is linear. Elongated projections are generally used for the thicker sheet gages.

On thin sheet, an annular projection of small diameter may be used instead of a round projection. The annular projection has greater stiffness to resist collapse when electrode force is applied.

Machined or Forged Parts

ANNULAR PROJECTIONS ARE frequently used on forged parts to carry heavy loads and for applications that require a pressure-tight joint around a hole between two parts. Such preparation also produces a high-strength weld when a large stud or boss is welded to thin sheet metal. Figure

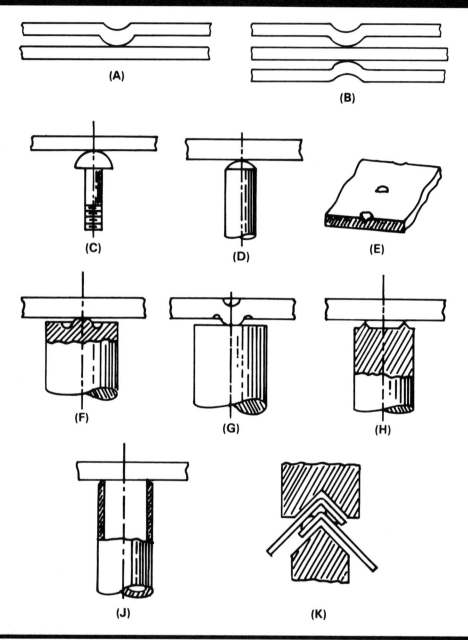

Figure 17.26–Examples of Various Projection Designs.

17.29 shows two applications of annular projections. The summit of the circular ridge should be rounded, particularly with heavy sections, to improve heat balance. Relief, as shown in Figure 17.29(C), should be provided at the base of the projection, for the upset metal to fill as the projection collapses. This will assure a tight joint without a gap, as shown in Figure 17.29(D).

Various designs of weld fasteners are available commercially for projection welding applications. Typical examples are shown in Figure 17.30. Projection designs and their number depend upon the application.

HEAT BALANCE

THE FOLLOWING FACTORS affect heat balance:

(1) Projection design and location
(2) Thickness of the sections

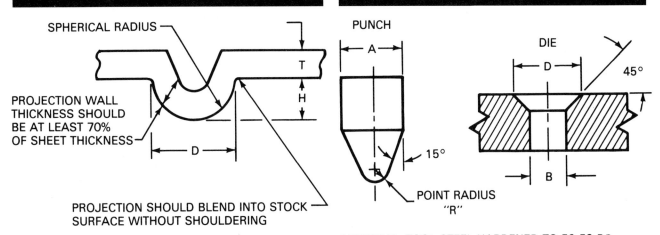

SPHERICAL RADIUS

T

H

PROJECTION WALL
THICKNESS SHOULD
BE AT LEAST 70%
OF SHEET THICKNESS

D

PROJECTION SHOULD BLEND INTO STOCK
SURFACE WITHOUT SHOULDERING

Figure 17.27–Basic Design of a Projection Placed in Steel Sheet

PUNCH

A

15°

POINT RADIUS
"R"

DIE

D

45°

B

MATERIAL: TOOL STEEL HARDENED TO 50-52 RC

Figure 17.28–Basic Design of a Punch and Die to Form Projections of the Type Shown in Figure 17.27 in Sheet Steel

(3) Thermal and electrical conductivities of the metal being welded

(4) Heating rate

(5) Electrode alloy

The distribution of heat in the two sections being projection welded together must be reasonably uniform to obtain strong welds, as in spot welding. The major portion of the heat develops in the projections during projection welding. Consequently, heat balance is generally easier to obtain in projection welding than in spot welding; how-ever, it may be complicated by simultaneously making multiple projection welds. Uniform division of welding current and electrode force is necessary to obtain even heating of all projections. Since the current paths through the projections are in parallel, any variation in resistance between the projections will cause the current to be distributed unequally.

Projections must be designed to support the electrode force needed to obtain good electrical contact with the

Table 17.3
Punch and Die Dimensions for Spherical Dome Projections (Refer to Figure 17.28)

| Thickness (T) | Projection | | Punch | | Die | |
	Height, H Within 2%	Diameter, D Within 5%	Diameter A	Point Radius, R Within 0.002	Hole Diameter, B Within 0.005	Chamber Diameter
0.022-0.034	0.025	0.090	0.375	0.031	0.076	0.090
0.036-0.043	0.035	0.110	0.375	0.047	0.089	0.110
0.049-0.054	0.038	0.140	0.375	0.047	0.104	0.130
0.061-0.067	0.042	0.150	0.375	0.062	0.120	0.150
0.077	0.048	0.180	0.375	0.062	0.144	0.180
0.092	0.050	0.210	0.500	0.078	0.172	0.210
0.107	0.055	0.240	0.500	0.078	0.196	0.240
0.123	0.058	0.270	0.500	0.094	0.221	0.270
0.135	0.062	0.300	0.500	01.09	0.250	0.300
0.153	0.062	0.330	0.500	0.125	0.270	0.330
0.164	0.068	0.350	0.500	0.141	0.297	0.360
0.179	0.080	0.390	0.500	0.156	0.328	0.390
0.195	0.084	0.410	0.500	0.156	0.338	0.410
0.210	0.092	0.440	0.500	0.187	0.358	0.440
0.225	0.100	0.470	0.500	0.187	0.368	0.470
0.245	0.112	0.530	0.500	0.187	0.406	0.530

All dimensions are in inches.

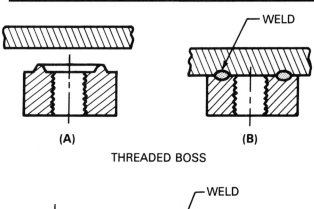

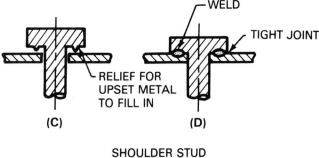

Figure 17.29–Application of Projection Welding Using Annular Projections

mating part, and to collapse when heated. With multiple projections, slight variations in projection heights can affect heat balance. This may occur as a result of wear of the projection-forming punches.

Heat balance in dissimilar thickness materials is maintained by placing the projection in the thicker of the pieces to be welded. The size of the projection is based upon the requirements for heating the thinner section. Similarly, to maintain heat balance in materials of dissimilar conductivity, the projection is located in the piece of higher conductivity (lower resistivity). The choice of electrode alloy determines the conductivity of the electrode, which can also affect heat balance.

WELDING CYCLE

Welding Current

THE CURRENT FOR each projection is generally less than that required to produce a spot weld in the same thickness of that same metal. The projection will heat rapidly and excessive current will melt it and result in expulsion; however, the current must be at least high enough to create fusion before the projection has completely collapsed.

For multiple projections, the total welding current will approximately equal the current for one projection multiplied by the number of projections. Some adjustment may be required to account for normal projection tolerances, part designs, and the impedance of the secondary circuit.

Weld Time

WELD TIME IS about the same for single or multiple projections of the same design. Although a short weld time may be desirable from a production standpoint, it will require correspondingly higher amperage. This may cause overheating and metal expulsion. In general, longer weld times and lower amperages are used for projection welding than those for spot welding.

In some cases, multiple impulse welding may be advantageous to control heating rate. This is helpful with thick sections and with metals of low thermal conductivity.

Electrode Force

THE ELECTRODE FORCE used for projection welding will depend upon the metal being welded, the projection design, and the number of projections in the joint. The force should be adequate to flatten the projections completely when they reach welding temperature, and so to bring the workpieces in contact. Excessive force will prematurely collapse the projections and the weld nuggets will be ring-shaped, with incomplete fusion in the center.

The welding machine must be capable of mechanically following the work with the electrodes as the projections collapse. Slow follow-up will permit metal expulsion before the workpieces are together.

The sequence of events during the formation of a projection weld is shown schematically in Figure 17.31. In Figure 17.31(A), the projection is shown in contact with the mating sheet. In Figure 17.31(B), the current has started to heat the projection to welding temperature. The electrode force causes the heated projection to collapse rapidly and then fusion takes place as shown in Figure 17.31(C). The completed weld is shown in Figure 17.31(D).

ELECTRODES AND WELDING DIES

THE AREAS OF parts to be joined are frequently flat except for the projections. In such cases, large flat-faced electrodes are used. When the surfaces to be contacted are contoured, the electrodes are fitted to them. With such electrodes, the electrode force can be applied without distorting the parts, and the welding current can be introduced without overheating the contact areas.

For a single projection, the electrode face diameter should be at least twice the diameter of the projection. With multiple projections, the electrode face should extend a minimum of one projection diameter beyond the boundry of the projection pattern.

The best electrode material is one that is sufficiently hard (to minimize wear) but does not crack or cause surface burning of the part. If burning or cracking is encoun-

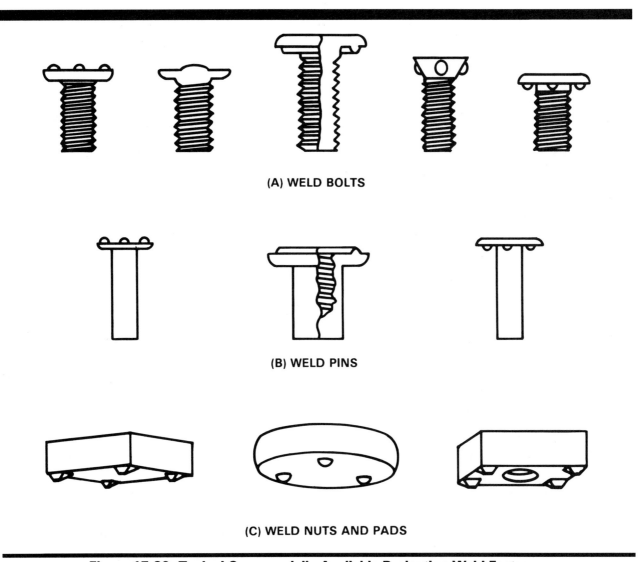

(A) WELD BOLTS

(B) WELD PINS

(C) WELD NUTS AND PADS

Figure 17.30–Typical Commercially Available Projection Weld Fasteners

tered, a softer alloy of higher conductivity should be used. With multiple projections, electrode wear can upset the balance of welding current and electrode force on the projections. Then the strength and quality of the welds may become unacceptable.

Electrodes for large production requirements often have inserts of Resistance Welder Manufacturers Association (RWMA) Group B material at the points of greatest wear. In some cases, it is more economical and equally satisfactory to use one-piece electrodes of RWMA Group A, Class 3 alloy.

Welding electrodes and locating dies for projection welding are usually combined. With the proper dies, it is possible to attain accuracy with projection welding equal to that of any other assembly process. The welding dies should meet the following requirements:

(1) Provide accurate positioning of the parts
(2) Permit rapid loading and unloading
(3) Have no alternative path for the welding current
(4) For ac welding, be made of nonmagnetic materials
(5) Be properly designed for operator safety

The dies must be mounted solidly on the welding machine. The parts are mated in one die and all the welds are made at once with one operation of the machine. One part may be located in relation to the other by punching holes in one, with semipunchings in the other to match. The projections can usually be embossed or forged in the same operation.

In some designs, insulated pins or sleeves may be used in the electrode or dies to position and align the parts. Simple examples are shown in Figures 17.32 and 17.33.

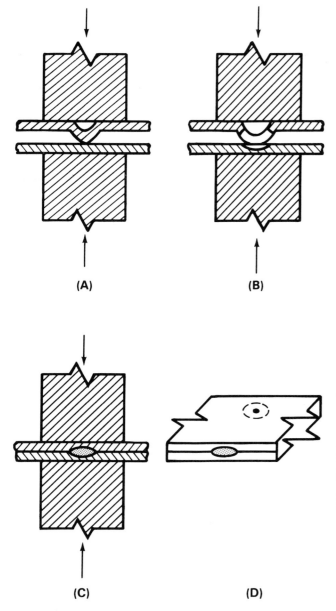

(A)   **(B)**

(C)   **(D)**

Figure 17.31—Sequence of Events During the Formation of a Projection Weld

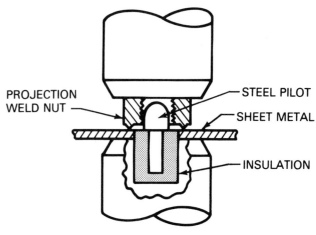

Figure 17.32—Use of an Insulating Pin to Locate a Weld Nut

retainer through the side of the electrode, holding a bolt for welding. Vacuum may also be used to hold small parts in the upper electrode or die when part design permits.

The success of projection welding operations in production with respect to the electrodes depends largely upon the proper selection of materials, proper installation, and proper maintenance. If the dies are correctly designed and constructed, the installation is next in importance. First, the platens of the welding machine must be parallel to each other and perpendicular to the motion of the ram. The platens should also be smooth, clean, and free of nicks and pit marks. If they are not, the platens should be removed and machined smooth and flat before installation of the

When the small part of an assembly can be placed on the bottom and the large part on top, it is a simple matter to hold the small part in a recessed lower electrode such as shown in Figure 17.34. When it is desired to locate a small part on top of a larger part, a problem exists. Sometimes the small part can be located and held by a removable device and then welded with a flat upper electrode. Parts that nest into the upper electrode may be held by spring clips attached to the electrode. Figure 17.35 shows a spring-loaded

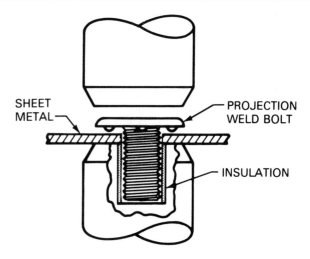

Figure 17.33—Positioning a Weld Bolt with an Insulating Sleeve

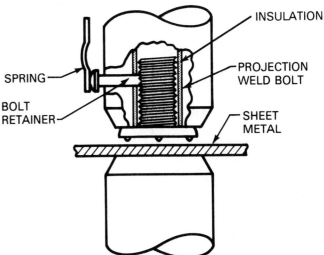

**Figure 17.35—Holding a Bolt in the Upper
Electrode with a Spring Retainer**

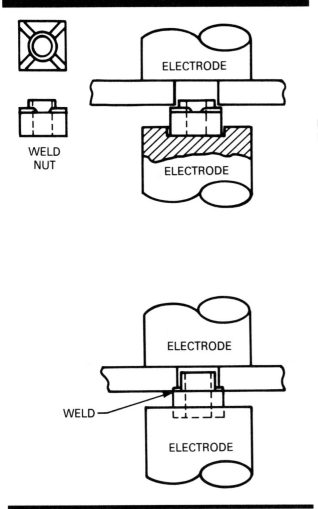

**Figure 17.34—Use of a Recessed Electrode to
Position a Weld Nut**

dies. The check for parallelism of the platens should be made under intended operating forces. This can best be done by placing a steel block with smooth parallel faces between the platens, applying the intended electrode force, and then checking for gap with feeler gages.

The next step is to check the bases of the die blocks. They must be clean, smooth, flat, and free from burrs and nicks. If not, they should be machined.

The dies are then installed on the machine. Most machines have tee slots at right angles to one another in the two platens to permit universal alignment of the dies. After the dies are properly lined up, they should be clamped securely to the platens. With the work in place in the dies, the position of the ram or knee of the machine should be adjusted for the proper stroke, including the necessary allowance for upset of the projections.

If the tips of the projections are in one plane and of uniform height, the setup is ready for trial welds. Nonuni-

formity of current or force on the projections may be caused by the following:

(1) Shunting of current through locators
(2) Unequal secondary circuit path lengths
(3) Excessive play in the welding head
(4) Too much deflection in the knee of the machine

The use of shims between die components or between dies and platens should be avoided. If shims must be used, they should be only clean, annealed, pure copper sheets, of sufficient area to carry the secondary current.

If projections are located on curved or angled surfaces, accurate templates should be provided for checking the dies. Note that when curved parts are welded, or two or more parts are welded to others, the mill tolerances for the metal thicknesses involved may cause problems. These tolerances must be provided for in the design of the parts and the arrangement of the projections.

JOINT DESIGN

LAP JOINT DESIGNS for projection welding are similar to those for spot welding. In general, joint overlap and edge distances for projection welding can be less than for spot welding. Most applications use multiple projections where the minimum distance between projections should be twice the projection diameter.

Part design at the joint location may be significantly limited because the welding electrodes normally contact several projections simultaneously. The electrodes must be mounted rigidly on the welding machine, and the supporting members must be strong enough to minimize deflection when electrode force is applied. Press type welding machines are commonly used for projection welding applications.

Fit-up is important with multiple projection welding. Each projection must be in contact with the mating surface to accomplish a weld. Uniformity of projection heights is a factor in good fit-up. The welding dies must be carefully designed and accurately manufactured to mate with the parts at the weld locations. They should not need to deform the parts to obtain good fit-up.

Where surface marking of one part must be minimized, the projections should be placed in the other part. A large, flat electrode on the show side of the joint should prevent electrode marking, although slight shrinkage may occur at each projection weld. This may be visible after some finishing operations.

When projection welds are used to attach other fasteners such as weld nuts and bolts, they must contain a sufficient number of projections to carry the design load. The design should be proven by applicable mechanical testing. Production quality control should be programmed to ensure that weld quality does not drop below the design standards.

CROSS WIRE WELDING

General Principles

RESISTANCE WELDING OF crossed wires is, in effect, a form of projection welding. In practice, it usually consists of welding a number of parallel wires at right angles to one or more other wires or rods. There are many specific ways to perform the welding operation, depending upon production requirements, but the final finished product is essentially the same regardless of the method used. Figure 17.36 shows a section of a typical cross wire weld.

Crossed-wire products include such items as stove and refrigerator racks, grills of all kinds, lamp shade frames,

poultry equipment, wire baskets, fencing, grating, and concrete reinforcing mesh.

Wire racks may be welded in a press-type projection welding machine or in special automatic indexing machines, with hopper feed and a separate gun for each weld.

Concrete reinforcing mesh is made on continuous machines. The stay wires are fed either from wire reels on the side of the machine or from magazines of cut wire. The welded mesh is either rolled into coils, like fencing, or cut into mats and then stacked and bundled.

As in spot and projection welding, the wire or rod should be clean and free from scale or rust, dirt, paint, heavy grease, or other high-resistance coatings. Plated or galvanized wire or rods may be used, but the coating at the weld will be destroyed.

Wire Materials

LOW CARBON STEEL wire is the wire most commonly welded. Typical machine settings for cross wire welding of this type of material are tabulated in Table 17.4. Next in importance are stainless steel and copper-nickel wires. Copper-nickel alloy wires will require about the same weld time and amperage as carbon steel wires, and about twice the electrode force. Stainless steel wires will require about the same weld time also, but 60 percent of the amperage, and 2.5 times the electrode force.

Welding Technique

NORMALLY, CROSS WIRE welds are not dressed after welding. Therefore, the major consideration may be appearance, with strength secondary in importance for some applications.

In setting up the welding machine, consideration must be given to the following:

(1) Design strength
(2) Appearance
(3) Welding electrodes
(4) Electrode force
(5) Weld time
(6) Welding current (heat)

The particular application will determine which is most important, strength or appearance, when setting up for a particular crossed wire welding application. It is normally assumed that high-strength welds with an acceptable appearance are desired.

The required electrode force, welding current, and weld time depend greatly upon the amount that the wires or rods are to be compressed together. This condition is called *setdown*. It is the ratio of the decrease in joint height to the diameter of the smaller wire. Weld strength generally increases as "setdown" percent increases.

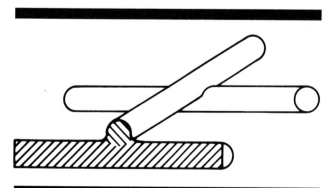

Figure 17.36—Section of a Typical Cross Wire Weld

The welding electrodes must be of the proper material and shape, with provision for water cooling. RWMA Class II alloy electrodes usually have acceptable life, although electrode facings of harder alloys are sometimes used for special applications. Although flat electrodes are commonly used for cross wire welding, certain advantages are gained by shaping them to mate with the wires or rods being welded. Shaped electrodes provide better contact between the electrode and the work.

The electrode force depends upon the wire diameter, the specified setdown, the desired appearance, and the weld-design strength. The electrode force will affect the appearance of the weld. The values given in Table 17.4 will produce welds with good appearance. Lower weld strengths than those shown in the table will result if higher forces are used without decreasing the weld time and increasing the welding current.

The weld time needed will depend upon the diameter of the wire to be welded. For best results, the values shown in the table should be used.

The welding current depends upon the diameter and the specified setdown. It should be slightly less than that which will result in spitting or expulsion of hot metal.

Table 17.4
Conditions for Cross Wire Welding of Low Carbon Steel Wire

Wire Diameter in.	Cold Drawn Wire				Hot Drawn Wire			
	Weld Time, Cycles	Electrode Force, lbs.	Weld Current, A	Weld Strength, lbs.	Weld Time, Cycles	Electrode Force, lbs.	Weld Current, A	Weld Strength, lbs.
	15% Setdown				15% Setdown			
1/16	5	100	600	450	5	100	600	350
1/8	10	125	1800	975	10	125	1850	750
3/16	17	360	3300	2000	17	360	3500	1500
1/4	23	580	4500	3700	23	580	4900	2800
5/16	30	825	6200	5100	30	825	6600	4600
3/8	40	1100	7400	6700	40	1100	7700	6200
7/16	50	1400	9300	9600	50	1400	10000	8800
1/2	60	1700	10300	12200	60	1700	11000	11500
	30% Setdown				30% Setdown			
1/16	5	150	800	500	5	150	800	400
1/8	10	260	2650	1125	10	260	2770	850
3/16	17	600	5000	2400	17	600	5100	1700
1/4	23	850	6700	4200	23	850	7100	3000
5/16	30	1450	9300	6100	30	1450	9600	5000
3/8	40	2060	11300	8350	40	2060	11800	6800
7/16	50	2900	13800	11300	50	2900	14000	9600
1/2	60	3400	15800	13600	60	3400	16500	12400
	50% Setdown				50% Setdown			
1/16	5	200	1000	550	5	200	1000	450
1/8	10	350	3400	1250	10	350	3500	900
3/16	17	750	6000	2500	17	750	6300	1800
1/4	23	1240	8600	4400	23	1240	9000	3100
5/16	30	2000	11400	6500	30	2000	12000	5300
3/8	40	3000	14400	8800	40	3000	14900	7200
7/16	50	4450	17400	11900	50	4450	18000	10200
1/2	60	5300	21000	14600	60	5300	22000	13000

$$\text{Setdown \%} = \frac{\text{Decrease in joint height}}{\text{Diameter of smaller wire}} + 100$$

METALS WELDED

PROPERTIES INFLUENCING WELDABILITY

THE FOLLOWING PROPERTIES of metals have a bearing on their resistance weldability:

(1) Electrical resistivity
(2) Thermal conductivity
(3) Thermal expansion
(4) Hardness and strength
(5) Oxidation resistance
(6) Plastic temperature range
(7) Metallurgical properties

Electrical Resistivity

WORKPIECE RESISTIVITY IS probably the most important welding property, from an RW standpoint, since the heat generated by the welding current is directly proportional to resistance. More current is required to generate heat for a metal of low resistivity than one of high resistivity. A metal such as pure copper is difficult to resistance weld, because of its low electrical resistivity. In addition, current shunting through adjacent welds is more significant in metals of low resistivity than in those of high resistivity. Therefore, metals of high electrical resistivity are considered more weldable than those of low resistivity. High currents also require large transformers and power lines which increase equipment costs.

Thermal Conductivity

THERMAL CONDUCTIVITY IS important because part of the heat generated during resistance welding is lost through conduction into the base metal. This loss must be overcome by greater power input. Therefore, metals of high heat conductivity are less weldable than those of low conductivity. Thermal conductivity and electrical conductivity of the various metals closely parallel one another. Aluminum, for instance, is a good conductor of both heat and electricity, while stainless steel is a poor conductor of both.

Thermal Expansion

THE COEFFICIENT OF thermal expansion is a measure of the change in dimensions that takes place with a temperature change. When the coefficient of thermal expansion is large, warping and buckling of welded assemblies can be expected.

Hardness And Strength

THE HARDNESS AND strength of metals are important to resistance welding. Soft metals can be indented easily by the electrodes. Hard, strong metals require high electrode forces, which in turn require electrodes with high hardness and strength to prevent rapid deformation of the electrodes. Metals that retain their strength at elevated temperatures may require the use of welding machines capable of applying a forging force to the weld.

Oxidation Resistance

ALL COMMONLY USED metals oxidize in air, some more readily than others. The surface oxide normally has high electrical resistance. Surface oxide films generally reduce the resistance weldability of metals. In spot and seam welding, they can cause surface flashing, pickup of metal on the electrode, and poor surface appearance. If the oxide film thickness varies from one part to another, inconsistent weld strength may result.

Aluminum alloys form surface oxides rapidly. Therefore, welding must be done within a short time after deoxidation cleaning to avoid significant variations in surface contact resistance. In contrast, preweld deoxidation cleaning is usually not necessary for stainless steel after being cleaned at the mill prior to packaging and shipping. Whether preweld deoxidation cleaning is necessary depends upon the amount of oxide present and how it will affect weld properties. Surface resistance measurements may be used to confirm cleanliness. In any case, all mill scale, heavy oxide from prior heat treatment, and extraneous material, such as paint, drawing compounds, or grease, should be removed prior to resistance welding.

Plastic Temperature Range

IF THE METAL melts and flows in a narrow temperature range, the welding variables must be more closely controlled than with a metal having a wide plastic temperature range. This property may have considerable bearing on the welding procedure and equipment. Aluminum alloys have narrow plastic ranges and require precise control of welding current, electrode force, and electrode follow-up during welding; projection welding of aluminum is not done commercially. Low carbon steel has a wide plastic range; it is easily resistance welded.

Metallurgical Properties

WITH RESISTANCE WELDING, a small volume of metal is heated to its forging or melting temperature in a short time. The heated metal is then cooled rapidly by the electrodes and surrounding metal. Cold-worked metal will be annealed in the areas exposed to this thermal cycle. In contrast, the rapid cooling will cause hardening in some steels. High carbon steel may harden so rapidly that the welds crack. A tempering cycle following the weld cycle is needed to avoid this cracking. For optimum mechanical properties in the weld region, the heat treatable alloys must be properly postweld heat-treated.

SPOT, SEAM, AND PROJECTION WELDING **571**

LOW CARBON STEEL

LOW CARBON STEELS generally contain less than 0.25 percent carbon. The overall resistance weldability of these steels is good. Their electrical resistivity is average. Hardenability is low. Welds with good strength can be obtained over a wide range of current, electrode force, and weld time settings.

HARDENABLE STEELS

MEDIUM CARBON STEELS may contain from 0.25 to 0.55 percent carbon; high carbon steels may contain 0.55 to 1.0 percent carbon. Low alloy steels contain up to 5.5 percent total alloying elements including cobalt, nickel, molybdenum, chromium, vandium, tungsten, aluminum, and copper.

Alloying additions produce certain desirable properties in steels. The steels may respond to heat treatment, and they may be hard and brittle unless a postheat tempering cycle is employed. With seam welding, that requires that travel stop after each weld nugget is formed to apply the postheat tempering cycle. Special controls are available to perform this function on standard machines.

In general, hardenable steels are less weldable, because of their hardenability, than low carbon steel.

STAINLESS STEELS

STAINLESS STEELS CONTAIN relatively large amounts of chromium or chromium and nickel as alloying elements. They are divided into three groups: martensitic, ferritic, and austenitic types. Whether a stainless steel is hardenable depends upon the amounts of carbon, chromium, and nickel present.

Ferritic and Martensitic Types

THESE STEELS MAY be hardenable (martensitic types) or nonhardenable (ferritic types). Both types have poor resistance weldability. When resistance welding the hardenable types, the precautions given for high carbon and low alloy steels should be followed. The nonhardenable types have low ductility and a characteristic coarse-grained structure in the weld region. These steels are generally not suitable for applications where a ductile weld is required. With the martensitic types, a postweld heat treatment improves weld ductility. However, postweld heat treatment of the ferritic types is not beneficial.

Austenitic Type

THERE ARE A number of austenitic stainless steels, each having suitable properties for particular uses. The ones most common contain 18 percent chromium, 8 percent nickel, and approximately 0.10 percent carbon. The non-stabilized ones are susceptible to carbide precipitation if heated for an appreciable time between 800 to 1600°F;

with short weld times they can be resistance welded without producing harmful carbide precipitation.

These alloys require less current than is required for low carbon steels, since their electrical resistances are about seven times greater. Relatively high electrode forces are needed because of their high strengths at elevated temperatures. Austenitic stainless steels have higher coefficients of thermal expansion than carbon steels. As a result, seam welded assemblies may warp excessively. Distortion may be reduced by using welding schedules that lower the total heat input.

NICKEL-BASE ALLOYS

GENERALLY, NICKEL-BASE ALLOYS are readily joined by resistance welding. However, the cast precipitation-hardenable nickel-base alloys, e.g., Alloy 713C, with low ductility, are normally difficult to resistance weld without cracking. High electrode forces are needed because of the high strength of nickel-base alloys at elevated temperatures. These alloys are subject to embrittlement by sulfur, lead, and other low-melting-point metals when exposed to them at high temperatures. Oils, grease, lubricants, marking materials, and other foreign materials which might contain sulfur or lead must removed from the parts prior to welding, or weld cracking may occur. Pickling prior to welding will only be necessary if a significant amount of oxide is present, recognized by surface discoloration.

Pure nickel can be welded rather easily. Some mechanical sticking of electrodes may be experienced because of the high electrical conductivity of nickel. A restricted dome electrode with 170 degree cone angle is recommended for spot welding.

Monel 400 is an alloy approximately two-thirds nickel and one-third copper. It has higher electrical resistivity and strength than low carbon steel. Therefore, somewhat lower welding current and higher electrode force are required for this alloy than for low carbon steel.

Monel K-500, which can be age-hardened at 1000°F (538°C), has higher electrical resistivity and strength but lower thermal conductivity than Monel 400. Therefore, lower welding currents but higher electrode forces are required for Monel K-500 than for Monel 400. Monel K-500 will crack in the age-hardened condition if subjected to appreciable tensile stress at 1100°F (595°C); spot, seam, and projection welding should be done on annealed material.

Inconel 600 contains approximately 78 percent nickel, 15 percent chromium, and 7 percent iron. It also has higher electrical resistivity and strength but lower thermal conductivity than Monel 400. Therefore, lower welding currents and higher electrode forces are required for this alloy than for Monel 400. Inconel 600 can be readily resistance welded using procedures similar to those for stainless steels.

Inconel X-750, Inconel 718, and Inconel 722 are age-hardenable alloys. They possess high strengths at elevated temperatures, and have high electrical resistances. Relatively low welding currents and high electrode forces are needed for these alloys. Projection welding can be readily accomplished with machines having adequate force capacity. These Inconels should be welded in the solution annealed condition.

COPPER ALLOYS

COPPER ALLOYS HAVE a wide range of weldability that varies almost inversely with their electrical resistance. When the resistance is low, they are difficult to weld; when the resistance is high, they are rather easy to weld. Machines having adequate current capacity and moderate forces are necessary. Because of the narrow plastic range of these alloys, machines with low inertia heads should be used to provide a faster follow-up of the upper electrode to maintain pressure on the joint, to prevent metal expulsion. The machines should be capable of accurate control of welding current, time, and electrode force, because of the sensitivity of these alloys to variations in welding conditions. Shorter welding times are recommended to prevent metal expulsion and welding the electrode to the work. Fusing of the electrodes with the workpiece can be reduced by using electrodes faced with a refractory metal.

Copper-zinc alloys (brasses) become easier to weld with increasing zinc content, because the electrical resistivity increases. The red brasses are difficult to weld, while the brasses with high zinc content can be welded throughout a range of welding conditions, even though the required energy input is high compared with that for carbon steel.

Copper-tin alloys (phosphor bronze), copper-silicon alloys (silicon bronze), and copper-aluminum (aluminum bronze) are relatively easy to weld because of their relatively high electrical resistance. These alloys, particularly phosphor bronze, have a tendency to be hot short, which may result in cracking in the weld.

ALUMINUM AND MAGNESIUM ALLOYS

ALL COMMERCIAL ALUMINUM and magnesium alloys that are produced in the form of sheet and extrusions may be spot and seam welded, provided the thicknesses involved are not too great. Proper welding equipment, correct surface preparation, and suitable welding procedures are necessary to produce satisfactory welds.

Aluminum and magnesium alloys have high thermal and electrical conductivities. Therefore, high welding currents and short welding times are needed. Machines with low inertia heads should be used for spot and seam welding

because these alloys soften rapidly at welding temperature. Rapid acceleration of the welding head is necessary to maintain contact between the electrodes and work to prevent expulsion. Projection welding of aluminum and magnesium is not done commercially because they are plastic in narrow temperature ranges.

TITANIUM ALLOYS

TITANIUM AND ITS alloys can be readily resistance welded. Resistance welding is facilitated by their relatively low electrical and thermal conductivities. Although titanium and titanium alloys are very sensitive to embrittlement, caused by reaction with air at fusion welding temperatures, they can be resistance welded without inert-gas shielding: during resistance welding, the molten weld metal is completely surrounded by the base metal, thus protecting it from contamination; furthermore the welding time is short.

Many plated and coated steels can be spot, seam, or projection welded, but the weld quality usually is affected by the composition and thickness of the coating. Coatings on steel are usually applied for corrosion resistance, decoration, or a combination of these. Welding procedures should assure reasonable preservation of the coating function as well as produce welds of adequate strength. Strength requirements usually require machine settings similar to those for bare carbon steel. Adjustments to compensate for the coating will be determined by a number of factors, including its effect on contact resistance, acceptable electrode indentation, tendency of the coating to alloy with the base metal, and tendency of the electrode to weld to the work.

Coating thickness is the most important variable affecting the weldability of these steels. When coating thickness presents problems in welding, better quality welds can often be obtained by decreasing the coating thickness. A weld nugget of the desired size may be obtained without too much disturbance of the outside surfaces by using higher welding current, greater electrode force, and shorter weld time than for the same thickness of bare steel. However, it is difficult to prevent alloying and metal pickup around the periphery of the electrode face, particularly with low-melting-point coatings such as lead, tin, and zinc. Short welding times, good tip maintenance, and attention to electrode cooling are the best preventive measures.

WELDING SCHEDULES

When setting up for welding a particular metal and joint design, a schedule must be established to produce welds that meet the design specifications. Previous experience can provide a starting point for the initial setup. If the application is a new one, reference to published information

on the welding of the material by the designated process will serve as a guide for the initial setup.

Sample welds should be made and tested while changing one process variable at a time within a range, to establish an acceptable value for that variable. It may be necessary to establish the effect of one variable at several levels of another. For example, weld or heat time and electrode force may be evaluated at several levels of current. Visual examination and destructive test results can be used to select an appropriate welding schedule. Finally, first production parts, or simulations thereof, should be welded and destructively tested. Final adjustments are then made to the welding schedule to meet design or specification requirements.

Starting schedules for many commercial alloys may be available from the equipment manufacturer. Some may also be found in the following publications:

(1) AWS C1.1, Recommended Practices for Resistance Welding

(2) AWS C1.3, Recommended Practices for Resistance Welding Coated Low Carbon Steels

(3) AWS D8.5, Recommended Practices for Automotive Portable Gun Resistance Spot Welding

(4) AWS D8.7, Recommended Practices for Automotive Weld Quality, Resistance Spot Welding

(5) Resistance Welding Manual, 4th Ed., Resistance Welder Manufacturers Association, 1989

(6) Metals Handbook, Vol. 6, 9th Ed., ASM International, 1983.

WELD QUALITY

THE WELD QUALITY required depends primarily upon the application. In applications such as aircraft and space vehicles, the weld quality must meet the requirements of rather stringent specifications. In other applications, such as automobiles, the requirements are less stringent. Generally, the quality of resistance spot, seam, and projection welds is determined based on the following:

(1) Surface appearance
(2) Weld size
(3) Penetration
(4) Strength and ductility
(5) Internal discontinuities
(6) Sheet separation and expulsion

Unfortunately, weld nugget size and penetration, two factors with the strongest influence on weld strengths, cannot be determined by nondestructive inspections. In addition, the commonly used destructive metallographic examination and the tensile shear test of sample welds each have inherent limitations. The designer should be aware of these shortcomings when considering resistance spot, seam, or projection welding for an application.

Some success has been achieved in application of monitoring or adaptive controls, e.g., those based on measuring the thermal expansion of the developing weld nugget and surrounding base metal during heating and melting, to assure producing consistently acceptable resistance welds. Such successes may compensate for the lack of nondestructive inspectability of weld nugget size and penetration, as well as the inherent limitations of destructive testing of sample welds.

SURFACE APPEARANCE

NORMALLY, THE SURFACE appearance of a spot, seam, or projection weld should be relatively smooth. It should be round or oval in the case of a contoured workpiece, and free from surface fusion, electrode deposit, pits, cracks, excessive electrode indentation, or any other condition that indicates improper electrode maintenance or operation. Table 17.5 lists some of the more common undesirable surface conditions, their causes, and the effects on weld quality.

WELD SIZE

THE DIAMETER OR width of the fused zone must meet the requirements of the appropriate specifications or design criteria. Table 17.6 lists the required diameter of fused zone for various workpiece thicknesses. In the absence of such requirements, either accepted shop practices or the following general rules should be used.

(1) Spot welds that are reliably reproduced under normal production conditions should have a minimum nugget diameter of 3.5 to 4 times the thickness of the thinner outside part.

(2) The individual nuggets in a leak-tight seam weld should overlap a minimum of 25 percent. The diameter of nugget should be at least 3.5 to 4 times the thickness of the thinner outside part.

(3) Projection welds should have a nugget diameter equal to or larger than the diameter of the original projection.

There is a maximum limit to the nugget size of a spot, seam, or projection weld. Since this limit is usually controlled by the part configuration, cost, or practicality of making the weld, each user should establish this limit based on the design requirements and prevailing shop practices.

Table 17.5
Undesirable Surface Conditions for Spot Welds

Type	Cause	Effect
1. Deep electrode indentation	Improperly dressed electrode face; lack of control of electrode force; excessively high rate of heat generation due to high contact resistance (low electrode force)	Loss of weld strength due to reduction of metal thickness at the periphery of the weld area; bad appearance
2. Surface fusion (usually accompanied by deep electrode indentation)	Scaly or dirty metal; low electrode force; misalignment of work; high welding current; electrodes improperly dressed; improper sequencing of pressure and current	Undersize welds due to heavy expulsion of molten metal; large cavity in weld zone extending through to surface; increased cost of removing burrs from outer surface of work; poor electrode life and loss of production time from more frequent electrode dressings
3. Irregular shaped weld	Misalignment of work; bad electrode wear or improper electrode dressing; badly fitting parts; electrode bearing on the radius of the flange; skidding; improper surface cleaning of electrodes	Reduced weld strength due to change in interface contact area and expulsion of molten metal
4. Electrode deposit on work (usually accompanied by surface fusion)	Scaly or dirty material; low electrode force or high welding current; improper maintenance of electrode contacting face; improper electrode material; improper sequencing of electrode force and weld current	Bad appearance; reduced corrosion resistance; reduced weld strength if molten metal is expelled; reduced electrode life
5. Cracks, deep cavities, or pin holes	Removing the electrode force before welds are cooled from liquidus; excessive heat generation resulting in heavy expulsion of molten metal; poorly fitting parts requiring most of the electrode force to bring the faying surfaces into contact	Reduction of fatigue strength if weld is in tension or if crack or imperfection extends into the periphery of weld area; increase in corrosion due to accumulation of corrosive substances in cavity or crack

Table 17.6
Typical Required Minimum Weld Nugget Size (Diameter) for Various Sheet Thicknesses

Nominal Thickness of Thinner Sheet		Nugget Size		Nominal Thickness of Thinner Sheet		Nugget Size	
in.	(mm)	in.	(mm)	in.	(mm)	in.	(mm)
0.001	(0.03)	0.010	(0.25)	0.036	(0.90)	0.160	(3.81)
0.002	(0.05)	0.015	(0.38)	0.040	(1.00)	0.160	(4.06)
0.003	(0.08)	0.020	(0.50)	0.045	(1.10)	0.170	(4.32)
0.004	(0.10)	0.030	(0.76)	0.050	(1.20)	0.180	(4.57)
0.005	(0.12)	0.035	(0.89)	0.056	(1.40)	0.190	(4.82)
0.006	(0.16)	0.040	(1.02)	0.063	(1.60)	0.200	(5.08)
0.007	(0.18)	0.045	(1.14)	0.071	(1.80)	0.210	(5.33)
0.008	(0.20)	0.050	(1.27)	0.080	(2.00)	0.225	(5.72)
0.010	(0.25)	0.060	(1.52)	0.090	(2.30)	0.240	(6.10)
0.012	(0.30)	0.070	(1.78)	0.100	(2.50)	0.250	(6.35)
0.016	(0.40)	0.085	(2.16)	0.112	(2.80)	0.260	(6.60)
0.018	(0.45)	0.090	(2.29)	0.125	(3.20)	0.280	(7.11)
0.020	(0.50)	0.100	(2.54)	0.140	(3.60)	0.300	(7.62)
0.022	(0.56)	0.105	(2.68)	0.160	(4.10)	0.320	(8.13)
0.025	(0.65)	0.120	(3.05)	0.180	(4.60)	0.340	(8.64)
0.028	(0.70)	0.130	(3.30)	0.190	(4.80)	0.350	(8.89)
0.032	(0.80)	0.140	(3.56)				

PENETRATION

PENETRATION IS THE depth to which the weld nugget extends into the pieces being welded. Generally, the acceptable minimum penetration is 20 percent of the thickness of the thinnr outside piece. If penetration is less than 20 percent, the weld is said to be cold, because the heat generated in the weld zone or joint interface was too small. Normally, the acceptable maximum penetration is 80 percent of the thickness of the thinner outside piece. Excessive penetration, e.g., 100 percent, will result in expulsion, deep indentation, and rapid electrode deterioration. Figure 17.37 shows normal, excessive, and insufficient penetration.

STRENGTH AND DUCTILITY

STRUCTURES JOINED BY spot, seam, and projection welds are usually designed so that the welds are loaded in shear when the parts are exposed to tension or compression loading. In some applications, the welds are loaded in tension, or a combination of tension and shear, but only where the direction of loading is normal to the plane of the joint. A seam weld may also be subjected to peeling action.

The strength requirements for spot and projection welds are normally specified in pounds (kilograms) per weld. For seam welds, the strength is usually specified in pounds per inch (kg/m) of joint length. These requirements are for a given sheet thickness. The strength of spot and projection welds increases as the nugget diameter increases, even though the average unit stress decreases. The unit stress decreases because of the increasing tendency for failure to occur at the edge of the nugget as its size increases. In low carbon steel, for example, the calculated average shear stress in good welds at fracture will vary from 10 to 60 ksi (69 to 414 MPa). Low values apply to relatively large welds, and high values to relatively small welds. In both instances, the actual tensile stress in the sheet at the weld periphery is at or near the ultimate tensile strength of the base metal. For this reason, the shear strength of circular welds tends to vary linearly with nugget diameter.

Single spot and projection welds are not strong in torsion where the axis of rotation is perpendicular to the plane of the welded parts. The torsional strength tends to vary with the cube of the nugget diameter. Little torsional deformation occurs with low ductility welds prior to failure. Angular displacements may vary from five to 180 degrees depending upon weld metal ductility. Torsion is normally used to shear welds across the interface to measure the nugget diameter, where the base metal is sufficiently thick.

The standard methods of measuring ductility, such as those that measure the percent elongation or reduction of area in a tensile test, are not adaptable to spot, seam, and projection welds. Hardness testing is the closest thing to ductility testing for these welds. It should be noted that although for a given alloy ductility decreases with increasing hardness, different alloys of the same hardness do not necessarily possess the same ductility.

Another method of indicating the ductility of a spot or projection weld is to determine the ratio of its direct tension-shear strength.[4] A weld with good ductility has a high ratio; a weld with poor ductility has a low ratio. Where this ratio is specified, 0.25 is usually the minimum for hardenable steel welds after tempering.

Various methods are available to minimize the hardening effect of rapid cooling in the welds. The following are some of these methods:

(**1**) Use long weld times to put heat into the work.
(**2**) Preheat the weld area with a preheat current.
(**3**) Temper the weld and heat-affected zones with a temper current at some interval after the weld time.
(**4**) Furnace anneal or temper the welded assembly.

INTERNAL DISCONTINUITIES

INTERNAL RESISTANCE WELD discontinuities include cracks, porosity, or spongy metal, large cavities, and, in some coated metals, metallic inclusions. Generally, these discontinuities will have no detrimental effect on the static or fatigue strength of a resistance weld if they are located entirely in the central portion of the weld nugget. This is true because the stresses are essentially zero in the central portion of the weld nugget. On the other hand, no defects should occur at the periphery of a weld, where the load stresses are highly concentrated. The high stresses at the weld periphery can be attributed to the high stress concen-

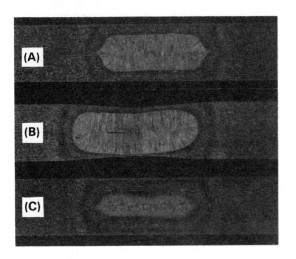

Figure 17.37–Penetration in Spot Welds

4. These tests are described in *Welding Handbook*, Vol. 1, 8th Ed. 390-394.

tration factor[5] associated with the overlapping joint geometry. Since high stress concentration can greatly reduce the fatigue strength or life of a metal, resistance spot, seam, and projection welding are not generally used for applications where the joint is subject to high cyclic load stresses.

Spot, seam, and projection welds in metal thicknesses of approximately 0.040 in. (1 mm) and greater may have small shrinkage porosity in the center of the weld nugget, as illustrated in Figure 17.38(A). This porosity is less pronounced in some welds than in others due to the difference in forging action of the electrodes on the hot metal. Porosity or cavities that result from heavy expulsion of molten metal, as shown in Figure 17.38(B), are much larger than shrinkage cavities. A certain number of expulsion cavities are usually expected in production welding of various metals. Heavy expulsion is a result of improper welding conditions.

Internal defects in spot, seam, and projection welds are generally caused by low electrode force, high welding current, poor fit-up, or inadequate overlap. They are also caused by excessive welding speeds or removing the electrode force too soon after welding current stops. When these conditions occur, the weld nugget is not properly forged during cooling.

When cracklike indications are observed in the heat-affected zone at low magnification, these indications should be examined at higher magnification to determine whether they are actual cracks or coring. As shown in Figure 17.39, a cored area is filled with material that has a dendritic structure. Coring appears to result from incipient melting or back filling of heat-affected zone cracks by the molten metal, based on the dendritic structure. Coring does not appear to affect the serviceability of the welded joint, based on the service experience of various resistance welded nickel-base alloy jet engine components, such as nozzles and combustor housings.

---

5. A detailed of discussion of stress concentration factors can be found in R. E. Peterson's *Stress Concentration Factors*, John Wiley and Sons, New York, 1974.

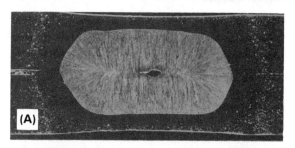

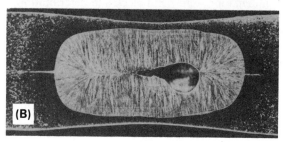

Figure 17.38—Shrinkage Cavities in Spot Welds

SHEET SEPARATION

SHEET SEPARATION OCCURS at the faying surfaces due to the expansion and contraction of the weld metal and the forging effect of the electrodes on the hot nugget. The amount of separation varies with the thickness of the sheet, increasing with greater thickness.

Excessive sheet separation results from the same causes as surface indentation, to which it is related. Improperly dressed electrode faces act as punches under high electrode force, which tends to decrease the joint thickness, upset the weld metal radially, and force the sheets up around the electrodes. Excessive sheet separation is illustrated in Figure 17.40 (note that one sheet is laminated).

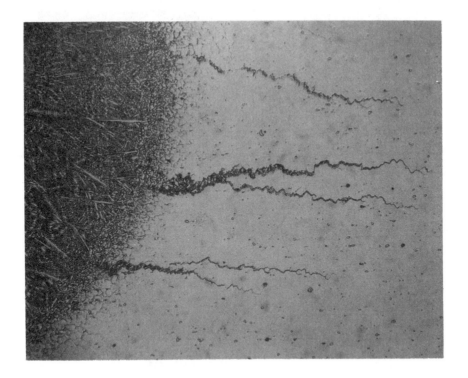

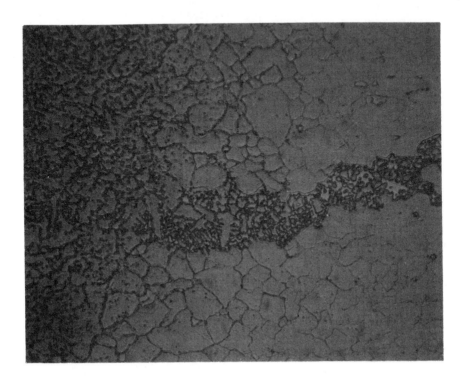

Figure 17.39–Coring in an Inconel 718 Seam Weld (Area Filled with Dendritic Material)

Figure 17.40–Excessive Sheet Separation (One Sheet Laminated)

SAFETY

SPOT, SEAM, AND projection welding may involve hazardous situations that can be avoided by taking the proper precautions outlined in the following section.

MECHANICAL

Guarding

INITIATING CONTROLS ON welding equipment, such as push buttons or switches, should be arranged or guarded to prevent the operator from inadvertently activating them.

In some multiple-gun welding machine installations, the operator's hands can be expected to pass under the point of operation. These machines should be effectively guarded by a suitable device, such as proximity-sensing gates, latches, blocks, barriers, or dual hand controls.

Stop Buttons

ONE OR MORE emergency stop buttons should be provided on all welding machines, with a minimum of one at each operator position.

PERSONAL EQUIPMENT

THE PROTECTIVE EQUIPMENT needed is dependent upon the particular welding application. The following equipment is generally needed for resistance welding:

(1) Eye protection, in the form of face shields or hardened lens goggles; face shields are the preferred form of protection

(2) Skin protection, provided by nonflammable gloves and clothing with the minimum number of pockets and cuffs in which hot or molten particles can lodge

(3) Protective footwear

ELECTRICAL

RESISTANCE WELDING EQUIPMENT should be designed to avoid accidental contact with parts of the system that are electrically hazardous. High-voltage components must have adequate electrical insulation and be completely enclosed. All doors, access panels, and control panels of resistance welding machines must be kept locked or interlocked, to prevent access by unauthorized persons. The interlocks must effectively interrupt power and discharge all high-voltage capacitors into a suitable resistive load when the door or panel is open. Additionally, a manually operated switch or suitable positive device should be provided to assure complete discharge of all high-voltage capacitors.

All electrical equipment must be suitably grounded and the transformer secondary may be grounded or provided with equivalent protection. External weld-initiating control circuits should operate at low voltage for portable equipment.

Additional information on safe practices for resistance welding is contained in ANSI Z49.1, *Safety in Welding and Cutting* (latest edition).

SUPPLEMENTARY READING LIST

Adams, T. "Nondestructive evaluation of resistance spot welding variables using ultrasound." *Welding Journal* 64 (6): 27-28; June 1985.

Aidun, D. K. and Bennett, R. W. "Effect of resistance welding variables on the strength of spot welded 6061-T6 aluminum alloy." *Welding Journal* 64(12): 15-25; December 1985.

Anon "Flexible controller helps 'Turn the Corner' in resistance welding." *Welding Journal* 62(11): 68-69; November 1983.

Bowers, R. T., et al, "Electrode geometry in resistance spot welding." *Welding Journal* 69(2): 455; February 1990.

Brown, B. M. "A comparison of ac and dc resistance welding of automotive steels." *Welding Journal* 66(1): 18-23; 1987.

Chang, H. S., and Cho, H.S. "R study on the shunt effect in resistance spot welding." *Welding Journal* 69(8): 308s-317s; August 1990.

Cho, H. S., and Cho, Y. J. "A study of thermal behavior in resistance spot welds." *Welding Journal* 68(6): 236s; June 1989.

Dickinson, D. W., et al, "Characterization of spot welding behavior by dynamic electrical parameter monitoring." *Welding Journal* 59(6): 170s-176s; June 1980.

Gedeon, S. A. "Measurement of dynamic electrical and mechanical properties of resistance spot welds." *Welding Journal* 66(12): 378s-382s; December 1987.

Gedeon, S. A., Schrock, D., LaPointe, J., Eagar, T. W. "Metallurgica l and process variables affecting the resistance spot weldability of galvanized sheet steels." SAE Technical Paper Series No. 840113. Warrendale, PA, 1988.

Gould, J. E. "An examination of nugget development during spot welding, using both experimental and analytical techniques." *Welding Journal* 66(1): 1s-5s; January 1987.

Hain, R. "Resistivity testing of spot welds challenges ultrasonics." *Welding Journal* 67(5): 46-50; 1988.

Hall, P. M., and Hain, W.R. "Nondestructive monitoring of spot weld quality using a four-point probe." *Welding Journal* 66(5): 20-24; May 1987.

Han, Z., et al, "Resistance spot welding: a heat transfer study." *Welding Journal* 68(9): 363s-368s; September 1989.

Howe, P. and Kelley, S. C. "Coating-weight effect on the resistance spot weldability of electrogalvanized sheet steels." *Welding Journal* 67(12): 271s-275s; December 1988.

——. "A comparison of the resistance spot weldability of bare, hot-dipped, galvannealed, and electrogalvanized DQSK sheet steels." SAE Technical Paper Series No. 880280. Warrendale, PA, 1988.

Kanne, R. "Solid-state resistance welding of cylinders and spheres." *Welding Journal* 65(5): 33-38; 1986.

Kim, E. W., and Eagar, T. W. "Measurement of transient temperature response during resistance spot welding." *Welding Journal* 68(8): 303s-307s; August 1989.

Kimichi, M. "Spot weld properties when welding with expulsion - A comparative study." *Welding Journal* 63(2): 58s-63s; 1984.

Lane, C. T., et al, "Cinematography of resistance spot welding of galvanized sheet steel." *Welding Journal* 66 (9): 260s-264s; 1987.

Nied, H. A. "The finite element modeling of the resistance spot welding process." *Welding Journal* 63(4): 123s-132s; April 1984.

Savage, W. F., et al, "Static contact resistance of series spot welds." *Welding Journal* 56(11): 365s-370s; November 1977.

——. "Dynamic contact resistance of spot welds." *Welding Journal* 57(2): 43s-50s; February 1978.

Sawhill, J. M. et al, "Spot weldability of Mn-Mo-Cb, V-N, and SAE 1008 steels." *Welding Journal* 56(7): 217s-224s; July 1977.

CHAPTER 18

FLASH UPS PER WE

insipient melting

PREPARED BY A COMMITTEE CONSISTING OF:

P. Dent, Chairman
Grumman Aircraft Systems

J. C. Bohr
General Motors

R. G. Gasser
Ferranti/Sciaky, Incorporated

J. M. Gerken
Lincoln Electric Corporation

D. L. Hallum
Bethlehem Steel Corporation

J. W. Lee
Textron Lycoming

R. B. McCauley
McCauley Associates

D. H. Orts
Armco, Incorporated

G. W. Oyler
Welding Research Council

W. T. Shieh
General Electric Company

K. C. Wu
Pertron/Square D

WELDING HANDBOOK COMMITTEE MEMBER:
A. F. Manz
A. F. Manz Associates

CHAPTER **18**

FLASH, UPSET, AND PERCUSSION WELDING

FLASH WELDING

FLASH, UPSET, AND percussion welding constitute a family of welding processes used to join parts of similar cross section by making a weld simultaneously across the entire joint area, without adding filler metal. Upset force is applied at some point before, during, or after the heating cycle to bring the parts into intimate contact. The method of heating and time of force application distinguish these three welding processes. Percussion welding may also be used to join the tip or end of a small part to a flat surface.

DEFINITION AND GENERAL DESCRIPTION

FLASH WELDING (FW) is a resistance welding process that produces a weld at the faying surfaces of a butt joint by a flashing action and by the application of pressure after heating is substantially completed. The flashing action, caused by the very high current densities at small contact points between the workpieces, forcibly expels material from the joint as the workpieces are slowly moved together. The weld is completed by a rapid upsetting of the workpieces.

Two parts to be joined are clamped in dies (electrodes) connected to the secondary of a resistance welding transformer. Voltage is applied as one part is advanced slowly toward the other. When contact occurs at surface irregularities, resistance heating occurs at these locations. High amperage causes rapid melting and vaporization of the metal at the points of contact, and then minute arcs form. This action is called "flashing". As the parts are moved together at a suitable rate, flashing continues until the faying surfaces are covered with molten metal and a short length of each part reaches forging temperature. A weld is

then created by the application of an upset force to bring the molten faying surfaces in full contact and forge the parts together. Flashing voltage is terminated at the start of upset. The solidified metal expelled from the interface is called "flash".

PRINCIPLES OF OPERATION

THE BASIC STEPS in a flash welding sequence are as follows:

(1) Position the parts in the machine.
(2) Clamp the parts in the dies (electrodes).
(3) Apply the flashing voltage.
(4) Start platen motion to cause flashing.
(5) Flash at normal voltage.
(6) Terminate flashing.
(7) Upset the weld zone.
(8) Unclamp the weldment.
(9) Return the platen and unload.

Figure 18.1 illustrates these basic steps. Additional steps such as preheat, dual voltage flashing, postheat, and trimming of the flash may be added as the application dictates.

Flashing takes place between the faying surfaces as the movable part is advanced toward the stationary part. Heat is generated at the joint and the temperature of the parts increases with time. Flashing action (metal loss) increases with part temperature.

A graph relating part motion with time is known as the flashing pattern. In most cases, a flashing pattern should show an initial period of constant velocity motion of one part toward the other to facilitate the start of flashing. This

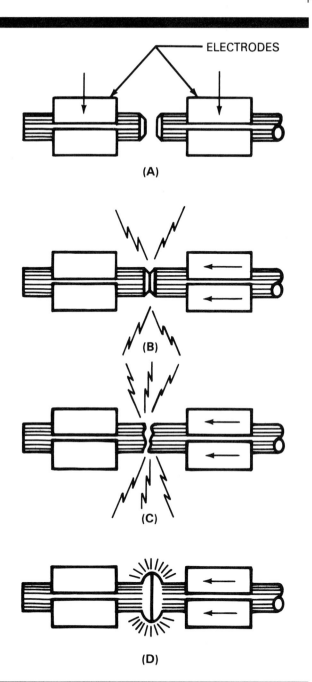

ELECTRODES

(A)

(B)

(C)

(D)

**Figure 18.1–The Basic Steps in Flash Welding:
(A) Position and Clamp the Parts; (B) Apply
Flashing Voltage and Start Platen Motion; (C)
Flash; (D) Upset and Terminate Current**

linear motion should then merge into an accelerating motion which should closely approximate a parabolic curve. This pattern of motion is known as *parabolic flashing*.

To produce a strong joint with uniform upset, the temperature distribution across the joint should be uniform and the average temperature of the faying surface should be the melting temperature of the metal. Once these conditions are reached, further flashing is not necessary.

The steepness of the temperature gradient corresponding to a stable temperature distribution is a function of the part acceleration during parabolic flashing. In general, the higher the rate of part acceleration, the steeper is the stable temperature gradient produced. Thus, the shape of the temperature distribution curve in a particular application can be controlled by appropriate choice of flashing pattern. Since the compressive yield strength of a metal is temperature sensitive, the behavior of the metal during the upsetting portion of the welding cycle is markedly dependent upon the flashing pattern. Therefore, the choice of flashing pattern is extremely important for the production of sound flash welds. The minimum flashing distance is the amount of flashing required to produce a stable temperature distribution. From a practical standpoint, the flashing distance should be slightly greater than the minimum acceptable amount, to ensure that a stable temperature distribution is always achieved.

Upset occurs when a stable temperature distribution is achieved by flashing and the two parts are brought together rapidly. The movable parts should be accelerated rapidly so that the molten metal on the flashing surfaces will be extruded before it can solidify in the joint. Motion should continue with sufficient force to upset the metal and weld the two pieces together.

Upset current is sometimes applied as the joint is being upset to maintain temperature by resistance heating. This permits upset of the joint with lower force than would be required without it. Upset current is normally adjusted by electronic heat control on the basis of either experience or welding tests.

ADVANTAGES AND LIMITATIONS

BUTT JOINTS BETWEEN parts with similar cross section can be made by friction welding and upset welding, as well as by flash welding. The major difference between friction welding and upset and flash welding is that the heat for friction welding is developed by rubbing friction between the faying surfaces, rather than from electrical resistance. Upset welding is similar to flash welding except that no flashing action occurs.

Listed below are some important advantages of flash welding:

(1) Cross sectioned shapes other than circular can be flash welded: for example, angles, H sections, and rectangles. Rotation of parts is not required.

(2) Parts of similar cross section can be welded with their axes aligned or at an angle to each other, within limits.

(3) The molten metal film on the faying surfaces and its ejection during upset acts to remove impurities from the interface.

(4) Preparation of the faying surfaces is not critical except for large parts that may require a bevel to initiate flashing.

(5) Rings of various cross sections can be welded.

(6) The heat-affected zones of flash welds are much narrower than those of upset welds.

The following are some limitations of the process:

(1) The high single-phase power demand produces unbalance on three-phase primary power lines.

(2) The molten metal particles ejected during flashing present a fire hazard, may injure the operator, and may damage shafts and bearings. The operator should wear face and eye protection, and a barrier or shield should be used to block flying sparks.

(3) Removal of flash and upset metal is generally necessary and may require special equipment.

(4) Alignment of workpieces with small cross sections is sometimes difficult.

(5) The parts to be joined must have almost identical cross sections.

FLASH WELDING APPLICATIONS

Base Metals

MANY FERROUS AND nonferrous alloys can be flash welded. Typical metals are carbon and low alloy steels, stainless steels, aluminum alloys, nickel alloys, and copper alloys. Titanium alloys can be flash welded, but an inert gas shield to displace air from around the joint is necessary to minimize embrittlement.

Dissimilar metals may be flash welded if their upsetting characteristics are similar. Some dissimilarity can be overcome with a difference in the initial extensions between the clamping dies, adjustment of flashing distance, and selection of welding variables. Typical examples are welding of aluminum to copper or a nickel alloy to steel.

Typical Products

THE AUTOMOTIVE INDUSTRY uses wheel rims produced from flash welded rings that are formed from flat cold-rolled steel stock. The electrical industry uses motor and generator frames produced by flash welding plate and bar stock previously rolled into cylindrical form. Cylindrical transformer cases, circular flanges, and seals for power transformer cases are other examples. The aerospace industry uses flash welds in the manufacture of landing gear struts, control assemblies, hollow propeller blades, and rings for jet engines and rocket casings.

The petroleum industry uses oil drilling pipe with fittings attached by flash welding. Several major railroads are using flash welding to join relatively high carbon steel track. In many cases, welding is done in the field using welding machines and portable generating equipment mounted on railroad cars.

Miter joints are sometimes used in the production of rectangular frames for windows, doors, and other architectural trim. These products are commonly made of plain carbon and stainless steels, aluminum alloys, brasses, and bronzes. Usually the service loads are limited, but appearance requirements of the finished joints are stringent.

EQUIPMENT

Typical Machines

A TYPICAL FLASH welding machine consists of six major parts:

(1) The machine bed which has platen ways attached

(2) The platens which are mounted on the ways

(3) Two clamping assemblies, one of which is rigidly attached to each platen to align and hold the parts to be welded

(4) A means for controlling the motion of the movable platen

(5) A welding transformer with adjustable taps

(6) Sequencing controls to initiate part motion and flashing current

Flash welding machines may be manual, semi-automatic, or fully automatic in their operations; however, most of them are either semi-automatic or fully automatic. With manual operation, the operator controls the speed of the platen from the time that flashing is initiated until the upset is completed. In semi-automatic operation, the operator usually initiates flashing manually and then energizes an automatic cycle that completes the weld. In fully automatic operation, the parts are loaded into the machine and the welding cycle is then completed automatically. The platen motion of many small flash welding machines is provided mechanically by a cam that is driven by an electric motor through a speed reducer. Large machines may be hydraulically or pneumatically operated. Equipment for flash welding is discussed in Chapter 19.

Operating personnel should be given instructions on how to operate the machinery in a safe manner. Hands must be kept clear of moving machinery, and contact with electrically charged surfaces must be avoided.

Controls and Auxiliary Equipment

ELECTRICAL CONTROLS ON flash welding machines are integral types designed to sequence the machine, control the welding current, and precisely control the platen position during flashing and upsetting. Silicon controlled rectifier (SCR) contactors are widely used on machines drawing up to 1200 A from the power lines. Ignitron contactors are common on larger machines.

Preheat and postheat cycles are normally controlled with electronic timers and phase-shift heat controls. Timers for these functions may be initiated manually or automatically in proper order during the welding period.

Dies

FLASH WELDING DIES, compared to spot and seam welding electrodes, are not in direct contact with the welding area. Dies may be considered work holding and current conducting clamps. Since the current density in these dies is normally low, relatively hard materials with low electrical conductivity may be used. However, water cooling of the dies may be necessary in high production to avoid overheating.

There are no standardized designs for these dies since they must fit the contour of the parts to be welded. The size of the dies depends largely upon the geometry of the parts to be welded and the mechanical rigidity needed to maintain proper alignment of parts during upsetting. The dies are usually mechanically fastened to the welding machine platens.

Electrode contact area should be as large as practical to avoid local die burns. The contact surfaces may be incorporated in small inserts attached to larger dies for low cost replacement and convenient detachment for redressing. A facing insert of RWMA[1] Group B material which is brazed to the die is frequently used for maximum wear resistance.

If the parts are backed up so that the clamping dies do not need to carry the upset force, clamping pressures need only be sufficient to provide good electrical contact. If the work cannot be backed up, it may be necessary to use serrated clamp inserts. In this case, the inserts are usually made of hardened tool steel.

Flash welding dies tend to wear but do not mushroom. As wear takes place, the contact area may decrease and cause local hot spots (die burns). The dies should be kept clean. Flash and dirt will tend to embed in the dies and cause hot spots and die burns. All bolts, nuts, and other die-holding devices should be tight. Additional information on flash welding dies and materials is given in Chapter 19.

Fixtures and Backups

THE FUNCTIONS OF fixtures for flash welding are (1) to rapidly and accurately locate two or more parts relative to each other, (2) to hold them in proper location while they are being welded, and (3) to permit easy release of the welded assembly. A fixture is either fastened to the machine or built into it. Parts are loaded directly into the fixture and welded.

Resistance welding processes are very rapid compared to other methods of joining. If maximum production is to be attained, fixtures must be easily loaded and unloaded. The

following factors should be considered when designing a fixture:

(1) Quick-acting clamps, toggles, and other similar devices should be employed. Sometimes ejector pins are used to facilitate removal of the finished assembly.

(2) The fixture must be designed so that welding current is not shunted through any locating devices. This may require insulation of pins and locating strips.

(3) Nonmagnetic materials are usually preferred, because any magnetic material located in the throat of the machine will increase the electrical impedance and limit the maximum current which the machine can deliver.

(4) The operator should be able to load and unload the parts safely. This may require the use of swivel devices or slides, so that the fixture can be moved out of the machine. A guard should swing in place to prevent the operator from reaching between the platens. The guard can also act as the flash shield.

(5) A fixture must provide for movement of the parts as they are being clamped in the dies.

(6) All bearings, pins, slides, etc., should be protected from spatter and flash.

Backups are needed if the clamping dies cannot prevent slippage of the parts when the upsetting force is applied. Slippage usually occurs when the section of the part in the die is too short for effective clamping, or the part is unable to withstand sufficient clamping force without damage.

A backup often consists of a steel bracket that can be bolted in various positions to the platen. Brackets can have either fixed or adjustable stops against the parts.

WELDING PROCEDURES

EVERY WELDING OPERATION involves numerous variables that affect the quality of the resulting weld. For this reason, a welding procedure should be developed that prescribes the settings for the welding variables to ensure consistent weld quality. Flash welding involves dimensional, electrical, force, and time variables. The dimensional variables are shown in Figures 18.2 A and B and 18.3. The paths of the movable platen and the faying surfaces during flashing and upsetting are also shown in Figure 18.3. The current, force, and time variables are shown in Figure 18.4. Most operations do not involve all of the variables shown. A simple flash welding cycle involves flashing at one voltage setting followed by upset.

Joint Design

THREE COMMON TYPES of welds made by flash welding are shown in Figure 18.5. Several basic design rules for flash welding are as follows:

---

1. Resistance Welder Manufacturers Association

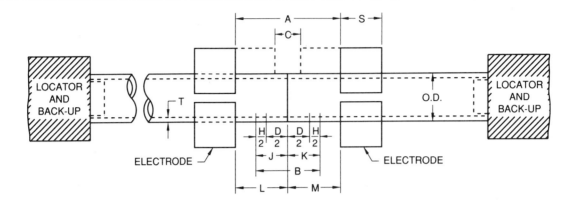

T = TUBE WALL OR SHEET
    THICKNESS
A = INITIAL DIE OPENING
B = MATERIAL LOST
C = FINAL DIE OPENING

D = TOTAL FLASH-OFF
H = TOTAL UPSET
J = K = MATERIAL LOST
    PER PIECE
L = M = INITIAL EXTENSION
    PER PIECE

O.D. = OUTSIDE DIA.
    OF TUBING
S = MINIMUM NECESSARY
    LENGTH OF ELECTRODE
    CONTACT

Figure 18.2A–Flash Welding of Tubing and Flat Sheets (See Table 18.5 for Recommended Data)

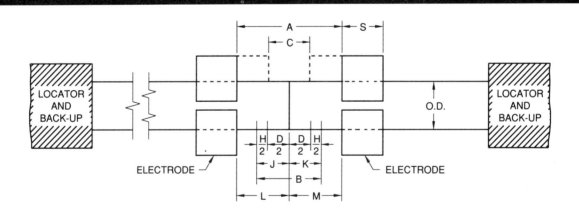

O.D. = OUTSIDE DIA. OF ROUNDS OR MINIMUM DIMENSION OF OTHER SECTIONS

A = INITIAL DIE OPENING
B = MATERIAL LOST
C = FINAL DIE OPENING
D = TOTAL FLASH-OFF

H = TOTAL UPSET
J = K = MATERIAL LOST
    PER PIECE
L = M = INITIAL EXTENSION
    PER PIECE

S = MINIMUM NECESSARY
    LENGTH OF ELECTRODE
    CONTACT

Figure 18.2B–Flash Welding of Solid Round, Hex, Square and Rectangular Bars (See Table 18.6 for Recommended Data)

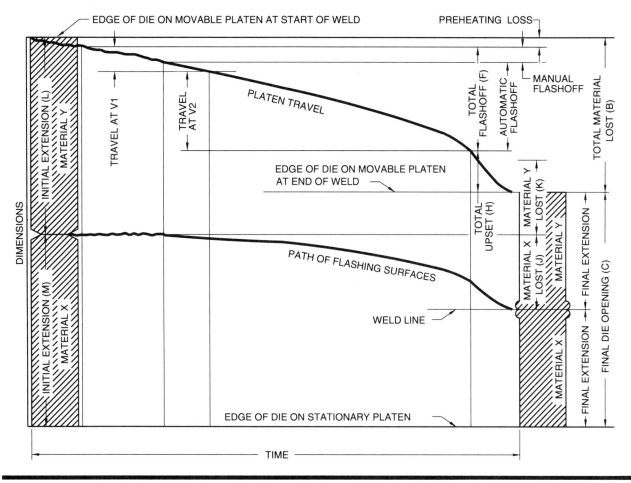

Figure 18.3–Flash Welding Dimensional Variables and Motions

(**1**) The design should provide for an even heat balance in the parts so that the ends to be welded will have nearly equal compressive strengths at the end of the flashing time.

(**2**) The metal lost during flashing (flash loss) and upset must be included in the initial length when designing the part. With miter joints, the angle between the two members must be taken into account in the design.

(**3**) The parts must be designed so that they can be suitably clamped and held in accurate alignment during flashing and upset, with the joint perpendicular to the upset force direction.

(**4**) The end preparation should be designed so that the flash material can escape from the joint, and that flashing starts at the center or the central area of the parts.

In general, the two parts to be welded should have the same cross section at the joint. Bosses may have to be machined, forged, or extruded on parts to meet this requirement.

In the flash welding of extruded or rolled shapes with different thicknesses within the cross-section, the temperature distribution during flashing will vary with section thickness. This tendency can often be counteracted by proper design of the clamping dies, provided the ratio of the thicknesses does not exceed about 4 to 1.

The recommended maximum joint lengths for several thicknesses of steel sheet are given in Table 18.1. The maximum diameters for steel tubing of various wall thicknesses are listed in Table 18.2. The limits can be exceeded in some cases using special procedures and equipment.

When flash welding rings, there is a ratio of circumference to cross-sectional area below which shunting of current becomes a problem. That power loss can be high. The minimum ratio will depend upon the electrical resistivity of the metal to be welded. With metals of high resistivity, such as stainless steel, the ratio can be lower than with low resistivity metals, such as aluminum.

When heavy sections are welded, it is often advisable to bevel the end of one part to facilitate the start of flashing.

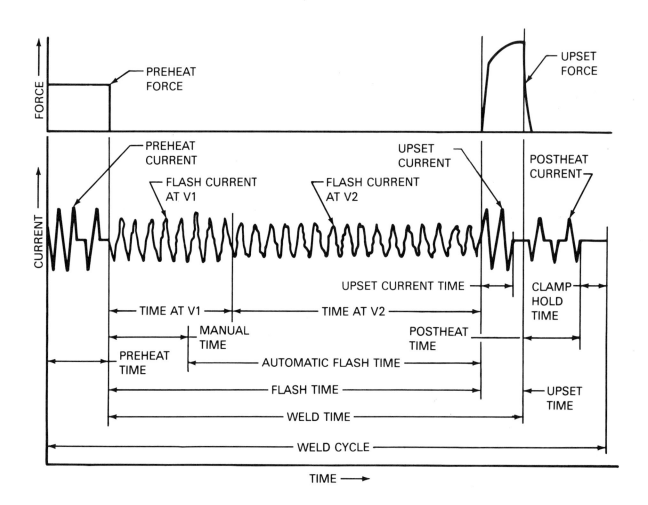

Figure 18.4—Flash Welding Current, Force, and Time Variables

Table 18.1
Recommended Maximum Joint Lengths of Flat Steel Sheet for Flash Welding

Sheet Thickness,		Max. Joint Length,		Sheet Thickness,		Max. Joint Length,	
in.	mm	in.	mm	in.	mm	in.	mm
0.010	.25	1.00	25	0.060	1.5	25.00	635
0.020	.50	5.00	125	0.080	2.0	35.00	890
0.030	.75	10.00	250	0.100	2.5	45.00	1145
0.040	1.0	15.00	375	0.125	3.2	57.00	1450
0.050	1.3	20.00	500	0.187	4.8	88.00	2235

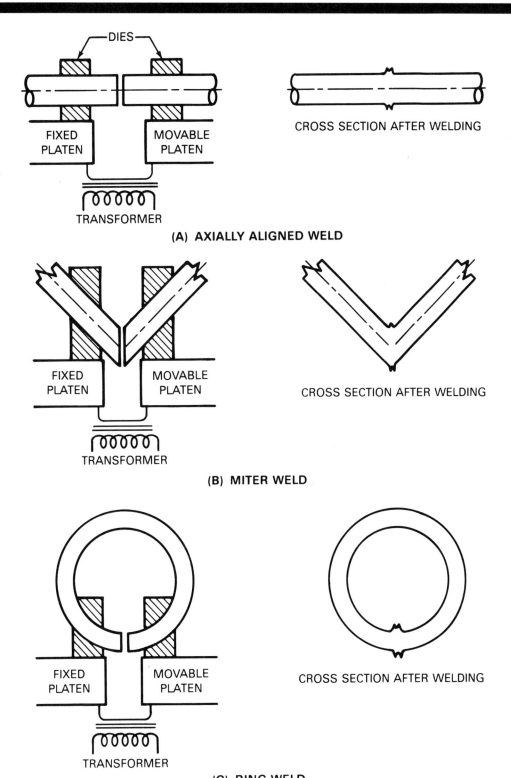

DIES

FIXED PLATEN

MOVABLE PLATEN

TRANSFORMER

CROSS SECTION AFTER WELDING

(A) AXIALLY ALIGNED WELD

FIXED PLATEN

MOVABLE PLATEN

TRANSFORMER

CROSS SECTION AFTER WELDING

(B) MITER WELD

FIXED PLATEN

MOVABLE PLATEN

TRANSFORMER

CROSS SECTION AFTER WELDING

(C) RING WELD

Figure 18.5–Common Types of Flash Welds

Such beveling may eliminate the necessity for preheating or initially flashing at a voltage higher than normal. Suggested dimensions for beveling plate, rod, and tubing are shown in Figure 18.6.

Heat Balance

IN AXIALLY ALIGNED joints, when the two parts to be welded are of the same alloy and cross section, the heat generated in each of the parts during the weld cycle will be the same, provided the physical arrangement for welding is uniform. Flash loss and upset loss will also be equal in each part. In general, the heat balance between two parts of the

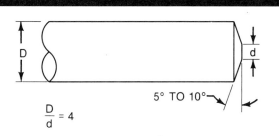

(A) RODS AND BARS OF 0.25 in. AND LARGER

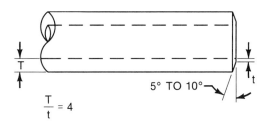

(B) TUBING OF 0.188 in. WALL AND LARGER

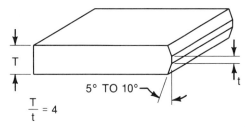

(C) FLAT PLATE OF 0.188 in. AND THICKER

NOTE: BEVEL ONLY ONE PIECE WHEN D IS 0.25 in. OR LARGER AND T IS 0.188 in. OR GREATER.

Figure 18.6–End Preparation for One Part to Facilitate the Flashing of Large Sections

same alloy will be adequate if their respective cross-sectional areas do not differ by more than normal manufacturing tolerances.

When flash welding two dissimilar metals, the metal loss during flashing may differ for each metal. Such behavior can be attributed to differences in electrical and thermal conductivities or melting temperatures, or both. To compensate for this, the extension from the clamping die of the more rapidly consumed part should be greater than that of the other part. In the case of aluminum and copper, the extension of the aluminum part should be twice that of the copper part.

Flash welding of nonaligned sections (miter joints) may produce a joint with varying properties across it because of heat unbalance across the joint. Since the faying surfaces are not perpendicular to their respective part lengths, the volume of metal decreases across the joint to a minimum at the apex. Consequently, flashing and upset at the apex may vary significantly from that which occurs across the remainder of the joint.

Miter joints between round or square bars should have a minimum included angle of 150 degrees. At smaller angles, the weld area at the apex will be poor quality because of the lack of adequate backup metal. Satisfactory miter joints may be made between thin rectangular sections in the same plane with an included angle as small as 90 degrees, provided the width of the stock is greater than 20 times its thickness. If service loading produces a tensile stress at the apex, the outside corner should be trimmed to remove the poor quality joint area.

Surface Preparation

SURFACE PREPARATION FOR flash welding is of minor importance and in most cases none is required. Clamping surfaces usually require no special preparation unless excessive scale, rust, grease, or paint is present. The abutting surfaces should be reasonably clean to accomplish electrical contact. Once flashing starts, dirt or other foreign matter will not seriously interfere with the completion of the weld.

Initial Die Opening

THE INITIAL DIE opening is the sum of the initial extensions of the two parts, as shown in Figures 18.2 (A) and (B) and 18.3. The initial extension for each part must provide for metal loss during flashing (flash loss) and upset, as well as some undisturbed metal between the upset metal and the clamping die. Initial extensions for both parts are determined from available welding data or from welding tests. The initial die opening should not be too large, otherwise nonuniform upset and joint misalignment may occur.

Alignment

IT IS IMPORTANT that the parts to be welded are properly aligned in the welding machine so that flashing on the faying surfaces is uniform. If the parts are misaligned, flashing will occur only across opposing areas and heating will not

Table 18.2
Recommended Maximum Diameters of Steel Tubing for Flash Welding

Wall Thicknesses,		Max. Tubing Diameter,		Wall Thicknesses,		Max. Tubing Diameter,	
in.	mm	in.	mm	in.	mm	in.	mm
0.020	0.5	0.50	13.0	0.125	3.2	4.00	102
0.030	0.8	0.75	19.0	0.187	4.7	6.00	152
0.050	1.3	1.25	32.0	0.250	6.4	9.00	230
0.062	1.6	1.50	38.0				
0.080	2.0	2.00	51.0				
0.100	2.5	3.00	76.0				

be uniform when upset, thus the parts will tend to slip past each other, as illustrated in Figure 18.7. Alignment of parts should be given careful consideration in designing of the machine, the parts to be welded, and the tooling for welding them. This is especially true when the ratio of the width to thickness of sections is large.

Material Loss

THE FINAL LENGTH of the welded assembly will be less than the sum of the lengths of the original workpieces because of flash and upset losses. These losses must be established for each assembly and then added to the workpiece length so that the welded assembly will meet design requirements. Changes in welding procedures may require modification of workpiece lengths.

Gas Shielding

IN SOME APPLICATIONS, displacement of air from the joint area by an inert or reducing gas shield may improve joint quality by minimizing contamination by oxygen, nitrogen, or both. However, gas shielding cannot compensate for improper welding procedures, and it should only be used when required by the application.

Argon or helium is particularly effective when flash welding reactive metals, such as titanium. At high temperatures, these metals are embrittled when they are exposed to air. Dry nitrogen may be effective with stainless and heat-resisting steels.

The value of a protective atmosphere depends upon the effectiveness of the shield design. The flash loss material may deposit on the gas shielding apparatus and interfere with its operation. Provisions for platen movement must be provided in the design.

If gas cylinders are used to provide the gas shielding, they should be protected from damage by plant traffic. Cylinder storage racks should have securing devices. If the gas shield is provided by a piping system, the piping should be properly labeled.

Preheating

DURING PREHEATING, THE parts are brought into contact under light pressure and then the welding transformer is energized. The resistance heating effect of high density current flow heats the metal between the dies. The temperature distribution across the joint during preheating approximates a sinusoidal waveform with the peak temperature point at the interface.

Three useful functions may be served by a preheating operation:

(1) It raises the temperature of the parts which, in turn, makes flashing easier to start and maintain.

(2) It produces a temperature distribution with a flatter gradient which persists throughout the flashing operations. This, in turn, distributes the upset over a longer length than is the case when no preheat is employed.

(3) It may extend the capacity of a machine and permit the joining of larger cross sections than would be otherwise possible. However, there is one possible drawback to preheating. Since preheating is often a manual operation, even when the machine is capable of flash welding automatically, the reproducibility of the preheating operation is largely a function of operator skill.

Welding

MOST COMMERCIAL FLASH welding machines are operated automatically. The welding schedule is established for the particular operations by a series of test welds that are evaluated for quality. The machine is then set up to repro-

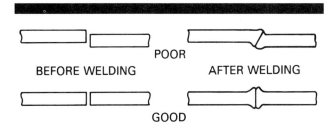

Figure 18.7–Effect of Poor Alignment on Joint Geometry

duce the qualified welding schedule for the particular application.

The operator may load and unload the machine and observe the welding cycle for consistency of operation. In some applications, automatic feed and ejection devices may be incorporated on the machine.

Postheating

STEELS WITH EXTREMELY high alloy or carbon content may crack if the weld is cooled too rapidly. In some cases, this condition may be avoided by preheating large parts, which will decrease the subsequent cooling rate. Postheating the joint in the welding machine by resistance heating or by immediately placing the weldment in a furnace operating at the desired temperature may prevent cracking when preheating is ineffective.

A postheat cycle may be incorporated in a flash welding machine using an electronic timer and phase-shift heat control. The postheat timer can be initiated at the end of upset or after a time delay. The desired temperature can be attained by adjustment of the heat control. However, heat will be transferred from the weldment to the clamping dies during the postheat. This must be considered in designing the die and in material selection, and water cooling may be necessary.

Flash Removal

IT IS FREQUENTLY necessary to remove the flash material from the welded joint. In some cases, this is done only for the sake of appearance. A joint is somewhat stronger in tension if the flash is not removed, because of the larger cross section provided by upset material. However, the notch effect at the weld line may then cause a reduction of fatigue strength. The notched portion of the upset material should be removed, but the balance may be left in place when the design of parts indicates that reinforcement is beneficial.

It is generally easier to remove the flash immediately after welding while the metal is still hot. This can be done by a number of methods, including machining, grinding, high-speed wheels, die trimming, oxy-fuel gas cutting, high-speed sanding, and pinch-off type clamping dies. With some alloy steels, flash removal with cutting tools is often difficult because of their hardness. In these cases, either grinding or oxyfuel gas cutting is usually employed.

With soft metals such as aluminum and copper, the flash may be almost sheared off using pinch-off dies. These dies have sharp tapered faces which cut almost through the metal as upsetting takes place. The final die opening is small. The partially sheared flash is then easily removed by other means. The joint can then be smoothed by filing or grinding.

PROCESS VARIABLES

Flashing Voltage

FLASHING VOLTAGE IS determined by the welding transformer tap setting. It should be selected to be as low as possible consistent with good flashing action. Electronic phase-shift heat control is not an effective means for reducing the flash voltage. The secondary voltage wave form produced by this means is incompatible with good flashing action.

Changes in flashing voltage should be made only by changing the tap setting of the transformer. One system for providing two voltage ranges uses two primary contactors, each of which is connected to separate transformer taps. One contactor is energized to provide a high secondary voltage (V_1 in Figures 18.3 and 18.4) during the initial stages of flashing. The high voltage assists in starting the flashing action. The other contactor is energized after a predetermined time in the flashing operation to provide a normal secondary voltage (V_2 in Figures 18.3 and 18.4). The first contactor is de-energized at the same time. The best flashing action is achieved with this arrangement.

Flashing Time

FLASHING IS CARRIED out over a time interval to obtain the required flash loss of metal. The time required will be related to the secondary voltage and the rate of metal loss as flashing progresses. Since a flashing pattern is generally parabolic, the variables are interrelated. In any case, smooth flashing action for some minimum flashing distance during some time interval is necessary to produce a sound, strong weld.

Upset

IN THE PRODUCTION of a satisfactory flash weld, the flash and upset variables must be considered together since they are interrelated. The upset variables include the following:

(1) Flashing voltage cutoff
(2) Upset rate
(3) Upset distance
(4) Upset current magnitude and duration

Flashing Voltage Cutoff

FLASHING VOLTAGE SHOULD be terminated at the moment that upset of the weld commences. Adjustments should be made during actual welding tests to ensure that voltage termination does not take place before the faying surfaces make full contact.

Upset Rate

UPSET IS INITIATED by increasing the acceleration of the parts to bring the faying surfaces together quickly. The molten metal and oxides present on the surfaces are forced out of the joint as this occurs. The hot weld zone is upset. The upset rate must be sufficient to expel the molten metal before it solidifies and to produce the optimum upset while the metal has adequate plasticity.

The welding machine must apply a force to the movable platen to properly accelerate the part and overcome the resistance of the parts to plastic deformation. The force required depends upon the cross-sectional area of the joint, the yield strength of the hot metal to be welded, and the mass of the movable platen. Table 18.3 gives the approximate minimum upset pressures for flash welding typical alloys. These values may be used for a first approximation in determining the welding machine size required to flash weld a particular joint area on one of these alloys.

Upset Distance

THE MAGNITUDE OF the upset distance must be sufficient to accomplish two actions:

(1) The oxides and molten metal must be expelled from the faying surfaces.

(2) The two faying surfaces must be brought into intimate metal-to-metal contact over the entire cross section.

The amount of upset required to obtain a sound flash weld depends upon the metal and the section thickness. If the flashing conditions produce relatively smooth flashed sur-

faces, smaller upset distances than needed for roughly flashed surfaces will be satisfactory for most metals. Some heat-resistant alloys may require upset distances as large as 1 to 1.25 times the section thickness. Satisfactory welds are made in aluminum with upset distances about 50 percent greater than those employed with steels of similar thicknesses. Typical flash welding dimensions including upset distances and material losses are shown in Tables 18.5 and 18.6. These data are for low and medium strength forging steels.

Upset Current

AS DISCUSSED UNDER postheating above, in some cases the weld zone may tend to cool too rapidly after flashing is terminated. This may result in inadequate upset or cold cracking of the upset metal. The joint temperature can be maintained during upset by resistance heating with current supplied by the welding transformer. The current magnitude is commonly controlled electronically.

Normally, upset current would be terminated at the end of upset. If the flash is to be mechanically trimmed immediately after welding, upset current may be maintained for an additional period to achieve the desired temperature for trimming.

WELD QUALITY

Effect of Welding Variables

WELD QUALITY IS significantly affected by the specific welding variables selected for the application. Table 18.4 indicates the effects of several variables on quality when they are excessive or insufficient in magnitude. Each variable is considered individually, although more than one can produce the same result. Commmon defects found in flash welds are discussed below.

Base Metal Structure

METALLURGICAL DISCONTINUITIES THAT often originate from conditions present in the base metal can usually be minimized by specifying necessary qualities in the materials selected. The inherent fibrous structure of wrought mill products may cause anisotropic mechanical behavior. An out-turned fibrous structure at the weld line often results in some decrease in mechanical properties as compared to the base metal, particularly in ductility.

The decrease in ductility caused by flash welding is normally insignificant except in two cases:

(1) The base material may be extremely nonhomogeneous. Examples are severely banded steels, alloys with excessive stringer type inclusions, and mill products with seams and cold shuts produced during the fabrication process.

(2) The upset distance may be excessive.

Table 18.3
Upsetting Pressures for Various Classes of Alloys

Strength Classification	Examples	Upset Pressure ksi	MPa
Low forging	SAE 1020, 1112, 1315 and those steels commonly designated as high strength low alloy	10	69
Medium forging	SAE 1045, 1065, 1335, 3135, 4130, 4140, 8620, 8630	15	103
High forging	SAE 4340, 4640, 300M, tool steel, 12% Cr and 18-8 stainless steel, titanium, aluminum	25	172
Extra high forging	Materials exhibiting extra high compressive strength at elevated temperature such as A286, 19-9 DL, nickel- and cobalt-based alloys	35	241

Table 18.4
Effect of Variables on Flash and Upset Weld Quality

	Voltage	Rate	Time	Current	Distance or Force
Excessive	Deep craters are formed that cause voids and oxide inclusions in the weld; cast metal in weld.	Tendency to freeze.	Metal too plastic to upset properly.	Molten material entrapped in upset; excessive deformation.	Tendency to upset too much plastic metal; flow lines bent perpendicular to base metal.
Insufficient	Tendency to freeze; metal not plastic enough for proper upset.	Intermittent flashing, which makes it difficult to develop sufficient heat in the metal for proper upset.	Not plastic enough for proper upset; cracks in upset.	Longitudinal cracking through weld area; inclusions and voids not properly forced out of the weld.	Failure to force molten metal and oxides from the weld; voids.

When the upset distance is excessive, the fibrous structure may be completely reoriented transverse to the original structure.

Oxides

ANOTHER SOURCE OF metallurgical discontinuities is the entrapment of oxides at the weld interface. Such defects are rare since proper upset should expel any oxides formed during the flashing operation.

Flat Spots

FLAT SPOTS ARE metallurgical discontinuities that are usually limited to ferrous alloys. Their exact cause is not certain. They appear on a fractured surface through the weld interface in the form of smooth, irregularly shaped areas.

There is excellent correlation between the location of flat spots and localized regions of carbon segregation in steels. In many cases, the cooling rates associated with flash welds are rapid enough to produce brittle, high carbon martensite at areas on the flashing interface where the carbon content happens to be greater than the nominal composition of the alloy. Microhardness tests and metallographic examination have confirmed the presence of high carbon martensite in the region surrounding a "flat spot" in almost every case, even in plain carbon steels. Furthermore, steels with banded microstructures appeared significantly more susceptible to this type of defect than unbanded steels.

Die Burns

BURNS ARE DISCONTINUITIES produced by local overheating of the base metal at the interface between the clamping die and the part surface. They can usually be avoided by keeping the parts clean and mating them properly with the dies.

Voids

VOIDS ARE USUALLY the result of either insufficient upset or excessive flashing voltage. Deep craters produced on the faying surfaces by excessive flashing voltage may not be completely eliminated during upset. Such discontinuities are usually discovered when the welding procedure is being qualified. They are readily avoided by decreasing the flashing voltage or increasing the upset distance. Figures 18.8 (A) and (B) show the appearance of flash welds with satisfactory and unsatisfactory upset.

Cracking

THE TYPE OF discontinuity known as a crack may be internal or external. It may be related to the metallurgical characteristics of the metal. Alloys that exhibit low ductility over some elevated temperature range may be susceptible to internal hot cracking. Such alloys, known as "hot-short" alloys, are somewhat difficult to flash weld, but usually can be successfully welded with the proper conditions.

Cold cracking may occur in hardenable steels. It can usually be eliminated by welding with conditions that moderate the weld cooling rate, coupled with post-welding heat treatment as soon as possible after welding.

Insufficient heating prior to or during upset is the usual cause of cracking in the external upset metal, as shown in Figure 18.8 (C). This can be eliminated by resistance heating during upset.

Mechanical Discontinuities

MECHANICAL DISCONTINUITIES INCLUDE misalignment of the faying surfaces prior to welding, and nonuniform upset during welding. These discontinuities are easily detected by visual inspection. Misalignment of the parts is corrected by adjustment of the clamping dies and fixtures. Nonuniform upset may be caused by part misalignment, insufficient clamping force, or excessive die opening at the start of upset. The latter can be corrected by decreasing the initial die opening and then adjusting the welding schedule, if necessary.

Table 18.5
Data for Flash Welding of Tubing and Flat Sheets* (See Fig. 18.2A for Assembly of Parts)

Thickness in.	Initial Die Opening in.	Material Lost in.	Final Die Opening in.	Total Flash Loss in.	Total Upset in.	Material Loss Per Piece in.	Initial Extension Per Piece in.	Flash Time Seconds	O.D. in.	Minimum Length of Electrode Contact	
										Within Backup	Without Backup
0.010	0.110	0.060	0.050	0.040	0.020	0.030	0.055	1.00	0.250	0.375	1.00
0.020	0.215	0.115	0.100	0.080	0.035	0.058	0.108	1.50	0.312	0.375	1.00
0.030	0.325	0.175	0.150	0.125	0.050	0.088	0.163	2.00	0.375	0.375	1.50
0.040	0.430	0.230	0.200	0.165	0.085	0.115	0.215	2.50	0.500	0.375	1.75
0.050	0.530	0.280	0.250	0.205	0.075	0.140	0.265	3.25	0.750	0.500	2.00
0.060	0.620	0.330	0.290	0.240	0.090	0.165	0.310	4.00	1.000	0.750	2.50
0.070	0.715	0.385	0.330	0.280	0.105	0.193	0.358	5.00	1.50	1.000	3.00
0.080	0.805	0.435	0.370	0.315	0.120	0.218	0.403	6.00	2.00	1.250	**
0.090	0.885	0.475	0.410	0.345	0.130	0.238	0.443	7.00	2.50	1.750	**
0.100	0.970	0.520	0.450	0.375	0.145	0.260	0.485	8.00	3.00	2.000	**
0.110	1.060	0.570	0.490	0.410	0.160	0.285	0.530	9.00	3.50	2.250	**
0.120	1.140	0.610	0.530	0.440	0.170	0.305	0.570	10.0	4.00	2.500	**
0.130	1.225	0.650	0.575	0.470	0.180	0.325	0.613	11.0	4.50	2.750	**
0.140	1.320	0.700	0.620	0.510	0.190	0.350	0.660	12.0	5.00	2.750	**
0.150	1.390	0.730	0.660	0.530	0.200	0.365	0.695	13.0	5.50	3.000	**
0.160	1.470	0.770	0.700	0.560	0.210	0.385	0.735	14.0	6.00	3.250	**
0.170	1.540	0.800	0.740	0.580	0.220	0.400	0.770	15.0	6.50	3.500	**
0.180	1.620	0.840	0.780	0.610	0.230	0.420	0.810	16.0	7.00	3.750	**
0.190	1.690	0.870	0.820	0.630	0.240	0.435	0.845	17.0	7.50	4.000	**
0.200	1.760	0.900	0.860	0.650	0.250	0.450	0.880	18.0	8.00	4.250	**
0.250	2.010	1.010	1.000	0.730	0.280	0.505	1.005	24.0	8.50	4.500	**
0.300	2.245	1.120	1.125	0.810	0.310	0.560	1.123	30.0	9.00	4.750	**
0.350	2.460	1.210	1.250	0.880	0.330	0.605	1.230	36.0	9.50	5.000	**
0.400	2.640	1.290	1.350	0.930	0.360	0.645	1.320	42.0			
0.450	2.780	1.350	1.430	0.970	0.380	0.675	1.390	48.0			
0.500	2.910	1.410	1.500	1.020	0.390	0.705	1.455	54.0			
0.550	3.040	1.465	1.575	1.055	0.410	0.733	1.520	60.0			
0.600	3.135	1.505	1.630	1.085	0.420	0.753	1.568	66.0			
0.650	3.245	1.555	1.690	1.125	0.430	0.778	1.623	73.0			
0.700	3.360	1.610	1.750	1.160	0.450	0.805	1.680	80.0			
0.800	3.525	1.675	1.850	1.210	0.465	0.838	1.763	92.0			
0.900	3.660	1.730	1.930	1.250	0.480	0.865	1.830	104.0			
1.000	3.800	1.800	2.000	1.300	0.500	0.900	1.900	116.0			

Notes:

* Data based on welding without preheat, and for two pieces of same welding characteristics, using constant acceleration of flash rate.

** Not recommended without use of backup.

TESTING AND INSPECTION

NONDESTRUCTIVE EVALUATION OF flash welded joints is complicated by several factors including the flash, the variation in thickness for bars, and other factors. Fortunately, one of the major advantages of flash welding is that it can be highly mechanized and automated. Therefore, a consistent quality level is readily maintained after satisfactory welding conditions are established. The fact that no filler metal is employed means that the strength of the weld is primarily a function of the base metal composition and properties. Consequently, properly made flash welds should exhibit satisfactory mechanical properties.

In commercial practice, both destructive and nondestructive tests are employed to ensure maintenance of the desired quality level in critical flash welded products. The process control procedure usually includes the following:

(1) Material certification
(2) Qualification of welding procedure
(3) Visual inspection of the product
(4) Destructive testing of random samples

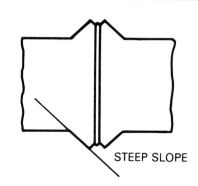

(A) SATISFACTORY HEAT AND UPSET

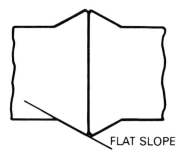

(B) INSUFFICIENT HEAT OR UPSET OR BOTH

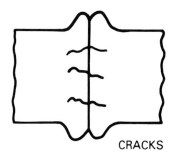

(C) CRACKS DUE TO INSUFFICIENT HEAT

Figure 18.8–Visual Indications of Flash Weld Quality

When the product is used in a critical application, the above procedure is supplemented by other tests such as magnetic particle and dye penetrant examination. When the welded joint is subsequently machined, routine measurement of the hardness of the weld area may also be specified. In addition, specifications may require proof testing of flash welded products.

Material Certification

SINCE MATERIAL DEFECTS may cause flash weld discontinuities, each lot of raw material should be carefully inspected upon delivery to ensure that it meets specifications. Certified chemical analysis, mechanical property tests, macroetch examination, and magnetic particle inspection may be applicable.

Procedure Qualification

EACH NEW COMBINATION of material and section size to be flash welded normally requires qualification of a new welding procedure. This usually involves welding a number of test specimens that duplicate the material, section size, welding procedure, and heat treatment to be used in producing the product. All of these specimens are visually inspected for cracks, die burns, misalignment, and other discontinuities. Where required, weld hardnesses are measured. To verify weld strength, a tensile specimen should be machined from a test weld using the entire welded cross section where possible. The test results should be compared to the base metal properties and design requirements.

All pertinent welding conditions used in producing the qualification test should be recorded. The production run is then made using the qualified welding procedure.

Nondestructive Inspection

EACH COMPLETED WELD in the production run should be visually examined for evidences of cracks, die burns, misalignment, or other external weld defects. Where specified, magnetic particle or fluorescent penetrant inspection is performed on random samples to assist in detecting flaws not visible to the unaided eye. In critical applications, random radiographic examination may also be specified.

Destructive Testing

DEPENDING UPON THE size of the production run, a specified number of randomly chosen parts may be selected for destructive testing of the welds. The results of these destructive tests must all meet the same criteria specified in the welding procedure qualification test. Additional tests are required if any fail. A report of the results of all destructive tests is then prepared to certify the maintenance of the required average quality level for the lot.

Bend Tests.   Notched bend tests may be used to force a fracture to occur along the weld interface for visual examination. A bend test may be useful as a qualitative means for establishing a welding schedule. However, such tests are not generally used for specification purposes.

Tension Tests.   Where strength testing is required, the tension test specimen should be machined to include the entire welded cross-section of the flash weld, as applied for Procedure Qualification above.

WELDING OF STEEL

TYPICAL DATA FOR the flash welding of steel tubing and flat sheets are given in Table 18.5. For welding solid round, hexagonal, square, and rectangular steel bars, data are given in Table 18.6. Both tables are applicable to steels of low and medium forging strength. They give the recommended dimensions for setting up a flash welding machine to weld the various sections. Total flashing time is based on welding without preheating.

When setting up a schedule, the dimensional variables and flashing time are selected from the tables. The welding machine is adjusted to the lowest secondary voltage at which steady and consistent flashing can be obtained. The secondary voltages available are dependent upon the electrical design of the welding transformer.

The upsetting force used for a particular application depends upon the alloy and the cross-sectional area of the joint. The selection of equipment for steels should be based on the values of recommended upset pressures given in Table 18.3. Such values are based on welding without preheat.

Table 18.6
Data for Flash Welding of Solid Round, Hex, Square and Rectangular Bars* (See Fig. 18.2(B) for Assembly of Parts)

Thickness in.	Initial Die Opening in.	Material Lost in.	Final Die Opening in.	Total Flash Loss in.	Total Upset in.	Material Loss Per Piece in.	Initial Extension Per Piece in.	Flash Time Seconds	O.D. in.	Minimum Length of Electrode Contact Within Backup	Minimum Length of Electrode Contact Without Backup
0.050	0.100	0.050	0.050	0.040	0.010	0.025	0.050	1.00	0.250	0.375	1.00
0.100	0.182	0.082	0.100	0.062	0.020	0.041	0.091	1.50	0.312	0.375	1.00
0.150	0.270	0.120	0.150	0.090	0.030	0.060	0.135	2.00	0.375	0.375	1.50
0.200	0.350	0.150	0.200	0.110	0.040	0.075	0.175	2.50	0.500	0.375	1.75
0.250	0.430	0.180	0.250	0.130	0.050	0.090	0.215	3.25	0.750	0.500	2.00
0.300	0.510	0.210	0.300	0.150	0.060	0.105	0.255	4.00	1.000	0.750	2.50
0.350	0.600	0.250	0.350	0.180	0.070	0.125	0.300	5.00	1.50	1.000	3.00
0.400	0.685	0.285	0.400	0.205	0.080	0.143	0.343	6.00	2.00	1.250	**
0.450	0.770	0.320	0.450	0.230	0.090	0.160	0.385	7.00	2.50	1.750	**
0.500	0.850	0.350	0.500	0.250	0.100	0.175	0.425	8.00	3.00	2.000	**
0.550	0.940	0.390	0.550	0.280	0.110	0.195	0.470	9.00	3.50	2.250	**
0.600	1.025	0.425	0.600	0.305	0.120	0.213	0.513	10.0	4.00	2.500	**
0.650	1.100	0.450	0.650	0.325	0.125	0.225	0.550	11.0	4.50	2.750	**
0.700	1.180	0.480	0.700	0.350	0.130	0.240	0.590	12.0	5.00	2.750	**
0.750	1.260	0.510	0.750	0.375	0.135	0.255	0.630	13.0	5.50	3.000	**
0.800	1.340	0.540	0.800	0.400	0.140	0.270	0.670	14.0	6.00	3.250	**
0.850	1.420	0.570	0.850	0.425	0.145	0.285	0.710	15.0	6.50	3.500	**
0.900	1.500	0.600	0.900	0.450	0.150	0.300	0.750	16.0	7.00	3.750	**
0.950	1.580	0.630	0.950	0.475	0.155	0.315	0.790	17.0	7.50	4.000	**
1.000	1.660	0.660	1.000	0.500	0.160	0.330	0.830	18.0	8.00	4.250	**
1.050	1.740	0.690	1.050	0.525	0.165	0.345	0.870	20.0	8.50	4.500	**
1.100	1.820	0.720	1.100	0.550	0.170	0.360	0.910	22.0	9.00	4.750	**
1.150	1.900	0.750	1.150	0.575	0.175	0.375	0.950	24.0	9.50	5.000	**
1.200	1.980	0.780	1.200	0.600	0.180	0.390	0.990	27.0			
1.250	2.060	0.810	1.250	0.625	0.185	0.405	1.030	30.0			
1.300	2.140	0.840	1.300	0.650	0.190	0.420	1.070	33.0			
1.400	2.300	0.900	1.400	0.700	0.200	0.450	1.150	36.0			
1.500	2.460	0.960	1.500	0.750	0.210	0.480	1.230	42.0			
1.600	2.620	1.020	1.600	0.800	0.220	0.510	1.310	49.0			
1.700	2.780	1.080	1.700	0.850	0.230	0.540	1.390	57.0			
1.800	2.940	1.140	1.800	0.900	0.240	0.570	1.470	66.0			
1.900	3.100	1.200	1.900	0.950	0.250	0.600	1.550	77.0			
2.000	3.260	1.260	2.000	1.000	0.260	0.630	1.630	92.0			

Notes:

* Data based on welding without preheat, and for two pieces of same welding characteristics, using constant acceleration of flash rate.

** Not recommended without use of backup.

UPSET WELDING

DEFINITION

UPSET WELDING (UW) is a resistance welding process that produces coalescence over the entire area of faying surfaces, or progressively along a butt joint, by the heat obtained from the resistance to the flow of welding current through the area where those surfaces are in contact. Pressure is used to complete the weld.

PRINCIPLES OF OPERATION

WITH THIS PROCESS, welding is essentially done in the solid state. The metal at the joint is resistance heated to a temperature where recrystallization can rapidly take place across the faying surfaces. A force is applied to the joint to bring the faying surfaces into intimate contact and then upset the metal. Upset hastens recrystallization at the interface and, at the same time, some metal is forced outward from this location. This tends to purge the joint of oxidized metal.

PROCESS VARIATIONS

UPSET WELDING HAS two variations:

(1) Joining two sections of the same cross section end-to-end (butt joint)
(2) Continuous welding of butt joint seams in roll-formed products such as pipe and tubing.

The first variation can also be accomplished by flash welding and friction welding. The second variation is also done with high frequency welding.

BUTT JOINTS

Metals Welded

A WIDE VARIETY of metals in the form of wire, bar, strip, and tubing can be joined end-to-end by upset welding. These include:

(1) Carbon steels
(2) Stainless steels
(3) Aluminum alloys
(4) Brass
(5) Copper
(6) Nickel alloys
(7) Electrical resistance alloys

Sequence of Operations

THE ESSENTIAL OPERATIONAL steps to produce an upset welded butt joint are as follows:

(1) Load the machine with the parts aligned end-to-end

(2) Clamp the parts securely
(3) Apply a welding force
(4) Initiate the welding current
(5) Apply an upset force
(6) Cut off the welding current
(7) Release the upset force
(8) Unclamp the weldment
(9) Return the movable platen and unload the weldment

The general arrangement for upset welding is shown in Figure 18.9. One clamping die is stationary and the other is movable to accomplish upset. Upset force is applied through the moveable clamping die or a mechanical backup, or both.

Joint Preparation

FOR UNIFORM HEATING, the faying surfaces should be flat, comparatively smooth, and perpendicular to the direction of the upsetting force. Prior to welding, they should be cleaned to remove any dirt, oil, oxidation, or other materials that will impede welding.

The contact resistance between the faying surfaces is a function of the smoothness and cleanliness of the surfaces and the contact pressure. This resistance varies inversely with the contact pressure, provided the other factors are constant. As the temperature at the joint increases, the

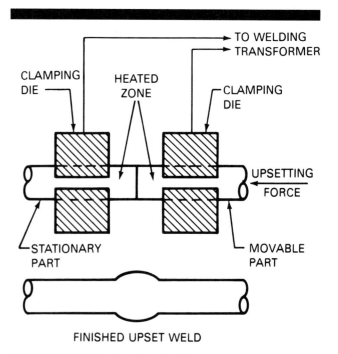

Figure 18.9–General Arrangement for Upset Welding of Bars, Rods, and Pipes

contact resistance changes, but it finally becomes zero when the weld is formed. Upset welding differs from flash welding in that no flashing takes place at any time during the welding cycle.

Generally, force and current are maintained throughout the entire welding cycle. The force is kept low at first to promote high initial contact resistance between the two parts. It is increased to a higher value to upset the joint when the welding temperature is reached. After the prescribed upset is accomplished, the welding current is turned off and the force is removed.

Equipment

EQUIPMENT FOR UPSET welding is generally designed to weld a particular family of alloys, such as steels, within a size range based on cross-sectional area. The mechanical capacity and electrical characteristics of the machine are matched to that application. Special designs may be required for certain aluminum alloys to provide close control of upset force.

Electric current for heating is provided by a resistance welding transformer. It converts line power to low voltage, high current power. No-load secondary voltages range from about 0.4 to 8 V. Secondary current is controlled by a transformer tap switch or by electronic phase shift.

Basically, an upset welding machine has two platens, one of which is stationary and the other movable. The clamping dies are mounted on these platens. The clamps operate either in straight line motion or through an arc about an axis, depending upon the application. Force for upset butt welding is produced generally by a mechanical, pneumatic, or hydraulic system.

Heat Balance

THE UPSET PROCESS is generally used to join together two pieces of the same alloy and same cross-sectional geometry. In this case, heat balance should be uniform across the joint. If the parts to be welded are similar in composition and cross section but of unequal mass, the part of larger mass should project from the clamping die somewhat farther than the other part. With dissimilar metals, the one with higher electrical conductivity should extend farther from the clamp than the other. When upset welding large parts that do not make good contact with each other, it is sometimes advantageous to interrupt the welding current periodically to allow the heat to distribute evenly into the parts.

Applications

UPSET WELDING IS used in wire mills and in the manufacture of products made from wire. In wire mill applications, the process is used to join wire coils to each other to facilitate continuous processing. The process also is used to fabricate a wide variety of products from bar, strip, and tubing. Typical examples of mill forms and products that have been upset welded are shown in Figure 18.10. Wire and rod from 0.05 to 1.25-in. (1.27 to 31.75 mm) diameter can be upset welded.

Weld Quality

BUTT JOINTS CAN be made that have about the same properties as the unwelded base metal. With proper procedures, welds made in wires are difficult to locate after they have passed through a subsequent drawing process. In many instances, the welds are then considered part of the continuous wire.

Upset welds may be evaluated by tension testing. The tensile properties are compared to those of the base metal. Metallographic and dye penetrant inspection techniques are also used.

A common method for evaluating a butt weld in wire is a bend test. A welded sample is clamped in a vise with the weld interface located one wire diameter from the vise jaws. The sample then is bent back and forth until it breaks in two. If the fracture is through the weld interface and shows complete fusion, or if it occurs outside the weld, the weld quality is considered satisfactory.

CONTINUOUS UPSET BUTT WELDING

General Description

IN THE MANUFACTURE of continuously welded pipe or tubing by upset welding, coiled strip is fed into a set of forming rolls. These rolls progressively form the strip into cylindrical shape. The edges to be joined approach each other at an angle and culminate in a longitudinal vee at the point of welding. A wheel electrode contacts each edge of the tube a short distance from the apex of the vee. Current from the power source travels from one electrode along the adjacent edge to the apex, where welding is taking place, and then back along the other edge to the second electrode. The edges are resistance-heated by this current to welding temperature. The hot edges are then upset together by a set of pinch rolls to consummate a weld.

Equipment

FIGURE 18.11 SHOWS a typical tube mill that uses upset welding to join the longitudinal seam. Figure 18.11 (A) shows the steel strip entering the strip guide assembly and the first stages of the forming section. The heat regulator, located behind the forming section, can be adjusted either manually or by phase-shift heat control. Figure 18.11 (B) shows a rotary type oil-cooled welding transformer. This welding equipment includes (1) a dressing tool assembly for dressing the welding electrodes without removing them from the welding machine, and (2) a scarfing tool assembly that removes the upset metal after welding. In the third step, the welded tube enters the straightening and sizing section, shown in Figure 18.11 (C). Following this, the tubing is cut to the desired length.

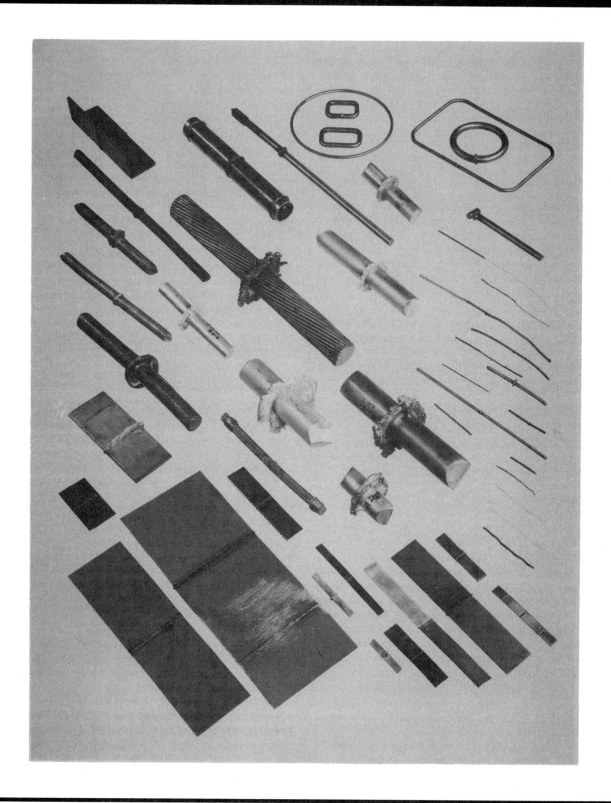

Figure 18.10–Typical Mill Forms and Products Joined by Upset Welding

**(A) STRIP GUIDE ASSEMBLY AND FIRST STAGES
OF THE FORMING SECTION**

(B) ROTARY TYPE OIL-COOLED WELDING TRANSFORMER

(C) STRAIGHTENING AND SIZING SECTION

Figure 18.11–Typical Tube Mill Using Welding for Joining the Longitudinal Seam

Welding can be done using either ac or dc power. Alternating current machines may be operated on either 60 Hz single-phase power or on power of higher frequency produced by a single-phase alternator. Direct current machines are powered by a three phase transformer-rectifier unit.

Welding Procedures

AS THE FORMED tube passes through the zone between the electrodes and the pinch rolls, there is a variation in pressure across the joint. If no heat were generated along the edges, this pressure would be maximum at the center of the squeeze rolls. However, since heat is generated in the metal ahead of the squeeze roll center line, the metal gradually becomes plastic and the point of initial edge contact is slightly ahead of the squeeze roll axes. The point of maximum upset pressure is somewhat ahead of the squeeze roll centerline.

The current across the seam is distributed in inverse proportion to the resistance between the two electrodes. This resistance, for the most part, is the contact resistance between the edges to be welded. Pressure is effective in reducing this contact resistance. As the temperature of the joint increases, the electrical resistance will increase and the pressure will decrease. A very sharp thermal gradient caused by the resistance heating at the peaks of the ac cycle produces a "stitch effect". The stitch is normally of circular cross section, lying centrally in the weld area and parallel to the line of initial closure of the seam edges. It is the hottest portion of the weld. The stitch area is molten while the area between stitches is at a lower temperature. The patches of molten metal are relatively free to flow under the influence of the motor forces (current and magnetic flux) acting on them. Consequently, they are ejected from the stitch area. If the welding heat is excessive, too much metal is ejected and pinhole leaks may result. With too little heat, the individual stitches will not overlap sufficiently, resulting in an interrupted weld.

The longitudinal spacing of the stitches must have some limit. The spacing is a function of the power frequency and the travel speed of the tube being welded. With 60 Hz power, the speed of welding should be limited to approximately 90 ft/min (0.45 m/sec). To weld tubing at higher speeds than this requires welding power of higher frequency. Typical welding speeds using various sizes of 180 Hz power sources for steel tubing of several wall thicknesses are shown in Table 18.7.

It is desirable to close the outside corners of the edges first as the formed tube moves through the machine so that the stitches will be inclined forward. This condition is known as an inverted vee. The advantages of using an inverted vee are twofold: (1) the angle deviation from the vertical reduces the forces tending to expel any molten metal in the joint, and (2) the major portion of the solid upset metal is extruded to the outside where it is easily removed. The tubing is normally formed so that the included angle of the vee is about 5 to 7 degrees.

Table 18.7
Typical Seam Welding Speeds for Steel Tubing Using 180Hz Power Sources

Wall Thickness, in.	Speed, ft/min.			
	125kVA	200kVA	300kVA	500kVA
0.050	150	200	--	--
0.065	110	140	200	--
0.083	72	105	145	--
0.095	--	85	115	--
0.109	--	66	90	--
0.125	--	50	70	140
0.134	--	--	60	125
0.156	--	--	--	85

Surface Burns

AS IN SPOT and seam welding, the current that provides the welding heat must enter the stock through electrode contacts. The resistance of these contacts must be kept to a minimum to avoid resistance heating sufficient to result in surface burns on the tube. Burns are actually surface portions of the tube that are heated to their fusion or melting point. They tend to stick to or embed themselves in the face of the wheel electrode. If large steel particles become embedded in the electrode face, the contact resistance will increase and cause more severe burning. This action continues to build up with each revolution of the electrode. To stop burning, the operation must be interrupted to clean or replace the electrode.

To eliminate burns, the area of contact and the pressure between the electrode and the tube must be adequate. As a rule of thumb, each electrode should have sufficient contact area so that the current density will be less than 50 000 A/in.^{2} (32 A/m^2). The relative shapes of the formed tube and the electrode should ensure that the maximum contact pressure occurs next to the seam.

Without the aid of some backup support, electrode contact pressure is limited by the ability of the tube to support the forces being applied. The maximum permissible pressure in the welding throat is a function of the yield strength of the metal and the ratio of tube diameter to wall thickness (D/t ratio). In extreme cases where the D/t ratio is high, a backup mandrel must be used to prevent distortion of the tube wall and misalignment of the joint.

INSPECTION AND TESTING

UPSET WELDS CAN be inspected and tested in the same manner as flash welds. In general, the quality requirements for upset welds are not so stringent as those specified for flash welds. The process normally cannot produce welds with the consistency available with flash welding.

PERCUSSION WELDING

DEFINITION AND GENERAL DESCRIPTION

PERCUSSION WELDING IS a joining process that produces coalescence with an arc resulting from a rapid discharge of electrical energy. Pressure is applied percussively during or immediately following the eletrical discharge.

In general, "percussion welding" is the term used in the electronics industry for joining wires, contacts, leads, and similar items to a flat surface. On the other hand, if the item is a metal stud that is welded to a structure for attachment purposes, it is called *capacitor discharge stud welding*[2].

In application of the process, the two parts are initially separated by a small projection on one part, or one part is moved toward the other. At the proper time, an arc is initiated between them. This arc heats the faying surfaces of both parts to welding temperature. Then, an impact force drives the parts together to produce a welded joint. There are basically two variations of the percussion process: capacitor discharge and magnetic force.

Although the steps may differ in certain applications because of process variations, the essential sequence of events in making a percussion weld is as follows:

(1) Load and clamp the parts into the machine.
(2) Apply a low force on the parts or release the driving system.
(3) Establish an arc between the faying surfaces (1) with high voltage to ionize the gas between the parts or (2) with high current to melt and vaporize a projection on one part.
(4) Move the parts together percussively with an applied force to extinguish the arc and consummate a weld.
(5) Turn off the current.
(6) Release the force.
(7) Unclamp the welded assembly.
(8) Unload the machine.

Percussion welding is similar to capacitor discharge stud welding. The differences between the two processes lie in the applications and the type of power source. Percussion welding may be used to join equal cross sections of wires, rods, and tubes. Welding current is supplied by a capacitor storage bank in these applications.

The process may also be used to weld wires or contacts to large flat areas with power from a capacitor bank or a transformer.

PRINCIPLES OF OPERATION

WELDING HEAT IS generated by a high current arc between the two parts to be joined. The current density is very high, and this melts a thin layer of metal on the faying surfaces in a few milliseconds. Then the molten surfaces are brought together in a percussive manner to complete the weld.

There are two process variations that differ in the type of power supply, method of arc initiation, and work drive motion.

Capacitor Discharge Percussion Welding

WITH THE CAPACITOR discharge method, power is furnished by a capacitor storage bank. The arc is initiated by the voltage across the terminals of the capacitor bank (charging voltage) or a superimposed high voltage pulse. Motion may be imparted to the movable part by mechanical or pneumatic means.

Magnetic Force Percussion Welding

FOR MAGNETIC FORCE welding, power is supplied by a welding transformer. The arc is initiated by vaporizing a small projection on one part with high current from the transformer. The vaporized metal provides an arc path. The percussive force is applied to the joint by an electromagnet that is synchronized with the welding current. Magnetic force percussion welds are made in less than one half cycle of 60 Hz. Consequently, the timing between the initiation of the arc and the application of magnetic force is critical.

ADVANTAGES OF PERCUSSION WELDING

THE EXTREME BREVITY of the arc in both versions of percussion welding limits melting to a very thin layer on the faying surfaces. Consequently, there is very little upset or flash on the periphery of the welded joint (but enough to remove impurities from the joint). Heat-treated or cold-worked metals can be welded without annealing them.

Filler metal is not used and there is no cast metal at the weld interface. A percussion welded joint usually possesses higher strength and conductivity than does a brazed joint. Unlike brazing, no special flux or atmosphere is required.

A particular advantage of the capacitor discharge method is that the capacitor charging rate is easily controlled and low compared to the discharge rate. The line power factor is better than with a single-phase ac machine. Both these factors give good operating efficiency and low power line demand.

Percussion welding can tolerate a slight amount of contamination on the faying surfaces because expulsion of the thin molten layer tends to carry any contaminants out of the joint.

2. See Chapter 9, Stud Welding.

Figure 18.12–Typical Electrical Contacts Joined by Magnetic Force Percussion Welding

LIMITATIONS

THE PERCUSSION WELDING process is limited to butt joints between two like sections and to flat pads or contacts joined to flat surfaces. In addition, the total area that can be joined is limited since control of an arc path between two large surfaces is difficult.

Joints between two like sections can usually be accomplished more economically by other processes. Percussion welding is usually confined to the joining of dissimilar metals not normally considered weldable by other processes, and to the production of joints where avoidance of upset is imperative.

Another limitation of this process is that two separate pieces must be joined. It cannot be used to weld a ring from one piece.

APPLICATIONS

Weldable Metals

THE MAGNETIC FORCE method is primarily used for joining electrical contacts to contactor arms. Combinations include copper to copper, silver-tungsten to copper, silver oxide to copper, and silver-cadmium oxide to brass. Areas from 0.040 to 1.27 in^2. (26 to 820 mm^2) are being welded in production. Some metal loss occurs at the weld interface, and in most instances some flash must be removed from the periphery of the weld. Figure 18.12 shows several contact designs welded by this process. Figure 18.13 shows a section through a typical weld.

The capacitor discharge method is usually employed to produce the following types of joints:

(1) Butt joints between wires or rods

(2) Lead wire ends to flat conductors or terminals
(3) Contacts to relay arms

The wire is usually made of copper and may be solid or stranded, bare or tinned. The rods are usually copper, brass, or nickel-silver. Other alloys such as steel, alumel, chromel, aluminum, and tantalum may be welded to themselves or to other materials. The method is also applicable to reactive, refractory, and dissimilar metal welds, because the short weld time limits contamination of the reactive metals and the formation of low-strength intermetallic zones in the joints.

Industrial Uses

COMPANIES USING PERCUSSION welding are mainly those in the electrical contact or component field. Large contact assemblies for relays and contactors are usually made in magnetic force percussion welding machines. Such machines can be automated for high production.

Figure 18.13–Photomacrograph of a Section Through a Silver Contact (Top) Welded to a Brass Terminal (Bottom)

Hand held capacitor discharge equipment may be used to weld wires to pins. This is particularly applicable to aerospace equipment that is subject to shock and vibration. The process is also used to weld electronic components to terminals.

EFFECT ON METALS WELDED

Heat Effects

A PERCUSSION WELD is made in an extremely short time. It may take milliseconds when using magnetic force welding. Because of this short time, the heat-affected zones of percussion welds are shallow, usually less than 0.010 in. (0.25 mm). There is little oxidation of mating surfaces and a minimum of alloying between dissimilar metals. Since the heat-affected depth is so small, heat-treated metals may be welded without softening them. The heat input is so concentrated and of such short duration that heat-sensitive components near the weld area are not affected by the welding cycle.

Heat balance between parts is usually not a factor of concern. Since percussion welding is essentially a dc process, polarity of the two parts involved may be important in some cases, as in arc welding.

Metal Loss

THE METAL LOSS that occurs during a percussion weld is not as great as in arc stud welding. The loss varies with the area of the weld and the type of welding machine. Metal loss can generally be ignored for parts to be joined by capacitor discharge percussion welding. However, it should be considered in magnetic force percussion welding.

Flash

FLASH IS THE metal that is expelled at high velocity from the weld interface during a percussion weld. It can damage adjacent tooling and may affect accuracy of assembly. Any flash attached to the weld joint should be removed so that it will not cause a problem in service.

MAGNETIC FORCE PERCUSSION WELDING

Welding Machines

MAGNETIC FORCE MACHINES use a low voltage power supply (20 to 35 volts from a transformer), a projection type arc starter, and an electromagnetic system to produce the weld force. A unit generally consists of a modified press type resistance welding machine with specially designed transformer, controls, and tooling. Figure 18.14 shows a typical machine used to weld the type of part shown in Figure 18.12. An air cylinder provides the initial force to bring the parts together.

Magnetic force percussion welding machines usually have an independent power source for the electromagnet so that the force magnitude and time of application can be varied with respect to the initiation of welding current. This is accomplished by using two separate transformers, one for welding power and one for the electromagnet power. The acceleration of the force member can be controlled by adjusting the magnitude of the electromagnet current, thereby providing a duration control for arc time.

Since welding is done during 1/2 cycle of 60 Hz, current is unidirectional. In some cases, the polarity of the two parts may have some effect on weld quality. In general, the same conditions that prevail in dc arc welding are also in effect in percussion welding with respect to polarity. The current is always passing through the transformer in the same direction and the core can become partially saturated. Consequently, the electrical controls should provide a low amplitude 1/2-cycle pulse in the opposite direction to deflux the transformer and electromagnet. This can be done during the loading time.

Joint Design

FOR WELDING TWO flat surfaces together, a projection similar to that for resistance welding must be formed on one piece as shown in Figure 18.15. Its diameter and height must be developed for each application. The diameter must be large enough to support the initial force applied to the parts but too small to carry the welding current. The height determines the gap between the faying surfaces and, thus, the initial arc voltage. When large area contacts are welded, two projections may be required.

The surfaces to be joined must be flat and parallel during welding so that arcing will occur over the entire area. Areas that are not melted will probably not weld when impacted together.

Voltage And Current

IT IS NECESSARY to establish and maintain the desired magnitude of voltage and current for the required weld area. These are determined by the projection desired, the capacity of the welding transformer, and the impedance of the secondary circuit. The transformer should have low impedance with secondary voltages higher than those commonly used in resistance welding.

Arc Time

ARC TIME CAN be considered as the time beginning with the explosion of the projection and ending when the two parts come together and the arc is extinguished. The timing between the initiation of the arc and the application of magnetic force is very critical.

Figure 18.14—Magnetic Force Percussion Welding Machine

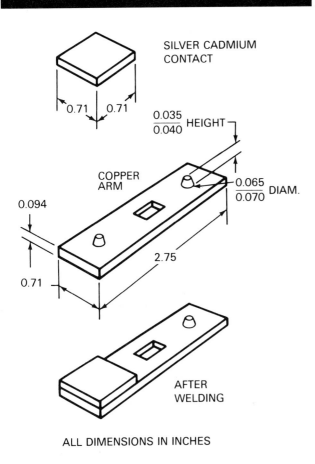

SILVER CADMIUM
CONTACT

0.71 | 0.71

$\frac{0.035}{0.040}$ HEIGHT

COPPER
ARM

0.094

$\frac{0.065}{0.070}$ DIAM.

0.71

2.75

AFTER
WELDING

ALL DIMENSIONS IN INCHES

Figure 18.15–Typical Design of a Magnetic Force Percussion Welded Contact Assembly

The arc time is a function of:

(1) Magnitude of magnetic force
(2) Timing of the magnetic force with relation to welding current
(3) Inertia or mass of the moving parts in the force system
(4) Height of the projection
(5) Magnitude of the welding current and the diameter of the projection

Acceleration of the movable head is directly proportional to the magnetic force applied and inversely proportional to the mass. The acceleration of the movable head

with the two transformer system can be controlled by adjusting the magnitude of the force current, which thereby provides a duration control for arc time, within limits.

CAPACITOR DISCHARGE PERCUSSION WELDING

TWO TYPES OF machines are presently used. One uses a high voltage, low capacitance system. Charging voltages range from 1 to 3 kV. With this system, wire end preparation is not critical since the applied potential is sufficient to ionize the air in the gap and start the arc.

The other system uses a low voltage, high capacitance energy source. This has the advantages of a safe working voltage (about 50V), a simple power supply, and low weld spatter. In some designs, the high voltage power is discharged through a transformer of low voltage output.

A low voltage system requires 600V arc starting circuit and special wire end preparation. Once the air gap is ionized with the 600V (low amperage) circuit, the arc is sustained by the 50V circuit. The arc initiation circuit does no appreciable melting.

One type of low voltage machine consists of a hand held gun and a portable power supply. The gun is designed to weld wires to terminals by holding a small flat or square terminal in one set of stationary jaws and the wire to be welded in a set of movable jaws. When the gun is triggered, springs move the wire toward the terminal at a high velocity. A feather edge on the end of the wire greatly improves arc starting. The arc is initiated at the point of contact of the wire and terminal. The welding current melts the feather edge on the wire faster than the wire is moving toward the, terminal. The arc spreads over the wire area and melts a layer about 0.002 to 0.003-in. (0.050 to 0.076 mm) thick in each part. The arc is extinguished after about 150 to 600 microseconds as the two parts come in contact.

Another version of a portable, low voltage welding machine employs a high frequency pulse to initiate the arc. This feature eliminates the need for a special shape on the wire end. The machine uses an electromechanical actuator to accelerate the wire and to provide the necessary forging force. One version of this machine is shown in Figure 18.16.

Low voltage semiautomatic and automatic machines are used to weld assemblies similar to the one shown in Figure 18.17. Component leads are usually tinned annealed copper. Terminals may be brass, tinned brass, or nickel-silver alloys. Wires and leads of 0.006 to 0.102-in. (0.2 to 2.6 mm) diameter can be welded to terminals and plates of various thicknesses above 0.006 in. (0.2 mm) thick.

Controls for capacitor discharge equipment usually include those for welding voltage, capacitance, and high frequency voltage when it is used. Control of the motion mechanism is also provided.

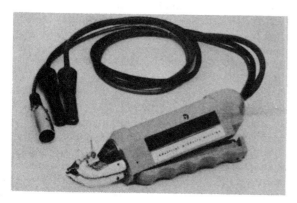

Figure 18.16–A Portable Capacitor Discharge Percussion Welding Power Supply and Hand-Held Gun

lographic examination will show the weld interface and the widths of the heat-affected zones. In the case of dissimilar metals, it may reveal the degree of alloying at the interface. Microhardness tests on a metallographic section may indicate the effect of welding on the base metal.

Welded joints may be tested in tension, bending, or shear, depending upon the joint design. The effect of vibration may be important in some applications. The test method should be designed to qualify the welding procedures and weld joint properties for the intended applications.

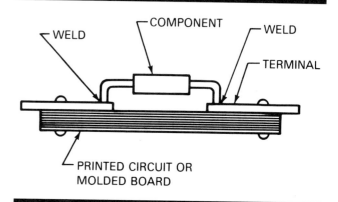

Figure 18.17–A Typical Percussion Welded Electronic Assembly

WELD QUALITY

THE QUALITY OF percussion welds can be determined by metallographic examination and mechanical tests. Metal-

SAFETY

MECHANICAL

THE WELDING MACHINE should be equipped with appropriate safety devices to prevent injury to the operator's hand or other parts of the body. Initiating devices, such as push buttons or foot switches, should be arranged and guarded to prevent them from being actuated inadvertently.

Machine guards, fixtures, or operating controls should prevent the hands of the operator from entering between the work-holding clamps or the parts to be welded. Dual hand controls, latches, presence-sensing devices, or any similar device may be employed to prevent operation in an unsafe manner.

ELECTRICAL

ALL DOORS AND access panels on machines and controls should be kept locked or interlocked to prevent access by unauthorized personnel. When the equipment utilizes capacitors for energy storage, the interlocks should interrupt the power and discharge all the capacitors through a suitable resistive load when the panel door is open. A manually operated switch or other positive device should also be provided in addition to the mechanical interlock or contacts. Use of this device will assure complete discharge of the capacitors.

A lock out procedure should be followed prior to working with the electrical or hydraulic systems.

SAFETY CONSIDERATIONS FOR PERSONNEL

FLASH GUARDS OF suitable fire resistant material should be provided to protect the operator from sparks and avoid fires. In addition, personal eye protection with suitable shaded lenses should be worn by the operator.

When the welding operations produce high noise levels, operating personnel should be provided with ear protection.

Metal fumes produced during welding operations should be removed by local ventilating systems.

Additional information on safe practices for welding may be found in the American National Standard Z49.1, Safety in Welding and Cutting (latest edition), available from the American Welding Society.

SUPPLEMENTARY READING LIST

Anon. "Union Pacific used flash welding to take clickity-clack out of its tracks." *Welding Journal* 55(11): 961-962; November 1976.

Cueman, M. K., and Williamson, R. "Process model for percussion welding." *Welding Journal* 68(9): 372s-376s; September 1989.

Holko, Kenneth H. "Magnetic force upset welding dissimilar thickness stainless steel tee joints." *Welding Journal* 49(9): 427-439s; September 1970.

Kotecki, D. J., Cheever, D. L., and Howden, D. G. "Capacitor discharge percussion welding; microtubes to tube sheets." *Welding Journal* 53(9): 557-560; September 1974.

MIL-W-6873, Military Specification, Welding; Flash, Carbon and Alloy Steel.

Petry, K. N., et al., "Principles and practices in contact welding." *Welding Journal* 49(2): 117-126; February 1970.

Savage, W. F. "Flash welding: the process and application." *Welding Journal* 41(3): 227-237; March 1962.

———. "Flash welding: process variables and weld properties." *Welding Journal* 41(3): 109s-119s; March 1962.

Sullivan, J. F. and Savage, W. F. "Effect of phase control during flashing on flash weld defects." *Welding Journal* 50(5): 213s-221s; May 1971.

Thompson, E. G. "Attachment of thermocouple instrumentation to test components by all-position percussion welding." *Welding Journal*. 61(6): 31-33; June 1982.

Turner, D. L., et. al., "Flash butt welding of marine pipeline materials." *Welding Journal* 61(4): 17-22; April 1982.

RESISTANCE WELDING EQUIPMENT

PREPARED BY A COMMITTEE CONSISTING OF:

P. Dent, Chairman
Grumman Aerospace Corporation

J. C. Bohr
General Motors

R. G. Gasser
Ferranti/Sciaky, Incorporated

J. M. Gerken
Lincoln Electric Corporation

D. L. Hallum
Bethlehem Steel Corporation

J. W. Lee
Textron Lycoming

R. B. McCauley
McCauley Associates

D. H. Orts
Armco, Incorporated

G. W. Oyler
Welding Research Council

W. T. Shieh
General Electric Company

K. C. Wu
Pertron/Square D

WELDING HANDBOOK COMMITTEE MEMBER:
A. F. Manz
A. F. Manz Associates

CHAPTER 19

RESISTANCE WELDING EQUIPMENT

INTRODUCTION

THE SELECTION OF resistance welding equipment is usually determined by the joint design, construction materials, quality requirements, production schedules, and economic considerations. Standard resistance welding machines are designed to meet the requirements of Bulletin No. 16, Resistance Welders Manufacturers Association (RWMA). These machines are capable of welding a wide variety of alloys and component sizes. Specially designed, complex resistance welding equipment may be necessary to meet the economic requirements of mass production or the quality requirements of military specifications.

A resistance welding machine has three principal elements:

(1) An electrical circuit consisting of a welding transformer and a secondary circuit with electrodes that conduct the current to the work
(2) A mechanical system consisting of a machine frame and associated mechanisms to hold the work and apply the welding force
(3) The control equipment to initiate and time the duration of current; it also may control the current magnitude as well as the sequence and the time of other parts of the welding cycle

With respect to electrical operation, resistance welding machines are classified in two basic groups: direct energy and stored energy. Machines in both groups may be designed to operate on either single-phase or three-phase power.

Most resistance welding machines are the single-phase direct energy type. This is the type of machine most commonly used because it is the simplest and least expensive in initial cost, installation, and maintenance. The mechanical systems and secondary circuit designs are essentially the same for all types of welding machines, but transformer designs and control systems can differ considerably.

A single-phase welding machine has a larger volt-ampere (kVA) demand than a three-phase machine of equivalent rating. The demand of a single-phase machine causes unbalance on a three-phase power line. Also, its power factor is relatively low because of the inherent inductive reactance in the welding circuit of the machine. Single phase demand may not be a problem if the welding machine is a small part of the total line load or a number of single-phase welding machines are connected to balance the load on the three phases of the power line.

A three-phase direct energy machine draws power from all three phases of the power line. The inductive reactance of the welding circuit is low because direct current is used for welding. Consequently, the required secondary circuit voltage for a given welding current is reduced; thus the kVA demand of a three-phase machine is lower than that of an equivalent (equal current) single-phase machine. This is a definite advantage where a large capacity machine is needed and power line capacity is limited.

The principle of a stored energy machine is to accumulate and to store electrical energy and then to discharge it to make the weld. The energy is normally stored in a capacitor bank. Single-phase power is generally used for small bench model stored energy machines. The power demand is low because storage time is relatively long in comparison to the weld time.

SPOT AND PROJECTION WELDING MACHINES

ROCKER ARM TYPE

THE SIMPLEST AND most commonly used spot welding machine is the rocker arm design, so called because of the rocker movement of the upper horn. A horn is essentially an arm or extension of an arm of a resistance welding machine which transmits the electrode force and, in most cases, the welding current. This type of machine is readily adaptable for the spot welding of most weldable metals. Three methods of operation are available: (1) air, (2) foot, and (3) motor.

Air-operated machines, such as the one in Figure 19.1, are the most popular. With air operation, the welding cycle is generally controlled automatically with a combination control unit. These machines can operate rapidly and are easily set up for welding.

Foot-operated machines are best suited for miscellaneous sheet metal fabrication, particularly for short production runs where consistent weld quality is not required.

Motor-operated machines are normally used where compressed air is not readily available.

Standard rocker arm machines are generally available with throat depths of 12 to 36 inches and transformer capacities of 5 to 100 kVA. The general construction of these machines is the same with all three types of operation.

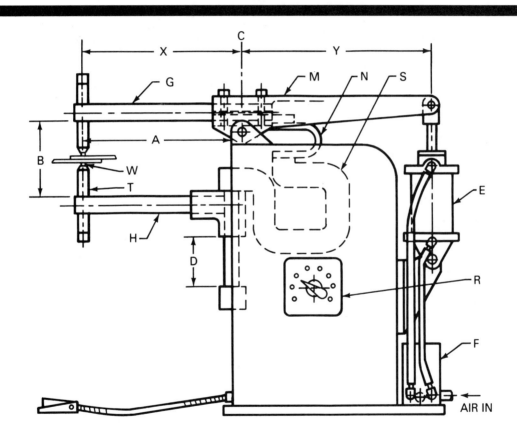

A — THROAT DEPTH	H — LOWER HORN
B — HORN SPACING	M — ROCKER ARM
C — CENTERLINE OF ROCKER ARM	N — SECONDARY FLEXIBLE CONDUCTOR
D — LOWER ARM ADJUSTMENT	R — CURRENT REGULATOR (TAP SWITCH)
E — AIR CYLINDER	S — TRANSFORMER SECONDARY
F — AIR VALVE	T — ELECTRODE HOLDER
G — UPPER HORN	W — ELECTRODE

Figure 19.1–Air-Operated Rocker Arm Spot Welding Machine

Electrode Position

THE TRAVEL PATH of the upper electrode is an arc about the fulcrum of the upper arm. The electrodes must be positioned so that both are in the plane of the horn axes. Also, the two horns should be parallel when the electrodes are in contact with the work. Even with parallel horns, electrode skidding can occur if the electrode holders or horns are not sufficiently rigid. Skidding can be reduced by changing to more rigid electrode holders, adjusting the position of the electrodes, or providing support to the lower horn. Because of the radial motion of the electrode, these machines are not recommended for projection welding.

Mechanical Design

THE MACHINE FRAME houses the transformer and tap switch and supports the mechanical and electrical components.

For air-operated machines, the stroke of the air cylinder must be proportioned to the required electrode spacing. Its diameter must be proportioned to the required electrode and lever arm ratio Y/X, as shown in Figure 19.1. For a given cylinder diameter, the available welding force will decrease as the throat depth is increased, maintaining the original fulcrum point. Electrode spacing can be set by adjusting the position of the electrodes in the horns. In most cases, however, it is desirable to use a double-acting air cylinder with adjustable stroke.

The force exerted by a piston is equal to the product of its surface area and the air pressure applied to that area. Most industrial air systems are operated at 80 psi (550 kPa) minimum, and cylinder size is generally determined on this basis.

Electrode force is the product of the piston force and the lever arm ratio Y/X. Consequently, it is in direct proportion to the air pressure as controlled by a pressure regulator. Air pressures below 20 psi (140 kPa) should not be used because of possible erratic and inconsistent behavior of the air cylinder.

With foot and motor operated machines, the air cylinder is replaced by a stiff spring. The spring is compressed by a foot operated lever arm or a motor-driven cam as it exerts a force on the end of the rocker arm. The amount of force is determined by the stiffness of the spring and the compression distance.

PRESS TYPE

PRESS TYPE MACHINES are recommended for all projection welding operations and many spot welding applications. With this type of machine, the movable welding head travels in a straight line in guide bearings or ways. These bearings must be of sufficient proportions to withstand any eccentric loading on the welding head.

Standard press type welding machines, as defined by the RWMA, are available with capacities of 5 to 500 kVA and throat depths up to 54 inches (Figure 19.2). Nonstandard units, such as magnetic force and bench types, are widely used for the manufacture of radios, instruments, electrical components, and jewelry.

Press type machines are classified according to their use and method of force application. They may be designed for spot welding, projection welding, or both. Force may be applied by air, hydraulic, or electromagnetic systems, or manually when used with small bench units.

A few general guidelines for the selection of a machine of this type are as follows:

(1) Hydraulic operation is not normally used on machines rated below 200 kVA because of the higher cost as compared to air operation. It is also not recommended for use with projection welding because of its slower "follow-up" characteristic compared to that of air. The follow-up of a welding machine is the ability of the force mechanism to react to the dynamic changes that occur during a weld and to maintain the proper clamping pressure.

(2) Air operation may be used on all sizes of machines. When high forces are required, however, air cylinders and valves will be quite large, operation will be slow, and air consumption will be high. When all factors are taken into consideration, most machines of 300 kVA and under are air operated, and machines of 500 kVA and above are hydraulically operated. In between, they may be operated by either method.

Frequently a high KVA rating system may be supplied to meet a high duty cycle requirement and not a high KVA demand. These machines can be equipped for air operation.

Fast electrode follow-up is particularly important when spot welding or projection welding relatively thin sections, particularly those of aluminum and other nonferrous metals. Air operation provides much faster follow-up than does hydraulic operation because of the compressibility of air. With hydraulic operation, follow-up must occur by liquid motion and is therefore determined by the capacity of the pump. A pre-charged air accumulator, commonly referred to as a surge tank, is sometimes used with the system to further improve the air operated follow-up system.

Fast follow-up on large machines (even air operated machines) is often achieved with the use of a spring system mounted below the guiding arrangement for the ram. This spring system allows the upper electrode to follow-up the weld independent of the inertia of the ram. Also, bellows-type air systems are used in conjunction with precision timers and dumping valves that allow the ram to follow-up independent of the friction and air limitations of the air cylinder. While the bellows system is costly, it has the advantage of using the precision guidance of the ram system, and this is especially important when welding multiple projections.

When welding precious or dissimilar metals, the synchronization of heat and pressure must be precise and con-

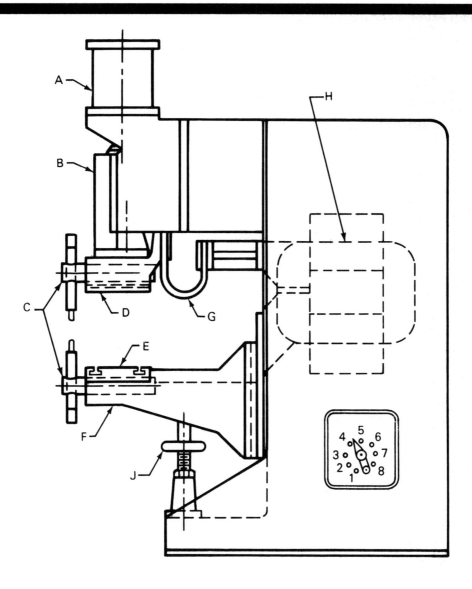

A — AIR OR HYDRAULIC CYLINDER
B — RAM
C — SPOT WELDING ATTACHMENT
D — UPPER PLATEN
E — LOWER PLATEN

F — KNEE
G — FLEXIBLE CONDUCTOR
H — TRANSFORMER SECONDARY
J — KNEE SUPPORT

Figure 19.2–Press Type Combination Spot and Projection Welding Machine

sistent. One answer is to use an electromagnetic force system assisted by a smaller air cylinder. This combination provides a workable system for precisely controlling the exact time when force is applied to the work.

The magnetic force builds up in synchronization with the weld current. This force, combined with the initial clamping force, assures that when maximum heating occurs there will be proper follow-up of the welder head. As the current decays towards zero, the magnetic force also decreases and in the end, the weld is held by the clamping force. The magnetic force builds to a peak on both the positive and negative half cycles of current.

GENERAL CONSTRUCTION

STANDARD PRESS TYPE welding machines are designed and built on the *unit principle* for economy in manufacture. The same frame size is used with two or three transformers of different kVA ratings and with a range of throat depths. A typical press type welding machine is shown in Figure 19.2.

Projection welding machines have platens on which dies, fixtures, and other tooling are mounted. In most cases, the platens are a direct part of the secondary circuit. The platens have flat surfaces and usually have standard T-slots on which to bolt attachments.

Machines designed for spot welding are equipped with horns and electrode holders. A combination unit will have both platens and horns. Such a machine will have one throat depth as a projection welding machine and a greater throat depth as a spot welding machine. The platens, the ram, and the force cylinder are all on the same center line. The distance from this center line to the face of the secondary plate is the depth of the projection welding throat. On standard machines with horns, the spot welding electrodes are located six inches or more from the face. This is true whether or not platens are used.

On projection and combination machines, the lower platen is mounted to, or may be a part of, a knee which can be adjusted vertically. The knee may be made of copper, bronze, steel, or cast iron.

MECHANICAL DESIGN

Air Operated Machines

THESE MACHINES ARE usually the direct acting type where the electrode force is exerted by the air cylinder through the ram. Four general types of double-acting air cylinders are employed. These are illustrated in Figure 19.3. In all cases, air for the pressure stroke enters at port A and exhausts at port B. For the return stroke, the air enters at Port B and exhausts at port A.

Figure 19.3A shows a fixed-stroke cylinder with stroke adjustment. The stroke adjuster K limits the travel of piston P and the electrode opening.

An adjustable-stroke cylinder with a dummy piston is shown in Figure 19.3B. The dummy piston R is attached to the adjusting screw K which positions this piston. Chamber L is connected to port A through the hollow adjusting screw. The stroke of the force piston P is adjusted by the position of the dummy piston R above it. This cylinder design responds faster than a fixed-stroke cylinder because the volume L above piston P can be made smaller than that of a fixed-stroke cylinder of the same size.

The adjustable-stroke cylinder can be modified to provide a retraction feature. This feature can accommodate additional electrode opening for loading and unloading the machine, or for electrode maintenance. See Figure 19.3C. With the adjustable-retractable stroke cylinder, a third port C is connected to chamber H above the dummy piston R. If air is admitted to chamber H at a pressure slightly higher than the operating pressure in chamber L, piston R will move down to a position determined by the adjustable stop X. This determines the UP position for piston P and the electrode opening for welding. When the air from chamber H is exhausted to atmosphere, piston P will lift piston R with it until stop X contacts the cylinder head. This will increase the electrode opening for loading and unloading the machine. Readmission of air to chamber H will return pistons P and R to welding position when the pressure in chamber H is slightly higher than that in chamber M. Flow control valves or cushions are usually used to control the operating speed of an air cylinder.

Figure 19.3D shows a diaphragm type cylinder. In this design, separate cylinders are used to retract the entire cylinder and ram to allow work piece loading. The deflection of the diaphragm by the pressure differential on either side of it provides the electrode movement. This system responds very rapidly due to its inherent low friction and inertia, providing fast follow-up of the electrodes as the weld nugget is formed. Dual electrode force is easily attained by alternately pressurizing and depressurizing chamber B while chamber A is held at a constant pressure.

Hydraulic Machines

WITH THESE MACHINES, a hydraulic cylinder is used in place of an air cylinder. The designs for hydraulic cylinders are similar to those for air-operated cylinders. Refer to Figure 19.3. Hydraulic cylinders are generally smaller in diameter than air cylinders because higher pressures can be developed with a liquid system.

In the simplest type of hydraulic system, a constant speed motor drives a constant pressure, constant delivery pump. The output pressure of the pump is controlled by an adjustable relief valve. Liquid delivery is controlled with a four-way valve of design similar to that employed in an air system. Auxiliary devices include a sump, a filter, a heat exchanger, a gauge, and sometimes an accumulator.

Portable Type

A TYPICAL PORTABLE spot welding machine consists of four basic components:

(1) A portable welding gun or tool
(2) A welding transformer and, in some cases, a rectifier
(3) An electrical contactor and sequence timer
(4) A cable and hose unit to carry power and cooling water between the transformer and welding gun

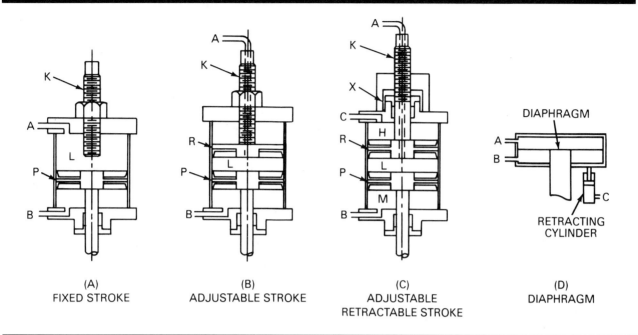

(A) FIXED STROKE

(B) ADJUSTABLE STROKE

(C) ADJUSTABLE RETRACTABLE STROKE

(D) DIAPHRAGM

Figure 19.3–Typical Air Cylinder Designs for Air-Operated Press Type Welding Machines

A typical portable welding gun consists of a frame, an air or hydraulic actuating cylinder, hand grips, and an initiating switch. The unit may be suspended from an adjustable balancing unit.

There are two basic types of air or hydraulically operated guns. One is the scissor type which is analogous to a rocker arm spot welding machine. The other is a "C" type, so-called because of its shape. This type has action similar to a press type spot welding machine.

The design of a gun is influenced by the electrode force required. To minimize the size and weight of a gun, a hydraulic cylinder is commonly used to provide forces greater than 750 pounds. However, air cylinders supplying up to 1500 pounds are sometimes used for simplicity of equipment.

Transformers for portable guns should produce open-circuit secondary voltages that are two to four times greater than those of transformers for stationary machines. The higher voltages are needed because of the cable added between the transformer and the gun. The introduction of this cable into the secondary circuit has three fundamental effects:

(1) It increases the total impedance on an ac or frequency converter control. Therefore, considerably higher secondary voltage is required in a gun welder to produce a given secondary current than is required on a stationary type welder.

(2) It increases the resistance component of the impedance so that the power factor is much higher than in a stationary type welder. On a dc machine, the increased resistance lowers the available current unless voltage is proportionately increased.

(3) It minimizes the effect of the impedance in the weld pieces upon both the current output of the welder and the power factor of the load. For power calculations, maximum welding amperes, kVA demand, and power factor may be assumed to be the same as their short circuit values.

Another type of gun currently in use is called a transgun. Transguns have transformers mounted directly to a self-equalizing force system and offer several advantages. They are significantly more compact than the transformer described previously. They also have power factors that can exceed 85 percent. The work, however, is the primary resistance component of the secondary circuit and must be taken into account when rated short circuit currents are used to size the transformer.

An air-hydraulic booster is a piston device for transforming air pressure into high hydraulic pressure. The pressure increase is proportional to the ratio of the area of the hydraulic piston to the area of the pneumatic piston. The booster provides the necessary hydraulic pressure to the gun cylinder.

A combination control is required to operate a portable gun unit. It consists of a primary contactor and a sequence timer. If an electronic tube contactor is used, the control is usually mounted separately, but as close to the transformer as possible. If the contactor is a solid state device, the compactness of this unit permits mounting the control directly on the transformer.

MULTIPLE SPOT WELDING TYPE

A MULTIPLE SPOT welding machine is a special purpose unit designed to weld a specific assembly. This type of machine should be considered when the production requirements and the number of spot or projection welds on an assembly are so large that welding with a single point machine is uneconomical. The principal advantages of these machines are:

(**1**) A number of welds can be made at the same time.

(**2**) Part dimensions and weld locations can be reasonably consistent.

(**3**) The equipment can be very reliable and easy to maintain.

Welding Station Design

MULTIPLE SPOT WELDING machines have a number of transformers, usually of dual secondary design. Figure 19.4 shows typical standardized components that are used in designing a wide range of multiple spot welding machines. Force is applied directly to the electrode through a holder by an air or hydraulic cylinder.

To make welds on close centers, the cylinder diameter must be small. This can be accomplished with tandem or triple pistons on the same shaft. A 2 in. (51 mm) diameter triple piston cylinder can develop 500 pounds of force (2224 N) at 60 psi (414 KPa). This force is adequate for spot welding two 0.030 in. (0.76 mm) thicknesses of cold rolled steel. Closer spot weld spacing may be obtained by using hydraulic cylinders of smaller diameter. Another

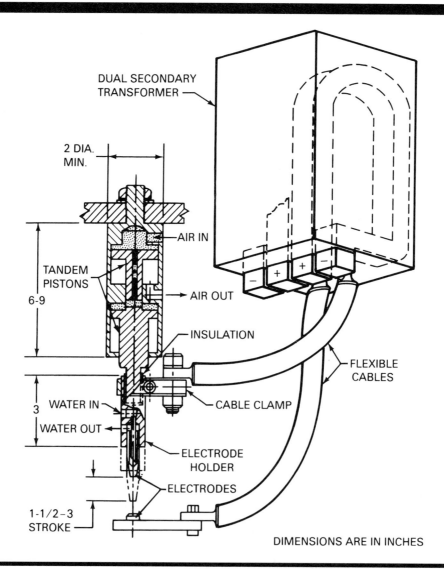

Figure 19.4–Basic Components of a Multiple Spot Welding System

method, generally of last resort, is to use offset electrodes. This does, however, induce eccentric loading on the cylinder. When this type of electrode is used, the stroke of the cylinder should be kept to a minimum. The combination cylinder and electrode holder assemblies are commonly referred to as welding guns.

A welding transformer with two insulated secondaries can power two separate welding circuits, though there can be no individual current control of each circuit. There can also be significant variance in the output of each secondary, even when great care is taken to provide duplicate secondary loops. When critical welding must be accomplished, it is recommended that separate transformers be used for each circuit.

The dual-secondary type of transformer is noted for its compact design and narrow width. If desired, only one of the dual secondaries need be used at one time. For higher secondary voltage, the two secondaries can be connected in series to feed one secondary circuit. To increase the welding current available to a single circuit, the secondaries can be connected in parallel. Welding guns and transformer units of this type can be designed to spot weld two sheets of cold rolled carbon steel up to 0.125 in. (3.2 mm) thick.

For most applications, the lower electrode is a piece of solid copper alloy with one or more electrode alloy inserts that contact the part to be welded. It is normally water cooled to remove heat. The inserts generally are designed with large contact areas to resist wear. Pointed electrodes are not normally used against the "show" side of the work to avoid marking.

Self-equalizing gun designs are often used where standard electrodes are needed on both sides of the weld to obtain good heat balance, or where variations in parts will not permit consistent contact with a large, solid lower electrode. The same basic welding gun is used for these designs but it is mounted on a special "C" frame similar to that for a portable spot welding gun. The entire assembly can move as electrode force is applied at the weld locations.

Machine Designs

MULTIPLE TRANSFORMER MACHINES are used extensively in the manufacture of formed sheet metal products. Because of their broad usage and requirements, many designs of multiple transformer machines are available. The machines may be designed as welding stations in large, high production, automated assembly lines, or they may be used independently. Independent machines may be loaded and unloaded either manually or automatically. They are commonly interfaced with robots for both welding and material handling purposes.

In many instances, a single weld control can be used to initiate all of the weld sequences. In such a case, up to six separate primary contactors are signaled in series fashion; i.e., each is fired after the previous one has completed its full weld sequence. This type of firing sequence is referred to as cascade firing. Some of these controls also have the ability to serve as PLCs (Programmable Logic Control) but are generally limited to small applications.

There are several advantages to using this type of weld control. The primary advantage is the economic saving of purchasing a single control. Also, a single unit control aids in machine maintenance and trouble shooting. There is, however, the disadvantage of lost cycle time waiting for completion of each weld.

In designing the machine for a particular weldment, a number of factors must be considered. These include:

(1) Shape, size, and complexity of the part
(2) Consistency of the parts being joined
(3) Part composition and thickness
(4) Required weld appearance
(5) Production rate requirements
(6) Available equipment (presses, frames, and dial tables)
(7) Changeover time for different assemblies
(8) Economic factors including initial cost, labor to operate, and maintenance.

ROLL SPOT AND SEAM WELDING MACHINES

A ROLL SPOT or seam welding machine is similar in principle to a spot welding machine, except that wheel-shaped electrodes are substituted for the electrode tips used in spot welding. Both roll spot and seam welding can be performed on the same type of machine.

The essential elements of a standard seam welding machine are as follows:

(1) A main frame that houses the welding transformer and tap switch
(2) A welding head consisting of an air cylinder, a ram, and an upper electrode mounting and drive mechanism

(3) The lower electrode mounting and drive mechanism, if used
(4) The secondary circuit connections
(5) Electronic controls and contactor
(6) Wheel electrodes
(7) Wheel bearings — current carrying type

The main frame, tap switch, ram, and air cylinder are essentially the same as those of a standard press type spot or projection welding machine. The transformer is normally heavier duty, due to the continuous nature of seam welding versus spot welding. Hydraulic cylinders are seldom used on seam welding machines because the electrode force requirements are not usually high.

To provide for electrode wear, either an adjustable connection is used between the ram and the piston rod or an adjustable-stroke air cylinder is employed. In addition, the position of the lower electrode and its mounting arrangement are sometimes adjustable. This adjustment is used to position the work at a proper height for convenient operation.

Most seam welding of thin gages is done using continuous drive systems. With thick gages, intermittent drive systems must be used to maintain electrode force on the weld nugget as it solidifies. The thickness range that can be welded with each drive system will depend upon the metal being joined.

The majority of continuous drive mechanisms use a constant speed, ac electric motor with a variable speed drive. The speed range depends upon the drive design and the electrode diameter. Good flexibility may also be obtained with a constant torque, variable speed dc drive.

TYPES

THERE ARE THREE general types of seam welding machines:

(1) Circular, where the axis of rotation of each electrode is perpendicular to the front of the machine; this type is used for long seams in flat work and for circumferential welds, such as welding the heads into containers; such a machine is shown in Figure 19.5

(2) Longitudinal, where the axes of rotation of the electrodes are parallel to the front of the machine; this type is used for such applications as the welding of side seams in cylindrical containers and short seams in flat work

(3) Universal, where the electrodes may be set in either the circular or longitudinal position; this is accomplished with a swivel type upper head in which the electrode and its bearing can be rotated 90 degrees about a vertical axis; two interchangeable lower arms are used, one for circular operation and the other for longitudinal operation

ELECTRODE DRIVE MECHANISMS

Knurl or Friction Roller

THE KNURL OR friction roller drive has either the upper or the lower electrode, or both, driven by a friction wheel on the periphery of the electrode. When these friction rolls have knurled teeth, they are known as knurls or knurl drives. Knurl or friction roller drive will maintain a constant welding speed as the electrode diameter decreases from wear.

A knurl drive is commonly used on machines for seam welding galvanized steel, terne plate, scaly stock, or other materials where the electrodes are likely to pick up surface

Figure 19.5–A Standard Circular Seam Welding Machine with a Special Fixture

material from the parts being welded. The knurl drive wheel tends to break up the material on the electrode face. Where the nature of the work permits, both electrodes should be knurl-driven to provide a more positive drive and lessen the possibility of skidding.

A knurl drive may also function to control the shape of the contact face of the wheel electrode. This can be accomplished by using knurlers designed with a radius in the wheel contact area, or by using a flat knurler designed with side cutters that constantly trim the wheel contact face to a specific width.

Gear Drive

WITH THIS METHOD, the electrode shaft is driven by a gear train powered by a variable speed drive. Only one electrode should be driven to avoid skidding. Otherwise, a differential gear box is necessary. This type of drive is generally less desirable than a knurl drive because the welding speed decreases as the electrode wears. This can be overcome by gradually increasing the drive speed.

The most important applications for a gear-driven machine are the welding of aluminum and magnesium and the fabrication of small diameter containers. Standard seam welding machines are designed with some minimum distance between electrode centers for each machine size. If one of the electrodes must be small to fit inside a container, the other must be correspondingly larger to maintain the required center distance. If the ratio of the two electrode diameters exceeds about 2 to 1, the smaller electrode should be driven and the large one should idle to minimize electrode skidding.

SPECIAL PURPOSE MACHINES

SPECIAL PURPOSE MACHINES are available for specific applications. Such machines can be generally grouped as traveling electrode type, traveling fixture type, and portable seam welding machines.

Traveling Electrode Type

WITH THIS TYPE of machine, the seam to be welded is clamped or otherwise positioned on a fixed mandrel or shoe of some type and the ram and wheel electrode are moved along the seam. The mandrel or shoe is the lower electrode. The ram and electrode are moved by an air or hydraulic cylinder or by a motor-driven screw. Sometimes two upper electrodes operating in series are used side by side or in tandem. Figure 19.6 shows a typical traveling electrode machine.

Traveling Fixture Type

IN THE TRAVELING fixture type, the upper electrode remains in a fixed position. The fixture and work are moved under the electrode by a suitable driving system. Multiple electrodes can also be used to advantage with this type of machine, such as the one shown in Figure 19.7.

Portable Type

PORTABLE SEAM WELDING guns may be used for work that is too large and bulky to be fed through a standard machine. The gun consists of a pair of motor-driven wheel electrodes and bearings, together with an air cylinder and associated mechanism for applying the electrode force. Welding current is supplied in the same manner as for portable spot welding. A variable speed dc drive may be used where a wide range of welding speeds is desirable. The motor and speed reducer are mounted directly on the welding gun frame.

Figure 19.6–Traveling Electrode Seam Welding Machine

COOLING

ONE REQUIREMENT IN seam welding is the proper cooling of the machine, the electrodes, the current-carrying bearings, and other components of the secondary circuit. Temperature rise in these components causes an increase in electrical resistance in the secondary circuit. This results in lower welding current. Therefore, proper cooling is necessary to maintain control of the resistance and current in the secondary circuit. Cooling the work is also important in most applications to minimize warpage from the local heating. Water jets spraying on both the work and the welding electrodes are usually satisfactory. Welding under water may be done in special cases.

Another method of cooling the weldment is a water mist that removes heat by evaporation. A mist is produced by mixing air and water in proper proportions in a nozzle.

Figure 19.7—Traveling Fixture Seam Welding Machine with Two Electrodes in Tandem

FLASH AND UPSET WELDING MACHINES

FLASH AND UPSET welding machines are similar in construction. The major difference is the motion of the movable platen during welding and the mechanisms used to impart the motion. Flash welding is generally preferred for joining components of equal cross section end-to-end. Upset welding is normally used to weld wire, rod, or bar of small cross section and to continuously join the seam in pipe or tubing. Flash welding machines usually have a much larger capacity than upset welding machines.

FLASH WELDING MACHINES

General Construction

A STANDARD FLASH welding machine consists of a main frame, a stationary platen, a movable platen, clamping mechanisms and fixtures, a transformer, a tap switch, elec-

trical controls, and a flashing and upsetting mechanism. The stationary platen is generally fixed in position, although some designs provide a limited amount of adjustment for electrode and work alignment. The movable platen is mounted on ways on the frame and connected to the flashing and upsetting mechanism. Both platens are usually of cast or fabricated steel, although some small welding machines may have cast bronze, cast iron, or copper platens. The platens are connected to the transformer secondary. Electrodes that hold the parts and conduct the welding current to them are mounted on the platens. The transformer and tap switch are generally located within or immediately behind the frame with short, heavy-duty copper leads to the platens.

The depth of the frame and, consequently, the width of the platens depend upon the size of the parts to be welded as well as the clamping mechanism design. Upsetting force should be aligned as nearly as possible with the geometric center of the parts to minimize machine deflection. Dual flashing and upsetting cylinders or cams are sometimes used with wide platens to provide uniform loading or clearance for long pieces to extend over the mechanism.

Transformer and Controls

A FLASH WELDING transformer is essentially the same as those used for other types of single-phase resistance welding machines. A tap switch in the primary circuit is normally used to adjust flashing voltage. An autotransformer is sometimes used to extend the adjustment range of the secondary voltage. The primary power to the transformer is switched with an electronic contactor. Phase-shift heat control may be initiated with the contactor to provide low power for preheating or postweld heat treating in the machine.

Phase shift heat control should never be used to control the secondary voltage during flashing; only voltage tap selection should be used for adjusting the voltage during flashing. If heat control is used during flashing, there are periods of time when no secondary voltage is present, followed by an instantaneous voltage which can be quite high. The result can be deep cratering and entrapped oxides in the weld zone.

With ignitron contactors, auxiliary load resistors must be connected in parallel with the transformer primary for proper operation of the ignitrons.

Programming of secondary current for preheating prior to flashing and postheating of the completed weld in the machine can be done with appropriate controls.

Flashing and Upsetting Mechanisms

IN THE OPERATION of a flash welding machine, the parts are moved together using a predetermined travel pattern. This movement must be carefully controlled to produce consistently sound welds. After the appropriate flashing time, the pieces are rapidly brought into contact and upset. The upsetting action must be accurately synchronized with the termination of flashing.

The type of mechanism used for flashing and upsetting will depend upon the size of the welding machine and the application requirements. Some mechanisms permit the faying surfaces to be butted together under pressure and then preheated. After the appropriate temperature is reached, the pieces are separated and then the flashing and upsetting sequence is initiated. The movable platen may be actuated with a motor-driven cam or with an air or hydraulic cylinder.

Motor-operated machines use an ac or dc motor with a variable speed drive, which in turn drives a rotary- or wedge-shaped cam. The cam is designed to produce a specific flashing pattern. It may contain an insert block to upset the joint at the end of flashing. The speed of the cam determines the flashing time. The platen may be moved directly by the cam or through a lever system. The motor may operate intermittently for each welding cycle or continuously. With continuous operation, the drive is engaged through a clutch on the output shaft of the speed reducer. The motor speed may be electronically controlled to produce a specific flashing pattern. A typical motor-operated flash welding machine is shown in Figure 19.8.

A motor-driven flashing cam may be used in combination with an air or hydraulic upsetting mechanism, particularly on larger machines. Such a combination provides adjustment of upset speed, distance, and force independently of the flashing pattern. Current is synchronized with the mechanical motion of the platen by limit switches or electronic sequence controls.

Medium and large flash welding machines use hydraulically operated flashing and upsetting mechanisms. These machines are capable of applying high upsetting forces for large sections. They are accurate in operation and are readily set up for a wide range of work requirements. A large hydraulic flash welding machine is shown in Figure 19.9. A servo system is used to control the platen motion for flashing and upsetting. The servo system may be actuated by a pilot cam mechanism, or by an electrical signal generated from the secondary voltage or the primary current. Choice of operating mode depends upon the application. The control may be programmed to include preheating and postheating. An accumulator is generally required to provide an adequate volume of hydraulic fluid from the pumping unit during upsetting.

Electro-hydraulic servo systems are generally of two designs. In one design, the servo valve meters the fluid directly to the hydraulic cylinder for position control. With the other design, the servo valve meters the fluid to a small control cylinder that operates a follower valve on a separate hydraulic system. The first design is simple and straightforward, but the second system has two distinct advantages. First, it has two separate hydraulic circuits for improved valve life. Second, the speed of response is fast, and control of the platen position is accurate.

Clamping Mechanisms and Fixtures

SEVERAL DESIGNS OF clamping mechanisms are available to accommodate different types of parts. These designs may be grouped generally as operating in either the vertical or the horizontal position. In special cases, the mechanisms may be mounted in other positions.

Vertical Clamping. The movement of the electrode may be in a plane perpendicular to the platen ways. The electrode may move either through a slight arc or in a straight line. If operating through an arc, a clamping arm pivots about a trunnion. This design is generally known as the "alligator" type. A machine with this type of clamping arrangement is shown in Figure 19.10. Clamping force may be applied by an air or hydraulic cylinder operating directly

Figure 19.8—Automatic Motor-Operated Flash Welding Machine

Figure 19.9—Automatic Hydraulically Operated Flash Welding Machine with Horizontal Clamping

Figure 19.10—An Automatic Flash Welding Machine with Vertical Alligator Type Clamping

or through a leverage or cam-operated mechanism. Vertical clamping is commonly used for bar stock and other compact sections.

Horizontal Clamping. With this design, the motion of the electrodes is parallel to the platen ways and generally in a straight line, as shown in Figure 19.9. The major advantage of this type of clamping mechanism is that the secondary of the welding transformer can be connected to both halves of the electrodes for uniform transfer of welding current into the work. This arrangement is highly desirable for welding parts with large cross sections. Clamping

force can be applied with one of the mechanisms described for vertical clamping.

Fixtures. Fixtures may be used to support and align the parts for welding as well as to back up the parts to prevent slippage of the electrodes during upsetting. They are usually adjustable to accommodate the geometry and length of the parts. The design must be sturdy to withstand the upsetting force without deflecting. When the parts can be supported, the clamping force on the electrodes can be limited to that needed to ensure good electrical contact and maintain satisfactory joint alignment.

UPSET WELDING MACHINES

UPSET WELDING MACHINES are quite similar to flash welding machines in principle, except that no flashing mechanism is required. A typical upset welding machine, such as the one in Figure 19.11, consists of a main frame that

houses a transformer and tap switch, electrodes to hold the parts and conduct the welding current, and means to upset the joint. A primary contactor is used to control welding current.

Figure 19.11–An Air-Operated Automatic Upset Welding Machine

The simplest type of upset welding machine is manually operated. In this machine, the pieces to be welded are clamped in position in the electrodes. A force is exerted on the movable platen with a hand-operated leverage system. Welding current is applied, and when the abutting parts reach welding temperature, they are compressed together to accomplish the weld. The current is manually shut off at the proper time during the welding cycle. The work is then removed from the electrodes. A limit switch or a timing device may be used to terminate the welding current automatically after the weld has upset a predetermined length.

Automatic machines may use springs or air cylinders to provide upset force. Either device can provide uniform force consistently. Spring or air-operated machines are particularly adapted for welding nonferrous metals having narrow plastic ranges.

There are three standard sizes of upset welding machines, rated at 2, 5, and 10 kVA. Normal upset forces are 12, 70, and 120 lb, respectively. Larger units are also available.

Upset welding is used extensively for welding of small wires, rods, and tubes in the manufacture of items such as chain links, refrigerator and stove racks, automotive seat frames, and for joining coils of wire for further processing. The upset welding process is often selected for applications where the upset is not objectionable in the context of the design. It is best adapted for joints between parts with relatively small cross section where uniformity of welding current is not a problem.

RESISTANCE WELDING CONTROLS

THE PRINCIPAL FUNCTIONS of resistance welding controls are to (1) provide signals to control machine actions, (2) start and stop the current to the welding transformer, and (3) control the magnitude of the current. There are three general groups of controls: timing and sequencing controls, welding contactors, and auxiliary controls.

TIMING AND SEQUENCE CONTROLS

Sequence Weld Timers

A SEQUENCE WELD timer is a device to control the sequence and duration of the elements of a complete resistance welding cycle. It may also control other mechanical movements of the machine such as driving or indexing mechanisms. Sequence weld timers are used on spot, seam, and projection welding machines.

The four basic steps in any spot, seam, or projection welding cycle are as follows:

(1) Squeeze time
(2) Weld time
(3) Hold time
(4) Off time

Squeeze time is the interval between the initial application of electrode force on the work and the first application of current. Weld time is the duration of welding current with single impulse welding. Hold time is the period during which the electrode force is maintained on the weld after current ceases. Off time is the period during which the electrodes are retracted from the work during repetitive welding. During off time, the work is moved to the next weld location.

A multiple impulse weld timer provides for a number of current pulses with an interval between them. It controls the duration of each pulse, called heat time, as well as the interval between them, or cool time. The sum of the heat and cool times is known as the weld interval.

Timers and combination controls now almost exclusively use synchronous precision phase controls for the welding functions. Non-synchronous controls are obsolete and are now rarely encountered in the workplace.

Single-phase and three-phase resistance welding controls are similar, except for the firing sequence of the electronic switch elements and the techniques of electronic heat control. The timing and control functions are nearly the same, but the terminology used may vary between the two types of equipment.

Synchronous Precision Controls

THIS TYPE OF control uses synchronous precision timers for accurate timing of all periods of current. The timer closes the primary circuit of the welding transformer at precisely the same point (electrical angle) with respect to the ac line voltage. Another distinction of a synchronous precision timer is that accuracy is absolute and equal to the set value. A synchronous precision control always contains an electronic heat control unit.

Control of the exact time when the primary circuit is closed is vital for precise results. This is necessary not only to control the heat obtained, but also because conduction, which does not begin at the same time in each half of a cycle, can cause saturation of the welding transformer.

Classification of Sequence Weld Timers

SEQUENCE WELD TIMERS are classified by the Resistance Welder Manufacturers Association according to the functions they control.

(1) Types 1AS and A1A control the weld time only.

(2) Type 1BS controls heat and cool times for multiple impulse welding operations.

(3) Type A3B covers sequence timers that control squeeze, weld, hold, and off times.

(4) Type A3C is similar to type A3B except that a squeeze delay or initial squeeze time is provided to account for the electrode travel time to contact the work. This type of timer is used for high-speed repetitive welding.

(5) Type A5B, which is also similar to type A3B, is designed for multiple impulse welding applications. This type controls heat, cool, and weld interval times instead of weld time.

(6) Type 7B is a sequence timer used in conjunction with a type 1AS weld timer to control squeeze, weld, hold, and off times.

(7) Type 9B is similar to type 7B except that it is used in conjunction with a type 1BS weld timer.

In the above designations, S indicates a synchronous precision timer, and the prefix A indicates an absolute cycle timer.

Time Mechanisms

SEVERAL TYPES OF timers have been developed to control the duration of various functions during the welding cycle. The availability of inexpensive microprocessors and associated digital circuits has led to their use in most, if not all, of the welding controls now being manufactured. Many older designs used RC (resistor-capacitor) timers for determining intervals.

Digital counters, with or without microprocessor control, provide accurate measurement and control of welding cycles or even parts of cycles (as in heat controls). These counters may be used to time conduction intervals or other actions associated with the welding process.

Some operations, such as postheating of flash or upset welds, are not critical with respect to timing accuracy. Pneumatic or motor-operated timers may be suitable for these applications. Timing ranges may vary from a few seconds to several minutes.

CONTACTORS

A CONTACTOR IS used to close and open the primary power line to the welding transformer. The term *contactor* is actually a misnomer; it is a carry-over from the mechanical (magnetic) contactors which were originally used to control welding transformer conduction in nonsynchronous welding controls. Modern welding controls typically use SCR (silicon controlled rectifier) switch assemblies, made up of a pair of inverse-paralleled devices which act as the switching element or *contactor*. In this arrangement, one SCR conducts during the positive portion of the conduction cycle, and the other during the negative portion. In single-phase equipment, only one set of SCRs is needed in one of the primary lines, as shown in Figure 19.12. With a three-phase frequency converter machine, one set is required in each leg of the transformer for a total of three sets as shown in Figure 19.13.

SCR contactor components are usually assembled in a package resembling a ceramic *hockey puck*, with anode and cathode connections at the faces of the puck and the gate leads exiting through the side of the puck insulator. Water cooled blocks of copper are used on one or on both faces of the pucks, and an insulated tensional bolt and compression springs or washers are used with the copper cooling blocks. SCR switches of this construction are available with continuous current ratings of thousands of amperes and considerably higher current ratings at lower duty cycles. Blocking voltage ratings of 2,500V or more are also available.

Firing of the SCRs is accomplished by applying a current pulse to the gate-cathode junction of the SCR, which is for-

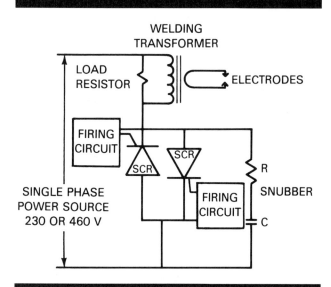

Figure 19.12–Single-Phase Welding Machine with a Pair of SCR Contactors

ward biased at the time that conduction is desired. In single-phase equipment, it is acceptable to fire both SCRs simultaneously, as only the device which is forward biased will conduct. Typically, firing pulses are delivered to the SCRs via pulse transformers which provide voltage isolation for the gating circuitry. The pulses are usually about 1 to 3 amperes in magnitude, with rise times of 1 to 2 microseconds and a total duration of 100 microseconds or less. Because of the low forward voltage drop of an SCR (a few volts), it is possible to control the welding transformer conduction over virtually the entire range of 0 to 100 percent.

SCRs are susceptible to spurious firing by line voltage spikes. They are also susceptible to the rate of rise of the voltage spikes, not just to the spike magnitude. For this reason a series-connected resistor and capacitor (RC) assembly, called a "snubber", is usually connected in parallel

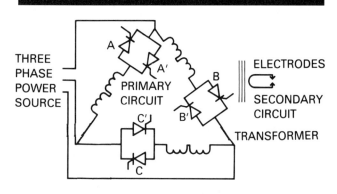

Figure 19.13–Three-Phase Welding Machine with SCR Contactors

with the SCRs. Properly snubbed, SCRs are durable and reliable switches.

Some older control designs use pairs of ignitron tubes triggered either by small SCRs or by thyratron tubes. Although extremely rugged devices, ignitrons have restrictions in mounting orientation because of their internal mercury pools. Also, because of their mercury content, ignitrons pose some personnel hazards. Ignitrons have relatively high holding current levels and forward voltages. As they age, both the minimum holding current levels and forward voltage drops of ignitrons increase. Even when new, ignitrons are not capable of handling the full range of conduction. Typically, their forward voltage drop limits the conduction range available to 20 percent to 100 percent (at 440 or 475V levels).

AUXILIARY CONTROLS

Heat Control

COARSE ADJUSTMENT OF the heat or current output of a welding machine can be accomplished with adjustable taps of the welding transformer. The tap switch changes the ratio of transformer turns for major adjustment of welding current. Precise control is accomplished using electronic heat control.

In electronic heat control circuits, the firing time of the SCRs relative to the start of each half cycle can be delayed to produce the desired heat setting. Referring to Figure 19.14, if the SCR firing pulses are produced 180 degrees out of phase with the ac supply, the SCRs will not conduct and no heat will be produced. As the out-of-phase or delay angle of the firing pulses is decreased, the SCRs begin to fire late in the half cycle and the rms value of the welding transformer primary voltage will be low. As the delay angle is further decreased, the SCRs will fire earlier, and will conduct current for a greater part of the half cycle. The rms current will increase. When the delay angle equals the power factor of the load, 100 percent rms primary current will be conducted to the welding transformer. Figure 19.14 illustrates this concept for welding machines with four different power factors. The higher the power factor (lower angle), the wider the range of heat control.

The reduction in heat or energy varies as the square of the current. Thus, if the rms current can be varied from 100 to 20 percent, the heat will vary from 100 to 4 percent. SCRs allow control of the heat over the entire range from zero to 100 percent, an ability which is not practical with ignitrons.

Automatic heat control is normally the basis of all auxiliary controls that change the welding amperage during a welding sequence. These include current and voltage regulators as well as upslope, downslope, and temper controls.

To minimize variations in welding current, the heat control should be operated as near to full heat as possible. At low settings, a small change of the dial setting can significantly change the rms current. Line voltage disturbances,

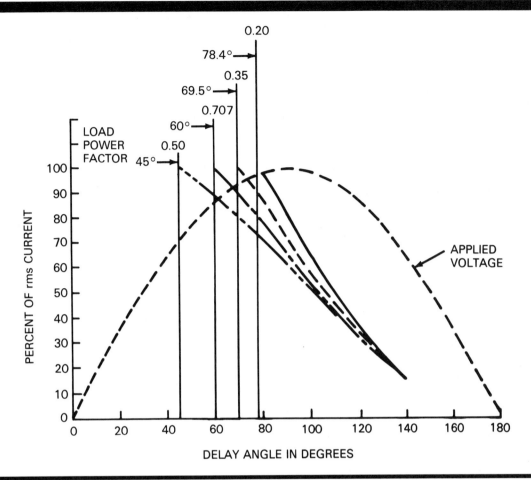

Figure 19.14–The Relationship Between Percent of RMS Current and Firing Delay Angle for Different Power Factors

such as the operation of another welding machine, can sufficiently distort the line voltage wave form to produce such a change. Major changes in welding transformer output should be made by changing the tap switch.

The power demand is always greater when heat control is used to adjust the magnitude of the welding current. In general, the kVA demand with heat control follows a linear relationship with current. For example, if the welding current is adjusted by heat control to 80 percent of its maximum value, the kVA demand will be about 80 percent of its maximum. However if the welding current is reduced to 80 percent of its maximum value by changing the transformer tap switch, the kVA demand will be only about 64 percent of maximum.

Upslope and Downslope Controls

UPSLOPE CONTROL IS used to start the welding current at some low value and control its rate of rise to some maximum value during a period of several cycles. It is frequently used to minimize or prevent the expulsion of molten metal from between the faying surfaces when welding coated steels and some nonferrous metals, particularly aluminum.

Downslope control is used to decrease the welding current from maximum to a lower value called the *postheat current*. The gradual decrease in current reduces the cooling rate of the weld. It may be useful when welding hardenable steels to minimize the cooling rate and the cracking tendency.

Upslope and downslope of welding current are illustrated in Figure 19.15. The accepted nomenclature for the various parts of a welding current cycle are also shown.

Quench and Temper Control

THE QUENCH AND temper control is a device that applies a temper cycle to the completed weld after a quench period during which no current is applied. In each case, the time period is adjustable. Temper current magnitude is normally adjustable with heat control.

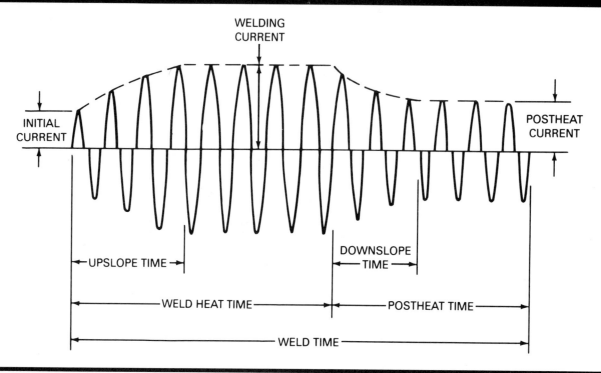

Figure 19.15—Welding Current with Upslope and Downslope Features

Heat control is frequently used when spot welding hardenable steels in thickness ranges from 0.016 to 0.125 in. (0.4 to 3.2 mm). After the weld is made, the weld cools rapidly and martensite is formed. Current pulses are then applied to reheat the weld zone and thus temper the martensite. Although this cycle cannot duplicate furnace heat treatment, it usually will prevent weld cracking.

Forge Delay Control

THIS CONTROL WILL initiate a forging force at a definite time interval after the start of weld time or weld interval. It is used to apply two levels of force to a weld, namely, a welding force and a forging force. Obviously, the welding machine must be designed to perform a dual force function.

Dual force is used when spot welding certain aluminum alloys. The principle is to produce the weld and then apply a high force during cooling to avoid the formation of cracks. It is common to downslope the welding current to retard the cooling rate during the application of forging force.

Electronic Current Regulator

AN ELECTRONIC CURRENT regulator is designed to maintain a constant welding current under changing conditions. This device will make corrections for either line voltage fluctuations or impedance changes caused by insertion of magnetic material into the throat of the welding machine. It first compares the primary current, as measured by a current transformer or other device (feedback signal), to a previously adjusted satisfactory level (command signal); then it varies the phase-shift heat control network to make these signals equal but opposite.

Electronic Voltage Regulator

IF HEAT CONTROLS are not operated too close to their maximum heat settings, many are capable of dynamically adjusting the SCR firing angle to maintain the desired heat level when variations in the AC line voltage (sags and surges) occur. Since heat also varies as the square of voltage (assuming impedance remains constant), a 10 percent drop in line voltage will result in a 19 percent reduction in heat. The welding transformer tap switch should then be set so that the desired weld is obtained at a control heat setting of 81 percent or less, if a 10 percent line sag is to be expected. Reaction time for heat controls employing this compensation feature can be less than one cycle.

Load Distribution Control

A LOAD DISTRIBUTION control is used with resistance welding machines that have two or more transformers. This control distributes the electrical power demand by energizing the welding transformers in sequence on one or more phases. Reconnection is normally provided to energize the transformers simultaneously on two or more phases.

This control generally contains several single-function timers to control mechanical functions, such as squeeze and hold timers, acting over two or more weld periods. In addition, it has a contactor for each transformer. The weld timers are functional but are weld safe; that is, the termination of weld time is not dependent upon conduction of a single electronic device. Accessories such as heat control and upslope control are sometimes added to this type of control.

A less expensive version of this control uses only one ignitron or SCR contactor and a series of magnetic contactors. The ignitron or SCR contactor switches the primary current on and off. The magnetic contactors connect the welding transformers in succession to the contactor circuit during a nonconductive period.

MONITORING AND ADAPTIVE CONTROLS

THERE ARE A number of factors that affect the consistency of resistance spot welds during a production run. These include line voltage variations, electrode deterioration, changes in surface resistance, shunt paths, and variations in the force system. There are several systems available to monitor specific welding variables or actions that occur during the welding cycle. If the monitor detects a fault, it can do one or more of the following:

(1) Turn on an alarm or signal light
(2) Document the information
(3) Reject or identify the faulty part
(4) Interrupt the process until the problem is corrected
(5) Alter time or current for the next weld
(6) Change a variable during the weld cycle to ensure a good weld.

Variables that affect process stability and weld consistency include weld time, welding current, impedance, welding energy, and electrode force. Physical changes that take place in the weld zone are temperature, expansion and contraction, electrical resistance, and, in some cases, metal expulsion.

Monitoring devices can compute either weld energy or impedance by measuring welding voltage, current, resistance, or time. When the computed value falls outside acceptable limits, the unit can notify the operator or automatically adjust one or more of the variables prior to the next weld.

Several systems of adaptive feedback have been developed which are intended to make consistent, acceptable welds. These adaptive feedback systems, whether used singly or in combination, have certain limitations. As examples of these limitations, they may require frequent calibration, work only on single-point welders, or add significantly to machine maintenance. While the systems described below are available, they are not widely used in industry.

In aerospace industries, electrode indentation is a limiting factor for acceptable welds. It has been reported that there is a relationship between electrode indentation and weld strength. Therefore, by controlling electrode indentation, welds of consistent strength with acceptable electrode indentation can be obtained. A welding control based on this principle has been developed.

A relationship between nugget expansion and weld strength has also been established. Instruments have been developed to control nugget expansion. They do this by increasing or decreasing the welding current in real time with reference to a baseline nugget/time expansion curve. The object, of course, is to obtain consistent, acceptable welds. Such a feedback control can compensate for any shunting effect, even in aluminum alloys.

In the automotive industry, portable welding guns are widely used. With a specially designed welding gun, adaptive feedback control can be achieved using the electrode indentation method. However, nugget expansion feedback control is difficult to achieve in a portable gun.

For this reason, the resistance method and the acoustic emission analysis method are used to improve the performance of a portable gun. In the resistance method, a resistance/time curve for a good weld is established. If and when the resistance/time curve of a subsequent weld deviates from this baseline curve, thereby indicating expulsion is imminent, welding current is terminated.

The acoustic method detects metallurgical actions such as melting, expulsion, solidification, phase transformation, and cracking by the acoustic waves they emit, each with a distinguishable wave form and amplitude. By detecting such acoustic waves at the threshold of expulsion, welding current can be terminated to obtain a strong weld.

As an alternative to either taking immediate action (terminating current) or passive monitoring (notifying the operator), some controls are capable of analyzing the data from many welds and detecting trends. Trend analysis allows the control to compensate for a lowering of the weld strength and slow deviations from the desirable weld results by varying process conditions to maintain high quality welds.

ELECTRICAL CHARACTERISTICS

SINGLE-PHASE EQUIPMENT

THE TYPICAL ELECTRICAL system of a single-phase resistance welding machine consists of (1) a transformer, (2) a tap switch, and (3) a secondary circuit including the electrodes.

The welding transformer, in principle, resembles any other iron-core transformer. The primary difference is that its secondary circuit has only one or two turns. Stationary machines usually have single turn secondaries. Portable gun welding transformers may have two turns that can be connected in series or parallel, depending upon the output requirements.

Transformer Rating

RESISTANCE WELDING TRANSFORMERS are normally rated on the basis of temperature rise limitations of the components. The standard rating in kVA is based on the ability of a transformer to produce that power at a 50 percent duty cycle without exceeding design limitations. This means that a transformer can produce its rated power for a total time of 30 seconds during each minute of operation without exceeding temperature limitations, if it is being properly cooled.

Duty cycle is the percentage of time that the transformer is actually "ON" during a one minute integrating period. For 60 Hz power, it can be expressed by the formula:

$$\text{percent duty cycle} = \frac{\text{welds/min x weld time in cycles}}{(60 \text{ cycles/sec}) (60 \text{ sec/min})} \times 100$$

For example, if a machine is producing 30 welds per minute with a weld time of 12 cycles (60 Hz), its operating duty cycle is:

$$\frac{30 \times 12}{3600} \times 100 = 10 \text{ percent}$$

If a welding transformer is operated at less than 50 percent duty cycle, it can be operated at a power level higher than its thermal rating.

The maximum permissible kVA input for a standard resistance welding transformer at a particular duty cycle can be determined using the following equation:

$$kVA_i = 7.07 \ kVA_r/(DC)^{1/2} \tag{19.1}$$

where
$$kVA_i = \text{maximum input power}$$
$$kVA_r = \text{standard power rating at 50 percent duty cycle}$$
$$DC = \text{operating duty cycle, percent}$$

For example, a welding transformer rated at 100 kVA may be operated at 141 kVA at 25 percent duty cycle without overheating.

Tap Switches

TAP SWITCHES ARE devices for connecting various primary taps on the transformer to the supply lines. They are usually rotary type and designed for flush mounting in an opening in the machine frame, or, in some cases, directly on the transformer. The switches are designed to accommodate the arrangements of the transformer taps. Straight rotary designs are normally used with 4, 6, or 8 tap transformers. To provide a range of secondary voltages, taps are placed at various turns on the primary winding. These taps are connected to the tap switch, and thus the turns ratio of the transformer can be changed to produce different secondary voltages (Figure 19.16). In addition, there may be a series-parallel switch that connects two sections of the primary in series or in parallel. This provides a wider range of secondary voltages.

Most switch handles have locking buttons so that the contacts are centered in each operating position. In addition, some switches have an OFF position which acts as a disconnect. A tap switch should not be operated while the transformer is energized; otherwise, arc-over between points will damage the contact surfaces of the tap switch.

Ac Secondary Circuit

THE GEOMETRY OF the secondary circuit (loop), the size of the conducting components, and the presence of magnetic material in the loop will affect the electrical characteristics of the welding machine. Available welding current and kVA demand will be influenced by the impedance of the secondary circuit.

The electrical impedance of an ac welding machine should be minimized to permit the delivery of the required welding current at minimum kVA demand. The electrical impedance will be smaller when:

(1) The throat area of the welding machine is decreased.
(2) The electrical resistance of the secondary circuit is decreased.
(3) The sizes of the secondary conductors are increased.
(4) The amount of magnetic material in or near the throat of the machine is decreased.

Power Factor Correction

Series Capacitors. Welding machines of good design effectively minimize the impedance of the secondary cir-

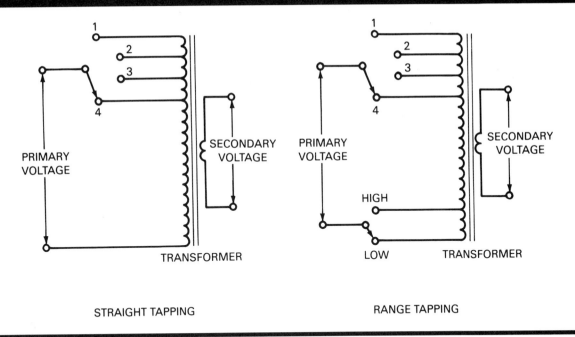

STRAIGHT TAPPING

RANGE TAPPING

Figure 19.16—Rotary Tap Switches Used to Provide a Range of Secondary Voltages

cuit. However, the size of the work to be welded and associated fixturing may require a large throat depth or throat height. This requirement may add considerable inductance to the secondary circuit. The increased inductance causes a reactive voltage drop which, in turn, decreases the power factor. To compensate for this, a higher secondary voltage is required and the necessary electrical kVA demand will be increased.

Low power factor and intermittent, high electrical demand are not desirable to the electric utility, which must maintain a stable power supply to other customers. One method of reducing line kVA demand and improving power factor is the use of series capacitors in the primary circuit. A specific amount of capacitance can be connected in series with the transformer of a welding machine to neutralize the inductance of the machine and improve the power factor. This, in turn, will reduce the demand from the power line.

This power factor correction method will increase the voltage applied to the welding machine transformer. High voltage insulation is therefore required. A transformer tap switch is not used because it changes the series resonant condition. The welding current is changed with phase-shift heat control or a tapped autotransformer.

The resistance of the secondary circuit limits the current in any high power factor system. Since the metal being welded has resistance, the welding current may vary significantly with slight changes in metal thickness or cleanliness. This may affect weld consistency and quality, particularly with alloys of high resistance.

Voltages appearing across the welding machine transformer and the series capacitors are higher than the electrical supply voltage. Therefore, special high voltage electrical control panels are normally required. A protective over-voltage device, a discharge resistor, and a contact to ground are generally provided for safe operation and maintenance.

Three-phase welding systems have largely replaced single-phase series capacitor installations. A welding machine with a high power factor is generally less troublesome than a series capacitor installation.

Shunt Capacitors. Shunt capacitors are seldom used with resistance welding equipment. The initial high inrush of current may actually increase the line demand. However, shunt capacitors may be preferred to series capacitors if the welding time is comparatively long, as in non-interrupted resistance seam welding.

Dc Secondary Circuit

ONE METHOD OF decreasing impedance losses in the secondary circuit is to rectify the secondary power to dc. Single-phase dc resistance welding machines have a center-tapped secondary and a full-wave silicon diode rectifier. With this system, the kVA rating of a machine need not be increased much to provide for a larger throat area. For a given size and application, the kVA demand of a dc machine will be significantly lower than that of an ac machine. The reason for this is the high power factor of about

90 percent for dc machines, compared to 25 to 30 percent for ac machines.

Secondary dc power is particularly useful for portable gun welding applications. The impedance loss in the cable connecting the gun and transformer is much lower with dc than with ac. This, in turn, decreases the kVA demand and the required size of the welding transformer. It is also advantageous for spot and seam welding operations, during which the amount of magnetic material in the machine throat increases or decreases as welding proceeds.

DIRECT ENERGY THREE-PHASE EQUIPMENT

Frequency Converter Type

TWO TYPES OF frequency converter systems exist: (1) the classic half-wave system illustrated in Figure 19.13 and (2) the full-wave type which uses three phase input to a rectifier to supply a low frequency converter. Both of these systems perform in a similar fashion, but the full-wave type uses a large core single-phase transformer while the half-wave type uses a large core three-phase transformer.

This type of machine has a specially designed transformer with three primary windings, each of which is connected across one of the three input phases. There is one secondary winding which is interleaved among the primary windings and connected to the secondary conductors.

Referring to Figure 19.13, these transformer primary windings are connected to the power lines by three electronic contactors. Ignitron tubes or SCRs may be used as contactors. A welding control causes contactors A, B, and

C to conduct in sequence. With the correct sequence and conduction time, current is passed through the three primary windings in the same direction. This causes unidirectional current in the secondary circuit. Contactors A, B, and C are then shut off at the end of a preselected time. Contactors A', B', and C' are caused to conduct next, with the correct sequence and conduction time, and current will be in the opposite direction through the primary windings and the secondary circuit. This action effectively applies a reversing "dc" voltage to the primary windings.

The maximum duration of unidirectional primary current is governed primarily by the size of the transformer and its saturation characteristics. It is common practice to have two maximum dc pulse lengths. One is a short time of about 5 cycles (60 Hz) for high current applications, and the other is usually 10 cycles with welding current limited to about 50 percent of maximum. Specially designed massive transformers may permit the use of high current for the longer time period.

Figure 19.17 shows a typical current-force diagram for this type of machine. Programming may be provided for other functions, such as preheat current, pre-compression force, and temper current. Single- or multiple-impulse welds may be made.

Dc Rectifier Type

A THREE-PHASE DC rectifier type welding machine is similar to the single-phase type in that each welding transformer powers a rectifier bank. The output of the rectifiers is fed into the welding circuit. Some machines use half-wave rectification, as shown in Figure 19.18(A). In this case, the transformer secondary is wye connected. Other machines,

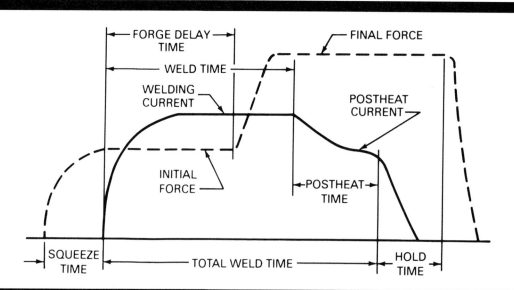

Figure 19.17–Typical Current-Force Diagram for Frequency Converter or DC Rectifier types of Three-Phase Spot Welding Machines

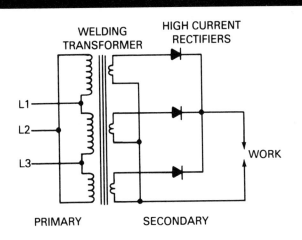

(A) HALF-WAVE RECTIFIER

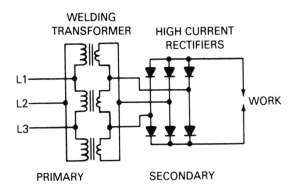

(B) FULL-WAVE RECTIFIER

Figure 19.18–Electrical Arrangements for Three-Phase DC Rectifier Welding Machines

particularly earlier versions, have full-wave rectification with the transformer secondary connected in delta arrangement, as shown in Fig. 19.18(B).

Welding current is controlled by electronic heat control, sometimes in conjunction with a transformer tap switch. The design of the primary circuit and control varies among equipment manufacturers. The secondary current output of a three-phase machine is much smoother than that of a single-phase machine. In addition, power demand is balanced on the input line.

The three-phase rectifier consists of silicon diodes mounted on water-cooled conductors. The arrangement of conductors and diodes is electrically symmetrical. The impedance of each diode circuit must be similar so that the diodes will share the load (current) equally. The diodes themselves must have similar electrical characteristics. Diodes have long life if properly applied and used. Welding current may be provided continuously as long as the thermal rating of the machine is not exceeded.

A variation of this scheme uses a rectifier in the primary to convert the ac power to dc, and a pulse width modulated power supply to generate a high frequency input to the welding transformer. The output of the welding transformer is then rectified to a smooth low ripple dc for welding. The advantage of this type of circuit is the reduction in size and weight of the welding transformer. This is particularly beneficial when using transguns for robotic welding.

A typical current-force diagram for this type of machine is similar to that of Figure 19.17. In addition, programming may be provided for other functions such as preheating, upslope, downslope, and tempering. Single- or multiple-impulse welds may be made.

STORED ENERGY EQUIPMENT

EQUIPMENT OF THE stored energy type is usually found in small units suitable for bench mounting. They are powered from a single-phase line. Many designs of welding heads or portable tongs that are connected to their power units with cables are available. They are used for a wide variety of applications, including assembly of small electrical components of nonferrous alloys, and the spot welding of foils.

Electrode force may range from a few ounces to several pounds. Calibrated springs are used in a manual force system to apply the electrode force. Stored energy is used to produce the welding current pulse. The welding current amplitude, duration, and wave shape are determined by the electrical characteristics of the power source, including capacitance, reactance, resistance, and capacitor voltage. Welding times are often substantially shorter than one half cycle of 60 Hz.

Figure 19.19 shows a typical foot-operated bench welding machine of the stored energy type, with a maximum electrode force of either 8 or 20 lb., depending upon the spring size. Electrode force is applied by actuating a foot pedal mounted beneath the welding head. A typical power source is rated at 40 watt-seconds, has 600 microfarads of capacitance, and can be adjusted for welding outputs as indicated by the curves in Figure 19.20. Larger machines are available.

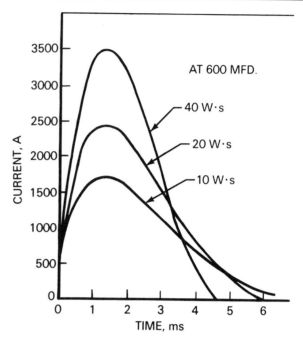

Figure 19.20–Typical Time-Current Wave Forms of a 40 Watt-Second Stored Energy Spot Welding Machine

Figure 19.19–A Bench-Mounted Stored Energy Spot Welding Machine

ELECTRODES AND HOLDERS

THE PERISHABLE TOOLS used in resistance welding are the electrodes, which may be in the form of a wheel, roll, bar, plate, clamp, chuck, or some modification of these. Most spot welding applications use electrode holders or adaptors for mounting the electrodes in the machine.

A welding electrode may perform one or more of the following functions:

(1) Conduct welding current to the parts
(2) Transmit a force to the joint
(3) Fixture or locate the parts in proper alignment
(4) Remove heat from the weld or adjacent part

The electrode design should always provide sufficient mass to transmit the required welding force and current, and provide adequate cooling when needed. High production applications sometimes involve thick sections that require special electrode designs. If it is necessary to compromise the design, that may affect electrode life, weld quality, production rate, or all three. Consequently, selection of the electrode material is very important for good performance.

ELECTRODE MATERIALS

RESISTANCE WELDING ELECTRODE materials are classified by the RWMA.[1] They are divided into three groups: A, copper-base alloys; B, refractory metal compositions, and C, specialty materials. In addition to these materials, there are a number of proprietary alloys available from the various electrode manufacturers. Table 19.1 gives the minimum properties for copper-base alloys to meet the various RWMA classification requirements. The specific alloy compositions are not specified, and they will vary among the manufacturers.

Group A: Copper-Base Alloys

THE COPPER-BASE ALLOYS are divided into five classes. Class 1 alloys are general purpose material for resistance welding

1. Standard electrode materials are described in ANSI/RWMA Bulletin No. 16, Resistance Welding Equipment Standards, Resistance Welder Manufacturers Association, Philadelphia, Pennsylvania.

Table 19.1
Minimum Properties for RWMA Electrode Materials

Group A Copper-Base Alloys	Proportional Limit Tension, psi			Hardness, Rockwell B			Conductivity, %[a]			Ultimate Tensile Strength, psi			Elongation, % in 2 in. or 4 diameters		
	Class 1	Class 2	Class 3	Class 1	Class 2	Class 3	Class 1	Class 2	Class 3	Class 1	Class 2	Class 3	Class 1	Class 2	Class 3
Rod Diam., In.															
Round Rod Stick (Cold Worked)															
Up to 1	17,500	35,000	50,000	65	75	90	80	75	45	60,000	65,000	100,000	13	13	9
Over 1 to 2	15,000	30,000	50,000	60	70	90	80	75	45	55,000	59,000	100,000	14	13	9
Over 2 to 3	15,000	25,000	50,000	55	65	90	80	75	45	50,000	55,000	95,000	15	13	9
Thickness, In.															
Square, Rectangular, and Hexagonal Bar Stock (Cold Worked)															
Up to 1	20,000	35,000	50,000	55	70	90	80	75	45	60,000	65,000	100,000	13	13	9
Over 1	15,000	25,000	50,000	50	65	90	80	75	45	50,000	55,000	100,000	14	13	9
Thickness, In.															
Forgings															
Up to 1	20,000	22,000[b]	50,000	55	65	90	80	75	45	60,000	55,000	94,000	12	13	9
Over 1 to 2	15,000	21,000[b]	50,000	50	65	90	80	75	45	50,000	55,000	94,000	13	13	9
Over 2	15,000	20,000[b]	50,000	50	65	90	80	75	45	50,000	55,000	94,000	13	13	9
Castings															
All	--	20,000	45,000	--	55	90	--	70	45	--	45,000	85,000	--	12	5

Group A Copper-Base Alloys	Proportional Limit Tension, psi	Hardness, Rockwell	Conductivity, %[a]	Ultimate Tensile Strength, psi	Elongation, % in 2 in. or 4 Diameters
Class 4 Alloys					
Cast	60,000	33C	18 (Average)	90,000	0.5
Wrought	85,000	33C	20 (Average)	140,000	1.0
Class 5 Alloys, cast					
Type H	16,000	88B	12	70,000	2
Type S	12,000	65B	15	65,000	12

Group B Refractory Metals	Hardness, Rockwell	Conductivity, %[a]	Ultimate Compression Strength, psi
Class 10 - Rods, bars, and inserts	72B	45	135,000
Class 11 - Rods, bars, and inserts	94B	40	160,000
Class 12 - Rods, bars, and inserts	98B	35	170,000
Class 13 - Rods, bars, and inserts	69B	30	200,000
Class 14 - Rods, bars, and inserts	85B	30	---

a. International Annealed Copper Standard.

b. Hot worked and heat treated but not cold worked.

applications. They may be used for spot and seam welding electrodes where electrical and thermal conductivities are of greater importance than mechanical properties. Other applications are seam welding machine shafts and welding fixtures. This alloy class is recommended for spot and seam welding electrodes for aluminum, brass, bronze, magnesium, and metallic coated steels, because Class 1 alloys have high electrical and thermal conductivity.

Class 1 alloys are not heat treatable. Their strength and hardness are increased by cold working. Therefore, they have no advantage over unalloyed copper for castings, and are rarely used or fabricated in this form.

Class 2 alloys have higher mechanical properties but somewhat lower electrical and thermal conductivities than Class 1 alloys. Class 2 alloys have good resistance to deformation under moderately high pressures, and are the best general purpose alloys. This alloy class is suitable for high production spot and seam welding of clean mild and low alloy steels, stainless steels, low conductivity copper-base alloys, and nickel alloys. These materials comprise the bulk of resistance welding applications.

Class 2 alloys are also suitable for shafts, clamps, fixtures, platens, gun arms, and various other current-carrying structural parts of resistance welding equipment. Class 2 alloys are heat treatable and may be used in both wrought and cast forms. Maximum mechanical properties are developed in wrought form by cold working after heat treatment.

Class 3 alloys are also heat treatable, but have higher mechanical properties and lower electrical conductivity than Class 2 alloys. The chief application for spot or seam welding electrodes made of this alloy is for welding heat resistant alloys that retain high strength properties at elevated temperatures. Welding of these alloys requires high electrode force, which in turn requires a strong Class 3 electrode alloy. Typical heat resistant alloys are some low alloy steels, stainless steels, and nickel-chromium-iron alloys.

Class 3 alloys are especially suitable for many types of electrode clamps and current-carrying structural members of resistance welding machines. Their properties are similar in both the cast and wrought conditions, because they develop most of their mechanical attributes from heat treatment.

Class 4 alloys are age-hardenable types that develop the highest hardness and strength of the Group A copper alloys. Their low conductivity and tendency to be hot-short make them unsuitable for spot or seam welding electrodes. They are generally recommended for components that have relatively large contact area with the part. These include flash and projection welding electrodes and inserts. Other applications are part backup devices, heavy-duty seam welding machine bearings, and other machine components where resistance to wear and high pressure are important.

Class 4 alloys are available in both cast and wrought forms. Because of their high hardness after heat treatment, they are frequently machined in the solution-annealed condition.

Class 5 alloys are available principally in the form of castings with high mechanical strength and moderate electrical conductivity. They are recommended for large flash welding electrodes, backing material for other electrode alloys, and many types of current-carrying structural members of resistance welding machines and fixtures.

Group B: Refractory Metal Compositions

THESE MATERIALS CONTAIN a refractory metal in powder form, usually tungsten or molybdenum. They are made by the powder metallurgy process. Their chief attribute is resistance to deformation in service. They function well for achieving heat balance when two different electrode materials are needed to compensate for a difference in thicknesses or composition of alloys being welded.

Class 10, 11, and 12 compositions are mixtures of copper and tungsten. The hardness, strength, and density increase and the electrical conductivity decreases with increasing tungsten content. They are used as facings or inserts where exceptional wear resistance is required in various projection, flash, and upset welding electrodes. It is difficult to establish guidelines for the application of each grade. The electrode design, welding equipment, opposing electrode material, and workpiece composition and condition are some of the variables that should be considered in each case.

Class 13 and Class 14 are commercially pure tungsten and molybdenum, respectively. They are generally considered to be the only electrode materials that will give good performance when welding nonferrous metals that have high electrical conductivity. The welding of braided copper wire or copper and brass wires to themselves or to various types of terminals are typical uses for Class 13 and 14 materials.

Group C: Other Materials

A NUMBER OF unclassified copper alloys and other materials may be suitable for resistance welding electrodes. Suitability of a particular material for electrodes will depend upon the application. Although most requirements are met by materials meeting RWMA standards, there are cases where other materials will function as well or better. For example, steel may be used for flash welding electrodes for certain aluminum applications.

Dispersion-strengthened copper is an unclassified material that may be used for electrodes. It is high purity copper that contains small amounts of submicroscopic aluminum oxide uniformly distributed in the matrix. The aluminum oxide significantly strengthens the copper matrix and raises the recrystallization temperature of cold worked material. The high recrystallization temperature of wrought material provides excellent resistance to softening and mushrooming of electrodes when the contacting surfaces are heated. This significantly contributes to long electrode life. The mechanical properties and electrical conductivity of dispersion strengthened copper bars meet the requirements for RWMA Group A, Class 1 and 2 alloys, but they are not so classified.

SPOT WELDING ELECTRODES

A SPOT WELDING electrode has four features:

(**1**) The face
(**2**) The shank
(**3**) The end or attachment
(**4**) Provision for cooling

Face

THE FACE OF the electrode is that portion which contacts the work. Its design is influenced by the composition, thickness, and geometry of the parts to be welded. In turn, the electrode face geometry determines the current and

pressure densities in the weld zone. Figure 19.21 shows the standard RWMA electrode face and taper designs. The radius, dome, and flat-faced contours are those most commonly used. The flat-faced electrode is used to minimize surface marking or to maintain heat balance.

The face may be concentric to the axis of the electrode, as in Figures 19.21(A),(B),(C),(E), and (F); eccentric or offset, as in Figure 19.21(D); or at some angle to the axis, as in Figure 19.22. So-called offset electrodes with eccentric faces are used to make a weld near a corner or in other less accessible areas. This is illustrated in Figure 19.23. A facing of Group B material may be brazed to a shank of a Group A alloy to produce composite electrodes for special applications as shown in Figure 19.24.

Shank

THE SHANK OF an electrode must have sufficient cross-sectional area to support the electrode force and carry the welding current. The shank may be straight, as in Figure 19.21, or bent, as in Figure 19.25. The standard shank diameters are shown in Figure 19.21.

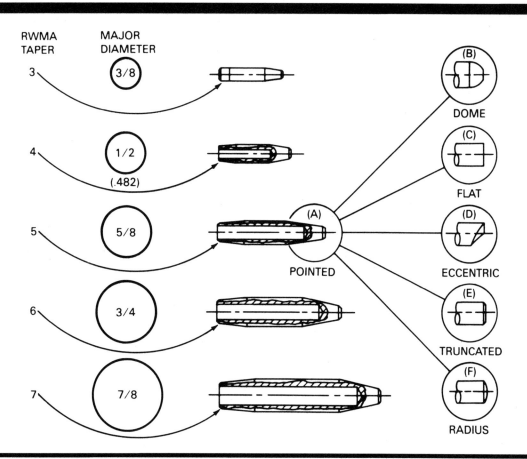

Figure 19.21–Standard RWMA Spot Welding Electrode Face and Taper Designs

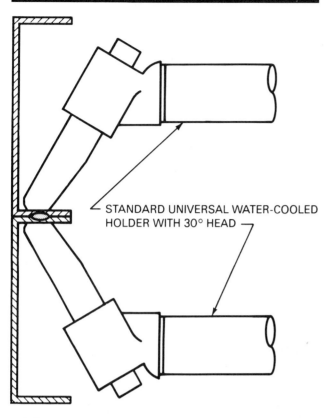

Figure 19.22—Special Spot Welding Electrodes with the Faces Angled at 30 Degrees

Attachment

THE METHOD OF attaching the shank end to the holder is usually one of three general types: tapered, threaded, or straight-shank.

RWMA tapered attachments use the Jarno taper as the standard. This taper offer the following advantages:

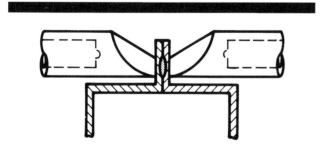

Figure 19.23—An Application of Type D Offset Spot Welding Electrodes

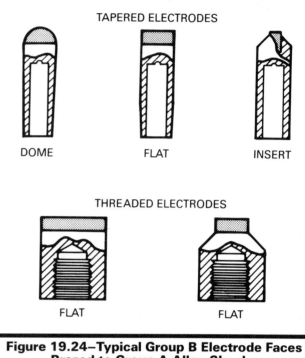

Figure 19.24—Typical Group B Electrode Faces Brazed to Group A Alloy Shanks

(1) The taper number multiplied by 1/8 in. gives the nominal major diameter. For example, RWMA No. 5 taper has 5/8 in. diameter.

(2) The taper numbers progress in sequence from 3 to 7.

(3) The RWMA taper is a uniform 0.600 in./ft for all sizes.

The electrode diameter and taper length increase as the taper number increases. Longer tapers can support higher electrode forces, but there is a maximum force that should be used with each electrode size. Recommended maximum electrode forces for the various sizes are given in Table 19.2.

Threaded attachments are used where high welding forces would make removal of tapered electrodes difficult, or where electrode position is critical. Typical threaded electrodes are shown in Figure 19.26.

Straight-shanked electrodes are used to transmit high welding forces, especially the 3/4 and 7/8 in. diameters. The base of the electrode bears against the holder socket. The water seal is an "O" ring in a recessed groove in the holder. The electrode is mechanically held in place by a coupling or collar, as shown in Figure 19.27.

Cooling

WHEREVER PRACTICAL, SPOT welding electrodes should have an internal cooling passage extending close to the welding face. This passage should be designed to accom-

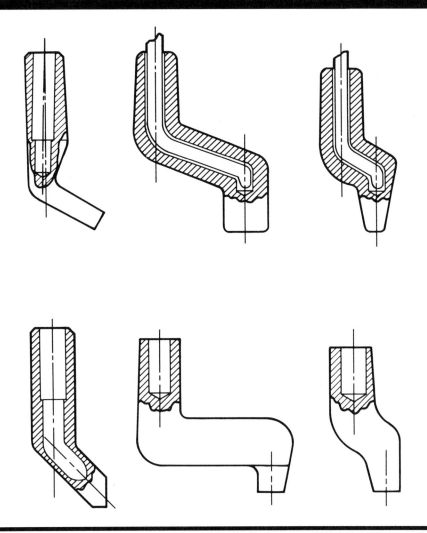

Figure 19.25–Typical Single and Double Bent Spot Welding Electrodes

modate a water inlet tube and to provide for water flow out around the tube. The tube should be positioned to direct the cooling water against the inside of the tip of the electrode. In most cases, the tube is a component of the electrode holder. An exception is the case of bent elec-

trodes. Where internal cooling is not practical, external cooling of the electrodes by immersion, flooding, or attached cooling coils should be considered.

Two-Piece Electrodes

TWO-PIECE OR CAP-AND-ADAPTOR electrodes are available with both male and female caps, as shown in Figure 19.28. They are available with straight and bent shanks. Use of this electrode design is a matter of economics. Tip maintenance costs may be lower because only the cap needs to be replaced when worn down. On the other hand, the resistance of the cap-to-adaptor interface may contribute to electrode heating and wear. Their use should be evaluated for each application, compared to the one-piece design.

Table 19.2
Recommended Maximum Electrode Force for Standard Spot Welding Electrodes

Taper No.	Shank Diameter in.	Face Diameter In.	Maximum Electrode Force, lb
4	0.482	0.19	800
5	0.625	0.25	1500
6	0.750	0.28	2000
7	0.875	0.31	2400

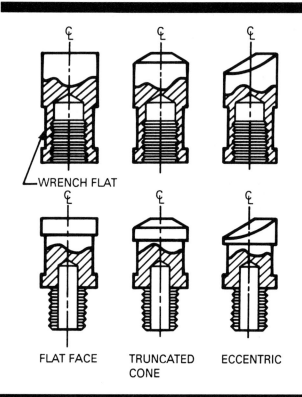

Figure 19.26—Typical Threaded Spot Welding Electrodes

WRENCH FLAT

FLAT FACE   TRUNCATED CONE   ECCENTRIC

Method of Manufacture

STRAIGHT ELECTRODES ARE machined from cold worked rods. Bent electrodes may be produced by cold forming of straight electrodes, by forging, or by casting. Forging or casting is normally used where the required shape cannot

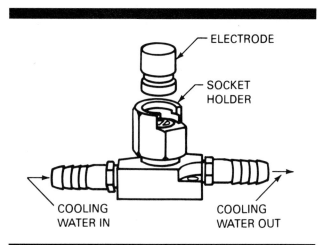

ELECTRODE

SOCKET HOLDER

COOLING WATER IN   COOLING WATER OUT

Figure 19.27—Straight Shanked Electrode with Socket Holder

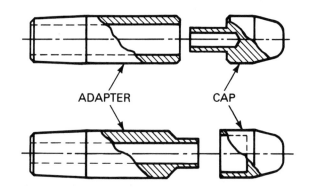

ADAPTER   CAP

Figure 19.28—Male and Female Designs of Two-Piece Spot Welding Electrodes

be produced by cold forming. Most bent electrodes are cold formed because they have distinct advantages over the others, including the following:

(1) The physical and mechanical properties of cold-drawn rod
(2) Placement of a water tube in the cooling hole prior to forming
(3) Lower manufacturing costs

Maintenance

A SPOT WELDING electrode has a specific face area in contact with the work. In use, this area will grow by mushrooming and the current and pressure densities will decrease at the same time. As a result, the weld will become smaller. In addition, the electrodes tend to pick up metal from the parts being welded. A small amount of pick up may not be harmful, but a considerable amount will cause the electrodes to overheat and mushroom faster.

It is not possible to predict how many welds can be made with a given setup before redressing of the electrodes is necessary. A periodic check of the weld quality as well as the electrode shape will help in determining the number of welds or assemblies that can be made before redressing. Then a schedule of electrode redressing should be set up as preventive maintenance to maintain weld quality.

A minor amount of redressing of electrodes in the machine is permissible using a plastic or metal paddle contoured on both sides to match the electrode face contour. The paddle is wrapped with fine abrasive cloth. The electrodes are brought against the abrasive cloth under a light load. The paddle is then rotated to redress the electrode faces.

Where a major amount of redressing is necessary, the electrode should be removed from the machine and refaced on a lathe. Alternatively, major redressing of the electrode may be done in the machine with a manual or power-operated dressing tool.

A file should never be used for redressing electrodes in the machine because the resulting electrode faces may be irregular in size and contour. Poorly dressed electrodes will reduce the quality of weld.

The following suggestions may be helpful in correctly using spot welding electrodes:

(**1**) Use standard electrodes and holders wherever possible

(**2**) Use the proper electrode material recommended for the application

(**3**) Use adequate water cooling and circulate it in the correct direction in the electrodes

(**4**) Align the electrodes properly; electrodes should not skid against the parts or be out of alignment when they are in contact with the parts

(**5**) Use only rawhide or rubber mallets for tapping electrodes into position and only ejector type holders or the proper tools for removing electrodes from the machine

(**6**) See that the machine is set up properly; the electrodes must contact the parts with minimum impact before current flows, and must remain in contact until termination of the current.

Specifications and Identification

SPOT WELDING ELECTRODES are covered by two standards:

(**1**) ANSI/RWMA Bulletin No. 16, Resistance Welding Standards, published by the Resistance Welder Manufacturer Association.

(**2**) AWS D8.6/SAE HS-J1156, Standard for Automotive Resistance Spot Welding Electrodes, published by the American Welding Society and the Society of Automotive Engineers.

These standards provide a code system for the various standard electrode designs. The code identifies the nose style, alloy class, shank size, and length. Methods are also given to identify bent electrode shapes, special-faced electrodes, and cap electrodes.

Straight electrodes are identified by a letter followed by four numbers with the following meanings:

(**1**) The letter indicates the nose style, as shown in Figure 19.22.

(**2**) The first digit indicates the Group A alloy class, as shown in Table 19.1.

(**3**) The second digit indicates the taper.

(**4**) The third and fourth digits indicate the overall length in 0.25-in. units.

For single bent electrodes, two digits are placed ahead of the letter to indicate the bend angle in degrees. For single

and double bent electrodes, two additional digits are added to indicate the offset distance in 0.062-in. units.

ELECTRODE HOLDERS

ELECTRODES ARE MOUNTED on a spot welding machine by means of electrode holders. Various holder designs permit positioning the electrodes properly with respect to the work. The holders are clamped to the arms of the welding machine. Most of them have provisions for conducting cooling water to the electrodes, and some have an ejector mechanism for easy removal of the electrodes.

There are three fundamental holder designs: straight, offset, and universal or adjustable offset. These three basic types are available in standard sizes and designs for use with standard spot welding electrodes. Similar design principles are generally employed for special holders, with or without adaptors, for use with a great variety of special or standard electrodes.

The three types of standard holders are available as nonejector and ejector types. Straight electrode holders of both types are shown in Figure 19.29. With the ejector type, the electrode is removed by striking the ejector head or button with a hammer. With the nonejector type, the electrode taper is released by rotating the electrode with a wrench. Holders are available in different lengths and several diameters.

Offset and universal holders are produced with 90° and 30° heads, as shown in Figure 19.30. Low inertia holders which incorporate a spring for rapid follow-up are also available.

Multiple electrode holders are available for producing two or more spot welds simultaneously in parallel. These holders have spring, mechanical, or hydraulic force equalizing systems. The lower electrode may be a flat block that opposes all upper electrodes or individual electrodes mounted in a block. Since the welds are made by parallel circuits, the proper division of current to each weld will depend upon the relative resistances of the paths. The path of lowest impedance will conduct more current than the others, and weld size may vary with the current magnitude.

PROJECTION WELDING ELECTRODES

PROJECTION WELDING ELECTRODES must have flat surfaces that are larger than the projection diameter. It is common practice to use large, flat electrodes or rectangular bar stock.

Projection welding electrodes usually consist of an internally water-cooled holder with replaceable inserts at the projection locations. These inserts may be threaded electrodes or pieces of Group A or B electrode materials pressed or otherwise secured in the holder. An example of this design is shown in Figure 19.31.

Since the area of contact between each electrode and the adjacent part is larger than in spot welding, current and

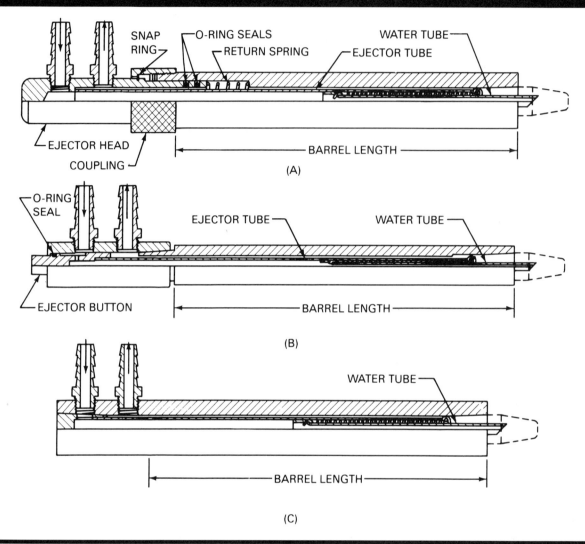

Figure 19.29–Typical Straight Spot Welding Electrode Holders: (A) and (B) Ejector Types, (C) Nonejector Type

pressure densities are lower. Therefore, electrode deterioration from wear, deformation, or pickup is not nearly as rapid as with spot welding. The electrodes do, however, eventually become pitted or deformed at the projection weld locations. When this deterioration interferes with proper electrode contact or weld quality, the electrodes or inserts must be redressed or replaced.

Selecting the best combination of opposing electrode materials for good heat balance will minimize deterioration. Regular cleaning of the electrodes to remove grease, dirt, flash, or other contamination will also prolong electrode life.

Multiple projection welding electrodes can be designed to compensate automatically for height variations or wear. Such equalizing electrodes generally employ some hydraulic or mechanical method to provide automatic floating or equalizing features.

SEAM WELDING ELECTRODES

SEAM WELDING ELECTRODES are wheels or disks. The five basic considerations are face contour, width, diameter, cooling, and method of mounting. The diameter and width of the wheel are usually dictated by the thickness, size, and shape of the parts. The face contour depends upon the requirements for current and pressure distribution in the weld nugget and the type of drive mechanism. The four basic face contours in common use are flat, single-bevel, double-bevel, and radius, as shown in Figure 19.32.

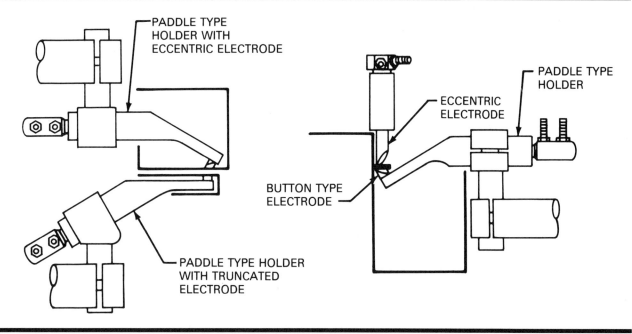

Figure 19.30–Various Combinations of Electrodes and Holders

The electrodes are usually cooled by either flooding or directing jets of water on both the electrodes and the work from top and bottom. Where those methods of cooling are unsatisfactory, the electrodes and shafts should be designed for internal cooling.

Cooling by simple flooding alone is not always adequate. A steam pocket may develop at the point where the electrode meets the work, which will keep cooling water from the immediate area. When flood cooling is unsuitable, water mist or vapor cooling may be effective.

A seam welding electrode is attached to the shaft with a sufficient number of bolts or studs to withstand the driving torque. The contact area with the shaft must be great enough to transmit the welding current with minimum heat generation.

Peripheral drive mechanisms, such as knurl or friction drives running against the electrode, require adequate work clearance. A knurl drive will mark the electrode face, which in turn will mar the surface of the weld. However, a knurl drive wheel tends to clean surface pickup from the electrode face.

Although the work and the drive method may require flat-faced electrodes with or without beveled edges, those electrodes are more difficult to set up, control, and maintain than radius-faced electrodes. In addition, radius faces give the best weld appearance.

Seam welding electrodes, like spot welding electrodes, have a predetermined area of contact with the parts which must be held within limits if consistent weld quality is to be maintained. Only minor dressing or touch up with light

abrasives should be attempted with the electrode in the machine. Wheel dressers may be used for continuous electrode maintenance. Machining in a lathe is the preferred method of redressing an electrode to its original shape.

Precautions must be taken to prevent foreign materials from becoming embedded in the electrode wheel or work. Rough faces do not improve traction. Welding should be stopped while electrodes are still on the work.

FLASH AND UPSET WELDING ELECTRODES

FLASH AND UPSET welding electrodes are not usually in direct contact with the weld area as are spot and seam welding electrodes. They function as work-holding and current-carrying clamps, and are often referred to as such. They are normally designed to contact a large area of the workpiece, and the current density in the contact area is relatively low. Accordingly, relatively hard electrode materials with low conductivity can give satisfactory performance.

Since the electrodes must conform to the parts to be welded, there are no standard designs. Two important requirements are that the materials have sufficient conductivity to carry the current without overheating, and that the electrodes be rigid enough to maintain work alignment and minimize deflection.

The electrodes are mechanically fastened to the welding machine platen. They can be solid, one-piece construction of one of the RWMA Group A electrode materials in Classes 1 through 5. Service life can sometimes be in-

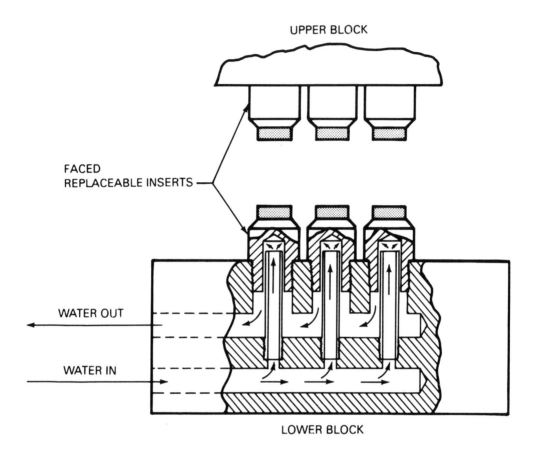

Figure 19.31–Typical Projection Welding Multiple Electrode Construction

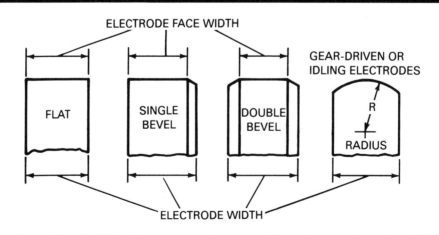

Figure 19.32–Seam Welding Wheel Face Contours

creased by using Class 2, 3, and 5 material with replaceable inserts of Class 3, 4, or one of the Group B materials at the wear points.

A varying amount of wear inevitably occurs, and this may result in decreased contact area and localized burning of the work. For good service, the electrodes should be kept cool, clean, and free of dirt, grease, flash, and other foreign particles. An antispatter compound may help prevent flash adherence. All fasteners and holding devices should be tight and properly adjusted, and their gripping surfaces should be properly maintained to avoid work slipping during welding.

POWER SUPPLY

POWER DEMAND FROM the line depends upon the welding method and the design of the welding machine. An adequate power source is one of the prerequisites for high-production resistance welding. A major part of the power source system for an industrial plant is within the plant itself. That part consists of the power source transformers and conductors.

POWER SOURCE TRANSFORMERS

IN CONSIDERING THE installation of a resistance welding machine, it is necessary to determine if the plant supply is adequate. This includes the kVA rating of the power source transformer and the size of the power source conductors. The power source transformer is connected to a 2300-, 4800-, 7500- or 13 000-volt primary feeder and produces 230 or 460 V power. It should not be confused with the welding transformer mounted in the welding machine. The power source conductors are the leads between the power source transformer and the welding machine.

The adequacy of the power source transformer and conductors is governed by two factors: the permissible voltage drop and the permissible heating. The permissible voltage drop is the determining factor in the majority of installations, but consideration must also be given to heating.

The size of the power source transformer for single-point welders should at least equal the value of the KVA demand during welding. Power source transformers have an impedance that is generally in the area of 5 percent. This means that at their KVA rating, the voltage drop on the secondary will be 5 percent. Further, the power conductors between the power source transformers and the welder will generally be sized to have no more than a 5 percent drop. This will add up to a 10 percent voltage drop at the welder, which is the maximum that most machinery manufacturers recommend for their products.

To determine the size of the power source transformer required to serve a welding machine on the basis of voltage drop, it is first necessary to determine the maximum permissible voltage drop specified by the machine manufacturer. Normally, it should not be greater than 5 percent. When the same power transformer is used with two or more machines, the voltage drop caused by one machine will be reflected in the operation of the second. Then it is advisable to confine the total voltage drop to not more than 10 percent for consistent weld quality. Voltage drop should be measured at the machine location. The percentage voltage drop is calculated by the following formula:

$$\text{Voltage drop, percent} = \frac{\text{(No-load voltage) - (Full-load voltage)}}{\text{No-load voltage}} \times 100$$

BUS OR FEEDER SYSTEM

IN GENERAL, THE bus or feeder from the transformer to the machines should always be as short as possible and of low reactance design to minimize the voltage drop in the line. The simplest and most economical power line consists of insulated wires taped together in a conduit. When only two or three machines are to be served at a common location, this construction is economical and effective. Bus duct construction that permits easy tap connections at frequent intervals along its length is desirable in production plants where manufacturing layouts are continually changing.

Systems are available to interlock two or more machines to prevent simultaneous firing and the accompanying excessive voltage drop. Any scheme of interlocking will cause some curtailment in production. However, this can be minimized with a voltage monitoring interlock, set to operate only when the voltage drops below a preset value.

Table 19.3
Equivalent Continuous Loading of Resistance Welding Machines

Type of Welding	Equivalent Continuous Load, Percent of Sum of Name-Plate Ratings
Spot, projection (single-impulse)	20
Spot, projection (multiple-impulse)	40
Flash, multipoint spot, or projection	20
Seam	70

INSTALLATION

RESISTANCE WELDING MACHINES should be connected to the power line according to electrical codes and the recommendations of the machine manufacturer. The size of the primary cable should be appropriate for both thermal and voltage drop considerations.

Because many control units contain phase-shift heat control, the control power source must be in phase with the welding power source. The control power source should be fused separately from the welding power source.

Enclosed fusible isolation switches are frequently used for the power or welding circuit. These switches seldom have adequate interrupting capacity for safe disconnection under load. For emergency disconnecting purposes, a cir-

cuit breaker should be used. The rating of the breaker in carrying capacity should be sufficient to carry the maximum demand of the machine when its welding circuit is shorted. That may be from two to four times the machine nameplate rating. One of the advantages of a circuit breaker is that a push button can be installed on the welding machine. In an emergency the operator can quickly open the circuit by hitting this button.

When fuses are used, their size should be that recommended by the machine manufacturer. Manufacturers normally provide wiring diagrams which include recommended fuse ratings. The fuses should function for any normal demand or operation of the machine. The purpose of fuses is to interrupt a short circuit in the electrical system.

SAFETY

RESISTANCE WELDING PROCESSES are widely used in high production operations, especially in the automobile and appliance industries. These processes include projection, spot, seam, flash, upset, and percussion welding in a wide range of machine types. The main hazards which may arise with the processes and equipment are as follows:

(1) Electric shock due to contact with high voltage terminals or components

(2) Eye injury or fires caused by ejection of small particles of molten metal from the weld

(3) Crushing of some part of the body between the electrodes or other moving components of the machine

(4) Welding fumes from the parts themselves or from oil, lubricant, or other material on the parts

MECHANICAL

Guarding

INITIATING DEVICES ON welding equipment, such as push buttons and switches, should be arranged or guarded so as to prevent the operator from inadvertently activating them.

In some multiple-gun welding machine installations, the operator's hands can be expected to pass under the point of operation. These machines should be effectively guarded by suitable devices such as proximity-sensing devices, latches, blocks, barriers, or dual hand controls.

All non-portable, single-ram welding machines should be equipped with one or a combination of the following:

(1) Machine guards or fixtures which prevent the operator's hands from passing under the point of operation

(2) Dual-hand controls, latches, proximity-sensing devices, or any similar mechanism which prevents operation of the ram while the operator's hands are under the point of operation

All chains, gears, operating linkages, and belts associated with the welding equipment should be protected in accordance with ANSI Standard B15.1, Safety Standard for Mechanical Power Transmission Apparatus (latest edition).

Static Safety Devices

ON PRESS TYPE, flash, and upset welding machines, static safety devices such as pins, blocks, or latches should be provided to prevent movement of the platen or head during maintenance or setup for welding. More than one device may be required, but each device should be capable of sustaining the load.

Portable Welding Machines

Support Systems. All suspended portable welding gun equipment, with the exception of the gun assembly, should have a support system that can withstand the total shock load in the event of failure of any component of the system. The system should be fail safe. The use of adequate devices such as cables, chains, or clamps is considered satisfactory.

Movable Arm. Guarding should be provided around the mounting and actuating mechanism of the movable arm of a welding gun if it can cause injury to the operator's hands. If suitable guarding cannot be achieved, two handles should be used. Each handle should have an operating switch connected in series so both handles must be actuated to energize the machine. These handles must be located at safe distances from any shear or pinch points on the gun.

Stop Buttons

ONE OR MORE emergency stop buttons should be provided on all welding machines, with a minimum of one at each operator position.

Guards

EYE PROTECTION AGAINST expelled metal particles must be provided by a guard of suitable fire-resistant material, or by the use of approved personal protective eye wear. The use of safety glasses with side shields is recommended in all work areas. The variations in resistance welding operations are such that each installation must be evaluated individually. For flash welding equipment, flash guards of suitable fire-resistant material must be provided to control flying sparks and molten metal.

ELECTRICAL

Voltage

ALL EXTERNAL WELD initiating control circuits should operate on low voltage. It should not be more than 120 V for stationary equipment and 36 V for portable equipment.

Capacitors

RESISTANCE WELDING EQUIPMENT and control panels containing capacitors involving high voltages must have adequate electrical insulation and be completely enclosed. All enclosure doors must be provided with suitable interlock switches, and the switch contacts must be wired into the control circuit.

The interlocks must effectively interrupt power and discharge all high voltage capacitors into a suitable resistive load when the door or panel is open. In addition, a manually operated switch or suitable positive device should be provided to assure complete discharge of all high voltage capacitors.

Locks and Interlocks

ALL DOORS, ACCESS panels, and control panels of resistance welding machines must be kept locked or interlocked. This is necessary to prevent access by unauthorized persons.

Grounding

THE WELDING TRANSFORMER secondary should be grounded by one of the following methods:

(**1**) Permanent grounding of the welding secondary circuit
(**2**) Connection of a grounding reactor across the secondary winding with a reactor tap to ground

As an alternative on stationary machines, an isolation contactor may be used to open all of the primary lines.

The grounding of one side of the secondary windings on multiple spot welding machines can cause undesirable transient currents to flow between transformers. This can happen when either multiphase primary supplies or different secondary voltages, or both, are used for the several guns. A similar condition may also exist with portable spot welding guns, when several units are used on the same fixture or assembly or on another that is nearby. Such situations require use of a grounding reactor or isolation contactor.

INSTALLATION

ALL EQUIPMENT SHOULD be installed in conformance with ANSI/NFPA No. 70, National Electric Code (latest edition). The equipment should be installed by qualified personnel under the direction of a competent technical supervisor. Prior to its production use, the equipment should be inspected by competent safety personnel to ensure that it is safe to operate.

Additional information on safe practices for resistance welding equipment may be found in ANSI Z49.1, Safety in Welding and Cutting (latest edition).

SUPPLEMENTARY READING LIST

Anon. "Railcar repair shop cuts costs with unique installation of welding equipment." *Welding Journal* 62(8): 51-55; August 1983.

Anon. "Resistance welding electrodes do their own part holding." *Welding Journal* 62(2): 43-47; February 1983.

Beemer, R. D. and Talbo, T. W. "Analyzer for non-destructive process control of resistance welding." *Welding Journal* 49(1): 9s-13s; January 1970.

Blair, R. H. and Blakeslee, R. C. "Half-wave and full-wave resistance welding power supplies." *Welding Journal* 50 (3): 174-6; March 1971.

Dilay, W. and Zulinski, E. "Evolution of the silicon-controlled rectifier for resistance welding." *Welding Journal* 51(8): 554-9; August 1972.

Johnson, K. I., ed., *Resistance Welding Control and Monitoring*. Cambridge, England: The Welding Institute, 1977.

Mollica, R. J. "Adaptive controls automate resistance welding." *Welding Design and Fabrication* 51(8): 70-72; August 1978.

Nadkarni, A. V. and Weber, E.P. "A new dimension in resistance welding electrode materials." *Welding Journal* 56(1): 331s-338s; November 1977.

Parker, F. "The logic of dc resistance welding." *Welding Design and Fabrication* 49(12): 55-58; December 1976.

Sherbondy, G. M. and Motto, J. W. Jr. "Current ratings of power semiconductors." *Welding Journal* 51(6): 393-400; June 1972.

Weber, E.P. et al. "The application of dispersion strengthened copper for resistance welding electrodes." *Welding Journal* 58(8): 34-40; August 1979.

HIGH-FREQUENCY WELDING

PREPARED BY A COMMITTEE CONSISTING OF:

H. N. Udall, Chairman
Thermatool Corp.

E. D. Oppenheimer
Consultant

W. C. Rudd
Consultant

WELDING HANDBOOK COMMITTEE MEMBER:
G. N. Fischer
Fischer Engineering Co.

CHAPTER 20

HIGH-FREQUENCY WELDING

INTRODUCTION

GENERAL DESCRIPTION

HIGH-FREQUENCY WELDING includes those processes in which the coalescence of metals is produced by the heat generated from the electrical resistance of the work to high frequency current, usually with the application of an upsetting force to produce a forged weld.

There are two processes that utilize high-frequency current to produce the heat for welding: *high frequency resistance welding* (HFRW), and *high frequency induction welding* (HFIW), sometimes called *induction resistance welding*. The heating of the work in the weld area and the resulting weld are essentially identical with both processes. With HFRW, the current is conducted into the work through electrical contacts that physically touch the work. With HFIW, the current is induced in the work by coupling with an external induction coil. There is no physical electrical contact with the work.

With low frequency (50 Hz – 360 Hz), direct current or "square wave" resistance welding, much higher currents are required to heat the metal and large electrical contacts must be placed very close to the desired weld area. The voltage drop across the weld is very low, and the current flows along the path of least resistance from one electrode to the other. With high-frequency welding, by contrast, the current is concentrated at the surface of the part. The location of this concentrated current path in the part can be controlled by the relative position of the surfaces to be welded and the location of the electrical contacts or induction coil. Heating to welding temperature can be accomplished with a much lower current than with low frequency or direct current resistance welding.

Although the welding process depends upon the heat generated by the resistance of the metal to high-frequency current, other factors must also be considered for successful high-frequency welding. Because the concentrated high-frequency current heats only a small volume of metal just where the weld is to take place, the process is extremely energy efficient, and welding speeds can be very high. Maximum speeds are normally limited by mechanical considerations of materials handling, forming and cutting. Minimum speeds are limited by material properties and weld quality requirements.

The fit of the surfaces to be joined and the manner in which they are brought together is important if high-quality joints are to be produced. Flux is not usually used but can be introduced to the weld area in an inert gas stream. Inert gas shielding of the welding area is generally needed only for joining reactive metals such as titanium and also for certain stainless steel products. Typical high-frequency welding applications are shown in Figure 20.1.

HISTORY

IN THE LATE 1940's and early 1950's, emphasis was on the development of procedures and equipment for high-frequency induction butt end welding of pipes and tubes. Successful welding operations were conducted using 10 kHz motor-generator power sources equipped with induction coils that could be opened for removal of the finished work. The first mobile installation for joining pipe in the field was utilized by a utility company in the streets of New York City. The unit produced welded butt joints in pipe up to 12 in. (305 mm) in diameter with wall thicknesses up to 5/16 in. (8 mm). With this process, pipe ends were pressed together and induction heated to forge welding temperature in approximately 60 seconds.

In 1949, the first high-frequency induction welding system for the continuous "forge welding" of the longitudinal seam in small steel tubes was developed. This used a 10 kHz motor generator with split return induction coil over the seam similar to those now used for continuous normalizing of the welded seam in pipe. This unit was successful

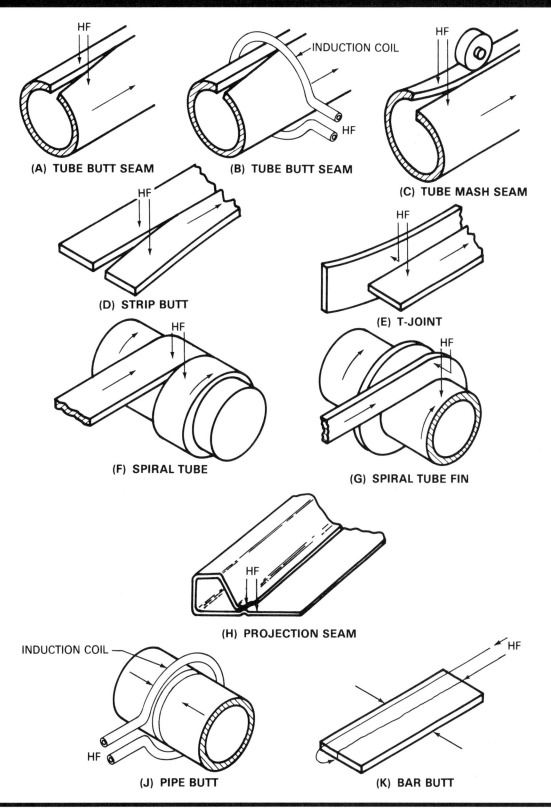

Figure 20.1–Basic High-Frequency Welding Applications

and operated for many years, but it had the disadvantage of heating a large portion of the tube. The unit was eventually replaced by a 400 kHz HFRW system.

In 1952, Thomas Crawford made tests on the continuous induction welding of the longitudinal seam of various metal tubes including some tubes with cables inside. The tests were made at a frequency of 400 kHz, and employed an induction coil surrounding the tube. Induction welding of tube and pipe has become very successful for smaller diameters up to about 6 in. (150 mm). It is now becoming more popular for larger diameters, up to 20 in. (500 mm) in some installations.

In 1952, W. C. Rudd and Robert Stanton invented a 400 kHz process for welding a large variety of continuous joints. The Rudd and Stanton process introduced the welding current directly into the work by means of sliding contacts. The sliding contacts permitted the production of butt, lap, and tee joints in pipe, strip, and structural products. Successful tests were made in 1954 on butt welding the longitudinal seam in aluminum tubing. This led to the first commercial installation in 1955, for the welding of aluminum irrigation tubing, and the mill operated for nearly twenty years.

Subsequently, over 3000 installations of various forms of the system have been installed around the world. A large variety of metals have been welded.

High-frequency welding is an automated process, and is not adaptable to manual welding.

ADVANTAGES AND LIMITATIONS

HIGH-FREQUENCY WELDING processes offer several advantages over low frequency and direct current resistance welding processes. One characteristic of the high-frequency processes is that they can produce welds with very narrow heat-affected zones. The high-frequency welding current tends to flow only near the surface of the metal because of the "skin effect" and along a narrow controlled path because of the "proximity effect". These effects are described in Principles of Operation later in the chapter. The heat for welding, therefore, is developed in a small volume of metal along the surfaces to be joined. A narrow heat-affected zone is generally desirable because it tends to give a stronger welded joint than with the wider zone produced by many other welding processes. With some alloys the narrow heat-affected zone and absence of cast structure may eliminate the need for postweld heat treatment to improve the metallurgical characteristics of the welded joint.

The shallow and narrow current flow path results in extremely high heating rates and therefore high welding speeds and low-power consumption. A major advantage of the continuous high-frequency welding process is its ability to weld at very high speeds. For instance, a 1/2 in. (12 mm) wall steel pipe with a diameter ranging from 8 to 48 in. (200 to 1200 mm) can be HFRW welded at speeds over 100 fpm (30 m/s) with a 600 kW output welder. Smaller tube diameters such as 1 to 2 in. (25 to 50 mm) with light walls in the range of .025 to 0.065 in. (0.6 to 1.7 mm) can be induction welded at speeds ranging from 200 to 800 fpm (60 to 240 m/min.) using welders with 100 kW to 400 kW output.

High-frequency welding can also be used to weld very thin wall tubes. Wall thicknesses down to less than 0.005 in. (0.13 mm) are presently being welded on continuous production mills. The process is equally adaptable to large diameter pipes with wall thicknesses up to 1.0 in. (25 mm).

The process is adaptable to many metals including low carbon and alloy steels, ferritic and austenitic stainless steels, and many aluminum, copper, titanium, and nickel alloys.

Because the time at welding temperature is very short and the heat is localized, oxidation and discoloration of the metal as well as distortion of the part are minimal. Materials that would normally be damaged from prolonged exposure to heat can be welded with high-frequency welding processes. For example, electrical cable sheathing with the cable inside is welded with HFIW.

High-frequency welding power sources have a balanced three-phase input power system. Using conventional vacuum tube welding power sources, as much as 60 percent of the energy is converted into useful heat in the work. Solid-state high-frequency power sources, which have even higher efficiencies, are now becoming available.

As with any process, there are also limitations. The equipment operates in the radio frequency range, and therefore special care must be taken in its installation, operation, and maintenance to avoid radiation interference in the plant's vicinity.

As a general rule, the minimum speed in carbon steel is about 25 fpm (0.125 m/s). For products which are only required in small quantities, the process may be uneconomical unless the technical advantages justify the application.

Because the process utilizes localized heating in the joint area, proper fit-up is important. Equipment is usually incorporated into mill or line operation and must be fully automated. The process is limited to the use of coil, flat, or tubular stock with a constant joint symmetry throughout the length of the part. Any disruption in the current path or change in the shape of the vee can cause significant problems. Also, special precautions must be taken to protect the operators and plant personnel from the hazards of high-frequency current.

FUNDAMENTALS OF THE PROCESS

HIGH-FREQUENCY CURRENT in metal conductors tends to flow at the surface of the material at relatively shallow depth. This is commonly called the *skin effect*. The effect of frequency on the depth of penetration for several metals is shown in Figure 20.2. Penetration is also a function of temperature as indicated in Figure 20.2. For example, the depth of current penetration in steel at 1470°F (800°C) is about 0.03 in. (0.8 mm) at 450 kHz and nearly 0.22 in.

(5.5 mm) at 10 kHz. At room temperature the penetration in steel is about 0.002 and 0.010 in. (0.05 and 0.25 mm) at 450 and 10 kHz respectively.

The high-frequency current path at the surface of the workpiece is controlled by the nearness of its own return flow path. This phenomenon, called *proximity effect*, is illustrated in Figure 20.3.

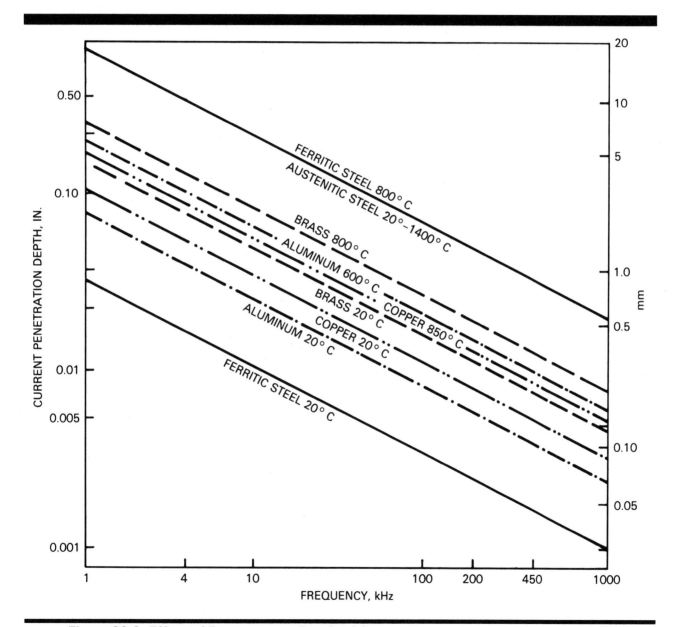

Figure 20.2—Effect of Frequency on Depth of Current Penetration Into Various Metals at Selected Temperatures

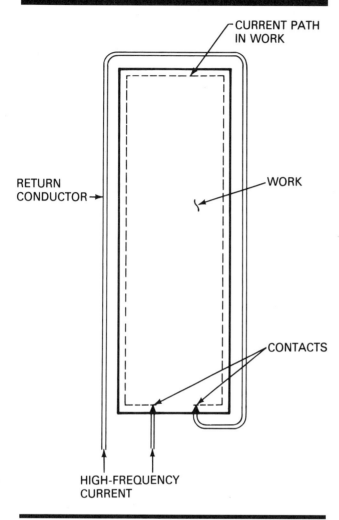

Figure 20.3–Restriction of the Flow Path of High-Frequency Current by the Proximity Effect of the Return Conductor

Both the skin effect and proximity effect become more pronounced with increasing frequency. Therefore, the effective resistance of the current path in the work increases as the frequency increases because as frequency increases the current is confined to a shallower and narrower path. This current concentration is advantageous because extremely high heating rates and high temperatures can be achieved in a localized area. Moreover, these high temperature concentrations can be positioned where they are needed at the surfaces to be welded.

Current patterns (high-temperature concentrations) in the work piece at frequencies of 60 Hz and 10 000 Hz are illustrated in Figure 20.4. In Figure 20.4(A), the return conductor is positioned parallel and close to the plate. The 60 Hz current flowing in the steel plate travels in opposite phase to the current in the adjacent proximity conductor.

In this case, the size and shape of the proximity conductor have a negligible effect on the distribution of the current in the steel plate. As a result, the current flows fairly uniformly throughout the plate cross section.

When a 10 000 Hz current is applied to the same system, as shown in Figure 20.4(B), the current in the work is confined to a relatively narrow band immediately beneath the proximity conductor. This narrow band is the path of lowest inductive reactance to the current in the plate. The shape and magnetic surroundings of the proximity conductor have considerable effect on the distribution of the current in the work, but have no effect on the depth of current penetration.

Two round proximity conductors at different distances from the work are illustrated in Figure 20.4(C). The closer

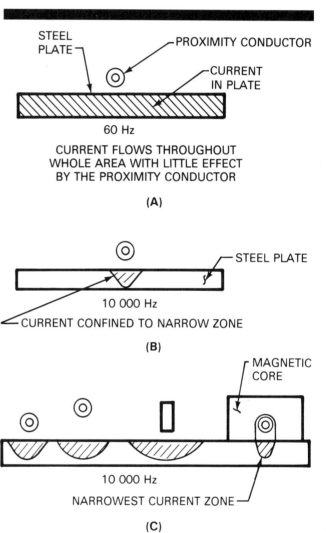

Figure 20.4–Current Depth and Distribution Adjacent to Various Proximity Conductors

proximity conductor develops a more confined current path. A rectangular proximity conductor with the narrow edge at the same distance from the work as the close round conductor exhibits a broader current distribution in the workpiece. If a magnetic core is placed around the proximity conductor, the current is further confined and heating takes place directly beneath the proximity conductor as shown.

If the two conductors with currents flowing in opposite directions are sheets placed edge to edge in a plane with a small gap between them, the proximity effect will cause the two adjacent edges to heat. The skin effect will confine the current to a shallow depth at those edges. This is the situation which occurs during the butt welding of the longitudinal seam of tube and pipe using either the HFIW or HFRW process.

Almost all high-frequency welding techniques employ some force to bring the heated metals into intimate contact during coalescence. During the application of force, an upset or flash occurs in the weld area. In many cases, the flash is removed after welding.

PROCESS VARIATIONS

HIGH-FREQUENCY INDUCTION WELDING

Tube Seam Welding

HIGH-FREQUENCY INDUCTION welding is generally used to weld continuous seam tube and pipe. The tube is formed from metal strip in a continuous roll-forming mill and enters the weld area with the edges to be welded slightly separated. In the weld area, they converge in a "vee" until they touch at what is referred to as the *weld point*. An induction coil, typically made of copper tubing, or copper sheet with attached cooling tubes, encircles the tube slightly ahead of the weld point. A current is induced by this induction coil which flows both around the tube immediately underneath the coil and along the tube edges between the induction coil and the weld point. This is illustrated in Figure 20.5.

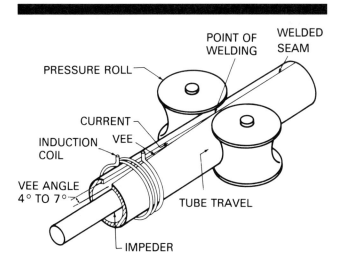

Figure 20.5–Joining a Tube Seam by High-Frequency Induction Welding

The high-frequency current follows in a localized path down one side of the vee consisting of the two edges being welded and back along the other due to the skin and proximity effects. The edges are resistance heated to a shallow depth. Welding speed and power level are adjusted so that the two edges are at welding temperature when they reach the weld point. At that point, pressure rolls forge the hot edges together and upset them to produce a weld. Hot metal containing impurities from the faying surfaces of the joint is squeezed out in both directions. The upset metal is normally trimmed off flush with the base metal. In the case of very thin wall tubes where the wall thickness is of the order of two times the depth of current penetration or less, a magnetic core called an *impeder* is required inside the tube. An impeder will limit the current flowing around the inside circumference of the tube which detracts from the current available to heat the metal at the vee and reduces the efficiency of the process.

The impeder is normally made of a ferrite material, and it must be cooled to prevent its temperature rising to the curie point where it becomes nonmagnetic. As the pipe or tube gets larger, the losses around the outside circumference of the tube become larger relative to the heating in the vee, making the process progressively less efficient. Therefore, induction welding is more efficient for smaller sizes of tube and pipe, and high frequency resistance welding becomes more competitive for larger sizes.

Butt Welding of Hollow Pieces

HFIW OF INDIVIDUAL pieces can be done only when the induced current can circulate in a closed circuit path. A typical application is the welding of butt joints between sections of pipe or tubing. A narrow induction coil is placed around the joint. High-frequency current in the coil induces a circulating current concentrated in the area of the butted pipe joint which is heated very rapidly. When the metal reaches welding temperature, it is upset to pro-

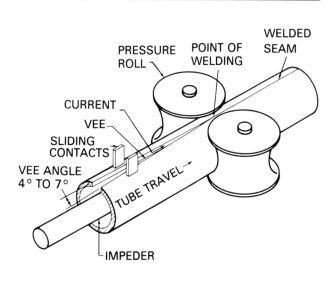

Figure 20.6–Joining a Tube Seam by High-Frequency Resistance Welding

duce a forge weld. The placement of the coil in this application is shown in Figure 20.1(J). This process is used for welding high-pressure boiler tube with diameters of from about 1 to 3 in. (25 to 75 mm) and with wall thicknesses up to about 0.375 in. (10 mm). It has also been used for joining pipe up to about 12 in. (300 mm) in diameter in the field. Welding times range from 10 to 60 seconds per joint.

HIGH-FREQUENCY RESISTANCE WELDING

Continuous Seam Welding

THIS APPLICATION IS generally similar to high-frequency induction welding, particularly when welding tube or pipe as shown in Figure 20.6. In this case, however, the high frequency current is introduced into the work with a pair of small sliding contacts placed on either side of the seam to be welded ahead of the weld point. The welding current travels directly from one contact along one edge of the welding vee to the weld point and back along the opposite edge to the other contact. The edges are forced together by weld pressure rolls at the weld point. When welding large diameter tube there is virtually no current flowing around the outside circumference of the tube and, therefore, the efficiency of the process does not decrease with increasing tube diameter. A current can flow around the inside circumference of the tube and, for smaller diameter tubes, an impeder is used to minimize this current flow.

Continuous HFRW seam welding is not confined to the welding of closed shapes such as tube and pipe. It can be

adapted to many other products. Some of these are illustrated in Figure 20.1(D), (E), (F), and (H). High-frequency welded beams are produced from low carbon or high-strength low alloy materials. The high-strength low alloy steel beams are used in commercial vehicle and trailer frames where their high-strength-to-weight ratio is particularly valuable. Nonferrous beams have been produced for special applications. High frequency welded finned tubes are widely used in heat exchange applications.

Finite Length Welding

TECHNIQUES ARE AVAILABLE for butt welding the ends of two strips together. This is done by passing a high-frequency current through the joint area. The current is introduced at each end of the joint by small contacts and is confined to the area of the joint by a proximity conductor. Generally, a magnetic core is used to assist in narrowing the current path, as shown in Figure 20.4 (C). By selection of the proper frequency, the depth of penetration of the current can be adjusted to heat the joint through its thickness. When the joint reaches welding temperature, forging force is applied, and the hot metal is upset. Joints are made in this fashion at rates as high as 1000 joints per hour. The use of a proximity conductor in the HFRW of strip butt joints is illustrated in Figure 20.7.

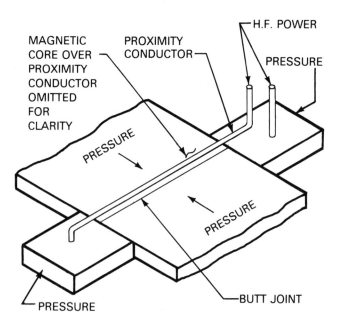

Figure 20.7–Joining Strips Together Using High-Frequency Welding

EQUIPMENT

POWER SOURCES

EXCEPT FOR A few special applications, such as tube and end and strip butt welding, vacuum tube oscillators with output power ranging from 50 to 1200 kW at frequencies ranging from 200 to 500 kHz are used for high-frequency welding. A basic circuit for a typical high frequency oscillator is shown in Figure 20.8.

Units can be manufactured for special line voltages, but typical input voltages are either 460 V, 60 Hz, 3-phase, or 380 V, 50 Hz, 3-phase. After the incoming circuit breaker and contactor, there is a 3-phase SCR (thyristor) voltage regulator. The regulator is designed both to maintain a preset voltage when the line voltage varies and also to enable the preset voltage to be controlled either automatically or by the welder operator. In automatic operation, control is dependent upon variables such as the weld temperature or the mill speed.

The plate transformer converts this controlled voltage to a high voltage which is then rectified to provide the direct current required for the oscillator circuit. The filter choke and filter capacitor reduce the ripple in the dc to an acceptable level, typically less than one percent. The oscillator circuit converts the direct current to high-frequency alternating current for the output transformer. The output transformer converts the high voltage-low current power to the low voltage-high current power required for welding.

Vacuum tube oscillators inherently have high output impedance (high voltage-low current) and must be fed into high impedance loads. The inductors and the workpiece contact circuits in high-frequency welding are low impedance (low voltage-high current) loads. An impedance matching output transformer is required to transfer energy efficiently from the oscillator to the work. Power is transferred most efficiently from the high-frequency generator to the work when the impedance of the work circuit matches the impedance of the generator. For induction welding, variable impedance matching transformers are often used in which the primary winding can be moved relative to the secondary winding to match the relatively wide impedance range typical of induction applications.

The secondary winding of the impedance matching transformer is in series with the induction coil or the contacts and workpiece. This forms the low voltage-high current welding system. The connecting leads should have the lowest possible impedance to obtain high efficiency and to minimize the voltage drop in the leads. This may be achieved by using short, wide leads made of flat copper plate separated by approximately 1/16 in. (1.6 mm) of insulation. Power losses in poorly designed leads or incorrectly matched transformers can seriously degrade the performance of a high-frequency welding system.

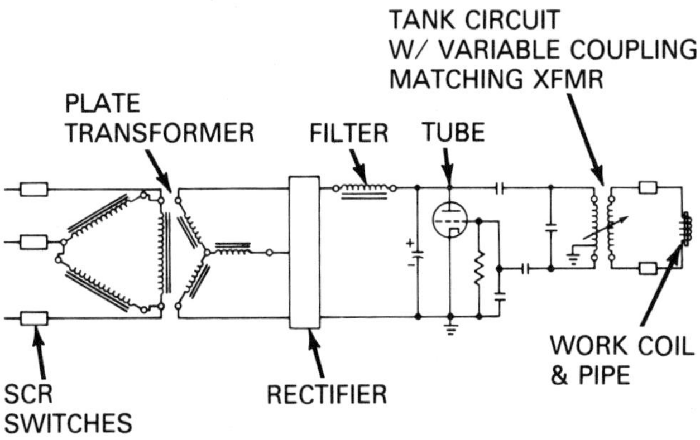

SOLID-STATE WELDING POWER SOURCES

FULLY TRANSISTORIZED SOLID-STATE welding power sources have now been developed which are expected to displace vacuum tube units in the future. A number of units are in operation under production conditions for tube welding applications.

The efficiency of a typical vacuum tube unit is between 50 and 65 percent depending on its age, design, and operating conditions. Solid-state power sources are smaller in size and have already demonstrated efficiencies of over 80 percent. Higher efficiencies can be expected as this technology develops. Economies result from a significant decrease in power consumption and a reduction in cooling water consumption. In addition, the incoming wiring and switchgear required for these units is smaller. The quality of welds made by these units has been found to be comparable to those made by vacuum tube units.

INDUCTION COILS

AN INDUCTION COIL, also called an *inductor,* is generally fabricated of copper tubing, copper bar, or copper sheet. It is normally water cooled. The highest efficiency is obtained when the induction coil completely surrounds the workpiece. The coil may have one or more turns as required by the application. The strength of the magnetic field which induces the heating current in the workpiece diminishes rapidly as the distance between the coil and the workpiece is increased. Typical spacing between coil and workpiece ranges from about 1/8 in. (3 mm) for small di-

Figure 20.8–Schematic Circuit for a Typical High-Frequency Oscillator

Figure 20.9–Typical Inductor Coils

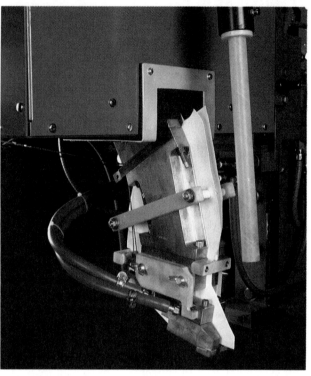

Figure 20.10–Typical Welding Contact Assemblies

ameter products up to 1 in. (25 mm) for large diameters. Some typical induction coils are shown in Figure 20.9.

CONTACTS

THE HIGH-FREQUENCY current transfer contacts are usually made of a copper alloy or of hard metallic or ceramic particles in a copper or silver matrix. The contacts are silver brazed to heavy water-cooled copper mounts. Replacements can be made by exchanging the mount and the contact tip assembly. Contact tip area ranges from 0.25 to 1 in.^{2} (160 to 650 mm^2) depending upon the current to be carried. Welding currents are usually in the range of 500 to 5000 A. Consequently, both internal and external cooling for the contact tip and mount is required.

The force of the contact tip against the work is usually in the range of 5 to 50 pounds (20 to 220 N) for a continuous welding system. It depends upon the contact size, the surface condition of the part, the contact material being used, and the current required. Welding current is determined by the thickness of the part being joined and the welding speed. Contact life is dependent upon a number of factors including the contact material, the contact pressure, the material being joined and the welding current. Contact life can be as low as 1000 ft (300 m) under very severe conditions of high current and poor workpiece surface condition such as heavy wall steel pipe, to over 300 000 ft (90 000 m) for light wall nonferrous materials. A typical contact system showing the flexible adaptors with pneumatic pressure regulation is shown in Figure 20.10.

IMPEDERS

WHEN WELDING TUBE and pipe with both the HFIW and HFRW processes, current can flow on the inside surface of the tube as well as on the outside surface. This current flows in parallel with the welding current and results in a substantial power loss at the joint edges. Because the lost power does not heat the joint edges, the weld temperature cannot be reached unless the welding speed is reduced or the power increased. To minimize this loss, an impeder is placed inside the tube in the weld area. The impeder increases the inductive reactance of the current path around the inside wall of the tube. The higher inductive reactance reduces the undesirable inside surface current. Thus, higher welding speeds are attainable for a given power input.

Impeders are typically made of one or more ferrite bodies and are usually cooled with water or mill coolant to keep their operating temperature below the Curie point where they lose their magnetic properties. Impeders are particularly important when a mandrel must run through the tube in the weld area in order to perform an inside weld bead treatment such as inside scarfing or bead rolling. Without impeders, such a mandrel, even though it should always be made of a nonmagnetic material such as austenitic stainless steel, reduces the inductive reactance of the

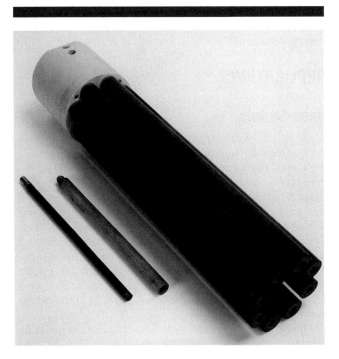

Figure 20.11–Typical Impeders

current path around the inside wall of the tube. The mandrel also reduces efficiency because it is induction heated due to the voltage induced in it by this same inside current. Impeders must therefore be placed on top of the mandrel immediately below the weld vee area or preferably completely surrounding the mandrel.

The impeder should extend from a point at or slightly upstream of the apex of the welding vee to a point at least 1-1/2 tube diameters upstream of the welding contact or the upstream edge of the induction coil in order to obtain the maximum beneficial effect. Some typical impeder arrangements are shown in Figure 20.11. Impeders are generally not needed when welding large diameter tube and pipe with the HFRW process.

CONTROL DEVICES

Input Voltage Regulators

HIGH-SPEED, HIGH-FREQUENCY seam welding requires accurate control of the weld power level. Short-term transient power fluctuations can result in intermittent weld imperfections, and a long-term drift in the power level will lead to less than optimum welding conditions. Thus, it is important that the power be automatically and continuously regulated. Today, virtually all high-frequency generators use silicon controlled rectifiers (thyristors) for power regulation. Their control circuits are designed to continuously correct for changes in input voltage, to regulate the

output power, and to rapidly turn off the power in response to overload or fault conditions. For continuous operation, it is essential that the welding power be essentially free from line frequency ripple, particularly when welding thin wall nonferrous metals at high speed. Excessive fluctuations in power can cause intermittent lack of fusion along the weld seam called *stitching*. Filters are used on the rectifier output to reduce the dc voltage ripple to one percent or less.

Speed Power Control

IN ORDER TO maintain proper welding conditions at different mill speeds and especially to minimize scrap when the mill is started and stopped, the weld power can be automatically adjusted as a function of mill speed. This system is most effective when welding low carbon steel tube, and can virtually eliminate any unwelded seam when the mill is stopped and restarted. The system will also reduce scrap when welding stainless and alloy steels and nonferrous materials, but typically a small unwelded length of product results when stopping and starting a mill running these materials.

Weld Temperature Control

VARIATIONS IN WELD temperature can be caused by variations in strip thickness and welding speed as well as deterioration of the impeder. These variations can be minimized by the addition of a weld temperature control. This reads the output of an optical pyrometer aimed at the weld vee and automatically adjusts the weld power to maintain a constant preset value.

MECHANICAL EQUIPMENT

MECHANICAL EQUIPMENT IS required in both the continuous and finite length high-frequency welding processes. The weld edges must be precisely aligned mechanically, and this alignment must be maintained as the upset pressure is applied to consummate the weld. The condition of the edges to be welded is also important. A mill-slit edge is normally satisfactory, if the edges are not damaged during shipment from the slitting mill. For both precision thin wall tubing and high-quality heavy wall tubing, the strip edges should be trimmed on the welding line. The edges can be trimmed with a stationary cutting tool or with a milling cutter. When welding large diameter pipe using single strand rolled to width material, the edges are often slit off in line immediately ahead of the forming section of the pipe mill.

In continuous seam welding, the edges to be joined are brought together to form a vee. The included angle of the vee is about four to seven degrees. If the vee angle is too small, arcing across the vee may occur, and it is difficult to maintain the vee apex at a constant location. If the vee is too wide, mechanical and thermal control of the edges may be lost and they may tend to buckle. The optimum vee angle depends on the tooling design and the base metal. Variations in the angle and length of the vee will cause variations in the weld quality.

The edges of the vee should be parallel to each other in the plane perpendicular to the weld travel. If the edges are closer on the inside as they approach the apex, then the inside edges will draw more current due to the proximity affect and become overheated relative to the outside edges. This will cause excessive upset inside the pipe or insufficient upset at the outside, or both, leading to difficulty in removing the inside upset and possible weld defects at the outside.

Both the outside and the inside flash can be removed using single point tools arranged closely behind the weld point. The inside flash may be left as welded or rolled smooth for some products.

ACCESSORY EQUIPMENT

FOR CERTAIN BASE metals, such as medium and high carbon or alloy steels, the products are postweld heat-treated. Low carbon steels may be annealed or stress relieved to restore ductility which was reduced during forming, welding, and sizing.

In most cases, only the weld zone is heat-treated. This is called *seam annealing*, but generally the seam is normalized and not fully annealed. Seam annealing is performed in line by induction heating immediately following the upset removal operation. A special linear inductor is used to seam anneal at a frequency of 1 to 3 kHz depending on the wall thickness of the pipe. Low frequencies are more efficient on wall thicknesses over 3/8 in. (10 mm), but they are extremely noisy. For this reason, tubes less than 1/2 in. (13 mm) thick are often seam annealed at frequencies between 2.5 and 3 kHz.

Some applications require the complete tube to be heat-treated. This heat treatment may be performed in line after welding and sizing using induction heating. The induction heating frequency used depends on the base metal, diameter, wall thickness, and the required temperature. Medium frequencies between 1 kHz to 10 kHz are usually used although higher frequencies may be needed for small diameter tubing. An inert atmosphere may be provided during heating and cooling to prevent surface oxidation. Tubes may also be heat-treated off-line.

CONSUMABLES

AS A GENERAL rule, there are no true consumables used in HFIW, and only the welding contacts are consumed in HFRW. As mentioned earlier, the contacts must be replaced from time to time. Typical contact life is between 1000 ft (300 m) to 300 000 ft (91 km) of tube or pipe. Filler metal is not used in any high-frequency welding process currently in production.

A flux or inert gas may be used when welding titanium, some grades of stainless steel, or brass tubing, but these are special situations.

APPLICATIONS

Base Metals

ALMOST ALL ENGINEERING metals and alloys can and have been joined by high-frequency welding. The exceptions are metals that cannot be hot formed, are unstable at elevated temperatures, or have unsatisfactory properties that cannot be restored by mechanical or thermal postweld treatments. Such a material would be cast iron. Reactive metals can be protected with an inert atmosphere. Inert shielding may be unnecessary because the weld cycle time is very short. High-conductivity metals such as pure copper are satisfactorily welded, and dissimilar metals can be readily joined. However, in dissimilar metals, the weld temperature is limited to the lower melting point of the two metals.

Metals with large nonmetallic inclusions, with large grain size, and with damaged or contaminated faying surfaces can be difficult or impossible to weld satisfactorily. Uniformity of dimensions, strength, and electromagnetic and thermophysical properties are desirable for consistent high quality welding.

JOINT DESIGN

THE FAYING SURFACES of butt, tee, and lap joints should be parallel in the plane perpendicular to the weld travel. This will assure uniform surface heating in the weld vee.

Unequal heating of the faying surfaces of tee welds is unavoidable. The thinner member will typically be at a higher temperature.

The direction of the upset should be perpendicular to the faying surfaces. Shearing forces during upset usually result in voids, contamination, or hot-tearing.

Metallurgical Considerations

BEFORE UPSETTING, A small layer of molten metal forms on one or both faying surfaces. Below the molten layer a heat-affected zone forms in which metallurgical changes occur. The weld thermal cycle is brief, and therefore the metallurgical reactions may not be completed resulting in unusual metastable structures.

Materials which can be hardened by heating and quenching have hard weld zones that may require postweld heat treatment. Work-hardened base metals are softened in the narrow heat-affected zone. Precipitation hardened materials may be partially annealed or overaged.

The joint upsetting process which occurs downstream of the weld apex not only forces most of the molten metal and the contaminants out of the joint, but also hot-works the adjacent metal. This may result in grain refinement and improved mechanical properties immediately next to the bond plane. The upset also creates a sharp rotation of the base metal so that laminar inclusions in the base metal adjacent to the bond plane may become substantially parallel to the bond plane resulting in heat-affected-zone discontinuities known as *hook cracks*.

Weld discontinuities are principally the result of significant, thin, flat nonmetallics at the weld interface. The discontinuities are usually caused by inadequate heating or inadequate upset. Inadequate heating or upset can be constant or variable due to unstable operation. If enough weld power is available, weld quality generally improves with increasing weld speed.

Faying surfaces with mechanical damage or containing excessive contaminants are a common cause of defects.

Typical Uses

BY FAR THE greatest number of high-frequency welding machines are used to make pipe and tube. There are over 40 American Society for Testing Materials standards covering all types of tubes and pipes. Some of these products require welds made only by high-frequency welding, and others specify high-frequency welding as well as other methods. There are also a number of American Petroleum Institute (API) specifications for line pipe and other oil country goods which allow high-frequency welding.

A 1000 kW induction welder in operation on a heavy wall structural tube mill is illustrated in Figure 20.12. The formed open seam tube proceeds from the right of Figure 20.12 through the induction coil. The edges are heated between the induction coil and the weld pressure roll assembly, and the hot-welded seam cools rapidly beyond the weld pressure rolls. The weld bead removal stand is hidden behind the structure supporting the upper weld pressure rolls. The weld temperature control pyrometer is mounted on this structure with its air-purged sight tube pointing to the weld point.

Longitudinal butt welded tube and pipe is made from several metals in many sizes. At the small end of the spectrum is automotive radiator tube made of either aluminum or brass. Radiator tube can be as small as 3/8 in. (10 mm) diameter and is induction welded in production in wall thicknesses as low as 0.0045 in. (0.11 mm). At the opposite end of the spectrum is API steel line pipe. Line pipe as large as 48 in. (1.2 m) diameter and 1 in. (25 mm) thick has been produced using the contact method. Among the metals welded by HFRW and HFIW are low carbon, high-strength low alloy, high-strength carbon and alloy steels, stainless steels, aluminum and aluminum alloys, copper and copper alloys, some nickel and titanium, and other nonferrous alloys.

Most tubing is produced in a round shape but tubes can be rolled to other shapes such as square or rectangular. Complex, roll-form shapes, which in some cases require

two longitudinal welded seams, are also produced. There are also a number of specialized applications such as the welding of tapered lamp posts and the welding of a metal sheath around an electrical cable. Thin walled, lap-welded tubing may also be high-frequency welded.

Helically wound pipe and tube can also be HFRW using lap or butt joints. Corrugated culvert pipe in diameters between 2 ft (600 mm) to 8 ft (2400 mm) are helically wound. Other operations may be integrated with the HFW mill either before or after welding. For example, hole patterns may be punched in the strip prior to forming and welding. A flying cutoff can be synchronized with the hole pattern providing a finished tubular part complete with the required hole pattern and needing no subsequent drilling or punching.

Zinc or aluminum coated tubing can be welded from precoated base metals. After the weld has been made and the outside weld bead removed, a metal spray can recoat the weld area to provide a complete coating. Alternatively, steel strip can be welded and then cleaned, full-body induction heated, and passed through a molten zinc bath to produce a galvanized tube directly from uncoated strip.

Tubes can also be hot-or cold-stretch reduced in line after welding. Induction heating is generally used to heat the tube prior to hot stretch reducing. Alternatively, the tubes can be cold stretch reduced in line and subsequently stress relieved by induction heating prior to coiling or cutting to length.

A.P.I. line pipe and some other tubular specifications require that the weld seam be normalized after welding. This heat treatment eliminates any untempered martensite in the weld heat-affected zone. The in line heat treatment is performed using a special linear inductor that induction heats the weld area to a width of between one and three times the wall thickness of the tube. The heated area is then allowed to air cool prior to final water quenching, sizing, and cutoff. The entire tube may also be fully heat-treated either in line or off line after cutoff.

Structural shapes such as I and H beams as well as tee sections can be high-frequency welded from flat strip using specially constructed structural welding mills. A typical range of structural member sizes is shown in Table 20.1. Structural shapes are made commercially from low carbon

Table 20.1
Typical Range of Sizes for High Frequency
Welded Structural Steel Shapes

	In.	mm
Section Height	3 to 20	75 to 500
Section Width	2 to 12	50 to 300
Web Thickness	.08 to .40	2 to 10
Flange Thickness	.12 to .50	3 to 12

Note: Tooling can be provided for I, T, and H Sections. The sections may be symmetrical or assymetric. The flanges of I and H Sections may be of different widths or thicknesses.

Figure 20.12–1000 kW Induction Welder

and high-strength low alloy steels. Stainless steels, titanium, and aluminum have also been welded on a laboratory scale. For some special applications, stiffening ribs can be formed into the web of the beam immediately after welding to increase its resistance to buckling.

Two different geometries of finned tubing can also be fabricated by high frequency welding. In one, the fin is helically wound on edge onto a tube and simultaneously welded to the surface of the tube. The fin and tube may be either of the same material or of different materials. In the other, fins are welded longitudinally to tubes for the manufacture of boiler water walls.

Tubes can be welded to strip or sheet metal for solar absorber plates or freezer liners. Two strips can be welded together with a butt or lap joint. These can be the same or different metals.

In addition to these continuous processes, high-frequency welding can also be used for making finite length joints between the ends of two tubes or the ends of two strips. Lower frequencies between 3 and 10 kHz are usually used to weld finite length metal weldments. Examples of a few products that can be fabricated by high frequency welding are shown in Figure 20.13.

Mechanical Properties

HIGH-FREQUENCY WELDS are autogenous hot-forged welds where most molten metal on the faying surfaces immedi-

ately prior to the forging operation is expelled leaving little or no cast structure in the joint area. The mechanical properties of the subsequent welded joints depend upon the inherent strength of the base metal after rapid heating, forging, and cooling. Low carbon and high-strength, low alloy steels and nonheat treatable aluminum weldments can be used in the as-welded condition. The joint is as strong as the base metal. Medium and high carbon steels may form martensite in the weld heat-affected zone, and therefore these weldments are almost always postweld heat-treated. Heat treatable grades of aluminum are softened by the rapid heating and quenching that occurs during welding. Postweld heat treatment is usually required to restore strength to the weld zones. Steel structural shapes are usually made either from low carbon or high-strength, low alloy steels, and therefore these sections are used in the as-welded condition.

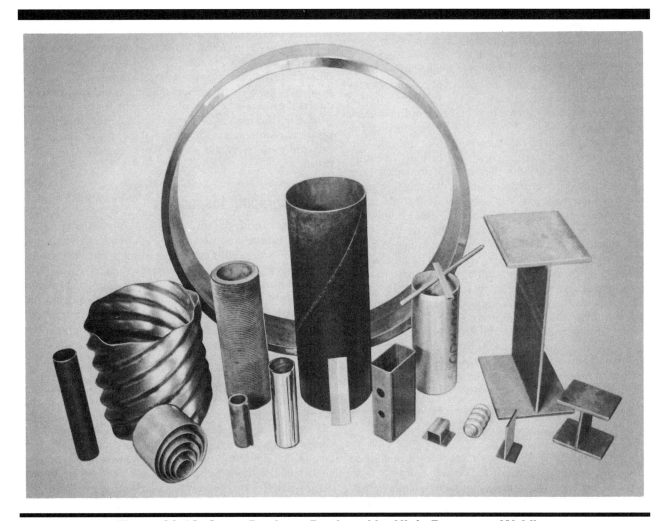

Figure 20.13–Some Products Produced by High-Frequency Welding

INSPECTION AND QUALITY CONTROL

PROCESS CONTROL

VIRTUALLY ALL HIGH-FREQUENCY welding systems are continuous mills. Fully automated mills contain strip accumulators to allow continuous operation even when the start of a new coil of strip is being welded to the end of the previous coil. The only reason for a mill with an accumulator to stop would be a fault condition or a planned shutdown. Thus, the best method of quality assurance in a high-frequency mill is by means of process control.

Theoretically, if the mill has been set up to make a satisfactory product and if every possible mill variable is monitored and remains constant, then the product quality will remain constant. However some variables are either not known or not properly understood. Others may be known but cannot be effectively monitored or controlled. However, the most important variables are known and can be controlled, and by doing so, the most consistant product quality is achieved.

Basic process control has always been practiced by tube mill operators. Meters provide information on such variables as mill speed, weld power, and individual drive motor current. The operator reads the information and adjusts the variable when it deviates from the standard practice value. Operators also observe the condition of the outside bead as it is removed and infer from this inspection the probable weld quality.

Today, however, there is increasing availability of equipment either to monitor items that have not previously been monitored or to better monitor those that have. In most cases, this equipment provides a visual output to the operator and an electronic output to a supervisory computer. As the number of items monitored increases, the need for a supervisory computer also increases because operators cannot effectively manage many variables. The supervisory computer, on the other hand, can be programmed to continuously scan all the available process sensors and to communicate with the operator only when one or more of these variables falls outside preset limits. As mill speeds increase, automatic process monitoring becomes more necessary. Failure to correct unsatisfactory welding variables may lead to large scrap losses at high mill speeds. Automated process control is therefore becoming increasingly common on modern high-speed mills.

Visual and Dimensional Inspection

A TYPICAL SPECIFICATION for a high frequency welded tube will include acceptance criteria for outside diameter, wall thickness, ovality, straightness, and general appearance. Other high frequency welded products are also required to meet specified dimensional tolerances. In most cases, the dimensional checks are done manually on a small sample of the total amount of the product produced. For critical products, noncontact gauging systems based on ultrasonics, lasers, or similar techniques can be used to provide continuous measurement of wall thickness and outside diameter.

Product Testing Procedures

TEST REQUIREMENTS FOR tubular products are described in the ASTM A450/A450M, *Standard Specification for General Requirements for Carbon, Ferritic Alloy and Austenitic Alloy Steel Tubes*. This specification covers mandatory and nonmandatory requirements for a large variety of tubular products. The nonmandatory requirements would be mandatory if they are specified in the product specification or the purchase documents. ASTM A450/A450M covers requirements for chemical and mechanical testing, product dimensional tolerances, hydrostatic testing and nondestructive testing.

ASTM A769/A769M, *Standard Specification for Electric Resistance Welded Steel Shapes*, covers the requirements for structural shapes such as I beams and T sections produced by high frequency welding. This specification covers the intended classes of application for the structural products, the manufacturing, chemical, and mechanical property requirements, dimensional tolerances, test methods and frequency, and requirements for inspection and testing.

Metallographic Inspection

THIS IS USED both for the metallographic examination of the base material and also of the high frequency weld. It is a common practice for evaluating weld quality. Transverse weld cross sections are most generally used. A typical weld cross section in a small diameter steel tube is shown in Figure 20.14. The outside upset has been removed, but the inside flash has not been removed. A large diameter, heavy wall steel tube in the as-welded condition with both the outside and inside upset removed is shown in Figure 20.15. This high quality hydraulic cylinder tubing is subsequently heat-treated and drawn over a mandrel.

Metallographic inspection is also valuable for determining the cause of failure in mechanical tests such as flattening tests. A cross section of flattening-test failure in API J55 well casing is shown in Figure 20.16. The failure path shown in Figure 20.16 is close to but not on the weld interface. The fracture follows lines parallel to and at right angles to the flow lines in the upset area which correspond to planes parallel to the surface of the material prior to upsetting.

The cause of failure can be traced, in this case, to a lack of ductility in the through-thickness direction of the hot-

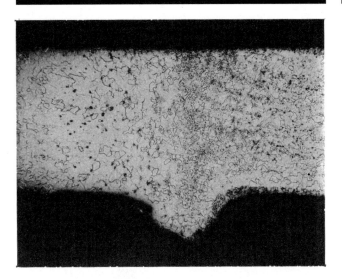

Figure 20.14—Cross Section Through High-Frequency Weld in 1.25 in. (32 mm) Diameter, 0.043 in. (1.1 mm) Wall Structural Tubing (Magnification 50x)

rolled material. A longitudinal cross section of the base metal (parallel to the rolling direction) is shown in Figure 20.17. Numerous nonmetallic inclusions lying in planes parallel to the surfaces of the material are evident. These nonmetallics result in weakness in the through thickness direction. Low ductility in the through-thickness direction is common in hot-rolled metals. It is improved by reducing the elongated nonmetallic inclusions in the metal.

Nondestructive Inspection

REQUIREMENTS FOR NONDESTRUCTIVE inspection of HFW tubes are provided in ASTM A450/A450M, *Standard Specification for General Requirements for Carbon, Ferritic Alloy and Austenitic Alloy Steel Tubes*, and other documents referenced by that specification.

Ultrasonic testing is described in ANSI/ASTM E213, *Standard Recommended Practice for Ultrasonic Inspection of Metal Pipe and Tubing*, and ANSI/ASTM E273, *Standard Method for Ultrasonic Inspection of Longitudinal and Spiral Welds of Welded Pipe and Tubing*.

Eddy-current testing procedures are described in ANSI/ASTM E309, *Standard Recommended Practice for Eddy-Current Examination of Steel Tubular Products Using Magnetic Saturation*, and ASTM E426, *Standard Recommended Practice for Electromagnetic (Eddy Current) Testing of Seamless and Welded Tubular, Austenitic Stainless Steel and Similar Alloys*. For smaller diameter tubes, coils encircling the tube are normally used, and for larger diameters, sector coils located over the weld seam can be used. A typical system consists of an exciting coil which

Figure 20.15—High-Frequency Welded Hydraulic Cylinder Tubing 11.25 in. Diameter, 0.65 in. Wall (As-Welded) (Magnification 6x)

induces eddy currents into the tube and a sensor coil that reads the resulting magnetic flux created by the induced currents. The exciting and sensor coils are typically packaged together as a single unit. A discontinuity in the welded seam will disturb the normal current flow pattern. The disturbed current will induce a magnetic field that differs from that produced in tube without a discontinuity. The sensor coil detects the difference.

Flux-leakage testing is described ANSI/ASTM E570, *Standard Recommended Practice for Flux Leakage Examination of Ferromagnetic Steel Tubular Products*. The tube is first magnetized to a level approaching its magnetic saturation. Discontinuities cause a leakage of the magnetic flux which is found by a magnetic detector.

In all of these methods, calibration standards are prepared by testing tubes of the same size and material as the one to be inspected. These test tubes contain known discontinuities such as drilled holes or transverse, tangential, or longitudinal notches, and they are used to simulate the

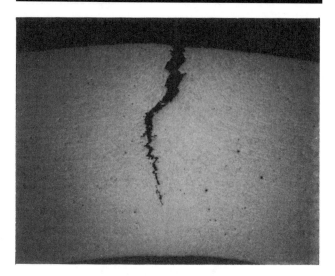

Figure 20.16–Cross Section of High-Frequency Weld in API J-55 Well Casing 4.5 in. O.D x 0.189 Wall Showing Flattening Crush Test Failure Near Bond Plane (Magnification 15x)

Figure 20.17–Longitudinal Section Through the Base Metal of the Pipe Shown in Figure 20.16.

type of discontinuity which may occur in the welding process. When a signal exceeding the magnitude of that required by the calibration procedures occurs, a marking system, typically a paint spray, identifies the location of the potential defect. In many cases, because of the high speed of the high-frequency welding process, it is difficult to discriminate between real weld defects and other discontinuities which may not be cause for rejection; therefore, the marked areas may subsequently be retested off line to verify the on line test.

Weld testing procedures such as x-ray, magnetic particle, and liquid penetrant are generally not applicable to high-frequency welding.

Ultrasonic testing, eddy-current testing, flux-leakage testing, or all three, may also be performed after subsequent processing such as stretch reduction, drawing, or cold expansion has been performed. Such subsequent pro-

cessing may enlarge a discontinuity and make it easier to detect.

WELD PROCEDURE AND PERFORMANCE QUALIFICATIONS

WELDING PROCEDURES FOR high-frequency welding depend on the design of the mechanical equipment which forms and upsets the material, the type of product being produced, and the process being used—HFIW or HFRW. Because of the wide variety of mechanical equipment used together with the large number of product types produced, there are no standard welding procedures published on this subject. General guidelines are available from the equipment manufacturers, but the actual procedures used for any particular product are usually developed by the mill operator management.

SAFETY

THE HEALTH AND safety of the welding operators, maintenance personnel, and other personnel in the area of the welding operations must be considered in establishing operating practices. Design, construction, installation, operation, and maintenance of the equipment, controls, power supplies, and tooling should conform to the requirements of Federal (OSHA), State, and local safety regulations, as well as those of the company.

Voltages in high-frequency generators range from 400 to 30 000 volts and can be lethal. Proper care and safety precautions should be taken while working on high-frequency generators and their control systems to prevent injury. Modern units are equipped with safety interlocks on access doors and automatic safety grounding devices that prevent operation of the equipment when the access doors are open. The equipment should never be operated with panels or high-voltage covers removed or with interlocks and grounding devices blocked. The new fully transistorized units will use significantly lower voltages, usually less than 1500 V, but such voltages are still dangerous, and the same safety practices must be observed.

The high voltage-high frequency leads should be encased in grounded metal ducts both for safety and to minimize E.M.I. (electromagnetic interference) radiation. Low-voltage induction coils and contact systems should always be properly grounded for operator protection. High-frequency currents are more difficult to ground than low-frequency currents, and grounding lines should be short and direct to minimize inductive impedance. Care should be taken to prevent the high-frequency magnetic field around the coil and leads from heating adjacent metal parts of the mill by induction.

Personnel injuries from direct contact with high-frequency voltages, especially at the upper range of welding frequencies, tend to produce severe local tissue damage. However, fatalities are unlikely because the current flows on the surface of the victim's body.

SUPPLEMENTARY READING LIST

Brown, G. H., Hoyler, C. N. and Bierwith, R. A. *Theory and applications of radio frequency heating.* New York: D. Van Nostrand Co., Inc., 1957.

Dailey, R. F. "Induction welding of pipe using 10,000 cycles." *Welding Journal* 44(6): 475-479; June 1965.

Haga, H., Aoki, K., and Sato, T. "Welding phenomena and welding mechanisms in high frequency electric resistance welding." *Welding Journal* 59(7): 208s-212s July 1980.

Haga, H. et al. *Intensive study for high quality ERW pipe*, Document Number 3101, ERW-01-81-0. Nippon Steel Corporation, 1981.

Harris, S. G. "Butt welding of steel pipe using induction heating." *Welding Journal* 40(2): 57s-65s; February 1961.

Johnstone, A. A., Trotter, F. J., and a'Brassard, H. F. "Performance of the thermatool high frequency resistance welding process. *British Welding Journal* 7(4): 238-249; April 1960

Koppenhofer, R. L. et al. "Induction-pressure welding of girth joints in steel pipe." *Welding Journal* 39(7): 685-691; July 1960.

Martin, D. C. "High frequency resistance welding." Bulletin No. 160. Welding Research Council, April 1971.

Oppenheimer, E. D. "Helical and longitudinally finned tubing by high frequency resistance welding." ASTME Tech Paper AD67-197. Dearborn, MI: Society of Manufacturing Engineers, 1967.

Oppenheimer, E. D., Kumble, R. G. and Berry, J. T. "The double ligament tensile test: its development and application." *Journal of Engineering Materials and Technology.* 107-112, April 1975.

Osborn, H. B., Jr. "High frequency continuous seam welding of ferrous and non-ferrous tubing." *Welding Journal* 35(12): 1199-1206; December 1956.

Rudd, W. C., "High frequency resistance welding." *Welding Journal* 36(7): 703-707; July 1957.

———. "High frequency resistance welding." *Metal Progress* 239-40, 244; October 1965.

———. "Current penetration seam welding - a new high speed process." *Welding Journal* 46(9): 762-766; September 1967.

Udall, H. N. "Metallographic techniques - their contribution to quality high frequency welded products." Proceedings of 1986 International Conference - Tomorrow's Tube, 10-12 June 1986. International Tube Association, 1986.

Udall, H. N. and Berry, J. T. "High frequency welding of HSLA steel structurals." *Metal Progress* 112(3): August 1977.

Udall, H. N., Berry, J. T., and Oppenheimer, E. D. "A high speed welding system for the production of custom designed HSLA structural sections." Proceedings of International Conference on Welding of HSLA (Microalloyed) Structural Steels in Rome, Italy 9-12 Nov. 1976. American Society for Metals, 1978.

Wolcott, C. G. "High frequency welded structural shapes." *Welding Journal* 44(11): 921-926; November 1965.

ELECTRON BEAM WELDING

PREPARED BY A COMMITTEE CONSISTING OF:

D. E. Powers, Chairman
PTR - Precision Technologies, Inc.

J. D. Ferrario
Ferranti Sciaky, Inc.

G. K. Hicken
Sandia National Laboratories

J. F. Hinrichs
A. O. Smith Corporation

J. O. Milewski
Los Alamos Scientific Laboratories

T. M. Mustaleski
Martin Marietta Energy Systems, Inc.

WELDING HANDBOOK COMMITTEE MEMBER:
L. J. Privoznik
Westinghouse Electric Corporation

ELECTRON BEAM WELDING

INTRODUCTION

PROCESS HISTORY

SINCE ELECTRON BEAM welding (EBW) was initially used as a commercial welding process in the late 1950's, the process has earned a broad acceptance by industry. First employed by the nuclear industry, and then shortly thereafter by the aircraft and aerospace industries, the process was quickly recognized as having the capacity for enhancing both the quality and reliability of the highly critical parts used by these industries. The process also reduced the manufacturing costs.

During this initial period of commercial application, the process was limited strictly to operation in a high vacuum chamber. However, a system was soon developed that required a high vacuum only in the beam generation portion. This permitted the option of welding in either a medium vacuum chamber or a nonvacuum environment. This advancement led to its acceptance by the commercial automotive and consumer product manufacturers. As a consequence, EBW has been employed in a broad range of industries worldwide. Since the late 1960's the process has provided both very shallow and extremely deep single-pass autogenous welds with a minimal amount of thermal distortion of the workpiece.

PROCESS OVERVIEW

EBW IS A fusion joining process that produces coalescence of materials with heat obtained by impinging a beam composed primarily of high-energy electrons onto the joint to be welded. Electrons are fundamental particles of matter, characterized by a negative charge and a very small mass. For EBW they are raised to a high-energy state by being accelerated to velocities in the range of 30 to 70 percent of the speed of light.

Basically, an electron beam welding gun functions in much the same manner as a TV picture tube. The primary difference is that a TV picture tube uses a low-intensity electron beam to continuously scan the surface of a luminescent screen, and thereby produces a picture. An electron beam welding gun uses a high-intensity electron beam to continuously bombard a weld joint, which converts that energy to the level of heat input needed to make a fusion weld.

In both of these cases, the beam of electrons is created in much the same manner, using an electron gun that typically contains some type of thermionic electron emitter (normally referred to as the gun "cathode" or "filament"), a biasing control electrode (normally referred to as the gun "grid" or "grid cup"), and an anode. Various supplementary devices, such as focus and deflection coils, are also provided to focus and deflect this beam. In EBW, the total beam generating system (gun and electron optics) is called either the electron beam gun/column assembly, or simply the electron beam gun column.

FUNDAMENTALS OF THE PROCESS

PRINCIPLES OF OPERATION

THE HEART OF the electron beam welding process is the electron beam gun/column assembly, a simplified representation of which is shown in Figure 21.1. Electrons are generated by heating a negatively charged emitting material to its thermionic emission temperature range, thus causing electrons to "boil off" this emitter or cathode and be attracted to the positively charged anode. The precisely configured grid or bias cup surrounding the emitter provides the electrostatic field geometry that then simultaneously accelerates and shapes these electrons into the beam. The beam then exits the gun through an opening in the anode.

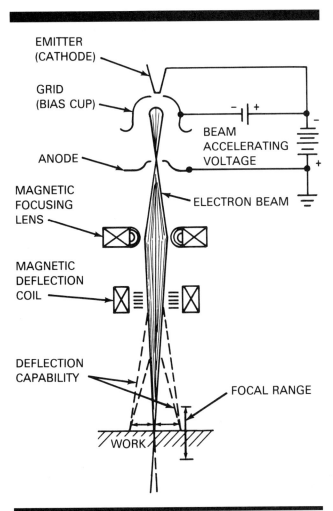

Figure 21.1–Simplified Representation of a Triode Electron Beam Gun Column

In a diode (cathode-anode) gun, this beam-shaping electrode and the emitter are both at the same electrical potential, and together are referred to as the cathode. In a triode (cathode-grid-anode) gun, the two are at different potentials; consequently the beam-shaping grid can be biased to a slightly more negative value than the emitter, in order to control beam current flow. In this case, the emitter alone is called the cathode (or filament) and the beam-shaping electrode is called the bias (or grid) cup. Since, in both cases, the anode is incorporated into the electron gun, beam generation (acceleration and shaping) is accomplished completely independent of the workpiece.

Upon exiting the gun, this beam of electrons accelerates to speeds in the range of 30 to 70 percent of the speed of light when gun operating voltages are in the range of 25kv to 200 kv. The beam then continues on toward the workpiece. Once the beam exits from the gun, it will gradually broaden with distance traveled, as illustrated in Figure 21.1. This divergance results from the fact that all the electrons in the beam have some amount of radial velocity, due to their thermal energy, and in addition, all experience some degree of mutual electrical repulsion. Therefore, in order to counteract this inherent divergence effect, an electromagnetic lens system is used to converge the beam, which focuses it into a small spot on the workpiece. The beam divergence and convergence angles are relatively small, which gives the concentrated beam a usable focal range, or "depth-of-focus", extending over a distance of an inch or so, as illustrated in Figure 21.1.

In practice, the rate of energy input to the weld joint is controlled by the following four basic variables:

(1) The number of electrons per second being impinged on the workpiece (beam current)
(2) The magnitude of velocity of these electrons (beam accelerating voltage)
(3) The degree to which this beam is concentrated at the workpiece (focal beam spot size)
(4) The travel speed with which the workpiece or electron beam is being moved (welding speed)

The maximum beam accelerating voltages and currents that can be achieved with commercially available electron beam gun/column assemblies vary over the ranges of 25 to 200 kV and 50 to 1000 mA, respectively, and the electron beams produced by these systems can generally be focused to diameters in the range of 0.01 to 0.03 in. (0.25 to 0.76 mm).

The resulting beam power levels and power densities attainable from these units can reach values as high as 100 kW and 10^7W/in.^{2} (1.55 x 10^4 W/mm^2), respectively. Such power densities are significantly higher than those possible with arc welding processes.

The potential welding capability of an electron beam system is indicated by the maximum power density that the system is capable of delivering to the workpiece. This comparison factor depends upon the maximum beam power (current x voltage) and the minimum focal spot size attainable with the system. At this writing, electron beam welding systems with beam power levels up to 300 kW and power densities in excess of $10^8 W/in.^2$ (1.55×10^5 W/mm^2) have been built, but these are not yet commercially available.[1]

At power densities on the order of $10^5 W/in.^2$ ($1.55 \times 10^2 W/mm^2$), and greater, the electron beam is capable of instantly penetrating into a solid workpiece or a butt joint and forming a vapor capillary (or "keyhole") which is surrounded by molten metal. As the beam advances along the joint, molten metal from the forward portion of the keyhole flows around its periphery and solidifies at the rear to form weld metal. In most applications, the weld penetration formed is much deeper than it is wide, and the heat-affected zone produced is very narrow. For example, the width of a butt weld in 0.5 in. (13 mm) thick steel plate may be as small as 0.030 in. (0.8 mm) when made in a vacuum. This stands in remarkable contrast to the weld zone produced in arc and gas welded joints, where penetration is achieved primarily through conduction melting.

Since the EB weld results from a keyhole that the beam forms, the angle of incidence with which the beam impinges on the surface of a workpiece can affect the final angle at which the keyhole (and thus the resulting weld zone) is produced with respect to that surface.

An electron beam can be readily moved about by electromagnetic deflection. This allows specific beam spot motion patterns (circles, ellipses, bow tie shapes, etc.) to be generated on the surface of a workpiece when an electronic pattern generator is used to drive the deflection coil system, as illustrated in Figure 21.2. This deflection capability can, in certain instances, also be used to provide beam travel motion. In most instances, however, deflection is used to adjust the beam-to-joint alignment, or to apply a deflection pattern. This deflection modifies the average power density being input to the joint, and results in a change in the weld characteristics achieved. However, as previously noted, care must always be taken when using any type of beam deflection to ensure that the beam angle of incidence does not adversely affect the final weld results. It especially must not cause part of the weld joint to be missed.

PROCESS VARIATIONS

THREE BASIC MODES of electron beam welding are now used. These are high vacuum (EBW-HV), medium vacuum (EBW-MV), and nonvacuum (EBW-NV). The principal difference between these process modes is the ambient pressure at which welding is done. With the high-vacuum (often referred to as "hard vacuum") mode, welding is done in the pressure range of 10^{-6} to 10^{-3} torr.[2] For medium vacuum, the pressure range is 10^{-3} to 25 torr. Within this range, the pressure span from about 10^{-3} to 1 torr is often called a "partial" or "soft" vacuum, and from about 1 to 25 torr, a "quick" vacuum. Nonvacuum electron beam welding is done at atmospheric pressure, and thus is sometimes called "atmospheric" EBW. In all cases, the electron beam gun pressure must be held below 10^{-4} torr for stable and efficient operation.

High vacuum and medium vacuum welding are done inside a vacuum chamber. This imposes an evacuation time penalty to create the "high purity" environment. The medium vacuum welding machine retains most of the advantages of high vacuum welding, with shorter chamber evacuation times, resulting in higher production rates. Nonvacuum EB welding, although it incurs no pumpdown time penalty, is not suitable for all applications because the welds it produces are generally wider and shallower than equal power EB welds produced in a vacuum.

With medium-vacuum operation, the beam is generated in high vacuum and then projected into a welding chamber operating at higher pressure. This is accomplished by providing an orifice below the beam generation column that is large enough to pass the beam, but still small enough to impede any significant back diffusion of gases into the gun chamber.

In nonvacuum electron beam welding equipment, the beam is generated in high vacuum and then projected through several specially designed orifices, separating a series of differentially pumped chambers, before finally emerging into a work environment that is at atmospheric pressure. For nonvacuum electron beam welding directly in the atmosphere, beam accelerating voltages of greater than 150 kV are normally required. However, if the atmospheric pressure environment around the weldment is a gas such as helium, beam accelerating voltages lower than 150 kV can sometimes be employed.

Figure 21.3 shows the three basic modes of electron beam welding. A fixed electron beam gun column is shown mounted on the exterior of the high- and medium-vacuum enclosures to illustrate these two modes. A mobile electron beam gun column may also be mounted on the interior of high- and medium-vacuum enclosures, as illustrated in Figure 21.4. This is commonly employed to provide a higher degree of gun column motion capability.

High Vacuum Welding

HIGH VACUUM (10^{-3} torr or lower) is the required environment for all electron guns. Thus, even though special

---

1. The Welding Research Institute of Osaka University in Japan has a 300 kw electron beam welding machine which they are presently using to investigate single-pass EB joining of extremely heavy sections.

2. A torr is the accepted industry term for a pressure of one millimeter of mercury. Standard atmospheric pressure can be expressed as 760 torr or 760 mm of mercury.

Figure 21.2—Beam Deflection Capability of an Electron Beam Column as Shown by a "Bow Tie" Pattern on a Workpiece

ELECTRON BEAM MODES OF OPERATION

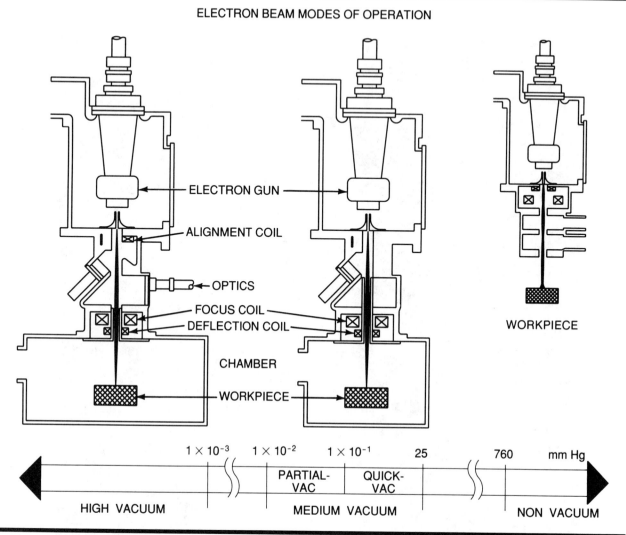

Figure 21.3–The Basic Modes of Electron Beam Welding, with a Corresponding Vacuum Scale

methods allow the beam to enter higher pressure environments, the gun itself will not operate effectively at pressures much greater than 10^{-4} torr.

The principal advantages of electron beam welding in high vacuum are as follows:

(1) Maximum weld penetration and minimum weld width can be achieved, thereby producing a minimum of weld shrinkage and distortion. A high depth-to-width ratio is achieved because of the high-energy density of the beam and the resultant keyhole mode of melting.

(2) Maximum weld metal purity is possible due to the relatively clean environment provided by a high vacuum.

(3) The relatively long gun-to-work distances possible in a hard vacuum improve the operator's ability to observe the welding process and to weld normally inaccessible joints.

Since the electrons in the beam would be scattered by collisions with any residual gas molecules that may be present in the beam's path, and because the frequency with which these collisions occur varies directly with both the concentration of gas molecules and the total distance traveled, the use of a high-vacuum environment minimizes scatter (particularly when long beam travel distances must be employed).

The high vacuum minimizes exposure of the hot weld zone to oxygen and nitrogen contamination, and concurrently causes gases evolved during welding to move rapidly away from the weld metal, thereby improving weld metal purity. For this reason, high vacuum welding is better suited for welding highly reactive metals than the medium and nonvacuum process variations.

Production of high vacuum requires pumping times which significantly limit production rates. This pumpdown limitation can be offset somewhat if a number of

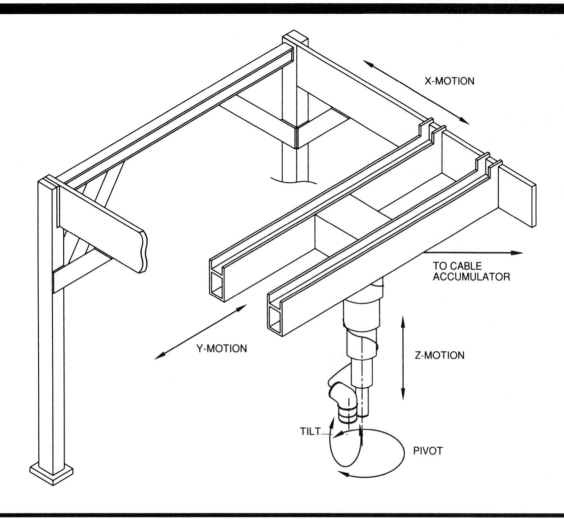

Figure 21.4–Mobile EB Gun / Column Multi-Axis Motion System

assemblies are welded in a single load and the chamber volume is small. The number of parts that can be welded, per batch load, will be limited by the chamber size employed. As a result, high vacuum welding is generally more suitable when relatively low production rates are involved. Various types of "air-to-air" part transfer schemes have been developed which allow parts to be moved in and out of a high-vacuum region without needing to vent it. These procedures make it possible to use high vacuum EBW on certain high-production joining applications, such as welding bimetallic saw blades.

Medium Vacuum Welding

A PRINCIPAL FEATURE of medium vacuum welding is the ability to weld without pumping the welding chamber to very low pressure (high vacuum). When the chamber is small, the pumping time required may be a matter of only a few seconds, which is of major importance in economical

processing. This makes medium vacuum welding ideally suited for use in the mass production of parts that involve repetitive welding tasks, and where a welding chamber of minimum volume can be used. For example, gears can be successfully welded to shafts in their final machined or stamped condition, without subsequent finishing to maintain close dimensional tolerances. Such an operation is shown in Figure 21.5.

Because medium vacuum welding is done at pressures with a significant (100 ppm) concentration of air, this mode of EBW is less desirable than high vacuum EBW for reactive metals. In addition to requiring specialized postweld heat treatment, many refractory metals require an ultrapure welding environment.

This higher concentration of air also scatters the beam electrons, enlarging the beam diameter and decreasing the power density. This results in welds that are slightly wider, more tapered, and less penetrating than similar type welds produced under high-vacuum conditions.

Figure 21.5—Electron Beam Welding a Gear in Medium Vacuum

Nonvacuum Welding

THE MAJOR ADVANTAGE of nonvacuum welding is that the work need not be enclosed in a vacuum chamber. Elimination of chamber evacuation time results in higher production rates and a lower cost per piece. Also, the size of the weldment is not limited by the size of the chamber.

These advantages, however, are gained at the expense of not being able to achieve the depth-to-width ratios, weld penetrations, and gun-to-work distances attainable in a vacuum. The welding atmosphere is not as "pure" as it is with high and medium vacuum welding, even when inert gas shielding is employed. Although the use of a workpiece vacuum enclosure is not required, some type of radiation shielding must still be provided to protect personnel from the X-rays generated when the electron beam strikes the work.

Operating conditions for nonvacuum welding differ from the other two variations. Beam dispersion increases rapidly at ambient pressure, as shown in Figure 21.6. The nonvacuum gun-to-work distance, even when a helium gas

environment is being employed, should be less than about 1.5 inches (38 mm). This restriction requires that the weld joint should not be shielded from the electron beam by the shape of the workpiece.

The depth of penetration achieved in nonvacuum electron beam welding is affected by beam power level, travel speed, gun-to-work distance, and the ambient atmosphere through which the beam passes. Figure 21.7 shows weld penetration as a function of travel speed for three different beam power levels. Note the increase in travel speed to be gained for a given penetration as the power level is increased.

Nonvacuum electron beam welding appears to demonstrate more efficient penetration at power levels above 50 kW. This result is attributed to a decreased gas density produced by local heating of the air by the high-energy electron beam.

The graph in Figure 21.8 shows the effect of the ambient atmosphere, gun-to-work distance, and travel speed on the weld penetration. Penetration is greater with helium, which is lighter than air, and lower with argon, which is heavier than air. For a given penetration and gun-to-work distance, higher travel speeds can be achieved in a helium shielding gas.

Many materials have been welded successfully using the nonvacuum technique. They include carbon, low alloy, and stainless steels; high-temperature alloys; refractory alloys; and copper and aluminum alloys. Some of these metals can be welded directly in air while others require inert gas protective atmospheres to avoid excessive oxygen and nitrogen contamination.

With 60 kW nonvacuum equipment, it is possible to produce single-pass welds in many metals 1 in. (25.4 mm) thick, at relatively high speeds. Figure 21.9 is a cross section of a nonvacuum weld in 3/4 in. (19 mm) Type 304 stainless steel plate.

ADVANTAGES AND LIMITATIONS

ELECTRON BEAM WELDING has unique performance capabilities. The high-quality environment, high-power densities, and outstanding control solve a wide range of joining problems.

The following are advantages of electron beam welding:

(1) The EBW directly converts electrical energy into beam output energy. Thus the process is extremely efficient.

(2) Electron beam weldments exhibit a high depth-to-width ratio. This feature allows for single-pass welding of thick joints.

(3) The heat input per unit length for a given depth of penetration can be much lower than with arc welding. The resulting narrow weld zone results in low distortion, and fewer deleterious thermal effects.

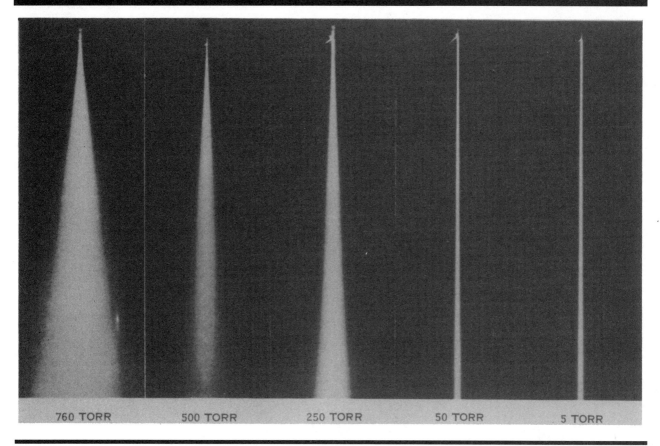

Figure 21.6–Electron Beam Dispersion Characteristics at Various Pressures

(4) A high-purity environment (vacuum) for welding minimizes contamination of the metal by oxygen and nitrogen.

(5) The ability to project the beam over a distance of several feet in vacuum often allows welds to be made in otherwise inaccessible locations.

(6) Rapid travel speeds are possible because of the high melting rates associated with this concentrated heat source. This reduces welding time and increases productivity and energy efficiency.

(7) Reasonably square butt joints in both thick and relatively thin plates can be welded in one pass without filler metal addition.

(8) Hermetic closures can be welded with the high- or medium-vacuum modes of operation while retaining a vacuum inside the component.

(9) The beam of electrons can be magnetically deflected to produce various shaped welds; and magnetically oscillated to improve weld quality or increase penetration.

(10) The focused beam of electrons has a relatively long depth of focus, which will accommodate a broad range of work distances.

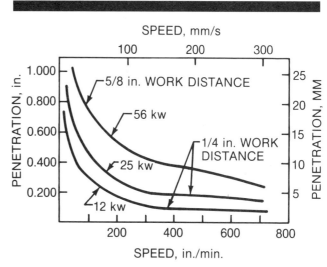

Figure 21.7–Effect of Travel Speed on Penetration of Nonvacuum Electron Beam Welds in Steel (175 kV in Air)

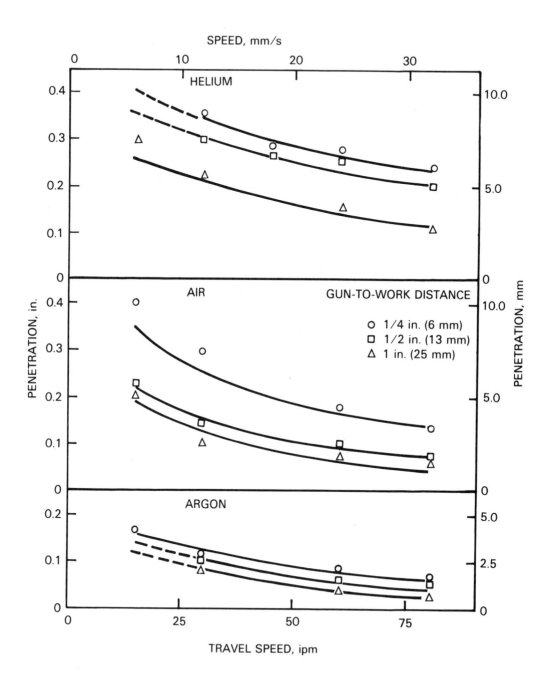

Figure 21.8–Penetration Versus Travel Speed for Nonvacuum Electron Beam Welds in AISI 4340 Steel in Helium, Air, and Argon with Three Gun-to-Work Distances (175 kV, 6.4 kW)

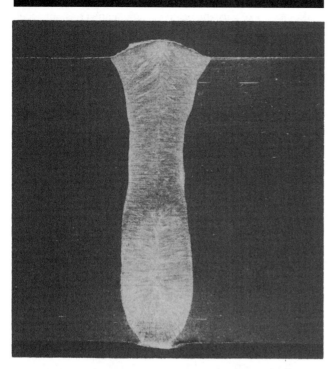

Figure 21.9–A Cross Section of a Nonvacuum Electron Beam Weld in 3/4 in. (19 mm) Stainless Steel Plate Made in Air With 12 kW of Power

(**11**) Full penetration, single-pass welds with nearly parallel sides, and exhibiting nearly symmetrical shrinkage, can be produced.

(**12**) Dissimilar metals and metals with high thermal conductivity such as copper can be welded.

Some of the limitations of electron beam welding are as follows:

(**1**) Capital costs are substantially higher than those of arc welding equipment. Depending on the volume of parts to be produced, however, the final "per piece" part costs attainable with EBW can be highly competitive.

(**2**) Preparation for welds with high depth-to-width ratio requires precision machining of the joint edges, exacting joint alignment, and good fit-up. In addition, the joint gap must be minimized to take advantage of the small size of the electron beam. However, these precise part-preparation requirements are not mandatory if high depth-to-width ratio welds are not needed.

(**3**) The rapid solidification rates achieved can cause cracking in highly constrained, low ferrite stainless steel.

(**4**) For high and medium vacuum welding, work chamber size must be large enough to accommodate the assembly operation. The time needed to evacuate the chamber will influence production costs.

(**5**) Partial penetration welds with high depth-to-width ratios are susceptible to root voids and porosity.

(**6**) Because the electron beam is deflected by magnetic fields, nonmagnetic or properly degaussed metals should be used for tooling and fixturing close to the beam path.

(**7**) With the nonvacuum mode of electron beam welding, the restriction on work distance from the bottom of the electron beam gun column to the work will limit the product design in areas directly adjacent to the weld joint.

(**8**) With all modes of EBW, radiation shielding must be maintained to ensure that there is no exposure of personnel to the x-radiation generated by EB welding.

(**9**) Adequate ventilation is required with nonvacuum EBW, to ensure proper removal of ozone and other noxious gases formed during this mode of EB welding.

EQUIPMENT

HIGH-VACUUM, MEDIUM-VACUUM, AND nonvacuum EBW equipment employs an electron beam gun/column assembly, one or more vacuum pumping systems, and a power supply. High- and medium-vacuum equipment operates with the work in an evacuated welding chamber. Although nonvacuum work does not need to be placed in a chamber, a vacuum environment is necessary for the electron beam gun column.

All three basic modes can be performed using so-called high-voltage equipment, i.e., equipment using gun columns with beam accelerating voltages greater than 60 kV. Nonvacuum electron beam welding performed directly in air requires beam accelerating voltages greater than 150 kV.

High vacuum and medium vacuum welding can also be performed with so-called low-voltage equipment (i.e., equipment with gun columns that employ beam accelerating voltages of 60 kV and lower). Because high-voltage gun columns are generally fairly large, they are usually mounted on the exterior of the welding chamber, and are either fixed in position or provided with a limited amount of tilting or translational motion, or both. Low-voltage gun columns are usually small. Some units are "fixed" externally. Others are internally mounted "mobile" units capable of being moved about, with up to five axes of combined translational motion, in a manner similar to that illustrated in Figure 21.4.

ELECTRON BEAM GUNS

IN GENERAL, ELECTRON beam welding guns are operated in a space-charge-limited condition. When a gun is operated in this condition, the beam current produced at any accelerating voltage is proportional to the 3/2 power of the accelerating voltage ($I = KV^{3/2}$), where the constant of proportionality, K, is a function of gun geometry.

Besides acceleration voltage, a broad range of conditions must be satisfied if an electron gun is to deliver the required power and power density.

Optimum gun performance depends upon gun configuration, emitter characteristics, total power capabilities, and focusing provisions. For a given metal and joint thickness, characteristically narrow welds can be made if (1) sufficient beam power is available to permit rapid travel speed, and (2) the beam power density is great enough to develop and continuously maintain a vapor hole to the depth of penetration required.

An electron beam gun generates, accelerates, and collimates the electrons into a directed beam. The gun components can logically be divided into two categories: (1) elements that generate free electrons (the emitter portion), and (2) elements that produce a useful beam, or electrodes that accelerate and form the electrons into a beam. The emitter may be either (1) a directly (resistance) heated wire or ribbon type filament, or (2) a rod or disc type filament indirectly heated by an auxiliary source, such as electron bombardment or induction heating. The specific emitter design chosen will affect the characteristics of the final beam spot produced on the work.

Only self-accelerated guns, similar to the Pierce and Steigerwald telefocus gun configurations, are used for electron beam welding. They have superior focusing and power capabilities, and also permit placing the gun anode and workpiece at earth ground potential.

The Pierce gun was originally designed as a diode capable of producing a rapidly converging beam with the primary focal point close to the anode, and having a beam divergence that was uniform thereafter. The Steigerwald telefocus gun was originally designed as a triode which produced a gradually converging beam, with its primary focal point some distance from the anode. Current Pierce and Steigerwald gun designs are modifications of the original designs.

When a change in beam current is desired at a given accelerating voltage in a diode gun, the cathode-to-anode spacing in the gun must be changed. This changes the proportionality constant, K, of the gun. Several spacers are supplied by the manufacturer to provide a wide range of operating conditions. Each spacer has a range of beam power, beam spot size, and control sensitivity.

Diode guns control the beam current at a given voltage by controlling the power to, and thus the temperature of, the electron emitter. Electron emission is related to both emitter temperature and accelerating voltage.

The triode-type gun is similar to the diode gun except that the beam-forming electrode ("grid cup") is biased with a variable negative voltage relative to the emitter. In this type of gun, the emitter is simply referred to as the *filament*. This makes it easy to vary the beam current at any constant accelerating voltage. Thus, both the accelerating voltage and beam current can be varied independently within limits.

The ability to control beam current with a bias voltage allows rapid changes in beam current. Electronic switching circuits permit users to repetitively pulse the beam current. Beam current pulsing helps to minimize heat input to the workpiece, while still achieving deep weld penetrations. This capability is not as extensively used for welding as it is for drilling. Accurate control of the beam current slope rates is extremely useful in many welding applications, particularly on circular welds at the start/finish overlap region.

All gun/column assemblies employ an electromagnetic lens to focus the electron beam into a small spot on the workpiece. Electromagnetic deflection coils usually oscillate the beam in a repetitive or nonrepetitive fashion. These deflection coils are generally positioned immediately below the electromagnetic focusing lens, and are used to deflect the electron beam from its normal axial position. Two sets of deflection coils at 90° will trace classical lissajous figures on the work surface. Sinusoidal beam deflection perpendicular to the direction of welding will broaden the weld bead to simplify manual tracking of weld seams. Circular and elliptical deflection tends to reduce weld porosity. More complex deflection patterns may also be employed, both to enhance weld penetration and to improve weld quality.

The modified Pierce and Steigerwald guns used today weld with both "high" and "low" beam accelerating voltages. Similar power levels are available with low- and high-voltage guns. Beam power is the product of the accelerating voltage and the beam current. Therefore, operation at high voltage requires less beam current than operation at low voltage for an equivalent beam output power. In high- or medium-vacuum applications, both low- and high-voltage equipment will produce similar-quality welds in most metals. However, differences in the weld cross section produced at the same beam power will exist, because one system operates with a low voltage and high current and the other with high voltage and low current.

POWER SUPPLIES

Electron Gun Power Supplies

THE GUN POWER source used for an electron beam welding machine is an assembly of at least one main power supply and one or more auxiliary power supplies. It produces high-voltage power for the gun and auxiliary power for the emitter and beam control. Depending upon whether the gun is a diode type or a triode type, the high-voltage power supply consists of two or more of the following components:

(1) A main high-voltage dc power supply that provides the constant beam accelerating voltage and total beam current

(2) An emitter (filament) power supply with either ac or dc output

(3) A bias electrode power supply that impresses a voltage between the emitter and the bias electrode (grid cup) to control beam current

The main high-voltage power supply and auxiliary-beam- generation power supplies are frequently placed together in a common oil-filled tank. A high-purity, electrical-grade transformer oil serves both as an electrical insulating medium and as a heat transfer agent to carry heat from the electrical components to the tank walls. The components are typically supported from the cover plate of the tank so that they can be removed from the tank with the cover plate; however, removal of these components is rarely required.

Another high-voltage insulating material occasionally used in EBW power supplies is sulfur hexafluoride gas at pressures up to 45 psi. A power supply with this gas insulation is considerably more compact and lighter than an oil-insulated unit of the same rating. The components in both the main high-voltage power supply and the auxiliary supplies are primarily transformers, diodes (rectifiers), capacitors, and resistors. Electron tube diodes were initially used in the main power supply by some manufacturers, but solid-state diodes, usually silicon, are presently used. Cost, regulation, physical size, ability to absorb transients of voltage and current, and thermal capability are just some of the considerations which affect the choice of components and insulating medium.

Main High-Voltage Power Source

THIS POWER UNIT converts line input power to high-voltage dc power for the electron beam gun. Power ratings commercially available are in the range of 3 to 100 kw, but up to 300 kw is readily attainable. Units are designed for a particular electron beam gun type (high or low voltage). Some typical machine ratings that are commercially available are shown in Table 21.1.

The maximum allowed voltage ripple in the dc output varies, depending upon the desired quality of beam focus. Excessive voltage ripple will produce an undesirable beam current ripple that may affect weld quality. Voltage ripple is usually controlled below one percent.

The inherent decrease in output voltage with increasing load is typically in the range of 15 to 20 percent. Various controls and regulators compensate for this voltage decrease and minimize the effects of line voltage variations, as well as the effects of temperature and other factors that influence the stability of the output voltage. Sophisticated controls eliminate all of the effects mentioned and maintain a stable accelerating voltage to within 1 percent of the selected value. Other less costly controls eliminate only some of the effects, but they are adequate for less critical applications. Some of the controls used, in approximate order of increasing sophistication and performance, include:

Table 21.1
Typical Electron Beam Welding Machine Ratings

Rating, kW	Output	
	kV, max	mA, max
3	30	100
3	60	50
6	30	200
7.5	150	50
15	60	250
15	150	100
25	175	144
25	150	167
30	60	500
35	200	175
45	60	750
60	175	345
100	100	1000

(1) Line voltage regulator (constant-voltage transformer)

(2) Servo-operated variable transformer with feedback from the high-voltage output

(3) Motor-generator with electronic exciter and feedback from the high-voltage output

(4) Electronic regulation with both current and voltage feedback available

Emitter Power Supply

DIRECTLY HEATED WIRE or ribbon filaments (emitters) are the most common. Filaments may be hairpin shaped or may be a more complex shape. The current that heats the filament can be ac or dc, but dc is preferred because the magnetic field created by the filament heating current may affect the beam's direction. The cyclic nature of ac heating currents causes the beam spot to oscillate with a small, but significant, amplitude about a fixed point.

Since the magnitude of magnetic effects will increase with the heating current, filtering must be used even with dc heating currents to reduce any ripple present to three percent.

Current and voltage ratings of a filament power supply depend upon the type and size of the directly heated filament. For 0.020 in. (0.5 mm) diameter tungsten wire filaments, the supply would be rated for 30A at 20V. Ribbon-type filaments have a much larger emitting area than wire type filaments. Ribbons require power supplies rated for higher currents and lower voltages (30 to 70A at 5 to 10V). High dc filament heating currents will produce a certain amount of initial beam deflection, but this is a static feature which can usually be corrected by beam alignment devices normally built into the gun/column assembly.

Indirectly heated emitters are also used. Here, an auxiliary bombardment or inductive-type heat source indirectly heats the gun emitter to electron emission temperatures. The power supply for driving the auxiliary electron gun used to heat a disc emitter by bombardment would be rated for 100 to 200 mA at several kilovolts. The power supply for indirectly heating a "bolt type" emitter, heated in a radial mode, would be rated for 2 to 3 amps at 400 to 600 volts.

Bias Voltage Supply

THE BIAS VOLTAGE supply for a triode style gun is usually designed to give complete control of beam current from zero to maximum. To do this, the dc power supply applies a variable voltage on the beam-shaping electrode (grid cup), making it negative with respect to the emitter. A voltage in the range of 1500 to 3000 V is needed to cut off the beam current. For maximum beam current, the voltage is 100 to 300 V. Here again the bias supply must also have no greater than one percent beam current ripple. Various electronic input power control devices are used to provide pulsing, ramping, etc. of the beam current.

Some electron beam equipment uses a self-biasing system. The bias voltage is derived partially from the main accelerating voltage through a voltage divider, and partially from the voltage across a series resistor in the main power circuit. This system does not have a separate bias supply.

Electromagnetic Lens and Deflection Coil Power Supplies

THE ELECTROMAGNETIC LENS (focusing lens shown in Figure 21.1) is generally powered by a solid-state constant-current power supply. The strength of the magnetic field varies with the current flowing through the coil. The current provided to the coil must remain constant, to produce a beam spot with consistent size, even when the voltage drop across the coil changes with temperature variations.

Beam deflection coils (Figure 21.1) are also powered by solid-state devices. Two sets of coils at 90° are usually placed at the base of the gun column for x and y deflection of the beam. Programming of the power sources for the two sets of coils can provide beam movement along either axis singularly or both axes simultaneously. Complex geometric beam patterns (circle, ellipse, square, rectangle, hyperbola, etc.) can be produced by electronic control. The ripple on the dc input to both the deflection coils and the electromagnetic lens must be low to minimize adverse effects of beam instability on weld quality.

VACUUM PUMPING SYSTEMS

VACUUM PUMPING SYSTEMS are required to evacuate the electron beam gun chamber, the work chamber for high- and medium-vacuum modes, and the orifice assembly used on the beam exit portion of the gun/column assemblies for medium vacuum and nonvacuum welding. Two basic types of vacuum pumps are used. One is a mechanical piston or vane type, to reduce pressure from 1 atmosphere to about 0.1 torr. For medium vacuum welding, these mechanical pumps are generally operated in conjunction with a Roots-type blower, another kind of mechanical pump. The other is an oil-diffusion- type pump used to reduce the pressure to 10^{-4} torr or lower. Sequencing these pumps to produce the needed vacuum can be accomplished by manual or automatic operation of valves in the system. Commercial electron beam welding equipment has standarized on automatic valve sequencing.

The vacuum system for an electron beam gun chamber consists of a mechanical roughing pump and a diffusion pump. A similar system is used to evacuate a high-vacuum-mode work chamber. Both systems may be ducted to their chambers through a water-cooled or liquid nitrogen-cooled (optically dense) baffle, if necessary, to prevent any oil from backstreaming out of the diffusion pump into the gun or work areas. Today, however, most diffusion pumps have some form of integrated cold cap to capture backstreaming pump oil, so baffles are rarely needed. In addition, turbomolecular pumps and cryogenic pumps are presently being used, which entirely eliminate any possibility that pump oil will enter the gun or chambers. A combination of diffusion and mechanical pumping is shown in Figure 21.10.

The vacuum pumping system can be mounted on the same base as the vacuum chamber and connected to it with rigid ducting. The primary exception to this rule is the mechanical roughing pump, which must be connected to the system with a flexible tube in order to minimize any chance of its vibration being transmitted to the welding chamber. A large diameter vacuum valve isolates the diffusion pump from the chamber during the roughing portion of the pumping cycle. A small mechanical pump keeps the isolated diffusion pump at low pressure.

The roughing and diffusion pumping periods are normally controlled by automatic sequencing of pneumatic or electric vacuum valves. Automatic evacuation cycles are accomplished with pressure-sensing relays that activate the appropriate valves in the preprogrammed sequence. The control units are designed to protect the vacuum system in case of accidental pressure rise in the chamber.

For medium vacuum welding, the work chamber is evacuated with a mechanical vacuum pumping system of high capacity. The types and sizes of mechanical pumps used in the system will depend upon the work chamber size, the work load, and the desired production rate. Automatic evacuation cycles speed production.

With nonvacuum welding equipment, the electron beam gun chamber is evacuated with a combination mechanical-diffusion pumping system. The various pressure stages in the orifice assembly through which the electron beam exits from the gun/column assembly are pumped with a series of mechanical vacuum pumps.

In vacuum EBW, the evacuation process and its rate depend upon the capabilities of the pumps, the work and fixturing load, the size of the chamber, and the total leakage rate of the system. The total leakage rate is the increase

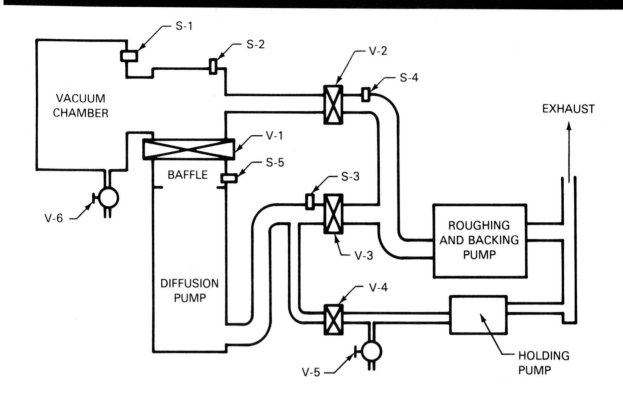

VALVES

V-1 — HIGH VACUUM
V-2 — ROUGHING
V-3 — BACKING
V-4 — HOLDING
V-5 — VACUUM RELEASE
V-6 — VACUUM RELEASE

VACUUM SENSORS

S-1 — ION TYPE
S-2 — THERMOCOUPLE TYPE
S-3 — THERMOCOUPLE TYPE
S-4 — THERMOCOUPLE TYPE
S-5 — ION TYPE

Figure 21.10–Vacuum Pumping System for High-Vacuum Operation

in chamber pressure per unit of time attributed to both real leaks and virtual leaks in the system.

Real leaks are actual holes or voids in the chamber capable of passing air or gas. A *virtual leak* is the semblance of a real leak somewhere in the vacuum system; in actuality, this type of leak results from the outgassing of absorbed or occluded gases on the interior surfaces of the system, when under vacuum. For satisfactory system operation, no in-leakage (real leaks) should be detectable with a helium mass spectrometer leak detector having a sensitivity of 1 x 10^{-4} standard mm^3/s of helium.[3] In addition, a pressure rate-of-rise test should be conducted to ensure that no detrimental virtual leaks are present. This test is conducted by isolating the chamber to be tested from the pumping system (without exposing it to atmosphere) immediately after doing a four-hour preparatory pump down of the chamber. A customary limiting value for a rate-of-rise test is in the range of 1 to 2 x 10^{-2} torr (10 to 20 micron) per hour, averaged over a 10-hour test period.

LOW-VOLTAGE SYSTEMS

WORK CHAMBERS OF low-voltage systems are usually made of carbon steel plate. The thickness of the plate is designed to provide adequate x-ray protection and the structural strength necessary to withstand atmospheric pressure. Lead shielding may be required in certain areas of the chamber to ensure total radiation tightness of the system.

The weldment inside the chamber may be observed by direct viewing through leaded glass windows. However, the effectiveness of this technique depends upon the dis-

3. See ASTM B498 for a description of this method of leak testing.

tance between the operator and the weld joint, and upon the shape of the workpiece. When direct viewing is difficult, an optical viewing system may be provided to give the operator a magnified view of the weld seam. It can be used for setup operations, inspection of the weld, alignment of the weld joint with respect to the electron beam, and positioning of the gun to center the sharply focused electron beam on the weld seam.

Closed-circuit television is also used, to provide another method of viewing. It may have both its light source and television camera mounted in a readily serviceable location outside of the chamber, or both items may be located inside the chamber. An optical protection system of some type is normally employed to shield the viewing equipment from metal spatter and metal vapor deposition. The closed-circuit television system permits continuous monitoring of welds and minimum operator exposure to the intense light from the weld.

HIGH-VOLTAGE SYSTEMS

HIGH-VOLTAGE ELECTRON BEAM systems operate above 60 kV, and generally are designed to operate at voltages between 100 kV and 200 kV (with beam powers up to 100 kW). The electron beam gun/column assembly of a high-vacuum welding machine is housed in an external vacuum chamber, mounted on either the top or side of the welding chamber, as illustrated by Figure 21.11. This mounting may involve either a stationary or sliding style of seal. In the latter case, the motion of the gun/column assembly will normally be limited to a single (x or y) axis of motion. Any other required axis of motion must be provided by the weldment, for example, the type of x, y, z, or rotary motion of the workpiece that is commonly employed.

The external location of the high-voltage gun reduces its maneuverability, but provides ready access to the gun components for service. This arrangement also provides the operator with a view of the beam spot and the weld through optics that are relatively coaxial with the electron beam (Figure 21.11). A view of the back side of a $10 bill through such a slightly oblique (but still effectively coaxial) optical system is shown in Figure 21.12. Direct viewing may also be provided through lead glass windows in the chamber walls.

Work chambers for this equipment are usually welded carbon steel boxes of ribbed design, externally clad with lead for x-ray protection of personnel.

Nonvacuum electron beam welding machines do not require a vacuum chamber around the workpiece. The electron beam gun column may be fixed atop an x-radiation shielding box containing the workpiece and travel carriage. Another arrangement is to place both this gun column and the workpiece in an x-ray shielded room, making both capable of transverse motion, and then operating the equipment remotely from outside the room.

SEAM TRACKING METHODS

WITH ELECTRON BEAM welding, as with other automatic welding processes, the relative positions of the beam spot and the weld joint must be established accurately prior to starting to weld. This relationship must then be accurately maintained throughout the entire welding cycle. This total requirement is somewhat complicated in electron beam welding because (1) the beam spot is very small and produces a relatively narrow weld bead; (2) welding is performed at relatively high travel speeds; and (3) the workpiece is contained in a vacuum chamber or radiation enclosure, making continuous observation difficult. As previously mentioned, most high vacuum electron beam systems are equipped with a viewing system that permits the operator to observe the weld joint and the beam spot. The initial correct position of the electron beam in relation to the joint can easily be established with a viewing system. On medium vacuum and nonvacuum systems, where operator viewing is not normally provided, this initial beam-to-joint alignment is accomplished through precise handling (tooling and fixturing) of the part.

For welding long or slightly irregular joints, a means for automatically maintaining proper beam-to-joint alignment is desirable. Optical viewing of welding and manual correction for deviations in the joint path is, at best, difficult, although some equipment is used in this fashion.

Two methods are used to maintain beam position along a nonlinear joint. The first involves programming by analog or continuous path numerical controls. This method is applicable where parts have been machined precisely to a required contour and are accurately positioned for welding.

The second method uses an adaptive electromechanical control. This control has a tracking device that follows the weld joint and signals the control to adjust the work or gun position to keep the beam on the joint. Stylus and contour seam tracking devices employ the same electrical circuitry and may be quickly interchanged to accomplish various tracking requirements.

The stylus-type "seam tracking" system has a probe, or stylus, that rides in the joint. Lateral (cross seam) movements of the probe, resulting from a change in joint position, are converted to electrical signals by a transducer. The electrical signals drive a positioning servomotor that moves the work or gun to maintain the preset alignment. The electrical signals from this system define a right-error, a left-error, and the null or correct gun position. Alternatively, the electron beam itself can be deflected electronically to accommodate changes in the location of the joint.

A contour-type system involves a simple modification of the stylus-type seam tracking system which permits edge welding of certain types of assemblies using the weldment as a cam. A preloaded ball-type stylus rides against the edge of the weldment as the work is rotated or driven linearly. As a proximity control to accommodate changes in vertical position of the weld joint, the stylus is used to maintain

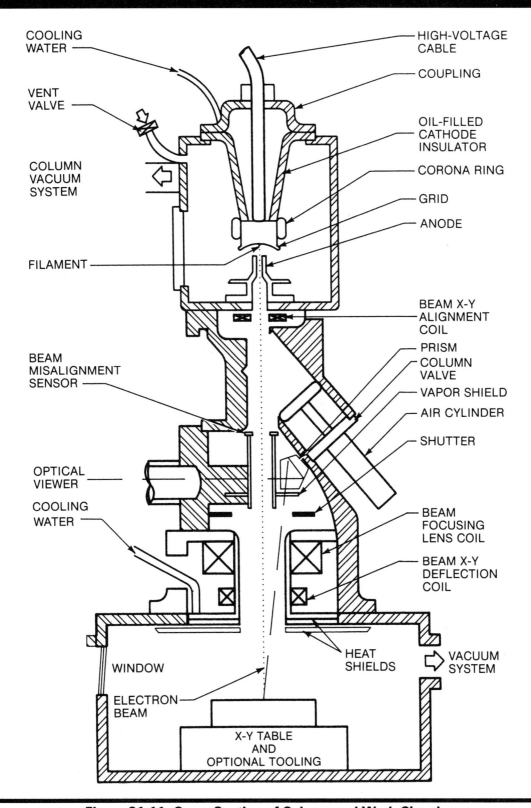

Figure 21.11–Cross Section of Column and Work Chamber

Figure 21.12–The Optical System for an Electron Beam Column Provides Magnification for Viewing Fine Work (This Example Shows a Partial View of the Reverse Side of a U.S. $10.00 Bill to Demonstrate the Capability)

a constant gun-to-work distance by feeding the tracking signal to a servomotor drive on the z axis.

An electronic joint-finding system is also available (Figure 21.13) which can be used for both "seam locating" and "seam tracking" functions. This unit taps into the same electron beam used for welding as a means for sensing the joint position. It finds the joint location by recognizing the absence of rays reflected by the joint in the midst of rays from the work surface, thereby eliminating any need to calibrate an auxiliary joint sensing device.

Once the vacuum chamber has been closed and evacuated for welding, a finely focused electron beam is aimed at the weld joint and scanned back and forth across it. This action produces a secondary electron emission or backflow that can be "collected" and continuously monitored. As the beam is traversed back and forth, the magnitude of secondary electron backflow being measured will decrease each time the beam passes across the joint. Thus, if the monitor signal is displayed as a visible oscilloscope trace, the resulting discontinuity in the oscilloscope line trace will indicate where the joint is with respect to the beam column centerline. Consequently, this method provides an easy means for initially aligning the beam column centerline with the joint.

By initially scanning the entire joint in discrete steps, probable misalignments can be anticipated and corrected during welding.

The stylus, contour, and electronic seam tracking devices are often used in conjunction with a record-and-playback tape system, where the joint can be traced and its location recorded. Then the joint is welded using playback programmed control of the beam or work position.

On systems equipped with CNC controls, programmed periodic scans by the welding beam across the joint during welding indicate the exact location of the joint, which the system repositions. The CNC control assumes on-line, "real time" seam tracking.

WORK-HANDLING EQUIPMENT

THE RESPONSE OF work-handling mechanisms must be accurate and well defined to maintain the relative positions of the electron beam and the weld joint during the entire welding operation. Their design and manufacture should follow good machine tooling practices. Ruggedness, repeatability, smoothness, accuracy, and suitability for operation in a vacuum (if so required) are prime requirements. Also, the magnetic susceptibility of the materials must be considered. Since travel speed affects weld geometry, this variable must be controlled accurately and be repeatable. In general, electric motor drives having an accuracy of about +2 percent of set speed are adequate.

Most electron beam welding machines provide standard mechanisms for linear and rotary motion of the workpiece relative to the electron beam. Horizontal linear motion is usually provided by movement of a work table or by movement of the electron gun. Rotary motion about a vertical axis is achieved with a motor-driven horizontal rotary table. Chambers can be equipped with external platform devices that allow the work table (and any work-handling mechanisms) to be withdrawn from the vacuum chamber, for ease of loading and fixturing the parts.

Figure 21.14 shows an x-y work table on its external platform. An adjustable (0 to 90 degree) rotary work positioner and tailstock have been mounted on the table. Rotary motion about a horizontal axis can be provided using the rotary positioner and tailstock. The positioner is power-driven. Simple linear, rotary, and circular joints can be aligned with the electron beam using these mechanisms.

It is often desirable to weld circular joints in several parts during a single loading of the chamber. In this case, the components are arranged on an eccentric table attached to a motor driven horizontal rotary table. The eccentric table holding the parts can position each piece in turn under the electron beam. The circular weld motion is made with the eccentric table. Programmed indexing of the eccentric table from piece to piece can be added.

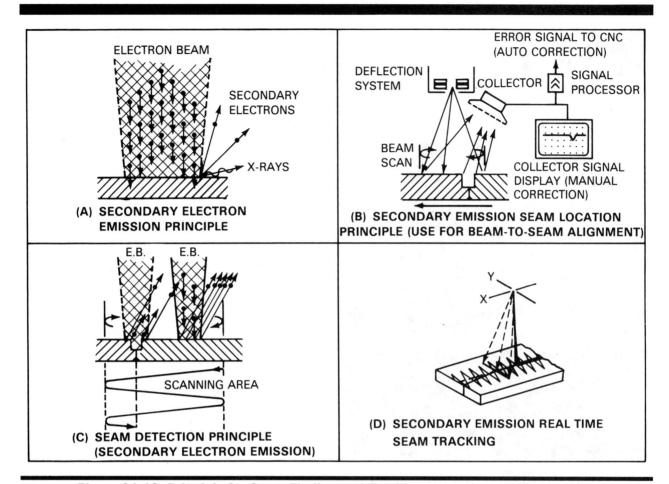

Figure 21.13–Principle for Seam Finding and Tracking Using Secondary Emission

Multiple-spindle rotary fixtures are also frequently used when making circumferential welds in a group of similar parts. The parts are again batch loaded, and then successively indexed into welding position by a motor drive. The joint on each part can be positioned for welding by linear movements of the work table on which the rotary fixture is mounted. It is possible to automate the entire operation. An example of a special purpose rotary fixture is shown in Figure 21.15.

CONTROLS

SINCE ALL THE operating variables of an electron beam welding system are directly controllable, the process is readily adaptable to computer numerical control (CNC). Movement of the workpieces or the gun, as well as electron beam deflection, can be preprogrammed in any combination. The beam current itself is also programmable. Thus, the beam can easily be changed from one discrete level to another, or changed at a specified rate. This ease of "upslope/downslope" control, and the capability for producing various beam deflection patterns, enhance the capacity of the process to produce extremely high-quality welds. Other variables, including the accelerating voltage, beam focus, emitter power, chamber pressure, as well as other auxiliary functions, can also be part of the program for control or monitoring. Electron beam welding systems perform computerized contour welding of intricately-shaped parts, where beam power and travel speed must be varied as a function of position along the weld path, as well as parts requiring multipass weld programs.

MEDIUM-VACUUM EQUIPMENT

EQUIPMENT FOR MEDIUM vacuum electron beam welding is basically a modification of standard high-vacuum equipment. An "aperture tube" (an orifice that allows beam passage, but impedes gas flow) is added into the gun column

Figure 21.14–Table With X-Y Drive and Rotary Positioner from Electron Beam Work Chamber

Figure 21.15–A Loaded Rotary Welding Fixture in Location Beneath the Electron Beam Gun

assembly, thereby allowing the separately pumped gun region to remain under high vacuum when the chamber is operated at a medium vacuum level. As on high-vacuum equipment, a column valve is used to isolate and maintain the gun region under high vacuum during chamber venting, and the aperture which is added helps to maintain a vacuum of 10^{-4} torr or better in the gun region during beam operation, while still allowing the beam to impinge on a workpiece located in a medium-vacuum environment. Thus, on medium-vacuum equipment, the chamber is cyclically vented as new parts are loaded and rapidly repumped down to some medium-vacuum welding level without the gun region being exposed to atmosphere. This permits high-volume part production. Both low- and high-voltage EBW systems are produced for medium vacuum welding.

General purpose medium-vacuum systems, such as the one shown in Figure 21.16, are used advantageously in short production runs. However, most medium-vacuum units are especially tooled for specific part assemblies. Figure 21.17 illustrates typical medium- vacuum tooling concepts. In each case, the work chamber and tooling are an integral assembly, specifically designed for a single-part design.

Various medium-vacuum welding systems are used for high- production applications. For example, a machine with a single welding station and multiple-loading stations can have a production capability in the region of 200 parts per hour. A dual welding station machine, on the other hand, could increase that production capability up to 500 parts per hour. The production rates in the final analysis are dependent upon the design of the parts.

Another method for achieving high production with medium-vacuum equipment is shown in Figure 21.18. Here, a sliding seal is used to provide intermediate vacuum zones before and after the separately pumped medium-vacuum welding chamber. This method maintains a series of continuously pumped vacuum zones which eliminate the need for evacuation time, thus enabling the high- production capability of a dial feed table to be fully utilized

Figure 21.16—A General Purpose Medium Vacuum Electron Beam Welding Machine

and allowing production rates in excess of 500 parts per hour to be attained.

NONVACUUM EQUIPMENT

A BEAM OF electrons passing through a gas is primarily scattered by the shell electrons of the gas atoms or molecules. As the gas pressure increases, scattering becomes more severe (Figure 21.6). This produces a noticeable broadening of the beam profile and a decrease in beam-power density, but not necessarily a loss in total beam power.

An electron beam must be generated in high vacuum. To weld with the beam at atmospheric pressure, the beam is passed through a series of chambers or stages operating at progressively higher pressures. In addition, the electron velocity (accelerating voltage) must be high enough to minimize the scattering effect of the atmosphere.

A series of chambers operating at successively higher pressures is obtained by staging; i.e., differentially pumping a number of chambers. A series of apertures is pro-

vided to permit the electron beam to pass through the wall of one chamber into the next, while restricting the gas flow in the opposite direction. This orifice and pumping system must be designed to maintain the atmospheric- to high-vacuum gradient required. The electron beam must be accelerated through a high voltage. If the last stage is in air, this voltage must be a minimum of 150 kV in order to provide a practical working distance between the final orifice and the workpiece. The beam power level used, and the gas comprising the atmosphere through which the beam eventually passes, can greatly influence the useful working distance.

Figure 21.19 shows a conventional nonvacuum electron beam gun/column assembly, complete with orifice system. The electron gun shown is typical of those used with the other modes of electron beam welding, and is capable of operating at accelerating voltages in the range of 150 to 200 kV. Beam current, and thus the power, is controlled by the voltage on the bias electrode of the gun. The beam is focused by an electromagnetic lens to the minimum diameter of the orifice system shown at the bottom of Figure 21.19. It emerges from the vacuum environment, into air at atmospheric pressure, through the lower orifice. Inert gas shielding can be added, if desired. The workpiece is placed near the lower orifice.

During operation, a high vacuum is continuously maintained in the upper gun area by using an oil diffusion or turbomolecular pump on this region. Lesser vacuum levels are maintained in the interim pressure stages by mechanical pumps. In most cases, the work is moved horizontally in front of the gun column, but the entire gun column can be moved if desired. As with the high-vacuum and medium-vacuum modes, the gun can be placed in either a vertical or a horizontal position, and the welding area must be shielded to protect personnel from the X-radiation produced during welding. Health hazards from this radiation are discussed at the end of this chapter under "Safety Precautions".

Another type of nonvacuum electron beam welding gun unit features a gas-filled, high-voltage power supply that can be mounted directly on the gun/column assembly during operation, both of which can then be traversed along a weld joint during operation.

As with sliding-seal style medium-vacuum equipment, the time for evacuation of the work area is eliminated, and thus production rates in excess of 500 parts per hour are readily attainable with the nonvacuum EBW mode. In addition, since the workpiece need not be enclosed in a chamber to be evacuated, part size and part condition requirements are greatly alleviated.

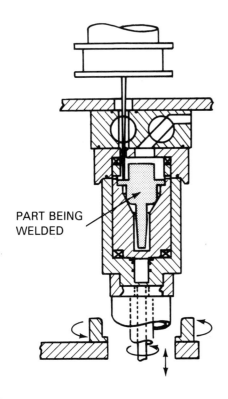

PART BEING
WELDED

(A) BALL JOINT ASSEMBLY

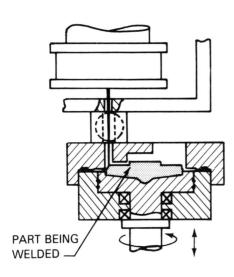

PART BEING
WELDED

(B) DIAPHRAGM ASSEMBLY

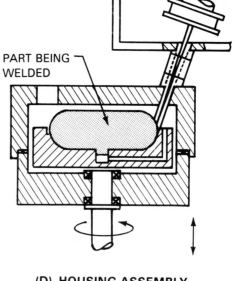

PART BEING
WELDED

(C) CARRIAGE ASSEMBLY

PART BEING
WELDED

(D) HOUSING ASSEMBLY

**Figure 21.17–Typical Tooling Concepts in Special Purpose Medium Vacuum Electron Beam
Welding Machines**

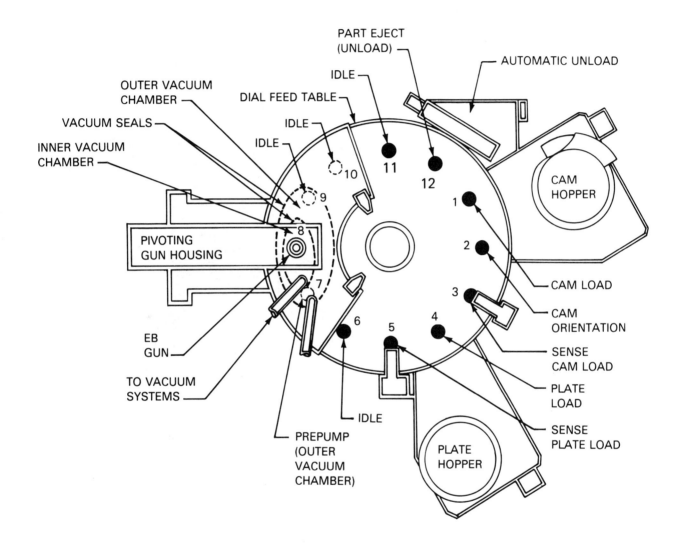

Figure 21.18–A Medium Vacuum Electron Beam Welding System with a Prepumping Zone for Continuous Part Feed Capability

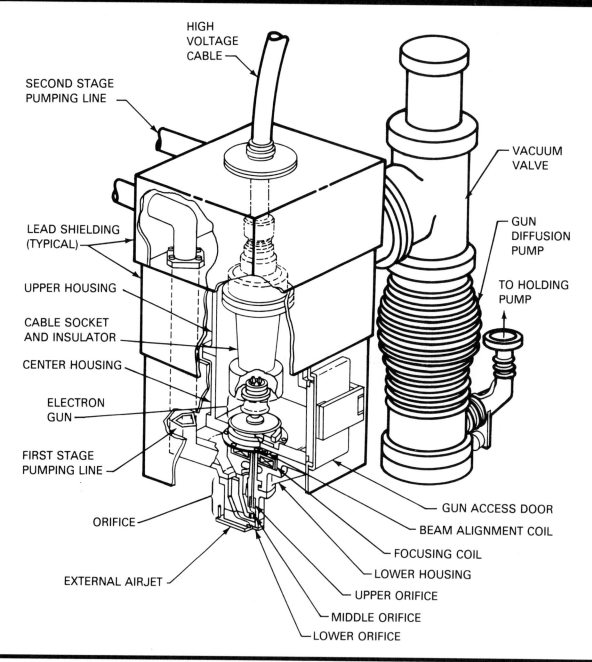

Figure 21.19–A Nonvacuum Electron Beam Gun Column Assembly

CHARACTERISTICS OF WELDS

THE ELECTRON BEAM welding process produces weld metal geometries that differ significantly from those made by conventional arc welding processes. Typical transverse cross sections through electron beam and gas tungsten arc welds are compared in Figure 21.20. The geometry of a typical electron beam weld exhibits a weld depth-to-width ratio that is very large in comparison to that of an arc weld. This feature results from the high-power density of the electron beam. The beam is concentrated in a small area, and the beam power density can exceed the power densities available in arc welding by several orders of magnitude.

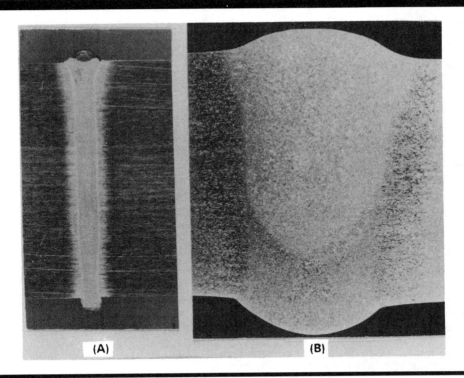

Figure 21.20–Comparison of Electron Beam (left) and Gas Tungsten Arc (right) Welds in 1/2 in. Thick Type 2219 Aluminum Alloy Plate

The high depth-to-width ratios of electron beam welds account for two important advantages of the process. First, relatively thick joints can be welded in a single pass. Thick weld joints, which require multiple-pass arc welding procedures, can be welded in a single pass by electron beam welding procedures, in considerably less time. An example of this is illustrated by Figure 21.21, which shows the cross section of a single-pass electron beam weld made in 4-in. thick carbon steel with a beam power of 33 kW and a travel speed of about 5 in./min (2 mm/sec). Second, for a given thickness, the rate (travel speed) at which welding can be accomplished is much greater than can be attained with arc welding. In turn, the electron beam welding process introduces less distortion and fewer thermal effects than arc welding.

High-vacuum welds with depth-to-width ratios of 50:1 are possible in a number of alloys. The welding of heavy sections in a single pass is practical using a square-groove butt joint. In aluminum, plates up to 18 in. (460 mm) thick have been welded in this manner.

Some problems remain with joining thick sections, but production applications in steel extend to over 6 in. (152 mm). The medium-vacuum mode sacrifices some of the penetration capability achievable in the high-vacuum mode, and in the partial (soft)-vacuum region of the medium-vacuum mode approximately five percent less penetration can be experienced. In the nonvacuum mode, the maximum penetration attainable in steel is presently under 2 in. (51 mm).

The ability to produce welds with these characteristics depends upon the process mode, whether high, medium, or nonvacuum. In all cases, it is highly dependent upon the beam spot size and the total beam power. Figure 21.22 is a representative plot showing how penetration decreases with increasing ambient pressure, due to the beam spreading brought about by increased pressure. (NOTE: The data are normalized relative to data achievable under high-vacuum conditions). The spread shown in this data plot is indicative of the fact that operating variables other than pressure (such as beam voltage, distance traveled, etc.) will also affect the penetration achieved at any given ambient pressure. The final depth-to width ratio achieved is also critically dependent upon base metal physical properties, especially melting point, heat capacity, thermal diffusivity, and vapor pressure.

EB welding, particularly in its high-vacuum mode, is an excellent tool for welding dissimilar metals and different masses, and for repair welding components impossible to salvage with other processes. Depending on the joint thickness and material being welded, the low total heat input to the workpiece can noticeably minimize weld joint distortion. In general, the high- and medium- vacuum modes are the most advantageous, although the nonvacuum mode offers some notable advantages over conventional arc welding processes.

Because a high-power density beam produces welds that are not controlled by thermal conduction, metals of significantly different thermal conductivities can be welded together. A joint between two pieces of different thickness has unequal heat losses into the thinner and thicker parts, but that is not a significant problem. In thin metals or low-

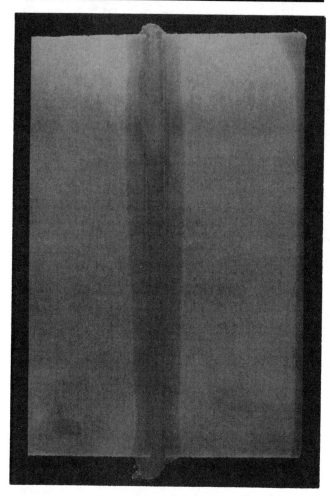

Figure 21.21–Single-Pass Electron Beam Weld in a 4 in. Thick Carbon Steel Section

melting metals, fusion can be accomplished without significant need for backup heat sinks that would be required for arc welding. For unequal masses, the beam energy is usually concentrated on the thicker section, and the power is adjusted for penetration through the thin section.

Arc welding procedures often require preheating thick sections of a metal having high thermal conductivity, such as aluminum or copper. Little or no preheat is required for electron beam welding those metals, because of the high-power density available.

Reactive and refractory metals are detrimentally affected by very small amounts of atmospheric contaminants, such as oxygen, nitrogen, and hydrogen. These metals may be electron beam welded without introducing these contaminants. Such metals include tungsten, molybdenum, columbium, tantalum, zirconium, and titanium. The high-vacuum mode is most suitable for joining these metals. The other two modes have decreasing weld perfor-

mance capabilities, although they still may be satisfactorily applied in select cases.

Although electron beam welding is a high-power density process, it is also a low-energy process. The total energy input needed to weld a joint of a given thickness is considerably less than that required by more conventional arc welding processes. Two advantages result from the low-energy input: first, it minimizes distortion and reduces the size of the weld heat-affected zone; second, the high cooling rates of narrow electron beam welds can avoid metallurgical reactions, such as phase changes. However, the fundamental rules of metallurgy regarding cooling rates and the resulting microstructure still apply. Nevertheless, the weld metal will have mechanical properties normally associated with the bulk properties of the base metal.

Another aspect of electron beam welding involves the control of the process. As the process is pushed to the limit of its capabilities, the operating variables that influence the final results require much greater control. Accurate control of the electron beam process has permitted a high degree of reliability. At present, incorporation of minicomputers and microprocessors offers additional control capability over the welding conditions.

Control of the welding environment can control the composition of the weld metal. Electron beam welding in a high vacuum permits gases to escape and high vapor pressure metals to evaporate. This produces a more refined melt zone material, but may also cause the loss of certain alloying elements. At the other pressure extreme, nonvacuum electron beam welding in air may increase the nitrogen and oxygen content of the weld metal.

The narrow weld metal shape produced by this process results in less distortion in weldments. In this type of weld, the weld metal is essentially parallel-sided except where the electron beam first impinges on the top surface of the abutted members. See Figure 21.20(A). Contraction of the metal during cooling is fairly uniform through the joint. When weld metal has a characteristic V-shape, as in arc welding, there is significant warpage from the unequal thermal contraction at top and bottom of the joint.

Since the electron beam can penetrate through extremely thick sections, beveling or chamfering one or both edges of abutting members is not necessary. However, for such welds, tighter machining tolerances are normally required for high vacuum electron beam welding than for arc welding.

In medium vacuum welding, weld metal cross sections are similar to those of high vacuum welding, but the depth-to-width ratios are somewhat smaller.

With nonvacuum electron beam welding, the weld bead produced may be nearly as wide as a typical gas tungsten arc weld. That weld metal may possess all the characteristics produced by the more conventional welding processes, because insufficient power or large gun-to-work distances were employed. However, at high power and speed, depth-to-width ratios on the order of 5 to 1 are feasible with the nonvacuum welding mode.

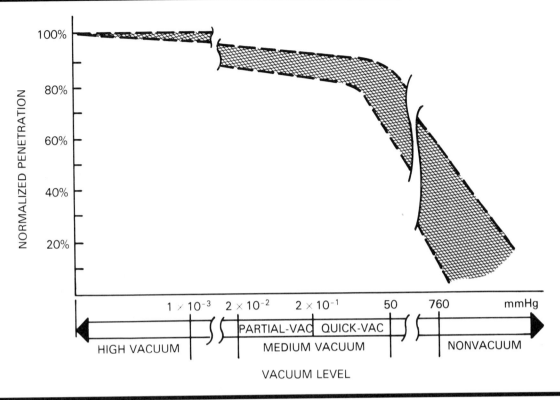

Figure 21.22—Penetration as a Function of Operating Pressure

WELDING PROCEDURES

JOINT DESIGNS

BUTT, CORNER, LAP, edge, and T-joints can be made by electron beam welding using square-butt joints or seam welds. Fillet welds are difficult to make and are not generally attempted. Typical electron beam weld joint designs are shown in Figure 21.23. Modifications of these designs are frequently made for particular applications.

Square butt joint welds require fixturing to maintain fit-up and alignment of the joint. They can be self-aligning if a rabbet joint design is used. The weld metal area can be increased using a scarf joint, but fit-up and alignment of the joint are more difficult than with a square butt joint weld. Edge, seam, and lap fillet welds are primarily used to join sheet gage thicknesses.

JOINT PREPARATION AND FIT-UP

WHEN NO FILLER wire is added, the fit-up of parts must be more precise than for arc welding processes, because poor fit up would result in a lack of fill of the weld joint. The beam must impinge on and melt both members simulta-neously, except for seam welds where the beam penetrates through the top sheet. Underfill or incomplete fusion will result from poor fit-up, and lap joints which are not clamped sufficiently will burn through.

A metal-to-metal fit between parts is desirable but difficult to obtain. The acceptable gap for a particular application will depend upon the process mode employed, the type of base metal, the thickness and configuration of the joint, and the required weld quality. Thus, while sheet sections being welded with the vacuum mode may require a fit-up of less than 0.004 in. (0.1 mm), plate sections being welded with the nonvacuum mode may tolerate a fit-up more than five times greater. Aluminum alloys can tolerate somewhat larger gaps than steel. Beam deflection or oscillation with high and medium vacuum welding, to widen the fusion zone, and with nonvacuum welding, may permit larger gaps. Consequently, the maximum acceptable joint gap and the tolerance for each particular application should be determined and qualified in order to avoid unnecessary joint preparation costs.

In general, roughness of the faying surfaces is not critical so long as the surfaces can be properly cleaned to remove

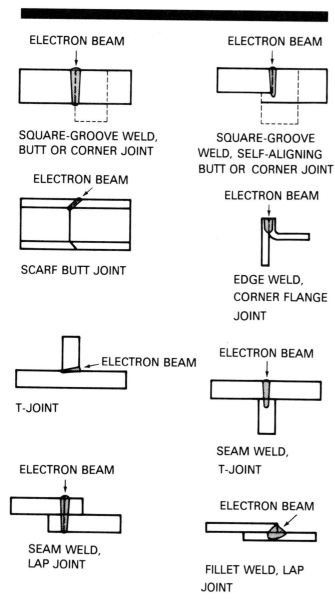

Figure 21.23—Typical Joint Designs for Electron Beam Welds

any contamination. Burrs on the sheared edges of sheet are not detrimental unless they separate the faying surfaces of lap joints.

CLEANING

CLEANLINESS IS A prime requisite for high-quality welding. The cleanliness level will depend upon the end use of the welded product. Contamination of the weld metal may cause porosity or cracking, or both, as well as a deterioration of mechanical properties. Improper cleaning of the components to be welded may lengthen chamber evacua-

tion time excessively, depending on the vacuum mode being employed.

Acetone and methylethylketone were, for many years, considered to be excellent solvents for cleaning electron gun components and workpiece parts. However, these chemicals are now considered possible toxic substances, and many facilities are presently using pure alcohol instead. Chlorinated hydrocarbon solvents should definitely not be used because of their detrimental effect on the operation of high-voltage equipment. If a vapor degreaser must contain a chlorinated hydrocarbon solvent for heavy degreasing tasks, the parts must be thoroughly washed in pure alcohol afterward. An alternative would be to degrease in a fluorocarbon type solvent. After final cleaning, the joint area should not be touched by hand or tools.

Surface oxides and other forms of contamination that solvents will not dissolve should be removed by mechanical or chemical means. Flat surfaces of soft metals, such as magnesium, aluminum, and copper, can be scraped by hand. Machining without coolant is preferred for all but very hard metals, where grinding must be used. Surfaces that are not prepared by machining should be chemically cleaned. Grit blasting and grinding are not recommended for soft metals, including soft steels, because the grit may be embedded in the surfaces. Wire brushing is not generally recommended, because it also tends to embed contaminants in the metal surface.

Nonvacuum welding will generally require less stringent precleaning than vacuum welding.

FIXTURING

ELECTRON BEAM WELDING can be accomplished by manually or automatically controlling the system's functional operation. The parts must be fixtured to align the joint, unless the design is self-fixturing, and then either the assembly or the electron beam gun column must be moved to accomplish the weld.

Where practical, self-fixturing joints should be used. A pressed or shrink fit can position circular parts for welding. However, these methods require close tolerance machining, which may not be economical for high-production welding.

Fixturing for electron beam welding need not be as strong and rigid as that required for automatic arc welding. The reason is that electron beam welds are generally made with much lower power than arc welds. Therefore, stresses in the weldment caused by thermal gradients extend over a smaller volume of metal. However, fixturing used for EBW must not introduce magnetic effects that adversely affect the beam.

The close joint fit-up and alignment required for electron beam welds generally call for fixturing made to the same tolerances. Copper chill blocks plated with nickel to avoid contamination can be used to remove heat from the joint.

Work tables and rotating positioners should have smooth and accurate motion at the required travel speeds. All fixturing and tooling should be made of nonmagnetic metals to prevent magnetic deflection of the beam. All magnetic metals should be demagnetized before welding them.

The entry and exit of the electron beam tends to produce underfill at both ends of the welded joint. To minimize or eliminate this defect, tabs of the same metal as the workpieces should be fitted tightly against both ends of the joint so that the beam can be initiated on the starting tab, traversed along the weld joint, and terminated on the runoff tab. These tabs can later be removed flush with the ends of the workpiece.

FILLER METAL ADDITIONS

WHEN FAYING SURFACES of butt joints are fitted together with acceptable tolerances, filler metal is not normally needed to obtain a full thickness weld. As welding progresses along the joint, weld metal flows from the leading edge to the trailing edge of the vapor hole. Thermal contraction, as the weld progressively freezes, usually produces a welded joint free of underfill, when proper welding procedures are used. Certain joint designs use the thermal contraction of the weldment to produce an autogenous weld from multiple weld passes; such weld procedures use a narrow tapered joint gap and a lower power density beam to produce full penetration welds. Such welds tend to exhibit few of the defects sometimes encountered with single-pass autogenous welds.

However, for some applications it is desirable or necessary to add filler metal to obtain an acceptable welded joint. Filler metal may be needed to obtain certain physical or metallurgical characteristics in the weld metal. Weld metal characteristics that may be altered or improved by filler metal addition include ductility, tensile strength, hardness, and crack resistance. For example, preplacing a thin aluminum shim in the joint can produce a deoxidizing action in mild steel, which will reduce porosity.

When filler metal is added to the joint for metallurgical purposes, wire feed is not employed exclusively. The dilution obtained from a dissimilar filler metal added as wire at the joint surface does not occur uniformly from top to bottom of the weld. For a single pass weld in heavy plate, filler metal may take the form of a thin shim. The presence of the filler shim requires that beam oscillation or a large diameter spot be used to melt the shim and the base metal on both sides of the joint. This is not the case with thin metal weldments, where filler wire can be added at the surface and dilution will occur throughout the entire joint. Typical examples of filler metal additions for metallurgical reasons are the welding of Type 6061 aluminum alloy using Type 4043 aluminum filler metal, and the welding of beryllium using aluminum or silver filler metal.

Filler metal may be added at the surface to fill the joint during a second pass after the penetration pass has been made. This is done to obtain a full thickness weld in thick plate.

Filler wire feeding equipment is usually either a modified version of that used for gas tungsten arc welding, or a unit specially designed for use in a vacuum chamber. Filler wire diameters are generally small, 0.030 in. (0.8 mm) and under, because the wire feeder must uniformly feed the wire into the leading edge of a small molten weld pool. The wire feeding nozzle should be made of a heat-resistant metal.

For welding in a vacuum chamber, the filler wire drive motor must be sealed in a vacuum-tight enclosure or otherwise designed for use in a vacuum. Outgassing from an open motor will greatly increase the work chamber evacuation time. Provisions must be made for adjusting the wire feed nozzle to position the wire with respect to the electron beam and to the weld joint over the entire length of the joint.

SELECTION OF WELDING VARIABLES

THE RATE OF energy input to the workpiece during EBW is commonly express in joules per inch, or joules per second.[4] The formula for this expression is:

$$\text{Energy input, J/in. (J/mm)} = \frac{E \times I}{S} = \frac{P}{S} \qquad (21.1)$$

where:
E = beam accelerating voltage, V
I = beam current, A
P = beam power, W or J/s
S = travel speed, in./s (mm/s)

Data for welding various thicknesses of a specific material can be plotted to permit interpolation of welding variables for that material over the range of values covered by the data. A curve relating energy input with thickness for a particular family of alloys can be determined from a few tests to establish the welding conditions for untested metal thicknesses. Figure 21.24 shows several such curves. These graphs are particularly useful to determine starting point conditions. Three factors make this possible: (1) electron beam welding machine settings are usually regulated by closed-loop servocontrols that ensure stability and reproducibility; (2) the adjustment of each variable is independently controlled, to permit flexibility in selection; (3) assuming the vacuum level and work distance being used are

---

4. Energy input to the weld from a heat source is discussed in Chapter 2, *Welding Handbook*, Vol. 1, 8th Ed., p. 33.

held constant, there are only four basic variables to adjust: accelerating voltage, beam current, travel speed, and beam focus. Beam deflection may constitute a fifth variable, if an oscillatory beam motion is employed. These variables combine to make the process of establishing the welding schedule relatively simple.

Once the required energy input per unit length is determined for a given metal thickness, the travel speed can be selected and the required welding power defined, or vice versa. The beam voltage and current can now be selected to produce the required power.

The beam size selected will be dependent upon the desired weld bead geometry. To maintain a selected beam spot diameter at the surface of the workpiece, it is necessary to correspondingly increase the focus coil current as the accelerating voltage is increased, since the beam spot size is a dependent function of the accelerating voltage. Many electron beam welding units automatically perform this compensation task. If the accelerating voltage is maintained constant but the gun-to-work distance is increased, a corresponding decrease in focus coil current is necessary to maintain a selected beam spot diameter at the surface of the workpiece.

Changes in individual welding variables will affect the penetration and bead geometry in the following manner:

(1) Accelerating voltage: as the accelerating voltage is increased, the depth of penetration achievable will also increase. For long gun-to-work distances or the production of narrow, parallel-sided welds, the accelerating voltage should be increased and the beam current decreased to obtain maximum focal range (see Figure 21.1).

(2) Beam current: for any given accelerating voltage, the penetration achievable will increase with beam current.

(3) Travel speed: for any given beam power level, the weld bead will become narrower and the penetration will decrease as the travel speed is increased.

(4) Beam spot size: sharp focus of the beam will produce a narrow, parallel-sided weld geometry because the effective beam power density will be maximum. Defocusing the beam, either by overfocusing or by underfocusing, will increase the effective beam diameter and reduce beam power density This, in turn, will tend to produce a shallow or V-shaped weld bead. These effects are shown in Figure 21.25.

Underfocusing is often used for heavy section welding in order to produce the highest possible effective aspect ratio. However, care should be taken to ensure that the depressed beam focal point does not produce a weldment having a large "nail head" or a bottle shape, since both conditions lead to weld cracking.

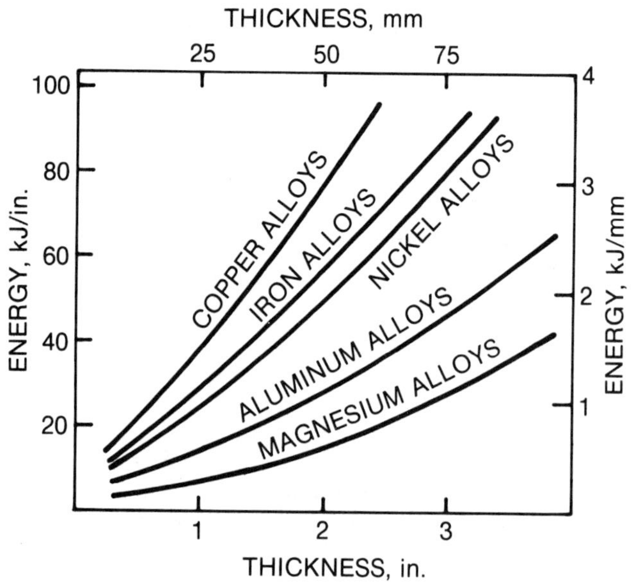

Figure 21.24–High Vacuum Electron Beam Welding Energy Requirements for Complete Penetration Welds in Several Metals as a Function of Joint Thickness

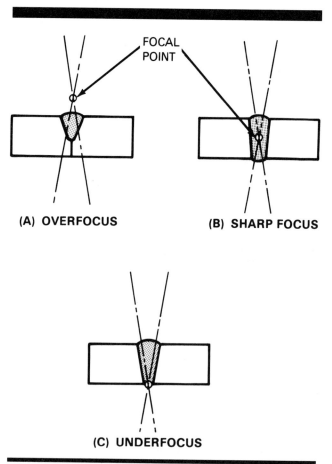

Figure 21.25–Effect of Electron Beam Focusing on Weld Bead Geometry and Penetration

METALS WELDED

IN GENERAL, METALS and alloys that can be fusion welded by other welding processes can also be joined by electron beam welding.[5] This includes similar and dissimilar metal combinations that are metallurgically compatible. The narrow weld metal geometry and thin heat-affected zones, especially in the high-vacuum mode, produce joints with better mechanical properties and fewer discontinuities than arc welded joints. However, electron beam welds in alloys that are subject to hot cracking or porosity will often contain such discontinuities. The weldability of a particular alloy or combination of alloys will depend upon the metallurgical characteristics of that alloy or combination, and the part configurations, joint design, process variation, and chosen welding procedure.

STEELS

Rimmed and Killed Steels

IN INGOT-CUT RIMMED steel, the chemical reaction that occurs between carbon and oxygen to form carbon monoxide gas (CO) will occur in the molten weld pool. As a result, violent weld pool action, spatter, and porosity in the solidified weld metal are expected with this type of steel.

Electron beam welds in rimmed steel can be improved if deoxidizers, such as manganese, silicon, or aluminum, are incorporated through filler metal additions. Deoxidizers can also be added locally to the joint area by painting, spraying, and shim inserts.

Running a low-power (low penetration) weld pass, before doing a required high-power (high penetration) pass, can often reduce the violence of the weld puddle action. Welding of rimmed steel can be improved by the careful selection of electron beam welding conditions, such as slow travel speed, to produce a wide and shallow weld cross section. The gases need time to escape from the molten weld metal. Then a weld of reasonable quality can be obtained. Various beam deflection patterns may help encourage gas escapement, and thus be effective in reducing weld porosity.

Continuous cast mild steels are silicon-aluminum killed, and therefore porosity is not a problem.

Hardenable Steels

THICK SECTIONS OF hardenable steels may crack when electron beam welded without preheat. Very rapid cooling in the fusion and heat-affected zones will result in the formation of brittle martensite. The combination of a hard, brittle microstructure and residual stresses can create cracks. Cracking can be prevented by preheating. Preheat can be supplied with a defocused electron beam in many applications, relying on careful programming and monitoring to achieve the proper preheat temperature.

Stainless Steels

Austenitic Stainless Steels. The high cooling rates typical of electron beam welds help to inhibit carbide precipitation in stainless steels because of the short time that the weld zone is in the sensitizing temperature range. However, the high cooling rate may cause cracking in highly constrained, low ferrite grades of material.

Martensitic Stainless Steels. Although these steels can be welded in almost any heat-treated condition, a hard martensitic heat-affected zone will result. Hardness and susceptibility to cracking increase with increasing carbon content and cooling rate. Rapid cooling can be prevented by preheating the base materials before welding.

Precipitation-Hardening Stainless Steels. These steels can, in general, be electron beam welded to produce good mechanical properties in the joint. The semiaustenitic types, such as 17-7PH[6] and PH14-8 Mo[6], can be welded as readily as the 18-8 types of austenitic stainless steel. The weld metal becomes austenitic during welding and remains austenitic during cooling. In the more martensitic types, such as 17-4 PH[6] and 15-5 PH[6], the low carbon content precludes formation of hard martensite in the weld metal and heat-affected zone. However, not all combinations of PH alloys can be welded without some cracking. Some precipitation-hardening stainless steels, such as 17-10P and HNM,[6] have poor weldability because of their high phosphorus content.

ALUMINUM ALLOYS

IN GENERAL, ALUMINUM alloys that can be readily welded by gas tungsten arc and gas metal arc welding can be electron beam welded. Two problems that may be encountered in some alloys are hot cracking and porosity.

The nonheat treatable series of aluminum alloy (1xxx, 3xxx, and 5xxx) can be electron beam welded without difficulty. Welded joints will possess mechanical properties similar to annealed base metal.

The heat treatable alloys (2xxx, 6xxx, and 7xxx) are crack sensitive to varying degrees when electron beam welded. Some may also be prone to weld porosity. Alumi-

---

5. The weldability of various metals is covered in Volume 4, *Welding Handbook*, 7th Ed.

6. Trademarks

num alloy Types 6061-T6 and 6066-T6, which are difficult alloys to join by other processes, can be successfully welded by the electron beam process. Best results with these alloys are obtained by incorporating small amounts of Type 4043 aluminum filler metal or 718 aluminum brazing foil in the weld.

As-welded joints in 1.5 in. (38 mm) thick Type 7075-T651 aluminum alloy exhibit lower mechanical properties than unwelded plate. The low weld properties are caused by overaging in the heat-affected zone. Postweld solution treating and aging will produce heat-treated properties in the welded joint. At high travel speeds, weld porosity may result from vaporization of certain elements in this alloy, the loss of which may change the weld metal properties. This effect should be taken into consideration before welding 7075. The high zinc content of Type 7075 aluminum alloy is responsible for vapor formation. At low travel speed, the vapor escapes to the surface before the weld metal solidifies.

Zinc-free aluminum alloys can be welded at higher speeds without developing severe porosity. It is advantageous to weld thermally hardened aluminum alloys at high travel speed to minimize the width of the softer weld and heat-affected zones.

TITANIUM AND ZIRCONIUM

TITANIUM AND ZIRCONIUM absorb oxygen and nitrogen rapidly at welding temperatures, and this reduces their ductility. Acceptable levels of oxygen and nitrogen are quite low. Therefore, these materials and their alloys must be welded in an inert environment. High vacuum electron beam welding is best for both metals, but medium vacuum and nonvacuum welding with inert gas shielding may be acceptable for some titanium applications. Most zirconium applications require that welding be performed in a vacuum or an inert gas environment, to preserve the corrosion resistance of the metal.

REFRACTORY METALS

ELECTRON BEAM WELDING is an excellent process for joining the refractory metals, because the high-power density allows the joint to be welded with minimum heat input. This is especially important with molybdenum and tungsten, because fusion and recrystallization raise the ductile-to-brittle transition temperatures of these metals above room temperature. The short time at temperature associated with electron beam welding minimizes grain growth and other reactions that raise transition temperatures.

The refractory metals rhenium, tantalum, vanadium, and niobium are readily welded. Molybdenum and tungsten are difficult to weld.

Electron beam welding joins molybdenum and tungsten successfully provided the joints are not restrained during welding. Thin sections are easy to handle, and it may be better to fabricate a composite structure by joining thin welded sections rather than to weld a single thick section. Freedom from impurities such as oxygen, nitrogen, and carbon is important. Alloys of metals containing rhenium are better suited for welding than the pure metals because they remain more ductile at lower temperatures.

DISSIMILAR METALS

WHETHER TWO DISSIMILAR metals or alloys can be welded together successfully depends upon their physical properties, such as melting points, thermal conductivities, atomic sizes, and thermal expansions. Weldability is usually predicted by empirical experience in this area. A generalization about weldability can be made by examining the alloy phase diagram of the metals to be joined. If intermetallic compounds are formed by the metals to be joined, the weld will be brittle.

Information on the relative weldability of some dissimilar metals is given elsewhere in the Handbook.[7] However, the available information about each particular application must be reviewed with regard to joint restraint and service environment. The problem of metallurgical incompatibility can sometimes be solved by the use of a filler metal shim or by welding each of the materials to a compatible transition piece. Examples are given in Table 21.2. Table 21.3 presents a summary of the weldability of various metal combinations derived from phase diagram information and accumulated practical experience.

Often the electrical couple formed at the interface produced when two dissimilar materials are being welded can induce an electromagnetic force (EMF) at elevated temperatures. If large circulating currents and magnetic fields are produced, they may cause the electron beam to be deflected from the joint centerline in medium to heavy section weldments. This undesirable effect may be corrected by broadening the beam spot, by providing a slight bias to the beam's angle of impingement, or by both techniques.

7. *Welding Handbook*, Vol. 4, 7th Ed.

Table 21.2
Examples of Filler Metal Shims for Electron Beam Welding

Metal A	Metal B	Filler Shim
Tough pitch copper	Tough pitch copper	Nickel
Tough pitch copper	Mild steel	Nickel
Hastelloy X [a]	SAE 8620 steel	321 stainless steel
304 stainless steel	Monel [a]	Hastelloy B *
Inconel 713 [a]	Inconel 713 [a]	Udimet 500 *
Rimmed steel	Rimmed steel	Aluminum

* Tradenames

Table 21.3
Weldability of Dissimilar Metal Combinations

	Silver	Aluminum	Gold	Beryllium	Cobalt	Copper	Iron	Magnesium	Molybdenum	Columbium	Nickel	Platinum	Rhenium	Tin	Tantalum	Titanium	Tungsten
Aluminum	2																
Gold	1	5															
Beryllium	5	2	5														
Cobalt	3	5	2	5													
Copper	2	2	1	5	2												
Iron	3	5	2	5	2	2											
Magnesium	5	2	5	5	5	5	3										
Molybdenum	3	5	2	5	5	3	2	3									
Columbium	4	5	4	5	5	2	5	4	1								
Nickel	2	5	1	5	1	1	2	5	5	5							
Platinum	1	5	1	5	1	1	1	5	2	5	1						
Rhenium	3	4	4	5	1	3	5	4	5	5	3	2					
Tin	2	2	5	3	5	2	5	5	3	5	5	5	3				
Tantalum	5	5	4	5	5	3	5	4	1	1	5	5	5	5			
Titanium	2	5	5	5	5	5	5	3	1	1	5	5	5	5	1		
Tungsten	3	5	4	5	5	3	5	3	1	1	5	1	5	3	1	2	
Zirconia	5	5	5	5	5	5	5	3	5	1	5	5	5	5	2	1	5

1. Very desirable
   (solid solubility in all combinations)

2. Probably acceptable
   (Complex structures may exist)

3. Use with caution
   (Insufficient data for proper evaluation)

4. Use with extreme caution
   (No data available)

5. Undesirable combinations
   (Intermediate compounds formed)

APPLICATIONS

ELECTRON BEAM WELDING is primarily used for two distinctly different types of applications: high precision and high production.

High-precision applications require that the welding be accomplished in a high-purity environment (high vacuum) to avoid contamination by oxygen, nitrogen, or both, and with minimum heat effects and maximum reproducibility. These types of applications are mainly in the nuclear, aircraft, aerospace, and electronic industries. Typical products include nuclear fuel elements, special alloy jet engine components, pressure vessels for rocket propulsion systems, and hermetically sealed vacuum devices.

High-production applications take advantage of the low-heat input and the high reproducibility and reliability of electron beam welding if a high-purity environment is not required. These relaxed conditions permit welding of components in the semifinished or finished condition using both medium and nonvacuum equipment. Typical examples are gears, frames, steering columns, and transmission and drive-train parts for automobiles; thin-wall tubing; bandsaw and hacksaw blades; and other bimetal strip applications. Figure 21.26 shows a bimetallic strip welding machine where individual strips are fed continuously into and out of the weld chamber through a series of pressure zones.

Nonvacuum electron beam welding has found its major application in high-volume production of parts whose size or composition precludes their effectively being welded in a vacuum. One example of this is the automotive industry, where nonvacuum welding is employed for many applications. A nonvacuum electron beam welded torque converter assembly is shown in Figure 21.27. The manufacture of welded tubing is another example. Integrated EB welding machine/tube mill units have been built to weld copper or steel tubing continuously at speeds up to 100 ft/min. (500 mm/s).

Figure 21.26–Electron Beam Welder Designed for Joining Bimetallic Strip

Figure 21.27–Torque Converter Assembly Welded with a Nonvacuum Electron Beam Welder (Welds Made to Hold Vanes in Place)

WELD QUALITY

TO PRODUCE WELDS that meet the requirements of specifications set forth in the welding industry, it is necessary to control three factors that are primarily responsible for electron beam weld quality: (1) joint preparation, (2) welding procedure, including provisions for keeping the beam on the seam, and (3) characteristics of the metals being welded. The first two of these are covered in other sections of this chapter.

The third factor relates to the physical and mechanical properties of the metals being welded, as well as to their metallurgical characteristics. Weld discontinuities of metallurgical origin include cracking and porosity.

Weld zones constitute regions of different microstructures within the base metal structure. Unlike a cast ingot, weld metal grains usually grow from partially melted grains at the fusion line. The phenomenon is called *epitaxial solidification*. The nature of the weld metal structure is controlled by the size and orientation of the base metal grains, and by the thermal gradients in and the shape of the weld pool.

The nature of the stress resulting from fusion welding is also important. Metal immediately adjacent to the moving weld pool is first heated; it expands against the restraining forces of the surrounding cold base metal; then it cools and contracts. In the process, that metal is plastically deformed (upset) during the heating cycle and restrained in tension during cooling. Residual tensile and compressive stresses surround the weld zone, often resulting in warpage of the welded assembly.

In considering these factors, it would appear that electron beam welding offers the following unique characteristics for controlling the weld joint properties.

(1) Base metal recrystallization and grain growth can be minimized.
(2) Beam oscillation and travel speed can be used to control the shape of temperature gradients in the weld pool.
(3) Low-heat inputs result in low thermal stresses in the base metal and, hence, low distortion.

Residual stresses are symmetrically distributed due to the characteristic two-dimensional symmetry (parallel sides) of the electron beam weld zone.

Unfortunately, it is not always possible to realize the full potential of the process, since the weldability of a metal is ultimately controlled by metallurgical factors. For this reason, electron beam welds may exhibit most of the common discontinuities associated with fusion welding. A possible exception is hydrogen-induced cold cracking of carbon steel weldments because normally there is no source of hydrogen in an autogenous high vacuum electron beam setup.

One type of discontinuity sometimes found in partial joint penetration welds is large voids at the bottom of the weld metal. Typically, a large number of these voids will be aligned and will appear as linear porosity rather than scattered porosity. When the weld barely penetrates through the joint, root porosity will appear as a lack of fill, accompanied by spatter on the back side of the weld.

Another occurrence peculiar to the vacuum mode of welding is the release of trapped air through the molten weld metal. This sometimes creates a defect. It happens during an attempt to weld a gas-filled container that is not properly vented to vacuum.

Other discontinuities are generally the same as those found in other types of fusion welds. Electron beam weld discontinuities include the following:

(1) Porosity
(2) Shrinkage voids
(3) Cracking
(4) Undercutting
(5) Underfill
(6) Missed joints
(7) Lack of fusion

The probability of encountering these discontinuities is more pronounced when welding thick sections. Knowledge of the causes of the discontinuities and means for avoiding them are essential for the production of high-quality welds. As an example, in welding thick sections in the horizontal position, holes and porosity can be avoided by tilting the beam axis a few degrees out of the plane of welding. Equally important is a reliable nondestructive testing method, such as ultrasonic inspection, to determine the presence of certain types of defects that are not detectable by radiography.

The narrow weld beads created by electron beam welding make radiographic inspection difficult. Certain joint designs incorporate a feature called a *radiographic window*. As shown in Figure 21.28, this provides a void within the joint that is easily resolved by radiographic technique, when not completely consumed by the weld. This window

can be located at any position in the joint, and its absence in the radiograph after welding assures that penetration to that depth has been achieved.

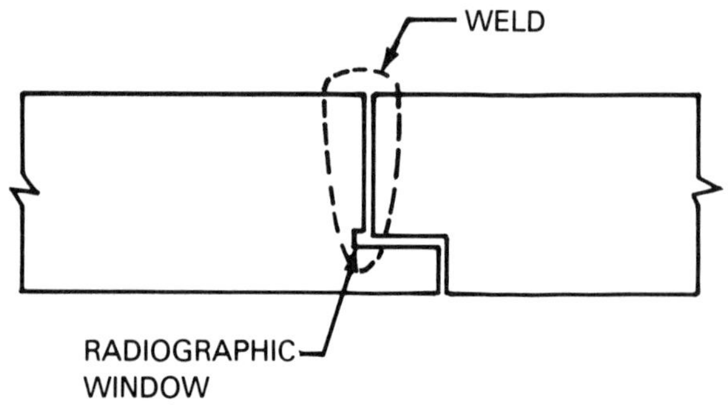

POROSITY AND SPATTER

POROSITY IN ELECTRON beam welds is caused by the evolution of gas as the metal is melted by the beam. The gas may form as a result of (1) the volatilization of high vapor pressure elements in the alloy, (2) the escape of dissolved gases, or (3) the decomposition of compounds such as oxides and nitrides. Copper-zinc alloys (brasses) and aluminum-magnesium alloys are difficult to electron beam weld because metal vapors evolve. Both zinc and magnesium have low boiling points. Dissolved gases and compounds are likely to be present in alloys originally melted in air or under protective gas atmospheres.

Spatter is caused by the same factors as porosity. The rapid evolution of gas or metal vapor causes the ejection of drops of molten weld metal that scatter over the work surface and within the chamber. Spatter and porosity can even occur in vacuum remelted alloys when a residual phase volatilizes under the intense heat of the electron beam.

An effective means of preventing porosity and spatter is to weld only vacuum melted or fully deoxidized metals. When gas-emitting metals or high vapor pressure alloys must be welded, special techniques are required to minimize porosity. Filler metal containing a deoxidizer may be added when welding metals that are not completely deoxidized. Slow welding speed will provide time for gas bubbles to escape from the molten metal.

An oscillatory beam deflection may be effective in reducing porosity. In extreme cases, remelting the joint a second or third time will reduce it. However, these techniques reduce joint strength in age-hardening alloys that are heat-treated prior to welding.

Figure 21.28–Radiographic Window Feature Often Used to Simplify Reading of Weld Joint Radiographs

SHRINKAGE VOIDS

SHRINKAGE VOIDS MAY occur between dendrites near the center of the weld metal. These voids are characterized by irregular outlines of porosity. Shrinkage voids usually occur in alloys having high volumetric shrinkage on solidification. In electron beam welds where the bond lines are essentially parallel, solidification proceeds uniformly from the base metal to the center of the weld. When solidification shrinkage of the metal is great, voids will form if the face and root surfaces freeze before the center of the weld. An example of shrinkage voids in an electron beam weld in 15-7Mo PH stainless steel is shown in Figure 21.29. Low travel speed or beam oscillation may minimize or eliminate shrinkage voids by increasing the volume of molten metal and decreasing the solidification rate. However, these conditions will generally widen the fusion zone.

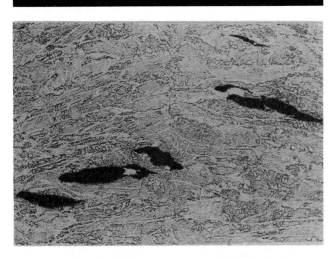

Figure 21.29–Shrinkage Voids in an Electron Beam Weld

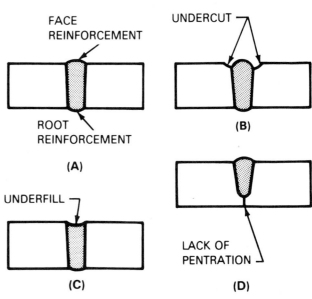

Figure 21.30–Correct (A) and Incorrect (B, C, and D) Electron Beam Weld Geometries

CRACKING

HOT OR COLD cracks may form in electron beam welds in alloys that are subject to these types of cracking. Hot cracking is generally intergranular and cold cracking is transgranular. Hot cracks form in a low-melting grain boundary phase during solidification of the weld metal. Cold cracks form after solidification as a result of high internal stresses produced by thermal contraction of the metal during cooling. A crack originates at some imperfection or point of stress concentration in the metal and propagates through the grains by cleavage.

Hot cracking may be minimized by welding at high travel speeds with minimum beam energy. Cold cracking may be overcome by redesigning the joint to eliminate points of stress concentration. Quench-hardenable steels should be preheated to a suitable temperature to control the formation of martensite in the weld zone.

UNDERCUTTING

ELECTRON BEAM WELDS with good bead geometry have essentially parallel bond lines with a uniform crown or buildup of weld metal on the top surface, as shown in Figure 21.30(A). Undercutting refers to grooves produced in the base metal at the edges of the weld bead, as shown in Figure 21.30(B). Undercut occurs when the weld metal does not wet the base metal. Undercutting is promoted by very high travel speeds, improper cleaning procedures, or beam asymmetry (it usually occurs on one side only). Alloy additions that reduce surface tension or increase fluidity, such as aluminum additions to carbon steel welds, have a beneficial effect.

Undercutting on the top surface of the weld can sometimes be filled by making a "cosmetic pass". This is usually performed at lower power levels relative to the penetration pass, and can be made more effective by beam deflection or a "defocused" beam to widen the top of the bead. Certain joint designs provide extra metal above the desired "finished surface" of the weldment which is machined off after the weld. Undercut is removed during the machining operation.

UNDERFILL

FULL PENETRATION WELDS can develop either a uniform or irregular root surface, depending on the weld variables and material. The width of the root surface is dependent upon the welding conditions. In thick sections of metal, such as 3-in. (76 mm) stainless steel, the face and root surface shapes are dependent upon the surface tension supporting the column of molten metal as it is being transported along the weld joint. At low welding speeds there will be a relatively large mass of molten weld metal, and the bead will tend to sag due to insufficient surface tension and the force of gravity. This will form an extremely heavy root reinforcement and the weld face may show severe underfill (concavity), as shown in Figure 21.30(C). Various techniques can eliminate this condition. These include the use of a backing strip, a step joint, or welding in the horizontal or the vertical position.

Excessive sagging of the root surface usually results when the beam energy is too high or the molten weld metal is too wide. This can be reduced by proper adjustment of the welding variables. If underfilling persists at the best beam operating conditions, filler metal must be added to fill the groove. A number of techniques are effective in providing the required filler metal. One is to place a narrow strip over the face of the joint and then weld through it. The thickness of the strip must be slightly greater than the depth of any undercut, so that the undercut will be entirely in the strip. Filler metal wire may similarly be added to the leading edge of the weld as it is being made, or during a subsequent smoothing pass made with a defocused beam. On circular welds, beam power ramping (upslope and downslope) may be used to minimize defects in the overlap area.

LACK OF PENETRATION

THERE ARE NUMEROUS applications of electron beam welding in which full penetration of the joint is not required. These applications generally involve seal welds or welds subjected to shearing forces only. In these cases, the sharp notch at the root of the weld is acceptable. However, when a welded joint must support a transverse tensile stress at the root of the weld, full joint penetration is required. Lack of penetration may be caused by low beam power, high travel speed, or improper focusing of the beam. This condition is shown in Figure 21.30(D).

MISSED JOINTS

WHEN A SMALL diameter electron beam is used to make a long joint in a thick section, the beam axis must be in the same plane as the joint faces and remain aligned with the joint along its entire length of travel. Otherwise, the possibility of missing the joint at some location is great. Even when the beam is properly aligned with the joint, electrostatic or magnetic forces can cause beam deflection, resulting in portions of the joint being missed. An electrostatic field can be generated by the accumulation of an electrical charge on an insulated surface, such as the glass in the vacuum chamber windows. The electron beam will be deflected away from or toward the charged surface if the beam passes close to it.

Residual magnetism in a ferromagnetic base metal or in the fixturing can cause unexpected beam deflection. For example, a steel part may be magnetized during grinding if it is held by a magnetic chuck, and the residual magnetism in the part will cause the beam to deflect and miss the joint. This can be avoided by demagnetizing all ferromagnetic parts before welding, and by using nonmagnetic materials for fixturing.

Unexpected beam deflection can occur when welding dissimilar metals, especially when one is ferromagnetic. An example of this is shown in Figure 21.31, a weld between a nonmagnetic nickel-base alloy and a magnetic maraging steel. Residual or induced magnetism in the steel deflected the electron beam and caused lack of fusion at the root of the joint. If dissimilar materials are to be welded in production, it is important that test welds be made and examined to determine whether beam deflection will occur.

Figure 21.31–Beam Deflection When Welding Dissimilar Metals

The occurrence of missed joints can be verified by using joint designs which include witness lines and radiographic windows. Witness lines are scribed parallel to the joint on the face or root side of the joint, or both. The weld lies between these lines, and its position relative to the joint can be determined by postweld examination. See Figure 21.32.

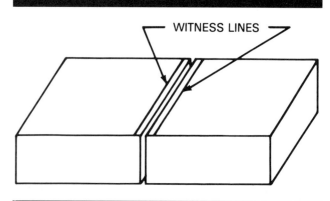

WITNESS LINES

Figure 21.32–Witness Lines Scribed Parallel to Joint

LACK OF FUSION

LACK OF FUSION occurs mostly in partial penetration welds. However, it can also occur near the root of full penetration welds made with insufficient beam power. Figure 21.33 shows an example of this in an electron beam weld in a titanium alloy.

Lack of fusion generally can be avoided by using properly adjusted welding variables. There are circumstances, however, where partial penetration welding is required. One example is the welding of circular joints where the beam power and penetration must be decreased as the end of the weld overlaps the start, to avoid crater formation. A partial penetration weld is formed in the overlap and lack of fusion can occur. Another example is the welding of thick sections. Two partial penetration weld passes, one from each side of the joint, may be required when the metal thickness is too great to be penetrated in a single pass. The second pass must reach the root of the first pass.

Welding with a slightly defocused beam and low travel speed (to compensate for the lower energy density) is effective in eliminating lack of fusion. Beam oscillation, either circular or transverse, is sometimes effective. Preheating is helpful because it reduces the thermal gradients at the root of the weld. Lack of fusion is difficult to locate nondestructively because it is similar to fine cracks and usually can not be detected with x-ray inspection. Penetrant tests are ineffective because the unfused area does not usually extend to the surface.

Ultrasonic testing is the only nondestructive test method that can detect lack of fusion in electron beam welds. Experienced nondestructive test personnel are required to perform the test and interpret the results. Even then, the test method is not suitable for many applications. Since certain joint designs can be easier to inspect ultrasonically, NDT personnel should be consulted prior to designing the joint on critical assemblies.

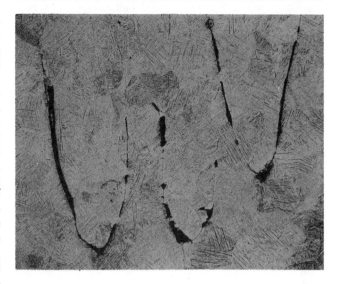

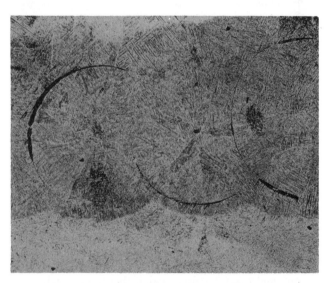

Figure 21.33–Lack of Fusion or Spiking in Vertical (top) and Horizontal (bottom) Sections Through an Electron Beam Weld in a Titanium Alloy

SAFETY PRECAUTIONS

SINCE ELECTRON BEAM welding machines employ a high-energy beam of electrons, the process requires users to observe several safety precautions not normally necessary with other types of fusion welding equipment. The four primary potential dangers associated with electron beam equipment are electric shock, x-radiation, fumes and gases, and damaging visible radiation. In addition to the potential dangers associated with welding specific materials (beryllium, etc.), there may also be a potential danger associated with collateral materials (solvents, greases, etc.) used in operating the equipment. Precautionary measures should be taken to assure that all required safety procedures are strictly observed. ANSI/AWS F2.1, *Recommended Safe Practices for Electron Beam Welding and Cutting*, and ANSI/ASC Z49.1, *Safety in Welding and Cutting* (latest editions) give the general safety precautions that must be taken.

ELECTRIC SHOCK

EVERY ELECTRON BEAM welding system operates with a voltage level high enough to cause fatal injury, regardless of whether the system is referred to as being a "low voltage" or a "high voltage" unit. Manufacturers of electron beam equipment, by meeting various underwriter requirements, attempt to ensure that their machines are well-insulated against the dangers of contact with the high voltage. However, all precautions required when working around high voltages should still be observed when working with EBW machines.

X-RADIATION

THE X-RAYS GENERATED by an electron beam welding machine are produced when electrons, traveling at a high velocity, collide with matter. The majority of these x-rays are produced when the electron beam impinges upon the workpiece. Substantial amounts are also produced when the beam strikes gas molecules or metal vapor in both the gun column and work chamber. Underwriters Laboratories and Federal regulations have established firm rules for permissible x-radiation exposure levels, and producers and users of equipment must observe these rules.

Approximately 1 in. (25 mm) thick steel, when used for manufacturing the work part vacuum enclosure, will be sufficient to satisfy the x-ray shielding required for beam systems using accelerating voltages of up to 60 kV, assuming proper design. For units with higher beam accelerating voltages, either a much thicker steel or use of a lead covering on top of steel is needed to satisfy the x-ray shielding requirements for these units. Leaded glass windows are employed in both high and low voltage electron beam systems. In general, commercial shielded vacuum chamber walls and leaded glass windows provide sufficient radiation protection for operators.

In the case of nonvacuum systems, some type of radiation enclosure must be provided to assure the safety of the operator. Thick walls of high-density concrete (or some other similar material) may be selected instead of steel and lead, especially if a large radiation enclosure is required. Special safety precautions should be imposed to prevent personnel from accidently entering or being trapped inside these enclosures when equipment is in operation.

A complete x-ray radiation survey of the electron beam equipment should always be made at the time of installation and at regular intervals thereafter. These surveys should be performed by personnel trained in the proper procedures for doing a radiation survey, and knowledgeable about the use of radiation survey equipment, in order to assure initial and continued compliance with all radiation regulations and standards applicable to the site where the equipment is installed.

FUMES AND GASES

IT IS UNLIKELY that the very small amount of air left in a high vacuum electron beam chamber would be sufficient to produce ozone and oxides of nitrogen in harmful concentrations. However, nonvacuum and medium vacuum electron beam systems are capable of producing these by-products, as well as other types of airborne contaminants, in concentrations well above acceptable levels.

Adequate area ventilation should be employed to reduce concentrations of any airborne contaminants around the equipment below the maximum allowable exposure levels. Proper exhausting techniques should be employed to maintain residual concentrations in the chamber or enclosure below these same limits.

VISIBLE RADIATION

DIRECT VIEWING OF visible radiation emitted by the molten weld metal can be harmful to eyesight. In the presence of intense light sources, proper eye protection is necessary. Optical viewing should be done through filters in accordance with ANSI Z87.1, *Occupational and Educational Eye and Face Protection* (latest edition).

SUPPLEMENTARY READING LIST

Baujat, V. and Charles C. "Submarine hull construction using narrow groove GMAW." *Welding Journal* 69(8): 31-35; August 1990.

Bench, F. K. and Ellison, G. W. "EB welding of 304L stainless steel with cold wire feed." *Welding Journal* 53(12): 763-766; December 1974.

Ben-Zvi, I., Bogart, L., and Turneaure, J. P. "Simple device for controlling 100 percent penetration in electron beam welds." *Welding Journal* 51(12): 842-843; December 1972.

Bibly, M. J., Burbridge, G., and Goldak, J. A. "Cracking in restrained EB welds in carbon and low alloy steels." *Welding Journal* 54(8): 253s-258s; August 1975.

———. "Gases evolved from electron beam welds in plain carbon steels." *Welding Journal* 51(12): 844-847; December 1972.

Caroll, M. J., and Powers, D. E. "Automatic joint tracking for CNC-programmed electron beam." *Welding Journal* 64(8): 34-38; August 1985.

Dietrich, W. "Investigation into electron beam welding of heavy sections." *Welding Journal* 57(9): 281s-284s; September 1978.

Dixon, R. D., Milewski, J. O. and Fetzko, S. "Electron beam welding data acquisition system using a personal computer." *Welding Journal* 66(4): 41-46; April 1987.

Dixon, R. D. and Pollard, L. Jr. "Effect of accurate voltage control on partial penetration EB welds." *Welding Journal* 53(11): 495s-497; November 1974.

Farrell, W. J. and Ferrario, J. D. "A computer-controlled, wide-bandwidth deflection system for EB welding and heat treating." *Welding Journal* 66(10): 41-49; October 1987.

Fink, J. H. "Analysis of atmospheric electron beam welding." *Welding Journal* 54(5): 137s-143s; May 1975.

Gajdusek, E., "Advances in nonvacuum electron beam technology." *Welding Journal* 59(7): 17-21; July 1980.

Hinrichs, J. F., et al., "Production electron beam welding of automotive frame components." *Welding Journal* 53 (8): 488-493; August 1974.

King, J. F., David, S. A., Sims, J. E. and Nasreldin, A. M. "Electron beam welding of heavy-section 3Cr - 1.5 Mo alloy." *Welding Journal* 65(7): 39-47; July 1986.

Lubin, B. T. "Dimensionless parameters for the correlation of electron beam welding variables." *Welding Journal* 47 (3): 140s-144s; March 1968.

Mayer, R., Dietrich, W., and and Sundermeyer, D. "New high-speed beam current control and deflection systems improve electron beam welding applications." *Welding Journal* 56(6): 35-41; June 1977.

Metzbower, E. A. "Laser beam welding: thermal profiles and HAZ hardness." *Welding Journal* 69(7): 272s; July 1990.

Metzger, G. and Lison, R. "Electron beam welding of dissimilar metals." *Welding Journal* 55(8): 230s-240s; August 1976.

Murphy, J. L., Mustaleski, T. M. and Watson, L. C. "Multipass autogenous electron beam welding." *Welding Journal* 67(9): September 1988.

Murphy, J. L. and Turner, P. W. "Wire feeder and positioner for narrow groove electron beam welding." *Welding Journal* 55(3): 181-190; March 1976.

Mustaleski, T.M., McCaw, R. L. and Sims, O. E. "Electron beam welding of nickel - aluminum bronze." *Welding Journal* 67(7): 53-59; July 1988.

O'Brien, T. B., el al., "Suppression of spiking in partial penetration EB welding." *Welding Journal* 53(8): 332s-338s; August 1974.

Patterson, R. A., et al. "Titanium aluminide: electron beam weldability." *Welding Journal* 69(1): 39s; January 1990.

Powers, D. E. and Colegrove, R. K. "A new mobile EB gun/column assembly." *Welding Journal* 65(9): 47-51; September 1986.

Powers, D. E. and LaFlamme, G. R. "EBW vs. LBW - A complete look at the cost and performance traits of both processes." *Welding Journal* 67(3): 25-31; March 1988.

Privoznik, L. J., Smith, R. S., and Heverly, J. S. "Electron beam welding of thick sections of 12 percent Cr turbine grade steel." *Welding Journal* 50(8): 567-572; August 1971.

Sandstrom, D. J., Bucken, J. F., and Hanks, G. S. "On the measurement and interpretation and application of parameters important to electron beam welding." *Welding Journal* 49(7): 293s-300s; July 1970.

Schumacher, B. W. "Atmospheric EB welding with large standoff distance." *Welding Journal* 52(5): 312-314; May 1973.

Schwartz, M. M., "Electron beam welding." *Bulletin 196* New York: Welding Research Council, July 1974.

Tews, P., et al., "Electron beam welding spike suppression using feedback control." *Welding Journal* 55(2): 52s-55s; February 1976.

Tong, H. and Geidt, W. H. "A dynamic interpretation of electron beam welding." *Welding Journal* 49(6): 259s-266s; June 1970.

Weber, C. M. and Funk, E. R. "Penetration mechanism of partial penetration electron beam welds." *Welding Journal* 51(2): 90s-94s; February 1972.

Weidner, C. W. and Schuler, L. E. "Effect of process variables on partial penetration electron beam welding." *Welding Journal* 52(3): 114s-119s; March 1973.

LASER BEAM WELDING

PREPARED BY A COMMITTEE CONSISTING OF:

D. E. Powers, Chairman
PTR - Precision Technologies, Incorporated

R. F. Duhamel, Co-Chairman
United Technologies Industrial Lasers

P. Anthony, Co-Chairman
Rofin-Sinar, Incorporated

D. A. Belforte
Belforte Associates

K. W. Carlson
Westinghouse Laser Center

L. S. Derose
Texcel Incorporated

D. Elza
Coherent General

D. Gustaferri
Ferranti Sciaky, Incorporated

A. Lingenfelter
Lawrence Livermore National Laboratory

R. W. Walker
Laser Consulting Services

WELDING HANDBOOK COMMITTEE MEMBER:
R. M. Walkosak
Westinghouse Electric Corporation

CHAPTER **22**

LASER BEAM WELDING

FUNDAMENTALS OF THE PROCESS

DEFINITION AND GENERAL DESCRIPTION

A LASER IS a device that produces a concentrated coherent light beam by stimulated electronic or molecular transitions to lower energy levels. Laser is an acronym for light amplification by stimulated emission of radiation. Coherent means that all the light waves are in phase.

In practice, a laser device consists of a medium placed between the end mirrors of an optical resonator cavity. When this medium is "pumped" (i.e., excited) to the point where a population inversion occurs, a condition wherein the majority of active atoms (or molecules) in this medium are put into a higher-than-normal energy state, a source of coherent light that can then reflect back and forth between the end mirrors of the cavity will be provided. This results in a cascade effect being induced which will cause the level of this coherent light to reach a threshold point (i.e., the point at which the gain in light amplification being produced begins to exceed any losses in light that might simultaneously be occurring), thereby allowing the device to start to emit a beam of laser light.

From an engineering standpoint, a laser is an energy conversion device that simply transforms energy from a primary source (electrical, chemical, thermal, optical, or nuclear) into a beam of electromagnetic radiation at some specific frequency (ultraviolet, visible, or infrared). This transformation is facilitated by certain solid, liquid, or gaseous mediums which, when excited on either a molecular or atomic scale (by various techniques), will produce a very coherent and relatively monochromatic (i.e., exhibiting a fairly singular frequency) form of light — a beam of laser light. Because they are coherent and monochromatic, both low-power and high-power laser light beams have a very low divergence angle. Thus they can be transported over relatively large distances before being highly concentrated (through the use of either transmissive or reflective-type focusing optics) to provide the level of beam power density needed to do a variety of material processing tasks such as welding, cutting, and heat treating.

The first laser beam was produced in 1960 using a ruby crystal pumped by a flash lamp. Solid-state lasers of this type produce only short pulses of light energy, and at repetition frequencies limited by heat capacity of the crystal. Consequently, even though individual pulses do exhibit instantaneous peak power levels in the megawatt range, pulsed ruby lasers are limited to low average power output levels. Both pulsed and continuously operating solid-state lasers, capable of welding and cutting thin sheet metal, are currently commercially available. Many of the latter utilize neodymium-doped, yttrium aluminum garnet (Nd-YAG) crystal rods to produce a continuous, monochromatic beam output in the 1 to 2 kW power range.

Electrically pumped, pulsed and continuous wave (CW) gas lasers of the ac, dc and rf excited variety have also been developed. Thus carbon dioxide (CO_2) lasers, with beam power outputs of up to 25 kW, are commercially available today, and are in use for a wide variety of industrial material processing tasks. Such lasers are capable of providing full penetration, single-pass welds in steel up to 1-1/4 in. (32 mm) thick.

PRINCIPLES OF OPERATION

LASER BEAM WELDING (LBW) is a fusion joining process that produces coalescence of materials with the heat obtained from a concentrated beam of coherent, monochromatic light impinging on the joint to be welded. In the LBW process, the laser beam is directed by flat optical elements, such as mirrors, and then focused to a small spot (for high-power density) at the workpiece using either reflective focusing elements or lenses. LBW is a noncontact process, and thus requires that no pressure be applied. Inert gas shielding is generally employed to prevent oxidation of the molten puddle, and filler metal may occasionally be used.

As described above, the lasers predominantly being used for industrial material processing and welding tasks are the 1.06 μm wavelength YAG laser and the 10.6 μm wavelength CO_2 laser, with the active element most commonly employed in these two varieties of lasers being the neodymium (Nd) ion, and the CO_2 molecule (respectively).

Solid-State Lasers

SOLID-STATE LASERS utilize an impurity in a host material as the active medium. Thus the neodymium ion (Nd^{+++}) is used as a "dopant", or purposely added impurity in either a glass or YAG crystal, and the 1.06 μm output wavelength is dictated by the neodymium ion. The lasing material, or host, is in the form of a cylinder about 6 in. (150 mm) long by 0.375 in. (9 mm) in diameter. Both ends of the cylinder are made flat and parallel to very close tolerances, then polished to a good optical finish and silvered to make a reflective surface. The crystal is excited by means of an intense krypton or xenon lamp. A simplified schematic arrangement of the rod, lamp, and mirrors is shown in Figure 22.1.

The selection of the host material for the neodymium ion depends on several factors. These include the ability to produce large quantities of good optical quality rods (i.e., having an acceptable hardness and polishability factor), with acceptable levels of thermal conductivity, fluorescent lifetime, efficiency, and optical absorption bands. All of these factors influence the ability of the system to emit reasonable amounts of energy in a single pulse, and successful materials are those from which large amounts of energy can be extracted. Since the YAG crystal possesses all of the ideal characteristics outlined, it makes an excellent host material.

The output characteristics of Nd:YAG lasers depend on the excitation method, which may be either continuous or repetitively pulsed in nature. In continuous operation, the laser is excited with either xenon lamps for power levels up to 10 W or krypton lamps for power levels in the range of 100 W and greater. For repetitively pulsed lasers, the output characteristics depend on lamp configuration. The most common configuration is shown in Figure 22.1. Table 22.1 gives the characteristics of Nd:YAG lasers and offers some idea of the capability for trade-offs between the average power, pulse energy, pulse duration, and pulse repetition rates for such lasers.

The relatively narrow frequency band exhibited by Nd:YAG lasers facilitates continuous wave operation at room temperature, making the continuous wave Nd:YAG laser second only to the CW gas lasers in terms of continuous wave power generation. However, its considerably lower overall efficiency capability (i.e., typically two percent versus ten percent for gas lasers) results in a lower power output.

In the pulsed mode, the active medium of a YAG laser is pumped intermittently, instead of continuously, by employing a pulsed power supply to drive the flashlamp.

Figure 22.2 shows the time relationship of the flashlamp and laser output pulses of a typical pulsed solid-state laser. The beginning of the flashlamp pulse establishes a population inversion in the active medium. When the loop gain reaches 1.0, lasing begins and continues as a series of closely spaced spikes for the duration of the flashlamp pulse. These spikes are produced by gain switching in the

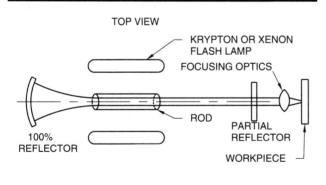

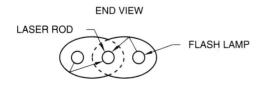

Figure 22.1–Schematic View of a Solid-State Laser

Table 22.1
Output of a Nd: YAG Laser

Continuous wave operation	
Average power	<1000 W (multimode)
	<20 W (TEM$_{00}$)
Divergence	1-20 mrad
Beam diameter	0.04-0.4 in.
Pulse Length of 0.1 TO 20 ms	
Output energy	<500 J/pulse (multimode)
	5J/pulse (TEM$_{00}$)
Repetition rate	200 Hz
Divergence	10 mrad (multimode)
	3 mrad (TEM$_{00}$)
Beam diameter	0.2-0.4 in.
Pulse length of 0.1 to 1 μs (repetitive switch)	
Output energy	1 mJ/pulse
Repetition rate	50-100 kHz
Average power	10-100 W
Peak power	10-50 kW

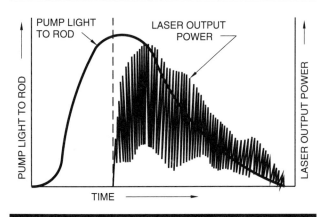

Figure 22.2–Output of a Typical Pulsed Solid-State Laser Compared to Pump Light Input to Rod as a Function of Time

Table 22.2
Output of a Nd-Glass Laser (for a Pulse Length of 1 to 10 ms)

Output energy	20 J/pulse (multimode)
Repetition rate	10 Hz
Divergence	5-10 mrad
Beam diameter	0.2-0.4 in.

active medium. The gain rises quickly to a high value because of the intense pumping level. This results in a high loop gain and a high-intensity standing wave in the optical cavity. This quickly depletes the population inversion for that particular wavelength, and lasing stops. Thus, the laser switches itself off momentarily by using up all of its gain.

Because of the spiking in the output, the peak power of a pulsed solid state laser tends to be difficult to determine, and tends to vary from pulse to pulse, even though the overall energy and duration of each pulse may remain constant.

For these reasons, specifications of pulsed solid state lasers usually do not include the maximum output power. Instead, pulse energy and pulse duration are specified. Peak output power may be approximated by dividing the energy of the output pulse by pulse duration as with other pulsed lasers.

Solid-state laser pulse durations vary from as short as 50 μs to as long as 50 ms, with the usual pulse duration employed being about 1 ms. Only Nd:YAG systems are capable of pulse durations much greater than 2 ms, and some Nd:YAG laser drillers use pulse durations of 5 to 8 ms.

Glass also has certain desirable characteristics as a laser host material. One is that large pieces of high optical quality can be fabricated into a variety of sizes and shapes, ranging from fibers with diameters of a few microns to rods with diameters on the order of 4 in. (100 mm) and lengths of up to 6 1/2 ft (2 m). However, since the thermal conductivity of glass is lower than that of most crystalline hosts, cooling it will present a problem that can limit the maximum repetition rate at any given pulse energy level. Also, the emission lines of ions in glass are broader than those in crystalline materials. This raises the threshold of the glass for laser action, because a higher population inversion is required to achieve the same gain. The output characteristics of Nd-glass lasers suitable for laser welding are given in Table 22.2.

Gas Lasers

THE ELECTRIC DISCHARGE style CO_2 gas lasers are the most efficient type currently available for high power LB material processing. These lasers employ a gas mixture of primarily nitrogen and helium containing a small percentage of carbon dioxide, and an electric glow discharge is used to pump this laser medium (i.e., to excite the CO_2 molecule). Gas heating produced in this fashion is controlled by continuously flowing the gas mixture through the optical cavity area, and thus CO_2 lasers are usually categorized according to the type of gas flow system they employ: slow axial, fast axial, or transverse.

Slow Axial Flow. The slow axial flow (SAF) is the simplest CO_2 gas laser. Gas flow is in the same direction as the laser resonator's optical axis and electric excitation field, or gas discharge path, as shown in Figure 22.3. The axial flow of gas is maintained through the tube to replenish molecules depleted from the effects of the gas discharge being used for excitation, which causes the CO_2 to be reduced to $CO + O$ by electron bombardment. Catalytic devices are employed in the recycling gas flow path to help accomplish a recombination effect.

Cooling of the laser gas is achieved by conduction through the walls of the discharge tubes to a liquid coolant in the cooling mantle, and some form of external heat exchanger system is then used to dissipate the heat being continuously extracted in this manner.

A mirror is located at each end of the discharge tubes to complete the resonator cavity. Typically, one mirror is totally reflective (rear mirror) and the other is partially reflective and partially transmissive (output coupler). Slow axial flow resonators are capable of generating laser beams with a continuous power rating of approximately 80 watts for every meter of discharge length. A folded tube configuration is used for achieving output power levels of 50 to 1000 watts, maximum, rather than simply extending the length of the resonator cavity to attain these power levels.

Fast Axial Flow. Fast axial flow (FAF) lasers have a similar arrangement of components to that of the slow axial flow laser described above except that, in the case of the FAF laser, a Roots blower or turbo pump is used to circulate the laser gas at high speed through the discharge region and corresponding heat exchangers. See Figure 22.4.

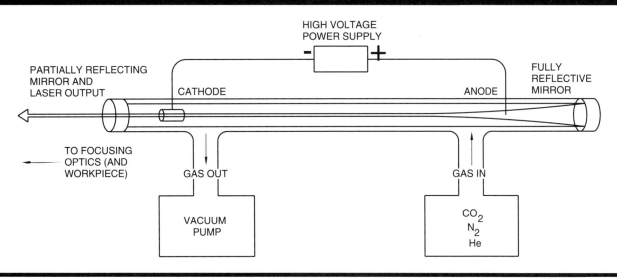

Figure 22.3–Schematic Diagram of a Slow Axial Flow Laser

Within the confines of the laser itself, cooling is enhanced by forcing the hot laser gas through gas-to-liquid heat exchangers. This gives a much higher rate of heat extraction than is available with slow-axial flow lasers, and provides the capacity for achieving output power levels of greater than 2 kW per meter of discharge length. Fast axial flow (FAF) lasers with CW output power levels of between 500 and 6000 watts are currently available. As with SAF lasers, most FAF lasers can readily be pulsed.

Transverse Flow.   Transverse flow lasers operate by continuously circulating gas across the resonator cavity axis by means of a high speed fan-type blower, while maintaining an electric discharge perpendicular to both the gas flow direction and the laser beam's optical axis. Because the volume of the resonator is large relative to its length, mirrors can be placed at each end to reflect the beam several times through the discharge region before transmission through an output coupler. A transverse flow laser is shown in Figure 22.5. The ability to achieve a long optical path within a short resonator structure allows transverse flow lasers to be both compact and capable of generating high output powers. Transverse flow lasers with CW output power levels of between 1 and 25 kW are currently available.

BEAM DELIVERY AND FOCUSING OPTICS

LASER BEAMS MUST be focused to a small diameter to produce the high-power density required for welding. This focusing is accomplished with transmitting optics (lenses) or reflective optics (mirrors). See Figures 22.6 and 22.7. Minimum spot size can be varied by optics design and choice of focal length. For a given laser beam, the final focused spot size attainable will be directly proportional to the focal length employed. Thus the resultant power density achieved will vary inversely proportional to the square of the focal length, while the depth of focus attained will vary directly with focal length. Therefore, laser beams being focused with short focal length optics require that greater precision be observed (in maintaining the lens-to-workpiece distance) than when longer focal length optics are employed.

The shortest practical focal length to use for CO_2 laser welding is approximately 5 in. (125 mm). This is because of

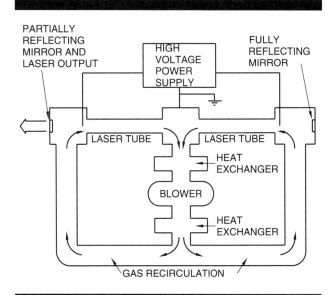

Figure 22.4–Schematic View of a Fast Axial Flow Laser

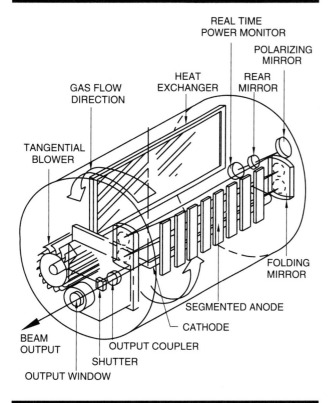

Figure 22.5–Schematic Diagram of a Transverse Flow Laser

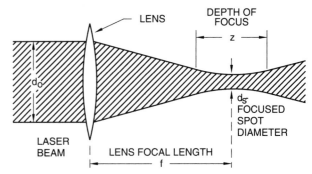

Figure 22.6–Focusing a Gaussian Beam With a Simple Lens

the adverse effects which the weld spatter and vapor produced during processing can have on the focusing optics. Since the spot size at the focal plane varies inversely with the diameter of the beam incident upon the focusing optic element, a beam expander may be used to increase the beam's diameter prior to focusing, thus allowing longer focal lengths to be employed, without having to sacrifice power density.

Referring to Figure 22.6, the ratio of the focal length of the optics to the beam diameter (f/d_o) is referred to as the F number (F#). The focused laser beam's spot size (i.e., the focused beam spot diameter d_s) will vary directly proportional to the laser beam's wavelength and the F# of the focusing system, as shown in equation 22.1.

$$d_s = K(F\#)\lambda = K (f/d_o)\lambda \qquad (22.1)$$

where d_s is the focused spot diameter, K is a quality measure which specifies the focusability of the laser beam (a factor discussed later under the beam quality section) and λ is the laser beam wavelength. The power density of the laser beam at focus is inversely proportional to the beam diameter squared. The smaller the F# used for a particular system, the smaller the focused spot diameter and the greater the power density.

Focusing Systems

LOW-POWER SOLID-STATE LASER systems usually employ transmissive style optics (lenses) to focus the beam on the workpiece, while higher power gas lasers generally employ reflective style optics (mirrors) for this purpose. These mirrors are usually made of metal, and are water cooled to withstand high incident powers. They may be either bare or coated. In comparison to transmissive optics, these mirrors are less sensitive to soiling from weld spatter and fumes and are easier to maintain in production. Highly polished, bare copper mirrors are commonly employed, but gold-coated mirrors will provide the highest reflectivity and thus the least amount of beam attenuation. However, they are expensive and susceptible to surface damage.

Molybdenum-coated mirrors, while still expensive, have good reflectivity and are less susceptible to fume and spatter damage. Thus, a laser system may use gold-coated mirrors to transmit the laser beam to the work station, but then employ molybdenum-coated mirrors within the work station. Several different types of reflective style focus heads used for high-power laser welding are shown in Figure 22.7.

The primary advantage of metal mirror-type optics is that they may be water cooled by flowing liquid through passages beneath the reflecting surface or around the periphery. Thus, in comparison to transmissive-type optics, reflective optics allow a more efficient cooling capability and thereby provide more repeatable results with time.

In recent years the number of vendors of optics systems has increased because of the increased demand for this product. Mirrors and mirror focusing systems can be purchased in the many different configurations needed to satisfy most manufacturing requirements. Many of these can be purchased directly off the shelf, while special orders require a modest lead time. In addition, there are many facilities available to repolish or refurbish mirrors at reasonable prices.

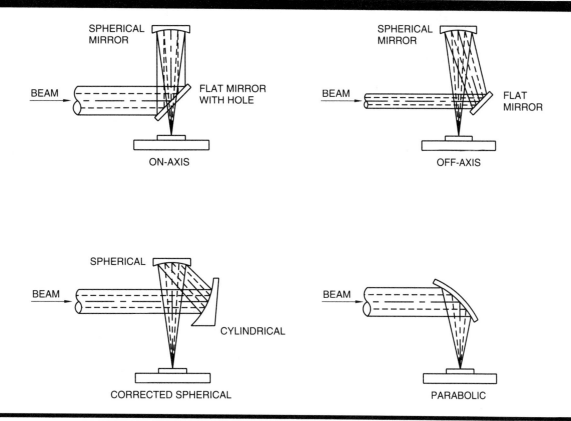

Figure 22.7–Multikilowatt Laser Beam Focusing Heads

Beam Quality

LASER BEAM QUALITY is a measure of focusability. It is a function of the beam's transverse mode, and the extent of aberrations and divergence introduced by the optics. The radiant energy oscillating from one end of the laser resonant cavity to the other forms an intense electromagnetic field. This field can assume many different cross-sectional shapes called *transverse electromagnetic* (TEM) *modes*, which establish the laser beam's radial energy distribution. This TEM mode is expressed as TEM_{mn}, where the subscripts "m" and "n" specify the transverse modal lines across the emerging beam's cross-section. Thus the beam, in cross-section, is segmented into two distinct planes (at right angles to each other) as illustrated in Figure 22.8, and then the number of energy density modes (or "valleys") in each of these directions is expressed as a subscript. The notation TEM_{oo} is, therefore, representative of the lowest order mode (or purest beam), and the power distribution across this beam is Gaussian. A pure TEM_{oo} beam is the highest "absolute quality" beam attainable, and thus is the most focusable. Although the TEM_{oo} mode is indicative of the highest quality beam obtainable, it may not be the most ideal mode to employ for welding — depending on the specific weld task to be performed.

At laser output powers in the multikilowatt range, the ability to generate a high-quality output beam can be limited by several factors. Non-uniformities in the lasing media, a phenomena which is more prevalent in transverse flow units than in axial flow designs, will affect beam quality. Thermally induced changes of diffraction in materials used as output couplers and windows can also significantly affect the output beam's quality and result in a degradation of weld performance. This phenomenon is commonly referred to as "thermal focusing" or "thermal lensing". This phenomenon can also be produced by the presence of freon or some other heavy molecular element somewhere along the beam's transmission path. In an effort to preserve process consistency and maintain a reasonable lifetime of laser optics, which are consumable items, many laser manufacturers resort to using either a higher order mode beam than TEM_{oo}, thus inducing less thermal distortion due to the refractive index of transmissive components, or employ a laser design that has no transmissive components at all. Figure 22.8 illustrates a variety of different type modes that can be generated by the CO_2 lasers being employed in industry today.

The TEM_{oo} mode and some of the higher order modes are usually produced using a stable oscillator configuration similar to that depicted in Figure 22.3. The unstable oscil-

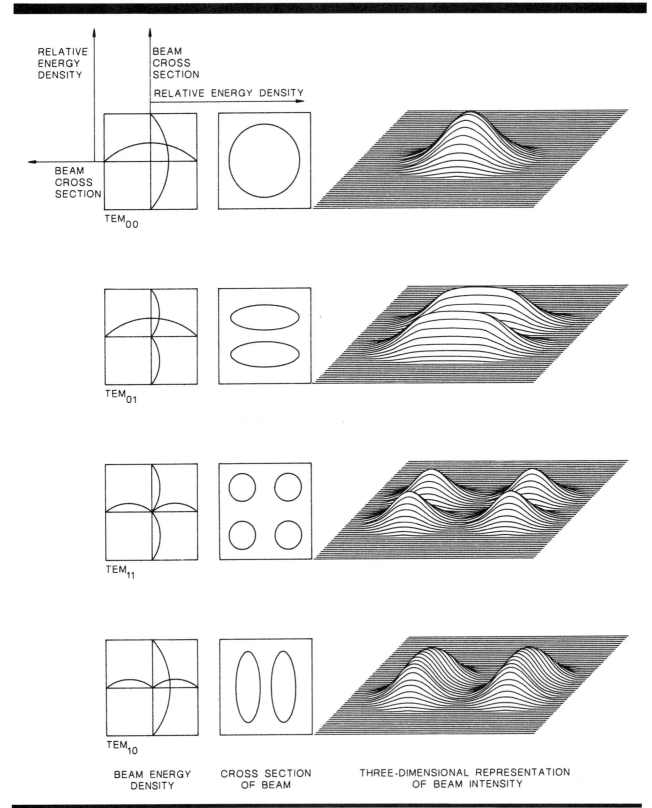

RELATIVE
ENERGY
DENSITY

BEAM
CROSS
SECTION

RELATIVE ENERGY DENSITY

BEAM
CROSS
SECTION

TEM_{00}

TEM_{01}

TEM_{11}

TEM_{10}

BEAM ENERGY
DENSITY

CROSS SECTION
OF BEAM

THREE-DIMENSIONAL REPRESENTATION
OF BEAM INTENSITY

Figure 22.8–Beam Cross Sections for Four Different TEM Modes

lator configuration depicted in Figure 22.9, on the other hand, illustrates another means for generating the multikilowatt level laser beams which are focusable to power densities sufficiently high enough to produce deep penetration welding. The intensity profile of the laser beam produced with the unstable oscillator configuration shown will be annular. Multikilowatt lasers employing this method, for use in production welding at power levels up to 25 kW, are readily available. Some of these lasers use no transmissive elements and, instead, employ an aerodynamic window to transmit the beam from the reduced pressure environment in the laser cavity to atmospheric pressure. Above 6 kilowatts, aerodynamic windows are almost mandatory for production usage because solid windows have too short an operating life.

The most significant unstable oscillator parameter affecting beam quality (i.e., focusability) is its magnification M, which is defined as the ratio of the near field annular output beam's outer diameter, (OD) to inner diameter, (ID). This is illustrated in Figure 22.9. As this magnification increases, both focusability and welding performance improve. Improved welding performance of M = 4 versus M = 2 laser cavity optics is shown in Figure 22.10. This effect is also illustrated in Figure 22.11, where fusion zone cross sections of welds made with M = 2, M = 3, and M = 4 laser cavity optics are compared.

Beam Polarization

LASER WELDING SPEED has been proven to be dependent on the alignment of the plane of the polarization of the incident laser beam on the workpiece relative to the direction of motion of the welding beam. Highest welding speeds with the narrowest weld bead geometries result when the

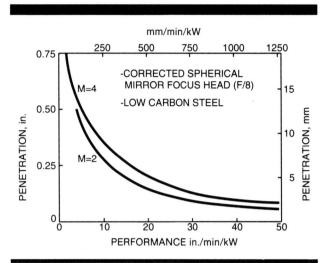

Figure 22.10–Welding Performance Comparison for M=2 and M=4 Laser Cavity Optics

polarization plane is coincident with the direction of welding. Conversely, welding in a direction perpendicular to the plane of polarization results in the lowest welding speed. Due to this effect, "circular polarization" (i.e., equal in all directions) of the laser beam is often employed to give consistent results regardless of the orientation of the welding direction with respect to the polarization plane of the laser output.

The output of a laser is frequently characterized as being either "randomly polarized" or "linearly polarized". In the case of the former, the design of the resonator in the laser allows the polarization plane to "drift" in a random fashion. This often happens at relatively high speed. Welds made with such lasers generally do not exhibit any polarization effects.

Linearly polarized output beams are produced by laser resonator designs where the plane of polarization is "locked" and cannot rotate. This characteristic must be noted in the design of welding systems and a choice made whether to employ circular polarization optics in the beam delivery system. In typical production welding systems, where the beam delivery optics are fixed and the workpiece is manipulated under the focused beam, the orientation of the plane of polarization of the incident beam in relationship to the direction of welding must be maintained constant in order to guarantee repeatable results.

Beam Switching

THE ABILITY TO transmit the laser beam directly through the atmosphere, over distances of up to several meters in air, makes it feasible to use a single laser source for servicing several different work stations. In production, a single

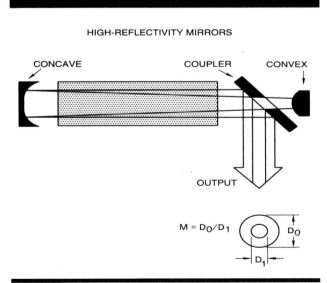

Figure 22.9–Unstable Oscillator Laser Cavity Optics

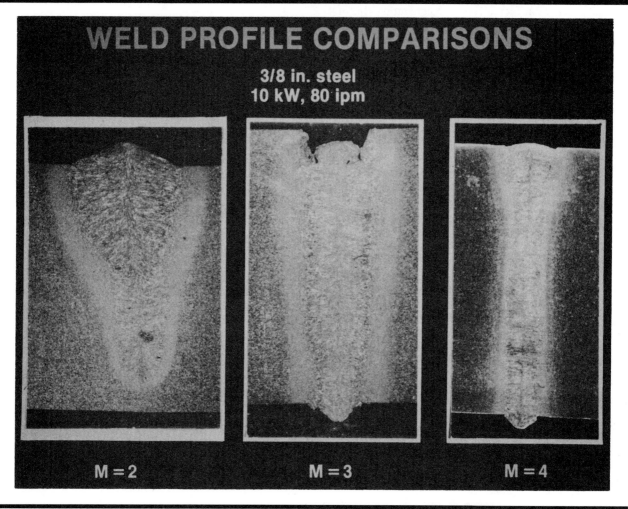

Figure 22.11–Comparison of Bead-on-Plate Weld Profiles in 3/8 in. Thick Carbon Steel Using M=2, M=3, and M=4. Welding Conditions: 10 KW, 80 in./min.

laser may often be used to service several work stations by employing beam switching mirrors.

The primary advantage of operating in this manner is that it helps increase the laser beam's usage factor. While the laser beam is processing material in one station, parts can be prepared for processing in another. Then, when the processing task in the first station is completed, the beam can immediately be switched to the second station.

Robotic Style Beam Delivery

WHILE LASER WELDING was initially performed in much the same manner as electron beam welding, by manipulating the joint to be welded under a stationary beam, it is currently quite common to use robotic style devices to manipulate the laser beams over stationary joints. Conse-

quently, highly flexible laser beam motion systems having the capability to perform three-dimensional welding tasks are readily available today.

Two different concepts are currently employed to provide this robotic style of LBW capability. The first involves mounting beam-focusing optics on the end of an articulated mechanical arm, and using a series of beam-directing mirror combinations (situated in the various joints of the arm) to direct the beam to these optics. The other is to mount the beam-focusing optics on a Z-travel axis, suspended from an X-Y gantry style motion system, and then use a series of singular beam-directing mirrors to direct the beam to them.

Both these robotic style laser concepts are presently employed in industry to accomplish a variety of welding tasks.

EQUIPMENT

High-Power Lasers

HIGH-POWER (I.E., 6 kW and greater) electric discharge gas lasers generally employ three basic sub-assemblies: the optical cavity, the gas flow loop, and the electric discharge system, which includes the power supply and electrodes. The optical cavity is usually formed by precise location of water-cooled metal mirrors of specific radius of curvature and spacing which determines the beam mode. The mirrors are mounted on a truss or similar structure which is designed to have minimum distortion due to temperature variations. The volume encompassed within the optical cavity is normally maintained at reduced pressure.

A typical system might operate at one tenth of an atmosphere. The gas flow loop contains the laser gas, comprised of approximately 95 percent helium and nitrogen and the remainder carbon dioxide. The gas is driven around an enclosed loop by a large axial vane or similar pump. The gas is electrically excited in the laser cavity where the optical power is produced; unused power in the form of heat is removed by a gas-to-water heat exchanger located just downstream of the laser cavity, and the gas is recirculated to be electrically excited again. The electric discharge system contains a high voltage dc, ac, or rf (radio frequency) power supply which is connected to an electrode array to provide excitation to the gas throughout the laser cavity volume. Sometimes ballast resistors are placed in series with the electrodes to provide a smoother electrical discharge for the CW laser beam.

In addition to the above, high-power industrial lasers utilize many ancillary systems, all usually controlled by a CNC, so that the laser may be operated quickly and easily. Control systems automate start up and laser operation with a minimum of operator induced functions so that operator skill may be minimized.

Some of the ancillary systems that may be found in an industrial laser include: a gas supply system where either premixed gas or separate bulk gases are supplied to the laser; a vacuum system to maintain the laser pressure at its operating level or to evacuate the laser cavity when a fresh mixture of gas is required; a bulk water supply for heat exchanger and mirror cooling; a solid or aerodynamic window to transport the high power beam from the reduced pressure laser cavity to the work environment at atmospheric pressure; power meters to monitor beam power; and a beam shuttering system which delivers the beam to the workstation on demand via operator or workstation CNC controls.

Industrial carbon dioxide lasers are available with up to 25 kW of focusable power which can produce penetrations of up to 1.25 in. (32 mm) in low carbon steel. Deep penetration laser welds can have depth to width ratios in the ten to one range, similar to those formed with electron beams. Laser welds, however, are formed without the need

of vacuum chambers or protection from x-rays (see Figure 22.12). Welding is normally done inside an enclosure which protects the operator from stray (reflected laser beam) radiation. High-power industrial lasers occupy a reasonable size floor space on the shop floor, have a relatively high "availability" time, and are not excessively expensive to operate. They normally use all reflective, water-cooled optic elements (requiring a minimum of periodic maintenance) and are ruggedly built to withstand the rigors of the manufacturing environment. An artist's rendition of a typical 25 kilowatt laser is shown in Figure 22.13, and Figure 22.14 shows a laser similar to this integrated into a finished production system.

The beam quality of high-power industrial lasers, although not as good or as focusable as near-Gaussian low power lasers, provides the focusing capability needed for "keyhole" welding. High-power industrial lasers, however, do not normally have the pulsing capability desirable for some applications. Cathode maintenance may be periodically required in some high power lasers. When solid windows are used to transmit the beam from the laser cavity to the ambient environment, limited life and cost of this consumable item also must be considered.

PROCESS ADVANTAGES

MAJOR ADVANTAGES OF laser beam welding include the following:

(1) Heat input is close to the minimum required to fuse the weld metal; thus, metallurgical effects in heat-affected zones are reduced, and heat-induced workpiece distortion is minimized.

(2) Single pass laser welding procedures have been qualified in materials of up to 1 1/4 in. (32 mm) thick, thus allowing the time to weld thick sections to be reduced and the need for filler wire (and elaborate joint preparation) to be eliminated.

(3) No electrodes are required; welding is performed with freedom from electrode contamination, indentation, or damage from high resistance welding currents. Because LBW is a noncontact process, distortion is minimized and tool wear essentially eliminated.

(4) Laser beams are readily focused, aligned, and directed by optical elements. Thus the laser can be located at a convenient distance from the workpiece, and redirected around tooling and obstacles in the workpiece. This permits welding in areas not easily accessible with other means of welding.

(5) The workpiece can be located and hermetically welded in an enclosure that is evacuated or that contains a controlled atmosphere.

(6) The laser beam can be focused on a small area, permitting the joining of small, closely spaced components with tiny welds.

Figure 22.12—Laser Weld Being Made on 1/8 in. (3.2 mm) Thick Type 304 Stainless Steel

(7) A wide variety of materials can be welded, including various combinations of different type materials.

(8) The laser can be readily mechanized for automated, high-speed welding, including numerical and computer control.

(9) Welds in thin material and on small diameter wires are less susceptible to burn-back than is the case with arc welding.

(10) Laser welds are not influenced by the presence of magnetic fields, as are arc and electron beam welds; they also tend to follow the weld joint through to the root of the workpiece, even when the beam and joint are not perfectly aligned.

(11) Metals with dissimilar physical properties, such as electrical resistance, can be welded.

(12) No vacuum or X-ray shielding is required.

(13) Aspect ratios (i.e., depth-to-width ratios) on the order of 10:1 are attainable when the weld is made by forming a cavity in the metal, as discussed later under the section on keyhole welding.

(14) The beam can be transmitted to more than one work station, using beam switching optics, thus allowing beam time sharing.

PROCESS LIMITATIONS

LASER BEAM WELDING has certain limitations when compared to other welding methods, among which are the following:

(1) Joints must be accurately positioned laterally under the beam and at a controlled position with respect to the beam focal point.

(2) When weld surfaces must be forced together mechanically, the clamping mechanisms must ensure that the final position of the joint is accurately aligned with the beam impingement point.

(3) The maximum joint thickness that can be laser beam welded is somewhat limited. Thus weld penetrations of

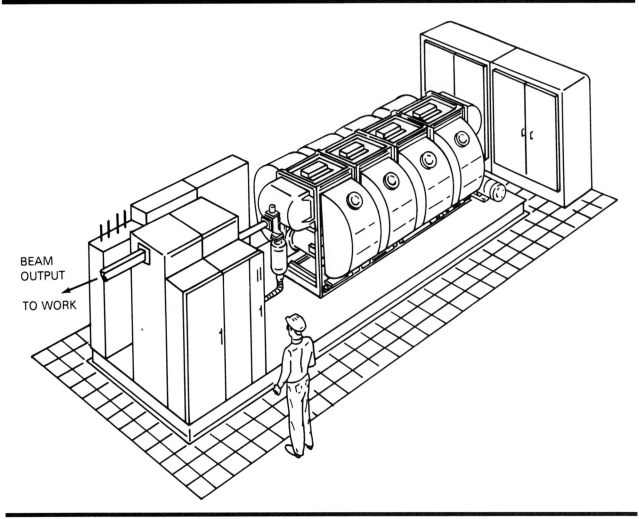

BEAM
OUTPUT

TO WORK

Figure 22.13—Artist's Conception of a 25 KW CO2 Laser

much greater than 0.75 in. (19 mm) are not presently considered to be practical production LBW applications.

(4) The high reflectivity and high thermal conductivity of some materials, such as aluminum and copper alloys, can affect their weldability with lasers.

(5) When performing moderate-to-high power laser welding, an appropriate plasma control device must be employed to ensure weld reproducibility is achieved.

(6) Lasers tend to have a fairly low energy conversion efficiency, generally less than 10 percent.

(7) As a consequence of the rapid solidification characteristic of LBW, some weld porosity and brittleness can be expected.

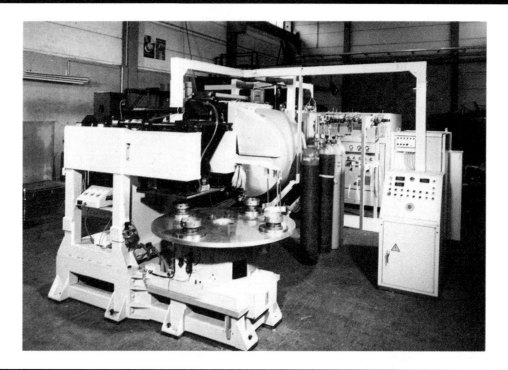

Figure 22.14—Production System Showing a CO_2 Laser Combined With a Rotary Work Table

CHARACTERISTICS OF THE WELD

KEYHOLE OR DEEP PENETRATION WELDING

WHEN BEAM POWER densities on the order of 1×10^6 W/in.^{2} (1.55×10^3 W/mm^2) or greater are achieved, deep penetration beam welding is accomplished by a keyhole energy transfer mechanism. At this power density level, the energy input of the impinging beam is so intense that it cannot be removed by the normal conduction, convection or radiation processes. Thus the area upon which the beam is being impinged melts and vaporizes. Power densities associated with the transition from conduction welding to keyhole or deep penetration beam welding are shown in Figure 22.15. This keyhole welding phenomena is common to both laser beam and electron beam welding, indicating that it is primarily a function of power density, and not dependent upon wavelength.

When the material at the interaction point melts and vaporizes, the vapor recoil pressure creates a deep cavity or "keyhole" as shown in Figure 22.16. This keyhole is a vapor column surrounded by a thin cylinder of molten metal. When the workpiece moves relative to the beam, the vapor

pressure of the metal sustains the keyhole, and the molten metal surrounding the keyhole flows opposite to the weld direction where it rapidly solidifies, forming a narrow fusion zone or weld. Depth-to-width ratios in the range of ten to one can be obtained when laser welding in the keyhole mode. The narrow, high depth-to-width ratio fusion zone formed by laser welds at atmospheric pressure are similar to electron beam welds made in vacuum.

Plasma Suppression

THE INTENSE HEAT generated by the laser beam melts the workpiece, and some of the liquid metal is vaporized into a gaseous state. A fraction of this gas is ionized by the high energy beam and becomes a plasma. The presence of this plasma is detrimental because it tends to absorb and attenuate the laser beam.

Since this plasma can cause the beam to be severely attenuated, failure to control or significantly suppress it will result in reduced welding performance, diminished depth of penetration and sometimes a collapse of the keyhole. Fortunately, plasma can be suppressed by blowing it away

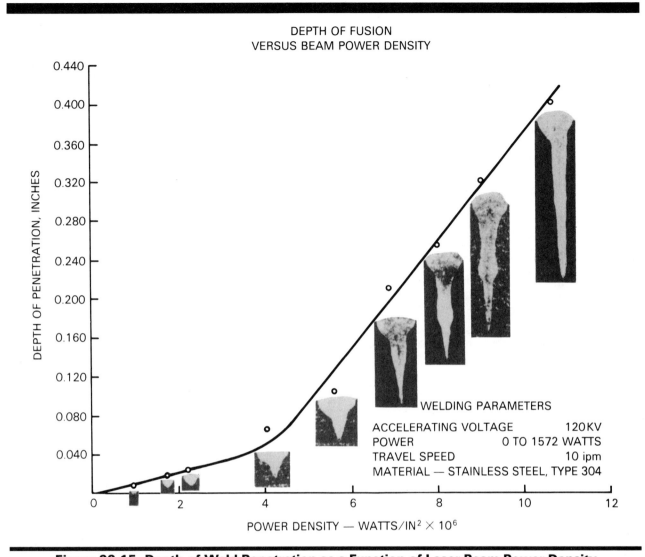

DEPTH OF FUSION
VERSUS BEAM POWER DENSITY

WELDING PARAMETERS

ACCELERATING VOLTAGE	120KV
POWER	0 TO 1572 WATTS
TRAVEL SPEED	10 ipm
MATERIAL — STAINLESS STEEL, TYPE 304	

Figure 22.15–Depth of Weld Penetration as a Function of Laser Beam Power Density

with a stream of gas having a component of velocity transverse to the laser beam axis.

For laser power less than 5 kilowatts, helium, argon or various mixtures of the two may be effectively utilized to suppress plasma. Nitrogen and carbon dioxide may also be used effectively provided that they are compatible with the molten metal and vapor. For higher powers, however, helium is preferred because it is less likely to break down under the influence of the high energy density laser beam.

Flow rates or flow velocities must be adjusted to suppress the plasma but should not be so great as to disturb the molten pool. Many plasma suppression jet designs have been used effectively. One of the simplest techniques is simply to flow the gas across the beam impingement point as shown in Figure 22.17. For production welding, the size of the plasma suppression jet can be minimized and

precisely aligned to reduce volume flow. For experimental and development work, a larger plasma suppression jet may be useful, because small changes in process parameters such as work distance can be evaluated independent of precise plasma suppression jet location.

Auxiliary Gas Shielding

WHEN THE WELD metal must be shielded until it cools below the oxidation temperature, auxiliary inert gas shielding can be provided using shielding hardware compatible with the laser welding process. As with more conventional welding processes, the main objective is to provide an inert environment over that length of the workpiece which is at or above the oxidation temperature without disturbing the molten weld puddle. In some cases it is also necessary to

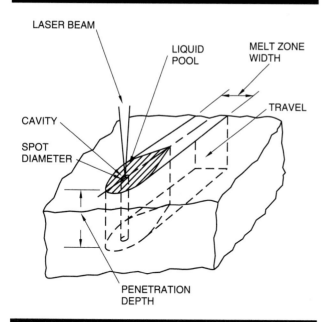

Figure 22.16–Schematic View of Keyhole Welding

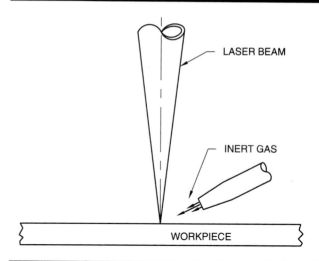

Figure 22.17–Plasma Suppression Using a Transverse Jet of Inert Gas

provide underbead shielding. Inert shielding also can be obtained by placing the workpiece in a dry box which can be purged and backfilled with inert gas.

Energy Absorption

EFFECTIVE LASER BEAM welding depends upon absorption of beam energy by the workpiece. However, shiny metal surfaces at room temperature have high reflectivity for laser light, particularly at a wavelength of 10.6 μm. For example, absorption of low intensity 10.6 μm CO_2 laser beam light is as low as forty percent for stainless steel and one percent for polished aluminum or copper. Absorption levels are higher for Nd:YAG and ruby laser beams. Fortunately, the absorption in most metals increases with temperature, and surface temperature increases rapidly at the beam impingement point when the metal is exposed to high-power density laser radiation. At power densities on the order of 10^6 W/in.^{2} $(1.55 \times 10^3$ W/mm^2), a threshold value of absorption is experienced for most steels and superalloys where the absorption level is approximately 90 percent. For aluminum and copper, this threshold occurs at intensities of approximately 1×10^7 W/in.^{2} $(1.55 \times 10^4$ W/mm^2) and for tungsten at 1×10^8 W/in.^{2} $(1.55 \times 10^5$ W/mm^2).

SHALLOW PENETRATION WELDING AND CONDUCTION EFFECTS

WHEN SUBKILOWATT LASER output power levels and welding speeds approach the limit of penetration dictated by

the lower laser output power being employed, the effects of thermal conduction become more prevalent than the deep welding effects described above.

The result of this transition is generally seen as a broadening of the weld bead, from the deep keyhole, high aspect ratio weld shape to the characteristic "wineglass" shape. See Figure 22.18.

Shallow penetration welds performed with a pulsed laser such as closure welds for hermetic sealing in electronic enclosures, batteries, etc., are typically conduction welds, with limited keyhole formation. Surface plasma generation aids in transferring the energy of the laser beam into the workpiece, giving the desired melt zone required to form the closure weld.

THIN SECTION WELDING

IN SOME APPLICATIONS, the welding speed may be reduced in order to obtain a wider weld through conduction effects. This method is commonly applied in sheet metal butt welding applications where fit-up tolerance forces the laser into accepting lower welding speeds to ensure the reliability and repeatability of the process.

Laser welding is an excellent process for welding thin sections. Stainless steel as thin as 0.0001 in. (0.0025 mm) has been successfully welded by pulsed lasers. Pulsed Nd:YAG and pulsed CO_2 laser welding machines are especially suited for most thin section welds. As with other processes, full penetration welds are preferable to partial penetration welds. Distinct advantages and disadvantages of pulsed welding include:

Advantages:

(1) Small fusion and heat-affected zones
(2) Low heat input

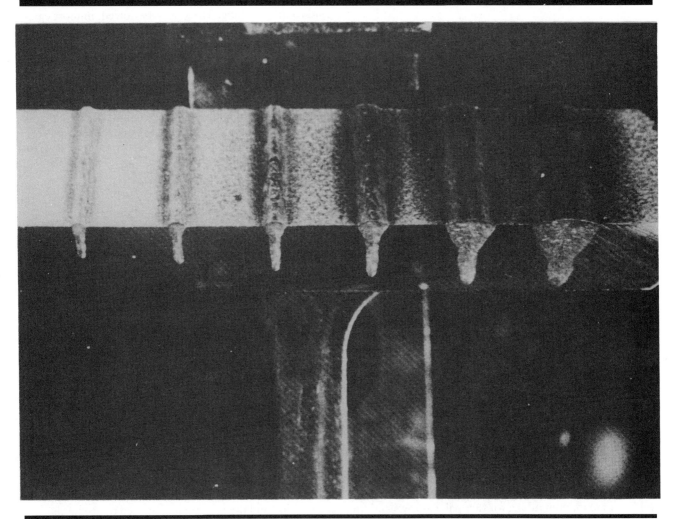

Figure 22.18–Variation of Weld Penetration as Travel Speed Changes Using a Constant Output Power

(3) Ability to precision tack weld a joint

(4) Unique properties of heat transfer with lasers

Disadvantages:

(1) Extremely high cooling rates

(2) Sensitivity to material chemistry

(3) Problems coupling with high reflectivity materials

The first two advantages of laser welding are fairly self-explanatory. The ability to precision tack weld a joint is aided by the pulsed mode of welding. A single pulse may be used to tack the joint before the final weld is made. The heat-transfer properties during laser welding differ dramatically from welding processes that depend on electrical conductivity to make welds. These properties allow laser welding of many materials which are not electrically con-

ducting. Materials transparent to light of a specific wavelength can, on the other hand, be used for hold-down fixturing during laser welding. See Figure 22.19.

Fixturing

FIXTURING IS EXTREMELY important for thin section laser welding. Tolerances must be held closely to maintain joint fitups without allowing either mismatch or gaps. Standing-edge joints are preferable for thin section welds since the actual weld joint cross section is enlarged. Butt joints are difficult to design for welding, and distortion during welding may cause joint mismatching or gaps.

The disadvantages of laser welding for thin sections are generally related to material cracking or laser coupling problems. Cracking is typically caused by high cooling rates, which may lead to undesirable brittle phases in some

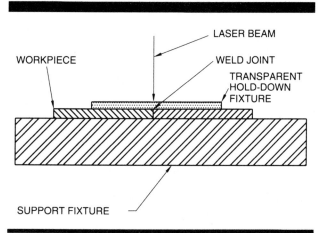

Figure 22.19–Use of Transparent Hold-Down Fixturing For Flat Thin Section Welding

materials, or by material chemistry problems, which may lead to hot cracking. These may be solved, in most cases, by either preheating or using a different laser wavelength to reduce high cooling rates, or by changing to a more suitable material in the case of hot cracking.

Coupling problems occur when materials (such as copper, aluminum, and silver) are highly reflective to the laser wavelength. The problem is usually solved in one of three ways: changing to a different wave length laser, etching or painting the surface to reduce reflectivity, or using a keyhole mode weld in which the energy density is great enough to overcome reflectivity.

APPLICATIONS

LASER BEAM WELDING is being used for an extensive variety of applications such as in the production of automotive transmissions and air conditioner clutch assemblies. In the latter application, laser welding permits the use of a design that could not otherwise be manufactured. The process is also being used in the production of relays and relay containers and for sealing electronic devices and heart pacemaker cases. Other applications include the continuous welding of aluminum tubing for thermal windows and for refrigerator doors.

Successful laser welding applications include welding transmission components (such as synchro gears, drive gears and clutch housings) for the auto industry. These annular and circumferential-type rotary welds need from 3 to 6 KW of beam power, depending on the weld speed being employed, and require penetrations which typically do not exceed .125 in. (3.2 mm). Materials welded are either carbon or alloy steels. In some cases, such as the gear teeth, they have been selectively hardened before welding. There are many advantages to laser welding such assemblies. The low heat input provided by the laser does not cause any affect to the prehardened zones adjacent to the weld. Also, this low heat input produces a minimal amount of distortion so that precision stampings can often be welded to finished dimensions. Since the ease of automation and high weld-speed capability of the laser process makes it ideal for automotive-type production, a number of these systems have recently been installed in the automotive industry.

Figure 22.20A shows a fully automated, 3 KW system employed for welding clutch assemblies. This system incorporates a beam switch to use one laser for sequentially welding at two separate but in-line work stations. While welding in one station, parts are loaded into the other station, thus helping to maximize the production capability of this dual-station system. As illustrated by Figure 22.20B, individual hub and housing components are brought to each station and then assembled, pressed to proper dimensions and welded, all under the control of a central unit. Weld speed is 90 ipm for these assemblies.

Figure 22.21 shows another transmission component weld, which involves welding a threaded annular boss onto a circular ring. Here a 2.5 kW (CO_2) laser is used to provide an 0.187 in. (4.75 mm) deep weld at 60 ipm, employing helium shield gas.

Figure 22.22(A) shows a recuperative plate pair for a heat exchanger which is joined by welds made around each of the air holes [Figure 22.22(B)] in the cutout pattern shown. The material of the these plates is 0.008 in. (0.2 mm) thick Inconel 625, and welding is done using a CO_2 laser rated for 750 W (continuous output power), operating in an enhanced pulse mode (i.e., 1.5 millisecond pulse length and 200 pulse per second repetition rate). Weld type is a lap weld, made at 120 ipm. A 5 in. (127 mm) focol length lens was used.

Figure 22.23 shows the cross section of a 416 stainless steel cap welded onto a 310 stainless steel body, using a 750 W CO_2 laser at 45 ipm weld speed. Penetration of weld into the body component is 0.050 in. (1.27 mm) deep.

Figure 22.20A–Production Welding System for Automotive Transmission Components

METALS WELDED

LASER BEAM WELDING can be used for joining most metals to themselves as well as dissimilar metals that are metallurgically compatible. Low carbon steels are readily weldable, but when the carbon content exceeds 0.25 percent, martensitic transformation may cause brittle welds and cracking. Pulsed welding helps minimize the tendency for cracking. Fully killed or semi-killed steels are preferable, especially for structural applications, because welds in rimmed steel may have voids. Steels having high amounts of sulfur and phosphorus may be subject to hot cracking during welding. Also, porosity may occur in free machining steels containing sulfur, selenium, cadmium, or lead.

Difficulty has been encountered in welding carburized or nitrided steels. Welds in these alloys are generally porous and exhibit cracks. Occasionally nickel shims are added to these metals and some alloy steels to increase toughness. Aluminum in small quantities has also been added to joints of rimmed steel to reduce porosity caused by entrapped gases.

Many stainless steels are considered good candidates for laser welding. The low thermal conductivity of these metals permits forming narrower welds and deeper penetrations than possible with carbon steels. Stainless steel of the 300 series, with the exception of free machining Types 303 and 303Se and stabilized Types 321 and 347, are readily weldable. Welds made in some of the 400 series stainless steels can be brittle and may require post weld annealing. Many heat resistant nickel and iron based alloys are being welded successfully with laser beams. Titanium alloys and other refractory alloys can be welded in this way, but an inert atmosphere is always required to prevent oxidation.

Copper and brass are often welded to themselves and other materials with specialized joint designs used for conduction welding. Aluminum and its weldable alloys can be joined for partial penetration assembly welds and are commonly joined by pulsed conduction welds for hermetically sealed electronic packages. Joint designs must retain aluminum in tension.

Refractory metals such as tungsten are often conduction welded in electronic assemblies, but require higher power than other materials. Nickel-plated Kovar is often used in sealing welds for electronic components, but special care is required to ensure that the plating does not contain phosphorous, which is usually found in the electroless nickel

Figure 22.20B–Production Welding System for Automotive Transmission Components

plating process commonly used for Kovar parts that are to be resistance welded.

Dissimilar metal joints are commonly encountered in conduction welds where the twisting of conductors forms a mechanical support that minimizes bending of potentially brittle joints. Dissimilar metals having different physical properties (reflectivity, conductivity and melting points) are often joined in the welding of conductors. Special techniques such as adding extra turns of one material to the joint as opposed to the other may be required to balance the melting characteristics of the materials. Some of these concepts can also be applied to structural and assembly welds, but the possibilities are much more limited.

JOINT DESIGN

JOINTS DESIGNED FOR laser welding must meet the criteria of the manufacturing engineer, and strength and safety specifications must be considered. Joints must be accessible to a focused laser beam and must be economical when considering machining operations before and after welding. A good joint design can enhance a laser welding production system because tooling design, manufacture, and maintenance are affected by weld joint design. An optimum weld joint design may facilitate assembly of a part before welding. The weld joint also should be easy to inspect.

A variety of joints is applicable to the laser welding process. Joints used in laser welded construction are normally designed for structural, assembly, sealing, or similar purposes. Some types of joints used in laser welded construction are shown in Figure 22.24. Butt, corner, T-lap, as well as variations and combinations are applicable to the laser welding process.

Butt Joint

BUTT JOINT GEOMETRIES may be annular, circumferential or linear. Joint cleanliness must be maintained and as with any welding operation, rust and scale must be removed so as not to inhibit fusion zone integrity. An important consideration in laser joint preparation is fit-up. In some cases, gaps of 3 percent of metal thickness can be tolerated. However, underfill occurs if gapping is too extreme.

When employing butt joint design for laser processing, hold-down tooling should be considered, especially if the application is to be repeated in high-volume production. Butt joints are conducive to automated, high volume production welding operations, but intimate contact must be assured through fixturing design and dimensional control of the parts. In the case of annular butt joints, subassemblies can be manufactured to include an interference fit, allowing assembly to take place independent of the weld process. Preassembly can simplify the overall design of a laser welding system. A separate press or assembly station

Figure 22.21–Cross Section of a Laser Beam Weld Joining a Boss to a Ring. A 2.5 KW CO$_2$ Laser Produced a Travel Speed of 60 in./min (25 mm/sec). Penetration was 0.187 in. (4.8 mm)

also adds another measure of quality control to the production system, as part size tolerances can be gauged before the joining process.

Butt joints [Figure 22.24(A)] are applicable in structural, assembly, and sealing applications. Single-pass penetration of 1.25 in. (31.8 mm) thick butt joints has been demonstrated with 25 kW high-power CO$_2$ lasers using the keyhole penetration technique. A majority of successful laser welding applications in the automotive transmission industry utilize single-pass, full penetration welds of 0.090 to 0.200 in. (2.3 to 5.1 mm) requiring 5 kW to 9 kW.

Keyhole penetration welding is easily performed with sharply focused laser beams of 1 kW or greater. Assembly or structural joints requiring limited or partial penetration also may be welded with lower powered lasers. At lower power levels, keyhole formation does not take place, but weld puddle creation is accomplished via conduction from the material surface similar to the more conventional welding processes.

T-JOINT

A LASER BEAM can be aimed at the root of an accessible T-joint [Figure 22.24(C)]. At an optimum directed angle, a focused beam may follow the gap between the intersecting workpieces. Depending upon plate thickness and laser power used, fusion takes place at the interface between

workpieces. The stress load is transferred from one member to another primarily through the root. If a fillet is formed, it will act to reduce stress. However, laser formed fillets are not usually as pronounced as typical arc welding fillets.

CORNER JOINT

CORNER JOINTS [FIGURE 22.24(B)] are often used in laser welding assembly and sealing applications. The use of a corner joint design is limited by plate or sheet thickness. The thinner the material, the less the power that is required, but the greater the reliability on focused spot location that is needed.

The corner joint has the advantage of accessibility where tooling and fixturing are important to maintain fit-up integrity.

LAP JOINT

LAP JOINTS [FIGURE 22.24(D)] are typically used in sheet metal assembly applications. The focused laser beam can be impinged onto the top surface, causing a weld penetration into or through one or more material sheets in contact. This type of weld is sometimes referred to as a burn-through lap weld.

(A)

(B)

Figure 22.22—Heat Exchanger Recuperator Plate Pair Joined by Laser Beam Welds Around Each of the Air Holes. Material is 0.008 in. (0.2 mm) Thick Inconel 625

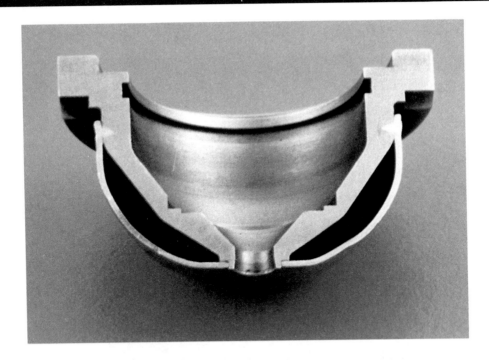

Figure 22.23–Cross Section of a Laser Beam Weld Joining a 416 Stainless Steel Cap to a 310 Stainless Steel Body

Intimate contact is not necessarily required in the burn-through lap weld because molten metal will bridge limited gaps between layers. In some applications using coated steels, a gap may be advantageous because it would avoid entrapment of outgassing products from the coating. The gap tolerance has a relatively narrow band which is related to workpiece thickness and beam spot size. For example, for 0.030 to 0.040 in. (0.8 to 1.0 mm) thick material welded with laser beam of approximately 0.020 in. (0.5 mm) diameter, a gap of approximately 0.003 in. (0.08 mm) is required to prevent porosity from outgassing. However, if the gap exceeds approximately 0.006 in. (0.15 mm), incomplete fusion of the joint may occur.

As with any lap joint weld, the fusion zone interface is the stress carrier. To increase the interface area, a laser beam may be directed in a circular or linear pattern (by moving the beam delivery optics). Although laser lap weld joints are not as sensitive to fit-up variances as other joints described above, part fit-up must generally be maintained using suitable fixtures or tooling.

Laser lap welds are usually characterized by a slight reinforcement at the fusion zone faces. For burn-through lap welds, a slight reinforcement of the root may also be achieved. When the bottom member of the assembly is not penetrated, a deformation is usually observed at the bottom surface. Finally, the lap joint is less dependent on the focused laser beam location tolerance than are butt, corner, or T-joints.

JOINT PREPARATION

Cleaning

ALL LASER WELDED joints must be free of rust, scale, lubricants, or other contaminants. Parts cleaning systems are easily integrated with laser processing production facilities via materials handling conveyors. The type of cleaning system used must comply with local and state laws.

Gap Tolerance

THE SENSITIVITY OF lap welds to gap was discussed above. Butt, T, and other similar joints also are sensitive to gaps. The gap tolerance for these joints is dependent on material thickness, weld speed, beam diameter, and beam quality. Normally, gap tolerance increases with material thickness. However, as the gap increases, the reinforcement normally associated with line-on-line fit up of laser welds decreases. When the gap is too large for weld bead reinforcement, underfill will occur. As the gap continues to increase, underfill will become more severe until complete lack of fusion occurs. This condition is characterized by the beam actually channeling through the gap without being absorbed by the workpiece at the mating surfaces.

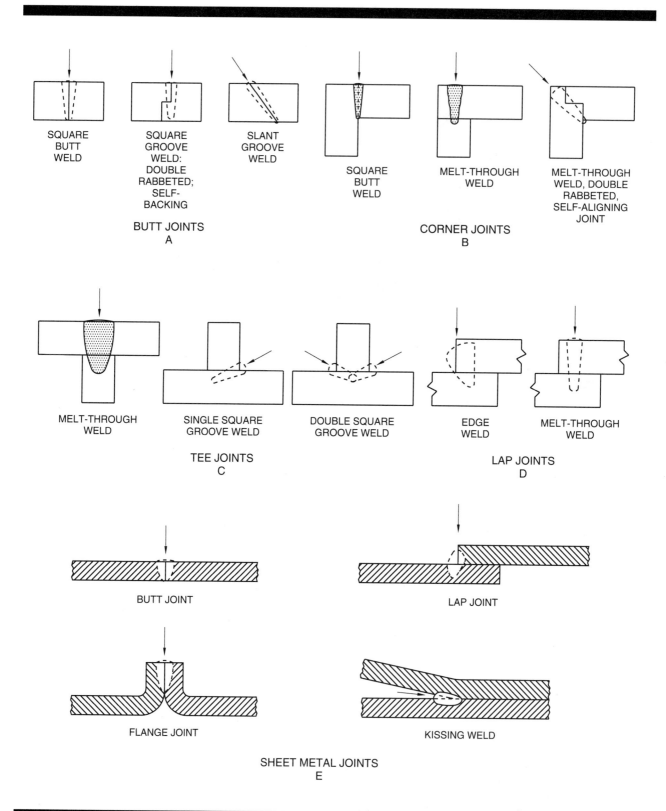

SQUARE
BUTT
WELD

SQUARE
GROOVE
WELD:
DOUBLE
RABBETED;
SELF-
BACKING

SLANT
GROOVE
WELD

**BUTT JOINTS
A**

SQUARE
BUTT
WELD

MELT-THROUGH
WELD

MELT-THROUGH
WELD, DOUBLE
RABBETED,
SELF-ALIGNING
JOINT

**CORNER JOINTS
B**

MELT-THROUGH
WELD

SINGLE SQUARE
GROOVE WELD

DOUBLE SQUARE
GROOVE WELD

EDGE
WELD

MELT-THROUGH
WELD

**TEE JOINTS
C**

**LAP JOINTS
D**

BUTT JOINT

LAP JOINT

FLANGE JOINT

KISSING WELD

**SHEET METAL JOINTS
E**

Figure 22.24–Joint Configurations for Laser Welds

For laser beams employed to weld materials up to 0.5 in. (12.7 mm) thick, a typical gap tolerance may be given as 3 percent of material thickness. However, the minimum gap specified should be based on the factors mentioned above as well as the required fusion zone geometry.

Mismatch

LASER BEAMS ARE considerably more tolerant to mismatch than gap. Mismatch as great as half the material thickness can be tolerated. However, joint specification and integrity should be used to designate the mismatch tolerance.

SAFETY

MISUSE OF LASER equipment can result in permanent damage to the eyes and skin of both operators and nearby personnel. In addition, specific precautionary measures are needed to avoid other potential hazards sometimes associated with using lasers such as dangers related to servicing high-voltage power sources, and harmful fumes that can be released when laser processing certain materials. These hazards can, in some instances, be far more significant than the beam-related hazards which are usually considered the most important.

Detailed laser safety information can be found in ANSI publication Z-136.1[1] and the Federal Performance Standard for Laser Products: FDA-Title 21, CFR, Section 1040.[2] Reference to these documents is strongly recommended.

Detailed training[3] is recommended for those working with lasers, including those on the technical support staff and technicians.

BEAM-RELATED HAZARDS

BOTH THE ANSI and FDA standards divide all lasers into four major classes with some sub classes which define the potential beam-related hazards associated with each type. These categories can be summarized as follows:

Class 1 - Denotes exempt lasers or laser systems that cannot, under normal operating conditions, produce a hazard. This would include, for example, bar code reading lasers found at grocery store check-out counters.

Class 2 - Denotes low power visible lasers or laser systems which, because of natural human aversion response to bright light, do not normally present a hazard but which may produce a hazard if viewed directly for extended periods.

Class 3A - Denotes lasers or laser systems that would not produce a hazard under normal conditions if viewed for

only momentary periods with the unprotected eye, but may present a hazard if viewed using some form of light collecting optics.

Class 3B - Denotes lasers or laser systems that can produce a hazard if viewed directly, including intrabeam (i.e., direct) viewing of specular (i.e., concentrated) reflections. Except for higher power Class 3B lasers, this class will not usually produce a hazardous diffuse reflection.

Class 4 - Denotes lasers or laser systems that can produce a hazard not only if the beam or specular reflections are viewed directly, but also from direct viewing of diffuse reflections. In addition to eye damage, the beam and its reflections may also produce both skin and fire hazards.

Control Measures

CONTROL MEASURES CENTER on enclosing as much as possible of the beam path and baffling the target area to reduce the chance of hazardous reflections. Care should be taken to employ dark filters to reduce the level of visible light to a comfortable level. Robotic systems should be designed and installed to limit laser beam traverse so as not to direct the beam at personnel.[4]

NONBEAM-RELATED HAZARDS

THE POTENTIAL NONBEAM-RELATED hazards associated with using a laser include items like electrical shock, toxic gases and other occupational hazards, the proper safety precautions for which are clearly defined in ANSI/ASC Z49.1, Safety in Welding and Cutting.[5] The general safety requirements expressed therein as well as those provided by OSHA's General Industry Safety Standards,[6] should be strictly adhered to at all times.

1. American National Standard for the Safe Use of Lasers, ANSI Z136.1 (latest edition).
2. Center for Devices and Radiological Health, Food and Drug Administration, Title 21, CFR-Section 1040: Federal Performance Standard for Laser Products (latest edition).
3. R. J. Rockwell, Jr., Controlling Laser Hazards, *Laser Applications*, 5 (9); 93-99: 1986 Sept.
4. Sliney and Wolbarsht, *Safety With Lasers and Other Optical Sources*, New York, Plenum, 1980.
5. American National Standard for Safety in Welding and Cutting, ANSI/ASC Z49.1 (latest edition).
6. Code of Federal Regulations, Title 29, Part 1910 (in its entirety): Occupational Safety and Health Standards (latest edition).

SUPPLEMENTARY READING LIST

Anthony, P. "Choosing the right CO_2 laser." *Industrial Laser Review* 4(2): July 1989.

Banas, C. M. "High power laser welding - 1978." *Optical Engineering* 17(3): 2410-16; May-June 1978.

Brown, C. and Banas, C. M. "High-power laser beam welding in reduced-pressure atmospheres." *Welding Journal* 65(7): 48-53; July 1986.

Crafer, R. C. "Improved welding performance from a 2kW axial flow CO_2 laser welding machine." *Advances in Welding Process, 4th Int. Conf., Harrogate, England, 9-11 May 1978*, Cambridge, England: The Welding Institute, 1978.

Duhamel, R. F. "Effect of laser optics on welding performance." *ICALEO* (Santa Clara, CA), November 1988.

Harry, J. E. "Industrial lasers and their applications." New York: McGraw-Hill, 1974.

Holbert, R. K., Mustaleski, T. M., and Frye, L. D. "Laser beam welding of stainless steel sheet." *Welding Journal* 66(8): 21-25; August 1987.

Jon, M. C. "Noncontact acoustic emission monitoring of laser beam welding." *Welding Journal* 64(9): 43-48; September 1985.

Mazumder, J. and Steen, W. M. "Laser welding of steels used in can making." *Welding Journal* 60(6): 19-25; June 1981.

Morgan-Warren, E. J. "The application of laser welding to overcome joint asymmetry." *Welding Journal* 58(3): 76s-82s; March 1979.

Powers, D. E. and LaFlamme, G. R. "EBW vs. LBW - A comparative look at the cost and performance traits of both processes." *Welding Journal* 67(3): 25-31; March 1988.

Ram, V., Kohn, G., and Stern, A. "CO_2 laser beam weldability of zircaloy 2." *Welding Journal* 65(7): 33-37: July 1986.

Rupp, E. W. "Water cooling of laser: design considerations and techniques." *Laser and Applications* 91, March 1985.

Russo, A. J., et al. "Thermocapillary flow in plused laser beam weld pools." *Welding Journal* 69(1):23s; January 1990.

Schwartz, M. M. "Laser welding and cutting." *Welding Research Council Bulletin* New York: No. 167; November 1971.

Seretsky, J. and Ryba, E. R. "Laser welding dissimilar metals: titanium to nickel." *Welding Journal* 55(7): 208s-11s; July 1976.

Sharp, C.M. and Nilsen, C. J. "High speed laser beam welding in the can making industry." *Welding Journal* 67 (1): 25-28; January 1988.

Sherwell, J. R. "Design for laser beam welding." *Welding Design and Fabrication* 50(6): 106-10; June 1977.

Yessik, M. and Schmaty, D. J. "Laser processing in the automotive industry." *SME Paper* #MR74-962; 1974.

FRICTION WELDING

PREPARED BY A COMMITTEE CONSISTING OF:

C. A. Johnson, Chairman
Naval Weapons Center

G. E. Beatty
Chance Collar Co.

G. A. Knorovsky
Sandia National Laboratories

D. L. Kuruzar
Manufacturing Technology, Inc.

H. W. Seeds
Saginaw Div. General Motors Corp.

D. E. Spindler
Manufacturing Technology, Inc.

J. S. Thrower
General Electric Co.

R. N. Vecchiarelli
Cindex Industries, Inc.

WELDING HANDBOOK COMMITTEE MEMBER:

J. C. Papritan
Ohio State University

FRICTION WELDING

DEFINITION AND PROCESS VARIATIONS

FRICTION WELDING (FRW) is a solid-state welding process that produces a weld under compressive force contact of workpieces rotating or moving relative to one another to produce heat and plastically displace material from the faying surfaces. While considered a solid-state welding process, under some circumstances a molten film may be produced at the interface. However, even then the final weld should not exhibit evidence of a molten state because of the extensive hot working during the final stage of the process. Filler metal, flux, and shielding gas are not required with this process. The basic steps in friction welding are shown in Figure 23.1.

First, one workpiece is rotated and the other is held stationary as shown in Figure 23.1(A). When the appropriate rotational speed is reached, the two workpieces are brought together and an axial force is applied, as in Figure 23.1(B). Rubbing at the interface heats the workpiece locally and upsetting starts, as in Figure 23.1(C). Finally, rotation of one of the workpieces stops and upsetting is completed, as in Figure 23.1(D).

The weld produced is characterized by a narrow heat-affected zone, the presence of plastically deformed material around the weld (flash), and the absence of a fusion zone.

ENERGY INPUT METHODS

THERE ARE TWO methods of supplying energy in friction welding. Direct drive friction welding, sometimes called *conventional friction welding*, uses a continuous input. Inertia friction welding, sometimes called *flywheel friction welding*, uses energy stored in a flywheel.

Direct Drive Welding

IN DIRECT DRIVE friction welding, one of the workpieces is attached to a motor driven unit, while the other is restrained from rotation. The motor driven workpiece is rotated at a predetermined constant speed. The workpieces to be welded are moved together and then a friction weld-

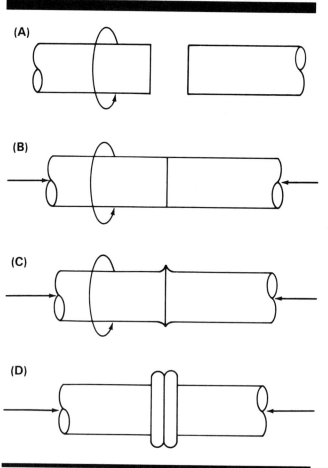

(A)

(B)

(C)

(D)

Figure 23.1–Basic Steps in Friction Welding

ing force is applied. Heat is generated as the faying surfaces (weld interface) rub together. This continues for a predetermined time, or until a preset amount of upset takes place. The rotational driving force is discontinued, and the rotating workpiece is stopped by either the application of a braking force or by its own resistance to rotation. The friction welding force is maintained or increased (forge force) for a predetermined time after rotation ceases. The relationship of direct drive friction welding parameter characteristics is shown in Figure 23.2.

Inertia Drive Welding

IN INERTIA FRICTION welding, one of the workpieces is connected to a flywheel, and the other is restrained from rotating. The flywheel is accelerated to a predetermined rotational speed, storing the required energy. The drive motor is disengaged and the workpieces are forced together by a friction welding force. This causes the faying surfaces to rub together under pressure. The kinetic energy stored in the rotating flywheel is dissipated as heat, through friction at the weld interface, as the flywheel speed decreases. An increase in friction welding force may be applied (forge force) before rotation stops. The forge force is maintained for a predetermined time after rotation ceases. The relationship of inertia friction welding parameter characteristics appears in Figure 23.3.

TYPES OF RELATIVE MOTION

WITH MOST FRICTION welding applications, one of the two workpieces is rotated about an axis of symmetry with the faying surfaces perpendicular to that axis. This means that in the normal case, one of the two workpieces must be circular or tubular in cross-section at the joint location. Typical arrangements for single and multiple welding operations are shown in Figures 23.4 (A) through (E).

Figure 23.4(A) depicts the conventional and most commonly used mode in which one workpiece rotates while the other remains stationary. Figure 23.4(B) shows another mode in which both workpieces are rotated, but in opposite directions. This procedure would be suitable for producing welds where very high relative speeds are needed. Figure 23.4(C) shows a third mode where two stationary workpieces push against a rotating piece positioned between them. This setup might be desirable if the two end parts are long or are of such an awkward shape that rotation would be difficult or impossible by the other modes.

A similar situation, shown in Figure 23.4(D), involves two rotating pieces pushing against a stationary piece at the middle. The same principle can be applied to the making of two welds back to back at the same time with one rotating spindle at the center, as shown in Figure 23.4(E), for the purpose of improving productivity.

Additional forms of friction welding are rather unique. Radial, orbital and angular reciprocating friction welding

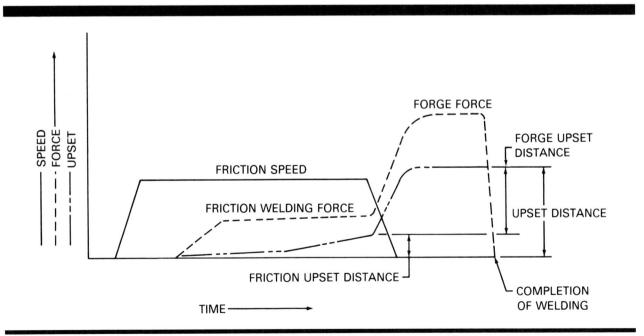

Figure 23.2–Direct Drive Friction Welding Parameter Characteristics

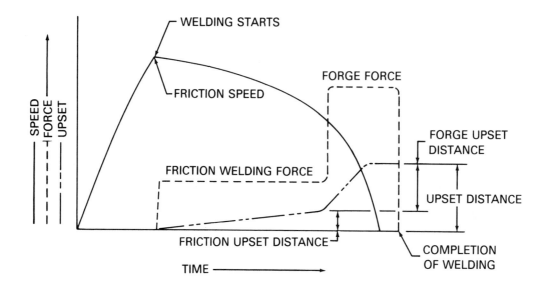

Figure 23.3–Inertia Friction Welding Parameter Characteristics

and friction surfacing are special cases using rotational motion. Linear reciprocating friction welding, as suggested by the name, uses a straight-line motion. These variations of friction welding are described below.

Radial

THIS PROCESS VARIATION can be used to join circular sections where it is undesirable to rotate the parts to be joined. It is also used to weld collars to shafts and tubes. As illustrated in Figure 23.5, the applied force on the rotating band is perpendicular to the axis of rotation. The collar is rotated and compressed as it is heated. An internal expanding mandrel supports the pipe walls and prevents penetration of upset metal into the bore of the pipe.

Orbital

IN THIS PROCESS variation, illustrated in Figure 23.6, one part rotates (or orbits) around the other. Neither part actually rotates around its axis. Consequently, the parts being joined do not need to be circular or tubular in cross-section. This variation is especially useful when part-to-part angular orientation is necessary.

Friction Surfacing

THIS PROCESS VARIATION uses rotational motion of one of the parts, but at the same time adds a relative motion in a direction perpendicular to the axis of rotation. This process is used to deposit material in a solid-state mode to a variety of configurations from flat plates to circular or cylindrical shapes. This variation is shown in Figure 23.7(A).

Angular Reciprocating

THIS PROCESS, SHOWN in Figure 23.7(B), is used primarily for joining plastics. It employs a cyclic reversing rotational motion in which the moving part is rotated through a given angle which is less than one full rotation.

Linear Reciprocating

THIS PROCESS USES a straight-line back and forth motion between the two parts to be joined. An advantage of this variation is that rotational symmetry of the parts to be joined is not necessary. This variation is illustrated in Figure 23.7(C).

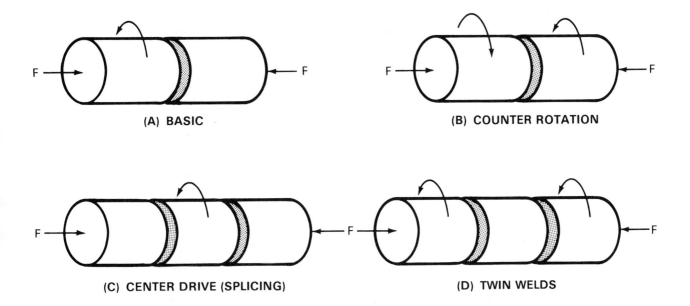

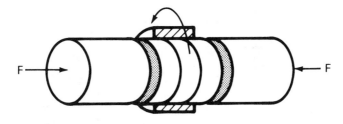

Figure 23.4–Typical Arrangements of Friction Welding

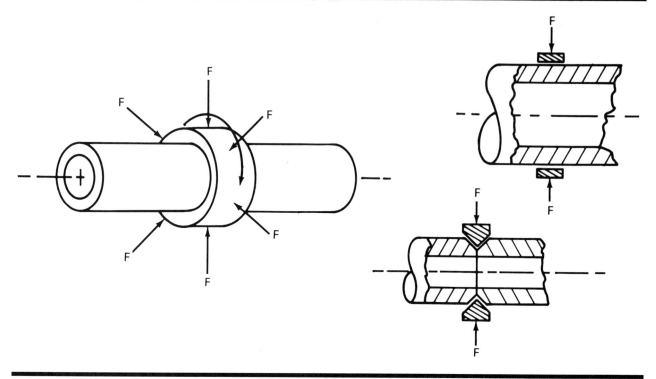

Figure 23.5–Radial Friction Welding

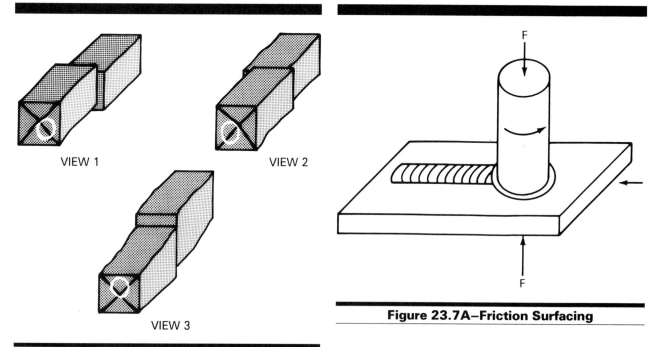

VIEW 1

VIEW 2

VIEW 3

Figure 23.6–Schematic View of Orbital Friction Welding. Three Consecutive Views Taken at 120 Degree Intervals

Figure 23.7A–Friction Surfacing

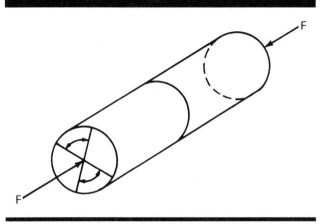

Figure 23.7B—Angular Reciprocating Friction Welding

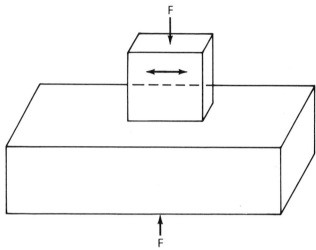

Figure 23.7C—Linear Reciprocating Friction Welding

PROCESS CHARACTERIZATION

WHILE SPECIFIC DETAILS of the bonding process are unclear, the welding cycle can be divided into two stages: the rubbing or friction stage, and the upsetting or forging stage. The welding heat is developed during the first stage, and the weld is consolidated and cooled during the second stage.

FRICTION STAGE

WITH IDENTICAL WORKPIECES, the joining mechanism occurs in steps. When the pieces make contact, rubbing takes place between the faying surfaces, and strong adhesion takes place at various points of contact. The unit pressure is high. At some points, the adhesion is stronger than the metal on either side. Shearing takes place and metal is transferred from one surface to the other. As rubbing continues, the torque and interfacial temperature both increase. The sizes of transferred fragments grow until they become a continuous layer of plasticized metal. If a liquid film forms, it occurs at this point. During this period, the torque peaks and decreases to some minimum value, which remains reasonably constant as metal is heated and forced from the interface and axial shortening continues.

FORGING STAGE

TOWARD THE END of the heating process, forging pressure is applied to the workpiece to cause axial shortening. This upset results in the flash shown in Figure 23.1(D). Comparing Figures 23.2 and 23.3, it can be seen that the latter part of the direct drive and the inertia friction welding processes is very similar with respect to axial shortening (upset), speed and pressure. As the speed decreases, a second torque peak occurs when the interface bonds and cools from its maximum temperature. The torque then decreases as the RPM drops to zero.

The bonding mechanism with dissimilar metals is more complex. A number of factors, including physical and mechanical properties, surface energy, crystal structure, mutual solubility, and intermetallic compounds, may play a role in the bonding mechanism. It is likely that some alloying will occur in a very narrow region at the interface as a result of mechanical mixing and diffusion. The properties of this layer may have a significant effect on overall joint properties. Mechanical mixing and interlocking may also contribute to bonding. The complexity makes prediction of weldability of dissimilar metals very difficult. Suitability of a particular combination should be established for each application with a series of tests designed for that purpose.

RELATIONSHIP BETWEEN VARIABLES

Speed

THE FUNCTION OF rotation is to produce a relative velocity at the faying surfaces. From a weld quality standpoint, speed is not generally a critical variable; that is, it can vary within a fairly broad tolerance band and still provide sound welds. For steels, the tangential velocity should be in the range of 250 ft/min (1.3 m/s). This is true for both solid and tubular workpieces. Tangential speeds below 250 ft/min produce very high torques that cause work clamping problems, non-uniform upset, and metal tearing. Production machines are normally designed to operate with speeds of 300 to 650 rpm. For example, a spindle speed of 600 rpm can be used to weld steel products of 2 to 4 in. (50 to 101 mm) diameters (310 to 620 ft/min or 1.6 to 3.2 m/s).

High rotational speeds and the lower heat inputs associated with them, see Figure 23.8, can be used to weld hardenable steels. Longer heating time preheats the metal to control cooling rates and avoid quench cracking. Con-

versely, for certain dissimilar metal combinations, low velocities (and their shorter heating times) can minimize the formation of brittle intermetallic compounds. In practice, however, heating time (for a given amount of upset) is usually controlled by varying the friction welding pressure.

Pressure

THE EFFECTIVE PRESSURE ranges are also broad for heating and forging, although the selected pressures should be reproducible for any specific operation. The pressure controls the temperature gradient in the weld zone, the required drive power, and the axial shortening. The specific pressure depends upon the metals being joined and the joint geometry. Pressure can be used to compensate for heat loss to a large mass, as in the case of tube-to-plate welds.

Heating pressure must be high enough to hold the faying surfaces in intimate contact to avoid oxidation. For a set spindle speed, low pressure limits heating with little or no axial shortening. High pressure causes local heating to high temperature and rapid axial shortening. With mild steel, the rate of axial shortening is approximately proportional to heating pressure, as illustrated in Figure 23.9. It also

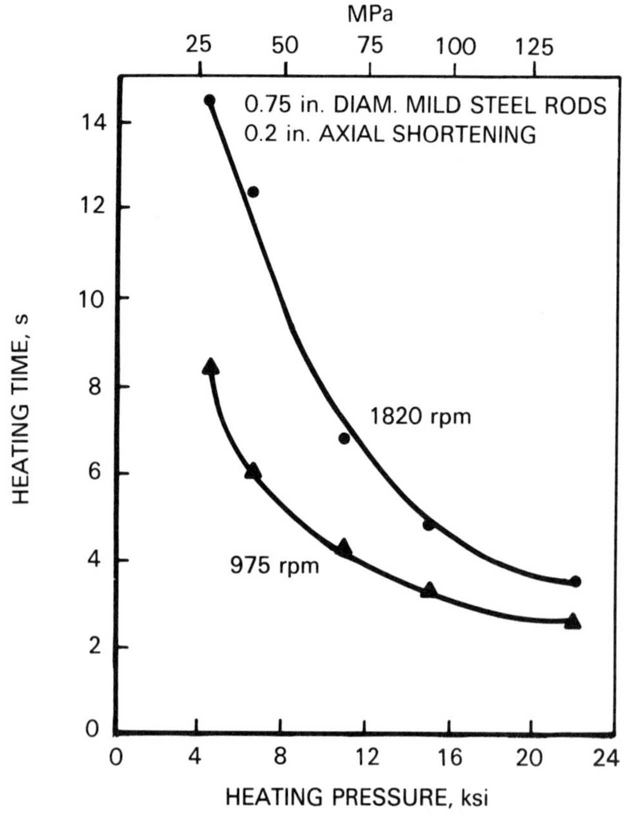

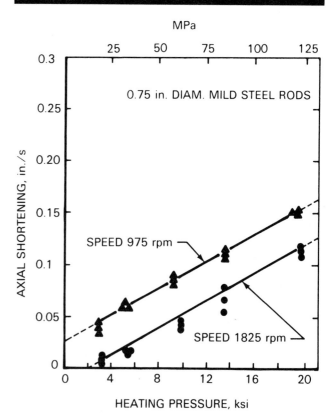

Figure 23.8–Relationship Between Heating Time and Heating Pressure for Mild Steel with Continuous Drive Friction Welding

Figure 23.9–Relationship Between Axial Shortening and Heating Pressure for Mild Steel With Continuous Drive Friction Welding

shows that for a given pressure during the heating phase, axial shortening is greater at low speed than at high speed. Joint quality is improved in many metals, including steels, by applying an increased forging force at the end of the heating period.

For steels, a wide range of pressures is applicable for making sound welds. In the case of mild steel, heating pressures of 4500 to 8700 psi (31 to 60 MPa) and forging pressures of 11 000 to 22 000 psi (76 to 152 MPa) are acceptable. Commonly used values are 8000 and 20 000 psi (55 and 138 MPa), respectively. High, hot-strength alloys, such as stainless steels and nickel base alloys, will require higher forging pressures.

If a "preheat" effect is desired in order to achieve a slower cooling rate, a pressure of about 3000 psi (21 MPa) can be applied for a brief period at the initiation of the weld cycle. The pressure is then increased to that required for welding.

Heating Time

FOR A PARTICULAR application, heating time is determined during setup or from previous experience. Excessive heating time limits productivity and wastes material. Insufficient time may result in uneven heating as well as entrapped oxides and unbonded areas at the interface. Uneven heating is typical of friction welds in barstock. Near the center of a rotating bar, the surface velocity may be too low to generate adequate frictional heating. Hence, thermal diffusion from the outer portion of the faying surface must take place in order to insure a sound bond overall.

Heating time can be controlled in two ways. The first is with a suitable timing device that stops rotation at the end of a preset time. Preheat and forging functions can be incorporated with heating time using a sequence timer.

The second method is to stop rotation after a predetermined axial shortening. This method is set to consume a sufficient length to assure adequate heating prior to upsetting. Variations in surface condition can be accommodated without a sacrifice in weld quality.

In summary, for a given axial shortening when welding mild steel, the heating time will be significantly influenced by heating pressure and speed. Heating time is reduced at a decreasing rate as heating pressure is increased. It also decreases with speed at the same heating pressure.

INERTIA FRICTION WELDING

WITH INERTIA FRICTION welding, the speed continuously decreases with time during both the friction and forging stages. This contrasts to direct drive friction welding where the friction stage occurs at constant speed. Speed is decreasing during the forging phase of both processes. Throughout the friction stage the thickness of the plasticized layer is related to the rubbing speed. As the speed decreases at the end of the friction stage, the generation of

heat decreases, the thickness of the hot plasticized layer decreases, and the torque peaks as the weld enters the forging stage. The axial pressure forces the hot metal from the joint. During this time, the rate of axial shortening increases and then stops as the joint cools.

Relationship Between Variables

THERE ARE THREE welding variables with this method: moment of inertia of the flywheel, initial flywheel speed, and axial pressure. The first two variables determine the total kinetic energy available to accomplish welding. The amount of pressure is generally based on the material to be welded and the interface area. The energy in the flywheel at any instant during the welding cycle is defined by the equation:

$$E = \frac{I S^2}{C} \tag{23.1}$$

where

E = Energy, ft-lb (J)
I = Moment of inertia, lb-ft^2 (kg-m^2)
S = Speed, rpm
C = 5873 when the moment of inertia is in lb-ft^2
C = 182.4 when the moment of inertia is in kg-m^2

For mathematical modeling and parameter calculations, the derived value of "Unit Energy" is defined by the following equation:

$$E_u = \frac{E}{A} \tag{23.2}$$

where

E_u = Unit Energy, ft-lb/in.^{2} (J/mm^2)
E = energy, ft-lb (J)
A = Faying Surface Area, in.^{2} (mm^2)

Unit energy can be used to scale or extrapolate data from one material, size or geometry to another. This can often serve as a first approximation.

With a particular flywheel system, the energy in the flywheel is determined by its rotational speed. If the mass of the flywheel is changed, the available energy at any particular speed changes. Therefore, the capacity of an inertia welding machine can be modified by changing the flywheel within the limits of the machine capability.

During welding, energy is extracted from the flywheel, and its speed decreases. The total time for the wheel to come to rest depends on the average rate at which the energy is being removed and converted to heat.

The shape of the heat-affected zone can be adjusted by varying the flywheel moment of inertia, heating pressure, and speed. Also, the heat input can be adjusted to control the width of the heat-affected zone and the cooling rate of the weldment. The effect of flywheel energy, heating pressure, and tangential velocity on the heat pattern and flash formation of welds in steel are shown in Figure 23.10.

Flywheel Effect

THE MOMENT OF inertia of the flywheel depends upon its section shape, diameter, and mass. For a specific application and initial speed, the energy available for welding can be increased by changing to a flywheel with a larger moment of inertia. The product of flywheel moment of inertia and the square of its initial velocity varies inversely for a given total energy requirement.

The amount of upset near the end of the welding cycle depends upon the remaining energy in the flywheel as well as the heating or forging pressure. For low carbon steel, forging usually starts at a peripheral velocity of about 200 ft/min (1.0 m/s). Large flywheels can prolong the forging or upsetting phase. If the flywheel is too small, the upset may be insufficient to consolidate the weld and eject impurities from the interface.

For a given initial velocity and heating pressure, a larger flywheel will increase the available energy. The effect of this is shown in Figure 23.10(A). As the available energy is increased, the amount of plasticized metal becomes greater as do the upset and flow of metal from the interface. The heating pattern remains fairly uniform, but the excessive energy wastes metal in the form of flash.

Velocity

THE INSTANTANEOUS TANGENTIAL velocity varies directly with the radius and rotational speed according to the following equation:

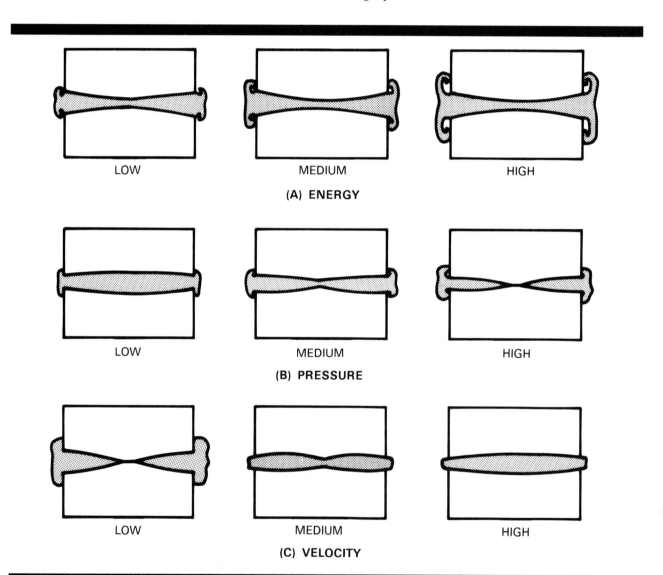

(A) ENERGY

(B) PRESSURE

(C) VELOCITY

Figure 23.10—Effect of Welding Variables on the Heat Pattern at the Interface and Flash Formation of Inertia Welds

$$V_t = K\,r\,s \tag{23.3}$$

where

$V_t =$ tangential velocity, ft/min
$r =$ radius, in. (m)
$s =$ instantaneous speed, rpm
$K = 0.52$ when r is in inches
$K = 0.1$ when r is in meters

With a rotating solid rod, the velocity varies linearly from zero at the center to a maximum at the periphery. This contrasts with the behavior of a thin wall tube where the change in velocity across the faying surface is minor. Hence, the energy required for welding a rod and a tube of the same alloy and equal faying surface area will be different.

For each metal, there is a range of peripheral velocities that produces the best weld properties. For welding solid bars of steel, the recommended initial peripheral velocity of the workpiece ranges from 500 to 1500 ft/min (2.5 to 7.5 m/s); however, welds can be made at velocities as low as 300 ft/min (1.5 m/s). If the velocity is too low, whether at the required energy level or not, the heating at the center will be too low to produce a bond across the entire interface and the flash will be rough and uneven. This is illustrated in Figure 23.10(C). At medium velocities of 300 to 800 ft/min (1.5 to 4.1 m/s), the heating pattern in steel has an hourglass shape at the lower end of the range and gradually flattens at the upper end of the range. At initial velocities above 1200 ft/min (6.1 m/s) for steel, the weld becomes rounded and is thicker at the center than at the periphery.

Heating Pressure

THE EFFECT OF varying heating pressure is generally opposite to that of velocity. As Figure 23.10(B) shows, welds made at low heating pressure resemble welds made at high velocity with regard to the formation and appearance of weld upset and heat-affected zones. Excessive pressure produces a weld that lacks good bonding at the center and has a large amount of weld upset, similar to a weld made at low velocity. The effective heating pressure range for a solid bar of medium carbon steel is 22 000 to 30 000 psi (152 to 207 Mpa).

ADVANTAGES AND LIMITATIONS

FRICTION WELDING, LIKE any welding process, has its specific advantages and disadvantages.

ADVANTANGES

THE FOLLOWING ARE some advantages of friction welding:

(1) No filler metal is needed.
(2) Flux and shielding gas are not required.
(3) The process is environmentally clean; no arcs, sparks, smoke or fumes are generated by clean parts.
(4) Surface cleanliness is not as significant, compared with other welding processes, since friction welding tends to disrupt and displace surface films.
(5) There are narrow heat-affected zones.
(6) Friction welding is suitable for welding most engineering materials and is well suited for joining many dissimilar metal combinations.
(7) In most cases, the weld strength is as strong or stronger than the weaker of the two materials being joined.
(8) Operators are not required to have manual welding skills.

(9) The process is easily automated for mass production.
(10) Welds are made rapidly compared to other welding processes.
(11) Plant requirements (space, power, special foundations, etc.) are minimal.

LIMITATIONS

SOME LIMITATIONS OF the process are as follows:

(1) In general, one workpiece must have an axis of symmetry and be capable of being rotated about that axis.
(2) Preparation and alignment of the workpieces may be critical for developing uniform rubbing and heating, particularly with diameters greater than 2 in. (50 mm).
(3) Capital equipment and tooling costs are high.
(4) Dry bearing and nonforgeable materials cannot be welded.
(5) If both parts are longer than 3 ft (1 m), special machines are required.
(6) Free-machining alloys are difficult to weld.

FRICTION WELDING VARIABLES

MATERIALS WELDED

FRICTION WELDING CAN be used to join a wide range of similar and dissimilar materials, including: metals, some metal matrix composites, ceramics, and plastics. Some combinations of materials that have been joined according to the literature and equipment manufacturers' data are indicated in Figure 23.11. This figure should only be used

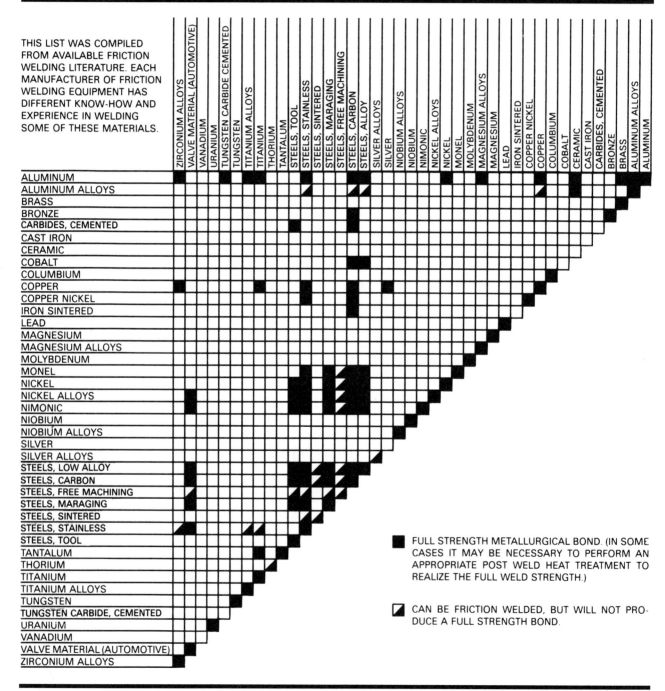

THIS LIST WAS COMPILED FROM AVAILABLE FRICTION WELDING LITERATURE. EACH MANUFACTURER OF FRICTION WELDING EQUIPMENT HAS DIFFERENT KNOW-HOW AND EXPERIENCE IN WELDING SOME OF THESE MATERIALS.

■ FULL STRENGTH METALLURGICAL BOND. (IN SOME CASES IT MAY BE NECESSARY TO PERFORM AN APPROPRIATE POST WELD HEAT TREATMENT TO REALIZE THE FULL WELD STRENGTH.)

◪ CAN BE FRICTION WELDED, BUT WILL NOT PRODUCE A FULL STRENGTH BOND.

Figure 23.11–Material Combinations Weldable by Friction Welding

as a guide. Specific weldability may depend upon a number of factors including specific alloy compositions, applicable process variation, component design, and service requirements.

In principle, almost any metal that can be hot forged and is unsuitable for dry bearing applications can be friction welded. Some metals may require postweld heat treatment to remove the effect of the severe deformation or quench hardening at the weld interface. Free-machining types of alloys should be welded with caution because redistribution of inclusions may create planes of weakness in the weld zone. Such welds exhibit low strength, decreased ductility, and reduced notch toughness.

In general, a consequence of reorienting the inclusion population into the weld plane is that the ductility and toughness across the joint will tend to approach the wrought short transverse properties of the materials being welded. If these properties are critical, it is essential to use microstructurally clean materials.

There are a number of dissimilar metal combinations that have marginal weldability. These may involve combinations that have high and low thermal conductivities, a large difference in forging temperatures, or the tendency to form brittle intermetallic compounds. Examples are aluminum alloys to both copper and steel and titanium alloys to stainless steel.

The metallurgical structures produced by friction welding are generally those resulting from elevated temperature deformation. Time at temperature is short, and the temperatures achieved are generally below the melting point. With nonhardenable metals such as mild steel, changes in properties are negligible in the weld zone. On the other hand, with hardenable steels, structural changes may occur in the heat-affected zone. They should be welded with a relatively long heating time to achieve a slower cooling rate and preserve toughness.

The interface structures of dissimilar metal combinations are significantly affected by the particular welding conditions employed. The longer the welding time, the greater the consideration that must be given to diffusion across the interface. Proper welding conditions will usually minimize undesired diffusion or intermetallic compound formation. The interface between an aluminum and carbon steel friction weld is shown in Figure 23.12. A very narrow diffusion zone is apparent.

In some cases, joints between dissimilar metals will show a mechanical mixing at the interface. Such action in a joint between Type 302 stainless steel and tantalum is shown in Figure 23.13.

JOINT DESIGN

THE NATURE OF friction welding suggests that the joint face of at least one workpiece must be essentially round, except in the cases of orbital and linear reciprocating friction welding. The rotated workpiece should be somewhat balanced in shape because it is revolved at relatively high

Figure 23.12–Interface of a Friction Weld Between Aluminum (top) and Carbon Steel (bottom) (x1000)

speed. Preparation of surfaces to be joined is not normally critical except in the case of alloys with distinct differences in mechanical or thermal properties, or both.

The basic joint designs for combinations of bars, tubes, and plates are illustrated in Figure 23.14. When bars or tubes are welded to plates, most of the flash comes from the bar or tube. This is true because there is less mass in the smaller section and the heat penetrates deeper into it. This effect can be usefully employed in joints between dissimilar metals with widely different mechanical or thermal properties. The material of lower forging strength or lower thermal conductivity should have a larger cross-sectional area.

Conical joints are usually designed with the faces at 45 to 60 degrees to the axis of rotation, as shown in Figure 23.15. For low-strength metals, large angles are preferred to support the axial thrust required to produce adequate

Figure 23.13–Interface of a Friction Weld Between Tantalum (top) and Type 302 Stainless Steel (bottom) (x200 and reduced 66 percent)

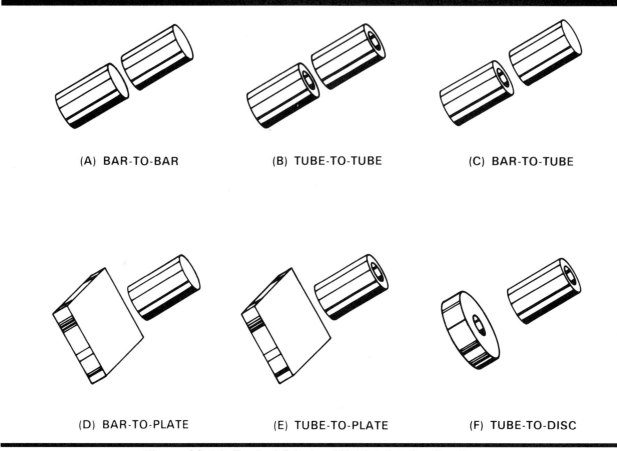

(A) BAR-TO-BAR   (B) TUBE-TO-TUBE   (C) BAR-TO-TUBE

(D) BAR-TO-PLATE   (E) TUBE-TO-PLATE   (F) TUBE-TO-DISC

Figure 23.14–Typical Friction Weld Joint Applications

heating pressure. Certain applications may require the use of conical joints; however, experience has shown that a butt weld (perpendicular) geometry is superior. The butt geometry leads to less residual stress and distortion.

For applications where the flash cannot be removed but must be isolated for cosmetic or functional reasons, clearance for it can be provided in one or both workpieces (flashtraps). Two flashtrap configurations are illustrated in Figures 23.16A and B.

FRICTION WELDING MACHINES

A TYPICAL FRICTION welding machine consists of the following components:

(1) Head
(2) Base
(3) Clamping arrangements
(4) Rotating and upsetting mechanisms
(5) Power supply
(6) Controls
(7) Optional monitoring devices

This is true for both process variations. However, the machines for each variation differ somewhat in design and method of operation.

Equipment is currently available (circa 1989) from 200 lb (890 N) to 275 tons (250 metric tons) maximum forge force in direct drive friction welders and up to 2250 tons (2040 metric tons) maximum forge force in inertia friction welders. This translates to parts ranging from 0.06 in (1.5 mm) diameter barstock to 24 in. (600 cm) diameter tubes. The faying surface area ranges from .003 to 250 in.^{2} (2 to 160 000 mm^2). A given machine can generally make welds which have a faying area range of eight to one. For example, a 30 ton (27 metric tons) maximum forge force machine can make welds in barstock ranging from 0.5 in. (13 mm) diameter to slightly less than 1.5 in. (38 mm) diameter. The above information is based on manufacturers recommendations for mild steel.

Direct Drive Welding Machines

WITH DIRECT DRIVE friction welding machines, one of the workpieces to be welded is clamped in a vise. The other workpiece is held in a centering chuck that is mounted on

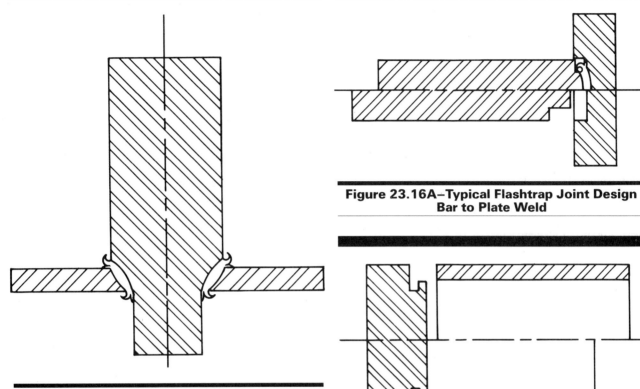

Figure 23.15–Typical Conical Weld Joint Design

Figure 23.16A–Typical Flashtrap Joint Design Bar to Plate Weld

Figure 23.16B–Typical Flashtrap Joint Design

a rotatable spindle. The spindle is driven by a motor through a single- or variable-speed drive.

To make a weld, the rotating workpiece is thrust against the stationary workpiece to produce frictional heat at the contact surfaces as illustrated in Figure 23.17. The combination of speed and pressure raises the contact surfaces to a suitable temperature and deformation (upset) occurs. Rotation is then stopped and the pressure is maintained or increased to further upset the interface and complete the weld. A typical weld cycle is illustrated in Figure 23.3.

The machine spindle can be driven directly by a motor and allowed to stop under its natural deceleration characteristics and the retarding torque exerted by the weld. In

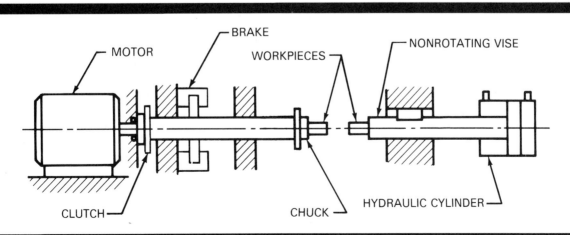

Figure 23.17–Basic Arrangement of a Direct Drive Welding Machine

practice, however, a clutch is normally used between the motor and spindle so that the motor can run continuously. The spindle can be engaged when required for the welding operation. This also conserves the starting energy that would be consumed if the motor is started for each weld.

A common practice is to include a fast-acting brake on the spindle. The function of the brake is to rapidly terminate rotation at the end of a specified heating time or after a preset axial shortening of the weldment. This feature provides good control of overall weldment length and broadens the acceptable range of welding variables for critical applications.

Two variables are used to control the friction heating phase. These are axial shortening and heating time. Under distance setting control, in axial shortening, the friction heating phase continues until a given part length is achieved. This is used to compensate for variations in preweld part length. The time setting control mode is intended to provide repeatable energy input. It is also possible to combine both options; a preweld distance can be set after which control changes to a time based mode (or vice versa).

For critical applications, where the workpieces would normally be of uniform length before welding, the time mode is preferred. In all cases, a minimum loss in length must occur between the components to ensure the removal of contaminants at the interface and a resulting sound weld.

There are a number of variables associated with this method:

(1) Rotation speed (rpm)
(2) Preheat pressure
(3) Preheat distance or time
(4) Friction pressure
(5) Friction distance or time
(6) Braking time (includes delay and rate)
(7) Forge Pressure
(8) Forge time (includes delay and rate)

This list is comprehensive, and not all machines or weld schedules will require every setting.

Inertia Welding Machines

WITH AN INERTIA welding machine, a flywheel is mounted on the spindle between the drive and the rotating chuck, as shown in Figure 23.18. The flywheel, spindle, chuck, and workpiece are accelerated to a selected speed corresponding to a specific energy level. When that speed is attained, driving is stopped and the flywheel and workpiece are allowed to spin freely. The two workpieces are then brought together and a specific axial thrust is applied. The kinetic energy of the flywheel is transferred to the weld interface and converted to heat. As a result, the flywheel speed decreases and finally comes to rest. Simultaneously, the tangential velocity is decreasing to zero with time in an essentially parabolic mode. Heating time is only a matter of seconds.

In the majority of applications, inertia friction welding uses a single axial thrust to produce heating and forging pressure. However, the machines are normally capable of applying more than one level of thrust. When forging pressure is used, it is triggered at a selected speed setting near the end of the cycle. A typical weld cycle is shown in Figure 23.3. This multiple force technique can also be used to provide a preheat effect before welding occurs, as with the direct drive method.

A distance setting control mode can also be achieved in inertia friction welding. Energy input is varied by adjusting

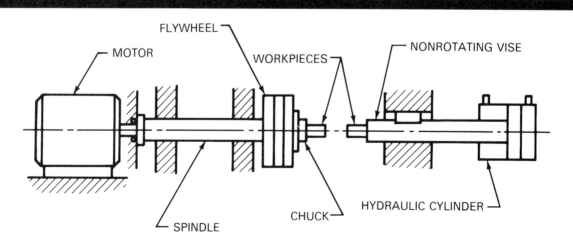

Figure 23.18–Basic Arrangement of an Inertia Welding Machine

RPM depending upon initial part lengths. An experimental correlation between energy and upset must be established before employing this mode.

The variables that control the weld quality are as follows:

(**1**) Total moment of inertia

(**2**) Weld speed (initial rpm)
(**3**) Weld pressure
(**4**) Upset speed (rpm at which upset pressure is applied)
(**5**) Upset pressure

Most welds are performed by varying weld speed and pressure only.

WELDING PROCEDURES

SURFACE PREPARATION AND FIT-UP

AS WITH ANY welding process, surface preparation can affect weld quality. Weld quality and consistency will be best when the faying surfaces are free of dirt, oxide or scale, grease, oil, or other foreign materials. In addition, the faying surfaces should mate together with very little gap.

In noncritical applications, some contamination and nonuniform contact of the faying surface may be tolerated. This is true if sufficient axial shortening is used to account for the gap and to extrude sufficient plasticized metal from the interface in order to carry away any contaminants. Sheared, flame cut, or sawed surfaces may be used with adequate axial shortening, provided the surfaces are essentially perpendicular to the axis of rotation. If the surfaces are not perpendicular, joint mismatch could result. For best practice, the squareness should be within 0.010 in./in. (0.01 mm/mm) of joint diameter.

Thick layers of mill scale should be removed from steel workpieces prior to welding to avoid unstable heating. A thin layer of rust may not be detrimental with adequate axial shortening.

Center projections left by cutoff tools are not harmful. However, pilot holes or concave surfaces should be avoided since they may entrap air or impurities at the interface.

For dissimilar metal welds between materials with large differences in hot-forging behavior, surface cleanliness of both workpieces is critical. Additionally, the squareness of the harder material is critical. Examples include steel to aluminum, steel to copper, and copper to aluminum.

TOOLING AND FIXTURES

ALL GRIPPING DEVICES used for holding the workpieces must be reliable. Slippage of a workpiece in relation to the chuck results either in a poor weld or in damage to the gripping device or the workpiece.

The gripping mechanism of the chucking devices must be rigid and resist the applied thrust. The extension of the workpiece from the device should be as short as practical to minimize deflection, eccentricity, and misalignment.

Grip diameter must be as large or larger than the diameter of the weld interface, otherwise the workpiece may shear at the grips. Serrated gripping jaws are recommended for maximum clamping reliability.

There are two basic types of tooling: rotating and nonrotating. The machines in Figure 23.19 are equipped with both types. Each type, in turn, is either manual or power operated. As a rule, manually actuated tooling is used only for small quantity production.

Rotating tooling must be well balanced, have high strength, and provide good gripping power. Collet chucks meet these requirements and are most frequently used.

The most commonly used nonrotating gripping device is a vicelike fixture with a provision for absorbing the thrust. This device permits reasonable tolerance in the stationary workpiece diameter and yet maintains concentricity with the other piece in the collet chuck. More accurate devices may be used where concentricity is very critical.

Mating of faying surfaces and concentricity of the workpieces depend upon the accuracy of manufacture, projecting length from the clamping fixture, and the rigidity of the tooling.

HEAT TREATMENT

PRIOR HEAT TREATMENT of the workpieces generally has little effect on the ability to friction weld specific alloys. However, it may affect the mechanical properties of the heat-affected zone and the gripping of the workpieces.

Postweld heat treatment is often employed to produce the desired properties in the base metal, the welded joint, or both. A postweld anneal may be used to soften or stress relieve the joint. This heat treatment improves the ductility.

In the case of dissimilar metal welds, a postweld heat treatment should not contribute to the formation or expansion of an intermetallic layer at the interface which may lower joint ductility or strength. The postweld heat treatment should be evaluated for the application by destructive testing.

(A) DIRECT DRIVE MACHINE

(B) INERTIA DRIVE MACHINE

Figure 23.19–Typical Friction Welding Machines

WELD QUALITY

Weld quality is dependent upon the proper selection of material (type and quality) and welding variables. Good welds can be made between like metals with a wide range of speeds, pressures and times. Dissimilar material combinations are more critical with respect to welding parameters.

JOINT DISCONTINUITIES

DISCONTINUITIES CHARACTERISTIC OF fusion welds, such as gas porosity and slag inclusions, are not encountered in friction welding. However, other types of discontinuities may occur. These are associated with improper surface preparation, incorrect welding conditions, defective material or combinations of the above.

Discontinuities at the center of a weld may occur for various reasons, such as welding conditions not creating sufficient heating at the center for coalescence. Inertia welds made with the same speed and inertial mass but with a decreasing heating pressure (axial shortening) from left to right are shown in Figure 23.20. Two cross sections, shown in Figures 23.20(E) and (F), exhibit center defects due to insufficient pressure. Lack of center bonding may also occur in continuous drive friction welds when inadequate speed, heating time or heating pressure is used.

Concave faying surfaces that prevent uniform contact during the early stages of welding can limit center heating and entrap oxides. A weld with a discontinuity that resulted when a center hole for machining operations was not removed prior to welding is shown in Figure 23.21.

PROCESS MONITORING

The advent of reliable computerized data acquisition and analysis systems has revolutionized process monitoring in friction welding. Microprocessor controlled welding machines are capable of maximizing both output and quality. Particularly useful is their ability to document each weld and manipulate data for statistical process control (SPC) purposes.

Factors which are documented include: friction and forge pressures, speed, upset and time. Other parameters such as torque and energy may also be monitored in specific cases.

INSPECTION AND TESTING

INSPECTION AND TESTING are applied both to input materials and resulting weldments. Rather than relying upon destructive testing to guarantee quality, in-process monitoring and nondestructive inspection are increasingly being employed. Depending upon the quality level needed, this may range from simple visual inspection and mechanical tests to the latest advances in nondestructive techniques. A photograph of a friction welded automotive "halfshaft" is shown in Figure 23.22(A). Peak temperature is used as a process control, and the inframetric image of the weld is shown in Figure 23.22(B).

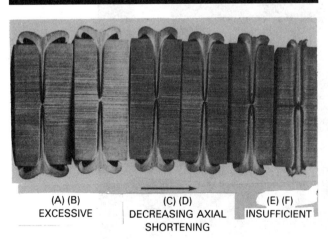

(A) (B)
EXCESSIVE

(C) (D)
DECREASING AXIAL
SHORTENING

(E) (F)
INSUFFICIENT

The center defects in (e) and (f) are indicated with arrows.

Figure 23.20–Effect of Axial Shortening on Bond Joint of Friction Welds

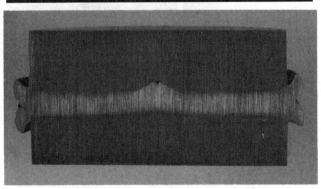

Figure 23.21–Discontinuity at the Center of a Friction Weld Caused by a Prior Center Hole

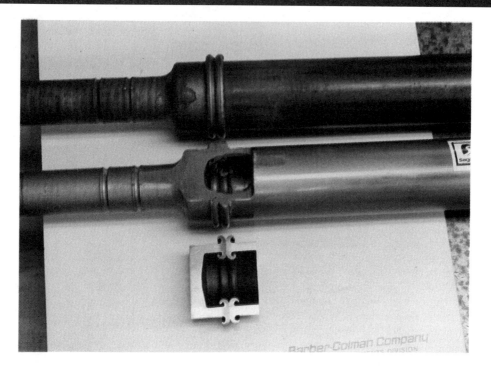

(A) AUTOMOTIVE "HALFSHAFT" FRICTION WELD JOINT

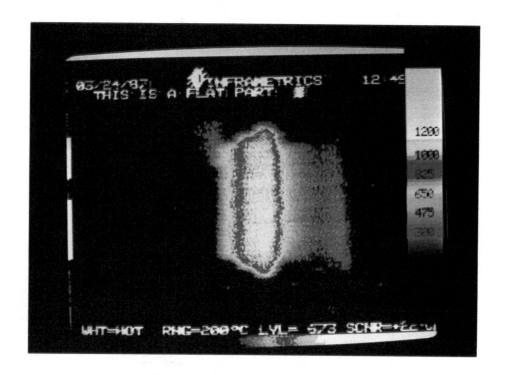

(B) INFRAMETRIC IMAGE

Figure 23.22–Inframetric Imagerty Used to Measure Peak Weld Temperature

APPLICATIONS

FRICTION WELDED PARTS in production applications span the aerospace, agricultural, automotive, defense, marine, and oil industries. Everything from tongholds on forging billets to critical aircraft engine components are friction welded in production.

Automotive parts which are manufactured by friction welding include gears, engine valves, axle tubes, driveline components, strut rods and shock absorbers. Hydraulic piston rods, track rollers, gears, bushings, axles and similar parts are commonly friction welded by the manufacturers of agricultural equipment. Friction welded aluminum/copper joints are in wide usage in the electrical industry. Stainless steels are friction welded to carbon steel in various sizes for use in marine drive systems and water pumps for home and industrial use. Friction welded assemblies are often used to replace expensive castings and forgings.

Some typical applications, including automotive, aircraft and medical, are shown in Figures 23.23 through 23.28.

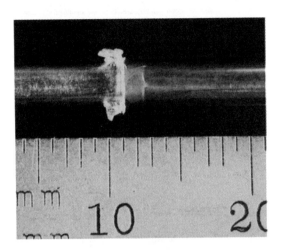

(A) AS-WELDED — NOTE FLASH ON 4043 ONLY

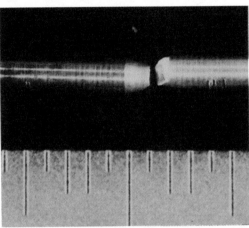

(C) REDUCED SECTION TENSILE TEST SHOWING FAILURE AWAY FROM THE INERTIA WELD

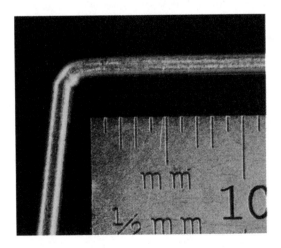

(B) REDUCED SECTION BEND TEST

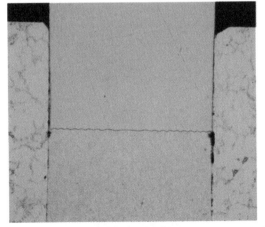

(D) METALLOGRAPHIC SECTION SHOWING FRICTION WELD INTERFACE

Figure 23.23–An Inertia Welded Transition Joint Between OFHC Copper and 4043 Aluminum to Facilitate Simultaneous Solderability and Weldability of a Ground Pin

(A) CAMSHAFT FORGING FRICTION WELDED TO TIMING GEAR

(B) FINISHED SPINDLES WELDED TO BRACKETS FOR TRAILER

Figure 23.24—Typical Steel Automotive Applications

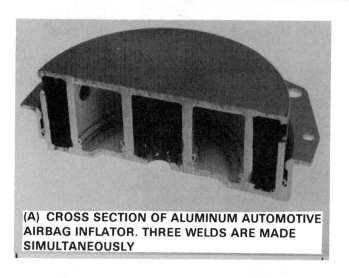

(A) CROSS SECTION OF ALUMINUM AUTOMOTIVE AIRBAG INFLATOR. THREE WELDS ARE MADE SIMULTANEOUSLY

(B) FITTINGS INERTIA WELDED TO AN ALUMINUM CANISTER

Figure 23.25—Two Typical Aluminum Automotive Applications

WELDS

(A) A JET ENGINE COMPRESSOR WHEEL FABRICATED BY FRICTION WELDING

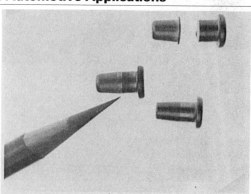

(B) TITANIUM-TO-TITANIUM ALLOY AIRCRAFT RIVETS

Figure 23.26—Typical Aircraft Applications

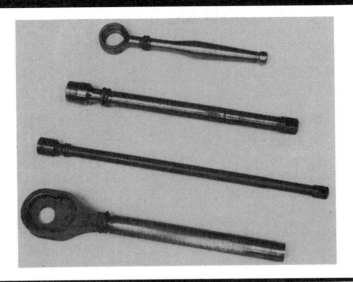

Figure 23.27–Hand Tools Inertia Welded From Forgings

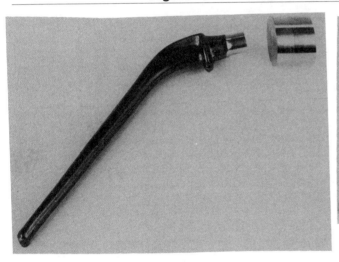

(A) RAW STOCK

(B) WELDED AND FINISH MACHINED

Figure 23.28–Cobalt Alloy Hip Replacement Prothesis

SAFETY

FRICTION WELDING MACHINES are similar to machine tool lathes in that one workpiece is rotated by a drive system. They are also similar to hydraulic presses in that one workpiece is forced against the other with high loads. Safe practices for lathes and power presses should be used as guides for the design and operation of friction welding machines. Typical hazards include high noise levels, high rotational speeds, and flying particles.

Machines should be equipped with appropriate mechanical guards and shields as well as two-hand operating switches and electrical interlocks. These devices should be designed to prevent operation of the machine when the work area, rotating drive, or force system is accessible to the operator or other personnel.

Operating personnel should wear appropriate eye and ear protection and safety apparel commonly used with machine tool operations. Ear protection should be provided to guard against high noise levels produced during friction welding. In any case, applicable OSHA standards should be strictly observed.

The machine manufacturers literature should be studied for complete safety precautions.

SUPPLEMENTARY READING LIST

Baeslack III, W. A. and Hagey, K. S. "Inertia friction welding of rapidly solidified powdered metallurgy aluminum." *Welding Journal* 67(7): 139-S; July 1988.

Bangs, S. "Inertia welding for fuel mandrels." *Welding Design & Fabrication.* 37-39; June 1986.

Bell, R.A., Lippold, J.C., and Adolphson, D.R. "An evaluation of copper -stainless steel inertia friction welds." *Welding Journal* 63(11): 325-S; November 1984.

Dawes, C.J. "An examination of orbital friction welding using axial offset." *Welding Institute Res Bull* 12(6): 161-167; 1971.

Dickson, G.R., et al. "Experiments on friction welding some nonferrous metals." International Conference on the Welding and Fabrication of Nonferrous Metal, 1972 May 2-3, 41-53. Eastbourne, Cambridge, England: The Welding Institute, 1972.

Dinsdale, W.O., Dunkerton, S.B. "The impact properties of forge butt welds in carbonmanganese steels," Part I: Continuous Drive Friction Welds. Welding Institute Research Report, 159/1981, September 1981.

Dinsdale, W.O. and Dunkerton, S.B. "The impact properties of forge butt welds in carbonmanganese steels," Part II: Orbital Friction and Inertia Welds. Welding Institute Research Report, 160/1981, September 1981.

Dunkerton, S.B. "Properties of 25 mm diameter orbital friction welds in three engineering steels." Welding Institute Research Report, 272/1985, April 1985.

———. "Toughness properties of friction welds in steels." *Welding Journal* 65(8): 193-S; April 1986.

Eberhard, B.J., Schaaf Jr., B.W., and Wilson, A.D. "Friction weld ductility and toughness as influenced by inclusion morphology." *Welding Journal* 62(7): 171-S; 1983.

Ellis, C.R.G. "Continuous drive friction welding of mild steel." *Welding Journal* 51(4): 183s-197s; April 1972.

Ellis, C.R.G. and Needham, J.C. *Quality control in friction welding,* IIW Document III-460-72. (Available from) Miami, Florida: American Welding Society, 1972.

Ellis, C.R.G. and Nicholas, E.D. "A quality monitor for friction welding." Advances in Welding Processes, 3rd International Conference, 1974 May 7-9, Harrogate, England, 14-18. Cambridge, England: The Welding Institute, 1974.

Ellis, C.R.G. "Recent industrial developments in friction welding." *Welding Journal* 54(8): 582-589; August 1975.

Forster, P.B. "Heat under power (HUP) friction welding." Advances in Welding Processes, 3rd International Conference, 1974 May 7-9, Harrogate, England. Cambridge, England: The Welding Institute, 1974

Jessop, T.J. "Friction welding of dissimilar metal combinations: aluminum and stainless steel." Welding Institute Research Report, P/73/75, November 1975.

Jessop, T.J., et al. "Friction welding dissimilar metals." Advances in Welding Processes, 4th International Conference, 1978 May 9-11, Harrogate, England, 23-36. Cambridge, England: The Welding Institute, 1978.

Kuruzar, D.L. "Joint design for the friction welding process." *Welding Journal* 58(6): 31-5; June 1979.

Kyusojin, A., et al. "Study on mechanism of friction welding in carbon steels." *Bulletin of the JSME* 23(182): August 1980.

Lebedev, V.K., et al. "The inertia welding of low carbon steel, *Avt Svarka* (7): 18-22; 1980.

Lippold, J.C. and Odegard, B.C., "Technical note: microstructural evolution during inertia friction welding of austenitic stainless steels." *Welding Journal* 64(12): 327-S; December 1985.

Murti, K.G.K., and Sundaresan, S. "Thermal behavior of austenitic-ferri tic joints made by friction welding." *Welding Journal* 64(12): 327-S; December 1985.

Needham, J.C. and Ellis, C.R.G. "Automation and quality control in friction welding." The Welding Institute Research Bulletin 12(12), 333-9 (Part 1), 1971 December; 13(2), 47-51 (Part 2), 1972 February.

Nessler, C.G., et al. "Friction welding of titanium alloys." *Welding Journal* 50(9): 379s-85s; September 1971.

Nicholas, E.D. "Radial friction welding." Advances in Welding Processes, 4th International Conference, 1978 May 9-11, Harrogate, England, 37-48. Cambridge, England: The Welding Institute, 1978.

———. "Radial friction welding." *Welding Journal* 62(7): 17-29; July 1983.

Nicholas, E.D. and Thomas, W.M. "Metal deposition by friction welding." *Welding Journal* 65(8): 17; August 1986.

Nicholas, E.D. "Friction Welding: state of the art." *Welding Design and Fabrication* 50(7): 56-62; July 1977.

Nicholas, E.D. "Friction welding noncircular sections with linear motion: a preliminary study." Welding Institute Research Report, 332/1987, April 1987.

Ruge, J., Thomas, K., and Sundaresan, S. "Joining copper to titanium by friction welding." *Welding Journal* 65(8): 28; November 1986.

Sassani, F. and Neelam, J.R. "Friction welding of incompatible materials." *Welding Journal* 67(11): 264-S; November 1988.

Searl, J. "Friction welding noncircular components using orbital motion." *Welding and Metals Fabrication* 39(8): 294-297; August 1971.

Tumuluru, M.D. "A parametic study of inertia friction welding for low alloy steel pipes." *Welding Journal* 63 (9): 289-S; September 1984.

Vill, V.I. *Friction welding of metals.* Translated from Russian. (published by) Miami, Florida: American Welding Society, 1962.

Wang, K.K. "Friction welding." Bulletin 204. New York: Welding Research Council, April 1975.

Wang, K.K. and Linn, W. "Flywheel friction welding research." *Welding Journal* 53(6): 233s-41s; June 1974.

Wang, K.K. and Rasmussen, G. "Optimization of inertia welding process by response surface methodology." *Trans—Asme, Journal Engrg. Ind.* 94, Series B (4): 999-1006; November 1972.

Wang, K.K., Reif, G.R., and OH, S.K. "In-process quality detection of friction welds using acoustic emission techniques." *Welding Journal* 61(9): 312-S; September 1982.

Yashan, D., Tsang, S., Johns, W.L., and Doughty, M.W. "Inertia friction welding of 1100 aluminum to type 316 stainless steel." *Welding Journal* 66(8): 27; August 1987.

EXPLOSION WELDING

PREPARED BY A COMMITTEE CONSISTING OF:

V. D. Linse, Chairman
Edison Welding Institute

P. I. Temple
Detroit Edison Co.

WELDING HANDBOOK COMMITTEE MEMBER:
P. I. Temple
Detroit Edison Co.

CHAPTER 24

EXPLOSION WELDING

FUNDAMENTALS OF THE PROCESS

DEFINITION AND GENERAL DESCRIPTION

EXPLOSION WELDING IS a solid-state welding process that produces a weld by high velocity impact of the workpieces as the result of controlled detonation. The explosion accelerates the metal to a speed at which a metallic bond will form between them when they collide. The weld is produced in a fraction of a second without the addition of filler metal. This is essentially a room temperature process in that gross heating of the workpieces does not occur. The faying surfaces, however, are heated to some extent by the energy of the collision, and welding is accomplished through plastic flow of the metal on those surfaces.

Welding takes place progressively as the explosion and the forces it creates advance from one end of the joint to the other. Deformation of the weldment varies with the type of joint. There may be no noticeable deformation in some weldments, and there is no measureable loss of metal. Welding is usually done in air, although it can be done in other atmospheres or in a vacuum where circumstances dictate. Most explosion welding is done on sections with relatively large surface areas, although there are applications for sections with small surface areas as well.

PRINCIPLES OF OPERATION

A TYPICAL ARRANGEMENT of the components for explosion welding is shown in Figure 24.1. Fundamentally, there are three components:

(1) Base metal
(2) Prime or cladding metal
(3) Explosive

The base component remains stationary as the prime component is welded to it. The base component may be supported by a backer or an anvil, particularly when it is relatively thin. The base component by itself or in combination with the backer should have sufficient mass to minimize distortion during the explosion welding operation.

The prime component usually is positioned parallel to the base component; however, for special applications it may be at some small angle with the base component. In the parallel arrangement, the two are separated by a specified spacing, referred to as the *standoff distance*. In the angular arrangement, a standoff distance may or may not be used at the apex of the angle. The explosion locally bends and accelerates the prime component across the standoff distance at a high velocity so that it collides at an angle with and welds to the base component. This angular collision and welding front progresses across the joint as the explosion takes place.

The explosive, almost always in granular form, is distributed uniformly over the top surface of the prime component. The force which the explosion exerts on the prime component depends upon the detonation characteristics and the quantity of the explosive.

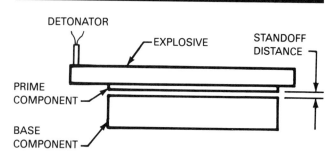

Figure 24.1–Typical Component Arrangement for Explosion Welding

A buffer layer such as neoprene material may be required between the explosive and the prime component to protect the surface of that component from erosion by the detonating explosive.

Explosive Detonation

THE ACTION THAT occurs during explosion welding is illustrated in Figure 24.2. The manner in which the explosive is detonated is extremely important. Detonation must take place progressively across the surface of the prime component. The speed of the detonation front establishes the velocity at which the collision progresses over the joint area. This is known as the *collision velocity* and is one of the important variables of the process. The selection of an explosive that will produce the required detonation velocity is of utmost importance in consistently obtaining good welds. Moreover, the explosive must provide uniform detonation so the collision velocity will be uniform from the start to the finish of the weld.

Prime Component Velocity and Angle

AS THE DETONATION front moves across the surface of the prime component, both the intense pressure in the front and the pressure generated by the expanding gases immediately behind the front accelerate the prime component to a certain angle and velocity. This angle and velocity depend upon the type and amount of explosive, the thickness and mechanical properties of the prime component, and the standoff distance employed.

Collision, Jetting, and Welding

THE FOLLOWING ARE important interrelated variables of the explosion welding process:

(1) Collision velocity
(2) Collision angle
(3) Prime component velocity

The intense pressure necessary to make a weld is generated at the collision point when any two of these variables are within certain well defined limits. These limits are determined by the properties of the particular metals to be joined. Pressure forces the surfaces of the two components into intimate contact and causes localized plastic flow in the immediate area of the collision point. At the same time, a jet is formed at the collision point, as shown in Figure 24.2. The jet sweeps away the original surface layer on each component, along with any contaminating film that might be present. This exposes clean underlying metal which is required to make a strong metallurgical bond. Residual pressures within the system are maintained long enough after collision to avoid release of the intimate contact of the metal components and to complete the weld.

NATURE OF THE BOND

THE INTERFACE BETWEEN the two components of an explosion weld is almost always wavy on a microscale, the wave size being dependant on the collision conditions employed in making the weld. A typical wavy explosion weld interface is shown in Figure 24.3. Most welds with a wavy

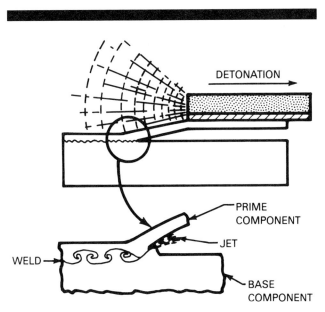

Figure 24.2–Action Between Components During Explosion Welding

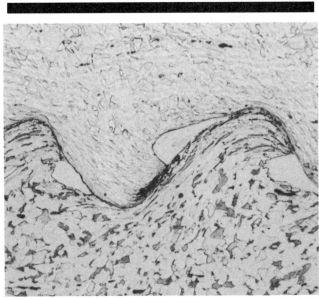

Figure 24.3–Typical Wavy Interface Formed Between Two Explosion Welded Components (Stainless Steel to Mild Steel)

interface contain small pockets of jet material which normally are located on the front and back slopes of the waves. This material is composed of some combination of the two parent metals, and partial or complete melting of the material generally occurs. The pockets will be ductile when the metal combinations can form solid solutions, but they may be brittle or may show discontinuities in those combinations that form intermetallic compounds. Pockets of the latter material may not be detrimental, if they are very small. Good welding practices will produce small pockets.

Large pockets, on the other hand, occur with excessive collision conditions (prime component velocity, collision velocity, and collision angle) or may even produce a continuous melted layer. The large pockets and the continuous melted layer often contain a substantial number of shrinkage voids and other discontinuities that reduce strength and ductility. They are usually detrimental to the soundness and serviceability of the weld. For this reason, welding practices that produce an excessively large wave size or a continuous melted layer must be avoided.

On certain occasions, a flat weld interface can be formed when the collision velocity is below some critical value for the particular combination of metals being welded. Welds of this type usually possess satisfactory mechanical properties but as a rule are not sought in practice. Small variations in the collision conditions which produce a flat weld interface can result in lack of bonding.

EXPLOSIVE MATERIAL PROPERTIES

EXPLOSIVES USED FOR the explosion welding process are almost always granular, and their composition is usually based around ammonium nitrate as the primary ingredient. This allows them to detonate in the velocity range of 6500 to 9800 ft/s (2000 to 3000 m/s) which is normally required to produce the collision point conditions necessary for optimal welding conditions. In general, the detonation velocity of an explosive will depend upon its composition, thickness, and packing or loading density.

PARALLEL AND PRESET ANGLE STANDOFFS

TWO TYPES OF standoffs can be employed for explosion welding: parallel or preset angle. The use of the preset angle approach is normally restricted to small areas or short length welds such as tube-to-tubesheet welding, lap welding between sheet or tube components, or other specialized small area welds. The parallel or constant standoff is used for all larger area welding which constitutes the majority of the explosion welding applications. For other than flat plate cladding operations, the standoff distance geometry and amount of explosive must be allowed for in the design of the prewelding components.

STANDOFF DISTANCE

THE STANDOFF DISTANCE employed in the explosion welding setup will have some influence on the interfacial wave size. Increasing the standoff distance increases the collision angle between the prime and base components (see Figure 24.2) up to the limiting dynamic bend angle to which the particular explosive loading being used is capable of accelerating the cladding component. The interfacial wave size correspondingly increases with the increasing angle of collision.

In general terms, the standoff distance in a parallel welding setup will normally be between one half and one times the thickness of the cladding component being accelerated by the explosive. In an angular arrangement, the preset angle will typically range between one and eight degrees.

JOINT QUALITY

THE QUALITY OF an explosion weld will depend upon the nature of the interface and the effect the process has on the properties of the metal components. The properties of the metal include strength, toughness, and ductility. The effect of welding on these properties can be determined by comparing the results of tension, impact, bending, and fatigue tests of welded and unwelded materials. Standard ASTM testing procedures may be used.

The quality of the bond can be determined by destructive and nondestructive tests. Since the size of test samples is limited by the thickness of the components and the weld is planar and in essence has no thickness, special destructive tests are used for evaluation of the bond. The tests should reflect the conditions the weld will have to endure in service.

NONDESTRUCTIVE INSPECTION

DUE TO THE nature of explosion welds, nondestructive inspection is restricted almost totally to the ultrasonic method. Radiographic inspection is only applicable to welds between metals with significant differences in density and an interface with a large wavy pattern.

Ultrasonic Inspection

ULTRASONIC INSPECTION IS the most widely used nondestructive method for the examination of explosion welds. It will not determine the strength of the weld, but it will indicate weld soundness. Pulse-echo techniques are normally used for clad steels in pressure vessels.[1] An ultrasonic frequency in the range of 2.5 to 10 MHz usually is adequate. Allowance needs to be made for the differences in acoustical impedance of various metals.

The ultrasonic instrument should be calibrated on standard samples containing both bonded and known unbonded areas which will provide a display signal amplitude of 50 to 75 percent of full screen height for the bonded area. Unbonded areas reflect the signal before it can complete the circuit. This shows up in the height of the signal at the appropriate location on the display scope. C-Scan recordings can be made to give a permanent record of the results of the examination.

For large clad plates where scanning 100 percent of the surface area is not necessary, the examination can be carried out on a rectangular grid pattern laid out on the plate. Unbonded areas which are detected should be investigated to determine whether they are small enough to be acceptable or are so large or so numerous that they are unacceptable. The size and number of unbonded areas that can be permitted in a clad plate depend upon the intended service for the plate. Clad plates for heat exchangers sometimes require over 98 percent bond, and limits are placed on the size and number of unbonded areas that are permitted.

Radiographic Inspection

RADIOGRAPHY CAN BE used to inspect explosion welds in metals that have significantly different densities and a wave size sufficiently large to be resolvable on a radiograph. Radiographs are marked to identify the plate and the precise location of the area they represent. The radiographs are taken perpendicular to the surface from the side with the high density metal. The film must be in intimate contact with the surface on the low density side. Radiographs can delineate a wavy interface as uniformly spaced light and dark lines. The number of waves per unit length is then counted and the weld quality correlated through previous destructive testing to the wave size. Further, those areas in which no wave patterns are delineated would indicate either a flat weld interface or no weld at all.

DESTRUCTIVE TESTING

DESTRUCTIVE TESTING IS used to determine the strength of the weld and the effect of the process on the properties of the base metals. Standard testing techniques can be used, but specially designed tests sometimes are required to determine bond strength for some configurations.

Clad Plates

THE REQUIREMENTS FOR carbon steel plates clad with copper, stainless steel, or nickel alloys are covered in appropriate ANSI/ASTM standards.[2] These Standards primarily use simple bend and shear tests to determine the strength of the composite.

Chisel Test

THE CHISEL TEST is widely used to determine the integrity of the bond in an explosion weld. The test is performed by driving a chisel into and along the interface. The ability of the interface to resist separation by the force of the chisel provides an excellent qualitative measure of the bond strength. If the weld is not good, failure will occur along the interface in advance of the chisel point. If the weld is good, either the chisel will cut through the weaker of the two parent metals or fracture will occur in one of the two parent metals away from the weld interface.

Tension-Shear Test

THIS TEST IS designed to determine the shear strength of the weld. The specimen configuration is shown in Figure 24.4. Equal thicknesses of the two components are preferred. The length of the shear zone, "d", should be selected so that little or no bending will occur in either component. Failure should occur by shear, parallel to the weld line. If failure occurs through one of the base metals, the shear strength of the weld is obviously greater than the strength of the base metal. In any event, the results are useful for comparison purposes only, using a common test specimen.

Tension Test

A SPECIAL "RAM" or ring rupture tension test can be used for evaluation of the tensile strength of explosion welds. As shown in Figure 24.5, the specimen is designed to subject the weld interface to a tensile load. The cross-sectional area of the specimen is the annulus between the outside and inside diameters. The typical specimen has a short gage length which is intended to cause failure at or immediately adjacent to the weld interface. If failure occurs in one of

---

1. See ANSI/ASTM A578, *Standard Specification for Straight Beam Ultrasonic Inspection of Plain and Clad Steel Plates for Special Applications* (latest edition).

2. See ANSI/ASTM standard specifications A263, A264, A265 and B432 (latest editions).

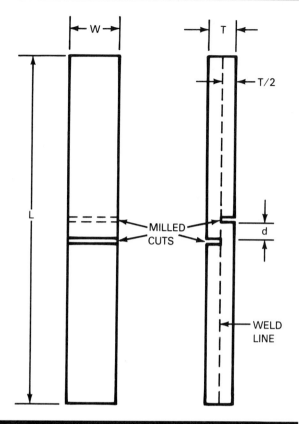

Figure 24.4–Tensile Shear Test Sample Configuration

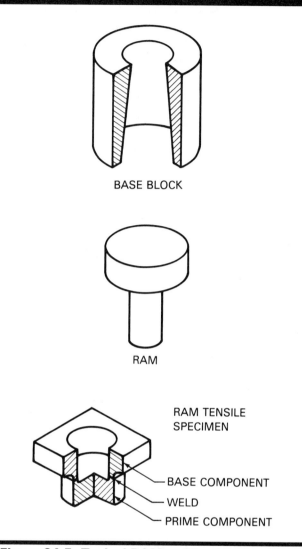

Figure 24.5–Typical RAM or Ring Rupture Test Sample Configuration

the base metals, the test shows that the weld is stronger than the base metal.

The test is conducted by placing the specimen on the base block with the ram in the hole. A compressive load is then applied through the ram and base. Load at failure is recorded.

Metallographic Examination

METALLOGRAPHY CAN PROVIDE useful information about the quality of explosion welds. The section for metallographic examination should be taken so that the interface can be examined on a plane parallel to the direction of detonation and normal to the surfaces of the welded components. A well-formed, well-defined wave pattern is generally indicative of a good weld. Depending on the combination of materials being evaluated, the amplitude and frequency of the wave can vary somewhat without significant influence on the strength of the weld. Small, isolated pockets of melt resulting from the vortices of the jet are usually not detrimental to the weld quality. Large melt pockets containing voids or even microcracks in the swirls

indicate that the collision angle and energy were too high and the weld is poor.

Excessive collision conditions with metals such as titanium, high strength nickel alloys, and martensitic steels can produce strain bands emanating from the interface wave slopes as a result of localized shear. Proper welding conditions must be employed to minimize the occurrence of these bands and their potential detrimental affect on the performance of the clad product.

Samples for metallographic examination should be taken from an area that is representative of the entire weld. Edge effects may result in areas of less than optimum weld quality along the edges of a weld. Samples taken from such locations would not be representative of the rest of the weld.

WELDING PROCEDURES

TYPES OF JOINTS

EXPLOSION WELDING IS limited to joints that overlap or have faying surfaces. In the case of cladding, the surfaces of both components have the same geometry, and one component overlays the other. In transition joints in pipe or tubing and tube-to-tube sheet joints, an overlapping joint configuration is usually used. The overlap and weld in such joints should be long enough to insure that it will not fail in service by shear along the interface.

SURFACE PREPARATION

THE SURFACES TO be joined should be clean and free of gross imperfections to produce welds of consistent soundness, strength, and ductility. The smoothness required depends upon the metals to be joined. In general, a surface finish of 150 microinches or better is required to obtain high quality welds.

FIXTURING AND BACKUP

FOR CONSISTENT QUALITY, the welding conditions should be uniform over the entire area to be joined. These include stand off distance with parallel components or initial angle with inclined components and sufficient rigidity or support for the base component. For cladding with a relatively thick prime component, spacers or supports for providing the required standoff distance are usually placed around the outer edges of the cladding plate where edges effects will normally be removed. Where the prime or cladding component is so thin that deflection due to its own weight combined with the weight of the explosive on top of it will cause a problem in maintaining the necessary standoff distance, additional gapping support may be required in the central areas. Typically, light weight materials such as small foam or balsa wood blocks are strategically placed under the middle areas of the cladding plate. They are normally consumed in the welding process and have minimal effect on the resulting weld.

During the cladding of plates with thicker base or backer components, the base is typically placed directly on packed sand or earth. If the base component is relatively thin or subject to more extensive deformation during the explosion welding process, it should be supported uniformly on a more rigid, massive anvil to minimize deflection. When cladding pipe or tubing, an internal or external mandrel normally is required to back up the base component.

CAPABILITIES AND LIMITATIONS

ONE ATTRIBUTE OF the explosion welding process is its ability to join a wide variety of similar and dissimilar metals. The dissimilar metal combinations range from those that are commonly joined by other welding processes, such as carbon steel to stainless steel, to those that are metallurgically incompatible for fusion welding or diffusion bonding processes, such as aluminum or titanium to steel.

The process can be used to join components of a wide range of sizes. Surface areas ranging from less than 1 in.^{2} (6.5 cm^2) to over 400 ft^2 (37m^2) can be welded. Since the base component is stationary during welding, there is no upper limit on its thickness. The thickness of the prime component may range from 0.001 to 1.25 in. (0.25 to 31.8 mm) or more depending on the material.

Geometric configurations that can be explosion welded are those which allow a uniform progression of the detonation front and, hence, the collision front. These include flat plates as well as cylindrical and conical structures. Welds may also be made in certain complex configurations, but such work requires thorough understanding and precise control of the process.

APPLICATIONS

METALS WELDED

AS A GENERAL rule, any metal can be explosion welded if it possesses sufficient strength and ductility to withstand the deformation required at the high velocities associated with the process. Metals that will crack when exposed to the shock associated with detonation of the explosive and the collision of the two components cannot be explosion welded. Metals with elongations of at least five to six percent [in a 2 in. (51 mm) gage length] and Charpy V-notch impact strengths of 10 ft-lb (13.6J) or better can be welded by this process. In special cases, metals with low ductility can be welded by preheating them to a slightly elevated temperature at which point they will have adequate impact resistance; however, the use of explosives in conjunction with elevated temperature components requires special safety considerations. The commercially significant metals and alloys that can be joined by explosion welding are given in Figure 24.6.

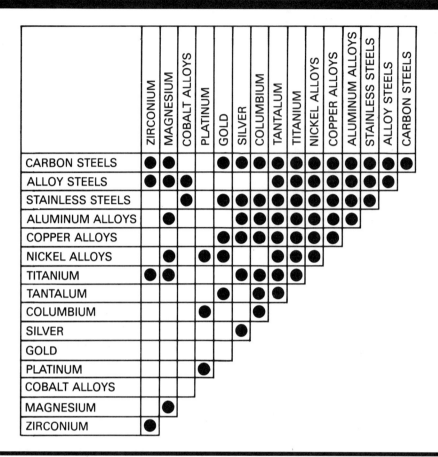

Figure 24.6—Commercially Significant Metals and Alloys that can be Joined by Explosion Welding

	ZIRCONIUM	MAGNESIUM	COBALT ALLOYS	PLATINUM	GOLD	SILVER	COLUMBIUM	TANTALUM	TITANIUM	NICKEL ALLOYS	COPPER ALLOYS	ALUMINUM ALLOYS	STAINLESS STEELS	ALLOY STEELS	CARBON STEELS
CARBON STEELS	●	●			●	●	●	●	●	●	●	●	●	●	●
ALLOY STEELS	●	●	●					●	●	●	●	●	●	●	
STAINLESS STEELS			●		●	●	●	●	●	●	●	●	●		
ALUMINUM ALLOYS		●				●	●	●	●	●	●	●			
COPPER ALLOYS					●	●	●	●	●	●	●				
NICKEL ALLOYS		●		●	●				●	●					
TITANIUM	●	●					●	●	●						
TANTALUM					●		●	●							
COLUMBIUM			●				●								
SILVER							●								
GOLD															
PLATINUM					●										
COBALT ALLOYS															
MAGNESIUM		●													
ZIRCONIUM	●														

While explosion welding does not produce changes in the bulk properties, it can produce some noteable changes in the mechanical properties and hardness of metals, particularly in the immediate area of the weld interface as indicated in Figure 24.7. In general, the severe localized plastic flow along the interface during welding increases the hardness and strength of the material in this region. Accordingly, the ductility decreases. Such effects may be erased by a postweld heat treatment as shown in Figure 24.7. However, the particular heat treatment applied should be one that will not reduce the ductility of the weld by unfavorable diffusion or the formation of brittle intermetallic compounds at the interface.

CLADDING

Plate

THE CLADDING OF flat plate constitutes the major commercial application of explosion welding. A typical clad plate is shown in Figure 24.8. It is customary to supply explosion clad plate in the as-welded condition because the hardening which occurs immediately adjacent to the interface usually does not significantly affect the bulk engineering properties of the plate. Despite this, some service requirements may demand postweld heat treatment. Clad plates usually are distorted somewhat during explosion welding and must be straightened to meet standard flatness specifications (Figure 24.9). Straightening is usually done with a press or a roller leveler.

Pressure vessel heads and other components can be made from explosion clad plates by conventional hot or cold forming techniques (see Figures 24.10, 24.11, and 24.12). Hot forming must take into account the metallurgical properties of the materials and the possibility that undesirable diffusion may occur at the interface. Compatible alloy combinations, such as stainless and carbon steel, may be formed by methods traditionally used for clad materials. Incompatible combinations such as titanium and steel, on the other hand, may require special procedures to limit the formation of undesirable intermetallic compounds at the interface. Titanium clad steel, for instance, should be hot formed at temperatures no higher than 1400°F (760°C) to prevent the formation of undesirable intermetallics which could lead to brittle failure of the bond.

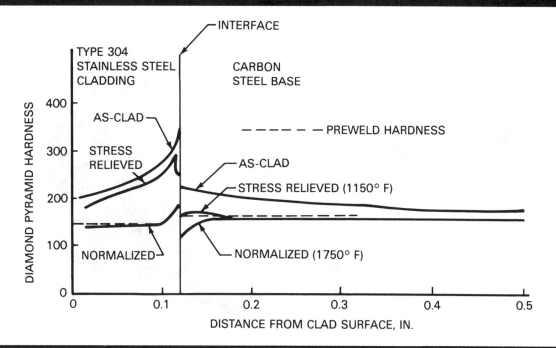

Figure 24.7—Hardness Profile Across Stainless Steel to Carbon Steel Clad Plate, As-Welded and After Heat Treatment

Figure 24.8–As-Explosion Clad Flat Plate Consisting of 13/16-Inch 304L Stainless Steel Clad on 8-Inch-Thick SA 516-70 Steel (DuPont photograph)

Figure 24.9–Explosion Clad Titanium (1/4-Inch) Steel (1-3/4-Inch) Tube Sheet Blanks Following Post Welding Flattening (Explosive Fabricators, Inc. photograph)

Figure 24.10–180 Inch Diameter Dome of 3/16 Inch Type 410 Stainless Steel on 3 Inch Thick A387 Steel Formed From Explosion Clad Plate (Explosive Fabricators, Inc. photograph)

Figure 24.11—Titanium (1/4-Inch) Clad Steel (1 1/4-Inch) Condenser Sheet (Explosive Fabricators, Inc. photograph)

Figure 24.12—Finished Vessel Fabricated From Explosion Clad Plate (Explosive Fabricators, Inc. photograph)

Reducing the thickness of clad plate by rolling (termed *conversion rolling*) provides a convenient and economical means of producing bimetal sheets of proper thickness for subsequent processing.

Cylinders

EXPLOSION WELDING CAN be used to clad cylinders on their inside or outside surfaces. One application of this is the internal cladding of steel forgings with stainless steel to make nozzles for connection to heavy-walled pressure vessels. Clad nozzles with inside diameters of 1/2 to 24 in. (13 to 610 mm) and lengths up to 3 ft (900 mm) have been made. A typical internally clad cylinder is shown in Figure 24.13.

Transition Joints

FUSION WELDED JOINTS between two incompatible metals are difficult or impossible to make. Some of those that can be made exhibit low strength and ductility. Transition joints produced by explosion welding may provide a solution to that problem. Many such joints can be cut from a single large clad plate. Conventional fusion welding practices may then be used to attach the members of the transition joint to their respective similar metal components. Care must be taken, however, to limit the installation temperature and subsequently the service temperature at the weld interface to a level suitable for the materials combination in the joint.

Electrical

ALUMINUM, COPPER, AND steel are the metals most commonly used in electrical systems. Joints between them frequently are necessary to take advantage of the special properties of each. Such joints must be sound if they are to conduct high amperages efficiently, minimize power losses, and avoid overheating of the member in service. Transition joints cut from thick explosion welded plates of aluminum and copper or aluminum and steel provide efficient conductors of electricity. This concept is routinely used in the fabrication of anodes for the primary aluminum industry.

Temperature limits for transition joints between aluminum and steel are 500°F (260°C) or less for long-term service. Copper-aluminum joints should be limited to 300°F (150°C). High quality transition welds are unaffected by thermal cycling below these temperatures. Short term exposure (10 to 15 minutes) during attachment welding, for example, may reach 550 to 600°F (290 to 315°C) with aluminum to steel and 400 to 500°F (200 to 260°C) with aluminum to copper without harm.

In the presence of an electrolyte such as salt water, aluminum and steel form a galvanic cell. In a mechanical connection, crevice corrosion in the joint can become a severe problem. A welded transition joint is metallurgically bonded, and there is no crevice in which the electrolyte can act. Structural transition joints are used to attach alu-

Figure 24.13–Steel Nozzle Internally Clad With 1/4-Inch Thick Inconel® 600 (The internal bore size is 9 inches.)

minum superstructures to the steel decks of naval vessels and commercial ships.

Tubular

TUBULAR TRANSITION JOINTS in various configurations can be machined from thick clad plate. The interface of the explosion weld is perpendicular to the axis of the tube in this case. Examples of a variety of transition joints machined from explosively clad plate are shown in Figure 24.14. While the majority of explosively welded tubular transition joints are aluminum to steel, other metal combinations for this type of joint include titanium to stainless steel, zirconium to stainless steel, zirconium to nickel-base alloys, and copper to aluminum.

Joints can also be fabricated directly by explosion welding in an over-lapping or telescoping style similar to a cylindrical cladding operation. They offer the advantage of a long overlap and frequently require little or no machining

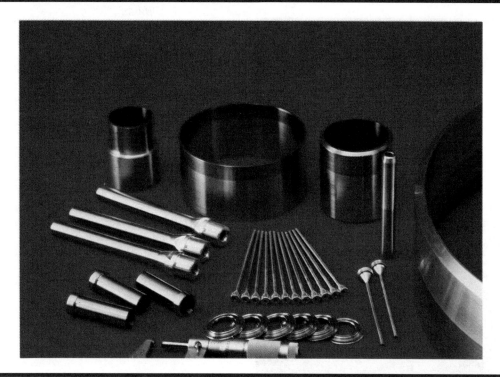

Figure 24.14–Examples of Aluminum to Steel, Titanium to Aluminum, and Titanium to Stainless Steel Tubular Transition Joints Machined from Explosion Clad Plate (Explosion Fabricators, Inc. photograph)

after welding. Typical direct explosion welded tubular transition joints are shown in Figure 24.15.

OTHER

Heat Exchangers

EXPLOSION WELDING MAY be used to make tube-to-tube sheet joints in heat exchanger fabrication. The process is essentially a version of short length internal cylinder cladding with a small explosive charge used to make the joint. In most instances, the weld is located near the front of the tube sheet and has a length of approximately 1/2 in. (13 mm) or three to five times the thickness of the wall of the tube, whichever is greater. Points that must be considered in determining whether explosion welding is suitable for particular tube-to-tube sheet application include the diameter of the tube, the ratio of the wall thickness to the diameter of the tube, the thickness of the ligament between the holes in the tube sheet, and the thickness of the tube sheet. Tubes may be welded individually or in groups. The number in any group is controlled by the quantity of explosive that can be detonated safely at any one time.

Most applications of explosion welding in tube-to-tube sheet joints involve tube diameters in the range of 1/2 to

1.5 in. (13 to 38.1 mm). Metal combinations include steel to steel, stainless steel to stainless steel, copper alloy to copper alloy, nickel alloy to nickel alloy clad steel, and both aluminum and titanium to steel.

Feed Water and Heat Exchanger Tube Plugging

EXPLOSION WELDING CAN be used for plugging leaking tubes in heat exchangers. Electric utilities and petro-chemical companies use the process because it is quick, easy, and reliable. Although the process appears simple, only qualified, trained technicians should implement it. An explosive handling permit is required.

Two examples of tube plugs are shown in Figures 24.16 and 24.17. All plugs are completely assembled by the manufacturer, ready for installation. Materials that match the tube or general purpose nickel alloy material can be used.

The actual welding process is identical with those described previously. Following insertion of the plug into the tube and detonation, the welding occurs automatically. Preparation for field use of this process, however, requires careful attention. This is due to the tubes being plugged having corrosion or process fluids in the tubes.

Figure 24.15—Explosion Welded 12-Inch-Diameter 3003 Aluminum to A106 Grade B Steel Tubular Transition Joint (Battelle photograph)

Preparation of the tubes for plugging requires the following steps:

(1) Remove all fluid from the tube. This is best accomplished by blowing air through the tube.

(2) If water or fluid reappears, a rubber plug may be inserted into the tube 6 to 8 in. (150 to 200 mm) deep to keep the fluid away from the weld area.

(3) The tube ID must be cleaned to a bright shiny surface. Use a carbide burr or other abrasive to clean the surface if wire brushing does not remove oxides or corrosion products. The tube should be cleaned to a depth of 3 1/2 to 4 in. (90 to 100 mm).

(4) Inspect the tube for grooves. These must be removed or reduced to only a few thousandths in depth.

Note: If the tubesheet has enough depth, some of the tube may be removed by drilling to the tubesheet diameter and the plug inserted directly in the tubesheet.

(5) Ensure the surface is clean and dry when the explosive plug is inserted.

The actual explosive weld can now be completed by detonating the inserted plug. Since this requires explosives, only specially trained and licensed technicians perform this step of the process. Once the area has been cleared of smoke and explosion cases, the plug is ready for inspection. Depending upon the availability of equipment and quality requirements, various testing methods may be used. Following a visual examination the plug may be tested using pneumatic pressure, hydrostatic pressure, or a helium gas sniffer. Careful attention to avoid plug ejection paths must be avoided during any pressure test. Repairs to a plug which failed testing requires removal of the plug and restoration of the surface to clean, dry conditions without grooves for rewelding using another plug.

Pipeline Welding

IN THE EARLY 1980's, the procedure for joining lengths of large diameter gas and oil transmission pipelines by explosion welding was commercialized as a field procedure. The first commercial application of this procedure was employed to join a 3.7 mile (6 km) long section of 42 in. (1067 mm) diameter line in 1984. The approach involved the use of balanced external and internal explosive charges (see Figure 24.18) to achieve the required short overlap or telescope type weld without the requirement for any support tooling and to allow the weld to be made quickly and economically.

Buildup and Repair

EXPLOSION WELDING MAY be used for buildup and repair of worn components. It is particularly applicable to the repair of inside and outside surfaces of cylindrical components. The worn area is clad with an appropriate thickness of metal and machined to the proper dimensions. In some instances, such as bearing surfaces, the repair can be made with a material that is superior to the original material.

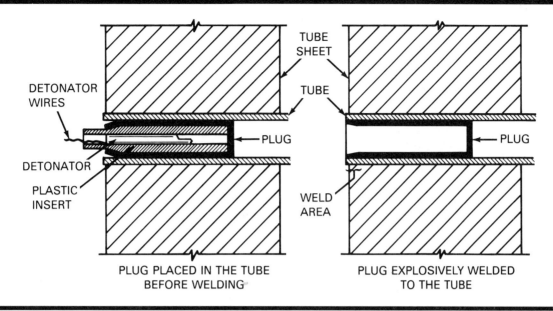

DETONATOR WIRES

DETONATOR

PLASTIC INSERT

TUBE SHEET

TUBE

PLUG

PLUG

WELD AREA

PLUG PLACED IN THE TUBE BEFORE WELDING

PLUG EXPLOSIVELY WELDED TO THE TUBE

Figure 24.16–Explosively Welded Plug

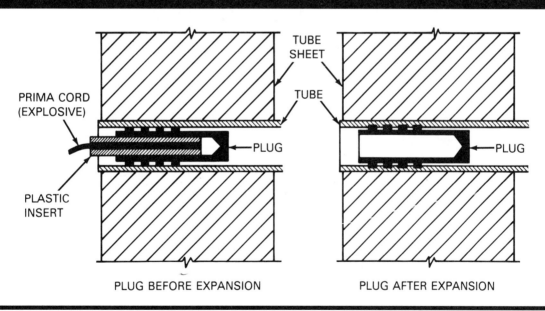

PRIMA CORD (EXPLOSIVE)

PLASTIC INSERT

TUBE SHEET

TUBE

PLUG

PLUG

PLUG BEFORE EXPANSION

PLUG AFTER EXPANSION

Figure 24.17–Explosively Expanded Plug

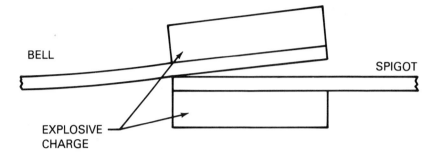

(A) GENERAL ARRANGEMENT OF PIPE ENDS AND EXPLOSIVES BEFORE WELDING

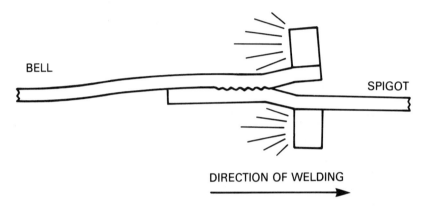

(B) DURING EXPLOSION WELDING

(C) FINAL JOINT CONFIGURATION AFTER WELDING

Figure 24.18–Schematic of Explosion Welding of Girth Joint in Pipe

SAFETY

EXPLOSIVES AND EXPLOSIVE devices are a part of explosion welding. Such materials and devices are inherently dangerous. Safe methods for handling them do exist. However, if the materials are misused, they can kill or injure anyone in the area and destroy or damage property.

Explosive materials should be handled and used by competent people who are experienced in that field. Handling and safety procedures must comply with all applicable federal, state, and local regulations. Federal jurisdiction over the sale, transport, storage, and use of explosives is through the U.S. Bureau of Alcohol, Tobacco, and Firearms; the Hazardous Materials Regulation Board of the U.S. Department of Transportation; the Occupational Safety and Health Agency; and the Environmental Protection Agency. Many states and local governments require a blasting license or permit, and some cities have special explosive requirements.

The Institute of Makers of Explosives provides educational publications to promote the safe handling, storage, and use of explosives. The National Fire Protective Association provides recommendations for safe manufacture, storage, handling, and use of explosives.

Personnel working in the vicinity of the explosion welding operation should be provided with eye protection (safety glasses) to guard against flying particles. They should also have ear protection to guard against the noise of explosions. Warning signs should be installed to warn people to wear eye and ear protection and to keep away from detonation areas.

SUPPLEMENTARY READING LIST

American Society of Mechanical Engineers. "High energy rate fabrication." Proceedings: 8th International Conference, San Antonio, Texas, 17-21 June 1984, Eds. Bermon, I. and Schroeder, J. W. New York: American Society of Mechanical Engineers, 1984.

Bilmes, P., Gonzlez A. C., and Cuyas, J. C. "Barrier interlayers in explosive cladding of aluminum to steel." *Metal Construction* 20(3): 113-114; March 1988.

Blazynski, T. Z. *Explosive welding, forming and compaction.* England: Applied Science Publishers Ltd, 1983.

Chadwick, M. D. and Jackson, P. W. "Explosive welding in planar geometries." *Explosive Welding Forming and Compaction*, Ed. T. Z. Blazynski, 219-287. England: Applied Science Publishers Ltd, 1983.

Cleland, D. B. "Basic consideration for commercial explosive cladding processes." *Explosive Welding Forming and Compaction*, Ed. T. Z. Blazynski, 159-188. England: Applied Science Publishers Ltd, 1983.

Crossland, B. "Review of the Present State-of-the-Art in Explosive Welding." *Metals Technology*, January 1976.

El-Sobky, H. "Mechanics of explosive welding." *Explosive Welding Forming and Compaction*, Ed. T. Z. Blazynski, 189-217. England: Applied Science Publishers Ltd., 1983.

Fujita, M. "An investigation of the combined underwater (explosive) bonding and forming process." *High Energy Rate Fabrication*, Vol. 70. Proceedings: ASME Winter Meeting, Phoenix, Arizona, 14-19 November 1982, 29-37. New York: American Society of Mechanical Engineers, 1982.

Holtzman, A. H. and Cowan, G. R. "Bonding of metals with explosives." Bulletin 104. New York: Welding Research Council, April 1965.

Jamieson, R. M., Loyer, A., and Hauser, W. D. "High impact girth welds in large diameter pipes." *Steels for Line Pipe and Pipeline fittings*, 342-453. England: The Metals Society and the Welding Institute, 1981.

Johnson, T. E. and Pocalyko, A. "Explosive welding for the 80's." *High Energy Rate Fabrication*, Vol 70. Proceedings: ASME Winter Meeting, Phoenix, Arizona, 14-19 November 1982, 63-82. New York: American Society of Mechanical Engineers, 1982.

Justice, J. T. "Explosion welding proven for large-diameter gas lines." *Oil and Gas Journal* 84(34): 44-50; August 1986.

Justice, J. T. and O'Beirne, J. J. Paper presented at Pipeline Engineering Symposium, New Orleans, 23-27 February 1986. New York: American Society of Mechanical Engineers (ASME), 1986.

————. "Explosion welding of a large diameter gas transmission pipeline." Proceedings: Pipeline Engineering Symposium, 23-27 February 1986, New Orleans, Ed. E. J. Seiders, 1-3. New York: American Society of Mechanical Engineers (ASME), 1986.

Linse, V. D. *The application of explosive welding to turbine components,*" 74-GT-85. New York: American Society of Mechanical Engineers, 1974.

Linse, V. D. and Lalwaney, N. S. "Explosive welding." *Journal of Metals* 36(5): May 1984.

Longstaff, G. and Fox, E. A. "Fabrication and plugging of tubes to tubesheet joints using "Impact" explosive welding technique." *High Energy Rate Fabrication*, Vol. 70. Proceedings: ASME Winter Meeting, Phoenix, Arizona, 14-19 November 1982, 39-53. New York: American Society of Mechanical Engineers, 1982.

Patterson, R. A. "Explosion bonding: aluminum-magnesium alloys bonded to austenitic stainless steel." *High Energy Rate Fabrication*, Vol. 70. Proceedings: ASME Winter Meeting, Phoenix, Arizona, 14-19 Nov. 1982, 15-27. New York: American Society of Mechanical Engineers, 1982.

Tatsukawa, I. "Interfacial phenomena in explosive welding of Al-Mg alloy/steel and Al-Mg alloy/titanium/steel." *Japan Welding Society* 17(2): 110-116; Oct 1986.

CHAPTER 25

ULTRASONIC WELDING

PREPARED BY A COMMITTEE CONSISTING OF:

J. L. Jellison, Chairman
Sandia National Laboratories

C. E. Albright
Ohio State University

J. Devine
Sonobond Ultrasonics

G. Harmon
National Bureau of Standards

G. A. Knorovsky
Sandia National Laboratories

V. H. Winchell II
Motorola Phoenix

WELDING HANDBOOK COMMITTEE MEMBER:
J. C. Papritan
Ohio State University

CHAPTER 25

ULTRASONIC WELDING

FUNDAMENTALS

DEFINITIONS AND GENERAL DESCRIPTION

ULTRASONIC WELDING (USW) is a solid-state welding process that produces a weld by local application of high-frequency vibratory energy while the workpieces are held together under pressure. A sound metallurgical bond is produced without melting of the base material.

Typical components of an ultrasonic welding system are illustrated in Figure 25.1. The ultrasonic vibration is generated in the transducer. This vibration is transmitted through a coupling system or sonotrode,[1] which is represented by the wedge and reed members in Figure 25.1. The sonotrode tip is the component that directly contacts one of the workpieces and transmits the vibratory energy into it. The clamping force is applied through at least part of the sonotrode, which in this case is the reed member. The anvil supports the weldment and opposes the clamping force.

Ultrasonic welding is used for applications involving both monometallic and bimetallic joints. The process is used to produce lap joints between metal sheets or foils, between wires or ribbons and flat surfaces, between crossed or parallel wires, and for joining other types of assemblies that can be supported on an anvil.

This process is being used as a production tool in the semiconductor, microcircuit, and electrical contact industries; for fabricating small motor armatures; in the manufacture of aluminum foil; and in the assembly of aluminum components. It is receiving acceptance as a structural joining method by the automotive and aerospace industries. The process is uniquely useful for encapsulating materials such as explosives, pyrotechnics, and reactive chemicals that require hermetic sealing but cannot be processed by high temperature joining methods.

---

1. The sonotrode is the acoustical equivalent of the electrode and its holder used in resistance spot or seam welding.

PROCESS VARIATIONS

THERE ARE FOUR variations of the process, based on the type of weld produced. These are spot, ring, line, and continuous seam welding. Furthermore, two variants of ultrasonic spot welding are used in microelectronics.

Spot Welding

IN SPOT WELDING, individual weld spots are produced by the momentary introduction of vibratory energy into the workpieces as they are held together under pressure between the sonotrode tip and the anvil face. The tip vibrates in a plane essentially parallel to the plane of the weld interface, perpendicular to the axis of static force application. Spot welds between sheets are roughly elliptical in shape at the interface. They can be overlapped to produce an essentially continuous weld joint. This type of seam may contain as few as 5 to 10 welds per inch. Closer weld spacing may be necessary if a leak-tight joint is required.

Ring Welding

RING WELDING PRODUCES a closed loop weld which is usually circular in form but may also be square, rectangular, or oval. Here, the sonotrode tip is hollow, and the tip face is contoured to the shape of the desired weld. The tip is vibrated torsionally in a plane parallel to the weld interface. The weld is completed in a single, brief weld cycle.

Line Welding

LINE WELDING IS a variation of spot welding in which the workpieces are clamped between an anvil and a linear sonotrode tip. The tip is oscillated parallel to the plane of the

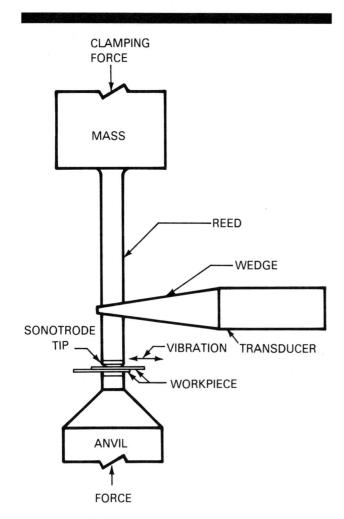

Figure 25.1—Wedge-Reed Ultrasonic Spot Welding System

weld interface and perpendicular to both the weld line and the direction of applied static force. The result is a narrow linear weld, which can be up to 6 in. in length, produced in a single weld cycle.

Continuous Seam Welding

IN CONTINUOUS SEAM welding, joints are produced between workpieces that are passed between a rotating, disk-shaped sonotrode tip and a roller type or flat anvil. The tip may traverse the work while it is supported on a fixed anvil, or the work may be moved between the tip and a counter-rotating or traversing anvil. Area bonds may be produced by overlapping seam welds.

MICROMINIATURE WELDING

THERE IS A consistency in the mechanism of ultrasonic joining throughout its various types of applications includ-

ing microelectronic wire bonding. In microelectronic applications, the wire diameter normally varies from 0.001 to 0.020 in. (25 to 500 µm), with the highest volume usage occurring in the 0.001 to 0.002 in. (25 to 50 µm) range. The significance of the material flow induced by ultrasonic energy can easily be observed by placing the bonding wedge on the wire to be joined to a substrate using the full bonding force, but with no ultrasonic energy being applied. With this condition, the flow of the wire is hardly noticeable.

Gradual application of ultrasonic power over time results in increased wire flow and a gradual bonding action between the outer surface of the deformed wire and the substrate material. The center of the bond area remains undisturbed, confirming the stable relationship between the bonding members. The vibratory action effectively removes surface contaminants to expose fresh material for welding. Using a scanning electron microscope as a diagnostic tool, the disruption of surface contaminants can readily be observed below the heel of the weld on the substrate.

Reflected sound and light are used to reveal the movements of the bonding tool. The magnitude of the wedge movement increases with increasing ultrasonic energy. By gradually increasing the power and the time for a given load on the wedge, characteristics of the ultrasonic wire bond interface can be observed. For a given combination of wire and substrate materials, a "window" or range of power (bond tool movement), time, and machine load variables can be determined which will provide acceptable weld strength values. There is a trade-off between reducing the wire strength because of deformation and strength of the bond interface itself. For reliable bonding, the wire itself should always be weaker than the bond interface.

MICROMINIATURE THERMOSONIC WELDING

MICROELECTRONIC WIRE JOINING represents a growing volume of industrial activity. Millions of wire bonds are performed daily. Because of the volume and the importance of the reliability of the products, wire bonding continues to evolve.

Ultrasonic wire bonding in the early 1970's was predominately the joining of aluminum wires to aluminum metallized bond pads on semiconductor devices, and joining wires to either aluminum-clad or gold-plated leads on the package. From purely ultrasonic and purely thermocompression[2] types of solid-state bonding has emerged today's thermosonic bonding. Thermosonic welding involves ultrasonic welding with heated substrates. Interface temperatures of 215 to 400°F (100 to 200°C) are normally used. It

---

2. Thermocompression welding is a deformation welding process in which fresh metal surfaces are exposed for welding by mechanical disruption of surface films. It is typically done at temperatures ranging from 215 to 660°F (100 to 350°C).

is now the most popular method of wire joining. Billions of wire joints were produced with this process each year during the late 1980's.

With the continually evolving improvements in wire bonding have come increasingly improved bonding machines and processes, bonding wedge designs and materials, bonding wire materials, and techniques for measuring bond quality. Ultrasonic wire bonding has been transformed from its "art" form in the 1960's to a common production technology in 1990.

MECHANISM OF THE PROCESS

ULTRASONIC WELDING INVOLVES complex relationships between the static clamping force, the oscillating shear forces, and a moderate temperature rise in the weld zone. The magnitudes of these factors required to produce a weld are functions of the thickness, surface condition, and the mechanical properties of the workpieces.

STRESS PATTERNS

IN ALL TYPES of ultrasonic welding, static clamping force is applied perpendicular to the interface between the workpieces. The contacting sonotrode tip oscillates approximately parallel to this interface. Combined static and oscillating shear forces create dynamic internal stresses in the workpieces at the faying surfaces, resulting in elastoplastic deformation.

Photoelastic stress models reveal significant aspects of these stress patterns. With applied static force only, the stress pattern is symmetrical about the axis of force application. With the superimposition of a lateral force, such as that occurring during one-half cycle of vibration, the force shifts in the direction of this lateral force, and shear stress is produced on that side of the axis. When the direction of the lateral force is reversed, as in the second half of the vibratory cycle, the shear stress shifts to the opposite side of the axis. During welding, the shear stress changes direction thousands of times per second.

As long as the stresses in the metal are below the elastic limit, the metal deforms only elastically. When the stresses exceed their threshold value, highly localized interfacial slip occurs, with no gross sliding. This action tends to break up and disperse surface films and permits metal-to-metal contact at many points. Continued oscillation breaks down surface asperities so that the contact area can grow until a physically continuous weld area is produced. At the same time, atomic diffusion occurs across the interface, and the metal recrystallizes to a very fine grained structure having the properties of moderately cold-worked metal.

Temperature Developed in Weld Zone

ULTRASONIC WELDING OF metals at room temperature produces a localized temperature rise from the combined effects of elastic hysteresis, localized interfacial slip, and plastic deformation. However, similar metals do not melt at the interface when the clamping force, power, and weld time are set correctly. Sections examined with both optical and electron microscopy have shown phase transformation, recrystallization, diffusion, and other metallurgical phenomena, but no evidence of melting.

Interfacial temperature studies made with very fine thermocouples and rapidly responding recorders show a high initial rise in temperature at the interface, followed by a leveling off. The maximum temperature achieved is dependent upon the welding machine settings. Increasing the power raises the maximum temperature achieved. Increased clamping force increases the initial rate of temperature rise, but lowers the maximum temperature achieved. Thus, it is possible to control the temperature profile, within limits, by appropriate adjustment of machine settings.

The interface temperature rise is also related to the thermal properties of the metal being welded. Generally, the temperature produced in a metal of low thermal conductivity, such as iron, is higher than that in a metal of high thermal conductivity, such as aluminum or copper.

Temperature measurements during welding of metals having a wide range of melting temperatures show that the maximum temperature in the weld is approximately 35 to 50 percent of the absolute melting temperature of the metal, when suitable welding machine settings are used.

Energy Delivered to the Weld Zone

THE FLOW OF energy through an ultrasonic welding system begins with the introduction of 60 Hz electrical power into a frequency converter. This device converts the applied frequency to that required for the welding system, which is usually in the range of from 10 to 75 kHz. The high-frequency electrical energy is conducted to one or more transducers in the welding system, where it is converted to mechanical vibratory energy of the same frequency. The vibratory energy is transmitted through the sonotrode and sonotrode tip into the workpiece. Some of the energy passes through the weld zone and dissipates in the anvil support structure.

Power losses occur throughout the system, in the frequency converter, transducer, sonotrode, and the interfaces between these components. However, with a well-designed system, as much as 80 to 90 percent of the input power to the converter may be delivered into the weld zone.

For practical usage, the power required for welding is usually measured in terms of the high-frequency electrical power delivered to the transducer. This power can be monitored continuously and provides a reliable average value to associate with equipment performance as well as with weld quality. The product of the power in watts and welding time in seconds is the energy, in watt-seconds or joules, used in welding.

Power Requirements and Weldability

THE ENERGY REQUIRED to make an ultrasonic weld can be related to the hardness of the workpieces and the thickness of the part in contact with the sonotrode tip. Analysis of data covering a wide range of materials and thicknesses has led to the following empirical relationship, which is accurate to a first approximation:

$$E = K(HT)^{3/2} \qquad (25.1)$$

where:

E = electrical energy, W·s (J)
K = a constant for a given welding system
H = Vickers hardness number
T = thickness of the sheet in contact with the sonotrode tip, in. (mm)

The constant "K" is a complex function that appears to involve primarily the electromechanical conversion efficiency of the transducer, the impedance match into the weld, and other characteristics of the welding system. Different types of transducer systems should have substantially different K values.

The above relationship has not been verified for welds used in microelectronics. This relationship predicts values that are about two orders of magnitude too high for microelectronic welds.

Figure 25.2 shows the relationship between the energy required for sound spot welds and the hardness of various sheet thicknesses of any weldable metal, based on the above equation. It provides a convenient first approximation of the minimum electrical input energy required (to produce sound welds) for a ceramic transducer type spot welding machine based on the Vickers hardness of the metal and the sheet thickness. Similar data can be derived for ring, line, and seam welds. For seam welds, the energy would be expressed in terms of the unit length of seam.

WELDING MICROELECTRONIC DEVICES

SMALL ULTRASONIC WELDS used in microelectronics are made with highly deformable materials such as aluminum, copper, and gold. The wire diameters are usually 25 to 50 μm. The welding process is complex but can be summarized as follows: Wire-to-substrate interfacial motion may

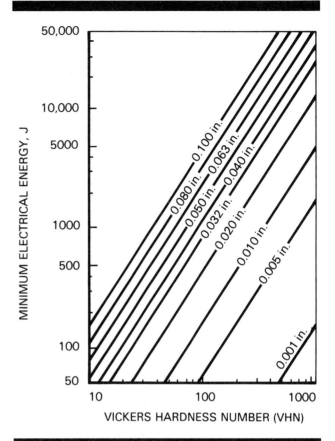

Figure 25.2—The Relationship Between the Minimum Electrical Energy Required for Ultrasonic (ceramic transducer) Spot Welding and Metal Hardness for Several Sheet Thicknesses

occur during the first few milliseconds, resulting in only minimal temperature rise [120 to 210°F (50 to 100°C)]. After this time, small microwelds form along or just inside the perimeter of the mated surfaces, and it is assumed that this interfacial motion slows and ceases as these microwelds grow. At this point, the interfacial motion progresses into the bonding tool-to-wire interface, and more ultrasonic energy is absorbed into the weld area. The microwelds join together and grow toward the center as shown in Figure 25.3, generally leaving the center unwelded.

The clamping force then deforms the wire and this process sweeps aside brittle surface oxides and contaminants, leaving clean surfaces in contact. Little deformation takes place in the center of the weld, so this area is often left unwelded, as shown in Figure 25.4. Transmission and scanning electron micrographs taken along the interface of monometallic welds have variously shown grain boundaries, no grain boundaries, debris zones of oxides and contaminants, and numerous crystallographic defects. Gold-alumi-

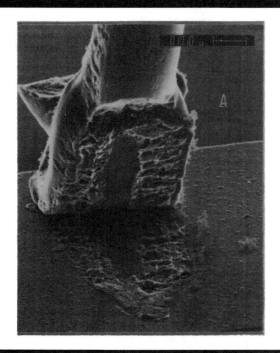

Figure 25.3–Typical Ultrasonic Bond Between a 25 µm Diameter Aluminum, 1% Silicon Wire and an Aluminum Substrate Using 50 ms Weld Time and 25 Kg Load (Note: The Wire has Been Lifted up so the Weld Pattern May be Seen. The Center is Unwelded).

Figure 25.4–Example of the Initial Stage of a Bond Between an Aluminum, 1% Silicon Ribbon (12 µm by 37µm Cross Section) and an Aluminum Substrate (Note: The Ribbon has been Removed, Revealing Microwelds Beginning to Form at the Perimeter of the Joint Where Deformation is Largest.)

num and other bimetallic welds made at room temperature do not show the formation of intermetallic compounds, but a clear boundary, similar to a grain boundary, is always observed.

The connections to essentially all semiconductor chips are welded in the same manner. The exceptions are a few types of devices that are specially prepared with solder bumps on their bond pads. Aluminum-to-aluminum cold ultrasonic welds are made between fine aluminum - 1 percent silicon wire, 25 to 50 µm in diameter, to various aluminum alloy (e.g., alloys with 1 percent Si, 1 to 2 percent Cu, etc.) bonding pads on semiconductor chips. Larger diameter wire, up to about 0.03 in. (0.75 mm), supplied in the fully annealed condition, is used to connect power devices that require higher currents.

The vast majority of interconnections to integrated circuits are made by thermosonic welding of gold wire. Thermocompression (solid state) welding usually requires interfacial temperatures between 575 and 750°F (300 and 400°C). This can damage plastics used to attach the chip to its package, whereas thermosonic welding can keep the interface temperature as low as 300°F (150°C). That lower temperature does not damage the plastics. The weld is matured with ultrasonic energy which, in combination with this temperature, can be kept small enough to prevent damage to the semiconductor chip.

Currently, gold wires are ultrasonically welded to semiconductor chips. However, there is a large effort to use copper wires in place of gold. Thermosonic ball bonding would then be used to produce the copper wire welds. If successful, copper should replace much of the gold wire in the future.

Tape automated bonding (TAB) is used to connect semiconductor devices where several hundred connections per device are required. For this technology, small leads of tin- or gold-plated copper ribbon are attached to plastic tape for rigidity. The leads are placed over a chip, which was designed with raised bonding pads. These tape-supported leads are then attached to the chip by either thermocompression or liquid interface bonding all at one time, or individually by modified automated wire welding equipment. In the latter case, the welds may be made by thermosonic, laser-heated thermocompression, laser-heated thermosonic, or liquid interface methods. This is a rapidly evolving technology.

An example of an ultrasonic transducer (60 kHz excitation) used in microelectronics welding including the capillary (sonotrode) is illustrated in Figure 25.5.

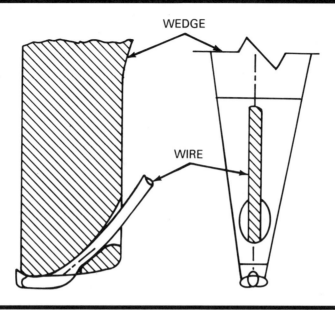

Figure 25.5—Ultrasonic Bonding Wedge With Wire

PROCESS ADVANTAGES AND DISADVANTAGES

ULTRASONIC WELDING HAS advantages over resistance spot welding in that little heat is applied during joining and no melting of the metal occurs. Consequently, no cast nugget or brittle intermetallics are formed. There can be no arc, and there is no tendency to expel molten metal from the joint.

The process permits welding thin to thick sections, as well as joining a wide variety of dissimilar metals. Welds can be made through certain types of surface coatings and platings.

Ultrasonic welding of aluminum, copper, and other high thermal conductivity metals requires substantially less energy than does resistance welding.

The pressures used in ultrasonic welding are much lower, welding times are shorter, and thickness deformation is significantly lower than for cold welding.

A major disadvantage is that the thickness of the component adjacent to the sonotrode tip must not exceed relatively thin gages because of the power limitations of present ultrasonic welding equipment. The range of thicknesses of a particular metal that can be welded depends upon the properties of that metal.

In addition, ultrasonic welding of metal is limited to lap joints. Butt welds cannot be made in metals because there is no effective means of supporting the workpieces and applying clamping force. However, ultrasonic butt welds are made in some polymer systems.

WELDABLE METALS

MOST METALS AND their alloys can be ultrasonically welded. Figure 25.6 identifies some of the monometallic and bimetallic combinations that currently can be welded on a commercial basis. Blank spaces in the chart indicate combinations that have not been attempted or are not known to have been successfully welded.

Various metals differ in weldability according to their composition and properties. The metals considered more difficult to weld are those that require either high power or long weld times, or both, and those that incur operational problems such as tip sticking or short tip life.

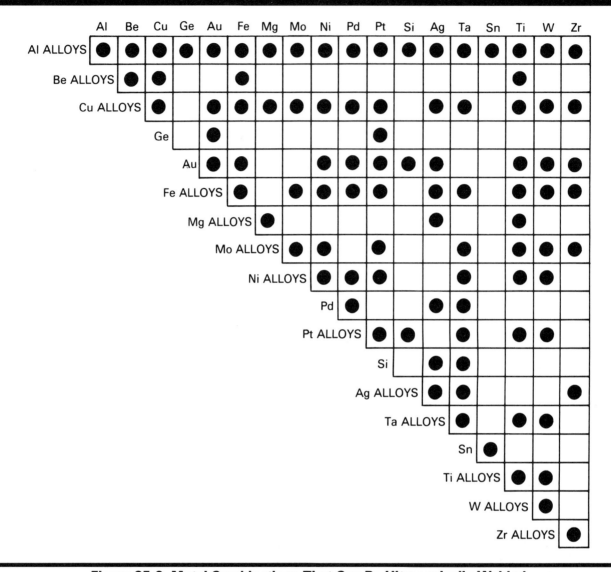

Figure 25.6–Metal Combinations That Can Be Ultrasonically Welded

ALUMINUM ALLOYS

ALL COMBINATIONS OF aluminum alloys form a weldable pair. They may be joined in any available form: cast, extruded, rolled, forged, or heat-treated. Soft aluminum cladding on the surface of these alloys facilitates welding. Aluminum can be welded to most other metals including germanium and silicon, the primary semiconductor materials.

COPPER ALLOYS

COPPER AND ITS alloys, such as brass and gilding metal, are relatively easy to weld. High thermal conductivity is not a deterrent factor in ultrasonic welding, as it is in fusion welding. Surface condition is an especially important variable in ultrasonic welding of copper alloys.

IRON AND STEEL

SATISFACTORY WELDS CAN be produced in iron and steel of various types, such as ingot iron, low carbon steels, tool and die steels, austenitic stainless steels, and precipitation-hardening steels. The power requirements are higher than for aluminum and copper.

PRECIOUS METALS

THE PRECIOUS METALS, including gold, silver, platinum, palladium, and their alloys, can be ultrasonically welded without difficulty. Most precious metals can be satisfactorily welded to other metals and to germanium and silicon.

REFRACTORY METALS

THE REFRACTORY METALS, including molybdenum, columbium, tantalum, tungsten, and some of their alloys, are among the most difficult metals to weld ultrasonically. Thin foil thicknesses of these metals can be joined if they are relatively free from contamination and surface or internal defects.

OTHER METALS

NICKEL, TITANIUM, ZIRCONIUM, beryllium, magnesium, and many of their alloys can be ultrasonically welded in thin gages to themselves and to other metals. Metal foils and wires are readily joined to thermally sprayed metals on glass, ceramics, or silicon. Such welds are particularly useful in the semiconductor industry. Typical combinations are shown in Table 25.1.

MULTIPLE-LAYER WELDING

MULTIPLE-LAYER WELDING IS feasible; for example, as many as 20 layers of 0.001 in. (25 μm) thick aluminum foil can be joined simultaneously with either spot welds or continuous seam welds. Several layers of dissimilar metals can also be welded together.

LIMITATIONS

THERE IS AN upper limit to the thickness of any metal that can be ultrasonically welded effectively, because the power output of available equipment is limited. For a readily weldable metal, such as Type 1100 aluminum, the maximum thickness in which reproducible high-strength welds can be produced is approximately 0.10 in. (2.5 mm). The present upper limit of harder metals is in the range of 0.015 to 0.040 in. (0.4 to 1.0 mm). This limitation applies only to the member of the weldment that is in contact with the welding tip; the other member may have greater thickness.

Table 25.1
Metal Wire and Ribbon Leads That May be Ultrasonically Welded to Thin Metal Surfaces on Nonmetallic Substrates

Substrate	Metal Film	Lead Material	Diameter or Thickness Range, in.
Glass	Aluminum	Aluminum wire	0.002-0.010
	Aluminum	Gold wire	0.003
	Nickel	Aluminum wire	0.002-0.020
	Nickel	Gold wire	0.002-0.010
	Copper	Aluminum wire	0.002-0.010
	Gold	Aluminum wire	0.002-0.010
	Gold	Gold wire	0.003
	Tantalum	Aluminum wire	0.002-0.020
	Chromel	Aluminum wire	0.002-0.010
	Chromel	Gold wire	0.003
	Nichrome	Aluminum wire	0.0025-0.020
	Platinum	Aluminum wire	0.010
	Gold-platinum	Aluminum wire	0.010
	Palladium	Aluminum wire	0.010
	Silver	Aluminum wire	0.010
	Copper on silver	Copper ribbon	0.028
Alumina	Molybdenum	Aluminum ribbon	0.003-0.005
	Gold-platinum	Aluminum wire	0.010
	Gold on molybdenum-lithium	Nickel ribbon	0.002
	Copper	Nickel ribbon	0.002
	Silver on molybdenum-manganese	Nickel ribbon	0.002
Silicon	Aluminum	Aluminum wire	0.010-0.020
	Aluminum	Gold wire	0.002
Quartz	Silver	Aluminum wire	0.010
Ceramic	Silver	Aluminum wire	0.010

Extremely thin sections can be welded successfully. For example, fine wires less than 0.0005 in. (0.01 mm) in diameter and thin foils of 0.00017 in. (0.004 mm) thickness have been welded.

Where a weld is difficult to achieve with available power levels, good quality joints might be made by inserting a foil of another metal in between the two workpieces. Three examples of this are (1) 0.0005 in. (0.01 mm) thick nickel or platinum foil has been used between molybdenum components; (2) beryllium foil has been welded to AISI Type 310 stainless steel using an interleaf of thin Type 1100-H14 aluminum foil; and (3) the weldable range of Type 2014-T6 aluminum alloy has been extended by using a foil interleaf of Type 1100-O aluminum.

For some metals, use of abrasive or textured tips and anvils will decrease the required clamping force and the welding power. This may permit welding thicker sections with a particular machine size.

Generally, ultrasonic welding can be used to join a metal within the thickness limitations of the process provided there is the following:

(1) Adequate joint overlap
(2) Access for the sonotrode tip to contact the parts
(3) An avenue for anvil support and clamping force application

APPLICATIONS

ELECTRONIC COMPONENTS

THE MOST IMPORTANT application of the USW process is the assembly of miniaturized electronics components. Fine aluminum and gold lead wires are attached to transistors, diodes, and other semiconductor devices. Wires and ribbons are bonded to thin films and microminiaturized circuits. Diode and transistor chips are mounted directly on substrates. Reliable joints with low electrical resistance are produced without contamination or thermal distortion of the components.

ELECTRICAL CONNECTIONS

ELECTRICAL CONNECTIONS OF various types are effectively made by ultrasonic welding. Both single and stranded wires can be joined to other wires and to terminals. The joints are frequently made through anodized coatings on aluminum, or through certain types of electrical insulation. Other current-carrying devices, such as electric motors, field coils, harnesses, transformers, and capacitors may be assembled with ultrasonically welded connections. A typical example is the field coil assembly for automotive starter motors shown in Figure 25.7. Ultrasonic welds are used here for joining aluminum ribbon to itself, to copper ribbon, to consolidated stranded copper wire, and to copper terminals.

For the starter motor armature of Figure 25.8, two wires are welded simultaneously into each slot of the barrel commutator. The entire process is accomplished automatically at rates up to 180 complete armatures per hour. Armatures for small motors in appliances, hand tools, fans, computers, and other electrical devices are assembled in a similar manner.

Thermocouple junctions involving a wide variety of dissimilar metals can be produced by this means.

FOIL AND SHEET SPLICING

BROKEN AND RANDOM lengths of aluminum foil are welded in continuous seams by foil rolling mills. Highly reliable splices, capable of withstanding annealing operations, are made rapidly in foils up to 0.005 in. (0.13 mm) thick and 72 in. (180 cm) wide. The splices are almost undetectable after subsequent working operations. Alumi-

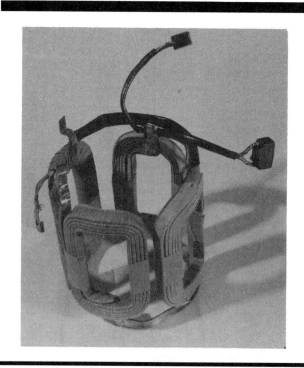

Figure 25.7–Field Coil Assembled by Ultrasonic Welding

Figure 25.8–Starter Motor Armature With Wires Joined in Commutator Slots by Ultrasonic Welding

SEQUENCE:
1. AS-RECEIVED CONTAINER
2. FLANGE FORMED
3. COVER WELDED TO THE FLANGE
4. COVER TRIMMED
5. CYLINDER REDRAWN

Figure 25.9–Cylinder Closure by the Flange-Weld-Redraw Technique

num and copper sheet up to about 0.020 in. (0.5 mm) thick can be spliced together by ultrasonic welding using special processing and tooling.

ENCAPSULATION AND PACKAGING

ULTRASONIC WELDING IS used for a wide variety of packaging applications that range from soft foil packets to pressurized cans. Leak tight seals are produced by ring, seam, and line welding.

The process is useful for encapsulating materials that are sensitive to heat or electrical current. Such materials may be primary explosives, slow-burning propellants and pyrotechnics, high-energy fuels and oxidizers, and living-tissue cultures.

Ultrasonic welding can be accomplished in a protective atmosphere or vacuum, and therefore instrument parts, ball bearings, and other items that must be protected from dust or contamination are frequently joined by the process. This capability also permits encapsulation of chemicals that react with air.

Ring welds up to about 1.5 in. (38 mm) diameter can be produced, but these welds are limited to thin sections of aluminum or copper. Straight cylindrical containers are often welded with a flanging and re-forming technique such as that shown in Figure 25.9. The cylinder ends are flared to form a 90° flange. The covers are ultrasonically welded to the flange, and then the welded flange is subsequently reformed to the original cylindrical geometry.

Line welding is used for packaging with one or more straight line seams, such as sealing the ends of squeeze tubes. Square or rectangular packets are produced by intersecting line welds on each of the four edges. Continuous seam welding is used to seal packages that cannot be accommodated with ring or line welding.

STRUCTURAL WELDING

ULTRASONIC WELDING PROVIDES joints of high integrity for structural applications within the limitations of weldable sheet thickness. The process is being used to assemble aircraft secondary structures, such as the helicopter access door in Figure 25.10. This assembly consists of inner and outer skins of aluminum alloy joined by multiple ultrasonic spot welds. Individual ultrasonic welds had 2.5 times the minimum average strength requirements for resistance spot welds in the same metals and thicknesses. Assembled doors sustained loads 5 to 10 times the design load without weld failure in air load tests. Significant savings in fabrication and energy costs were evident when compared to those of adhesive bonding.

In another application, small clips are attached to cylindrical reactor fuel elements with ultrasonic spot welds. Eight clips are attached to each element, and production rates of about 200 elements per hour are achieved in a semiautomatic setup.

SOLAR ENERGY SYSTEMS

ULTRASONIC WELDING HAS reduced fabrication costs for some solar energy conversion and collection systems. Systems for converting solar heat to electricity frequently involve photovoltaic modules of silicon cells which are connected by aluminum connectors. An ultrasonic seam welding machine, operating at speeds up to 30 feet per minute, joins all connectors in a single row in a fraction of the time required for hand soldering or individual spot

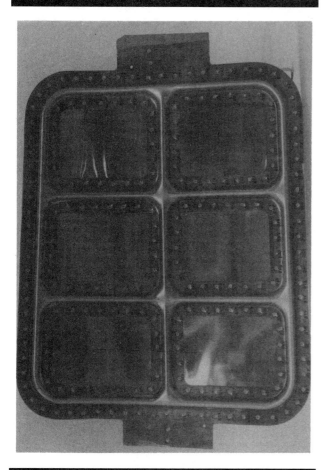

Figure 25.10–Ultrasonically Welded Helicopter Access Door

welding. After all connections are made on one side of the assembly, the process is repeated on the opposite side.

Solar collectors for hot-water heating systems may consist of copper or aluminum tubing attached to a collector plate. An automated ultrasonic system makes successive spot welds spaced on 1-in. (25 mm) centers between the plate and tubing as the assembly is passed beneath the welding tip. A 36 in. (1 m) tube can be welded to a plate in about 2 minutes, at an energy cost of about 0.3 cent. Fabrication costs are lower than those of soldering, resistance spot welding, or roll welding.

OTHER APPLICATIONS

APPLICATIONS IN OTHER areas have also been successful. Continuous seam welding was used to assemble components of corrugated heat exchangers. Strainer screens were welded without clogging the holes. Beryllium foil windows for space radiation counters were ultrasonic ring welded to stainless steel frames to provide a helium leak-tight bond.

Currently, pinch-off weld closures in copper and aluminum tubing, which are used with tubes in refrigeration and air conditioning, are produced with special serrated bar tips and anvils. Aluminum foil, surrounding fiberglass insulated ducting, is overlapped and welded with a traversing-head seam welding machine.

EQUIPMENT

GENERAL DESCRIPTION

AN ULTRASONIC WELDING machine consists of the following components:

(1) A frequency converter to provide electrical power at the design frequency of the welding system
(2) A transducer-sonotrode system to convert this power into elastic vibratory energy and deliver it to the weld zone
(3) An anvil which serves as a support for the workpieces
(4) A force application mechanism
(5) Either a timing device to control the weld interval in spot, ring, and line welding, or a rotating and translating mechanism for seam welding

(6) Appropriate electrical, electronic, and hydraulic or pneumatic controls

Vibratory Frequency

ULTRASONIC WELDING CAN be accomplished over a broad frequency range from less than 0.1 to about 300 kHz. However, the frequencies used for welding machines are usually in the range of 10 to 75 kHz.

An ultrasonic welding machine is designed to operate at a single frequency. There is no critical frequency for welding of specific metals or thicknesses. Due to the practical fundamentals of transducer-sonotrode design, it is expedient to build both light, low-power machines that operate at high frequencies, and heavy, high-power machines that operate at low frequencies. For example, welding machines in the power range of about 1200 to 8000 W operate with frequencies in the 10 to 20 kHz range. Conversely, small machines joining fine wires may have a power capacity of only a few watts and an operating frequency in the range of about 40 to 75 kHz.

A machine is designed to operate at some nominal frequency that may actually vary about 1 percent above or below the design frequency due to manufacturing variations. Adjustments are provided to tune the equipment to its optimum operating frequency.

Transducer-Sonotrode System

BOTH MAGNETOSTRICTIVE AND piezoelectric types of transducers are used in ultrasonic welding systems. Magnetostrictive materials have the property of changing length under the influence of varying magnetic flux density. Such transducers, which usually consist of a laminated stack of nickel or nickel alloy sheets, are rugged and serviceable for continuous duty operation but have low electromechanical conversion efficiency. Piezoelectric ceramic materials, such as lead zirconate titanate, are capable of changing dimensions under the influence of an electrical field. These materials have a conversion efficiency of more than twice that of magnetostrictive transducers. When they are operated at high duty cycles, both types of transducers must be cooled to prevent overheating and a loss of transduction characteristics.

The sonotrode system is designed to operate at the resonant frequency of the transducer and usually to provide gain in the amplitude of the delivered vibration. Sonotrode materials are selected to provide low energy losses and high fatigue strength under the applied static and vibratory stresses. A titanium alloy and stainless steel are the most commonly used sonotrode materials.

For high reliability, the various joints in the system must have high integrity and excellent fatigue life. Because of the high-frequency vibration, brazed, welded, and mechanical junctions have been used, but most current welding machines use mechanical joints for ease of interchangeability.

Transducer-sonotrode systems usually have acoustically designed mounting arrangements to ensure maximum efficiency of energy transmission when static force is applied through the system. These force-insensitive mounts prevent any shift of the resonant frequency of the system, and minimize loss of vibratory energy into the supporting structure.

Anvil

THE ANVIL, IN addition to supporting the workpiece, provides the necessary reaction to the applied clamping force. Its geometry is seldom critical except that it must not permit the workpiece to vibrate in compliance with the applied frequency.

Clamping Mechanism

THE STATIC LOAD is always applied normal to the plane of the weld interface. The means for applying this load depend upon the overall design of the welding machine. With larger units, hydraulic systems are satisfactory. Intermediate size units may incorporate pneumatically actuated or spring loaded systems. Miniature welding machines that require very small clamping loads may be spring actuated or dead-weight loaded. Such mechanical devices are suitable for production applications where frequent adjustments are not required.

Frequency Converter

THE FUNCTION OF the frequency converter is to change electrical line power of 50 or 60 Hz to the design frequency of the welding system in an oscillator stage, and then to amplify the output power in an amplifier stage. The output of such a system is the high-frequency electrical power of the ultrasonic transducer in the welding head.

Most ultrasonic welding systems use a solid-state frequency converter of the silicon-controlled rectifier (SCR) or transistor type. Transistors can operate efficiently at high frequencies, but their power-handling capability is low. At high power levels, multiple units must be used, resulting in a more complex but less reliable circuit. SCRs can handle more power per device and are normally useful at frequencies below 20 kHz. SCRs cannot be turned off by a control signal but require commutating circuitry, which adds to cost and reduces efficiency. The overall efficiency of SCR and transistor converters is approximately the same.

Automatic frequency control is a standard feature on frequency converters. Both SCR and transistor circuits can be either controlled by free-running oscillators or operated in a self-excited mode through positive feedback derived from the load. The self-excited mode is preferable because such systems automatically track the mechanical resonance of the loaded transducer, which assures optimum load matching under all conditions.

Some ultrasonic welding machines also incorporate constant amplitude control. By keeping the transducer mechanical amplitude constant, regardless of load, transducer dissipation is held at a safe level for all loading conditions, and the sonotrode tip amplitude is constant.

In most ultrasonic welding machines, the frequency converter and the welding head are separate assemblies connected by lightweight cables. In some of the low-power units, the converter is attached to the welding head.

TYPES OF EQUIPMENT

Spot Welding Machines

ULTRASONIC SPOT WELDING machines all operate on the same basic principle. They deliver single pulses of high-frequency vibration to the weld interface to produce a spot weld. Two different systems are used to apply the clamping force and transmit the ultrasonic energy to the workpiece: the wedge-reed system and the lateral drive system. Spot welding machines range in size from about 10 W to 8000 W capacity. Figure 25.11 indicates the capabilities of various machines for welding some common metals.

Wedge-Reed System. In the wedge-reed system, the transducer drives a wedge-shaped coupler in longitudinal vibration, as shown in Figure 25.1. The wedge is rigidly attached to the reed and produces flexural vibration in it. The sonotrode at the end of the reed undergoes essentially lateral vibration in a plane parallel to that of the weld inter-

face. For very high powers, two or more transducers may be used to drive the wedge.

Some systems have a movable head and a fixed anvil, with force being applied through the reed. Other systems have a fixed head and a movable anvil, and force is applied through the anvil. In both cases, the reed is acoustically designed for force insensitivity so that the vibration is not damped during force application. Figure 25.12 shows a typical machine of the wedge-reed type with a movable head. A 1500 watt machine with a movable anvil and stationary wedge-reed system is shown in Figure 25.13.

Lateral Drive System. In this system, the sonotrode tip is attached to a lateral sonotrode which vibrates longitudinally to produce tip motion parallel to the weld inter-

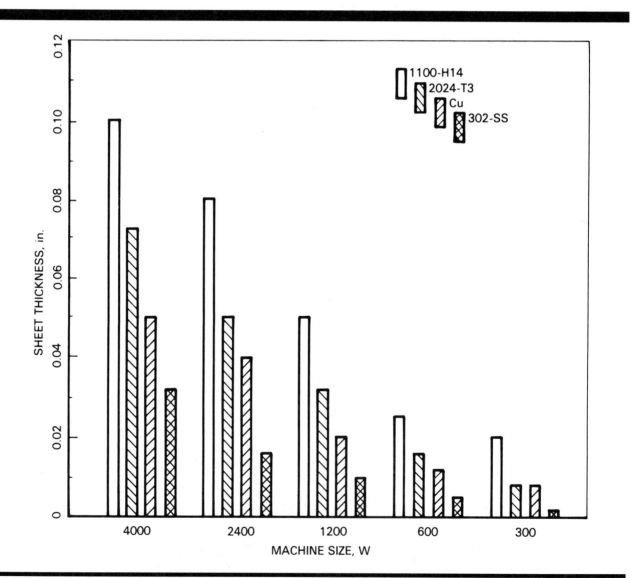

Figure 25.11–Capacities of Several Ultrasonic Spot Welding Machines for Joining Selected Metals

Figure 25.12–Typical Wedge-Reed Ultrasonic Spot Welding Machine with Movable Head. Machine Shown has 4500 Watt Rating

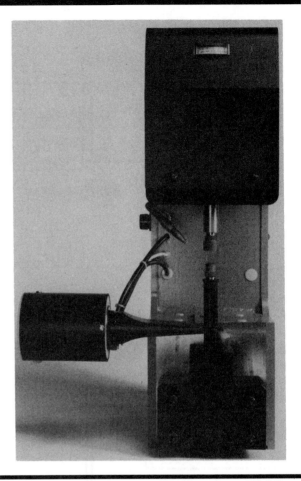

Figure 25.13–Typical 1500 Watt Spot Welding Machine With Moveable Anvil and Stationary Wedge-Reed Drive System

face. Two representative arrangements are illustrated in Figure 25.14. A system used in low-power welding machines is shown in Figure 25.14(A). Clamping force at the tip results when a bending moment is applied to the sonotrode, indicated by the vertical arrows on the figure. This lateral sonotrode is surrounded by a metal sleeve which has a flange located at a vibratory node to isolate the acoustic system from the applied force.

In the system shown in Figure 25.14(B), the lateral sonotrode is mounted through resilient metal diaphragms located at vibratory antinodes. These diaphragms isolate the vibration from the force application system.

A typical 100 W welding machine incorporating a lateral drive system is shown in Figure 25.15. Low-power systems, such as those used for lead wire welding in semiconductor devices, may be incorporated in a micropositioner. They may also include automatic wire feed, work manipulation, positioning, and microscopic observation.

Ring Welding Machines

RING WELDING MACHINES use a torsionally driven coupling arrangement as shown in Figure 25.16. The axial driving coupling members are tangent to the torsional coupler. They vibrate out of phase to produce a torsional vibratory displacement of the sonotrode tip in a plane parallel to the weld interface.

Ring welding machines are available in sizes ranging from 100 W to about 4000 W capacity. They are capable of making circular ring welds in diameters ranging from approximately 0.040 to approximately 2.5 in. (1 to 60 mm).

Line Welding Machines

THESE MACHINES USUALLY incorporate an overhung, lateral drive, transducer-sonotrode system. The couplers extend beyond the sonotrode tip, and the clamping force is

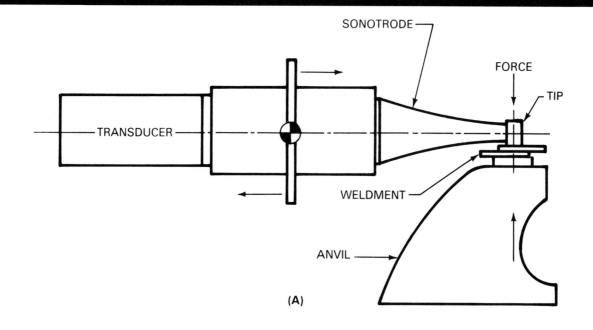

(A)

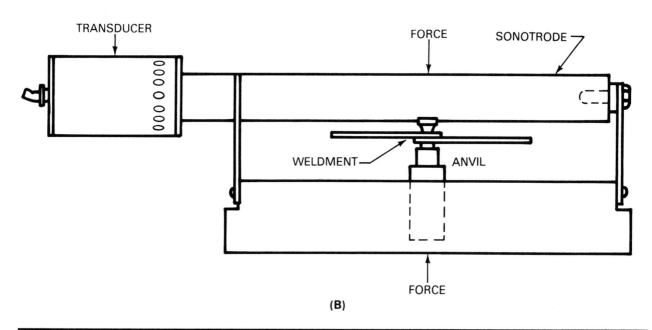

(B)

Figure 25.14–Two Types of Lateral Drive Spot Welding Systems

applied to the overhung portion of the sonotrode. This design eliminates high bending movements in the coupler when clamping forces are high.

A cluster of multiple transducer-sonotrode units attached to a single sonotrode tip is used for weld lengths greater than about 1 in. (25 mm). An array for making 5 in. (125 mm) long line welds incorporates five such units with an interconnecting tip. Hydraulic cylinders apply force to each coupler, thereby equalizing the force for each increment of weld length.

Continuous Seam Welding Machines

CONTINUOUS SEAM WELDING machines usually incorporate lateral drive transducer-sonotrode systems and antifriction bearings. The entire transducer-sonotrode-tip assembly is

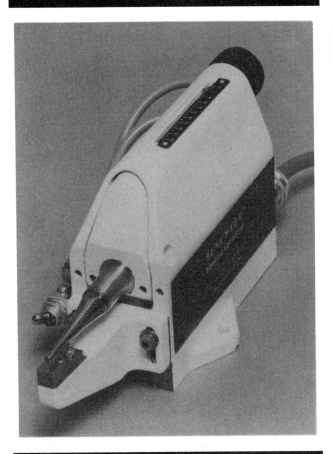

Figure 25.15—Typical 100 Watt Spot Welding Machine With Lateral Drive System

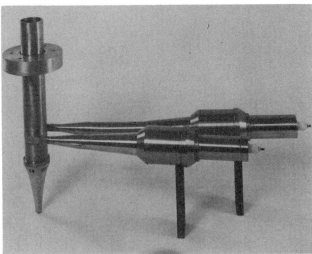

Figure 25.16—Transducer-Sono trode System for Ring Welding Machines

which the disk tip rotates as it traverses across the stationary workpiece. A typical unit is shown in Figure 25.17. A 100 W system of this type is used in aluminum foil mills for splicing coils. Machines of higher power capacity are used for splicing thicker materials.

rotated by a motor drive. The rotating tip rolls on the work along the desired path. Three arrangements are used for handling the work: a roller-roller system, a traversing anvil system, and a traversing head system.

Roller-Roller System. In this arrangement, the workpiece is driven between a rotating disk tip and a counter-rotating anvil. A compact 100 W machine is used for thin foil materials. Heavier materials are welded with a 2000 W unit capable of welding rates up to 450 ft/min (135 m/min).

Traversing Anvil System. In this system, the rotating transducer-sonotrode system is mounted in a fixed position. The workpiece is located on an anvil that traverses laterally under the rotating coupler disk.

Traversing Head System. The predominant type of seam welding machine is the traversing head system in

Figure 25.17—Typical Traversing Head, Continuous Seam Ultrasonic Welding Machine

SONOTRODE TIPS AND ANVILS

Tip and Anvil Geometry

WELDING IS MOST effectively accomplished when the sonotrode tip and anvil are contoured to accommodate the specific geometry of the parts being joined.

For spot welding of flat sheets, the tip is contoured to a spherical radius of about 50 to 100 times the thickness of the sheet adjacent to the tip. The anvil face is usually flat. This provides a friction-type drive in which slippage can occur between the tip and the top sheet or between the anvil and the bottom sheet as well as at the weld interface.

A positive type of drive is illustrated in Figure 25.18, which shows the arrangement for welding a small rib to a tubular member. The sonotrode tip is contoured to mate with the rib so that they are locked together. With this drive, maximum energy is delivered at the weld interface.

When joining a wire to a flat surface, the tip is preferably grooved to match the wire so that the wire is not excessively deformed during welding. With small wires, such as those used for connections to semiconductor devices, the tips must be precise in dimensions and finish. For joining two wires together, the anvil may be grooved to support and position both wires. The tip may fit into the groove and contact the upper wire.

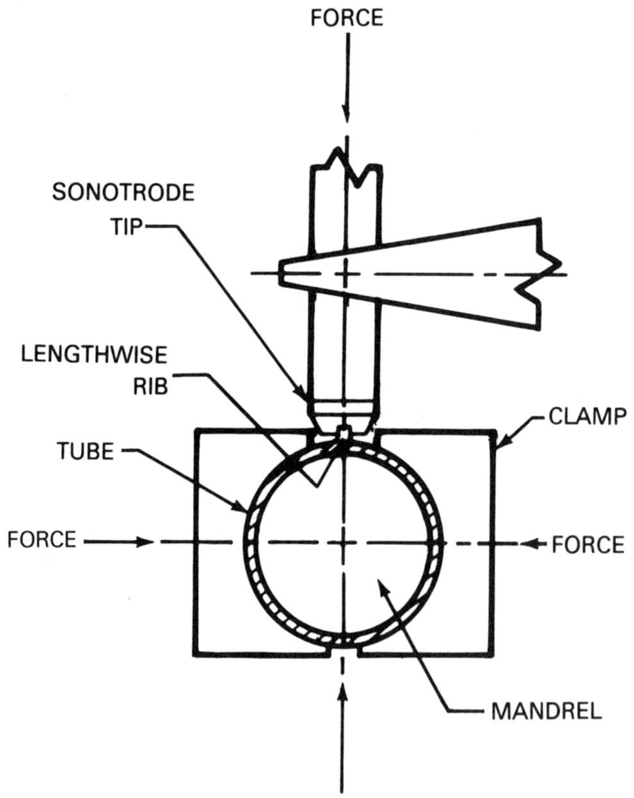

Ring welding tips are usually hollow members having the shape of the desired welds: circular, elliptical, square, or rectangular. The wall thickness is determined by the desired weld width, and the edge of the tip contacting the work is convex. Anvils may be flat or appropriately contoured to mate with the workpiece. In welding a lid to a cylindrical container, the anvil is usually recessed to accommodate the container, with the flange in contact with the anvil surface.

Line welding tips have narrow, elongated shapes with the contacting surface any desired width up to about 0.1 in. (2.5 mm). The anvil is designed to accommodate the workpiece. For line welding of can side seams, for example, the anvil consists of a cylindrical mandrel around which the can body blank is wrapped and supported with clamping jaws, as shown in Figure 25.18.

Continuous seam welding tips are resonant disks. For welding flat surfaces, the disks are machined with a convex edge. The edge of the disk may also be contoured to mate with the workpiece; for example, the entire periphery of a disk can be grooved to permit continuous seam welding of a rib to a cylinder, with an arrangement similar to that shown in Figure 25.18.

Tip Materials

AS IN RESISTANCE spot welding, wear of the sonotrode tip and anvil depends upon the properties and geometry of the parts being welded. Tips made of high-speed tool steel are generally satisfactory for welding relatively soft materials, such as aluminum, copper, iron, and low carbon steel. Tips of hardenable nickel-base alloys usually provide good service life with hard, high-strength metals and alloys. The material used for the sonotrode tip is also satisfactory for the anvil face.

Frequently, longer tip life and more effective welding are possible using tips and anvils with rough faces because they tend to prevent gross slippage between themselves and the workpieces. The roughening may be accomplished with electrical discharge machining (EDM) or by sandblasting to a finish of about 200 microinch. Abrasive tips usually permit the use of lower powers and clamping forces than are required with smooth tips.

Tip Maintenance

WHEN SONOTRODE TIPS begin to show wear, erosion, or material pickup, they may be reconditioned by cleaning and burnishing. Light sanding with 400-grit silicon carbide paper is usually sufficient.

If the wear is excessive, the tips should be replaced. Most welding machines have mechanically attached tips to simplify replacement.

Figure 25.18–Welding Array for Joining Small Longitudinal Ribs to Cylinders

The tip may tend to stick to the weld surface occasionally, particularly if improper machine settings are used. The sticking may be alleviated by increasing the clamping force or decreasing the welding time. With some materials, application of a lubricant, such as a faint trace of very dilute soap solution, to the surfaces being joined will reduce sticking. If these measures are not adequate, tip sticking may usually be eliminated by welding with a tip having an insert of tungsten carbide.

CONTROLS

THE BASIC CONTROLS for an ultrasonic welding machine are relatively simple. They consist of a master switch for introducing line power and controls to adjust clamping force, power, welding time, and sometimes resonance. Appropriate readout is normally provided for all adjustments.

The welding cycle is generally controlled automatically and is usually actuated by dual palm buttons or a foot switch. The automatic cycle consists of lowering the sonotrode tip or raising the anvil, application of clamping force, introduction of the ultrasonic pulse, and retraction of the tip or anvil.

Other controls and indicators are included on some welding machines to monitor operation of the equipment or to provide flexibility in use. Means may be provided for adjustment of the following:

(1) Sonotrode stroke length
(2) Speed of sonotrode advance and retraction from the weldment
(3) Speed of traverse for continuous seam welding machines
(4) Height of anvil to provide clearance for the workpiece
(5) Anvil position, particularly on ring welding machines, where precise alignment of tip and anvil are essential to ensure uniform contact around the periphery of the weld

Weld quality control monitors may also be included. One such device is a weld power meter which indicates the power delivered into the weld. A substantial change in load power indicates a faulty weld, which may be due to changes in part dimensions or surface finish, improper assembly of parts, or machine malfunction. On some machines, the high and low limits of acceptable power can be set, and deviations from this range can be used to trigger a visual or audible signal to alert the operator or to actuate a reject mechanism.

A recent development permits control of distance, either by controlling the distance the part is compressed or the final height of the part above an arbitrary datum, such as the anvil surface.

Microprocessor controlled machines permit selection of the controlling variables. Factors to be controlled are time, energy input, and distance. Most machines of this type are equipped with ports suitable for connection with a printer or computer. With appropriate software, statistical process control tests can be performed.

Another type of weld quality monitor is based on a constant energy principle. This system automatically adjusts welding time so that a predetermined amount of energy is delivered to each weld. When the energy cannot be delivered within the preset time, an alarm is activated.

Automated equipment may also include frequency counters, weld counters, material handling actuators, indexing mechanisms, and other devices to minimize operator functions and maximize production rates.

AUTOMATED PRODUCTION EQUIPMENT

SEVERAL FEATURES OF ultrasonic welding equipment make it particularly adaptable to automated or semiautomated production lines, namely:

(1) The welding head can be readily interfaced with other automatic processing equipment. It can be mounted on any rigid structure and in any position with the tip contacting the work from any direction.
(2) The frequency converter may be located as far as 150 ft (46 m) away from the welding head.
(3) Welding times are usually a fraction of a second, and production rates are limited primarily by the speed of the work-handling equipment.
(4) The process does not involve extensive heating of the equipment or the workpiece.
(5) In automatic filling and closing lines, accidental spillage of the contents on the weld interface usually will not significantly affect weld quality.

MICROELECTRONIC WELDING EQUIPMENT

MICROELECTRONIC WELDERS (WIRE bonders) are high-speed automated machines. Typically, they are capable of joining 6 to 8 wires/sec (12 to 16 welds). There are two basic welding process types, thermosonic ball bonding and ultrasonic wedge bonding. Both processes use ultrasonic energy; however, the tooling, operations, and materials are different. Ultrasonic energy for both process types is generated by a stacked piezo-electric sandwich structure attached to a horn (transducer). The transducer design is tapered to provide mechanical gain. The system resonates at 60 kHz. Phase locked loop circuits latch the electronic and mechanical systems for optimized output. Recent trends include the use of high-speed signal analysis to measure changes in system impedance during welding and to achieve real-time control of ultrasonics during bond formation.

Motion control is accomplished through software controlled servo systems. The thermosonic ball bonder requires three axes of motion for the welding head manipulator, plus one for the positioner. The wedge bonder requires a rotational axis. The servos are required to position the tool with an accuracy of ±2.5 microns (63.5 μm) in all three axes.

Figure 25.19 shows a ball bonding tool with an unbonded ball in position to begin the bonding process. The ball is formed by a spark discharge that melts the tip of the wire. The heat-affected zone above the ball is fully annealed by the discharge. The wire feeds through the capillary allowing the welding head to travel on the surface towards the crescent bond. In ball bonding, it is possible for the crescent bond to be located in any direction with respect to the ball bond. High-purity gold wire (99.99 percent gold) is the predominant bonding wire. Microalloying of the residual 100 ppm impurities must be carefully controlled to insure acceptable ball formation and wire loop shape control. Application of the principal welding variables: weld force, weld time and ultrasonic energy, is software controlled. Welding temperature is controlled independently.

Figure 25.5 shows an ultrasonic wedge bonding tool. Typically, wedge bonding is used with aluminum - 1 percent silicon wire in a room temperature process. However, it is also used, with heat, for bonding gold wire in microwave packages. As shown in Figure 25.5, the wire is positioned under the wedge, normal to the wedge axis. The addition of a rotational axis allows axial alignment of first

and second bonds either by rotating the weldhead or by rotating the work.

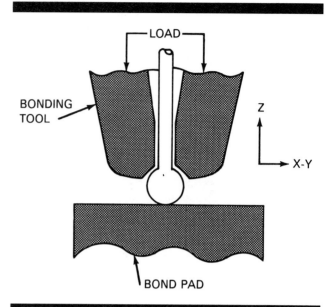

Figure 25.19–Ball Bonding Tool With Unbonded Ball

JOINING PROCEDURES

JOINT DESIGN

JOINT DESIGNS FOR ultrasonic welding are less restrictive than for some other types of welding. Edge distance is not critical. The only restriction is that the sonotrode tip should not crush or gouge the sheet edge. Welds in structural aluminum alloys of several thicknesses have shown the same strength at both 1/8 and 3/4 in. (3 and 19 mm) from the edge. Weight and material savings are achieved by using the minimum acceptable overlap.

Ultrasonic welding places no restrictions on spot spacing or row spacing with any of the four types of welds. Consecutive or overlapped welds have no effect on the quality of previously made welds, except perhaps under resonance conditions described below.

Ring welding offers unique capabilities for hermetic sealing, as indicated by the joint designs in Figure 25.20. Ring welds may also be preferred to spot welds for structural applications. The rings provide relatively uniform stress distribution with less stress concentration, less tendency toward cracking, and generally no parts resonance (see below).

CONTROL OF PARTS RESONANCE

SOMETIMES THE ENTIRE workpiece may be excited to vibration by the ultrasonic welding system. When this occurs, inferior welds may result, previously made welds may fracture, or cracks may be generated in the workpiece. Any of several remedies may be applied singly or in combination.

Resonance vibration may be eliminated by altering the workpiece dimensions or the orientation of the workpiece in the welding machine. Damping of vibration in thin sections can frequently be accomplished by applying pressure-sensitive tape to the part. Clamping of masses to the workpiece or clamping into a comparatively massive fixture usually suffices in even the most difficult cases.

SURFACE PREPARATION

A GOOD SURFACE finish contributes to the ease with which ultrasonic welds are made. Some of the more readily weldable metals, such as aluminum, copper, or brass, can be welded in the mill finish condition if not heavily oxidized, and may require only the removal of surface lubricants

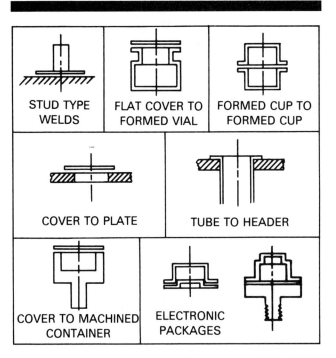

Figure 25.20–Typical Ultrasonic Ring Welding Applications

with a detergent. Normally, thin oxide films do not inhibit welding since they are disrupted and dispersed during the process.

Metals that are heavily oxidized or contain surface scale require more careful surface preparation. Mechanical abrasion or descaling in a chemical etching solution is usually necessary to provide a clean surface for welding. Once the surface scale is removed, the elapsed time before welding is not critical as long as the materials are stored in a noncorrosive environment.

It is possible to weld metals through certain surface films, coatings, or insulations, but somewhat higher ultrasonic energy levels are required. Some types of films cannot be penetrated and must always be removed prior to welding.

SPECIAL WELDING ATMOSPHERES

ULTRASONIC WELDING USUALLY does not require special atmospheres. With some metals, the process may produce discoloration of the surface in the vicinity of the weld. When such a surface is undesirable, it can be minimized by inert gas protection, such as small jets of argon impinging around the tip contact area. For packaging applications in which sensitive materials must be protected from contamination, welding can be accomplished in a chamber filled with inert gas.

PROCESS VARIABLES

THE VARIABLES OF ultrasonic power, clamping force, and welding time or speed are established experimentally for a specific application. Once determined, they usually require no adjustment unless there are alterations to the equipment, such as sonotrode tip changes or changes in the workpiece.

ULTRASONIC POWER

THE POWER SETTING may be indicated in terms of the high-frequency power input to the transducer, or the load power (the power dissipated by the transducer-sonotrode-workpiece assembly). As previously noted, the power requirement varies with the material and thickness of the workpiece adjacent to the sonotrode tip.

The minimum effective power for a given application can be established by a series of tests from which a threshold curve for welding is plotted. Details of this procedure are described later.

CLAMPING FORCE

AN ULTRASONIC WELDING machine usually provides a fairly broad range of clamping forces. Table 25.2 shows typical ranges for machines of various power capacities. The clamping-force range of machines with hydraulic or pneumatic force systems can be modified by changing the pressure cylinder.

The function of clamping force is to hold the workpieces intimately together. Excessive force produces needless surface deformation and increases the required welding power. Insufficient force permits tip slippage that may cause surface damage, excessive heating, or poor welds. Clamping force for a specific application is established in conjunction with ultrasonic power requirements.

WELDING TIME OR SPEED

THE INTERVAL DURING which ultrasonic energy is transmitted to the workpieces in spot, ring, or line welding is usually within the range of 0.005 second for very fine wires to about 1 second for heavier sections. The need for a long welding time indicates insufficient power. High power and a short welding time will usually produce welds that are superior to those achieved with low power and a long welding time. Excessive welding time causes poor surface appearance, internal heating, and internal cracks.

Table 25.2
Typical Clamping Force Ranges for Ultrasonic Welding Machines of Various Power Capacities

Machine Power Capacity, W	Approximate Clamping Force Range, lbf
20	0.009-0.39
50-100	0.5-15
300	5-180
600	70-400
1200	60-600
4000	250-3200
8000	800-4000

The same factors of power and unit time are significant in continuous seam welding. With available equipment, the travel speed for hard, thick metals may be as low as 5 ft/min. (1.5 m/min). Thin aluminum, 0.00l-in. thick, can be welded at speeds up to about 500 ft/min (150 m/min).

FREQUENCY ADJUSTMENT

ADJUSTMENT OF THE frequency converter output to match the operating frequency of the welding system is necessary for good performance. A system has a given nominal frequency, but the best operating frequency may vary with changes in the sonotrode tip, the workpiece, or the clamping force. The method of frequency adjustment varies with different types of frequency converters. After the setting is established for a specific welding setup, usually no further adjustment is necessary.

INTERACTION OF WELDING VARIABLES

FOR A GIVEN application, there is an optimum clamping force at which minimum vibratory energy is required to produce acceptable welds. This condition can be established by plotting the threshold curve. This curve, illustrated in Figure 25.21, defines the conditions of best dynamic coupling between the sonotrode tip and the workpiece and, thus, the minimum energy to produce strong welds.

The technique consists of making welds at selected power and clamping force settings and conducting cursory evaluation of weld quality as noted on Figure 25.21. For ductile, thin sheets and fine wires, a useful criterion of successful bonding is the ability to pull a nugget from one of the workpieces when peel tested. Welds in hard or brittle metals may be evaluated on the basis of weld strength or evidence of material transfer when peel tested. The threshold curve is normally derived as follows:

(1) The welding time is set at a reasonable value. One-half second is a good starting point for most metals. For very thin metals, a shorter weld time is usually chosen.

(2) Welding is started at low values of clamping force and power, and a series of test welds is made with incrementally increasing values of clamping force at a fixed power level. The welds are evaluated and the results plotted as in Figure 25.21, indicating acceptable and unacceptable welds.

(3) This procedure is then repeated at other values of ultrasonic power until an inverted bell-shaped curve is obtained.

These data will generate a curve separating the acceptable from the unacceptable welds. Welding is ordinarily done using the clamping force value for minimum acceptable power and a power level somewhat above the minimum. The product of the selected power and weld time is the total energy required. If welding time is decreased, then power must be increased accordingly. The threshold curve is a practical and efficient method for determining proper machine settings for all types of ultrasonic welds.

POWER-FORCE PROGRAMMING

CERTAIN MATERIALS, SUCH as the refractory metals and alloys, are more effectively welded when power-force programming is used. Weld strength is higher and cracking of the weld metal is minimized when these programming techniques are used.

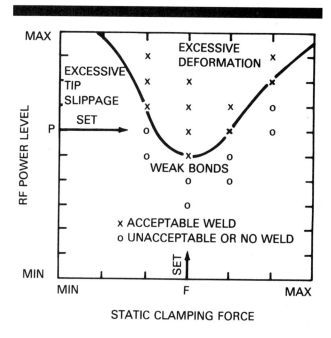

OPTIMUM CONDITIONS FOR MINIMUM POWER ARE CLAMPING FORCE F AND POWER P

Figure 25.21—Typical Threshold Curve Relating RF Power and Clamping Force

Power-force programming involves incremental variations in power and clamping force during the welding cycle. The cycle is initiated at low power and high clamping force. After a brief interval, the power is increased and the force reduced. The cycle is accomplished automatically with special logic circuitry.

WELD QUALITY

INFLUENCING FACTORS

VARIATIONS IN WELD quality may result from several factors which are generally associated with the workpieces and the welding machine or its settings. Weld quality is ordinarily not affected by normal manufacturing variations in metal parts. Metals that meet the specification requirements can usually be consistently welded without varying machine settings. Problems are sometimes encountered, however, if close tolerances are not held. For example, nickel, copper, and gold platings on metal surfaces frequently have thickness variations that affect weld quality. Surfaces for ring welding must be flat and parallel to ensure uniform welding around the periphery.

If there is any change in the workpieces during a production operation, the welding schedule must usually be adjusted to accommodate such change. Variations in weld quality during production runs have been traced to unauthorized changes in metal alloy, geometry, or surface finish. Wires such as magnet wires that are lubricated to facilitate coil winding may be ultrasonically welded without cleaning. However, a change in the type of lubricant may cause unacceptable welds unless machine settings are appropriately adjusted.

Uniform quality welding also depends upon the mechanical precision of the welding machine. Lateral deflection of the sonotrode or looseness of the anvil can produce unacceptable aberrations in the welds. Erratic weld quality may result from the use of a force-sensitive machine if power is lost and the frequency shifts as clamping force is applied. Sonotrode tips must be acoustically designed and precision ground to the desired contour. Their surfaces must be properly maintained to ensure reproducible welds.

PHYSICAL AND METALLURGICAL PROPERTIES

ULTRASONIC WELDS HAVE distinctive characteristics when examined both internally and externally.

Surface Appearance

THE SURFACE OF the work at a weld location is usually roughened slightly by the combined compressive and shear forces. The roughness can be minimized with adjustments in the machine settings and with careful sonotrode tip maintenance. The surface contour depends primarily on tip geometry. Spot welds usually leave an elliptical impression because of the linear displacement of the tip. A weld impression is larger in soft, ductile metals, such as aluminum, than in hard metals of the same thickness with appropriate adjustment in machine settings. Spot size can be increased by using a larger tip radius.

The actual weld area does not necessarily duplicate the surface impression, except in thin sheet. Sometimes, spot welds have unwelded areas in the center. This condition can usually be eliminated by decreasing the tip radius or reducing the clamping force.

Thickness Deformation

A WELD MAY show some thickness deformation because of the applied clamping force. Such deformation in sheet metals is usually less than 20 percent of the total joint thickness, even with soft metals. With contoured parts such as wires, deformation is somewhat greater unless the tip contour mates with the workpiece. Deformation may exceed 50 percent in fine wires.

Microstructural Properties

METALLOGRAPHIC EXAMINATION OF ultrasonic welds in a wide variety of metals shows that a number of different phenomena occur as a result of the vibratory energy introduced into the weld zone. The following are three important ones:

(1) Interfacial phenomena, such as interpenetration and surface film disruption
(2) Working effects, such as plastic flow, grain distortion, and edge extrusion
(3) Heat effects, such as recrystallization, precipitation, phase transformation, and diffusion

Ultrasonic welding is usually accompanied by local plastic deformation along the faying surfaces, by interdiffusion or recrystallization at the interface, and by interruption and displacement of oxide or other barrier films. Surface films, which are broken up by the stress reversals and plastic deformations that occur along the interface, may be displaced in the vicinity of the interface or may simply be interrupted in continuity in random areas within the bond zone. The actual behavior of such films depends upon several factors, including the machine settings, the properties of the film and the base metal, and the temperature achieved at the interface.

The temperature effect is significant in welding certain metals. Recrystallization of the metal frequently occurs in the weld nugget. Sufficient heat may be generated in certain alloys that exhibit precipitation behavior or phase transformation to induce these effects. Although diffusion may occur across the interface, the extent of atom movements is limited by the short weld time.

More than one of the above effects may be apparent in the same weld, and different effects may occur in welds in the same metal produced at different machine settings.

Several typical examples are illustrated in Figure 25.22. An extreme example of interpenetration across an interface is shown in Figure 25.22(A), in which a Kovar[3] foil has intruded into as much as 75 percent of the thickness of a nickel foil. A gold plate on the surface of the Kovar has been dispersed throughout the highly worked region. Interfacial ripples in a nickel-to-molybdenum weld, shown in Figure 25.22(B), illustrate the plastic flow that occurs locally. Entrapped oxide is indicated by the dark patches on the extreme right of the figure.

The weld between two sheets of arc-cast molybdenum, Figure 25.22(C), shows very little interpenetration, and the bond line is thin. Figure 25.22(D) illustrates the surface oxide film dispersion that may occur during welding of aluminum sheet. General plastic flow along the interface is observed in the Type 2024-T3 aluminum alloy weld of Figure 25.22(E) where the metal has recrystallized to a fine grain size.

Evidence of recrystallization has been observed in ultrasonic welds in several structural aluminum alloys, beryllium, low carbon steel, and other metals, even though they were not in the cold-worked condition prior to welding. For instance, in the Type 2020 aluminum alloy weld in Figure 25.22(F), mutual deformation of the surfaces and subsequent recrystallization are evident. In Figure 25.22(G), the elevated temperature during welding resulted in recrystallization of prior cold-worked nickel.

Still another effect of interfacial heating is illustrated in Figure 25.22(H), which shows a weld in solution treated and aged nickel base alloy. In the aged condition, a precipitate normally appears throughout the grains and in the grain boundaries. In the vicinity of this interface, the oxide scale is dispersed and the grain boundaries appear to stop short of the interface, indicating that the precipitate was dissolved during welding. An example of alloying that may occur in the bond between ferrous metals of different carbon content is shown in Figure 25.22(J).

---

3. Low-expansion iron-base alloy containing 29 percent nickel and 17 percent cobalt.

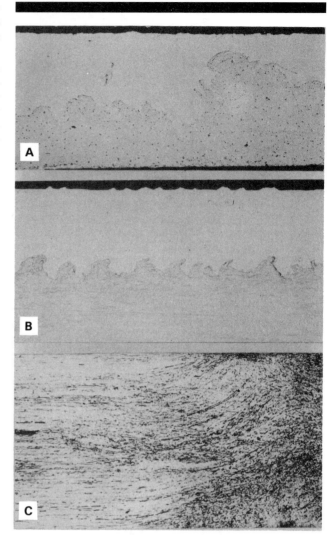

(A) 0.003-in. NICKEL FOIL (TOP) TO 0.003-in. GOLD-PLATED KOVAR FOIL (× 150)

(B) 0.005-in. NICKEL FOIL (TOP) TO 0.020-in. MOLYBDENUM SHEET (× 100)

(C) 0.008-in. ARC-CAST MOLYBDENUM SHEET TO ITSELF (× 70)

Figure 25.22–Photomicrographs of Typical Ultrasonic Welds

MECHANICAL PROPERTIES

A VARIETY OF mechanical tests may be used to evaluate weld quality. The property most frequently tested is shear strength. In addition, data are reported on tensile strength,

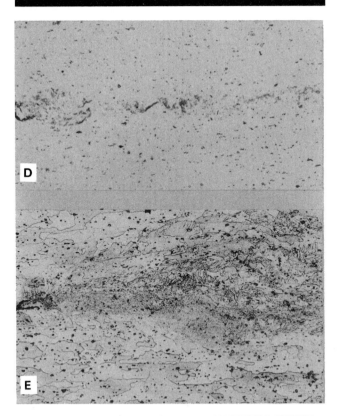

(D) 0.012-in. TYPE 1100-H14 ALUMINUM SHEET TO ITSELF (× 250)

(E) 0.032-in. TYPE 2024-T3 ALUMINUM ALLOY SHEET TO ITSELF (× 75)

Figure 25.22 (Continued)–Photomicrographs of Typical Ultrasonic Welds

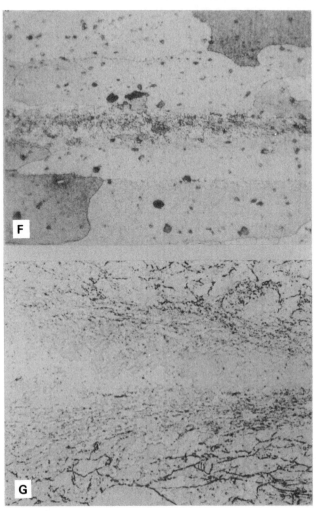

(F) 0.040-in. TYPE 2020 ALUMINUM ALLOY SHEET TO ITSELF (× 375)

(G) 0.014-in. HALF-HARD NICKEL SHEET TO ITSELF (× 250)

Figure 25.22 (Continued)–Photomicrographs of Typical Ultrasonic Welds

microhardness, corrosion resistance, and hermetic sealing properties. All available information indicates that the ultrasonic technique, properly applied, produces welds of acceptable strength and integrity.

Shear Strength

SHEAR TESTS ARE usually conducted on simple lap joints containing single or multiple spot welds or predetermined lengths of seam or line welds. For convenience, test specimen preparation and testing procedures essentially duplicate those used for resistance spot and seam welds. Microshear tests have also been developed for thermosonic ball bonds.

Figure 25.23 shows the increase in shear strength with sheet thickness for single-spot specimens in two aluminum alloys. Usually in the thin gages of aluminum sheet, and often in the intermediate gages, failure occurs by fracture of the base metal or by tear-out of the weld button rather than by shear of the weld itself. Similar data for several stainless steel and nickel alloys are shown in Figure 25.24 and for several refractory metals and alloys in Figure 25.25.

Typical spot-weld strengths in a variety of metals are summarized in Table 25.3. Of particular interest is the low

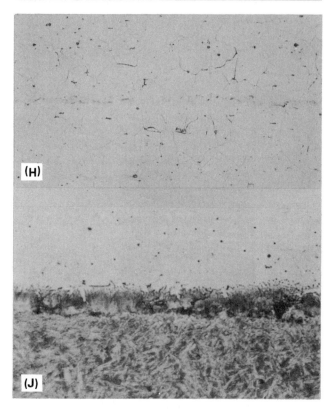

(H) 0.012-in. SOLUTION HEAT-TREATED AND AGED INCONEL X-750 SHEET TO ITSELF (× 75)

(J) 0.032-in. DIE STEEL (0.9% C) (TOP) TO 0.032-in. INGOT IRON (× 500)

Figure 25.22 (Continued)–Photomicrographs of Typical Ultrasonic Welds

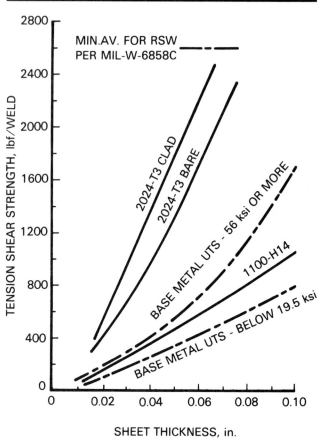

Figure 25.23–Typical Shear Strengths of Ultrasonic Spot Welds in Aluminum Alloy Sheet

variability associated with the strength data. In most instances this is less than 10 percent.

Line welds and seam welds show approximately the same strength as the base metal, at least with thin gages. As an example, spot seam welds in structural aluminum alloys have shown strengths equivalent to 85 to 95 percent of the ultimate tensile strength of the material under both shear and hydrostatic tests. Line welds in 0.00l-in. type 5052-H16 aluminum alloy average 85 to 92 percent of the base metal strength. Continuous seam welds in thin gage 1100 aluminum show 88 to 100 percent joint efficiency.

Elevated temperature tests on welded specimens of several metals and alloys indicate that weld strength is no lower than that of the base material at the same temperature.

Tensile Strength

TENSILE TESTS ON welds in selected metals indicate tensile strengths usually within the range of 20 to 40 percent of the shear strength. With resistance welds, the ratio of direct tension to shear strength is usually taken as a criterion of weld ductility. The significance of this ratio for ultrasonic welds has not been established.

CORROSION RESISTANCE

THE CAST NUGGET of a resistance spot weld is frequently the site of localized corrosion attack when the weldment is exposed to an unfavorable environment. This is not true of ultrasonic welds. Weld specimens in aluminum alloys and stainless steels that have been exposed in boiling water, sodium chloride solutions, and other corrosive materials have shown no preferential attack in the weld.

However, when dissimilar metals are welded, the possibility of galvanic corrosion at the weld nugget must be recognized.

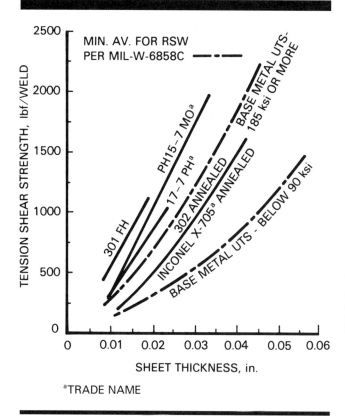

Figure 25.24—Typical Shear Strengths of Ultrasonic Spot Welds in Stainless Steel and Nickel-Base Alloys

ᵃTRADE NAME

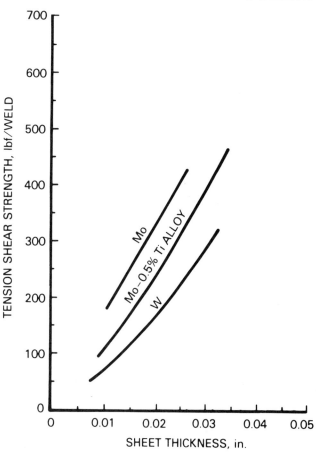

Figure 25.25—Typical Shear Strengths of Ultrasonic Spot Welds in Several Refractory Metals and Alloys

QUALITY CONTROL

Nondestructive Evaluation

THE WELD POWER monitor, discussed previously, provides an effective means of monitoring weld quality at the time the weld is made. The operator can immediately detect an improper cycle and reject the part, or logic may be provided for automatic rejection of a part that is made at an unsatisfactory level.

A number of postweld nondestructive techniques are also available. Ultrasound, radiography, and infrared radiation techniques may be used in specific applications. If hermetic sealing is the primary requirement of the weld, helium leak tests are effective.

Destructive Testing

AN APPROACH USED for some applications involves destructive testing of randomly selected specimens during a production run. For relatively thin ductile sheet, a peel test will indicate adequate weld strength if failure occurs by nugget tear-out or fracture of the base metal. Metallographic sectioning for examination provides a reliable indication of weld quality, but it is slow and expensive.

For most applications, shear testing is the most practical destructive test. Figure 25.26 shows typical variations in shear strength of random spot weld samples in 0.040-in. (1.0 mm) Type 2024-T3 aluminum alloy, produced with a specific machine setting for a number of days at different times of the day. The maximum, average, and minimum-strength values for each set of weld samples are shown. The horizontal lines indicate the mean value and standard deviation range for the entire group. The process began to show poor control on the seventh and eighth days. Control was restored on the ninth day by making the appropriate amplitude adjustments.

Table 25.3
Typical Shear Strengths of Ultrasonic Spot Welds in Several Alloys

Metal	Alloy or Type	Sheet Thickness, in.	Mean Shear Strength With 90% Confidence Interval, lbf
Aluminum	2020-T6	0.040	1240 ± 50
	3003-H14	0.040	730 ± 40
	5052-H34	0.040	750 ± 30
	6061-T6	0.040	800 ± 40
	7075-T6	0.050	1540 ± 90
Copper	Electrolytic	0.045	850 ± 20
Nickel	Inconel X-750 (*)	0.032	1520 ± 100
	Monel K-500 (*)	0.032	900 ± 60
	Rene 41 (*)	0.020	380
	Thoria dispersed	0.025	910
Steel	AISI 1020	0.025	500 ± 20
	A-286	0.015	680 ± 70
	AM-350	0.008	310 ± 20
	AM-355	0.008	380 ± 70
Titanium	8% Mn	0.032	1730 ± 200
	5% Al-2.5% Sn	0.028	1950 ± 120
	6% Al-4% V	0.040	2260 ± 180

* Trade Names

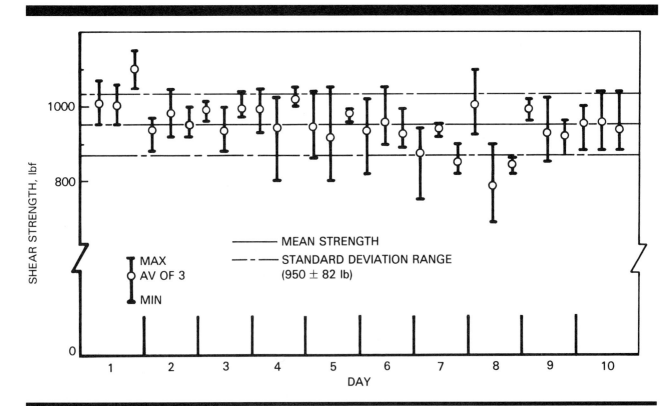

Figure 25.26–Typical Variance in Ultrasonic Weld Shear Strength in 0.040 in. Type 2024-T3 Aluminum Alloy

SAFETY

THE INTENT OF this section is to outline any probable hazards specifically associated with operating ultrasonic welding equipment. Thus, no attempt will be made here to discuss all of the potential hazards associated with welding and cutting processes in general, which are covered in detail in ANSI/ASC Z49.1 (latest edition), *Safety In Welding and Cutting*, nor all the OSHA regulations established to protect personnel working on (or around) various forms of industrial machinery and welding equipment. Strict conformance to the general requirements outlined in such applicable documents should be adhered to at all times.

The operator may require both eye and ear protection, depending on the specific application and equipment. Safety glasses are recommended for all workplace environments. Sound levels in the work area should be monitored to determine if a need for ear protection exists.

Most ultrasonic welding equipment is designed with interlocks and other safety devices to prevent operation under unsafe conditions. Nevertheless, consideration must be given to the health and safety of the operators, maintenance personnel, and other personnel in the area of the welding operations. Good engineering practice must be followed in the design, construction, installation, operation, and maintenance of equipment, controls, power supplies, and tooling to assure conformance to Federal safety laws (OSHA), state safety laws, and safety standards of the using company.

With high-power equipment, high voltages are present in the frequency converter, the welding head, and the co-axial cable connecting these components. Consequently, the equipment should never be operated with the doors open or housing covers removed. Door interlocks are usually installed to prevent introduction of power to the equipment when the high-voltage circuitry is exposed. The cables are shielded fully and present no hazard when properly connected and maintained.

Because of hazards associated with application of clamping force, the operator should never place hands or arms in the vicinity of the welding tip when the equipment is energized. For manual operation, the equipment is usually activated by dual palm buttons that meet the requirements of the Occupational Safety and Health Administration (OSHA). Both buttons must be pressed simultaneously to actuate a weld cycle, and both must be released before the next cycle is initiated. For automated systems in which the weld cycle is sequenced with other operations, guards should be installed for operator protection. Such hazards can be further minimized by setting the welding stroke to the minimum that is compatible with workpiece clearance.

Ring welding machines may be used for closure of containers filled with detonable materials. While no instance is known of premature ignition of such materials during ultrasonic welding, adequate provisions should always be made for remote operation by placing the welding machine either in a separate room from the control station or behind an explosion-proof barrier.

SUPPLEMENTARY READING LIST

anon. "Ultrasonic welding sees growing use in small motor assembly." *Welding Journal* 57(9): 41-43; September 1978.

anon. "Ultrasonic welding of silver electrical contacts." *Welding Journal* 59(5): 41,42; May 1980.

Avila, A. J. "Metal bonding in semiconductor manufacturing - a survey." *Semiconductor Products and Solid-State Technology* 7(11): 22-26; 1964.

Chang, U. I. and Frisch, J. "An optimization of some parameters in ultrasonic metal welding." *Welding Journal* 53(1): 24s-35s; January 1974.

Devine, J. "Joining electric contacts ultrasonics works fast." *Welding Design and Fabrication*, March 1980.

———. "Joining metals with ultrasonic welding." *Machine Design*, September 20, 1984.

Dzierwa, R. "The welding proliferation." *Appliance*, June 1988.

Estes, C. L. and Turner, P. W. "Ultrasonic closure welding of small aluminum tubes." *Welding Journal* 52(8): 359s-369s; August 1973.

Harman. G. G. and Keedy, K. O. "An experimental model of the microelectronic ultrasonic wire bonding mechanism." 10th Annual Proceedings Reliability Physics, 49-56. Las Vegas, NV, April 5-7 1972.

Hazlett, T. H. and Ambekar, S. M. "Additional studies of interface temperature and bonding mechanisms of ultrasonic welds." *Welding Journal* 49(5): 196s-200s; May 1970.

Hulst, A. P. and Lasance, C. "Ultrasonic bonding of insulated wire." *Welding Journal* 57(2): 19-25; February 1978.

Jones, J. B. "Ultrasonic welding." Proceedings of the CIRP International Conference on Manufacturing Technology, 1387-1410. Ann Arbor, MI, September 1967.

Jones, J. B. et al., "Phenomenological considerations in ultrasonic welding." *Welding Journal* 40(4): 289s-305s; April 1961.

Joshi, K. C. "The formation of ultrasonic bonds between metals." *Welding Journal* 50: 840-848; 1971.

Kelly, T. J. "Ultrasonic welding of Cu-Ni to steel." *Welding Journal* 60(4): 29-31; April 1981.

Kirzanowski, J. E. "A transmission electron microscopy study of ultrasonic wire bonding." Proceedings, 39th IEEE Electronic Components Conference, 450-455. Houston, TX, May 22-24, 1989, (Modified version to be published in IEEE Transactions on CHMT-12. No. 4, 1989).

Koziarski, J. "Ultrasonic welding: engineering, manufacturing and quality control problems." *Welding Journal* 40(4): 349-358; April 1961.

Langenecker, B., "Effects of ultrasound on deformation characteristics of metals." IEEE Trans. Sonics and Ultrasonics, SU-13, 1-8; 1966.

Littleford, F. E. "Welding electronic devices by ultrasonics." *Industrial Electronics* 6(3): 123-126; 1976.

Meyer, F. R. "Assembling electronic devices by ultrasonic ring welding." *Electronic Packaging and Production* 16 (7): 27-29; 1976.

————. "Ultrasonic welding process for detonable materials." *National Defense* 60(334): 291-293; 1976.

————. "Ultrasonics produces strong oxide-free welds." *Assembly Engineering* 20(5): 26-29; 1977.

Shin, S. and Gencsoy, H. T. "Ultrasonic welding of metals to nonmetallic materials." *Welding Journal* 47(9): 398s-403s; September 1968.

Yeh, C. J., Libby, C. C., and McCauley, R. B. "Ultrasonic longitudinal mode welding of aluminum wire." *Welding Journal* 53(6): 252-260; June 1974.

DIFFUSION WELDING AND BRAZING

PREPARED BY A COMMITTEE CONSISTING OF:

M. M. Schwartz, Chairman
Sikorsky Aircraft

J. M. Gerken
Lincoln Electric Company

WELDING HANDBOOK COMMITTEE MEMBER:
J. R. Condra
E. I. DuPont de Nemours

CHAPTER **26**

DIFFUSION WELD-ING AND BRAZING

FUNDAMENTALS OF THE PROCESSES

DEFINITIONS AND GENERAL DESCRIPTIONS

DIFFUSION WELDING (DFW) is a solid state welding process that produces a weld by the application of pressure at elevated temperature with no macroscopic deformation or relative motion of the workpieces. A filler metal may be inserted between the faying surfaces. Terms which are sometimes used synonymously with diffusion welding include diffusion bonding, solid-state bonding, pressure bonding, isostatic bonding, hot press bonding, forge welding, and hot pressure welding.

Several kinds of metal combinations can be joined by diffusion welding:

(**1**) Similar metals may be joined directly to form a solid-state weld. In this situation, required pressures, temperatures, and times are dependent only upon the characteristics of the base metals and their surface preparation.

(**2**) Similar metals can be joined with a filler metal in the form of a thin layer of a different metal between them. In this case, the filler metal may promote more rapid diffusion or permit increased microdeformation at the joint to provide more complete contact between the surfaces. This filler metal may be diffused into the base metal by suitable heat treatment until it no longer remains a separate layer.

(**3**) Two dissimilar metals may be joined directly where diffusion-controlled phenomena occur to form a metallic bond. The mechanisms are similar to those in category (1) above, with the added effects that dissimilar metals create.

(**4**) Dissimilar metals may be joined with a third metal; i.e., a filler metal, between the faying surfaces to enhance weld formation either by accelerating diffusion or permitting more complete initial contact in a manner similar to category (2) above.

Diffusion brazing (DFB) is a process that forms liquid braze metal by diffusion between dissimilar base metals or between base metal and filler metal preplaced at the faying surfaces. The process is used with the application of pressure. The filler metal may be diffused into the base metal to the extent that a distinct layer of brazing filler metal does not exist in the joint after the diffusion brazing cycle is completed. The joint properties approach those of the base metal. The process is sometimes called liquid phase diffusion bonding, eutectic bonding, or activated diffusion bonding.

Diffusion welding and diffusion brazing are similar in that a filler metal may be used with both processes. However, melting takes place by diffusion at the faying surfaces during the early stage of diffusion brazing. If diffusion at the interface continues with sufficient time at elevated temperature, any distinct layer of filler metal will finally disappear. Then the joint properties are nearly the same as those of the base metal.

If a filler metal is used and it does not melt, or alloy with the base metal to form a liquid phase, the process is diffusion welding. The purpose of the filler metal is to aid metallic bonding, particularly during the first stage of diffusion welding. It helps to eliminate voids at the interface that result when two rough surfaces are mated together. By proper selection, the filler metal will soften at welding temperature and flow under pressure to fill the interface voids. Also, it will diffuse into the base metal and produce a joint with acceptable properties for the application. The filler metal may be a diffusion aid, but it is not a brazing filler metal.

DIFFUSION WELDING PRINCIPLES

AS ILLUSTRATED IN Figure 26.1, metal surfaces have several general characteristics:

(**1**) Roughness

(**2**) An oxidized or otherwise chemically reacted and adherent layer

(**3**) Other randomly distributed solid or liquid products such as oil, grease, and dirt

(**4**) Adsorbed gas or moisture, or both

Two necessary conditions must be met before a satisfactory diffusion weld can be made:

(**1**) Mechanical intimacy of faying surfaces.

(**2**) The disruption and dispersion of interfering surface contaminants to permit metallic bonding.

For conventional diffusion welding without a diffusion aid, a three-stage mechanistic model, shown in Figure 26.2, adequately describes weld formation. In the first stage, deformation of the contacting asperities occurs primarily by yielding and by creep deformation mechanisms to produce intimate contact over a large fraction of the interfacial area. At the end of this stage, the joint is essentially a grain boundary at the areas of contact with voids between these areas. During the second stage, diffusion becomes more important than deformation, and many of the voids disappear as grain boundary diffusion of atoms continues. Simultaneously, the interfacial grain boundary migrates to an equilibrium configuration away from the original weld interface, leaving many of the remaining voids within the grains. In the third stage, the remaining voids are eliminated by volume diffusion of atoms to the void surface (equivalent to diffusion of vacancies away from the void). The stages overlap, and mechanisms that may dominate

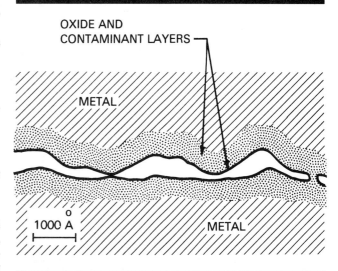

Figure 26.1–Characteristics of a Metal Surface Showing Roughness and Contaminants Present

one stage also operate to some extent during the other stages.

This model is consistent with several experimentally observed trends:

(**1**) Temperature is the most influential variable since it, together with pressure, determines the extent of contact area during stage one and it alone determines the rate of diffusion that governs void elimination during the second and third stages of welding.

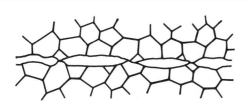

(A) INITIAL ASPERITY CONTACT

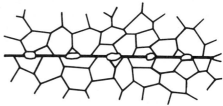

(B) FIRST STAGE DEFORMATION AND INTERFACIAL BOUNDARY FORMATION

(C) SECOND STAGE GRAIN BOUNDARY MIGRATION AND PORE ELIMINATION

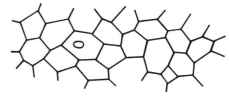

(D) THIRD STAGE VOLUME DIFFUSION PORE ELIMINATION

Figure 26.2–Three-Stage Mechanistic Model of Diffusion Welding

(2) Pressure is necessary only during the first stage of welding to produce a large area of contact at the welding temperature. Removal of pressure after this stage does not significantly affect joint formation. However, premature removal of pressure before completion of the first stage is detrimental to the process.

(3) Rough initial surface finishes generally adversely affect welding by impeding the first stage and leaving large voids that must be eliminated during the later stages of welding.

(4) The time required to form a joint depends upon the temperature and pressure used; time is not an independent variable.

This model is not applicable to diffusion brazing or hot pressure welding processes where intimate contact is achieved through the use of molten filler metal and bulk deformation, respectively.

At the same time that intimate contact is being achieved as described above, various intervening surface films must be disrupted and dispersed so that metallic bonds can form. During initial faying surface contact (stage 1), the films are locally disrupted and metal-to-metal contact begins at places where the surfaces move together under shear. The subsequent steps in the process involve thermally activated diffusion mechanisms that complete film dispersion and eliminate voids to achieve intimate metal contact (stages 2 and 3).

The barrier film is largely an oxide. Proper cleaning methods reduce the other components of film to negligible levels. Two actions tend to disrupt and disperse the oxide film. The first is solution of the oxide in the metal; the second is spheroidization or agglomeration of the film. Oxides decompose and the decomposition products are dissolved in titanium, tantalum, columbium, zirconium, and other metals in which interstitial elements are highly soluble. If the oxide is relatively insoluble in the metal, as is the case for aluminum, the disruption action for the trapped film is spheroidization. This leaves a few oxide particles along the weld interface. However, if the weld is properly made, these particles are no more detrimental than inclusions normally present in most metals and alloys.

Both decomposition and spheroidization of the oxides require diffusion. Decomposition results in the diffusion of interstitial atoms of oxygen into the metal and spheroidization by diffusion as a result of the excessive surface energy of the thin films. The time of solution of a film of thickness X is proportional to X^2/D, where D is the diffusion coefficient. The film must be kept very thin if diffusion welding times are to be within acceptable limits. Spheroidization occurs more rapidly if the oxide films are thin. Hence, control of the film thickness after cleaning and during heating to welding temperature is a critical factor in diffusion welding.

Once actual metal-to-metal contact is established, the atoms are within the attractive force fields of each other and a high-strength joint is generated. At this time, the joint resembles a grain boundary because the metal lattices on each side of the line have different orientations. However, the joint may differ slightly from an internal grain boundary because it may contain more impurities, inclusions, and voids that will remain if full asperity deformation has not occurred. (Stage 2 in the model for achieving intimate contact is not yet complete.) As the process is carried to completion, this boundary migrates to a more stable non-planar configuration, and any remaining interfacial voids are eliminated through vacancy diffusion.

An intermediate filler metal is of significant practical importance in many systems, although the mechanisms so far described do not consider its use. When a filler metal is used or dissimilar base metals are welded, the diffusion of each metal into the other must be considered to develop a complete understanding of the DFW process.

DIFFUSION BRAZING PRINCIPLES

DIFFUSION BRAZING PRODUCES joint properties that are significantly different from those of conventional brazed joints. The main objective of the process is to produce joints having mechanical properties approaching those of the base metal in applications where other joining processes are unacceptable. Some examples are the following:

(1) Cast nickel-base superalloys for high temperature service, and beryllium alloys
(2) Some dissimilar metal combinations
(3) Assemblies where a combination joining and heat treating cycle is desirable to minimize distortion
(4) Elevated temperature applications, such as high strength titanium alloys in aircraft
(5) Large, complicated assemblies where it is economical to produce many strong joints simultaneously

Two approaches to diffusion brazing are used. One utilizes a brazing filler metal that has a chemical composition approximately the same as the base metal but with a lower melting temperature. Melting temperature is suppressed by adding certain alloying elements to the base metal composition or to a similar alloy composition. For example, the melting temperature of a nickel-base high-temperature alloy can be lowered by a small addition of silicon or boron. In this case, the brazing filler metal melts and wets the base metal faying surfaces during the brazing cycle. This approach is sometimes called activated diffusion bonding or transient liquid phase bonding.

The second approach uses a filler metal that will alloy with the base metal to form one or more eutectic or peritectic compositions. When the brazing temperature is slightly higher than the eutectic or peritectic temperature, the filler metal and base metal will alloy to produce a low-melting composition. The filler metal itself does not melt, but a low-melting alloy is formed *in situ*. This method is also known as eutectic brazing. An example is the diffusion brazing of titanium alloys with copper.

With either approach, the assembly is held at brazing temperature for a sufficient time for diffusion to produce a nearly uniform alloy composition across the joint. As this takes place, the melting temperature of the braze metal and the strength of the joint increase. The brazing time depends upon the degree of homogeneity desired, the thickness of the initial filler metal layer, and the brazing temperature. The relationship of heating rate to brazing temperature may also be important. A low heating rate will allow more solid-state diffusion to take place, and more filler metal will be required to provide sufficient liquid to fill the joint. Conversely, if a large quantity of filler metal and fast heating are used, the molten metal may run out of the joint and erode the base metal. The thick joint so formed will require a longer diffusion time to achieve a suitable composition gradient across the joint.

The composition of the braze metal may be important with respect to response to subsequent heat treatment. This is particularly true for metals that undergo phase transformation during heating and cooling. Alloy composition will determine the transformation temperature and rate of transformation. Therefore, the phase morphology and mechanical properties of the joint can be controlled by the joint design and the brazing cycle.

ADVANTAGES AND LIMITATIONS

DIFFUSION WELDING AND brazing have a number of advantages over the more commonly used welding and brazing processes, as well as a number of distinct limitations on their applications. Some of the advantages of the two processes are as follows:

(1) Joints can be produced with properties and microstructures very similar to those of the base metal. This is particularly important for lightweight fabrications.

(2) Components can be joined with minimum distortion and without subsequent machining or forming.

(3) Dissimilar alloys can be joined that are not weldable by fusion processes or by processes requiring axial symmetry, such as friction welding.

(4) A large number of joints in an assembly can be made simultaneously.

(5) Members with limited access can be joined.

(6) Large joint members of base metals that would require extensive preheat for fusion welding can be more readily joined. An example is thick copper.

(7) Defects normally associated with fusion welding are not encountered.

Following are some important process limitations:

(1) The thermal cycle is normally longer than that of conventional welding and brazing processes.

(2) Equipment costs are usually high, and this can limit the maximum size of components that can be produced economically.

(3) The processes are not adaptable to a high production rate, although a number of assemblies may be joined simultaneously.

(4) Adequate nondestructive inspection techniques for quality assurance are not available, particularly those that assure design properties in the joint.

(5) Suitable filler metals and procedures have not yet been developed for all structural alloys.

(6) The faying surfaces and the fit of joint members generally require greater care in preparation than for conventional hot pressure welding or brazing processes. Surface smoothness may be an important factor in quality control in the case of diffusion brazing.

(7) The need to simultaneously apply heat and a high compressive force in the restrictive environment of a vacuum or protective atmosphere is a major equipment problem with diffusion welding.

SURFACE PREPARATION

THE FAYING SURFACES of joint members to be diffusion welded or diffusion brazed must be carefully prepared before assembly. Surface preparation involves more than cleanliness. It also includes (1) the generation of an acceptable finish or smoothness, (2) the removal of chemically combined films (oxides), and (3) the cleansing of gaseous, aqueous, or organic surface films. The primary surface finish is obtained ordinarily by machining, abrading, grinding, or polishing.

One property of a correctly prepared surface is its flatness and smoothness. A certain minimum degree of flatness and smoothness is required to assure uniform contact. Conventional metal cutting, grinding, and abrasive polishing methods are usually adequate to produce the needed surface flatness and smoothness. A secondary effect of machining or abrading is the cold work introduced into the surface. Recrystallization of the cold worked surfaces increases the diffusion rate across the weld or braze interface.

Chemical etching (pickling), commonly used as a form of preweld preparation, has two effects: the first is the favorable removal of nonmetallic surface films, usually oxides; the second is the removal of part or all of the cold worked layer that forms during machining. The need for oxide removal is apparent because it prevents metal-to-metal contact.

Degreasing is a universal part of any procedure for surface cleaning. Alcohol, acetone, detergents, and many other cleaning agents may be used. Frequently, the recommended degreasing technique is intricate and may include multiple rinse-wash-etch cycles using several solutions. Because some of these cleaning solvents are toxic or flammable, proper safety precautions should always be followed.

Heating in vacuum may also be used to obtain clean surfaces. The usefulness of this method depends to a large extent upon the type of metal and the nature of its surface films. Organic, aqueous, or gaseous adsorbed layers can be removed by vacuum heat treatment at elevated temperature. Most oxides do not dissociate during a vacuum heat treatment, but it may be possible to dissolve adherent oxides in some metals at elevated temperature. Some metals that may dissociate oxides and dissolve the resulting oxygen, at an elevated temperature, are zirconium, titanium, tantalum, and columbium. Cleaning in vacuum usually requires subsequent vacuum or inert atmosphere storage and careful handling to avoid the recurrence of surface contamination.

Many factors enter into selecting the faying surface treatment. In addition to those already mentioned, the specific welding or brazing conditions may affect the selection. With higher temperature or pressure, it becomes less important to obtain extremely clean surfaces. Increased atomic mobility, surface asperity deformation, and solubility of impurity elements all contribute to the dispersion of surface contaminants. With lower temperature or pressure, better prepared and preserved surfaces are more important.

Preservation of the clean faying surface is necessary following the surface preparation. One requirement is the effective use of a protective environment during diffusion welding or brazing. A vacuum environment provides continued protection from contamination. A pure hydrogen atmosphere will minimize the amount of oxide formed and it will reduce existing surface oxides of many metals at elevated temperature. However, it will form hydrides with titanium, zirconium, columbium, and tantalum that may be detrimental. High-purity argon, helium, and sometimes nitrogen can be used to protect clean surfaces at elevated temperature. Many of the precautions and principles applicable to brazing atmospheres can be applied directly to diffusion brazing or welding.[1]

---

1. Brazing atmospheres are discussed in Chapter 12 of this volume.

DIFFUSION WELDING

PROCESS CONDITIONS

Temperature

TEMPERATURE IS AN important diffusion welding process condition for a number of reasons:

(1) It is readily controlled and measured.
(2) In any thermally activated process, an incremental change in temperature will cause the greatest change in process kinetics when compared to most other process conditions.
(3) Virtually all the mechanisms are temperature dependent.
(4) Elevated temperature physical and mechanical properties, critical temperatures, and phase transformations are important reference points.
(5) Temperature must be controlled to promote or avoid certain metallurgical factors, such as allotropic transformation, recrystallization, and solution of precipitates.

Kinetic theory provides a means for understanding the quantitative effects of temperature in diffusion welding. Diffusivity can be expressed as a function of temperature as:

$$D = D_o e^{-Q/kT} \tag{26.1}$$

where
  D = diffusion coefficient at temperature T

  D_o = a constant of proportionality
  Q = activation energy for diffusion
  T = absolute temperature
  k = Boltzmann's constant

From this, it is apparent that the diffusion-controlled processes vary exponentially with temperature. Thus, relatively small changes in temperature produce significantly large changes in process kinetics.

In general, the temperature at which diffusion welding will take place is above $0.5\,Tm$, where Tm is the melting temperature of the metal in degrees Kelvin or degrees Rankine. Many metals and alloys can best be diffusion welded at temperatures between 0.6 and $0.8\,Tm$. For any specific application the temperature, pressure, time, and faying surface preparation are interrelated.

Time

TIME IS CLOSELY related to temperature in that most diffusion controlled reaction rates vary with time. The diffusion length, x, is the average distance traveled by migrating atoms during diffusion. It can be approximated as:

$$x = C(Dt)^{1/2} \tag{26.2}$$

where
  x = diffusion length
  D = diffusion coefficient at T
  t = time
  C = a constant

Thus, diffusion reactions progress with the square root of time (longer times become less and less effective), whereas they progress exponentially with temperature, as was previously shown.

Experience indicates that increasing both the time and the pressure at welding temperature increases joint strength up to a limit. Beyond this point no further gains are achieved. This illustrates that time is not a quantitatively simple condition. The simple relationship that describes the average distance traveled by an atom does not reflect the more complex changes in micro structure that result in the formation of a diffusion weld. Although atom motion continues indefinitely, micro structural changes tend to approach equilibrium. An example of similar behavior is the recrystallization of metals.

In a practical sense, time may vary over a very broad range, from seconds to hours. Production factors influence the practical time for diffusion welding. An example is the time necessary to provide the heat and pressure.

When the welding equipment has thermal and mechanical (or hydrostatic) inertia, welding times are long because of the impracticality of suddenly changing the conditions. When there are no inertial problems, welding time may be as short as 0.3 min, as is the case when joining thoria-dispersed nickel to itself. On the other hand, it may be as long as 4 hours, as when joining columbium to itself with zirconium as a filler metal.

Pressure

PRESSURE IS AN important factor. It is more difficult to deal with as a quantitative value than either temperature or time. Pressure affects several aspects of the process. The initial phase of metallic bond formation is certainly affected by the amount of deformation induced by the pressure applied. This is the most obvious single effect and probably the most frequently and thoroughly considered. Higher pressure invariably produces better joints when the other variables are fixed, within the limits of the welding range. The most apparent reasons for this effect are the greater faying surface deformation and asperity collapse. The greater deformation may also lower recrystallization temperature and accelerate the process of recrystallization at the welding temperature.

The welding equipment and the joint geometry place practical limitations on the magnitude of welding pressure. The pressure needed to achieve a good weld is closely related to the temperature and time. Pressure has additional significance when dissimilar metal combinations are considered. From economic and manufacturing aspects, low welding pressure is desirable. High pressure requires more costly equipment, better controls, and generally involves more complex production procedures.

The pressures and temperatures employed are largely interdependent, but the pressure need not exceed the yield stress of the base metal or filler metal at the welding temperature. Thus, unless retaining dies are used, the pressure is usually kept slightly below the yield stress at the welding temperature. The temperature and pressure are normally selected to produce a weld in an acceptable time.

Metallurgical Factors

IN ADDITION TO the process conditions, there are a number of metallurgically important factors to be considered. Two factors of particular importance with similar metal welds are phase transformation and microstructural factors that tend to modify diffusion rates. Phase transformation (allotropic transformation) occurs in some metals and alloys. Steels are the most familiar of these, but titanium, zirconium, and cobalt also undergo phase transformation. During phase transformation, the metal is very plastic, and this promotes rapid faying surface deformation at lower pressures. Diffusion rates are generally higher during transformation, and also during recrystallization.

Another means of increasing diffusion is alloying or, more specifically, introducing elements with high diffusivity at the faying surfaces. The function of a high-diffusivity element is to accelerate void elimination. In addition to simple diffusion acceleration, these alloying elements may have secondary effects. The elements should have reasonable solubility in the base metal, but should not form stable compounds. Alloying should not promote melting at the weld interface.

When using a diffusion activator, it is desirable to hold the weldment at the diffusion temperature either during or after the welding process to reduce the high concentration of the element at the weld interface. If this is not done, the high concentration may produce metallurgically unstable microstructures. This is particularly important for joints that will be exposed to elevated temperature service.

It is sometimes advantageous to use some form of filler metal between the faying surfaces. One purpose of a filler metal is to provide a layer of soft metal between the faying surfaces. A soft metal layer permits plastic flow to take place at lower pressures than would be required without it during the first stage of welding. See Figure 26.2. After the joint is formed, the diffusion of alloying elements from the base metal into the filler metal reduces the compositional gradient across the joint.

Filler metals may be necessary or advantageous in certain applications in order to:

(1) Reduce welding temperature
(2) Reduce welding pressure
(3) Reduce process time
(4) Increase diffusivity
(5) Scavenge undesirable elements

Filler metals can be applied in many forms. They can be electroplated, condensed, or sputtered onto the faying surface, or they can be in the form of foil inserts or powder. The thickness of the filler metal should not exceed 0.010 in. (0.25 mm).

Generally, the filler metal is a purer version of the base metal. For example, unalloyed titanium often is used as an filler metal with titanium alloys, and nickel is sometimes used with nickel-alloys. An exception to this rule is the use of silver as filler metal in the diffusion welding of aluminum.

Aluminum alloys are among the most difficult metals to diffusion weld, because of the rapid formation of a stable oxide film on bare aluminum surfaces. Most aluminum diffusion welding is done at high temperatures. Lower temperature is required if foil interlayers or electroplated coatings are used as filler metals, and still lower pressures and deformations are needed in the presence of a transient liquid phase. However, these methods must be controlled, otherwise they can produce brittle intermetallic phases and low weld strengths. Significant quantities of silver can dissolve in aluminum at 896-986°F (480-530°C), and silver oxides are unstable above 392°F (200°C). Thus, if the diffusion temperature is 896-968°F (480-530°C), silver oxides will not form, and the silver will dissolve in the aluminum base metal.

A diffusion welding application of aluminum is illustrated in Figure 26.3. Any aluminum alloy can be welded by combining a silver coating on a surface clad with aluminum, and in practice this could have the advantage of a single welding procedure for all aluminum alloys.

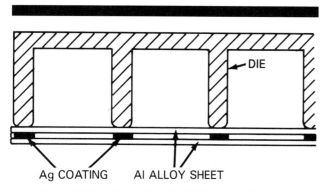

(A) SETUP PRIOR TO DIFFUSION WELDING

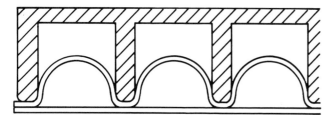

(B) ASSEMBLY AFTER WELDING AND FORMING

Figure 26.3–Fabrication of Diffusion Welded Aluminum Assembly Using Silver as a Filler Metal

Filler metals containing rapidly diffusable elements can also can be used. For example, beryllium can be used with nickel alloys to decrease diffusion time. A properly selected diffusion aid will not melt at welding temperature or form a low melting eutectic with the base metal. An improperly chosen diffusion aid can:

(1) Decrease the temperature capability of the joint
(2) Decrease the strength of the joint
(3) Cause microstructural degradation
(4) Result in corrosion problems at the joint

PROCESS VARIATIONS

Conventional Diffusion Welding

CONVENTIONAL DIFFUSION WELDING involves the application of pressure and heat to accomplish a weld along the entire length of one or more joints simultaneously. Filler metal may or may not be used. Pressure may be applied using gas pressure or a press (mechanical or hydraulic). Heat may be applied by any convenient means but electrical resistance heaters are the most common source. Forming parts to shape is done prior to or after welding using equipment designed for that purpose.

Continuous Seam Diffusion Welding

CONTINUOUS SEAM DIFFUSION welding (CSDW) joins components by "yield-controlled diffusion welding." With this process variation, the parts are positioned with tooling and then fed through a machine with four rollers. The top and bottom rollers are made of molybdenum and function much like resistance seam welding wheels. The two side rollers are used to maintain the shape of the components. The wheels and parts are heated to the desired temperature by electrical resistance. A special control system monitors part temperature. Welding temperature is usually between 1800 and 2000°F (982 and 1090°C) for titanium and between 2000 and 2200°F (1090 and 1200°C) for nickel-base superalloys. The hot wheels apply pressure in the range of 1 to 20 ksi (7 TO 138 MPa) on the seam. The actual pressure depends upon the metal being joined, the joint design, the temperature, and the welding speed. An application of this process could be the joining of two flanges to a web to form a structural beam.

Combined Forming and Welding

TWO PROCESS VARIATIONS take advantage of the superplastic properties of certain metals or alloys. Some alloys can deform or flow significantly at elevated temperatures under very small applied loads without necking or fracture. Titanium and its alloys exhibit this superplastic behavior in the temperature range of 1400 to 1700°F (760 to 925°C). Complex shapes can be formed using moderate gas pressures; then the shapes can be diffusion welded, or vice versa.

One of these process variations is called creep isostatic pressing (CRISP). It is a two-step process combining creep or superplastic forming of titanium sheet structures with hot isostatic pressing to produce a diffusion welded structure.

Inherent in the CRISP process variation is the mating of two external skins. First, one skin is creep formed by gas pressure to the contour of a die. Then shaped inserts are located on the skin and a second skin is creep formed by gas pressure over the first skin and inserts. Diffusion welding of the formed sheets and inserts is achieved by hot isostatic pressing in an autoclave. This method eliminates the need for precision machined die sets and close dimensional tolerances in parts.

Another process variation takes advantage of the same properties of titanium and its alloys described previously; however, the welding is performed under low pressure conditions. This variation is called "superplastic forming diffusion welding" (SPF/DW). Since superplastic forming and diffusion welding of selected titanium alloys can be accomplished using the same temperature, the two operations can be combined in a single fabrication cycle. The welding is accomplished under low pressure conditions.

For titanium alloys that exhibit superplastic properties, SPF/DW considerably extends the range of low cost and structurally efficient titanium aerospace components that can be manufactured. SPF/DW titanium parts may be substituted for conventionally fabricated aluminum alloy components.

Recent developments in the SPF/DW of high strength aluminums and metal matrix composites have stimulated work in the field of diffusion welding of aluminum.

The superplastic forming of the sheet may be done first, followed by welding, or the steps can be reversed. The order depends upon the design of the component. Forming is done first if this is required to bring the faying surfaces of the joint together for welding. If the faying surfaces are in contact, welding is the first process and then the part is formed to final shape. A suitable nonmetallic agent may be used to prevent welding in selected areas.

Superplastic forming of Ti-6%Al-4%V alloy sheet can be done by the application of low pressure argon at 1700°F (925°C) in a sealed die. Gas pressure of about 150 psi (1035 kPa) is used for both forming and welding. Preparation of titanium alloy sheets is usually limited to degreasing and acid etch.

EQUIPMENT AND TOOLING

A WIDE VARIETY of equipment and tooling is employed for diffusion welding. The only basic requirement is that pressure and temperature must be applied and maintained in a controlled environment. Various types of equipment have been developed, each with its special advantages and disadvantages. There are numerous variations of a given type of equipment or approach depending upon the specific ap-

plication. A general description of three types of diffusion welding equipment follows.

Isostatic Gas Pressure

THE PRESSURE FOR welding can be applied uniformly to all joints in an assembly using gas pressure. It is important that all air be removed from the assembly prior to welding. The assembly itself may be evacuated and sealed by fusion welding, if this is possible. Otherwise, the assembly must be sealed in a thin, gas-tight envelope which is evacuated and sealed. Electron beam welding in vacuum is a convenient process for evacuating and sealing in one operation.

Gas pressure is applied externally against the evacuated assembly at welding temperature. Very high pressures can be applied using an autoclave, but the assembly must be capable of withstanding the applied pressure without macrodeformation. Some designs may require internal support tooling with provisions for removing it after welding.

The primary component of hot isostatic equipment is a cold wall autoclave, which can be designed for gas pressures up to 150 ksi (1035 MPa) and for part temperature in excess of 3000°F (1650°C). A typical autoclave is shown in Figure 26.4. Work to be welded is placed in the heated cavity. Internal water cooling is usually provided to maintain a low wall temperature. Openings on each end provide access to the vessel cavity. Utilities and instrumentation are brought into the vessel through high pressure fittings located in the end closures. The high temperatures are produced with an internal heater. Resistance heaters of various designs are used. Alumina or silica insulation is used to reduce heat losses to the cold wall. Temperature is monitored and controlled by thermocouples located throughout the furnace and vessel. Pressurization is achieved by pumping inert gas into the autoclave with a multiple-stage piston-type compressor. Temperature and pressure are controlled independently, and any combination of heating and pressurizing rates can be programmed. Autoclaves are pressure vessels and should be designed to meet applicable code requirements.

The most important consideration is the gas-tight envelope or container in which the workpieces are contained. If a leak develops in the container, pressure cannot be applied to or maintained on the joint. Sufficient gas pressure is applied so that local plastic flow will occur at the faying surface and all void space will be filled as a result of local deformation. With proper conditions, essentially no macrodeformation or changes in part dimensions will occur during welding.

The chief advantage of this technique is the ability to handle complex shapes. It is also well suited to batch operations where large quantities of relatively small assemblies can be welded simultaneously. The major drawbacks are the capital equipment costs and the size limitations imposed by the internal dimensions of the autoclave. Opera-

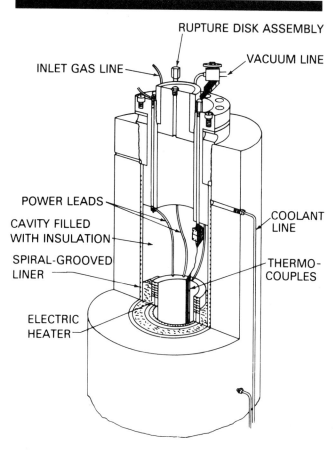

INLET GAS LINE

RUPTURE DISK ASSEMBLY

VACUUM LINE

POWER LEADS

CAVITY FILLED
WITH INSULATION

SPIRAL-GROOVED
LINER

ELECTRIC
HEATER

COOLANT
LINE

THERMO-
COUPLES

Figure 26.4–A Typical High Temperature, Cold-Wall Autoclave

tional equipment ranges up to 36 in. (92 cm) inside diameter and inside lengths of up to 108 in. (275 cm).

The gas pressure process variables are selected to suit the base metals. Usually joints are made at the highest possible pressure to minimize the temperature needed. This method is well suited for welding brittle metals or metals to ceramics and cermets because the isostatic pressure eliminates tensile stresses in the materials.

Presses

A COMMON APPROACH for diffusion welding employs a mechanical or hydraulic press. The basic requirements for the press are (1) sufficient load and size capacity, (2) an available means for heating, and (3) the maintenance of uniform pressure for the required time. It is often necessary to provide a protective atmosphere chamber around the weldment. Press equipment that can be adapted to diffusion welding applications, as shown in Figure 26.5, is often available in a manufacturing or development organization.

There is no standard press design for diffusion welding. Some units provide a vacuum or an inert atmosphere around the parts. Radiant, induction, and resistance types of heating are used. One advantage of a press is the ease of operation and the excellent process control available. One disadvantage is the practical limitation of press size when large weldments are considered. Presses do not lend themselves to high production rates, or batch operations.

Some of the limitations on size can be overcome by operating in a large forming or forging press without an inert atmosphere chamber. Heated platens apply both heat and pressure to the components. The platens may be metallic or ceramic, depending upon the temperature and pressure employed. Castable ceramics are particularly useful because contours can easily be accommodated without extensive machining. Heating elements can be cast into a ceramic die to provide uniform heat during welding. Close tolerances must be maintained between the die and the part so that uniform pressure will be applied to the joint. This is a major problem with press type equipment. It is difficult to maintain uniform pressure on the joint, and variations in weld quality can result.

Tooling requirements vary with application. If no lateral restraint is provided, excessive upsetting may occur during welding. In such cases, lower pressure or temperature is usually required. Heated dies are required and die materials can be a problem. The die must be able to withstand both the temperature and the pressure and must be compatible with the base metal. Interaction between the part and the die can be controlled by stopoff agents and sometimes by oxidizing the die surface. Atmosphere protection is often achieved by sealing the parts in evacuated metal cans that are designed to conform to the die shape.

Retorts can be used in conjunction with presses for diffusion welding of titanium. Tooling blocks and spacers of Type 22-4-9 stainless steel may be used to fill any voids between the titanium workpieces to maintain their shapes. Presses with side and end restraining jacks can exert up to 2 ksi (13.8 MPa) pressure on the retort in all directions. In actual production, the completed assembly pack (retort, heating pads, and insulation) is heated before it is placed in the press. Large structures may require a preheat of as long as 40 hours. Several packs may be in assembly and preheat at one time. The actual time in the press will vary from 2 to 12 hours, depending upon the shape of the structure and the mass of titanium. The assembly pack is cooled to room temperature, dismantled, and the retort is then cut open. This approach is quite slow and is not readily adapted to high production rates.

Resistance Welding Machines

RESISTANCE WELDING EQUIPMENT may be used to produce diffusion spot welds between sheet metal parts. In general, modification of standard equipment is not necessary to achieve successful diffusion welds. The interface is resis-

Figure 26.5–Diffusion Welding Vacuum Hot Press

tance heated under pressure with this equipment. The cycle is designed to avoid melting of metal at the interface. Weld times are generally less than 1 second.

As in standard resistance welding, selection of a suitable electrode material is important. The electrodes must be electrical conductors, possess high strength at welding temperatures, resist thermal shock, and resist sticking to the parts. There is no universal electrode material because of potential interaction with the workpiece. Therefore,

each combination must be carefully evaluated from a metallurgical compatibility standpoint.

In some applications, a small chamber surrounding the electrodes is used to provide an inert atmosphere or vacuum during welding.

One advantage of this type of equipment is the speed at which diffusion welds can be made. Each weld is made in a very short time; however, only a small area is welded in each cycle and many welds are needed to join a large area.

Tooling

A NUMBER OF important considerations must be observed when selecting tooling materials. The main criteria are the following:

(1) Ease of operation
(2) Reproducibility of the welding cycle
(3) Operational maintenance
(4) Weld cycle time
(5) Capital cost

Furthermore, the materials must be capable of maintaining their proper positions and shapes throughout the heating cycle.

Suitable fixture materials are limited when welding temperatures are above 2400°F (1320°C). Only the refractory metals and certain nonmetallic materials have sufficient creep strength at such high temperatures. For example, only tantalum and graphite may be suitable fixturing for diffusion welding of tungsten. Ceramic materials are suitable for fixtures provided they are completely outgassed prior to welding.

Fixtures can be designed to take advantage of the difference in thermal expansion between the base metals and the fixture material. It is possible to generate some, if not all, of the pressure required for welding by appropriate selection of the fixture material, base metal and the clearances between the fixture and part. These principles have been used to join Type 2219 aluminum alloy tubing to Type 321 stainless steel. A precise method was devised to apply the correct welding pressure to the tubular assembly. The reproducible uniform welding pressure was developed by taking advantage of the difference in the thermal expansions of low alloy steel and stainless steel.

DIFFUSION BRAZING

PROCESS VARIABLES

DIFFUSION BRAZING IS similar to conventional brazing. Various methods of heating, atmospheres, joint designs, and equipment can generally be used interchangeably. With diffusion brazing, the filler metal, the brazing temperature, and the brazing time are selected to produce a joint with physical and mechanical properties almost identical to those of the base metal. To do this, it is necessary to completely diffuse the braze metal into the base metal.

Temperature and Heating Rate

THE TEMPERATURE CYCLE used for diffusion brazing depends upon the base metal and the design of the brazing system. When the filler metal composition is similar to that of the base metal, the assembly must be heated to the melting temperature of the filler metal as in conventional brazing. As the brazing alloy melts, it wets the base metal and fills the voids in the joint; then the temperature can be maintained or reduced to solidify the braze metal.

Diffusion brazing forms a filler metal *in situ* during brazing. The metals are generally selected to form a molten eutectic that flows and fills the voids in the joint at brazing temperature. The brazing temperature is somewhat higher than the eutectic that flows and fills the voids in the joint at brazing temperature. The brazing temperature is somewhat higher than the eutectic temperature. For example, a plating of copper on a silver base metal faying surface will form a eutectic when heated to 1500°F (815°C). The eutectic melting temperature is 1435°F (780°C).

In systems where several eutectic and peritectic reactions take place at different temperatures, both the brazing temperature and the heating rate are important. Although a liquid phase can form at the lowest eutectic temperature, diffusion rates will be faster at higher temperatures. The heating rate will determine whether a molten eutectic is formed. If the heating rate is too low, solid-state diffusion will prevent the formation of a molten eutectic. The voids at the faying surface will not be filled by the braze metal.

The maximum brazing temperature may be established by the characteristics of the base metal: for example, incipient melting in most metals is not desirable. Brazing temperature may also be limited by the effect of temperature on the final metallurgical structure or by the heat treatment requirements for the weldment.

After brazing is completed and the braze metal solidified, a high temperature is maintained while solid state diffusion takes place.

Time

THE DURATION OF the diffusion brazing cycle will depend on (1) the brazing temperature, (2) the diffusion rates of the filler metal and the base metal at brazing temperature, and (3) the maximum concentration of filler metal permissible at the joint. The alloy composition at the joint may influence the response to heat treatment or the resulting mechanical properties of the joint. Therefore, the joint must be held at high temperature for some minimum time to reduce the concentration of filler metal to an acceptable value.

Pressure

CONVENTIONAL BRAZING REQUIRES little or no pressure across the joint. In some cases, fixturing may be necessary to avoid excessive pressure. This is particularly so when the molten filler metal is to flow into the joint by capillary action. When the filler metal is placed in the joint before brazing, excessive pressure may force low melting constituents to flow out of the joint before brazing temperature is achieved. In that case, the molten filler metal may not be sufficiently fluid to fill interface voids.

Metallurgical Factors

THE METALLURGICAL EVENTS that transpire during diffusion brazing are similar to those that occur during diffusion welding. An additional factor is the variation in chemical composition across the joint. Compositional variations can significantly affect the response of a particular alloy to heat treatment. For metals that exhibit an allotropic transformation, the chemical composition affects both the transformation temperature and the rate of transformation. Thus, the response to heat treatment across a diffusion brazed joint varies with the local chemical composition. For example, copper stabilizes the beta phase in titanium and decreases the beta-to-alpha transition temperature.

Filler Metals

THE FILLER METAL is a metal that will alloy with the base metal to form a molten alloy at some elevated temperature. A eutectic must form that melts at a temperature compatible with the metallurgy and properties of the base metal. The filler metal may be in powder, foil, or wire form, or it

may be plated onto the surface of the base metal. Close control of the amount of filler metal in the joint is essential for consistent results.

Application of pure metals and simple alloys by electroplating or vapor deposition can be accurately controlled. Films of desired thickness can be deposited on the faying surfaces. However, these processes are not always economical. Metal foil or wire formed into suitable shapes are better for many applications.

In the case of nickel- and cobalt-base alloys, elements commonly added to brazing filler metal to depress the melting temperature also increase alloy hardness and brittleness. Consequently, these filler metals can only be produced as powders. Powders are a problem when precise amounts of filler metal are required. Boron in the range of 2.0 to 3.5 percent is used in nickel-base filler metals. Boron can be diffused into the surfaces of nickel alloy foil or wire shapes to produce filler metal preforms. These preforms provide good control of filler metal placement for diffusion brazing applications.

EQUIPMENT AND TOOLING

THE EQUIPMENT AND tooling used for diffusion brazing are essentially the same as those used for conventional brazing. If furnace brazing is used, the entire cycle can be done in the same equipment or in a dedicated furnace. In some cases it may be more economical and convenient to braze with one piece of equipment and then follow with a diffusion heat treatment with other equipment. For example, the brazing could be done with resistance welding or induction heating equipment, and the diffusion heat treatment could be performed in a furnace.

APPLICATIONS

A WIDE VARIETY of similar and dissimilar metal combinations may be successfully joined by diffusion welding and brazing. Most applications involve titanium, nickel, and aluminum alloys, as well as several dissimilar metal combinations. The mechanical properties of the joint depend on the characteristics of the base metals. For example, the relatively low creep strength and the solubility of oxygen at elevated temperature contribute to the excellent properties of titanium alloy diffusion weldments.

Nickel-base heat-resistant alloys are difficult to join because their creep strengths are high, requiring high pressures for diffusion welding. In addition, a thin, stable oxide film interferes with metal to metal contact because, unlike titanium, the oxygen is not soluble in the nickel. These factors contribute to poor solid-state weldability of

these nickel-base heat-resistant alloys. This problem can be overcome by the use of a relatively soft filler metal that provides more intimate contact.

Base metals strengthened by cold working will be irreversibly softened by the joining heat treatment. However, heat treatable alloys may be rehardened during the joining heat treatment or may be hardened with a postweld heat treatment.

TITANIUM ALLOYS

MANY DIFFUSION WELDING and brazing applications involve titanium alloy components, the majority of which are Ti-6%Al-4%V alloy.[2] The popularity of the processes with titanium alloys stems from the following factors:

(1) Titanium is readily joined by both processes without special surface preparation or unusual process controls.

(2) Diffusion welded or brazed joints may have better properties for some applications than fusion welded joints.

(3) Most titanium structures or components are used principally in aerospace applications where weight savings or advanced designs, or both, are more important than manufacturing costs, within limits.

A number of well-established diffusion welding and brazing methods are available for joining titanium alloys. Welding can be accomplished using pressures in the range of several hundred to several thousand psi. High pressures are used in conjunction with low welding temperatures and when the assembly is welded in a retort. Inserts may be used to maintain the required dimensions. When welding at higher temperatures without an enclosure surrounding the joint members, maximum pressure is usually limited by the allowable deformation in the weldment, and this pressure must be determined empirically. Pressures of 300 to 500 psi (2070 to 3450 kPa) work well in many cases. In some applications, total weldment deformation and deformation rate instead of pressure are used for process control during welding.

Titanium Diffusion Welding

WELDING TEMPERATURE IS probably the most influential condition in determining weld quality; it is set as high as possible without causing irreversible damage to the base metal. For the commonly used alpha-beta type titanium alloys, this temperature is about 75 to 100°F (24 to 38°C) below the beta transus temperature. Thus, Ti-6%Al-4%V alloy with a beta transus of approximately 1825°F (996°C) is normally diffusion welded between 1700 and 1750°F (925 and 955°C).

The time required to achieve high weld strength can vary considerably with other factors, such as mating surface roughness, welding temperature, and pressure. Welding times of 30 to 60 minutes should be considered a practical minimum, with 2 to 4 hours being more desirable.

Faying surface finish and preweld cleaning procedure are two other important considerations. Although the general rule that a smooth faying surface makes welding easier still applies, parts with relatively rough (milled or lathe-turned) faying surfaces can be successfully diffusion welded as long as welding temperature, time, and pressure are adjusted to accommodate such rough finishes. Freshly machined faying surfaces only need to be degreased with a suitable solvent prior to welding. Hydrocarbon and chlorinated solvents should not be used because of safety considerations.

A preferred cleaning method is acid cleaning in a nitric and hydrofloric acid solution. Any residue remaining from cleaning must be removed by thorough rinsing.

Several industries have taken advantage of the benefits of the diffusion welding process, particularly the aerospace industry with its high usage of titanium alloys. The engine mount of the space shuttle vehicle was designed to have 28 diffusion welded titanium parts, ranging from large frames to interconnecting box tubes. This structure is capable of withstanding three million pounds of thrust. Eight-inch (203 mm) square tubes with 0.75 in. (19 mm) thick wall were fabricated by diffusion welding in lengths up to 180 in. (457 cm).

The use of diffusion welding in the gas turbine industry reached a milestone with the production application of a Ti-6%Al-4%V component for an advanced, high-thrust engine. This application marks the first production use of diffusion welding in a rotating engine component.

Titanium Diffusion Brazing

CONTINUOUS SEAM DIFFUSION brazing has been used to produce stiffened skins fabricated as an integral one-piece structure. An example is shown in Figure 26.6. One of the first applications of this method was the fabrication of curved Ti-6%Al-4%V alloy I-beams used as structural members to support boron-aluminum composite on a fighter airplane. These beams were made from 0.025 in. (0.64 mm) sheet.

Superplastic forming/diffusion brazing of titanium parts is also used. An augmentor flap fabricated by the process is shown in Figure 26.7.

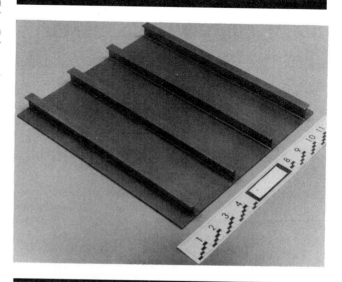

Figure 26.6–A Titanium Alloy Stiffened Sheet Structure Fabricated by Continuous Seam Diffusion

---

2. The weldability of titanium alloys is discussed in the *Welding Handbook*, Vol. 4, 7th Ed. 433-487.

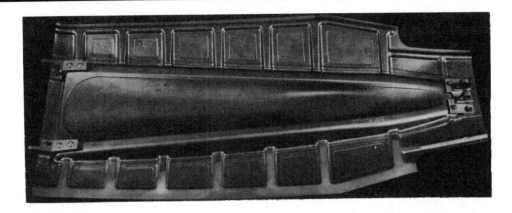

Figure 26.7–Augmentor Flap for a Jet Engine Fabricated by Superplastic Forming/Diffusion Brazing of Titanium. (The structure features two and three sheet construction and weighs three pounds)

Conventional diffusion brazing techniques are also used for joining titanium alloys. Brazing times, temperatures, and preweld cleaning procedures are much the same as for diffusion welding. Pressure may just be sufficient to hold the joint members in contact, and faying surface finish requirements are not as stringent.

The faying surfaces of the titanium alloy are electrolytically plated with a thin film of either pure copper or a series of elements, such as copper and nickel. When heated to the brazing temperature of 1650 to 1700°F (900 to 925°C), the copper layer reacts with the titanium alloy to form a molten eutectic at the braze interface. The brazement is then held at the brazing temperature for at least 1.5 hours. The assembly may also be given a subsequent heat treatment at the brazing temperature for several hours, to reduce the composition gradient in the braze metal. Diffusion brazed joints made with a copper filler metal and a cycle of 1700°F (925°C) for 4 hours exhibited tensile, shear, unnotched fatigue, and stress corrosion properties equal to those of the base metal. However, they had slightly lower notch fatigue and corrosion fatigue properties, and significantly lower fracture toughness. A typical photomicrograph of a diffusion brazed T-joint between Ti-6%Al-4%V and Ti-3%Al-2.5%V alloys is shown in Figure 26.8. A Widmanstatten structure formed at the braze interface because the plated filler metal stabilized the beta phase.

Diffusion brazing is being used to fabricate light-weight cylindrical cases of titanium alloys for jet engines. In this application, the titanium core is plated with a very thin layer of copper and nickel that reacts with the titanium to form a eutectic. During brazing in a vacuum of 10^{-5} torr, a eutectic liquid forms at 1650°F (900°C). This liquid performs the function of a brazing filler metal between the core and face sheets. The eutectic quickly solidifies due to rapid diffusion at the braze interface.

In the past, the copper-nickel filler metal was electro-deposited on the edge of the core in a lamellar fashion. Currently joints are produced using a homogeneous, thin copper-nickel foil as the filler metal. The use of foil has the advantage of allowing more precise control of the filler metal thickness and composition. In addition, the use of foil eliminates several complicated steps that are required in the plating process. Finally, as the foil is a homogeneous layer, it produces its available liquid all at once as soon as the ternary eutectic point is reached. This situation is an improvement over the stepwise formation of liquid that is produced from the electro- plated method.

Diffusion brazed assemblies are held at the brazing temperatures for one to four hours to reduce the composition gradient at the braze interface by diffusion. A typical diffusion brazed titanium alloy honeycomb structure is shown in Figure 26.9.

NICKEL ALLOYS

Diffusion Welding of Nickel Alloys

MANY NICKEL ALLOYS, specifically the high-strength heat-resistant alloys, are more difficult to diffusion weld than most other metals. These alloys must be welded at temperatures close to their melting temperatures, and because of their high-temperature strengths, relatively high pressures are required. In addition, extra care must be taken in preparing the faying surfaces to be welded to ensure cleanliness and mutual conformity. Surface oxides that form on these alloys are stable at high temperatures and will not dissolve or diffuse into the base metal. During welding, the ambient atmosphere must be carefully controlled to prevent faying surface contamination.

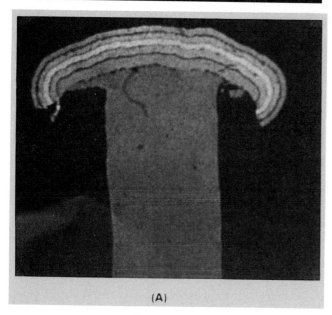

(A)

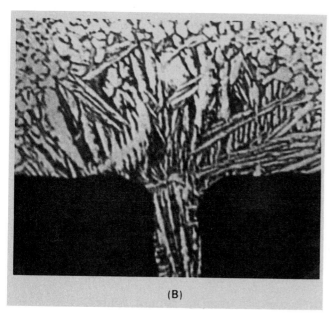

(B)

Figure 26.8—A Diffusion Brazed T-Joint Between Ti-6%Al-4%V and Ti3%Al-2.5%V Alloys

Pure nickel or a soft nickel alloy is commonly used as filler metal when diffusion welding nickel alloys. These filler metals, generally from 0.0001 to 0.001 in. (2.5 to 25 μm) thick, serve several functions. Their relatively low yield strength allows surface conformity to take place at relatively low welding pressures. More important, they are used during welding to prevent the formation of stable precipitates, such as oxides, carbides, or carbonitrides, at

the weld interface. The diffusion welding time must be adequate to allow sufficient interdiffusion to occur at the weld interface.

Welding conditions for some diffusion welded nickel base heat-resistant alloys are shown in Table 26.1.

The pressure required for satisfactory welding is influenced strongly by the geometry of the joint members. Therefore, the required pressure for each application must be determined empirically.

The significance of filler metal and its composition was demonstrated by a series of diffusion welds in a wrought and cast proprietary nickel alloy. Welds were made without filler metal and then with 0.0002 in. (5 μm) both pure nickel and Ni-35%Co alloy. The welding conditions were the same as those listed for this alloy in Table 26.1.

The microstructure of the welds in wrought proprietary alloy are shown in Figure 26.10. With no filler metal, fine Ti (C,N) and NiTiO3 precipitates formed at the weld interface during welding and pinned the interfacial boundary, causing very poor weld mechanical properties. The nickel filler metal consisted of an electroplated layer on each surface. These layers probably welded together early in the cycle, and no precipitates were present to interfere with welding. Subsequent diffusion and grain boundary movement resulted in much improved mechanical properties. The pure nickel filler metals, however, resulted in preferential diffusion of aluminum and titanium into the nickel. This led to the formation of excessive amounts of the strengthening precipitate Ni3(Al, Ti) in the joint. The use of a nickel- 35% cobalt alloy filler metal prevented the diffusion of aluminum and titanium and resulted in a homogeneous joint.

Diffusion Brazing of Nickel Alloys

NICKEL-BASE HEAT RESISTANT alloys can be diffusion brazed using two variations of the process. The variations differ primarily in the thermal cycle to accomplish diffusion. Both methods produce high-strength joints that resemble the base metal in both structure and mechanical properties.

The First Variation With the first variation, a filler metal of 0.001 to 0.004 in. (0.025 to 0.1 mm) thickness, is used. The joint members are held together under slight pressure [under 10 psi (69 kPa)] and heated to the brazing temperature [typically 2000 to 2200°F (1090 to 1200°C)] in vacuum or an argon atmosphere. At brazing temperature, the filler metal melts, filling the voids between the faying surfaces with a thin, molten layer. While the parts are held at the brazing temperature, rapid diffusion of alloying elements occurs between the braze metal and the base metal. This change of composition at the braze interface causes the braze metal to isothermally solidify, thus forming solid braze metal while still at the brazing temperature. After isothermal solidification occurs, the joint microstructure generally resembles that of the base metal except for some compositional and structural variations.

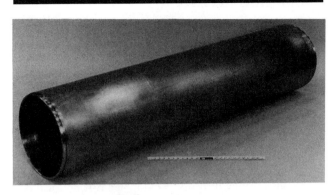

Figure 26.9–Diffusion Brazed Titanium Alloy Honeycomb Structure (81 in. long, 6 lb total weight)

Two single crystal components can be joined together by diffusion brazing to form a joint without grain boundaries and with the same crystal orientation as the base material. The two components have to have the same crystal orientation. A braze alloy of essentially the same composition as the base metals to be joined but with melting point depressents is required. The assembly is then heated to a temperature above the liquidus of the braze alloy and held at this temperature to allow the melting point depressent to diffuse from the filler metal into the base alloy. The base metal then solidifies isothermally. The solidification grows epitaxially from the base metal surfaces, and because the base metals are single crystal with the same orientation, the solidifying braze grows as a single crystal with the same orientation as the base metal.

A diffusion brazed joint between two single crystals is shown in Figure 26.11. This joint was brazed with B-Ni2 foil (Ni-7%Cr-3%Fe-4.5%Si-3.2%B) for 16 hours at 2100°F (1150°C) followed by 12 hours at 2275°F (1245°C). Notice the absence of grain boundaries in the joint area. The fractured surface of a similarlly brazed joint is shown in Figure 26.12(A). The microstructure across the fracture of the tensile specimen is shown in Figure 26.12

(B). The joint was brazed using a nickel base foil containing 15% Cr and 4% B at 2150°F (1175°C) for 16 hours plus 2275°F (1245°C) for 22 hours. The tensile strength at 2000°F (1095°C) was 46.6 ksi (321 MPa). The reduction in area was 12.5%.

At this stage the joint has good properties, although not fully equivalent to those of the base metal. By permitting the brazement to remain at the brazing temperature for a longer time, the braze metal can be homogenized both in composition and structure until it is essentially equivalent to the base metal.

The Second Variation   The second variation involves joining nickel-base components with a specially designed brazing filler metal that completely melts at some elevated temperature below the incipient melting point of the base metal. Subsequent to this, the brazed component is given a diffusion heat treatment to homogenize the brazing filler with the base metal. This is followed by an appropriate aging heat treatment designed for the base metal.

A brazing filler metal contains melting point depressants, such as silicon, boron, manganese, aluminum, titanium, and columbium. The filler metal contains sufficient amounts of depressants so that the resultant alloy is molten at a temperature that does not impair the properties of the base metal. Ideally, brazing is accomplished at the normal solution heat treating temperature for a given base metal. Figure 26.13 shows a diffusion brazed joint made in a wrought proprietary nickel alloy using the first procedure described. A filler metal of 0.003 in. (0.08 mm) thick Ni-15%Cr-15%Co-5%Mo-3%Be was used in the joint with a processing cycle of 2140°F (1170°C) for 24 hours in vacuum. A microprobe chemical analysis across a joint showed a uniform chemical composition, essentially that of the base metal. Stress rupture tests at 1600 and 1800°F (870 and 980°C) showed that the diffusion brazed joints had essentially the same properties as the base metal.

Diffusion brazed joints produced at lower temperatures and with shorter time may not be uniform in composition. As a result, some elevated temperature mechanical properties of the joints may be lower than those of the base metal, particularly under stress-rupture conditions.

Table 26.1
Typical Diffusion Welding Conditions for Some Nickel-Base Alloys

Base Metal*	Filler Metal	Welding Temp.		Pressure,		Time, h
		°F	°C	psi	kPa	
Inconel 600	Ni	2000	1090	100-500	690-3450	0.5
Hastelloy X	Ni	2050	1120	100-500	690-3450	4
Wrought Udimet 700	Ni-35%Co	2140	1170	1000	6900	4
Cast Udimet 700	Ni-35%Co	2175	1190	1200	8275	4
Rene 41	Ni-Be	2150	1180	1550	10690	2
Mar-M 200	Ni-25%Co	2200	1205	1000-2000	6900-13800	2

* Tradenames

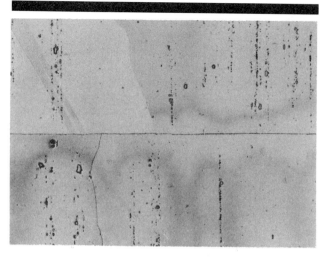

(A)

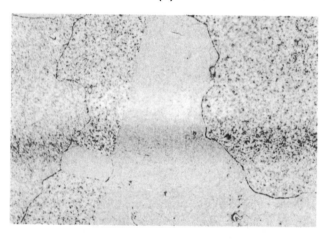

(B)

(C)

Figure 26.10–Diffusion Welds in a Wrought Alloy with (A) No Interlayer, (B) A Nickel Interlayer, and (C) A Ni-35% Co Alloy Interlayer (x250)

ALUMINUM ALLOYS

ALUMINUM ALLOYS CAN be successfully diffusion welded as long as some means is employed to avoid, disrupt, or dissolve the tenacious surface oxide. A wide range of temperatures, pressures, and times may be utilized: for example, with Type 6061 aluminum alloy, welding conditions as divergent as 725°F (385°C) and a pressure of 3800 psi (26 MPa) for several hours or 1000°F (538°C) and a pressure of 1000 psi (7 MPa) for one hour have been satisfactory. However, the main boundary condition is the melting point of the base metal. Welding is normally carried out in vacuum or inert gas although aluminum-boron fiber composites can be diffusion welded in air. If no local deformation of the parts can be tolerated, the faying surfaces should be coated with a thin layer of silver or gold-copper alloy by electrolytic or vapor deposition. The coating will prevent surface oxidation during welding.

Aluminum and aluminum alloys can be diffusion brazed using a copper filler metal. Sound, strong joints can be produced in aluminum by limiting the copper thickness to 2×10^{-5} in. (0.5 µm) and restricting the brazing temperature to between 1030 and 1060°F (554 and 571°C). The time at temperature should not exceed 15 minutes at the lower temperature limit or 7 minutes at the upper limit. Type A356.0 aluminum-7% silicon casting alloy can be diffusion brazed by electroplating one of the joint members with copper that will form a eutectic with the aluminum and silicon in the casting alloy when heated to 975°F (524°C).

To ensure optimum joint properties, copper thickness, brazing temperature, and brazing time must be selected to promote isothermal solidification during brazing and thereby prevent the formation of the compound $CuAl_2$. Proper balancing of these conditions results in strong joints that can withstand quenching from the solution temperature required for heat treating Type A356.0 alloy to the T61 condition. Electroplating the cover sheets with 1.5 to 2.0×10^{-5} in. (0.38 to 0.5 µm) of copper and holding between 980 and 1000°F (527 and 538°C) for one hour are satisfactory conditions. After quenching and aging, the joint strength will equal that of the casting itself. Microstructurally, the brazed joint will be indistinguishable from the casting.

STEELS

STEELS ARE NOT normally diffusion welded because they are more easily joined by conventional brazing or fusion welding processes for most applications. Diffusion welding may be utilized successfully for specialized applications

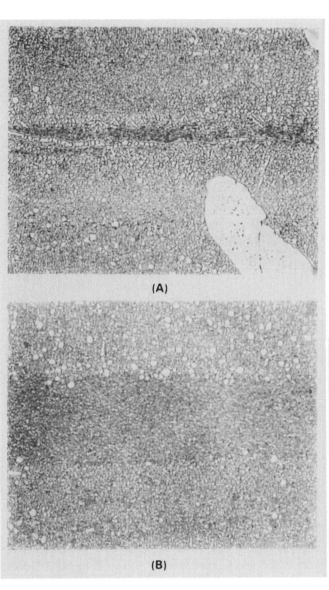

(A)

(B)

Figure 26.11–Diffusion Brazed Single Crystal. (Brazing Temperature was 16 Hours at 2100°F Followed by a Diffusion Heat Treatment of 22 Hours at 2275°F)

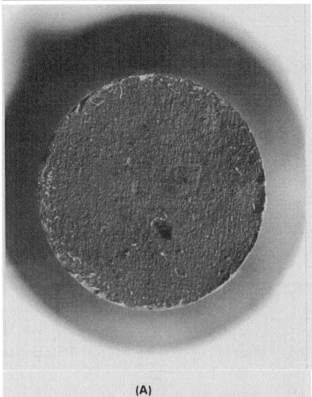

(A)

(B)

Figure 26.12–Fracture Surface (8X) (A) and Microstructure (100X) (B) of Single Crystal Brazement

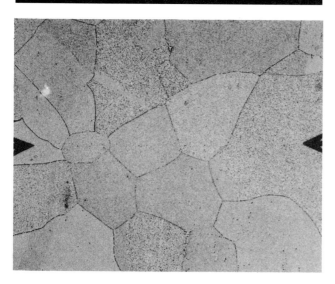

Figure 26.13–A Diffusion Brazed Joint in Wrought Nickel Alloy

where high-quality joints are required between large, flat surfaces. For example, low carbon steels have been welded without a filler metal over a wide range of conditions. Two sets of conditions that produced excellent welds in AISI 1020 steel are 1800 to 2200°F (982 to 1204°C) with a pressure of 1 ksi (7 MPa) for 1 to 15 minutes and 2000 to 2200°F (1093 to 1204°C) with a pressure of 5 psi (35 kPa) for 2 hours. Welding can be accomplished either in a protective atmosphere or in air, provided the joint is first seal welded around the periphery to exclude air.

Stainless steels can be diffusion welded using conditions similar to those used for carbon steel; however these steels are normally covered by a thin adherent oxide that must be removed prior to welding. This can be accomplished either by welding at high temperatures in dry hydrogen or by copper plating the faying surfaces after anodic cleaning. Copper oxide on the plating is relatively easy to reduce in hydrogen during heating to welding temperature. For illustrative purposes, sound welds were made in AMS 5630 martensitic stainless steel at 2000°F (1093°C) with a pressure of 100 psi (690 kPa) for 1.5 hour using a 0.0001 in. (2.5 μm) thick copper filler metal.

DISSIMILAR METAL COMBINATIONS

DIFFUSION WELDING IS particularly well suited for joining many dissimilar metal combinations, especially when the melting points of the two base metals differ widely or when they are not metallurgically compatible. In such cases, conventional fusion welding is not practical because it would either result in excessive melting of one of the metals or in the formation of a brittle weld metal. Diffu-

sion welding is also suitable when the high temperatures of fusion welding would cause an alloy to become brittle or lower its strength drastically, as is the case with some refractory metal alloys. Filler metals are sometimes used to prevent the formation of brittle intermetallic phases between certain metal combinations.

When determining conditions and filler metal requirements for diffusion welding a particular dissimilar metal combination, the effects of interdiffusion between the two base metals must be considered. Interdiffusion can cause certain problems as a result of the following metallurgical phenomena:

(1) An intermediate phase or a brittle intermetallic compound may form at the weld interface. Selection of an appropriate filler metal can usually prevent such problems.

(2) Low melting phases may form. Sometimes this effect is beneficial.

(3) Porosity may form due to unequal rates of metal transfer by diffusion in the region adjacent to the weld interface. This is known as Kirkendall porosity. Proper welding conditions or the use of an appropriate filler metal, or both, may prevent this problem.

One problem that often exists and that is not unique to diffusion welding is the difference in the thermal expansion characteristics of the two base metals. Simply stated, any combination of dissimilar metals that is heated and cooled during welding or brazing will develop shear stresses in the joint if the coefficients of thermal expansion are not identical. The severity of the problem will vary depending upon the temperature span, the difference between the expansion coefficients, the size and shape of the joint members, and the nature of the weld formed between them. This becomes a design problem, in part, since distortion can result. Cracking can result when the joint strength or ductility, or both, are low and the stresses are high.

Many dissimilar combinations can be formed but result in brittle intermetallic phases, and in some cases the reaction proceeds very rapidly due to the formation of a liquid phase at the welding temperature. Although these combinations are brittle, useful joints can be attained by allowing for it in the component design. A combination of Zircaloy 2 with type 304 stainless steel is a good example of the situation where a strong useful joint can be made despite the presence of brittle phases. Figure 26.14 shows the joint designs employed for joining type 304 stainless tube to Zircaloy 2.

The conical tapered joint shown in Figure 26.14(A) uses the differential expansion of the two materials to provide the required pressure. The joint shown in Figure 26.14(B) requires a longitudinal force to maintain pressure during the diffusion welding process.

Joints between stainless steel and Zircaloy 2 tubing of 7/8 in. (22.2 mm) diameter and 1/8 in. (3.2 mm) wall can withstand from 12 000 to 17 000 psi (83 to 117 MPa) internal pressure when tested hydraulically. The fracture initiates by longitudinal splitting of the Zircaloy tube. Similar joints have withstood 100 pressure cycles between 100 and 3500 psi (690 to 24 000 kPa) at 500°F (260°C) and 200 temperature cycles between 100 and 600°F (38 and 316°C) without failure.

Representative conditions used for diffusion welding some dissimilar metal combinations are presented in Table 26.2. Often the temperature and the time used for a particular combination are selected as part of the necessary heat treatment for one of the alloys to develop design properties for a specific application.

Several recent Japanese successes are shown in Table 26.3

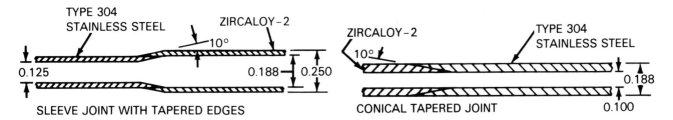

TYPE 304 STAINLESS STEEL   ZIRCALOY-2

10°

0.125   0.188 — 0.250

SLEEVE JOINT WITH TAPERED EDGES

(A) BONDING PRESSURE IS APPLIED BY EXPANSION OF THE STAINLESS STEEL AGAINST THE ZIRCALOY.

ZIRCALOY-2   TYPE 304 STAINLESS STEEL

10°

0.188

CONICAL TAPERED JOINT   0.100

(B) BONDING PRESSURE IS APPLIED BY AN AXIAL COMPRESSIVE FORCE FROM AN INTERNAL BAR.

Figure 26.14—Joint Design Used to Overcome Existence of Brittle Phases

Table 26.2
Diffusion Welding Conditions for Some Dissimilar Metal Combinations

Base Metal Combinations	Filler Metal	Temperature, °F	°C	Time,h	Pressure[a], ksi	MPa	Atmosphere
Cu to Al	—	950	510	0.25	1	7	Vacuum
Cu to 316 Stainless Steel	Cu	1800	982	2		a	Vacuum
Cu to Ti	—	1560	849	0.25	0.7	5	Vacuum
Cu to Cb-1%Zr	Cb-1%Zr	1800	982	4		a	Vacuum
Cu-10%Zn to Ti-6%Al-6%V-2%Sn	—	900	482	8		a	Vacuum
4340 Steel[c] to Inconel 718[b]	—	1730	943	4	29	200	Vacuum
Nickel 200 to Inconel 600[b]	—	1700	927	3	1	7	Not Reported
Pyromet X-15[b] to T-111 Ta alloy	Au-Cu	1100	593	4	30	207	Not Reported
Cb-1%Zr to 316 Stainless Steel	Cb-1%Zr	1800	982	4		a	Vacuum
Zircaloy-2 to 304 Stainless Steel	—	1870-1900	1021-1038	0.5		a	Vacuum

a. Pressure is applied with differential thermal expansion tooling.

b. Tradenames

c. Outgassing of the 4340 steel at 10(-5) torr (130 mPa) and 1850°F (1010°C) for 24 hours prior to welding was critical to the formation of satisfactory welds.

Table 26.3
Recent Applications of Diffusion Joining in Japanese Industries

Products	Materials	Reason for Adoption	Previous Method	Product Records
F-15 Fighter Fitting	Ti-6Al-4V	Cost reduction	Machining from forging or plate	more than 1000 parts
Impeller for Liquid Fuel Rocket	Ti-5Al-2.5Sn	Higher quality	—	100 parts
Jewelry	18K gold alloy	Higher quality	Brazing	400 million yen per year
Electrode	Cu-316L stainless steel	New type only possible by DJ	—	2200 parts
Tube sheet	Cupro Ni-mild steel-316L ss	Cost reduction	Rolling or explosive bonding	1000 parts
Chock liner for steel rolling mill	Brass-mild steel	Cost reduction	Solid Cu alloy	100 parts
Cooling plate for cyclotron accelerator	Cu-316Lss	New type only possible by DJ	—	50 parts
Continuous casting mold	wear resistant material - Cu alloy - 304 ss	New type only possible by DJ	—	20 parts

INSPECTION

ESTABLISHING THE QUALITY of a diffusion welded or brazed joint is difficult with current nondestructive examination procedures. This is due to the nature of the joints. Usually, little or no porosity exists if the joint is made with properly developed procedures. The main defect in a diffusion weld is lack of grain growth across the original interface. Efforts to distinguish intimate contact without grain growth across the interface from a perfect bond have not been successful.

Radiography, eddy current, and thermal methods are relatively unsatisfactory for inspection of most applications. Dye penetrant methods have been found useful for edge inspections.

Ultrasonic examination has proved the most useful for internal inspection, especially if a hairline separation exists. The sensitivity varies with the metal being tested, the ultrasonic frequency, the skill of the operator, and the degree of sophistication of the equipment. In general, defects of less than 0.1 in. (2.5 mm) diameter are difficult to locate and a practical limit of about 0.04 in. (1.0 mm) exists. With special methods and very sophisticated equipment, it has been reported that defects equivalent to 0.005 in. (0.13 mm) diameter can be detected in some metals. These testing approaches cannot be considered routine and they only work under special conditions.

Various types of diffusion weld defects representing possible production defects for superplastic form-ing/diffusion brazing processes have been evaluated in numerous titanium specimens. These specimens were used to evaluate ultrasonic techniques with regard to defect resolution. Table 26.4 summarizes the results. Higher frequency, focusing probes appear to produce the best results, but investigations are continuing.

Ultrasonic inspection cannot differentiate between complete intimate contact and an actual diffusion weld. Only metallographic examination can assure complete welding or brazing. Because this is a destructive test, it cannot always be performed on the part in question. Fortunately, diffusion welding and brazing are reproducible when good process control is exercised. Random destructive sampling coupled with ultrasonic examination will provide a high confidence level. This approach has been successfully used in production.

A reliable nondestructive examination method for inspection of diffusion welded structures is desireable. However, the present conventional nondestructive examination procedures and equipment do not adequately differentiate between acceptable and unacceptable diffusion welded or brazed joints. Therefore it is necessary to supplement the nondestructive tests with destructive tests. Conventional radiography is not suitable for detecting the extremely small defects involved in diffusion welding, but use of x-ray micro-focus techniques coupled with digital image enhancement offer an improvement in resolution.

Table 26.4
Detectability of Diffusion Brazing Defects with Various Ultrasonic Testing Methods

Diffusion Brazing Defects	Ultrasonic Testing Methods			
	Transmission-focussing, non-focussing probes 5 and 10 MHz	Pulse/echo, non-focussing probes 10 up to 20 MHz	Pulse/echo non-focussing probes 30 up to 110 MHz	Pulse/echo point-focussing probes > 30 MHz
Course-dispersive macro defects (single defect size w = 200μm; φ = 3 mm)	D	WD	WD	WD
Course-dispersive macro defects (single defect size w = 5μm; φ> 5 mm)	PD	D	WD	WD
Fine-dispersive macro defects (single defect size w = 3μm; 0,1 <φ< 5 mm)	PD	PD	WD	WD
Micro defect configurations (single defect size w = 1μm; φ< 15μm)	ND	ND	PD	D

ND   not detected
PD   partially detected
D   detected
WD   well detected

SAFE PRACTICES

HAZARDS ENCOUNTERED WITH diffusion welding and brazing are similar to those associated with other welding and cutting processes. Personnel must be protected against hot materials, gases, fumes, electrical shock, and chemicals.

The operation and maintenance of diffusion welding and brazing equipment should conform to the provisions of American National Standard Z49.1, Safety in Welding and Cutting. This standard provides detailed procedures and instructions for safe practices which will protect personnel from injury or illness, and property and equipment from damage by fire or explosion arising from diffusion welding.

It is essential that adequate ventilation be provided so that personnel will not inhale gases and fumes generated while diffusion welding and brazing. Some filler metals and base metals contain toxic materials such as cadmium, beryllium, zinc, mercury, or lead, which are vaporized during brazing.

Solvents such as chlorinated hydrocarbons, and cleaning compounds such as acids and alkalies, may be toxic or flammable or cause chemical burn when present in the brazing environment.

Requirements for the purging of furnaces or retorts that will contain a flammable atmosphere are also given in the standard. In addition, to avoid suffocation, care must be taken with atmosphere furnaces to insure that the furnace is purged with air before personnel enter it.

SUPPLEMENTARY READING LIST

Adam, P. and Steinhauser, L., Bonding of superalloys by diffusion welding and diffusion brazing. MTU, pages 9-1 to 9-6, 61st Meeting of Structural and Materials Panel of AGARD, Oberammergau, Germany, 11-13 September 1985, AGARDCP-398, July 1986.

Army Material Development and Readiness Command, "Improved fabrication of fluidic laminates: fineblanking and semisolid-state diffusion bonding promise to improve fluidic components." *Journal Vol-U8202*, 1 page, October 1981.

Arvin, G. H. et al, "Evaluation of superplastic forming and co-diffusion bonding of Ti-6Al-4V titanium alloy expanded sandwich structures." Rockwell International, NASA-CR-165827, NA-81-185-1, Cont. No.-NASI-15788, 134 pages, May 1979 - December 1980.

Blackburn, L. B. "Effect of LID[R] processing on the microstructural and mechanical properties of Ti-6Al-4V and Ti-6Al-2Sn-4Zr-2Mo titanium foil-gauge materials." NASA Langley Research Center, NASA TP-2677, 24 pages, April 1987.

Blair, W. "Fabrication of titanium multiwall thermal protection system (TPS) curved panel." Rohr Industries, Inc., NASA-CR-165754, Cont. No.-NASI-15646, 29 pages, August 1981.

Boire, M. and Jolys, P. "Application du soudage par diffusion associe au formage superplastique (SPF/DB) a la realization de structures en toles minces de TA6V." Aerospatiale, pages 10-4 to 10-12, 61st Meeting of Structural and Materials Panel of AGARD, Oberammergau, Germany, 11-13 September 1985, AGARD-CP-398, July 1986.

Calderon, P. D. et al, "An investigation of diffusion welding of pure and alloyed aluminum to type 316 stainless steel." *Welding Journal* 64(4): 1045-1125; April 1985.

Dini, J. W. et al, "Use of electrodeposited silver as an aid in diffusion welding." *Welding Journal* 63(1): 285-345; January 1984.

Doherty, P. E. and Harraden, D. R. "New forms of filler metal for diffusion brazing." *Welding Journal* 56(10): 37-39; October 1977.

Dunkerton, S. B. and Dawes, C. J. "The application of diffusion bonding and laser welding in the fabrication of aerospace structures." The Welding Institute, pages 3-1 to 3-12, 61st Meeting of Structural and Materials Panel of AGARD, Oberammergau, Germany, 11-13 September 1985, AGARD-CP-398, July 1986.

Duvall, D. S., Owczarski, W. A., and Paulonis, D. F. "TLP* Bonding: a new method for joining heat resistant alloys." *Welding Journal* 53(4): 302-14; April 1974.

Elmer, J. W. et al, "The behavior of silver-aided diffusion-welded joints under tensile and torsional loads." *Welding Journal* 67(7): 1575; July 1988.

Godziemba-Malisqewski, J. "Thermal surge in diffusion welding-generati on, inrush characteristic, and effects." *Welding Journal* 66(6): 1745; June 1987.

Isserow, S. "Diffusion welding of copper to titanium by hot isostatic pressing (HIP)." AMMRC, Final Rept. AMRRC-TR-80-85, Journal Vol-U8111, 14 pages, December 1980.

Kamat, G. R. "Solid state diffusion welding of nickel to stainless steel." *Welding Journal* 67(6): 44; June 1988.

Kapranos, P. and Priestner, R. "NDE of diffusion bonds." University of Manchester/UMIST, *Metals and Materials* 194-198; April 1987.

Leodolter, W. "Tool sealing arrangement and method." Department of Air Force and McDonnell Douglas, Pat-App1-6-300767, Filed 10 15 pages, September 1981.

Lison, R. and Stelzer, J. F. "Diffusion welding of reactive and refractory metals to stainless steel." *Welding Journal* 58(10): 3065-3145; October 1979.

McQuilkin, F. T. "Feasibility of SPF/DB titanium sandwich for LFC wings." Rockwell International, NASA-CR-165929, Cont. No. - NASI-16236, 62 pages, June 1982.

Moore, T. J. and Glasgon, T. K. "Diffusion welding of MA6000 and a conventional nickel-base superalloy." *Welding Journal* 64(8): 2195-2265; August 1985.

Morley, R. A. and Caruso, J. "Diffusion welding of 390 aluminum alloy hydraulic valve bodies." *Welding Journal* 59(8): 29-34; August 1980.

Munir, Z. A. "A theoretical analysis of the stability of surface oxides during diffusion welding of metals." *Welding Journal* 62(12): 3335-3365; December 1983.

Naimon, E. R. et al, "Diffusion welding of aluminum to stainless steel." *Welding Journal* 60(11): 17-20; November 1981.

Niemann, J. T. and Garrett, R. A. "Eutectic bonding of boron-aluminum structural components." *Welding Journal* Part 1, 53(4): 175s-84s; April 1974; Part 2, 53(8): 351s-9s; August 1974.

Niemann, J. T. and White, G. W. "Fluxless diffusion brazing of aluminum castings." *Welding Journal* 57(10): 285s-91s; October 1978.

Norris, B. and Gojny, F. "Joining processes used in the fabrication of titanium and Inconel honeycomb sandwich structures." Rohr Industries, Inc., lst SAMPE International Metals Symposium, August 18-20, 1987, Cherry Hill, N. Jersey, 8 pages.

O'Brien, M., Rice, C. R., and Olson, D. L. "High strength diffusion welding of silver coated base metals." *Welding Journal* 55(1): 25-27; January 1976.

Owezarski, W. A. and Daulonis, D. F. "Application of diffusion welding in the USA." *Welding Journal* 60(2): 22-33; February 1981.

Partridge, P. G., Harvey, J. and Dunford, D. V. "Diffusion bonding of Al-alloys in the solid state." Royal Aircraft Establishment, pages 8-1 to 8-23, 61st Meeting of Structures and Materials Panel of AGARD, Oberammergau, Germany, 11-13 September 1985, AGARD-CP-398, July 1986.

Rosen, R. S. et al, "The properties of silver-aided diffusion welds between uranium and stainless steel." *Welding Journal* 65(4): 835; April 1986.

Sheetz, H. A., Coppa, P. L. and Devine, J. "Ultrasonically activated diffusion bonding for fluidic control assembly." Sonobond Corp., Cont. No. - DAAA21-76-C0186, RLCD CR-79005, Final Report February 1979, 161 pages.

Schwartz, M. M. "Diffusion brazing titanium sandwich structures." *Welding Journal* 57(9): 35-8; September 1978.

———. *Metals Joining Manual* McGraw-Hill Book Co., 490 pages, ISBN 0-07-055720-9, September 1979.

Sharples, R. V. and Bucklow, I. A. "Diffusion bonding of aluminum alloys to titanium." *The Welding Institute* 7836.01/85/448.3, 307/1986, 15 pages, July 1986.

Signes, E. G. "Diffusion welding of steel in air." *Welding Journal* 47(12): 571s-4s; December 1968.

Stephen, D. and Swadling, S. J. "Diffusion bonding in the manufacture of aircraft structure." British Aerospace pages 7-1 to 7-17, 61st Meeting of Structures and Materials Panel of AGARD, Oberammergau, Germany, September 11-13 1985, AGARD-CP-398, July 1986.

Sullivan, P. G. "Elevated temperature properties of boron/aluminum composites." Nevada Engineering and Technology Corp., NASA-CR-159445, Cont. No. - NAS320079, Final Report 26 January 1976 - 26 January 1977, November 1978, 116 pages.

Tanzer, H. J. "Fabrication and development of several heat pipe honeycomb sandwich panel concepts." Hughes Aircraft Co., Cont. No. - NASI-16556, NASA CR-165962, 55 pages, June 1982.

Tobor, G. and Elze, S. "Ultrasonic testing techniques for diffusion-bonded titanium components." MBB, pages 11-1 to 11-10, 61st Meeting of Structures and Materials Panel of AGARD, Oberammergau, Germany, 11-13 September 1985, AGARDCP-398, July 1986.

Weisert, E. D. and Stacher, G. W. "Fabricating titanium parts with SPF/DB process." *Metal Progress* 111(3): 32-7; March 1977.

Wells, R. R. "Microstructural control of thin-film diffusion brazed titanium." *Welding Journal* 55(1): 20s-8s; January 1976.

Wigley, D. A. "The structure and properties of diffusion assisted bonded joints in 17-4 PH, type 347, 15-5 PH and Nitronic 40 stainless steels." Southampton University, NASA-CR-165745, Cont. No. - NASI-16000, 31 pages, July 1981.

Wilson, V. E. "Superplastic formed and diffusion bonded titanium landing gear component feasibility study." Rockwell International, Cont. No. - F33615-79-C3401, TR-80-3081, Final Report March 1979 - July 1980, 80 pages, July 1980.

Witherell, C. E. "Diffusion welding multifilament superconducting components." *Welding Journal* 57(6): 153s-60s; June 1978.

ADHESIVE BONDING OF METALS

PREPARED BY A COMMITTEE CONSISTING OF:

P. R. Khaladkar, Chairman
E. I. DuPont de Nemours and Company

R. Hartshorn
3M Company

D. Zaluca
Johnson Wax Company

WELDING HANDBOOK COMMITTEE MEMBER:
J. R. Condra
E. I. DuPont de Nemours and Company

CHAPTER **27**

ADHESIVE BONDING OF METALS

FUNDAMENTALS OF THE PROCESS

DEFINITIONS AND GENERAL DESCRIPTION

ADHESIVE BONDING IS a materials joining process in which a nonmetallic adhesive material is placed between the faying surfaces of the parts or bodies, called *adherends*. The adhesive then solidifies or hardens by physical or chemical property changes to produce a bonded joint with useful strength between the adherends. During some stage of processing, the adhesive must become sufficiently fluid to wet the faying surfaces of the adherends.

Adhesive[1] is a general term that includes such materials as cement, glue, mucilage, and paste. Although natural organic and inorganic adhesives are available, synthetic organic polymers are usually used to join metal assemblies. Various descriptive adjectives are applied to the term adhesive to indicate certain characteristics, as follows:

(1) Physical form: liquid adhesive, tape adhesive
(2) Chemical type: silicate adhesive, epoxy adhesive, phenolic adhesive
(3) Materials bonded: paper adhesive, metal-plastic adhesive, can label adhesive
(4) Application method: hot-setting adhesive, sprayable adhesive

Although adhesive bonding is used to join many nonmetallic materials, only the bonding of metals to themselves or to nonmetallic structural materials is covered in this chapter.

Adhesive bonding is similar to soldering and brazing of metals in some respects but a metallurgical bond does not take place. The surfaces being joined are not melted, although they may be heated. An adhesive in the form of a liquid, paste, or tacky solid is placed between the faying

surfaces of the joint. After the faying surfaces are mated with the adhesive in between, heat or pressure, or both, are applied to accomplish the bond.

An adhesive system must have the following characteristics:

(1) At the time the bond is formed, the adhesive must become fluid so that it wets and comes into close contact with the surface of the metal adherends.
(2) In general, the adhesive cures, cools, dries, or otherwise hardens during the time the bond is formed or soon thereafter.
(3) The adhesive must have good mutual attraction with the metal surfaces, and have adequate strength and toughness to resist failure along the adhesive-to-metal interface under service conditions.
(4) As the adhesive cures, cools, or dries, it must not shrink excessively. Otherwise, undesirable internal stresses may develop in the joint.
(5) To develop a strong bond, the metal surfaces must be clean and free of dust, loose oxides, oil, grease, or other foreign materials.
(6) Air, moisture, solvents, and other gases which may tend to be trapped at the interface between the adhesive and metal must have a way of escaping from the joint.
(7) The joint design and cured adhesive must be suitable to withstand the intended service.

PRINCIPLES OF OPERATION

FOR WETTING TO occur, the surface free energy of the adherend must be greater than that of the adhesive. This is usually the case for metallic adherends and polymeric adhesives; however, contaminants adsorbed on the metal can lower the surface free energy and prevent the formation of a good adhesive bond. Contaminants can be removed from

---

1. Terms relating to adhesives are defined in ANSI/ASTM D907.

the surface by washing with solvent or by abrasion. The latter treatment, using grit blasting, abrasive paper, or abrasive pads, is frequently used to prepare metal surfaces before bonding.

In addition to the surface energetics, an adhesive must have low viscosity during the bond forming process in order to spread readily over the adherend surface. The higher the viscosity, the greater the probability that the adhesive will not completely wet the surface and will entrap gases, liquids, or vapors in the bondline. This tendency can be reduced by the application of pressure during the cure process.

PROCESS ADVANTAGES

ADHESIVE BONDING HAS several advantages for metals joining when compared to resistance spot welding, brazing, soldering, or mechanical fasteners such as rivets or screws.

Bonding Dissimilar Materials

IT IS POSSIBLE to bond dissimilar metals with minimal galvanic corrosion in service, provided the adhesive layer maintains electrical isolation between the metals. Many types of adhesive formulations are flexible enough to permit the bonding of dissimilar metals with widely different coefficients of thermal expansion. Such possibilities depend, of course, upon the size of the pieces and the degree

of joint strength required. A single adhesive may be used for joining a number of dissimilar metal combinations in a single assembly.

Adhesive bonding also makes it possible to join metals to nonmetallic materials such as various types of plastics. Figure 27.1 shows the bonding of rigid urethane foam to sheet steel with a free-flowing, room-temperature-curing, epoxy adhesive in the fabrication of insulating panels.

Bonding Thin Gage Metals

VERY THIN METALS parts can be adhesively bonded. Three examples of this are: (1) multiple layers of thin metal sheets can be bonded together to form electric motor laminates; (2) various metal foils can be joined to themselves or to other materials; and (3) thin-gage metal sheets may be used as sandwich panel skins.

Low Processing Temperatures

THE TEMPERATURES USED for the heat curing of most adhesives are between 150 and 350°F (65 and 176°C), temperatures that are below the normal soldering range. Room-temperature-curing formulations are available that provide sturdy structural bonds for service temperatures up to 180°F (82°C) under humidity conditions that do not exceed 70 percent relative humidity. High performance room-temperature-curing epoxy adhesives are available

Figure 27.1–Rigid Urethane Foam Bonded to Sheet Metal

that maintain good strength up to 300°F (150°C). They can be used to join heat-sensitive components without damage. Adhesive bonding should be considered when high temperature joining operations would cause metallurgical or structural damage to the parts.

Many adhesives provide suitable performance at temperatures above their curing temperature, which is not the case with metal solders.

Combination Bonding and Sealing

THE ADHESIVE THAT bonds the components may also serve as a sealant or coating to provide protection from oils, chemicals, moisture, or a combination of these. In Figure 27.2, a room-temperature-curing adhesive is being applied to seal the ends of an antenna circuit in the handle of a marine radio. The same adhesive is also used to bond neoprene tubing and plastic-coated lead-in wires to aluminum at three points within the radio housing.

Thermal and Electrical Insulation

ADHESIVES CAN PROVIDE thermal or electrical insulating layers between the two surfaces being joined. For example, almost all mass-produced printed circuits use adhesive bonding. In this application, the adhesive used to bond the copper conductor to the base material has electrical characteristics similar to those of the base material. An adhesive may also serve as an insulator between adjacent conductors.

The addition of certain metallic or carbon fillers to adhesive formulations can make them electrically conductive. Some testing before acceptance should be done under simulated service conditions because corrosion may occur in some metal structures bonded with electrically conductive adhesives that are exposed to moisture. Metal powder additives can improve the thermal conductivity of adhesives.

Uniform Stress Distribution

JOINTS CAN BE designed to distribute the load over a relatively large bonded area to minimize stress concentrations. In wall panel construction, for example, metal skin sheets are bonded to metal or paper honeycombs, foamed polystyrene, or other core materials.

Smooth Surface Appearance

ADHESIVES CAN ENSURE smooth, unbroken surfaces without protrusions, gaps, or holes. A typical example is the vinyl-to-metal laminate widely used in the production of television cabinets, housing for electronic equipment, and automotive trim. Figure 27.3 shows adhesive-bonded truck

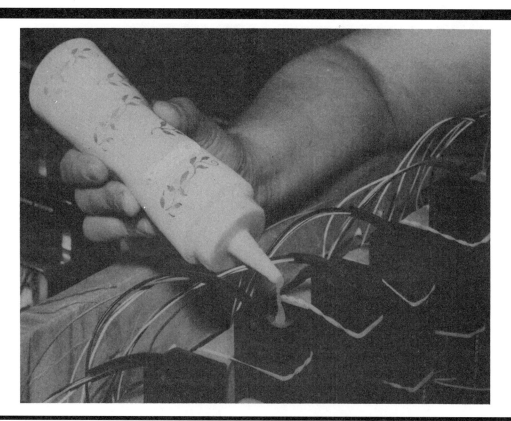

Figure 27.2–Application of an Epoxy Adhesive to Seal the End of a Marine Radio Antenna

door panels where broad, smooth areas are required. Hood and roof stiffeners on automobiles are adhesive bonded rather than resistance spot welded to the panels to avoid marks that would be susceptible to rusting and might require filling, grinding, and polishing prior to painting.

Good Vibration and Sound Damping

THE ABILITY OF FLEXIBLE adhesives to absorb shock and vibration gives the joints good fatigue life and sound-dampening properties. The use of adhesives rather than rivets has increased joint fatigue life by a factor of ten or more in some applications. A specific example is the improved fatigue life of adhesive bonded helicopter rotor blades.

A combination of adhesives and rivets for joints in very large aircraft structures has increased the fatigue life of joints from 2×10^5 cycles for rivets alone to more than 1.5×10^6 cycles for the bonded and riveted joints. The large bonded area also dampens vibration and sound.

Weight Savings

ADHESIVE BONDING MAY permit significant weight savings in the finished product by utilizing lightweight fabrications. Honeycomb panel assemblies, used extensively in the aircraft industry and the construction field, are excellent examples of lightweight fabrications. Typical panels are shown in Figure 27.4. Not only is the honeycomb core material bonded to the metal face sheets, the honeycomb core itself is generally adhesive bonded. Although weight reduction can be important in the function of the product, it may also provide considerable cost savings in packing, shipping, and installation labor.

Simplification of Design

ADHESIVES OFTEN PERMIT design simplification. In Figure 27.5, aluminum die cast pump sections have been adhesive bonded to a steel core. Previously, the part was cast as one piece of steel, but blow holes in the casting resulted in an excessive number of rejects. Redesigning to an adhesive-bonded assembly reduced the number of rejects to near zero.

Figure 27.3–Truck Door Panels of Aluminum Adhesive Bonded to Chipboard

Figure 27.4–Adhesive Bonded Honeycomb Panels

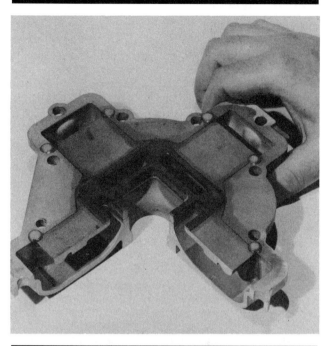

Figure 27.5—Aluminum Die Cast Pump Sections Adhesive Bonded to a Steel Core

PROCESS LIMITATIONS

ADHESIVE BONDING HAS certain limitations which should be considered in its applications, the most important of which are listed below.

Low Peel Strength

ADHESIVES WILL NOT support high peel loads above 250°F (120°C), not even those adhesives that have high tensile and shear strengths at temperatures as high as 300°F (150°C). For applications where high peel strength is essential, some mechanical reinforcement may be necessary.

Operational Temperature Ceiling

ADHESIVES, INCLUDING EPOXY-PHENOLICS that are designed for low creep at elevated temperatures, have an operational temperature ceiling of about 500°F (260°C). Some new high temperature adhesives derived from heat stable polyamides, polybenzimidazoles, and related compounds show promise for use at temperatures up to 700°F (371°C), but they are costly and difficult to process.

Equipment and Processing Costs

CAPITAL INVESTMENT FOR equipment and tooling to process components may be high when large bonding areas and special service requirements are involved. The benefits of an adhesive-bonded joint must be balanced against the cost of autoclaves, presses, tooling, and other special equipment needed to perform the bonding operation.

Process control costs may be higher than those for other joining processes. Surfaces must be properly cleaned, treated, and protected from contamination prior to bonding if the best bond durability is required. Surface preparation can range from a simple solvent wipe to multi-step cleaning, etching, anodizing, rinsing, and drying procedures which must be very carefully followed in critical structural bonding applications. Control of ambient temperature and humidity may also be necessary.

Curing Time

TO DEVELOP FULL strength, joints must be fixtured and cured at temperature for some time. On the other hand, mechanical fasteners provide design strength immediately and usually do not require extensive fixturing.

Bond Testing Procedures

NONDESTRUCTIVE INSPECTION METHODS normally used for other joining methods are not generally applicable to the evaluation of adhesive bonds. Both destructive and nondestructive testing must be used with process controls to establish the quality and reliability of bonded joints.

Limited Service Conditions

SERVICE CONDITIONS MAY be restrictive. Many adhesive systems degrade rapidly when the joint is both highly stressed and exposed to a hot, humid environment.

ADHESIVES

GENERAL DESCRIPTION

ADHESIVES MAY BE either thermosetting or thermoplastic. The principal ingredients in most adhesive formulations following are:

(1) A synthetic resin system
(2) An elastomer or flexibilizer
(3) Inorganic materials

Thermosetting Adhesives

THERMOSETTING RESINS ARE the most important materials on which metal adhesive formulations are based. Their properties can be modified for specific applications by the addition of modifying agents and fillers. Thermosetting adhesives harden or cure by chemical reactions that occur with the addition of a hardener or catalyst. Heat, pressure, radiation, or other energy can accelerate the curing rate. Once they cure, these adhesives cannot be remelted, and a broken joint cannot be rebonded by heating. Depending upon composition, thermosetting adhesives may soften or weaken at high temperature and ultimately decompose. Thermosetting resins, or adhesives, are frequently referred to as *structural adhesives*.

Thermoplastic Adhesives

THERMOPLASTIC RESINS ARE long chain molecular compounds that soften upon heating and harden upon cooling. They undergo no chemical change upon heating, so the cycle can be repeated. However, they will oxidize and decompose at excessively high temperatures. Many thermoplastic resins can also be softened at room temperature with organic solvents. They harden again as the solvent evaporates. Limited resistance to heat, solvents, and load-induced stresses makes thermoplastic resins generally unsuitable as structural adhesives. However, some thermoplastic resins or elastomers are combined with thermosetting resins such as epoxies and phenolics for improved flexibility, peel strength, and impact resistance.

Flexibilizers or elastomers are added to adhesive formulations to add resiliency, improve peel strength, and increase resistance to shock and vibration.

Inorganic materials are added as fillers to improve the mechanical and physical properties of the adhesives. Fillers can add greatly to the stability of bonded joints by reducing shrinkage and thermal expansion and by increasing the modulus of elasticity of the adhesive.

TYPES OF ADHESIVES

THE TYPES OF polymeric adhesives used to bond metals are listed in Table 27.1.

Solvent Type Adhesives

THE SOLVENT TYPE contact adhesives are predominantly elastomeric thermoplastics produced as solutions. They achieve their bond strength upon removal of the solvent. The liquid adhesive is applied to the adherend surfaces and, for metal adherends, time is allowed for the solvent to evaporate. The adhesive-coated surfaces are then joined under contact pressure. Sometimes heat is applied to fuse the coated surfaces after drying.

Hot-Melt Adhesives

THE HOT-MELT ADHESIVES are thermoplastics. After the adherends have been coated with adhesive and mated, heat

Table 27.1
Types of polymeric adhesives used to bond metals

Solvent

    Neoprene
    Nitrile
    Urethane (thermoplastic)
    Block copolymer
    Styrene-butadiene

Hot Melt

    Ethylene vinyl acetate
    Block copolymer
    Polyester
    Polyamide

Pressure Sensitive

    Block Copolymer
    Acrylic

Chemically Reactive

    Epoxy
    Phenolic
    Structural Acrylic
    Anaerobic
    Cyanoacrylate
    Urethane

and pressure are applied to the assembly. The joint is then cooled to solidify the adhesive and achieve a bond. These adhesives are not normally used for structural applications.

Pressure Sensitive Adhesives

PRESSURE SENSITIVE ADHESIVES are formulations that instantly provide a relatively low-strength bond upon the brief application of pressure. They may be applied to any clean, dry surface. Since they are capable of sustaining only very light loads because of retention of their flow characteristics, they are not considered structural adhesives. However, in recent years a number of pressure sensitive transfer tapes and double sided tapes have become available that exhibit high shear strengths. They are used in the construction of trucks and trailers.

Chemically Reactive Adhesives

THE CHEMICALLY REACTIVE adhesives consist primarily of thermosetting resins in liquid and solid forms, including films and tapes. They are activated either by the addition of a catalyst or hardener, or by the application of heat. Bond strength is achieved from the chemical reaction that takes place during the cure. Catalysts or hardeners may be incorporated by the adhesive manufacturer or may be added by the user just prior to application. Such formulations generally must be used within a prescribed period of time after mixing to avoid premature setting.

STRUCTURAL ADHESIVES

THE ULTIMATE OBJECTIVE of a structural adhesive is to create a bond that is as strong as the materials it joins. Since this goal is not always attainable, a structural adhesive can be defined as one that is used to transfer required loads between adherends in a structure for its life expectancy when exposed to its service environment.

Although structural bonding has been successfully used for many aerospace applications since the 1950's, concerns about the long-term durability of structural adhesive bonds have limited the widespread use of this joining method. The combination of stress, even as low as 20 percent of the initial adhesive strength, and exposure to hot, humid environments, can cause significant degradation of bond performance, sometimes leading to failure. Many of the factors affecting the durability of adhesive joints are known and include the type of adhesive, the nature of the adherends, the surface preparation before bonding, and the service conditions.

There are several mechanisms by which the performance of an adhesive can be affected by the presence of moisture. The mechanical properties can change as water is absorbed due to the plasticizing action of water. Swelling stresses can lead to the formation of crazes and microcracks. In unfavorable situations, hydrolysis of the adhesive can occur. Water can also displace the adhesive from the metal surface and thus induce interfacial disbonding. Finally, water may hydrate and weaken the metal oxide surface layer of the adherend. The relative importance of each of these factors is, in general, unknown, as are the details of the various mechanisms. For these reasons it is not usually possible to predict the durability of a given adhesive joint.

Phenolic Resins

PHENOLIC RESINS ARE modified with thermoplastics or elastomers for structural adhesive applications. These modified phenolics are available as solutions in organic solvents and also as films, both supported and unsupported. Such adhesives feature high peel strengths, and tensile and shear strengths in the range of 3000 to 5000 psi (20 to 34 MPa).[2]

Epoxy Resins

THE EPOXY RESINS combine the properties of excellent wetting action, low shrinkage, high tensile strength, toughness, and chemical inertness to produce adhesives noted for their strength and versatility. Unlike phenolic adhesives, epoxies do not form volatile products during curing. They can be applied in liquid form without a solvent carrier. Because of this, volatile entrapment is minimized.

Only low pressure is necessary to maintain intimate contact between the adherends during bonding, resulting in greatly simplified equipment requirements.

Epoxy adhesives are available as free-flowing liquids, films, powders, stocks, pellets, and mastics. This variety of forms permits considerable latitude in the selection of application technique and equipment. Fillers or plasticizers may be added to minimize stresses that can develop when the adhesive and adherends have different coefficients of thermal expansion.

The wide choice of hardeners available for epoxy formulations offers curing cycles ranging from a few seconds at elevated temperatures to several minutes or hours at room temperature. However, the heat-resistant formulations require high temperature cures.

Unmodified epoxy-based adhesives show high shear and tensile strengths, but tend to be brittle and thus perform poorly in cleavage and peel. High peel strength can be obtained by flexibilizing the epoxy resin, but this reduces the modulus of the adhesive. Specially formulated "toughened" epoxy adhesives, in which a modifying rubber is present as a distinct phase evenly distributed throughout the cured adhesive, are now available. These adhesives have high shear strengths together with high peel strength and impact resistance. The phase-separated structure imparts high fracture toughness, or resistance to crack propagation, to the adhesive.

Anaerobic Adhesives

ANAEROBIC ADHESIVES ARE shelf-stable, ready-to-use formulations that cure at room temperature. Their cure is inhibited by the presence of air (oxygen) in the package and during application. Once the joint is assembled and air excluded from the liquid adhesive, curing begins. The major use of these adhesives is for sealing and securing threaded fasteners, although formulations are available for applications requiring high bond strength.

Cyanoacrylate Adhesives

CYANOACRYLATE ADHESIVES ARE also shelf-stable and cure rapidly at room temperature when placed in contact with most surfaces. The cure is catalyzed by traces of basic compounds present on the surface. Water, which is adsorbed on most materials, often acts as the effective catalyst. Cyanoacrylates work best when the surfaces are well matched, so that a controlled, even bond is possible. Until recently these adhesives found only limited use in industrial applications because of their relatively poor moisture and heat resistance. However, some of the newer adhesives appear to perform better in these respects.

Acrylic Adhesives

ACRYLIC STRUCTURAL ADHESIVES cure by a chain-growth free-radical mechanism that allows cure times from about one minute to a few hours. The adhesives are generally

---

2. The mechanical properties stated for these adhesive systems pertain to bonded structures which have not been stressed prior to testing and are stored in a low or normal relative humidity condition; i.e., less than 70 percent relative humidity.

supplied as two components that are mixed just prior to the assembly of the joint. Some formulations are available that require no mixing; cure is initiated on contact of the adhesive with a surface that has been coated with a special primer/activator.

Acrylics adhere well to a variety of metals and engineering plastics and are tolerant of surface contamination. Formulations are available that form strong bonds to metals whose surfaces have not been cleaned of mill oils and drawing compounds. However, the most durable bonds are produced when clean surfaces are used.

An epoxy-phenolic resin cured with dicyandiamide is an adhesive that performs well in the 400 to 500°F (205 to 260°C) temperature range. It will produce a bond having good shear and creep properties at 500°F (260°C), but relatively poor peel resistance when cured in a representative cycle of 2 hours at 350°F (177°C) and 150 psi (1 MPa) pressure. Better "all-around" strength properties are often obtained by using this adhesive in conjunction with a specially formulated primer having a high peel strength. The elastomer-modified phenolics, which require curing at temperatures from 300 to 350°F (149 to 177°C) and pressures up to 100 psi (690 kPa), offer good resistance to heat and water in service.

Other classes of high-temperature resistant polymers have been evaluated as high-temperature adhesives. Of these the polyamides are available as structural adhesives; however, applications for such adhesives are limited at present owing to the difficult processing conditions.

FORMS AND APPLICATION METHODS

INDUSTRIAL ADHESIVES ARE available in a number of forms:

(1) Liquids, ranging in viscosity from free-flowing to thick syrups
(2) Pastes
(3) Mastics
(4) Solids
(5) Powders
(6) Supported and unsupported films

The method used for application of a particular adhesive should be selected after careful consideration of these factors:

(1) Available forms of the selected adhesive
(2) Methods available for applying the various forms
(3) Joint designs and order of assembly
(4) Production rate requirements
(5) Equipment costs

Adhesives may be applied with rollers, brushes, caulking guns, trowels, spray guns, or by dipping. The form of the adhesive and method of application must be compatible.

Liquid Adhesives

FOR SMALL ASSEMBLY work, liquid adhesives are commonly applied by brush, short-napped paint roller, or dipping. The more viscous liquids are often applied by trowel, extrusion gun, or plastic squeeze bottle. Polyethylene nozzles of bottle or tube containers should not be rubbed across prepared surfaces such as etched aluminum. This action may deposit a waxlike coating to which the adhesive will not adhere. Small applicators, resembling a ballpoint pen or a hypodermic needle, may be used to deposit very narrow glue lines for spot application. The silk screen process is also useful for applying an adhesive to selected areas. Automatic dispensing machines, which simplify the proportioning and the mixing of many two-component formulations, are also available.

For large areas such as curtain wall panels, liquid adhesive may be applied by spraying, flow coating, roller coating, or troweling. Depending upon the application method, consideration must be given to the viscosity and working (pot) life of the adhesive formulation.

Paste and Mastic Adhesives

PASTE AND MASTIC adhesives may be applied with a smooth or serrated trowel, knife roller, or an extrusion device.

Some paste formulations contain a thixotropic additive that inhibits sag or flow during application and cure. This feature permits their use on vertical or overhead surfaces and may eliminate or significantly reduce the need for special cleanup operations.

Solid Adhesives

ONE METHOD FOR applying a solid adhesive is to first heat the substrate to a temperature slightly above the melting point of the adhesive and then add the adhesive to the hot surface as it melts. Some rod and powder forms of epoxy adhesives are applied in this manner. Specially developed flame spray guns can also be used, but this method may require powders with a particle size within narrow tolerances. Also, care must be taken to avoid overheating the adhesive during application.

Film Adhesives

ADHESIVES IN FILM or tape form are extremely simple to use and produce a bond line having relatively constant thickness and coating weight per unit area. These are important factors in most bonding applications. Films made from adhesives that are thermoplastic, thermosetting, or pressure sensitive are supplied in rolls or sheets that can be blanked or cut to the required shape with scissors. Film adhesives are particularly useful for bonding large areas such as honeycomb sandwich panels. Special films are available for use as adhesive backing on items such as nameplates and decals.

Generally, the film adhesive is placed in position between the adherends and activated with heat or solvent. In the case of pressure-sensitive film, pressure is applied to accomplish bonding. For applications demanding high strength at elevated temperatures, the films usually require both heat and pressure to create the bond.

Solvent reactive and pressure sensitive films are intended primarily for bonding large sheets where only nominal pressure is required. These types of films do not need special heating equipment. They are particularly useful for short-run production and for bonding parts at room temperature.

Duplex bonding films, which combine the properties of elastomeric adhesives and epoxies, are sometimes used for honeycomb sandwich constructions. The elastomeric side, usually a nitrile phenolic, is bonded to the facing to provide peel resistance. The epoxy side forms fillets around the cell walls, and this increases the effective bonding area and the resulting joint strength.

PRIMERS

CERTAIN SERVICE CONDITIONS may require the use of a primer for improved corrosion resistance, flexibility, shock resistance, or peel resistance of the adhesive bond. Primers may also be used to wet or penetrate the substrate or to protect a treated surface of a substrate prior to the application of adhesive.

Most primers are low-viscosity solutions commonly applied by spraying. Brush applications may be satisfactory when relatively small areas are to be coated. In some cases, roll coating or dipping may be employed. Several primer coats may be required to build up the desired thickness, particularly if the adherends are porous. Air drying and full or partial curing are generally necessary prior to further processing.

ADHESIVE SELECTION

THE SELECTION OF the proper adhesive for production depends basically upon the answers to four key questions:

(1) What materials are being joined?
(2) What are the service requirements?
(3) What method of adhesive application is most suitable?
(4) Are bonding costs competitive with other joining methods?

The service requirements of the completed assembly must be studied thoroughly. Several factors to be considered in describing the bonding application:

(1) Type of loading
(2) Operating temperature range
(3) Chemical resistance
(4) Weather and environmental resistance
(5) Flexibility
(6) Differences in thermal expansion rates

(7) Odor or toxicity problems
(8) Color match

Once the service requirements are known, adhesive systems with good durability potential can be selected. The desired form of the adhesive and method of application can then be chosen based on availability of production equipment and scheduling requirements.

In adhesive selection, the tendency to over-design must be avoided. Requirements for higher strength or greater heat resistance than is actually required for the specific application may exclude from consideration many formulations that are adequate for the job, and perhaps less costly or easier to handle in production.

In adhesive selection for a specific application, there are certain physical properties of the adhesive itself and mechanical properties of bonded joints that should be considered. These properties pertain to the behavior of an adhesive from the time that it is made until the bond is accomplished, as well as its performance in service. Table 27.2 lists some of the properties and the applicable ASTM Standards.

For most applications several candidate adhesives will usually be selected based on the above considerations. The final choice is generally made after a test program to determine the suitability of the adhesives for the particular application. Testing may involve various laboratory specimens or a prototype of the complete assembly. In either case, testing should include some estimate of the long-term durability of the adhesive bond in the service environment. The adhesive supplier will generally provide help in the selection and evaluation of adhesives.

Table 27.2
ASTM Standards for Determining Adhesive Properties

Property	ASTM Designation(s)
Physical	
Aging	D1151, D1183, D2918, D2919, D3236, D3762
Chemical	D896
Corrosivity	D3310, D3482
Curing rate	D1144
Flow properties	D2183
Storage life	D1337
Viscosity	D1084
Volume resistivity	D2739
Mechanical	
Cleavage strength	D1062, D3433
Creep	D1780, D2293, D2294
Fatigue	D3166
Flexural strength	D1184
Impact strength	D950
Peel	D903, D1781, D1876, D2918, D3167
Shear	D1002, D2182, D2295, D2919, D3528, D3983
Tensile	D897, D1344, D2095

JOINT DESIGN

TO INCORPORATE AS many of the advantages of adhesive bonding as possible, joint design should be a part of the early stages of product planning. If bonding is being considered as part of a redesign program, structural adhesives should not be substituted directly for other joining methods. The joint should be redesigned to take advantage of adhesive bonding.

Although the primary objective is a strong assembly capable of meeting service requirements, proper joint design can often lead to other cost-saving benefits. Through good design, it may be possible to achieve satisfactory results with an economical adhesive formulation, to utilize a simple bonding process, and to minimize the quality control steps needed to ensure reliability.

Joint design often influences the form and characteristics of the bond line. The design must provide space for sufficient adhesive and a means of getting the adhesive into the joint area.

When considering a new design or a redesign for adhesive bonding, three rules should be observed:

(1) The design loading should produce shear or tensile loading on the joint; cleavage or peel loading should be minimized.

(2) The joint design should ensure that the static loads do not exceed the adhesive-plastic strain capacity.

(3) If it is anticipated that the joint will be exposed to low cyclic loads, the joint overlap should be increased sufficiently to minimize the possibility of creep in the adhesive.

These rules may be difficult to achieve in practice. Some stress concentrations are unavoidable, and it is difficult to design a joint that will be stressed in one mode only.

The four main types of loading are illustrated in Figure 27.6. An adhesive bonded joint performs best when loaded in shear: that is, when the direction of loading is parallel to the plane of the faying surfaces. With thin-gage metal bonds, joint designs can provide large bond areas in relation to the metal cross-sectional area. This makes it possible to produce joints that are as strong as the metal adherends.

The relationship between joint strength and overlap length for a double-lap shear joint is shown in Figure 27.7. The joint strength and overlap distance are proportional up to some limit (point A on Figure 27.7). Then the unit increase in strength decreases as the overlap distance increases. Beyond some overlap (point B), the failure load does not change significantly with overlap distance.

SHEAR LOADING

FIGURE 27.8 INDICATES the shear stress distribution across lap joints caused by load P with short, medium, and long overlaps. With short overlap, Figure 27.8(A), the shear

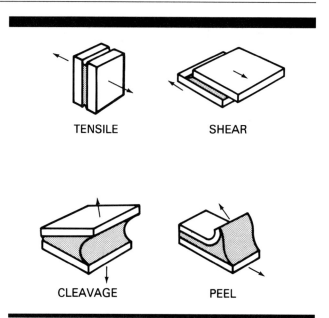

TENSILE **SHEAR**

CLEAVAGE **PEEL**

Figure 27.6–Four Principal Types of Loading

stress is uniform along the joint. In this case, the joint can creep under load with time and failure may occur prematurely. When the overlap exceeds some value, the adhesive at the ends of the joint carries a larger portion of the load than the adhesive at the center. Therefore the shear stress at the center is lower, as shown in Figure 27.8(B), and the likelihood of creep is decreased. With long overlap, Figure 27.8(C), the portion of the joint overlap that sees low shear stress is a greater percentage of the total, and creep poten-

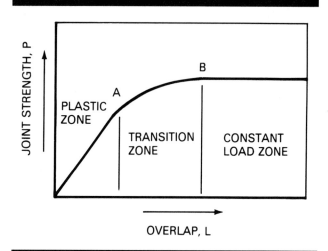

Figure 27.7–Relationship Between Joint Strength and Overlap in Shear

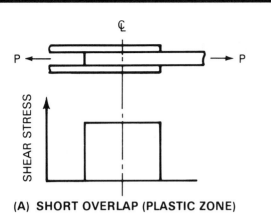

(A) SHORT OVERLAP (PLASTIC ZONE)

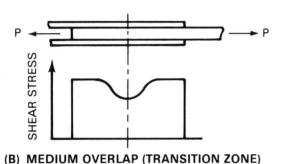

(B) MEDIUM OVERLAP (TRANSITION ZONE)

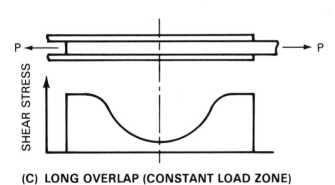

(C) LONG OVERLAP (CONSTANT LOAD ZONE)

Figure 27.8–Change in Shear Stress Distribution with Overlap for Constant Load, P

tial is minimized. The joint overlap for minimum creep will depend upon the mechanical properties of the base metal, the adhesive properties and thickness, the type of loading, and the service environment.

PEEL LOADING

DIFFICULTIES MAY ARISE when cleavage or peel-type loading is present. Cleavage loading produces nonuniform stress across the joint, and this causes failure to initiate at the edge of the adhesive. Obviously, such a joint is considerably weaker than the same bonded area under uniform shear or tensile stress.

The situation is even more critical when the adhesive is subject to peel-type loading. A very narrow line of adhesive at one edge of the joint must withstand the load. Peel loading produces failure at only a fraction of the tensile load needed to rupture a bond of the same area.

As noted earlier, undirectional loading is rarely accomplished. Most joints are subjected to loads which combine cleavage or peel loading with tension or shear stress in the bond. One example is a straight butt joint that is designed to be stressed strictly in tension but is subjected to a slight bending moment that creates a cleavage load. Another example is a single lap joint that is designed to withstand expected shear stress but sees cleavage or peel loads when the joint rotates slightly as the load forces tend to align, as shown in Figure 27.9. These problems can usually be minimized by selecting an adhesive designed to carry the type of loading expected and by employing the proper joint design.

Several of the more common types of joints for sheet metal are shown in Figure 27.10. A butt joint design, shown in Figure 27.10(A), is not recommended. Cleavage loading may develop if the applied loading is eccentric. A scarf joint [Figure 27.10(B)] is a better design because the bonded area can be greater than with a butt joint. Cleavage stress concentrations at the edges are minimized by the tapered edges of the adherends. Although widely used in wood bonding, this configuration is difficult to make in metals with regard to alignment and pressure application during curing.

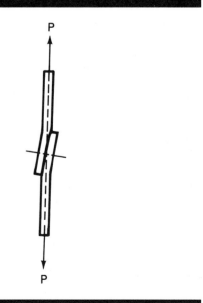

Figure 27.9–Lap Joint Rotation as a Result of Loading

The single lap joint, Figure 27.10(G), is probably the most commonly used type and is adequate for many applications. The beveled lap joint, Figure 27.10(H), has less stress concentration at the edges of the bond because of the beveled edges. The thin edges of the adherends deform as the joint rotates under load, and this minimizes peel action.

If joint strength is critical and the components are thin enough to bend under load, a joggle lap joint, Figure 27.10 (I), is better, The load is aligned across the joint and parallel with the bond plane, thus minimizing the possibility of cleavage loading.

If sections to be bonded are too thin to permit edge tapering, a double-strap joint, Figure 27.10(E), will give good results. The best design is the beveled double-strap joint, Figure 27.10(F), with its tapered straps.

Adhesive bonding can be used to advantage on extruded, cast, or machined components. The butt joints shown in Figure 27.11 can easily be incorporated into machined or extruded shapes that are to be assembled by adhesive bonding. The tongue and groove joint not only aligns the load-bearing interfaces with the plane of shear stress but also provides good resistance to bending. The landed-scarf tongue and groove joint offers production advantages. Its configuration automatically aligns the parts to be mated, controls the length of joint, and establishes the thickness of the glue line. It is a good design for an assembly that will see high compressive forces, and it offers a clean appearance.

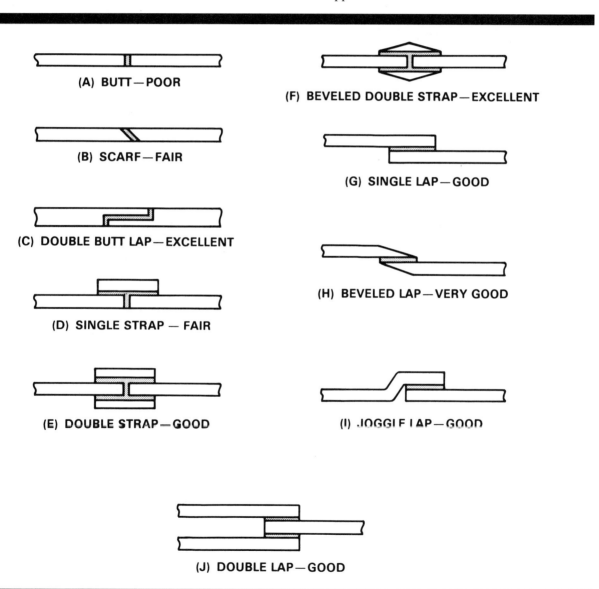

(A) BUTT—POOR

(B) SCARF—FAIR

(C) DOUBLE BUTT LAP—EXCELLENT

(D) SINGLE STRAP — FAIR

(E) DOUBLE STRAP—GOOD

(F) BEVELED DOUBLE STRAP—EXCELLENT

(G) SINGLE LAP—GOOD

(H) BEVELED LAP—VERY GOOD

(I) JOGGLE LAP—GOOD

(J) DOUBLE LAP—GOOD

Figure 27.10—Adhesive Bonded Joint Designs for Sheet Metal

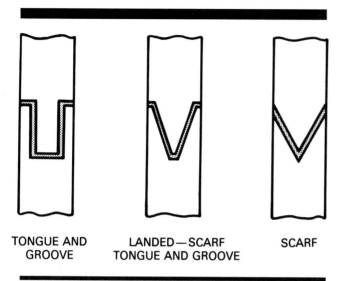

TONGUE AND
GROOVE

LANDED—SCARF
TONGUE AND GROOVE

SCARF

Figure 27.11–Adhesive Bonded Butt Joint Designs for Machined or Extruded Shapes

Corner and T-joint designs are shown in Figure 27.12. The use of beveled or tapered reinforcing members requires a cost analysis to determine if the improved joint properties are justifiable. Joints requiring machined slots or complex corner fittings are seldom of interest in sheet metal designs.

Adhesive bonding is also useful for tube joints, examples of which are shown in Figure 27.13. Large bonded areas give very strong joints with clean appearance. Processing may be complicated with some designs. During assembly of designs shown in Figures 27.13(A) and (B), the adhesive may be pushed out of the joint. The design shown in Figure 27.13(C) partially overcomes this problem. Adhesive in the corners is forced into the joint by a positive pressure filling action during assembly. Tapered or scarfed tubular joint designs as illustrated in Figures 27.13(D), (E), and (F), will produce a positive pressure on the adhesive during assembly to completely fill the gap, but they are costly to produce. The design pictured in Figure 27.13(G) shows a tubular sleeve joint that can be filled by injecting the adhesive under a positive pressure through a hole in the sleeve. This technique results in completely filled and bonded joints at reasonable fabrication costs.

With the availability of computers and the development of analytical mathematical models, it is now possible to optimize joint design by taking into account the geometry of the adherends and the properties of the adhesive. The thick adherend shear test (ASTM D 3983) provides the shear modulus, the elastic shear stress limit, and the asymptotic shear stress of the adhesive which are useful in this respect.

It may be possible to eventually include the effects of environmental exposure into the mathematical modeling of joint performance. This will allow the durability of adhesively bonded structures to be predicted.

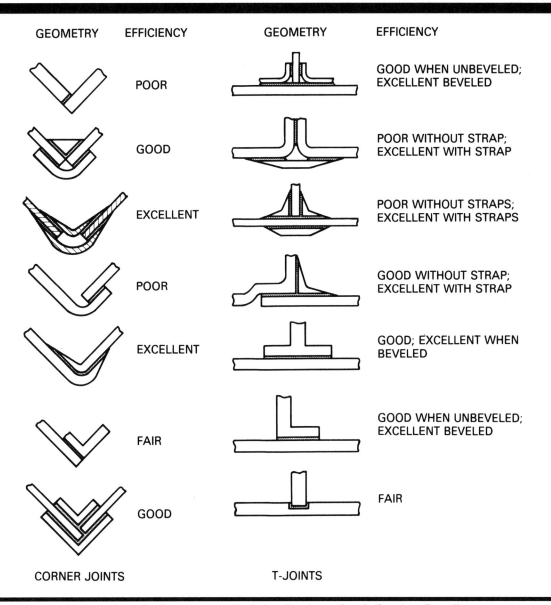

GEOMETRY	EFFICIENCY	GEOMETRY	EFFICIENCY
	POOR		GOOD WHEN UNBEVELED; EXCELLENT BEVELED
	GOOD		POOR WITHOUT STRAP; EXCELLENT WITH STRAP
	EXCELLENT		POOR WITHOUT STRAPS; EXCELLENT WITH STRAPS
	POOR		GOOD WITHOUT STRAP; EXCELLENT WITH STRAP
	EXCELLENT		GOOD; EXCELLENT WHEN BEVELED
	FAIR		GOOD WHEN UNBEVELED; EXCELLENT BEVELED
	GOOD		FAIR

CORNER JOINTS   T-JOINTS

Figure 27.12—Corner and T-Joint Designs for Adhesive Bonding

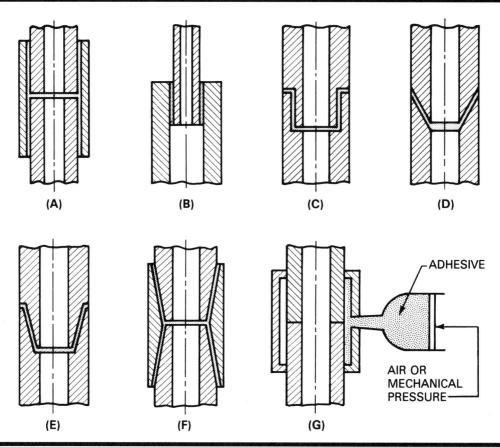

Figure 27.13—Tubular Joint Designs for Adhesive Bonding

SANDWICH CONSTRUCTION

A MAJOR USE of structural adhesive bonding is in sandwich construction. First used in the aircraft industry to meet the demand for a high stiffness-to-weight ratio, sandwich panels are now widely used throughout industry. Their characteristics make them equally valuable in the fabrication of walls, truck bodies, refrigerators, cargo pallets, and a great many other commercial applications

The sandwich skins, or facings, behave approximately as membranes stabilized by a lightweight core material that transmits shear between the skins. Basically, a high stiffness-to-weight ratio is achieved by placing the two load carrying skins as far from the neutral axis as possible with the lightweight core bonded to them. In a sense, the face sheets perform the same functions as the flanges of an I-beam, and the core performs as the beam web.

The selection of skin and core materials for sandwich construction is dictated by service requirements and economic considerations, including the cost of fabrication and materials. Many different skin and core materials may be used as sandwich components. Sheets of metal, plastic, wood, and fire-resistant inorganics are all commonly used as skin materials. Core materials are of the following three basic types:

(1) Solid types such as hardwood, balsa, or metal
(2) Honeycomb or corrugated types made of various materials, typically metal foils, resin-impregnated paper, or reinforced plastics
(3) Open- or closed-cell foamed materials such as polystyrene, polyurethane, polyisocyanurate, and glass

Metal foil honeycomb cores are available with and without perforations in the cell walls. The perforations permit equalization of pressure in the completed panel and also vent gases produced by some types of adhesives during bonding. Cores are also available in truss and waffle configurations that are produced by corrugating and folding or by pressing the material to shape between matched dies.

Expanded plastic cores may be supplied in a preformed state that requires a separate bonding step, or the plastic may be foamed in place and thus simultaneously formed and bonded to the skins of the sandwich.

Special consideration should be given to selecting the material combination that will produce the optimum composite structure for a specific application. For example, a honeycomb core between thin metal skins can provide high strength and low weight; or a foamed plastic core can be used to assure high thermal insulation with almost any skin material.

Several basic types of adhesives with very similar properties are used in sandwich manufacture. These basic formulations are modified to meet specific environmental conditions. A careful study of the design, environmental factors, and material properties should be made before selecting the adhesive.

SURFACE PREPARATION

FREQUENTLY, THE WEAKEST link in an adhesive-bonded joint is the interfacial bond between the adhesive and the adherents. Exactly the right adhesive may have been selected for the job, the joints may have been designed properly, and correct application and cure procedures and equipment used. However, if the adherents are not properly cleaned and prepared to receive the adhesive, the bond will show less than optimum performance and the environmental resistance will usually be significantly reduced. Surfaces should be cleaned by procedures that ensure that the bond between the adhesive and metal surfaces is as strong as the adhesive itself. Failure should occur in the adhesive rather than at the bond line when the joint is tested under simulated service conditions.

The degree of surface preparation depends chiefly upon the nature of the material and, to some extent, on the service requirements, bonding cycle, and the probable nature of contaminants. For some less critical applications, a solvent wipe or washing in a detergent solution may prove to be adequate. Care should be taken, however, to remove all cleaning agents from the surface by rinsing and drying thoroughly prior to the application of the adhesive. However, for the best joint performance, the surfaces must be prepared using procedures that will provide the best bond between the adhesive and adherent.

It is equally important to avoid recontamination of the clean surfaces during processing. Components should be handled with clean gloves, tongs, or hooks, and all contact with the bonding area should be avoided. Priming, bonding, or a combination of these should be accomplished as soon after surface cleaning as possible. In the interim, parts should be stored in a clean, dry place.

METAL PREPARATION

METAL SURFACES MAY be cleaned by chemical means or mechanical abrasion. These are the two basic methods, but some variation in procedures is advised by various adhesive manufacturers. Metal faying surfaces should be free from oxide scale, deep scratches, burrs, and other irregu-larities. Cleaning should be done after machining, heat treating, welding, sand-blasting, abrading, deburring, polishing, or similar treatment that might leave foreign material on the metal surface.

The surface preparation procedure should be reproducible, easily controlled, and production oriented if it is to be economical. In addition, it should satisfy the following requirements:

(1) Remove all contaminants from the surface
(2) Make the surface chemically receptive to the adhesive or primer and give satisfactory wetting characteristics
(3) Prevent poorly adhering or low-strength compounds from forming on the adherents
(4) Remove minimum amounts of metal
(5) Avoid embrittlement or corrosion, or forming a surface prone to environmental attack

Surface preparation methods for aluminum alloys, stainless steels, carbon steels, magnesium alloys, titanium alloys, and copper alloys are given in ASTM D2651, *Standard Recommended Practice for Preparation of Metal Surfaces for Adhesive Bonding*. The best corrosion-resistant surface preparation for aluminum alloys is stated in SAE *Aerospace Recommended Practice* (ARP) No. 1524.

To obtain optimum strength characteristics on aluminum, surface preparation is usually done in steps that include:

(1) Vapor degreasing
(2) Drying
(3) Chemical cleaning such as an alkaline precleaning and rinse followed by an acid etch
(4) Anodizing
(5) Careful rinsing in clean water
(6) Air drying, forced drying, or a combination of these

The vapor degreasing solution should be checked periodically for oily contaminants and decomposition products, and the solvent changed when necessary. Solvent degreasing is rarely recommended, but if it is used, the adherents should only be wiped with clean cloths or disposable tissues.

The recommended composition of the bath for chemical cleaning varies from one manufacturer to another. It depends upon the kind of metal being treated. Forced drying in an oven after water rinsing is preferred since this reduces the possibility of recontamination from dust and impurities when air drying at room temperature. Forced drying temperatures must be selected judiciously. Temperatures that are too high can adversely affect the surface condition.

Mechanical abrading is sometimes used to clean metal surfaces and increase the effective bonding area by roughening. Grinding, filing, wire brushing, sanding, and abrasive blasting are some methods. Abrasive cleaning is not usually as effective as chemical treatment, but it may be adequate in certain applications. It is best to clean the surface with solvent before abrading so that any contaminants are not ground into the surface. It is also desirable to clean the surface with solvent after the abrading to remove contaminants and debris.

For magnesium, one of the corrosion-inhibiting pretreatments developed specifically for this metal should be used. It is possible to clean magnesium mechanically, but care must be taken to prevent the exposure of any magnesium dust to an open flame.

Mechanical pretreatments should not be used for structural magnesium joints or for joints that will be subjected to severe environments. Some treatments for magnesium have been found to be effective only with certain adhesives. Certain corrosion-inhibiting pretreatments produce a weak surface layer that fails before the adhesive; these are not satisfactory for structural applications. If such pretreatments are used, the joints should be reinforced with mechanical fasteners.

A thorough check of treating solution composition should be made periodically. Special consideration should be given to the rate of metal processing over a given period of time. Failure to control the concentration of strong acid or alkali solutions may result in excessive metal loss. Aluminum and magnesium alloys are more reactive than stainless steel and titanium alloys, and exposure of these metals to high concentrations of acid or alkali may affect adhesion characteristics with certain kinds of adhesives. The time between composition checks depends upon the rate of exhaustion of the treating solution.

PREPARATION OF OTHER MATERIALS

RIGID PLASTICS CAN be lightly sanded to reduce gloss and remove mold-release compounds. Then they should be wiped and flushed with an oil-free solvent. Certain types of plastics, such as fluorocarbon isomer and polyethylene, are difficult to bond and may require chemical treatment.

Glass is easily cleaned by wiping with a suitable solvent. Joint durability can be greatly enhanced, particularly in moist environments, by first cleaning with a laboratory glassware cleaning solution or with a 30 percent hydrogen peroxide solution and then priming the surfaces with a silane finish.

INSPECTION OF PREPARED SURFACES

THE AFFINITY OF a clean metal surface for water is the most common test for a chemically clean surface. It is called the *water-break test*. If clean water from the rinse bath spreads smoothly over the metal surface as it runs off, it indicates that the surface is clean. If it collects in droplets, there is probably a thin film of oil on the surface. The oil should be removed completely, and the water-break test repeated.

If water drops placed on a flat, dry, treated surface spread rapidly and uniformly, this indicates that the surface is free of oil or grease. If the contact angle of the water drop is low (10 degrees or less), the surface has been cleaned adequately to remove greases, oils, and other nonpolar contaminants. When contact angle measurement is used as a quality control device for cleanliness, inspection should be made immediately after the metal has been dried. If the water remains in droplet form, the surface has not been suitably prepared.

ASSEMBLY AND CURE

THE PROCEDURES AND equipment used for assembly and cure of bonded components depend upon:

(1) The type of adhesive used
(2) The type, size, and configuration of the assembly
(3) The service requirements of the completed assembly

ASSEMBLY

DRYING TIME IS an important factor when solvent-dispersed adhesives are used. Since this time varies with different formulations, it is essential that the adhesive manufacturer's recommendations be followed. Solvent evaporation rate may be increased by moderate heating with infrared lamps, a hot-air oven, or other methods.

If there is sufficient porosity in one component to allow the solvent to escape, the parts may be mated during the drying time. In any case, the assembly must not be heated for curing until the solvent has evaporated. It is also essential that coated parts be mated before the tack range of the adhesive has expired.

Parts may be mated immediately after they are coated with chemically reactive adhesives. Mating may be delayed, but it must be done before the adhesive starts to "body" or thicken excessively.

FIXTURING

PROVISION SHOULD BE made for positioning the components for mating and holding them in place while the adhesive cures or sets. Assembly fixtures are frequently used for positioning. They may be simple jigs or self-contained equipment with provision for applying pressure or heat, or both. The fixture design depends upon the amount of heat and pressure needed to cure the adhesive and the size and configuration of the assembly.

Fixturing is particularly important when a contact adhesive is used. Care should be taken to align the parts accurately before they are mated since a strong bond is created instantly upon contact of the two coated surfaces. The use of release sheets, untreated kraft paper for example, is often helpful to avoid premature contact. Positioning may not be so critical with some formulations of less aggressive tack if the assembly can be slightly adjusted after mating without damage to the bond.

The fixture should properly position the parts to meet assembly tolerances and glue line thickness requirements. It should be lightweight for ease of handling and heat transfer. A heavy fixture presents a large heat sink which may retard heating and cooling rates. This may be detrimental for some adhesive systems. Nevertheless, the fixture must be strong enough to maintain dimensions under the curing conditions for the assembly. The expansion rate of the fixture material should nearly match that of the assembly to minimize part distortion and subsequent stressing of the adhesive.

Pressure-sensitive tape may be used to hold parts in position if it can withstand the curing temperature. Tapes are particularly useful with epoxy formulations that cure at room temperature or slightly warmer and require only moderate pressures.

Adhesive bonding may be combined with resistance welding or mechanical fasteners to improve the load carrying capacity of the joint. The adhesive is applied to the adherents first. Then the components are joined together with spot welds or mechanical fasteners to hold the joints rigid while the adhesive cures. Figure 27.14 illustrates typical design combinations. These techniques significantly reduce or eliminate fixturing requirements and decrease assembly time when compared to conventional adhesive bonding methods.

PRESSURE APPLICATION

WITH CERTAIN ADHESIVE formulations, it is necessary to apply and maintain adequate pressure during cure to:

(1) Produce a uniformly thin glue line over the entire bonded area for optimum strength characteristics
(2) Facilitate flow or spreading of viscous adhesives
(3) Counteract any internal pressure caused by the release of volatiles
(4) Overcome minor imperfections in the faying surfaces
(5) Compensate for solvent loss and dimensional changes

Pressure may be applied to the joint by several methods which include the following:

(1) Dead weights such as bags of sand or shot
(2) Mechanical devices such as clamps, wedges, bolts, springs, and rollers
(3) Inflated tubes

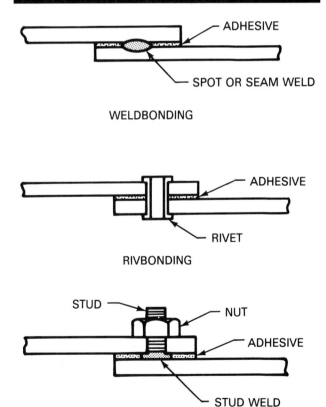

Figure 27.14–Adhesive Bonding in Combination with Resistance Welds and Mechanical Fasteners

(**4**) Air pressure bearing on the assembly located in a flexible, evacuated bag

(**5**) Mechanical or hydraulic presses

(**6**) Autoclaves

Inflated tubes are used in conjunction with a rigid backing fixture. When inflated, the tube presses uniformly along the bond line. Ambient air pressure is adequate for some applications. It is applied by enclosing the assembly in a thin, air-tight bag and then evacuating the bag. Autoclaves are used in a similar fashion in that the assembly is placed in a thin, gas-tight bag vented to ambient pressure. The bagged assembly is placed in an air-tight chamber which then is pressurized to several atmospheres. The pressure forces the bag to conform to the part and transmits the pressure to the assembly.

Phenolic-based adhesives generally require curing pressures in the 300 psi (2070 kPa) range, although adequate bonds may be attained with pressures as low as 50 psi (345 kPa).

Flat panels coated with a neoprene contact bond adhesive are generally mated by passing them through rollers under as much pressure as the components will withstand without crushing. A weighted hand roller or other pressure device can also be used.

For sandwich panel fabrication, the upper pressure limit is governed by the compressive strength of the core material. The lower limit depends upon the minimum requirements of the adhesive formulation. For sandwich panels containing solid inserts or edgings, special fixtures may be used to apply higher pressure at the specific locations.

Throughout the curing cycle, pressure should be as uniform and constant as possible over the entire bond area. If necessary, irregular surfaces can be built up with pads of compressible material. In some cases, soft rubber pads are used to compensate for variations in the dimensions of sheet material and fixtures. The mass of such materials should be minimized to avoid heat sink and insulating effects. Matched tooling is not often used for curved panels because of the high cost. A better method is to use a male or female tool in conjunction with the vacuum bag or autoclave technique.

CURING TEMPERATURE

SINCE VARIATIONS IN the thermal conductivity of the components influence the amount of heat transmitted to the adhesive layer, curing temperature should be measured at the glue line. Otherwise, the adhesive may not develop the desired properties for the application because of improper curing temperature.

Most phenolic-based structural adhesives require curing at elevated temperatures, generally from about 300 to 400°F (150 to 205°C), for periods ranging from 0.5 to 2 hours. Many one-part epoxy adhesives can be cured at temperatures as low as 250°F (120°C). A great number of two-component epoxy systems cure at room temperature; however, their properties are generally better when they are cured at elevated temperatures.

When neoprene contact-bond adhesives are used, the adhesive-coated surfaces are frequently heated during the drying cycle and mated under pressure while still warm. When design requirements are not so stringent, the adhesive may be dried and the components mated at room temperatures. Joint properties tend to be more variable, however, than those obtained when the hot contact bonding procedure is used.

As a general rule, curing time decreases as the curing temperature is increased within limits. Even the epoxies designed to cure at room temperature will cure faster when heated to moderately elevated temperatures. Curing time may be reduced from a number of hours to several minutes by heating. On the other hand, some room-temperature-curing adhesives will not cure properly below 60°F (16°C). This may be an important factor in field applications.

In some instances, longer curing times at elevated temperatures will improve the bond strength of the joint for service above room temperature. Post-curing of the bonded joint without pressure can also improve the heat resistance of the bond.

OVENS

OVENS ARE A widely used, inexpensive method for heat curing when only moderate pressure or simple positioning of the parts is required. They can be heated by gas, electricity, or steam. Adhesives that give off flammable vapor or solvent during cure should not be exposed to open flame or electrical elements. Ovens should be vented, temperature controlled, and fitted with an air circulating fan for uniformity of heat throughout. Infrared lamps and ovens are commonly used for contact-bond rapid-drying neoprene formulations.

HEATED PRESSES

HYDRAULIC PLATEN PRESSES are frequently used for applying heat and pressure to flat assemblies. These are usually heated by electric heating elements, high-pressure steam, hot water, or some other heat-exchanging fluid.

When the work is placed in the press at temperatures below about 150°F (73°C), it is called "cold entry". Entry at the adhesive curing temperature is known as "hot entry". In general, adhesives that release volatiles perform better when cold entry is employed. Certain adhesives are also affected by the rate of temperature rise or heat input. These factors influence the chemical reactions, the flow, and the density of cured adhesives of the volatile releasing types. For example, cold entry and a rate of temperature rise of less than 10°F (6°C) per minute result in better shear strength at elevated temperatures for certain nitrile-phenolic film adhesives. Other adhesives, such as epoxy-phenolics, require either stepped heat input or release of pressure (breathing) at specific temperatures to allow volatiles to escape. Nonvolatile adhesives, such as epoxies, are not affected to any great extent by entry temperature or by the rate of heat input.

Large autoclaves are used for bonding aircraft assemblies and other extremely large parts. The typical operating range of such autoclaves is 200 psi (1380 kPa) maximum pressure and 350°F (180°C) maximum temperature. Pressure is generally provided by compressed air, and curing temperature is achieved with steam heated tubes or electrical elements.

QUALITY CONTROL

TESTING

ADHESIVE BONDED JOINTS are inspected and tested to determine their quality and performance under the specific loading and environmental conditions they will see in service. Based on the test results, quality requirements can be established. Inspection methods and procedures can be specified to assure that quality. The advantages and limitations of inspection and testing procedures must be understood in order to apply them successfully. There are a number of military and industry standard specifications for testing adhesive bonded joints (Table 27.2).

Testing may also consist of accelerated, simulated, or actual use tests of the end product devised by the individual manufacturer or an industry group. For this reason, industry associations may be good sources of information on testing procedures.

If an adhesive is to be used with a metal for which no performance data exist, or if it is to be used in an unusual environment, it should be subjected to some testing. Single overlap shear specimens can be used to evaluate the compatibility of a metal surface condition with an adhesive system and to evaluate the effect of any unusual environmental exposure. If an adhesive is to be used in a structural joint under stress in a certain environment, test joints should be simultaneously subjected to both the stress and the environmental conditions expected in service.

PROCESS CONTROL AND QUALITY ASSURANCE

GOOD PROCESS CONTROL usually requires inspection of all cleaning and processing equipment, evaluation of all materials, and control of storage time and conditions.

Adhesives and primers should be evaluated to assure conformance to the requirements of the design and the user's or adhesive manufacturer's specifications. Certified test reports from the manufacturer may be acceptable in lieu of actual performance tests.

Periodic tests should be performed to determine that cleaning, mixing, and bonding procedures are adequately controlled. Lap shear tests are generally satisfactory for control of mixing, priming, and bonding. Peel tests should be performed to ascertain the adequacy of cleaning procedures. The climbing drum method, described in ASTM D1781, as well as the crack extension (wedge) test, ASTM D3762, may also be used for this purpose.

The crack extension (wedge) test is designed for rapid screening of adhesive joint durability in a controlled humidity and temperature environment. The test specimen design for aluminum alloys is shown in Figure 27.15. One or more specimens are cut from an adhesive bonded panel. The wedge is forced between the adherents and bends them apart. This separates the adhesive and produces cleavage loading at the apex of the separation. The location of the apex of the sheet separation is recorded.

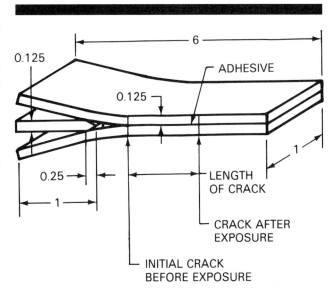

DIMENSIONS ARE IN INCHES

Figure 27.15–Crack Extension (Wedge) Specimen Designed for Aluminum Alloys

The wedged specimens are then exposed at 120°F (49°C) to an air environment of 95 to 100 percent relative humidity for 60 to 75 minutes. The water should contain less than 200 parts per million total solids. The distance that the apex moved during exposure is measured two hours after exposure.

The test is used for surface preparation process control by comparing test results with a maximum acceptable increase in adhesive crack length. It is also used for adhesive durability characteristics and surface preparation procedures. The test was originally designed for adhesive bonded aluminum. However, it may be suitable for other metals with design modifications to account for differences in stiffness and yield strength.

The frequency of testing will depend upon the volume of parts produced and the requirements for the application. However, many manufacturers who employ adhesive bonding in critical applications perform suitable quality control tests at least daily to ensure that the process is within specifications. Any production parts rejected for dimensional reasons or structural damage should be destructively inspected for joint quality.

EVALUATION OF FABRICATED PARTS

AFTER THE MECHANICAL and processing properties of an adhesive system have been determined through destructive laboratory testing, the ability of manufacturing departments to duplicate these properties should be established. Therefore, rather complete testing of the first part or the first few parts of the production run is recommended.

Test loads should be applied in the same manner in which the part will be loaded in use. However, actual loading conditions are often difficult to simulate. In cases involving multidirectional loads, design loads may be applied in each plane individually. The part can then be loaded to failure in the most critical load path to determine if it meets the minimum design strength.

When it is impractical to load a completed part for test because of geometry or difficulty in mounting, many companies fabricate test specimens that are either an integral part of the assembly or separate panels processed in the same manner as the part. Mechanical properties of such specimens closely represent the actual strength of the part. This procedure can provide close control over materials and processing equipment.

NONDESTRUCTIVE INSPECTION

THERE ARE SEVERAL nondestructive inspection methods other than visual that may apply to adhesive bonding:

(1) Ultrasonic
(2) Acoustic impact (tapping)
(3) Liquid crystals
(4) Birefringent coatings
(5) Radiography
(6) Holography
(7) Infrared
(8) Proof test
(9) Leak test

Methods which can be used for a specific application will depend upon one or more of the following factors:

(1) Design and configuration of the structure
(2) Materials of construction
(3) Types of joints
(4) Material thicknesses
(5) Type of adhesive
(6) Accessibility of the joints

In some cases, it may be necessary to incorporate features in the component design or the adhesive to utilize an inspection process. For example, a filler may be required in the adhesive to increase its thermal or electrical conductivity or its density. To determine the applicability of a particular inspection method, the manufacturers of the particular type of equipment should be consulted.

SAFE PRACTICES

ADEQUATE SAFETY PRECAUTIONS must be observed with adhesives. Corrosive materials, flammable liquids, and toxic substances are commonly used in adhesive bonding. Therefore, manufacturing operations should be carefully supervised to ensure that proper safety procedures, protective devices, and protective clothing are being used. All federal, state, and local regulations should be complied with, including OSHA Regulation 29CFR 1900.1000, *Air Contaminants*. The material safety data sheet of the adhesive should be carefully examined before the adhesive is handled to ensure that the appropriate safety precautions are being followed.

GENERAL REQUIREMENTS

Flammable Materials

ALL FLAMMABLE MATERIALS such as solvents should be stored in tightly sealed drums and issued in suitably labeled safety cans to prevent fires during storage and use. Solvents and flammable liquids should not be used in poorly ventilated, confined areas. When solvents are used in trays, safety lids should be provided. Flames, sparks, or spark-producing equipment must not be permitted in the area where flammable materials are being handled. Fire extinguishers should be readily available.

Toxic Materials

SEVERE ALLERGIC REACTIONS can result from direct contact, inhalation, or ingestion of phenolics and epoxies as well as most catalysts and accelerators. The eyes or skin may become sensitized over a long period of time even though no signs of irritation are visible. Once a worker is sensitized to a particular type of adhesive, allergic reactions may keep that individual from working close to it. Careless handling of adhesives by production workers may expose others to toxic materials if proper safety rules are not observed: for example, co-workers may touch tools, door knobs, light switches, or other objects contaminated by careless workers.

For the normal individual, proper handling methods that eliminate skin contact with the adhesive should be sufficient. It is mandatory that protective equipment, protective creams, or both be used to avoid skin contact with certain types of formulations.

Factors to be considered in determining the extent of precautionary measures to be taken include:

(1) The frequency and duration of exposure
(2) The degree of hazard associated with the specific adhesive
(3) The solvent or curing agent used
(4) The temperature at which the operations are performed
(5) The potential evaporation surface area exposed at the work station

All these elements should be evaluated in terms of the individual operation.

PRECAUTIONARY PROCEDURES

A NUMBER OF measures are recommended in the handling and use of adhesives and auxiliary materials.

Personal Hygiene

PERSONNEL SHOULD BE instructed in the proper procedures to prevent skin contact with solvents, curing agents, and uncured base adhesives. Showers, wash bowls, mild soaps, clean towels, refatting creams, and protective equipment should be provided.

Curing agents should be removed from the hands with soap and water. Resins should be removed with soap and water, alcohol, or a suitable solvent. Any solvent should be used sparingly and be followed by washing with soap and water. In case of allergic reaction or burning, prompt medical aid should be obtained.

Work Areas

AREAS IN WHICH adhesives are handled should be separated from other operations. These areas should contain the following facilities in addition to the proper fire equipment:

(1) A sink with running water
(2) An eye shower or rinse fountain
(3) First aid kit
(4) Ventilating facilities

Ovens, presses, and other curing equipment should be individually vented to remove gases and vapors. Vent hoods should be provided at mixing and application stations.

Protective Devices

PLASTIC OR RUBBER gloves should be worn at all times when working with potentially toxic adhesives. When contaminated, the gloves must not contact objects that others may touch with their bare hands. Contaminated gloves should be discarded or cleaned using procedures that will remove the particular adhesive. Cleaning may require solvents, soap and water, or both. Hands, arms, face, and neck should be coated with a commercial barrier ointment or cream. This type of material may provide short term protection and facilitate removal of adhesive components by washing.

Full face shields should be used for eye protection whenever the possibility of splashing exists, otherwise glasses or goggles should be worn. In case of irritation, the eyes should be immediately flushed with water and then promptly treated by a physician.

Protective clothing should be worn at all times by those who work with the adhesives. Shop coats, aprons, or coveralls may be suitable and they should be cleaned before reuse.

SUPPLEMENTARY READING LIST

Cagle, C. V. *Handbook of Adhesive Bonding*. New York: McGraw-Hill, 1973.

Cotter, J. L. and Hockney, M. G. D. "Metal joining with adhesives." *International Metallurgical Reviews* 19: 103-115; 1974.

De Lollis, N. J. "Adhesives for metals - theory and technology." New York: Industrial Press Inc., 1970.

Hartshorn, S. R., ed. "Structural adhesives: chemistry and technology." New York: Plenum Press, 1986.

Katz, I. "Adhesive materials, their properties and usage." Long Beach, CA: Foster Publishing Co., 1971.

Kinloch, A. J., ed. "Durability of structural adhesives." New York: Applied Science Publishers, 1983.

———. "Structural adhesives: developments in resins and primers." London: Applied Science Publishers, 1986.

———. "Adhesion and adhesives: science and technology." London: Chapman and Hall, 1987.

Landrock, A. H. "Adhesives technology handbook." Park Ridge, NJ: Noyes Publications, 1985.

Minford, J. D. "Evaluating adhesives for joining aluminum." *Metals Engineering Quarterly*, November 1972.

———. "Aluminum adhesive bond permanence." *Treatise on Adhesion and Adhesives*, Vol. 5. New York: Marcel Dekker 45-137; 1981.

Patrick, R. L., ed. *Treatise on Adhesion and Adhesives*, Vol. 4. New York: Marcel Dekker, 1976.

Rogers, N. L. "Surface preparation of metals for adhesive bonding." Applied Polymer Symposium No. 3, *Structural Adhesive Bonding*. New York: Interscience Publishers; 327-340; 1966.

Schneberger, G. L., ed. "Adhesives in manufacturing." New York: Marcel Dekker, 1983.

Shields, J. *Adhesives handbook*, 3rd ed. London: Butterworths, 1984.

Skeist, I., ed. *Handbook of adhesives*, 2nd ed. New York: Van Nostrand Reinhold, 1976.

Snogren, R.C. *Handbook of surface preparation*. New York: Palmerton Publishing Co., 1974.

THERMAL SPRAYING

PREPARED BY A COMMITTEE CONSISTING OF:

E. R. Sampson, Chairman
Tafa, Incorporated

R. A. Douty
Westinghouse Electric Corporation

J. O. Hayden
Hayden Corporation

J. E. Kelly
Eutectic Corporation

R. A. Sulit
Integrated Systems Analysts, Incorporated

WELDING HANDBOOK COMMITTEE MEMBER:
C. W. Case
Inco Alloys International

CHAPTER 28

THERMAL SPRAYING

INTRODUCTION

GENERAL HISTORY

ALTHOUGH THERMAL SPRAYING has been in use since the early part of the 20th Century, many of the early applications were concerned mainly with reclamation. Since 1960, there has been a dramatic expansion in the number and diversity of thermal spraying processes, methods, and materials.

Technological advances in equipment, process variations, and materials and forms (wire, rod, cord, or powder) now available have resulted in a multiplicity of new and potential applications. Only the present range and scope of the process variations and applications are reviewed in this chapter. The information will be useful as a guide to the equipment and consumables available, and as a reference for selecting the process variation suitable for each application. Attention is paid to setting variables once thermal spraying is selected as the production or repair process.

PROCESS DESCRIPTION

THERMAL SPRAYING (THSP) is a group of processes in which finely divided metallic or nonmetallic surfacing materials are deposited in a molten or semimolten condition on a substrate to form a thermal spray deposit. The surfacing material may be in the form of wire, rod, cord, or powder. The spray material is heated to its plastic or molten state by an oxyfuel gas flame, electric arc, plasma, or by detonation of an explosive gas mixture. The hot material is propelled from the spray gun to the substrate in a gas stream. Most metals, cermets, oxides, and hard metallic compounds can be deposited by one or more of the process variations. The process is sometimes called "metallizing", "metal spraying", or "flame spraying". A schematic view of a wire flame spray system is shown in Figure 28.1.

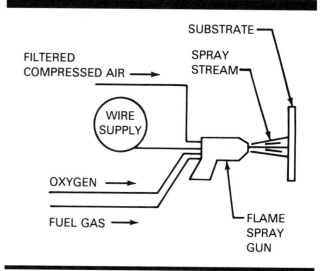

Figure 28.1–Schematic View of Wire Flame Spray System

The substrate is usually roughened before spraying, generally by grit blasting using aluminum oxide or chilled iron grit.

When the molten particles strike the substrate, they flatten and form thin platelets that conform to irregularities of the part geometry and to each other. The platelets rapidly cool and solidify. Successive layers are built up to the desired thickness by the impingement of particles upon the substrate, building up, particle by particle, into a lamellar structure, as shown in Figure 28.2.

The bond between the substrate and the coating material may be mechanical, metallurgical, chemical, or a combination of these. A post spray heat treatment of the coating may be required to increase the bond strength, by

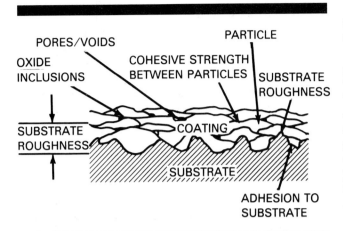

Figure 28.2–Typical Coating Cross Section to Illustrate Lamellar Structure of Oxides and Inclusions

diffusion or by chemical reaction between the spray deposit and the substrate.

The density of the coating deposit will depend upon the material type, method of deposition, the spraying conditions, and post spray processing. The density is generally 85 to 90+ percent of the filler metal density. The properties of the deposit depend upon such factors as porosity, the cohesion between deposited particles, adhesion to the substrate including interface integrity, and chemistry of the coating material.

PROCESS SELECTION

THE ANTICIPATED SERVICE conditions should be considered in choosing the thermal spray process, procedures, consumables, and quality requirements. This chapter covers basic knowledge about thermal spray processes and potential applications. Practical application procedures, to the extent that is possible, are also given. For more detailed information refer to the AWS publication *Thermal Spraying: Practice, Theory and Application.*

ADVANTAGES AND APPLICATIONS

THERMAL SPRAYING PROCESSES and procedures are specialized, yet find wide use in both manufacturing and maintenance applications. The nature of thermal spraying processes is inherently synergistic. Many components and variables are involved, which, when working together and properly applied, produce an effect far greater than they would individually. Yet each component and variable must be separately and jointly understood, to permit the selection and operation of a suitable process. The background of this chapter will help the user tailor that process to particular applications.

The end use of sprayed coatings determines the properties needed in the final coating, the type of consumable employed, and the kind of equipment needed.

Manufacturing

THERMAL SPRAYING IS used extensively in the manufacture of original equipment components. For example, the aerospace industry has developed hundreds of applications. In addition, marine, mining, food, automotive, petroleum, electrical power generation, thermal processing, chemical processing and electronic applications use thermally sprayed coatings to achieve results that no substrate by itself can provide.

Maintenance

EXISTING INDUSTRIAL FACILITIES save hundreds of millions of dollars annually through the use of thermal spraying for repair and maintenance. This includes not only in-plant but also on-site applications, to coat structures and equipment parts. Repair of components, where thermal spraying is applicable, is both economical and time saving. When corrosion or wear, or both, are encountered, thermal spraying should be considered. The use of sprayed coatings, often impregnated with sealers, has received worldwide acceptance by industry for such applications. In many cases, the thermal spray application ends up making the component better than new.

LIMITATIONS

APPLICATION ENGINEERS NEED to be aware of the nature of thermally sprayed coatings, as contrasted with fusion welding, and evaluate them accordingly. For example, thermally sprayed wear coatings usually should not be selected over welded overlays if high impact resistance or resistance to aggressive liquid corrosion is needed in the end-use of the component. For these applications, consideration should be given to fused coatings which have a true metallurgical bond. In addition, the engineer needs to consider the effect of part geometry on coating quality and buildup. In cases where fused coatings cannot be used, successful results have been obtained by applying sealers, selected for a specific environment, over a sprayed coating.

Thermal spraying embodies a group of processes, as does fusion welding, and selection should proceed in the normal fashion. For example, capital investment for plasma spraying equipment is 10 times more expensive than arc spraying. Careful consideration should to be given to equipment and process costs.

The heterogenous structure of sprayed deposits creates factors unique to thermal spraying:

(1) Microhardness is lower than that exhibited by the original spray consumable.

(2) Bond strength is mechanical, metallurgical, or a combination of these, and can be modified in a number of ways.

(3) Deposit densities are less than 100 percent.

(4) Shrinkage stresses may be a factor affecting coating bond strength in certain configurations and applications. Low shrinkage materials should be selected for difficult part geometries.

(5) Thermally sprayed deposits usually have some porosity, but sealers can be used to minimize coating penetration by corrosive media.

SPRAY CONSUMABLES

THE SPRAY MATERIALS are in the form of wire (both solid and cored), rod, cord (a continuous length of powder-filled plastic tubing), or powder. Cord spraying is primarily used in Europe. Many metals, oxides, ceramics, intermetallic compounds, some plastics, and certain types of glass can be deposited by one or more of the processes.

PROCESS VARIATIONS

THERMAL SPRAYING PROCESSES can be categorized under two basic groups, according to the methods of heat generation. Group I uses combustible gases as the heat source, while Group II uses electrical power. See Table 18.1. Additional heat is generated at impact during hypersonic flame spraying, as the spray material gives up its kinetic energy. This is discussed further in the section on Hypersonic Spraying.

GROUP I: COMBUSTION

Subsonic Flame Spraying

IN SUBSONIC FLAME spraying, the spray material is fed into and melted by an oxyfuel gas flame. Whether the material is in the form of wire, rod or powder, molten particles are propelled onto the substrate by the force of the flame.

A wide variety of materials in these forms can be sprayed with the flame. Materials that cannot be melted with an oxyfuel gas flame, and those that burn or become severely oxidized in the oxyfuel flame, cannot be flame sprayed.

Flame spray accessories in the form of air jets and air shrouds are available to change the flame characteristics. These accessories can be used to adjust the shape of the flame and the velocity of the sprayed materials.

Materials are deposited in multiple layers, each of which can be as thin as 0.0005 in. (130 μm) per pass. The total thickness of material deposited will depend upon several factors including the following:

(1) Type of surfacing material and its properties
(2) Condition of the workpiece material including geometry
(3) Service requirements of the coated product
(4) Postspray treatment of the coated product

Hypersonic Flame Spraying

DETONATION AND CONTINUOUS flame guns are two types of hypersonic spray guns.

The detonation gun operates on principles significantly different from other flame spray methods. This method repeatedly heats and projects charges of powder onto a substrate by rapid successive detonations of an explosive mixture of oxygen and acetylene in the gun chamber.

The continuous flame hypersonic guns used in the United States use a propylene-oxygen flame. Overseas operators prefer ethylene, hydrogen, and propane as fuel gases. The powder is brought to the torch using a nitrogen carrier. The torch is designed to confine the powder in the center of the flame. The particles leave the gun at velocities generally in excess of mach 4. This speed is far greater than achieved in most other spray methods. The kinetic energy released by impingement upon the substrate contributes additional heat that promotes bonding, high density, and appreciable hardness values.

GROUP II: ELECTRIC

Arc Spraying

THE SPRAY MATERIALS used with arc spraying, commonly called "electric arc spraying", are metals and alloys in wire form, and powders contained in a metal sheath (cored wire). Two continuously fed wires are melted by an arc operating between them. The molten metal is atomized and propelled onto a substrate by a high-velocity gas jet, usually air. Recent work has been done using other gases.

Table 28.1
Basic Groups of Thermal Spraying

Group I Combustion	Group II Electrical
1. Flame	1. Arc
a. Subsonic	2. Plasma arc
b. Hypersonic	3. Induction coupled plasma

This method is restricted to spraying consumables that can be produced in continuous wire form.

Plasma Spraying

PLASMA SPRAYING IS a thermal spraying process in which a nontransferred plasma arc gun is used to create an arc plasma that melts and propels the surfacing material to the substrate.

The term *nontransferred arc* means that the plasma arc is contained within the gun, and that the substrate is not part of the electric circuit. The arc is maintained between a tungsten cathode and a constricting nozzle which serves as the anode. An inert gas or a reducing gas, under pressure, enters the annular space between the anode and cathode, where it becomes ionized, producing temperatures up to 30 000°F (17 000°C). The hot plasma gas passes through the nozzle as a high velocity jet. The surfacing material, in powder form, is injected into the hot gas stream, where it becomes molten and is propelled onto the substrate.

Vacuum Plasma Spraying

VACUUM SPRAYING IS a variation of plasma spraying which is performed in a vacuum chamber. The advantage of the process is the elimination of oxides from the deposit. This is especially advantageous in aircraft engine applications. The cost of this apparatus is about ten times that of standard plasma spray equipment. Operating costs are also higher.

Induction Coupled Plasma Spraying

INDUCTION COUPLED PLASMA equipment is used to create an ultra high temperature arc region 2 in. (50 mm) in diameter by 6 in. (150 mm) long, into which powders are injected. The powder is heated along a substantially longer path than that within a comparable plasma spray gun. The longer powder residence time makes possible the use of larger particles, assures the melting of the particles, and results in a more consistent sprayed coating.

Because of the size of the equipment, this system has limited torch movement and portability.

NATURE OF SPRAYED COATINGS

SUCCESS IN THE use of thermally sprayed coatings relies on careful adherence to specific process procedures. This is a basic rule of thermal spraying, and deviation from the standards for a particular application, or inattention to detail, especially preparation, will produce an unreliable result.

Sprayed coating systems have four basic components: substrate type, bond coats as necessary, coating structure, and finish.

SUBSTRATES

SUBSTRATES ON WHICH the thermally sprayed coatings are applied include metals, oxides, ceramics, glass, most plastics, and wood. All spray materials cannot be applied to all substrates, since some require special techniques or are temperature sensitive.

Substrate preparation is required for every thermal spraying process, and is virtually the same for each process. Two important steps are: (1) cleaning the surface to eliminate contamination that will inhibit the bonding of the coating to the substrate, and (2) roughening the substrate surface to create minute asperities or irregularities (anchor teeth), which provide a greater effective surface area to enhance coating adhesion and bond strength.

Attention must also be paid to part geometries (no sharp edges where the coating ends), and base material (affected by cleaning agents, grit type, and blasting pressure).

BOND COATS

THE BOND BETWEEN the coating and the substrate may be mechanical or metallurgical. Adhesion is influenced by a combination of: (1) coating material, (2) spray particle size, (3) substrate condition and geometry, (4) degree of surface roughness, (5) surface cleanliness, (6) surface temperature before, during, and after spraying, (7) particle impact velocity, (8) type of base material, and (9) spray angle.

COATING STRUCTURE

THE STRUCTURE AND chemistry of coatings sprayed in ambient air are different from those of the same material in the wrought or presprayed form.

The differences in structure and chemistry are due to the incremental nature of the coating, and its reaction with the process gases and the atmosphere surrounding the coating material while in the molten state. For example, when air or oxygen is used as the process gas, oxides of the spray material are formed while the particles are in transit and become a part of the coating.

Metal coatings tend to be porous and brittle, and to differ in hardness from the original consumable material. The "as-sprayed" structures of coatings will be similar in their lamellar nature, but will exhibit varying characteristics, depending on the particular spraying process used, process variables, techniques employed, and the nature of the spray material applied.

The coating density will vary with the particle velocity, the heat source temperature of the spray process, and the amount of air used. The density also varies with the type of powder, its mesh size, spray rate, standoff distance, and method of injection.

Microscopic examination is the only means of quality evaluation for porosity.

The average particle impact velocities for several thermal spray processes are shown in Figure 28.3.

The nature of the bond in the "as-sprayed" condition can be modified by post spray thermal treatment. Modification is by diffusion, chemical reaction, or both, between the coating and the substrate.

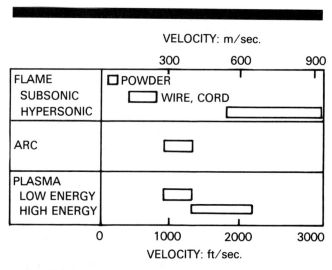

Figure 28.3—Average Particle Impact Velocities for Various Thermal Spray Processes

THERMAL SPRAY EQUIPMENT

FLAME SPRAY EQUIPMENT

A TYPICAL FLAME spraying arrangement consists of the following:

(1) Spray gun
(2) Spray material and the associated feeding equipment
(3) Oxygen and fuel gas supplies, pressure regulators, and flowmeters
(4) A compressed air source and control unit, when required
(5) Workpiece holding device
(6) Gun or workpiece handling device for semi-automatic or automatic processing, as required
(7) Air cooling ring, air jets, or siphon

The gun design depends upon the type of material to be sprayed and its physical form (wire, rod, or powder). When automated, the gun or the workpiece, or both, are driven by mechanisms designed to produce the desired deposit configuration.

The four fuel gases most commonly used for flame spraying are: acetylene, propane, methylacetylene-propadiene (MPS), and propylene.[1] Acetylene in combination with oxygen produces the highest flame temperature. The distinct characteristics of an oxyacetylene flame make it easy to adjust the stoichiometry to produce oxidizing,

neutral, or reducing conditions. The significant changes in flame appearance are not so evident with the other three gases. Hydrogen, which is used occasionally, and propane, are suitable for flame spraying metals with low melting points such as aluminum, tin, zinc, and babbitt metal. Table 28.2 lists heat source temperatures for various fuel gases.

Gas Controls

OXYGEN AND FUEL gas flowmeters are used to provide good control of the gas ratio and flame intensity. Their use permits higher spraying rates than with valve control of gas flows. Since the molten particles are exposed to oxygen, an oxide film forms on them, even when a reducing gas mixture is used. The thickness of the oxide film does not vary greatly with changes in the fuel-gas-to-oxygen ratio.

**Table 28.2
Heat Source Temperatures**

Source	Temperature, °F	Temperature, °C
Acetylene, oxygen	5625	3100
Arcs and plasmas	4000-15000	2200-8300
Hydrogen, oxygen	4875	2690
MPS, oxygen	5200	2870
Natural gas, oxygen	4955	2735
Propane, oxygen	4785	2640

---

1. Properties of these and other fuel gases are discussed in Chapter 14, Oxyfuel Cutting, Welding Handbook, Vol. 2, 8th Ed.

Compressed Air Supply

THE CLEANLINESS AND dryness of the compressed air, when used to atomize and propel the molten surfacing material, is important in producing a quality deposit. Oil or water in the compressed air will cause fluctuations in the flame, produce poor or irregular atomization of the spray material, reduce bond strength, and affect the quality of the deposit. Aftercoolers or a desiccant dryer and chemical filters should be installed between the air source and the spray unit. Accurate regulation of the air pressure is important for uniform atomization.

WIRE FLAME SPRAYING EQUIPMENT

WITH WIRE FLAME spraying, the metal wire to be deposited is normally supplied to the gun continuously from a coil or spool. In some cases, cut lengths of metal rods are used.

A cross section of a typical wire flame spray gun is shown in Figure 28.4. The gun consists essentially of two subassemblies: a drive unit which feeds the wire, and a gas head which controls and mixes the flows of fuel gas, oxygen, and compressed air. The principles of operation of all wire type gas guns are similar. Commercial equipment for wire flame spraying is shown in Figure 28.5.

The wire drive unit consists of a motor and drive rolls. They may be air or electrically powered, with adjustable speed controls. Speed controls may be mechanical, electromechanical, electronic, or pneumatic.

The wire is fed through a central orifice in the nozzle, where it is melted by a coaxial flame. The flame is surrounded by a coaxial stream of compressed gas, usually air, to shear the molten material into droplets and propel it onto the substrate. In special applications, inert gas may be used instead of air. Various sizes of nozzles and air caps are used to accommodate different wire sizes. The arrangement of the oxyfuel gas jets and compressed air orifices differs with the various manufacturers, as do the mechanisms for feeding the wire through the flame.

If the wire feed rate is excessive, the wire tip will extend beyond the hot zone of the flame and not melt or atomize properly. This produces very coarse deposits. If the feed is too slow, the metal will severely oxidize, and the wire may fuse to the nozzle. Such deposits have high oxide content.

Wire spraying units vary in size. Small hand-held units are manipulated in much the same manner as paint spray guns. They are often used to apply protective coatings of aluminum or zinc to large objects such as tanks, ship hulls, and bridges. Larger units are usually designed to be mechanically manipulated for spraying moving parts.

CERAMIC ROD FLAME SPRAYING EQUIPMENT

CERAMIC ROD FLAME spraying is similar to wire flame spraying. Straight lengths of ceramic rod are successively fed into the flame by driven plastic rollers in the gun.

The bond between a ceramic deposit and the substrate is mechanical in nature. The semi-molten particles deform and take the shape of the prepared surface. Proper surface preparation is therefore a prerequisite for a firmly bonded deposit.

The equipment for ceramic rod spraying is similar to wire spraying equipment (Figure 28.1). This equipment requires greater care in adjusting the spraying variables than does wire spraying equipment because of the higher melting points and lower thermal conductivities of ceramics as compared to metals.

Some ceramics applied by this technique are:

(**1**) Alumina-titania
(**2**) Alumina
(**3**) Zirconia

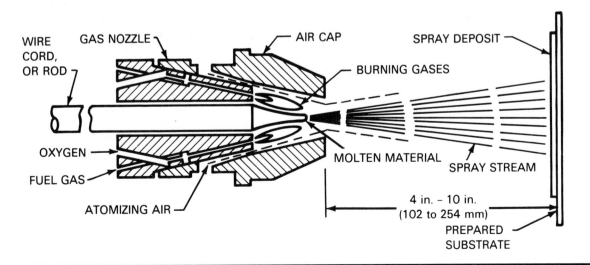

Figure 28.4–Cross Section of Typical Wire, Rod, or Cord Flame Spray Gun

conditions, with due consideration of the following factors:

(1) Thermal, electrical, and chemical characteristics
(2) Melting point
(3) Adherence or bond strength
(4) Density
(5) Cost

Some of the important characteristics of ceramic spray deposits are:

(1) Good adherence to a variety of substrate materials
(2) Economically applied in controlled thicknesses
(3) Good physical and chemical properties
(4) Low thermal and electrical conductivities
(5) High wear resistance
(6) Good finishing characteristics.

POWDER FLAME SPRAYING

IN POWDER FLAME spraying, the material to be sprayed is supplied to the gun in powder form from a hopper. The hopper may be remote from the gun or mounted onto it. The powder may be aspirated or carried into the flame by an air feed system, by the oxygen stream, or by gravity. The powder is melted by the flame and propelled onto the substrate by either a compressed air jet or the combustion gases. A hypervelocity powder flame spray torch is shown in operation in Figure 28.6.

In all thermal spraying processes, the powder particle velocity feed rate affects the structure and the deposit efficiency of the coating. If the raw material is not properly heated, deposit efficiency will rapidly decrease, and the coating will contain trapped, unmelted particles. If the particle velocity is too low, some powder will be volatilized, resulting in coating deterioration and higher operating costs.

Powder flame spraying equipment is simpler and less costly than plasma spray equipment. However, the spray rate with flame spraying is lower. The equipment is designed for easy portability.

A special case is a powder flame spray gun which is similar to an oxyacetylene welding torch. Powder to be sprayed is metered into the gas stream before it leaves the tip. Compressed air is not used. The torch can be used for preheating or fusing spray deposits, when powder is not being injected into the gas stream.

Metals, ceramics, and ceramic-metal mixtures can be flame sprayed by the powder method. The metals are usually hard alloys designed for specific wear or corrosion resistant applications. Very hard metallic compounds, such as carbides and borides, can be blended with metal powders to form a composite, wear-resistant coating. The degree of melting of the particles of spray powder depends upon both the melting point of the material and the time

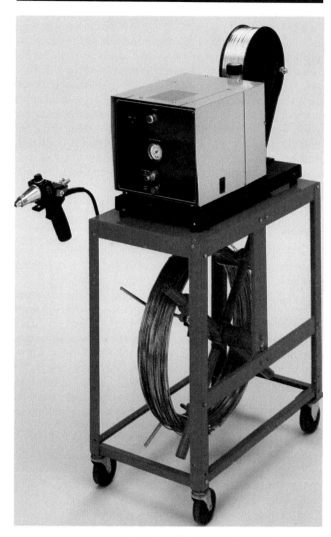

Figure 28.5—Oxyfuel Gas Wire Spray Equipment Capable of Spraying Wires Ranging from Low Melting Alloys (Babbitt) to Higher Melting Point Steels. Aluminum Wire Shown on Top Spool and Carbon Steel Wire on the Bottom Reel.

(4) Rare earth oxides
(5) Zirconium silicate
(6) Magnesium zirconate
(7) Barium titanate
(8) Chromium oxide
(9) Magnesia-alumina
(10) Mullite
(11) Calcium titanate

Each ceramic surfacing material has specific characteristics, economics, advantages, and limitations. The material is selected to provide specific properties for the service

Figure 28.6—Hyper Velocity Oxyfuel Gas Spray Gun (Note Diamond Pattern Resulting from Super Sonic Outlet Velocity. Shown Spraying Tungsten Carbide Powder)

that the particles are exposed to the heat of the flame (called the dwell time). Powders with low melting points will become completely molten, and those with high melting points, such as ceramics, may melt only on the particle surface.

Due to the lower particle velocities and lower temperatures obtained, the coatings produced by powder flame spraying generally have lower adhesion strength, higher porosity content, and lower overall corrosion strength than coatings produced by other spray processes.

The powder feedstock may be pure metal, an alloy, a composite, a carbide, a ceramic, or any combination of these. The process is used to apply "self fluxing" metallic alloy coatings. These materials contain boron and silicon, which serve as fluxing agents, and oxidation is minimized. Fusion or metallurgical bonding to a metal substrate is accomplished by heating the deposit to its melting temperature range. The fusing temperature is usually in excess of 1900°F (1040°C), and is accomplished with any heating source such as a flame, induction coil, or a furnace.

A small amount of gas is diverted to carry the powder from the hopper into the oxyfuel gas stream, where the powder is melted and carried by the flame onto the substrate.

Variations in the powder flame spraying process include compressed gas to feed powder to the flame, additional air jets to accelerate the molten particles, a remote powder feeder with an inert gas to convey powder through a pressurized tube into the gun, and devices for high speed powder acceleration at atmospheric pressure. Such refinements tend to improve flow rate, and sometimes to increase particle velocity, which enhances bond strength and coating density.

DETONATION SPRAYING

DETONATION SPRAYING IS accomplished with a specially designed gun shown in Figure 28.7. The detonation gun is different from other combustion spraying devices. It uses the energy of explosions of oxygen-acetylene mixtures, rather than a steady flame, to blast powdered particles onto the surface of the substrate. The resulting deposit is extremely hard, dense, and tightly bonded.

The detonation gun, schematically shown in Figure 28.8, consists of a long barrel into which a mixture of oxygen, fuel gas, and powdered coating material, suspended in nitrogen, is introduced. The oxygen-acetylene mixture is ignited by an electric spark several times per second, creating a series of controlled detonation waves (flame fronts) which accelerate and heat the powder particles as they move down the barrel. Exit particle velocities of approximately 2,500 ft/sec. (760m/sec.) are produced. After each ejection of powder, nitrogen purges the unit prior to successive detonations. Multiple detonations per second build up the coating to the specified thickness.

Figure 28.7–Detonation Flame Spraying Equipment

Temperatures above 6000°F (3315°C) are achieved within the detonation gun, while the substrate temperature is maintained below 300°F (150°C) by a carbon dioxide cooling system.

Coating thicknesses range between 0.002 and 0.02 in. (50 and 500 μm). The process produces a sound level in excess of 150 decibels, and is housed in a sound isolating room. The actual coating operation is completely automatic and remotely controlled. The high particle impingement velocity results in a strong bond with the substrate. Excellent finishes are achievable. The porosity content of the coating is low.

CONTINUOUS COMBUSTION SPRAYING

EQUIPMENT FOR THE continuous combustion spraying process is similar to that for the subsonic flame spray process, in that a fuel gas such as propylene is burned with oxygen to provide a heat source. The powder to be sprayed is entrained in a nitrogen carrier gas and injected axially into the torch. The nozzle on the hypersonic gun restricts gas flow and results in exit velocities up to 3,000 ft/s (900 m/s).

Flame sprayed deposits produced by the hypersonic gun are similar to those produced by detonation spraying. Because of the high impingement velocities, the sprayed particles are very tightly bonded to the substrate.

ARC SPRAYING

THE ARC SPRAY process uses an arc between two wires (feedstock). They are kept insulated from each other and automatically advance to meet at a point within an atomizing gas stream. A potential difference of 18 to 40 volts applied across the wires initiates an arc as they converge, melting the tips of both wires. An atomizing gas, usually compressed air, is directed across the arc zone, shearing off molten droplets which form the atomized spray.

The velocity of the gas through the atomizing nozzle can be regulated over a range of 800 to 1100 ft/min. (4.0 to 5.5 m/s) to control deposit characteristics desired. Molten metal particles are ejected from the arc at the rate of several thousand particles per second.

In comparison with wire flame spraying, the quantity of metal oxides is better controlled and spray rates are higher in wire arc spraying. Thus wire arc spraying is often more economical.

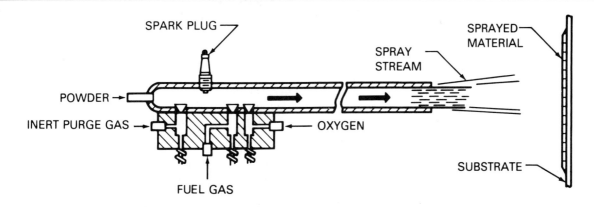

Figure 28.8–Schematic Arrangement of an Oxygen-Fuel Gas Detonation Gun

The wire control unit consists of two reel (or coil) holders, which are insulated from each other, and connected to the spray gun with flexible insulated wire guide tubes. Wire sizes range from 1/16 to 1/8 in. (1.6 to 3.2 mm).

Arc Equipment

A WIRE ARC spray gun is shown schematically in Figure 28.9. A welding type power supply is required to maintain the arc between the two wires.

The arc temperatures exceed the melting point of the spray material. During the melting cycle, the metal is superheated to the point where some volatilization may occur, especially with aluminum and zinc. The high particle temperatures produce metallurgical interactions or diffusion zones, or both, after impact with the substrate. These lo-

calized reactions form minute weld spots with good cohesive and adhesive strengths. Thus the coatings develop excellent bond strengths.

The wire arc spray process can deposit as little as one lb/hr. Higher deposition rates than those possible with other spray processes are also available with arc spraying. Factors controlling the rate of application are the current rating of the power source and the permissible wire feed rate to carry the available power.

Direct current constant potential power sources are normally used for wire arc spraying; one wire is positive (anode) and the other is negative (cathode). The tip of the cathode wire is heated to a higher temperature than the tip of the anode wire and melts at a faster rate. Consequently, the particles atomized from the cathode are much smaller than those from the anode wire when the two wires are of the same diameter.

The power source, providing a voltage of 18 to 40 volts, permits operation over a wide range of metals and alloys. The arc gap and spray particle size increase with a rise in voltage. The voltage should be kept at the lowest possible level, consistant with good arc stability, to provide the smoothest coatings and maximum coating density.

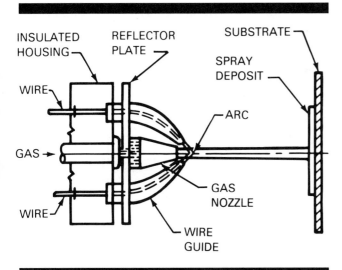

Figure 28.9–Schematic View of a Wire Arc Spray Gun

Systems Operations

WIRE ARC SPRAY systems can be operated from a control console or from the gun. The control console will have the switches and regulators necessary for controlling and monitoring the operating circuits that power the gun and control the spray procedure, namely the following:

(1) A solid-state direct current power source, usually of the constant voltage type
(2) A dual wire feeding system
(3) A compressed gas supply with regulators and flowmeters built into the control assembly
(4) Arc spray gun and pertinent console switching

Controls arc provided for the atomizing gas pressure and gas flow rate, wire feed speeds, and arc power requirements. On most equipment, switches are provided at the spray gun to energize the wire feed and the atomizing gas flow.

Energy and labor costs are lower for wire arc spraying because of its higher deposition rate, lower maintenance, low gas costs, and higher deposition efficiencies. One adverse effect of the high energy state of the atomized particles is their tendency to change composition through oxidation or vaporization, or both. The nature of these effects is complex but they can be minimized by judicious wire selection.

The arc spray method is less versatile than flame or plasma methods, because powders and nonconductive materials cannot be used. Particle velocities and temperatures are generally higher than those with wire flame spraying, but lower than those with plasma spraying.

Higher strength bonds can be achieved with some materials by spraying the first pass using high arc voltage, a low gas flow rate, and a short gun-to-work distance. This is called the bond coat mode. These conditions will produce coarse, hot particles that will adhere well to the substrate. To avoid overheating of the substrate, traversing speed across the substrate is rapid, especially in manual spraying, where travel speed is not automatically controlled.

After the first pass has been applied over the entire surface, subsequent spraying is done using standard gas pressure, the lowest possible arc voltage consistant with good arc stability, and the normal spray gun to work distance. These conditions ensure the following:

(1) Fine spray particle size
(2) Minimum loss of alloy constituents
(3) A concentrated spray pattern
(4) High melting rate

PLASMA ARC SPRAYING

THE TERM "PLASMA ARC" is used to describe a family of metal working processes used in spraying, fusion welding and surfacing, and cutting. They all use a constricted arc to provide high density thermal energy. Arc constriction is accomplished by forcing the electric arc through an orifice. During heating, the accompanying gas is partially ionized, producing a plasma. In plasma spraying, a nontransferred arc is established between an electrode and a constricting nozzle. The substrate is not part of the electrical circuit.

Turbine and rocket engine components are exposed to extreme service conditions. Existing engineering materials are not adequate without a protective thermally sprayed coating. In many cases, the spray coating consists of ceramic oxides and carbides which require temperatures higher than those possible with flame and arc processes. The plasma spray process evolved to meet these needs. The plasma spray process also stimulated the evolution of a new family of materials and application techniques for a greatly expanded range of industrial applications. Plasma spraying supplements the older processes of flame and wire arc spraying.

The process uses powdered materials in a plasma (hot ionized or dissociated gas) as the heat source. Plasma generators provide controllable temperatures of from 4000 to 15 000°F (2200 to 8300°C). These temperatures will melt most substances.

In the plasma spray process, a gas or gas mixture is passed through an electric arc between a coaxially aligned tungsten alloy cathode and an orifice within a copper anode. The process is illustrated in Figure 28.10. The gas passing through the orifice is ionized. The temperature of the ionized plasma is much higher than that obtained with a combustion flame.

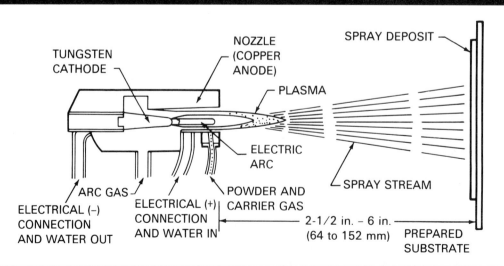

Figure 28.10–Sectional View of Plasma Arc Spraying Torch

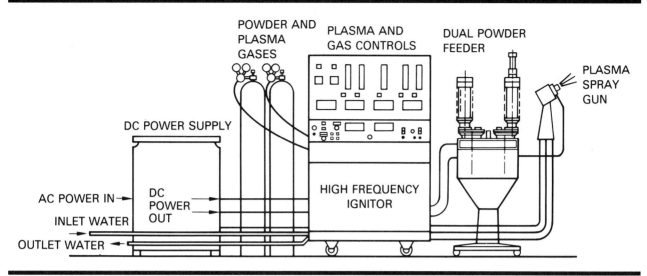

Figure 28.11–Complete Plasma Spray System

EQUIPMENT

A PLASMA SPRAY unit will consist of a plasma gun, power source, powder, powder feed system, and associated fixturing and traversing devices. This is shown schematically in Figure 28.11.

Torch Design

SEVERAL TYPES OF plasma spray guns are available. An 80 kw plasma spray gun is shown in Figure 28.12. In each instance, the arc is generated between an electrode and a water-cooled chamber (nozzle) into which a plasma gas is injected. The gas expands in the heat of the arc, is accelerated, and exits through a nozzle in a cone shaped configuration.

As shown in Figure 28.10, the rear electrode may be fixed or adjustable, but it must be aligned coaxial with the nozzle or front electrode. Flowmeters are used to control the flow of gas through the gun. Several nozzle configurations can be used to accommodate various plasma gases and to spray different types of powders.

Quality deposits require introducing powder at the proper point in the arc plasma and at the correct feed rate. Since the particles are in the plasma for very short times, slight variations in the location of the feed point may significantly change the amount of heat transmitted to the powder.

Current gun designs have power capacities of from 40 to 100 KW. Direct current of 100 to 1100 A is used at 40 to 100 V. High power is necessary when spraying with high particle velocities. Particle velocity is an important variable with respect to bond strength and deposit density and integrity.

Power Supply

POWER SUPPLIES FOR plasma spraying should have the following characteristics:

(1) Constant current dc output
(2) Variable open-circuit and load voltages
(3) Variable current control
(4) Low ripple
(5) Good regulation
(6) Arc starting capability

Rectifier type, solid-state units generally meet the above requirements. Units are easily operated in parallel for high-power operations. In general, they resemble arc welding power sources.

Powder Feed Devices

POWDER FEED MECHANISMS are of three types: aspirator, mechanical, and localized fluid bed. Mechanical feed is the most popular type. It uses the metering action of a screw or wheel to deliver powder at a constant rate to a mixing chamber. The powder is introduced into the carrier gas stream in the mixing chamber.

Units are available to cover a wide range of spray rates. The range of a particular design is determined by the specific gravity of the surfacing material. Modifications are usually available to meet specific spray rate requirements.

System Control

A COMPLETE SYSTEM, including the spray unit, can be operated from a control console. The console provides adjustment of the plasma gas flow rates, plasma current, starting

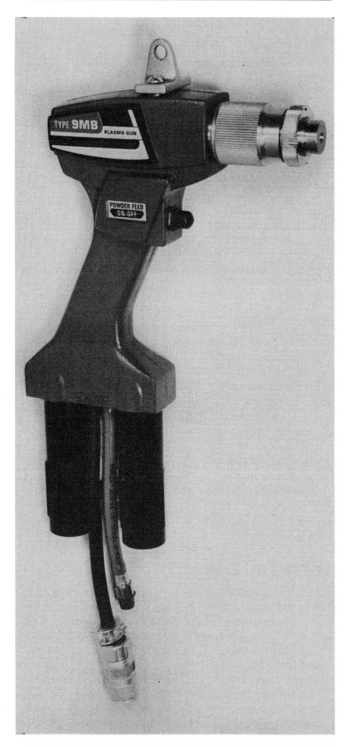

Figure 28.12–Photograph Shows an 80kw Nontransferred Plasma Gun Used to Spray Powders

and stopping functions, and, in some cases, operation of the powder feed unit. These functions are common to all plasma spray systems.

GASES

GASES SERVE THREE purposes in plasma spray systems: as primary plasma gas; as secondary gas mixed in small volumes with the plasma gas; and as the powder carrying gas, usually the same as the primary gas.

Monatomic and diatomic gases can be used for plasma spraying. Argon and helium are the two most frequently used monatomic gases. With monatomic gases, it is possible to attain powder heating rates sufficiently high for most applications. Plasmas generated with polyatomic gases have greater heat contents. They not only release ionization energy but also the energy of molecular recombination. The choice of gas affects the quality of the plasma. The gases should be welding grade with low moisture and oxygen contents.

The four gases commonly used for plasma spraying and their important characteristics are as follows:

(1) Nitrogen is widely used because it is inexpensive, diatomic, and permits high spraying rates and deposit efficiencies. Nozzle life will be shorter than with monatomic gases, but this factor may be offset by the lower cost of this gas.

(2) Argon provides a high velocity plasma. It is used to spray materials that would be adversely affected if hydrogen or nitrogen were used. Carbides and high temperature alloys are most commonly sprayed with argon, especially in aircraft applications.

(3) Hydrogen may be used as a secondary gas in amounts from 5 to 25 percent, with nitrogen or argon. Hydrogen additions raise the arc voltage, and thus the power and heat content of the arc. Hydrogen may have a detrimental affect on certain metals which tend to absorb hydrogen when in a molten condition.

(4) Helium is usually used as a secondary gas mixed with argon, especially when titanium is the substrate. It will also tend to raise the arc voltage.

PLASMA SURFACING

IT IS NECESSARY to use plasma spraying equipment for powders with melting points above 5000°F (2800°C). Because this method is capable of depositing refractory metals and ceramics, it can also deposit powdered materials that are normally applied by flame spraying, but it does so at a higher rate.

A partial list of surfacing materials applied by this method is shown in Table 28.3. Many commercial compositions are proprietary and designed for specific applications.

Plasma sprayed ceramic coatings exhibit higher densities and hardness than flame sprayed deposits. High density

Table 28.3
Materials Commonly Applied by Plasma Spraying

Metals	Carbides[a]	Oxides	Cermets
Aluminum	Chromium carbide	Alumina	Alumina-nickel
Chromium	Titanium carbide	Chromium oxide	Alumina-nickel
Copper	Tungsten carbide	Magnesia	aluminide
Molybdenum		Titania	Magnesia-nickel
Nickel		Zirconia	Zirconia-nickel
Nickel-chromium			Zirconia-nickel
alloys			aluminide
Tantalum			
Tungsten			

a. Normally combined with a metal powder that serves as a binder.

plasma sprayed deposits can be thinner in some cases, but may be more susceptible to cracking. Deposition procedures can be designed to overcome differences in coefficients of thermal expansion of the ceramic coating and the metal substrate. This can be achieved by spraying mixtures of the ceramics and a suitable metal in various proportions to produce graded (layered) deposits.

CONTROLLED ATMOSPHERE PLASMA SPRAYING

PLASMA ARC SPRAYING lends itself to controlled atmosphere applications. Temperature regulation of both the substrate and atmosphere are more precise in controlled atmospheres. This results in lower oxidation of the sprayed materials and less porosity in the sprayed deposit. It also produces closer control of the composition and morphology of the sprayed coating. This results in greater structural homogeneity, absence of oxide, increased hardness, and a thicker deposit capability. These benefits are produced at a higher deposition rate.

When spraying in an inert gas atmosphere chamber, improvements are achieved only with considerable capital equipment cost. The need for improved coating properties must be weighed against the additional expense of the equipment.

INDUCTION COUPLED PLASMA TORCH

THE INDUCTION COUPLED plasma torch generates a plasma by producing a conductive load (arc) within the inductive

field by an ignition system. The inductive field then couples to the conductive gas as it would to an iron bar. See Figure 28.13.

Plasma stability, power conversion efficiency, and maximum heat content are all related to the gas flow pattern, and this pattern varies with different gases. Since there are no electrodes, continuous operation on reactive as well as inert gases is possible without torch deterioration. These gases include air, argon, nitrogen and oxygen.

Control of the plasma effluent is obtained by varying the plasma gas and its flow rate, power input to the induction coil, and the design of the exit nozzle. Gas velocity can be varied from a few feet per second to over 10,000 ft/s (3000 m/s) by changing the exit nozzle size.

This heat source has been used to spray intermetallic powders such as titanium aluminide with excellent results. Lack of an electrode, which might deteriorate during operation, eliminates that potential source of contamination and results in a purer deposit.

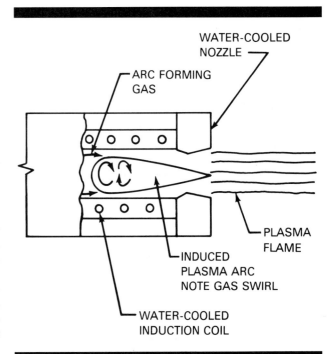

Figure 28.13–Schematic View of Induction Coupled Plasma Torch

FUSED SPRAY DEPOSITS

GENERAL DESCRIPTION

A FUSED SPRAY deposit is a self-fluxing alloy deposited by thermal spraying, which is subsequently heated to coalescence within itself and with the substrate. The materials wet the substrate without the addition of a fluxing agent, provided the substrate is properly cleaned and prepared to receive it. The materials are powdered nickel or cobalt alloys, and they may be applied by powder flame spraying or by plasma spraying.

The application of a fused deposit involves four operations:

(1) Surface preparation
(2) Spraying the self-fluxing alloy
(3) Fusing the coating to the substrate
(4) Finishing the coating to meet surface and dimensional requirements

Fused coatings are dense and nearly porosity free. The alloy compositions can result in hardness levels greater than 50 Rc. Coating thickness is limited to those ranges which can be heated to melting temperature without spalling. Self fluxing coatings are limited to applications where the effects of fusing temperatures and any distortion can be tolerated. Thick coatings of dissimilar metals can be applied in multiple passes. For optimum results, the surface to be coated should be cleaned of all oxide residues after each fusing stage or layer.

A finishing operation is not always required if the as-fused surface is suitable for the application. Centrifuge screw flutes, buffing fixtures, and process piping are examples of components that may be used in the as-fused condition. Pump packing sleeves, pump plungers, piston rods, and process rolls are examples of surfaced machine parts that require a subsequent finishing operation on fused deposits.

A properly sprayed and fused deposit will be nearly homogeneous, metallurgically bonded to the substrate, and have no open or visible porosity. It will have higher hardness than an equivalent mechanically bonded deposit, and will withstand pressures and environments better than nonfused deposits.

SELF-FLUXING ALLOYS

MOST SELF-FLUXING ALLOYS fall into two general groups: Nickel-chromium-boron-silicon alloys and Cobalt-chromium-boron-silicon alloys.

In some cases tungsten carbide or chromium carbide particles are blended with an alloy from one of the above groups.

The boron and silicon additions are crucial elements that act as fluxing agents and as melting point depressants.

They permit fusing at temperatures compatible with steels, certain chromium-iron alloys, and some nickel base alloys.

The hardness of fused coatings will range from 20 to 60 Rc, depending upon alloy composition. Hardness is virtually unaffected by the thermal spraying procedures since there is almost no dilution with the base metal.

Selection of an alloy composition for a particular application should be based on certain considerations, including the following:

(1) Fusion temperature of the alloy and thermal effects on the base metal
(2) Relative difference in the coefficients of thermal expansion of the base metal and alloy deposit
(3) Service requirements of the part
(4) Finish requirements of the fused deposit and available finishing equipment

EQUIPMENT

IN ADDITION TO cleaning, blasting, thermal spraying, and work-handling equipment, some device or method is needed to fuse the sprayed deposit. Fusing may be done with an oxyfuel gas torch, in a furnace, or by induction heating.

Fusing Torches

A FUSING TORCH can have a single or multiple jet tip, depending upon the mass of the workpiece. The fusing gas is usually oxy-acetylene, and a neutral or reducing flame is used. Other fuel gases may be used except for cobalt-base alloys, where an oxyacetylene reducing flame is recommended. A combination spraying and fusing torch is available for applying these types of deposits. The coating is alternately deposited and fused. This type of equipment is particularly suited for repair work, but not for large workpieces nor for production work.

Fusing Furnaces

SPRAY DEPOSITS CAN be fused by placing the coated workpiece in an atmosphere furnace operating at the fusing temperature. Argon, dry hydrogen, or vacuum atmospheres may be used. Furnace fusing is advantageous for high production applications, intricate part geometries, or parts with significant variations in section thickness.

BASE METALS

FUSED THERMAL SPRAYED deposits can be applied to a wide variety of metals. However, varying degrees of skill, technique, and procedures are required. Some base metals are easier to surface than are others. Those which can be readily sprayed with one or more self-fluxing alloys and then fused are as follows:

(**1**) Carbon and low alloy steel with less than 0.25 percent carbon

(**2**) AISI 300 series stainless steels, except Types 303 and 321

(**3**) Certain grades of cast iron

(**4**) Nickel and nickel alloys that are free of titanium and aluminum

Metals that require special procedures to avoid undesirable metallurgical changes are carbon and low alloy steels with more than 0.25 percent carbon, and AISI 400 series stainless steels, except Types 414 and 431. Types 414, 431, and the precipitation hardening stainless steels are not recommended as base metals for self-fluxing alloys.

Cracking of some types of fused sprayed deposits on hardenable steels can be avoided by isothermal annealing of the parts from the fusing temperature. The isothermal anneal prevents the formation of martensite in the substrate material. Fused deposits with a hardenable steel composition and hardnesses above 25 Rc will likely crack when the steel transforms to martensite. Surface cracking results from the rapid expansion that takes place during the transformation. However, there are applications in which cracks in the fused deposit are not detrimental to service requirements.

FUSING

FUSING A SPRAYED deposit is accomplished by heating the workpiece to a temperature range dependent on the particular self-fluxing alloy. The fusing temperatures of nickel-chromium-boron-silicon alloys range from 1875 to 2150°F (1025 to 1175°C). The cobalt-chromium-boron-silicon alloys fuse in the range of 2150 to 2250°F (1175 to 1230°C). The actual fusing temperature depends upon the composition of the alloy.

The most common method of fusing is with one or more oxyfuel gas heating torches, using a reducing flame. A typical torch fusing operation is shown in Figure 28.14. First, the torch is directed on the workpiece, which is heated to a dull red color, about 1400 to 1600°F (760 to 870°C). Then the torch is moved across the spray deposit to gradually increase the surface temperature until the de-

Figure 28.14–Fusing a Deposit on a Large Roll with Oxyacetylene Torches

posit shows a glossy or greasy appearance. This indicates that the deposit has fused. Overheating should be avoided to prevent flow of the molten alloy. The temperature of the workpiece and deposit must be maintained as uniform as possible.

The fusing operation may also be done with other methods of heating, including furnace and induction heating. With these processes, heating is done in a neutral or reducing atmosphere, to avoid oxidation of both the deposit and the base metal.

POST-TREATMENTS

SEALING

SEALING OF SPRAYED deposits is performed to lengthen the service life or prevent corrosion of the substrate, or both. Sprayed deposits of aluminum or zinc may be sealed with vinyl coatings, either clear or aluminum pigmented. The sealer may be applied to fill only subsurface pores in the deposit, or both subsurface pores and surface irregularities. The latter technique will provide a smooth coating to resist industrial atmospheres. The vinyl coatings may be applied with a brush or spray gun.

Sealing is also used on coated machine parts. Where the spray deposit will be exposed to acids, it is recommended that the surface be sealed with either a high melting point wax sealer or a phenolic plastic solution. Spray deposits on high-pressure hydraulic rams, pump shafts, and similar parts should be sealed with air-drying phenolics, to prevent the seepage of liquid through the coating around the packing. Pressure cylinders of all types are reclaimed by thermal spraying. Prior to finish grinding, the cylinder bore is sealed with a phenolic. This prevents grinding wheel particles from embedding in the pores of the sprayed metal and causing premature wear.

Epoxies, silicones, and other similar materials are used as sealants for certain corrosive conditions. Vacuum impregnations with plastic solutions is also possible.

DIFFUSING

A THIN LAYER of aluminum may be diffused into a steel or silicon bronze substrate at 1400°F (760°C). The diffused layer can provide corrosion protection against hot gases up to 1600°F (870°C). After depositing the aluminum, the part can be coated with an aluminum pigmented bitumastic sealer or other suitable material, to prevent oxidation of the aluminum during the diffusion heat treatment. There are similar aircraft applications with diffusion temperatures dependent upon the base material to which the aluminum is applied.

SURFACE FINISHING

TECHNIQUES FOR SURFACE finishing of thermal spray deposits differ somewhat from those commonly used for metals. Most sprayed deposits are primarily mechanically bonded to the substrates, except for fused coatings. Excessive pressure or heat generated in the coating during the finishing operation can cause damage such as cracking, crazing, or separation from the substrate.

Since the composition of an as-sprayed deposit is an aggregation of individual particles, improper finishing techniques can dislodge particles singly or in clusters. This may cause a severely pitted surface. The deposited particles should be cleanly cut and not pulled from the surface. Even so, a totally finished surface will probably not be shiny but may have a matte finish due to porosity of the deposit.

The selection of a finishing method depends on the type of deposit material, its hardness, and the coating thickness. Consideration should be given to the properties of the substrate material as well as dimensional and surface roughness requirements. Spray deposits of soft metals are usually finished by machining, especially those applied to machine components. A good finish is obtained using high cutting speeds and carbide tools for such applications. More often, however, sprayed deposits are finished by grinding, particularly the hardfacing and ceramic coatings.

Various other finishing methods are occasionally used. These include buffing, tumbling, burnishing, belt polishing, lapping, and honing.

Machining

TUNGSTEN CARBIDE TOOLS are commonly used for machining sprayed metal deposits and fused coatings. Proper tool angles play a critical role in the success of machining these coatings. The surface speeds and the depth of cut are of equal importance. Improper tool angle and tool pressure can result in excessive surface roughness and the destruction of the bond between the coating and the substrate.

A cutting tool with a slightly rounded tip and a rake angle of three degrees should be used. On outside circumferences, the tip of the tool should be set three degrees below center; in bores, this should be three degrees above center. This will help to limit the stress on the deposit. Peripheral speed should not exceed 75 ft/min. (400 mm/sec.). The feed should be slow with light cuts for best surface finish.

Special cutting tools, such as oxide-coated carbide, cubic boron-nitride, ceramics, cermets, and diamonds may be used to machine very hard metal, ceramic, and cermet deposits. In many cases, machining with these tools is replacing grinding of intricate shapes and large pieces. Machining of flat deposits requires extreme care at the corners and edges to avoid damage. Depth-of-cut and feed rate should be low.

Grinding

Metal Deposits. Wet grinding is the preferred method. Large, wide wheels are used, and the required amount of stock can be removed with one operation. Wet grinding permits closer tolerances than does dry grinding. Grinding wheel manufacturers can recommend wheel types and grinding procedures for various metal deposits, using a particular type of grinding machine.

If it is necessary to grind metal deposits dry, as is done with portable grinders mounted on a lathe, the major amount of materials should be removed first by machining. Then the deposit is ground to the required finish and dimensions.

Wheels used for dry grinding operations may be either aluminum oxide or silicon carbide, depending upon the metal to be ground. The factors to be considered in selecting a wheel for a spray deposit are similar to those for grinding the same metal in wrought or cast form. The grinding technique should be designed to minimize heat buildup in the deposit. The structure of the wheel should be as open as possible and the grain size as coarse as possible, consistent with the finish requirements. The wheel should be narrow, the infeed light, and the traverse as fast as possible without spiraling.

When grinding equipment is not available, metal deposits can be machined to within 0.002 to 0.006 in. (50 to 150 µm) of final size. Then they can be finished to size with a belt polishing unit. Close tolerances and fine finishes are possible with belt polishing, by proper selection of abrasive type and grit size.

Grinding Fused Deposits. Because most fused deposits are designed for hardfacing purposes, grinding is usually the most economical method for finishing them.

Although most fused deposits can be machined with the proper type of cutting tool, close tolerance work is difficult because of rapid tool wear and the large amount of heat generated. Dry grinding may be suitable for some operations, but, in this case also, heat and fast wheel wear make close tolerance difficult. Wet grinding can produce close tolerance parts, fine finishes, and economical stock removal rates. Nickel-base alloys are best ground with silicon carbide grinding wheels, and cobalt-base alloys with aluminum oxide grinding wheels.

Grinding wheel manufacturers should be consulted for recommendations of the appropriate type for the job. Good practice usually suggests a coarse wheel, consistent with finish requirements; an open structure or soft bond; as large a wheel as possible; and good wheel dressing techniques. Surface finish of fused coatings can often be improved after grinding by polishing with fine grit belts.

Grinding Ceramic Deposits. The as-sprayed surface finish of flame sprayed ceramics is, in general, more coarse than 150 µm. Many applications require a better finish, and this can be accomplished by grinding. Although the individual particles of a ceramic deposit have extreme hardness, the deposit can be finished by conventional grinding techniques on standard equipment. However, it is necessary to use the proper grinding wheel, in some cases a diamond wheel, and to follow correct procedures. General recommendations for grinding ceramic deposits are available from grinding wheel manufacturers.

Flood cooling should be employed during grinding. Water containing a rust inhibitor is best. Water-soluble coolants are likely to stain light-colored ceramic deposits.

Other Finishing

OTHER METHODS OF surface finishing are sometimes used for as-sprayed and fused deposits. These include:

(1) Manual buffing or polishing
(2) Abrasive tumbling
(3) Honing
(4) Lapping

As-sprayed or machined deposits may be buffed or polished manually with abrasive stones, cloth, or paper. Abrasive tumbling of small parts will polish the surface by removing the "high spots." An abrasive medium, cleaners, and usually a liquid are vibrated or rotated in a drum in which the parts are finished.

Honing is done with abrasive stones mounted in a loading device. The part normally moves in one direction or rotates while the stones are oscillated under pressure, transverse to the work motion. Lapping is done with a fine, loose abrasive mixed with a vehicle, such as water or oil. The mixture is spread on lapping shoes or plates that are then rubbed against the spray deposit. The lap rides against the deposit, and their relative movements are continually changed.

QUALITY CONTROL

A PROPERLY DESIGNED quality control program can ensure consistent quality in thermal sprayed deposits. Proper quality control consists of more than just the examination of the workpiece after the spraying is complete. Each step in the operation should be monitored by an inspector. This includes not only the spraying and fusing steps, but also the preparation of the substrate and the various stages of handling and storage of the workpiece between operations. In addition, the quality of the spray materials must be controlled. Since bond strength and spray deposit soundness are difficult to determine by nondestructive techniques, the procedures for accomplishing each step of

the thermal spraying operation should be documented. The procedures should be qualified by appropriate destructive tests of sample parts.

In general, sprayed deposits are inspected visually for quality and soundness. With fused deposits, lack of bonding may be detected by localized torch heating of the suspected area. Lack of bonding will be indicated by a hot spot or spalling of the deposit material. Ultrasonic techniques may also be used to detect lack of bonding. Penetrant or magnetic particle inspection can detect surface porosity and cracks. Magnetic particle inspection can be used only on ferromagnetic spray deposits.

PROPERTIES

THE QUALITY AND the properties of thermal sprayed deposits are largely determined by the size, temperature, and velocity of the spray droplets as they impinge on the substrate, and the degree of oxidation of both the droplets and the substrate during spraying. These factors will vary with the method of spraying and the procedures employed.

Metals and alloys deposited by the thermal spray process do not retain their original chemical composition unless special techniques are used. Their properties may change significantly depending upon the spray method used. With plasma and other arc methods, appreciable amounts of low melting point constituents may be lost by vaporization. Oxidation of the droplets may also be significant when air is used as the propellant.

The physical and mechanical properties of a spray deposit normally differ greatly from those of the original material. The deposit structure is lamellar and nonhomogeneous. Its cohesion is generally the result of mechanical interlocking, some point to point fusion, and sometimes oxide to oxide bonding. The tensile strengths of these structures are low compared to those of the same materials in wrought or cast form. Sometimes the compressive strength is quite high but the ductility is low. Deposits from wire or rod are less dense than the original material. In any case, spray deposits should be considered as a separate and distinct form of fabricated material.

Oxide spray deposits tend to retain their physical properties with only modest losses. In many cases, the deposit will have a crystalline structure. Alpha alumina may deposit with a metastable gamma structure. The chemical compositions of reactive type ceramics, such as carbides, silicides, and borides, normally change when the materials are sprayed in air with the flame or plasma methods.

MICROSTRUCTURE

THE MICROSTRUCTURE OF a transverse section through a flame sprayed metal deposit will show a heterogeneous mixture of layered metal particles (white), metal oxide inclusions (gray), and pores (black). A photomicrograph of a transverse section through a flame sprayed deposit of 0.80 percent carbon steel is shown in Figure 28.15. The light layered particles are bonded to one another by chemical and mechanical interactions. A photomicrograph of a transverse section through a copper deposit and its substrate at the bond line is shown in Figure 28.16. The roughness of the prepared substrate surface is apparent.

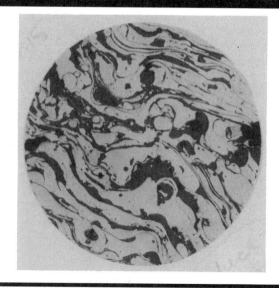

Figure 28.15–Transverse Section Through a Flame Sprayed AISI 1080 Steel Deposit (x500 Reduced on Reproduction)

Figure 28.16–Transverse Section Through a Thermal Sprayed copper Deposit (Top) and the Substrate (Bottom) (x500 Reduced on Reproduction)

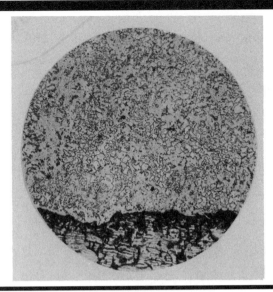

Figure 28.18–Microstructure of a Fused Coating of a Self-Fluxing Nickel-Chromium Alloy (Top) on a Substrate (Bottom) (x250 Reduced on Reproduction)

The microstructure of the polished and etched surface of the 0.80 percent carbon steel deposit is shown in Figure 28.17. It has an emulsified appearance because the flattened steel particles (light) are separated by the oxide (gray).

As-sprayed, self-fluxing alloy deposits are similar in appearance to any typical metal deposit, except that there is significantly less oxide. These materials are oxidation resis-

tant in nature. After fusing, the deposit will have a cast structure with some porosity and inclusions. The microstructure of a fused nickel-chromium self-fluxing alloy deposit is shown in Figure 28.18. The roughness of the prepared substrate is also evident.

HARDNESS

THE HETEROGENEOUS STRUCTURES of spray deposits generally have a lower macrohardness than the original rod or wire supplied to the gun. However, the hardness of individual deposit particles (microhardness) may be much higher than that of the overall deposit. The hardness test should be selected to give the overall deposit hardness or the particle hardness. The thickness of the deposit must also be considered in selecting the type of test. If the deposit is too thin, the indenter may penetrate through it and into the substrate. This would obviously give a false reading.

The Brinell and Rockwell hardness tests can be used to determine the hardness of fairly thick metallic deposits. Superficial Rockwell and Vickers hardness tests are suitable for thin metallic deposits. Requirements for various hardness tests are covered in the appropriate ASTM Standards. Table 28.4 relates the minimum spray deposit thickness to the various Rockwell hardness tests.

Hardness tests with diamond indenters are not entirely satisfactory for determining the true hardness of heterogeneous spray deposits, but they can be used for spot checks and shop guides. Microhardness tests can be used to determine the hardness of individual particles. Since the deposited particles are relatively thin, hardness impressions

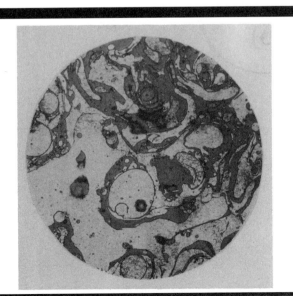

Figure 28.17–Section Parallel to the Surface of a Flame Sprayed Deposit of AISI 1080 Steel (x500 Reduced on Reproduction)

Table 28.4
Minimum Deposit Thickness for Rockwell Hardness Tests

Rockwell Scale	Minimum Thickness	
	in.	mm
15N	0.015	0.38
30N	0.025	0.64
45N	0.035	0.89
A	0.040	1.0
B	0.060	1.5
C	0.070	1.8
D	0.050	1.3

Table 28.5
Typical Bond Strength of Self-Bonding Materials*

Material	Bonding strength, psi (kPa)		
	Plasma Sprayed	Flame Sprayed	
		Powder	Wire
Columbium	2400 (16500)	-	-
Molybdenum	3200 (22100)	3600 (24800)	3300 (22800)
Nickel-aluminide	3000 (20700)	2750 (19000)	3150 (21700)
Tantalum	2750 (19000)	-	-

* Applied to a smooth, unprepared metal surface.

should be taken on a transverse section. The Knoop indentation hardness test is best suited for this.

BOND STRENGTH

THE STRENGTH OF the bond between a spray deposit and the substrate depends upon many factors, including the following:

(1) Substrate material and its geometry
(2) Preparation of the substrate surface
(3) Spray angle to substrate
(4) Preheat
(5) Bond layer material and its application method and procedures
(6) Deposit material and its application method and procedures
(7) Thickness of deposit
(8) Post spraying thermal treatment

A standard test for determining the bond or cohesive strength of thermal spray deposits is described in ASTM C633, Standard Test Method for Adhesion or Cohesive Strength of Flame Sprayed Coatings.

In this test, each specimen is an assembly of a coated substrate block and a loading block. The flat end of the substrate block is prepared, and the deposit is applied. Then, the deposit is machined or ground flat and uniform in thickness using procedures appropriate for the deposited material. The loading block is then adhesive bonded to the flat deposit surface to produce a tension specimen. The specimen is loaded in tension at a constant rate using a self-aligning device. The maximum load is recorded. From this, the bond strength or the cohesive strength of the deposit can be calculated, depending upon the fracture location. The bond strengths of several self-bonding spray materials are presented in Table 28.5.

This test method is limited to deposit thicknesses greater than 0.015 in. (0.4 mm), because adhesive bonding agents tend to infiltrate porous deposits. If the bonding agent penetrates to the substrate, it will affect the test results.

DENSITY

THERMAL SPRAYED DEPOSITS have densities less than 100 percent of the filler metals because they are porous and contain some oxide. The densities of the flame sprayed deposits and the original wire for several metals are given in Table 28.6.

Porosity in spray deposits consists of isolated and sometimes interconnected pores. It is difficult to determine the amount accurately. However, it can be estimated by several methods. The simplest one is to superimpose a grid over the microstructure (photomicrograph) of a prepared surface and then count the number of grid squares occupied by pores. Other methods include water or toluene immersion, and paraffin absorption. Because of lack of total interconnection of the pores, however, no method is perfect.

The porous nature of spray deposits can be used to advantage, especially for bearing surfaces. The porosity permits oil retention and provides an escape for foreign material from actively loaded areas. Where corrosion is a factor, porosity is a disadvantage. It limits the use of deposits to those that are anodic to the base material, unless special overcoatings of paint or sealers are used.

SHRINKAGE

SPRAY DEPOSITS CONTRACT upon cooling. The amount of shrinkage varies widely with different materials and spray-

Table 28.6
Comparison of the Densities of Flame Sprayed Metal Deposits and the Wire

Metal	Density, lb/in.^{3} (kg/m^3)	
	Flame Sprayed Deposit (Wire)	Wire
Type 1100 Aluminum	0.087 (2408)	0.098 (2713)
Copper	0.271 (7501)	0.324 (8968)
Molybdenum	0.326 (9024)	0.369 (10214)
AISI 1025 steel	0.244 (6754)	0.284 (7861)
Type 304 stainless steel	0.249 (6892)	0.290 (8027)
Zinc	0.229 (6839)	0.258 (7141)

ing methods, but it will not be the same as that of the original material in cast or wrought form. Contraction sets up tensile stresses in the deposit as well as shear stresses across the bond between the deposit and substrate. These stresses tend to crack or spall the deposit. Surface prepara-tion, selection of material, and coating thickness are im-portant factors in preventing these problems. Metals hav-ing low coefficients of thermal expansion should be used where possible, especially for thick deposits and buildup of internal surfaces.

APPLICATIONS

CORROSION AND OXIDATION PROTECTION

THERMAL SPRAY DEPOSITS can provide protection against many types of corrosive attack on iron and steel. Zinc, aluminum, stainless steel, bronze, hard alloys, and ceram-ics are used as surfacing materials. Service conditions de-termine both the material type and its application proce-dures. Undercoatings for organic materials, such as paints and plastic finishes, can be applied by this process. A thick layer of zinc or aluminum can protect steel against oxida-tion and provide a strong bond for an organic coating.

Nickel, nickel-copper alloys, stainless steels, and bronzes are some metals that are cathodic to steel. They should be used as a deposit on steel only if they are made impermeable to corrosive agents by sealing. The sealer is likely to entrap air bubbles in pores of the spray deposits. The component should not be heated because expansion of the air bubbles may rupture the sealer.

Hard alloy deposits are often used on machine compo-nents such as pump plungers, pump rods, hydraulic rams, packing sections of steam turbine shafts, and valves. When sealed, these materials provide both corrosion and wear resistance.

Several different materials may be used to give oxidation protection, the choice depending on the operating temper-ature. For applications up to 1600°F (870°C), the part can be aluminized by depositing a thin layer of aluminum. The aluminum is then diffused into the surface by a suitable heat treatment. For temperatures above 1600°F (870°C), a nickel-chromium alloy deposit may be used. This is fol-lowed by a coating of aluminum. Often, this combination deposit is then covered with an aluminum pigmented bitu-mastic sealer. The part is then diffusion heat-treated in a furnace, or it is placed directly in service if the operating temperature is above 1600°F (870°C). Such deposits are sometimes used for cyanide pots, furnace kiln parts, an-nealing boxes, and furnace conveyors.

Zirconia and alumina ceramics are sometimes used for thermal barrier layers. When the workpiece will be ex-posed to thermal cycling, a bond layer of nickel aluminide or nickel-chromium alloy may help to minimize thermal stresses in the ceramic deposits.

WEAR RESISTANCE

IN THE MECHANICAL field, thermal spray hardfacing mate-rials can be used to combat many types of wear. The ability of metal spray deposits to absorb and maintain a film of lubricant is a distinct advantage in many applications. Spray deposits often give longer life than the original sur-faces, except where severe conditions of shock loading or abrasion are encountered. A low cost base metal can be protected on just the areas of wear with a high quality, wear resistant deposit.

An arc spray gun being used to deposit a nickel alumi-nide bonding material on the I.D. of a four inch steel cylin-der is shown in Figure 28.19. Compressed air shears off molten droplets at a right angle to the axis of the spray gun.

Some metal deposits, such as nickel-copper alloys, nickel, and stainless steel, are virtually impervious to pene-tration by corrosives when they are applied in sufficient thickness and are exposed only to moderate pressure. These surfaces can be vacuum impregnated with various phenolic or vinyl solutions or with fluorocarbon resins for high-pressure operation. For applications where extreme wear or corrosion resistance, or both, are encountered, fused spray coatings may be used.

ELECTRICAL CHARACTERISTICS

THE ELECTRICAL RESISTANCE of a metal spray deposit may be 50 to 100 percent higher than that of the same metal in cast or wrought form. This should be taken into consider-ation in the design of spray deposits for electrical conduc-tors. Such applications include spraying of copper on elec-trical contacts, carbon brushes, and glass in automotive fuses, as well as silver or copper contacts.

In the field of electrical insulation, various ceramic de-posits can be used for insulators. Magnetic shielding of electrical components may be provided with deposits of zinc or tin zinc applied to electronic cases and chassis. Condenser plates can be made by spraying aluminum on both sides of a cloth tape.

FOUNDRY

CHANGES IN CONTOUR of expensive patterns and match plates can be readily accomplished by the application of thermal spray deposits followed by appropriate finishing. Patterns and molds can be repaired with wear resistant de-posits. Blow holes in castings that appear during machining can be filled to salvage the parts.

Figure 28.19–Wire Arc Spray Gun Used to Deposit Nickel Aluminum Bonding Material on the I.D. of a 4-in. (100 mm) Diameter Steel Cylinder. Compressed Air Shears Off Droplets at a Right Angle to Axis of Spray Gun.

BRAZING AND SOLDERING

THERMAL SPRAYING IS frequently used for the preplacement of soldering or brazing filler metals. The usual practice is to apply the filler metal using standard thermal spraying techniques.

AIRCRAFT AND MISSILES

THERMAL SPRAYING IS used for air seals and wear resistant surfaces to prevent fretting and galling at elevated tempera-

tures. Deposits of alumina and zirconia are used for thermal insulation.

A robot set up for plasma arc spraying of a bond coat on a part from the hot section of a gas turbine engine is shown in Figure 28.20. The bond coat is part of a thermal barrier coating system.

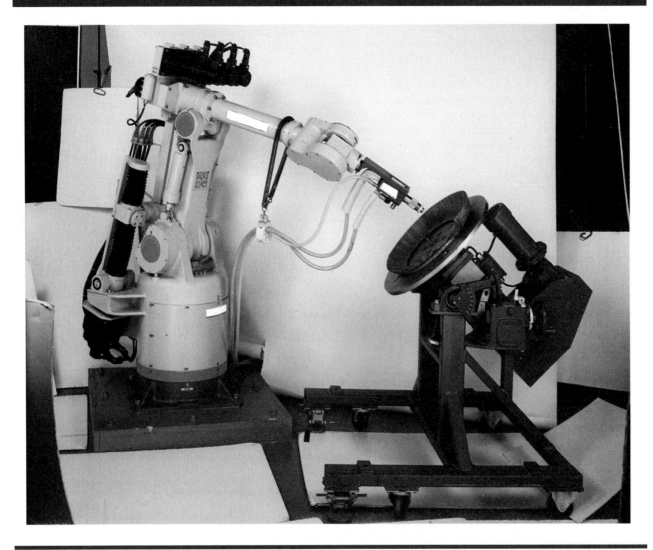

Figure 28.20–Robotic Spraying of Bondcoat on a Part from the Hot Section of a Gas turbine Engine

SAFETY

THE POTENTIAL HAZARDS to the health and safety of personnel involved in thermal spraying operations and to persons in the immediate vicinity can be grouped as follows:

(1) Electrical shock
(2) Fire
(3) Gases
(4) Dust and fumes
(5) Arc radiation
(6) Noise

These hazards are not unique to thermal spraying methods. For example, flame spraying has hazards similar to those associated with the oxyfuel gas welding and cutting processes. Likewise, arc spraying and plasma spraying are similar in many respects to gas metal arc and plasma arc welding, respectively. However, thermal spraying does generate dust and fumes to a greater degree than the welding processes do.

GAS SYSTEMS

LOCAL, STATE, AND federal regulations relative to the storage of gas cylinders should be investigated and complied with. Safe storage, handling, and use of gas cylinders are described in ANSI Z49.1, *Safety in Welding and Cutting*, and CGA P-1, *Safe Handling of Compressed Gases*. Improper storage, handling, and use of these cylinders constitute safety hazards in thermal spraying operations. Oil or grease must not be used on oxygen equipment; only special oxidation resistant lubricants may be used.

Acetylene pressures in excess of 15 psi (103kPa) are dangerous and should not be used. When acetylene pressure of 15 psi is too low for the application, another fuel gas should be used. Alloys containing more than 67 percent copper or silver must not be used in acetylene systems, because dangerous explosive compounds may be formed.

FLAME SPRAY GUNS

FLAME SPRAY GUNS must be maintained in accordance with the manufacturer's recommendations. Each operator should be familiar with the operation of the flame spray gun and should read the instruction manual thoroughly before using it.

A friction lighter, a pilot light, or arc ignition should be used to ignite the fuel gas. Matches are not safe. A flame spray gun and its hoses should not be hung on gas regulators or cylinder valves because of the danger of fire or explosion.

PLASMA AND ARC SPRAYING EQUIPMENT

PLASMA AND ARC spraying use equipment where high voltages and amperages present a hazard. Operators should be thoroughly instructed and trained in the operation of the unit. They should be familiar with the operating and safety recommendations, and at the same time observe proper safety precautions for electrical equipment.

The plasma spraying equipment itself should be kept in a condition safe to operate. Exposed electrodes of plasma guns should be grounded or adequately insulated. Periodic inspections should be made of cables, insulation, hoses, and gas lines. Faulty equipment must be repaired or replaced immediately. The entire system, including the power supply, must be shut down before repairing any part of the power supply, console, or gun.

Arc spray guns should be cleaned frequently according to the manufacturer's operation manual to prevent the accumulation of metal dust. If the arc spray gun is suspended on a cable, the suspension hook must be insulated or grounded. Contact between any ungrounded portion of the plasma or arc spray gun and the spray chamber must be avoided.

FIRE PREVENTION

FINELY DIVIDED AIRBORNE solids, especially metallic dusts, must be treated as explosives. To minimize danger from dust explosions, adequate ventilation should be provided to spray booths. A wet collector of the water-wash type is recommended to collect the spray dust. Bag or filter type collectors are not recommended. Good housekeeping in the work area should be maintained to avoid accumulation of metal dusts, particularly on rafters, tops of booths, and in floor cracks.

Paper, wood, oily rags, and other combustibles in the spraying area can cause a fire and should be removed before the equipment is operated.

PROTECTION OF PERSONNEL

THE GENERAL REQUIREMENTS for the protection of thermal spray operators are the same as for welders, set forth in ANSI Z49.1, *Safety in Welding and Cutting*; ANSI Z87.1, *Practices for Occupational and Educational Eye and Face Protection*; ANSI Z88.2, *Practices for Respiratory Protection*; and ANSI Z89.1, *Safety Requirements for Industrial Head Protection*.

Eye Protection

HELMETS, HAND HELD shields, face shields, and goggles should be used to protect the eyes, face, and neck during all thermal spraying operations. These are described in ANSI Z87.1 and Z89.1. Safety goggles should be worn at all times. Helmets, hand held shields, and goggles must be equipped with suitable filter plates to protect the eyes from excessive ultraviolet, infrared, and intense visible radiation. A guide for the selection of the proper filter shade number is shown in Table 28.7.

Respiratory Protection

MOST THERMAL SPRAYING operations require that respiratory devices be used by the operator. The nature, type, and magnitude of the fume and gas exposure determine which

Table 28.7
Recommended Eye Filter Plates for Thermal Spraying Operations

Operation	Filter Shade Numbers
Wire flame spraying (except molybdenum)	5
Wire flame spraying of molybdenum	5 to 6
Flame spraying of metal powder	5 to 6
Flame spraying of exothermics or ceramics	5 to 8
Plasma and arc spraying	9 to 12
Fusing operations	5 to 6

respiratory protective device should be used. The selection of these devices should be in accordance with ANSI Z88.2, *Practices for Respiratory Protection*. This standard contains descriptions, limitations, operational procedures, and maintenance requirements for standard respiratory devices. All devices selected should be of a type approved by U.S. Bureau of Mines, National Institute for Occupational Safety and Health, or other approving authority for the purpose intended.

Ear Protection

EAR PROTECTORS OR properly fitted soft rubber ear plugs must be worn to protect the operator from the high intensity noise from the thermal spray gun. Such protection should reduce the noise level to below 80 decibels. Cotton wads are not recommended for ear protection, as they are ineffective against high-intensity noise. Federal, State, and local codes should be followed for noise protection requirements.

Protective Clothing

APPROPRIATE PROTECTIVE CLOTHING requirements for a thermal spraying operation will vary with the size, nature, and location of the work to be performed. When working in confined spaces, flame resistant clothing as well as leather or rubber gauntlets should be worn. Clothing should be fastened tightly around the wrists and ankles to keep dusts from contacting the skin.

For work in the open, ordinary clothing such as overalls and jumpers may be used. However, open shirt collars and loose pocket flaps are potential hazards. High-top shoes are recommended and cuffless trousers should cover the shoe tops.

The intense ultraviolet radiation of plasma arc spraying can cause skin burns through normal clothing. When using this process, the clothing should provide protection against such radiation. For exposure to more intense radiation, leather welder capes are necessary. Protection against radiation with arc spraying is essentially the same as that used for electric arc welding.

SUPPLEMENTARY READING LIST

Anon. "Paperboard plant beats high replacement part costs with thermal spraying." *Welding Journal* 56(1): 34; June 1977.

Anon. "Thermal spraying saves 75% of replacement cost for hydraulic press ram." *Welding Journal* 56(8): 41; August 1977.

Anon. "Flame spraying cuts coal-processing equipment maintenance costs." *Welding Journal* 59(5): 39-40; May 1980.

Anon. "Nonskid deck surface is arc sprayed in place." *Welding Journal* 60(4): 37; April 1981.

Anon. "Thermal spray coating adheres to plastics." *Welding Journal* 65(1): 55; January 1986.

American Welding Society. AWS C2.16-78, Guide for Thermal Spray Operator and Equipment Qualification. Miami, FL: American Welding Society, 1978.

———. Thermal Spraying Practice, Theory and Application. Miami, FL: American Welding Society, 1985.

———. C2.2-67, Recommended Practices for Metallizing With Aluminum and Zinc for Protection of Iron and Steel. Miami, FL: American Welding Society, 1967.

Clark, W.P. "The development of thermal spray hard surfacing." *Welding Journal* 60(7): 27-29; July 1981.

Cullison, A. "Thermal spraying sparks artists imagination." *Welding Journal* 64(7): 58-60; July 1985.

Hermanek, F. J. "Determining the adhesive/cohesive strength of thin thermally sprayed deposits." *Welding Journal* 57(11): 31-35; November 1978.

Hermanek, F. J. "Thermal Conductivity and thermal shock qualities of zirconia coatings on thin gage Ni-Mo-C metal." *Metal Progress*, 97(3): 104; March 1970.

Ingham, H. S., Jr. and Fabel, A. J. "Comparison of plasma flame spray gases." *Welding Journal* 54(2): 101-5; February 1975.

Irons, G. C. "Laser fusing of flame sprayed coatings." *Welding Journal* 57(12): 29-32; December 1978.

Longo, F. N. "Use of flame sprayed bond coatings." *Plating* 61(10): 306-11; October 1974.

Longo, F. N. and Durmann, G. J. "Corrosion prevention with thermal sprayed zinc and aluminum coatings." *Welding Journal* 53(6): 363-70; June 1974.

Papers of the Eighth International Thermal Spray Conference, Miami Beach, FL: September 27 to October 1, 1976; American Welding Society, 1976.

Phelps, H. C. "Fuel gas additive helps solve problem of rejects during thermal spraying operation." *Welding Journal* 56(7): 32-35; July 1977.

Sulit, R.A. et al "Thermal spray repair for naval machinery at SIMA." *Welding Journal* 67(12): 31; December 1988.

OTHER WELDING PROCESSES

PREPARED BY A COMMITTEE CONSISTING OF:

L. Heckendorn, Chairman
Toledo Scale Co.

H. R. Castner
Edison Welding Institute

WELDING HANDBOOK COMMITTEE MEMBER:
J. R. Hannahs
Midmark Corporation

OTHER WELDING PROCESSES

THERMIT WELDING

FUNDAMENTALS OF THE PROCESS

Definition

THERMIT[1] WELDING (TW) is a process that produces coalescence of metals by heating them with superheated molten metal from an aluminothermic reaction between a metal oxide and aluminum. Filler metal is obtained from the liquid metal. The process had its beginning at the end of the 19th century when Hans Goldschmidt of Goldschmidt AG West Germany (Orgotheus Inc. USA) discovered that the exothermic reaction between aluminum powder and a metal oxide can be initiated by an external heat source. The reaction is highly exothermic, and therefore, once started, it is self-sustaining.

Principles of Operation[2]

THE THERMOCHEMICAL REACTION takes place according to the following general equation:

Metal oxide + aluminum (powder) →
aluminum oxide + metal + heat

The reaction can only be started and completed if the oxygen affinity of the reducing agent (aluminum) is higher than the oxygen affinity of the metal oxide to be reduced. The heat generated by this exothermic reaction results in a liquid product consisting of metal and aluminum oxide. If the density of the slag is lower than that of the metal, as in the case of steel and aluminum oxide, they separate immediately. The slag floats to the surface and the molten steel drops into the cavity to be welded.

Typical thermochemical reactions and the thermal energies produced are as follows:

$$3Fe_3O_4 + 8\ Al \rightarrow 9Fe + 4Al_2O_3: H = 3350kJ$$
$$3FeO + 2\ Al \rightarrow 3Fe + Al_2O_3: H = 880kJ$$
$$Fe_2O_3 + 2\ Al \rightarrow 2Fe + Al_2O_3: H = 850kJ$$
$$3CuO + 2\ Al \rightarrow 3Cu + Al_2O_3: H = 1210kJ$$
$$3Cu_2O + 2\ Al \rightarrow 6Cu + Al_2O_3: H = 1060kJ$$

In the above reactions, aluminum is the reducing agent. Theoretically, the elements magnesium, silicon, and calcium can be used as well; but, for general applications, magnesium and calcium have found limited use. Silicon is often used in Thermit mixtures for heat treatment, but it is rarely used in welding. In some cases, an aluminum-silicon alloy is used as the reducing agent.

The first of the reactions above is the one most commonly used as a basis of mixtures for Thermit welding. The proportions of such mixtures are usually about three parts by weight of iron oxide to one part of aluminum. The theoretical temperature created by this reaction is about 5600°F (3100°C). Additions of nonreacting constituents, as well as heat loss to the reaction vessel and radiation, reduce this temperature to about 4500°F (2480°C). This is about the maximum temperature that can be tolerated, since aluminum vaporizes at 4530°F (2500°C). On the other hand, the maximum temperature should not be much lower because the aluminum slag (Al_2O_3) solidifies at 3700°F (2040°C).

The heat loss depends very much upon the quantity of Thermit being reacted. With large quantities, the heat loss per pound of Thermit is considerably lower and the reaction more complete when compared with small quantities of Thermit.

---

1. *Thermit* is the term commonly used to identify this welding process even though it is a registered trademark. More detailed metallurgical information is presented in the Thermit welding chapter, ASM International's *Metal Handbook*, Vol. 6, 9th Ed., 1985.
2. See also Chapter 2, "Physics of welding", Vol. 1, 8th Ed., 38-39.

Alloying elements can be added to the Thermit compound in the form of ferroalloys to match the chemistry of the parts to be welded. Other additions are used to increase the fluidity and lower the solidification temperature of the slag.

The Thermit reaction is nonexplosive and requires less than one minute for completion, regardless of quantity. To start the reaction, a special ignition powder or ignition rod is required; both can be ignited by a regular match. The ignition powder or rod will produce enough heat to raise the Thermit powder in contact with the rod to the powder's ignition temperature, which is about 2200°F (1200°C).

The parts to be welded should be aligned properly; the faces to be joined should be free of rust, loose dirt, moisture, and grease. A proper gap must be provided between the faces, the size depending upon the width of the joint. Wider joints normally require a larger gap. A mold, which may be built up on the parts or premanufactured to conform to the parts, is placed around the joint to be welded.

To fabricate a butt joint, the joint faces should be preheated sufficiently to promote complete fusion between the Thermit deposit and the base metal. Even though it is called a welding process, Thermit welding resembles metal casting where proper gates and risers are needed to:

(1) Compensate for shrinkage during solidification
(2) Eliminate typical defects that appear in castings
(3) Provide proper flow of the molten steel
(4) Avoid turbulence as the metal flows into the joint

APPLICATIONS

Rail Welding

THE MOST COMMON application of the process is the welding of rail sections into continuous lengths. It is an effective means of minimizing the number of bolted joints in the track structure. In coal mines, the main haulage track is often welded to minimize maintenance and to reduce excessive coal spillage caused by uneven track. Crane rails are usually welded to minimize joint maintenance and vibration of the building as heavily loaded wheels pass over the joint.

Thermit mixtures are available for all types of rail steels. The majority of rails are C-Mn steels, but Cr, Cr-Mo, Cr-V, Cr-Mn, and Si alloy rail steels are manufactured abroad. Addition of rare earth metals or alloys may decrease the amount of sulfur and phosphorous in the weld deposit, resulting in an improvement in mechanical properties.

Welding with Preheat. Premanufactured molds of split design are generally used for welding standard rail sizes. The mold should be aligned so that its center coincides with the center of the gap between the rail ends. The rail ends are preheated in the range of 1100 to 1800°F (600

Figure 29.1–Preheating Rail Ends With a Gas Torch Prior to Thermit Welding

to 1000°C) with a gas torch flame directed into the mold as shown in Figure 29.1. A refractory-lined crucible containing the Thermit charge is positioned above the mold halves after preheating is completed. The charge is then ignited, and the molten steel pours into the joint. In most procedures, the metal is fed into the middle of the joint gap (center-poured); in other procedures, the metal enters the sidecup of the mold at the outer leg of the rail base and rises vertically in the center of the joint.

A self-tapping seal (thimble) is used in the bottom of the crucible. A few seconds after the Thermit reaction is complete, the molten metal melts the seal and pours out of the bottom of the crucible into the gap between the two rail sections. The lower density liquid slag floats to the top of the Thermit metal in the crucible. It does not reach the mold cavity until all of the molten steel has entered and filled both the cavity between the rail sections and the mold itself. The slag remains on top of the weld and solidifies there. When the metal has solidified, the mold halves are removed and discarded. The excess metal is removed by hand grinding or by hydraulic or manual shearing devices.

Preheating times and temperatures may be reduced by using a larger Thermit charge. The heat dissipated into the work during welding has to be provided by a larger mass of molten steel.

Welding without Preheat. The self-preheating method is designed to eliminate the variables associated with torch preheating and the equipment needed to perform that operation. The rail ends are preheated by a portion of the molten metal produced by the Thermit reaction. The crucible and mold are a one-piece design as shown in Figure 29.2. The molds, commonly known as *shell molds*, are premanufactured of sand, bonded with phenolic resins. They are very light, nonhygroscopic, and moisture-free with a long shelf life. After the Thermit reaction is completed, the molten steel automatically flows from the crucible into the joint rather than passing through the atmosphere, as is the case with a separate crucible.

Figure 29.3, a section through a mold, shows the shape of the cavity in which the molten filler metal flows. There is a hollow chamber in the mold underneath the weld area that receives the first molten metal, allowing it to preheat the rail ends. This metal is called the *preheat metal*. By the time the chamber is filled, sufficient molten metal should have passed over the rail ends to preheat them to the re-

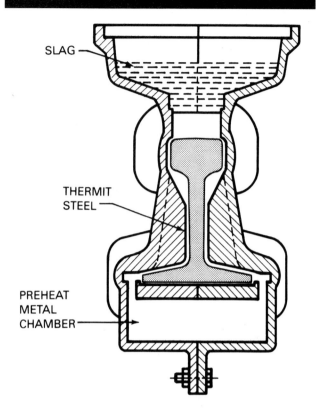

Figure 29.3—Section Through a Mold-Crucible Having a Preheat Metal Chamber

quired temperature to assure complete fusion with the base metal. Thermit portions for this process are about twice the size of those used for the external preheat method.

The heat-affected zones in the adjacent rail sections are considerably smaller than when external preheating is used. Figure 29.4 shows a typical section through a Thermit rail weld made with the self-preheating process.

Repair Welding

REPAIR WELDS ARE normally nonrepetitive, and therefore, premanufactured molds are not used. A mold must be made for each weld so that it will conform to the shape of the part.

Preparing the Joint. The pieces to be joined should be properly positioned in contact and aligned for welding. Firm marks should then be made on the pieces outside the area to be covered by the mold box. They will be used to reposition the pieces after the grooved faces are prepared for welding, thus maintaining the original part dimensions. The metal may then be cut with a cutting torch along the line of fracture to provide a parallel-sided gap. The gap

Figure 29.2—Combination Mold-Crucible in Position for Thermit Welding Without Preheat

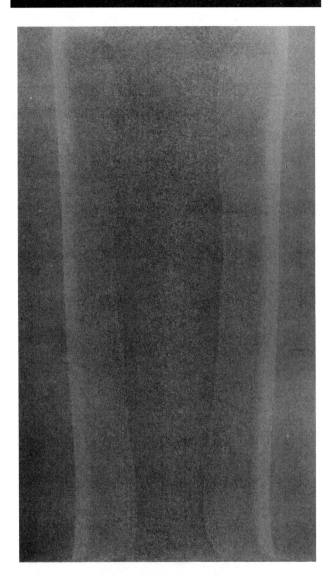

Figure 29.4—Photomacrograph of a Vertical Section Through a Typical Thermit Rail Weld

Applying the Mold. When a single large weld is to be made, a wax pattern is used to shape the mold cavity at the joint, similar to the investment (lost-wax) casting process. The wax is placed in the gap and on the surfaces of the parts to produce the exact shape desired for the finished weld, including the collar of weld reinforcement. A sand mold is then built up around the pattern using a suitable mold box to contain the mold sand.

Wood patterns of pouring and heating gates and risers are positioned within the mold as it is being rammed. Where two pieces of the same size are to be welded together, the heating gate is centered directly on the wax pattern. If unequal sections are being welded together, the heating gate is directed toward the larger section to provide somewhat uniform heating of the two parts. Where there are one or more high points on a joint of complex cross section, riser gates will be required at all of them. The top of the mold is hollowed out to provide a basin for the slag produced by the thermit reaction. The mold must be adequately vented to facilitate the escape of moisture and other gases during the preheating and welding process. Finally, the wood patterns are removed. Figure 29.5 shows a section through a Thermit weld mold with the crucible in position and ready for welding.

The quality of the molding sand requires special attention. It must have a high melting temperature, high permeability, and adequate shear strength. The sand should be free of clay components with low melting points.

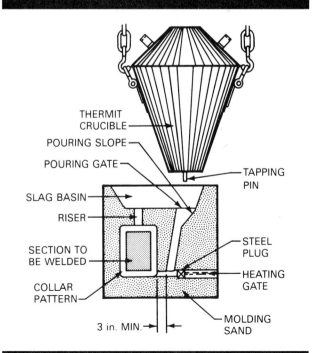

Figure 29.5—Cross Section of a Typical Thermit Mold for Repair Welding With External Preheat

width depends upon the size of the section to be welded as shown in Table 29.1. All loose oxide and slag from torch cutting, as well as dirt and grease, should be removed from the workpieces where the mold will be located.

To allow for contraction of the weld during cooling, the pieces are initially spaced 1/16 to 1/4 in. (2 to 6 mm) further apart than their original positions, using the markers on the pieces for reference. The exact increase depends upon the size of the weld and the gap length. The amount of contraction allowance required can be judged quite accurately with experience.

Preheating.   Preheating is accomplished by directing a gas flame into the chamber through the heating gate. A torch designed specifically for the purpose may burn propane, natural gas, kerosene, or gasoline.

The initial purpose of preheating is to remove the wax. The heat is applied gradually, and the torch is frequently removed from the heating gate to allow the melted wax to flow out. After the wax has been removed, the heat is gradually increased to preheat the faces of the base metal and thoroughly dry the mold. The mold must be completely dried to avoid weld porosity generated by residual moisture in the molding sand. Preheating is continued until the ends of the parts to be welded are cherry red in color, an indication that their temperature is between 1500°F and 1800°F (800 and 1000°C).

Upon completion of the preheating, the heating gate must be blocked. A short length of steel rod of appropriate diameter is pushed into the gate against a shoulder and then backed with molding sand.

Charging the Crucible.   The Thermit reaction, as in the case of rail welding with preheat, takes place in a refractory-lined, cone-shaped crucible as shown in Figure 29.6. A hard refractory (magnesite) stone at the bottom of the crucible holds a replaceable refractory orifice or thimble. The thimble is plugged by inserting a tapping pin through it and then placing a metal disk on top of the pin.

The disk is covered with a layer of refractory sand. The Thermit mixture should be placed in the crucible in a manner that will not dislodge the sand layer.

Low carbon steel punchings are sometimes added to the Thermit mixture to augment the metal produced. The quantity of Thermit mixture required for a joint can be calculated by the following equation:

$$X = \frac{E}{0.5 + 0.01S} \tag{29.1}$$

where:

X = quantity of Thermit required
E = quantity of molten steel required to fill the gap, including 10 percent for losses
S = percent of steel punchings to be included in the charge

The quantity X will be in lb when E is in lb and kg when E is in kg.

Approximately 25 lb (11.5 Kg) of Thermit mixture is required for each pound of wax in the pattern.

Making the Weld.   The reaction can be initiated by two methods: (1) starting powder that can be ignited by a match or regular gas striker, or (2) an ignition rod.

After the reaction is complete and the action of the molten metal subsides, the crucible is tapped by striking the

Table 29.1
Examples of Thermit Weld Dimensions and Mold Requirements

Section Size or Diam., in.	Gap, in.	Collar, in.	Risers		Pouring Gates		Heating Gates		Connecting Gates		Thermit Req'd[a] lb.
			No.	Diam., in.	No.	Diam., in.	No.	Diam., in.	No.	Diam., in.	
						Rectangular Sections					
2x2	7/16	1-1/2x7/16	1	3/4	1	3/4	1	1-1/4			6
2x4	9/16	1-5/16x9/16	1	3/4	1	1	1	1-1/4			12
4x4	11/16	2-5/8x11/16	1	1	1	1	1	1-1/4			25
4x8	7/8	3-7/16x7/8	1	1	1	1	2[b]	1-1/4			50
8x8	1-1/8	4-5/8x1-1/8	1	1-3/4	1	1-1/4	2[b]	1-1/4	—	—	125
8x12	1-1/4	5-1/2x1-1/4	1	1-3/4	1	1-1/4	1	1-1/4	1	1-1/4	175
12x12	1-7/16	6-1/2x1-7/16	1	2-1/2	1	1-1/2	2[b]	1-1/2	1	1-1/2	300
12x18	1-11/16	7-3/4x1-11/16	1	2-1/2	1	1-1/2	2[b]	1-1/2	1	1-1/2	500
16x16	1-3/4	8-15/16x1-3/4	1	2-3/4	2	2	2	1-1/2	2	1-1/2	700
16x24	2	9-15/16x2	1	2-3/4	2	2	2	1-1/2	2	1-1/2	1150
24x24	2-5/16	11-13/16x2-5/16	2	2-1/2	2	2	2	1-3/4	2	1-3/4	1875
24x36	2-5/8	14-1/8x2-5/8	2	2-1/2	2	2	2	2	4	2	3125
						Round Sections					
2	7/16	1-3/8x7/16	1	3/4	1	3/4	1	1/1/4			5
4	5/8	2-3/8x5/8	1	1	1	1	1	1-1/4			25
8	1	4-3/16x1	1	1-1/2	1	1-1/4	1	1-1/4			75
12	1-5/16	5-7/8x1-5/16	1	1-3/4	1	1-1/2	1	1-1/112	1	1-1/2	200
16	1-5/8	7-1/2x1-5/8	1	2	1	1-1/2	1	1-1/2	1	1-1/2	425

a. Thermit required includes provision for a 10% excess of steel in slag basin for a single pour and a 20% excess for a double pour.

b. Includes one separate back heating gate.

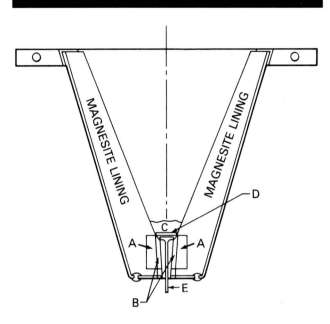

A - MAGNESITE STONE
B - MAGNESITE THIMBLE
C - PLUGGING MATERIAL
D - METAL DISC
E - TAPPING PIN

Figure 29.6—Cross Section of a Thermit Crucible

tapping pin with a sharp upward blow. The molten steel flows into the mold and fills the joint.

The mold is stripped away after the weld metal has solidified. Whenever possible, the entire weldment should be annealed to stress relieve it.

If required, the collar around the weld may be removed by machining or grinding. Risers and the gate are removed with an oxyfuel gas cutting torch.

Applications of Repair Welding. Thermit welding is employed in the marine field for repair of heavy sections of ferrous metal such as broken stern frames, rudder parts, shafts, and struts.

Broken necks, pinions, and pinion teeth of sheet and plate rolls are replaced with entirely new pieces, cast or forged slightly oversize to permit machining. They are Thermit welded to the main section.

Badly worn wobblers on the ends of steel mill rolls may similarly be replaced with a sufficiently tough Thermit metal deposit that is machinable. The method is particularly applicable for repairs involving large volumes of metal, where the heat of fusion cannot be raised satisfactorily or efficiently by other means or where fractures or voids in large sections require a large quantity of weld metal.

Thermit welding can be used for the repair of ingot molds at significant savings over replacement: the bottom of the mold can be cut off and completely rebuilt with Thermal metal, or an eroded cavity in the bottom can be filled with Thermit metal. The first method of repair is more sophisticated and requires larger quantities of Thermit, but the life of the ingot mold will be more than doubled. The latter type of repair has to be repeated after every second or third pour.

With large dredge cutters, the blades may be Thermit welded to a center ring. Quantities up to several thousand pounds are poured at one time. In this case, Thermit welding is a production tool rather than a repair method.

Reinforcing Bar Welding

THERMIT WELDING WITHOUT preheat is one way of splicing concrete reinforcing steel bars. Continuous reinforcing bars permit the design of concrete columns or beams smaller in section than when the bars are not welded together.

Two premanufactured mold halves are positioned at the joint in the aligned bars and sealed to them with adhesive compound and sand to avoid loss of molten metal. The arrangements for horizontal and vertical welding are shown in Figure 29.7. A closure disk is located in a well at the base of the Thermit crucible section of the mold. The Thermit powder is placed in the crucible and the reaction initiated. After completion of the reaction, the molten steel melts through the closure disk and fills the gap between the bars. The initial molten steel entering the mold chamber preheats the bar ends as it flows over them into a preheat metal chamber. The molten steel fills the joint gap and completes the weld. Reinforcing bars can be welded by this process in any position with properly designed molds. Thermit welded reinforcing bar specimens tested in tension and bending are shown in Figure 29.8.

As an alternative to welding, reinforcing bars can be joined end-to-end by depositing Thermit metal between a steel sleeve and the enclosed bars. The joint is primarily mechanical. Arrangements for horizontal and vertical connections are shown in Figure 29.9 The sleeve is placed around the abutted bars. Its inside diameter is somewhat larger than the diameter of the bars to provide space for the cast steel. Both the inner surface of the sleeve and the surfaces of the bars are serrated. A graphite or CO_2—cured sand mold is mounted over an orifice in the sleeve through which the molten steel flows into the annular space between the bars and sleeve. The Thermit mix is placed above a metal closure disk within the mold.

Upon ignition of the Thermit powder, molten steel is produced. It melts the metal disk, flows into the annular space between the bars and the sleeve, and then solidifies. About five minutes are required to make a joint with this technique.

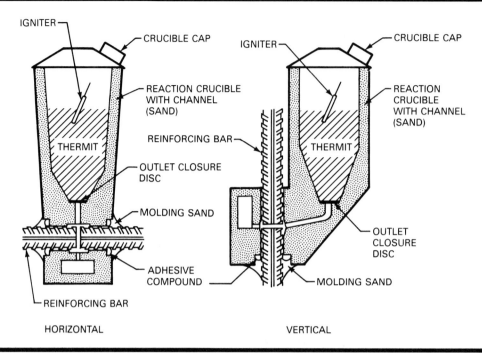

Figure 29.7–Thermit Welding Concrete Reinforcing Steel Bars

Electrical Connections

A THERMIT MIXTURE of copper oxide and aluminum is used for the welding of joints in copper conductors. The reaction between the two materials produces superheated molten copper and slag in one to five seconds. Other metals in the form of slugs or powder can be added to produce alloys for particular applications of the process.

The process is primarily used for welding copper bars, cables, and wires together, as well as copper conductors to steel rails for grounding. For the latter application, a graphite mold is clamped to the rail section at the joint. As soon as the Thermit reaction is complete, the molten copper melts the disk and flows into the joint cavity. It solidifies in a few seconds and creates a weld between the base metal and the copper cable. Then the mold is removed. It can be used again after removing the slag from the reaction chamber.

Heat Treatment of Welds

THE DEVELOPMENT OF high alloy steels and the application of the Thermit process for welding them created a need for special types of Thermit mixtures that produce heat only. Molten metal is not produced. This particular type of Thermit is designed to create sufficient heat for heat-treating purposes.

Using special binders, the Thermit mixture itself is formed to the configuration of the parts to be heat-treated. It keeps its exact shape during and after the

Figure 29.8–Thermit Welded Reinforcing Steel Bars After Tensile and Bend Testing

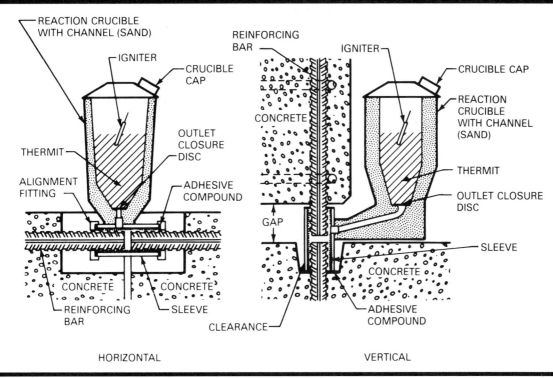

Figure 29.9–Thermit Sleeve Joint for Reinforcing Bars

Thermit reaction. The maximum temperature produced in the part can be adjusted by the design of the Thermit compound. Figure 29.10 shows Thermit blocks in place for heat treating a rail section.

SAFETY

THE PRESENCE OF moisture in the Thermit mix, in the crucible, or on the workpieces can lead to rapid formation of steam when the Thermit reaction takes place. Steam pressure may cause violent ejection of molten metal from the crucible. Therefore, the Thermit mix should be stored in a dry place, the crucible should be dry, and moisture should not be allowed to enter the system before or during welding.

The work area should be free of combustible materials that may be ignited by sparks or small particles of molten metal. The area should be well ventilated to avoid the buildup of fumes and gases from the reaction. Starting powders and rods should be protected against accidental ignition.

Personnel should wear appropriate protection against hot particles or sparks. This includes gloves, full face shields with filter lenses for eye protection, and headgear. Safety boots are recommended to protect the feet from hot sparks. Clothing should not have pockets or cuffs that might catch hot particles.

Figure 29.10–Bonded Thermit Blocks in Place for Heat Treating a Rail Section

Preheating should be done taking safety precautions applicable to all oxyfuel gas equipment and operations. For additional information, refer to American National Standard Z49.1, *Safety in Welding and Cutting.*[3]

---

3. Available from the American Welding Society.

COLD WELDING

FUNDAMENTALS OF THE PROCESS

Definition and General Description

COLD WELDING (CW) is a solid-state process in which pressure is used at room temperature to produce coalescence of metals with substantial deformation at the weld. A characteristic of the process is the absence of heat, either applied externally or generated by the welding process itself. A fundamental requisite for satisfactory cold welding is that at least one of the metals to be joined is highly ductile and does not exhibit extreme work-hardening. Both butt and lap joints can be cold welded. Typical joints are shown in Figure 29.11.

Materials Welded

METALS WITH FACE-CENTERED cubic (FCC) lattice structure are best suited for cold welding, provided they do not work-harden rapidly. Soft tempers of metals such as aluminum and copper are most easily cold welded. It is more difficult to weld cold-worked or heat-treated alloys of these metals. Other FCC metals that may be cold welded readily are gold, silver, palladium and platinum.

The joining of copper to aluminum by cold welding is a good application of the process, especially where aluminum tubing or electrical conductor grade aluminum is joined to short sections of copper to provide transition joints between the two metals. Such cold welds are characterized by substantially greater deformation of the aluminum than the copper because of the difference in the yield strengths and work-hardening behaviors of the two metals.

Dissimilar Metal Welds

NUMEROUS DISSIMILAR METALS may be joined by cold welding whether or not they are soluble in one another. In some cases, the two metals may combine to form intermetallic compounds. Since cold welding is carried out at room temperature, there is no significant diffusion between dissimilar metals during welding. The alloying characteristics of the metals being joined do not affect the manner in which the cold welding operation is carried out. However, interdiffusion at elevated temperatures can af-

fect the choice of postweld thermal treatments and the performance of the weld in service.

Welds made between metals that are essentially insoluble in each other are usually stable. Diffusion can form an intermetallic compound at elevated service temperatures. In some cases, this intermetallic layer can be brittle and cause a marked reduction in the ductility of the weld. Such welds are particularly sensitive to bending or impact loading after an intermetallic layer has formed.

The rate at which intermetallic compounds form depends upon the specific diffusion constants for the particular metals in the weld as well as the time and temperature of exposure. Thus, bimetal cold weldments require careful consideration of the diffusion couple and the service environment. For example, a layered structure forms at the interface in an aluminum-copper weldment at elevated temperatures as shown in Figure 29.12. The layered structure contains a brittle Al-Cu intermetallic compound that weakens the weldment. Figure 29.13 shows how rapidly the thickness of the diffused zone increases at high service temperatures. Mechanical tests have shown that the strength and ductility of the joint decrease when the thickness of the interfacial layer exceeds about 0.002 in. (0.05 mm). Consequently, aluminum-copper cold welds should be used only in applications where service temperatures are low and peak temperatures seldom, if ever, exceed 150°F (65°C).

Metallurgical Structure

IN BUTT JOINTS, the lateral flow of metal between the dies during upset produces a cross-grained structure adjacent to the interface of the weld, as shown in Figure 29.14. This cross-grained material is essentially a narrow transverse section in the weldment. The presence of this section is not important in metals that are essentially isotropic, such as aluminum and some aluminum alloys. In nonisotropic metals, fatigue or corrosion resistance may be substantially lower at the welded joint.

Surface Preparation

COLD WELDING REQUIRES that clean metal faces come into intimate contact for a satisfactory joint. Proper surface preparation is necessary to assure joints of maximum strength. Dirt, absorbed gas, oils, or oxide films on the

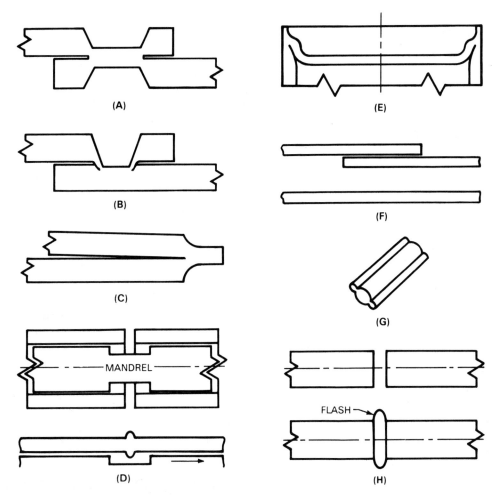

(A) LAP WELD, BOTH SIDES INDENTED; (B) LAP WELD, ONE SIDE INDENTED; (C) EDGE WELD, BOTH SIDES INDENTED; (D) BUTT JOINT IN TUBING, BEFORE AND AFTER WELDING; (E) DRAW WELD; (F) LAPPED WIRE, BEFORE AND AFTER WELDING; (G) MASH CAP JOINT; (H) BUTT JOINT IN SOLID STOCK, BEFORE AND AFTER WELDING.

Figure 29.11–Contour of Typical Cold Welded Joints

surface interfere with metal-to-metal contact and must be removed to obtain strong welds.

The best method of surface preparation for lap welds is wire brushing at a surface speed of about 3000 ft/min (15.2 m/s). A motor-driven rotary brush of 0.004 in. (0.1 mm) diameter stainless steel wire is commonly used. Softer wire brushes may burnish the surface; coarser types may remove too much metal and roughen the surface. The surfaces should be degreased prior to brushing to avoid contamination of the wire brushes. It is important that the clean surface not be touched with the hands because grease or oil on the faying surfaces impairs the formation of a strong joint. Welding should take place as soon as practical after cleaning to avoid interference with bonding

from oxidation. In the case of aluminum, for example, welding should be done within about 30 minutes.

Chemical and abrasive cleaning methods have not proved satisfactory for cleaning surfaces to be joined by cold welding. The residue from chemical cleaning or abrasive particles embedded in or left on the surface may prevent the formation of a sound weld.

EQUIPMENT

PRESSURE FOR WELDING may be applied to overlapped or butted surfaces with hydraulic or mechanical presses, rolls, or special manual or pneumatically operated tools. A hand tool of the toggle cutter type is suitable for very light work.

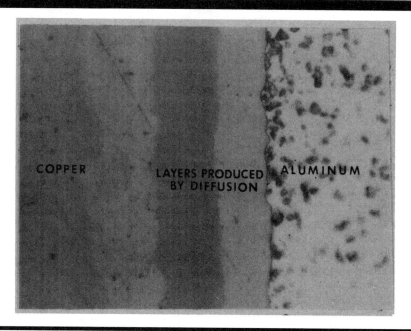

Figure 29.12–Layered Structure in an Aluminum-Copper Cold Weld After Exposure at 500°F (260°C) for 60 Days

A common manually operated press, as shown in Figure 29.15, may be used for medium size work. Heavy work requires power operated machines. The rate of pressure application does not usually affect the strength or quality of the weld.

Regardless of how pressure is applied, the proper indentation for lap welds is important. The indentation may take the form of a narrow strip, a ring, or a continuous seam. Typical weld indentor configurations used for cold welding are presented in Figure 29.16. The selection of indentation configuration is largely determined by the desired appearance and performance characteristics.

The bar type indentation, Figures 29.16(A) and (B), causes metal deformation along both of its sides. Indentation in the form of a ring, Figure 29.16(C), may cause undesirable curvature of the sheet surfaces. A ring weld may

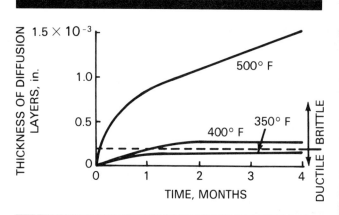

Figure 29.13–Change in Thickness of the Diffusion Layer in Aluminum-Copper Cold Welds With Time at Three Elevated Temperatures

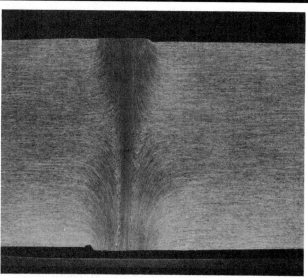

Figure 29.14–Transverse Flow Lines in a Cold Welded Butt Joint

Figure 29.15–A Manually Operated Cold Welding Machine

have a smooth convex dome of metal within the ring. This dome is formed when the metal is forced from between the dies as pressure is applied. Continuous seam welding, Figure 29.16(E), can be employed in the manufacture of thin-wall tubing or lap welds in sheet.

Symmetrical dies that indent both sides of the joint, Figures 29.11(A) and (C), are generally used. If one surface must be free of indentation, a flat plate or anvil may be used on one side to produce the weld shown in Figure 29.11(B). Thinner gages of sheet metal or wire can be cold welded using simple dies mounted in hand-operated tools.

Draw welding is a form of lap welding used to seal containers. Both the lid and the can are flared before welding. The components are placed in a close fitting die. A punch forces the components into the die which cold welds the flared metal as it is drawn down over the punch. Figure 29.11(E) illustrates such a joint.

Dies are usually subjected to high pressures and should be made of tool steel hardened to about 60 Rockwell C. Pressures of from 150 to 500 ksi (1000 to 3400 MPa) are required to weld aluminum depending upon alloy composition and temper. Copper requires pressures that may be two to four times greater than those required for alumi-

num. Aluminum can be cold welded to copper using specially designed dies to compensate for the difference in yield strengths.

BUTT JOINTS

COLD WELDING IS commonly used to produce butt joints in wire, rod, tubing, and simple extruded shapes of like and unlike metals. Figures 29.11(D) and (H) illustrate butt joints in tubing and solid forms. The weld can be as strong as the base metal when correct procedures are used.

Preparation for Welding

A SHORT SECTION is usually sheared from the ends of the parts to be joined to expose fresh clean surfaces. The shear should be designed to produce square ends so that the parts do not deflect from axial alignment as welding force is applied. During shearing, a thin film of a particular metal being cut can accumulate on the shear blades. If the shear is then used to cut a different metal, accumulated metal on the blade may transfer to the cut surface and inhibit welding. Therefore, the shear blades should be cleaned before shearing parts of another metal for cold welding.

It is not usually necessary to degrease parts to be cold welded before shearing if the residual film of lubricant is very thin. However, degreasing may be necessary if there is a heavy oil film on the metal to avoid contamination of the cut surfaces and to prevent the parts from slipping in the clamping dies. The best practice is to clean before shearing.

Welding Procedure

THE PARTS SHOULD be positioned in the clamping dies with sufficient initial extension of each part between the dies to ensure adequate upset to produce a satisfactory weld. However, extension of the parts should not be excessive or the parts may bend instead of upsetting. The upsetting force will cause the parts to bend or assume an S-shaped curve, as shown in Figure 29.17, if the initial die opening is too large. The ends can deflect and slide past one another when force is applied if the projecting length of each part exceeds about twice the thickness or diameter of the parts. In other words, the initial opening between the dies should be no greater than four times the diameter or thickness of the parts. This distance is the maximum total upset that can be used to effect welding. The minimum upset distance varies with the alloy being joined.

The welding dies should firmly grip the parts to prevent slippage when the upset force is applied. Any slippage will reduce the amount of upset. For a firm grip, the dimensions of the parts are critical so that the dies can nearly close to hold each part securely. The allowable tolerance depends on (1) the design of the die and die holder, and (2) the gripping surface finish. Deep knurling on the gripping

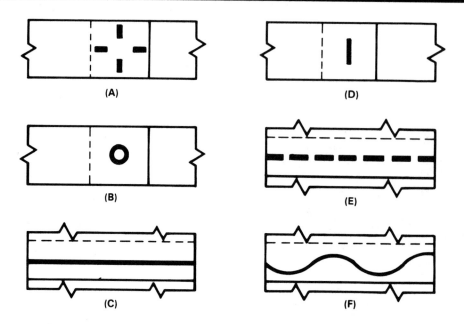

(A) AND (B), BAR TYPE; (C) RING TYPE; (D), (E), AND (F), INTERMITTENT AND CONTINUOUS
SEAM TYPES

Figure 29.16–Typical Lap Weld Indentor Configurations Used in Cold Welding

surfaces will indent into the part. The allowable tolerance for round parts is about 3 percent of the part diameter.

Somewhat wider tolerances are permissible for rectangular-shaped parts because the dies usually bear on only two sides. The gap between the closed grips must, however, be small to obtain uniform upsetting of metal. It should be no more than about 10 percent of the part thickness.

The application of upset force causes the metal between the dies to upset laterally as illustrated in Figure 29.18. This lateral flow of metal:

(**1**) Breaks up the oxide film present on the abutting surfaces and carries most of it away from the joint surface
(**2**) Enables oxide-free metal on one side of the interface to achieve intimate contact with oxide-free metal on the other side
(**3**) Provides the energy that enables the contacting surfaces to achieve a metallurgical bond with one another

Thus, all requirements needed to form a metal bond are fulfilled, and a metallurgical union forms. The flash formed by the lateral flow of metal can be pinched off by the dies as they close together.

Cold welds are usually insensitive to the rate of upsetting of the metal, within limits. Regardless of upset speed, welding will take place if there is sufficient upset.

Multiple Upset

THE AMOUNT OF upset required to produce a full-strength weld in some alloys sometimes exceeds that which can be provided in one step because of a limitation on part extension. If an initial upset will produce a bond of sufficient strength to hold the parts together, additional upset can be applied to produce a full-strength weld by repositioning the weldment in the dies. Surface preparation prior to welding is relatively unimportant when a multiple upset-

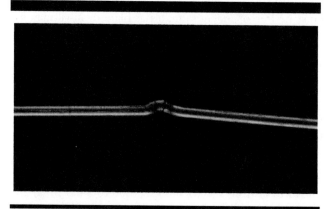

Figure 29.17–Bending Produced During Upset From Excessive Projecting Lengths

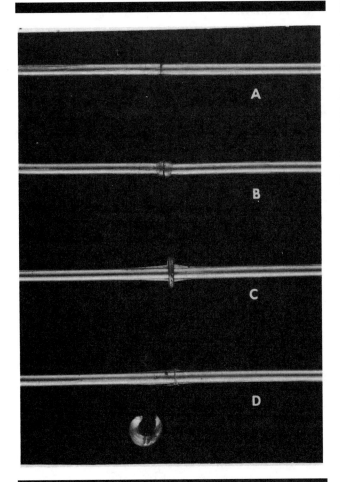

Figure 29.18–States of Upset During Butt Cold Welding

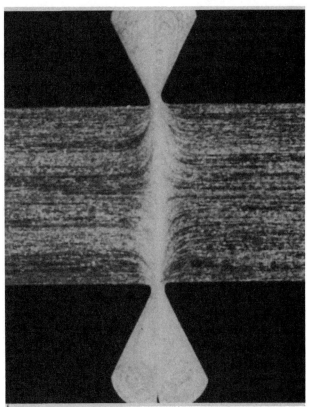

Figure 29.19–Single Upset Cold Weld in Type 1100 Aluminum Wire

ting technique is used. This technique will completely displace contaminants from the interface.

Single upset welding is not normally used for welding butt joints in wires smaller than 3/16 in. (4.8 mm) in diameter. A cross section of a single upset butt joint in type 1100 aluminum wire is shown in Figure 29.19. Compare this weld with the multiple upset weld in Type 1100 aluminum alloy illustrated in Figure 29.20. The multiple upset, offset-flash technique is commonly used in wire drawing to splice 0.025 to 0.128 in. (0.64 to 3.25 mm) diameter wires as well as those aluminum alloy wires that cannot be welded effectively with single upset. Figure 29.21 illustrates the various stages involved in making a multiple upset butt weld between strips.

Offset Welds

FIGURE 29.22 ILLUSTRATES a cold weld being made in a die designed to produce an offset flash. This technique will produce a discontinuous flash that is easy to remove as well as a weld joint that is at an angle to the wire axis. The weld joint being at an angle to the axis will be less influenced by discontinuities in the weld.

Application

COLD WELDED BUTT joints are used in the manufacture of aluminum, copper, gold, silver, and platinum wire. The most common use is to join successive reels of wire for continuous drawing to smaller diameters. Butt joints are also used to repair breaks in the wire that occur during the drawing operation. Diameters ranging from 0.0025 to 0.50 in. (0.06 to 12.7 mm) have been welded successfully. The aluminum alloys that have been welded with good results include Types EC, 1100, 2319, 3003, 4043, all of the 5000 series, 6061, and 6201. With most of these alloys, the as-welded wire can be drawn successfully to smaller diameters after removal of the flash. Cold welded joints in wire of very high-strength alloys, such as Types 2014 and 7178, usually must be annealed to prevent breaks during subsequent drawing operations. Aluminum alloys that contain lead and bismuth (Types 2011 and 6262) are difficult to cold weld.

Figure 29.20–Multiple Upset Cold Weld in Type 1100 Aluminum Wire Using an Offset Flash Technique

Where welding is permitted by ASTM Specifications, cold welding is used to join successive lengths of wire to permit stranding of long lengths of multiple-strand electrical conductors. The weld flash is removed, and the weld is dressed with a file or suitable abrasive to obtain a smooth, uniform appearance.

Welds in annealed wire of any of the weldable aluminum alloys exhibit tensile strengths exceeding 95 percent of the base metal. In cold-worked Type EC or 5005 aluminum wire and heat-treated Type 6201 wire, weld efficiencies of 92 to 100 percent are attained. In bend testing, the welded joint can be bent or twisted about half as many times before failure as unwelded wire of the same alloy.

Seven-strand No. 4 AWG aluminum alloy conductors with a cold weld in one strand have shown the same breaking strength as similar conductors without welded strands. Types EC, 5005, and 6201 aluminum alloys were used in making these tests.

For copper wire, work hardening at the cold weld increases the metal strength to that of the drawn wire.

LAP WELDING

Procedures

LAP WELDS CAN be used for joining aluminum sheet or foil to itself and also to copper sheet or foil. Pressure is applied to the lapped parts by dies that indent the metal and cause it to flow at the interface. This pressure ranges between 150 and 500 ksi for aluminum, depending upon the compressive yield strength of the alloy being welded. Excellent lap welds can be produced in nonheat-treatable aluminum alloys, such as Types EC, 1100, and 3003. However, aluminum alloys containing more than three percent magnesium, the 2000 and 7000 series of wrought alloys, and castings are not readily lap welded.

Table 29.2 gives recommended deformations and typical joint efficiencies for lap welds in several common aluminum alloys. For most alloys, joint strength is maximum when deformations between 60 and 70 percent are used. It is apparent that the intrinsic strength of the weld may increase at deformations exceeding 70 percent, but the overall strength of the assembly is decreased. Lap welds exhibit good shear and tensile strengths but have poor resistance to bending or peel loading.

Equal deformations can be achieved when lap welding aluminum to copper or welding dissimilar aluminum alloys together by using dies with the bearing areas approximately in inverse proportion to the compressive yield strength of each metal.

Applications

COMMERCIAL APPLICATIONS OF lap welding are shown in Figures 29.23 and 29.24. They include packaging as well as electrical applications. The latter is probably the field in which lap welding finds the greatest application. It is especially useful in the fabrication of electrical devices in which

**Table 29.2
Lap Type Cold Welds in Selected Aluminum Alloys**

Alloy	Temper	Recommended Deformation, %	Joint Efficiency, %
3003	0	50	85
3003	H14	70	70
3003	H16	70	60
3003	H18	60	55
3004	0	60	60
3004	H34	55	40
5052	0	60	65
5052	H34	60	45
6061	T6	60	50
7075 Alclad	T6	40	10

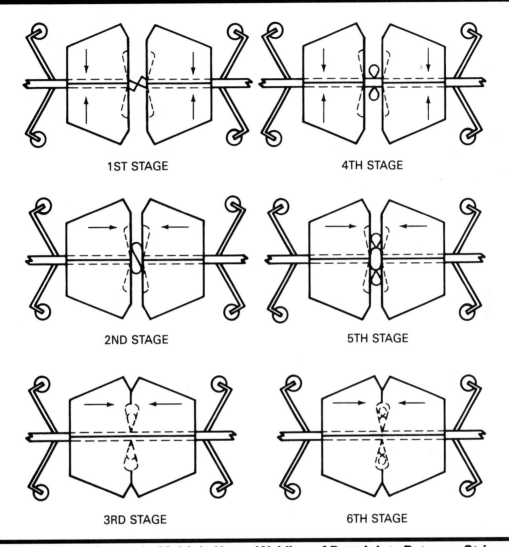

1ST STAGE

4TH STAGE

2ND STAGE

5TH STAGE

3RD STAGE

6TH STAGE

Figure 29.21–Stages in Multiple Upset Welding of Butt Joints Between Strips

a transition from aluminum windings to copper terminations is required. The range of electrical applications covers large distribution transformers to small electronic devices.

A variation of cold lap welding is the sealing of commercially pure aluminum, copper, or nickel tubing by pinching it in two between two dies. The tubing is placed transversely between two radius-faced linear dies. As the dies are forced together, the tubing is pinched flat against itself. As the force on the dies is increased, the metal between the dies is upset and extruded from between them as in lap welding. The force is increased until the tube walls are cold welded together and then finally parted in two across the midpoint of the weld. As with sheet, the interior of the tubing must be clean to accomplish a leak-free weld across the flattened tube.

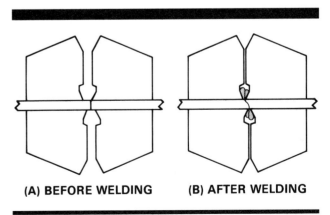

(A) BEFORE WELDING   (B) AFTER WELDING

Figure 29.22–Cold Welding of Wire With an Offset Flash Technique

Figure 29.23–Application of Cold Welding in the Manufacture of an Electrical Component

Figure 29.24–Application of Cold Welding in the Manufacture of Industrial Packaging

The die face radius and width must be designed to accomplish cold welding and ultimately cut the tubing in two. The opposing dies must be carefully aligned for welding. The face radius is the key to successful cold welding and must be determined experimentally for the metal and tube wall thickness.

HOT PRESSURE WELDING

DEFINITION

HOT PRESSURE WELDING (HPW) is a solid-state welding process that produces coalescence of metals with heat and application of pressure sufficient to produce macro deformation of the work pieces. Vacuum or other shielding media may be used.

PRESSURE GAS WELDING (PGW)

Definition and General Description

PRESSURE GAS WELDING[4] is an oxyfuel gas welding process which produces a weld simultaneously over the entire faying surfaces. The process is used with the application of pressure. No filler metal is used. The two variations of the process are the closed joint and open joint methods. In the closed joint method, the clean faces of the parts to be joined are abutted together under moderate pressure and heated by gas flames until a predetermined upsetting of the joint occurs. In the open joint method, the faces to be joined are individually heated by the gas flames to the melting temperature and then brought into contact for upsetting. Both methods are easily adapted to mechanized operation. Pressure gas welding can be used for welding low and high carbon steels, low and high alloy steels, and several nonferrous metals and alloys.

In the closed joint method, since the metal along the interface does not reach the melting point, the mode of welding is different from that of fusion welding. Generally speaking, welding takes place by the action of grain growth, diffusion, and grain coalescence across the interface under the impetus of high temperature [about 2200°F

---

4. More detailed information is presented in Chapter 22, *Welding Handbook*, Section 2, 5th Ed., 1963.

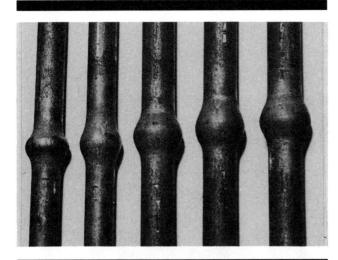

Figure 29.25–Typical Pressure Gas Welds in 1 in. (25 mm) and 1-1/4 in. (32 mm) Diameter Steel Bars

(1200°C) for low carbon steel] and upsetting pressure. The welds are characterized by a smooth-surfaced bulge or upset, as shown in Figure 29.25, and by the general absence of cast metal at the weld line.

In the open joint method, the joint faces are melted, but the molten metal is squeezed from the interface to form a flash when the joint is upset. These welds resemble flash welds in general appearance.

Principles of Operation

Closed Joint Method. The faces to be welded are butted together under initial pressure to assure intimate contact. The metal at the joint is then heated to welding temperature with a gas flame. Finally, the metal is upset sufficiently to produce a weld.

Heating is generally done with water-cooled, multiflame oxyacetylene torches. These torches are designed to generate sufficient heat and distribute it uniformly throughout the entire section to be welded. For sections over 1 in. (25 mm) thick, it is advisable to heat the joint uniformly from all sides as shown in Figure 29.26.

Solid or hollow round sections, such as shafts or piping, are usually welded with circular ring torches. The torch head may be a split type for easy loading and removal of work from the welding machine. A typical head of this type for welding 2.5 in. (63.5 mm) diameter tubing with a 0.25 in. (6.4 mm) wall thickness is shown in Figure 29.27. More elaborate heating heads are required for more complicated shapes. They should conform to the shape of the part to provide uniform heating.

Figure 29.26–Torch Arrangement for Pressure Gas Welding a Type 321 Stainless Steel Ring With 1-7/8 by 2-5/8 in. (50 X 70 mm) Cross Section

Figure 29.27–Split Annular Torch for Pressure Welding Piping, Tubing or Solid Rounds

The apparatus for pressure gas welding must be designed to apply the desired pressure and maintain alignment during welding. Provision for maintaining uniform pressure is essential.

The quality and type of end preparation of the parts to be welded depend upon the type of steel. In general, the abutting ends should be machined or ground to a smooth, clean surface. Freedom from oil, rust, grinding dust, and other foreign material is of great importance.

The geometry of the abutting faces depends upon the application and the alloy. Some control of the shape of the upset metal can be obtained by beveling one or both of the parts. Figure 29.28 illustrates typical joint preparations for pressure gas welding and the effect of beveling on the shapes of the completed welds.

For illustration, assume that two 5 in. (125 mm) diameter by 1/4 in. (6.4 mm) wall steel pipes are to be pressure gas welded end-to-end using a butt joint. The general procedures are as follows: A split torch head that will provide small oxyacetylene flames for the full circumference of the joint is selected. The head is mounted in the same plane as the interface with provision for axial oscillation. The abutting ends of the pipe are beveled to an included angle of 6 to 10 degrees with a smooth, clean finish. The pipes are placed in the machine and aligned. Then a force of 5850 lbs (2650 Kg) is applied to produce a low compressive pressure of 1500 psi (10 MPa).

While this force is maintained, the torch flames are oscillated axially a short distance across the weld joint. As the joint heats up, the metal will upset. The joint faces will close together preventing oxidation at higher temperatures. As the metal temperature increases, the compressive strength of the steel decreases and the joint begins to upset uniformly. At this time, the metal is at welding temperature throughout its full thickness and an upsetting pressure of 4000 psi (28 MPa) is applied until the weld zone is upset for a distance of 0.188 in. (4.8 mm). The torches may then be extinguished and the assembly removed from the machine.

There are a few variations of this basic procedure, the principal one being the sequence of pressure application. These variations are introduced to meet special require-

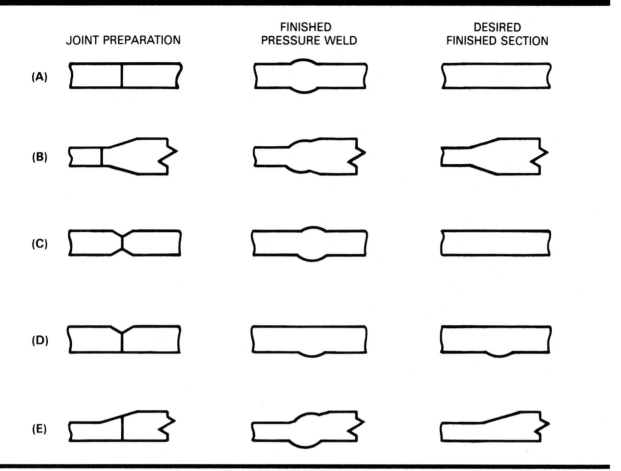

| JOINT PREPARATION | FINISHED PRESSURE WELD | DESIRED FINISHED SECTION |

(A)

(B)

(C)

(D)

(E)

Figure 29.28–Joint Designs for Pressure Gas Welding

ments of certain metals such as high carbon steels, high chromium steels, and nonferrous metals. For example, the constant pressure method is recommended for welding high carbon steel parts.

Another variation of the basic method is applicable to high chromium steels and some nonferrous metals. A high initial pressure in the range of 6000 to 10 000 psi (40 to 70 MPa) is applied before heating is started and is maintained until the metal in the weld zone starts to upset. This high pressure forces the joint faces together to prevent oxidation. Pressure is then decreased until welding temperature is reached when the high pressure is again applied to upset the joint.

Examples of typical pressure cycles used for pressure gas welding several metals are given in Table 29.3. Table 29.4 gives the average dimensions of closed joint pressure welds in parts of various thicknesses.

The quality of a weld depends to an important degree upon proper upsetting during the welding operation. The upset distance or shortening of the weld zone increases with metal thickness. Recommended amounts of upset are given in Table 29.4. These values are usually mea-

sured from fixed points on the parts or the welding machine.

Open Joint Method. Machines for open joint pressure gas welding must provide more accurate alignment and be of rugged construction to withstand rapidly applied upset forces. Machines similar to those used for flash welding are suitable.

The most satisfactory heating head is a flat, multiflame type burner, such as the one shown in Figure 29.29, that produces a uniform flame pattern conforming to the cross section of the members to be welded. Good alignment of the heating head with the faces of the joint is important to minimize oxidation and to obtain uniform heating and subsequent upsetting. A removable spacer block can be used during alignment.

Saw cut surfaces are satisfactory for welding since the ends are thoroughly melted before the weld is consummated. A thin layer of oxide on the joint faces has little effect on weld quality, but major amounts of foreign substances, such as rust or oil, should be removed before welding.

Table 29.3
Typical Upset Pressure Cycles for Pressure Gas Welds

Type of metal	Method	End pressure, psi (MPa)		
		Initital	Intermediate	Final
Low carbon steel	Closed joint	500-1,500 (3-10)	. . .	4,000 (28)
High carbon steel	Closed joint	2,700 (19)	. . .	2,700 (19)
Stainless steel	Closed joint	10,000 (69)	5,000 (34)	10,000 (69)
Monel alloy	Closed joint	6,500 (45)	. . .	6,500 (45)
Steel (carbon and alloy)	Open joint	. . .	. . .	4,000-5,000 (28-34)

The general procedure for open joint pressure gas welding is to align the parts with a suitable torch tip properly spaced between the joint faces (Figure 29.29). When thin sections are welded, the torch tip is placed just outside the joint with the flames directed at the joint faces. The tip is designed to heat the full cross section of the faces. The flames are maintained in this position until a molten film entirely covers both faces. The torch is then withdrawn, and the parts are rapidly brought together with a force that will produce a constant pressure of 4000 to 5000 psi (28 to 35 MPa) at the interface. This step is shown in Figure 29.30. Pressure is maintained until upsetting of the metal ceases. The total upset is controlled by both the applied pressure and the temperature of the hot metal. It is not preset on the equipment.

Equipment

Machines. The apparatus for pressure gas welding comprises:

(1) Equipment for applying upsetting force
(2) Suitable heating torches and tips designed to provide uniform and controlled heating of the weld zone
(3) Necessary indicating and measuring devices for regulating the process during welding

The complexity of the machine depends upon the configuration and size of the parts to be welded and the degree to which the process is mechanized. In most cases, it is advisable to use special heating torches and tips as well as special apparatus for gripping and applying force to the parts.

Figure 29.31 illustrates a simple, manually operated gas pressure welding machine capable of welding bars and tubes up to 3 in. (76 cm) in diameter.

Auxiliary Equipment. Some auxiliary equipment is necessary. The gas supply must be adequate for the maximum flow requirement, and the gas regulators must be capable of maintaining a uniform flame adjustment. Quick-acting gas shut-off valves are very desirable. In many instances, needle valves are advantageous for fine adjust-

Table 29.4A
Joint Dimensions of Pressure Gas Welds, Squared End Preparation, Closed Joint Method

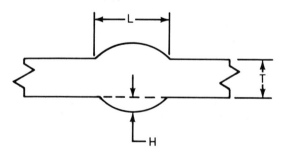

Metal Thickness (T), in.	Length of Upset (L), in.	Approx. Upset Hgt (H), in.	Total Upset, in.
1/8	3/16 - 1/4	1/16	1/8
1/4	5/16 - 1/2	3/32	1/4
3/8	9/16 - 5/8	1/8	5/16
1/2	3/4 - 7/8	3/16	3/8
3/4	1-1/16 - 1-3/16	1/4	1/2
1	1-1/4 - 1-1/2	3/8	5/8

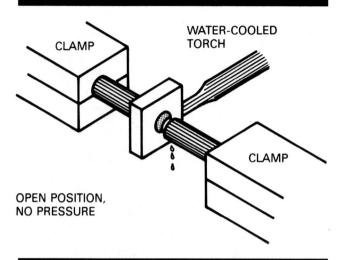

Figure 29.29–Torch and General Setup for Open Joint Pressure Gas Welding

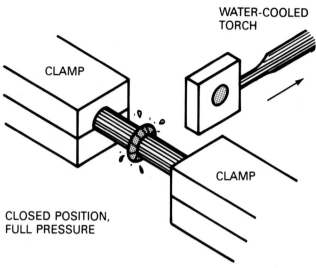

Figure 29.30–An Open Joint Pressure Gas Weld as Upsetting Starts

ment of the flame. The best control of gas flow and heat input is obtained when the pressure gages are located close to the torch. This permits the operator to check gas pressures readily. Flowmeters may be used to assure uniform gas flow.

An ample supply of water is needed for cooling the torches and, in some cases, the clamps and parts of the press. Adequate jigs for aligning and supporting the parts are needed. Automatic control units for regulating the upset force and heating cycles and then terminating the operation can be incorporated in a machine.

Applications

Metals Welded. Pressure gas welding has been successfully applied to plain carbon, low alloy, and high alloy steels, and to several nonferrous metals, including nickel-copper, nickel-chromium, and copper-silicon alloys. It has been very useful for joining dissimilar metals.

Table 29.4B
Joint Dimensions of Pressure Gas Welds, Squared End Preparation, Closed Joint Method

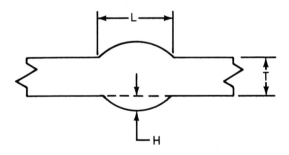

Metal Thickness (T), mm	Length of Upset (L), mm	Approx. Upset Hgt (H), mm	Total Upset, mm
3	5-6	2	3
6	8-13	2	6
10	14-16	3	8
13	19-22	5	10
19	27-30	6	13
25	32-38	10	16

Figure 29.31–A Small Pressure Gas Welding Machine for Rods and Tubes Up to 3 in. Diameter

Rails. The first commercial application of this process was the welding of railroad rails, and thousands of joints were made. However, this process has largely been replaced by the flash welding process. When pressure gas welding is used, the closed butt method is usually employed with equipment specifically designed for this application.

The rail ends are carefully prepared by power sawing and then are cleaned. The rails are gripped by special clamps and a force applied to produce about 2800 psi (20 MPa) on the joint. The joint is then heated with specially shaped heating tips or heads that are oscillated automatically across the joint until the metal reaches welding temperature. Adequate pressure is then applied to produce the required upset. Typical rail welding equipment is shown in Figure 29.32. Most of the upset metal on the ball and edges of the base is removed by oxygen cutting. The welded joint is then indexed or moved to the next position, where the weld zone is normalized to refine the grain size and restore normal hardness. Finally, the weld is ground to rail contour, examined by magnetic particle inspection, and oiled for protection against rusting. Rails that have been welded, normalized, and inspected in this manner have given satisfactory service under both heavy and fast traffic for extended time periods.

Other Applications. Pressure gas welding has been largely superseded by other welding processes. The basic elements of the process assisted in the development of similar processes, such as flash and friction welding, that use other sources of energy. Automatic welding of pipe, a former application, is now accomplished using automatic gas metal arc welding.

Properties and Heat Treatment

IN GENERAL, PRESSURE gas welding has a minimum effect on the mechanical and physical properties of the base metals. Because of the relatively large mass of hot metal in the weld zone, its cooling rate is usually quite low.

In the closed-joint method, the maximum temperature of the metal is below the temperature at which overheating and rapid grain growth occur. In the open-joint method, the melted metal film is squeezed out of the joint during upset. These characteristics are advantageous for welding high carbon steels and some nonferrous alloys that are hot-short or affected by overheating.

Another important factor is the absence of deposited metal. The entire weld zone is base metal, and, hence, it responds to heat treatment in the same manner. This, of course, includes the effect of the heat of welding on the

Figure 29.32–Pressure Gas Welding Railroad Rails

corrosion resistance of welded stainless steels. If unimpaired corrosion resistance is desired, stabilized stainless steel must be used or the welded assembly must be given a stabilization heat treatment after welding.

Pressure gas welds in low carbon steels seldom require heat treatment or stress relief since the heat-affected zone in such steels is usually normalized and relatively stress free. Pressure gas welding has been used with low alloy and high carbon steels for fabricating assemblies subject to high service stresses, and postweld heat treatment was necessary. Heat treatment may frequently be done with the same heating heads used for welding.

In rails, for example, the annealed zone on each side of the weld may be too soft. To overcome this problem, the weld zone can be heated to normalizing temperature using heating heads and then air cooled to restore the desired hardness. Similarly, heat treatment with the welding flame may be suitable for developing desired mechanical properties in welded joints in some low alloy steels such as those used for oil well drilling tools. Such heat treatment, which is essentially a normalizing operation, will refine the grain size in the weld zone and improve ductility and toughness.

For highly hardenable steels, annealing or slow cooling after the welding operation may be necessary to prevent hardening or surface cracking in the weld zone. To develop optimum properties in welds in heat-treatable steels, furnace heat treatment is commonly used.

Weld Quality

Mechanical Properties.[5]   Since there is no deposited metal in pressure welds, the mechanical properties of welds will depend upon the composition of base metals, the cooling rate, and the quality of the weld. When dissimilar steels are joined, the properties of the welded joint will be more nearly those of the weaker member.

Metallurgical Structure.   The location of the original interface in pressure gas welds in plain carbon steels and many alloy steels is very difficult to detect in a metallo-

5. Mechanical properties of typical pressure gas welds are given in Chapter 22, "Pressure gas welding," *Welding Handbook*, Section 2, 5th ed., 1963.

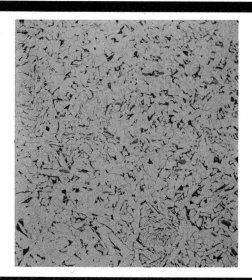

Figure 29.33–Photomicrograph of a Pressure Gas Weld in 1020 Steel, As-Welded

graphic cross section using normal etchants. It is possible to locate the weld line with special polishing and etching techniques. A typical photomicrograph of a pressure gas weld in steel is shown in Figure 29.33. Although not apparent, the interface extends vertically through the center of the photomicrograph.

Quality Control

Process Control. Successful pressure gas welding by the closed joint method requires positive and continuous control of the variables that influence the quality of a weld. The variables include the following:

(1) Degree of roughness and cleanliness of end preparation
(2) Pressure cycle
(3) Alignment of the parts
(4) Welding cycle time
(5) Performance of the heating torches
(6) Desired upset or shortening
(7) Cooling time in the machine after upset

Examination of the pressure-upsetting cycle can indicate if the welding conditions are conforming to the prescribed procedure. With constant heat input into the weld zone, constant width of heated zone, and uniform pressure sequence, the entire cycle of heating and upsetting should be completed with a variation in welding time of not more than 10 percent. On this basis, if a weld requires an unduly long or short time, the conditions that prevailed during welding should be evaluated. A large time variation from nominal would indicate that (1) some factor other than

time was controlled improperly, and (2) the weld might be of questionable quality. Malfunctioning of the pressure system or the heating heads and slipping of the parts in the clamps are examples of the conditions that might cause poor quality welds.

Autographic or other records of the following variables have proved of value in maintaining good control:

(1) Pressure cycle
(2) Total time or times for certain stages of the procedure
(3) Gas flow rates
(4) Total upset distance

Some conditions of importance in the closed joint method are not so important with the open joint method. With the open joint method, the cleanliness of the joint faces is not critical except for excessive amounts of foreign matter. Melting of the faces offsets a need for thorough surface preparation. The amount of upset or shortening is not necessarily constant, and, therefore, it is not an index of weld quality. However, due attention must be given to the pressure cycle and the performance of the heating torches.

Inspection. The first inspection is usually a visual one to evaluate the following general characteristic:

(1) Presence or absence of excessive melting
(2) Contour and uniformity of upset
(3) Position of the weld line with respect to the midpoint of the upset zone

If there is no appreciable variation from an accepted standard and the controls were adequate, it may usually be concluded that the pressure weld is of normal quality.

In many highly stressed assemblies, added assurance of weld consistency and quality may be needed. Sample welds selected either at random or at fixed intervals should be destructively tested. This procedure will serve as a positive and continuous check on the welding cycle and process controls as well as the properties of the welded assemblies.

Magnetic particle inspection can be used for nondestructive inspection of pressure welded rails. The nick-break test can be used as a convenient quality check for soundness.[6] Fracture of a sample weld along the weld line will show the extent of metallic bonding, grain size, and evidence of overheating of the faces. Changes in the welding cycle can be checked quickly by this test. Experience has proved that when the nick-break tests show satisfactory crystalline fracture throughout the weld cross section, all other tests will usually prove satisfactory.

6. For a description of this test, refer to *Welding Inspection*, 113–114. Miami: American Welding Society, 1980 or API Standard 1104, Para 2.63, 15th ed., 1980, Supplement 1, 1982.

Proof testing may be used as an alternative to destructive testing. The test is designed to disclose defective welds and pass acceptable welds. A welded joint is subjected to either a tensile or a bending load, or both, to produce a maximum tensile stress just below the yield strength of the metal. A poor quality weld will fail in this test.

FORGE WELDING (FOW)

Fundamentals of the Process

FORGE WELDING[7] IS a solid-state welding process that produces a weld by heating the work pieces to welding temperature and applying blows sufficient to cause permanent deformation at the faying surfaces. Forge welding was the earliest welding process and the only one in common use until well into the nineteenth century. Blacksmiths used this process. Pressure vessels and steel pipe were among the industrial items once fabricated by forge welding.

The process finds some application with modern methods of applying the heat and pressure necessary to achieve a weld. The chief present day applications are in the production of tubing and clad metals.

Principles of Operation

THE SECTIONS TO be joined by forge welding may be heated in a forge, furnace, or by other appropriate means until they are very malleable. A weld is accomplished by removing the parts from the heat source, superimposing them, and then applying pressure or hammer blows to the joint.

Heating time is the major variable that affects joint quality. Insufficient heat will fail to bring the surfaces to the proper degree of plasticity, and welding will not take place. If the metal is overheated, a brittle joint of very low strength may result. The overheated joint is likely to have a rough, spongy appearance where the metal is severely oxidized. The temperature must be uniform throughout the joint interfaces to yield a satisfactory weld.

Process Modes

Hammer Welding. In hammer welding, coalescence is produced by heating the parts to be welded in a forge or other furnace and then applying pressure by means of hammer blows. Manual hammer welding is the oldest technique. Pressure is applied to the heated members by repeated high velocity blows with a comparatively light sledge hammer. Modern automatic and semiautomatic hammer welding is accomplished by blows of a heavy

power-driven hammer operating at low velocity. The hammer may be powered by steam, hydraulic, or pneumatic equipment.

The size and quantity of parts to be fabricated will determine the choice of either manual or power-driven hammer welding. This process may still be used in some maintenance shops, but it largely has been replaced by other welding processes.

Die Welding. This is a forge welding process where coalescence is produced by heating the parts in a furnace and then applying pressure by means of dies. The dies also shape the work while it is hot.

Metals Welded

LOW CARBON STEELS are the metals most commonly joined by forge welding. Sheets, bars, tubing, pipe and plates of these materials are readily available.

The major influences on the grain structure of the weld and heat-affected zone are the amount of forging applied and the temperature at which the forge welding takes place. A high temperature is generally necessary for the production of a sound forge weld. Annealing can refine the grain size in a forge welded steel joint and improve joint ductility.

Thin, extruded sections of aluminum alloy are joined edge-to-edge by a forge welding process with automatic equipment to form integrally stiffened panels. The panels are used for lightweight truck and trailer bodies. Success of the operation depends upon the use of correct temperature and pressure, effective positioning and clamping devices, edge preparation, and other factors. Although the welding of aluminum for this application is called forge welding, it could be classified as hot pressure welding because the edges to be joined are heated to welding temperature and then upset by the application of pressure.

Joint Design

THE FIVE JOINT designs applicable to manual forge welding are the lap, butt, cleft, jump, and scarf types shown in Figure 29.34. The joint surfaces for these welds are slightly rounded or crowned. This shape ensures that the center of the pieces will weld first so that any slag, dirt, or oxide on the surfaces will be forced out of the joint as pressure is applied. Lap, pin, and butt joints used for automatic forge welding are shown in Figure 29.35.

Scarfing is the term applied to the preparation of the workpieces of forge welding. Similarly, the prepared surface is referred to as a scarfed surface. Each workpiece to be welded must be upset sufficiently for an adequate distance from the scarfed surface to provide metal for mechanical working during welding.

---

7. More detailed information is presented in Chapter 61, *Welding Handbook*, Section 3B, 6th Ed., 1971.

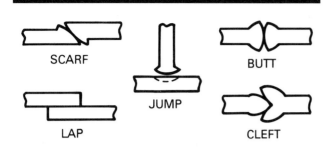

Figure 29.34–Typical Joint Designs Employed for Manual Forge Welding

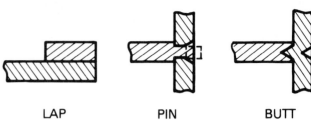

Figure 29.35–Typical Joint Designs Used for Automatic Forge Welding

Fluxes

IN THE FORGE welding of certain metals, a flux must be used to prevent the formation of oxide scale. The flux and the oxides present combine to form a protective coating on the heated surfaces of the metal. This coating prevents the formation of additional oxide and lowers the melting point of the existing oxide.

Two commonly used fluxes for steels are silica sand and borax (sodium tetraborate). Flux is not required for very low carbon steels (ingot iron) and wrought iron because their oxides have low melting points. The flux most commonly used in the forge welding of high carbon steels is borax. Because it has a relatively low fusion point, borax may be sprinkled on the metal while it is in the process of heating. Silica sand is suitable as a flux in the forge welding of low carbon steel.

CARBON ARC WELDING

DEFINITION AND GENERAL DESCRIPTION

CARBON ARC WELDING (CAW) is a process in which an arc is established between a nonconsumable carbon (graphite) electrode and the work, or between two carbon electrodes. The latter is a variation known as twin carbon arc welding. Two other variations known as shielded and gas carbon arc welding no longer have commercial significance.

Although carbon arc welding has been superseded to a great extent by other welding processes, there are many applications for which it can be used to good advantage. In operation, the carbon arc is used only as a source of heat. In this respect, it resembles the gas tungsten arc welding process.

Figure 29.36 shows the carbon arc welding process. The arc stream usually develops a temperature of from 7000 to 9000°F (3870 to 4980°C), depending upon the amount of current used. Because the electrode burns off very slowly, it does not have an appreciable effect on the composition of the deposited metal, provided filler metal is added.

PRINCIPLES OF OPERATION

THE APPLICATION OF heat and the filler metal feed are controlled separately in carbon arc welding. With some welds, the carbon arc is used to fuse the edges together without the addition of filler metal. A carbon arc can produce the high heat needed for welding metals that have high thermal conductivity, such as copper.

In carbon arc welding, direct current electrode negative (straight polarity) should be used. The arc is formed between the tip of the carbon electrode and the base metal. Welding current is adjusted to provide sufficient heat to melt the base metal and welding rod uniformly as welding progresses. Recommended current ranges for carbon and graphite electrodes are given in Table 29.5. Amperages are recommended on the basis of maximum electrode life. Higher amperages can be used, but the electrode will be consumed faster.

The properties of welds made with the carbon arc in mild steel may be adequate for noncritical applications. The process does not provide as much shielding from the atmosphere as the shielded metal arc or gas metal arc welding processes.

EQUIPMENT

CARBON ELECTRODES RANGE in size from 1/8 to 7/8 in. (3.2 to 22 mm) diameter. Baked carbon electrodes last longer than graphite electrodes. Figure 29.37 shows typical air-cooled carbon electrode holders. Water-cooled holders are available for use with the larger sizes of electrodes, or adapters can be fitted to regular holders to permit ac-

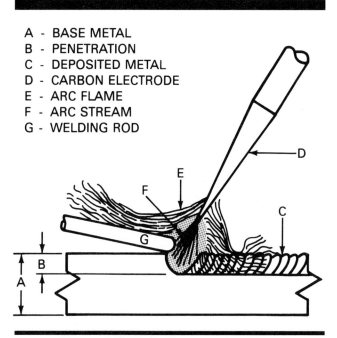

A - BASE METAL
B - PENETRATION
C - DEPOSITED METAL
D - CARBON ELECTRODE
E - ARC FLAME
F - ARC STREAM
G - WELDING ROD

Figure 29.36–Carbon Arc Welding

commodation of the larger electrodes. Direct current welding machines of either the rotating or rectifier type are excellent power sources for the carbon arc welding process.

WELDING TECHNIQUE

THE WORKPIECES MUST be free from grease, oil, scale, paint, and other foreign matter. The two pieces should be

Table 29.5
Recommended Current Ranges for Carbon and Graphite Electrodes[a]

Electrode Diameter		Current, A[b]	
in.	mm	Carbon Electrodes	Graphite Electrodes
1/8	3.2	15-30	15-35
3/16	4.8	25-55	25-60
1/4	6.4	50-85	50-90
5/16	7.9	75-115	80-125
3/8	9.5	100-150	110-165
7/16	11.1	125-185	140-210
1/2	12.7	150-225	170-260
5/8	15.9	200-310	230-370
3/4	19.0	250-400	290-490
7/8	22.2	300-500	400-750

a. Recommended with regard to maximum electrode life. Where electrode cost is not a factor, higher amperages may be used.

b. Direct current electrode negative (straight polarity)

clamped tightly together with no root opening. They may be tack welded together.

Carbon electrodes, 1/8 to 5/16 in. (3.2 to 8 mm) diameter may be used depending upon the current required for welding. The end of the electrode should be prepared with a long taper to a point. The diameter of the point should be about half that of the electrode. For steel, the electrode should protrude about 4 to 5 in. (100 to 125 mm) from the electrode holder.

A carbon arc may be struck by bringing the tip of the electrode into contact with the work and immediately withdrawing it to the correct length for welding. In general, an arc length between 1/4 and 3/8 in. (6 and 10 mm) will be best. If the arc length is too short, there is likely to be excessive carburization of the molten metal resulting in a brittle weld.

When the arc is broken for any reason, it should not be restarted directly upon the hot weld metal as this is likely to cause a hard spot in the weld at the point of contact. The arc should be started on cold metal to one side of the joint and then quickly returned to the point where welding is to be resumed.

When the joint requires filler metal, the welding rod is fed into the molten weld pool with one hand while the arc is manipulated with the other. The arc is directed on the surface of the work and gradually moved along the joint, constantly maintaining a molten pool into which the welding rod is added in the same manner as in gas tungsten arc welding. Progress along the weld joint and the addition of welding rod must be timed to provide the size and shape of weld bead desired. Welding vertically or overhead with the carbon arc is difficult because carbon arc welding is essentially a puddling process. The weld joint should be backed up, especially in the case of thin sheets, to support the molten weld pool and prevent excessive melt-through.

For outside corner welds in 14 to 18 gage steel sheet, the carbon arc can be used to weld the two sheets together without a filler metal. Such welds are usually smoother and more economical to make than shielded metal arc welds made under similar conditions.

METALS WELDED

THE CARBON ARC can be used for welding steels and nonferrous metals. It can also be used for surfacing.

Steels

THE PRINCIPLE USE of carbon arc welding of steel is making edge welds without the addition of filler metal. This is done chiefly in thin gage sheet metal work, such as tanks, where the edges of the work are fitted closely and fused together using appropriate flux.

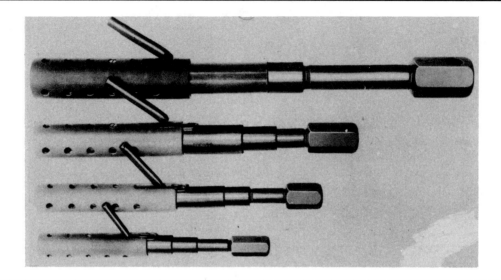

Figure 29.37–Typical Air-Cooled Carbon Electrode Holders

Galvanized sheet steel can be braze welded with the carbon arc.[8] A bronze welding rod is used. The rod is placed in the arc so that the zinc is not burned off the steel sheet. The arc should be started on the welding rod or a starting block. Low current, a short arc length, and rapid travel speed should be used. The welding rod should melt and wet the galvanized steel.

Cast Iron

IRON CASTINGS MAY be welded with the carbon arc and a cast iron welding rod. The casting should preheated to about 1200°F (650°C) and slowly cooled if a machinable weld is desired.

Copper

STRAIGHT POLARITY SHOULD always be used for carbon arc welding of copper. Reverse polarity will produce carbon deposits on the work that inhibit fusion.

The work should be preheated in the range of 300 to 1200°F (150 to 650°C) depending upon the thickness of the parts. If this is impractical, the arc should be used to locally preheat the weld area. The high thermal conductivity of copper causes heat to be conducted away from the point of welding so rapidly that it is difficult to maintain welding heat without preheating.

A root opening of 1/8 in. (3.2 mm) is recommended. Best results are obtained at high travel speeds with the welding rod held within the arc. A long arc length should be used to permit carbon from the electrode to combine with oxygen to form CO. This will provide some shielding of the weld metal.

---

8. Refer to Chapter 12 of this volume for a discussion of braze welding.

TWIN CARBON ARC WELDING

WITH A TWIN carbon torch, the arc heat can be used for welding, brazing, surfacing, or soldering operations as well as for preheating or postheating the work. The heat is produced by an arc between two carbon electrodes; the work is not part of the electrical circuit. Twin carbon arc welding is used principally for maintenance operations. A twin carbon arc torch, shown in Figure 29.38, has two adjustable arms in which the carbon electrodes are clamped. To maintain a constant distance between the electrodes (arc length) as they are consumed, adjustment of electrode position can usually be made while operating the torch.

Small ac arc welding machines are normally used with the twin carbon arc. Copper coated carbon electrodes are generally used in 0.250 to 0.375 in. (6.4 to 9.5 mm) diameter. The current should never be set so high that the copper coating is burned away over 0.5 in. (12.7 mm) ahead of the arc. Only enough current should be used to cause the filler material to flow freely on the work. This will avoid consuming carbons too rapidly.

SAFETY

SAFETY PROCEDURES AND equipment normally used with other arc welding processes should also be used with this one. This includes welding helmets with appropriate filter lenses, protective clothing, and gloves. Adequate ventilation should be provided. The requirements of ANSI Z49.1, Safety in Welding and Cutting, latest edition, and appropriate federal, state, and local regulations should be followed when carbon arc welding.

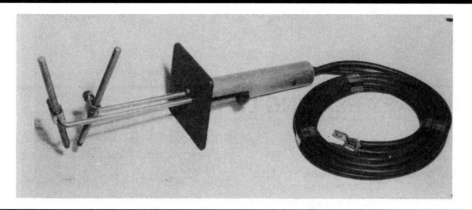

Figure 29.38—A Twin Carbon Arc Welding Torch

BARE METAL ARC WELDING

BARE METAL ARC welding[9] (BMAW) is an arc welding process that uses an arc between a bare or lightly coated electrode and the weld pool. Neither shielding nor pressure is used, and filler metal is obtained from the electrode.

The chief disadvantage of welding with a bare electrode is that the molten filler and weld metal are exposed to the atmosphere. The molten metal transferring across the arc and the molten weld metal are both subjected to oxidation and nitrification. As a result, the molten metal oxidizes rapidly, and the weld metal is likely to have unsatisfactory fusion with the base metal. Formation of porosity in the

weld will have a detrimental effect upon the strength and ductility of the welded joint. Nitrogen in the form of nitrides tends to cause high hardness and poor ductility. Water vapor dissociates in the arc to produce hydrogen, which may cause hydrogen embrittlement of some metals and underbead cracking in some steels.

Extruded covered electrodes have largely replaced bare and lightly covered electrodes on the market today. There is still, however, a considerable tonnage of bare electrodes manufactured and used. Most of the bare wires manufactured today are either coiled or spooled for use with the gas shielded welding processes.

9. More information is presented in Chapter 61, *Welding Handbook*, Section 3B, 6th Ed., 1971.

ATOMIC HYDROGEN WELDING

ATOMIC HYDROGEN WELDING (AHW) is an arc welding process that uses an arc between two metal tungsten electrodes in a shielding atmosphere of hydrogen and without the applicaton of pressure. Shielding is obtained from the hydrogen. Filler metal may or may not be added.

In this process, the arc is maintained entirely independent of the work or parts being welded. The work is a part of the electrical circuit only to the extent that a portion of the arc comes in contact with the work, at which time a voltage exists between the work and each electrode.

Historically, atomic hydrogen welding was the forerunner of the gas shielded arc welding processes. At that time, it was the best process for welding of metals other than carbon and low alloy steels. With the advent of low-cost inert gases, the gas shielded arc welding processes have largely replaced atomic hydrogen welding.

Hydrogen in its normal state is diatomic. Each molecule consists of two atoms. When an arc is established in hydrogen between two electrodes, the temperature in the arc stream reaches approximately 11 000°F (6090°C) and the molecular hydrogen dissociates into its atomic form. In the process of dissociation, a large amount of heat is absorbed from the arc by the hydrogen. The heat is subsequently liberated on recombination of the hydrogen atoms at the surface of the work. A sudden decrease in the temperature of the hydrogen as it strikes a relatively cold surface (weld area) is accompanied by a rapid release of heat as the hydrogen atoms recombine to the molecular form. By varying the distance between the arc stream and the surface, the available energy can be varied over a wide range. The hydrogen also cools the electrodes and protects both the electrodes and the metal from oxidation.

Atomic hydrogen welding had one unique advantage: the ability to control heat input over a very wide range by manipulating the arc. It was widely used for tool and die repair and similar operations where very precise metal buildup with accurate alloy control was necessary.

SUPPLEMENTARY READING LIST

THERMIT WELDING

Ailes, A. S. "Modern applications of Thermit welding." *Weld. Met. Fab* 32(9): 335-43, 414-19; 1964.

Cikara, M. "Repair of rails by Thermit welding and some observations on the testing of welded joints." *Welding and Allied Processes in Maintenance and Repair Work*, 318-34. New York: Elsevier Pub. Co., 1961.

Fricke, H. D. "Thermit welding." *ASM International's Metals Handbook*, Vol. 6, 9th Ed., 1985.

Frick, H. D., Guntermann, H., and Jacoby, N. "Thermit welding process for rails of special quality." *ETR.* 25(4): 1976 (in German).

Guntermann, H. "The applications of the Thermit process in areas besides rail welding." *ZEV-Glaser Annalen*, 1975 (in German).

————. "Thermit butt joints for concrete-steel construction." *Maschinernmarket* 75 (75): 1969 (in German).

Jacoby, N. "Special processes of the thermit welding technique." *Der Eisenbahningenieur.* No. 3, 1977 (in German).

Kubaschewski, E., Evans, L. L., and Alcock, C. B. *Metallurgical thermochemistry*, 4th Ed. London-New York: Pergamon Press, 1967.

Rossi, B. E. *Welding engineering.* New York: McGraw-Hill, 1954.

COLD WELDING

Jellison, James L. and Zanner, Frank J. "Solid-state welding." *ASM International's Metals Handbook*, Vol. 6, 9th Ed., 1985.

Houldcraft, P. T. *Welding process technology*, 217-21. London: Cambridge University Press, 1977.

Milner, D. R. and Rowe, G. W. "Fundamentals of solid phase welding." *Metallurgical Review* 28(7): 433-80; 1962.

Mohamed, H. A. and Washburn, J. "Mechanism of solid-state pressure welding." *Welding Journal* 54(9): 302s-10s; September 1975.

Tylecote, R. F. *The solid-state welding of metals.* New York: St. Martin's Press, 1968.

HOT PRESSURE WELDING

Bryant, W. A. "A method for specifying hot isostatic pressure welding parameters." *Welding Journal* 54(12): 433s-35s; December 1975.

Guy, A. G. and Eiss, A. L. "Diffusion phenomena in pressure welding." *Welding Journal* 36(11): 473s-80s; November 1957.

Hastings, D. C. "An application of pressure welding to fabricate continuous welded rails." *Welding Journal.* 34 (11): 1065-69; November 1955.

Jellison, James L. and Zanner, Frank J. "Solid-state welding." *ASM International's Metals Handbook*, Vol. 6, 9th Ed., 1985.

Lage, A. P. "Application of pressure welding to the aircraft industry." *Welding Journal* 35(11): 1103-09; November 1956.

Lessmann, G. G. and Bryant, W. A. "Complex rotor fabrication by hot isostatic pressure welding." *Welding Journal.* 51(12): 606s-14s; December 1972.

McKittrick, E. S. and Donalds, W. E. "Oxyacetylene pressure welding of high-speed rocket test track." *Welding Journal* 38(5): 469-74; May 1959.

Metzger, G. E. "Hot pressure welding of aluminum alloys." *Welding Journal* 57(1): 37-43; January 1978.

WELDING HANDBOOK
INDEX OF MAJOR SUBJECTS

Z

INDEX